Lecture Notes in Artificial Intelligence 814

Subseries of Lecture Notes in Computer Science
Edited by J. G. Carbonell and J. Siekmann

Lecture Notes in Computer Science
Edited by G. Goos and J. Hartmanis

Alan Bundy (Ed.)

Automated Deduction — CADE-12

12th International Conference on
Automated Deduction
Nancy, France, June 26 - July 1, 1994
Proceedings

Springer-Verlag
Berlin Heidelberg New York
London Paris Tokyo
Hong Kong Barcelona
Budapest

Series Editors

Jaime G. Carbonell
School of Computer Science, Carnegie Mellon University
Schenley Park, Pittsburgh, PA 15213-3890, USA

Jörg Siekmann
University of Saarland
German Research Center for Artificial Intelligence (DFKI)
Stuhlsatzenhausweg 3, D-66123 Saarbrücken, Germany

Volume Editor

Alan Bundy
Department of Artificial Intelligence, The University of Edinburgh
80 South Bridge, Edinburgh EH1 1HN, United Kingdom

CR Subject Classification (1991):I.2.3, F.4.1-2

ISBN 3-540-58156-1 Springer-Verlag Berlin Heidelberg New York
ISBN 0-387-58156-1 Springer-Verlag New York Berlin Heidelberg

CIP data applied for.

This work is subject to copyright. All rights are reserved, whether the whole or part of the material is concerned, specifically the rights of translation, reprinting, re-use of illustrations, recitation, broadcasting, reproduction on microfilms or in any other way, and storage in data banks. Duplication of this publication or parts thereof is permitted only under the provisions of the German Copyright Law of September 9, 1965, in its current version, and permission for use must always be obtained from Springer-Verlag. Violations are liable for prosecution under the German Copyright Law.

© Springer-Verlag Berlin Heidelberg 1994
Printed in Germany

Typesetting: Camera ready by author
SPIN: 10131285   45/3140-543210 - Printed on acid-free paper

Preface

This volume contains the papers presented at the Twelfth International Conference on Automated Deduction (CADE-12) held on June 26–July 1, 1994, at the Palais des Congrès de Nancy, France, sponsored by CADE Inc. and hosted by INRIA and CRIN; many other organizations provided co-sponsorship. The CADE conferences are the major forum for the presentation of new research in all aspects of automated deduction.

There were 177 papers submitted to CADE-12: 152 regular papers, 23 system descriptions and 2 problem sets. Of these 177 papers, 67 were accepted: 46 regular papers, 19 system descriptions and 2 problem sets (some papers switched categories during refereeing). In addition, the CADE-12 programme contained 3 invited talks, a banquet speech, a panel discussion, 7 workshops and 8 tutorials. The number of papers submitted and accepted are both records for CADE conferences. The field of automated deduction is clearly growing. This growth has been achieved with no dilution of the very high standards in the field. In particular, the quality of papers accepted by CADE-12 is at least as high as at previous CADEs. This volume presents many of the most important research results in automated deduction since CADE-11.

Much of the growth is accounted for by the spread of interest in the field around the globe. CADE-12 received papers from: Australia, Austria, Belgium, Canada, France, Germany, Italy, Japan, Norway, China, Poland, Portugal, Spain, Sweden, The Netherlands, UK, USA and Ukraine.

Since CADE-11, the CADE conference series has been put on a firmer footing with the incorporation of CADE. CADE Inc. will sponsor all future CADE conferences. It will protect the officers of CADE from financial liability and provide initial funding for CADE conferences as well as providing a holding fund for any surpluses. CADE Inc. is a sub-corporation of the Association of Automated Reasoning. It will be run by a Board of Trustees consisting of the current and past four CADE Programme Chairs plus a Secretary and Treasurer. I am particularly grateful to Neil Murray for agreeing to hold the joint post of Secretary and Treasurer of CADE Inc.

I would like to thank the many people who made CADE-12 possible. I am grateful: to the Programme Committee and the additional referees they recruited for reviewing a record number of papers to a very high standard in a short time scale; to Claude Kirchner and his Local Arrangements Committee for their energy and dedication in organising an excellent conference; to my fellow Trustees for their advice and encouragement, especially to Secretary/Treasurer Neil Murray; to the Invited Speakers, John Slaney, Ursula Martin and Bob Constable, the Banquet Speaker, Richard Platek, and the Panel Organiser, Bob Boyer; to the tutorial and workshop organisers; to Janice Lonsdale for help in organising the Programme Committee Meeting at Imperial College; and last, but by no means least, to Ian Green, Millie Tupman, Carole Douglas and the members of my research group at Edinburgh, who helped with the myriad administrative tasks involved in programme chairing a conference.

Alan Bundy
CADE-12 Programme Chair

Herbrand Award for Distinguished Contributions to Automated Reasoning
Recipient: Professor W. W. Bledsoe

We are delighted to announce that the second Herbrand Award for Distinguished Contributions to Automated Reasoning will be given at CADE-12 to Professor Woody Bledsoe, of the University of Texas at Austin. The Herbrand Award was initiated at CADE-11 to honour an individual or (a group of) individuals for exceptional contributions to the field of Automated Deduction. The winner is selected by the Trustees, the Programme Committee and previous winners. At CADE-11 the Award was made to Dr. Larry Wos. Woody Bledsoe has made numerous contributions to the field of Automated Deduction including natural proof systems, interactive systems, decision procedures, analogical reasoning and applications to set theory, analysis and program verification. He has been a source of inspiration to generations of researchers at Austin and throughout the world.

Nomination for this award can be made at any time to the CADE chair. A nomination should include a letter (of up to 2000 words) from a principal nominator describing the nominee's contributions, along with two other letters (of up to 2000 words) of endorsement.

Previous CADEs

CADE-1 Argonne National Laboratory, USA, 1974 (IEEE Transactions on Computers **C-25**, #8).

CADE-2 Oberwolfach, Germany, 1976.

CADE-3 MIT, USA, 1977.

CADE-4 University of Texas at Austin, USA, 1979.

CADE-5 Les Arcs, France, 1980 (Springer-Verlag LNCS **87**).

CADE-6 Courant Institute, New York, USA, 1982 (Springer-Verlag LNCS **138**).

CADE-7 Napa, California, USA, 1984 (Springer-Verlag LNCS **170**).

CADE-8 University of Oxford, UK, 1986 (Springer-Verlag LNCS **230**).

CADE-9 Argonne National Laboratory, USA, 1988 (Springer-Verlag LNCS **310**).

CADE-10 Kaiserslautern, Germany, 1990 (Springer-Verlag LNAI **449**).

CADE-11 Saratoga Springs, USA, 1992 (Springer-Verlag LNAI **607**).

CADE Inc. trustees

Alan Bundy (Chair)	University of Edinburgh
Neil Murray (Secretary/Treasurer)	State University of New York at Albany
Deepak Kapur (CADE-11)	State University of New York at Albany
Mark Stickel (CADE-10)	SRI International
Ewing Lusk (CADE-9)	Argonne National Laboratory
Jörg Siekmann (CADE-8)	Universität Kaiserslautern

Programme chair

Alan Bundy   University of Edinburgh

Assistant to programme chair

Ian Green   University of Edinburgh

Programme committee

Luigia Carlucci Aiello	Università di Roma "La Sapienza"
Jürgen Avenhaus	Universität Kaiserslautern
David Basin	Max-Planck-Institut für Informatik, Saarbrücken
Wolfgang Bibel	Technische Hochschule Darmstadt
Francois Bry	Ludwig-Maximilians-Universität München
Ricardo Caferra	LIFIA-IMAG
Edmund Clarke	Carnegie Mellon University
Anthony Cohn	University of Leeds
William Farmer	The MITRE Corporation
Alan Frisch	University of York
Dov Gabbay	Imperial College
Harald Ganzinger	Max-Planck-Institut für Informatik, Saarbrücken
Stephen Garland	MIT
Fausto Giunchiglia	IRST & University of Trento
Mike Gordon	University of Cambridge
Elsa Gunter	AT&T Bell Laboratories
Masami Hagiya	University of Tokyo
Ryuzo Hasegawa	Institute for New Generation Computer Technology
Larry Henschen	Northwestern University
Larry Hines	University of Texas at Austin
Steffen Hölldobler	Technische Universität Dresden
Douglas Howe	AT&T Bell Laboratories
Matt Kaufmann	Computational Logic Inc.
Hélène Kirchner	CRIN-CNRS & INRIA Lorraine
Alexander Leitsch	Technische Universität Wien
Ursula Martin	University of St. Andrews
David McAllester	MIT
William McCune	Argonne National Laboratory
Michael McRobbie	The Australian National University
Tobias Nipkow	Technische Universität München
Hans-Jürgen Ohlbach	Max-Planck-Institut für Informatik, Saarbrücken

Programme committee (contd.)

William Pase — *ORA Canada*
Alberto Pettorossi — *Università di Roma Tor Vergata*
Frank Pfenning — *Carnegie Mellon University*
David Plaisted — *UNC Chapel Hill & MPI für Informatik, Saarbrücken*
Uday Reddy — *University of Illinois at Urbana-Champaign*
Michael Rusinowitch — *CRIN-INRIA*
Taisuke Sato — *Electrotechnical Laboratory*
Ken Satoh — *Fujitsu Laboratories Ltd.*
Natarajan Shankar — *SRI International*
John Slaney — *The Australian National University*
Alan Smaill — *University of Edinburgh*
Gert Smolka — *DFKI & Universität des Saarlandes, Saarbrücken*
Andrei Voronkov — *Uppsala University*
Richard Waldinger — *SRI International*
Lincoln Wallen — *University of Oxford*
Christoph Walther — *Technische Hochschule Darmstadt*

Referees

Alessandro Armando
Leo Bachmair
Jonas Barklund
Peter Barth
Antje Beringer
Bob Boyer
Stefan Brüning
Bruno Buchberger
Marco Cadoli
Alessandro Cimatti
Pierangelo Dell'Acqua
Jörg Denzinger
Francesco M. Donini
Uwe Egly
Norbert Eisinger
Amy Felty
Chris Fermüller
Hiroshi Fujita
Ulrich Furbach
Stefan Gerberding
Jürgen Giesl
Enrico Giunchiglia
John Gooday
Mark Grundy
Ursula Hans
Hirotaka Hara
Robert Hasker
Kouji Iwanuma
Jürgen Janas

Deepak Kapur
Hans Kleine Büning
Eckhard Koch
Christoph Kreitz
Sentot Kromodimoeljo
Reinhold Letz
Craig MacNish
Paolo Mancarella
M. Martelli
Ron van der Meyden
J Moore
Kunaki Mukai
Daniele Nardi
Paliath Narendran
Nicola Olivetti
Eugenio Omodeo
Uwe Petermann
William Pierce
Fiora Pirri
Katalin Prasser
Christian Prehofer
Maurizio Proietti
Martin Protzen
Thomas Rath
Horst Reichel
Michael M. Richter
Mark Saaltink
Ko Sakai

Gernot Salzer
Manfred Schmidt-Schauß
Peter H. Schmitt
Josef Schneeberger
Johann Schumann
Roberto Sebastiani
Luciano Serafini
Ute Sigmund
Don Simon
Mark Stickel
Jürgen Stuber
Sakthi Subramanian
Christian Suttner
Toshihisa Takagi
Kazuko Takahashi
Yukihide Takayama
Michael Thielscher
Paul Thistlewaite
Yozo Toda
Paolo Traverso
Ralf Treinen
Haruyasu Ueda
Adolfo Villafiorita
Jörg Würtz
Uwe Waldmann
Emile Weydert
Matt Wilding
Jia-Huai You
Bill Young

Invited speakers

Robert L. Constable	*Cornell University*
Ursula Martin	*University of St. Andrews*
Richard Platek	*Odyssey Research Associates & Cornell University*
John Slaney	*The Australian National University*

Local arrangements committee

Claude Kirchner (Chair)

Anne-Lise Charbonnier	Miki Hermann
Eric Domenjoud	Armelle Savary

Assistance with refereeing process

Josie Bundy	Weiru Liu
Ian Frank	Raul Monroy
Ian Gent	Santiago Negrete
Ian Green	Julian Richardson
Andrew Ireland	Alan Smaill
Ina Kraan	Geraint Wiggins

Sponsor

CADE Inc.

Host

INRIA Lorraine & CRIN

Co-sponsors

Department of Artificial Intelligence, University of Edinburgh	Institut National Polytechnique de Lorraine
CCL Esprit Working Group	Ministère de l'Enseignement Supérieur et de la Recherche
Computational Logic Inc.	
CNET	Sun Microsystems
CNRS-SPI	Université de Nancy I
Conseil Général de Meurthe et Moselle	Université de Nancy II
Conseil Régional de Lorraine	Ville de Nancy
District de l'agglomération Nancéenne	Ville de Villers lès Nancy

With the cooperation of

CEBEA	Nogema
Hewlett Packard	OCÉ
IBM	Summer Systems

The International Joint Conference on Artificial Intelligence, Inc.

Table of contents

Invited talk
The crisis in finite mathematics: Automated reasoning as cause and cure 1
   John Slaney (The Australian National University)

Heuristics for induction
A divergence critic 14
   Toby Walsh (INRIA Lorraine)

Synthesis of induction orderings for existence proofs 29
   Dieter Hutter (DFKI Saarbrücken)

Lazy generation of induction hypotheses 42
   Martin Protzen (Technische Hochschule Darmstadt)

Experiments with resolution systems
The search efficiency of theorem proving strategies 57
   David A. Plaisted (University of North Carolina at Chapel Hill)

A method for building models automatically: Experiments with an extension of OTTER 72
   Christophe Bourely, Ricardo Caferra and Nicolas Peltier (LIFIA-IMAG)

Model elimination without contrapositives 87
   Peter Baumgartner and Ulrich Furbach (Universität Koblenz)

Implicit vs. explicit induction
Induction using term orderings 102
   Francois Bronsard, Uday S. Reddy and Robert W. Hasker (The University of Illinois at Urbana-Champaign)

Mechanizable inductive proofs for a class of $\forall \exists$ formulas 118
   Jacques Chazarain and Emmanuel Kounalis (Université de Nice Sophia Antipolis)

On the connection between narrowing and proof by consistency 133
   Olav Lysne (University of Oslo)

Induction
A fixedpoint approach to implementing (co)inductive definitions 148
   Lawrence C. Paulson (University of Cambridge)

On notions of inductive validity for first-order equational clauses 162
   Claus-Peter Wirth and Bernhard Gramlich (Universität Kaiserslautern)

A new application for explanation-based generalisation within automated deduction 177
   Siani Baker (University of Cambridge)

Heuristics for controlling resolution

Semantically guided first-order theorem proving using hyper-linking 192
*Heng Chu and David A. Plaisted
(University of North Carolina at Chapel Hill)*

The applicability of logic program analysis and transformation to theorem proving 207
D.A. de Waal and J.P. Gallagher (University of Bristol)

Detecting non-provable goals 222
Stefan Brüning (Technische Hochschule Darmstadt)

Panel discussion

A mechanically proof-checked encyclopedia of mathematics: Should we build one? Can we? 237
Robert S. Boyer (University of Texas at Austin & Computational Logic, Inc.), N.G. de Bruijn (Eindhoven University of Technology), Gérard Huet (INRIA Rocquencourt) and Andrzej Trybulec (Warsaw University in Białystok)

ATP problems

The TPTP problem library 252
Geoff Sutcliffe (James Cook University), Christian Suttner and Theodor Yemenis (Technische Universität München)

Unification

Combination techniques for non-disjoint equational theories 267
Eric Domenjoud, Francis Klay and Christophe Ringeissen (CRIN-CNRS & INRIA Lorraine)

Primal grammars and unification modulo a binary clause 282
Gernot Salzer (Technische Universität Wien)

Logic programming applications

Conservative query normalization on parallel circumscription 296
Kouji Iwanuma (Yamanashi University)

Bottom-up evaluation of datalog programs with arithmetic constraints 311
Laurent Fribourg and Marcos Veloso Peixoto (LIENS Paris)

On intuitionistic query answering in description bases 326
Veronique Royer (ONERA, Chatillon) and J. Joachim Quantz (Technische Universität Berlin)

Applications

Deductive composition of astronomical software from subroutine libraries .. 341
Mark Stickel, Richard Waldinger (SRI International), Michael Lowry, Thomas Pressburger and Ian Underwood (NASA Ames Research Center)

Proof script pragmatics in IMPS 356
   William M. Farmer, Joshua D. Guttman, Mark E. Nadel and F. Javier Thayer (The MITRE Corporation)

A mechanization of strong Kleene logic for partial functions 371
   Manfred Kerber and Michael Kohlhase (Universität des Saarlandes)

Special-purpose provers

Algebraic factoring and geometry theorem proving 386
   Dongming Wang (Institut IMAG)

Mechanically proving geometry theorems using a combination of Wu's method and Collins' method 401
   Nicholas Freitag McPhee (University of Minnesota), Shang-Ching Chou and Xiao-Shan Gao (The Wichita State University)

Str+ve and integers 416
   Larry M. Hines (University of Texas at Austin)

Banquet speech

What is a proof? 431
   Richard Platek (Odyssey Research Associates & Cornell University)

Invited talk

Termination, geometry and invariants 432
   Ursula Martin (University of St. Andrews)

Rewrite rule termination

Ordered chaining for total orderings 435
   Leo Bachmair (SUNY at Stony Brook) and Harald Ganzinger (Max-Planck-Institut für Informatik, Saarbrücken)

Simple termination revisited 451
   Aart Middeldorp (University of Tsukuba) and Hans Zantema (Utrecht University)

Termination orderings for rippling 466
   David A. Basin (Max-Planck-Institut für Informatik, Saarbrücken) and Toby Walsh (INRIA Lorraine)

ATP efficiency

A novel asynchronous parallelism scheme for first-order logic 484
   David B. Sturgill and Alberto Maria Segre (Cornell University)

Proving with BDDs and control of information 499
   Jean Goubault (Bull Corporate Research Center)

Extended path-indexing 514
   Peter Graf (Max-Planck-Institut für Informatik, Saarbrücken)

Invited talk

Exporting and reflecting abstract metamathematics 529
Robert L. Constable (Cornell University)

AC unification

Associative-Commutative deduction with constraints 530
Laurent Vigneron (CRIN-CNRS & INRIA Lorraine)

AC-superposition with constraints: No AC-unifiers needed 545
Robert Nieuwenhuis and Albert Rubio (Technical University of Catalonia)

The complexity of counting problems in equational matching 560
Miki Hermann (CRIN & INRIA Lorraine) and Phokion G. Kolaitis (University of California at Santa Cruz)

Higher-order theorem proving

Representing proof transformations for program optimization 575
Penny Anderson (INRIA, Unité de Recherche de Sophia-Antipolis)

Exploring abstract algebra in constructive type theory 590
Paul Jackson (Cornell University)

Tactic theorem proving with refinement-tree proofs and metavariables 605
Amy Felty and Douglas Howe (AT&T Bell Laboratories)

Higher-order unification

Unification in an extensional lambda calculus with ordered function sorts and constant overloading 620
Patricia Johann and Michael Kohlhase (Universität des Saarlandes)

Decidable higher-order unification problems 635
Christian Prehofer (Technische Universität München)

Theory and practice of minimal modular higher-order E-unification 650
Olaf Müller (Forschungszentrum Informatik) and Franz Weber (Bertelsmann Distribution)

General unification

A refined version of general E-unification 665
Rolf Socher-Ambrosius (Max-Planck-Institut für Informatik, Saarbrücken)

A completion-based method for mixed universal and rigid E-unification 678
Bernhard Beckert (Universität Karlsruhe)

On pot, pans and pudding or how to discover generalised critical pairs 693
Reinhard Bündgen (Universität Tübingen)

Natural systems

Semantic tableaux with ordering restrictions 708
Stefan Klingenbeck and Reiner Hähnle (University of Karlsruhe)

Strongly analytic tableaux for normal modal logics 723
Fabio Massacci (Università di Roma "La Sapienza")

Reconstructing proofs at the assertion level 738
Xiaorong Huang (Universität des Saarlandes)

Problem sets

Problems on the generation of finite models 753
Jian Zhang (Academia Sinica)

Combining symbolic computation and theorem proving: Some problems
of Ramanujan 758
Edmund Clarke and Xudong Zhao (Carnegie Mellon University)

System descriptions

SCOTT: Semantically constrained OTTER 764
*John Slaney (The Australian National University), Ewing Lusk and
William W. McCune (Argonne National Laboratory)*

PROTEIN: A *PRO*ver with a *T*heory *E*xtension *IN*terface 769
Peter Baumgartner and Ulrich Furbach (Universität Koblenz)

DELTA — A bottom-up preprocessor for top-down theorem provers 774
Johann M. Ph. Schumann (Technische Universität München)

SETHEO V3.2: Recent developments 778
*Ch. Goller, R. Letz, K. Mayr and Johann M. Ph. Schumann
(Technische Universität München)*

KoMeT 783
*W. Bibel, Stefan Brüning, U. Egly and T. Rath (Technische Hochschule
Darmstadt)*

Ω-MKRP: A proof development environment 788
*Xiaorong Huang, Manfred Kerber, Michael Kohlhase, Erica Melis, Dan
Nesmith, Jörn Richts and Jörg Siekmann (Universität des Saarlandes)*

leanT^{A}P: Lean tableau-based theorem proving 793
Bernhard Beckert and Joachim Posegga (Universität Karlsruhe)

FINDER: Finite domain enumerator 798
John Slaney (The Australian National University)

Symlog: Automated advice in Fitch-style proof construction 802
Frederic D. Portoraro (University of Toronto)

KEIM: A toolkit for automated deduction 807
*Xiaorong Huang, Manfred Kerber, Michael Kohlhase, Erica Melis, Dan
Nesmith, Jörn Richts and Jörg Siekmann (Universität des Saarlandes)*

Elf: A meta-language for deductive systems 811
Frank Pfenning (Carnegie Mellon University)

EUODHILOS-II on top of GNU Epoch 816
*Takeshi Ohtani, Hajime Sawamura and Toshiro Minami
(Fujitsu Laboratories Ltd.)*

Pi: An interactive derivation editor for the calculus of partial inductive
definitions 821
    Lars-Henrik Eriksson (Swedish Institute of Computer Science)

Mollusc: A general proof-development shell for sequent-based logics 826
    Bradley L. Richards (Swiss Federal Institute of Technology), Ina Kraan
    (University of Zurich), Alan Smaill and Geraint A. Wiggins (University
    of Edinburgh)

KITP-93: An automated inference system for program analysis 831
    T.C. Wang and Allen Goldberg (Kestrel Institute)

SPIKE: A system for sufficient completeness and parameterized inductive
proofs 836
    Adel Bouhoula (CRIN & INRIA Lorraine)

Distributed theorem proving by *Peers* 841
    Maria Paola Bonacina (University of Iowa) and William W. McCune
    (Argonne National Laboratory)

Author index 847

The Crisis in Finite Mathematics: Automated Reasoning as Cause and Cure

John Slaney

Australian National University
Canberra 0200 Australia

1 Mathematics in Trouble

When electronic technology first became available to mathematicians working on finite structures, they welcomed the computer as the most useful tool for their purposes since old envelopes were invented. It could be used not only for arithmetic—as we all now use pocket calculators—but for a host of other operations such as generating structures from their descriptions, plotting graphs and suchlike visualisation aids, and, most pleasing of all, filling in the boring parts of those proofs (common in the field) which contain lines like 'There are 124 cases to consider...'. Of course, there was constant speculation as to whether "real" theorems would ever be proved purely mechanically, and if so whether human mathematics was on the verge of technological obsolescence, but generally—at first—such an idea was classed with those of machine-generated speech, self-focussing cameras, and satellite navigation, as blatant science fiction. To be sure, some doubted the reliability of mechanically computed results, and even refused to regard theorems as proved unless by human beings. Well, the Luddites, like the poor, are always with us. On the whole, however, the new tool was grasped eagerly and applied vigorously, and the field blossomed.

That was the state of things until roughly the mid-1970s. By then, computational techniques were in widespread use in fintite group theory, design theory, graph theory and many related studies. Of course, *applied* mathematics, statistics and the like already relied heavily on computation, and are unthinkable in their modern form without it, but the present talk is concerned rather with pure mathematics. There it seemed twenty years ago that the new machines were fitting neatly into the old disciplines, much to the benefit of the latter. The theory of finite structures is hard, and it is hard because they are complicated. Computers are better than we are at handling a certain amount of certain kinds of complexity, so they have found a niche in that kind of work from which they cannot be moved.

There was, however, a cloud on the horizon. For a proof to compel belief in what is proved, it should lie open to inspection. To be sure, nothing is totally sceptic-proof, and no mathematical process is immune to mistakes, but at least in mathematics as traditionally practiced the support for a theorem is a recipe for a train of thought, each step in which is available for tracing. Now where the proof involves automated reasoning essentially, in that human capacities are inadequate to the task of following the steps, this standard concept of mathematical proof is violated. Taking the working out of 124 cases on trust because of the behaviour of a machine is unlike doing so because your graduate students claim to have checked them, even though on the score of reliability computers have it over graduate students by a wide margin. With time, the level of computer involvement increased: for one thing, the hardware became more powerful and more readily available; for another, mathematicians themselves became computer literate and began to find computer-aided thought more natural; moreover, once the "easy" cases of a problem had fallen to a few minutes' cpu time, it seemed reasonable to spend a few hours on the next ones, then a few days... and now computations taking months are quite routine; finally, increasingly sophisticated and powerful automated reasoning software offered more in functionality and performance. This process of increasing reliance on mechanical assistance in turn increased the significance of the gaps in proofs where computer-generated fragments outstripped human verification capabilities. The cloud on the horizon grew alarmingly. The inevitable storm broke in 1977, following publication of the celebrated solution to the four-colour problem.[1]

It is not my present purpose to recapitulate the philosophical debate of that period, which tended to polarise around what seems to me the rather uninteresting question of the merits of the formal logical definition of proof as against human hand-waving. What should be noted is that the issue has not gone away, despite the waning interest in that particular theorem. On the contrary, it has deepened to the point where many mathematicians are seriously unsure of the attitude they should take to their own discipline. The disquiet has even spilled over into the popular press. In 1991 the New York Times carried an article by Gina Kolata headed *Computers Still Can't Do Beautiful Mathematics*. This reported the action of Fan Chung, editor of the *Journal of Graph Theory*, in making a test case of a paper by McKay and Radziszowski on some finite cases of Ramsey's theorem. She challenged the editorial board with the question: 'Is an important result with a computer-assisted proof acceptable? ... how should the paper be evaluated knowing that it is almost impossible for the referee to verify its correctness?' The replies she received revealed no universal agreement except on the point that the question is seriously worrying. Now it is of course unfair to score cheap philosophical points off a newspaper article, but that particular one is recommended as an illustration of most of the confusions it is possible to commit with regard to its subject (starting, as Larry Wos immediately pointed out, with its blatantly false headline). Still more recently, *Scientific American* devoted a full-length article to 'The Death Of Proof'[8] noting many examples of 'experimental mathematics' including another theorem of McKay and Radziszowski[7] that the largest simple graph containing no 5-clique and no 4-independent set has 24 nodes (a solution to what is popularly called the 5,4 Party Problem). The main emphasis of the *Scientific American* article is on the blurring of the distinction between mathematics and the empirical sciences. I shall return to the issues in the final section below; for the present it is necessary only to note the articles as symptoms of a field in crisis.

---

[1] See [1]. The computational proof has been independently reproduced many times.

2 A Case Study

It is instructive, and valuable for the purposes of the argument, to illustrate with a concrete example. A very suitable one is provided by a current research project in which I have the good fortune to be involved, concerning the existence problems for a range of finite algebras.[2] This section outlines the problem set and the recent computational attack on it by members of the Automated Reasoning community.

The objects whose existence is in doubt are quasigroups certain special kinds. A quasigroup is simply a groupoid (a set on which is defined a binary operation) such that the equations $ax = b$ and $xa = b$ have unique solutions for all elements a and b. That is, the "multiplication table" of a quasigroup is a Latin square; each row and each column is a permutation of the elements. Most of the problems of recent interest concern idempotent quasigroups, which are those satisfying the condition $xx = x$.

A quasigroup $\langle S, \circ \rangle$ has six conjugates corresponding to the six permutations of the variables x, y and z in the expression $x \circ y = z$ as follows. The following are all equivalent to $x \circ y = z$:

$x \circ_{123} y = z$ $\qquad$ $x \circ_{132} z = y$ $\qquad$ $z \circ_{312} x = y$

$y \circ_{213} x = z$ $\qquad$ $y \circ_{231} z = x$ $\qquad$ $z \circ_{321} y = x$

We refer to the quasigroup $\langle S, \circ_{ijk} \rangle$ as the (i,j,k)-conjugate of $\langle S, \circ \rangle$.

Two quasigroups $\langle S, \circ \rangle$ and $\langle S \star \rangle$ are said to be orthogonal if there are no a, b, c and d in S, $a \neq b$ and $c \neq d$ such that $a \circ c = b \circ d$ and $a \star c = b \star d$. A well-known fact, first conjectured by Euler, is that there are no orthogonal quasigroups of order 6. There are also none of order 2, but they exist for all other orders.

It sometimes happens that a quasigroup is orthogonal to one (or more) of its own conjugates. Such a quasigroup, orthogonal to its (i,j,k)-conjugate, is called an (i,j,k)-conjugate-orthogonal Latin square, or (i,j,k)-COLS. If it is idempotent, it is a COILS rather than just a COLS, and where it is of order v it is an (i,j,k)-COLS(v) or an (i,j,k)-COILS(v). In general, the problem of discovering such objects is highly nontrivial, though they do exist for almost all orders.[3]

Our recent research has concentrated on eight problem classes which we have dubbed QG1–QG8. The existence problem of order v for QGi we call QG$i.v$. The default is to impose idempotence as an extra condition; where this is not done, we append 'ni' (non-idempotent). There is no significance to the numbering of 1...8, and nor do we think that these are the only problems in the area usefully approached by computational means. The problems are as follows.

---

[2] The research to the end of 1992 is reported in [3] and that to mid-1993 in [9]. For the immediate mathematical background see [2].

[3] The exact spectra for the various (i,j,k)-COLS and COILS are given in [2] and in [9].

QG1 Determine the spectrum of (3,2,1)-COILS. The only open problem is order 12. This appears to be very difficult. We have not succeeded in exhausting the search space for any problem larger than QG1.9.

QG2 Determine the spectrum of (3,1,2)-COILS. Of the two open problems, QG2.10 and QG2.12, one has been closed as a corollary to our results for QG4 (below). There exists a (3,1,2)-COILS(12). Order 10 remains open.

QG3 Determine the spectrum of [idempotent] models satisfying the equation $xy.yx = x$ known as Schröder's second law. All models of this equation are (2,1,3)-COLS. The one previously open problem, again order 12, was solved positively by the generation of idempotent quasigroups satisfying the equation.

QG4 Determine the spectrum of Stein's third law $xy.yx = y$ in [idempotent] quasigroups. Again all models are (2,1,3)-COLS, again order 12 was the open problem, and again we solved it positively. These results for QG3 and QG4 are mathematically significant, as they complete the spectra of several associated structures in design theory, for details of which see [2].

QG5 Determine the spectrum of the equation $(yx.y)y = x$. Especially, investigate its spectrum in idempotent quasigroups. Here there are many open problems. Our new results are all negative: there is no solution of order 10 or 14, and there is no idempotent solution of order 9, 10, 12, 13, 14, 15 or 16. Order 18 is now the smallest open case.

QG6 Investigate the spectrum of the equation $xy.y = x.xy$. Is there a solution of order 9, 12 or 17? Are all solutions of orders congruent to 0 or 1 (mod 4)? It is easy to see that all models are idempotent, so there is no separate non-idempotent problem for QG6. We found a model of order 9, and obtained negative results for orders 7, 10, 11, 12, 14 and 15. Order 17 remains an open problem.

QG7 Investigate the spectrum of the equation $(xy.x).y = x$.[4] Is it restricted to orders congruent to 1 (mod 4)? Also, is there a model of order 33? Here again we obtained only negative results, for all orders up to 15 except for those congruent to 1 (mod 4). As in the case of QG6, all models are idempotent. Order 33 is too big for our methods at present.

QG8 Investigate the spectrum of Stein's first law $x.xy = yx$, or its conjugate equivalent $(x.xy)y = x$ which is more convenient for our computational approach. Before we started, there were 23 open problems—the smallest QG8.15 and the largest QG8.126. Now there are 22, since we have a negative result for order 15.

A number of programs have been used, differing from each other in several respects. Among the researchers and programs involved are the following.

- In 1990 Jian Zhang (Beijing) used a finite domain constraint satisfaction method to solve QG5.9 which was then an open problem. He has since used the same software to corroborate results of other researchers.

---

[4] [3] uses the equation $yx.y = x.yx$ instead. Its spectrum is known to be the same.

- Masayuki Fujita (ICOT/MRI, Tokyo) applied the ICOT group's theorem prover MGTP-G to quasigroup problems starting in 1991. His early work is summarised in [3]. The QG problems provided a significant spur to the development and refinement of MGTP-G, which differs from most of the other programs used in that it does not represent the problems as ground (propositional) ones but in terms of first order clauses from which it reasons by hyper-resolution after case-splitting on appropriate instances. Parallel processing on ICOT's Parallel Inference Machines was quite important to the success of MGTP. MGTP solved the open problems QG5.10, QG5.12, QG5.10ni, QG6.7, QG6.9, QG6.10, QG6.11, QG6.12, QG7.7, QG7.8 and QG7.10.
- The author's model-generating program FINDER resembles that of Jian Zhang in that it is dedicated to finite constraint satisafaction problems. FINDER has also been made much more efficient as a result of experiences with the QG problems. Its basic inference mechanism is case analysis to deal with the positive clauses in the problem representation, followed by unit resolution to reduce the search space. It also deduces certain extra information from failed branches in the search tree. It is quite fast, partly because of careful coding in a low-level language (C), but is actually the least efficient of the successful programs used. Open problems first closed by FINDER include QG3.12, QG7.11, QG7.14 and QG7.15, and it has confirmed most of the results of the other researchers.
- Mark Stickel (SRI) and Hantao Zhang (Iowa) have collaborated on a series of implementations of the Davis-Putnam algorithm for propositional SAT problems. Of all researchers in the program, they have been the most vigorous in pursuing new results. Stickel's DDPP (Discrimination Tree based Davis-Putnam Prover) and the considerably faster LDPP (Linear-list-based Davis-Putnam Prover) are written in LISP, while Zhang has implemented a very similar algorithms in his C program SATO (Satisfiability Testing Optimized) which has been parallelised using the P4 package from Argonne.[5] The hardest problems we have solved, including the monstrous QG5.14ni, were done by SATO. DDPP solved the previously open QG4.12, QG5.13, QG5.14, QG5.15 and QG7.12. SATO has added QG5.14ni, QG5.16, QG6.14.and QG6.15.
- Mark Wallace and Micha Meier (ECRC, Munich) have investigated some of the QG problems using the constraint logic programming system ECLIPSE. They obtained interesting effects, including the first complete traversal of the search space of QG1.9. The constraint solver used for these problems within ECLIPSE is based on arc consistency.
- Ryuzo Hasegawa (ICOT, Tokyo) has experimented with a constraint propagation method coded in Prolog, and also with an extension of MGTP to do constraint logic programming. He reports extremely encouraging results, especially on QG5, where his programs achieve the smallest search space, in terms of the number of branches, of any we have seen.

There is insufficient time and space to give a detailed presentation of any of

---

[5] Ewing Lusk and William McCune helped to make this work possible.

the programs. For a good account of the work of Zhang and Stickel, see [10]. For some more remarks on this and on MGTP-G and FINDER, see [9]. It is worth noting that some GSAT experiments were performed on the QG problems by Toby Walsh and Ian Gent (Edinburgh) in order to discover whether they were appropriate for hill climbing methods or the like. The results were quite negative, even where many solutions exist in the search space. This is interesting, if disappointing. We have yet to understand fully why it happens.

The new results are of two radically different kinds, though generated by the same search methods. In some cases, such as QG3.12, QG4.12 and QG6.9, positive solutions were found. In most cases, however, such as the open problems from QG5 and QG7, our results were negative. There we can only report that we looked everywhere but found nothing. It is this latter kind of result that is of interest for the purposes of the present talk. Where a structure is found, it can be displayed and, being rather small, it can be checked—by hand if necessary, or by machine if you value your sanity—for satisfaction of the defining conditions. This is not possible in the negative cases, some of which resulted from the equivalent of some cpu months of computation and involved the closing of many millions of subcases.

3   Searching and Logic

It is apparent that the algorithms used in searching, such as the Davis-Putnam algorithm, can function as propositional logic theorem provers. It is worth emphasising that, for a large class of these algorithms the moves made in conducting a search correspond precisely to very familiar sorts of logical inference. To indicate briefly how this is the case, we begin with an outline of constraint satisfaction problems (CSPs) over finite domains.

A CSP is defined by a finite set N of nodes, a finite set $D(n)$ called the *domain* of n associated with each node n and a set of constraints. We shall think of the constraints negatively as sets C of pairs $\langle n, d \rangle$ where n is a node and $d \in D(n)$, such that if $\langle n, d_1 \rangle$ and $\langle n, d_2 \rangle$ are in C then $d_1 = d_2$. Intuitively, the nodes are variables describing a state, the domains are the possible values these can take, and a negative constraint is an assignment of values which may not all be made together in any consistent state. The problem is to enumerate the consistent states, or perhaps just to check for the existence of a consistent state. Formally, a state is a function mapping each node into its domain, and a state is consistent if it has no constraint as a subset.

There are many algorithms for solving CSPs, and it is no part of the purpose of this talk to survey the voluminous literature devoted to them.[6] What is to the point is to indicate a certain *rapprochement* between the fields of CSP and Automated Deduction. In general, CSP algorithms have two parts. They work with some notion of the consistency of a state space which reduces to consistency as defined above in the case that every domain is a singleton. Then one part

---

[6] For a general introduction, see the article on Constraint Satisfaction in the *Encyclopedia of Artificial Intelligence*.

of the algorithm removes values from domains until the state space becomes consistent; it may also refine the constraints in various ways. The second aspect of the algorithm is some technique for backtracking from one state subspace to another, constructing the search tree as it does so. Often, as in our programs MGTP and FINDER, this is a matter of choosing a node with more than one value in its domain, splitting the search space by setting the domain to the singleton of each value in turn and in each case recursively searching the resulting subspace. In the case of DDPP, the technique is to choose one value from the domain of the node and split by asserting first that that is the only value for the node and then that that value is not among the possibilities for the node.

To represent a CSP as a propositional logic problem in clause form is very easy. We need constants to name the nodes and the values in their domains. This is readily arranged, for instance by numbering everything and using natural numerals as names. An atomic formula is the assertion that a particular node n has a particular value d, which we may write $p(n,d)$. Now the specification of the domains just consists of a set of positive clauses, one for each node:

$$p(1, v_1^1) \vee \ldots \vee p(1, v_{k_1}^1)$$
$$\vdots$$
$$p(m, v_1^m) \vee \ldots \vee p(m, v_{k_m}^m)$$

A negative constraint
$$\{\langle n_1, d_1 \rangle, \ldots, \langle n_k, d_k \rangle\}$$
on the other hand, is represented logically as a negative clause:

$$\neg p(n_1, d_1) \vee \ldots \vee \neg p(n_k, d_k)$$

A consistent state is then exactly a model of this set of clauses, in the normal logical sense. Note that all of the clauses are ground and that the domain specification clauses are non-Horn.[7]

In the logical recension of the CSP, the operation of dividing the search space and recursively searching the resulting subspaces is exactly the familiar logical technique of case-splitting on positive clauses in order to search for a model. Manthey and Bry's SATCHMO, for example, does exactly that. In other words, CSP algorithms may be construed formally as constructing a semantic tableau, branching on positive disjunctions as required. On such an account, the consistency phase of a CSP search corresponds to the application of α rules to extend a branch without dividing it.

In the cases of FINDER and DDPP, constraint propagation is a matter of appealing to constraints of cardinality 2 in order to remove possible values from the domains of nodes. That is, where we have a constraint

$$\{\langle n_1, d_1 \rangle, \langle n_2, d_2 \rangle\}$$

---

[7] Some of our programs, including FINDER and the Davis-Putnam implementations, allow positive literals to occur in constraints, but this refinement need not concern us for the purposes of the present explanation.

and have either assumed or deduced that the value assigned to n_1 is d_1, clearly any state in which d_2 is also assigned to n_2 is inconsistent, so we may remove d_2 from the domain of n_2 until such time as the search backtracks again to this point. Constraint refinement is more of the same in a sense: given a constraint

$$\{\langle n_1, d_1 \rangle, \langle n_2, d_2 \rangle \ldots, \langle n_k, d_k \rangle\}$$

for some $k > 2$, where again n_1 is given the value d_1, it is clear that within the subspace to be searched next

$$\{\langle n_2, d_2 \rangle \ldots, \langle n_k, d_k \rangle\}$$

is a constraint of cardinality $k - 1$.

It is evident that this simple move of constraint refinement coresponds exactly on the logical reading to a step of unit resolution:

$$\frac{\neg p(n_1, d_1) \vee \ldots \vee \neg p(n_k, d_k) \quad p(n_1, d_1)}{\neg p(n_2, d_2) \vee \ldots \vee \neg p(n_k, d_k)}$$

The constraint propagation inference is just two unit resolution steps linked together thus:

$$\frac{p(n_1, d_1) \vee \ldots \vee p(n_1, d_k) \quad \neg p(n_1, d_1) \vee \neg p(n_2, d) \quad p(n_2, d)}{p(n_1, d_2) \vee \ldots \vee p(n_1, d_k)}$$

The other types of consistency reasoning, such as MGTP's hyper-resolution, clearly also correspond to logical inferences of fairly standard sorts. Arc consistency, as used in many constraint solvers, is based on unit-resulting negative hyper-resolution linked to binary resolution:

$$\frac{p(n_1, d) \vee P \\ p(n_2, d_1) \vee \ldots \vee p(n_2, d_k) \\ \{\neg p(n_1, d) \vee \neg p(n_2, d_i) : 1 \leq i \leq k\}}{P}$$

FINDER, as noted, also adds constraints as the search progresses, but this process too corresponds to a simple propositional inference: to negative hyper-resolution in fact:

$$\frac{p(n, d_1) \vee \ldots \vee p(n, d_k) \\ \{\neg p(n, d_i) \vee P_i : 1 \leq i \leq k\}}{P_1 \vee \ldots \vee P_k}$$

Hence in every case the search tree really can be construed as a tableau-like proof in which the α rules are variants of resolution. This fact is perhaps rather obvious, especially in this community, but it is worth labouring the point because we want to milk it a little for its philosophical import.

4 Automated Reasoning and Proof

Armed with a clear example of a series of computer-generated results, we may now return to the difficult question of the relationship between Automated Reasoning and the crisis in finite mathematics. The quasigroup theorems afford a better illustration than better known ones such as the four colours or the nonexistence of a projective plane of order 10, since they were generated purely by computation from a very simple problem axiomatisation, rather than by clever human mathematics merely assisted by computation. If it is found useful to focus even more narrowly on a single result, let this be Hantao Zhang's demonstration that there is no model of QG5.14, even without the assumption of idempotence. The computation took the equivalent of around 16 months of Sparc-2 cpu time distributed over many machines during a 5-month period. As SATO explores some 300,000 branches per hour, the search tree may be calculated to contain in the region of 3.5×10^9 branches.

One natural reaction to such essentially computational proofs is to see them as forcing a significant change to the concept of proof, in that they require explicit reference at least to an algorithm and maybe also to its implementation. That is, it is natural to regard the verification of the program as part of the proof. If this natural thought is right, then the nature of mathematics is indeed affected by the contribution from Automated Reasoning since theorems of quasigroup theory (in our example) require for their support part of the formal theory of algorithms and computation. Kolata's New York Times article [4] expresses it clearly: 'In a sense, the program itself becomes part of the proof, and few mathematicians are equipped to check its reliability.'

Now program verification is very important for several reasons, some of which will be revisited below, but the suggested contribution to mathematical proof is not one of them. It is my view that the above 'natural reaction' rests on a conceptual confusion about the nature of proof and its relation to practice.

As we all learned in our undergraduate logic training, a finitary proof (and infinitary ones are not in question here) is a finite tree whose nodes are formulae of some language, whose arcs are instances of rules of inference and whose leaves are axioms of the theory in which the proof takes place. Now a proof in this formal sense is an abstract object. Rarely do mathematicians exhibit such proofs in full. Instead they sketch them. Often they say things like '... and the other cases are similar' or 'verification is left as an exercise for the reader' or 'for $k < 3$ the result is trivial'. What this kind of demonstration amounts to is not a formal proof but a reason for thinking that a formal proof exists. Since the formal proof is an abstract object, its existence does not imply natural instantiation, either on paper or in anyone's mind.

Another common response, to be distinguished from the above 'natural reaction', is to regard computer-generated results not as proofs but as reports of experiments which yield evidence for the existence of proofs. This at least seems to have the epistemology approximately right. The evidence is empirical, just like any other experimental data, though the proof, if one does exist, is necessarily valid and its conclusion necessarily true just like any other true proposition of pure mathematics. My difficulty with this response is not just that it severely understates the case, but that it misses a significant point. There is a clear distinction to be drawn between what really are instances of mathematical evidence and what are just long proofs. This distinction is notably *not* drawn in the *Scientific American* article of last October. The work of McKay and Radziszowski on the Party Problem offers a clear example. They produced a computer demonstration, in a few cpu months, that $R(5,4)$ is 25. They have also investigated $R(5,5)$, the smallest number of people you have to invite to a party in order to ensure either a clique of 5 who are all acquainted or a set of 5 who are total strangers to each other. It is known that $43 \leq R(5,5) \leq 49$. Their work strongly suggests that $R(5,5) = 43$ since about 2500 (5,5,42)-graphs have been generated by simulated annealing from random starting points, and all have proved to be among the 328 known such graphs, none of which can be extended to a (5,5,43)-graph. McKay estimates the probability of this happening by chance if $R(5,5) = 44$ at less than 0.0006. Thus the experimental results would be very unlikely if $R(5,5)$ were 44, and almost inconceivable if it were greater. In the light of the evidence, no-one would rationally bet against the theorem at everyday odds, yet we do not count it as proof and do not regard the problem as solved.

It seems to me that the difference between genuinely experimental mathematics and merely computer-generated mathematics is crucial. There is something non-constructive about experimental evidence for the existence of a proof, based on detecting effects of the theorem. This is quite apart from the point that such evidence is inconclusive. In contrast, the argument from failed search is perfectly constructive, since if the procedure followed was correct then a proof in the formal sense was actually instantiated in the sequence of states of the machine. In a case such as the quasigroup one, belief in the proof is the basis for belief in the theorem, whereas in the 5,5 Party Problem case the probabilistic evidence for the proposition is the ground for thinking that it has a proof. The difference is in direction of conceptual priority, not in degree of support. This difference, however, points to a genuine ground for disquiet about proofs such as ours. Where a proof is used to underpin belief in the proposition proved, we cannot thus gain knowledge of the proposition unless our belief in the correctness of the proof is well founded; and unless the proof is accessible to us, this is not obviously the case. The normal point of exhibiting a proof in support of a theorem is to allow the inference to that theorem to be carried out by tracing the proof steps. In this respect unsurveyably long proofs *are* different from short ones. There is still an available inference to the theorem that there is no Bennett quasigroup of order 14, but it is an inference starting from our belief in the correctness of Zhang's SATO program together with the observed result of the computation. Thus while program correctness is no part of the proof, belief in it certainly may be part of the grounds for belief in the existence of the proof and therefore in the theorem proved.

For that reason, it is valuable to confirm our results by searching with independent programs. Almost all of the QG results have been obtained by two or more of our programs, instantiating significantly different algorithms in different languages, written by different people and run independently on different machines. Of course, part of this corroboration process is to check not only that there is agreement on the negative results, but also that the positive results are the same.

For that reason, too, program verification is important not just to end users with peculiar applications such as Air Traffic Control but to pure mathematics itself. As [5] points out, verification of the code does not itself suffice to establish the result beyond all possible doubt, for hardware malfunctions, noise on networks and even unfortunate strikes by cosmic rays can cause output or lack of it even where the software is perfect. Moreover, an error of that sort in a computer proof may be extremely difficult to detect in comparison with the much more common errors made by human beings. However, as I have been at pains to stress, program verification does not contribute to the proof of a theorem such as our quasigroups example, so the possibility of error that it leaves open does not contaminate everything. Program verification in these cases contributes rather to the level of confidence which it is appropriate to have in the results of computations; and if this is increased to the point where the probability of damage by cosmic rays is seriously relevant then it is about as good as anyone could reasonably want. Verification of nontrivial programs, however, is very difficult. There is no prospect in the near future that FINDER's 7000 lines of C code will be completely verified, though the chances of verifying DDPP's few pages of Lisp are perhaps rather better. Even there, though, the program is quite experimental and subject to change as new heuristics and similar ideas are tried, and realistically we do not expect to see a satisfactory proof of correctness in the kind of detail that would contribute greatly to the present exercise. Fortunately, I have a Modest Proposal as to a better way forward.

Since the search follows the steps of a simple, if tedious, proof, it costs little to have the searching program dump a trace of the proof as it goes. This proof would be too long for human evaluation, but could be verified directly by passing it through an independent proof checker. Not every step would have to be verified, as most of the resolution inferences do not participate in branch closure and therefore do not have to be recorded. Moreover, since the mechanisms and heuristics for selecting clauses for case analysis and so forth do not affect the validity of the proof, they do not have to be recorded, checked or duplicated in the checker. Thus the proof checker can be a much simpler and much more obviously correct piece of software than the searcher. The checking process would be another very long computation in one of the difficult cases, but would be composed of very simple steps such as verifying that what is claimed to be an instance of unit resolution, negative hyper-resolution or whatever really is one. It does seem that the branching and backtracking of the search tree would have to be reproduced in the checker, though the latter does not have to decide how the branching is to be done but only to follow it, noting whether it is done in accordance with the rules.

This, then, is the present Modest Proposal: that we verify the proofs rather than the programs which produce them. After all, it does not actually matter to the mathematicians who are our clients how many bugs our programs contain, so long as the proofs are correct. In the same way, fortunately, no-one needs to verify the brain processes of a mathematician—a theorem prover (with additional functionality) instantiated in meat—in order to evaluate its output. It appears to be much easier to verify proofs than programs. Well, at least in principle, and not too rarified a principle at that, we know how it is done. Of course, those who object to computational results on ideological or mystical grounds, as not having received the breath of life from a warm and cuddly graph theorist in contrast to an emotionless workstation, will continue to object to computer verification of the results. Well, they will have to shop in a different store. For those who only want overwhelming evidence that a proof has actually been found, I think we have just the product.

It seems that the difficulty of checking proofs as long as that of QG5.14ni is caused by the complexity of the branching structure, not by the moves which extend branches. That is, the β rules, not the α rules, cause the main problem. Here, though, there are possibilities for interesting research. It is clear from examination of the behaviour of machines performing the QG searches that many branches of the search tree have very similar structure. This suggests that there is a great deal of symmetry in the search and therefore in the proof. Techniques for symmetry detection and its application to proof compression may therefore yield large rewards in the proof checking exercise. Moreover, understanding of the symmetries in the proofs could lead to a better understanding of those in the search spaces, leading in turn to more efficient searches for future proofs to check. I consider symmetry to be one of the most important topics of current research in ground theorem proving, and one which holds out the most hope for really significant advances in finite constraint satisfaction as well as in proof verification. More generally, the quasigroup problems and similar ones in graph theory and the like offer a splendid test set for all manner of proof compression methods. In urging this community to dedicate research effort to proof checking, therefore, I am not necessarily recommending the substitution of tedium for excitement. On the contrary, I am indicating an opportunity for solid achievement which will be of serious value both to Automated Reasoning and to the fields of mathematics which it now serves. Please see to it.[8]

---

[8] The material in this paper is similar to, and in places closely follows, that in [9] which contains a rather fuller account of the QG problems and of the programs used to attack them. Since last year, when [9] was written, I have come to a somewhat clearer view of the conceptual issues surrounding computational proof, so the present exercise represents another attempt to explore them, this time in the rather more relaxed format of an invited address. I am personally grateful to Larry Wos for prompting me to write up the quasigroups work, and to CADE for the invitation to present this talk.

References

1. K. Appel & W Haken, *Proof of the 4-Color Theorem*, **Discrete Mathematics**, 16 (1977), pp. 179–180.
2. B. Benhamou & L. Sais, *Theoretical Study of Symmetries in Propositional Calculus and Applications*, **Proc. 11th International Conference on Automated Deduction** (1992), pp. 281–294.
3. F. Bennett & L. Zhu, *Conjugate-Orthogonal Latin Squares and Related Structures*, **Contemporary Design Theory: A Collection of Surveys**, ed. J. Dinitz & D. Stinson, New York, Wiley, 1992.
4. M. Fujita, J. Slaney & F. Bennett, *Automatic Generation of Some Results in Finite Algebra*, **Proc. 13th International Joint Conference on Artificial Intelligence** (1993), pp. 52–57.
5. G. Kolata, *Computers Still Can't Do Beautiful Mathematics*, **New York Times, Week in Review** §E, Sunday, July 14, 1991, p.4.
6. C. W. H. Lam, L. Thiel & S. Swiercz, *The Non-existence of Finite Projective Planes of Order 10*, **Canadian Journal of Mathematics** 41 (1989), pp. 1117–1123.
7. A. Mackworth, *Constraint Satidfaction*, **Encyclopedia of Artificial Intelligence** (ed. Shapiro) 1987, pp. 205–211.
8. R. Manthey & F. Bry, *SATCHMO: A Theorem Prover Implemented in Prolog*, **Proc. 9th International Conference on Automated Deduction** (1988), pp 415–434.
9. B. McKay & S. Radziszowski, $R(4,5) = 25$, **Journal of Graph Theory**, forthcoming.
10. J. Horgan, *The Death of Proof*, **Scientific American**, 237.4 (October 1993), pp. 74–82.
11. J. Slaney, M. Stickel & M. Fujita, *Automated Reasoning and Exhaustive search: Quasigroup Existence Problems*, **Journal of Computers in Mathematics with Applications**, special issue on Automated Reasoning, forthcoming.
12. H. Zhang & M. Stickel, *Implementing the Davis-Putnam Algorithm by Tries*, typescript, University of Iowa, 1994.

A Divergence Critic

Toby Walsh*
INRIA-Lorraine
615, rue du Jardin Botanique, B.P. 101
F-54602 Villers-les-Nancy, France
walsh@loria.fr

Abstract. Inductive theorem provers often diverge. This paper describes a critic which monitors the construction of inductive proofs attempting to identify diverging proof attempts. The critic proposes lemmas and generalizations which hopefully allow the proof to go through without divergence. The critic enables the system SPIKE to prove many theorems completely automatically from the definitions alone.

1 Introduction

Rippling is a powerful heuristic developed at Edinburgh for proving theorems involving explicit induction [6]. The essential idea behind rippling is to remove the "difference" between the induction conclusion and the induction hypothesis using a very goal directed form of rewriting. When augmented with a "difference matching" procedure to identify such differences, rippling has also proved useful in domains outside of explicit induction. For example, it has been used to sum series, to prove limit theorems, and to perform normalization [3, 15, 16]. In this paper, I describe an experiment to apply rippling and difference matching to a new domain, the problem of overcoming the divergence of a prover using implicit induction. The experiment is very successful. A critic has been implemented which is often able to identify diverging proof attempts and to propose lemmas and generalizations which will allow the proof to go through successfully. Although the critic is designed to work with SPIKE, a theorem prover which uses implicit induction [4], the critic should also work with other implicit and explicit induction provers.

Induction in SPIKE is performed by means of test sets. A test set is essentially a finite description of the initial model. It is a more powerful concept than the related notion of cover set since, in combination with a ground convergent rewrite system, a test set can be used to refute false conjectures. The basic idea in SPIKE is to instantiate induction variables in the conjecture to be proved with the members of the test set (using the `generate` rule), and then to use rewriting to simplify the resulting expressions (using the `simplify` rules). This rewriting uses any of the axioms, lemmas and induction hypotheses provided they are smaller (with respect to a well-founded relation) than the current conjecture. Although SPIKE has proved several challenging theorems without assistance (*eg.* the binomial theorem), its attempts to prove many theorems diverge without an appropriate generalization or the addition of a suitable lemma. The aim of the

---

* Current address: IRST, Location Panté di Povo, I38100 Trento, Italy. toby@irst.it

critic described in this paper is to identify when a proof attempt is diverging, and to speculate a lemma or generalization which would prevent this divergence. The identification of divergence is described in more detail in the next two sections, and the speculation of an appropriate generalization or lemma is described in the following three sections.

2 Divergence Analysis

Several properties of rewrite rules have been identified which give rise to divergence (*eg.* forwards and backwards crossed systems [8]). However, these properties fail to capture all diverging rewrite systems since the problem is, in general, undecidable. The divergence critic proposed here studies just the proof attempt looking for patterns of divergence; no attempt is made to analyse the rewrite rules themselves for structures which give rise to divergence. The advantage of this approach is that the critic need not know the details of the rewrite rules applied, nor the type of induction being performed, nor the control structure used by the prover. The critic can thus recognise divergence patterns arising from complex mutual or multiple inductions with little more difficulty than divergence patterns arising from simple straightforward inductions. The disadvantage of this approach is that the critic will sometimes identify a "divergence" pattern when none exists. Fortunately, such cases appear to be rare, and even when they occur, the critic usually suggests a lemma or generalization which gives a shorter and more elegant proof.

The divergence critic attempts to find term structure introduced by induction which is accumulating in an equation and which is preventing simplification. To do this, the critic partitions the sequence of diverging equations generated by SPIKE. This is necessary since several diverging sequences can be interleaved in the prover's output. Several heuristics are used to reduce the number of partitions considered. The main heuristic is parentage; that is, the critic partitions the sequence so that each equation in a partition is derived from the previous one. Other heuristics which can be used include: the function and constant symbols which occur in one equation occur in the next equation in the partition, and the weights of the equations in a partition form a simple arithmetic progression. The critic then attempts to find term structure introduced by induction which is accumulating at some position in the equation and which is causing divergence. To identify such accumulating term structure, the critic uses the difference matching procedure introduced in [2]. This is explained in more detail in the next section. To fix divergence, the critic then speculates a lemma or generalization of the theorem which moves this accumulating term structure out of the way.

3 An Example

To illustrate the essential ideas behind the critic's divergence analysis, consider SPIKE's attempt to prove the length-append theorem.

$$len(app(a, b)) = len(app(b, a))$$

SPIKE begins by applying the **generate** rule. This instantiates the induction variables with members of the test set, $\{nil, cons(h, t)\}$. Up to variable renaming, this gives 3 distinct equations, which are rewritten by the **simplify** rules using the definitions of *len* and *app* to give a simple identity and the following 2 equations,

$$len(app(b, nil)) = s(len(b))$$
$$len(app(a, cons(d, b))) = len(app(b, cons(c, a)))$$

Since no further simplification can be made, SPIKE performs another induction by applying the **generate** rule. After simplification, this gives,

$$len(app(b, cons(c, nil))) = s(s(len(b)))$$
$$len(app(a, cons(f, cons(d, b)))) = len(app(b, cons(e, cons(c, a))))$$

No further simplification can be performed so SPIKE again applies the **generate** rule. Unfortunately, the proof attempt will continue to diverge like this ad infinitum.

Using the parentage heuristic, the divergence critic partitions the equations produced into two sequences. The first sequence is,

$$len(app(a, b)) = len(app(b, a))$$
$$len(app(a, cons(d, b))) = len(app(b, cons(c, a)))$$
$$len(app(a, cons(f, cons(d, b)))) = len(app(b, cons(e, cons(c, a))))$$
$$\vdots$$

The second sequence is,

$$len(app(b, nil)) = s(len(b))$$
$$len(app(b, cons(c, nil))) = s(s(len(b)))$$
$$len(app(b, cons(d, cons(c, nil)))) = s(s(s(len(b))))$$
$$\vdots$$

The critic then attempts to find the accumulating term structure in each sequence which is causing divergence. In this case, in both sequences, cons functions are accumulating on the second argument of append. Since append is defined recursively on its first argument, SPIKE is unable to simplify such terms, and the proof diverges. To identify this accumulating term structure, the critic uses difference matching. This procedure annotates terms with wavefronts, boxes with holes which mark where the terms differ. For example, taking the first sequence, difference matching successive equations gives the annotated sequence,

$$len(app(a, b)) = len(app(b, a))$$
$$len(app(a, \boxed{cons(d, \underline{b})})) = len(app(b, \boxed{cons(c, \underline{a})}))$$
$$len(app(a, \boxed{cons(f, \underline{cons(d, b)})})) = len(app(b, \boxed{cons(e, \underline{cons(c, a)})}))$$

An annotation consists of a **wavefront**, a box, with a **wavehole**, an underlined term. The **skeleton** is formed by deleting everything that appears in the wavefront but not in the wavehole. The **erasure** of an annotated terms is formed by deleting the annotations but not the terms they contain. In the above sequence, the skeleton of every annotated equation is identical to the erasure of the previous equation in the sequence. Difference matching guarantees this; that is, difference matching s with t annotates s so that its skeleton matches t. More formally, s' is a **difference match** of s with t with substitution σ iff $\sigma(skeleton(s')) = t$ and $erase(s') = s$ where $skeleton(s')$ and $erase(s')$ build the skeleton and erasure of the annotated term s'.

Difference matching successive equations in the second sequence gives,

$$len(app(b, nil)) = s(len(b))$$
$$len(app(b, \boxed{cons(c, \underline{nil})})) = \boxed{s(s(len(b)))}$$
$$len(app(b, \boxed{cons(d, \underline{cons(c, nil)})})) = \boxed{s(s(s(len(b))))}$$

Note that similar term structure is accumulating on the lefthand side of this sequence as in the first sequence.

The critic then tries to speculate a lemma which moves the accumulating and nested term structure out of the way. In this case, the critic speculates a rule for moving a cons off the second argument of append. That is, the lemma,

$$len(app(a, cons(d, b))) = s(len(app(a, b)))$$

With this lemma, SPIKE is able to prove the len-app theorem without divergence. In addition, this lemma is sufficiently simple that SPIKE can prove it without assistance. The heuristics used by the critic to perform this lemma speculation are described in more detail in the next two sections.

The divergence analysis performed by the critic can be summarised as follows:

---

1. There is a sequence of equations $s_i = t_i$ to which the **generate** rule is applied ($i = 0, 1 \ldots$);
2. There exists (non trivial) G, H such that for each j, difference matching gives $s_j = G(U_j)$, and $s_{j+1} = G(\boxed{H(\underline{U_j})})$.

Preconditions to the divergence critic.

By "non-trivial" I wish to exclude unhelpful answers like $\lambda x.x$. H is thus the accumulating and nested term structure. For the first sequence of equations in the len-app example, H was $\lambda x.cons(y, x)$, G was $\lambda x.app(a, x)$, and U_0 was b. For simplicity, I have ignored the orientation of equations. In addition, the preconditions can be easily generalised to include multiple and nested annotations. This allows the critic to recognise multiple sources of divergence in the same equation. Techniques like those of [7] which identify accumulating term structure by most specific generalization cannot cope with divergence patterns that give rise to nested annotations.

The preconditions above leave the length of sequence undefined. If the sequence is of length 2, then the critic can be thought of as preemptive. That is, it will propose a lemma just before another induction is attempted and divergence begins. However, using such a short sequence risks identifying divergence when none exists. On the other hand using a long sequence is expensive to test and allows the prover to waste time on diverging proof attempts. Empirically, a good compromise appears to be to look for sequences of length 3. This is both cheap to test and reliable. To identify accumulating term structure, it also appears to be sufficient to use ground difference matching with alpha conversion of variable names. There exists a fast polynomial algorithm to perform such difference matching based upon the ground difference matching algorithm given in [3].

4 Lemma Speculation

One way of removing the accumulating and nested term structure is to apply a lemma, called a **wave rule**, which moves this difference to the top of the term leaving the skeleton unchanged. The hope is that the prover will then be able to cancel the difference with a similar difference on the other side of the equality.

For the len-app theorem, the divergence pattern suggests a lemma of the form,

$$len(app(a, \boxed{cons(d,\underline{b})})) = \boxed{F(len(app(a,b)))}$$

The only problem is to determine a suitable instantiation for F. Two heuristics are used for this: **cancellation** and **petering out**. The cancellation heuristic uses difference matching to identify term structure accumulating on the opposite side of the sequence which would allow cancellation to occur; failing that, it looks for suitable term structure to cancel against in a new sequence (the original sequence is usually a divergence pattern of a step case, whilst the new sequence is usually a divergence pattern of a base case). In the len-app example, the successor functions accumulating at the top of the righthand side of the second sequence of equations suggests that F be instantiated to $\lambda x.s(x)$. Thus, as required, the cancellation heuristic suggests the lemma,

$$len(app(a, \boxed{cons(d,\underline{b})})) = \boxed{s(len(app(a,b)))}$$

The other heuristic used is petering out. In moving the differences up to the top of the term, they may disappear altogether. The petering out heuristic uses regular matching to identify such situations. Petering out occurs, for example, in the speculation of the lemma,

$$sorted(\boxed{insert(y,\underline{x})}) = sorted(x)$$

This lemma is proposed in the analysis of the diverging proof attempt of the theorem $sorted(isort(x))$ where $isort$ is insertion sort and $insert(y,x)$ inserts the element y into the list x in order. In the case of petering out, F is instantiated to the identity function $\lambda x.x$.

All lemmas speculated are filtered through a conjecture disprover. When a confluent set of rewrite rules exists for ground terms, exhaustive normalization of some represenative set of ground instances of the equations is used to filter out non-theorems. For more sophisticated techniques for disproving conjectures see [12]. Alternatively, SPIKE itself could be used to filter out non-theorems.

The critic's lemma speculation can be summarized as follows (using the same variable names as the preconditions):

---

1. The critic proposes a lemma of the form,

$$G(\boxed{H(U_0)}) = \boxed{F(G(U_0))}$$

2. F is instantiated by the cancellation or petering out heuristics;
3. Lemmas are filtered through a conjecture disprover;
4. If several lemmas are suggested, the critic deletes any that are subsumed.

Postconditions for the divergence critic.

As before, this definition can be easily extended to deal with multiple and nested wavefronts. Note that as the lemma proposed moves the wavefronts to top of the term, it usually only introduces fresh divergence in the rare cases that cancellation or fertilization fails. This is unlikely since the cancellation and petering out heuristics attempt to ensure that cancellation or fertilization can take place.

5 Generalization

One major cause of divergence is the need to generalize. Although any lemma proposed by the critic is usually sufficient to fix divergence, attempting to prove the lemma itself can cause fresh divergence. In addition, several speculated lemmas can often be replaced by a single generalization, and a generalized lemma frequently leads to a shorter, more elegant and natural proof. The critic therefore attempts to generalize the lemma speculated, using the conjecture disprover to guard against over-generalization.

The main heuristic used for generalization is an extension of the primary term heuristic of Aubin [1]. The **primary terms** are those terms encountered as a term is explored from the root to the leaves ignoring non-recursive argument positions to functions. Consider, for example, the theorem,

$$len(app(a, b)) = len(a) + len(b)$$

where $+$ is defined recursively on its second argument. The primary terms of the righthand side are $\{len(a) + len(b), len(b), b\}$.

Analysis of SPIKE's diverging attempt to prove this theorem suggests the lemma,

$$\boxed{s(len(a))} + len(b) = \boxed{s(len(a) + len(b))}$$

A set of candidate terms for generalization is constructed by computing the intersection of the primary terms of the two sides of the equation. In this case, the intersection of the primary terms is $\{len(b), b\}$. The critic then picks members of this set to generalize to new variables. Picking b justs gives an equivalent lemma up to renaming of variables. Picking $len(b)$ gives the generalization,

$$\boxed{s(len(a))} + y = \boxed{s(len(a) + y)}$$

The reason for considering just primary terms is that the recursive definitions typically provide wave rules for removing differences which accumulate at these positions. In addition to primary terms, the divergence critic therefore also considers the positions of the waveholes in the lemma being speculated. The motivation for this extension is that the speculated lemma will allow differences to be moved from the wavehole positions; such positions are therefore also candidates for generalization.

For instance, + in the above example is considered to be "recursive" on both the first and second arguments since + is recursively defined on its second argument and the lemma being speculated has a wavehole on the first argument of +. The terms $\{len(a), a\}$ are therefore also included in the intersection set of candidate terms for generalization. Picking a to generalize gives, as before, an equivalent lemma up to renaming. Picking $len(a)$ gives the generalization,

$$\boxed{s(\underline{x})} + y = \boxed{s(\underline{x + y})}$$

The speculated lemma is now as general as is possible. This lemma allows the proof to go through without divergence.

The critic also has a heuristic for merging speculated lemmas. For instance, with the theorem $sorted(isort(x))$, the critic's divergence analysis and lemma speculation actually suggest the lemmas,

$$sorted(\boxed{insert(0, \underline{x})}) = sorted(x)$$
$$sorted(\boxed{insert(s(y), \underline{x})}) = sorted(x)$$

The critic identifies that $\{0, s(y)\}$ is a cover set for the natural numbers and merges these two lemma to the give the generalization,

$$sorted(\boxed{insert(y, \underline{x})}) = sorted(x)$$

6 Transverse Wave Rules

The lemmas speculated so far have moved differences directly to the top of the term where they are removed by cancellation or petering out. An alternative way of removing a difference is to move the difference onto another argument position where: either it can be removed by matching with a "sink", a universally quantified variable in the induction hypothesis; or it can be moved upwards by rewriting with the recursive definitions. Theorems involving functions with accumulators provide a rich source of examples where such lemmas prevent divergence.

Consider, for example, the theorem,

$$qrev(a, b) = app(rev(a), b)$$

where rev is naive list reversal and $qrev$ is tail recursive list reversal which builds the reversed list on its second accumulator argument. That is,

$$rev(nil) = nil$$
$$rev(cons(h, t)) = app(rev(t), cons(h, nil))$$
$$qrev(nil, r) = r$$
$$qrev(cons(h, t), r) = qrev(t, cons(h, x))$$

SPIKE's attempt to prove this theorem diverges generating the following sequence of equations to which the **generate** rule are applied,

$$qrev(a, b) = app(rev(a), b)$$
$$qrev(a, \boxed{cons(c, \underline{b})}) = app(\boxed{app(\underline{rev(a)}, cons(c, nil))}, b)$$
$$qrev(a, \boxed{cons(c, \underline{cons(d, b)})}) = app(\boxed{app(app(\underline{rev(a)}, cons(c, nil)), cons(d, nil))}, b)$$

$$\vdots$$

Divergence analysis of the righthand side of these equations identifies some accumulating term structure. Rather than move this term structure to the top of the term, it is much simpler to move it onto the second argument of the outermost append. The critic therefore proposes a **transverse** wave rule, which preserves the skeleton but moves the difference onto a different argument position. In this example, this is a lemma of the form,

$$app(\boxed{app(\underline{rev(a)}, cons(c, nil))}, b) = app(rev(a), \boxed{F(\underline{b})})$$

In moving the difference onto another argument position, the difference may change syntactically. The righthand side of the lemma is therefore only partially determined. To instantiate F, the critic uses two heuristics: **fertilization** and **simplification**. The fertilization heuristic uses matching to find an instantiation for F which enables immediate fertilization. In this case, matching against the induction hypothesis suggests,

$$app(\boxed{app(\underline{rev(a)}, cons(c, nil))}, b) = app(rev(a), \boxed{cons(c, \underline{b})})$$

The simplification heuristic uses matching to find an instantiation for F which enables the term to be simplified using one of the recursive definitions.

Finally the critic generalizes the lemma using the same primary term heuristic as before (augmenting recursive positions with wavehole positions). This gives the lemma,

$$app(\boxed{app(\underline{a}, cons(c, nil))}, b) = app(a, \boxed{cons(c, \underline{b})})$$

This is exactly the lemma needed by SPIKE to complete the proof. In addition, the lemma is simple enough to be proved by itself without divergence; this is not true of the ungeneralized lemma.

The actions of the critic can be summarized as follows,

Preconditions:

1. There is a sequence of equations $s_i = t_i$ to which the **generate** rule is applied ($i = 0, 1 \ldots$);
2. There exists (non trivial) G, H such that for each j, difference matching gives $s_j = G(U_j, Acc)$ and $s_{j+1} = G(\boxed{H(\underline{U_j})}, Acc)$.

Postconditions:

1. The critic proposes a lemma of the form,

$$G(\boxed{H(\underline{U_0})}, Acc) = G(U_0, \boxed{F(\underline{Acc})})$$

2. F is instantiated by the fertilization or simplification heuristics;
3. The lemma is generalized as much as possible;
4. Generalized lemmas are filtered through a conjecture disprover;
5. If several lemmas are suggested, the critic deletes any that are subsumed.

Speculation of transverse wave rules.

The preconditions and postconditions can be easily generalised to include multiple and nested annotations. The critic also uses an additional cancellation heuristic to generalize transverse wave rules. This heuristic attempts to cancel equal outermost functors where possible. For example, consider the theorem,

$$half(x + x) = x$$

From SPIKE's diverging proof attempt, the critic suggests the lemma,

$$half(\boxed{s(\underline{x})} + y) = half(x + \boxed{s(\underline{y})})$$

The cancellation heuristic deletes the equal outermost function. This gives the more general lemma,
$$\boxed{s(\underline{x})} + y = x + \boxed{s(\underline{y})}$$

7 Results

The critic described in the previous sections has been implemented in Prolog. It is successful at identifying divergence and proposing appropriate lemmas and generalizations for a large number of theorems. A few examples are given in Table 1. The times given are in seconds for the average of 10 runs on a Sun 4 running Quintus 3.1.1. For brevity, :: is written for infix cons, <> for infix append, [] for the empty list nil, and [x] for the list cons(x,nil). In addition, + is defined recursively on its second argument, even is defined by a s(s(x)) recursion, $even_m$ is defined by a mutual recursion with odd_m, and $rot(n, l)$ rotates a list l by n elements.

SPIKE's proof attempt diverges on each example when given the definitions alone. In each of the 25 cases, however, the critic is quickly able to suggest a lemma which overcomes divergence. When multiple lemmas are proposed (with the exception of 15) any one on its own is sufficient to fix divergence. In every case (except 9 and 19) the lemmas proposed are sufficiently simple to be proved automatically without introducing fresh divergence. In many cases, the lemmas proposed are optimal; that is, they are the simplest possible lemmas which fix divergence. In the cases when the lemma is not optimal, it is close to optimal.

Example 2 is a simple program verification problem taken from [7]. The second lemmas proposed in examples 3 and 5 are somewhat surprising; they are nevertheless just as good at fixing divergence as the first lemmas. Examples 7 and 8 demonstrate that the critic can cope with divergence in theories involving mutual recursion. In example 9, the proposed lemma is too difficult to be proved automatically. However, the divergence critic is able to identify the cause of this difficulty and propose a lemma which allows the proof to go through (example 11). In example 15, the critic identifies two separate divergence patterns. To overcome divergence, the first lemma plus one or other of the second and third are therefore needed. Example 19 is the only disappointment; the lemma proposed fixes divergence but is too difficult to be proved automatically, even with the assistance of the divergence critic. The problem seems to be that the example needs the introduction of a derived function like append which does not occur in the specification of the theorem. Examples 24 and 25 demonstrate that the critic can cope with divergence in theories containing conditional equations.

Divergence analysis is very quick in each case. The divergence pattern is recognized usually in less than a second. Most of the time is spent looking for generalizations and refuting over-generalizations using the conjecture disprover. Indeed, in the slowest example, less than 1% of the time is spent performing divergence analysis. Additional heuristics for preventing over-generalization and a more efficient implementation of the conjecture disprover would thus speed up the slower examples considerably.

No	Theorem	Lemmas speculated	τ/s
1	$s(x)+x=s(x+x)$	$s(x)+y=s(x+y)$ $s(x)+y=x+s(y)$	7.8
2	$dbl(x)=x+x \leftrightarrow$ $dbl(0)=0$, $dbl(s(x))=s(s(dbl(x)))$	$s(x)+y=s(x+y)$ $s(x)+y=x+s(y)$	8.2
3	$len(x<>y)=len(y<>x)$	$len(x<>z::y)=s(len(x<>y))$ $len(x<>z::y)=len(w::x<>y)$	3.6
4	$len(x<>y)=len(x)+len(y)$	$s(x)+y=s(x+y)$, $s(x)+y=x+s(y)$	7.2
5	$len(x<>x)=dbl(len(x))$	$len(x<>z::y)=s(len(x<>y))$ $len(x<>w::z::y)=s(len(x<>w::y))$	11.6
6	$even(x+x)$	$even(s(s(x))+y)=even(x+y)$	5.4
7	$even_m(x+x)$	$even_m(s(s(x))+y)=even_m(x+y)$ $odd_m(s(s(x))+y)=odd_m(x+y)$	28.4
8	$even_m(x) \rightarrow half(x)+half(x)=x$	$s(x)+y=s(x+y)$, $s(x)+y=x+s(y)$	6.0
9	$rot(len(x),x)=x$	$rot(len(x),x<>[y])=y::rot(len(x),x)$	2.4
10	$len(rot(len(x),x))=len(x)$	$len(rot(x,z<>[y]))=s(len(rot(x,z)))$	4.8
11	$rot(len(x),x<>[y])=y::rot(len(x),x)$	$(x<>[y])<>z=x<>y::z$ $rot(len(x),(x<>[y])<>z) =$ $y::rot(len(x),x<>z)$	86.3
12	$len(rev(x))=len(x)$	$len(x<>[y])=s(len(x))$	2.0
13	$rev(rev(x))=x$	$rev(x<>[y])=y::rev(x)$	1.2
14	$rev(rev(x)<>[y])=y::x$	$rev(x<>[y])=y::rev(x)$	16.0
15	$len(rev(x<>y))=len(x)+len(y)$	$len(x<>[y])=s(len(x))$ $s(x)+y=s(x+y)$, $s(x)+y=s(x+y)$	10.0
16	$len(qrev(x,[]))=len(x)$	$len(qrev(x,z::y))=s(len(qrev(x,y)))$	2.2
17	$qrev(x,y)=rev(x)<>y$	$(x<>[y])<>z=x<>y::z$	3.4
18	$len(qrev(x,y))=len(x)+len(y)$	$s(x)+y=s(x+y)$, $s(x)+y=x+s(y)$	12.0
19	$qrev(qrev(x,[]),[])=x$	$qrev(qrev(x,[y]),z)=y::qrev(qrev(x,[]),z)$	5.0
20	$rev(qrev(x,[]))=x$	$rev(qrev(x,[y]))=y::rev(qrev(x,[]))$	5.8
21	$qrev(rev(x),[])=x$	$qrev(x<>[y],z)=y::qrev(x,z)$	5.2
22	$nth(i,nth(j,x))=nth(j,nth(i,x))$	$nth(s(i),nth(j,y::x))=nth(j,nth(i,x))$	7.4
23	$nth(i,nth(j,nth(k,x))) =$ $nth(k,nth(j,nth(i,x)))$	$nth(s(i),nth(j,y::x))=nth(j,nth(i,x))$	7.6
24	$len(isort(x))=len(x)$	$len(insert(y,x))=s(len(x))$	2.0
25	$sorted(isort(x))$	$sorted(insert(y,x))=sorted(x)$ $sorted(insert(y,insert(z,x)))=sorted(x)$	114

Table 1. Some lemmas speculated by the Divergence Critic and time, τ taken.

The results are very pleasing. Using the divergence critic, the theorems listed (with the exception of 19) can all be proved from the definitions alone. For comparison, the NQTHM system [5] (perhaps the best known explicit induction theorem prover) when given just the definitions was unable to prove more than half these theorems.[2] Of course, with the addition of some simple lemmas, NQTHM is able to prove all these theorems. Indeed, in many cases, NQTHM needs the

---

[2] To be precise, NQTHM failed on 5, 6, 7, 9, 10, 11, 14, 16, 17, 19, 20, 21, 22, and 23.

same lemmas as those proposed by the divergence critic and required by SPIKE. This suggests that the divergence critic is not especially tied to the particular prover used nor even to the implicit induction setting. To test this hypothesis, I presented the output of a diverging proof attempt from NQTHM to the critic. I chose the commutativity of times as this is perhaps the simplest theorem which causes NQTHM to diverge. The critic proposed the lemma,

(EQUAL (TIMES Y (ADD1 X)) (PLUS Y (TIMES Y X))))

where TIMES and PLUS are primitives of NQTHM's logic recursively defined on their first arguments. This is exactly the lemma needed by NQTHM to prove the commutativity of times.

8 Related Work

Critics for monitoring the construction of proofs were first proposed in [9] for "proof planning". In this framework, failure of one of the proof methods automatically invokes a critic. Various critics for explicit induction have been developed that speculate missing lemmas, perform generalizations, look for suitable case splits, *etc* using heuristics based upon rippling similar to the ones described here [10]. There are, however, several significant differences. First, the divergence critic described here works in an implicit (and not an explicit) induction setting. Second, the divergence critic is not automatically invoked but must identify when the proof is failing. Third, the divergence critic is less specialized. These last two differences reflect the fact that critics in proof planning are usually associated with the failure of a particular precondition to a heuristic. The same divergence pattern can, by comparison, arise for many different reasons: the need to generalize variables apart, to generalize common subterms, to add a lemma, *etc*. Fourth, the divergence critic must use difference matching to annotate terms; in proof planning, terms are often already appropriately annotated. Finally, the divergence critic is less tightly coupled to the the theorem prover's heuristics. The critic can therefore exploit the strengths of the prover without needing to reason about the complex heuristics being used. For instance, the divergence critic has no difficulty identifying divergence in complex situations like nested or mutual inductions.

Divergence has been studied quite extensively in completion procedures. Two of the main novelties of the critic described here are the use of difference matching to identify divergence, and the use of rippling in the speculation of lemmas to overcome divergence. Dershowitz and Pinchover, by comparison, use most specific generalization to identify divergence patterns in the critical pairs produced by completion [7]. Kirchner uses generalization modulo an equivalence relation to recognise such divergence patterns [11]; meta-rules are then synthesized to describe infinite families of rules with some common structure. Thomas and Jantke use generalization and inductive inference to recognize divergence patterns and to replace infinite sequences of critical pairs by a finite number of generalizations

[13]. Thomas and Watson use generalization to replace an infinite set of rules by a finite complete set with an enriched signature [14].

Generalization modulo an equivalence enables complex divergence patterns to be identified. However, it is in general undecidable. Most specific generalization, by comparison, is more limited. It cannot recognize divergence patterns which give nested wavefronts like,

$$s^n(\boxed{s^n(\underline{x})+x}).$$

In addition, most specific generalization cannot identify term structure in waveholes. For example, consider the divergence pattern of example (20),

$$rev(qrev(x, nil)) = x$$
$$rev(qrev(x, \boxed{cons(y, \underline{nil})})) = \boxed{cons(y, \underline{x})}$$
$$rev(qrev(x, \boxed{cons(z, cons(y, nil))})) = \boxed{cons(z, \underline{cons(y, x)})}$$
$$\vdots$$

Most specific generalization of the lefthand side of this sequence gives the term $rev(qrev(x, z))$ (or, ignoring the first term in the sequence, $rev(qrev(x, cons(y, z)))$). Most specific generalization cannot, however, identify the more useful pattern, $rev(qrev(x, cons(y, nil))))$ which suggests the simpler lemma,

$$rev(qrev(x, cons(y, nil))) = cons(y, rev(qrev(x, nil)))$$

9 Future Work

There are many other types of divergence which could be incorporated into the divergence critic. Further research is needed to identify such divergence patterns, isolate their causes and propose ways of fixing them. This research may take advantage of the close links between divergence patterns and particular types of generalization. For instance, I am currently attempting to incorporate two other divergence patterns into the critic which arise when variables need to be renamed apart, and common subterms generalized in the theorem being proved. (Note that the current heuristics already rename variables apart and generalize common subterms in the speculated lemma.)

To illustrate a divergence pattern associated with the need to rename variables apart, consider the theorem,

$$app(x, app(x, x)) = app(app(x, x), x)$$

SPIKE's attempt to prove this theorem diverges giving the equations,

$$app(x, app(x, x)) = app(app(x, x), x)$$
$$app(x, \boxed{cons(y, app(x, \boxed{cons(y, \underline{x})}))}) = app(app(x, \boxed{cons(y, \underline{x})}), \boxed{cons(y, \underline{x})})$$

As before difference matching identifies the accumulating term structure. However, rather than try to speculate a lemma to move this difference out of the way, the critic generalizes the original theorem so that the difference is not introduced in the first place. In this case, it is sufficient to rename the first variable apart,

$$app(z, app(x, x)) = app(app(z, x), x)$$

SPIKE is able to prove this theorem without difficulty. The task of divergence analysis here is to isolate the variables to be renamed apart.

To illustrate a divergence pattern associated with the need to generalize common subterms consider the theorem,

$$len(rev(rev(x))) = len(rev(x))$$

SPIKE's attempt to prove this theorem diverges giving the equations,

$$len(rev(rev(x))) = len(rev(x))$$
$$len(rev(\boxed{app(\underline{rev(x)}, cons(y, nil))})) = len(\boxed{app(\underline{rev(x)}, cons(y, nil))})$$
$$\vdots$$

As before difference matching identifies the accumulating term structure. However, rather than try to speculate some lemma for moving this difference out of the way, we generalize the common subterm $rev(x)$. This has the effect of changing the wavefront from, $\boxed{app(\cdots, cons(y, nil))}$ to $\boxed{cons(y, \cdots)}$. The equation can then be simplified using the definitions of app and len.

10 Conclusions

This paper has described a critic which attempts to identify diverging proof attempts and to propose lemmas and generalizations which overcome the divergence. The critic has proved very successful; it enables the system SPIKE to prove many theorems from the definitions alone. The critic's success can be largely attributed to the power of the rippling heuristic. This heuristic was originally developed for proofs using explicit induction but has since found several other applications. To apply the rippling heuristic to the divergence of proofs using implicit induction required the addition of the difference matching procedure. This identifies accumulating term strucure which is causing the divergence. Lemmas and generalizations are then proposed to move this term structure out of the way.

Acknowledgments

This research was supported by a Human Captial and Mobility Postdoctoral Fellowship. I wish to thank: Adel Bouhoula and Michael Rusinowitch for their invaluable assistance with SPIKE; Pierre Lescanne for inviting me to visit Nancy; David Basin, Alan Bundy, Miki Hermann, Andrew Ireland, and Michael Rusinowitch for their helpful comments and questions; the members of the Eureca and Protheo groups at INRIA; and the members of the DReaM group at Edinburgh.

References

1. R. Aubin. *Mechanizing Structural Induction*. PhD thesis, University of Edinburgh, 1976.
2. D. Basin and T. Walsh. Difference matching. In D. Kapur, editor, *11th Conference on Automated Deduction*, pages 295–309. Springer Verlag, 1992. Lecture Notes in Computer Science No. 607.
3. D. Basin and T. Walsh. Difference unification. In *Proceedings of the 13th IJCAI*. International Joint Conference on Artificial Intelligence, Chambery, France, 1993.
4. A. Bouhoula, and M. Rusinowitch. Automatic Case Analysis in Proof by Induction. In *Proceedings of the 13th IJCAI*. International Joint Conference on Artificial Intelligence, Chambery, France, 1993.
5. R.S. Boyer and J.S. Moore. *A Computational Logic*. Academic Press, 1979. ACM monograph series.
6. A. Bundy, A. Stevens, F. van Harmelen, A. Ireland, and A. Smaill. Rippling: A heuristic for guiding inductive proofs. *Artificial Intelligence*, 62:185–253, 1993.
7. N. Dershowitz and E. Pinchover. Inductive Synthesis of Equational Programs. In *Proceedings of the 8th National Conference on AI*, pages 234–239. American Association for Artificial Intelligence, 1990.
8. M. Hermann. Crossed term rewriting systems. CRIN Report 89-R-003, Centre de Recherche en Informatique de Nancy, 1989.
9. A. Ireland. The Use of Planning Critics in Mechanizing Inductive Proof. In *Proceedings of LPAR'92*. Springer-Verlag, 1992. Lecture Notes in Artificial Intelligence 624.
10. A. Ireland and A. Bundy. Using failure to guide inductive proof. Technical report 613, Dept. of Artificial Intelligence, University of Edinburgh, 1992.
11. H. Kirchner. Schematization of infinite sets of rewrite rules. Application to the divergence of completion processes. In *Proceedings of RTA'87*, pages 180–191, 1987.
12. M. Protzen. Disproving conjectures. In D. Kapur, editor, *11th Conference on Automated Deduction*, pages 340–354. Springer Verlag, 1992. Lecture Notes in Computer Science No. 607.
13. M. Thomas and K.P. Jantke. Inductive Inference for Solving Divergence in Knuth-Bendix Completion. In *Proceedings of International Workshop AII'89*, pages 288–303, 1989.
14. M. Thomas and P. Watson. Solving divergence in Knuth-Bendix completion by enriching signatures. *Theoretical Computer Science*, 112:145–185, 1993.
15. T. Walsh, A. Nunes, and A. Bundy. The use of proof plans to sum series. In D. Kapur, editor, *11th Conference on Automated Deduction*, pages 325–339. Springer Verlag, 1992. Lecture Notes in Computer Science No. 607.
16. T. Yoshida, A. Bundy, I. Green, T. Walsh, and D. Basin. Coloured rippling: the extension of a theorem proving heuristic. Technical Report, Dept. of Artificial Intelligence, University of Edinburgh, 1993.

Synthesis of Induction Orderings for Existence Proofs

Dieter Hutter

German Research Center for Artificial Intelligence, DFKI
Stuhlsatzenhausweg 3
D-66123 Saarbrücken
Germany
E-mail: hutter@dfki.uni-sb.de

Abstract. In the field of program synthesis inductive existence proofs are used to compute algorithmic definitions for the skolem functions under consideration. While in general the recursion orderings of the given function definitions form the induction ordering this approach often fail for existence proofs. In many cases a completely new induction ordering has to be invented to prove an existence formula. In this paper we describe a top-down approach for computing appropriate induction orderings for existence formulas. We will use constraints from the knowledge of guiding inductive proofs to select proper induction variables and to compute a first outline of the proof. Based on this outline the induction ordering is finally refined and synthesized.

1 Introduction

In the field of program synthesis inductive existence proofs are used to compute algorithmic definitions for the skolem functions under consideration e.g. [MW80], [Bi92], [He92]. The problem arises to find an appropriate well-founded ordering for the generation of an induction formula. Since the induction ordering corresponds closely to the recursion ordering of the synthesized function selecting a wrong ordering results either in the failure of the proof or in a very complicated proof and thus, in a complicated definition of the synthesized program. All the approaches so far combine orderings of existing definitions to formulate the induction ordering (see [BM79] for details). But in many cases appropriate induction orderings are not based on definitions of the functions used to specify the theorem. Hence, proving existentially quantified formulas results in the synthesis of suitable function definitions with the help of which an appropriate induction scheme can be formulated.

Suppose for example, an algorithmic specifications for quotient and remainder (on natural numbers) has to be synthesized by proving the following formula:

$$\forall x, y : nat \, \exists u, v : nat \, \neg(y = 0) \rightarrow ((u \times y) + v = x \land v < y)$$

The specification uses $+$, $\times$, and $<$ which are all defined by structural recursion:

Definition of $+$:

$$\forall x, y : nat \, (x + 0) = x \quad \land$$
$$\forall x, y : nat \, (s(x) + y) = s(x + y)$$

Definition of $\times$:

$$\forall x, y : nat \, (x \times 0) = 0 \quad \land$$
$$\forall x, y : nat \, (s(x) \times y) = (y + (x \times y)) \quad \land$$

Definition of $<$:

$$\forall x, y : nat \, (x < 0) \leftrightarrow \textit{False} \quad \land$$
$$\forall x, y : nat \, (s(x) < 0) \leftrightarrow \textit{False} \quad \land$$
$$\forall x, y : nat \, (s(x) < s(y)) \leftrightarrow (x < y)$$

However, using structural induction the guidance of the proof renders more difficult and results in more complicated algorithms for quotient and remainder. An appropriate induction scheme would be to use the difference of x and y as an inductive predecessor of x. Introducing such a function *minus* we are able to define the quotient of two numbers in a rather easy way:

$$\forall x, y : nat \, (y = 0 \lor x < y) \rightarrow quot(x, y) = 0 \quad \land$$
$$\forall x, y : nat \, (\neg(y = 0) \land \neg x < y) \rightarrow quot(x, y) = s(quot(minus(x, y), y))$$

It will turn out that our approach will automatically synthesize exactly this definition of the quotient. During the proof a lemma will be postulated specifying the function *minus*. Furthermore, the synthesized definition of *minus* will be used to formulate the induction formula for the original theorem.

Our approach to compute an induction ordering for an arbitrary formula is divided into two tasks:

– We have to select the induction variables - i.e. the set of universally quantified variables we want to induce upon. Corresponding to the induction variables we also have to select a set of existentially quantified variables which we want to instantiate during the proof. A discussion of both problems is given in section 3.
– Based on the selected induction variables an appropriate induction ordering has to be computed. This task will be addressed in section 4.

2 A Closer Look to the Problem

Proving a formula $\forall x, y: nat \, \exists z: nat \, \Psi(x, y, z)$ by induction on x yields the following proof-obligations: [1].

Base case: $\forall x \, \Phi(x) \rightarrow (\forall y \, \exists z \, \Psi(x, y, z))$
Induction step: $\forall x \, \neg\Phi(x) \rightarrow (\forall y \, \exists z \, \Psi(p(x), y, z) \rightarrow \forall y \, \exists z \, \Psi(x, y, z))$

A common method to guide induction proofs is the technique of rippling and colouring terms [Bu88], [Bu90], [Hu90].

First, the syntactical differences between induction hypothesis and induction conclusion are shaded. We call such a shaded area a *context* or a *wave-front*. Hence, we obtain $\forall y \, \exists z \, \Psi(\boxed{p(x)}, y, z)$ as shaded hypothesis while $\forall y \, \exists z \, \Psi(x, y, z)$ denotes the conclusion. We obtain the following coloured induction step:

$$\forall x \, \neg\Phi(x) \rightarrow (\forall y \, \exists z \, \Psi(\boxed{p(x)}, y, z) \rightarrow \forall y \, \exists z \, \Psi(x, y, z))$$

Analogously, syntactical differences between both sides of equations or implications given in the database are shaded. In case of the definitions of $+$, $\times$, and $<$ we obtain the following coloured formulas:

$$\forall x, y: nat \; (x \boxed{+ 0}) = x \quad \wedge \tag{1}$$

$$\forall x, y: nat \; (\boxed{s(x)} + y) = \boxed{s}(x + y) \tag{2}$$

$$\forall x, y: nat \; (x \boxed{\times 0}) = 0 \quad \wedge \tag{3}$$

$$\forall x, y: nat \; (\boxed{s(x)} \times y) = \boxed{y +}(x \times y) \quad \wedge \tag{4}$$

$$\forall x, y: nat \; (\boxed{s(x)} < \boxed{s(y)}) \leftrightarrow (x < y) \tag{5}$$

Depending on the locations of the contexts or wave-fronts inside the white expressions we call these C(oloured)-equations (resp. C-implications) context-moving (wave-fronts on both sides) (e.g. 2, 4, or 5), context-creating or context-deleting (wave-fronts only on one side) (e.g. 1 or 3. Using these C-equations we are able to move, insert, or delete wave-fronts within the conclusion. E.g applying the C-equation 2 will move a wave-front s from the first argument of $+$ into the front of $+$ while using the C-equation 1 from left to right will create a wave-front $\boxed{+\ldots}$.

---

[1] For sake of readability we will use a human-oriented calculus: we allow to deduce Ψ from Ψ' if $\Psi \rightarrow \Psi'$ holds, and we prove a formula Ψ by deducing $True$. However, the approach has originally been developed in the context of resolution and paramodulation and is embedded in the further development of the INKA-system [Biundo et al. 86]. The extraction of algorithmic specifications out of existence proofs using the resolution calculus is described in [MW80].

Second, a context $\tilde{p}$ of the hypothesis is created inside the conclusion which is done by context-creating C-equations. Following this approach in general, more wave-fronts are created than intended, e.g. $\forall y\ \exists z\ \Psi(\tilde{s}(\tilde{p}(x)), y, z)$. [2] The strategies to get rid of these obstructing contexts are based on the observation that they are only permitted to occur at certain positions without preventing the use of the hypothesis. A tolerated position of contexts is at top-level of the formula since for the use of the hypothesis it is necessary that an instance of it occurs *inside* the conclusion. Furthermore, we may instantiate the all-quantified non-induction variables of the hypothesis and hence, wave-fronts may also occur in the positions of the corresponding variables - so-called *sinks* [Bu90] - in the conclusion.

Moving wave-fronts to top-level or into a sink we may also instantiate an existentially quantified variable z by a C-term $\tilde{s}(z'')$ in the conclusion (where z'' is also existentially quantified) which creates an additional wave-front inside the conclusion. In order to enable the use of the hypothesis we have to move also this wave-front towards top-level or towards some sink.

The question arises why existentially quantified variables should be instantiated if we only obtain additional wave-fronts which have to be moved right across the conclusion. The answer is that in many cases contexts cannot be moved or deleted independently.

Consider, for instance, a C-equation like

$$g(\tilde{s}(y), \tilde{s}(z)) = g(y, z).$$

In order to enable its application we need contexts in both arguments of g. Suppose, we induce on x and have to manipulate a conclusion

$$\exists z\ \Psi(g(x, z))$$

to enable the use of the corresponding hypothesis

$$\exists z\ \Psi(g(\tilde{p}(x), z)).$$

Blowing up the induction variable x by context-creating C-equations results in a formula

$$\exists z\ \Psi(g(\tilde{s}(\tilde{p}(x)), z)).$$

In order to lift the obstructing context $\tilde{s}$ towards top-level we have to instantiate z by $\tilde{s}(z')$ and obtain finally an instance

$$\exists z'\ \Psi(g(\tilde{p}(x), z'))$$

of the hypothesis.

---

[2] We use a light grey $\tilde{p}$ to denote wave-fronts which coincide with wave-fonts in the hypothesis while the darker grey $\tilde{s}$ denotes wave-fronts which have to be rippled away.

As we have seen in the example, the instantiation of existentially quantified variables is used to enable the move or the deletion of wave-fronts caused by blowing up induction variables. Hence, the selection of the induction ordering - and thus, the selection of the obstructing wave-fronts - and the instantiation of the existentially quantified variables, both are strongly connected and cannot be done independently. The wave-fronts which are caused by both processes have to fit into each other to enable the use of context-moving/removing C-equations. We will use this relation to synthesize induction orderings.

3 Selection of Induction Variables

In [BM79] induction schemes (and thus also induction variables) are suggested by each term occurring in a formula. Several heuristics like *merging* and *subsumption* are used to combine or select different induction schemes (see [St88] for details). But terms suggest always the same induction schemes regardless in which context they occur. E.g. $x * y$ proposes an induction on x according the structural ordering regardless whether this term occurs inside the first or second argument of $quot$. While in case of $...quot(\boxed{s(x)} * y, ...)...$ we are able to ripple the obstructing wave-front $\boxed{s(...)}$ in front of $quot$, rippling gets stuck in case of $...quot(..., \boxed{s(x)} * y)...$.

Thus, the selection of appropriate induction variables is determined by the possible ways contexts can be moved across the conclusion. Inducing on a variable which occurs inside a function for which there are no context-moving or context-removing C-equations will fail since moving the generated wave-front will be impossible. But, there may be inductive consequences of the given axioms which cause appropriate context-moving C-equations and which are not stated in the database explicitly. Formulating these theorems and proving them by induction could unblock the process.

Hence, on one hand fixing the induction ordering yields necessary C-equations to be stated as new lemmata. On the other hand, the given C-equations are used to formulate an induction ordering. We overcome this "deadlock" by a more abstract notion of C-equations. In order to calculate appropriate induction variables we abstract the notion of C-equations by removing the structure of the wave-fronts. Furthermore, we remove those parts of the skeleton which are not affected by wave-fronts, like variables or constants. Given the C-equations resp. C-implications

$$\boxed{s(x)} + y = \boxed{s(x+y)}$$
$$\boxed{s(x)} < \boxed{s(y)} \leftrightarrow x < y \text{ and}$$
$$\boxed{s(x)} = \boxed{s(y)} \leftrightarrow x = y$$

the following labeled fragments (Figure 1) are created in which dots denote the positions of wave-fronts on the left and the right side of the C-equation/C-implication. For each C-equation or C-implication in the database we compute its labeled fragment and add it to the database.

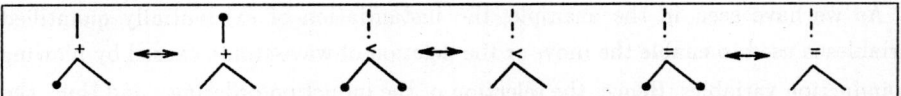

Fig. 1. Labeled fragments

In order to select appropriate induction variables and existentially quantified variables to be instantiated we search for a subset $\mathcal{V}$ of variables such that the following condition holds: marking each occurrence of a variable of $\mathcal{V}$ by a dot we have to remove or move all dots up to the predicate symbols with the help of the given labeled fragments. Then, the variables whose occurrences are marked with a dot determine the induction variables resp. the existentially quantified variables to be instantiated.

Consider again the example of the synthesis of quotient/remainder. Using $\mathcal{V} = \{x, u\}$ we are able to remove all dots while moving them towards the occurrences of the predicate symbols. Figure 2 illustrates this process. Selecting x as an induction variable would imply to select also y as an induction variable. This is especially caused by the labeled fragments of $<$ and $=$ which only enables us to label both arguments simultaneously. Hence, selecting y as an induction variable would imply to choose v as a variable to be instantiated and vice versa. Since also $=$ requires dots on both arguments x has also to be an induction variable because of $((u \times y) + v = x$. Thus, x and u are a minimal set of candidates for induction and instantiation.

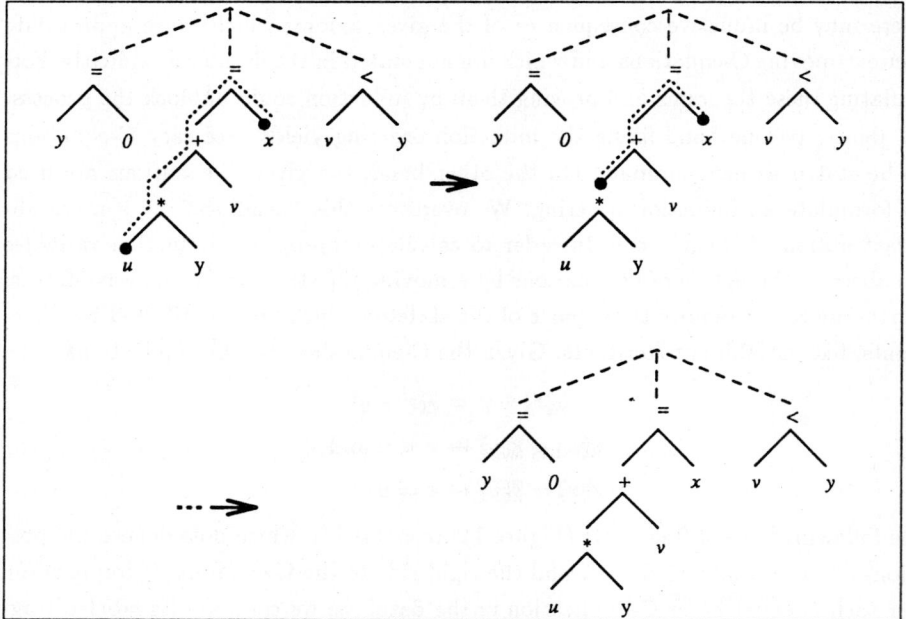

Fig. 2. Verifying $\{x, u\}$ as possible variable set

It turns out that this method of selecting induction variables is rather insensitive to missing lemmata. The reason is that in many cases these lemmata move wave-fronts into similar directions as the C-equations used to prove the lemmata.

4 Computing Induction Orderings

Based on the selected induction variables an appropriate induction ordering is computed. Again, remember that the wave-fronts have to be moved into dedicated positions and the restrictions for doing that will help us to formulate an appropriate induction ordering.

For example, consider our introductory example of synthesizing algorithms for quotient and remainder. Blowing up the induction variable x and instantiating the existentially quantified variable u creates wave-fronts which have to be moved towards top-level until they meet each other close to the equality symbol where they have finally to be deleted. Blowing up x restrains the possible instantiations of u and vice versa. In general there is rarely more than one C-equation to move or propagate a given wave-front into a specified direction. Hence, if we fix the context at some position on its way between x and u we fix also how to blow up x and how to instantiate u. Depending on the position where we fix the context we have different possibilities to formulate the wave-front. Using the wrong one can result in the failure of the proof. Hence, we have to determine the bottleneck on our way through the conclusion. Once we have fixed the context at a position in the formula we propagate it to the occurrences of the induction variables and the selected existentially quantified variables.

Depending on the position where we fix the context on the way between induction variables and existentially quantified variables we propose different methods to formulate an induction formula.

- A well-known approach fixes the wave-fronts close to the occurrences of the induction variables. The induction ordering is chosen according to the recursion ordering of the outer function symbols (see [BM79] for details).
- Fixing the wave-fronts at the occurrence of an existentially quantified variable they have to be propagated towards the occurrences of the induction variables during the proof attempt. Proving that the resulting wave-front at the occurrence of the induction variable x denotes a predecessor of x guarantees the soundness of the used induction scheme. We shall describe this approach in the next paragraph.
- Fixing the wave-front between the occurrences of the induction variables and top-level we have to propagate it into both directions, towards the induction variables as well as towards the existentially quantified variables. This task will be addressed in paragraph 4.2.

4.1 Synthesis of Induction Orderings

This section describes a method to synthesize an induction ordering by fixing the wave-front around the existentially quantified variable and propagating this context into the occurrences of the induction variables.

Let us illustrate this method in detail by the already mentioned quotient/remainder example.

$$\forall x, y : nat \, \exists u, v : nat \, \neg(y = 0) \rightarrow ((u \times y) + v = x \land v < y)$$

Due to the selection of the variable set $\mathcal{V}$ (see section 3) we know that we have to move the generated wave-fronts at x and u towards the equality predicate where both have to be deleted. Since x occurs on top of the right-hand side of $=$ moving wave-fronts of x towards $=$ is trivial regardless what wave-front is generated. Hence, the occurrence of x does not restrict the number of possible induction orderings. Since the wave-front generated by instantiating u has to be moved across $\times$ and $+$ we fix the wave-front at the occurrence of u using the C-equation obtained by the definition of $\times$. In a first step we are concerned with the separation of base case and induction step. In order to compute the conditions for the base case we instantiate the existentially quantified variables and simplify the obtained formula. The value of u is chosen according to the non-recursive case of the definition of $\times$ (since we use the recursive case for rippling). Hence, we instantiate u by 0 and symbolic evaluation yields:

$$\forall x, y : nat \, \exists v : nat \, \neg y = 0 \rightarrow (v = x \land v < y)$$

Using the equation $v = x$ we eliminate all occurrences of v by x which results in the following formula:

$$\forall x, y : nat \, y = 0 \lor x < y$$

Hence, $y = 0 \lor x < y$ forms the condition of the base case which is now already proved by the above transformation.

The case $\neg y = 0 \land \neg x < y$ forms the condition of the induction step. Starting the proof we do not know the predecessors of x which have to to be synthesized during the proof. Since we induce on x the induction hypothesis has the following form:

$$\exists u', v : nat \, \neg y = 0 \rightarrow ((u' \times y) + v = \boxed{pred(x)} \land v < y)$$

where $\boxed{pred(x)}$ denotes some (unknown) predecessor of x with respect to some well-founded ordering. Given the induction conclusion

$$\exists u, v : nat \, \neg y = 0 \rightarrow ((u \times y) + v = x \land v < y)$$

we start again with the instantiation of u. Since $\times$ is defined by structural recursion u is instantiated by $\boxed{s(u')}$ to enable the rippling process which yields the following formula:

$$\exists u', v\!:\!nat \ \neg y = 0 \rightarrow ((\boxed{s(u')} \times y) + v = x \wedge v < y)$$

The rest of the proof is almost done by standard rippling techniques: in order to move the created wave-front across the conclusion close to the occurrence of x we move it to the top-level of the equation with the help of the definition of $\times$ and the associativity law of $+$:

$$\exists u', v\!:\!nat \ \neg y = 0 \rightarrow \boxed{y+ ((u' \times y) + v)} = x \wedge v < y$$

In order to remove the wave-front on the left side we have to use the functionality axiom of the equality which postulates the task of blowing x up to a term like $\boxed{y + pred(x)}$. If we succeed we obtain a formula

$$\exists u', v\!:\!nat \ \neg y = 0 \rightarrow ((u' \times y) + v) = \boxed{pred(x)} \wedge v < y)$$

which is interpreted as an instance of a induction hypothesis provided $pred(x)$ denotes a predecessor of x wrt some well-founded ordering.

Remaining the problem to blow up x to a term $\boxed{y + pred(x)}$ we proceed with stating a lemma:

$$\forall x, y\!:\!nat \ \exists z\!:\!nat \ (\neg y = 0 \wedge \neg x < y) \rightarrow (y + z) = x$$

Proving this lemma our system comes up with an algorithmic specification of *minus*. The system automatically recognizes that in case of $\neg y = 0 \wedge \neg x = 0$ the function *minus* is argument bounded on the first argument [Wa88], i.e. $minus(x, y)$ is less than x according to a well-founded ordering (i.e. the count-ordering). Thus, the used scheme

$$\forall x, y\!:\!nat \ (y = 0 \vee x < y) \rightarrow \Psi(x, y) \wedge$$
$$(\neg y = 0 \wedge \neg x < y) \rightarrow (\Psi(minus(x, y), y) \rightarrow \Psi(x, y))$$
$$\rightarrow \forall x, y\!:\!nat \ \Psi(x, y)$$

is a sound induction axiom and the proof is completed.

As we have seen in the example this method synthesizes new induction orderings $\mathcal{R}$ by fixing the wave-fronts at the occurrences of the existentially quantified variables. Moving these wave-fronts into the positions of the selected induction variables their predecessors are computed. Thus the description of the predecessors wrt $\mathcal{R}$ is not restricted to functions given to the system but uses also function definitions which are synthesized during the proof. The induction ordering is extracted from the proofs of lemmata which are automatically stated while proving the original theorem.

4.2 Induction on Terms

Another alternative to compute appropriate induction orderings is to fix the wavefront between the occurrences of the induction variables and top-level of the formula. This approach corresponds to an induction on terms, i.e. we induce on the value of a non-variable subterm of the given theorem. Suppose, we want to prove a formula

$$\forall x \; \exists z \; \Psi(f(x), z)$$

by induction on x. Given a well-founded ordering $\mathcal{R}$ a new well-founded ordering $\mathcal{R}'$ is defined by

$$(y, x) \in \mathcal{R}' \text{ iff } (f(y), f(x)) \in \mathcal{R}$$

and used to formulate an induction scheme. The crucial point is how to denote the predecessor of x wrt $\mathcal{R}'$ since usually the predecessor of a term t wrt $\mathcal{R}$ is denoted by a term $g(t)$ where g is argument bounded under a condition $\Phi(x)$. Thus, in case $\Phi(f(t))$ holds $g(f(t))$ constitutes a predecessor of $f(t)$ wrt $\mathcal{R}$. Since $g(f(x))$ is no instance of a term $f(u)$ also

$$\exists z \; \Psi(g(f(x)), z)$$

is no admissible induction hypothesis unless $g(f(x))$ is in the range of f, i.e.

$$\exists y \; f(y) = g(f(x))$$

holds. Thus, all terms t' with $f(t') = g(f(t))$ are predecessors of x wrt $\mathcal{R}'$.

We use this property to formulate the following induction scheme:

$$\forall x \; (\Phi(f(x)) \rightarrow \exists z \; \Psi(f(x), z)) \wedge$$
$$(\neg \Phi(f(x)) \rightarrow \exists z \; \Psi(g(f(x)), z) \rightarrow \exists z \; \Psi(f(x), z))$$
$$\rightarrow \forall x \; \exists z \; \Psi(f(x), z)$$
$$\wedge \quad \forall x \; \neg\Phi(f(x)) \rightarrow \exists y \; f(y) = g(f(x))$$

Provided the induction variable x occurs only inside $f(x)$ within Ψ the scheme is proved to be sound [Hu91].

Let us demonstrate this approach by the following example. Suppose, we define the function *half* (division by 2) by

$$half(0) = 0$$
$$half(s(0)) = 0$$
$$\forall x : nat \quad half(s(s(x))) = s(half(x))$$

and we want to prove the following theorem:

$$\forall x, y : nat \; \exists z : nat \; quot(half(x), y) = half(quot(z, y))$$

Analyzing all possible ways to move context across the conclusion yields that we have to induce on x and that we have to instantiate z. The bottlenecks when moving the wave-fronts to top-level are the occurrences of $quot$ since there is only one context-moving C-equation concerning $quot$. The definition of $quot$ restricts admissible contexts movable across $quot$ to $\underline{minus(...)}$. Hence, the recursion ordering of $quot$ (applied to $half(x)$ and y) is used to obtain the following induction scheme:

$$\forall x, y : nat \, ((y = 0 \vee half(x) < y) \rightarrow \exists z : nat \, \Psi(half(x), y, z)) \wedge$$
$$((\neg y = 0 \wedge \neg half(x) < y)$$
$$\rightarrow (\exists u : nat \, half(u) = minus(half(x), y)) \wedge$$
$$\exists z : nat \, \Psi(minus(half(x), y), y, z)) \rightarrow \exists z : nat \, \Psi(half(x), y, z))$$
$$\rightarrow \forall x, y : nat \, \exists z : nat \, \Psi(half(x), y, z)$$

While the base case can easily be proved the induction step is split into two formulas. First, we verify the existence of a predecessor of x denoted by $minus(half(x), y)$:

$$\forall x, y : nat \, (\neg y = 0 \wedge \neg half(x) < y) \rightarrow \exists z : nat \, minus(half(x), y) = half(z)$$

which is an easy consequence of the given definitions. Second, we have to prove the actual induction step:

$$\forall x, y : nat \, (\neg y = 0 \wedge \neg half(x) < y) \rightarrow$$
$$(\exists z' : nat \, quot(\underline{minus(half(x), y)}, y) = half(quot(z', y)))$$
$$\rightarrow (\exists z : nat \, quot(half(x), y) = half(quot(z, y))$$

Opening up the definition of $quot$ the missing context is create and we obtain

$$\exists z : nat \, \underline{s(quot(\underline{minus(half(x), y)}, y))} = half(quot(z, y)))$$

as conclusion. The proof attempt succeeds if we find an instantiation $pred(z')$ of z such that $half(quot(pred(z'), y)) = \underline{s(half(quot(z', y)))}$ holds. We compute this instantiation by propagating the top-level context $\underline{s(...)}$ to the existentially quantified variable z':

$$\underline{s(half(quot(z', y)))}$$
$$= half(\underline{s(s(quot(z', y)))})$$
$$= half(\underline{s(quot(\underline{y+}z', y))})$$
$$= half(quot(\underline{y+(y+}z'), y))$$

provided we know that $\forall x, y : nat \, quot(y + x, y) = s(quot(x, y))$ holds.

The above deduction chain is used to instantiate z by $y+(y+z')$ in the conclusion

$$\exists z : nat \, \underline{s(quot(\underline{minus(half(x), y)}, y))} = half(quot(z, y)))$$

which results in

$$\exists z' : nat \; \boxed{s(quot(\boxed{minus(half(x),y)},y))} = \boxed{s(half(quot(z')))}$$

as a reformulated conclusion. Using the induction hypothesis also the second part of the theorem is proven.

Summing up, we induce on a term $half(x)$ with $minus(half(x),y)$ as its predecessor. It corresponds to an induction on x where in case of $\neg y = 0 \land \neg half(x) < y$ we use $minus(minus(x,y),y)$ as the predecessor of x.

Suppose, we want to prove a formula $\forall x \; \Phi(f(g(x)))$ by induction on x. While proving the induction step the task arises to move a context towards top-level:

$$\Psi(f(g(x))) \Rightarrow \Psi(f(g(\boxed{s(\underline{p(x)})}))) \Rightarrow \Psi(f(\boxed{s(g(\underline{p(x)}))})) \Rightarrow \Psi(f(g(\underline{p(x)})))$$

Now, induction on $g(x)$ splits this task into two parts:

$$\Psi(f(g(x))) \Rightarrow \Psi(f(\boxed{s(\underline{p'(g(x))})})) \Rightarrow \Psi(f(\underline{p'(g(x))})) \quad \text{and}$$

$$\forall x \exists u \; \underline{p'(g(x))} = g(u)$$

The advantages of this approach are twofold. The wave-fronts are closer to the top-level and thus, they are nearer to their target position. Additionaly, the form of the wave-fronts can be selected according to the available C-equations (resp. definition) of f in order to enable the rippling process. Proving the existentially quantified formula $\forall x \exists u \; \underline{p'(g(x))} = g(u)$ we compute an instantiation of u which denotes an appropriate predecessor of x. Using induction in order to prove the above formula we also use recursion in order to formulate the predecessor of x.

In some cases induction on terms can be simulated by a reformulation of the original theorem. Instead of proving

$$\forall x \quad \Psi(f(g(x)))$$

by induction on $g(x)$ an implication

$$\forall x, y \quad y = g(x) \to \Psi(f(y))$$

is proven by induction on y. Thus, the coloured induction step has the form

$$(\forall x' \; \underline{p'(y)} = g(x') \to \Psi(f(\underline{p'(y)}))) \to (\forall x \; y = g(x) \to \Psi(f(y)))$$

The task of manipulating the wave-fronts in $\Psi(f(y))$ is similar to the induction on $g(x)$ while modifying $y = g(x)$ in order to use $\underline{p'(y)} = g(x')$ corresponds to the proof of $\forall x \exists u \; \underline{p'(g(x))} = g(u)$.

5 Conclusion

We have presented a novel approach to compute induction orderings for existence proofs. We use labeled fragments of C-equations (and thus information of the rippling process) in order to select appropriate induction variables and existentially quantified variables to be instantiated. In contrast to [BM79] we consider not only the recursion analysis of the given definitions but also the possibilities of rippling given by additional lemmata in the database. Based on the computed outline of the existence proof the induction ordering is synthesized according to the possible ways wave-fronts can be moved across the conclusion. In contrast to other approaches the orderings are formulated with the help of recursively defined functions which are synthesized during the proof and which are unknown to the prover before. E.g. Middle-Out-Reasoning [He92] computes the predecessor of induction variables (and thus the induction ordering) by higher order unification. Hence, recursion is not available to specify the predecessor. On the other hand the presented techniques can be integrated into the MOR-approach yielding an improved approach of formulating and proving inductive formulas.

We have successfully tested the approach on several existence proofs, e.g. for synthesizing the inverse function of *log*, *half*, *quot*, and *reverse*. In the latter case the synthesized inverse function of *reverse* recurs on a function *butlast* which cuts off the last element of a list.

References

[Biundo et al. 86] Biundo et al. The Karlsruhe Induction Theorem Proving System. 8th CADE, LNCS 230, Springer, 1986

[Bi92] Biundo, S. Automatische Synthese rekursiver Programme als Beweisverfahren. Springer, IFB302, 1992

[BM79] Boyer, R.S. Moore J S. A Computational Logic. Academic Press, 1979

[Bu88] Bundy, A. et al. The Use of Explicit Plans to Guide Inductive Proofs. 9th CADE, LNAI 310, Springer-Verlag, 1988

[Bu90] Bundy, A. et al. Extensions to the Rippling-Out Tactic for Guiding Inductive Proofs. 10th CADE, LNAI 449, Springer, 1990

[He92] Hesketh, J. et al. Using Middle-Out Reasoning to Control the Synthesis of Tail-Recursive Programs. 11th CADE, LNAI 607, Springer, 1992

[Hu90] Hutter, D. Guiding Induction Proofs. 10th CADE, LNAI 449, Springer, 1990

[Hu91] Hutter, D. Mustergesteuerte Strategien für das Beweisen von Gleichheiten. Ph.D. thesis, University of Karlsruhe, 1991

[MW80] Manna, Z.; Waldinger, R. A Deductive Approach to Program Synthesis. ACM Transactions on Programming Languages and Systems, Vol. 2 No. 1, 1980

[St88] Stevens, A. A Rational Reconstruction of Boyer and Moore's Technique for Constructing Induction Formulas. Proceeding of ECAI88, 1988

[Wa88] Walther, C. Argument-Bounded Algorithms as a Basis for Automated Termination Proofs. 9th CADE, LNCS310, Springer, 1988

Lazy Generation of Induction Hypotheses

Martin Protzen

Fachbereich Informatik, Technische Hochschule Darmstadt,
Alexanderstr. 10, D-64283 Darmstadt, Germany
protzen@inferenzsysteme.informatik.th-darmstadt.de

Abstract: A novel approach for automating explicit induction is suggested. Analysis of successful induction proofs reveals that these proofs can be guided without reference to a specific induction axiom. This means that required induction hypotheses can be computed during the proof. We show that some instances of the generalized induction scheme provide induction hypotheses which cannot be obtained using common techniques. Our proposal also leads to an extension of known strategies to guide induction proofs. Criteria are developed which justify the soundness of the computed induction axioms without requiring additional proofs. The performance of the new technique is illustrated by some non-trivial problems for which necessary induction hypotheses are generated automatically.

1 Introduction

The degree of mechanization of an induction theorem prover strongly depends on the prover's ability to generate induction axioms which provide relevant induction hypotheses. Automated induction proofs following the explicit induction paradigm[1] have since the fundamental work of Aubin [Aubin 79] and of Boyer & Moore [Boyer & Moore 79] constructed inductions schemes using the technique of *Recursion Analysis:* (1) collect induction schemes suggested by recursive functions, (2) manipulate induction schemes (heuristically) and (3) compare and combine different induction schemes, cf. [Walther 94].

The key idea is to construct an induction that closely matches the recursion structure of some function appearing in the goal to prove. The motivation for this approach comes from an analysis of the role of induction in proofs, viz. to provide induction hypotheses such that the evaluation of the induction conclusion by unfolding function definitions introduces terms which then can be matched by the induction hypothesis. Recursion Analysis has been refined by several authors, the biggest improvement being the universal quantification of non-induction variables in induction hypotheses. In [Walther 92] the key idea is to provide as many induction hypotheses as possible before the proof. For this purpose, the union of suggested induction relations is computed. However, as the union of two well-founded relations is not necessarily well-founded, additional well-foundedness proofs are required, cf. [Walther 93]. All approaches based on Recursion Analysis have two deficiencies in common:

---

[1] as opposed to the implicit induction paradigm which evolved from the Knuth-Bendix Completion Procedure

- in some cases necessary induction hypotheses are not provided,
- the applicability of generated induction hypotheses is not guaranteed.

We claim that computing the hypotheses *before* the proof is not a solution to the problem and so the central idea for the lazy method is to postpone the generation of hypotheses until it is evident which hypotheses are required for the proof.

Consider e.g. the statement $\Psi[x, y]$: Perm(x y) = Perm (y x), where Perm(k l) returns True if the list k is a permutation of the list l. The recursive case of Perm(k l) is defined by the following conditional equation:

$$k \neq \text{nil} \rightarrow \text{Perm}(k\ l) = \text{Member}(\text{head}(k)\ l) \land \text{Perm}(\text{tail}(k)\ \text{delete}(\text{head}(k)\ l)),$$

where Member(n l) returns True if the number n is an element of the list l, delete(n l) deletes the first occurrence of n from l, and head(l) resp. tail(l) denote the first element of a list l resp. the list l without the first element.

Here the conventional approach fails to prove $\Psi[x, y]$ because all induction hypotheses are built corresponding to direct predecessors of the computation order of involved algorithms, cf. [Walther 94]. For our example, Recursion Analysis would suggest the induction hypotheses $\Psi[\text{tail}(x), y]$ or $\Psi[x, \text{delete}(\text{head}(x)\ y]$ or combinations thereof, and a proof using only these hypotheses is cannot be obtained[2]. But a proof is possible using the induction hypotheses $\Psi[\text{tail}(x), \text{tail}(y)]$, $\Psi[\text{tail}(x), \text{delete}(\text{head}(x)\ y)]$, $\Psi[\text{delete}(\text{head}(y)\ x), \text{tail}(y)]$ and $\Psi[\text{delete}(\text{head}(y)\ \text{tail}(x)), \text{delete}(\text{head}(x)\ \text{tail}(y))]$.

To mechanize induction proofs as far as possible a tool is needed which generates these required hypotheses. In this paper a method is proposed which computes hypotheses *during* the proof automatically when they are required and this is the reason to call the method *lazy*. The proposed approach can be used to extend known methods to guide induction proofs, e.g. the *rippling* method [Bundy et al. 93] or the *C-term* method [Hutter 90] by instantiating *induction* variables during the proof, although only to a limited extent. All hypotheses suggested by Recursion Analysis can still be obtained and additionally other ones, e.g. those mentioned above. The approach successfully generates induction hypotheses even for hard problems, e.g. the *Gilbreath Card Trick*, cf. [Huet 91].

The remainder of this paper is organized as follows: after introducing some basic concepts in section 2, we will illustrate and formalize the proposed method in section 3. The well-foundedness of generated induction axioms will be discussed in section 4.

2   Formal Preliminaries

Given a |w|-tuple of distinct variables $x^* \in \mathcal{V}_w$, we call a substitution σ a *substitution for* x^* iff for each substitution pair y/t in σ, all variables in t as well as y are contained in

---

[2] The statement is provable using two nested inductions but we claim that proofs involving fewer inductions should be preferred over proofs using more inductions.

x*. We assume a signature $\Sigma = \Sigma^c \cup \Sigma^d$, where Σ^c contains *constructor functions* (e.g. 0, succ for numbers and nil, add for linear lists) and Σ^d contains *defined functions* f, defined by a set of conditional equations of the form $\varphi_i \to f(x_1...x_n) = r_i$ where the *conditions* φ_i form a complete case analysis and exclude each other. Such equations are also called the *cases* of f. Additionally we demand that termination of each function in Σ^d has been verified. A function f is called *recursively defined* if some case of f is recursive, i.e. the the condition φ_i or the result r_i of the case contains a term $f(t_1...t_n)$. A term $f(t_1...t_n)$ is also called an f-term. Each n-ary constructor is associated with n *selector* functions which compute the inverse operations to their constructor, e.g. the constructors succ and add are associated with the selectors pred resp. head and tail. We also assume a *standard interpretation* M for the functions in Σ mapping each ground term from $\mathcal{T}(\Sigma)$ to a *constructor ground term* from $\mathcal{T}(\Sigma^c)$ such that $M(q)=q$ holds for $q \in \mathcal{T}(\Sigma^c)$. $M[\![x^*/r^*]\!](t)$ denotes the standard interpretation of a term t with variables x* under the variable assignment $\{x^*/r^*\}$. If the standard interpretation M satisfies a formula Ψ then Ψ is an *inductive truth* and we write $\Psi \in Th_{ind}$ for short. This means in particular that $\forall x. \Psi[x] \in Th_{ind}$ iff $\Psi(q) \in Th_{ind}$ for all $q \in \mathcal{T}(\Sigma^c)$. Further we assume an *axiom set* AX which contains all conditional equations which define the elements of Σ^d and also additional lemmata.

$\Psi[v_1...v_n]$ denotes that the v_i occur as subterms in the formula Ψ. If all v_i are distinct variables $\Psi[v_1/t_1...v_n/t_n]$ is an *instance* of Ψ and denotes the formula Ψ where all occurrences of v_i have been replaced by t_i. Subsequently we will abbreviate $\Psi[v_1/t_1...v_n/t_n]$ by $\Psi[t_1...t_n]$ if it is obvious from the context which variables have been replaced. $\Psi|_u$ denotes the subterm of Ψ at occurrence u, and $\Psi[u \leftarrow t]$ denotes the term which is obtained from Ψ by replacing the subterm at occurrence u with the term t.

As reasoning about relations plays a crucial role in this area we use the concept of *relation descriptions* from [Walther 92] to define relations on which induction proofs are based.

Definition An *atomic relation description* C for $x^* \in \mathcal{V}_w$ is a pair (φ, Δ), where φ, called the *range formula* of C, is a quantifier-free formula with at most the variables in x* as variables and Δ is a finite and non-empty set of partial substitutions[3]. A variable x is called an *induction variable* of C iff $x \in Dom(\delta)$ for some partial substitution $\delta \in \Delta$. An atomic relation description C defines a relation $<_C$ on the cartesian product $\mathcal{T}(\Sigma^c)_w$ of constructor ground terms by: $q^* <_C r^*$ iff $M[\![x^*/r^*]\!] \models \varphi$ and some $\delta \in \Delta$ exists such that $q_i = M[\![x^*/r^*]\!](\delta(x_i))$ for all $x_i \in Dom(\delta)$. Throughout this paper only atomic relation descriptions of the form $C = (\varphi, \{\delta\})$ will be used.

An *r-description* D for $x^* \in \mathcal{V}_w$ is a finite and non-empty set $\{C_1, ... , C_k\}$ of atomic r-descriptions for x*. A variable x is called an *induction variable* of D iff x is an induction variable for some atomic relation description $C \in D$. Each r-description D defines a relation $<_D$ on $\mathcal{T}(\Sigma^c)_w$ by $<_D := <_{C_1} \cup ... \cup <_{C_k}$. An r-description D is called *well-founded* iff $<_D$ is a well-founded relation.

---

[3]Partial substitutions δ differ from ordinary substitutions in that $\delta(x)$ is undefined for $x \notin Dom(\delta)$, whereas $\sigma(x)=x$ for an ordinary substitution σ, i.e. σ is total. Consequently substitution pairs x/x are meaningful for partial substitutions.

With each function $f \in \Sigma^d$ a relation description D_f is associated which includes for each recursive case $\varphi_i \rightarrow f(x_1...x_n) = r_i$ of f and each f-term $f(t_1...t_n)$ in φ_i or r_i the atomic relation description $(\varphi_i, \{\{x_1/t_1,...,x_n/t_n\}\})$. Note that all $C \in D_f$ are of the special form $(\varphi_i, \{\delta\})$. On relation descriptions the following operations are defined:

Definition An r-description D' for y* is called a *domain generalization* of an r-description D for x* upon the induction variable y of D (and D is called a *domain specialization* of D') iff D' is obtained by removing each substitution pair y/t from each partial substitution in each atomic r-description of D.
Given an atomic r-description $C = (\varphi, \{\delta_1, \delta_2, ..., \delta_n\})$ for $x^* \in \mathcal{V}_w$ and a unique variable renaming ν for the variables in x*, the atomic r-description $\nu(C) = (\nu(\varphi), \{\nu(\delta_1), \nu(\delta_2), ..., \nu(\delta_n)\})$ for $\nu(x^*)$ where each $\nu(\delta_i)$ is obtained from δ_i by replacing each substitution component x/t in δ_i with $\nu(x)/\nu(t)$ is called a *renamed variant* of C.

Since each well-founded relation description D represents a well-founded relation, we may uniformly associate an induction axiom with D: Given a formula ψ with free variables y* and an r-description $D = \{(\varphi_1, \Delta_1), ..., (\varphi_k, \Delta_k)\}$ for $x^* \in \mathcal{V}$, D is said to be in the *scope* of ψ iff $x^* \subset y^*$. If so, D is associated with $k+1$ so-called *induction formulas* $\psi_0, ..., \psi_k$ given as

$$\psi_0 = \forall y \, [\neg \varphi_1 \wedge ... \wedge \neg \varphi_k \rightarrow \psi] \quad \text{and}$$

$$\psi_i = \forall y \, [\varphi_i \wedge \bigwedge_{\delta \in \Delta_i} \forall u^* \, \psi[\delta'(y^*)] \rightarrow \psi]$$

for $1 \leq i \leq k$. Here the (total) substitution δ' is defined as $\delta'(y) = \delta(y)$, if $y \in \text{Dom}(\delta)$, and $\delta'(y) = u_y$ otherwise, where u_y is a fresh variable. The variables in $\text{Dom}(\delta)$ are called the *induction variables* of the induction axiom. Each subformula $\forall u^* \, \psi[\delta'(y^*)]$ is called an *induction hypothesis* and ψ is the *induction conclusion*. The variables $u \in u^*$, i.e. the variables introduced by δ', remain universally quantified in the induction hypotheses. Therefore the smaller $\text{Dom}(\delta)$ is, the more variables remain universally quantified in the induction hypotheses, and consequently the stronger the induction hypotheses are.

3 Generating Induction Hypotheses

To perform the manipulation of an induction conclusion, the technique of *rippling* [Bundy et al. 93] and *colouring terms* [Hutter 90] has been developed. There, in a first step the syntactical differences between an induction hypothesis and an induction conclusion are marked, e.g. $\forall y. \, \Psi[\underline{\text{pred}(x)}, y]$ is a hypothesis and $\Psi[x, y]$ is the conclusion for a proof of $\forall x, y. \, \Psi[x, y]$. Note that the differences are calculated with respect to a specific induction hypothesis, and so the presence of induction hypotheses prior to a proof is essential for this technique. Following [Hutter 90] syntactical differences will be underlined in this paper and will be called *context* while everything else which does not change during a proof will be called the *skeleton* of an equation[4].

---
[4]In [Bundy et al. 93] the notion of *wave fronts* resp. *holes* is used. A context roughly corresponds to a wave front without the hole.

Also, all theorems and defining equations from the axiom set have to be scanned and the syntactical differences between both sides of the equations are marked, e.g.
- (a) times(plus(x y) z) = plus(times(x z) times(y z))
- (b) times(plus(x y) z) = plus(times(x z) times(y z))
- (c) times(plus(x y) z) = plus(times(x z) times(y z))
- (d) x≠0 → plus(x y) = succ(plus(pred(x) y))
- (e) times(0 x) = x

Note that generally several marked versions, called *C(ontext)-equations*, can be derived from a single unmarked equation, each denoting a different manipulation of the context. In the example above (a) denotes a movement of the context plus(... y) out of the skeleton times(x z), (b) denotes a movement of plus(x ...) out of times(y z) while (c) denotes a movement of times(... z) into plus(x y). Using C-equations, contexts can be inserted (d), moved (a-c) or deleted (e) within the conclusion to obtain a match with an induction hypothesis.

In this paper the approach to mark contexts will be generalized. The generalization is based on the observation that the manipulation of the context does not rely on the actual context provided by the induction hypotheses *but on the position where the context appears*. Thus, if the induction conclusion Ψ can be transformed into a formula Ψ_1, e.g. by symbolic evaluation, the syntactical differences between Ψ_1 and the conclusion Ψ itself can be marked as context. If this context can be moved to positions where it is "tolerable" it is guaranteed that an induction hypothesis will be applicable. Hence, if successful, an induction hypothesis which was not known before the proof can be deduced and there is no need to generate induction hypotheses prior to the proof. To emphasize our point we take a closer look at what constitutes an induction hypothesis.

A formula Ψ' is an induction hypothesis for a formula Ψ if Ψ' is an *instance* of the induction conclusion Ψ, i.e. Ψ' is obtained from Ψ by substituting terms for some variables of Ψ, and secondly Ψ' is *smaller* than Ψ with respect to some well-founded relation. For an induction hypothesis Ψ' to be applicable to the conclusion Ψ we have to manipulate Ψ such that Ψ' appears as a subterm of the manipulated Ψ. Now we are able to formulate where context can be *tolerated*: (1) at top-level of Ψ, because to be applicable an induction hypothesis Ψ' has to occur as some subterm of Ψ, or (2) substituting for variables of the conclusion. We will illustrate tolerable and non tolerable positions for contexts with an example:

For x≠0 the induction conclusion $\Psi[x, y, z]$: times(plus(x succ(y)) z) = times(succ(plus(x y)) z) can be rewritten to

times(succ(plus(pred(x) succ(y))) z) = times(succ(succ(plus(pred(x) y))) z).

The context pred(...) is on tolerable positions (occupying positions corresponding to a variable of the induction conclusion) but the contexts succ(...) are not. A second rewrite step moves the contexts succ(...) to top level where they can be tolerated:

plus(z times(plus(pred(x) succ(y)))) = plus(z times(succ(plus(pred(x) y))))

Now the contexts plus(z ...) also are on tolerable positions (at top level of the formula) and the induction hypothesis Ψ[x/pred(x), y/y, z/z] can be applied.
Context substituted for variables of the conclusion is critical, because different occurrences of the same variable have to be substituted by equal terms (*both* occurrences of x in Ψ[x, y, z] have been substituted by pred(x) in Ψ[x/pred(x), y/y, z/z]) and the terms which have been substituted for the variables of the conclusion have to be smaller with respect to some well-founded order. If context does appear on a position *not* tolerable, then no induction hypothesis can be applied and therefore the context has to be moved. Different contexts at different occurrences of the same variable have to be equated, cf. section 3.3.

3.1 Instantiation of Induction Variables

We will illustrate by an example how a proof can be guided automatically and how appropriate induction hypotheses can be constructed at proof time.
The occurrence of universally quantified variables in induction hypotheses provides useful hypotheses for many induction proofs because these variables can be instantiated with arbitrary terms. We will explain how even induction variables (which are bound in induction hypotheses) can be instantiated - although only to a limited extent. Instantiation of induction variables will lead to an extension of existing guiding methods for induction proofs.

Let us consider a formula which arises as a subgoal in the correctness proof of the quicksort algorithm.

$$\Psi[n, l] : last(qsort(smaller(n\ l))) \leq n$$

Here, last(l) returns the last element of a non-empty list l, smaller(n l) returns a list of those members of the list l which are smaller than or equal to n, bigger(n l) returns a list of those members of the list l which are bigger than n and qsort(l) returns an ordered permutation of the list l. These functions are defined by conditional equations (for brevity we omit obvious cases which do not appear in the example):

(1) tail(l) ≠ nil → last(l) = last(tail(l))

(2) l ≠ nil ∧ ¬ head(l) ≤ n → smaller(n l) = smaller(n tail(l))
(3) l ≠ nil ∧   head(l) ≤ n → smaller(n l) = add(head(l) smaller(n tail(l)))

(4) l ≠ nil → qsort(l) = append(qsort(smaller(head(l) tail(l)))
                                  add(head(l) qsort(bigger(head(l) tail(l)))))

We start the proof by evaluating the subterm smaller(n l). This step will introduce contexts in the induction conclusion which have to be moved away in subsequent proof steps. The cases l=nil and l≠nil ∧ ¬head(l) ≤ n are easily reduced to True (the latter using the induction hypothesis $\Psi[n, \text{tail}(l)]$) and are omitted for the sake of brevity. Note that $\Psi[n, \text{tail}(l)]$ and $\Psi[\text{pred}(n), l]$ (and combinations thereof) are the only induction hypotheses suggested by Recursion Analysis.

Using (3) in case $l \neq \text{nil} \wedge \text{head}(l) \leq n$ we obtain

$\quad$ last(qsort(add(head(l) smaller(n tail(l)))))) $\leq n$.

The context add(head(l) ...) can be moved using (4)

$\quad$ last(app(qsort(smaller(head(l) smaller(n tail(l))))
$\quad\quad\quad\quad$ add(head(l) qsort(bigger(head(l) smaller(n tail(l))))))) $\leq n$

The last formula could also be marked in a different way but then no C-equation will be applicable. Using the lemma $y \neq \text{nil} \rightarrow \text{last}(\text{app}(x\ y)) = \text{last}(y)$ context can be deleted:

$\quad$ last(add(head(l) qsort(bigger(head(l) smaller(n tail(l))))))) $\leq n$

In the same way definition (1) $y \neq \text{nil} \rightarrow \text{last}(\text{add}(n\ y)) = \text{last}(y)$ is useful. The condition is established by a case split. Case qsort(...) = nil yields

$\quad$ head(l) $\leq$ n,

which is one of the conditions governing this case, and qsort(...) $\neq$ nil yields

$\quad$ last(qsort(bigger(head(l) smaller(n tail(l))))) $\leq$ n.

The only symbol still in an intolerable position (thus preventing the application of any induction hypothesis) is the symbol bigger. Moving bigger up over qsort is not promising because many additional contexts would be introduced. But bigger commutes over smaller, so using the lemma bigger(x smaller(y l)) = smaller(y bigger(x l)) we achieve

$\quad$ last(qsort(smaller(n bigger(head(l) tail(l))))) $\leq$ n.

This is the instance $\Psi[n, \text{bigger}(\text{head}(l)\ \text{tail}(l))]$ of the induction conclusion $\Psi[n, l]$. Of course, the well-foundedness of

$\quad D_1 = \{\ (l \neq \text{nil} \wedge \neg\ \text{head}(l) \leq n,\ \{\{n/n,\ l/\text{tail}(l)\}\}),$
$\quad\quad\quad (l \neq \text{nil} \wedge \text{head}(l) \leq n \wedge \text{qsort}(...) \neq \text{nil},\ \{\{n/n,\ l/\text{bigger}(\text{head}(l)\ \text{tail}(l))\}\})\}$

has still to be verified, cf. section 4.2.

Note that in the last step the direction of the movement of the context has been *reversed*. The rippling mechanism [Bundy et al. 93] does not allow context to be moved towards induction variables. But since the conventional approach does not provide the induction hypotheses which are obtained by moving context to induction variables, this kind of movement is not required in the rippling method. In our approach however, we are free to use any instance of the induction conclusion, provided the soundness of the induction axiom can be verified. This can be guaranteed for certain classes of instances for the induction variables, cf. section 4. Therefore, if rippling is chosen to implement the (*rewrite*) rule, rippling has to be extended by a strategy which enables to move context towards induction variables. This will lead to an increased branching rate for the proof guiding method, but for the benefit of additional possible proofs. The restriction in section 4.2 for context moved towards induction variables will also keep the additional branching rate low.

3.2 A Calculus for Induction Proofs

We will now formalize the approach taken in the previous section. The steps of an induction proof can be distinguished into three different categories:
- manipulation of induction conclusions using conditional C-equations from the axiom set,
- application of potential induction hypotheses;
- reasoning about the conditions which govern the respective proof branches or recognizing that the induction conclusion has been transformed into a tautology.

This distinction is reflected by the following calculus.

Objects of the calculus are triples of the form $< \Phi, D, \psi >$ where
Φ is a formula, the condition which governs the current proof branch,
D is a relation description denoting the applied induction hypotheses,
ψ is a formula, the intermediate result into which the original statement has been transformed.

$H = \rho \rightarrow t_1 = t_2$ is the statement which is currently under verification.

The calculus has three *inference rules*

$\quad$ (*rewrite*) $\quad$ applies conditional equations from the axiom set

$$\frac{< \Phi, D, \psi >}{< \Phi \wedge \sigma(\varphi), D, \psi[u \leftarrow \sigma(r_2)] > \ | \ < \Phi \wedge \neg \sigma(\varphi), D, \psi >}$$

if $\varphi \rightarrow r_1 = r_2 \in AX$ and σ is a substitution such that $\psi|_u = \sigma(r_1)$.

$\quad$ (*postulate*) $\quad$ postulates and applies a (potential) induction hypothesis

$$\frac{< \Phi, D, \psi >}{< \Phi \wedge \sigma(\rho), D \cup \{(\Phi \wedge \sigma(\rho), \{\sigma\})\}, \psi[u \leftarrow \sigma(t_2)] > \ | \ < \Phi \wedge \neg \sigma(\rho), D, \psi >}$$

if $H = \rho \rightarrow t_1 = t_2$ is the goal of the proof, σ is a substitution such that $\psi|_u = \sigma(t_1)$.

(*stop*) closes a branch of the proof

$$\frac{<\Phi, D, \psi>}{<\Phi, D, \text{TRUE}>}$$

if either the condition Φ is unsatisfiable, or $\Phi \rightarrow \psi$ is a tautology.

Deduction trees correlate the calculus to induction proofs
- $<\text{TRUE}, \{\}, H>$ is a deduction tree for H,
- if T is a deduction tree for ψ, then T* is also a deduction tree for ψ iff T* results from T by application of one of the rules above to a leaf of T,
- a branch of T is *closed* if it contains a leaf $<\Phi, D, \text{TRUE}>$, a deduction tree T is closed if all branches of T are closed.
- a deduction tree T for ψ represents an (induction) *proof* of ψ iff T is closed and the union D_{ind} of all relation descriptions D_i in the leaves of T is a well-founded relation description.

When applying the cases of a defined function, a case split according to the function definition should be performed (as long as the case split is complete and the conditions mutually exclude each other). This can easily be simulated by repeated application of (*rewrite*) with subsequent application of (*stop*) to proof branches with unsatisfiable condition sets.

The abstract formalization preserves the decomposition of an induction proof into the subproblems to select an appropriate induction relation and to manipulate the induction conclusion. Each of the rules will be implemented by a specialized calculus, e.g (*rewrite*) can be implemented by an extension of the rippling method while (*stop*) can be implemented by any first order calculus.

3.3 Equating Instances of Induction Variables

We resume the example from the introduction to demonstrate how the method can be used to equate different instances for different occurrences of the same variable of the induction conclusion (which also prevent the application of any induction hypothesis).

The functions Perm and delete are defined by the following conditional equations:

(5) $k = \text{nil} \wedge l = \text{nil}$ $\rightarrow \text{Perm}(k\ l) = \text{True}$
(6) $k = \text{nil} \wedge l \neq \text{nil}$ $\rightarrow \text{Perm}(k\ l) = \text{False}$
(7) $k \neq \text{nil}$ $\rightarrow \text{Perm}(k\ l) = \text{Member}(\text{head}(k)\ l) \wedge$
                                                          $\text{Perm}(\text{tail}(k)\ \text{delete}(\text{head}(k)\ l))$,

(8) $l = \text{nil}$ $\rightarrow \text{delete}(n\ l) = l$
(9) $l \neq \text{nil} \wedge n = \text{head}(l)$ $\rightarrow \text{delete}(n\ l) = \text{tail}(l)$
(10) $l \neq \text{nil} \wedge n \neq \text{head}(l)$ $\rightarrow \text{delete}(n\ l) = \text{add}(\text{head}(l)\ \text{delete}(n\ \text{tail}(l)))$

As a first step we evaluate both sides of the equation in Ψ[x, y]: Perm(x y) = Perm(y x). Non-recursive cases of Perm trivially yield True, and for x ≠ nil ∧ y ≠ nil the recursive case (7) yields

$\quad$ Member(head(x) y) ∧ Perm(tail(x) delete(head(x) y))
= Member(head(y) x) ∧ Perm(tail(y) delete(head(y) x))

Here all contexts are on tolerable positions but different occurrences of the same variable are instantiated by different terms, e.g. x is instantiated both by t_1=tail(x) and by t_2= delete(head(y) x). To obtain True both Perm-terms have to be eliminated. This can either be achieved by (a) applying the non-recursive cases (5), (6) of Perm, (b) by transforming the formula, such that either both Perm-terms are equal or (c) an instance of Ψ[x, y] can be used as an induction hypothesis. The first two possibilities (a) and (b) both get stuck immediately. But an instance of Ψ could be obtained if t_1=tail(x) and t_2=delete(head(y) x) (resp. tail(y) and delete(head(x) y)) were equal. Using x≠nil, x can be replaced by add(head(x) tail(x)) in t_2= delete(head(y) x), introducing t_1 as subterm of t_2. Thus to equate t_1 and t_2 the context delete(head(y) add(head(x) ...)) has to be deleted or moved. We continue with a case analysis suggested by the definition of delete: if head(x)=head(y) then Member(head(x) y) (resp. Member(head(y) x)) evaluates to True and using (9) we obtain

$\quad$ Perm(tail(x) tail(y)) = Perm(tail(y) tail(x)),

i.e. an instance Ψ[tail(x), tail(y)] of the induction conclusion Ψ[x, y]. This proof branch can be closed, provided that Ψ[tail(x), tail(y)] can be proven smaller than Ψ[x, y] with respect to some well-founded relation <.

The case head(x)≠head(y) needs a more detailed analysis. Still our aim is to equate the terms tail(x) and delete(head(y) add(head(x) tail(x))), i.e. the marked context has to be eliminated. Using (10) delete(head(y) add(head(x) tail(x))) can be rewritten to add(head(x) delete(head(y) tail(x))). Now add could be removed using (7) if add(head(x) delete(head(y) tail(x))) would occur as first argument of a Perm-term. But the arguments can be permuted in Perm(tail(x) delete(head(y) add(head(x) tail(x)))) using Ψ[delete(head(y) x), tail(y)] as induction hypothesis. Note that we have to apply the induction hypothesis from right to left. Otherwise, i.e. if we would apply the induction hypothesis Ψ[tail(y), delete(head(x) y)], the used induction relation would not be well-founded. Following the same line of argumentation tail(x) and delete(head(x) y) must also be permuted using the induction hypothesis Ψ[tail(x), delete(head(x) y)] and y must be replaced by add(head(y) tail(y)).

Thus the case head(x) ≠ head(y) yields

$\quad$ Member(head(x) y) ∧ Perm(add(head(y) delete(head(x) tail(y))) tail(x))
= Member(head(y) x) ∧ Perm(add(head(x) delete(head(y) tail(x))) tail(y))

which using (10) can be evaluated to

Perm(delete(head(x) tail(y)) delete(head(y) tail(x)))
  ∧ Member(head(x) y) ∧ Member(head(y) tail(x))
= Perm(delete(head(y) tail(x)) delete(head(x) tail(y)))
  ∧ Member(head(y) x) ∧ Member(head(x) tail(y))

But here another instance of the induction conclusion can be applied, and the proof is (nearly) complete. The context delete(head(y) ...) did not need to be moved because the same context had been introduced at the corresponding occurrence of tail(x).

So Perm(x y) = Perm(y x) can be proved using the relation description

D_2 = { (x≠nil ∧ y≠nil ∧ head(x)=head(y), {{x/tail(x), y/tail(y)}}),
  (x≠nil ∧ y≠nil ∧ head(x)≠head(y), {{x/tail(x), y/delete(head(x) y)}}),
  (x≠nil ∧ y≠nil ∧ head(x)≠head(y), {{x/delete(head(y) x), y/tail(y)}}),
  (x≠nil ∧ y≠nil ∧ head(x)≠head(y),
        {{x/delete(head(y) tail(x)), y/delete(head(x) tail(y))} }) }.

However, this induction proof is only *sound* if D_2 is well-founded, cf. section 4.2. Again the proof has been guided without referring to specific induction hypotheses. Instead the hypotheses have been generated during the proof.

4 Soundness of the Computed Induction Axiom

To guarantee the soundness of the computed induction axiom we have to verify the well-foundedness of the computed relation description. This can be seen as a separate proof but it is more efficient to consider also known well-founded relation descriptions. As it will turn out, requirements for computed induction hypotheses can be formulated such that the soundness of the computed induction axiom is automatically guaranteed. This is based on the observation, that in most proofs the instances of the variables introduced by symbolic evaluation as first proof step remain (nearly) unchanged throughout the proof. In section 4.2 we will formulate a requirement which entails the well-foundedness even if the instances introduced by symbolic evaluation are modified during the proof.

4.1 Unchanged Instances of Induction Variables

Our aim is to use well-founded relation descriptions which were computed before to justify the well-foundedness of the generated relation description D_{ind}. As we only accept terminating functions in Σ^d, we already have a large set of well-founded relation descriptions to compare D_{ind} with, viz. the relation descriptions D_f associated with each $f \in \Sigma^d$.

For our comparison we use the *containment principle* as defined in [Walther 92]. An r-description D_1 for $x^* \in \mathcal{V}_u$ is *contained* in an r-description D_2 for $y^* \in \mathcal{V}_v$, in symbols $D_1 \sqsubseteq D_2$, iff for all $(\varphi, \{\delta\}) \in D_1$

$$[\forall x^* \forall y^* \; \varphi \to \bigvee_{(\psi,\{\theta\}) \in D_2} (\psi \wedge \bigwedge_{y \in \text{Dom}(\theta)} \delta(y) = \theta(y))] \in \text{Th}_{\text{Ind}}.$$

Containment compares relation descriptions defined for different sets of variables, and $D_1 \sqsubseteq D_2$ entails $\to_{D_1} \subseteq \to_{D_2}$ where $\to_{D_1}$ and $\to_{D_2}$ are the relations on $\mathcal{T}(\Sigma^c)_w$ defined by D_1 resp. D_2 considered as relation descriptions for some sequence of distinct variables of x^* and y^*. Thus the well-foundedness of D_2 entails the well-foundedness of D_1.

We will now verify that the generated relation description D_{ind} is contained in (a renaming of) a relation description D_f *if no context has been moved towards induction variables*, i.e. $D_{\text{ind}} \sqsubseteq v(D_f)$.

Proof: as a first step in the proof of Ψ the cases $\varphi \to f(x_1, ..., x_n) = r$ of a function definition f will be applied. The corresponding nodes of the deduction tree are labelled with triples $< \sigma(\varphi), \{\}, \Psi[u \leftarrow \sigma(r)] >$ where σ is a matcher of $f(x_1, ..., x_n)$ to a subterm $f(s_1, ..., s_n)$ of the original statement Ψ. If a recursive case has been applied, r contains a subterm $f(t_1, ..., t_n)$, hence $\Psi[u \leftarrow \sigma(r)]$ contains a term $f(\sigma(t_1), ..., \sigma(t_n))$.

The term $f(\sigma(t_1), ..., \sigma(t_n))$ will remain unchanged in subsequent proof steps if no context is moved towards induction variables. Under this assumption, any induction hypothesis applied in this proof branch will instantiate the term $f(s_1, ..., s_n)$ from the induction conclusion by $f(\sigma(t_1), ..., \sigma(t_n))$ in the induction hypothesis, and these induction hypotheses will be applied in proof branches governed by the condition $\sigma(\varphi) \wedge \tau$, where τ is the conjunction of case conditions introduced by subsequent case splits. Hence the corresponding atomic relation description in D_{ind} will be $C = (\sigma(\varphi) \wedge \tau, \{\delta_{\text{ind}}\})$ where $\delta_{\text{ind}} = \{s_i / \sigma(t_i)\}$.

But by construction we have $(\varphi, \{\{x_1/t_1, ..., x_n/t_n\}\}) \in D_f$, or $(\sigma(\varphi), \{\theta\}) = (\sigma(\varphi), \{\{\sigma(x_1)/\sigma(t_1), ..., \sigma(x_n)/\sigma(t_n)\}\}) \in \sigma(D_f)$ if σ is a variable renaming of the induction variables $x_1, ..., x_n$ of D_f. Thus, if the terms s_i are all different variables, we have $\sigma(x_i) = s_i$ and $\sigma(x_i)/\sigma(t_i) \in \theta$, and hence $\delta_{\text{ind}}(u) = \theta(u)$ for all $u \in \text{Dom}(\theta)$.

As $\sigma(\varphi) \wedge \tau \to \sigma(\varphi) \in \text{Th}_{\text{ind}}$ holds as well, $D_{\text{ind}} \sqsubseteq \sigma(D_f)$ is verified, and thus the well-foundedness of $\sigma(D_f)$ entails the well-foundedness of D_{ind}.

Note that it is favorable to compare D_{ind} not with $\sigma(D_f)$ but with a domain generalization $\sigma(D_f')$ of $\sigma(D_f)$ because some requirements would weakened : (1) those terms $\sigma(t_i)$ introduced by evaluation of $f(s_1, ..., s_n)$ corresponding to non-induction variables x_i of D_f' do not have to remain unchanged during the proof, and (2) the substitution σ has only to be a variable renaming of the induction variables of D_f', i.e. the terms s_i in $f(s_1, ..., s_n)$ corresponding to non-induction variables x_i of D_f' do not have to be variables. Therefore it is desirable not only to prove termination for a function $f \in \Sigma^d$ but also to compute the "maximal" domain generalization(s) of D_f as e.g. the method described in [Walther 88] does.

For example, to prove $\Psi[x, y, z]$: times(plus(x succ(y) z) = times(succ(plus(x y)) z) the induction hypothesis $\Psi[x/\text{pred}(x), y/y, z/z]$ has been used. $D = (x \neq 0, \{\{x/\text{pred}(x), y/y,$

z/z}}) is well-founded because $D \subseteq \nu(D'_{plus})=(x{\neq}0, \{\{x/pred(x)\}\})$ can be verified.

This result is not surprising as Recursion Analysis would suggest the induction axiom associated with $\sigma(D'_f)$ if an induction heuristic would select $f(s_1, ..., s_n)$ as the subterm of Ψ to base the induction on.

4.2 Moving p-bounded Functions Towards Induction Variables

In section 3 we emphasized that sometimes other instances of induction variables have to be used than Recursion Analysis suggests. Therefore it is not surprising that the criterion just developed does not justify the soundness of the induction axioms of section 3. However, if we restrict the instances which we allow for the induction variables, a separate well-foundedness proof can still be avoided.

For this case we need the concept of *argument bounded functions* which has been developed to increase the degree of automatization for termination proofs, cf. [Walther 88]. A function $g \in \Sigma$ is called *p-bounded* if its result is always (count-) smaller or (count-) equal than its input on argument position p. A function g is called argument bounded iff it is p-bounded for some fixed argument position p. So we have for each p-bounded function g

$$\forall\ x^*.\ \#(g(x_1...x_n)) \leq \#(x_p) \in Th_{ind} \quad \text{for a function } \#: \mathcal{T}(\Sigma) \to \mathbb{N}.[5]$$

Now assume that D describes a relation $<_D$ over $\mathcal{T}(\Sigma)_w$ which satisfies the requirement

$$(*) \quad \forall\ x^*.\ (\delta(x_1)...\delta(x_n)) <_D (x_1...x_n) \to \Sigma_{i \in P}\ \#(\delta(x_i)) < \Sigma_{i \in P}\ \#(x_i)$$
$$\text{for some set } P \subseteq \{1...n\}.$$

With $\#(g(u_1...\delta(x_k)...u_n)) \leq \#(\delta(x_k))$ we have

$$\Sigma_{i \in P}\ \#(\delta'(x_i)) \leq \Sigma_{i \in P}\ \#(\delta(x_i)) < \Sigma_{i \in P}\ \#(x_i)$$

for $\delta'(x_i) = \delta(x_i)$ if $i \neq k$ and $\delta'(x_k) = g(u_1...\delta(x_k)...u_n)$.

Consequently, if the instance $\sigma(t_i)$ of an induction variable s_i (cf. section 4.1) in a well-founded r-description is replaced by a term $g(u_1...u_{p-1}\ \sigma(t_i)\ u_{p+1}...u_n)$ for a p-bounded function g, the obtained r-description is still well-founded.

Using this criterion we can prove the well-foundedness of

$$D_1 = \{\ (l{\neq}nil \wedge \neg\ head(l){\leq}n,\ \{\{n/n,\ l/tail(l)\}\}),$$
$$(l{\neq}nil \wedge head(l){\leq}n \wedge qsort(...){\neq}nil,\ \{\{n/n,\ l/bigger(head(l)\ tail(l))\}\})\}.$$

As bigger is a 2-bounded function, we obtain

---

[5]In [Walther 88] #(t) returns the *size* of a term t defined as the number of reflexive constructors in M(t).

$D_1^* = \{ \, (l \neq nil \land \neg \, head(l) \leq n, \{\{n/n, l/tail(l)\}\}),$
$\qquad (l \neq nil \land head(l) \leq n \land qsort(...) \neq nil, \{\{n/n, l/tail(l)\}\}) \}$

from D_1 by replacing the substitution pair l/bigger(head(l) tail(l)) by l/tail(l). D_1^* is well-founded because $D_1^* \sqsubseteq D_{smaller}$, $D_{smaller}$ is well-founded because the function smaller terminates, and with (*) the well-foundedness of D_1 is guaranteed.

From $D_2 = \{ \, (x \neq nil \land y \neq nil \land ..., \{\{x/tail(x), y/tail(y)\}\})$
$\qquad (x \neq nil \land y \neq nil \land ..., \{\{x/tail(x), y/delete(head(x) \, y)\}\}),$
$\qquad (x \neq nil \land y \neq nil \land ..., \{\{x/delete(head(y) \, x), y/tail(y)\}\}),$
$\qquad (x \neq nil \land y \neq nil \land ..., \{\{x/delete(head(y) \, tail(x)), y/delete(head(x) \, tail(y))\}\}) \, \}$

we can obtain

$D_2^* = \{ \, (x \neq nil \land y \neq nil \land ..., \{\{x/tail(x), y/tail(y)\}\})$
$\qquad (x \neq nil \land y \neq nil \land ..., \{\{x/tail(x), y/y\}\}),$
$\qquad (x \neq nil \land y \neq nil \land ..., \{\{x/x, y/tail(y)\}\}),$
$\qquad (x \neq nil \land y \neq nil \land ..., \{\{x/tail(x), y/tail(y)\}\}) \, \}$

because delete is 2-bounded. As $D_2^* \sqsubseteq D_{Perm}$ does not hold (this is prevented by the substitution pair x/x) the well-foundedness of D_2^* has to be verified in a separate proof, e.g using the procedure from [Walther 88]. If successful D_2 is well-founded, too.

As we have seen moving p-bounded symbols toward induction variables does not violate the well-foundedness of the computed induction axiom. Additionally this restriction can be used to control the branching rate of the (*rewrite*) rule. In practice, well-foundedness tests will be integrated into the proof, such that application of induction hypotheses which would violate the well-foundedness of D_{ind} can be avoided, cf. example in section 3.3.

5 Conclusion

A novel approach for automating explicit induction has been suggested. Information relevant to guide the proof can be obtained without reference to induction hypotheses computed before the proof, hence induction hypotheses required in the proof are generated "on the fly" at proof time. No precomputation of induction hypotheses is required. Following the proposed approach hypotheses can be generated which are not available using conventional techniques based on Recursion Analysis. Also requirements for induction hypotheses are developed which guarantee the soundness of the computed induction axiom without additional well-foundedness proof. These requirements can be tested by reusing well-founded relation descriptions which have been computed for functions at definition time. They can be established for a big class of non-trivial problems. The method is presently being integrated into the INKA system, an induction theorem prover under development at *Technische Hochschule Darmstadt* and *Universität des Saarlandes, Saarbrücken*. First experiments with the method seem promising, for instance the induction axiom for the *Gilbreath Card Trick* has successfully been generated.

Acknowledgements
I would like to thank Christoph Walther and my colleagues Stefan Gerberding, Jürgen Giesl and Thomas Kolbe for comments on earlier versions of this paper.

References

[Aubin 79] R. Aubin, Mechanizing Structural Induction. Theoretical Computer Science, vol. 9, 329-362, 1979.

[Boyer&Moore 79] R.S. Boyer, J S. Moore, A Computational Logic. Academic Press, 1979.

[Bundy et al. 93] A. Bundy et al., Rippling: A Heuristic for Guiding Inductive Proofs. Artificial Intelligence, vol. 62, no. 2, 185-253, August 1993.

[Huet 91] G. Huet, The Gilbreath Trick: A Case Study in Axiomatization and Proof Development in the COQ Proof Assistant, Technical Report 1511, INRIA, 1991.

[Hutter 90] D. Hutter, Guiding Induction Proofs. Proc. 10th Conf. on Automated Deduction, Kaiserslautern, 147-161, Springer LNAI vol. 449, 1990.

[Walther 88] C. Walther, Argument-Bounded Algorithms as a Basis for Automated Termination Proofs, Proc. 9th Conference on Automated Deduction, Argonne, Springer LNAI vol. 310, 1988. Revised version to appear as "On Proving the Termination of Algorithms by Machine" in AI Journal, 1994.

[Walther 92] C. Walther, Computing Induction Axioms, Proc. Conference on Logic Programming and Automated Reasoning, St. Petersburg, Springer LNCS vol. 624, 1992.

[Walther 93] C. Walther, Combining Induction Axioms by Machine, Proc. Int. Joint Conference on Artificial Intelligence, Chambery, 95-101, Morgan Kaufmann Publishers, 1993.

[Walther 94] C. Walther, Mathematical Induction. In: D.M. Gabbay, C.J. Hogger & J.A. Robinson (eds.), Handbook of Logic in Artificial Intelligence and Logic Programming, Vol. 2, Oxford University Press, 1994.

The Search Efficiency of Theorem Proving Strategies *

David A. Plaisted

Department of Computer Science
University of North Carolina at Chapel Hill
Chapel Hill, NC 27599-3175
e-mail: plaisted@cs.unc.edu
and
MPI fuer Informatik
Im Stadtwald
D-66123 Saarbruecken
and
Universitaet Kaiserslautern
Fachbereich Informatik
675 Kaiserslautern

Abstract. We analyze the search efficiency of a number of common refutational theorem proving strategies for first-order logic. We show that most of them produce search spaces of exponential size even on simple sets of clauses, or else are not sensitive to the goal. We also discuss clause linking, a new procedure that uses a reduction to propositional calculus, and show that it, together with methods that cache subgoals, have behavior that is more favorable in some respects.

1 Introduction

We are interested in the sizes of the search spaces produced by clause form refutational theorem proving strategies for first-order logic. This interest is different from that of most logicians who are interested in provability or the length of proofs. For some examples of the latter, see [CR79, Hak85, Urq87, Ede92]. The paper [Let93] studies how accurately the length of a derivation reflects the actual complexity of a proof. Our interest is in the search space size, that is, the number of proofs and partial proofs. This latter measure is more relevant for the efficiency of theorem provers than the size of a minimal proof. The paper [KBL93] shows that many refinements of resolution do not increase a certain measure of search space size by more than a factor of four, but does not compare refinements with one another. We demonstrate some surprising and little appreciated inefficiencies of many common strategies, which may help to explain their poor performance on some kinds of problems. We also discuss the clause linking method [LP92] and methods that cache subgoals and show that they overcome some of these limitations. We present some examples where resolution

---

* This research was partially supported by the National Science Foundation under grant CCR-9108904

has better performance. These analyses are interesting because they do not depend on particular machine architectures or data structures used to implement strategies, and are thus of a more universal nature. We only consider clause form refutational theorem proving methods for first-order logic; it would be interesting to extend this analysis to Hilbert-style, sequent-style, semantic tableau, and other methods. We emphasize Horn clauses, which are common in practice. We analyze the behavior of strategies on propositional Horn sets as well as giving some first-order clauses sets with a similar behavior.

We note that theorem proving methods based on term-rewriting, though often efficient, also have severe inefficiencies in propositional logic. This suggests that term-rewriting techniques alone do not suffice to obtain efficient theorem provers for general first-order logic. In addition, SLD resolution [Llo87] has similar inefficiencies. This may suggest the possibility of basing logic programming on other theorem proving strategies.

2   First order logic and refutational theorem proving

We assume the standard definitions of propositional and first-order logic. For a discussion of first-order logic and theorem proving strategies see [CL73, Lov78, Bun83, WOLB84]. We restrict our attention to clause form first-order refutational theorem proving. A *Horn clause* is a clause having at most one positive literal. Thus $\{\neg P, \neg Q, R\}$ is a Horn clause.

We define the operation of *unit simplification* as follows: Suppose we have a unit clause $\{L\}$ and another clause $\{M_1, ..., M_n\}$, where L and M_1 are complementary. Then, the clause $\{M_1, ..., M_n\}$ can be deleted and replaced by $\{M_2, ..., M_n\}$. This extends to first-order logic; in that case, we require that M_1 be an instance of the negation of L.

We also define *pure literal clause deletion* as follows: Suppose S is a set of clauses and C is a clause in S. Suppose L is a literal in C and there is no literal M in any clause of S such that L and the complement of M are unifiable. Then L is said to be *pure*. Also, in this case, $S - \{C\}$ is unsatisfiable iff S is. So, pure literal clause deletion is the operation of deleting such clauses from S.

We also define *subsumption*. In the propositional setting, a clause C subsumes D if C is a subset of D. In the first-order setting, we say that C subsumes D if C has a (substitution) instance that is a subset of D. Note that if C subsumes D, then C logically implies D. If C is derived, then one can often simplify clause sets by removing subsumed clauses D without losing completeness.

3   Search space formalism

We formalize theorem proving strategies as directed graphs. Formally, a *theorem proving strategy* is a 5-tuple $<S, V, i, E, u>$ where S is a set of states, i maps the input clauses to a set of states, E is a set of *edges* (pairs of states), and u maps S to $\{True, False\}$. Each state s is labeled with a set $label(s)$ of elements from some underlying set V of structures (such as clauses or chains). If an edge (s, t) is in E, this means that t is a possible successor state to s. Thus, (S, E)

is a directed graph. We require that no two distinct edges (s_1, t_1), (s_2, t_2) have $t_1 = t_2$. Thus the graph is a set of trees. Also, u is an unsatisfiability test; $u(s)$ is $True$ if the state s corresponds to a proof of unsatisfiability. We say such a strategy is *complete* if for all sets R of clauses, if R is unsatisfiable then there exists a path from some element of $i(R)$ to a state s such that $u(s)$ is $True$. We say such a strategy is *sound* if R is unsatisfiable whenever there is a path from some element of $i(R)$ to a state s such that $u(s)$ is $True$. A strategy is *linear* if for all s in S, there is a unique t in S such that there is an edge from s to t in E. The intention of this definition is that i and u are computable and of low complexity. Let F_M be the set of ordered pairs $\{(s, \{t : (s,t) \in E\}) : s \in S\}$ for a strategy M. Thus, $F_M(s)$ is the set of successors of a state s. We require that F_M be a function, in the sense that if the labels of s_1 and s_2 are the same then the sets of labels of their successors should also be the same. (Recall that each state is labeled with a set of elements of V.) Also, we intend that F_M should be computable and of low complexity. Often we omit V and write the strategy as a 4-tuple $< S, i, E, u >$.

As an example, we formalize resolution in this way. For this, we have the 5-tuple $< S, V, i, E, u >$ where each state in S is labeled with a finite set of clauses, V is the set of all clauses over some set of predicates and function symbols, $i(R) = \{R\}$ for all R, and (s, t) is in E if t is s together with all resolvents of clauses in s. Thus resolution is a linear strategy, in this formalism. Finally, $u(s) = True$ iff the empty clause is in $label(s)$. Now, resolution formalized in this way is complete, since if R is unsatisfiable, there is a resolution proof of the empty clause from R. Also, resolution is sound. In contrast, model elimination is not linear in this formalism. For model elimination, the labels of the states consist of single chains. Here $i(R)$ is a set of states, one for each clause in R, each state labeled with a singleton set containing a single chain. Also, (s, t) is in E if the chain in the label of t is obtained by a permissible operation (extension, reduction, or contraction) from the chain in the label of s. Thus, strategies that are conventionally thought of as linear, become non-linear in this framework, but strategies that are non-linear like resolution become linear in this framework.

4 Measures of search duplication

We now define some measures of search space duplication for such strategies. For this, we assume that R is a set of propositional clauses, for simplicity, although these ideas can be lifted to first-order logic. We can think of a search space $G = < S, V, i, E, u >$ as a function mapping a set R of clauses to a graph $G(R)$ representing the search space for R. For this, we define an *initial state* to be an element of $i(R)$ and a *final state* to be a state s such that $u(s) = True$. Thus the task of the theorem prover is to find a path from an initial to a final state. We say a state s is *reachable* from R if there is a path from some element of $i(R)$, to s. We are only interested in the nodes s that are reachable from R. Also, we are only interested in edges in E that occur on some such path. So, we define $S(R)$ to be the set of nodes reachable from R. We define $E(R)$ to be the set of edges in E that occur on some path of reachable states. Also, we define $G(R)$ to be the graph $< S(R), E(R) >$. Let $|T|$ be the number of elements in a set T. Then, we are interested to know how $|S(R)|$ depends on the length $c(R)$ of R, represented as a string of characters. For example, is $|S(R)|$ linear in $c(R)$, polynomial in $c(R)$, or exponential in $c(R)$? Also, we are interested in the structure of the states. Recall that S is a set of states, each labeled with a set of structures indicating lemmas or partial proofs. We are interested in how big these sets of structures can become, because this is a meaningful measure of search complexity. Thus, the most meaningful measure of search complexity is the sum, over all s in $S(R)$, of $|label(s)|$. Let us call this measure $||G(R)||$, and refer to it as the total duplication for R.

To further refine this measure, we consider three other measures: 1. The maximum length of a path in $G(R)$. 2. The maximum size of a subset of $S(R)$, no two elements of which are on the same path. 3. The maximum of $|label(s)|$ for all s in $S(R)$. We call the first, the duplication by *iteration*, the second, the duplication by *case analysis*, and the third, the duplication by *combination*. The intuition for this is that the length of a path represents the number of times that search must be iterated. Also, each path represents a case that must be considered in the search, so the second measure indicates the number of cases there are. The third measure concerns the sizes of the labels of the states. If the sizes of the labels are large, then there must be many elements of V in the same state label. However, in common propositional strategies, the elements of V are constructed from the predicates appearing in the input clauses. This means that there must be many combinations of these predicates, hence the term duplication by combination.

For each measure, we are interested in whether it is a constant, polynomial, or exponential in $c(R)$. We are also interested in the size of the total duplication $||G(R)||$. It is not difficult to show that $||G(R)||$ is bounded by the product of these three measures. To see this, we note that $G(R)$ is a tree. Each tree is a union of a set of paths from the root to a leaf. We can thus identify each state of $G(R)$ with a pair $(path, position)$ where the position tells the distance from the root. We thus have that the number of ordered pairs (s, v) such that $v \in label(s)$ is equal to the number of triples $(path, position, v)$ where $v \in label(s)$ for s the state corresponding to $(path, position)$. Thus the sum of such ordered pairs (s, v) is bounded by the product of the number of paths, the length of the longest path, and the number of elements in the largest label. But the sum of such ordered pairs is just $||G(R)||$ and the product is just the product of the three measures of duplication. This shows that the total duplication is bounded by the product of duplication by iteration, combination, and case analysis.

We say the duplication by iteration for R is *constant* if the ratio of the duplication by iteration to $c(R)$ is bounded. We say the duplication by iteration for R is *linear* if the ratio of the duplication by iteration to $c(R)$ is bounded. We say it is *polynomial* if the duplication by iteration is polynomial in $c(R)$. We say it is *exponential* if the duplication by iteration is exponential in $c(R)$. Similarly, we can define what it means for duplication by combination and case analysis to be constant, linear, et cetera. We also define in this way what it means for the total duplication to be linear, et cetera. We say a strategy has *polynomial behavior* if all three kinds of duplication are polynomial, or equivalently, if the total duplication is polynomial. We say a strategy has *exponential behavior* if the total duplication is exponential, or equivalently, if one of the three kinds of duplication is exponential.

If a strategy is linear, then a *round* is an edge in $E(R)$. The rounds are ordered; the *first* round is the edge of the form $(i(R), s)$, the second round is the edge of the form (s, t), and so on, so the edges are ordered by their distance from $i(R)$. Sometimes we use a similar terminology for non-linear strategies. It is often useful to discuss the behavior of the rounds in order to analyze a strategy.

5 Analysis of duplication for various strategies

We are interested in determining the degree of duplication for various strategies and their refinements. In this way we obtain the following chart. The chart is given for Horn clauses. This chart also shows, for each strategy, whether the strategy is goal sensitive. A strategy is *goal sensitive* for Horn clauses if each inference depends on a negative Horn clause; this means that some kind of backward chaining from the goal clauses is being done. In logic programming applications, one considers the negative clauses as goals or queries, and we adopt the same convention here. This seems to be true of many mathematical theorems as well as logic programs. Of course, there is no intrinsic reason why negative literals should be treated differently than positive literals in a more general context. If a strategy G is goal sensitive, then $||G(R)||$ will be empty for sets R of Horn clauses containing no all-negative clauses. We also indicate the search depth; this is the maximum length of a path in the search space from an initial to a final state. This indicates the depth at which a proof can be found. This differs from duplication by iteration; duplication by iteration considers essentially the maximum number of rounds of inference that can be done, whether or not a proof is found. This could conceivably be larger than the search depth.

We would like to emphasize that the functions in this chart are upper bounds, valid for *all* propositional Horn sets. In addition, the bounds are tight, meaning that there are propositional Horn sets for which these bounds are achieved.

Also, we are not considering which search method is used, whether depth-first, breadth-first, best-first, or some other search method. We only consider the total size of the search space. It's possible that a very good search method could lead to better bounds. However, we are not aware of any search method that can improve on the bounds given below. In particular, breadth-first search and depth-first iterative deepening [Kor85, ST85] should explore a portion of the search space having the same size as that indicated here. That is, if any of the bounds are exponential, these search methods will explore an exponential amount of the search space. Also, for theorem proving strategies having exponential search depth, any search method will explore an exponential amount of the search space.

The following abbreviations are used in this table: hyper-res means hyper-resolution, ord means ordering the literals, P_1-ded means P_1-deduction, 3-lit means 3-literal clauses, res means resolution, A-ord means A-ordering, neg means negative, g.o. means good ordering, b.o. means bad ordering, supp means support, ME means model elimination, lemm means lemmas, cach means caching, sprf means the simplified problem reduction format, mprf means the modified problem reduction format, clin means clause linking, f. means forward, b. means backward, and conn means a connection calculus.

Strategy	Search Depth	Combination	Iteration	Case Analysis	Goal Sensitive
hyper-res	linear	linear	linear	O(1)	no
hyper-res, ord	linear	linear	linear	O(1)	no
P_1-ded	linear	exp.	linear	O(1)	no
P_1-ded, 3 lit	linear	linear	linear	O(1)	no
P_1-ded, ord neg	linear	linear	linear	O(1)	no
res, A-ord	linear	exp.	linear	O(1)	no
all-neg res	linear	exp.	linear	O(1)	yes
all-neg res, g.o.	linear	exp.	?	O(1)	yes
all-neg res, b.o.	exp.	exp.	exp.	O(1)	yes
res, neg supp	linear	exp.	linear	O(1)	yes
ME	exp.	O(1)	exp.	exp.	yes
ME, unit lemm	linear	exp.	linear	O(1)	yes
ME, unit lemm, cach	linear	linear	linear	O(1)	yes
MESON	exp.	O(1)	exp.	exp.	yes
MESON, unit lemm, cach	linear	linear	linear	O(1)	yes
sprf, no cach	exp.	O(1)	exp.	exp.	yes
sprf, cach	linear	linear	linear	O(1)	yes
mprf, no cach	exp.	O(1)	exp.	exp.	yes
mprf, cach	linear	linear	linear	O(1)	yes
clin, f. supp	linear	linear	linear	O(1)	no
clin, b. supp,	linear	linear	linear	O(1)	yes
f. conn.	linear	linear	linear	O(1)	no
b. conn.	exp.	O(1)	exp.	exp.	yes

We can make some general observations about this table. The backward chaining strategies are goal-sensitive, but are mostly inefficient. Forward chaining strategies, though efficient for Horn clauses, are not goal-sensitive. All of the strategies that are goal sensitive have exponential duplication, except for the simplified problem reduction format with caching, the modified problem reduction format with caching, and clause linking with backward support. MESON and model elimination with caching and unit lemmas have this property, but the versions that are efficient on Horn clauses are not complete for general first-order clauses. A recent implementation of model elimination and unit lemmas with caching is described in [AS92]. Note that some refinements can be very damaging to a strategy. For example, ordering negative literals can severely degrade the performance of negative resolution.

5.1 Hard sets of clauses for the strategies

We now indicate how the above results were derived. For this we consider the sets of clauses S_n^1, S_n^2, and S_n^3 defined as follows. Note that S_n^1 is unsatisfiable but S_n^2 and S_n^3 are satisfiable.

Let S_n^1 be the set of $n+2$ clauses $\{\{\neg P_1, \neg P_2, ..., \neg P_n, P\}, \{P_1\}, \{P_2\}, ..., \{P_n\}, \{\neg P\}\}$. We sometimes write clauses in Prolog format; a clause $\{P, \neg P_1, ..., \neg P_n\}$ is written as $P : -P_1, ..., P_n$. A clause $\{\neg P_1, ..., \neg P_n\}$ is written as $: -P_1, ..., P_n$. Let S_n^2 be the following clauses, written in Prolog format for readability:

$$\begin{aligned} &goal\ clause &&: -P_{1,n} \\ &type\ 1\ clauses\ P_{i,j} &&: -P_{i+1,j}, P_{i,j-1},\ 1 \le i < j \le n \\ & P_{i,j} &&: -Q_{i+1,j}, Q_{i,j-1},\ 1 \le i < j \le n \\ & Q_{i,j} &&: -P_{i+1,j}, Q_{i,j-1},\ 1 \le i < j \le n \\ & Q_{i,j} &&: -Q_{i+1,j}, P_{i,j-1},\ 1 \le i < j \le n \\ &type\ 2\ clauses\ \ P_{i,i} &&: -P_{i,i+n/2},\ \ i \le n/2 \\ & Q_{i,i} &&: -Q_{i,i+n/2},\ \ i \le n/2 \\ & P_{i,i} &&: -P_{i-n/2,i},\ \ i > n/2 \\ & Q_{i,i} &&: -Q_{i-n/2,i},\ \ i > n/2 \end{aligned}$$

We can think of backward chaining theorem proving strategies on this set of clauses as ways of moving P-pebbles and Q-pebbles around on a graph. Initially, there is a P-pebble on the $(1,n)$ vertex. At each step, we are permitted to remove a pebble. If we remove a P pebble from vertex (i,j), we must either add two P pebbles or two Q pebbles to the two vertices $(i, j-1)$ and $(i+1, j)$ below. If we remove a Q pebble, we must add a P pebble and a Q pebble to these vertices. Note that the parity of the number of Q pebbles never changes unless some Q literal is generated in two or more ways.

Let S_n^3 be the following clauses, in Prolog format:

$$\begin{aligned} &goal\ clause &&: -P_0, Q_0 \\ &type\ 1\ clauses\ P_i &&: -P_{i+1}, P_{i+2},\ 0 \le i < 2n-2 \\ & P_i &&: -Q_{i+1}, Q_{i+2},\ 0 \le i < 2n-2 \\ & Q_i &&: -P_{i+1}, Q_{i+2},\ 0 \le i < 2n-2 \\ & Q_i &&: -Q_{i+1}, P_{i+2},\ 0 \le i < 2n-2 \\ &type\ 2\ clauses\ P_{2n-1} &&: -P_{n-1} \\ & P_{2n} &&: -P_n \\ & Q_{2n-1} &&: -Q_{n-1} \\ & Q_{2n} &&: -Q_n \end{aligned}$$

Here we can think of backward chaining strategies as methods of pebbling a graph. Whenever a pebble is removed from vertex i, pebbles must be added to vertices $i+1$ and $i+2$. As before, there are P pebbles and Q pebbles and the parity of the number of Q pebbles is preserved unless a Q pebble is generated in two ways.

We also introduce the set T_n^2 of clauses which is S_n^2 together with the unit clauses $P_{i,i}$ and $Q_{i,i}$ for $1 \leq i \leq n$. We introduce T_n^3 which is S_n^3 together with the unit clauses P_{2n-2}, P_{2n-1}, Q_{2n-2}, Q_{2n-1}. T_n^2 and T_n^3 are unsatisfiable, but easy if unit simplification is done.

We now indicate why these clauses sets are difficult for some strategies. In S_n^1, there are a large number of negative literals in the non-unit clause, and if an order for resolving them is not specified, many clauses can be generated with some of the negative literals deleted, since there are exponentially many orders for resolving the literals. This causes a problem for forward chaining methods that do not order the negative literals. In S_n^2, for each subgoal $P_{i,j}$ and $Q_{i,j}$, there is a choice of two clauses to resolve with it, each generating two more literals (subgoals). (This corresponds to the two ways of choosing P pebbles and Q pebbles.) These choices each generate more subgoals, each having two choices for a clause to solve it. Therefore, these choices can be made in many ways, generating many combinations of the $P_{i,j}$ and $Q_{i,j}$ for backward chaining methods. Also, for some methods, the same subgoal will be solved repeatedly. This set of clauses was chosen to neutralize the obvious methods of reducing the search space. The type 2 clauses were added so there would be no pure literals, to neutralize pure-literal-clause deletion. In S_n^3 and T_n^3, there are fewer clauses altogether and fewer subgoals at each level. The subgoals P_i and Q_i both depend on P_{i+1}, Q_{i+1}, P_{i+2}, and Q_{i+2}, for all $i < n - 1$.

Now, the strategies that have exponential behavior can often be made more efficient in simple ways, such as adding unit simplification. For example, T_n^2 and T_n^3 can be shown unsatisfiable in polynomial time in this way. However, unit simplification and subsumption do not help S_n^2 and S_n^3 because there are no positive unit clauses in the input. Also, these examples could be made slightly more complicated or lifted to first order logic and would still reveal the same poor behavior, even with unit simplification. Later we give such simple modifications to these sets of clauses. We think it is most illuminating initially to give the simplest examples demonstrating bad behavior.

5.2 Discussion of the behavior of individual strategies

We now discuss the strategies in turn, justifying the entries in the above chart. Often we identify a state with its label, and thus each state is considered as a set of elements of V, though this is not formally correct. First we consider hyper-resolution [Rob65].

Forward chaining strategies Hyper-resolution is equivalent to a sequence of resolutions that eliminate all the negative literals in a clause, by resolving with clauses that are all positive. We assume that subsumed clauses are deleted. After a linear number of rounds, a proof will be obtained or no additional inferences are possible (for Horn sets). P_1-deduction is the strategy that resolves two clauses only if one of them is positive [Rob65]. Assuming that subsumption is only tested after each round of resolution, P_1-deduction behaves worse than hyper-resolution on S_n^1. Any subset of the n subgoals (negative literals) can be generated, leading to 2^n combinations. Even with subsumption deletion, there are exponentially many clauses generated. Ordering the negative literals (so that only literals smallest in the ordering are resolved on) reduces the behavior to polynomial. If the input clauses are restricted to have 3 literals, then there are at most two subgoals per clause, and at most 4 subsets of these exist. Thus the duplication by combination is linear.

A-ordering Resolution with A-ordering [Sla67] is the strategy in which an ordering is specified on predicate symbols, and literals with predicates that are maximal in the ordering, are resolved on first. If subsumed clauses are deleted, we can simulate P_1-deduction and obtain similar complexity results on unsatisfiable clause sets. On satisfiable clause sets, after all the P_1-deductions are simulated, we will still not have a proof, and then A-ordering might have worse behavior than P_1-deduction. One can show that the search depth and the duplication by iteration for A-ordering are always linear, regardless of the ordering. Once a predicate P is resolved on it is in effect eliminated from the set of clauses and can never be resolved on again. After all predicates have been resolved on, the search stops. The duplication by combination for A-ordering can still be exponential, as shown by S_n^2. If we choose the A-ordering in S_n^2 so that the predicates $P_{i,j}$ are ordered by $j - i$, that is, $P_{i,j} > P_{k,l}$ if $j - i > l - k$, then exponentially many combinations of literals are generated. This is so because whenever we resolve on a literal $P_{i,j}$ we have two clauses to choose from, each generating a different combination of literals. The same is true for $Q_{i,j}$. Also, A-ordering is not goal-sensitive. Note that the first-order strategies based on term-rewriting techniques [HR91, BG90] generally reduce to A-ordering methods on clauses without equality. This shows that these methods also suffer from exponential search inefficiency and lack goal sensitivity. However, term-rewriting methods are often very efficient on pure equality problems.

General properties of clause sets We say that a set S of Horn clauses is *well-ordered* if there is a partial ordering $<$ on the predicate symbols so that if $P : -P_1, ..., P_n$ is a clause in S then $P_i < P$ for all i. Note that minimal unsatisfiable Horn sets are well-ordered. We call the minimal such ordering the *well-ordering* of the predicate symbols and abbreviate it by $<_S$. We say that a set S of Horn clauses is *deterministic* if for every predicate symbol P there is at most one clause C in S such that P appears positively in C, that is, $P \in C$. We note that minimal unsatisfiable Horn sets are both well-ordered and deterministic. Well-orderings of clause sets are useful for studying their search space behavior.

All-negative resolution All-negative resolution is like P_1-deduction with signs reversed: One of the parent clauses in a resolution must be all negative. As explained earlier, this strategy does poorly on S_n^2 and S_n^3, generating exponentially many combinations of the subgoals. However, the search depth for all-negative resolution is still linear, and there is no (i.e., constant) duplication by case analysis. To see that the search depth is linear, suppose that S is minimal unsatisfiable (and therefore well-founded). Each resolution involves a literal $\neg P$ from an all-negative clause C and a literal P from another clause D. Now, the effect of the resolution is to replace the literal $\neg P$ in C by the other literals $\neg Q$ in D such that $Q <_S P$. Therefore, if one always chooses a literal $\neg P$ to resolve such that P is maximal in $<_S$, each resolution will reduce the maximum predicates P in literals in the clause, and a proof will be found after a linear number of resolutions. If S is unsatisfiable but not minimal unsatisfiable, this reasoning can still be applied to a minimal unsatisfiable subset of S. If S is satisfiable, then we can still show that the duplication by iteration is linear, but the argument is a little more complicated, as follows.

Theorem 1. *Suppose S is a propositional Horn set containing n different predicate symbols. Let C be some clause generated from S by all-negative resolution. Then there is some clause C' generated from S by not more than n all-negative resolutions such that C' is a subset of C.*

Corollary 2. *After n rounds of all-negative resolution, all such C' will be generated, and so every clause C that can be generated by all-negative resolution will be subsumed by an already-generated clause. Then the search will stop, assuming that subsumed clauses are deleted. Thus the duplication by iteration is linear for all-negative resolution.*

All-negative resolution with ordering We can specify all-negative resolution with an ordering on the predicate symbols. This means that in an all-negative clause C, the predicate symbol P of C that is maximal in the ordering is the only one that is resolved on. One would expect that this would improve the behavior of the strategy, since fewer resolutions are possible. However, if an ordering on predicate symbols is specified, it can actually make the behavior much worse. It is only necessary to order the predicates so that those predicates P smaller in $<_S$ are resolved on first. This corresponds to moving the pebbles first that are farthest from the goal clause. For example, on T_n^3, this can cause each subgoal to be completely solved before working on the others, if we order the predicate symbols so that the P_j and Q_j with high j are resolved first. This can lead to exponential search depth (and duplication by iteration), and still allows exponential duplication by combination on T_n^2 and T_n^3.

A good ordering can lead to a linear search depth. We obtain a good ordering by resolving first on the predicate symbols P that are maximal in $<_S$. For example, on S_n^3 and T_n^3, this means that we resolve the P_j and Q_j with low j first. Then we obtain linear search depth and polynomial behavior. However, a good ordering cannot always reduce the duplication by combination to a polynomial amount. On S_n^2, there is exponential duplication by combination regardless of the ordering on predicate symbols chosen, but the proof is somewhat subtle. The reason for this is that we have to consider all possible orderings of predicate symbols and show that for all of them, the duplication by combination is exponential.

Theorem 3. *All-negative resolution with an ordering on the negative literals produces an exponential search space on S_n^2, regardless of the ordering used. This is still true if subsumed clauses are deleted.*

We note that this exponential bound applies also to all-negative resolution without an ordering on negative literals, and provides a rigorous proof for that case. Many of our other exponential lower bounds are based on this one, so we can also establish them. We don't know whether a good ordering can always lead to linear duplication by iteration for all-negative resolution with ordering. The problematic case is satisfiable clause sets that are not deterministic or well-founded. Another interesting open problem is whether all-negative resolution with a good ordering has exponential behavior on unsatisfiable Horn sets. We only showed this for S_n^2, which is satisfiable. We believe that even with a good ordering, the behavior is exponential. We believe also that A-ordering with a good ordering has exponential behavior on satisfiable clause sets. If true, these results would further indicate that sometimes even a good ordering cannot help the ordering strategies to perform well on easy problems.

SLD-resolution SLD-resolution is similar to all-negative resolution with ordering of the negative literals, and has similar complexity properties. Because of its importance for Prolog, we make some comments concerning it. The actual execution of Prolog programs is more restrictive than SLD-resolution; there is less flexibility in which literals can be resolved. Literals must be chosen for resolution in a last-in first-out manner. This means that the literal of an all-negative clause C chosen to resolve on must be one of the literals most recently added to C. Another difference between SLD-resolution and Prolog is that duplicate subgoals will be deleted from a clause but not from the Prolog execution.

Since Prolog is efficient in practice, we look for special cases where SLD-resolution performs well. For deterministic, well-ordered clause sets, there is an ordering that causes breadth-first SLD-resolution to have polynomial search depth and duplication by iteration and also polynomial duplication by combination. For such programs, A-ordering and all-negative resolution with a good ordering of the predicate symbols also have polynomial search depth and duplication by iteration and polynomial duplication by combination. For all these strategies, the good behavior is obtained by always resolving on the predicate symbols that are maximal in the well-ordering. For arbitrary unsatisfiable Horn sets, these results continue to hold if depth-first search is specified and the proper ordering of clauses and literals is used. To see this, suppose S is an unsatisfiable Horn set, and let T be a minimal unsatisfiable subset of S. Note that T is deterministic and well-ordered. We can order the clauses so that the clauses in T are used before those in S. We note that Prolog cannot always achieve this good behavior because last-in first-out SLD resolution does not always permit the desired ordering of literals and because a given subgoal may be solved repeatedly.

Remaining strategies We now consider the set-of-support strategy. This strategy [WRC65] initially chooses some subset of the input clauses as the *support set*. A clause is *supported* if it is in the support set, or if it is the resolvent of two clauses, at least one of which is supported. The support strategy restricts resolutions to those in which one of the parent clauses is supported. The behavior of resolution with the set-of-support restriction and a negative set of support is the same as all-negative resolution, for Horn clauses. If the set of support is chosen as the all positive clauses, then the behavior is like P_1-deduction except that additional resolvents can be generated.

We next consider model elimination. For Horn sets, assuming that a negative clause is chosen to start, model elimination [Lov69] and the MESON strategy [Lov78] behave essentially the same as SLD-resolution or all-negative resolution with an ordering on the negative literals. For technical reasons, however, we view the search space so that each state has only one chain, since these strategies are "input" strategies.

It is possible to use a lemma mechanism with model elimination and the MESON strategy. This makes use of the fact that when a negative literal $\neg P$ is eliminated by resolution, and all literals descending from $\neg P$ are also eliminated, then we have essentially derived a proof of P. This corresponds to a successful return from a call to the procedure P in Prolog. This means that any further occurrences of $\neg P$ can also be eliminated by the "lemma" P. That is, further calls to the procedure P will also return successfully, so the computation does not need to be repeated. This improves the behavior, as shown in the chart, but the behavior is still exponential.

We now consider caching. By this we mean that failures as well as successful returns from a procedure are remembered. If a procedure was called and failed before, then the computation does not have to be repeated when it is called again. This results in polynomial behavior.

We now consider the MESON strategy. This strategy has behavior like model elimination for Horn clauses, so the bounds are the same. For the MESON strategy, unit lemmas result in behavior like that of model elimination with unit lemmas. The MESON strategy with unit lemmas and caching has behavior like model elimination with unit lemmas and caching.

We consider two problem reduction formats. The simplified and modified problem reduction formats [Pla82, Pla88] simulate Prolog's back chaining mechanism, but are complete for first-order logic. These strategies have behavior much the same as that of model elimination or the MESON strategy, with and without caching.

The clause linking method [LP92] reduces first-order logic to propositional calculus, and then applies a Davis and Putnam-like procedure [DP60]. This reduction to the propositional calculus is done by successively instantiating the clauses using unification with literals of other clauses. A propositional decision procedure is then periodically applied to the resulting clauses. With backward support, this strategy is also goal sensitive. We refer to clause linking as a caching strategy, though it does not explicitly cache subgoals, because the deletion of duplicate instances has a similar effect. Thus we have five *caching strategies* in all.

As for connection calculi [Bib87], there are many of them. The connection calculi make use of connections between literals in (possibly) different clauses to control the search. The chart is only intended to show that they can be implemented to simulate forward reasoning, like hyper-resolution, or the backward chaining resolution strategies. It is also of course possible that connection calculi with behavior like that of the clause linking method exist.

5.3 Preventing unit simplifications

We now give modifications of these clause sets that still display the same exponential behavior, even for strategies used together with unit simplification. These are presented in the form of unsatisfiability-preserving transformations from clauses sets to clause sets.

The following transformation eliminates unit simplifications for back chaining methods: For each Horn clause $L : -L_1, ..., L_n$ where L and all L_i are positive literals, delete this clause and replace it by the clauses $L, P : -L_1, ..., L_n$ and $L : -P$, where P is a new predicate symbol. Note that the first of these is a non-Horn clause, since both L and P are positive literals. Let $U(S)$ be S transformed in this way.

The following transformation also prevents unit simplifications. If S is a set of propositional clauses, let $M(S)$ be a set of monadic first-order clauses with positive unit clauses P in S replaced by $P(a)$, and other clauses in S transformed by replacing positive literals P by $P(x)$ and negative literals $\neg P$ by $\neg P(x)$. Thus a clause $\{P, \neg Q, \neg R\}$ would be replaced by $\{P(x), \neg Q(x), \neg R(x)\}$, but $\{P\}$ would be replaced by $\{P(a)\}$.

5.4 Additional hard sets of clauses

Given a set S of propositional clauses, let $N(S, n)$ be S with clauses $P : -P_1, ..., P_m$ for $m \geq 1$ replaced by $P(x_1...x_n) : -P_1(x_1...x_n), ..., P_m(x_1...x_n)$. Also, a positive unit clause $\{P\}$ is replaced by $P(x_1...x_n) : -Q_1(x_1), ..., Q_n(x_n)$. In addition, negative clauses $: -P_1...P_m$ are replaced by $: -P_1(a, a, ..., a), ..., P_m(a, a, ..., a)$. Finally, the unit clauses $Q_i(a)$ and $Q_i(b)$ are added. These clauses generate exponential search spaces for forward chaining strategies because of the many possible combinations of a and b. Then the sets $N(T_n^2, n)$ and $N(T_n^3, n)$ are unsatisfiable first-order Horn sets that produce exponential behavior for all strategies considered, even with unit simplification, except the caching strategies.

If S is a set of clauses, let $\overline{S}$ be S with the signs of all predicate symbols changed, and predicate symbols systematically renamed to new predicate symbols. Note that this causes forward chaining strategies to behave like backward chaining strategies, and vice versa. Let $Sym(S)$ be $S \cup \overline{S}$. Now, $Sym(S_n^2)$ and $Sym(S_n^3)$ have exponential behavior for hyper-resolution, since negative hyper-resolution (all-negative resolution) is exponential for S_n^2 and S_n^3. $Sym(S_n^2)$ and $Sym(S_n^3)$ also have exponential behavior for all strategies (and refinements) discussed except clause linking and possibly the other caching strategies.

Consider $U(T_n^2)$ and $U(T_n^3)$. $U(T_n^2)$ has exponential behavior for all the back chaining methods even with unit simplification, except possibly the caching strategies, since T_n^2 does. $U(T_n^3)$ can be solved in polynomial time by all-negative resolution with a proper ordering of negative literals.

To defeat forward chaining methods, consider the clause sets $Sym(U(T_n^2))$ and $Sym(U(T_n^3))$. These still have exponential behavior for back chaining methods, except possibly the caching strategies. In addition, they have exponential behavior for forward chaining methods.

Adequacy We propose that all new strategies be analyzed theoretically as we have done, or by running them on some of the above-mentioned clause sets for all n up to say 50 (subject to time and space limitations!). The clause sets that seem most significant for this are S_n^2, S_n^3, T_n^2, T_n^3, $Sym(S_n^2)$, $Sym(S_n^3)$, $U(T_n^2)$, $U(T_n^3)$, $M(T_n^2)$, $M(T_n^3)$, $Sym(U(T_n^2))$, $Sym(U(T_n^3))$, $Sym(M(T_n^2))$, $Sym(M(T_n^3))$, $N(T_2^n, n)$, and $N(T_3^n, n)$. We say a strategy is *adequate* if it runs in polynomial time on all these clause sets, as well as propositional Horn sets, and satisfies the following additional requirements: It should be complete, goal-oriented, and natural. Natural means that the strategy is not specifically designed to do well on these clause sets. We don't know if clause linking is adequate, and know of no other strategy that might be adequate.

5.5 Discussion

The methods of this paper are in a way more discriminating than the results of Haken [Hak85], who showed that resolution is exponential for any refinement. This tends to suggest that all strategies are the same. Our methods discriminate between strategies more finely, and support the argument that some strategies are better than others.

There are some sets of clauses where the behavior of resolution is better than suggested above. For example, [Zam72, Zam89, Tam90, Tam91] have some first-order clause sets where a particular refinement of resolution similar to A-ordering is a decision procedure, but clause linking may generate an infinite search space. On the other hand, clause linking has a finite search space on clause sets containing variables and constant symbols but no function symbols. However, resolution can generate an infinite search space on these clause sets. For example, consider the following clause set: $\{X \neq Y, Y \neq Z, X = Z\}, \{a \neq b\}$. All-negative resolution generates an infinite search space on this clause set.

References

[AS92]   Owen Astrachan and M. Stickel. Caching and lemma use in model elimination theorem provers. In D. Kapur, editor, *Proceedings of the Eleventh International Conference on Automated Deduction*, 1992.

[BG90]   Leo Bachmair and Harold Ganzinger. On restrictions of ordered paramodulation with simplification. In Mark Stickel, editor, *Proceedings of the 10th International Conference on Automated Deduction*, pages 427–441, New York, 1990. Springer-Verlag.

[Bib87]   W. Bibel. *Automated Theorem Proving*. Vieweg, Braunschweig/Weisbaden, 1987. second edition.

[Bun83]   A. Bundy. *The Computer Modelling of Mathematical Reasoning*. Academic Press, New York, 1983.

[CL73]   C. Chang and R. Lee. *Symbolic Logic and Mechanical Theorem Proving*. Academic Press, New York, 1973.

[CR79]   S. A. Cook and R. Reckhow. The relative efficiency of propositional proof systems. *Journal of Symbolic Logic*, 44(1):36–50, March 1979.

[DP60]   M. Davis and H. Putnam. A computing procedure for quantification theory. *Journal of the Association for Computing Machinery*, 7:201–215, 1960.

[Ede92] E. Eder. *Relative Complexities of First-Order Calculi*. Vieweg, Braunschweig, 1992.

[Hak85] A. Haken. The intractability of resolution. *Theoretical Computer Science*, 39:297–308, 1985.

[HR91] J. Hsiang and M Rusinowitch. Proving refutational completeness of theorem-proving strategies: the transfinite semantic tree method. *J. Assoc. Comput. Mach.*, 38(3):559–587, July 1991.

[KBL93] H. Kleine Buening and T. Lettman. Search space and average proof length of resolution. Unpublished, 1993.

[Kor85] R. E. Korf. Depth-first iterative deepening: An optimal admissible tree search. *Artificial Intelligence*, 27:97–109, 1985.

[Let93] R. Letz. On the polynomial transparency of resolution. In *Proceedings of the 13th International Joint Conference on Artificial Intelligence*, pages 123–129, 1993.

[Llo87] J.W. Lloyd. *Foundations of Logic Programming*. Springer-Verlag, Berlin, 1987. 2nd edn.

[Lov69] D. Loveland. A simplified format for the model elimination procedure. *J. ACM*, 16:349–363, 1969.

[Lov78] D. Loveland. *Automated Theorem Proving: A Logical Basis*. North-Holland, New York, 1978.

[LP92] S.-J. Lee and D. Plaisted. Eliminating duplication with the hyper-linking strategy. *Journal of Automated Reasoning*, 9(1):25–42, 1992.

[Pla82] D. Plaisted. A simplified problem reduction format. *Artificial Intelligence*, 18:227–261, 1982.

[Pla88] D. Plaisted. Non-Horn clause logic programming without contrapositives. *Journal of Automated Reasoning*, 4:287–325, 1988.

[Rob65] J. Robinson. Automatic deduction with hyper-resolution. *Int. J. Comput. Math.*, 1:227–234, 1965.

[Sla67] J.R. Slagle. Automatic theorem proving with renameable and semantic resolution. *J. ACM*, 14:687–697, 1967.

[ST85] M.E. Stickel and W.M. Tyson. An analysis of consecutively bounded depth-first search with applications in automated deduction. In *Proceedings of the 9th International Joint Conference on Artificial Intelligence*, pages 1073–1075, 1985.

[Tam90] T. Tammet. The resolution program: able to decide some solvable classes. In *International Conference on Computer Logic, 1988*, pages 300–312, 1990. Springer Verlag LNCS 417.

[Tam91] T. Tammet. Using resolution for deciding solvable classes and building finite models. In *Baltic Computer Science*, pages 33–64, 1991. Springer Verlag LNCS 502.

[Urq87] A. Urquhart. Hard examples for resolution. *J. ACM*, 34(1):209–219, 1987.

[WOLB84] L. Wos, R. Overbeek, E. Lusk, and J. Boyle. *Automated Reasoning: Introduction and Applications*. Prentice Hall, Englewood Cliffs, N.J., 1984.

[WRC65] L. Wos, G. Robinson, and D. Carson. Efficiency and completeness of the set of support strategy in theorem proving. *Journal of the Association for Computing Machinery*, 12:536–541, 1965.

[Zam72] N.K. Zamov. On a bound for the complexity of terms in the resolution method. *Trudy. Mat. Inst. Steklov*, 128:5–13, 1972.

[Zam89] N.K. Zamov. Maslov's inverse method and decidable classes. *Annals of pure and applied logic*, 42:165–194, 1989.

A Method for Building Models Automatically. Experiments with an extension of OTTER

Christophe Bourely*, Ricardo Caferra*, and Nicolas Peltier*
LIFIA-IMAG 46, Avenue Félix Viallet 38031 Grenoble Cedex FRANCE
{bourely| caferra| peltier}@lifia.imag.fr, Phone: (33) 76.57-46.59 or -48.05

Abstract. A previous work on Herbrand model construction is extended in two ways. The first extension increases the capabilities of the method, by extending one of its key rules. The second, more important one, defines a new method for simultaneous search of refutations and models for set of equational clauses. The essential properties of the new method are given. The main theoretical result of the paper is the characterization of conditions assuring that models can be built. Both methods (for equational and non equational clauses) have been implemented as an extension of OTTER. Several running examples are given, in particular a new automatic solution of the ternary algebra problem first solved by Winker.
The examples emphasize the *unified approach* to model building allowed by the ideas underlying our method and the usefulness of using constrained clauses. Several problems open by the present work are the main lines of future work.

1 Introduction

It is trivial to say that the use of models or counterexamples is of great value in all aspects of reasoning. Very early researchers in Automated Deduction tried to incorporate such abilities into their systems. Two pioneer works deserves to be mentioned. The first one is the use of diagrams (models) as a powerful heuristic to prune the proof search in plane geometry in GTM (**G**eometric **T**heorem-proving **M**achine) [GHL83]. The second one is semantic resolution [SLA67] in which models play an essential role in the application of the resolution rule. Semantic resolution corresponds roughly, as Slagle pointed out, to the system used in GTM. It is worth citing a statement in [SLA67] page 695:

"*A disadvantage of semantic strategy is that the program must at present be given a model along with the theorem to be proved. One might hope that someday the program would devise its own models*".

Since then few works have been devoted to model building compared with those dedicated to refutational methods. At the same time, model construction with the help of automated theorem provers is considered to be one of the most outstanding successes in the field [BL84, WOS93]. Among these works the best

---

* This work has been partially supported by ESPRIT-BRA No 6471 "Medlar 2" and PRC-IA (MRE-CNRS, FRANCE)

known is surely Winker's proof of independence of axioms in ternary boolean algebra (see for example [WIN82, WOS93]). Despite this and other striking results Wos wrote recently [WOS93] page 17:

"I also note that our programs are still not effective for model generation".

Some more recent approaches must be mentioned. SATCHMO [MB88] is a specialized theorem prover that applies to a restricted class of (non equational) clauses. It is based on the use of a bottom up strategy, splitting of positive ground clauses and backtracking. Methods by Tammet and Fermüller and Leitsch [FLTZ93, FL92] apply to some classes of non equational clauses and rely on a resolution refinement and can *extract* models with finite domains from a saturated set of inferences. Slaney's method combines clever enumeration and backtracking techniques for building finite models. It has been used to prove some open results for certain classes of quasi-groups (see for example [FSB93]). A more detailed analysis of all these methods is done elsewhere [BCP94]. One common feature of almost all the approaches mentioned above is that they are able to build only *small finite* models and several of the used techniques are applicable *only* for this kind of models. The only exception is [FL92] which can build rather large finite models.

The present paper is an extension of previous work by the authors [CZ91, CZ92]. The method builds models automatically for sets of non equational clauses. Here similar ideas to those used for non equational sets of clauses are applied in extending the method to sets of equational clauses. Both methods allow the building of *finite* and *infinite* models. As the former our new method looks simultaneously for refutation and models. Four examples are given in the paper. Two of them have been taken from other approaches in order to allow a rough comparison between three different methods and to emphasize the *unified* approach of ours.

The plan of the paper is the following: In section 2 we recall very briefly the main features of the method for non equational clauses and give an extension of one of the key rules of the original method. Two running examples (in an extension of OTTER) show the capabilities of the method. Section 3 presents the method for equational clauses and gives its main theoretical properties. Section 4 is devoted to running examples, particularly to a new, fully automated solution of the ternary algebra problem first solved by S. Winker. Conclusion and the main subjects of future research are given in section 5.

2 The method for non equational clauses

In [CZ91, CZ92], methods have been introduced for simultaneous search for refutations and Herbrand models for first-order formulas without equality. The main idea of the approach is to set conditions *both to apply* inference rules and *to avoid* their application. The first conditions correspond to inference rules (or r-rules, "r" for refutation) in the usual sense and the latter correspond to the so called disinference rules (or mc-rules, "mc" for model construction). The r-rules are used in the refutation (proof) component of the method and the mc-rules are used in the model (or counterexample) construction component of the method. Intuitively it is easy to see that from a point of view of refutational (or proof) completeness "nothing is lost" because the mc-rules are *added* to the inference rules. From a technical point of view the key point is to consider *constrained formulas* (or c-formulas), i.e. couples $< formula, constraint >$. Constraints code either the conditions necessary to the application or the impossibility of application of the inference rules and denote the range of the variables of the formulas. Roughly speaking the method associates to each inference rule its disinference counterpart and introduces some essentially new rules, which are not inference rules in the usual sense. Constraints are represented by equational problems, in the sense of [CL89].

Equational problems are formulas containing only equalities and inequalities, connected by "$\wedge$" and "$\vee$", quantified in a particular way ($\exists^*\forall^*$).

A *solution* of an equational problem is a ground substitution of its free variables such that the corresponding logical formula is satisfied in the Herbrand universe. $\perp$ denotes problems without solution. Consequently $P \neq \perp$ means that P has solutions. Solutions of these equational problems define a kind of "dynamic sorts" for n-tuples of variables. These "sorts" are successively refined in order to produce pure literals which are then used to define Herbrand models of satisfiable sets of formulas. If we are looking for Herbrand models we need to solve P in the Herbrand universe (in this case $P \neq \perp$? is decidable).

It is neither possible nor useful to paste here large parts of results published elsewhere. Instead we recall two disinference rules representative enough to give a taste of the method (for technical details see [CZ91, CZ92]):

- The *GPL-rule*, ("**G**enerating **P**ure **L**iteral" rule).
Let S be a finite set of c-clauses. Let c be a c-clause in S and l be a literal in c. Let S' be the greatest subset of S such that each c-clause in S' contains at least one occurrence of a literal l^c complementary to l. The GPL rule computes constraints for c in order to prevent application of bc-resolution (i.e. binary resolution for c-clauses) upon l and l^c between the c-clause c and any of the c-clauses in S'.

Formally, let S be a set of c-clauses and $c : [\![l(\bar{t}) \vee c' : \mathcal{X}]\!]$ be a c-clause in S. The *GPL-rule* is defined as follows:

$$\frac{[\![l(\bar{t}) \vee c' : \mathcal{X}]\!] \quad S}{[\![l(\bar{t}) : \mathcal{X}_{pure}]\!]}$$

where $\mathcal{X}_{pure} = \bigwedge \{\forall \bar{y}.[\neg \mathcal{Y} \vee \bar{s} \neq \bar{t}] : [\![k : \mathcal{Y}]\!] \in S \text{ and } l^c(\bar{s}) \in k\} \wedge \mathcal{X}$ where $\bar{y}$ are the variables in $var(\mathcal{Y}) \cup var(k)$.

$S \cup [\![l(\bar{t}) : \mathcal{X}_{pure}]\!]$ is satisfiable if S is satisfiable.

Remark: GPL can be applied to self-resolving clauses, but the result (a unit clause with contradictory constraint) is useless.

- The one literal rule or unit *bc-dissubsumption rule*.

For a given unit c-clause c, the bc-dissubsumption rule keeps the c-clauses that cannot be subsumed by c and discards those subsumed by c.

Formally, let $c_1 : [\![l(\bar{s}) : \mathcal{X}]\!]$ be a unit c-clause and c_2: $[\![l(\bar{t}) \vee c' : \mathcal{Y}]\!]$ be a c-clause. The *unit bc-dissubsumption rule* is defined as follows (where $\bar{x}$ are the variables in $var(\mathcal{X}) \cup var(l(\bar{t}))$):

$$\frac{[\![l(\bar{t}) \vee c' : \mathcal{Y}]\!] \quad [\![l(\bar{s}) : \mathcal{X}]\!]}{[\![l(\bar{t}) \vee c' : \mathcal{Y} \wedge \forall \bar{x}.[\neg \mathcal{X} \vee \bar{s} \neq \bar{t}]]\!]}$$

The conclusion of this rule is denoted $bc\text{-}Dsub(c_1, c_2)$. c_2 is called the *ancestor* of the conclusion. c_2 can be removed from the set of c-clauses once the unit bc-dissubsumption rule has been applied.

2.1 Extending the GPL rule

It is not too difficult to allow generation of pure literals occurring in self-resolving clauses, as shown by the procedure below.

Procedure EGPL (Extended GPL)
    input: a finite set of c-clauses S
    output: a pure literal
begin
choose $L(t) \in C \in S$
Split S into two disjoints sets S_1 and S_2, such that S_2 contains all the self resolving c-clauses and $S_1 = S - S_2$ % Possibly $S_1 = \emptyset$ %
    repeat
    apply GPL on $L(t)$ (in S_1)
    $S_2' := \bigcup_{c_2 \in S_2} bc\text{-}Dsub([\![L(t) : \mathcal{X}_{pure}]\!], c_2)$
    if $\exists C' = [\![L^c(t') \vee D : W]\!] \in S_2'$ such that $\mathcal{X}_{pure} \wedge t = t' \wedge W \neq \bot$
    then $S_1 := S_1 \cup \{C''\}$; $S_2 := S_2 - \{C''\}$ (where C'' is the ancestor of C')
    else STOP
end

EGPL terminates and $S \cup [\![L(t) : \mathcal{X}_{pure}]\!]$ is satisfiable iff S is satisfiable (the proof comes down trivially to the one in [CZ92] pages 625-626).

Remark: The same procedure applies when dealing with set of equational clauses (but in this case EGPL will not always terminate).

2.2 Extending OTTER

The algorithms corresponding to the methods RAMC1 [CZ92] (including EGPL) and RAMCEC1 (see section 3) have been implemented in C in a SUN4 workstation. OTTER [McC90] has been adapted to the treatment of c-clauses and two modules have been added: one corresponding to model building rules of RAMC1 and RAMCEC1 and another one corresponding to constraint handling (simplification and solving). The module for constraint handling (one of the keys of the system) allows to treat problems of significant size: some rules are compiled and "good" data structures are used. Some simple heuristics have been incorporated.

2.3 Examples

Example 1.
1 [] P(x,x) : TRUE.
2 [] -P(x,y) | P(y,x) : TRUE.
3 [] P(x,y) | -P(f(x),f(y)) : TRUE.
4 [] -P(x,y) | P(f(x),f(y)) : TRUE.
5 [] -P(a,f(x)) : TRUE.

** KEPT: 6 [distautology,2] -P(x,y) | P(y,x) : (x != y).
Clause #2 is deleted.
** KEPT: 7 [dissubsumption,1,3] P(x,y) | -P(f(x),f(y)) : (y != x).
Clause #3 is deleted.
** KEPT: 8 [dissubsumption,1,4] -P(x,y) | P(f(x),f(y)) : (y != x).
Clause #4 is deleted.
** KEPT: 9 [egpl,8] -P(x,y) : (y != x).
Clause #5 is deleted by dissubsumption 9,5.
Clause #6 is deleted by dissubsumption 9,6.
Clause #7 is deleted by dissubsumption 9,7.
Clause #8 is deleted by dissubsumption 9,8.

1 [] P(x,x) : TRUE.
9 [egpl,8] -P(x,y) : (y != x).

Comments: This example shows that RAMC1 can build models for set of clauses belonging to non finitely controllable classes (a class is finitely controllable iff each satisfiable formula in the class has a finite model). It also shows the usefulness of the EGPL, because c-clause 9 cannot be generated by using GPL.

Example 2.
1 [] R(f(x),g(x),x) | -P(x,f(x)) : TRUE.
2 [] P(a,x) : TRUE.
3 [] R(f(a),g(a),a) : TRUE.
4 [] S(f(x),x,x) | L(x) : TRUE.
5 [] -S(f(x),x,a) : TRUE.
6 [] -S(f(x),x,b) : TRUE.
7 [] L(a) : TRUE.
8 [] L(b) : TRUE.

** KEPT: 9 [dissubsumption,8,4] S(f(x),x,x) | L(x) : (b != x).
Clause #4 is deleted.
** KEPT: 10 [dissubsumption,7,9] S(f(x),x,x) | L(x)
: (b != x) & (a != x). Clause #9 is deleted.
** KEPT: 11 [egpl,10] S(f(x),x,x) : (a != x) & (b != x).
Clause #10 is deleted by dissubsumption 11,10.
** KEPT: 12 [dissubsumption,3,1] R(f(x),g(x),x) | -P(x,f(x)) : (a != x).
Clause #1 is deleted.
** KEPT: 13 [egpl,12] R(f(x),g(x),x) : (a != x).
Clause #12 is deleted by dissubsumption 13,12.

```
7  []       L(a)           : TRUE.
8  []       L(b)           : TRUE.
2  []       P(a,x)         : TRUE.
5  []       -S(f(x),x,a)   : TRUE.
6  []       -S(f(x),x,b)   : TRUE.
3  []       R(f(a),g(a),a) : TRUE.
11 [egpl,10] S(f(x),x,x)   : (a != x) & (b != x).
13 [egpl,12] R(f(x),g(x),x) : (a != x).
```

Comments: This example is from [FLTZ93] page 169. It belongs to the AM class and is used by Tammet to illustrate its approach. AM class is the class for which Tammet's method is able to build models. AM is a finitely controllable class and does not contain formulas with equality. Several other examples in [FLTZ93] (for example those of pages 29, 153,...) are treated in a similar simple way.

3 The method for equational clauses

We have extended the method to equational clauses basically by applying the same idea used in GPL but to paramodulation instead of resolution: we shall obtain the *c-disparamodulation* rule. When equality appears in the clauses (and not only in the constraints), constraints must be handled in the context of equational theories and it is no more possible to use the rules in [CL89] valid only in the algebra of finite terms. The rules valid in any algebra must be used instead. In this case there is no general decision procedure for constraint solving. Some decidable cases, though rather simple from a constraint solving point of view can be very useful in model building (see section 4). Next section gives a flavor of the way in which constraints must be treated in the context of equational theories. The possibility of model building comes down to termination criteria, considered in section 3.2.

3.1 Solving constraints in a domain different from the Herbrand universe

Let S be $\{[\![P(x) : x \neq a]\!], [\![\neg P(x) : x \neq b]\!], [\![a = b : \top]\!]\}$.

The intuitive meaning of $[\![C : P]\!]$ is that it denotes the set of all ground instances σC with σ ground solution of P.

If we are looking for a Herbrand model we must solve constraints in the Herbrand universe (i.e. $H = \{a, b\}$) considering the empty theory. The solution of $x \neq a$ is $x = b$ and the one of $x \neq b$ is $x = a$. So we get $\{P(b), \neg P(a)\}$ and taking into account $a = b$ we conclude that S is unsatisfiable. But in domain $D = \{a\}$, S is satisfiable.

This example motivates the necessity of modifying handling constraint procedure used in [CZ92] in order to be able to build non Herbrand models. The result of applying domain-dependent rules (such as the explosion rule in [CL89]) will be of course different when considering a domain $D \neq H$. A theory, as for example $\{b = a\}$, can be incorporated in adding the demodulator (i.e. a rewrite rule) $b \rightsquigarrow a$ as a transformation rule in the constraint solving procedure. Therefore the constraint solving procedure in domain $D = \{a\}$ for S above will look like:

$x \neq a \rightsquigarrow x \neq a \wedge x = a \rightsquigarrow a \neq a \wedge x = a \rightsquigarrow \bot \wedge x = a \rightsquigarrow \bot$
$x \neq b \rightsquigarrow x \neq a \rightsquigarrow^* \bot$
$S \rightsquigarrow^* \{[\![P(x) : \bot]\!], [\![\neg P(x) : \bot]\!], [\![a = b : \top]\!]\} \rightsquigarrow^* \{[\![a = b : \top]\!]\}$
Then S is clearly satisfiable on the domain $D = \{a\}$.

3.2 The new rules: c-paramodulation and c-disparamodulation

The two key rules of the method are c-paramodulation and c-disparamodulation defined below (standard notations are used). Let $h_1 : [\![f(\bar{s}_1) = t_1 : \mathcal{X}]\!]$ and $h_2 : [\![u[f(\bar{s}_2)]_p = t_2 : \mathcal{Y}]\!]$ be two one-literal c-clauses.

- *c-paramodulation* from h_1 into h_2 is defined as:

$$\frac{[\![f(\bar{s}_1) = t_1 : \mathcal{X}]\!] \quad [\![u[f(\bar{s}_2)]_p = t_2 : \mathcal{Y}]\!]}{[\![u[p \leftarrow t_1] = t_2 : \mathcal{X} \wedge \mathcal{Y} \wedge \bar{s}_1 = \bar{s}_2]\!]}$$

- *c-disparamodulation* from h_1 into h_2 is defined as:

$$\frac{[\![f(\bar{s}_1) = t_1 : \mathcal{X}]\!] \quad [\![u[f(\bar{s}_2)]_p = t_2 : \mathcal{Y}]\!]}{[\![u[f(\bar{s}_2)]_p = t_2 : \mathcal{Y} \wedge (\forall \bar{v}. \neg \mathcal{X} \vee \bar{s}_1 \neq \bar{s}_2)]\!]}$$

where $\bar{v}$ is a variable vector defined by the variables in $Var(\mathcal{X}) \cup Var(\bar{s}_1)$.

After application of both rules on h_1 and h_2, h_2 can be deleted. The soundness and refutational completeness of a method using this rule are not too difficult to prove [BCP94]. The proofs are based on similar principles that those in [CZ92]. Intuitively: refutational completeness is preserved because "nothing is lost" for paramodulation.

3.3 The algorithm for model building

Procedure RAMCEC1: Refutation And Model Construction for Equational Clauses - version 1
   input: A finite set S of unit equational constrainted clauses
             A finite set Σ of functions symbols
   output: (if any) UNSATISFIABLE or SATISFIABLE & A MODEL
begin
   $S_1 := S; S_2 := \emptyset$
   while NotFound[$S_1 \cup S_2 := LookingForModel(S_1 \cup S_2)$] **do**
     either $DisParamodulation(S_1)$
     or $ConstraintHandling(S_1 \cup S_2)$
   end-while
   if Found $= \square$ (i.e. the empty c-clause)
     then UNSATISFIABLE
     else *Exhibit* Model
end-procedure % RAMCEC1 %

Procedure LookingForModel
   input: S
   output: $S_1 \cup S_2$ & Found = $(YES|NO|\Box)$
begin
   $S_1 := \emptyset; S_2 := \emptyset$
   for all $[\![C\ :\ P]\!] \in S$
     if $C \equiv \Box$ **then** $S_2 := S_2 \cup [\![C\ :\ P]\!]$ **else** $S_1 := S_1 \cup [\![C\ :\ P]\!]$
   end-for all
   Found := NO
   if $\Box \in S$ **then** Found := $\Box$
   if $TerminationCriteria\,(S_1)$ **then**
     if (**for all** $[\![\Box\ :\ P]\!] \in S_2$ $ConstraintSolving\,(P) = \bot$)
       then Found := YES
       else $S := S_1 \cup S_2$
     end-if
   end-if
end-procedure % LookingForModel %

Procedure DisParamodulation
   input: S_1
   output: $S'1$
begin
   choose $H : [\![f(\overline{x}) = s\ :\ \mathcal{X}]\!] \in S_1$
   for all $CCL \in S_1$ and PossibleChoices
     apply *disparamodulation* from H into CCL
   end-for all
end-procedure % DisParamodulation %

Procedure ConstraintHandling
   input: S
   output: S'
begin
   choose one c-clause in S and one rule in the following list:
     i.e. the list of rules allowing to simplify or solve constraints
     $[\![C\ :\ P]\!] \mapsto [\![C\ :\ P']\!]$, where $P \mapsto P'$ using a sound and preserving rule
     $[\![\Box\ :\ P]\!] \mapsto \Box$, if P has at least one solution in the domain
     $[\![t = u\ :\ \top]\!]$ then Add $(t \leadsto u)$ into *Demodulators*, if $u \prec t$
     $[\![C\ :\ P_1 \vee P_2]\!] \mapsto [\![C\ :\ P_1]\!] \wedge [\![C\ :\ P_2]\!]$
     etc...
end-procedure % ConstraintHandling %

3.4 Termination criteria and model building

Of course RAMCEC1 may not halt. Termination criteria are more complicated to find than for the non equational case. We give two termination criteria. If the first one is used the built model is called *explicit* (i.e. the values of the functions are already in the domain). If the second one is used the built model is called *implicit* (i.e. the values of the functions are computed by an algorithm). Theorem 1 and 4 and Corollary 2 below are the main theoretical results of the paper.

Explicit definition of the model

Let us consider the finite set of c-clauses $S_1 = \{C_i | C_i = [\![f_i^k(\overline{s}_i) = t_i \ : \ P_i]\!]\}$, where $f_i^k \in \Sigma$ (k is the arity), t_i and the s_l in $\overline{s}_i$ are terms and P_i is a constraint in a solved form (i.e. that from which a solution in the domain can always be computed). Let us define:

$D_0 = \Sigma_0$ (Σ_0 denotes a finite set of constants. If several constants are equal only one of them is included in Σ_0)

$(*): D_{n+1} = \{f^k(\overline{x}) \text{ where } \overline{x} \in D_n^k \text{ and } \not\exists \ C_i : [\![f^k(\overline{s}_i) = t_i \ : \ P_i]\!] \in S_1$
such that $(\sigma = mgu(\overline{x}, \overline{s}_i)$ and σ is a solution of $P_i))\}$

$D_\infty = \cup_{n=0}^\infty D_n$.

$(i): \forall i.$ (if σ is a solution in D_∞ of $\overline{x} = \overline{s}_i \wedge P_i$ then $\sigma t_i \in D_\infty$)
$(ii): \forall i \forall j.$ (if $i \neq j, f_i = f_j$ then $\overline{x} = \overline{s}_i \wedge \overline{x} = \overline{s}_j \wedge P_i \wedge P_j$ has no solution)

Remarks:

- condition (i) will allow to build the function (from D_∞^k into D_∞) assigned by the interpretation to f_i^k.
- condition (ii) assures that the model is really one and it means that disparamodulation cannot be applied on S_1.
- In order to see how tests corresponding to conditions (i) and (ii) work, the reader is referred to examples in section 4.

Theorem 1. *Let S_1 a set of c-clauses satisfying (i) and (ii) above. Then a model (possibly infinite) of S_1 can be built automatically.*

Proof. The proof is constructive. We build an interpretation $\mathcal{I}$ (i.e. we define a domain and assign functions to all function symbols in Σ).

- *Domain definition:*

The tests required by condition $(*)$ are decidable: the terms in $\overline{x}$ are ground and P_i is in a solved form, therefore D_n can be built for arbitrary n.

- *Assigning functions to function symbols:*

Let us consider $f^q \in \Sigma$ and $\overline{u} \in D_\infty^q$.
If $\exists C_i = [\![f^q(\overline{s}_i) = t_i \ : \ P_i]\!] \in S_1$ such that $\exists \sigma$ solution of $\overline{u} = \overline{s}_i \wedge P_i$ (note that C_i is *unique* by (ii)) then we define $f^q(\overline{u}) = \sigma t_i$. By condition (i) $\sigma t_i \in D_\infty$ and the function is well defined, else $f^q(\overline{u}) \in D_\infty$ (by $(*)$) and again the function is well defined.

$\mathcal{I}$ is a model of S_1 because (by construction) for all solution σ of $\overline{x} = \overline{s}_i \wedge P_i$ in D_∞, the instance of the clause $\sigma [f_i^k(\overline{s}_i) = t_i]$ is the unique one in $\mathcal{I}$. □

Corollary 2. *Let S be a set of unit equational clauses. Consider S_1 and S_2 as defined in RAMCEC1. If S_1 verifies (i) and (ii) and if for each c-clause $[\![\square \ : \ P]\!] \in S_2$, P can be reduced to $\bot$, then S is satisfiable.*

RAMCEC1 stops and the interpretations it builds are models of S.

Proof. (Trivial) As P in $[\![\square \ : \ P]\!]$ can be reduced to $\bot$, $[\![\square \ : \ P]\!]$ can be deleted from S_2. The model of S_1 built by RAMCEC1 (see Theorem 1) is also a model of S (because $S_2 = \emptyset$). □

Conjecture 3. *If RAMCEC1 can build models (M with domain D_∞) then a decision procedure for solving constraints in M can be explicitly given.*

Remark: In order to better understand the conjecture it should be added that when the theory is constructors-free the constraints can be transformed in the form "definition with constraints" [CL89], page 391 (as in Example 4 in section 4.2). When the theory is not constructors-free then the solved form will be "substitutions with exceptions" [COM91], page 21 (as in Example 3 in section 4.1).

Implicit definition of the model
Let us consider the finite set of c-clauses $S_1 = \{C_i | C_i = [\![f_i^k(\overline{s}_i) = t_i \; : \; P_i]\!]\}$. D_∞ is defined the same way as in explicit definition of the model, but now S_1 is assumed to satisfy (i') and (ii) ($\prec$ denotes an order on ground terms) where:
$(i') : \forall i. (if\ \sigma\ is\ a\ solution\ in\ D_\infty\ of\ \overline{x} = \overline{s}_i \wedge P_i\ then\ \sigma t_i \prec \sigma f_i(\overline{s}_i))$

Remark: (i') allows to compute the functions on D_∞ assigned by the interpretation to the function symbols of Σ.

Theorem 4. *Let S_1 be a set of c-clauses satisfying (i') and (ii) above. Then a model (possibly infinite) of S_1 can be built automatically.*

Proof. The proof is essentially the same as that of Theorem 1, except that (i') only proves that $t_\infty \in D_\infty$ can be computed in a finite number of steps, where $t_\infty =_\mathcal{I} f_i(\overline{u})$. □

Remark: A corollary similar to Corollary 2 could be stated for Theorem 4.

4 Effective building of infinite models for equational clauses

We have intentionally chosen a trivial example to show how the method works. The problem is to build a model (or detect unsatisfiability) for the formula:

$$\forall x \exists y \exists z. [f(y) = x \wedge f(z) = x \wedge y \neq z]$$

4.1 Illustrating the method

Example 3. The corresponding set of c-clauses is:

$$\{[\![f(g_1(x)) = x \; : \; \top]\!], [\![f(g_2(x)) = x \; : \; \top]\!], [\![g_1(x) \neq g_2(x) \; : \; \top]\!]\}$$

Extended OTTER gives:
```
1 [] (f(g1(x)) = x) : TRUE.
2 [] (f(g2(x)) = x) : TRUE.
3 [] (g1(x)) != g2(x)) : TRUE.

** KEPT: 4 [para_into,1,1] (x = y) : (g1(x) = g1(y)).
** KEPT: 5 [para_into,2,2] (x = y) : (g2(x) = g2(y)).
** KEPT: 6 [dis_para_into,1,2] (f(g2(x)) = x) : (all y. g2(x) != g1(y)).
** KEPT: 7 [para_into,1,2] (x = y) : (g1(x) = g2(y)).
** KEPT: 9 [para_into,7,3] (g1(x) != g2(y)) : (g1(x) = g2(y)).

** KEPT: 10 [solving,4] FALSE : (g1(x) = g1(y)) & (x != y).
Clause #4 is deleted.
** KEPT: 11 [solving,5] FALSE : (g2(x) = g2(y)) & (x != y).
Clause #5 is deleted.
** KEPT: 12 [solving,9] FALSE : (g1(x) = g2(y)).
Clause #9 is deleted.

1 [] (f(g1(x)) = x) : TRUE.
6 [dis_para_into,1,2] (f(g2(x)) = x) : (all y. g2(x) != g1(y)).
10 [solving,4] FALSE : (g1(x) = g1(y)) & (x != y).
11 [solving,5] FALSE : (g2(x) = g2(y)) & (x != y).
12 [solving,9] FALSE : (g1(x) = g2(y)).

t = x in D. % clause #1
t = x in D. % clause #6
** Cond 1 verified **
X = g1(x) & X = g2(y) & TRUE & (all z. g2(y) != g1(z)) |--> FALSE.
** Cond 2 verified **
g1(x) = g1(y) & x != y |--> FALSE. % clause #10
g2(x) = g2(y) & x != y |--> FALSE. % clause #11
g1(x) = g2(y) |--> FALSE. % clause #12
** Cond 3 verified **

Found = YES.
f: g1(x) -> x.
f: g2(x) -> x.
f: x -> f(x) if (all y z.x != g1(y) & x != g2(z)).
g1: x -> g1(x).
g2: x -> g2(x).
```

Comments: This particularly simple example is taken from [SHE77] page 32. In most cases of decidability, as Shelah points out, the existence of a model implies the existence of a finite model, but this is not the case for this formula. The following steps go into details of the domain building process.

$D_0 = \{a\}$

$D_1 = \{g_1(a); g_2(a); f(a)\}$

$D_n = \{g_1^{(p)}(f^{(q)}(a)); g_2^{(r)}(f^{(q')}(a)); f^{(n)}(a)\}$ where $p + q = r + q' = n$ and $p \geq 1$ and $r \geq 1$

$D_{n+1} = \{g_1(x)/x \in D_n\} \cup \{g_2(x)/x \in D_n\} \cup \{f(x)/x \in D_n \land x \neq g_1(u) \land x \neq g_2(v)\}$

$D_\infty = \{g_1^{(p)}(f^{(q)}(a)); g_2^{(r)}(f^{(q')}(a)); f^{(n)}(a)\}$ where $p \geq 1$ and $r \geq 1$

$D_\infty = \{g_1(x)/x \in D_\infty\} \cup \{g_2(x)/x \in D_\infty\} \cup \{f(x)/x \in D_\infty \land x \neq g_1(u) \land x \neq g_2(v)\}$

The methods stops given a model because conditions (i) and (ii) are verified and the constraints of S_2 can be reduced to $\bot$ in the considered model.

4.2 The ternary boolean algebra example revisited

The problem is to prove independence of (unnegated) axiom 5 with respect to axioms 1 to 4, otherwise stated: to build a model of 1-4 that is a countermodel of (unnegated) 5 ([WIN82, WOS93]). We give only some steps of the proof, but remarks a, b, c and d should help the reader to understand the method given in section 3. A detailed, commented proof will be included in [BCP94].

Example 4.

```
1 [] (f(x,x,y) = x) : TRUE.
2 [] (f(g(x),x,y) = y) : TRUE.
3 [] (f(x,y,g(y)) = x) : TRUE.
4 [] (f(x,y,f(z,u,v)) = f(f(x,y,z),u,f(x,y,v))) : TRUE.
5 [] (f(a,b,b) != b) : TRUE.

** KEPT: 6 [dis_para_into,1,5] (f(a,b,b) != b) : (b != a).
** KEPT: 7 [para_into,1,5] (a != b).
** KEPT: 8 [dis_para_into,1,2] (f(g(x),x,y) = y) : (x != g(x)).
** KEPT: 9 [para_into,1,2] (g(x) != x).
** KEPT: 11 [dis_para_into,8,10] (f(x,y,g(y)) = x)
 : (y != x) & (g(y) != x).
** KEPT: 19 [dis_para_into,15,17] (f(x,y,f(z,u,v)) = f(f(x,y,z),u,
f(x,y,v))) : (y != x) & (g(y) != x) & (f(x,y,v) != f(x,y,z)).
** KEPT: 20 [para_into,15,17] (f(x,y,f(z,u,v)) = f(x,y,z))
 : (y != x) & (g(y) != x) & (f(x,y,z) = f(x,y,v)).
** KEPT: 22 [dis_para_into,8,16] (x = f(x,y,x)) : (y != x) & (g(y) != x).

1 [] (f(x,x,y) = x) : TRUE.
6 [dis_para_into,1,5] (f(a,b,b) != b) : (b != a).
7 [para_into,1,5] (a != b).
8 [dis_para_into,1,2] (f(g(x),x,y) = y) : (x != g(x)).
9 [para_into,1,2] (g(x) != x).
22 [dis_para_into,8,16] (x = f(x,y,x)) : (y != x) & (g(y) != x).
11 [dis_para_into,8,10] (f(x,y,g(y)) = x) : (y != x) & (g(y) != x).
19 [dis_para_into,15,17] (f(x,y,f(z,u,v)) = f(f(x,y,z),u,f(x,y,v)))
 : (y != x) & (g(y) != x) & (f(x,y,v) != f(x,y,z)).
20 [para_into,15,17] (f(x,y,f(z,u,v)) = f(x,y,z))
 : (y != x) & (g(y) != x) & (f(x,y,z) = f(x,y,v)).
```

Model building
RAMCEC1 is a non deterministic procedure. Several strategies can be employed (and have been experimented) to build a model. The principle used here is to identify c-clauses which are either "big" or allow paramodulation into themselves. For such c-clauses the constraint is negated in order to delete them and to get interesting information for model building. The chosen c-clause is:

```
19 [dis_para_into,15,17] (f(x,y,f(z,u,v)) = f(f(x,y,z),u,f(x,y,v)))
 : (y != x) & (g(y) != x) & (f(x,y,v) != f(x,y,z)).
```

Negation of its constraint allows to test termination criteria on S_1:
```
1 []  (f(x,x,y) = x) : TRUE.
8 [dis_para_into,1,2] (f(g(x),x,y) = y) : (x != g(x)).
23 [solving,11] (f(x,y,z) = x) : (y != x) & (g(y) != x).
```

a) Domain construction:
$D_0 = \{a; b\}$
$D_1 = \{g(a); g(b)\}$
$D_n = \{g^{(n)}(a); g^{(n)}(b)\}$
$D_{n+1} = \{g(x)/x \in D_n\}$
$D_\infty = \{g^{(p)}(a); g^{(q)}(b)\}$
$D_\infty = \{g(x)/x \in D_\infty\}$

b) Conditions (i) and (ii) are verified on this domain?
```
t = x in D. % clause #1
t = y in D. % clause #8
t = x in D. % clause #23
** Cond 1 verified **
(X1 = x) & (X2 = x) & (X3 = y) & TRUE & (X1 = g(z)) & (X2 = z) &
(X3 = u) & (z != g(z)) |--> FALSE. % clause #1 with #8
(X1 = x) & (X2 = x) & (X3 = y) & TRUE & (X1 = z) & (X2 = u) & (X3 = v) &
(u != z) & (g(u) != z) |--> FALSE. % clause #1 with #23
(X1 = g(x)) & (X2 = x) & (X3 = y) & (x != g(x)) & (X1 = z) & (X2 = u) &
(X3 = v) & (u != z) & (g(u) != z) |--> FALSE. % clause #8 with #23
** Cond 2 verified **
```

c) Now we can build the interpretation on the domain D_∞:
```
f: x, x, y -> x.
f: g(x), x, y -> y.
f: x, y, z -> x if (y != x) & (g(y) != x).
g: x -> g(x).
```

d) Constraints in S_2 reduce to $\bot$?
```
(b != a) & (f(a,b,b) = b) |--> FALSE. % clause #6
(a = b) |--> FALSE. % clause 7
(g(x) = x) |--> FALSE. % clause #9
(y != x) & (g(y) != x) & (f(x,y,v) != f(x,y,z)) |--> FALSE. % clause #19
** Cond 3 verified **
```

Here RAMCEC1 stops.

Remark: It is easy to transform the obtained infinite model in a finite one. The cardinality of the domain must be greater or equal to 3. We can define for example $g(a) = b$ and $g(g(g(a))) = a$.

Comments: Several differences between the new solution given here and Winker's solution[WIN82, WOS93] should be remarked: our approach is systematic, model building is fully automated, we give an infinite model and a finite one (obtained by particularizing the former), techniques used are independent of the size of the model, model construction is monotone. A detail of the construction of a model by RAMCEC1 must be underlined: Winker's approach uses the hypothesis g(g(g(x)))=g(x) which is *independent* of the axioms, as proved by our model which is a model of axioms and a counter-model of g(g(g(x)))=g(x).

5 Conclusion and future work

We have presented a method for simultaneous search for refutations and (finite or infinite) models for sets of equational clauses. Hints of why it is refutational complete have been given and criteria assuring that the method will build models for satisfiable set of equational clauses have been identified. A running system designed as an extension of OTTER which implement the method (and a previous one) has been applied to several non trivial examples. In particular a new automatic solution for the independence of axioms in ternary boolean algebra has been found. The main ways of present and future research are:

1. We are presently studying an extension of GPL generating more that one pure literal in a c-clause. This extension enlarges the class of formulas for which models can be built.
2. Study of the classes for which the method is a decision procedure (some results are given in [CZ92]). In this respect we shall compare our approach to those in [BGW93, FLTZ93].
3. Existing ordering strategies can be incorporated to the purely deductive component of the method. Ordering strategies will very likely increase the power of RAMC1 and RAMCEC1.
4. Our method seems to be a good framework to simulate or to incorporate other approaches. A thorough comparison of different approaches are needed and we are working on it [BCP94]. As an example of this work we should mention that we have implemented Tammet's method in order to compare it to ours (some rough comparisons can be already done from the example in section 2). Of course theoretical comparison is also needed.
5. Extension of the class of formulas admitted as constraints will be considered [BCP94].
6. We are studying termination criteria broader than those in section 3.

References

[BCP94] Christophe BOURELY, Ricardo CAFERRA, and Nicolas PELTIER. Model building in automated deduction. Forthcoming, 1994.

[BGW93] Leo BACHMAIR, Harald GANZINGER, and Uwe WALDMANN. Superposition with simplification as a decision procedure for the monadic class with equality. In *Computational Logic and Proof Theory, KGC'93*, pages 83–96. Springer-Verlag, Lecture Notes in Computer Science 713, 1993.

[BL84] W.W. BLEDSOE and D.W. LOVELAND. *Automated Theorem Proving after 25 years*, volume 29 of *Contemporary Mathematics*. American Mathematical Society, Providence, RI, USA, 1984.

[CL89] Hubert COMON and Pierre LESCANNE. Equational problems and disunification. *Journal of Symbolic Computation*, 7:371–475, 1989.

[COM91] Hubert COMON. Disunification: a survey. In Jean-Louis Lassez and Gordon Plötkin, editors, *Computational Logic: Essays in Honor of Alan Robinson*. MIT Press, 1991.

[CZ91] Ricardo CAFERRA and Nicolas ZABEL. Extending resolution for model construction. In *Logics in AI, JELIA'90*, pages 153–169. Springer-Verlag, Lecture Notes in Artificial Intelligence 478, 1991.

[CZ92] Ricardo CAFERRA and Nicolas ZABEL. A method for simultaneous search for refutations and models by equational constraint solving. *Journal of Symbolic Computation*, 13:613–641, 1992.

[FL92] Christian G. FERMÜLLER and Alexander LEITSCH. Model building by resolution. In *Computer Science Logic, CSL'92*, pages 134–148. Springer-Verlag, Lecture Notes in Computer Science 702, 1992.

[FLTZ93] C. FERMÜLLER, A. LEITSCH, T. TAMMET, and N. ZAMOV. *Resolution Methods for the Decision Problem*. Lecture Notes in Artificial Intelligence 679. Springer-Verlag, 1993.

[FSB93] Masayuki FUJITA, John SLANEY, and Frank BENNETT. Automatic generation of some results in finite algebra. In *Proceedings of IJCAI'93*, pages 52–67. Morgan and Kaufmann, 1993.

[GHL83] H. GELERNTER, J.R. HANSEN, and D.W. LOVELAND. Empirical explorations of the geometry theorem-proving machine. In Jörg Siekmann and Graham Wrightson, editors, *Automation of Reasoning, vol. 1*, pages 140–150. Springer-Verlag, 1983. Originally published in 1960.

[MB88] Rainer MANTHEY and François BRY. Satchmo: A theorem prover implemented in PROLOG. In *Proc. of CADE-9*, pages 415–434. Springer-Verlag, Lecture Notes in Computer Science 310, 1988.

[McC90] William W. McCUNE. *OTTER 2.0 Users Guide*. Argonne National Laboratory, 1990.

[SHE77] Saharon SHELAH. Decidability of a portion of the predicate calculus. *Israel Journal of Mathematics*, 28(1-2):32–44, 1977.

[SLA67] James R. SLAGLE. Automatic theorem proving with renamable and semantic resolution. *Journal of the ACM*, 14(4):687–697, October 1967.

[WIN82] Steve WINKER. Generation and verification of finite models and counterexamples using an automated theorem prover answering two open questions. *Journal of the ACM*, 29:273–284, 1982.

[WOS93] Larry WOS. Automated reasoning. *Notices of the American Mathematical Society*, 40(1):15–26, January 1993.

Model Elimination Without Contrapositives

Peter Baumgartner and Ulrich Furbach

Universiät Koblenz
Institut für Informatik
Rheinau 1
56075 Koblenz, Germany

E-mail: {peter,uli}@informatik.uni-koblenz.de

Abstract. We present modifications of model elimination which do not necessitate the use of contrapositives. These restart model elimination calculi are proven sound and complete. The corresponding proof procedures are evaluated by a number of runtime experiments and they are compared to other well known provers. Finally we relate our results to other calculi, namely the connection method, modified problem reduction format and Near-Horn Prolog.

1 Introduction

This paper demonstrates that model elimination can be defined such that it is complete without the use of contrapositives. We believe that this result is interesting in at least two respects: it makes model elimination available as a calculus for non-Horn logic programming and it enables model elimination to perform proofs of mathematical theorems by case analysis.

Let us first explain what we mean by the term "without the use of contrapositives". In implementations of theorem proving systems usually n procedural counterparts $L_i \leftarrow \overline{L}_1 \wedge \cdots \wedge \overline{L}_{i-1} \wedge \overline{L}_{i+1} \wedge \cdots \wedge \overline{L}_n$ for a clause $L_1 \vee \cdots \vee L_n$ have to be considered. Each of these is referred to as a *contrapositive* of the given clause and represents a different entry point during the proof search into the clause. It is well-known that for Prolog's SLD-resolution one single contrapositive suffices, namely the "natural" one, selecting the head A of the clause $A \vee \neg B_1 \vee \cdots \vee \neg B_n$ as entry point. For full first-order systems the usually required n contrapositives are either given more *explicitly* (as in the SETHEO prover [LSBB92]) or more *implicitly* (as in the connection method by allowing to set up a connection with *every* literal in a clause [Ede92]). The distinction is merely a matter of presentation and will be given up for this paper. Now, by a system "without contrapositives" we mean more precisely a system which does not need all n contrapositives for a given n-literal clause.

Model elimination [Lov68] is a calculus, which is the base of numerous proof procedures for first order deduction. There are high speed theorem provers, like METEOR [AS92] or SETHEO [LSBB92]. The implementation of model elimination provers can take advantage of techniques developed for Prolog. For instance, Stickel's Prolog technology theorem proving system (PTTP, [Sti88]) uses Horn clauses as an intermediate language. Hence, it should be straightforward to use model elimination and PTTP in the context of **non-Horn logic programming**. Indeed this possibility is discussed in various papers; however it is discarded by some authors because of the necessity to use contrapositives (e.g. [Lov91, Pla88]). The argument is given by Plaisted [Pla88] explicitly: "In general, however, we feel that the need for contrapositives makes it difficult to view model elimination as a true programming language in the style of Prolog, since the user has less control over the search." Suppose, for example we are given an input clause[1]

$$prove(and(X, Y)) \leftarrow prove(X) \land prove(Y)$$

which can be used within a formalization of propositional calculus. A possible contrapositive is

$$\neg prove(X) \leftarrow \neg prove(and(X, Y)) \land prove(Y)$$

The procedural reading of this contrapositive is somewhat strange and leads to an unnecessary blowing-up of the search space; in order to prove $\neg prove(X)$ one has to prove $prove(Y)$ – a goal which is totally unrelated to $\neg prove(X)$ by introducing a new variable. Such observations had been the motivation for the development of calculi which need no contrapositives, e.g. Loveland's NearHorn-Prolog. Gabbay's N-Prolog [Gab85] when restricted to clause logic is general enough to be instantiated to both NearHorn-Prolog and problem reduction formats ([Pla88], see also Section 5 below; [RL92] contains a comparison of these).

Another motivation for the new calculi is as follows: in proving theorems such as "if $x \neq 0$ then $x^2 > 0$" a human typically uses case analysis according to the axiom $X < 0 \lor X = 0 \lor -X < 0$. This seems a very natural way of proving the theorem and leads to a well-understandable proof. Our modified model elimination procedure carries out precisely such a proof by case analysis. Experimental results with similar examples from calculus and a comparison with other proof procedures are presented in this paper.

The calculi we derive in this paper are a very small modification of model elimination and hence allow for Prolog implementation techniques. They are complete without the use of contrapositives and hence well-suited for logic programming and for theorem proving by "case analysis" or "splitting".

As a more theoretical contribution we will show that one of the NearHorn-Prologs, namely InH-Prolog ([LR89]), can be seen as one of our modified model elimination procedures.

As a final point, we discovered that the connection method ([Bib87, Ede92], [BF93] contains a comparative study) is complete without contrapositives and *without any change to the calculus*. This surprising result is due to a relaxed complementary-literal condition which subsumes the above-mentioned small change in model elimination.

This paper is organised as follows: in the following section we review the model elimination calculus we use as a starting point of our investigation. In section 3 we define various variants of this calculus and give soundness and completeness proofs of the weakest one. In section 4 we give some experimental results with a PTTP-implementation and in section 5 we discuss related work.

---

[1] Taken from [Pla88].

2 Review of Tableau Model Elimination

As a starting point we use a model elimination calculus that differs from the original one presented by [Lov68]; it is described in [LSBB92] as the base for the prover SETHEO. In [BF93] this calculus is discussed in detail by presenting it in a consolution style [Ede91] and comparing it to various other calculi. This model elimination manipulates trees by extension- and reduction-steps. In order to recall the calculus and to state a running example consider the clause set

$$\{\{P, Q\}, \{\neg P, Q\}, \{\neg Q, P\}, \{\neg P, \neg Q\}\},$$

A model-elimination refutation is depicted in Figure 1 (left side). It is obtained by successive fanning with clauses from the input set (*extension steps*). Additionally, it is required that every inner node (except the root) is complementary to one of its sons. An arc indicates a *reduction step*, i.e. the closing of a branch due to a path literal complementary to the leaf literal.

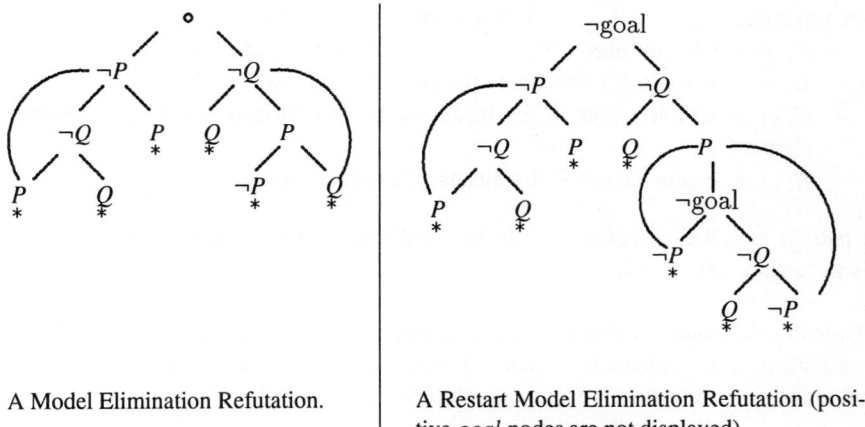

A Model Elimination Refutation. | A Restart Model Elimination Refutation (positive *goal*-nodes are not displayed).

Fig. 1. Model Elimination (left side) vs. Restart Model Elimination (right side) as defined in Section 3.

In the following we use a formal presentation of the calculus along the lines of [BF93]. Instead of trees we manipulate multisets of paths, where paths are sequences of literals.

A clause is a multiset of literals, usually written as the disjunction $L_1 \vee \ldots \vee L_n$. A *connection* in a set of clauses is a pair of literals, written as (K, L), which can be made complementary by application of a substitution. A *path* is a sequence of literals, written as $p = \langle L_1, \ldots, L_n \rangle$. L_n is called the *leaf* of p, which is also denoted by $leaf(p)$; similarly, the first element L_1 is also denoted by $first(p)$. '∘' denotes the append function for literal sequences. Multisets of paths are written with caligraphic capital letters.

Definition 1. (Tableau Model Elimination) Given a set of clauses S.

- The inference rule *extension* is defined as follows:

$$\frac{\mathcal{P} \cup \{p\} \qquad L \vee R}{\mathcal{R}}, \text{ where}$$

  1. $\mathcal{P} \cup \{p\}$ is a path multiset, and $L \vee R$ is a variable disjoint variant of a clause in S; L is a literal and R denotes the rest literals of $L \vee R$.
  2. $(leaf(p), L)$ is a connection with MGU σ.
  3. $\mathcal{R} = (\mathcal{P} \cup \{p \circ \langle K \rangle \mid K \in R\})\sigma$.

- The inference rule *reduction* is defined as follows:

$$\frac{\mathcal{P} \cup \{p\}}{\mathcal{P}\sigma}, \text{ where}$$

  1. $\mathcal{P} \cup \{p\}$ is a path multiset, and
  2. there is a literal L in p such that $(L, leaf(p))$ is a connection with MGU σ.

- A sequence $(\mathcal{P}_1, \ldots, \mathcal{P}_n)$ is called a *model elimination derivation* iff
  $\mathcal{P}_1$ is a path multiset $\{\langle L_1 \rangle, \ldots, \langle L_n \rangle\}$ consisting of paths of length 1, with $L_1 \vee \ldots \vee L_n$ in S (also called the *goal clause*), and
  $\mathcal{P}_{i+1}$ is obtained from $\mathcal{P}_i$ by means of an extension step with an appropriate clause C, or
  $\mathcal{P}_{i+1}$ is obtained from $\mathcal{P}_i$ by means of a reduction step.

The path p is called *selected path* in both inference rules. Finally, a *refutation* is a derivation where $\mathcal{P}_n = \{\}$.

Note that this calculus does not assume a special selection function which determines which path is to be extended or reduced next. Correctness and completeness of this calculus follows immediately from a result proved in [Bau92].

3 Restart Model Elimination Calculi

Let us now modify the calculus given above, such that no contrapositives are needed.

In order to get a complete calculus, we have to assume that there exists only one goal, i.e. a clause containing only negative literals, which furthermore does not contain variables. Without loss of generality this can be achieved by introducing a new clause $\leftarrow goal$ where $goal$ is a new predicate symbol and by modifying every purely negative clause $\neg B_1 \vee \cdots \vee \neg B_n$ to $goal \leftarrow B_1, \ldots, B_n$. In the following we will refer to clause sets S satisfying that property as clause sets in *goal-normal form*. Note that since the $goal$ literal is attached only to purely negative clauses the Horn status of the given clause set is not affected by the transformation. Furthermore, besides the $goal$-normal form we will only allow derivations which start with the goal clause $\leftarrow goal$.

Soundness of this transformation is evident. Completeness holds as follows:

Theorem 2. (Completeness of Model Elimination) *Let S be an unsatisfiable clause set in goal-normal form. Then there exists a tableau model elimination refutation of S with goal clause ← goal. Furthermore, if S is Horn then no reduction steps are required in this refutation.*

We are now ready to modify the calculus, such that no contrapositives are necessary. This will be done in three steps: as a base we define a *restart model elimination* by a small modification in the definition of tableau model elimination, with the result that no contrapositives are needed. Then we weaken this calculus by introducing a *selection function*, which determines which positive head literal can be used for an extension step. Finally, as a further weakening we introduce *strict restart model elimination* by disallowing reduction steps with a positive leaf literal.

A completeness proof is given for the weakest variant, i.e. strict restart model elimination with selection function. Completeness of the stronger variants follows from this result as a simple corollary.

Definition 3. (Restart Model Elimination) Assume the following line additionally given between conditions 1 and 2 in the definition of *extension* (Def. 1).

1a. if $leaf(p)$ is positive then let $p := p \circ \langle first(p) \rangle$ in conditions 2 and 3.

The calculus of *restart model elimination* consists of thus modified extension inference rule and of the reduction inference rule. An extension step with lengthening according to 1a above is also called a *restart*.

If clauses are written in a sequent style $A_1, \ldots, A_n \leftarrow B_1, \ldots, B_m$ then it is clear that, for syntactical reasons extension steps are possible only with head literals A's and not with B's from the body. Thus it is possible to represent clauses as above *without* the need of augmenting them with all contrapositives; only contrapositives with conclusions (i.e. entry points) stemming from the positive literals are necessary.

The price of the absence of contrapositives is that whenever a path ends with a positive literal, the root of the tree, i.e. the clause $\neg goal$ has to be copied and the path has to be lengthened with that literal. Note that there is only a restriction on the applicability of extension steps – reductions are still allowed when a path ends with a positive literal.

In Figure 1 (left side) there is one extension step, which is no longer allowed in restart model elimination, namely the extension of the path $\langle \neg Q, P \rangle$. Note that with our assumptions on goals, this path becomes $p = \langle \neg goal, \neg Q, P \rangle$ in restart model elimination. There is no reduction step possible and since $leaf(p)$ is positive, we lengthen p to $p' = \langle \neg goal, \neg Q, P, \neg goal \rangle$ in a restart step, which can finally be extended to the path multiset

$$\{\langle \neg goal, \neg Q, P, \neg goal, \neg P \rangle, \langle \neg goal, \neg Q, P, \neg goal, \neg Q \rangle\}$$

The complete restart model elimination refutation is depicted in Figure 1 (right side).

It is obvious that for reasons of efficiency a proof procedure based on this calculus must provide some refinements. For example, the use of lemmas or factoring might reduce the amount of redundancy introduced by restart steps. In the example from

Figure 1 (right side) the restart led to the newly introduced paths ending with $\neg P$ and $\neg Q$, respectively. When processing the tree from left to right it is obvious that solving $\neg P$ would be unnecessary, since there is already a closed subtree containing $\neg P$ as a root; thus a proof procedure would benefit extremely from the possibility of using lemmas or factoring [LMG93]. However, we are lucky, the same effect could be achieved by a reduction step. These and other topics concerning proof procedures are discussed in the following section.

Selection Function

Now we weaken the calculus by introducing a selection function on head literals.

Definition 4. (Selection Function) A *selection function* f is a function that maps a clause $A_1, \ldots, A_n \leftarrow B_1, \ldots, B_m$ with $n \geq 1$ to an atom $L \in \{A_1, \ldots, A_n\}$. L is called the *selected literal* of that clause by f. The selection function f is required to be *stable under lifting* which means that if f selects $L\gamma$ in the instance of the clause $(A_1, \ldots, A_n \leftarrow B_1, \ldots, B_m)\gamma$ (for some substitution γ) then f selects L in $A_1, \ldots, A_n \leftarrow B_1, \ldots, B_m$.

Now let f be a selection function, and assume the following line additionally given between conditions 2 and 3 in the definition of *restart* (Def. 3).

2a L is selected in C by f.

This modified calculus is called *(restart) model elimination with selection function*.

Assume there is a path p in our running example with $leaf(p) = \neg P$ and the selection function gives $f(P, Q \leftarrow) = Q$, then it is not allowed to perform an extension step with $P, Q \leftarrow$. Positive literals not selected by f can only be used within reduction steps. Note that the proof in Figure 1 (right side) is a proof with the above assumed selection function.

Strict Restart Model Elimination

As a further restriction we force the calculus to perform restarts whenever they are possible, i.e. if a leaf of a path is a positive literal it may not be used for a reduction step. Since for these leaves extension steps are possible only through a restart, we call this calculus strict restart model elimination. This restriction is motivated in several ways: first, a comparable restriction is formulated within Plaisted's modified problem reduction format ([Pla88], see also Section 5 below) and we would like to evaluate it within our framework; second, strict restart model elimination minimizes the search in ancestor lists, which occasionally results in shorter runtimes to find a proof. See also [Pla90] for restrictions on accessing ancestor lists within a non-restart calculus.

Definition 5. (Strict Restart Model Elimination) The inference rule *reduction* is modified by adding the following line after condition 2 to the definition of *reduction* (Def. 1).

3. and $leaf(p)$ is a negative literal.

Such reduction steps are called *strict reduction steps*. Strict restart model elimination is defined to be the same as restart model elimination, except that "reduction step" is replaced by "strict reduction step".

For the rest of this section we deal with soundness and completeness of restart model elimination. Soundness of the calculus follows immediately by showing that every restart model elimination proof can be mapped to a proof in the free variable semantic tableau calculus ([Fit90]), while completeness will be proven directly.

Theorem 6. **(Soundness)** *Restart model elimination is sound.*

Theorem 7. **(Completeness)** *Let f be a selection function and S be a clause set in goal-normal form. Then there exists a strict restart model elimination refutation of S with goal ← goal and selection function f.*

Since a strict reduction step is by definition also a reduction step we obtain as a corollary the completeness of the non-strict restart model elimination.

Since a selection function restricts the set of permissible derivations, completeness without selection function follows immediately from completeness with selection function.

Here we restrict to the proof on the ground level (Lemma 8 below). Although not quite trivial, lifting can be carried out by using standard techniques. In particular, by stability under lifting (Def. 4) it is guaranteed that the selection function will select on the first order level a literal whose ground instance was selected at the ground level.

Ground completeness reads as follows:

Lemma 8. **(Ground Completeness)** *Let f be a selection function and S be an unsatisfiable ground clause set in goal-normal form. Then there exists a strict restart model elimination refutation with selection function f of S with goal clause ← goal.*

Proof. Informally, the proof is by splitting the non-Horn clause set into Horn sets, assuming by completeness of model elimination refutations without reduction steps, and then assembling these refutations into the desired restart model elimination refutation. There, reduction steps come in by replacing extension steps with split unit clauses by reduction steps to the literals where the restart occurred.

For the formal proof some terminology is introduced: we say that a path multiset $\mathcal{P}$ "contains (an occurrence of) a clause $A_1, \ldots, A_n \leftarrow B_1, \ldots, B_m$" iff for some path p it holds $\{p \circ \langle A_1 \rangle, \ldots, p \circ \langle A_n \rangle, p \circ \langle \neg B_1 \rangle, \ldots, p \circ \langle \neg B_m \rangle\} \subseteq \mathcal{P}$ If we speak of "replacing a clause C in a derivation by a clause $C \cup D$" we mean the derivation that results when using the clause $C \cup D$ in place of C in extension steps. Also, the same literal $L \in C$ must be used to constitute the connection.

By a "derivation of a clause C" we mean a derivation that ends in a path multiset which contains (several occurrences of) the clause C.

Let $k(S)$ denote the number of occurrences of positive literals in S minus the number of definite clauses[2] in S ($k(S)$ is related to the *excess literal parameter* in [AB70]). Now we prove the claim by induction on $k(S)$.

---

[2] A *definite clause* is a clause containing exactly one positive literal.

Induction start ($k(S) = 0$): M must be a set of Horn clauses. By Theorem 2 there exists a model elimination refutation of S with goal $\leftarrow$ *goal* without reduction steps. Furthermore, for syntactical reasons, in every extension step only the single positive literal (and never a negative literal) of the extending clause can be selected. Thus, this refutation is also a strict restart model elimination refutation.

Induction step ($k(S) > 0$): As the induction hypothesis suppose the result to hold for unsatisfiable ground clause sets S' in goal-normal form with $k(S') < k(S)$.

Since $k(S) > 0$, S must contain a non-Horn clause $C = A_1, A_2, \ldots, A_n \leftarrow B_1, \ldots, B_m$ with $n \geq 2$. W.l.o.g. assume that A_1 is the literal selected by f in C. Now define n sets

$$S_1 := (S \setminus C) \cup \{A_1 \leftarrow B_1, \ldots, B_m\}$$
$$S_2 := (S \setminus C) \cup \{A_2\}$$
$$\vdots$$
$$S_n := (S \setminus C) \cup \{A_n\}$$

Every set S_i ($i = 1 \ldots n$) is unsatisfiable (because otherwise, a model for one of them would be a model for S). Furthermore, it holds $k(S_i) = k(S) - n + 1 < k(S)$. Thus, by the induction hypothesis there exist strict restart model elimination refutations R_i with goal clauses $\leftarrow$ *goal* of S_i, respectively.

Now consider R_1 and replace in R_1 every occurrence of the clause $A_1 \leftarrow B_1, \ldots, B_m$ by C. Call this derivation R'_1. Since A_1 is the sole positive literal in $A_1 \leftarrow B_1, \ldots, B_m$, A_1 must have been selected in the extension steps with that clause in R_1. Thus the corresponding extension steps in R'_1 with C are legal in the sense of the restriction to the selection function.

R'_1 is a derivation of, say, k_j occurrences of the positive unit clauses A_j ($j = 2 \ldots n$) from the input set S. Now every A_j can be eliminated from the input set S according to the following procedure: for $j = 2 \ldots n$ append R'_{j-1} k_j times with the refutation R_j, however with the first extension step in R_j being replaced by a restart step at one of the paths ending A_j. Note here that the restart step produces exactly the same goal clause $\leftarrow$ *goal* as is in R_j. Let R''_j be the refutation resulting from these k_j restart steps at leaf A_j. R''_j is a restart model elimination refutation of $S \cup \{A_j, \ldots, A_n\}$. In order to turn R''_j into a refutation of $S \cup \{A_{j+1}, \ldots, A_n\}$, replace every extension step with A_j in the appended refutations R_j by a reduction step to the positive path literal A_j. The resulting refutation R'_j is a desired strict model elimination refutation of $S \cup \{A_{j+1}, \ldots, A_n\}$.

Finally, R'_n is the desired strict restart model elimination refutation of S alone.

Regularity in Restart Theory Model Elimination.

Regularity means for ordinary model elimination that it is never necessary to construct a tableau where a literal occurs more than once along a path. Regularity tends to be one of the more useful refinements of model elimination. Unfortunately, regularity is *not* compatible to restart model elimination. This can be seen easily as the *goal*-literal is copied in restart steps, thus violating the regularity restriction. Hence at least the goal literal has to be excluded from the regularity restriction, because otherwise restart steps are impossible! But even with this exception completeness is lost in general, since after a restart step it might be necessary to repeat – in parts – a proof found so far up to the restart step.

However the following observations allows to define a somewhat weaker notion of regularity: First, the proof of Lemma 8 proceeds by splitting the input clause set into Horn sets and then assembles the existing non-restart refutations $R_1, \ldots R_n$ into a restart refutation R'_n. Since this assembling is done "blockwise" (the R_is are not interleaved among each other and keep their structure) some properties of the R_is carry over to their respective occurrence in R'_n. In particular, the *regularity* of the R_is carries over in this way. Hence we define a path as *blockwise regular (version 1)* iff every pair of occurrences of identical literals (unequal to $\neg goal$) is separated by at least one occurrence of the literal $\neg goal$. A derivation is is called *blockwise regular* iff every path in every of its path multisets is blockwise regular. From these considerations and definitions it follows with Lemma 8 that this restriction is complete. Since all the R_is are refutations of *Horn* clause sets this regularity restriction applies only to *negative* literals along a path. Thus we might derive a blockwise regular path $p = \neg goal \cdots A \cdots \neg goal \cdots A$. We wish to extend blockwise regularity to forbid such duplicate occurrences of positive literals, and say that a branch is *positive regular* iff all the positive literals occurring in it are pairwise distinct (not identical). Extending the preceding definition, we define a branch to be *blockwise regular (final version)* iff it is blockwise regular (version 1) and positive regular. Fortunately it holds:

Theorem 9. *Restart model elimination is complete when restricted to blockwise regular refutations (final version).*

4 PTTP without Contrapositives – Experimental results

The reader familiar with Stickel's PTTP-technique ([Sti88]) may have already noticed that the restart variant can be implemented very easily using the PTTP-technique. Indeed, we implemented the restart model elimination calculus in the theorem proving system PROTEIN ([BF94]). Both the implementation language and the target language for the compiled code is ECL^{i}PS^e, an enhanced Prolog-dialect.

We ran several examples known from the literature, and some new ones. We compared several versions of the prover, varying in strict restart model elimination vs. (non-strict) restart model elimination, and selection function vs. no selection function; the first four columns in Figure 3 contain the runtime results. Column 5 (**MPRF**) contains the results for the *Modified Problem Reduction Format* prover (see also the section on related work below). The option "nosave flag cleared" means that caching is enabled. The data, taken from [Pla88], were obtained on a SUN 3 workstation, whereas the other provers ran on a SUN Sparc 10/40. Hence, for normalisation the times for the MPRF prover were divided by 14.

The default flag settings in our provers include *ground reduction steps* (in reduction steps where no substitution is involved no further proof alternatives need to be explored) and *blockwise regularity* as defined at the end of Section 3. Other flags allow for the generation of *(Unit)-lemmas* (currently *all* lemma candidates are stored) and *factoring*. Unless otherwise noted, in Figure 3 the default flags are used.

For iterative deepening, the threshold was increased in each iteration by 1, and each extension step was uniformly charged with a cost of 1.

Furthermore, we also found it interesting to run standard model elimination provers which use contrapositives ("PTTP" and "Setheo", the right two columns in Figure 3). This demonstrate that in some cases, namely the examples from real-analysis, restart model elimination results in better performance, whereas in the usual benchmarks the results with restart based procedures are often in the same order of magnitude.

Now let us summarize the results. Compared among each other, each of the 4 versions of restart model elimination has its justification by dedicated examples. In the parameter space *non-strict restart* vs. *strict restart* model elimination we prefer as the default strategy the *non-strict* version, since whenever the *strict* version found a proof in reasonable time, the *non-strict* version did as well; but on most examples the *strict* version failed or behaved poorly, while the other version found a proof.

Example	Restart Model Elimination		Strict Restart Model Elimination		MPRF	ME PTTP	ME Setheo
	w/o Selection	w. Selection	w/o Selection	w. Selection			
Non-Obvious MSC/MSC006-1	**2.7**	∞	∞	∞	128 3.2[4]	0.3	0.5
Eder45	3.4 1.7[1]	1.8/10.1[3] 2.5[1]	0.5 0.7[1]	3.1/8.5[3] 0.9[1]		0.7	1.0
Steamroller	9.9	∞	6.3	∞	2246 14.5[4]	1.5	0.18
$x \neq 0 \rightarrow x^2 > 0$	0.7	0.8/46[3]	0.5	0.6/42[3]		2.4	0.8
Bledsoe 1 ANA/ANA003-4	∞	7.4[2]	∞	∞		∞	87
Bledsoe 2 ANA/ANA004-4	∞	32[2]	∞	∞		∞	∞
Natnum3	2.1	0.15	0.6	0.05		0.07	0.03
Wos 4 GRP/GRP008-1	20	∞	43	∞		22	13
Pelletier 48	1.1	4.1	∞	∞		5.9	0.2

Remarks: 1 – With (back) factoring. 2 – With lemmas.
3 – Depends from selected literal. 4 – "nosave" flag cleared.

Fig. 3. Runtime results (in seconds) for various provers.
Entries such as MSC/MSC006-1 refer to the respective TPTP-names [SSY94]. All examples were drawn from that problem library without modification.

In the parameter space *selection function* vs. *no selection function* we will not strictly prefer the one to the other. Note the extreme dependence on the "right" choice in the $x \neq 0 \rightarrow x^2 > 0$-example (this holds also for the Bledsoe examples), while in the Eder-example the right choice is not that crucial. On the other side, the Non-Obvious, Steamroller and Wos 4 examples obviously require the use of several contrapositives.

From this results we learn that the selection function should be carefully determined. Here, heuristics are conceivable such as "always select a biggest (in some ordering) head literal" in order to work in a decreasing direction. The selection function can even be determined dynamically within the bounds of Definition 4.

Currently, the user has to supply the selection function for a given input clause. This function is inherited to all instances of that clause. Here we see potential for further improvements.

But even at the moment our restart provers can well compete with the MPRF prover, which is according to our classification, closest to the strict restart prover with selection function (see also the next section).

Ordinary model elimination as implemented by ME-PTTP and Setheo is sometimes faster than restart model elimination. In the case of Setheo this may especially be due to the numerous refinements not present in the other provers. However there are many examples where restart model elimination finds a proof more quickly, which suggest to us that it is an interesting alternative to traditional model elimination.

5 Related Work

Connection Method. The *connection method* [Bib87] is an analytic calculus closely related to model elimination. Clause sets are called *matrices* there, and a *path through a matrix* is obtained by taking exactly one literal from every clause in the matrix. The method proceeds by systematically checking all paths through the matrix to contain complementary literals. If this is the case, a refutation has been found.

A somewhat higher-level formulation of the connection method can be found in [Ede92], and in [BF93] we showed that this connection method can, in steps, simulate model elimination. The converse, however, is not true for the following essential difference between the connection method and model elimination: in model elimination in extension steps a complementary pair of literals (called *connection*) must be established between the *leaf* literal where the extension occurred and some literal of the extending clause. In the connection method this restriction is dropped, and so every literal along the path (or even none) may be part of the connection.

This property is also the key for the observation stated in the introduction, namely that the connection method is complete without the use of contrapositives. In order to see this, recall that a restart step consists of copying the first literal of the path, followed by an extension step. Thus, copying is not necessary if the first literal in the path is accessible for the connection — as is the case in the connection method. Hence we get as a corollary to theorem 7, the completeness of strict restart model elimination with selection function:

Corollary 10. *The connection method is complete for input sets in goal-normal form, even if no contrapositives are used.*

Problem Reduction Formats. In [Pla88] two calculi named *simplified problem reduction format* and *modified problem reduction format* are described. They are goal-oriented, and neither of these needs contrapositives. We will discuss both of them.

The *simplified problem reduction format* (SPRF) is a variant of the Gentzen sequent calculus (see e.g. [Gal87]). A sequent is pair, written as $\Gamma \to L$ where Γ is a list of literals, and L is a literal. From the model elimination point of view a sequent $\Gamma \rightarrow L$ corresponds to the path $\Gamma \circ \langle \neg L \rangle$, i.e. the goal L is to be proven in the context (ancestor list) Γ.

Clauses are translated to inference rules operating on sequents; a clause $L \leftarrow L_1, \ldots, L_n$ is translated into the inference rule (where Γ is a variable)

$$\frac{\Gamma \rightarrow L_1 \quad \cdots \quad \Gamma \rightarrow L_n}{\Gamma \rightarrow L}$$

The interesting case is to see how model elimination restart steps can be mapped to derivations in SPRF. Suppose we have in a restart model elimination derivation a leaf $\neg p$ and wish to extend with the clause $p, q \leftarrow r$. After copying, the situation looks as depicted in Figure 2 (left side). This situation can be mirrored in SPRF by the partial proof in Figure2 (right side).

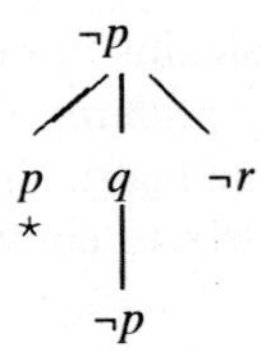

$$\frac{\dfrac{not(q) \rightarrow not(q) \quad \mathbf{not(q) \rightarrow r}}{not(q) \rightarrow p} R_1 \quad \mathbf{q \rightarrow p}}{\rightarrow p} \text{Split}$$

Fig. 2. Restart Model Elimination vs. Simplified Problem Reduction Format.

The *Split* rule is in effect the cut-rule, and R_1 stems from the clause $p, q \leftarrow r$. While the sequent $not(q) \rightarrow not(q)$ is an instance of an axiom, the boldface sequents are unproved. Note the close relationship to restart model elimination: the sequent $q \rightarrow p$ immediately corresponds to the goal $\neg p$ with ancestor list $\neg p \circ q$ in restart model elimination; it is even identical in strict restart model elimination, as negative ancestors need not be stored. For the other sequent $not(q) \rightarrow r$ note that the corresponding goal $\neg r$ in restart model elimination does not have the ancestor $not(q)$. If additional information – such as $not(q)$ – is considered as an advantage for proof finding, this is a shortcoming of restart model elimination. The situation however can easily be repaired either by an explicit change to the calculus, or by incorporating a more general *factoring* rule[3].

In this way, restart model elimination steps can be mapped to partial SPRF proofs. The converse, however, is not true. This is due to the fact that the splitting rule can be applied in every proof situation, i.e. to every sequent derived along a proof. In other words, a case analysis p or $\neg p$ can be carried even to goals totally unrelated to p.

Thus, in sum, the restart model elimination is more restricted than SPRF.

---

[3] Factoring means that a branch may be closed if its leaf is identical to some brother node of a predecessor of this leaf.

The *modified problem reduction format* (MPRF) avoids the problem of uncontrolled application of the splitting rule. This is formally carried out by an additional syntactical layer between "sequents" and "inference rules". As an essential difference, MPRF allows (in our terminology) for restart steps with *any* goal along the current path, not just with the goal literal as in restart model elimination. While this feature clearly increases the local search space, shorter proofs may be enabled. Another notable difference is that restart model elimination includes the negative literals along paths. As our experiments show this is often valuable information and should not be thrown away.

Near-Horn Prolog. As already mentioned in the introduction, there is a close relation to Loveland's Near-Horn Prolog, especially to the InH-Prolog variant from [LR89]. Instead of one tableau in our model elimination calculi, InH-Prolog deductions consist of a sequence of Prolog-like computations, called blocks. The activation of such blocks corresponds to our restart extension steps. If we agree that Prolog stepwise transforms a goal set G into the empty goal set, then the Prolog-like computations in InH-Prolog deal with triples of the form G # A { D }. Here, the list A is called *active heads* and the list D is called *deferred heads*. These components can easily be explained from the viewpoint of restart model elimination: the active heads A corresponds to the *positive* literals of the path in restart model elimination which was most recently selected for a restart step; consequently, since A is a left-ended stack the leftmost literal in A is the literal which caused the restart step. In the Prolog-like restart blocks every literal in A may be used in the role of a unit input clause ("cancellation step) in order to get rid of a goal literal. The deferred heads D correspond to the remaining positive leaf literals of the path multiset; they will cause new restart blocks (or restart steps) at a later time.

Let us compare our refutation from Figure 1 (right side) with the following InH-Prolog refutation. In this example no deferred head occurs.

```
?- GOAL
:- P,Q
:- Q,Q              % factoring to simplify presentation!
:- Q
:-        # P       % P from disjunctive clause Q,P <-
                    % is deferred, and the Block is finished
% restart:
?-GOAL # P
:- P,Q  # P         % cancellation (reduction)
:- Q    # P
:- P    # P         % cancellation
:-      # P         %
```

The cancellation steps in this derivation correspond to the two reduction steps in the right subtree of Figure 1 (right side). The derivation from the left subtree does not have a counterpart in the above InH-Prolog refutation, because of the factoring step we performed in the first block; this, of course, would have been possible in the restart model elimination refutation.

The reduction steps starting from positive leaf-literals have no counterpart in InH-Prolog - within a block there are only extension or cancellation steps. The latter correspond to reductions with a negative leaf-literal.

On the other side, the concept of a *strong cancellation pruning rule* of InH-Prolog has (so far) no counterpart in restart model elimination. By this rule, a certain class of refutations is discarded. Stated positively, and in the terminology of restart model elimination, only those refutations are acceptable in which a literal which caused a restart step is used in a (any) subsequent reduction step. Thus restart steps not relevant for the proof are filtered out. The completeness of this restriction can be seen again by analyzing the completeness proof of Lemma 8. In brief, a restart step applied to $A, B \leftarrow C$ causes by the splitting rule Horn refutations with $A \leftarrow C$ and B. Now, if the given clause set is supposed (without loss of generality) to be *minimal* unsatisfiable, then also the splitted sets contain minimal unsatisfiable subsets containing $A \leftarrow C$ and B, respectively. Hence these clauses must be used in the Horn refutations, and consequently, the restart step occurring at B must be followed by a reduction step to B.

Summarizing on all these considerations we conclude that InH-Prolog is very closely related to strict restart model elimination without selection function. As a consequence we see that our PTTP implementation can be seen as an implementation of InH-Prolog.

SLWV-Resolution. In [PCA91] a theorem prover that retains the procedural aspects of logic programming is defined. This so called SLWV resolution system is based on SL-resolution, a linear resolution format. SLWV saves contrapositives and uses case analysis as an additional inference rule. To this end the usual resolution step from SL-resolution is modified such that besides the current goal any ancestor is allowed to be expanded. In our terminology this would mean that every negative literal along a path can be copied in a restart step. As the authors of [PCA91] explain, this freedom clearly increases the search space when compared to Near-Horn Prolog in the case of near-Horn problems. As a further difference to our restart model elimination, SLWV-Resolution needs a completely new framework. Pereira et.al. had to redesign the PTTP-implementation technique for their prover, whereas we were able to implement restart model elimination by a small change of our existing prover.

References

[AB70]   R. Anderson and W. Bledsoe. A linear format for resolution with merging and a new technique for establishing completeness. *J. of the ACM*, 17:525–534, 1970.

[AS92]   Owen L. Astrachan and Mark E. Stickel. Caching and Lemmaizing in Model Elimination Theorem Provers. In D. Kapur, editor, *Proceedings of the 11th International Conference on Automated Deduction (CADE-11)*, pages 224–238. Springer-Verlag, June 1992. LNAI 607.

[Bau92] P. Baumgartner. A Model Elimination Calculus with Built-in Theories. In H.-J. Ohlbach, editor, *Proceedings of the 16-th German AI-Conference (GWAI-92)*, pages 30–42. Springer, 1992. LNAI 671.

[BF93]   P. Baumgartner and U. Furbach. Consolution as a Framework for Comparing Calculi. *Journal of Symbolic Computation*, 16(5), 1993.

[BF94]   P. Baumgartner and U. Furbach. PROTEIN: A *PRO*ver with a *T*heory *E*xtension *I*nterface. In *12th International Conference on Automated Deduction*. Springer, 1994. (in this volume).

[Bib87] W. Bibel. *Automated Theorem Proving*. Vieweg, 2nd edition, 1987.

[Ede91] E. Eder. Consolation and its Relation with Resolution. In *Proc. IJCAI '91*, 1991.
[Ede92] E. Eder. *Relative Complexities of First Order Languages*. Vieweg, 1992.
[Fit90] M. Fitting. *First-Order Logic and Automated Theorem Proving*. Texts and Monographs in Computer Science. Springer, 1990.
[Gab85] D. M. Gabbay. N-Prolog: An extension of Prolog with hypothetical implication II. logical foundations, and negation as failure. *The Journal of Logic Programming*, 2(4):251–284, December 1985.
[Gal87] J. Gallier. *Logic for Computer Science: Foundations of Automatic Theorem Proving*. Wiley, 1987.
[LMG93] R. Letz, K. Mayr, and C. Goller. Controlled Integrations of the Cut Rule into Connection Tableau Calculi. Journal of Automated Reasoning (to appear 1994), 1993.
[Lov68] D. Loveland. Mechanical Theorem Proving by Model Elimination. *JACM*, 15(2), 1968.
[Lov91] D. Loveland. Near-Horn Prolog and Beyond. *Journal of Automated Reasoning*, 7:1–26, 1991.
[LR89] D.W. Loveland and D.W. Reed. A near-horn prolog for compilation. Technical Report CS-1989-14, Duke University, 1989.
[LSBB92] R. Letz, J. Schumann, S. Bayerl, and W. Bibel. SETHEO: A High-Performace Theorem Prover. *Journal of Automated Reasoning*, 8(2), 1992.
[PCA91] L. M. Pereira, L. Caires, and J. Alferes. SLWV - A Theorem Prover for Logic Programming. AI Centre, Uninova, Portugal, July 1991.
[Pla88] D. Plaisted. Non-Horn Clause Logic Programming Without Contrapositives. *Journal of Automated Reasoning*, 4:287–325, 1988.
[Pla90] D. Plaisted. A Sequent-Style Model Elimination Strategy and a Positive Refinement. *Journal of Automated Reasoning*, 4(6):389–402, 1990.
[RL92] D. W. Reed and D. W. Loveland. A Comparison of Three Prolog Extensions. *Journal of Logic Programming*, 12:25–50, 1992.
[SSY94] G. Sutcliffe, C. Suttner, and T. Yemenis. The TPTP problem library. In *Proc. CADE-12*. Springer, 1994.
[Sti88] M. Stickel. A Prolog Technology Theorem Prover: Implementation by an Extended Prolog Compiler. *Journal of Automated Reasoning*, 4:353–380, 1988.

Induction using Term Orderings

Francois Bronsard[*][1], Uday S. Reddy[2], Robert W. Hasker[2]

[1] CRIN-INRIA-Lorraine and The University of Illinois at Urbana-Champaign
[2] The University of Illinois at Urbana-Champaign

Abstract. We present a procedure for proving inductive theorems which is based on explicit induction, yet supports *mutual induction*. Mutual induction allows the postulation of lemmas whose proofs use the theorems *ex hypothesi* while the theorems themselves use the lemmas. This feature has always been supported by induction procedures based on Knuth-Bendix completion, but these procedures are limited by the use of rewriting (or rewriting-like) inferences. Our procedure avoids this limitation by making explicit the implicit induction realized by these procedures. As a result, arbitrary deduction mechanisms can be used while still allowing mutual induction.

1 Introduction

"Classical" induction is Noetherian induction. Its use in reasoning about program properties was recognized early [10] and it has been used to implement many automated theorem provers, most notably, Nqthm [5; 6]. More recent work on this form of induction includes [3], [9], and [32].

A different induction method for equational systems was proposed by Musser [26] using the Knuth-Bendix *completion procedure* [22] and further developed by various researchers over the years [18; 15; 11; 20; 19; 1; 27; 23; 25; 16]

In the completion-based method, one is able to apply an induction hypothesis to complete an induction proof without ever testing that it is applied to a smaller instance. Because it contrasts sharply with classical induction, the method was called "inductionless induction" and formalized as "proof by consistency" [20; 1]. However, a simpler interpretation is possible. The completion-based method can also be viewed as Noetherian induction over *terms* where the well-founded order is implicit in the deduction mechanism. This seems to have been part of the folklore of term rewriting theory and was recently formalized by Reddy [27]. Based on this viewpoint, the method is now increasingly referred to as *implicit induction*.

As pointed out by Reddy [27], the completion-based induction method derives its power from its ability to carry out *mutual induction* proofs. It is able to adopt irreducible goals as lemmas and prove the theorem and the lemmas in a mutually recursive fashion, *i.e.*, the proof of the lemma and the theorem can use each other

---
[*] Work supported by Ministère des Affaires Etrangères. Address for correspondence: F. Bronsard, CRIN-INRIA-Lorraine, 54602, Nancy, France. Email: bronsard@loria.fr.

as inductive hypotheses. In contrast, the classical method requires the user to state the lemmas in advance so that a simultaneous induction proof can be carried out. In an illustrious demonstration of the power of mutual induction, Bouhoula and Rusinowitch [4] recently showed that the Gilbreath card trick theorem required only two user lemmas using an implicit induction prover while classical induction provers like Coq, Nqthm, and RRL required approximately fifteen to twenty lemmas!

In this paper, we propose a fundamental generalization of the completion-based method. Our point is that the power of the completion-based method owes to the fact that it uses *orderings on terms*. The fact that this ordering is implicit is purely incidental. Indeed, the ordering can be made *explicit* and this makes the method more general and flexible. But, the method is still different from classical induction because the ordering is on terms rather than values.

To see this distinction, consider the strict subterm order over terms. It is evidently well-founded, and we can carry out induction proofs based on this order. In contrast, classical induction uses orderings on values. So, all terms that denote the same value must be treated as equal and no distinctions can be made within them. For example, $x + 0 \succ x$ by the subterm order, and we can prove $P(x+0)$ using $P(x)$ in the term ordering method. But, the classical method cannot use such a distinction because $x + 0 = x$ and there can be no well-founded ordering $>$ satisfying $x + 0 > x$. Since any order on values also defines an order on terms, there exists infinitely many more term ordering proofs than classical proofs. So it is easier to find proofs with term orderings.

Thus, in our opinion, the implicit-explicit distinction (the subject of many heated debates) is entirely superfluous. The important issue is the use of term orderings and the powerful proof methods that they allow.

We make term orderings the core of our proposal. We formulate an explicit induction method based on term orderings (Sections 3 and 4) and show that it subsumes implicit methods (Section 5). We then show how to integrate completion-like forward inferences with the explicit term ordering method (Section 6). This last contribution is entirely novel. There are no term ordering induction methods at present which can use forward inferences (superposition, paramodulation, resolution, *etc.*) in their deduction mechanisms. It might seem paradoxical, but the completion-based induction method cannot use completion in its deduction mechanism (see [27] for a discussion of this). This weakness was our main motivation for proposing this new induction method based on explicit term orderings.

2 Preliminaries

2.1 Deduction Systems

Because our inductive proof procedure is applicable to many kinds of deduction systems, we use a fairly general notion of deduction system:

Definition 1. A (first-order) deduction system is specified by the following data:

- F, a finite, ranked alphabet of *function symbols*,
- P, a finite, ranked alphabet of *predicate symbols*,
- X, a countable set of variables,
- $\mathcal{L}$, a decidable set of *propositions* over F, P, and X, and
- $\vdash \subseteq (\mathbb{P}_R \mathcal{L} \times \mathcal{L})$, a recursively enumerable *deduction* relation that satisfies

$$\phi \in \mathcal{A} \Longrightarrow \mathcal{A} \vdash \phi$$
$$\mathcal{A} \vdash \phi \land \mathcal{A} \subseteq \mathcal{A}' \Longrightarrow \mathcal{A}' \vdash \phi$$
$$\mathcal{A} \vdash \phi \land \mathcal{A} \cup \{\phi\} \vdash \phi' \Longrightarrow \mathcal{A} \vdash \phi'$$
$$\mathcal{A} \vdash \phi \Longrightarrow \mathcal{A} \vdash \phi\theta \quad \text{where } \theta \text{ is a substitution}$$

($\mathbb{P}_R\mathcal{L}$ denotes recursive sets over $\mathcal{L}$.) We use the following conventions: $\mathcal{T}$ (and $\mathcal{T}_G$) denotes terms (and ground terms) over F and X; $\mathcal{P}$ (and $\mathcal{P}_G$) denotes atoms (and ground atoms) over P, F, and X; and $\mathcal{L}_G$ denotes ground (or closed) propositions.

Example 1. An *equational* deduction system has an empty set of predicate symbols. A proposition is a multiset $\{t, u\}$ of two terms, normally written as $t = u$ (or $u = t$). Let $\leftrightarrow_{\mathcal{A}}$ be the relation defined by $t[a\sigma/\alpha] \leftrightarrow_{\mathcal{A}} t[b\sigma/\alpha]$ whenever α is a position in t, $a = b$ is an equation in $\mathcal{A}$ and σ is a substitution. The deduction relation is

$$\mathcal{A} \vdash t = u \text{ iff } t \leftrightarrow^*_{\mathcal{A}} u$$

Example 2. A clausal deduction system has, as its propositions, multisets $\{l_1, \ldots, l_n\}$ of literals (*i.e.*, signed atoms). We normally write such a proposition as $l_1 \lor \ldots \lor l_n$ (or as $l_1 \lor \ldots \lor l_k \Leftarrow \neg l_{k+1} \land \ldots \land \neg l_n$). The deduction relation is

$$\mathcal{A} \vdash l_1 \lor \ldots \lor l_n \text{ iff } A \cup L \vdash_R \bot$$

where L is the skolemized negation of $l_1 \lor \ldots \lor l_n$ and $\vdash_R$ is the deduction mechanism built using the resolution and factoring inferences [30].

Example 3. A clausal deduction system with *equality* has, among its predicate symbols, a distinguished symbol "$=$" of arity 2. The deduction relation is the clausal deduction mechanism augmented with the paramodulation inference (or the appropriate axioms such that "$=$" is an equivalence relation) [29; 24].

2.2 Propositional Orderings

Briefly, we recall the basic notions. A *preorder* $\leqslant$ is a reflexive, transitive relation. The equivalence $\approx$ of a preorder is defined by $x \approx y \iff x \leqslant y \land y \leqslant x$. The *partial order* $<$ of $\leqslant$ is: $x < y \iff x \leqslant y \land x \not\approx y$. A partial order $<$ is said to be *well-founded* if there is no infinite descending sequence $x_1 > x_2 > \ldots$. A preorder is said to be well-founded if its partial order is well-founded.

Our inductive proof procedures rely on orderings over propositions instead of the conventional semantic orderings. We usually define such orderings as extensions of classical *term orderings* over $\mathcal{T}$. (See [12] for a survey.) If $\prec$ is a well-founded order on $\mathcal{T}$, the corresponding multiset order $\prec_{mul}$ is a well-founded order on multisets over $\mathcal{T}$ [13]. Another such extension is useful for our purposes:

Definition 2 (Max-extension). Let $<$ be a partial order on a set S, and let $\max(X)$ denote the multiset of $<$-maximal elements of the multiset X. The *max-extension* of $<$ is a preorder $\leqslant_{max}$ on multisets over S defined by:

$$X \leqslant_{max} Y \iff \max(X) <_{mul} \max(Y) \vee \max(X) = \max(Y)$$

It follows that $X \approx_{max} Y \iff \max(X) = \max(Y)$ and $X <_{max} Y \iff \max(X) <_{mul} \max(Y)$.

Lemma 3. *If $<$ is a well-founded order, its max-extension is well-founded.*

We use the max-extension preorder to define propositional orderings for various kinds of deductive systems:

Example 4.
1. For the equational deduction system (of Example 1), let $\prec$ be a well-founded order on $\mathcal{T}$. Then $\prec_{max}$ can be used as a well-founded preorder on equations. For example, $(x+0 = x) \prec_{max} (S(x)+0 = S(x))$ and $(x+0 = x) \prec_{max} (x+0 = 0+x)$ using any multiset path order for $\prec$.
2. For clauses (of Example 2), let $\prec$ be a well-founded order on atoms ($\mathcal{P}$). We extend it to literals by ignoring their sign and use $\prec_{max}$ as well-founded order on clauses.
3. For clauses with equality, let $\prec$ be a well-founded order on terms and non-equality atoms. The order $\prec_A$ on atoms is obtained by treating equality atoms $t_1 = t_2$ as binary multisets $\{t_1, t_2\}$ and other atoms a as unary multisets $\{a\}$, and comparing them by the max extension of $\prec$. The order on clauses is then $(\prec_A)_{max}$.

2.3 Induction

We take the following to be our operational definition of inductive consequence:

Definition 4 (Inductive Consequence). Given a deduction system, let $\mathcal{A}$ be a set of axioms. A proposition ϕ is an *inductive consequence* of $\mathcal{A}$, written $\mathcal{A} \vdash_{ind} \phi$, if and only if, for every ground instance of ϕ (with instantiations from $\mathcal{T}_G$), $\mathcal{A} \vdash \phi\theta$.

The above definition is classical. (See [31] for a discussion.) From a model-theoretic point of view it defines inductive consequences as those properties that hold in all reachable models (or, equivalently, term models). Sometimes, variants of Definition 4 may be preferred. For example, if some functions are only partially defined, it is more intuitive to restrict instantiations (θ) to constructor terms. In this case, the models of interest are the *constructor models*. In the context of a parameterized specification, we might allow instantiation only of variables of non-parameter sorts. This would allow models with arbitrary domains for the parameter sorts. The methods of this paper can be easily adapted to such variants.

Let $\phi[x]$ be a proposition with a single free variable x. To prove that $\phi[x]$ is an inductive consequence by conventional Noetherian induction,[3] we need a predicate $<$ in the alphabet P and an axiomatization, in $\mathcal{A}$, of $<$ as a partial order. Further, it must be known that $<$ is a well-founded order. (This may or may not be axiomatizable in the deduction system.) Then, $\mathcal{A} \vdash_{ind} \phi[x]$ if

$$\mathcal{A} \vdash (\forall y < x.\ \phi[y]) \Rightarrow \phi[x]$$

This places many requirements on the deduction system. In addition to those mentioned above, the language of the deduction system must include quantifiers and implication. This approach cannot be used, for instance, in equational or conditional equational systems (which lack predicate symbols) or in clausal systems (which lack quantifiers). Derived proof rules are often used to get around these limitations [5; 32].

A more widely applicable proof method, proposed in [27], uses *term orderings*:

Proposition 5 (Noetherian induction principle using term orderings).
Let $\prec$ be a well-founded order over $\mathcal{T}_G$ (a term ordering), $\mathcal{A}$ be a set of axioms, and $\phi[x]$ be a proposition. $\mathcal{A} \vdash_{ind} \phi[x]$ if, for all ground terms g,

$$\mathcal{A} \cup \{\phi[g'] \mid g' \prec g\} \vdash \phi[g]$$

There is a qualitative difference between the traditional Noetherian induction and Noetherian induction using term orderings. The conventional Noetherian induction relies on a *semantic* ordering $<$. In contrast, induction using term orderings relies on a *syntactic* ordering $\prec$ defined over $\mathcal{T}_G$. In the rest of the paper, we extend this notion to an induction principle based on propositional orderings, *i.e.*, well-founded orderings over $\mathcal{L}_G$ or $\mathcal{L}$. As we will show, such an extension offers significant advantages over traditional induction.

3 Induction Using Propositional Orderings

Given a well-founded order over ground propositions, *i.e.*, an order over $\mathcal{L}_G$, the Noetherian induction principle can be expressed as follows:

Proposition 6 (Noetherian induction with propositional orderings). *Let $\preccurlyeq$ be a well-founded preorder over $\mathcal{L}_G$, $\mathcal{A}$ be a set of axioms, and $\phi[x]$ be a proposition. $\mathcal{A} \vdash_{ind} \phi[x]$ if, for all ground terms g,*

$$\mathcal{A} \cup \{\phi[g'] \mid \phi[g'] \prec \phi[g]\} \vdash \phi[g]$$

---

[3] It should be noted that conventional induction is a "weaker" principle than inductive consequence. In particular, it is valid for certain unreachable models. We mention it here only by way of illustration.

Proof: Suppose the contrary. Then there is an instance $\phi[g]$ that is not provable. Since $\prec$ is well-founded, there is a minimal $\phi[g_0] \not\preceq \phi[g]$ that is not provable. That would mean all $\phi[g'] \prec \phi[g_0]$ are provable. The hypothesis of the theorem then implies $\mathcal{A} \vdash \phi[g_0]$ using the axioms of $\vdash$. This gives a contradiction. □

Notation. If ϕ and ψ are propositions, we write $\{\phi\}_{\prec \psi}$ (resp. $\{\phi\}_{\preceq \psi}$) to denote the set $\{\phi\theta \mid \phi\theta \prec \psi\}$ (resp. $\{\phi\theta \mid \phi\theta \preceq \psi\}$). Using this, Proposition 6 becomes

$$\mathcal{A} \vdash_{ind} \phi[x] \text{ if for all ground terms } g, \; \mathcal{A} \cup \{\phi[x]\}_{\prec \phi[g]} \vdash \phi[g] \; .$$

Proposition 6 is at the heart of our proposal in that all of the induction methods or inductive procedures presented in this paper ultimately rely on it. To some extent, we could say that these other methods are effective variants of the above principle.

Proposition 6 is not an effective method because it involves testing all ground instances, an infinite set. To develop an effective method, we consider finite sets of nonground propositions in $\mathcal{L}$ that can *cover* all ground cases. This is formalized in the following definition.

Definition 7 (cover sets). Let $\mathcal{A}$ be a set of axioms, ϕ a proposition and $\preceq$ a stable,[4] well-founded order over $\mathcal{L}$. A finite set of propositions $\{\psi_i\}_i$ is called a *cover set* for ϕ if, for all instances $\phi\alpha$, there is a proposition ψ_i and a substitution σ such that $\psi_i\sigma \preceq \phi\alpha$ and $\mathcal{A} \cup \{\psi_i\sigma\} \vdash \phi\alpha$.

We also say that a set of terms $\{t_i\}_i$ is a cover set if, for every ground term g, there is a t_i and a substitution σ such that $t_i\sigma \preceq g$ and $\mathcal{A} \vdash g = t_i\sigma$. Also, a set of substitutions $\{\sigma_i\}_i$ is a cover set for a proposition ϕ if $\{\phi\sigma_i\}_i$ is a cover set.

The notion of cover sets arose in the work on completion-based induction methods and its various extensions [1; 32; 23; 21]. The present notion is a generalization of that in [27]. Automatic methods for finding cover sets for equational deduction systems may be found in the above citations. Work remains to be done on finding similar methods for conditional equation systems and clausal deduction systems (see however [23]).

Example 5. 1. $\{\phi\}$ is always a cover set for ϕ (for any order $\preceq$).
2. Consider a deduction system with equality. If there exists $F_0 \subseteq F$ such that, for every ground term g, there exists $g_0 \in \mathcal{T}(F_0)$ such that $\mathcal{A} \vdash g = g_0$ and $g_0 \preceq g$, then $\{\phi[f(\bar{z})/x] : f \in F_0\}$ is a cover set for ϕ.
3. Consider a clausal deduction system. Suppose ψ is a proposition such that $\psi \prec \phi$ and $\mathcal{A} \vdash \psi\alpha \vee \neg\psi\alpha$ for all substitutions α. Then $\{\psi \vee \phi, \neg\psi \vee \phi\}$ is a cover set for ϕ.
4. Consider a clausal deduction system with equality. Suppose ϕ is a proposition and $\{\psi_i\}_i$ a set of consequences of ϕ; i.e., $\mathcal{A} \cup \phi \vdash \psi_i$ for all i. If the set $\{\psi_i\}_i$ completely defines ϕ, i.e., $\forall$ ground α, $\exists i$ such that $\mathcal{A} \cup \{\psi_i\alpha\} \vdash \phi\alpha$ and $\psi_i\alpha \preceq \phi\alpha$, then $\{\psi_i\}_i$ is a cover set for ϕ.

---

[4] An order is *stable* if it is preserved by substitution: $\phi \preceq \psi \implies \phi\theta \preceq \psi\theta$ for all substitutions θ.

Proposition 8 (Noetherian Induction with cover sets). *Let $\preccurlyeq$ be a stable, well-founded preorder over $\mathcal{L}$. $\mathcal{A} \vdash_{ind} \phi$ if*

1. *$\{\psi_i\}_i$ is a cover set for ϕ, and*
2. *for each proposition ψ_i, we have $\mathcal{A} \cup \{\phi\}_{\prec \psi_i \kappa} \vdash \psi_i \kappa$ where κ is a skolemizing substitution.*

Proof: Suppose the contrary. Then there is an instance $\phi\theta$ that is not provable. Since $\preccurlyeq$ is well-founded, there is a minimal $\phi\theta_0 \preccurlyeq \phi\theta$ that is not provable. Thus, there is an instance $\psi_j \alpha \preccurlyeq \phi\theta_0$ of a proposition ψ_j of the cover set that is not provable. Yet all $\phi\theta' \prec \psi_j \alpha$ are provable. This contradicts the hypothesis 2. (Note that $\mathcal{A} \cup \{\phi\}_{\prec \psi_j \kappa} \vdash \psi_j \kappa$ implies that $\mathcal{A} \cup \{\phi\}_{\prec \psi_j \alpha} \vdash \psi_j \alpha$ for all ground substitutions α.) □

Example 6. Consider the following equations defining natural number addition:

$$(N_1) \qquad x + 0 = x$$
$$(N_2) \qquad x + S(y) = S(x + y)$$

To prove the inductive theorem

$$(P) \qquad (x + y) + z = x + (y + z)$$

we use the multiset path order $\succ$ generated by the precedence $+ > S > 0$ and compare equations by $\succ_{max}$. Every ground term is equal to a smaller (or equal) term over 0 and S. So, we can instantiate z to 0 and $S(z')$ to obtain a cover set.

1. $(x + y) + 0 \leftrightarrow_{N_1} x + y \leftrightarrow_{N_1} x + (y + 0)$
2. $(x+y)+S(z') \leftrightarrow_{N_2} S((x+y)+z') \leftrightarrow_P S(x+(y+z')) \leftrightarrow_{N_2} x+S(y+z') \leftrightarrow_{N_2} x + (y + S(z'))$

To show that the use of the inductive hypothesis is valid, verify that

$$\{(x + y) + z', x + (y + z')\} \prec_{max} \{(x + y) + S(z'), x + (y + S(z'))\}$$

Example 7. Consider a clausal deduction system with equality containing a predicate $\geq$ denoting a total order, a predicate $\in$ denoting membership in a list, and a function max which returns the maximal element of a list. (We also assume the presence of the list constructors Nil and ".".) Assume the following axioms:

$$\begin{aligned}
\mathsf{max}(a.\mathsf{Nil}) &= a \\
\mathsf{max}(a.l) &= \mathsf{max}(l) \Longleftarrow l \neq \mathsf{Nil} \wedge \mathsf{max}(l) \geq a \\
\mathsf{max}(a.l) &= a \qquad\quad \Longleftarrow l \neq \mathsf{Nil} \wedge a \geq \mathsf{max}(l) \\
\mathsf{max}(l) &\in l \qquad\qquad\; \Longleftarrow l \neq \mathsf{Nil}
\end{aligned}$$

To verify the correctness of the definition of max, we wish to prove

$$P[l]: \qquad \mathsf{max}(l) \geq x \Longleftarrow l \neq \mathsf{Nil} \wedge x \in l$$

To order terms, we use the multiset path order with the precedence order max $>$ "$\in$" $>$ "$\geq$" $>$ "." $>$ Nil and extend it to clauses as described in Example 4. A cover set for this theorem is $\{P[\text{Nil}], P[a.\text{Nil}], (P[a.l'] \Longleftarrow l' \neq \text{Nil} \wedge \max(l') \geq a), (P[a.l'] \Longleftarrow l' \neq \text{Nil} \wedge a \geq \max(l'))\}$. The first two cases reduce to true trivially. The others simplify to

$$Q_1: \quad \max(l') \geq x \Longleftarrow \max(l') \geq a \wedge x \in (a.l') \wedge l' \neq \text{Nil}$$
$$Q_2: \quad a \geq x \quad \Longleftarrow a \geq \max(l') \wedge x \in (a.l') \wedge l' \neq \text{Nil}$$

Consider the proposition Q_1. Using the axioms for "$\in$", we can replace $x \in (a.l')$ separately by $x = a$ and $x \in l'$. The first case is trivial and the second is proven by applying the inductive hypothesis $P[l']$ (which is smaller than $P[a.l']$).

Similarly, proposition Q_2 splits into the trivial case $x = a$ and the case $x \in l'$. For $x \in l'$, we need to prove

$$a \geq x \Longleftarrow a \geq \max(l') \wedge x \in l' \wedge l' \neq \text{Nil}$$

This requires the following resolvent of the transitivity axiom and the inductive hypothesis $P[l']$:

$$z \geq x \Longleftarrow l' \neq \text{Nil} \wedge x \in l' \wedge z \geq \max(l')$$

This completes the proof of Q_2. The use of the inductive hypothesis is valid since we have used the (smaller) instance $P[l']$ in proving $P[a.l']$.

4 Mutual Induction

The major benefit of using propositional orderings rather than semantic orderings for inductive reasoning is that it allows proof by *mutual induction*, that is, proofs of multiple propositions which use each other in their proofs in a mutually recursive fashion. For example, suppose we have two propositions $\phi[x]$ and $\psi[y]$, a preorder $\preceq$, and we show that for all ground terms g,

$$\mathcal{A} \cup \{\phi[x]\}_{\prec \phi[g]} \cup \{\psi[y]\}_{\prec \phi[g]} \vdash \phi[g] \qquad (1)$$
$$\mathcal{A} \cup \{\phi[x]\}_{\prec \psi[g]} \cup \{\psi[y]\}_{\prec \psi[g]} \vdash \psi[g] \qquad (2)$$

Then, both $\phi[x]$ and $\psi[y]$ are inductive theorems of $\mathcal{A}$. For if a proof of $\phi[g]$ by (1) uses an instance $\psi[g']$ of $\psi[y]$, then $\psi[g'] \prec \phi[g]$. And if the proof of $\psi[g']$ by (2) uses an instance $\phi[g'']$ of $\phi[x]$, then $\phi[g''] \prec \psi[g']$. Thus, $\phi[g''] \prec \phi[g]$. Hence, such mutually inductive proofs, where the proof of $\phi[x]$ can use $\psi[y]$ and the proof of $\psi[y]$ can use $\phi[x]$, are valid.

The following procedure formalizes proof by mutual induction:

Definition 9 (Mutual induction procedure). Let $\preceq$ be a well-founded propositional ordering and $\mathcal{A}$ a set of axioms. The *mutual induction procedure* is defined by the following transition rules applied to pairs of the form (P, H) where P is the set of inductive propositions and H the set of inductive hypotheses:

Expand $(P \cup \{\phi\}, H) \Longrightarrow (P \cup \Psi, H \cup \{\phi\})$ if Ψ is a cover set of ϕ.
Delete $(P \cup \{\phi\}, H) \Longrightarrow (P, H)$ if $\mathcal{A} \cup (P)_{\preceq \phi\kappa} \cup (H)_{\preceq \phi\kappa} \vdash \phi\kappa$.
Lemma $(P, H) \Longrightarrow (P \cup P', H)$

A successful derivation is a sequence of transitions, $(P_0, \emptyset) \Longrightarrow \cdots \Longrightarrow (P_n, H_n)$, such that $P_n = \emptyset$.

We often use the following transition in place of *Delete*:

Simplify $\quad (P \cup \{\phi\}, H) \Longrightarrow (P \cup P', H) \quad$ if $\mathcal{A} \cup (P \cup P')_{\preceq \phi} \cup (H)_{\prec \phi} \vdash \phi$.

It is obtained by combining *Delete* with *Lemma*.

Example 8. Consider the equations (N_1), (N_2) given in Example 6 defining natural number addition. We want to prove the inductive theorem

$$(P_0) \qquad x + y = y + x$$

To prove this theorem, we instantiate y using the cover set $\{0, S(y')\}$ and reduce the resulting equations to normal form:

$$(P_1) \qquad (x + 0 = 0 + x) \quad \leftrightarrow_{N_1} \quad x = 0 + x$$
$$(P_2) \qquad (x + S(y') = S(y') + x) \leftrightarrow_{N_2} S(x + y') = S(y') + x$$

The normal forms can be proven as lemmas. For (P_1) the proof is straightforward and resembles Ex. 6. In the second case, we get

$S(0 + y') \leftrightarrow_{P_1} S(y') \leftrightarrow_{N_1} S(y') + 0$
$S(S(x') + y') \leftrightarrow_{P_2} S(S(y' + x')) \leftrightarrow_{P_0} S(S(x' + y')) \leftrightarrow_{P_2} S(S(y') + x') \leftrightarrow_{N_2} S(y') + S(x')$

Note the mutual use of the lemmas and the main theorem here. The proof of P_0 requires lemma P_2; in turn, the proof of P_2 needs P_0 in its last stage to show $y' + x' = x' + y'$. The use of the inductive hypothesis corresponding to P_0 is valid since

$$\{y' + x', x' + y'\} \prec_{max} \{S(S(x') + y'), S(y') + S(x')\} \ .$$

Theorem 10 (Soundness of Mutual Induction). *If* $(P_0, \emptyset) \Longrightarrow \cdots \Longrightarrow (\emptyset, H_n)$ *is a derivation of the mutual induction procedure, then* $\mathcal{A} \vdash_{ind} P_0$.

To prove this theorem, we rely on the following lemma:

Lemma 11. *Let* $(P_0, \emptyset) \Longrightarrow \cdots \Longrightarrow (P_i, H_i) \Longrightarrow \cdots \Longrightarrow (\emptyset, H_n)$ *be a derivation of the mutual induction procedure. For all* $i = 0, \ldots, n$, *the following invariants hold:*

$$\forall \phi \in \bigcup_{j=0..i} P_j, \ \forall \ ground \ \alpha, \ \mathcal{A} \cup (P_i)_{\preceq \phi \alpha} \vdash \phi \alpha \tag{3}$$

$$\forall \phi \in H_i, \ \forall \ ground \ \alpha, \ \mathcal{A} \cup (P_i)_{\preceq \phi \alpha} \vdash \phi \alpha \tag{4}$$

To simplify the argument, we introduce some terminology. We say that a set of propositions S *covers* Φ if, for every ground instance $\phi \alpha$ of a proposition in Φ, $\mathcal{A} \cup (S)_{\preceq \phi \alpha} \vdash \phi \alpha$. Invariant (3) states that each P_i covers the earlier P_j's, while (4) states that P_i covers H_i. Note that covering is a preorder on sets of propositions.

Proof. For $i = 0$, the invariants hold because P_0 covers itself and H_0 is empty. For $i > 0$, $(P_{i-1}, H_{i-1}) \Longrightarrow (P_i, H_i)$ is an instance of one of the transition rules. The rules *Expand* and *Lemma* are straightforward. For the *Expand* rule, (3) follows from Ψ being a cover set of ϕ, and (4) follows from applying the inductive hypotheses to show that $P \cup \Psi$ covers H. To show that *Delete* maintains (3), we need to show $\mathcal{A} \cup (P)_{\preccurlyeq \phi \alpha} \vdash \phi \alpha$. By the hypothesis of the rule, $\mathcal{A} \cup (P)_{\preccurlyeq \phi \alpha} \cup (H)_{\preccurlyeq \phi \alpha} \vdash \phi \alpha$ (since the order is stable). By inductive hypothesis (4), we have $\mathcal{A} \cup (P \cup \{\phi\})_{\preccurlyeq \phi' \alpha'} \vdash \phi' \alpha'$ for every ground instance $\phi' \alpha'$ of a proposition in H. Thus $P \cup \{\phi\}$ covers H. So, by the hypothesis of the rule, $\mathcal{A} \cup (P)_{\preccurlyeq \phi \alpha} \cup (P \cup \{\phi\})_{\preccurlyeq \phi \alpha} \vdash \phi \alpha$. This is equivalently written as $\mathcal{A} \cup (P)_{\preccurlyeq \phi \alpha} \cup \{\phi\}_{\preccurlyeq \phi \alpha} \vdash \phi \alpha$. By Proposition 6, $\mathcal{A} \cup (P)_{\preccurlyeq \phi \alpha} \vdash \phi \alpha$. Thus, P cover $P \cup \{\phi\}$. The invariant (4) now follows since P covers $P \cup \{\phi\}$ which, in turn, covers H. □

Invariant (3) applied to $i = n$ gives that $P_n (= \emptyset)$ covers P_0. Thus, $\mathcal{A} \vdash_{ind} P_0$, proving Theorem 10.

5 Implicit Induction

The origins of our induction method ultimately lie in the completion-based induction methods first described by Musser [26] and Huet and Hullot [18]. It is only fitting to pay our debt to this work by showing how our method generalizes it. The completion-based method has already been formulated as *term-rewriting induction* [27]. We first generalize this to deduction mechanisms other than term rewriting, e.g., conditional rewriting [7; 23; 2] and clausal deduction. This generalization is called *implicit induction*. We then show that implicit induction is a special case of our method.

Implicit induction is based on the notion of a "reductive" deduction relation:

Definition 12 (Reductive deduction relation). We call a deduction relation $\vdash_R$ *reductive* (with respect to a well-founded preorder $\preccurlyeq$) if

$$\mathcal{A} \vdash_R \phi \iff \mathcal{A}_{\preccurlyeq \phi \kappa} \vdash_R \phi \kappa$$

Example 9. Term rewriting [22] is a reductive deduction relation. A terminating rewrite system $\mathcal{A}$ is a set of rewrite rules with a well-founded reduction order $\prec$ on $\mathcal{T}$ such that, for each $(a \to b) \in \mathcal{A}$, $a \succ b$. Assume, without loss of generality, that $\prec$ includes the (strict) subterm order.

The rewrite relation $\to_\mathcal{A}$ is defined by $t[a\sigma/\alpha] \to_\mathcal{A} t[b\sigma/\alpha]$ where α is a position in t, $a \to b$ a rewrite rule in $\mathcal{A}$ and σ a substitution. Consider the term t in the equation $t = u$. It is easy to see that wherever $t \to_\mathcal{A} t'$ by an instance $a\sigma \to b\sigma$ of a rewrite rule, $\{t, u\} \succcurlyeq_{max} \{a\sigma, b\sigma\}$. More generally, given a rewrite proof $t \to^*_\mathcal{A} v \leftarrow^*_\mathcal{A} u$ of $t = u$, each instance of the rewrite rules used in the proof is smaller than $\{t, u\}$ by $\preccurlyeq_{max}$.

Similarly, conditional rewriting [28] is a reductive mechanism by the order on conditional equations. Likewise, reductive deduction for clauses [8] is a reductive mechanism by the order on clauses.

The implicit induction principle is a corollary of Proposition 6 where the deduction mechanism is restricted to reductive deduction.

Corollary 13 (Implicit induction). *Let $\vdash_R$ be a reductive deduction relation. $\mathcal{A} \vdash_{ind} \phi[x]$ by implicit induction if, for all ground terms g,*

1. *there exists a proposition[5] $\phi'[g]$ such that $\mathcal{A} \cup \{\phi'[g]\} \vdash_R \phi[g]$ and $\phi'[g] \prec \phi[g]$, and*
2. *$\mathcal{A} \cup \{\phi[x]\} \vdash_R \phi'[g]$.*

The main difference between Proposition 6 and this corollary is that we use $\{\phi[x]\}$ rather than $\{\phi[x]\}_{\prec \phi[g]}$ in (2). This is valid because we are using a reductive deduction relation. Thus, although $\{\phi[x]\}$ is apparently used without any qualifications, only instances $\phi[g'] \preccurlyeq \phi'[g]$ can be used in the proof. Since $\phi'[g] \prec \phi[g]$, only instance $\phi[g'] \prec \phi[g]$ are used in the proof of $\phi[g]$. Thus, by Proposition 6, $\mathcal{A} \vdash_{ind} \phi[x]$.

Definition 14 (Implicit Induction Procedure). Let $\mathcal{A}$ be a set of axioms and $\vdash_R$ be a reductive deduction relation. The *implicit induction procedure* is defined by the following transition rules:

Expand $(P \cup \{\phi\}, H) \Longrightarrow (P \cup \Psi, H \cup \{\phi\})$ where Ψ is a conditioned cover set of ϕ

Simplify $(P \cup \{\phi\}, H) \Longrightarrow (P \cup P', H)$ if $\mathcal{A} \cup P \cup P' \cup H \vdash_R \phi$

Lemma $(P, H) \Longrightarrow (P \cup P', H)$

(Ψ is a *conditioned* cover set of ϕ if, for every ground instance $\phi\alpha$, there exists $\psi \in \Psi$ and a substitution σ such that $\psi\sigma \prec \phi\alpha$ and $\mathcal{A} \cup \{\psi\sigma\} \vdash_R \phi\alpha$.)

The correctness of this procedure follows immediately from the correctness of the mutual induction procedure and the use of reductive deduction relations.

Instances of this procedure include the Hierarchical Induction procedure defined by Reddy [27] where $\vdash_R$ is the rewriting deduction mechanism, the inductive procedures for conditional equations of Kounalis and Rusinowitch [23], Bronsard and Reddy [7], and Bouhoula and Rusinowitch [4].

Critique of Implicit Induction. The limitation of the implicit induction procedure is that only *reductive* deduction relations can be used in the *Delete* transition. Such deduction mechanisms are, in general, incomplete. The benefit obtained in return is the ability to apply induction hypotheses without a test. This "benefit" is not worth the cost. See [14] for a similar sentiment.

While reductive deduction is generally incomplete, completeness can be recovered by using the appropriate *completion* procedure. Yet, paradoxically, the completion-based induction procedures, despite their name, do not allow completion inferences in the *Delete* rule. Instead, any conclusion ϕ of a completion

---

[5] By an abuse of notation, we write $\phi'[g]$ to indicate that $\phi'[g]$ is function of g, although g might not be a subterm of ϕ'.

inference is treated as a lemma that is potentially useful for the proof of the theorem, and an inductive proof of the lemma is attempted. This leads to wasteful computation. ϕ is a consequence of the inductive hypotheses and using it amounts to an indirect use of these hypotheses. This use may correspond to a valid use of the inductive hypotheses and, if so, no separate proof of ϕ is required. Reddy [27] pointed this out and gave an example showing that such needless lemmas can generate an *infinite* amount of computation. While such examples may be rare, they underscore the inefficiency of generating lemmas by completion.

Example 10. Recall the verification of the max function in Example 7. To prove the cover set instance $P[x, a.l'] \Longleftarrow l' \neq \mathsf{Nil} \wedge a \geq \mathsf{max}(l')$, we need the following resolvent with the transitivity property of $\geq$:

$$\frac{(P[x,l]:) \quad \mathsf{max}(l) \geq x \Longleftarrow l \neq \mathsf{Nil} \wedge x \in l \quad z \geq x \Longleftarrow z \geq u \wedge u \geq x}{(Q[z,x,l]:) \quad z \geq x \Longleftarrow l \neq \mathsf{Nil} \wedge x \in l \wedge z \geq \mathsf{max}(l)} \text{ Resolution}$$

Since $Q[a, x, l']$ is a consequence of $P[x, l']$ and $P[x, l'] \prec P[x, a.l']$ by the multiset path order, it seems reasonable to use $Q[a, x, l']$ in the proof of $P[x, a.l']$. However, implicit induction requires that $Q[a, x, l'] \prec P[x, a.l']$. This inequality does not hold, so the implicit induction proof fails.

6 Forward Inferences

We now outline a method by which completion-like mechanisms can be used in the *Delete* rule. This is the main technical contribution of this paper. The heart of the method is a subtle, yet simple, bookkeeping technique using which one can track dependencies between inductive hypotheses and their consequents.

The essence of the completion proof methods (for first-order reasoning) is to use a *backward* reductive deduction mechanism (such as rewriting) and a *forward* non-reductive deduction mechanism (such as superposition). By a "forward deductive mechanism", we mean an inference mechanism that operate on axioms independently of the goal proposition. Such inferences have the general form

$$\{\psi_1, \psi_2\} \vdash_F \psi$$

where ψ is obtained from ψ_1 and ψ_2 using some operation that typically involves unification. To capture this aspect, we make the following definition:

Definition 15. We call a deduction relation a *forward* deduction relation if, whenever $\mathcal{A} \vdash_F \psi$, there exists a finite $\Psi_0 \subseteq \mathcal{A}$ and a substitution θ such that $\Psi_0 \vdash_F \psi$ and $\Psi_0 \theta \alpha \vdash_F \psi \alpha$ for all ground substitutions α. We call the inference $\Psi_0 \vdash_F \psi$ a *canonical* inference with underlying substitution θ.

Using forward inferences in the *Delete* rule requires some care. Consider $\{\psi_1, \psi_2\} \vdash \psi$. If one of the premises ψ_1 is an induction hypothesis, there are restrictions on what instances of ψ_1 can be used in a proof. For example, if

ψ_1 is used in the proof of a formula ϕ then only the instances $\{\psi_1\}_{\prec\phi}$ can be used in the proof of ϕ. So, there must be some similar restriction on how ψ can be used as well. The restriction cannot simply be $\{\psi\}_{\prec\phi}$ because $\psi\theta \prec \phi$ does not guarantee that $\psi_1\theta \prec \phi$. Therefore, we must track the dependence between induction hypotheses and their consequences to ensure a valid proof. This motivates the following annotation:

Definition 16. A *tagged proposition* is a pair, written ψ^W, of a proposition ψ and a set of propositions $W = \{\omega_1, \ldots, \omega_n\}$.

The propositions that are derived by forward inferences from the inductive hypotheses have tags which record the instances of the inductive hypotheses used to derive them. Specifically, for every tagged proposition ψ^W, we ensure that

$$\forall \text{ ground } \alpha, \; \mathcal{A} \cup W\alpha \vdash \psi\alpha$$

A fresh inductive hypothesis has the form $\psi^{\{\psi\}}$ while an axiom ψ of $\mathcal{A}$ written as tagged proposition would have the form $\psi^\emptyset$.

If S is a set of tagged propositions, we use the notation

$$S_{\triangleleft\phi} = \{ \psi\theta \mid \psi^W \in S, \, \omega\theta \prec \phi \text{ for all } \omega \in W \}$$

to denote all instances of S derivable from the instances of hypotheses admissible for ϕ. Note that $\{\psi^{\{\psi\}}\}_{\triangleleft\phi} = \{\psi\}_{\prec\phi}$ and $\{\psi^\emptyset\}_{\triangleleft\phi} = $ all instances of ψ.

Definition 17. If $\vdash_F$ is a forward deduction relation, the corresponding *tagged* relation $\vdash_T$ for tagged propositions is the least relation satisfying the following condition: Whenever $\{\psi_1, \ldots \psi_n\} \vdash_F \psi$ is a canonical forward inference with a substitution θ,

$$\{\psi_1^{W_1}, \ldots, \psi_n^{W_n}\} \vdash_T \psi^{W_1\theta \cup \ldots \cup W_2\theta}$$

(That is, the conclusion inherits all the tags of the premises after instantiating them appropriately.)

Note that the definition gives an effective mechanism for transforming the inference rules of $\vdash_F$ to those of $\vdash_T$.

Example 10 (continued). The tagged version of the required resolution inference is:

$$\frac{P[x,l]^{\{P[x,l]\}} \quad z \geq x \Longleftarrow z \geq u \wedge u \geq x}{(Q[z,x,l])^{\{P[x,l]\}}} \text{ Resolution}$$

The tag $\{P[x,l]\}$ of the conclusion indicates that $Q[a,x,l']$ may be used in the proof of $P[x,a.l']$.

Definition 18 (Explicit Induction Procedure). Let $\mathcal{A}$ be a theory, $\vdash_R$ a reductive deduction relation, $\vdash_F$ a forward deduction relation and $\vdash_T$ the tagged version of $\vdash_F$. The *explicit induction procedure* is defined by the following transition rules applied to pairs of the form (P, H) where P is the set of inductive propositions and H the set of inductive hypotheses:

Expand $(P \cup \{\phi\}, H) \Longrightarrow (P \cup \Psi, H \cup \{\phi^{\{\phi\}}\})$ if Ψ is a cover set for ϕ.
Simplify $(P \cup \{\phi\}, H) \Longrightarrow (P \cup P', H)$ if $\mathcal{A} \cup P \cup P' \cup (H)_{\triangleleft \phi \kappa} \vdash_D \phi \kappa$.
Forward $(P, H) \Longrightarrow (P, H \cup \{\phi^W\})$ if $\mathcal{A} \cup H \vdash_T \phi^W$.
Lemma $(P, H) \Longrightarrow (P \cup P', H)$

Theorem 19 (Soundness). *If* $(P_0, \emptyset) \Longrightarrow \cdots \Longrightarrow (\emptyset, H_n)$ *is a derivation of the explicit induction procedure then* $\mathcal{A} \vdash_{ind} P_0$.

The proof of this is similar to that of Theorem 10 based on the following invariants:

$$\forall \phi \in \bigcup_{j=0..i} P_j, \ \forall \text{ ground } \alpha, \ \mathcal{A} \cup (P_i)_{\preccurlyeq \phi \alpha} \vdash \phi \alpha \tag{5}$$

$$\forall \phi^W \in H_i, \ \forall \text{ ground } \alpha, \ \mathcal{A} \cup (P_i)_{\preccurlyeq W \alpha} \vdash W \alpha \tag{6}$$

$$\forall \phi^W \in H_i, \ \forall \text{ ground } \alpha, \ \mathcal{A} \cup W \alpha \vdash \phi \alpha \tag{7}$$

Here, $\vdash$ is the closed union of $\vdash_R$ and $\vdash_F$. Note that (6) is an adaptation of the invariant (4) of the mutual induction procedure. Invariant (7) is new. It represents the semantics of tags.

7 Conclusion

We have developed a concept of induction using syntactic orders defined over propositions. Such an extension offers a powerful deduction paradigm: the capability to do proof by mutual induction. With this technique, different theorems can share their inductive hypotheses; this often allows simpler or more direct proofs. Our examples illustrate the power of this technique and its ease of use. Formalizing and developing this paradigm is one of the contribution of this paper.

We have formalized the technique of implicit induction as a mechanism of induction using propositional orders. The essence of an implicit induction procedure is the use of reductive deduction relations and, implicit in such procedure, the use of proposition orders to control deductions. As a result, explicit induction tests, to validate the use of the inductive hypotheses, becomes unnecessary. This continues the works of Hofbauer and Kutsche [17], Kounalis and Rusinowitch [23], and Reddy [27] to propose an interpretation of completion-based induction as classical Noetherian induction rather than the proof by consistency interpretation [20; 1].

The restriction to reductive deduction mechanisms, essential in implicit induction, limits the deductive capabilities of induction. We have described an effective induction procedure allowing both reductive and non-reductive inferences. This combines the advantages of implicit induction and explicit induction. This procedure is a proposed implementation of the technique of induction using propositional orders. It shows that with some simple bookkeeping mechanism one can divide the first-order deduction part of the procedure into combinations of reductive inferences and forward inferences. The resulting procedure has the simplicity and efficiency of implicit induction while lifting the restriction to reductive deductions.

Acknowledgements We would like to thank Sergeï G. Vorobyov for many valuable discussions and explanations, and Adel Bouhoula, Nachum Dershowitz, Michaël Rusinowitch, and Pierre Lescanne for fruitful discussions.

References

1. L. Bachmair. Proof by consistency in equational theories. In *Proc. 3rd LICS Symp., Edinburgh (UK)*, pages 228–233, 1988.
2. E. Bevers and J. Lewi. Proof by consistency in conditional equational theories. In S. Kaplan and M. Okada, editors, *2nd CTRS Workshop, LNCS*, vol. 516, pages 195–205. Springer-Verlag, 1991.
3. S. Biundo, B. Hummel, D. Hutter, and C. Walther. The Karlsruhe induction theorem proving system. In *8th CADE Conf., LNCS*, vol. 230. Springer-Verlag, 1986.
4. Adel Bouhoula and Michaël Rusinowitch. Automatic case analysis in proof by induction. In Ruzena Bajcsy, editor, *Proc. 13th IJCAI Conf., Chambéry (France)*, volume 1, pages 88–94. Morgan Kaufmann, August 1993.
5. R. S. Boyer and J. S. Moore. *A Computational Logic*. Academic Press, New York, 1979.
6. R. S. Boyer and J. S. Moore. A theorem prover for a computational logic. In M. E. Stickel, editor, *Proc. 10th CADE Conf., Kaiserslautern (Germany), LNCS*, vol. 449, pages 1–15. Springer-Verlag, 1990.
7. F. Bronsard and U. S. Reddy. Conditional rewriting in Focus. In S. Kaplan and M. Okada, editors, *2nd CTRS Workshop, LNCS*, vol. 516, pages 2–13. Springer-Verlag, 1991.
8. F. Bronsard and U. S. Reddy. Reduction techniques for first-order reasoning. In M. Rusinowitch and J. L. Rémy, editors, *3rd CTRS Workshop, LNCS*, vol. 656, pages 242–256. Springer-Verlag, 1992.
9. A. Bundy. A rational reconstruction and extension of recursion analysis. In *IJCAI*, 1989.
10. R. M. Burstall. Proving properties of programs by structural induction. *Computer Journal*, 12:41–48, 1969.
11. N. Dershowitz. Completion and its applications. In *Resolution of Equations in Algebraic Structures*, volume 2: Rewriting Techniques, pages 31–86. Academic Press, San Diego, 1989.
12. N. Dershowitz and J.-P. Jouannaud. Rewrite systems. In J. van Leeuwen, editor, *Handbook of Theoretical Computer Science B: Formal Methods and Semantics*, chapter 6, pages 243–320. North-Holland, Amsterdam, 1990.
13. N. Dershowitz and Z. Manna. Proving termination with multiset orderings. *Comm. ACM*, 22(8):465–476, August 1979.

14. S. J. Garland and J. V Guttag. Inductive methods for reasoning about abstract data types. In *ACM POPL Symp.*, pages 219–228. ACM, 1988.
15. J. A. Goguen. How to prove inductive hypotheses without induction. In *5th CADE Conf.*, LNCS, vol. 87, pages 356–372. Springer Verlag, Jul 1980.
16. B. Gramlich. Induction theorem proving using refined unfailing completion techniques. In *ECAI*, 1989. (also vailable as Technical Report SR-89-14, Universität Kaiserslautern, Germany.).
17. D. Hofbauer and R. D. Kutsche. Proving inductive theorems based on term rewriting systems. In J. Grabowski, P. Lescanne, and W. Wechler, editors, *Proc. 1st ALP Workshop*, pages 180–190. Akademie Verlag, 1988.
18. G. Huet and J.-M. Hullot. Proofs by induction in equational theories with constructors. *J. Comp. and System Sciences*, 25:239–266, 1982.
19. J.-P. Jouannaud and E. Kounalis. Automatic proofs by induction in equational theories without constructors. *Information and Computation*, 82:1–33, 1989.
20. D. Kapur and D. R. Musser. Proof by consistency. *Artificial Intelligence*, 31(2):125–157, February 1987.
21. D. Kapur, P. Narendran, and H. Zhang. Automating inductionless induction using test sets. *J. Symbolic Computation*, 11:83–112, 1991.
22. D. Knuth and P. Bendix. Simple word problems in universal algebras. In J. Leech, editor, *Computational Problems in Abstract Algebra*, pages 263–297. Pergamon Press, Oxford, 1970.
23. E. Kounalis and M. Rusinowitch. Mechanizing inductive reasoning. In *Proc. AAAI Conf.*, pages 240–245. AAAI Press and MIT Press, July 1990.
24. R. Kowalski. *Studies in the completeness and efficiency of theorem-proving by resolution*. PhD thesis, University of Edinburgh, 1970.
25. D. McAllester. Term rewriting induction. theorem-provers@ai.mit.edu electronic bulletin board, 1990.
26. D. R. Musser. On proving inductive properties of abstract data types. In *ACM POPL Symp.*, pages 154–162. ACM, 1980.
27. U. S. Reddy. Term rewriting induction. In M. Stickel, editor, *10th CADE Conf.*, volume 449 of *LNAI*, pages 162–177. Springer-Verlag, 1990.
28. Jean-Luc Rémy. *Etude des systèmes de Réécriture Conditionnels et Applications aux Types Abstraits Algébriques*. Th. Etat, INPL, Nancy (France), 1982.
29. G. A. Robinson and L. T. Wos. Paramodulation and first-order theorem proving. In B. Meltzer and D. Mitchie, editors, *Machine Intelligence 4*, pages 135–150. Edinburgh University Press, 1969.
30. J. A. Robinson. A machine-oriented logic based on the resolution principle. *J. ACM*, 12:23–41, 1965.
31. C.-P. Wirth and B. Gramlich. On notions of inductive validity for first-order equational clauses. In *12th CADE Conf.*, 1994.
32. H. Zhang, D. Kapur, and M. S. Krishnamoorthy. A mechanizable induction principle for equational specifications. In E. Lusk and R. Overbeek, editors, *9th CADE Conf.*, LNCS, vol. 310, pages 162–181. Springer-Verlag, 1988.

Mechanizable Inductive Proofs for a Class of ∀ ∃ formulas

Jacques CHAZARAIN and Emmanuel KOUNALIS

e-mail:{jmch, kounalis}@mimosa.unice.fr
Laboratoire d'Informatique I3S
Université de Nice Sophia Antipolis
Bat 4- 250 Av. Albert Einstein
06560 Valbonne, FRANCE

Abstract. We show how to prove formulas of the form $\forall x \exists y\, \Phi(x,y)$ in the initial model of an equational variety by using purely algebraic simplifications. This allows to tackle theorems whose proofs requires complicated noetherian induction. We also give a disproof theorem, which is not only useful to prove that a conjecture is false but is also used to shorten the proof search of a theorem. We show applications of the method to fragments of arithmetic.

1 Introduction

1.1 The Problem

A ∀∃-formula is a formula of the form $\forall x \exists y\, \Phi(x,y)$, where $x = x_1, \ldots, x_m$, $y = y_1, \ldots, y_n$ are variables and $\Phi(x,y)$ is any quantify-free first-order formula with equality as only predicate symbol. Reasoning about a ∀∃-formula may, for example, involves deciding if it is true in the class of all models of a given set of equations, or if it is true in the initial model of a given set of equations. Whereas, validity in all models of a ∀∃-formula is well-understood and there are several theorem-proving methods to accomplish the task, proofs of ∀∃-formulas in the initial model of a given set of equations turn out to be more complicated. The motivation for this paper is to present a systematic method for proofs in the initial model of an equational variety. In other words we shall deal with the following problem:

- Instance : Given a set E of equations and a formula $\forall x \exists y\, \Phi(x,y)$
- Question : Is $\forall x \exists y\, \Phi(x,y)$ valid in the initial model of E ?

Since the validity of universally quantified equations and more generally ∀∃-formulas in the initial model of a set E of equations is usually established by induction with respect to a noetherian relation on the structure of ground terms,we call them ∀∃-inductive theorems.

1.2 Background and Notations

We assume familiarity with the model and the proof theory of equational logic. For simplicity of notation, we assume have only one sort; all the results carry over to many-sorted case without difficulty. Let $T(F, X)$ denote the set of (first-order) terms built out of *function* symbols taken from a finite vocabulary F and a denumerable set X of *variables*. We assume that F contains at least one *constant* symbol. Thus, the set $T(F)$ of *ground* terms (variable-free), is non-empty. The term, $t\sigma$, which is the result of applying the *substitution* σ to a term t, is said to be an *instance* of t . If each term value of σ is ground, then σ is said to be a *ground substitution* . A set E of *equations* over $T(F, X)$ is a set of equalities of the form $l = r$, where l and r are in $T(F, X)$ and variables appearing in them are universally quantified .

A set E of equations defines the *variety $Vart(E)$*, that is, the class of algebras which are models of the equations considered as axioms. An equation $l = r$ is a logical consequence of E iff it is valid in all models of E. This semantics notion can be characterised in proof-theoretical terms. Let E be a set of equations. We denote by $\vdash_{equ}$ the smallest monotonic congruence that contains E. Then an equation $l = r$ is a logical consequence of E iff $E \vdash_{equ} l = r$. This is actually the well-known completeness Birkhoff's theorem.

The initial model $INIT(E)$ is defined to be the quotient of the algebra $T(F)$ by restriction of the congruence $\vdash_{equ}$ to ground terms. Initiallity of $INIT(E)$ means that there is a unique F- homomorphism from $INIT(E)$ to every algebra in $Vart(E)$. However, the validity of an equation $l = r$ in the initial model requires more than just equational reasoning: some kind of induction is necessary. Unfortunately, there is no simple proof theory which captures the semantic notion of the initial model. To overcome this problem, one uses formula schemata to formulate induction axioms which are used to prove formulas for validity in the initial model.

One way to reason with a set of equations E consists in compiling E into a rewrite system $\mathcal{E}$, i.e. into a set of rules $l \rightarrow r$ where $l = r$ belongs to E. It defines a rewrite relation $\longrightarrow$ on terms , i.e , $s \longrightarrow t$ if there is a rule $l \rightarrow r$ in $\mathcal{E}$, a position p in s and a substitution σ s.t $s/p = l\sigma$ and $t = s[p \leftarrow r\sigma]$. A *position p* in a term is a sequence of integers which defines the root of a subterm. Let $\longrightarrow^*$ the reflexive transitive closure of $\longrightarrow$. A term s is *irreducible* if there is no rule such that $s \longrightarrow t$ for some term t. The system $\mathcal{E}$ is *normalizing* if every ground term has an irreducible form. It is *ground confluent* if for all ground terms s such that $s \longrightarrow^* t_1$ and $s \longrightarrow^* t_2$, we can reduce the terms t_1 and t_2 to the same term.

1.3 Motivations and Overview of our Method

The need to be able to reason with $\forall\exists$-formulas is very important in many applications including number theory, program verification, and program synthesis. For example, the specification of a function $x \rightarrow y = f(x)$ satisfying a property in a theory E (integers, lists, algebraic data types,...) is often written down

$$E \vdash \forall x \exists y \, \Phi(x,y).$$

The constructive framework consists in proving the theorem in a logic with an algorithmic content and then extract a program for f from the proof. The extraction of program from a proof in certain logics can be automatic, but there is no systematic methods to find such a proof (and in fact, we know that an automatic proof is impossible in general). As a consequence, tools for proving theorems in the initial model of an equational variety are of singular importance.

Relation to Others Approach. As we pointed out above, one of the most powerful ways of proving properties of numbers, data structures, or programs is proof by induction. To accomplish the task, classical theorem proving provides explicit induction (more or less heuristic) to solve the problem (Burstall [6] , Boyer and Moore [4], Zhang and Kapur and Krishnamoorthy [15], Padawitz [20], Garlang and Guttag[11], Bundy et al.[5], Walter [23]). Musser [19] first suggested using completion to prove theorems in the standard model of an equational theory; such proofs normally requires structural induction. This approach, so-called inductionless- induction (since it tries to get rid of induction) has further extended by Huet and Hullot [12], Dershowitz [7], Jouannaud and Kounalis [13] , Kapur and Narendran and Zhang [14] , Fribourg [10], Bachmair [1], Reddy [21], Dershowitz and Reddy [9] etc...Wainer [22] has made a study of $\forall \exists$-formulas in arithmetic. However, one of the hardest problems in using either approaches is to find induction schemata which describe suitably the infinitely many premises needed for the proof of the theorem. In Kounalis and Rusinowitch [18] a new system for mechanizing proof of universal formulas has been proposed. It has been implemented in the system "SPIKE" [2], for extensions see [3].

We show here how to prove $\forall \exists$-formulas in the initial model by using purely algebraic simplification. Further, when we wish to carry out proofs of only universally quantified formulas, our method does not require the given set of equations to be turned into a terminating set of rules. More importantly, it is intended for both interactive and automatic program verification and synthesis. The presented technique combines the full power of classical induction and inductionless induction.

Overview. Let us outline our method in very general terms. To carry out a proof of $\forall x \exists y \, \Phi(x,y)$ in the initial model of a given set E of equations, we shall use the equations as rewrite rules to perform simplifications in the formulas. To achieve this task we impose directionality on the equations in E in such a way that the resulted rewrite system $\mathcal{E}$ to have the normalizing property.

Having associated to the set of equations E a normalizing rewrite system $\mathcal{E}$, our next step consists of computing a Test Set for the set $\mathcal{E}$. This is a finite set of terms $T_1, \ldots, T_k$ which, in essence, is a finite description of the initial model of E. The construction of a Test Set $TS(\mathcal{E})$ for $\mathcal{E}$ is decidable and can be performed in a relatively efficient way, (see [16]). The main purpose of the computation of

$TS(\mathcal{E})$ is to decompose automatically the proof of $\forall\,\exists$-formulas into a set of more simpler proofs.

Next, we replace variables of the conjecture to be proved with elements of $TS(\mathcal{E})$ and a simplification strategy is started. The simplification use rules from $\mathcal{E}$, and try to reduce to a smaller (w.r.t a well founded ordering) instance of the conjecture itself. This last point captures the notion of Induction Hypothesis in the proof by induction paradigm. However, most of the time several rounds are needed before getting the proof of a conjecture. The process of repeatedly breaking a proof down into simpler proofs must then continue until we eventually end up with proofs that can be trivial. At each round, the obtained formulas are in fact intermediate lemmas needed for the proof of the initial conjecture. Further, there is no hierarchy among the intermediate lemmas to be proved, and therefore no difficulty for the management of inductive hypothesis: every intermediate lemmas will be set into the initial set of conjectures.

As an informal example, consider the specification of the minus function in the theory of integers $\mathbb{Z}$:
$$\forall x\,\exists y \quad x+y=0 \qquad (*)$$
To prove (*) we assume given the following rules for addition in $\mathbb{Z}$:

$x+0 \to x \qquad x+S(y) \to S(x+y) \qquad x+P(y) \to P(x+y) \qquad S(P(x)) \to x$
$0+x \to x \qquad S(x)+y \to S(x+y) \qquad P(x)+y \to P(x+y) \qquad P(S(x)) \to x$

Here S stands for successor and P stands for predecessor. The computation of a test set gives $\{0, S(x), P(x)\}$. Now, we have to replace the universal variable x with each term of the test set. If we consider the case of $S(x')$ we get
$$\forall x'\,\exists y \quad S(x')+y=0$$
which reduces to
$$\forall x'\,\exists y \quad S(x'+y)=0. \qquad (**)$$
In order to prove (**), we replace the existential variable y with each term in the test set. For instance, using $P(y')$, we obtain
$$\forall x'\,\exists y' \quad S(x'+P(y'))=0$$
which reduces (using the rules for P and for SP) to
$$\forall x'\,\exists y' \quad x'+y'=0 \qquad (***)$$
which is the formula (*) modulo variable renaming. Since the term x' is strictly less than the term $S(x')$ under the subterm ordering, we may use (***) to achieve the proof of (**).

We will give a more systematic treatment of this example in section 2.3 .We give others examples in section 3 and in particular in section 3.3 where we show a case where an intermediate lemma is introduced.

The rest of this paper is organized as follows: in section 2 we show how to prove (or disprove) $\forall\,\exists$-Formulas. We first provide a definition of what it means for $\forall\,\exists$-formulas to be an inductive theorem. We then give our main results. In section 3 we present severals improvements with further examples.

2 Proofs of ∀∃-Formulas

In this section we develop the machinery for proving ∀∃-formulas in the initial model of an equational variety. We first provide the fundamental notions we will be dealing with and then set up the methods to operationalize their treatment. We finally illustrate the method on simple examples.

2.1 The Basics

As we have pointed out in the introductory remarks of this paper, an equation $l = r$ is valid in the initial model of an equational variety iff for every ground substitution σ, the equation $l\sigma = r\sigma$ is a logical consequence of E. That is

$$E \vdash_{ind} l = r \quad \text{iff} \quad E \vdash_{ind} l\sigma = r\sigma.$$

As a classical example, the formula $\forall x \forall y \quad x + y = y + x$ (*) is not deducible from the axioms :

$$x + 0 = x \quad \text{and} \quad x + S(y) = S(x + y)$$

However, every ground instance of (*) can be shown to be deducible from these axioms.

Let us first define what it means for a ∀∃-formula to be valid in the initial model of a set of equations.

Definition. 1 *Let E be a set of equations. A formula of the form $\forall x \exists y \, \Phi(x, y)$ is an inductive theorem of E if for all ground substitutions σ there exists a ground substitution η such that $E \vdash_{equ} \Phi(x\sigma, y\eta)$. This is denoted by $E \vdash_{ind} \forall x \exists y \, \Phi(x, y)$*

Clearly the major problem in automatizing definition 1 is the unbounded number of ground substitutions. The central idea consists of finding a "sufficiently representative" finite set of models for the set of all ground substitutions.

The following definition gives a suitable finite description of the ground substitutions to be considered for proofs in the initial model of a set E of equations, provided that E is transformed into a rewrite system $\mathcal{E}$.

There is a general definition of the test set [18] which is rather technical, so we shall only give the definition in the particular case of left-linear rewrite systems.

Definition. 2 *Let $\mathcal{E}$ be a left-linear rewrite system. A Test Set $TS(\mathcal{E})$ for $\mathcal{E}$ is a finite set of $\mathcal{E}$-irreducible terms such that:*

1. *For any $\mathcal{E}$-irreducible ground term s, there exists a term t in $TS(\mathcal{E})$ and a substitution σ such that $s = t\sigma$.*
2. *Every non-ground term in $TS(\mathcal{E})$ has infinitely many $\mathcal{E}$-irreducible ground instances,*
3. *Every non-ground term in $TS(\mathcal{E})$ contains variables which occur only at depth greater or equal than $depth(\mathcal{E})$, where $depth(\mathcal{E})$ is the maximum of the depth of the left hand sides of rules in $\mathcal{E}$.*

In practice, the rewrite systems are often left-linear and it will be the case in all our examples.

The Test Sets with which we deal in the present context possess themselves a well-defined mathematical structure. This, to a certain degree, mirrors the structure of initial model. The hypothesis 1) establishes a kind of completeness on the ground substitutions we need for proofs and 2), 3) give the conditions which are needed for proofs and disproofs. The construction of Test sets $TS(E)$ for a set of rules is decidable and can be performed in a relatively efficient way, [16, 17].

Remark :The variable names of terms of a test set are irrelevant , so we can use alpha conversion.

2.2 Main Theorem

We assume given an ordering $\ll$ on the terms which is well-founded on ground terms and such that for every terms s1 and s2 $\quad s_1 \ll s_2 \Rightarrow s_1\sigma \ll s_2\sigma \quad$ for any ground substitution σ. We know from [8] that such ordering can be extended to a multiset ordering.

We use the notation $x = (x_1, \ldots, x_n)$ to denote the sequence of variables $x_1, \ldots, x_n$. At the same time, we use the informal notation $\Phi(x/s, x/t)$ to denote the formula obtained from $\Phi(x, y)$ by applying the substitution $(x_1 \leftarrow s_1, \ldots, x_n \leftarrow s_n \quad y_1 \leftarrow t_1, \ldots, y_n \leftarrow t_n)$.

Theorem. 1 *Let E a set of equations defining a set $\mathcal{E}$ of rewrite rules , and let $TS(\mathcal{E})$ be a Test Set . We consider a finite set of $\forall\,\exists$-formulas :*

$$\forall x\,\exists y\,\Phi_i(x, y)\ (i = 1, \ldots, n).$$

If

(H1) Every ground term can be $\mathcal{E}$-rewrite to an irreducible form (normalizing property)

(H2) For all Φ_i , for all substitution σ with values in $TS(\mathcal{E})$,there exists a substitution η with values in $TS(\mathcal{E}) \cup X$ such that $\Phi_i(x\sigma, y\eta) \longrightarrow^* \Psi_i$ where Ψ_i is either
  - *(a) an equational tautology (say $z = z$) , or*
  - *(b) a formula of the form $\forall x'\,\exists y'\ y' = H(x')$ (explicit form)*
  - *(c) a formula $\Phi_k(x\tau, y/y')$ with a substitution τ such that $\tau \ll \sigma$ for the multiset ordering*

then, we get

$$E\ \vdash_{ind}\ \forall x\,\exists y\,\Phi_i(x, y)\ (i = 1, \ldots, n).$$

Proof.
For simplicity, we prove the theorem in the particular case of formulas with one universal variable et one existential variable.

We use noetherian induction, for the ordering $\ll$ on closed terms, to show that for any of closed term s the following property holds :

$Q(s) = $ (for any i there exists a closed term t s.t $\;E \vdash_{equ} \Phi_i(x/s, y/t)\;$).

We can assume s irreducible, since a reduced form of s exists by (H1) . Note that s satisfies the property Q since it is conserved by rewriting. Assume $Q(s')$ is true for any $s' \ll s$ let us prove $Q(s)$.

By definition of a test set, there is a test set term S and a ground substitution σ such that $S\sigma = s$.
By hypothesis (H2), we get for any i the existence of a term T such that

(*) $$\Phi_i(x/S, y/T) \longrightarrow^* \Psi_i.$$

By applying to (*) the ground substitution σ (we can always assume that T and S have disjoint variables names) we get :

$$\Phi_i(x/S\sigma, y/T) \longrightarrow^* \Psi_i\sigma.$$

There are three cases depending on the form of Ψ_i :

case 1 : Ψ_i is an equational tautology. Then $\Phi_i(x/S\sigma, y/T)$ is also one. Therefore, we can prove $Q(s)$ by taking for t any closed instance of T.

case 2 : Ψ_i is of the form $\forall x' \exists y'\; y' = H(x')$. This formula is obviously an equational theorem and therefore $\Phi_i(x/S\sigma, y/T)$ is one too. Hence, we can prove $Q(s)$ by taking for t the ground term $T(H(x\sigma))$.

case 3 : Ψ_i is of the form $\Phi_k(x/W, y/y')$, we note that the closed term $s' = W\sigma$ is strictly less than $s = S\sigma$ by (H2) and the stability of the ordering for substitutions. Therefore the induction hypothesis implies the existence of a closed term t' such that

$$E \vdash_{equ} \Phi_k(x/s', y/t').$$

By applying the ground substitution $\eta : y' \longrightarrow t'$ to the above rewriting, we get

$$\Phi_i(x/S\sigma, y/T\eta) \longrightarrow^* \Phi_k(x/W\sigma, y/y'\eta).$$

By definition of s' and t', we have

$$\Phi_k(x/W\sigma, y/y'\eta) = \Phi_k(x/s', y/t').$$

Hence $\Phi_i(x/S\sigma, y/T\eta)$ is an equational theorem . Therefore, we have proved $Q(s)$ by taking for t the closed term $T\eta$ $\square$.

Remark. It will be useful for section 3 to note that the theorem will be still valid if we extend the rewriting relation $\longrightarrow^*$ to any relation which keeps the the property of being an equational theorem.

Of course, we do not claim any completeness result, our theorem is only a sufficient condition for proving such inductive result, but as we shall see in the examples, it often works quite well.

Let us outline how the theorem above can be used to find a proof of an $\forall x \exists y\, \Phi$ formula : for each term $S_i(x')$ in the test set, we reduce the formula $\Phi(S_i(x'), y)$ to a formula $\Phi_1(x', y)$. If it is already in one of the case a), b) ,c) of the theorem we are done. If not, we replace the existential variable with each test set term $S_j(y')$ and reduce $\Phi_1(x', S_j(y'))$ to a formula $\Phi_2(x', y')$. If it is already in one of the case a), b) ,c) of the theorem we are done. Otherwise, we consider Φ_2 as a new lemma to prove by the same method, and so on... Of course, we can loop, but when we succeed, our theorem implies the validity of all intermediate lemmas Φ_i.

2.3 An Example

Let us now illustrate this with the very simple example given in the introduction. Assume we wish to prove the following $\forall\, \exists$-formula in the set $\mathbb{Z}$ of integers:

$$\forall x \exists y \quad x + y = 0. \tag{$*$}$$

Suppose that we are given the following facts about the function symbols $+$ to denote the addition function, S to denote the successor function, P to denote the predecessor function, and 0 to denote the zero function. The arrow indicates how to use an equation as a rewrite rule :

$$\begin{array}{llll}
x + 0 \to x & x + S(y) \to S(x+y) & x + P(y) \to P(x+y) & S(P(x)) \to x \\
0 + x \to x & S(x) + y \to S(x+y) & P(x) + y \to P(x+y) & P(S(x)) \to x
\end{array}$$

In order to prove (*), a Test set for the system $\mathcal{E}$ is first computed. In the present case $TS(\mathcal{E}) = \{0, S(x), P(x)\}$. Having computed a test set, universal variables of the formula to be proved are replaced with the elements of $TS(\mathcal{E})$. We get the following three formulas :

$$\exists y \quad 0 + y = 0$$
$$\forall x' \exists y \quad S(x') + y = 0$$
$$\forall x' \exists y \quad P(x') + y = 0.$$

These formulas, in turn, are reduced to

$$\exists y \quad y = 0 \tag{1}$$
$$\forall x' \exists y \quad S(x' + y) = 0 \tag{2}$$
$$\forall x' \exists y \quad P(x' + y) = 0 \tag{3}$$

using the rewrite rules in $\mathcal{E}$.

Thus, in order to prove (*) we try to reduce formulas (1),(2), (3) to formulas which satisfy one of the cases of the theorem 1.

Formula 1 is already in the explicit form pointed out by case 2 of theorem 1 . Let us now consider formula (2). By replacing the existential variable y with the test set elements : $\{0, S(y'), P(y')\}$. We get

$$\forall x' \quad S(x' + 0) = 0 \tag{2.1}$$

$$\forall x' \exists y' \ S(x' + S(y')) = 0 \tag{2.2}$$

$$\forall x' \exists y' \ S(x' + P(y')) = 0 \tag{2.3}$$

Formulas (2.1),(2.2) and (2.3) are in turn reduced to

$$\forall x' \quad S(x') = 0 \tag{2.1.r}$$

$$\forall x' \exists y' \quad S(S(x' + y') = 0 \tag{2.2.r}$$

$$\forall x' \exists y' \quad x' + y' = 0 \tag{2.3.r}$$

Now formula (2.3.r) is an instance of (*). Since x' in (2.3.r) is strictly smaller than $S(x')$ the subterm ordering, we use the case 3 of our theorem. This achieve the proof of formula (2). The proof of formula (3) is carried out by symmetry in a similar way. Therefore, (*) is true in the initial model of E.

2.4 A Disproof Theorem

It can be useful to have a criterion to verify whether a conjecture is false. The reason is that when we look for a proof, we may avoid some dead paths connected with false hypothesis.

We need a more restrictive hypothesis on the rewrite systems :

Theorem. 2 *Let $\mathcal{E}$ be a linear, ground confluent rewrite system. Consider the formula:* $\forall x \exists y \ s(x,y) = r(x,y)$. *We assume that:*

1. *there exists a test-set substitution σ such that for all test-set substitutions η $\quad s(x\sigma, y\eta)$ and $r(x\sigma, y\eta)$ are both $\mathcal{E}$-irreducible*
2. *there exists an existential variable position p in either $s(x,y)$ or $r(x,y)$ such that $s(x,y)/p$ and $r(x,y)/p$ are not unifiable (i.e, they do not share common instances).*

then $\forall x \exists y \ s(x,y) = r(x,y)$ is not an inductive theorem of E.

We omit the proof which is quite long and technical, but we give below a simple example of application. Suppose we wish to prove that the formula

$$\forall x \exists y \quad x + y = 0 \tag{0}$$

is not an inductive theorem in the natural integers $\mathbb{N}$.

We are given the following equations for the additive theory of the natural integers :

$$x + 0 = x$$

$$x + S(y) = S(x+y).$$

A test set is given by $\{0, S(x)\}$. We have now to prove the formulas

$$\exists y \quad 0 + y = 0 \qquad (1)$$

$$\forall x' \exists y \quad S(x') + y = 0. \qquad (2)$$

We immediately succeed with the first one by taking $y = 0$. The second formula reduces to the proof of

$$\forall x' \exists y \quad S(x' + y) = 0. \qquad (2.r)$$

By replacing x' with 0 we get

$$\exists y \quad S(0 + y) = 0$$

which reduces the proof to those of

$$\exists y \quad S(y) = 0. \qquad (3)$$

Now (3) verifies the requirements 1 and 2 of theorem 2 and therefore (3) is not an inductive theorem. If

$$\forall x' \exists y \quad S(x') + y = 0$$

was an inductive theorem, formula (3) would necessary be one too. Therefore, we have proved that (0) is not an inductive theorem in $\mathbb{N}$.

3 Improvements and Others Examples

We give some extensions of our method and apply them to some examples.

3.1 Logical Rules

In some case, we shall reduce equational formulas using not only rewriting rules but also "logical rules".
A logical rule in our context is an inductive theorem of the form $u_1 = u_2 \Rightarrow v_1 = v_2$. More precisely :

$$E \vdash_{ind} \quad \forall x \, (u_1 = u_2 \Rightarrow v_1 = v_2).$$

We shall use it to reduce equational formulas. A formula $e_1 = e_2$ which is an instance of $u1 = u2$ can be reduced to the corresponding instance of $v1 = v2$. This new kind of reduction keeps obviously the property that a formula $e1 = e2$ is an inductive theorem. The reduction which combines rewriting and logical reduction by a set of logical rules is still denoted by $\longrightarrow^*$. The main theorem remains without any change because a logical reduction keeps the property of being an equational theorem for ground terms. More precisely, given a logical rule and a ground substitutions σ we have the equational theorem

$$E \vdash_{equ} (u_1 = u_2)\sigma \Rightarrow (v_1 = v_2)\sigma.$$

Therefore, if $(u_1 = u_2)\sigma$ is an equational theorem then $(v_1 = v_2)\sigma$ is one too.

Let us give a theorem which is the main source of logical rules. We say that a term T is *regular* for a set of rewrite rules if no non-variable subterm of T unifies with the left-hand side of a rule.

Theorem. 3 *Let E be a set of equations. Assume that E can be oriented into a set $\mathcal{E}$ of rewrite rules which is normalizing and confluent on ground terms. Then for any regular term $T(x_1, \ldots, x_n)$ the following formulas are valid in the initial model of E:*

$$\forall x, y \ (\ T(x_1, \ldots, x_n) = T(y_1, \ldots, y_n) \Rightarrow x_i = y_i\ ) \quad \textit{for all} \ \ i \leq n.$$

Proof. To prove that

$$T(x_1, \ldots, x_n) = T(y_1, \ldots, y_n) \Rightarrow x_i = y_i \quad \text{for all} \ \ i \leq n$$

is valid in the initial model of E we must show that for any ground substitution σ the equality $T(x_1\sigma, \ldots, x_n\sigma) =_E T(y_1\sigma, \ldots, y_n\sigma)$ implies $x_j\sigma =_E y_j\sigma$.

By hypothesis, the ground terms $x_i\sigma$ and $y_i\sigma$ have an irreducible form, we denote them respectively by u_i and v_i. After reduction, we get the equality $T(u_1, \ldots, u_n) =_E T(v_1, \ldots, v_n)$. The regular hypothesis on the term T implies that the ground terms $T(u_1, \ldots, u_n)$ and $T(v_1, \ldots, v_n)$ are $\mathcal{E}$-irreducible. Therefore the unicity of the reduced form implies the identities $u_i = v_i$ for all $i \leq n$
□

Let us illustrate the use of these rules on an example. Consider the set (N) of rules for the natural integers $\mathbb{N}$:

$$0 + x \to x$$
$$S(x) + y \to S(x + y)$$

This rewrite system is normalizing and confluent on ground terms. The term $S(x)$ is regular, so the formula

$$S(u) = S(v) \Rightarrow u = v$$

is an inductive theorem of (N). For an application of the use of that kind of formulas, see the end of the next example.

3.2 Partial Evaluation of Boolean Formulas

To reduce formulas with propositional connectors as $\vee$ or $\wedge$, we can make use of the following rules that allows partial evaluation of Boolean expressions :

$X \vee True \rightarrow True \qquad True \vee X \rightarrow True$
$X \vee False \rightarrow X \qquad False \vee X \rightarrow X$
$X \wedge True \rightarrow X \qquad True \wedge X \rightarrow X$
$X \wedge False \rightarrow False \qquad False \wedge X \rightarrow False$

These rules keep the logical content of the propositional connectors. Therefore we can use it also to reduce our formulas without loosing the property of being an inductive theorem.

As an example, we consider again the case of the natural integer $\mathbb{N}$ and try to prove the formula

$$\forall x \forall y \exists z \quad (x + z = y \quad \vee \quad y + z = x) \tag{0}$$

which means that any two natural numbers are comparable.

Using the test set $\{0, S(x')\}$ we need to prove the following two formulas:

$$\forall y \exists z \quad (0 + z = y \quad \vee \quad y + z = 0) \tag{1}$$

$$\forall x' \forall y \exists z \quad (S(x') + z = y \quad \vee \quad y + z = S(x')) \tag{2}$$

After simplification of (1) and (2) we get

$$\forall y \exists z \quad (z = y \quad \vee \quad y + z = 0) \tag{1.r}$$

$$\forall x' \forall y \exists z \quad (S(x' + z) = y \quad \vee \quad y + z = S(x')) \tag{2.r}$$

Let us begin with (1.r). We can take for z the value $z = y$, so (1.r) reduce to true.

Now, we consider the formula (2.r) and we replace the variable y with the terms: $0, S(y')$. After simplifications, we get :

$$\forall x' \exists z \quad (S(x' + z) = 0 \quad \vee \quad z = S(x')) \tag{2.r.1}$$

$$\forall x' \forall y' \exists z \quad (S(x' + z) = S(y') \quad \vee \quad S(y' + z) = S(x')) \tag{2.r.2}$$

Taking $z = S(x')$, the first formula is true by the same reason as above . Using the logical rule $S(u) = S(v) \Rightarrow u = v$, the (2.r.2) formula may further be simplified, to

$$\forall x' \forall y' \exists z \quad (x' + z = y' \quad \vee \quad y' + z = x')$$

which is of the form (*) and therefore by the case 3 of theorem 1 we get that (0) is an inductive theorem.

3.3 Example with Mutual Induction

We illustrate an application of our main theorem in the case where mutual induction is needed. The euclidian division by 2 is usually specified by the formula:

$$\forall x \, \exists y, r \quad (x = y + (y + r) \quad \wedge \quad r \leq S(0) = True) \tag{0}$$

We assume given the following rewrite system for the definition of + in integers:

$$x + 0 \to x \qquad x + S(y) \to S(x + y)$$

$$0 + x \to x \qquad S(x) + y \to S(x + y).$$

The definition of $\leq$ is given by the rules :

$$0 \leq x \to True$$

$$S(x) \leq 0 \to False$$

$$S(x) \leq S(y) \to x \leq y.$$

A test set of the above rewrite system is : $\{0, S(x), True, False\}$.

Remark . In this example, we need to consider two sorts : integer and Boolean. Of course, the variables of integer sort can only be replaced with test set terms of the same sort : $0, S(x')$.

By replacing the universal variable x with the elements of the test set, we obtain the following two formulas :

$$\exists y, r \quad (0 = y + (y + r) \quad \wedge \quad r \leq S(0) = True) \tag{1}$$

$$\forall x' \, \exists y, r \quad (S(x') = y + (y + r) \quad \wedge \quad r \leq S(0) = True) \tag{2}$$

Case of formula (1). By replacing the existential variable y with 0 we get:

$$\exists r \quad (0 = 0 + (0 + r) \quad \wedge \quad r \leq S(0) = True).$$

By simplification,

$$\exists r \quad (0 = r \quad \wedge \quad r \leq S(0) = True).$$

If we substitue 0 for the existential variable r, we get
$(0 = 0 \quad \wedge \quad True = True)$, a tautology.

Case of formula (2).

$$\forall x' \exists y, r \quad (S(x') = y + (y + r) \quad \wedge \quad r \leq S(0) = True)$$

There is no simplification and it is also a $\forall \exists$- formula. Hence, we consider it as a new lemma. In this case the two new formulas, corresponding to the test set values for x' : 0 and $S(x'')$, are :

$$\exists y, r \quad (S(0) = y + (y + r) \quad \wedge \quad r \leq S(0) = True) \qquad (2.1)$$

$$\forall x'' \exists y, r \quad (S(S(x'')) = y + (y + r) \quad \wedge \quad r \leq S(0) = True) \qquad (2.2)$$

- Since (2.1) cannot be reduced, we susbstitue 0 for the existential variable y. We get after simplification:

$$\exists r \quad (S(0) = r \quad \wedge \quad r \leq S(0) = True.$$

Now, the first operand of $\wedge$ is in explicit form and gives $r = S(0)$, which succeeds also in the second operand.
- It remains to treat formula (2.2)

$$\forall x'' \exists y, r \quad (S(S(x'')) = y + (y + r) \quad \wedge \quad r \leq S(0) = True).$$

If we skip some dead end when $y = 0$ and substitute the test set value $S(y')$ for y, we get after simplifications:

$$\forall x'' \exists y', r \quad (S(S(x'')) = S(S(y' + (y' + r))) \quad \wedge \quad r \leq S(0) = True).$$

By using the logical reduction for S we get :

$$\forall x'' \exists y', r \quad (x'' = y' + (y' + r) \quad \wedge \quad r \leq S(0) = True).$$

Now this formula is the initial formula (modulo alpha renaming) .
So, our theorem shows that formula (0) and the auxiliary lemma (2) are inductive theorems.

The important point, is that the auxiliary lemma has been generated by our proof method, we do not have to guess it.

4 Conclusion

We have given a constructive method for performing inductive proofs of a class of $\forall \exists$ -formulas. The main advantage of this method, based on the test set tool, is that we do not have to explicit the inductions hypothesis. They are automatically generated for the main formula and also for the auxiliary lemmas. We give also a result concerning the disproof of theorem which is mainly used to avoid dead end in the proof search. An implementation of our method by S. Muller permits to prove automatically all the examples of this paper.

An important aspect of our process to find a proof is that it can also be used to generated a recursive definition of a Skolem function for an $\forall \exists$ - formula. We shall develop this point in a future work.

References

1. Bachmair, L. Proof by consistency in equational theories, Proc. 3th IEEE Symposium on Logic in Computer Science, Edinburgh.
2. Bouhala, A. Kounalis, E. and Rusinowitch, M. (1992). Automated Mathematical induction, Research Report of INRIA, Submitted.
3. Bouhala, A. and Rusinowitch, M. Automatic Case Analysis in Proof by Induction. Int. Joint Conf. on Artificial Intelligence Chambery, 1993.
4. Boyer, R.S. and Moore, J.S. A Computational Logic, Academic Press, New York.
5. Bundy, A. and van Harmelen, F. and Smaill A. and Ireland A. Extensions to the rippling-out tactic for guiding inductive proofs. Proc. 10th CADE, LNCS volume 449 (1989).
6. Burstall, R.M : Proving properties of programs by structural induction. Computer Journal , 12(1), pp.41-48, 1969.
7. Dershowitz, N. Termination of rewriting, Journal of Symbolic Computation 3 (1-2).
8. Dershowitz, N. and Manna, Z. Proving termination with multiset orderings. CACM 22(8) ,69-116 (1987).
9. Dershowitz, N. and Reddy, U. Deductive and Inductive Synthesis of Equational Programs. J. Symbolic Computation (1993) 15 , 467-494.
10. Fribourg, L. A Strong Restriction of the Inductive Completion Procedure, Proc. 13th ICALP, Rennes
11. Garland, S. J. and Guttag, J.V. Inductive methods for reasoning about abstract data types." Proc. ACM POPL Conference, (1988)
12. Huet, G. and Hullot, J.M. (1982). Proofs by induction in equational theories with constructors, Journal of Computer System Sciences 25 (2).
13. Jouanaud, J.P. and Kounalis, E. (1986 and 1989). Automatic Proofs by induction in equational theories without constructors, Proc. 1st IEEE Symposium on Logic in Computer Science. Full paper in Information and Control 82 (1989).
14. Kapur, D., Narendran, P. and Zhang, H. Proof by induction with test sets, Proc. 8th CADE, LNCS volume 230 (1986).
15. Kapur, D.,Krishnamoorthy. and Zhang, H. A mechanizable induction principle for equational specifications" , Proc. 9th CADE, LNCS volume 310 (1988).
16. Kounalis, E. Pumping lemmas for tree languages generated by rewrite systems, Proc. 15th Conference on Mathematical Foundations of Computer Science (MFCS 90), Banska Bystrica, LNCS 452, Springer-Verlag.
17. Kounalis, E. Testing for inductive-(co)-reducibility in rewrite systems, Proc. 15th Colloquium on Trees in Algebra and Programming (CAAP 90),LNCS 431, Full paper in Theoretical Computer Science, 1992.
18. Kounalis, E. and Rusinowitch, M. Mechanizing Inductive Reasoning, Proc. 8th National Conference on Artificial Intelligence (AAAI-90), Boston(USA). Also in the Bulletin of the European Association of Theoretical Computer Science (EATCS), n 41, June 1990.
19. Musser, D.R. On proving inductive properties of abstract data types, Proc. 7th POPL Conference, Las Vegas (1980).
20. Padawitz, P. , Computing with Horn Clauses, Springer-Verlag 1988.
21. Reddy, U. Term rewriting induction, Proc. 10th CADE, LNCS volume 449 (1989).
22. Wainer, S. Computability- Logical and Recursive Complexity. NATO ASI Series. Logic, Algebra and Computation. Edited by F. L. Bauer. Springer Verlag 1991.
23. Walther, G. Argument-bounded algorithms as bases for automated termination proofs, Proc. 9th CADE, LNCS volume 310 (1988).

On the Connection between Narrowing and Proof by Consistency

Olav Lysne

Department of Informatics
University of Oslo
Po. box 1080 Blindern
0316 OSLO
NORWAY
E-mail: Olav.Lysne@ifi.uio.no

Abstract. We study the connection between narrowing and a method for proof by consistency due to Bachmair, and we show that narrowing and proof by consistency may be used to simulate each other. This allows for the migration of results between the two process descriptions. We obtain decidability results for validity of equations in the initial algebra from existing results on narrowing. Furthermore we show that several results on completeness of position selection strategies for narrowing are special cases of a generalization of a result on covering sets presented by Bachmair.

1 Introduction

The seminal ideas for *narrowing* and *proof by consistency* emerged more or less simultaneously in the late seventies/early eighties. Although narrowing seems to have roots in an early work by Lankford, the first description of the systematic use of narrowing steps to find E-unifiers appears to be by Fay [7]. Since then many researchers have been elaborating the method, most of which have concentrated on reducing the search space [5, 6, 8, 12, 20, 23].

The idea of using the Knuth and Bendix completion process to prove properties of algebraic specifications is due to Musser [19]. The key observation is that the addition of an inconsistent equation to a specification leads to the coalescence of two equivalence classes containing ground terms. Under certain restrictions on the specification, such inconsistencies may be detected by using the Knuth and Bendix completion algorithm [15]. Extensions of the proof method have appeared regularly since then, both improving the methods by which inconsistencies may be found, and also extending the sets of specifications to which the method applies [10, 11, 13, 17]. Whereas the mentioned papers all are based on the completion process, Fribourg and Bachmair have presented methods for proof by consistency which do not apply completion [1, 8]. In this paper we shall concentrate on the approach of Bachmair.

There are several results indicating that there is a strong relation between proof by consistency and narrowing. The earliest ones of these perhaps stemming from the work of Fribourg, who used the same concept of *innermost positions* in both settings [8, 9]. Fairly recently it has been reported that *complete positions*, a notion developed by Fribourg in order to reduce the search space in proof by consistency, may be used to reduce the search space in narrowing as well [6].

In this paper we focus on the connection between narrowing and Bachmair's process for proof by consistency. The outline of the paper is as follows: Section 2 presents the preliminaries and basic notation. In section 3 we define the notions of unifiers and inconsistencies, and show how they are related. In section 4 we show how proof by consistency may be used to simulate an interleaving derivation of all possible narrowing paths, and in section 5 we modify the proof of refutational completeness of Bachmair's method in order to be able to transfer the concept of covering sets into narrowing. Section 6 is dedicated to examples of diffusion of results between narrowing and proof by consistency.

2 Basic Notions

We assume familiarity with the basic notions and notation of term rewriting, and only state the sightly less standardized notation.

If t is a term we shall let Vars(t) denote the set of variables occurring in t. The set of positions in a term t is denoted $P(t)$, and the set of non-variable positions in t is denoted $\bar{P}(t)$. The subterm of t at position i, written $t[i]$, is the subterm of t which has its root symbol at position i. The result of replacing the subterm of t at position i by the term u is written $t[i/u]$. By $t!_R$ we shall mean an (the) irreducible form of t wrt. the terminating (convergent) set R of rewrite rules.

We adopt the postfix notation on substitutions, thus a concatenated substitution, e.g. $\gamma\sigma$, corresponds to first applying γ and then σ. We let $\leq$ denote the preorder on substitutions such that $\sigma \leq \gamma$ iff there exists a substitution ρ such that $\sigma\rho = \gamma$. Having a set E of equations we extend the relation $\leftrightarrow^*_E$ to substitutions, such that $\sigma \leftrightarrow^*_E \gamma$ iff $x\sigma \leftrightarrow^*_E x\gamma$ for all variables x. When V is a set of variables we shall let $\sigma \leq_E \gamma[V]$ denote that there exists a substitution ρ such that $\forall x \in V | x\sigma\rho \leftrightarrow^*_E x\gamma$.

The initial algebra in the set of models for the set E of equations is denoted $\mathcal{I}(E)$, and the corresponding equivalence relation is $=_{\mathcal{I}(E)}$.

3 E-Unifiers and Inconsistencies

The problem generally approached by narrowing is that of unification modulo a set E of equations:

Definition 1. We say that a substitution σ is an E-unifier of the terms t and u iff $t\sigma \leftrightarrow^*_E u\sigma$. Furthermore Unif($t, u, E$) is the set of all E-unifiers of t and u, and GUnif(t, u, E) is the greatest subset of Unif(t, u, E) mapping all variables in Vars(t)∪ Vars(u) into ground terms.

Definition 2. Let S be a set of substitutions. We say that S is *a complete set of E-unifiers of the terms t and u* iff

1. $S \subseteq$ Unif(t, u, E)
2. $\forall \sigma \in$ Unif(t, u, E), $\exists \gamma \in S | \gamma \leq_E \sigma[\text{Vars}(t) \cup \text{Vars}(u)]$

S is a *ground* complete set of E-unifiers of the equation $t = u$ if 1. above holds, and requirement 2. is replaced by

3. $\forall \sigma \in \text{GUnif}(t, u, E), \exists \gamma \in S | \gamma \leq_E \sigma[\text{Vars}(t) \cup \text{Vars}(u)]$

Notice that we have two flavors of completeness. Whereas the the first should need no justification, ground complete sets of unifiers correspond to a complete set of solutions of an equation in the initial algebra. This notion of completeness is of special interest in logic programming languages with equality.

Example 1. Let E contain the following four equations:

$$\text{EQ}(x, x) = \text{T} \qquad \text{EQ}(0, s(x)) = \text{F}$$
$$\text{EQ}(s(x), 0) = \text{F} \qquad \text{EQ}(s(x), s(y)) = \text{EQ}(x, y)$$

The set $\{y \mapsto 0, y \mapsto s(z)\}$ is not a complete set of E-unifiers of the terms $\text{EQ}(y, y)$ and T because the empty substitution which also is an E-unifier of the two terms is not represented. The set is however ground complete when the language is restricted to the mentioned function symbols. This because it represents all ground unifiers, i.e. all ground solutions to the equation $\text{EQ}(y, y) = \text{T}$.

We shall now move our focus to proof by consistency. This is a proof method which is based on the observation that $t =_{\mathcal{I}(E)} u$ if and only if $t\sigma \leftrightarrow_E^* u\sigma$ for all ground substitutions σ. From this observation it follows that $t =_{\mathcal{I}(E)} u$ if and only if $\mathcal{I}(E)$ and $\mathcal{I}(E \cup \{t = u\})$ are isomorphic, i.e. if the addition of the equation $t = u$ to the specification does not make two $\leftrightarrow_E^*$ congruence classes both containing ground terms coalesce. Such a coalescence is often called an *inconsistency*. We shall, however, need to view inconsistencies as substitutions, thus we introduce the following notions:

Definition 3. A substitution σ is an E-inconsistency for the equation $t = u$ wrt. the set T of terms if there exist two ground terms $g, h \in T$ such that $t\sigma \leftrightarrow_E^* g$, $u\sigma \leftrightarrow_E^* h$ and $g \not\leftrightarrow_E^* h$. $\text{Inc}(t = u, E, T)$ denotes the set of all E-inconsistencies for $t = u$ wrt. T, and $\text{GInc}(t = u, E, T)$ is the greatest subset of $\text{Inc}(t = u, E, T)$ containing substitutions that map all variables in $\text{Vars}(t) \cup \text{Vars}(u)$ to ground terms.

Definition 4. Let S be a set of substitutions. We say that S is a complete set of E-inconsistencies for the equation $t = u$ wrt. T iff

1. $S \subseteq \text{Inc}(t = u, E, T)$
2. $\forall \sigma \in \text{Inc}(t = u, E, T), \exists \gamma \in S | \gamma \leq_E \sigma[\text{Vars}(t) \cup \text{Vars}(u)]$

S is a *ground* complete set of E-inconsistencies of the equation $t = u$ wrt. T if 1. above holds, and requirement 2. is replaced by

3. $\forall \sigma \in \text{GInc}(t = u, E), \exists \gamma \in S | \gamma \leq_E \sigma[\text{Vars}(t) \cup \text{Vars}(u)]$

Now we shall study how the concepts of E-unifiers and E-inconsistencies relate. Let T be the set of all ground terms. An interesting fact is that $\text{GInc}(t = u, E, T)$ and $\text{GUnif}(t, u, E)$ are inverses of each other in the sense that for every ground substitution σ either $\sigma \in \text{GInc}(t = u, E, T)$ or $\sigma \in \text{GUnif}(t, u, E)$, and no substitution is a member of both sets. We shall, however, use the following connection:

Lemma 5. Let E be a set of equations and t and u be two terms. Let EQ, T and F be a binary function and two constants respectively that are not referred to in neither E, t nor u. Furthermore let $E' = E \cup \{\text{EQ}(x,x) = \text{T}\}$. Then any (ground) complete set S of E' inconsistencies for the equation $\text{EQ}(t,u) = \text{F}$ wrt. the set $\{\text{T},\text{F}\}$, which does not have occurrences of any of the function symbols EQ, T and F is also a (ground) complete set of E-unifiers for t and u.

Proof. First we must prove that all elements in S are unifiers as well. Assume that $\sigma \in S$. Then we must have $\text{EQ}(t,u)\sigma \leftrightarrow^*_{E'}$ T and F$\sigma \leftrightarrow^*_{E'}$ F. Since $\text{EQ}(t,u)\sigma$ is identical to $\text{EQ}(t\sigma, u\sigma)$, and since the only equation in E' referring to EQ is $\text{EQ}(x,x) = \text{T}$, we must have $t\sigma \leftrightarrow^*_{E'} u\sigma$. From the requirements of the lemma we know that neither $t\sigma$ nor $u\sigma$ have occurrences of the symbols EQ and T, thus $t\sigma \leftrightarrow^*_E u\sigma$ and therefore $\sigma \in \text{Unif}(t,u,E)$.

It remains to prove that all unifiers are represented by S. Since all inconsistencies are represented by S it suffices to prove that every unifier is also an inconsistency. Assume $\sigma \in \text{Unif}(t,u,E)$. Then $t\sigma \leftrightarrow^*_E u\sigma$, thus $\text{EQ}(t,u)\sigma = \text{EQ}(t\sigma, u\sigma) \leftrightarrow^*_{E'} \text{EQ}(u\sigma, u\sigma) \leftrightarrow_{E'}$ T. Obviously F$\sigma \leftrightarrow^*_{E'}$ F, thus $\sigma \in \text{Inc}(\text{EQ}(t,u) = \text{F}, E', \{\text{T},\text{F}\})$. The ground case is analogous. □

This lemma invites us to show that the complete sets generated by the methods we consider do not contain references to the symbols EQ, T and F. We shall return to this problem later.

We could wish to strengthen the lemma by referring to $\text{Inc}(\text{EQ}(t,u) = \text{F}, E', T)$ instead of $\text{Inc}(\text{EQ}(t,u) = \text{F}, E', \{\text{T},\text{F}\})$, where T denotes the set of all ground terms. This would give more immediate correspondence between E-unifiers and the idea of inconsistencies in the initial algebra. It would, however, require E' to define EQ completely in the sense that all ground instances of $\text{EQ}(x,y)$ should be $\leftrightarrow^*_{E'}$-equivalent to either T or F. This is theoretically unproblematic, because Meseguer and Goguen have proved that this is always possible by adding a finite number of equations whenever E constitutes a finite convergent set of rewrite rules [18]. The problem of actually finding these equations is, however, non-trivial. So in order to get a better correspondence between the *process* of proof by consistency and the *process* of finding unifiers by narrowing we avoid this problem by focusing on inconsistencies wrt. the set $\{\text{T},\text{F}\}$.

4 Narrowing and Proof by Consistency

We shall in this section present a basic connection between the two processes we consider, and we start with a formal definition of narrowing.

Definition 6. Let R be a convergent set of rewrite rules. We say that the term t is *narrowable to the term u using the rewrite rule $l \to r$ at position i*, if $t[i]$ is not a variable, is unifiable with l with mgu. σ, and $u = t[i/r]\sigma$. This is written $t \leadsto_{[i,l \to r,\sigma]} u$.

It is customary to restrict the narrowing substitution σ to be *away* from the variables in the initial term t in the above definition. This in the sense that it only introduces subterms with fresh variables. We have chosen not to do this because it would introduce a lot of detail without having any clarifying effect on our results. One should however notice that the following theorem of Hullot [12] on the completeness of narrowing derivations only needs the subset of all narrowing derivations where all narrowing substitutions introduce only fresh variables:

Theorem 7 (Hullot). *Let R be a convergent set of rewrite rules, and let EQ be a binary function symbol not referred to in R, thus $R' = R \cup \{\text{EQ}(x,x) \to \text{T}\}$ is also convergent. Let S be the set of substitutions such that $\gamma \in S$ iff there exists a narrowing derivation $\text{EQ}(t,u) \leadsto_{[i_1,r_1,\sigma_1]} \text{EQ}(t_1,u_1) \ldots \leadsto_{[i_n,r_n,\sigma_n]} \text{T}$ where $r_1,\ldots,r_n \subseteq R'$, and $\sigma_1\sigma_2\ldots\sigma_n = \gamma$. Then S is a complete set of R-unifiers of the equation $t = u$.*

This theorem makes it possible to enumerate all E-unifiers by following all possible narrowing derivations simultaneously, thus we may say that the naive narrowing strategy of following all possible derivations is *complete*. The method has been refined and elaborated giving *basic* narrowing [12], *normal* narrowing [7], combinations of the two [20, 22] and different position selection strategies, e.g. [5, 6, 8, 23].

Having defined a process for searching for unifiers, we now move our focus to a process designed for search of inconsistencies. We present Bachmair's algorithm. The data structure consists of two sets of equations, L and C, such that C contains the equations to proved or disproved, and L is supposed to contain lemmas already proven valid in $\mathcal{I}(R)$. The set R of equations is supposed to constitute a convergent set of rewrite rules, and will be viewed as such. Let $CP(R,C)$ be the set of critical pairs emerging from superpositioning left hand sides of rules from R on nonvariable positions of either side of any equation from C, and let $\succ_R$ be a term ordering proving termination of R. These transition rules describe the actual process:

Deduction	$L, C \vdash L, C \cup \{s = t\}$	if $s = t \in CP(R,C)$
Induction	$L, C \vdash L \cup \{s = t\}, C$	if $s =_{\mathcal{I}(R)} t$
Deletion	$L, C \cup \{s = t\} \vdash L, C$	if $s =_{\mathcal{I}(R)} t$
Simplification	$L, C \cup \{s = t\} \vdash L, C \cup \{u = t\}$	if $s \succ_R u$ and $s \leftrightarrow^+_{R \cup L} u$ or $s \leftrightarrow_C u$ by $v = w$ where $s \rhd v$ and $v \succ_R w$

Here $\rhd$ is basically any well founded ordering on terms, but it is customary in such cases to assume the encompassment ordering, such that $t \rhd u$ if a subterm of t is an instance of u, and not vice versa.

Let us now show how proof by consistency contains narrowing. By $\vdash_{basic}$ we shall mean the transition system consisting of only the transition *Deduction* above. It is well known that the deductive power of proof by consistency is maintained by $\vdash_{basic}$ alone.

Theorem 8. *Let R be a convergent set of rewrite rules such that the function F is not referred to in R. For every set of narrowing derivations*

$$t \leadsto_{[i_1,r_1,\sigma_1]} \cdots \leadsto_{[i_n,r_n,\sigma_n]} t_1$$

$$\vdots$$

$$t \leadsto_{[j_1,s_1,\gamma_1]} \cdots \leadsto_{[j_m,s_m,\gamma_m]} t_k$$

wrt. R there exists a $\vdash_{basic}$ *derivation*

$$L, \{t = \text{F}\} \vdash_{basic} \cdots \vdash_{basic} L, C \cup \{t_1 = \text{F}, \ldots, t_k = \text{F}\}$$

wrt. R. and vice versa.

Proof. We first prove the implication from left to right by induction on the number of involved narrowing steps. If all narrowing sequences are empty the lemma obviously holds. Let us assume that the lemma holds for the actual situation stated, and let us extend one narrowing sequence by one step. We assume without loss of generality that the first sequence is extended by the step $t_1 \leadsto_{[i_{n-1}, l \to r, \sigma_{n+1}]} t_1[i_{n+1}/r]\sigma_{n+1}$. Now $t_1[i_{n+1}]$ is unifiable with l with most general unifier σ_{n+1}, thus the equation $t_1[i_{n+1}/r]\sigma_{n+1} = \text{F}$ is a member of $CP(R, C \cup \{t_1 = \text{F}, \ldots, t_n = \text{F}\})$. This means that there is a $\vdash_{basic}$ step

$$L, C \cup \{t_1 = \text{F}, \ldots, t_n = \text{F}\} \vdash_{basic} L, C \cup \{t_1 = \text{F}, \ldots, t_n = \text{F}, t_1[i_{n+1}/r]\sigma_{n+1} = \text{F}\}$$

For the reverse implication the base case is also trivial. When we extend by one $\vdash_{basic}$ transition we must have generated a critical pair by superpositioning a rewrite rule from R on a term u in an equation $u = \text{F}$. This because F is not referred to in R. We know that there is a narrowing derivation generating u from t, and this sequence may be extended in the wanted fashion by reversing the above argument. □

Example 2. Let R be the following
(1) $x + 0 \to x$
(2) $x + s(y) \to s(x + y)$
(3) $\text{EQ}(x, x) \to \text{T}$

Let a and b be variables, and consider the following two narrowing derivations
$\text{EQ}(a + a, a) \leadsto_{[1,(1),\{a \mapsto 0, x \mapsto 0\}]} \text{EQ}(0, 0) \leadsto_{[top,(3),\{x \mapsto 0\}]} \text{T}$
$\text{EQ}(a + a, a) \leadsto_{[1,(2),\{a \mapsto s(b), x \mapsto s(b), y \mapsto b\}]} \text{EQ}(s(s(b) + b), s(b))$

They are simulated by the following $\vdash_{basic}$ derivation
$L, \{\text{EQ}(a + a, a) = \text{F}\} \vdash_{basic}$
$L, \{\text{EQ}(a + a, a) = \text{F}, \text{EQ}(0, 0) = \text{F}\} \vdash_{basic}$
$L, \{\text{EQ}(a + a, a) = \text{F}, \text{EQ}(0, 0) = \text{F}, \text{T} = \text{F}\} \vdash_{basic}$
$L, \{\text{EQ}(a + a, a) = \text{F}, \text{EQ}(0, 0) = \text{F}, \text{T} = \text{F}, \text{EQ}(s(s(b), b) + s(b)) = \text{F}\}$

From the above proof and example we see that the substitution which is developed through a narrowing derivation may be developed in the same manner from the corresponding proof by consistency derivation using the generated most general unifiers in each transition step. In other words unification of the two terms t and u by narrowing may be simulated by finding the inconsistencies of the equations $\text{EQ}(t, u) = \text{F}$ in the fashion indicated by lemma 5. All unifiers of t and u correspond to a $\vdash_{basic}$ derivation creating the inconsistent equation $\text{T} = \text{F}$. This result may be extended to narrowing with interleaving reduction steps, by inserting *Simplification* steps in the proof. One must, however, restrict the *Simplification* rule such that it corresponds with the more restrictive reductions allowed in narrowing derivations.

We may not at this point assume that the refutational completeness results of proof by consistency with reduced search space automatically give us corresponding completeness results for narrowing. The reason for this is that whereas narrowing aims at finding a complete set of unifiers, proof by consistency targets at detecting the mere existence of inconsistencies. This problem will be addressed in the next section.

5 Conservative Covering Sets

We shall in this section study how Bachmair's method for proof by consistency may be used to generate complete sets of inconsistencies. This requires us to solve two problems. The first is to identify and alter the parts of the proof by consistency method where inconsistencies may be lost, and the second is that we must prove the wanted completeness result.

Let us first turn to the alteration of proof by consistency. The only form of rewriting which is legal in narrowing is reduction with the initial set R of rewrite rules. We therefore introduce the following reduction transition to proof by consistency:

$$\text{Reduction } L, C \cup \{s = t\} \vdash L, C \cup \{u = t\} \text{ if } s \to_R^+ u$$

We shall use the symbol $\vdash_{reduc}$ to denote transitions for proof by consistency with the *Deduction* and the *Reduction* rules alone. The *Induction* rule has lost its meaning when we only allow simplification by R, and *Deletion* is kept out for simplicity reasons.

In order to be able to prove the wanted completeness result we must give some introductory definitions. The key concept for detecting inconsistencies in all facets of proof by consistency is *ground reducibility*.

Definition 9. A term t is *ground reducible* with respect to a set R of rewrite rules if all of its ground instances are R-reducible. An equation $t = u$ is ground reducible wrt. R if for every ground instance $t\sigma = u\sigma$, either $t\sigma$ and $u\sigma$ are identical, or one of them is R-reducible.

It is well known that ground reducibility is decidable for terms as well as for equations [13, 14, 21].

Definition 10. Let R be a convergent set of rewrite rules, terminating by the ordering $\succ_R$. An equation $t = u$ is *provably inconsistent* wrt. $\mathcal{I}(R)$ if either $t \succ_R u$ and t is not ground reducible, $u \succ_R t$ and u is not ground reducible or the equation $t = u$ itself is not ground reducible.

All the three cases of the above definition separately guarantee that adding $t = u$ to R leads to the coalescence of two ground term $\mathcal{I}(R)$-congruence classes. Now if $\sigma \in \text{Inc}(t = u, R, T)$ where T is the set of all ground terms, there must exist an equational proof $t\sigma!_R \leftarrow_R^* t\sigma \leftrightarrow_{\{t=u\}} u\sigma \to_R^* u\sigma!_R$ such that $t\sigma!_R \neq u\sigma!_R$. We shall call such proofs *inconsistency witnesses*. Inspired by definition 10 we give the following definition

Definition 11. Let T be the set of all ground terms. An inconsistency $\sigma \in \text{Inc}(t = u, R, T)$ is *detectable* if there exists an inconsistency $\gamma \in \text{Inc}(t = u, R, T)$ such that $\sigma \leftrightarrow_R^* \gamma$ and the inconsistency witness corresponding to γ contains no R-steps, i.e. $t\gamma$ and $u\gamma$ are irreducible.

It is clear that an equation is provably inconsistent if and only if it has detectable inconsistencies. The reason for including $\leftrightarrow_R^*$ equivalent inconsistencies in the definition is connected to the fact that in a complete set of inconsistencies one substitution represents all of its $\leftrightarrow_R^*$ equivalents.

Now for the completeness proof. The development of our extended proof follows the same lines as that in [1]. In the proof of refutational completeness of Bachmair's procedure it is shown that the equation $t = u$ is successively replaced by equations that have inconsistencies which are gradually getting closer to being detectable. This proof requires a well founded ordering on inconsistency witnesses in which the cornerstone is a complexity function c:

$$c(t\sigma!_R \leftarrow_R^* t\sigma \leftrightarrow_{\{t=u\}} u\sigma \rightarrow_R^* u\sigma!_R) = \langle \{t\sigma\}, t, u\sigma \rangle \quad \text{if } t \succ_R u.$$
$$= \langle \{u\sigma\}, u, t\sigma \rangle \quad \text{if } u \succ_R t.$$
$$= \langle \{u\sigma, t\sigma\}, -, - \rangle \quad \text{otherwise.}$$

In the following we shall let $\succ_c$ denote the lexicographic combination of the multiset extension of $\succ_R$, $\rhd$, and $\succ_R$ itself. $\succ_c$ may now be viewed as an ordering on inconsistency witnesses, and is obviously well founded.

Now for the completeness. First we show that no inconsistencies are lost, or receding from being detected by any $\vdash_{reduc}$ step.

Lemma 12. *Let $\sigma \in \text{Inc}(t = u, R, T)$ where $t = u$ is an equation in the set C, and T is the set of all ground terms. Furthermore let $L, C \vdash_{reduc} L', C'$. Then there exists an equation $t' = u' \in C'$ such that $\sigma \in \text{Inc}(t' = u', R, T)$, and such that*

$$c(t\sigma!_R \leftarrow_R^* t\sigma \leftrightarrow_{\{t=u\}} u\sigma \rightarrow_R^* u\sigma!_R) \succeq_c c(t'\sigma!_R \leftarrow_R^* t'\sigma \leftrightarrow_{\{t'=u'\}} u'\sigma \rightarrow_R^* u'\sigma!_R)$$

Proof. If $t = u \in C'$ we are done, thus we need only consider the cases where C' is C with the equation $t = u$ replaced by $s = u$ such that $t \rightarrow_R^+ s$. We prove the lemma using the equation $s = u$ for $t' = u'$. Since $t\sigma \rightarrow_R^+ s\sigma$ and R is convergent, $t\sigma$ and $s\sigma$ must have the same R-minimal form. But then σ is an R-inconsistency of $s = u$ as well. It remains to prove that the inconsistency witness corresponding to σ and $s = u$ is less than the inconsistency witness corresponding to σ and $t = u$ wrt. $\succ_c$. This follows from the fact that $t\sigma \succ_R s\sigma$ by cases on whether $t \succ_R u$, $u \succ_R t$ or t and u incomparable. □

The natural question to ask at this point is whether the *Reduction* rule might be extended in order to regain the power of the *Simplification* rule, without losing the nice feature of lemma 12. What is clear is that we may not reduce equations in C by other equations in C without loosing inconsistencies. We can, however, obtain a similar result to the above by allowing to reduce by the inductive lemmas from L. This requires that instead of basing on the notion of R-inconsistencies we should build upon a notion of $\mathcal{I}(R)$-inconsistencies, where a substitution σ is a $\mathcal{I}(R)$-inconsistency for $t = u$ if $t\sigma \neq_{\mathcal{I}(R)} u\sigma$. This notion of inconsistencies is the dual of what Echahed calls IE-unifiers in [6], and allowing for reduction by lemmas in L also corresponds to *inductively normal narrowing* defined in the same paper.

Now we shall show that when generating $CP(R, \{t = u\})$, any inconsistency of the equation $t = u$ is maintained, partly as an inconsistency of a generated critical pair, and partly as the corresponding mgu. This is inspired by the way unifiers are preserved by all possible singleton narrowing steps from any given equation in theorem 7. In addition we clarify that the computation of critical pairs makes all inconsistencies closer to being detectable.

Definition 13. Let C be a set of pairs of equations and substitutions, and T be the set of all ground terms. Then C is a *conservative covering set for the equation* $t = u$ if for every undetectable inconsistency $\sigma \in \text{Inc}(t = u, R, T)$, there exists a pair $\langle t' = u', \gamma \rangle \in C$ and an R-inconsistency $\sigma' \in \text{Inc}(t' = u', R, T)$ such that $\gamma\sigma' \leftrightarrow^*_R \sigma$, and $c(t\sigma!_R \leftarrow^*_R t\sigma \leftrightarrow_{\{t=u\}} u\sigma \rightarrow^*_R u\sigma!_R) \succ_c c(t'\sigma'!_R \leftarrow^*_R t'\sigma' \leftrightarrow_{\{t'=u'\}} u'\sigma' \rightarrow^*_R u'\sigma'!_R)$. A *ground* conservative covering set is defined by replacing the above references to Inc by references to GInc.

Let $CCP(R, C)$ be $CP(R, C)$ such that to each critical pair is assigned its corresponding most general unifier.

Lemma 14. *Consider any equation $t = u$. $CCP(R, \{t = u\})$ is a conservative covering set for $t = u$.*

Proof. Let σ be an undetectable inconsistency in $\text{Inc}(t = u, R, T)$, and let us from σ create ρ such that for all variables x we have $x\rho = x\sigma!_R$. Obviously $\rho \leftrightarrow^*_R \sigma$, and ρ is also an undetectable R-inconsistency for $t = u$ (otherwise σ would have been detectable as well). We must show that there exists a pair $\langle e, \gamma \rangle \in CCP(R, \{t = u\})$ and an R-inconsistency σ' for e such that $\gamma\sigma' = \rho$. In addition we must show that the inconsistency witness of σ' is less than that of σ wrt. $\succ_c$.

Since ρ is not detectable, we may without loss of generality assume that there exists an inconsistency witness $t\rho!_R \leftarrow^*_R t\rho[i/r\kappa] \leftarrow_{\{l \rightarrow r\}} t\rho \leftrightarrow_{\{t=u\}} u\rho \rightarrow^*_R u\rho!_R$ for some substitution κ and $l \rightarrow r \in R$. Since $x\rho$ is irreducible for every variable x, there must be a critical overlap between the equational and the rewrite redex in the subproof $t\rho[i/r\kappa] \leftarrow_{\{l \rightarrow r\}} t\rho \leftrightarrow_{\{t=u\}} u\rho$. Therefore $\langle t[i/r]\gamma = u\gamma, \gamma \rangle \in CCP(R, \{t = u\})$, where γ is the mgu. of $t[i]$ and l. It is well known that the equation formed by the critical pair in such cases is able to replace the former two proof steps by one, thus we must have $t\rho[i/r\kappa] \leftrightarrow_{\{t[i/r]\gamma = u\gamma\}} u\rho$. There must therefore exist a substitution σ' such that both $t[i/r]\gamma\sigma' = t\rho[i/r\kappa]$ and $u\gamma\sigma' = u\rho$.

By a trivial argument considering mapping of irrelevant variables we get that $\gamma\sigma' = \rho$ from $u\gamma\sigma' = u\rho$. Since $t\rho[i/r\kappa]$ and $u\rho$ have distinct R-normal forms, σ' must also be an R-inconsistency for the equation $t[i/r]\gamma = u\gamma$.

It remains to show that the witness $t\rho!_R \leftarrow^*_R t\rho[i/r\kappa] \leftarrow_{\{l \rightarrow r\}} t\rho \leftrightarrow_{\{t=u\}} u\rho \rightarrow^*_R u\rho!_R$ is greater than $t\rho!_R \leftarrow^*_R t\rho[i/r\kappa] \leftrightarrow_{\{t[i/r]\gamma = u\gamma\}} u\rho \rightarrow^*_R u\rho!_R$ with respect to $\succ_c$. This follows from the fact that $t\rho$ rewrites by R to $t\rho[i/r\kappa]$ in the first of the two witnesses. □

By the above, and the fact that $\succ_c$ is well founded, it is obvious that fair proof by consistency with *Reduction* instead of *Simplification* will enumerate a complete set of inconsistencies for the input equations. This by for each emerging provably inconsistent equation considering the concatenation of the sequence of mgu.'s leading to it. From theorem 8 and the form of the *Reduction* rule it is also clear that fair $\vdash_{reduc}$ derivations may be used to simulate a fair execution of all possible narrowing derivations with interleaving simplification, and thus enumerate a complete set of R-unifiers for any equation. The following lemma will together with lemma 5 guarantee that the set of generated unifiers and inconsistencies are identical:

Lemma 15. *Let R be a convergent set of rewrite rules and t, u be two terms. Let EQ, T and F be a binary function and two constants respectively that are not referred to in neither R, t nor u. Furthermore let $R' = R \cup \{\text{EQ}(x,x) \to \text{T}\}$. Consider any $\vdash_{reduc}$ derivation wrt. R' starting from the situation $C = \{\text{EQ}(t,u) = \text{F}\}$. No mgu. computed in this derivation will contain references to any of the functions EQ, T and F.*

Proof. An mgu. violating this lemma could only appear if one of the two terms unified contained either EQ, T or F at a position not at the top. From the form of R' we know that this term would have to come from C. There is no such term in C initially, and by studying R' it is easy to see that this fact is maintained as an invariant through every possible $\vdash_{reduc}$ step. □

6 Relating Results

A lot of research on narrowing and proof by consistency has focused on reducing the search space. The main concept used for reducing the search space in proof by consistency is *complete positions* [9]. A position in a term t is said to be complete with respect to a set R of rewrite rules if $t[i]$ is not a variable, and $t[i]\sigma$ is reducible at the top position for every ground substitution σ which is such that $x\sigma$ is irreducible for every variable x. The concept of complete positions may be extended in the obvious way to complete sets of positions with respect to a subset of R [16]. It is clear that a term has a complete set of positions with respect to a subset of R if and only if it is ground reducible wrt. the same subset.

At this point we narrow our focus to the preservation of *ground inconsistencies*, and as for the modeling of narrowing, to generate *ground* complete sets of unifiers. The following is a generalization of a result stated in [1], the generalization being the preservance of *all* ground inconsistencies.

Lemma 16. *Let $t = u$ be an equation such that $t \succ_R u$, and such that P is a complete set of positions for t wrt. $R_C \subseteq R$. Furthermore let $D \subseteq CCP(R,C)$ be the set of critical pairs and mgu.'s emerging from positioning left hand sides from R_C on a position in P. Then D is a ground conservative covering set for $t = u$.*

Proof. Let σ be an arbitrary substitution in $\text{GInc}(t = u, R, T)$ where T is the set of all ground terms. Let us from σ create ρ such that $x\rho = x\sigma!_R$ for all variables x.

In the same way as in the proof of lemma 14 we may assume that there is an inconsistency witness on the form $t\rho!_R \leftarrow^*_R t\rho[i/r\kappa] \leftarrow_{\{l \to r\}} t\rho \leftrightarrow_{\{t=u\}} u\rho \to^*_R u\rho!_R$ for some κ. Since $t\rho$ is a ground term and P is a complete set of positions with respect to $R_C \subseteq R$, we may assume that the rule $l \to r \in R_C$ and that $i \in P$. The rest follows from the same lines of reasoning as the proof of lemma 14. □

By virtue of lemma 12 and lemma 16 it is now easy to see that a ground complete set of inconsistencies will eventually be found by a fair $\vdash_{reduc}$ derivation that generates critical pairs with complete sets of positions only. If we now move focus to how this result may be used in narrowing we get that narrowing restricted to ground complete sets of positions is ground complete due to lemma 5, lemma 15 and theorem 8:

Theorem 17. *Let R be a convergent set of rewrite rules, let EQ and T be function symbols not referred to in R and let R' be $R \cup \{\text{EQ}(x,x) \to \text{T}\}$. Let S be a set of substitutions such that $\gamma \in S$ iff there exists a narrowing derivation*

$$\text{EQ}(t, u) \leadsto_{[i_1, r_1, \sigma_1]} \text{EQ}(t_1, u_1) \ldots \leadsto_{[i_n, r_n, \sigma_n]} \text{T}$$

where $r_1 \ldots r_n \subseteq R'$, and $\sigma_1 \sigma_2 \ldots \sigma_n = \gamma$, and for every j such that $1 \leq j \leq n$, we have $i_j \in P_j$ and $r_j \in R_{Cj}$ where P_j is a fixed complete set of positions for $\text{EQ}(t_{j-1}, u_{j-1})$ wrt. $R_{Cj} \subseteq R'$ whenever it exists, and $P_j = \bar{P}(\text{EQ}(t_{j-1}, u_{j-1}))$ and $R_{Cj} = R'$ otherwise. Then S is a ground complete set of R-unifiers of the equation $t = u$.

Sketch of proof: Consider a fair $\vdash_{reduc}$ derivation starting from $L, \{\text{EQ}(t, u) = \text{T}\}$. By the above this derivation will generate a ground complete set of inconsistencies for $\text{EQ}(t, u) = \text{T}$ by only considering critical pairs from complete positions. From lemma 5 and lemma 15 we get that this ground complete set of inconsistencies is also a ground complete set of unifiers of the terms t and u, and from theorem 8 we also get that this derivation corresponds to a set of narrowing derivations restricted to complete sets of positions. □

Now let us take a look at existing results from narrowing. In [8] Fribourg introduced the *innermost narrowing strategy*, and proved that under certain conditions this strategy is ground complete. In [9] he himself showed that what he called innermost terms are actually ground reducible at the top position. Every position containing innermost terms is therefore complete wrt. R, so completeness of innermost narrowing is contained in theorem 17.

Echahed [5] proposed a syntactical restriction on the specification based on a notion of *sub-unifiability*. Based on his restriction he was able to prove any position selection strategy ground complete. The proof for this basically consists of showing that every narrowable position is complete wrt. the set of rewrite rules, thus this result is also contained in theorem 17. Echahed himself extended his result in [6] where he proved ground completeness of a position selection strategy choosing a single complete position wrt. R whenever such a position exists. Theorem 17 is an extension of this result in that it allows a *set* of positions, possibly divided between the two terms we are trying to unify, with respect to a subset of R.

In order to be able to benefit from results on narrowing in proof by consistency, we must first show that all derivations in proof by consistency may be simulated by a set of narrowing derivations. The following theorem is the dual of theorem 8.

Theorem 18. *Let R be a convergent set of rewrite rules, and let EQ be a binary function symbol not referred to in R. Let $t = u$ be an equation. For every $\vdash_{basic}$ derivation $L, \{t = u\} \vdash_{basic} \ldots \vdash_{basic} L, \{t_1 = u_1, \ldots, t_v = u_v\}$ wrt. R there exist narrowing derivations wrt. R on the form*

$$\mathrm{EQ}(t,u) \leadsto_{[p_1,r_1,\sigma_1]} \cdots \leadsto_{[p_n,r_n,\sigma_n]} \mathrm{EQ}(t_1,u_1)$$

$$\vdots$$

$$\mathrm{EQ}(t,u) \leadsto_{[q_1,s_1,\gamma_1]} \cdots \leadsto_{[q_m,s_m,\gamma_m]} \mathrm{EQ}(t_v,u_v)$$

and vice versa.

Proof. We only consider the equation $t_i = u_i$ for some i, $1 \leq i \leq v$. In the derivation

$$L, t = u \vdash_{basic} \cdots \vdash_{basic} L, \{t_1 = u_1, \ldots, t_v = u_v\}$$

we first remove all steps that do not contribute to the generation of the equation $t_i = u_i$, and then rename the terms $t_1 \ldots t_v, u_1 \ldots u_v$ in order to get continuous indices. We now have a derivation

$$L, t = u \vdash_{basic} \cdots \vdash_{basic} L, \{t_1 = u_1, \ldots, t_i = u_i\}$$

Since $\vdash_{basic}$ is purely additive we may assume that t equals t_1, u equals u_1 and $t_{j+1} = u_{j+i} \in CP(R, \{t_j = u_j\})$ for $1 \leq j < i$. We shall show for all j that $\mathrm{EQ}(t_j, u_j) \leadsto_{[k,l \to r,\sigma]} \mathrm{EQ}(t_{j+1}, u_{j+i})$ for some position k, rewrite rule $l \to r \in R$ and substitution σ.

We assume without loss of generality that for every j the subterm $t_j[h]$ is unifiable with l with mgu. σ, such that $t_{j+1} = t_j[h/r]\sigma$ and $u_{j+1} = u_j\sigma$. But this means that $\mathrm{EQ}(t_j, u_j) \leadsto_{[1h,l \to r,\sigma]} \mathrm{EQ}(t_j[h/r], u_j)\sigma$ and obviously $\mathrm{EQ}(t_j[h/r], u_j)\sigma = \mathrm{EQ}(t_j[h/r]\sigma, u_j\sigma) = \mathrm{EQ}(t_{j+1}, u_{j+1})$.

The reverse implication is proved by reversing the above argument in order to get a set of $\vdash_{basic}$ derivations, one for each narrowing derivation. These $\vdash_{basic}$ derivations may be merged in a straightforward manner. □

Example 3. Let again R be the following
(1) $x + 0 \to x$
(2) $x + s(y) \to s(x + y)$
(3) $\mathrm{EQ}(x,x) \to \mathrm{T}$

Let a and b be variables, and consider the following $\vdash_{basic}$ derivation
$L, \{a + a = a\} \vdash_{basic}$
$L, \{a + a = a, s(s(b) + b) = s(b)\} \vdash_{basic}$
$L, \{a + a = a, s(s(b) + b) = s(b), s(s(0)) = s(0)\}$

It is simulated by the following narrowing derivations and its two proper prefixes:
$\mathrm{EQ}(a+a, a) \leadsto_{[1,(2),\{a \mapsto s(b), x \mapsto s(b), y \mapsto b\}]} \mathrm{EQ}(s(s(b)+b), s(b))$
$\leadsto_{[11,(1),\{x \mapsto s(0), b \mapsto 0\}]} \mathrm{EQ}(s(s(0)), s(0))$

As for theorem 8, theorem 18 trivially extends to $\vdash_{reduc}$ derivations and narrowing with interleaving simplification steps.

Let us consider *basic* narrowing which was introduced by Hullot [12]. Informally, a narrowing derivation is basic if no narrowing position in the derivation is introduced by the narrowing substitution in a previous narrowing step. One of the most important results from Hullot's paper is that only basic narrowing derivations are needed in order to generate a complete set of unifiers. This is proved by means of the following theorem from Hullot. By $t \to_{[j,r]} u$ we denote that the term t rewrites to u in position j by the rule r:

Theorem 19 (Hullot). Let γ be a normalized substitution, and t be any term. For every rewrite sequence $t\gamma = t_1 \rightarrow_{[j_1,r_1]} \cdots \rightarrow_{[j_{n-1},r_{n-1}]} t_n$ there exists a basic narrowing derivation $t = u_1 \leadsto_{[j_1,r_1,\sigma_1]} \cdots \leadsto_{[j_{n-1},r_{n-1},\sigma_{n-1}]} u_n$ such that for each i, $1 \le i < n$ there exists a normalized substitution ρ_i such that $u_i \rho_i = t_i$ and $\sigma_1 \ldots \sigma_{i-1}\rho = \gamma$. And conversely for every narrowing derivation we can associate a rewrite derivation correspondingly.

The above theorem implies that for every term t and normalized substitution γ, there exists a basic narrowing derivation starting from t that will derive a term u which is a generalization of the minimal form of $t\gamma$. If we let t be $\text{EQ}(t', u')$ and γ be a normalized unifier of t' and u' get that the minimal form of $\text{EQ}(t', u')\gamma$ is T. Since the only generalization of T is T itself, the theorem implies completeness of narrowing. Let us now move our focus to proof by consistency again, and explain the fact that $\vdash_{basic}$ is refutationally complete in terms of theorem 18 and theorem 19.

If we want to disprove the equation $t = u$ in $\mathcal{I}(R)$, we will be looking for a normalized ground substitution γ such that $t\gamma!_R \ne u\gamma!_R$. If we are able to derive an equation $t' = u'$ which is a generalization of $t\gamma!_R = u\gamma!_R$, we will have an inconsistency witness by the equation $t' = u'$ with no rewrite steps, thus the equation $t' = u'$ must be provably inconsistent and γ must be detectable. Theorem 19 tells us that there exists a narrowing derivation issuing from $\text{EQ}(t,u)$ finding the term $\text{EQ}(t', u')$, thus theorem 18 gives that there is a corresponding $\vdash_{basic}$ derivation issuing from $L, t = u$ generating the provably inconsistent equation $t' = u'$.

We may now extend the notion of basic narrowing position to a notion of basic positions in $\vdash_{basic}$ derivations in the obvious way according to the correspondence in theorem 18. From the above, theorem 18 and theorem 19 we now immediately get

Proposition 20. *Refutational completeness is preserved in a $\vdash_{basic}$ derivation when only critical pairs from basic positions are computed.*

The direct significance of the above proposition is not so big, since Fribourg's notion of complete positions gives a more restrictive procedure than one relying on basic positions in general. What is perhaps more interesting is the generic nature of the justification of the proposition. It means that every narrowing strategy having the property that it will find a generalization of $t\gamma!_R$ for every term t and minimal substitution γ may be transformed into a strategy for proof by consistency giving a refutationally complete process. We may therefore e.g. conjecture that $\vdash_{basic}$ restricted to computing critical pairs on outermost positions is refutationally complete for constructor based systems. The corresponding proof for narrowing may be found in [23].

Hullot also proved that when all basic narrowing derivations starting from right hand sides in R terminate, then all basic narrowing derivations wrt. that particular R terminate. In the setting of proof by consistency this amounts to the following.

Proposition 21. *Let R be a finite convergent set of rewrite rules such that any basic narrowing derivation issuing from right hand sides in R terminate. Then $\vdash_{basic}$ restricteds to basic positions yields a decision procedure for validity in $\mathcal{I}(R)$.*

Proof. We know that the inference system $\vdash_{basic}$ is refutationally complete. Since there are only finitely many narrowing steps possible issuing from one and the same term wrt. a finite R, and all narrowing sequences terminate, König's lemma and lemma 18 gives termination of $\vdash_{basic}$. Termination without the encountering of inconsistencies obviously yields a proof. □

Again the most interesting part of this is the generic nature of the proof, thus every terminating narrowing strategy whose completeness is based on variants of theorem 19 yields a decidability result for the initial algebra, see e.g. [4]. There are however termination results for strategies whose completeness is not based upon the result of Hullot, e.g. [3], and the extraction of decisionality results from these will need some adaptation of the proofs.

7 Conclusion and Further Work

We have described how narrowing and Bachmair's method for proof by consistency may be simulated by each other, and thereby been able to let results migrate from one area to another. This has till now resulted in an extension of existing results for narrowing, new results on reducing search spaces for proof by consistency and also sufficient criteria on specifications for proof by consistency to be a decision algorithm for the corresponding initial algebra. We have also established a close connection between the existence of terminating complete narrowing strategies and decidability of equational validity in the underlying initial algebra.

What we consider to be the most important in this work is the possibility to let results migrate from one field to another more than it is the results we have chosen as examples of such migration. Further migration of results and evaluation of the new results obtained by migration is therefore an interesting path for further research.

Our focus has been on narrowing and proof by consistency in unconditional theories, but it seems reasonable to assume that there is a similar connection between narrowing in conditional theories and the corresponding extension of Bachmair's method [2]. Such a relation might open for interesting migration of results as well.

Acknowledgements: I would like to thank the anonymous referee who pointed out an error in the original version of lemma 5, and Henrik Linnestad for comments on a previous version of this paper.

References

1. L. Bachmair. Proof by consistency in equational theories. In *Proceedings 3rd IEEE Symposium on Logic in Computer Science, Edinburgh (UK)*, pages 228–233, 1988.
2. E. Bevers and J. Lewi. Proof by consistency in conditional equational theories. In *Proceedings Second International Workshop on Conditional and Typed Rewriting Systems*, volume 516 of *Lecture Notes in Computer Science*, pages 194–205. Springer-Verlag, 1990.
3. J. Chabin and P. Réty. Narrowing directed by a graph of terms. In *Proceedings 4th Conference on Rewriting Techniques and Applications, Como (Italy)*, volume 488 of *Lecture Notes in Computer Science*, pages 112–123, 1991.
4. J. Christian. Some termination criteria for narrowing and E-narrowing. In *Proceedings 11th International Conference on Automated Deduction, Saratoga Springs (NY, USA)*, volume 607 of *Lecture Notes in Artificial Intelligence*. Springer-Verlag, 1992.

5. R. Echahed. On completeness of narrowing strategies. *Theoretical Computer Science*, 72:133–146, 1990.
6. R. Echahed. Uniform narrowing strategies. In *Proceedings 3rd International Conference on Algebraic and Logic Programming, Pisa (Italy)*, volume 632 of *Lecture Notes in Computer Science*, pages 259–275. Springer-Verlag, 1992.
7. M. Fay. First-order unification in an equational theory. In *Proceedings of the 4th Workshop on Automated Deduction*, pages 161–167, 1979.
8. L. Fribourg. SLOG: A logic programming interpreter based on clausal superposition and rewriting. In *Proceedings of the 1985 Symposium on Logic Programming, Boston*, pages 172–184, 1985.
9. L. Fribourg. A strong restriction on the inductive completion procedure. In *Proceedings 13th International Colloquium on Automata, Languages and Programming*, volume 226 of *Lecture Notes in Computer Science*, pages 105–115. Springer-Verlag, 1986.
10. J. A. Goguen. How to prove inductive hypotheses without induction. In W. Bibel and R. Kowalski, editors, *Proceedings of the 5th Conference on Automated Deduction*, volume 87 of *Lecture Notes in Computer Science*, pages 356–373. Springer-Verlag, 1980.
11. G. Huet and J.-M. Hullot. Proofs by induction in equational theories with constructors. *Journal of Computer and System Sciences*, pages 239–266, 1982.
12. J.-M. Hullot. Canonical forms and unification. In *Proceedings 5th International Conference on Automated Deduction*, volume 87, pages 318–334. Springer-Verlag, 1980.
13. J.-P. Jouannaud and E. Kounalis. Automatic proofs by induction in theories without constructors. *Information and Computation*, 82(1):1–33, July 1989.
14. D. Kapur, P. Narendran, and H. Zhang. On sufficient-completeness and related properties of term rewriting systems. *Acta Informatica*, 24(4):395–415, 1987.
15. D. E. Knuth and P. B. Bendix. Simple word problems in universal algebras. In J. Leech, editor, *Computational Problems in Abstract Algebra*, pages 263–297. Pergamon Press, Oxford, 1970.
16. W. Küchlin. Inductive completion by ground proof transformation. In H. Aït-Kaci and M. Nivat, editors, *Resolution of Equations in Algebraic Structures*, volume 2 of *Rewriting Techniques*, chapter 7. Academic Press, 1989.
17. O. Lysne. Proof by consistency in constructive systems with final algebra semantics. In *Proceedings 3rd International Conference on Algebraic and Logic Programming, Pisa (Italy)*, volume 632 of *Lecture Notes in Computer Science*, pages 276–290. Springer-Verlag, 1992.
18. J. Meseguer and J. A. Goguen. Initiality, induction and computability. In M. Nivat and J. C. Reynolds, editors, *Algebraic Methods in Semantics*, chapter 14. Cambridge University Press, 1985.
19. D. L. Musser. On proving inductive properties in abstract data types. In *Proceedings of the 7th Annual ACM Symposium on Principles of Programming Languages*, pages 154–162, January 1980.
20. W. Nutt, P. Réty, and G. Smolka. Basic narrowing revisited. *Journal of Symbolic Computation*, 7:295–317, 1989.
21. D. Plaisted. Semantic confluence tests and completion methods. *Information and Control*, 65:182–215, 1985.
22. P. Réty. Improving basic narrowing techniques. In *Proceedings 2nd Conference on Rewriting Techniques and Applications, Bordeaux (France)*, volume 256 of *Lecture Notes in Computer Science*, pages 228–241. Springer-Verlag, 1987.
23. Jia-Huai You. Outer narrowing for equational theories based on constructors. In *Proceedings 15th International Colloquium on Automata, Languages and Programming, Tampere (Finland)*, volume 317 of *Lecture Notes in Computer Science*, pages 727–741. Springer-Verlag, 1988.

A Fixedpoint Approach to Implementing (Co)Inductive Definitions*

Lawrence C. Paulson
lcp@cl.cam.ac.uk

Computer Laboratory, University of Cambridge, England

Abstract. This paper presents a fixedpoint approach to inductive definitions. Instead of using a syntactic test such as 'strictly positive,' the approach lets definitions involve any operators that have been proved monotone. It is conceptually simple, which has allowed the easy implementation of mutual recursion and other conveniences. It also handles coinductive definitions: simply replace the least fixedpoint by a greatest fixedpoint. This represents the first automated support for coinductive definitions.

The method has been implemented in Isabelle's formalization of ZF set theory. It should be applicable to any logic in which the Knaster-Tarski Theorem can be proved. Examples include lists of n elements, the accessible part of a relation and the set of primitive recursive functions. One example of a coinductive definition is bisimulations for lazy lists.

1 Introduction

Several theorem provers provide commands for formalizing recursive data structures, like lists and trees. Examples include Boyer and Moore's shell principle [4] and Melham's recursive type package for the HOL system [11]. Such data structures are called **datatypes** below, by analogy with `datatype` definitions in Standard ML.

A datatype is but one example of an **inductive definition**. This specifies the least set closed under given rules [2]. The collection of theorems in a logic is inductively defined. A structural operational semantics [9] is an inductive definition of a reduction or evaluation relation on programs. A few theorem provers provide commands for formalizing inductive definitions; these include Coq [15] and again the HOL system [5].

The dual notion is that of a **coinductive definition**. This specifies the greatest set closed under given rules. Important examples include using bisimulation relations to formalize equivalence of processes [13] or lazy functional programs [1]. Other examples include lazy lists and other infinite data structures; these are called **codatatypes** below.

Not all inductive definitions are meaningful. **Monotone** inductive definitions are a large, well-behaved class. Monotonicity can be enforced by syntactic conditions such as 'strictly positive,' but this could lead to monotone definitions being rejected on the grounds of their syntactic form. More flexible is to formalize monotonicity within the logic and allow users to prove it.

This paper describes a package based on a fixedpoint approach. Least fixedpoints yield inductive definitions; greatest fixedpoints yield coinductive definitions. The package has several advantages:

- It allows reference to any operators that have been proved monotone. Thus it accepts all provably monotone inductive definitions, including iterated definitions.
- It accepts a wide class of datatype definitions, though at present restricted to finite branching.
- It handles coinductive and codatatype definitions. Most of the discussion below applies equally to inductive and coinductive definitions, and most of the code is shared. To my knowledge, this is the only package supporting coinductive definitions.
- Definitions may be mutually recursive.

The package is implemented in Isabelle [19], using ZF set theory [20, 21]. However, the fixedpoint approach is independent of Isabelle. The recursion equations are specified as introduction rules for the mutually recursive sets. The package transforms these rules into a mapping over sets, and attempts to prove that the

---

* J. Grundy and S. Thompson made detailed comments; the referees were also helpful. Research funded by SERC grants GR/G53279, GR/H40570 and by the ESPRIT Project 6453 'Types'.

mapping is monotonic and well-typed. If successful, the package makes fixedpoint definitions and proves the introduction, elimination and (co)induction rules. The package consists of several Standard ML functors [17]; it accepts its argument and returns its result as ML structures.[2]

Most datatype packages equip the new datatype with some means of expressing recursive functions. This is the main omission from my package. Its fixedpoint operators define only recursive sets. To define recursive functions, the Isabelle/ZF theory provides well-founded recursion and other logical tools [21].

Outline. Section 2 introduces the least and greatest fixedpoint operators. Section 3 discusses the form of introduction rules, mutual recursion and other points common to inductive and coinductive definitions. Section 4 discusses induction and coinduction rules separately. Section 5 presents several examples, including a coinductive definition. Section 6 describes datatype definitions. Section 7 presents related work. Section 8 draws brief conclusions.

Most of the definitions and theorems shown below have been generated by the package. I have renamed some variables to improve readability.

2 Fixedpoint operators

In set theory, the least and greatest fixedpoint operators are defined as follows:

$$\text{lfp}(D, h) \equiv \bigcap \{X \subseteq D \,.\, h(X) \subseteq X\}$$
$$\text{gfp}(D, h) \equiv \bigcup \{X \subseteq D \,.\, X \subseteq h(X)\}$$

Let D be a set. Say that h is **bounded by** D if $h(D) \subseteq D$, and **monotone below** D if $h(A) \subseteq h(B)$ for all A and B such that $A \subseteq B \subseteq D$. If h is bounded by D and monotone then both operators yield fixedpoints:

$$\text{lfp}(D, h) = h(\text{lfp}(D, h))$$
$$\text{gfp}(D, h) = h(\text{gfp}(D, h))$$

These equations are instances of the Knaster-Tarski Theorem, which states that every monotonic function over a complete lattice has a fixedpoint [6]. It is obvious from their definitions that lfp must be the least fixedpoint, and gfp the greatest.

This fixedpoint theory is simple. The Knaster-Tarski Theorem is easy to prove. Showing monotonicity of h is trivial, in typical cases. We must also exhibit a bounding set D for h. Frequently this is trivial, as when a set of 'theorems' is (co)inductively defined over some previously existing set of 'formulae.' Isabelle/ZF provides a suitable bounding set for finitely branching (co)datatype definitions; see §6.1 below. Bounding sets are also called **domains**.

The powerset operator is monotone, but by Cantor's Theorem there is no set A such that $A = \mathcal{P}(A)$. We cannot put $A = \text{lfp}(D, \mathcal{P})$ because there is no suitable domain D. But §5.5 demonstrates that $\mathcal{P}$ is still useful in inductive definitions.

3 Elements of an inductive or coinductive definition

Consider a (co)inductive definition of the sets $R_1, \ldots, R_n$, in mutual recursion. They will be constructed from domains $D_1, \ldots, D_n$, respectively. The construction yields not $R_i \subseteq D_i$ but $R_i \subseteq D_1 + \cdots + D_n$, where R_i is contained in the image of D_i under an injection. Reasons for this are discussed elsewhere [21, §4.5].

The definition may involve arbitrary parameters $\mathbf{p} = p_1, \ldots, p_k$. Each recursive set then has the form $R_i(\mathbf{p})$. The parameters must be identical every time they occur within a definition. This would appear to be a serious restriction compared with other systems such as Coq [15]. For instance, we cannot define the lists of n elements as the set listn(A, n) using rules where the parameter n varies. Section 5.2 describes how to express this set using the inductive definition package.

To avoid clutter below, the recursive sets are shown as simply R_i instead of $R_i(\mathbf{p})$.

---

[2] This use of ML modules is not essential; the package could also be implemented as a function on records.

3.1 The form of the introduction rules

The body of the definition consists of the desired introduction rules, specified as strings. The conclusion of each rule must have the form $t \in R_i$, where t is any term. Premises typically have the same form, but they can have the more general form $t \in M(R_i)$ or express arbitrary side-conditions.

The premise $t \in M(R_i)$ is permitted if M is a monotonic operator on sets, satisfying the rule

$$\frac{A \subseteq B}{M(A) \subseteq M(B)}$$

The user must supply the package with monotonicity rules for all such premises.

The ability to introduce new monotone operators makes the approach flexible. A suitable choice of M and t can express a lot. The powerset operator $\mathcal{P}$ is monotone, and the premise $t \in \mathcal{P}(R)$ expresses $t \subseteq R$; see §5.5 for an example. The 'list of' operator is monotone, as is easily proved by induction. The premise $t \in \text{list}(R)$ avoids having to encode the effect of $\text{list}(R)$ using mutual recursion; see §5.6 and also my earlier paper [21, §4.4].

Introduction rules may also contain **side-conditions**. These are premises consisting of arbitrary formulae not mentioning the recursive sets. Side-conditions typically involve type-checking. One example is the premise $a \in A$ in the following rule from the definition of lists:

$$\frac{a \in A \quad l \in \text{list}(A)}{\text{Cons}(a, l) \in \text{list}(A)}$$

3.2 The fixedpoint definitions

The package translates the list of desired introduction rules into a fixedpoint definition. Consider, as a running example, the finite set operator $\text{Fin}(A)$: the set of all finite subsets of A. It can be defined as the least set closed under the rules

$$\emptyset \in \text{Fin}(A) \qquad \frac{a \in A \quad b \in \text{Fin}(A)}{\{a\} \cup b \in \text{Fin}(A)}$$

The domain in a (co)inductive definition must be some existing set closed under the rules. A suitable domain for $\text{Fin}(A)$ is $\mathcal{P}(A)$, the set of all subsets of A. The package generates the definition

$$\text{Fin}(A) \equiv \text{lfp}(\mathcal{P}(A),\ \lambda X\,.\,\{z \in \mathcal{P}(A)\,.\,z = \emptyset \vee \\ (\exists a\,b\,.\,z = \{a\} \cup b \wedge a \in A \wedge b \in X)\})$$

The contribution of each rule to the definition of $\text{Fin}(A)$ should be obvious. A coinductive definition is similar but uses gfp instead of lfp.

The package must prove that the fixedpoint operator is applied to a monotonic function. If the introduction rules have the form described above, and if the package is supplied a monotonicity theorem for every $t \in M(R_i)$ premise, then this proof is trivial.[3]

The package returns its result as an ML structure, which consists of named components; we may regard it as a record. The result structure contains the definitions of the recursive sets as a theorem list called defs. It also contains, as the theorem unfold, a fixedpoint equation such as

$$\text{Fin}(A) = \{z \in \mathcal{P}(A)\,.\,z = \emptyset \vee \\ (\exists a\,b\,.\,z = \{a\} \cup b \wedge a \in A \wedge b \in \text{Fin}(A))\}$$

It also contains, as the theorem dom_subset, an inclusion such as $\text{Fin}(A) \subseteq \mathcal{P}(A)$.

---

[3] Due to the presence of logical connectives in the fixedpoint's body, the monotonicity proof requires some unusual rules. These state that the connectives $\wedge$, $\vee$ and $\exists$ preserve monotonicity with respect to the partial ordering on unary predicates given by $P \sqsubseteq Q$ if and only if $\forall x\,.\,P(x) \rightarrow Q(x)$.

3.3 Mutual recursion

In a mutually recursive definition, the domain of the fixedpoint construction is the disjoint sum of the domain D_i of each R_i, for $i = 1, \ldots, n$. The package uses the injections of the binary disjoint sum, typically Inl and Inr, to express injections $h_{1n}, \ldots, h_{nn}$ for the n-ary disjoint sum $D_1 + \cdots + D_n$.

As discussed elsewhere [21, §4.5], Isabelle/ZF defines the operator Part to support mutual recursion. The set $\text{Part}(A, h)$ contains those elements of A having the form $h(z)$:

$$\text{Part}(A, h) \equiv \{x \in A \,.\, \exists z \,.\, x = h(z)\}.$$

For mutually recursive sets $R_1, \ldots, R_n$ with $n > 1$, the package makes $n + 1$ definitions. The first defines a set R using a fixedpoint operator. The remaining n definitions have the form

$$R_i \equiv \text{Part}(R, h_{in}), \qquad i = 1, \ldots, n.$$

It follows that $R = R_1 \cup \cdots \cup R_n$, where the R_i are pairwise disjoint.

3.4 Proving the introduction rules

The user supplies the package with the desired form of the introduction rules. Once it has derived the theorem unfold, it attempts to prove those rules. From the user's point of view, this is the trickiest stage; the proofs often fail. The task is to show that the domain $D_1 + \cdots + D_n$ of the combined set $R_1 \cup \cdots \cup R_n$ is closed under all the introduction rules. This essentially involves replacing each R_i by $D_1 + \cdots + D_n$ in each of the introduction rules and attempting to prove the result.

Consider the Fin(A) example. After substituting $\mathcal{P}(A)$ for Fin(A) in the rules, the package must prove

$$\emptyset \in \mathcal{P}(A) \qquad \frac{a \in A \quad b \in \mathcal{P}(A)}{\{a\} \cup b \in \mathcal{P}(A)}$$

Such proofs can be regarded as type-checking the definition. The user supplies the package with type-checking rules to apply. Usually these are general purpose rules from the ZF theory. They could however be rules specifically proved for a particular inductive definition; sometimes this is the easiest way to get the definition through!

The result structure contains the introduction rules as the theorem list intrs.

3.5 The case analysis rule

The elimination rule, called elim, performs case analysis. There is one case for each introduction rule. The elimination rule for Fin(A) is

$$\frac{x \in \text{Fin}(A) \qquad [x = \emptyset] \quad \begin{matrix} \vdots \\ Q \end{matrix} \qquad [x = \{a\} \cup b \quad a \in A \quad b \in \text{Fin}(A)]_{a,b} \\ \vdots \\ Q}{Q}$$

The subscripted variables a and b above the third premise are eigenvariables, subject to the usual 'not free in ...' proviso. The rule states that if $x \in \text{Fin}(A)$ then either $x = \emptyset$ or else $x = \{a\} \cup b$ for some $a \in A$ and $b \in \text{Fin}(A)$; it is a simple consequence of unfold.

The package also returns a function for generating simplified instances of the case analysis rule. It works for datatypes and for inductive definitions involving datatypes, such as an inductively defined relation between lists. It instantiates elim with a user-supplied term then simplifies the cases using freeness of the underlying datatype. The simplified rules perform 'rule inversion' on the inductive definition. Section §5.3 presents an example.

4 Induction and coinduction rules

Here we must consider inductive and coinductive definitions separately. For an inductive definition, the package returns an induction rule derived directly from the properties of least fixedpoints, as well as a modified rule for mutual recursion and inductively defined relations. For a coinductive definition, the package returns a basic coinduction rule.

4.1 The basic induction rule

The basic rule, called induct, is appropriate in most situations. For inductive definitions, it is strong rule induction [5]; for datatype definitions (see below), it is just structural induction.

The induction rule for an inductively defined set R has the following form. The major premise is $x \in R$. There is a minor premise for each introduction rule:

- If the introduction rule concludes $t \in R_i$, then the minor premise is $P(t)$.
- The minor premise's eigenvariables are precisely the introduction rule's free variables that are not parameters of R. For instance, the eigenvariables in the Fin(A) rule below are a and b, but not A.
- If the introduction rule has a premise $t \in R_i$, then the minor premise discharges the assumption $t \in R_i$ and the induction hypothesis $P(t)$. If the introduction rule has a premise $t \in M(R_i)$ then the minor premise discharges the single assumption

$$t \in M(\{z \in R_i . P(z)\}).$$

Because M is monotonic, this assumption implies $t \in M(R_i)$. The occurrence of P gives the effect of an induction hypothesis, which may be exploited by appealing to properties of M.

The induction rule for Fin(A) resembles the elimination rule shown above, but includes an induction hypothesis:

$$\frac{x \in \text{Fin}(A) \quad P(\emptyset) \quad \begin{array}{c}[a \in A \quad b \in \text{Fin}(A) \quad P(b)]_{a,b} \\ \vdots \\ P(\{a\} \cup b)\end{array}}{P(x)}$$

Stronger induction rules often suggest themselves. We can derive a rule for Fin(A) whose third premise discharges the extra assumption $a \notin b$. The Isabelle/ZF theory defines the **rank** of a set [21, §3.4], which supports well-founded induction and recursion over datatypes. The package proves a rule for mutual induction and inductive relations.

4.2 Mutual induction

The mutual induction rule is called mutual_induct. It differs from the basic rule in several respects:

- Instead of a single predicate P, it uses n predicates $P_1, \ldots, P_n$: one for each recursive set.
- There is no major premise such as $x \in R_i$. Instead, the conclusion refers to all the recursive sets:

$$(\forall z . z \in R_1 \to P_1(z)) \land \cdots \land (\forall z . z \in R_n \to P_n(z))$$

Proving the premises establishes $P_i(z)$ for $z \in R_i$ and $i = 1, \ldots, n$.
- If the domain of some R_i is the Cartesian product $A_1 \times \cdots \times A_m$, then the corresponding predicate P_i takes m arguments and the corresponding conjunct of the conclusion is

$$(\forall z_1 \ldots z_m . \langle z_1, \ldots, z_m \rangle \in R_i \to P_i(z_1, \ldots, z_m))$$

The last point above simplifies reasoning about inductively defined relations. It eliminates the need to express properties of $z_1, \ldots, z_m$ as properties of the tuple $\langle z_1, \ldots, z_m \rangle$.

4.3 Coinduction

A coinductive definition yields a primitive coinduction rule, with no refinements such as those for the induction rules. (Experience may suggest refinements later.) Consider the codatatype of lazy lists as an example. For suitable definitions of LNil and LCons, lazy lists may be defined as the greatest fixedpoint satisfying the rules

$$\frac{}{\text{LNil} \in \text{llist}(A)} \qquad \frac{a \in A \quad l \in \text{llist}(A)}{\text{LCons}(a, l) \in \text{llist}(A)} \ (-)$$

The $(-)$ tag stresses that this is a coinductive definition. A suitable domain for llist(A) is quniv(A), a set closed under variant forms of sum and product for representing infinite data structures (see §6.1). Coinductive definitions use these variant sums and products.

The package derives an unfold theorem similar to that for Fin(A). Then it proves the theorem co_induct, which expresses that llist(A) is the greatest solution to this equation contained in quniv(A):

$$\frac{x \in X \quad X \subseteq \text{quniv}(A) \quad \overset{[z \in X]_z}{\vdots} \quad z = \text{LNil} \lor \left(\exists a\, l.\ z = \text{LCons}(a,l) \land a \in A \land l \in X \cup \text{llist}(A)\right)}{x \in \text{llist}(A)}$$

This rule complements the introduction rules; it provides a means of showing $x \in$ llist(A) when x is infinite. For instance, if $x = \text{LCons}(0, x)$ then applying the rule with $X = \{x\}$ proves $x \in$ llist(nat). (Here nat is the set of natural numbers.)

Having $X \cup \text{llist}(A)$ instead of simply X in the third premise above represents a slight strengthening of the greatest fixedpoint property. I discuss several forms of coinduction rules elsewhere [18].

5 Examples of inductive and coinductive definitions

This section presents several examples: the finite set operator, lists of n elements, bisimulations on lazy lists, the well-founded part of a relation, and the primitive recursive functions.

5.1 The finite set operator

The definition of finite sets has been discussed extensively above. Here is the corresponding ML invocation (note that cons(a, b) abbreviates $\{a\} \cup b$ in Isabelle/ZF):

```
structure Fin = Inductive_Fun
    (val thy        = Arith.thy addconsts [(["Fin"],"i=>i")]
     val rec_doms   = [("Fin","Pow(A)")]
     val sintrs     = ["0 : Fin(A)",
                       "[| a: A;  b: Fin(A) |] ==> cons(a,b) : Fin(A)"]
     val monos      = []
     val con_defs   = []
     val type_intrs = [empty_subsetI, cons_subsetI, PowI]
     val type_elims = [make_elim PowD]);
```

We apply the functor Inductive_Fun to a structure describing the desired inductive definition. The parent theory thy is obtained from Arith.thy by adding the unary function symbol Fin. Its domain is specified as $\mathcal{P}(A)$, where A is the parameter appearing in the introduction rules. For type-checking, the structure supplies introduction rules:

$$\frac{}{\emptyset \subseteq A} \qquad \frac{a \in C \quad B \subseteq C}{\{a\} \cup B \subseteq C}$$

A further introduction rule and an elimination rule express the two directions of the equivalence $A \in \mathcal{P}(B) \leftrightarrow A \subseteq B$. Type-checking involves mostly introduction rules.

ML is Isabelle's top level, so such functor invocations can take place at any time. The result structure is declared with the name Fin; we can refer to the Fin(A) introduction rules as Fin.intrs, the induction rule as Fin.induct and so forth. There are plans to integrate the package better into Isabelle so that users can place inductive definitions in Isabelle theory files instead of applying functors.

5.2 Lists of n elements

This has become a standard example of an inductive definition. Following Paulin-Mohring [15], we could attempt to define a new datatype $\text{listn}(A, n)$, for lists of length n, as an n-indexed family of sets. But her introduction rules

$$\text{Niln} \in \text{listn}(A, 0) \qquad \frac{n \in \text{nat} \quad a \in A \quad l \in \text{listn}(A, n)}{\text{Consn}(n, a, l) \in \text{listn}(A, \text{succ}(n))}$$

are not acceptable to the inductive definition package: listn occurs with three different parameter lists in the definition.

```
structure ListN = Inductive_Fun
  (val thy       = ListFn.thy addconsts [(["listn"],"i=>i")]
   val rec_doms  = [("listn", "nat*list(A)")]
   val sintrs    =
       ["<0,Nil>: listn(A)",
        "[| a: A;  <n,l>: listn(A) |] ==> <succ(n), Cons(a,l)>: listn(A)"]
   val monos     = []
   val con_defs  = []
   val type_intrs = nat_typechecks @ List.intrs @ [SigmaI]
   val type_elims = [SigmaE2]);
```

Fig. 1. Defining lists of n elements

The Isabelle/ZF version of this example suggests a general treatment of varying parameters. Here, we use the existing datatype definition of $\text{list}(A)$, with constructors Nil and Cons. Then incorporate the parameter n into the inductive set itself, defining $\text{listn}(A)$ as a relation. It consists of pairs $\langle n, l \rangle$ such that $n \in \text{nat}$ and $l \in \text{list}(A)$ and l has length n. In fact, $\text{listn}(A)$ is the converse of the length function on $\text{list}(A)$. The Isabelle/ZF introduction rules are

$$\langle 0, \text{Nil} \rangle \in \text{listn}(A) \qquad \frac{a \in A \quad \langle n, l \rangle \in \text{listn}(A)}{\langle \text{succ}(n), \text{Cons}(a, l) \rangle \in \text{listn}(A)}$$

Figure 1 presents the ML invocation. A theory of lists, extended with a declaration of listn, is the parent theory. The domain is specified as $\text{nat} \times \text{list}(A)$. The type-checking rules include those for 0, succ, Nil and Cons. Because $\text{listn}(A)$ is a set of pairs, type-checking also requires introduction and elimination rules to express both directions of the equivalence $\langle a, b \rangle \in A \times B \leftrightarrow a \in A \wedge b \in B$.

The package returns introduction, elimination and induction rules for listn. The basic induction rule, ListN.induct, is

$$\frac{x \in \text{listn}(A) \quad P(\langle 0, \text{Nil} \rangle) \quad \begin{array}{c} [a \in A \quad \langle n, l \rangle \in \text{listn}(A) \quad P(\langle n, l \rangle)]_{a,l,n} \\ \vdots \\ P(\langle \text{succ}(n), \text{Cons}(a, l) \rangle) \end{array}}{P(x)}$$

This rule requires the induction formula to be a unary property of pairs, $P(\langle n, l \rangle)$. The alternative rule, ListN.mutual_induct, uses a binary property instead:

$$\frac{P(0, \text{Nil}) \quad \begin{array}{c} [a \in A \quad \langle n, l \rangle \in \text{listn}(A) \quad P(n, l)]_{a,l,n} \\ \vdots \\ P(\text{succ}(n), \text{Cons}(a, l)) \end{array}}{\forall n\, l\,.\, \langle n, l \rangle \in \text{listn}(A) \rightarrow P(n, l)}$$

It is now a simple matter to prove theorems about $\text{listn}(A)$, such as

$$\forall l \in \text{list}(A)\,.\, \langle \text{length}(l), l \rangle \in \text{listn}(A)$$

$$\text{listn}(A)``\{n\} = \{l \in \text{list}(A)\,.\, \text{length}(l) = n\}$$

This latter result — here $r``X$ denotes the image of X under r — asserts that the inductive definition agrees with the obvious notion of n-element list.

Unlike in Coq, the definition does not declare a new datatype. A 'list of n elements' really is a list and is subject to list operators such as append (concatenation). For example, a trivial induction on $\langle m, l \rangle \in \text{listn}(A)$ yields

$$\frac{\langle m, l \rangle \in \text{listn}(A) \quad \langle m', l' \rangle \in \text{listn}(A)}{\langle m + m, \ l@l' \rangle \in \text{listn}(A)}$$

where $+$ here denotes addition on the natural numbers and @ denotes append.

5.3 A demonstration of rule inversion

The elimination rule, ListN.elim, is cumbersome:

$$\frac{x \in \text{listn}(A) \quad [x = \langle 0, \text{Nil} \rangle] \quad \begin{bmatrix} x = \langle \text{succ}(n), \text{Cons}(a, l) \rangle \\ a \in A \\ \langle n, l \rangle \in \text{listn}(A) \\ \vdots \\ Q \end{bmatrix}_{a,l,n}}{Q}$$

The ML function ListN.mk_cases generates simplified instances of this rule. It works by freeness reasoning on the list constructors: $\text{Cons}(a, l)$ is injective in its two arguments and differs from Nil. If x is $\langle i, \text{Nil} \rangle$ or $\langle i, \text{Cons}(a, l) \rangle$ then ListN.mk_cases deduces the corresponding form of i; this is called rule inversion. For example,

```
ListN.mk_cases List.con_defs "<i,Cons(a,l)> : listn(A)"
```

yields a rule with only two premises:

$$\frac{\langle i, \text{Cons}(a, l) \rangle \in \text{listn}(A) \quad \begin{bmatrix} i = \text{succ}(n) \\ a \in A \\ \langle n, l \rangle \in \text{listn}(A) \\ \vdots \\ Q \end{bmatrix}_n}{Q}$$

The package also has built-in rules for freeness reasoning about 0 and succ. So if x is $\langle 0, l \rangle$ or $\langle \text{succ}(i), l \rangle$, then ListN.mk_cases can similarly deduce the corresponding form of l.

The function mk_cases is also useful with datatype definitions. The instance from the definition of lists, namely List.mk_cases, can prove the rule

$$\frac{\text{Cons}(a, l) \in \text{list}(A) \quad \begin{matrix} [a \in A \quad l \in \text{list}(A)] \\ \vdots \\ Q \end{matrix}}{Q}$$

A typical use of mk_cases concerns inductive definitions of evaluation relations. Then rule inversion yields case analysis on possible evaluations. For example, the Isabelle/ZF theory includes a short proof of the diamond property for parallel contraction on combinators.

5.4 A coinductive definition: bisimulations on lazy lists

This example anticipates the definition of the codatatype $\text{llist}(A)$, which consists of finite and infinite lists over A. Its constructors are LNil and LCons, satisfying the introduction rules shown in §4.3. Because $\text{llist}(A)$ is defined as a greatest fixedpoint and uses the variant pairing and injection operators, it contains non-well-founded elements such as solutions to $\text{LCons}(a, l) = l$.

The next step in the development of lazy lists is to define a coinduction principle for proving equalities. This is done by showing that the equality relation on lazy lists is the greatest fixedpoint of some monotonic operation. The usual approach [23] is to define some notion of bisimulation for lazy lists, define equivalence to be the greatest bisimulation, and finally to prove that two lazy lists are equivalent if and only if they are equal. The coinduction rule for equivalence then yields a coinduction principle for equalities.

A binary relation R on lazy lists is a **bisimulation** provided $R \subseteq R^+$, where R^+ is the relation

$$\{\langle \text{LNil}, \text{LNil}\rangle\} \cup \{\langle \text{LCons}(a, l), \text{LCons}(a, l')\rangle . a \in A \land \langle l, l'\rangle \in R\}.$$

A pair of lazy lists are **equivalent** if they belong to some bisimulation. Equivalence can be coinductively defined as the greatest fixedpoint for the introduction rules

$$\frac{}{\langle \text{LNil}, \text{LNil}\rangle \in \text{lleq}(A)} \qquad \frac{a \in A \quad \langle l, l'\rangle \in \text{lleq}(A)}{\langle \text{LCons}(a, l), \text{LCons}(a, l')\rangle \in \text{lleq}(A)} \; (-)$$

To make this coinductive definition, we invoke CoInductive_Fun:

```
structure LList_Eq = CoInductive_Fun
  (val thy = LList.thy addconsts [(["lleq"],"i=>i")]
   val rec_doms = [("lleq", "llist(A) * llist(A)")]
   val sintrs   =
      ["<LNil, LNil> : lleq(A)",
       "[| a:A;  <l,l'>: lleq(A) |] ==> <LCons(a,l), LCons(a,l')>: lleq(A)"]
   val monos       = []
   val con_defs    = []
   val type_intrs = LList.intrs @ [SigmaI]
   val type_elims = [SigmaE2]);
```

Again, addconsts declares a constant for lleq in the parent theory. The domain of lleq(A) is llist(A) × llist(A). The type-checking rules include the introduction rules for lazy lists as well as rules for both directions of the equivalence $\langle a, b\rangle \in A \times B \leftrightarrow a \in A \land b \in B$.

The package returns the introduction rules and the elimination rule, as usual. But instead of induction rules, it returns a coinduction rule. The rule is too big to display in the usual notation; its conclusion is $x \in \text{lleq}(A)$ and its premises are $x \in X, X \subseteq \text{llist}(A) \times \text{llist}(A)$ and

$$[z \in X]_z$$
$$\vdots$$
$$z = \langle \text{LNil}, \text{LNil}\rangle \lor \left(\exists a l l' . z = \langle \text{LCons}(a, l), \text{LCons}(a, l')\rangle \land a \in A \land \langle l, l'\rangle \in X \cup \text{lleq}(A)\right)$$

Thus if $x \in X$, where X is a bisimulation contained in the domain of lleq(A), then $x \in \text{lleq}(A)$. It is easy to show that lleq(A) is reflexive: the equality relation is a bisimulation. And lleq(A) is symmetric: its converse is a bisimulation. But showing that lleq(A) coincides with the equality relation takes some work.

5.5 The accessible part of a relation

Let $\prec$ be a binary relation on D; in short, $(\prec) \subseteq D \times D$. The **accessible** or **well-founded** part of $\prec$, written acc($\prec$), is essentially that subset of D for which $\prec$ admits no infinite decreasing chains [2]. Formally, acc($\prec$) is inductively defined to be the least set that contains a if it contains all $\prec$-predecessors of a, for $a \in D$. Thus we need an introduction rule of the form

$$\frac{\forall y . y \prec a \to y \in \text{acc}(\prec)}{a \in \text{acc}(\prec)}$$

Paulin-Mohring treats this example in Coq [15], but it causes difficulties for other systems. Its premise does not conform to the structure of introduction rules for HOL's inductive definition package [5]. It is also unacceptable to Isabelle package (§3.1), but fortunately can be transformed into the acceptable form $t \in M(R)$.

The powerset operator is monotonic, and $t \in \mathcal{P}(R)$ is equivalent to $t \subseteq R$. This in turn is equivalent to $\forall y \in t . y \in R$. To express $\forall y . y \prec a \to y \in \text{acc}(\prec)$ we need only find a term t such that $y \in t$ if and only if $y \prec a$. A suitable t is the inverse image of $\{a\}$ under $\prec$.

The ML invocation below follows this approach. Here r is $\prec$ and field(r) refers to D, the domain of acc(r). (The field of a relation is the union of its domain and range.) Finally $r^{-\prime\prime}\{a\}$ denotes the inverse image of $\{a\}$ under r. The package is supplied the theorem Pow_mono, which asserts that $\mathcal{P}$ is monotonic.

```
structure Acc = Inductive_Fun
  (val thy        = WF.thy addconsts [(["acc"],"i=>i")]
   val rec_doms   = [("acc", "field(r)")]
   val sintrs     = ["[| r-`'{a}: Pow(acc(r));  a: field(r) |] ==> a: acc(r)"]
   val monos      = [Pow_mono]
   val con_defs   = []
   val type_intrs = []
   val type_elims = []);
```

The Isabelle theory proceeds to prove facts about acc($\prec$). For instance, $\prec$ is well-founded if and only if its field is contained in acc($\prec$).

As mentioned in §4.1, a premise of the form $t \in M(R)$ gives rise to an unusual induction hypothesis. Let us examine the induction rule, Acc.induct:

$$\frac{x \in \mathrm{acc}(r) \qquad [r^{-\prime\prime}\{a\} \in \mathcal{P}(\{z \in \mathrm{acc}(r) \,.\, P(z)\}) \quad a \in \mathrm{field}(r)]_a \\ \vdots \\ P(a)}{P(x)}$$

The strange induction hypothesis is equivalent to $\forall y \,.\, \langle y, a\rangle \in r \to y \in \mathrm{acc}(r) \land P(y)$. Therefore the rule expresses well-founded induction on the accessible part of $\prec$.

The use of inverse image is not essential. The Isabelle package can accept introduction rules with arbitrary premises of the form $\forall \mathbf{y} \,.\, P(\mathbf{y}) \to f(\mathbf{y}) \in R$. The premise can be expressed equivalently as

$$\{z \in D \,.\, P(\mathbf{y}) \land z = f(\mathbf{y})\} \in \mathcal{P}(R)$$

provided $f(\mathbf{y}) \in D$ for all $\mathbf{y}$ such that $P(\mathbf{y})$. The following section demonstrates another use of the premise $t \in M(R)$, where $M = \mathrm{list}$.

5.6 The primitive recursive functions

The primitive recursive functions are traditionally defined inductively, as a subset of the functions over the natural numbers. One difficulty is that functions of all arities are taken together, but this is easily circumvented by regarding them as functions on lists. Another difficulty, the notion of composition, is less easily circumvented.

Here is a more precise definition. Letting $\mathbf{x}$ abbreviate $x_0, \ldots, x_{n-1}$, we can write lists such as $[\mathbf{x}]$, $[y+1, \mathbf{x}]$, etc. A function is **primitive recursive** if it belongs to the least set of functions in $\mathrm{list}(\mathrm{nat}) \to \mathrm{nat}$ containing

- The **successor** function SC, such that $\mathrm{SC}[y, \mathbf{x}] = y + 1$.
- All **constant** functions $\mathrm{CONST}(k)$, such that $\mathrm{CONST}(k)[\mathbf{x}] = k$.
- All **projection** functions $\mathrm{PROJ}(i)$, such that $\mathrm{PROJ}(i)[\mathbf{x}] = x_i$ if $0 \leq i < n$.
- All **compositions** $\mathrm{COMP}(g, [f_0, \ldots, f_{m-1}])$, where g and $f_0, \ldots, f_{m-1}$ are primitive recursive, such that

$$\mathrm{COMP}(g, [f_0, \ldots, f_{m-1}])[\mathbf{x}] = g[f_0[\mathbf{x}], \ldots, f_{m-1}[\mathbf{x}]].$$

- All **recursions** $\mathrm{PREC}(f, g)$, where f and g are primitive recursive, such that

$$\mathrm{PREC}(f, g)[0, \mathbf{x}] = f[\mathbf{x}]$$
$$\mathrm{PREC}(f, g)[y + 1, \mathbf{x}] = g[\mathrm{PREC}(f, g)[y, \mathbf{x}], y, \mathbf{x}].$$

```
structure Primrec = Inductive_Fun
 (val thy       = Primrec0.thy
  val rec_doms  = [("primrec", "list(nat)->nat")]
  val sintrs    =
      ["SC : primrec",
       "k: nat ==> CONST(k) : primrec",
       "i: nat ==> PROJ(i) : primrec",
       "[| g: primrec; fs: list(primrec) |] ==> COMP(g,fs): primrec",
       "[| f: primrec; g: primrec |] ==> PREC(f,g): primrec"]
  val monos       = [list_mono]
  val con_defs    = [SC_def,CONST_def,PROJ_def,COMP_def,PREC_def]
  val type_intrs  = pr0_typechecks
  val type_elims  = []);
```

Fig. 2. Inductive definition of the primitive recursive functions

Composition is awkward because it combines not two functions, as is usual, but $m+1$ functions. In her proof that Ackermann's function is not primitive recursive, Nora Szasz was unable to formalize this definition directly [25]. So she generalized primitive recursion to tuple-valued functions. This modified the inductive definition such that each operation on primitive recursive functions combined just two functions.

Szasz was using ALF, but Coq and HOL would also have problems accepting this definition. Isabelle's package accepts it easily since $[f_0, \ldots, f_{m-1}]$ is a list of primitive recursive functions and list is monotonic. There are five introduction rules, one for each of the five forms of primitive recursive function. Let us examine the one for COMP:

$$\frac{g \in \text{primrec} \quad fs \in \text{list}(\text{primrec})}{\text{COMP}(g,fs) \in \text{primrec}}$$

The induction rule for primrec has one case for each introduction rule. Due to the use of list as a monotone operator, the composition case has an unusual induction hypothesis:

$$[g \in \text{primrec} \quad fs \in \text{list}(\{z \in \text{primrec} . P(z)\})]_{fs,g}$$
$$\vdots$$
$$P(\text{COMP}(g,fs))$$

The hypothesis states that fs is a list of primitive recursive functions satisfying the induction formula. Proving the COMP case typically requires structural induction on lists, yielding two subcases: either $fs = \text{Nil}$ or else $fs = \text{Cons}(f, fs')$, where $f \in \text{primrec}$, $P(f)$, and fs' is another list of primitive recursive functions satisfying P.

Figure 2 presents the ML invocation. Theory Primrec0.thy defines the constants SC, CONST, etc. These are not constructors of a new datatype, but functions over lists of numbers. Their definitions, which are omitted, consist of routine list programming. In Isabelle/ZF, the primitive recursive functions are defined as a subset of the function set $\text{list}(\text{nat}) \to \text{nat}$.

The Isabelle theory goes on to formalize Ackermann's function and prove that it is not primitive recursive, using the induction rule Primrec.induct. The proof follows Szasz's excellent account.

6 Datatypes and codatatypes

A (co)datatype definition is a (co)inductive definition with automatically defined constructors and a case analysis operator. The package proves that the case operator inverts the constructors and can prove freeness theorems involving any pair of constructors.

6.1 Constructors and their domain

Conceptually, our two forms of definition are distinct. A (co)inductive definition selects a subset of an existing set; a (co)datatype definition creates a new set. But the package reduces the latter to the former. A set having strong closure properties must serve as the domain of the (co)inductive definition. Constructing this set requires some theoretical effort, which must be done anyway to show that (co)datatypes exist. It is not obvious that standard set theory is suitable for defining codatatypes.

Isabelle/ZF defines the standard notion of Cartesian product $A \times B$, containing ordered pairs $\langle a, b \rangle$. Now the m-tuple $\langle x_1, \ldots, x_m \rangle$ is the empty set $\emptyset$ if $m = 0$, simply x_1 if $m = 1$ and $\langle x_1, \langle x_2, \ldots, x_m \rangle \rangle$ if $m \geq 2$. Isabelle/ZF also defines the disjoint sum $A + B$, containing injections $\text{Inl}(a) \equiv \langle 0, a \rangle$ and $\text{Inr}(b) \equiv \langle 1, b \rangle$.

A datatype constructor $\text{Con}(x_1, \ldots, x_m)$ is defined to be $h(\langle x_1, \ldots, x_m \rangle)$, where h is composed of Inl and Inr. In a mutually recursive definition, all constructors for the set R_i have the outer form h_{in}, where h_{in} is the injection described in §3.3. Further nested injections ensure that the constructors for R_i are pairwise distinct.

Isabelle/ZF defines the set $\text{univ}(A)$, which contains A and furthermore contains $\langle a, b \rangle$, $\text{Inl}(a)$ and $\text{Inr}(b)$ for $a, b \in \text{univ}(A)$. In a typical datatype definition with set parameters $A_1, \ldots, A_k$, a suitable domain for all the recursive sets is $\text{univ}(A_1 \cup \cdots \cup A_k)$. This solves the problem for datatypes [21, §4.2].

The standard pairs and injections can only yield well-founded constructions. This eases the (manual!) definition of recursive functions over datatypes. But they are unsuitable for codatatypes, which typically contain non-well-founded objects.

To support codatatypes, Isabelle/ZF defines a variant notion of ordered pair, written $\langle a; b \rangle$. It also defines the corresponding variant notion of Cartesian product $A \otimes B$, variant injections $\text{QInl}(a)$ and $\text{QInr}(b)$ and variant disjoint sum $A \oplus B$. Finally it defines the set $\text{quniv}(A)$, which contains A and furthermore contains $\langle a; b \rangle$, $\text{QInl}(a)$ and $\text{QInr}(b)$ for $a, b \in \text{quniv}(A)$. In a typical codatatype definition with set parameters $A_1, \ldots, A_k$, a suitable domain is $\text{quniv}(A_1 \cup \cdots \cup A_k)$. This approach using standard ZF set theory [22] is an alternative to adopting Aczel's Anti-Foundation Axiom [3].

6.2 The case analysis operator

The (co)datatype package automatically defines a case analysis operator, called R_case. A mutually recursive definition still has only one operator, whose name combines those of the recursive sets: it is called $R_1\_\ldots\_R_n\_\text{case}$. The case operator is analogous to those for products and sums.

Datatype definitions employ standard products and sums, whose operators are split and case and satisfy the equations

$$\text{split}(f, \langle x, y \rangle) = f(x, y)$$
$$\text{case}(f, g, \text{Inl}(x)) = f(x)$$
$$\text{case}(f, g, \text{Inr}(y)) = g(y)$$

Suppose the datatype has k constructors $\text{Con}_1, \ldots, \text{Con}_k$. Then its case operator takes $k + 1$ arguments and satisfies an equation for each constructor:

$$R\_\text{case}(f_1, \ldots, f_k, \text{Con}_i(\mathbf{x})) = f_i(\mathbf{x}), \qquad i = 1, \ldots, k$$

The case operator's definition takes advantage of Isabelle's representation of syntax in the typed λ-calculus; it could readily be adapted to a theorem prover for higher-order logic. If f and g have meta-type $i \Rightarrow i$ then so do $\text{split}(f)$ and $\text{case}(f, g)$. This works because split and case operate on their last argument. They are easily combined to make complex case analysis operators. Here are two examples:

- $\text{split}(\lambda x \,.\, \text{split}(f(x)))$ performs case analysis for $A \times (B \times C)$, as is easily verified:

$$\text{split}(\lambda x \,.\, \text{split}(f(x)), \langle a, b, c \rangle) = (\lambda x \,.\, \text{split}(f(x))(a, \langle b, c \rangle)$$
$$= \text{split}(f(a), \langle b, c \rangle)$$
$$= f(a, b, c)$$

- $\text{case}(f, \text{case}(g, h))$ performs case analysis for $A + (B + C)$; let us verify one of the three equations:

$$\text{case}(f, \text{case}(g, h), \text{Inr}(\text{Inl}(b))) = \text{case}(g, h, \text{Inl}(b))$$
$$= g(b)$$

Codatatype definitions are treated in precisely the same way. They express case operators using those for the variant products and sums, namely qsplit and qcase.

The package has processed all the datatypes discussed in my earlier paper [21] and the codatatype of lazy lists. Space limitations preclude discussing these examples here, but they are distributed with Isabelle.

7 Related work

The use of least fixedpoints to express inductive definitions seems obvious. Why, then, has this technique so seldom been implemented?

Most automated logics can only express inductive definitions by asserting new axioms. Little would be left of Boyer and Moore's logic [4] if their shell principle were removed. With ALF the situation is more complex; earlier versions of Martin-Löf's type theory could (using wellordering types) express datatype definitions, but the version underlying ALF requires new rules for each definition [7]. With Coq the situation is subtler still; its underlying Calculus of Constructions can express inductive definitions [10], but cannot quite handle datatype definitions [15]. It seems that researchers tried hard to circumvent these problems before finally extending the Calculus with rule schemes for strictly positive operators.

Higher-order logic can express inductive definitions through quantification over unary predicates. The following formula expresses that i belongs to the least set containing 0 and closed under succ:

$$\forall P \, . \, P(0) \wedge (\forall x \, . \, P(x) \rightarrow P(\text{succ}(x))) \rightarrow P(i)$$

This technique can be used to prove the Knaster-Tarski Theorem, but it is little used in the HOL system. Melham [11] clearly describes the development. The natural numbers are defined as shown above, but lists are defined as functions over the natural numbers. Unlabelled trees are defined using Gödel numbering; a labelled tree consists of an unlabelled tree paired with a list of labels. Melham's datatype package expresses the user's datatypes in terms of labelled trees. It has been highly successful, but a fixedpoint approach would have yielded greater functionality with less effort.

Melham's inductive definition package [5] uses quantification over predicates, which is implicitly a fixedpoint approach. Instead of formalizing the notion of monotone function, it requires definitions to consist of finitary rules, a syntactic form that excludes many monotone inductive definitions.

The earliest use of least fixedpoints is probably Robin Milner's datatype package for Edinburgh LCF [12]. Brian Monahan extended this package considerably [14], as did I in unpublished work.[4] LCF is a first-order logic of domain theory; the relevant fixedpoint theorem is not Knaster-Tarski but concerns fixedpoints of continuous functions over domains. LCF is too weak to express recursive predicates. Thus it would appear that the Isabelle/ZF package is the first to be based on the Knaster-Tarski Theorem.

8 Conclusions and future work

Higher-order logic and set theory are both powerful enough to express inductive definitions. A growing number of theorem provers implement one of these [8, 24]. The easiest sort of inductive definition package to write is one that asserts new axioms, not one that makes definitions and proves theorems about them. But asserting axioms could introduce unsoundness.

The fixedpoint approach makes it fairly easy to implement a package for (co)inductive definitions that does not assert axioms. It is efficient: it processes most definitions in seconds and even a 60-constructor datatype requires only two minutes. It is also simple: the package consists of under 1100 lines (35K bytes) of Standard ML code. The first working version took under a week to code.

In set theory, care is required to ensure that the inductive definition yields a set (rather than a proper class). This problem is inherent to set theory, whether or not the Knaster-Tarski Theorem is employed. We must exhibit a bounding set (called a domain above). For inductive definitions, this is often trivial. For

---

[4] The datatype package described in my LCF book [16] does *not* make definitions, but merely asserts axioms. I justified this shortcut on grounds of efficiency: existing packages took tens of minutes to run. Such an explanation would not do today.

datatype definitions, I have had to formalize much set theory. I intend to formalize cardinal arithmetic and the $\aleph$-sequence to handle datatype definitions that have infinite branching. The need for such efforts is not a drawback of the fixedpoint approach, for the alternative is to take such definitions on faith.

The approach is not restricted to set theory. It should be suitable for any logic that has some notion of set and the Knaster-Tarski Theorem. I intend to use the Isabelle/ZF package as the basis for a higher-order logic one, using Isabelle/HOL. The necessary theory is already mechanized [18]. HOL represents sets by unary predicates; defining the corresponding types may cause complications.

References

1. Abramsky, S., The lazy lambda calculus, In *Resesarch Topics in Functional Programming*, D. A. Turner, Ed. Addison-Wesley, 1977, pp. 65–116
2. Aczel, P., An introduction to inductive definitions, In *Handbook of Mathematical Logic*, J. Barwise, Ed. North-Holland, 1977, pp. 739–782
3. Aczel, P., *Non-Well-Founded Sets*, CSLI, 1988
4. Boyer, R. S., Moore, J. S., *A Computational Logic*, Academic Press, 1979
5. Camilleri, J., Melham, T. F., Reasoning with inductively defined relations in the HOL theorem prover, Tech. Rep. 265, Comp. Lab., Univ. Cambridge, August 1992
6. Davey, B. A., Priestley, H. A., *Introduction to Lattices and Order*, Cambridge Univ. Press, 1990
7. Dybjer, P., Inductive sets and families in Martin-Löf's type theory and their set-theoretic semantics. In *Logical Frameworks*, G. Huet, G. Plotkin, Eds. Cambridge Univ. Press, 1991, pp. 280–306
8. Farmer, W. M., Guttman, J. D., Thayer, F. J., IMPS: An interactive mathematical proof system, *J. Auto. Reas.* **11**, 2 (1993), 213–248
9. Hennessy, M., *The Semantics of Programming Languages: An Elementary Introduction Using Structural Operational Semantics*, Wiley, 1990
10. Huet, G., Induction principles formalized in the Calculus of Constructions, In *Programming of Future Generation Computers* (1988), Elsevier, pp. 205–216
11. Melham, T. F., Automating recursive type definitions in higher order logic, In *Current Trends in Hardware Verification and Automated Theorem Proving*, G. Birtwistle, P. A. Subrahmanyam, Eds. Springer, 1989, pp. 341–386
12. Milner, R., How to derive inductions in LCF, note, Dept. Comp. Sci., Univ. Edinburgh, 1980
13. Milner, R., *Communication and Concurrency*, Prentice-Hall, 1989
14. Monahan, B. Q., *Data Type Proofs using Edinburgh LCF*, PhD thesis, University of Edinburgh, 1984
15. Paulin-Mohring, C., Inductive definitions in the system Coq: Rules and properties, Research Report 92-49, LIP, Ecole Normale Supérieure de Lyon, Dec. 1992
16. Paulson, L. C., *Logic and Computation: Interactive proof with Cambridge LCF*, Cambridge Univ. Press, 1987
17. Paulson, L. C., *ML for the Working Programmer*, Cambridge Univ. Press, 1991
18. Paulson, L. C., Co-induction and co-recursion in higher-order logic, Tech. Rep. 304, Comp. Lab., Univ. Cambridge, July 1993
19. Paulson, L. C., Introduction to Isabelle, Tech. Rep. 280, Comp. Lab., Univ. Cambridge, 1993
20. Paulson, L. C., Set theory for verification: I. From foundations to functions, *J. Auto. Reas.* **11**, 3 (1993), 353–389
21. Paulson, L. C., Set theory for verification: II. Induction and recursion, Tech. Rep. 312, Comp. Lab., Univ. Cambridge, 1993
22. Paulson, L. C., A concrete final coalgebra theorem for ZF set theory, Tech. rep., Comp. Lab., Univ. Cambridge, 1994
23. Pitts, A. M., A co-induction principle for recursively defined domains, *Theoretical Comput. Sci.* (1994), In press; available as Report 252, Comp. Lab., Univ. Cambridge
24. Saaltink, M., Kromodimoeljo, S., Pase, B., Craigen, D., Meisels, I., An EVES data abstraction example, In *FME '93: Industrial-Strength Formal Methods* (1993), J. C. P. Woodcock, P. G. Larsen, Eds., Springer, pp. 578–596, LNCS 670
25. Szasz, N., A machine checked proof that Ackermann's function is not primitive recursive, In *Logical Environments*, G. Huet, G. Plotkin, Eds. Cambridge Univ. Press, 1993, pp. 317–338

On Notions of Inductive Validity for First-Order Equational Clauses[*]

Claus-Peter Wirth and Bernhard Gramlich

Universität Kaiserslautern, D-67663 Kaiserslautern, Germany
{wirth,gramlich}@informatik.uni-kl.de

Abstract. We define and discuss various conceivable notions of inductive validity for first-order equational clauses. This is done within the framework of constructor-based positive/negative conditional equational specifications which permits to treat negation and incomplete function definitions in an adequate and natural fashion. Moreover, we show that under some reasonable assumptions all these notions are monotonic w. r. t. consistent extension, in contrast to the case of inductive validity for initial semantics (of unconditional or positive conditional equations). In particular from a practical point of view, this monotonicity property is crucial since it allows for an incremental construction process of complex specifications where consistent extensions of specifications cannot destroy the validity of (already proved) inductive properties. Finally we show how various notions of inductive validity in the literature fit in or are related to our classification.

1 Introduction, Motivation, and Overview

Given some finite axiomatization R, e. g., by means of a set of equations or — more generally — of first-order formulas, one is often not only interested in those properties that are logical consequences of R, i. e. that hold in all models of R but also in properties that only hold in some specific intended model (or class of models) specified by R. Instead of restricting the class of models by some required property, one may also define notions of validity by saying that a formula $P(\mathbf{x})$ is to hold schematically in the sense that it is to hold for certain sets of instantiations for the variables $\mathbf{x}$ of $P(\mathbf{x})$.

Concerning the adequacy of a notion of inductive validity we think that there are at least three important criteria.

Coincidence with intuition: It should capture the intention of the human specifier as close as possible.

Monotonicity behaviour: Whenever we extend a specification in some consistent manner then previously valid formulas should still be valid w. r. t. the extended specification.

---
[*] This research was supported by "Deutsche Forschungsgemeinschaft, SFB 314 (D4)".

Operational feasibility: It should be operationally feasible in the sense that there are operational characterizations or at least sufficient operational criteria which provide the basis for corresponding theorem proving techniques.

For specifications with unconditional (or positive conditional) equations it is well-known how to obtain initial semantics and how to prove inductive theorems, i. e. equations that hold in the initial model (cf. e. g. [Bac88]). Consider for instance the following specification R over the natural numbers where addition ($+$) is defined in terms of zero (0) and successor (s).

$$R: \quad 0 + y = y, \; s(x) + y = s(x+y).$$

Here it is easy to show by standard techniques (cf. e. g. [BM79], [Bac88]) that

$$x + 0 = x$$

is an inductive theorem, i. e. holds in the initial model. If we now enrich the above specification by a subtraction operation ($-$) with

$$x - 0 = x, \; s(x) - s(y) = x - y,$$

yielding a new specification R', then, unfortunately, $x + 0 = x$ is no longer inductively valid in the enriched specification (to wit, substitute $0 - s(0)$ for x). Hence, initial semantics does not enjoy the above-mentioned monotonicity property w. r. t. consistent extension. The intuitive reason for that phenomenon is that by enriching our basic specification R as described above we have introduced new *junk terms* which should not be considered for verification purposes. This intention will be formally reflected in our approach by distinguishing between *constructor* and *general* variables which permits to refine the class of (ground) instances of a theorem that we would like to hold.

Another crucial problem with initial semantics has to do with the way that negative statements are interpreted. For instance, the negative equation

$$0 - s(0) \neq 0$$

holds in the initial model of R'. But if we now complete the partial definition of subtraction in some *consistent* manner, e. g., by adding

$$0 - s(y) = 0$$

yielding the new specification R'', then $0 - s(0) \neq 0$ does not hold any more in the initial model of R''. Of course, also validity of conditional statements or general first-order clauses is influenced by this phenomenon. One may argue now that in R' we are not yet sure whether $0 - s(0) \neq 0$ is to hold because it depends on the way we might (consistently) complete the definition of '$-$' later on in the specification process. Thus, the existence of negative literals in formulas to be proved opens up — or even necessarily entails — various ways of defining inductive validity, in particular if we want to guarantee some reasonable monotonicity property w. r. t. consistent extension. Pioneering papers along this line of reasoning are [KM87] and [KM86] (cf. also [Zha88], [ZKK88]). But whereas in these papers the specifications treated are systems of unconditional equations, and the formulas considered are pure equations, here we shall permit general equational first-order clauses as formulas and, moreover, as specifications we admit positive/negative conditional equational systems which naturally arise in many cases.

Before going into details, let us give a rough idea of our underlying specification formalism using constructor-based positive/negative conditional equations. The basic characteristics of our approach (as developed in [WG93]) may be summarized as follows: The set of function symbols is partitioned into *constructor* and *non-constructor* function symbols. The non-constructor function symbols can be used for (possibly partially) defining functions on a domain of discourse supplied by the constructor (ground) terms and called the *constructor sub-universe*. For such partial definitions of functions, variables ranging only over the constructor terms are useful because sometimes the specifier does not want to prescribe how the functions have to behave on objects that are *undefined* in the sense that they do not belong to the domain of discourse. Therefore, in addition to the usual *general variables* we also permit *constructor variables*. As axiomatizations we consider sets of positive/negative conditional equations of the form $\bigwedge_{i=1}^{m}(s_i = t_i) \wedge \bigwedge_{j=1}^{n}(u_j \neq v_j) \Rightarrow l = r$.

In general, specifications with such positive/negative conditional equations lack a unique minimal (i. e. an initial) model. In [Kap87] an operational semantics is developed which — under some reasonable assumptions — distinguishes one of the minimal models by extracting control information from the equations (considered as rewrite rules). In contrast to [Kap87] we use two syntactically expressible restrictions on the form of the equations/rules in order to obtain an appropriate semantics: Namely, the terms in the negative conditions must be *defined* and the *constructor rules* are required to have *Horn*-form and to be *constructor-preserving* (see below). For such positive/negative conditional rule systems a reduction relation can be defined which is monotonic w. r. t. consistent extension and which provides an operational characterization of a unique minimal model.

As formulas to be proved we consider (implicitly universally quantified) first-order clauses of the form $A_1 \wedge \ldots \wedge A_m \longrightarrow B_1 \vee \ldots \vee B_n$ where the A_i's and the B_j's are atoms over the predicate symbols '=' (equality) and 'Def' (definedness predicate) on terms (with general and constructor variables). Some interesting types of inductive validity of such formulas w. r. t. a positive/negative conditional rule system R are defined by restricting the class of models to be considered. Other interesting types of inductive validity are defined by means of *inductive substitutions*, i. e., substitutions that substitute constructor ground terms for constructor variables and leave general variables invariant. The latter invariance for general variables corresponds to the intuition that general variables should be permitted to range not only over the *junk* generated by general ground terms but also over additional *junk*, e. g., of non-constructor symbols which might be introduced later on. Indeed, permitting this additional *junk* is necessary for the intended monotonicity of the notions of inductive validity w. r. t. consistent extension of the specification.

A more detailed discussion of the subject (as well as missing proofs which we have to omit here due to lack of space) can be found in [WGKP93].

2 Preliminaries

2.1 Basic Notions and Notations

We assume familiarity with the basic notions and notations for syntax and semantics of (conditional) term rewriting systems (cf. e. g. [DJ90]). Due to lack of space and for the sake of readability we restrict our presentation to the one-sorted case (cf. [WGKP93] for the more general many-sorted case). We will consider terms over a signature $sig = (\mathsf{F}, \alpha)$ with sub-signature $cons = (\mathsf{C}, \alpha|_\mathsf{C})$ where the functions symbols from $\mathsf{C} \subseteq \mathsf{F}$ are *constructor* and those of $\mathsf{F} \setminus \mathsf{C}$ are *non-constructor* function symbols, and where α denotes the corresponding arity function (all symbols are assumed to have fixed arity). We consider two distinct types of variables corresponding to their intended usage later on. Namely, V_{SIG} and V_{CONS} denote the set of *general variables* and of *constructor variables*, respectively. The set V of all variables is given by $V := V_{\text{SIG}} \uplus V_{\text{CONS}}$. $\mathcal{T}_{\text{SIG}} := \mathcal{T}(sig, V_{\text{SIG}} \uplus V_{\text{CONS}})$ is the set of (variable-mixed) *general terms*, $\mathcal{T}_{\text{CONS}} := \mathcal{T}(cons, V_{\text{CONS}})$ the set of *pure constructor* terms, and $\mathcal{T}(cons, V_{\text{SIG}} \uplus V_{\text{CONS}})$ the set of (variable-mixed) *constructor terms*. $\mathcal{GT}(cons) := \mathcal{T}(cons, \emptyset)$ denotes the set of all *constructor ground terms* and $\mathcal{GT}(sig) := \mathcal{T}(sig, \emptyset)$ the set of all *ground terms*. To avoid problems with empty domains we assume that $\mathcal{GT}(cons)$ is non-empty. In order to avoid confusion note that we have $\mathcal{T}_{\text{CONS}} \subseteq \mathcal{T}_{\text{SIG}}$ but $V_{\text{CONS}} \cap V_{\text{SIG}} = \emptyset$. Furthermore, for $X = X_{\text{CONS}} \uplus X_{\text{SIG}}$, $X_{\text{CONS}} \subseteq V_{\text{CONS}}$, $X_{\text{SIG}} \subseteq V_{\text{SIG}}$, and some set $T = T_{\text{CONS}} \cup T_{\text{SIG}}$ we define $\mathcal{SUB}(X,T) := \{\sigma : X \to T \mid \forall \zeta \in \{\text{SIG}, \text{CONS}\} : \forall x \in X_\zeta : \sigma(x) \in T_\zeta\}$. For a term t, we shall use postfix-notation $t\sigma$ instead of $\sigma(t)$.

Instead of the usual *sig*-algebras we deal with *sig/cons*-algebras with *universe* $\mathcal{A}(\text{SIG})$ and *constructor sub-universe* $\mathcal{A}(\text{CONS}) \subseteq \mathcal{A}(\text{SIG})$ which are defined as usual except for the requirement $c^{\mathcal{A}}[\mathcal{A}(\text{CONS}) \times \ldots \times \mathcal{A}(\text{CONS})] \subseteq \mathcal{A}(\text{CONS})$ (for $c \in \mathsf{C}$). A *sig/cons*-homomorphism $h : \mathcal{A} \longrightarrow \mathcal{B}$ is a usual *sig*-homomorphism with the additional requirement $h[\mathcal{A}(\text{CONS})] \subseteq \mathcal{B}(\text{CONS})$. For $X \subseteq V$ we use $\mathcal{T}(X)$ to denote the term algebra over X and *sig/cons*/V (where $\mathcal{T}(X)(\text{SIG}) := \mathcal{T}(sig, X) \cap \mathcal{T}_{\text{SIG}}$ and $\mathcal{T}(X)(\text{CONS}) := \mathcal{T}(sig, X) \cap \mathcal{T}_{\text{CONS}}$). For a *sig/cons*-algebra $\mathcal{A}$ and a *sig*-congruence $\sim$ on $\mathcal{A}(\text{SIG})$, the factor algebra $\mathcal{B}$ of $\mathcal{A}$ modulo $\sim$ (denoted by $\mathcal{A}/\sim$) is given by $\mathcal{B}(\text{CONS}) := \{\sim[\{a\}] \mid a \in \mathcal{A}(\text{CONS})\}$ with $\mathcal{B}(\text{SIG})$ and $f^{\mathcal{B}}$ as usual (here, $\sim[\{a\}]$ denotes the $\sim$-congruence class of a). For a *sig/cons*-algebra $\mathcal{A}$, an $\mathcal{A}$-*valuation* κ of X is an element of $\mathcal{SUB}(X, \mathcal{A})$. For DUNNO $\in \{\text{SIG}, \text{CONS}\}$, $dunno \in \{sig, cons\}$ a *sig/cons*-algebra $\mathcal{A}$ is called DUNNO:*dunno*-term-generated if the following holds: $\forall a \in \mathcal{A}(\text{DUNNO}) : \exists t \in \mathcal{GT}(dunno) : \mathcal{A}(t) = a$. $\mathcal{A}$ is called *dunno*-term-generated if it is SIG:*dunno*-term-generated. For K a class of *sig/cons*-algebras, $\mathcal{A}$ a *sig/cons*-algebra, $X \subseteq V$ and $\kappa \in \mathcal{SUB}(X, \mathcal{A})$ we need the following definitions: $\mathcal{A}$ is *initial in* K if $\mathcal{A} \in K$ and for all $\mathcal{B} \in K$ there is a unique $h : \mathcal{A} \to \mathcal{B}$. $\mathcal{A}$ is *free for K over* X *w. r. t.* κ if for all $\mathcal{B} \in K$ and $\mu \in \mathcal{SUB}(X, \mathcal{B})$ there is a unique $h : \mathcal{A} \to \mathcal{B}$ with $\mu = \kappa h$. $\mathcal{A}$ is *free in K over* X *w. r. t.* κ if $\mathcal{A} \in K$ and $\mathcal{A}$ is free for K over X w. r. t. κ.

Note that our notion of *sig/cons*-algebras can be viewed as a very special (and simple) case of order-sorted structures, e. g., in the sense of [SNGM89].

2.2 Constructor-Based Positive/Negative Conditional Equational Specifications

Definition 1. (Positive/Negative Conditional Term Rewriting System)
A *positive/negative conditional term rewriting system* (*PNCTRS* for short) over *sig/cons*/V is a set of rules[2] of the form
$$\bigwedge_{i=1}^{m}(s_i = t_i) \wedge \bigwedge_{j=1}^{n}(u_j \neq v_j) \wedge \bigwedge_{k=1}^{p}(\text{Def } w_k) \quad \Rightarrow \quad l = r$$
with all s_i, t_i, u_j, v_j, w_k in $\mathcal{T}(sig, V_{\text{SIG}} \uplus V_{\text{CONS}})$.

Definition 2. (Semantics of PNCTRSs)
If R is a PNCTRS over *sig/cons*/V then a *sig/cons*-algebra $\mathcal{A}$ is said to be a *model of R* if we have:
$$\forall (P \Rightarrow l{=}r) \in R: \forall \kappa \in \mathcal{SUB}(V, \mathcal{A}): ((P \text{ is true w.r.t. } \mathcal{A}_\kappa) \Rightarrow \mathcal{A}_\kappa(l) = \mathcal{A}_\kappa(r))$$
where P is true w. r. t. $\mathcal{A}_\kappa$ if all its literals are true w. r. t. $\mathcal{A}_\kappa$, i. e., $\mathcal{A}_\kappa(u) = \mathcal{A}_\kappa(v)$ for $(u = v)$ in P, $\mathcal{A}_\kappa(u) \neq \mathcal{A}_\kappa(v)$ for $(u \neq v)$ in P, and $\mathcal{A}_\kappa(u) \in \mathcal{A}(\text{CONS})$ for $(\text{Def } u)$ in P.

Definition 3. (Minimal and Minimum Model)
Let the quasi-orderings $\lesssim_{\text{H}}$ and $\lesssim_{\text{CONS}}$ on *sig/cons*-algebras be defined by: $\mathcal{A} \lesssim_{\text{H}} \mathcal{B}$ if there exists a *sig/cons*-homomorphism from $\mathcal{A}$ to $\mathcal{B}$. $\mathcal{A} \lesssim_{\text{CONS}} \mathcal{B}$ if there exists a *cons*-homomorphism from $\mathcal{A}|_{C \uplus \{\text{CONS}\}}$ to $\mathcal{B}|_{C \uplus \{\text{CONS}\}}$.[3]
A *sig/cons*-algebra $\mathcal{A}$ is a *minimal model* of a PNCTRS R if it is minimal w. r. t. $\lesssim_{\text{H}}$ in the class M of all models of R. It is a *constructor-minimal model* of R if it is minimal w. r. t. $\lesssim_{\text{CONS}}$ in M. $\mathcal{A}$ is a *minimum model* of R if it is a least model of R w. r. t. $\lesssim_{\text{H}}$, and it is a *constructor-minimum model* of R if it is a least model of R w. r. t. $\lesssim_{\text{CONS}}$.

Now, every PNCTRS possesses minimal models, but not necessarily a minimum model. But as we shall see, under certain restrictions we are able to guarantee the existence of a constructor-minimum model which can be constructed as a quotient term algebra. To this end we have to define a reduction relation for PNCTRSs. This is only possible in a reasonable way — without assuming some termination ordering conditions as in [Kap87] — by imposing additional restrictions on the structure of the rules. To be precise, we require for every *constructor rule* of R, i. e., for every $(P \Rightarrow l{=}r) \in R$ with $l \in \mathcal{T}(cons, V_{\text{SIG}} \uplus V_{\text{CONS}})$:

---

[2] We shall always use '=' in equations or rules. The interpretation as '=' or '→' tacitly depends on the context.

[3] By the notation $\mathcal{A}|_{C \uplus \{\text{CONS}\}}$ we mean the *cons*-algebra of $\mathcal{A}$ with universe $\mathcal{A}(\text{CONS})$.

(1) There are no negative equations in P (*constructor rules have "Horn"-form*), and
(2) All terms in r and P are constructor terms, i. e., in $\mathcal{T}(cons, V_{\text{SIG}} \uplus V_{\text{CONS}})$, and all variables of r and P occur in l (*"constructor-preservation"*).

PNCTRSs satisfying these conditions are called *constructor-based*. In the rest of the paper all PNCTRSs are tacitly assumed to be constructor-based.

Now we are prepared to define the reduction relation.

Definition 4. (Reduction Relation)
Let R be a (constructor-based) PNCTRS over $sig/cons/V$ and $R_{\mathbf{C}}$ its subset of constructor rules.[4] For $X \subseteq V$ the *reduction relation* $\to_{R,X}$ on $\mathcal{T}(sig, X)$ induced by R is the smallest relation $\to$ satisfying $\to \cap (\mathcal{GT}(cons) \times \mathcal{T}_{\text{SIG}}) \subseteq \to_{R_{\mathbf{C}},X}$,[5] and $s \to t$ if $s, t \in \mathcal{T}(sig, X)$ and

$$\exists (P \Rightarrow l=r) \in R: \exists \sigma \in \mathcal{SUB}(V, \mathcal{T}(X)): \exists p \in \mathcal{POS}(s): \begin{pmatrix} s/p = l\sigma \land \\ t = s[p \leftarrow r\sigma] \land \\ P\sigma \text{ fulfilled w.r.t.} \to \end{pmatrix}$$

where "Q is fulfilled w. r. t. $\to$" is a shorthand for

$$\begin{pmatrix} \forall (u = v) \text{ in } Q : u \downarrow v \\ \land \forall (\text{Def } u) \text{ in } Q : \exists \hat{u} \in \mathcal{GT}(cons) : u \to^* \hat{u} \\ \land \forall (u \neq v) \text{ in } Q : \exists \hat{u}, \hat{v} \in \mathcal{GT}(cons) : u \to^* \hat{u} \updownarrow \hat{v} \leftarrow^* v \end{pmatrix}$$

The well-definedness of $\to_{R,X}$ can be established by a double closure construction, firstly for the constructor rules only (which is possible due to their required Horn-form), and secondly for general rules knowing that the reduction relation on constructor ground terms remains invariant (due to the required constructor-preservation property).

Note moreover that $\to_{R,X}$ is stable under substitutions from $\mathcal{SUB}(V, \mathcal{T}(X))$. Furthermore, for $X \subseteq Y$, confluence of $\to_{R,Y}$ implies confluence of $\to_{R,X}$.

In order to guarantee the existence of a constructor-minimum model of R we have to require that the reduction relation is (ground) confluent[6] and that the terms of the negative equations in the conditions of R are "defined".

Definition 5. (Def-Moderate PNCTRS)
A (constructor-based) PNCTRS is called Def-*moderate* (Def-MCTRS for short) if for each rule $(P \Rightarrow l=r) \in R$ and each negative condition $(u \neq v)$ in P we have that (Def u), (Def v) are in P, too.

Theorem 6. (*Existence and Characterization of a Constructor-Minimum Model*)
Let R be a Def-MCTRS over $sig/cons/V$ and K be the class of all constructor-minimal models of R. Moreover, let $X \subseteq V$ and κ be given by: $x \mapsto \leftrightarrow^*_{R,X} [\{x\}]$. Then the following holds:

---
[4] i. e., $R_{\mathbf{C}}$ is a positive conditional system consisting of those rules of R that do not involve non-constructor function symbols.
[5] This requirement ensures minimality of $\to$ on the constructor (ground) terms.
[6] Although many of our results do not depend on the (ground) confluence assumption, this property is desirable anyway, since otherwise, for instance the congruence induced by the reduction relation of some R need not yield a model of R in general.

- If $\to_{R,\emptyset}$ is confluent then $T(X)/\leftrightarrow^*_{R,X}$ is free for K over X w. r. t. κ.
- If $\to_{R,X}$ is confluent and $X \subseteq V_{\text{SIG}}$, then $T(X)/\leftrightarrow^*_{R,X}$ is a minimal model of R which is free in K over X w. r. t. κ and which is moreover a constructor-minimum model of R.

Corollary 7. *If R is a Def-MCTRS such that $\to_{R,\emptyset}$ is confluent then $\mathcal{G}T/\leftrightarrow^*_{R,\emptyset}$ is a minimal model of R which is initial in the class of all constructor-minimal models of R and which is the (up to isomorphism) unique $\lesssim_H$-minimum of all sig-term-generated constructor-minimal models of R.*

Hence, for confluent $\to_{R,\emptyset}$ the factor algebra $\mathcal{G}T/\leftrightarrow^*_{R,\emptyset}$ provides us with a constructive operational characterization of the intended unique minimal model. Moreover, the reduction relation is monotonic w. r. t. consistent extension in the following sense.

Theorem 8. *(Monotonicity of $\to_{R,X}$ w. r. t. Consistent Extension)*
Let R be a PNCTRS over $sig/cons/V$ and let R' be another PNCTRS over $sig'/cons'/V'$ with

$$\begin{array}{|c|c|c|} \hline sig' = (F', \alpha') & F \subseteq F' & R \subseteq R' \\ cons' = (C', \alpha'|_{C'}) & C = C' & X \subseteq X' \\ V' = V & \alpha \subseteq \alpha' & \\ \hline \end{array}.$$

Moreover assume that no new (i. e., of a rule from $R' \setminus R$) left-hand side is a constructor term. Then the following properties hold:

- *The reduction relations $\to_{R,X}$ and $\to_{R',X'}$ coincide for constructor terms, i. e.: $\forall s \in T(cons, X) : \forall t : (s \to^*_{R',X'} t \Leftrightarrow s \to^*_{R,X} t \Leftrightarrow s \to^*_{R_c,X} t)$.*
- *The reduction relation $\to_{R,X}$ is monotonic in X and R: $\to_{R,X} \subseteq \to_{R',X'}$.*

In [WG93] we have developed a couple of confluence criteria for PNCTRSs as well as some slightly extended "decreasingness"-notions which generalize known results for positive conditional rewrite systems (e. g., of [DOS88]). For related work on PNCTRSs see also [AB92], [Bec93].

3 Inductive Validity: Notions and Interrelations

Definition 9. (Syntax of Formulas)
Let $X \subseteq V$. The set of *formulas* (or *Gentzen clauses*) over sig, X is defined to be $\text{FORM}(sig, X) := \text{ATOM}(sig, X)^* \times \text{ATOM}(sig, X)^*$, where $\text{ATOM}(sig, X)$ is the set of *atoms* over the predicate symbols '=' and 'Def' on terms from $T(sig, X)$. A formula (Γ, Δ) will also be denoted by $\Gamma \longrightarrow \Delta$. For the special cases of $\longrightarrow \Delta$ and $(s = t) \longrightarrow$ we also write Δ and $(s \neq t)$, respectively.

Definition 10. (Validity of Formulas in Algebras)
Let $X \subseteq V$, $\mathcal{A}$ be a *sig/cons*-algebra, and $\kappa \in \mathcal{SUB}(X, \mathcal{A})$.
An atom $(u = v) \in \text{ATOM}(sig, X)$ is *true w. r. t. $\mathcal{A}_\kappa$* if $\mathcal{A}_\kappa(u) = \mathcal{A}_\kappa(v)$.
An atom $(\text{Def } u) \in \text{ATOM}(sig, X)$ is *true w. r. t. $\mathcal{A}_\kappa$* if $\mathcal{A}_\kappa(u) \in \mathcal{A}(\text{CONS})$.
A formula $(\Gamma \longrightarrow \Delta) \in \text{FORM}(sig, X)$ is *valid in $\mathcal{A}$* if $\forall \kappa \in \mathcal{SUB}(X, \mathcal{A})$:
$(\forall A \text{ in } \Gamma: (A \text{ is true w. r. t. } \mathcal{A}_\kappa) \Rightarrow \exists B \text{ in } \Delta: (B \text{ is true w. r. t. } \mathcal{A}_\kappa))$.

Next we introduce a notion for substitutions that replace constructor variables by constructor ground terms and leave the general variables invariant.

Definition 11. (Inductive Substitutions)
The set of *inductive substitutions* is defined by: INDSUB(V, *cons*)
:= $\{\tau \in \mathcal{SUB}(V, \mathcal{T}(V_{\text{SIG}})) \mid \tau[V_{\text{CONS}}] \subseteq \mathcal{GT}(cons) \land \tau|_{V_{\text{SIG}}} = \text{id}|_{V_{\text{SIG}}}\}$.

Definition 12. (Type-A / B' / B / C / D' / D / E Inductive Validity)
Let R be a PNCTRS over $sig/cons/V$, let M be the class of all models of R and K be the class of all constructor-minimal models of R. Then a formula $(\Gamma \longrightarrow \Delta) \in \text{FORM}(sig, V)$ is said to be
— *type-A inductively valid w. r. t. R*, denoted by $R \models^A_{ind} \Gamma \longrightarrow \Delta$, if
  $\forall \mathcal{A} \in M:$ $\quad \forall \tau \in \text{INDSUB}(V, cons):$ $\quad (\Gamma \longrightarrow \Delta)\tau$ is valid in $\mathcal{A}$.
— *type-B' inductively valid w. r. t. R*, denoted by $R \models^{B'}_{ind} \Gamma \longrightarrow \Delta$, if
  $\forall \mathcal{A} \in M:$ (($\mathcal{A}$ is CONS:*cons*-term-generated) $\Rightarrow$ $(\Gamma \longrightarrow \Delta)$ is valid in $\mathcal{A}$).
— *type-B inductively valid w. r. t. R*, denoted by $R \models^B_{ind} \Gamma \longrightarrow \Delta$, if
  $\forall \mathcal{A} \in K:$ $\quad \forall \tau \in \text{INDSUB}(V, cons):$ $\quad (\Gamma \longrightarrow \Delta)\tau$ is valid in $\mathcal{A}$.
— *type-C inductively valid w. r. t. R*, denoted by $R \models^C_{ind} \Gamma \longrightarrow \Delta$, if
  $\forall \mathcal{A} \in K:$ (($\mathcal{A}$ is CONS:*cons*-term-generated) $\Rightarrow$ $(\Gamma \longrightarrow \Delta)$ is valid in $\mathcal{A}$).
— *type-D' inductively valid w. r. t. R*, denoted by $R \models^{D'}_{ind} \Gamma \longrightarrow \Delta$, if
  $\forall \mathcal{A} \in K:$ (($\mathcal{A}$ is SIG:*cons*-term-generated) $\Rightarrow$ $(\Gamma \longrightarrow \Delta)$ is valid in $\mathcal{A}$).
— *type-D inductively valid w. r. t. R*, denoted by $R \models^D_{ind} \Gamma \longrightarrow \Delta$, if
  $(\Gamma \longrightarrow \Delta)$ is valid in $\mathcal{T}(V_{\text{SIG}})/\stackrel{*}{\leftrightarrow}_{R,V_{\text{SIG}}}$.
— *type-E inductively valid w. r. t. R*, denoted by $R \models^E_{ind} \Gamma \longrightarrow \Delta$, if
  $(\Gamma \longrightarrow \Delta)$ is valid in $\mathcal{GT}/\stackrel{*}{\leftrightarrow}_{R,\emptyset}$.

Type-A formulates the idea that (constructor) variables are meant to denote objects denoted by (constructor) ground terms. However, contrary to all other types, type-A does not restrict the models of the specification that have to be considered, but considers only instances of formulas obtained by inductive substitutions. Type-B' forbids junk in the constructor sub-universe (i. e. the domain of interest), which makes inductive substitutions redundant. Type-B forbids unnecessary confusion in the constructor sub-universe. Type-C combines both restrictions, which is appealing if one wants to prescribe a precise and fixed knowledge on the basic objects for computation. Type-D' corresponds to the philosophy that a partially defined function is to be interpreted as the set of all possible complete and consistent extensions. Type-D and E finally fix one specific unique minimal model which has neither junk nor confusion in the constructor sub-universe and which can be described constructively in terms of the factor algebra of the term algebra modulo the congruence induced by the reduction relation (provided the latter is confluent). Type-E uses the ground term algebra $\mathcal{GT}$, which is only adequate (cf. Theorem 17 and Example 14) when no general variables occur in the formula. Therefore, Type-D uses the term algebra $\mathcal{T}(V_{\text{SIG}})$ over V_{SIG}. In a sense, type-D means *inductive semantics* for V_{CONS} and *free semantics* for V_{SIG}. The notations of the types of inductive validity are motivated by the following result:

Lemma 13. *(From Type-A down to Type-E)*
Let R be a PNCTRS over $sig/cons/V$ and $\Gamma \longrightarrow \Delta \in \text{FORM}(sig, V)$.
(a) If $R \models_{ind}^{A} \Gamma \longrightarrow \Delta$, then $R \models_{ind}^{B} \Gamma \longrightarrow \Delta$ and $R \models_{ind}^{B'} \Gamma \longrightarrow \Delta$.
(b) If $R \models_{ind}^{B} \Gamma \longrightarrow \Delta$ or $R \models_{ind}^{B'} \Gamma \longrightarrow \Delta$, then $R \models_{ind}^{C} \Gamma \longrightarrow \Delta$.
(c) If $R \models_{ind}^{C} \Gamma \longrightarrow \Delta$, then $R \models_{ind}^{D'} \Gamma \longrightarrow \Delta$ and even $R \models_{ind}^{D'} \Gamma' \longrightarrow \Delta$ where Γ' results form Γ by deleting (some) Def-atoms.
(d) If R is a Def-MCTRS, $\rightarrow_{R,V_{SIG}}$ is confluent, and $R \models_{ind}^{C} \Gamma \longrightarrow \Delta$, then $R \models_{ind}^{D} \Gamma \longrightarrow \Delta$.
(e) If $R \models_{ind}^{D} \Gamma \longrightarrow \Delta$, then $R \models_{ind}^{E} \Gamma \longrightarrow \Delta$.
(f) If R is a Def-MCTRS, $\rightarrow_{R,\emptyset}$ is confluent, and $R \models_{ind}^{C} \Gamma \longrightarrow \Delta$, then $R \models_{ind}^{E} \Gamma \longrightarrow \Delta$.
(g) Under the restrictive condition of sufficient completeness of $\rightarrow_{R,\emptyset}$ (i. e.: $\forall s \in \mathcal{GT}(sig)$: $\exists t \in \mathcal{GT}(cons)$: $s \rightarrow_{R,\emptyset}^{*} t$) the following holds:
If R is a Def-MCTRS and $\rightarrow_{R,\emptyset}$ is confluent, then we get:
$R \models_{ind}^{D'} \Gamma \longrightarrow \Delta \Leftrightarrow R \models_{ind}^{E} \Gamma \longrightarrow \Delta$.

The basic characteristics of and the relations between these notions of inductive validity are illustrated in Figure 1 below where implications are indicated by arrows. Missing arrows indicate non-implications as shown by Example 14 below.

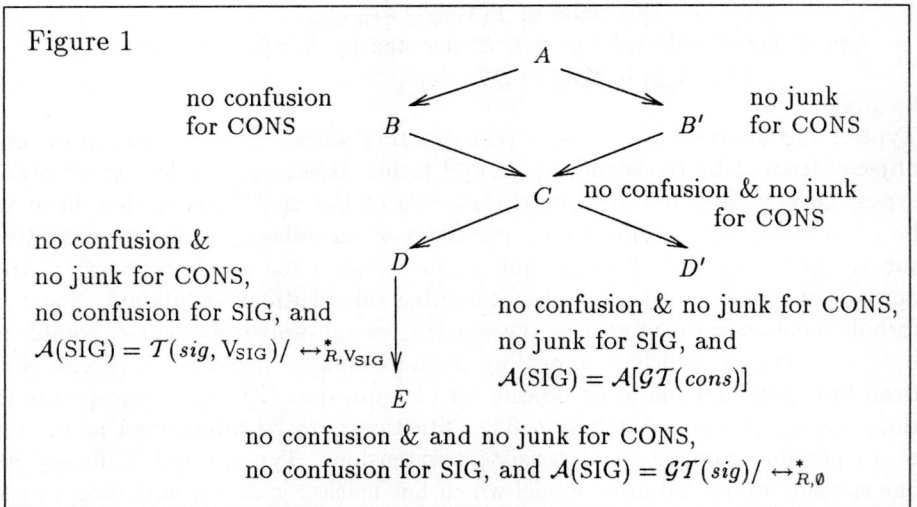

Figure 1

Example 14. *Let us return to the specification on natural numbers from section 1 and start with* $\mathbb{F} := \mathbb{C} := \{s, 0\}$ *and* $R := \emptyset$. *Then we have*
$$R \models_{ind}^{B} s(0) \neq 0 \text{ , but } R \not\models_{ind}^{A} s(0) \neq 0 \text{ ,}$$
since 0 and $s(0)$ are interpreted as distinct objects in any constructor-minimal model of R, and since $s(0) \neq 0$ does not hold for instance in the trivial model of

R which identifies everything. Furthermore, for the same reason we have
$$R \models_{ind}^{C'} s(0) \neq 0 \; , \; \text{but} \; R \not\models_{ind}^{B'} s(0) \neq 0 \; .$$
Keeping $C' := C$, let us add two non-constructor symbols $+$ and ω, yielding $F' := \{+, \omega\} \uplus C$, with the rules
$$R'\colon \; 0 + y = y, \quad s(x) + y = s(x + y),$$
where it does not matter whether x, y are from V_{SIG} or V_{CONS}. Then we have
$$R' \models_{ind}^{C} (\operatorname{Def} \omega) \longrightarrow (\omega + 0 = \omega) \; , \; \text{but} \; R' \not\models_{ind}^{B} (\operatorname{Def} \omega) \longrightarrow (\omega + 0 = \omega) \; ,$$
since if $(\operatorname{Def} \omega)$ is valid in some CONS:*cons-term-generated* model then ω can be denoted by some constructor ground term $s^i(0)$, and since for type-B inductive validity we also have to consider (*constructor-minimal*) models containing constructor objects that are not denoted by some term of the form $s^i(0)$, e. g. $\mathcal{GT}'/\leftrightarrow_{R',\emptyset}^{*}$ where $\mathcal{GT}'$ differs from $\mathcal{GT}$ only in the constructor sub-universe given as $\mathcal{GT}'(\text{CONS}) := \mathcal{GT}(\text{SIG})$. Note that $\mathcal{GT}'/\leftrightarrow_{R',\emptyset}^{*}$ is constructor-minimal indeed, due to $\mathcal{GT}'/\leftrightarrow_{R',\emptyset}^{*} \lesssim_{CONS} \mathcal{GT}/\leftrightarrow_{R',\emptyset}^{*}$. Furthermore, for the same reason we have
$$R' \models_{ind}^{B'} (\operatorname{Def} \omega) \longrightarrow (\omega + 0 = \omega) \; , \; \text{but} \; R' \not\models_{ind}^{A} (\operatorname{Def} \omega) \longrightarrow (\omega + 0 = \omega) \; .$$
Again keeping $C'' := C$, let us add a non-constructor symbol $-$, yielding $F'' := \{-\} \uplus C$, with the rules
$$R''\colon \; x - 0 = x, \quad s(x) - s(y) = x - y,$$
where it does not matter whether x, y are from V_{SIG} or V_{CONS}. Then we have
$$R'' \models_{ind}^{D} 0 - s(0) \neq 0 \; , \; \text{but} \; R'' \not\models_{ind}^{D'} 0 - s(0) \neq 0$$
because $0 - s(0) \neq 0$ does not hold in the SIG:*cons-term-generated* constructor-minimal model obtained by identifying $0 - y$ with 0. Furthermore,
$$R'' \models_{ind}^{D'} (0 - s(0)) - (0 - s(0)) = 0 \; , \; \text{but} \; R'' \not\models_{ind}^{E} (0 - s(0)) - (0 - s(0)) = 0$$
because a SIG:*cons-term-generated* model must satisfy $0 - s(0) = s^i(0)$ for some i, but $(0-s(0)) - (0-s(0)) \not\leftrightarrow_{R'',\emptyset}^{*} 0$ (i. e., $(0-s(0)) - (0-s(0)) = 0$ does not hold in the CONS:*cons-term-generated*, constructor-minimal model $\mathcal{GT}/\leftrightarrow_{R'',\emptyset}^{*}$ of R'').
Finally, keeping $C''' := C$ and choosing $F''' := \{+\} \uplus C$, for $X \in V_{SIG}$ we have $R' \models_{ind}^{E} X + 0 = X$ (which is not the case for the consistent extension via $F''' := F'$), but $R' \not\models_{ind}^{D} X + 0 = X$, since $\forall t \in \mathcal{GT}(sig)\colon t + 0 \leftrightarrow_{R',\emptyset}^{*} t$ but not $X + 0 \leftrightarrow_{R',V_{SIG}}^{*} X$.

While the example has shown that the reverse of each implication depicted in the figure does not hold in general, the following lemma gives sufficient conditions.

Lemma 15. *(From Type-E up to Type-A)*
Let R be a PNCTRS over $sig/cons/V$ and $\Gamma \longrightarrow \Delta \in \text{FORM}(sig, V)$.

(a) If $\Gamma \longrightarrow \Delta$ does not contain variables from V_{SIG}, then we get:
$$R \models_{ind}^{E} \Gamma \longrightarrow \Delta \quad \Rightarrow \quad R \models_{ind}^{D} \Gamma \longrightarrow \Delta \; .$$
(b) If R is a Def-MCTRS, $\rightarrow_{R,\emptyset}$ is confluent, and if for all (top level) terms u of atoms in Γ we have $R \models_{ind}^{D} (\operatorname{Def} u)$, then we get:
$$R \models_{ind}^{D} \Gamma \longrightarrow \Delta \quad \Rightarrow \quad R \models_{ind}^{C} \Gamma \longrightarrow \Delta \; .$$

(c) *If for each atom* (Def u) *in* Γ *we have* $R \models_{ind}^{C}$ (Def u),[7] *then we get:*
$$R \models_{ind}^{C} \Gamma \longrightarrow \Delta \quad \Rightarrow \quad R \models_{ind}^{B} \Gamma \longrightarrow \Delta \;.$$
If for each atom (Def u) *in* Γ *we have* $R \models_{ind}^{B'}$ (Def u),[7] *then we get:*
$$R \models_{ind}^{B'} \Gamma \longrightarrow \Delta \quad \Rightarrow \quad R \models_{ind}^{A} \Gamma \longrightarrow \Delta \;.$$
(d) *If no rule in R has a negative condition (like $u \neq v$), then we get:*
$$R \models_{ind}^{C} \Delta \quad \Rightarrow \quad R \models_{ind}^{A} \Delta \;.$$

Note that (by Lemma 13) Lemma 15 also permits to conclude from type-C to B' (via A), from type-B to A (via C), and from type-D' to C (via E, D).

Lemma 16. *(Operational Characterization of Type-D Inductive Validity)*
Let R be a PNCTRS over sig/cons/V and $\Gamma \longrightarrow \Delta \in \text{FORM}(sig, V)$.
Then $R \models_{ind}^{D} \Gamma \longrightarrow \Delta$ is equivalent to $\forall \tau \in \mathcal{SUB}(V, \mathcal{T}(V_{SIG}))$:
$$\begin{pmatrix} \forall (u{=}v) \text{ in } \Gamma: u\tau \leftrightarrow^{*}_{R,V_{SIG}} v\tau \;\wedge\; \forall(\text{Def }u) \text{ in } \Gamma: \exists \hat{u} \in \mathcal{GT}(cons): u\tau \leftrightarrow^{*}_{R,V_{SIG}} \hat{u} \\ \Rightarrow \\ \exists (u{=}v) \text{ in } \Delta: u\tau \leftrightarrow^{*}_{R,V_{SIG}} v\tau \;\vee\; \exists(\text{Def }u) \text{ in } \Delta: \exists \hat{u} \in \mathcal{GT}(cons): u\tau \leftrightarrow^{*}_{R,V_{SIG}} \hat{u} \end{pmatrix}$$

Finally we show that (under some reasonable assumptions) all defined notions of inductive validity are monotonic w. r. t. consistent extension.

Theorem 17. *(Monotonicity of Inductive Validity w. r. t. Consistent Extension)*
Let R be a PNCTRS over sig/cons/V and let R' be another PNCTRS over
sig'/cons'/V' with[8]
$$\begin{array}{|c|c|c|} \hline sig' = (F', \alpha') & F \subseteq F' & R \subseteq R' \\ cons' = (\mathbb{C}', \alpha'|_{\mathbb{C}'}) & C = C' & X \subseteq X' \\ V' = V & \alpha \subseteq \alpha' & \\ \hline \end{array}\;.$$
Moreover assume that

$$(*) \quad \rightarrow_{R',\emptyset} \text{ is confluent}$$

and that the following condition for the new left hand sides holds:

$$(**) \quad \forall (C \Rightarrow l{=}r) \in R' \setminus R: \; l \notin \mathcal{T}(cons, V_{SIG} \uplus V_{CONS}).$$

Then, for any clause $\Gamma \longrightarrow \Delta \in \text{FORM}(sig, V)$ we have:

(A) $R \models_{ind}^{A} \Gamma \longrightarrow \Delta \;\Rightarrow\; R' \models_{ind}^{A} \Gamma \longrightarrow \Delta$ *(even without assuming $(*)$, $(**)$).*
(B') $R \models_{ind}^{B'} \Gamma \longrightarrow \Delta \;\Rightarrow\; R' \models_{ind}^{B'} \Gamma \longrightarrow \Delta$ *(even without assuming $(*)$, $(**)$).*
(B) *If R' is Def-moderate then we get:* $R \models_{ind}^{B} \Gamma \longrightarrow \Delta \;\Rightarrow\; R' \models_{ind}^{B} \Gamma \longrightarrow \Delta$.
(C) *If R' is Def-moderate then we get:* $R \models_{ind}^{C} \Gamma \longrightarrow \Delta \;\Rightarrow\; R' \models_{ind}^{C} \Gamma \longrightarrow \Delta$.
(D') *If R' is Def-moderate then we get:* $R \models_{ind}^{D'} \Gamma \longrightarrow \Delta \;\Rightarrow\; R' \models_{ind}^{D'} \Gamma \longrightarrow \Delta$.
(D) *If for all (top level) terms u of atoms in Γ we have $R \models_{ind}^{D}$ (Def u),*
  then we get: $\quad R \models_{ind}^{D} \Gamma \longrightarrow \Delta \;\Rightarrow\; R' \models_{ind}^{D} \Gamma \longrightarrow \Delta$.
(E) *If $\Gamma \longrightarrow \Delta$ does not contain variables from V_{SIG},*
  and if for all (top level) terms u of atoms in Γ we have $R \models_{ind}^{E}$ (Def u),
  then we get: $\quad R \models_{ind}^{E} \Gamma \longrightarrow \Delta \;\Rightarrow\; R' \models_{ind}^{E} \Gamma \longrightarrow \Delta$.

---

[7] Even if this condition is not satisfied, the following equivalence transformation for type-B', C, D', D, and E validity (but not for A and B) may help to apply the Lemma: $\Gamma, (\text{Def }u), \Gamma' \longrightarrow \Delta$ is equivalent to $\Gamma, (x{=}u), \Gamma' \longrightarrow \Delta$ for a fresh (i. e. not occurring in Γ, u, Γ', Δ) constructor variable x.

[8] In a sorted framework one may even add new constructor symbols for new sorts and permit new rules with left-hand sides that are new (but not old) constructor terms.

Proof. Let $\mathcal{A}'$ be some $sig'/cons'$-algebra. We define the $sig/cons$-algebra $\mathcal{A}$ by $\mathcal{A} := \mathcal{A}'|_{F \uplus \{SIG, CONS\}}$. Now an "old" formula $\Gamma \longrightarrow \Delta \in \text{FORM}(sig, V)$ is valid in $\mathcal{A}$ if and only if it is valid in $\mathcal{A}'$. Its inductive instances $(\Gamma \longrightarrow \Delta)\tau$ do not differ for $\tau \in \text{INDSUB}(V, cons)$ and $\tau \in \text{INDSUB}(V', cons')$ due to $\mathcal{GT}(cons) = \mathcal{GT}(cons')$. Furthermore, if $\mathcal{A}'$ is CONS:$cons'$-term-generated (or even SIG:$cons'$-term-generated), then $\mathcal{A}$ is CONS:$cons$-term-generated (or even SIG:$cons$-term-generated). Thus for proving (A), (B'), [and (B), (C), (D'),] it suffices to show the following claim: (∗∗∗) If $\mathcal{A}'$ is a [constructor-minimal] model of R', then $\mathcal{A}$ is a [constructor-minimal] model of R.

Proof of (∗∗∗): Assume that $\mathcal{A}'$ is a model of R'. Then $\mathcal{A}$, as defined above, is a model of R since each rule from R can be translated into an "old" formula (on which $\mathcal{A}$ and $\mathcal{A}'$ do not differ). Let us now assume that R' is Def-moderate and that $\mathcal{A}'$ is constructor-minimal. Let $\mathcal{GT}'$ and $\mathcal{GT}(cons)$ denote the ground term algebra over $sig'/cons'$ and $cons$, respectively. By (∗) and Theorem 6, $\mathcal{GT}'/\leftrightarrow^*_{R',\emptyset}$ is a constructor-minimum model of R' w. r. t. $sig'/cons'$. Thus, $\mathcal{A}'$ must be a constructor-minimum model of R' w. r. t. $sig'/cons'$, too. Hence, there exists some $cons'$-homomorphism $h' : \mathcal{A}'|_{C \uplus \{CONS\}} \to (\mathcal{GT}'/\leftrightarrow^*_{R',\emptyset})|_{C \uplus \{CONS\}}$. By defining $h(a) := h'(a) \cap \mathcal{GT}(cons)$ (for $a \in \mathcal{A}(CONS)$), we get a $cons$-homomorphism $h : \mathcal{A}|_{C \uplus \{CONS\}} \to \mathcal{GT}(cons)/(\leftrightarrow^*_{R',\emptyset} \cap (\mathcal{GT}(cons) \times \mathcal{GT}(cons)))$. By confluence of $\to_{R',\emptyset}$ and Theorem 8 we get $(\leftrightarrow^*_{R',\emptyset} \cap (\mathcal{GT}(cons) \times \mathcal{GT}(cons))) \subseteq (\downarrow_{R',\emptyset} \cap (\mathcal{GT}(cons) \times \mathcal{GT}(cons))) \subseteq \downarrow_{R_C,\emptyset} \subseteq \leftrightarrow^*_{R_C,\emptyset} \subseteq \ker(\mathcal{B})$ for the kernel of any model $\mathcal{B}$ of R, and by the Homomorphism-Theorem we then get $\mathcal{A} \lesssim_{CONS} \mathcal{B}$. Thus, $\mathcal{A}$ is not only a model of R, but also a constructor-minimal one. ∎

Proof of (D): We use the operational version of type-D inductive validity given in Lemma 16. Let $\tau' \in \mathcal{SUB}(V', \mathcal{T}'(V_{SIG}'))$ and assume the condition of the implication of Lemma 16 to hold for this τ'. There exist $\tau \in \text{INDSUB}(V', cons')$ and $\sigma \in \mathcal{SUB}(V'_{SIG}, \mathcal{T}'(V_{SIG}'))$ such that $\tau' = \tau\sigma$ and $\tau|_V \in \mathcal{SUB}(V, \mathcal{T}(V_{SIG}))$. By assumption we know that for each atom $(u = v)$ in Γ the formulas (Def u) and (Def v) are valid in $\mathcal{T}(V_{SIG})/\leftrightarrow^*_{R,V_{SIG}}$. Thus there exist $\hat{u}, \hat{v} \in \mathcal{GT}(cons)$ with $u\tau \leftrightarrow^*_{R,V_{SIG}} \hat{u}$ and $v\tau \leftrightarrow^*_{R,V_{SIG}} \hat{v}$. By Theorem 8 we obtain $u\tau \leftrightarrow^*_{R',V_{SIG}'} \hat{u}$ and $v\tau \leftrightarrow^*_{R',V_{SIG}'} \hat{v}$. Together with $u\tau' \leftrightarrow^*_{R',V_{SIG}'} v\tau'$, which holds by assumption, this yields (exploiting stability of $\to_{R,X}$) $\hat{u} \leftrightarrow^*_{R',V_{SIG}'} u\tau\sigma \leftrightarrow^*_{R',V_{SIG}'} v\tau\sigma \leftrightarrow^*_{R',V_{SIG}'} \hat{v}$. Thus (since $\leftrightarrow^*_{R,X} \cap (\mathcal{GT}(cons) \times \mathcal{GT}(cons)) = \leftrightarrow^*_{R,\emptyset}$) we get $\hat{u} \leftrightarrow^*_{R',\emptyset} \hat{v}$ which, due to confluence of $\to_{R',\emptyset}$, implies $\hat{u} \downarrow_{R',\emptyset} \hat{v}$. Theorem 8 yields $\hat{u} \downarrow_{R,\emptyset} \hat{v}$, hence (by monotonicity of $\to_{R,X}$ in X (Theorem 8)) $u\tau \leftrightarrow^*_{R,V_{SIG}} \hat{u} \leftrightarrow^*_{R,V_{SIG}} \hat{v} \leftrightarrow^*_{R,V_{SIG}} v\tau$. Furthermore, for each atom (Def u) in Γ we have assumed that (Def u) is valid in $\mathcal{T}(V_{SIG})/\leftrightarrow^*_{R,V_{SIG}}$. Thus there exists some $\hat{u} \in \mathcal{GT}(cons)$ with $u\tau \leftrightarrow^*_{R,V_{SIG}} \hat{u}$. Because of $R \models^D_{ind} \Gamma \longrightarrow \Delta$ we can conclude that there is an atom $(u = v) \in \Delta$ with $u\tau \leftrightarrow^*_{R,V_{SIG}} v\tau$ or an atom (Def u) $\in \Delta$ with $u\tau \leftrightarrow^*_{R,V_{SIG}} \hat{u}$ for some $\hat{u} \in \mathcal{GT}(cons)$. Application of Theorem 8 yields $u\tau \leftrightarrow^*_{R',V_{SIG}'} v\tau$ or $u\tau \leftrightarrow^*_{R',V_{SIG}'} \hat{u}$, respectively. Finally, due to $\tau' = \tau\sigma$ (and stability of $\to_{R,X}$) we obtain $u\tau' \leftrightarrow^*_{R',V_{SIG}'} v\tau'$ or $u\tau' \leftrightarrow^*_{R',V_{SIG}'} \hat{u}$, respectively, as desired. Summarizing, we can conclude $R' \models^D_{ind} \Gamma \longrightarrow \Delta$ as was to be shown. ∎

Proof of (E): By Lemma 15(a), (D), Lemma 13(e). ∎

4 Discussion and Related Work

4.1 Advantages/Disadvantages

As shown above, all our notions of inductive validity have the desired monotonic behaviour w. r. t. consistent extension (under reasonable assumptions).

Concerning operational feasibility of the different types of inductive validity, the following can be said. Type-A can be approached by usual first-order theorem proving via induction schemes relying on the fact that only the validity of the inductive instances has to be considered. For type-B, C, and D we are and will be investigating inductive theorem proving techniques which are supported by a confluent rewriting relation. Lemma 16 allows us to develop powerful inference rules for type-D, some of which are not applicable for type-C (and B), which is no surprise since less formulas are of these types. Such an approach is more powerful than presenting an inference system for type-D only and then to approach type-C (and B) via Lemma 15(b) (and (c)). To see this, consider the specification with constructor constants a, b, c and non-constructor function symbol h with partial specification $h(a) = b$, $h(b) = a$. Now the formula $\text{Def}\, h(c) \longrightarrow h(h(h(c))) = h(c)$ is type-C valid which cannot be inferred via Lemma 15(b). It is, however, easy to show its type-C validity via a type-C equivalence transformation into $x = h(c) \longrightarrow h(h(h(c))) = h(c)$ (for $x \in V_{\text{CONS}}$) and subsequent case analysis (via a "covering set" of substitutions for x) followed by "contextual rewriting". Note that the formula $h(h(h(c))) = h(c)$ is of type-D' only. In fact, there are examples for type-D' that require inferences of this kind which are far more complicated since they may not allow for a finite argumentation. Therefore, we suspect that it will be difficult to develop a prover that can effectively show those type-D' inductively valid formulas that do not result from type-C valid formulas $\Gamma \longrightarrow \Delta$ by deleting Def-atoms in Γ (cf. Lemma 13(c)).

4.2 Special Cases

In the sequel we will assume R to be a Def-MCTRS and $\rightarrow_{R,V_{\text{SIG}}}$ to be confluent.

Let us first consider the special case that $cons = sig$ and that no variables from V_{SIG} occur in formulas. Note that, for $cons = sig$, our requirement of being constructor-based permits positive conditional equations only. In this case (disregarding formulas that involve Def-literals) type-A and B' validity are validity in all ($cons = sig$)-term-generated models. Furthermore, Type-B, C, D', D, and E coincide with validity in the unique minimal term-generated model (i. e. the initial model). Hence, we obtain classic initial semantics (for positive conditional equational specifications) as a special case of our general framework.

Still not permitting general variables in formulas, another important special case is that of sufficient completeness (cf. Lemma 13(g)), where type-C, D', D, and E again coincide, and where the model $\mathcal{GT}/\overset{*}{\leftrightarrow}_{R,\emptyset}$ (which establishes type-E) is $cons$-isomorphic to $\mathcal{GT}(cons)/\overset{*}{\leftrightarrow}_{R_{\mathbf{c}},\emptyset}$ (cf. Definition 4).

Let us finally restrict to positive formulas $\longrightarrow \Delta$. Here type-A and B' coincide. So do B, C, and D. Furthermore, these two groups coincide when no rule in R has a negative condition (cf. Lemma 15(d)).

4.3 Related Work

Let us now have a brief look at notions of inductive validity in the literature, most of which can be described as specializations within our framework. If we consider all symbols to be constructor symbols (and, consequently, forbid negative conditions in the rules), then we find type-A in [KR90] as well as in [BKR92]. The crucial idea of requiring the values of variables in formulas to be *defined*, i. e., to be substituted by constructor ground terms only, seems to appear first in [Zha88], [ZKK88]. The "constructor models" of [Zha88], [ZKK88], which are not models in the usual algebraic sense since functions are allowed to be partially interpreted, are consistently formalized in our framework by introducing the simple (order-sorted) notion of *sig/cons*-algebras. The notion of inductive validity of [ZKK88], [Zha88] can be described as type-A by implicitly interpreting all variables in rules as general variables and all variables in formulas as constructor variables. In [GS92] inductive validity is defined to be validity in the perfect model as introduced in [BG91]. This approach has a more general specification formalism permitting sets of first-order clauses instead of constructor-based PNCTRSs. The perfect model is determined as the least term-generated model w. r. t. some ground-total reduction ordering. The perfect model semantics, however, lacks the discussed monotonicity property. In [Pad90], [BL90], [KR88] and [BR93] we find the usual validity in the initial model which is like our type-E, assuming again all symbols to be constructor symbols and forbidding negative conditions in the rules. Kapur and Musser ([KM87], [KM86]) consider only unconditional equations and only congruences on ground terms (i. e. term-generated models) for validity, namely those that are maximally enlarged by random identification of undefined terms with defined ones (i. e. constructor ground terms) as long as this identification does not identify distinct constructor ground terms. Their intended congruence is then the intersection of all those maximally enlarged congruences. In [KM87] the maximal congruences are allowed to have some undefined terms left (accounting for the fact that the constructor-minimality requirement may forbid any further identification from some point on). The resulting "inductive models", however, lack the discussed monotonicity property, cf. [WGKP93] for an example. Therefore, in [KM86] the intersection is formed only over those congruences that have no undefined ground terms left. While we have no notion of inductive validity corresponding to that of [KM87], inductive validity in [KM86] coincides with type-D'. Finally, the problems due to not yet known or incompletely specified function symbols are also discussed in [Wal94], mainly along the lines of [KM86].

5 Conclusion

We have shown that considering inductive validity of first-order equational clauses instead of pure unconditional equations gives rise to various conceivable notions of inductive validity. Within the framework of constructor-based positive/negative conditional equational specifications (which provides an adequate unique model semantics) we have demonstrated that all these notions enjoy a desirable monotonic behaviour w. r. t. consistent extension, which is not the case for classic initial or perfect model validity.

Acknowledgements: We would like to thank Ulrich Kühler and Horst Prote for many valuable discussions and detailed criticisms on earlier versions of this paper, Jürgen Avenhaus and Klaus Becker for fruitful discussions, and Klaus Madlener for useful hints.

References

[AB92] J. Avenhaus, K. Becker. Conditional rewriting modulo a built-in algebra. SEKI-Report SR-92-11, FB Informatik, Univ. Kaiserslautern, 1992.

[Bac88] L. Bachmair. Proof by consistency in equational theories. In *Proc. 3rd IEEE Symposium on Logic in Computer Science*, pp. 228-233, 1988.

[Bec93] K. Becker. Semantics for positive/negative conditional rewrite systems. In M. Rusinowitch, J.L. Rémy, eds., *Proc. 3rd CTRS, LNCS 656*, pp. 213-225. Springer, 1993.

[BG91] L. Bachmair, H. Ganzinger. Perfect model semantics for logic programs with equality. In *Proc. 8th ICLP*, pp. 645-659. MIT Press, 1991.

[BKR92] A. Bouhoula, E. Kounalis, M. Rusinowitch. Automated mathematical induction. Rapport de Recherche 1663, INRIA, April 1992.

[BL90] E. Bevers, J. Levi. Proof by consistency in conditional equational theories. In S. Kaplan, M. Okada, eds., *Proc. 2nd CTRS, LNCS 516*, pp. 194-205. Springer, 1990.

[BM79] R. S. Boyer, J S. Moore. *A Computational Logic*. Academic Press, 1979.

[BR93] A. Bouhoula, M. Rusinowitch. Automatic case analysis in proof by induction. In *Proc. 13th IJCAI*, pp. 88-94, 1993.

[DJ90] N. Dershowitz, J.-P. Jouannaud. Rewrite systems. In J. van Leeuwen, ed., *Formal models and semantics, Handbook of Theoretical Computer Science*, vol. B, chapter 6, pp. 243-320. MIT Press, 1990.

[DOS88] N. Dershowitz, M. Okada, G. Sivakumar. Confluence of conditional rewrite systems. In S. Kaplan, J.-P. Jouannaud, eds., *Proc. 1st CTRS, LNCS 308*, pp. 31-44. Springer, 1988.

[GS92] H. Ganzinger, J. Stuber. Inductive theorem proving by consistency for first-order clauses. In M. Rusinowitch, J.-L. Rémy, eds., *Proc. of 3rd CTRS, LNCS 656*, pp. 226-241. Springer, 1993.

[Kap87] S. Kaplan. Positive/negative conditional rewriting. In S. Kaplan, J.-P. Jouannaud, eds., *Proc. 1st CTRS, LNCS 308*, pp. 129-143. Springer, 1987.

[KM86] D. Kapur, D. R. Musser. Inductive reasoning with incomplete specifications. Prelim. report. In *Proc. 1st LICS*, Cambridge, MA, 1986.

[KM87] D. Kapur, D. R. Musser. Proof by consistency. *Artificial Intelligence*, 31:125-157, 1987.

[KR88] E. Kounalis, M. Rusinowitch. On word problems in Horn theories. In E. Lusk, R. Overbeek, eds., *Proc. 9th CADE, LNCS 310*, pp. 527-535. Springer, 1988.

[KR90] E. Kounalis, M. Rusinowitch. Mechanizing inductive reasoning. In *Proc. 8th AAAI*, pp. 240-245. MIT Press, 1990.

[Pad90] P. Padawitz. Horn logic and rewriting for functional and logic program design. Technical Report MIP-9002, Univ. Passau, 1990.

[SNGM89] G. Smolka, W. Nutt, J.A. Goguen, J. Meseguer. Order-sorted equational computation. In H. Ait-Kaci, M. Nivat, eds., *Proc. CREAS*, vol. 2, Academic Press, 1989.

[Wal94] C. Walther. Mathematical induction. In *Handbook of Logic in Artificial Intelligence and Logic Programming*, vol. 2, Clarendon Press, 1994.

[WG93] C.-P. Wirth, B. Gramlich. A constructor-based approach for positive/negative-conditional equational specifications. In M. Rusinowitch, J.-L. Rémy, eds., *Proc. of 3rd CTRS, LNCS 656*, pp. 198-212. Springer, 1993. Revised and extended version to appear in J. Symbolic Computation.

[WGKP93] C.-P. Wirth, B. Gramlich, U. Kühler, H. Prote. Constructor-based inductive validity in positive/negative-conditional equational specifications. SEKI-Report SR-93-05, FB Informatik, Univ. Kaiserslautern, 1993.

[Zha88] H. Zhang. *Reduction, Superposition and Induction: Automated Reasoning in an Equational Logic*. PhD thesis, Rensselaer Polytech. Inst., Troy, NY, 1988.

[ZKK88] H. Zhang, D. Kapur, M.S. Krishnamoorthy. A mechanizable induction principle for equational specifications. In E. Lusk, R. Overbeek, eds., *Proc. 9th CADE, LNCS 310*, pp. 162-181. Springer, 1988.

A New Application for Explanation-Based Generalisation within Automated Deduction

S. Baker

Cambridge University, Cambridge CB2 3QG, UK
e-mail: Siani.Baker@cl.cam.ac.uk
tel: +44-223-334422

Abstract. Generalisation is currently a major theorem-proving problem. This paper proposes a new method of generalisation, involving the use of explanation-based generalisation within a new domain, which may succeed when other methods fail. The method has been implemented for simple arithmetical examples.

1 Introduction

Generalisation is a powerful tool in automated theorem proving with a variety of rôles, such as enabling proofs, defining new concepts, turning proofs for a specific example into ones valid for a range of examples and producing clearer proofs. Van der Waerden's account of how the proof of Baudet's conjecture was found [19] illustrates how generalisation lies at the heart of mathematical discovery, and how generalised theorems may be easier to prove than the original goal (because the induction hypothesis is also made stronger when the goal is strengthened by generalisation). If induction is blocked for an expression[1], generalisation may be used as a step to convert this expression into a new, more general, expression which may be proved by induction. In order to verify a goal Φ, one may instead prove its generalisation Ψ. Indeed, as discussed in Section 3, it might actually be *necessary* to adopt this approach in cases where Φ is a theorem of a system $\mathcal{S}$ but it is not the case that Φ may be proven by induction within $\mathcal{S}$ (although Ψ may be found such that Ψ is provable by induction within $\mathcal{S}$, and such that Ψ is a generalisation of Φ). Heuristics are needed for the suggestion of Ψ since there is no appropriate algorithm for finding suitable generalisations.[2] A suitable suggested generalisation must be just general enough to provide a proof by induction whilst not being too general, for not only may generalisations Ψ be computed such that it is not the case that Ψ is provable by induction, but overgeneralisations may also be computed (where Ψ is not a theorem of $\mathcal{S}$).

---

[1] Induction is said to be blocked if, after all available symbolic evaluation has been carried out, the induction conclusion is still not an instance of the induction hypothesis, and hence remains unprovable.

[2] Of course, an algorithm could be used which tried every possible option, but the search required might be infeasibly large, and the conclusion that there was no suitable generalisation would not be allowed: this trivial approach is not what is required.

Although generalisation is an important problem in theorem-proving, it has by no means been solved. It is important and still being investigated for reasons which relate to cut elimination and the lack of heuristics for providing cut formulae. A cut elimination theorem for a system states that every proof in that system may be replaced by one which does not involve use of the cut rule[3]. Uniform proof search methods can be used for logical systems, in sequent calculus form, where the cut rule is not used. In general, cut elimination holds for arithmetical systems with the ω-rule, but not for systems with ordinary induction. Hence in the latter, there is the problem of generalisation, since arbitrary formulae can be cut in. This makes automatic theorem-proving very difficult, especially as there is no easy or fail-safe method of generating the required cut formula.

When discussing generalisation I have so far referred to goal generalisation, which is the proof step described above which allows the postulation of a new theorem as a substitute for the current goal, from which the latter follows easily (for example, backward application of the $\forall$-elim rule of a natural deduction calculus). Yet generalisation may be carried out on proofs as well: it has a different emphasis, namely on providing a more general proof, and although goal generalisation may involve generalisation of the proof of a theorem in order to generalise the theorem itself, this need not be the case.

Generalisation of proofs has the advantage that more information may be available than for goal generalisation. The method proposed in this paper exploits this fact by enabling goal generalisation by means of carrying out explanation based generalisation on proofs. The problem of generalisation is tackled by use of an alternative (stronger) representation of arithmetic, in which proofs may be more easily generated. The "guiding proofs" in this stronger system may succeed in producing proofs in the original system when other methods fail (cf. Table 1). This method has been automated for simple arithmetical examples and results in the suggestion of an appropriate cut formula. More specifically, in order to carry out generalisations of the form $\forall x A(x)$ to $\forall x Q(x)$, individual proof instances of $A(x)$ are generated, and then generalised to a proof of the arbitrary case $A(r)$ using some inductive inference process (which presents algorithms to obtain a general pattern from individual instances). The latter proof can be generalised via explanation-based generalisation to the most general proof using the same rule set. The new proof can be shown to be of a linear form. Hence the formula for which this is a proof provides a suitable cut formula for $A(x)$, since it satisfies the properties of being a more general form of which A is a specific case, and such that induction may be performed upon it. Figure 1 represents the overall strategy for generalisation: a "general proof" (of $A(r)$) is provided by some means (one possible option being an automatic derivation using the ω-rule), and a suitable cut formula is suggested by inspection of this proof.

The following section presents ω-proofs, from which it is possible to read off appropriate cut formulae, and the semi-formal system of arithmetic in which they are derived. Section 3 discusses how the explanation-based generalisation method may be applied to these proofs in order to suggest cut formulae. Sec-

---

[3] See [18], for example.

A New Approach to Generalisation

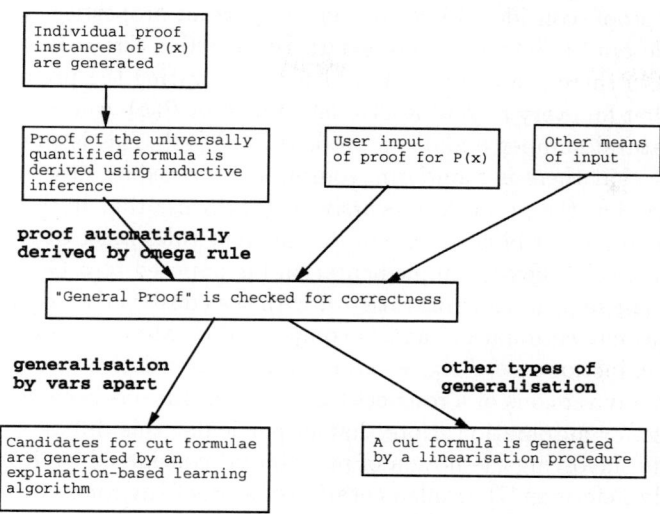

Fig. 1. Generalisation Strategy

tion 4 discusses how conventional inductive proofs are related to ω-proofs, and in particular how the explanation-based generalisation method linearises ω-proofs, and hence provides a cut formula which is provable by induction. Finally, conclusions are given.

2   The Constructive Omega Rule

In order to describe the generalisation method proposed, it is first necessary to provide a description of the 'stronger' system mentioned above. A suitable rule other than induction which might be added to Peano's axioms to form a system formalising arithmetic is the ω-rule:

$$\frac{A(0), A(\underline{1}) \ldots A(\underline{n}) \ldots}{\forall x A(x)}$$

where $\underline{n}$ is a formal numeral, which for natural number n consists in the n-fold iteration of the successor function applied to zero, and A is formulated within the language of arithmetic. This rule is not derivable in Peano Arithmetic (PA)[4], since for example, for the Gödel formula $G(x)$, for each natural number n, $PA \vdash G(\underline{n})$ but it is not true that $PA \vdash \forall x G(x)$. This rule together with Peano's axioms gives a complete theory — the usual incompleteness results do not apply since this is not a formal system in the usual sense.

---

[4] See for example [18] for a formalisation.

However, this is not a good candidate for implementation since there are an infinite number of premises. It would be desirable to restrict the ω-rule so that the infinite proofs considered possess some important properties of finite proofs. One suitable option is to use a **constructive ω-rule**. The ω-rule is said to be constructive if there is a recursive function f (generating the premises of the ω-rule) such that for every n, $f(n)$ is a Gödel number of $P(n)$, where $P(n)$ is defined for every natural number n and is a proof of $A(\underline{n})$ [18]. This is equivalent to the requirement that there is a uniform, computable procedure describing $P(n)$, or alternatively that the proofs are recursive (in the sense that both the proof-tree and the function describing the use of the different rules must be recursive) [20], which is the basis taken for implementation (as opposed to a Gödel numbering approach). The sequent calculus enriched with the constructive ω-rule (let us call it $PA_{c\omega}$) has cut elimination, and is complete [17]. Moreover, since the ω-rule implies the induction rule, $PA_{c\omega}+induction$ is a conservative extension of $PA_{c\omega}$. There are many versions of a restricted ω-rule; this one has been chosen because it is suitable for automation. Note that in particular this differs from the form of the ω-rule (involving the notion of provability) considered by Rosser [15] and subsequently Feferman [7]. Implementation of a proof environment with the constructive ω-rule is described in [2]; this provides a basis for the implementation of the generalisation method described in this paper. In the context of theorem proving, the presence of cut elimination for these systems means that generalisation steps are not required. In the implementation, although completeness is not claimed, some proofs that normally require generalisation can be generated more easily in $PA_{c\omega}$ than PA.

The constructive ω-rule may be used to enable automated proof of formulae, such as $\forall x \ (x+x) + x = x + (x+x)$, which cannot be proved in the normal axiomatisation of arithmetic without recourse to the cut rule. In these cases the correct proof could be extremely difficult to find automatically. However, it is possible to prove this equation using the ω-rule since the proofs of the instances $(0+0)+0 = 0+(0+0), (1+1)+1 = 1+(1+1), \ldots$ are easily found, and the general pattern determined by inductive inference. One motivation behind use of the ω-rule is that such an approach may be seen as an attempt to provide induction systems with the sub-formula property: systems which include the ordinary induction rule lose the sub-formula property (namely, that each formula in the derivation does not necessarily have to be composed of sub-formulae in the preceding derivation) due to the introduction of arbitrary "cut formula". Another motivation is that the resulting proofs may seem in some sense to be more intuitive than standard inductive proofs of the same theorems, in the sense of corresponding more closely to the way in which people convince themselves of the correctness of the proof. So, automated proof in such a system might be seen as a goal in itself, but the concern of this paper is rather how it is possible to use this system as a guide to the provision of difficult proofs in more conventional systems.

One way in which the constructive ω-rule may be put into effect is to require that there is an enumeration of the derivations which prove the premises — for example one could code proofs by numbers, by means of a primitive recursive function which generates them. But I have not used such a traditional representation; it was sufficient for my purposes to provide (for the nth case) a description for the ω-proof of A, namely $P(n)$, in a constructive way (in this case a recursive way), which captures the notion that each $P(n)$ is being proved in a uniform way (from parameter n). This is done by manipulating $A(\underline{n})$, where $\forall x A(x)$ is the sequent to be proved, and using recursively defined function definitions of PA as rewrite rules, with the aim of reducing both sides of the equation to the same formula. The recursive function sought is described by the sequence of rule applications, parametrised over n. In practice, the first few proofs will be special cases, and it is rather the correspondence between the proofs of $P(99)$, say, and $P(100)$, which should be captured.

$\underline{n} \equiv s^n(0)$
$USE\ (2)\ n\ TIMES\ ON\ LEFT\ ([1,1])$
$USE\ (1)\ ON\ LEFT\ ([2,2,1,1])$
$USE\ (2)\ n\ TIMES\ ON\ RIGHT\ ([2])$
$USE\ (1)\ ON\ RIGHT\ ([2,2,2])$
$USE\ (2)\ n\ TIMES\ ON\ LEFT\ ([1])$

$(\underline{n} + \underline{n}) + \underline{n} = \underline{n} + (\underline{n} + \underline{n})$
$(s^n(0) + s^n(0)) + s^n(0) = s^n(0) + (s^n(0) + s^n(0))$
$s^n(0 + s^n(0)) + s^n(0) = s^n(0) + (s^n(0) + s^n(0))$
$s^n(s^n(0)) + s^n(0) = s^n(0) + (s^n(0) + s^n(0))$
$s^n(s^n(0)) + s^n(0) = s^n(0 + (s^n(0) + s^n(0)))$
$s^n(s^n(0)) + s^n(0) = s^n(s^n(0) + s^n(0))$
$s^n(s^n(0) + s^n(0)) = s^n(s^n(0) + s^n(0))$

$EQUALITY$

Fig. 2. An ω-Proof of $\forall x\ (x+x)+x = x+(x+x)$

The ω-proof representation represents $P(n)$, the proof of the nth numerator of the constructive ω-rule, in terms of rewrite rules applied $f(n)$ or a constant number of times to formulae (dependent upon the parameter n). (For more technical details regarding the representation of ω-proofs, see [2].) As an example, the implementational representation of the ω-proof for $\forall x\ (x+x)+x = x+(x+x)$ takes the form given in Figure 2 (although it may be represented in a variety of ways) presuming that, within the particular formalisation of arithmetic chosen, one is given the axioms $0 + y = y$ (1) and $s(x) + y = s(x+y)$ (2).

By $s^n(0)$ is meant the numeral $\underline{n}$, ie. the term formed by applying the successor function n times to 0. The next stages use the axioms as rewrite rules from left to right, and substitution in the ω-proof, under the appropriate instantiation of variables, with the aim of reducing both sides of the equation to the same formula. The subpositions to which the rewrite rules are applied are given in parentheses, where positions are lists of integers representing tree co-ordinates in the syntax-tree of the expression, but in reverse order. The ω-proof represents, and highlights, blocks of rewrite rules which are being applied. Induction may be used (on the first argument) to prove the more general rewrite rules from one block to the next: for example, $\forall n\ s^n(x) + y = s^n(x+y)$ corresponds to n applications of axiom (2) above.

The processes of generation of a (recursive) ω-proof from individual proof instances, and the (metalevel) checking that this is indeed the correct proof have been automated (see [2]). Any appropriate inductive inference algorithm[5], such as Plotkin's least general generalisation [14], or that of Rouveirol [16], could be used to guess the ω-proof from the individual proof instances.

Note that such inductive inference algorithms for generating generalisations produce a proof for an arbitrary instance: it is suggested below how explanation-based generalisation can be applied using this information to find the proper induction formulae for inductive theorem provers. The problems involved in the inductive inference process differ from those involved in (goal) generalisation because inductive inference generalises a proof of $A(k)$ where k is some number to a proof of $A(n)$, whereas generalisation involves finding a formula C such that $C \vdash A$ and in addition such that C may be proven by induction. Inductive inference may be carried out by a relatively simple algorithm (such as updating a guess). However, there is no such known algorithm for goal generalisation.

In the implementation, the theorem under consideration is proved for a few cases, and then learning induction is used to guess the ω-proof from these cases; next it is verified that the guessed ω-proof is correct. (Verifying that the guessed ω-proof works for all n involves mathematical induction at the meta-level; however, the induction required for this differs from that required to prove the original theorem directly, and is usually simpler, so there is no circularity.) As an alternative, the user may bypass this whole stage by specifying the ω-proof directly. Further details of the algorithms and representations used, together with the correspondence between the adopted implementational approach and the formal theory of the system are described in [2].

The next section discusses how ω-proofs may be generalised.

3 Explanation-Based Generalisation Applied to ω-Proofs

There is a class of proofs which are provable in PA only using the cut rule but which are provable in $PA_{c\omega}$ [2, 11]. I shall consider whether the proof in $PA_{c\omega}$ suggests a proof in PA, and in particular, what the cut formula would be in a proof in Peano Arithmetic.

To illustrate the general principle, consider the simple example for $\Psi \equiv \forall x \; (x + x) + x = x + (x + x)$. The problem is to find Φ such that Φ should be a more general version of the goal Ψ, in order to prove $\Phi \vdash \Psi$, but on the other hand it should be suitable to give an inductive proof of Φ.

$$\frac{\Phi \vdash \forall x \; (x + x) + x = x + (x + x) \qquad \vdash \Phi}{\vdash \forall x \; (x + x) + x = x + (x + x)} CUT$$

Ordinary induction does not work on Ψ, primarily because the second, third and sixth terms in the step case may not be broken down by the rewrite rules

---

[5] Explanation-based learning permits learning from a single example, whereas inductive learning usually requires many examples to learn a concept.

corresponding to (1) and (2) above, and so fertilisation (substitution using the induction hypothesis) cannot occur. Hence it is necessary to use the cut rule.

In order to suggest a cut formula from an ω-proof, one method already proposed is to see what remains unaltered in the nth case proof, and then write out the original formula, but with the corresponding term re-named [3]. So, for the example in Figure 2, the variable corresponding to λ would be rewritten as y. In this case, this would give

$$\Phi \equiv \forall x \forall y \; (x+y) + y = x + (y+y).$$

Φ could then be proved by induction on x. Note that what is meant by 'unaltered' is defined by what is unaffected in structure by the rewrite rules. This procedure has been automated (all that is required is detection of the unaltered terms), and so the cut formula may be produced automatically. This method of generalisation will allow the proof of some theorems which pose a problem for other methods, such as $x \neq 0 \rightarrow p(x) + s(s(x)) = s(x) + x$, where p is the predecessor function (detailed comparisons of this 'unaltered term' method with other generalisation methods with regard to this example are given in [3]).

3.1 Generalisation using Explanation-Based Generalisation

A new development of this solution, which produces a more general generalisation, is to look at the rules of the ω-proof, and work out what the most general statement could be which was proved using these rules. This process has been applied in various other domains, and is the approach of explanation-based generalisation (denoted 'EBG' as an abbreviation). EBG is a technique for formulating general concepts on the basis of specific training examples, first described in [13]. The process works by generalising a particular solution to the most general possible solution which uses the rules of the original solution. It does this by applying these rules, making no assumptions about the form of the generalised solution, and using unification to fill in this form.

rules of ω-proof	generalised ω proof	instantiations
(2) n times at $[1,1]$	$fn0([s^n(X) + Y\|K]) = W$	original
(1) once at $[2,2,1,1]$	$fn0([s^n(X+Y)\|K]) = W$	
(2) n times at $[2]$	$fn0([s^n(Y)\|K]) = W$	$X = 0$
(1) once at $[2,2,2]$	$fn0([s^n(Y)\|K]) = s^n(P+Q)$	$W = s^n(P) + Q$
(2) n times at $[1]$	$s^n(Y+K) = s^n(Q)$	$P = 0, fn0 = +$
EQUALITY	$s^n(Y+K) = s^n(Y+K)$	$Q = Y + K$

Fig. 3. Illustration of Explanation-Based Generalisation on Rules of ω-Proof

The EBG method is applied in this instance to a new domain, namely that of ω-proofs. As an illustration of the method, let us apply explanation-based generalisation to Figure 2, to give the process shown in Figure 3. The right hand column is the instantiations of variables, which are finally to be filtered back up into the original expression using constraint back-propagation. Essentially what is happening is that each application of the rewrite rules of the original ω-proof is matched with the latest line of the new ω-proof to see the necessary form of a generalised ω-proof. If (2) is applied m times, this will match with the form $s^m(X) + Y \Rightarrow s^m(X + Y)$. Nothing more is supposed about the original form of the ω-proof than that it is of the form $U = W$. The rule application blocks on the left hand side of this figure are identical with those of the ω-proof given in Figure 2. The procedure is to form the most general ω-proof which could use those same rules to achieve equality. Hence, these same rewrite rules are applied at the specified subpositions to give a new ω-proof. In so doing the structure of U and W is revealed. For instance, the fact that rule (2) may be applied n times at subposition [1,1] of $U = W$ reveals that U must be of the form $fn0([s^n(X) + Y|K])$ (which represents some functor $fn0$ of as yet unknown arity with initial argument $s^n(X) + Y$ and additional arguments K) before the rule application, and of the form $fn0([s^n(X + Y)|K])$ afterwards. This process is repeated until all the given rules are exhausted. Finally, the left-hand side and the right-hand side of the ω-proof are unified (since the original proof resulted in equality). Throughout this process, information will have been obtained regarding the structure of some of the postulated variables in this new ω-proof, such as that presented in the final column of Figure 3. Feeding such variable instantiation information back to the original expression $U = W$ shows that it must be of the form $(\underline{n} + Y) + K = \underline{n} + (Y + K)$. This gives the most general generalisation as being $\forall x\ \forall y\ \forall z\ (x + y) + z = x + (y + z)$. By means of this method, the generalisations are found by recognising patterns, and a uniform approach for generalisation is provided.

Although the heuristic of replacing unaltered terms is suitable for implementation (and was successfully implemented), the method of explanation-based generalisation extends this idea to provide a uniform algorithm based on the underlying structure of the proof. The implementation of EBG, in which the previous "ω-rule" environment was extended to include generalisation tactics, follows the unification process described above, and thus subsumes the implementation of the heuristic method. [2] provides full details of the implementational environment, which will automatically suggest cut formulae for examples such as those of Table 1. Most of these examples involve generalisation of variables apart, or else generalisation of common subexpressions. In these cases both the heuristic of replacing unaltered variables and the explanation-based generalisation method work fairly straightforwardly. The method may also be applied to complicated examples containing nested quantifiers, etc., for the ω-rule applies to arbitrary sequents. $\forall xy\ (x + y) + x = x + (y + x)$ provides an instance of nested use of the ω-rule, which carries through directly. In the case of $\forall x\ even(x + x)$, the cut formula of $even(2.x)$ could possibly be extracted by a user from the form of the ω-proof (see [3]), which is an improvement over other generalisation methods. However, in some cases where an ω-proof may be provided, it is not clear what the cut formula might be.

3.2 Comparison with Related Work

These methods involving manipulation of the ω-proof should be compared with current generalisation methods. Of these, perhaps the most famous is that implemented by Boyer and Moore in their theorem-prover NQTHM [4]. The main heuristic for (goal) generalisation is that identical terms occurring on both the left and right side of the equation are picked for rewriting as a new variable (with certain restrictions). This may be a quick method if it happens to work, but may also entail the proofs of many lemmas, which might need to be stored in advance in anticipation of such an event in order to be more efficient. The problems inherent in Boyer and Moore's approach have led Raymond Aubin to extend their work in this field [1]. Aubin's method is to "guess" a generalisation by generalising occurrences in the definitional argument position, and then to work through a number of individual cases to see if the guess seems to work; if it does work, he will look for a proof; if it does not, then he will "guess" a different generalisation. However, Aubin's solution does not work in all cases. In particular, if a constructor such as a successor function appears in an original goal, together with individual variables, Aubin's method may result in overgeneralisation or indeed no solution at all (and the same applies to Castaing's very similar approach [5]). These methods are used as the basis for generalisation in many different theorem-proving systems. However, Hesketh's approach of directing generalisation by the failure of heuristics involved in proof search has proved more successful [10], although very often several alternative solutions have to be tested rather than a single solution being given in a uniform manner. Related work in proof generalisation has been carried out by Masami Hagiya, who has considered generalization of proofs in type theories. Hagiya approaches the problem of proof generalisation by synthesising a general proof from a concrete example proof by higher-order unification in a type theory with a recursion operator. Rather than the ω-rule, he uses ordinary recursion terms for representing inductive proofs: in order to make recursion terms more expressive, he has extended the calculus with implicit arithmetical inferences [9]. However, the degree of automation provided is not very high, and he has not addressed the issue of ensuring that the proofs from which one generalises are in the form required to produce a suitable result. Hence, he has not produced a generalisation tactic which automatically suggests cut formulae upon input of theorems.

The generalisation method proposed in this paper provides a uniform approach and does not have to check extra criteria, nor work through individual examples. Moreover, it is not possible to overgeneralise to a non-theorem (the method is sound but not complete — it does not always provide a solution, nor necessarily the best solution possible). Table 1 provides a comparison between cut formulae suggested by the various methods, and demonstrates that the EBG guiding method works well on a range of arithmetical examples.

By the EBG method, the most general generalisation of an expression is achieved, in a uniform manner. This is a new and desirable achievement: Hesketh

Methods	Selection of Types of Example	
	$x.(x+x) = x.x + x.x$	$(x+x) + x = x + (x+x)$
Boyer & Moore	fail	$y + x = x + y$
Aubin	$x.(y+y) = x.y + x.y$	1. $(x+y) + y = x + (y+y)$
Hesketh	$x.(y+y) = x.y + x.y$	2. $(x+y) + y = x + (y+y)$
Baker	$x.(y+z) = x.y + x.z$	$(x+y) + z = x + (y+z)$
	$x + s(x) = s(x) + x$	$even(x+x)$
Boyer & Moore	$x + y = y + x$	fail
Aubin	fail	fail
Hesketh	$x + s(y) = s(x) + y$ 3.	fail
Baker	$x + s(y) = s(x) + y$	$even(2.x)$ 4.
	$x \neq 0 \to p(x) + s(s(y)) = s(x) + y$	$x.(y.0) = y.(y.0)$
Boyer & Moore	fail	fail
Aubin	fail	fail
Hesketh	$x \neq 0 \to p(x) + s(s(y)) = s(x) + y$ 5.	fail
Baker	$x \neq 0 \to p(x) + s(s(y)) = s(x) + y$	ω-proof only 6.

1. Solution achieved by testing various possibilities after initial failure of method.
2. Solution achieved by pruning search space.
3. Probable result (this example was not considered in [10]).
4. Generalisation algorithm fails, but this cut formula is 'suggested' from the ω-proof.
5. Probable result (this example was not considered in [10]).
6. Generalisation algorithm fails, but user might be able to suggest a generalisation from the ω-proof.

Table 1. Cut Formulae Suggested Using Different Generalisation Methods

writes [10, p.271] that for her generalisation method "The generalisation from $\forall x.x + (x+x) = (x+x) + x$ will be found as $\forall x \forall y.x + (y+y) = (x+y) + y$ not $\forall x \forall y \forall z.x + (y+z) = (x+y) + z$ but I know of no theorem prover that can find this last generalisation in a principled way, ie. other than with just trial and error." Note also how if an ω-proof is provided, even without a generalisation being suggested, something has still been achieved, in the sense that a pattern might still emerge for the user. Thus the method may still be useful within a co-operative environment if it breaks down (cf. note 6., Table 1). This contrasts with alternative methods of generalisation, which do not provide much information if they fail. Moreover, because the suggested method explicitly exploits general patterns, it has a higher-level structure and thus greater potential for extension than other more special-purpose approaches.

The EBG method proposed in this section will succeed in the sense that there does exist some ω-proof such that a correct cut formula could be found by EBG (so long as inductive proof by generalisation apart, that is, generalisation by means of renaming some occurrences of the same variable in an expression, is possible). However, it will not necessarily work if generalisation apart is not appropriate for the example under consideration.

Note that it is possible to carry out explanation-based generalisation on proofs in other systems such as Peano arithmetic, in order to give a more general expression. However, if inductive proof is not possible in these systems (because induction is blocked) and hence application of the cut rule is needed in order to obtain a proof, then such a method will not work. However, it is possible that an ω-proof may be returned, and in this case explanation-based generalisation may be applied to the latter in order to obtain a solution.

The explanation-based generalisation method proposed in this paper can be used in a more general context for lemma generation, regardless of the way an ω-proof was obtained, so long as one can represent in the particular system of interest the notions of nth-successor, parametrised applications of rewrite rules and also the correctness check. Hence, although the learning device for cut formulae is not reliant upon the given proof system involving use of the ω-rule, in practice suitable proof environments such as HOL [8] would require extension, and moreover the user would have to input the proof directly unless some other method of generation could be provided. Hence, if the ω-proof were represented directly in higher-order logic, one would still be faced with the problems of application of rewrite rules a parametrised number of times, and a method of provision of the ω-proof (solved by inductive inference in the case of $PA_{c\omega}$).

4 Linearisation

This section briefly explains what linearisation is, and then discusses how conventional inductive proofs are related to ω-proofs in the context of EBG. It shall be shown that EBG linearises ω-proofs (and therefore, the initial formula of the generalised ω-proof may be proven by induction). When the EBG method (relying on generalisation of variables apart) is not appropriate, there are other ways of linearising ω-proofs.

The general form of a linear proof involving a single use of the constructive ω-rule is given in Figure 4,[6] with the proviso that there may be additional branches if the $P(i)$ consist of branching proof rules. Any proof which is of the form of Figure 4 (with additional leaves in the case of inductive proofs with branching in the step proof) shall be called *linear*, with the constraints that $A(\underline{k})$ is reduced to $A(\underline{k-1})$ in a uniform way for each k, and that the proof is cut-free. In the case of the constructive ω-rule operating on $\forall x\ A(x)$, the $A(i)$ are uniformly generated, and there will be a relationship between their proofs — otherwise, this may not be the case. An ω-proof is a parametrised version of an arbitrary subtree of a proof in $PA_{c\omega}$ of a universal statement; thus, a "linear" ω-proof will correspond to the kth subtree of Figure 4.

It shall be shown that ω-proofs may be turned into a linear form, which will suggest that induction can be done on the original formula (that is to say that this formula is a suitable cut formula). Hence it is necessary to recognise Q such that an original ω-proof for $P(x)$ may be transformed into a "linearised" ω-proof of form $Q(s(n))$ reducing in a uniform manner (via rewrite rule block i) to $Q(n)$

---

[6] Use of two inductions should be compared with P. Madden's transformations which linearise proofs defined by functions with two inductive variables [12].

$$\begin{array}{cccc}
& & & A(0) \\
& & & \vdots \text{ kth subtree} \\
& A(0) & A(k-1) & \\
\vdots P(0) & \vdots P(1) & \vdots P(k) & \\
A(\underline{0}) & A(\underline{1}) & \cdots \quad A(\underline{k}) & \cdots \\
\hline
& \forall x \, A(x) & & \text{constructive } \omega \text{ rule}
\end{array}$$

Fig. 4. Form of Linear Proof in $PA_{c\omega}$

to $Q(n-1)$ down to $Q(0)$, which reduces to an equality using a rewrite rule block j. The required proof of $\forall x \, P(x)$ in PA is:[7]

$$\cfrac{\forall x \, Q(x) \vdash \forall x P(x) \qquad \cfrac{\cfrac{\vdots j}{\vdash Q(0)} \quad \cfrac{\vdots i}{Q(r) \vdash Q(s(r))}}{\vdash \forall x \, Q(x)} ind(x)}{\vdash \forall x \, P(x)} cut$$

If the ω-proof is of a linear form, it is the case that $\forall x \, P(x)$ may be proved in this manner. The fact that a proof of $\Gamma, A(\underline{k}) \vdash A(s(\underline{k}))$ exists provided there is a derivation of $\Gamma \vdash A(s(\underline{k}))$ from $\Gamma \vdash A(\underline{k})$ (in the sense of there being a natural deduction proof of the former from the latter, possibly using additional axioms) forms an analogue of the deduction theorem.[8]

Theorem 1 Deduction Theorem for Sequents. *The sequent $\Gamma, A \vdash B$ is provable in $PA_{c\omega}$ if and only if there is a cut-free derivation in $PA_{c\omega}$ of*

$$\Gamma \vdash A$$
$$\vdots$$
$$\Gamma \vdash B$$

Proof. $\Rightarrow$: Suppose $\Gamma, A \vdash B$ is provable in $PA_{c\omega}$. The cut rule may be used to derive $\Gamma \vdash B$ from $\Gamma \vdash A$ and $\Gamma, A \vdash B$. Hence, given $\Gamma, A \vdash B$, then there is a derivation of $\Gamma \vdash B$ from $\Gamma \vdash A$. By the cut elimination theorem, there must be a cut-free derivation (in $PA_{c\omega}$) of $\Gamma \vdash B$ from $\Gamma \vdash A$.

$\Leftarrow$: Suppose that there is a proof in $PA_{c\omega}$ of $\Gamma \vdash B$ from $\Gamma \vdash A$. By using the same rules, there must be a proof of $\Gamma, A \vdash B$ from $\Gamma, A \vdash A$. However, $\Gamma, A \vdash A$ is an axiom. Hence, there is a proof in $PA_{c\omega}$ of $\Gamma, A \vdash B$. $\square$

Theorem 2 Linearisation Theorem. *Q can be proved using a single induction if and only if there is a linear ω-proof of Q*

---

[7] The stages i and j apply rewrite rules on a formula, rather than structural manipulations on a sequent.

[8] If $\Gamma, B \vdash_K A$, then $\Gamma \vdash_K B \to A$, for some logical system K [6, P127].

Proof. $\Rightarrow$: By Theorem 1.

$\Leftarrow$: *Given an ω-proof of the form $Q(s(k))$ reducing via rewrite steps $i(k)$ to $Q(k)$, and with such a repeated structure finally to $Q(0)$ (and equality via rewrite steps j), then $Q(s(\underline{k}))$ reduces to $Q(\underline{k})$ for all numerals such that $k \leq n$, where n was arbitrary — hence k is arbitrary. The rules $i(k)$ here form a repeated rule-block (parametrised over k): this is what has been described above as a linearisable ω-proof. By the soundness of the ω-proof representation with respect to $PA_{c\omega}$ (cf. [2]), there is a proof in $PA_{c\omega}$ of $\vdash Q(\underline{k})$ assuming $\vdash Q(s(\underline{k}))$, and by Theorem 1 there is a proof of $Q(k) \vdash Q(s(k))$, for each numeral k. By the form of the original linear proof, each proof segment $P(k)$ is obtained by taking the same shape of proof with a numerical parameter that is replaced by the appropriate value of k. So, there is a uniform correspondence between how the different instances of the numeral k are treated in each proof of $Q(k) \vdash Q(s(k))$, which suggests that the free variable version $Q(r) \vdash Q(s(r))$ is provable. The final lines of the ω-proof provide a proof of $Q(0)$, and therefore the prooftree in PA using induction may be completed.* □

The argument above suggests that if the EBG method linearises the ω-proof, then it also suggests a suitable cut formula. It remains to be shown that the EBG method does linearise the ω-proof. Recall that the ω-proof of the generalisation is the EBG proof with the final (instantiated) result. By doing EBG, the parts of a proof which should not be altered when using induction are renamed. This puts the ω-proof into a linear form, in that (for inital line $P(n)$) $P(n)$ reduces to $P(n-1)$, whereas it did not before. More specifically, if generalisation apart (involving differentiation between the same variable) is a suitable method of generalisation, the initial theorem $(P(n))$ must be of the form $Q(\mathbf{n})$ where $\mathbf{n}$ is a sequence of n's and each occurrence of n may be generalised differently (although there might be additional free variables in Q which are not being represented in this notation). Q may have an arbitrary number of argument positions, but for clarity let us consider the simple example when there are three, and when $Q(n_1, n_2, n_3)$ reduces in the ω-proof to $Q(n_1 - 1, n_2, n_3)$. The second and third arguments will renamed by the EGB process, thus giving a linear form. A corresponding argument applies given different argument positions of the altered variable, or varied arities of Q. Thus, EBG is a way of linearising the ω-proof, and if a solution is possible by generalisation of variables apart, then this method will find it (so long as a correct ω-proof is given initially). In conclusion, when a cut formula is suggested by the EBG method proposed, it will have been shown via linearisation of the ω-proof that induction may be successfully carried out upon it. Hence, it is not necessary to actually carry out the induction to know that the proof may be completed by using the cut rule and induction.

The EBG proof provides a most general generalisation of the initial formula which can be proved using the rules given in the ω-proof. However, in some cases (eg. $\forall l\ len(rev(l)) = len(l)$), such a generalisation will be the same, and still not provable by induction, since what is needed is the addition of a new structure (ie. an 'accumulator'). However, this additional structure is apparent in the ω-proof (for example being explicit in the rth line of the proof), and by means of linearising the ω-proof by looking for a repeated structure while allowing some generalising, a correct cut formula may again be suggested [2]. The ω-proof is put into a form such that there is a repeated rule block i: thus the rth line after r uses of i is $Q(r)$, such that $Q(r)$ reduces to $Q(r-1)$. In this way, correct cut formulae are suggested for many difficult examples [2]: in particular, one generalisation provided by this method of $\forall al\ len(append(rev(l),a)) = len(append(rev(a),l))$ (from $len(rev(l)) = len(l)$) is a better result (since it only requires one induction) than the only alternative suggestion provided for this example of $\forall al\ len(append(rev(l),a)) = len(append(a,l))$[9] (requiring two inductions).

5 Conclusions

In summary, an ω-proof is constructed using some inductive inference process (the individual proofs being easily generated using basic tactics), or obtained by some other means. Unification is used for the stage of constructing a generalised ω-proof and information about unification is fed through various stages of the construction using back propagation. The result is a proof of a generalisation of the original expression, which is linear in form, and hence suggests an appropriate cut formula. The approach also applies generally to other data-types. Not only is it the case that certain new structural patterns may be seen in the ω-proof which may guide generalisation, but also that the general representation of an arbitrary object of that type (eg. $s^n(0)$ for natural numbers, $[x_1, x_2, \ldots x_m]$ for lists, etc.) enables the structure of that particular data-type to be exploited, in the sense that rewrite rules may be used which would not otherwise be applicable.

The new method for generalisation which has been proposed is robust enough to capture in many cases what the alternative methods can do (in some cases with less work), plus it works on examples on which they fail. Implementation of this method has been carried out within the framework of an interactive theorem-prover with Prolog as the tactic language, in which the object-level logic is replaced by classical and constructive theories of arithmetic [2]. This approach works for theories other than arithmetic and logics other than a sequent version of the predicate calculus, and may rather be regarded as suggesting a general framework. So long as a procedure for constructing a proof for each individual of a sort is specified, universal statements about objects of the sort could be proved. Thus it appears that the approach described in this paper may be an aid to automated deduction, and provides a mechanism for guiding proofs in more conventional systems.

Acknowledgements I would like to acknowledge the help of Alan Smaill, and many others, from the Mathematical Reasoning Group in Edinburgh University.

---

[9] cf. [10].

References

1. Aubin, R.: Some generalization heuristics in proofs by induction. In G. Huet and G. Kahn, editors, *Actes du Colloque Construction: Amélioration et vérification de Programmes*. Institut de recherche d'informatique et d'automatique (1975)
2. Baker, S.: *Aspects of the Constructive Omega Rule within Automated Deduction*. PhD thesis, University of Edinburgh (1992)
3. Baker, S., Ireland, A., Smaill, A.: On the use of the constructive omega rule within automated deduction. In A. Voronkov, editor, *International Conference on Logic Programming and Automated Reasoning – LPAR 92, St. Petersburg*, Lecture Notes in Artificial Intelligence, Springer-Verlag **624** (1992) 214–225
4. Boyer, R.S., Moore, J.S.: Proving theorems about LISP functions. In N. Nilsson, editor, *Proceedings of the third IJCAI* (1973) 486–493
5. Castaing, J.: How to facilitate the proof of theorems by using induction-matching, and by generalisation. In A. Joshi, editor, *Proceedings of the ninth IJCAI* (1985) 1208–1213
6. Dummett, M.: *Elements of Intuitionism*. Oxford Logic Guides. Oxford Univ. Press, Oxford (1977)
7. Feferman, S.: Transfinite recursive progressions of axiomatic theories. Journal of Symbolic Logic **27** (1962) 259–316
8. Gordon, M: HOL: A proof generating system for higher-order logic. In G. Birtwistle and P.A. Subrahmanyam, editors, *VLSI Specification, Verification and Synthesis* Kluwer (1988)
9. Hagiya, M.: A Typed λ-Calculus for Proving-by-Example and Bottom-Up Generalisation Procedure. Algorithmic Learning Theory 93, Lecture Notes in Artificial Intelligence **744** (1993)
10. Hesketh, J.T.: *Using Middle-Out Reasoning to Guide Inductive Theorem Proving*. PhD thesis, University of Edinburgh (1991)
11. Kreisel, G.: On the Interpretation of Non-Finitist Proofs. Journal of Symbolic Logic **17** (1952) 43–58
12. Madden, P.: The specialization and transformation of constructive existence proofs. In Sridharan, editor, *Proceedings of the Eleventh International Joint Conference on Artificial Intelligence*, Morgan Kaufmann (1989)
13. Mitchell, T.M.: Toward combining empirical and analytical methods for inferring heuristics. Technical Report LCSR-TR-27, Laboratory for Computer Science Research, Rutgers University (1982)
14. Plotkin, G.: A note on inductive generalization. In D. Michie and B. Meltzer, editors, *Machine Intelligence* **5** Edinburgh University Press (1969) 153–164
15. Rosser, B.: Gödel-theorems for non-constructive logics. JSL **2** (1937) 129–137
16. Rouveirol, C.: Saturation: Postponing choices when inverting resolution. In *Proceedings of ECAI-90* (1990) 557–562
17. Shoenfield, J.R.: On a restricted ω-rule. Bull. Acad. Sc. Polon. Sci., Ser. des sc. math., astr. et phys. **7** (1959) 405–7
18. Takeuti, G.: *Proof theory*. North-Holland, 2 edition (1987)
19. Van der Waerden, B.L.: How the proof of Baudet's conjecture was found. In L. Mirsky, editor, *Papers presented to Richard Rado on the occasion of his sixty-fifth birthday*, Academic Press, London-New York (1971) 252–260
20. Yoccoz, S.:. Constructive aspects of the omega-rule: Application to proof systems in computer science and algorithmic logic. Lecture Notes in Computer Science **379** (1989) 553–565

Semantically Guided First-Order Theorem Proving using Hyper-Linking*

Heng Chu[1] and David A. Plaisted[2]

[1] Department of Computer Science
University of North Carolina
Chapel Hill, NC 27599-3175, USA
e-mail: chu@cs.unc.edu

[2] Department of Computer Science
University of North Carolina at Chapel Hill
Chapel Hill, NC 27599-3175, USA
e-mail: plaisted@cs.unc.edu
and
MPI fuer Informatik
Im Stadtwald
D-66123 Saarbruecken
and
Universitaet Kaiserslautern
Fachbereich Informatik
675 Kaiserslautern

Abstract. We present a new procedure, *semantic hyper-linking*, which uses semantics to guide an instance-based clause-form theorem prover. Semantics for the input clauses is given as input. During the search for the proof, ground instances of the input clauses are generated and new semantic structures are built based on the input semantics and a model of the ground clause set. A proof is found if the ground clause set is unsatisfiable. We give some results in proving hard theorems using semantic hyper-linking; no other special human guidance was given to prove those hard problems. We also show that our method is powerful even with a trivial semantics (that is, even with no guidance in the form of semantic information).

1 Introduction

Instance-based theorem proving is a direct application of Herbrand's theorem [13]. Gilmore [12] implemented a prover to generate instances of the input clauses by enumerating the Herbrand base. It suffers from inefficiencies due to the large number of instances. Lee and Plaisted [15] developed an instance-based theorem proving technique, *hyper-linking*, that uses unification to generate instances of the input clauses, and a propositional calculus prover to test the satisfiability

---

* This research was partially supported by the National Science Foundation under grant CCR-9108904

of the grounded instances. Lee has implemented a theorem prover `CLIN` in Prolog [14]. Many favorable results show that hyper-linking is a promising strategy and, unlike resolution, does not combine literals from different clauses.

Most current theorem provers apply inference rules based on syntactic features of the input data. In [29], it is argued that proofs should rely on the meaning (*semantics*) of the relevant concepts, not on the syntactic form used to represent the concepts. For human mathematicians, the use of examples or semantics is as important as the deductive abilities.

In this paper, we describe *semantic hyper-linking* to use semantics with hyper-linking to prove hard theorems. Semantics about the theorem and the axioms is given as input; the theorem prover applies inference rules based on the semantic and syntactic features of the input data. With no other human guidance, our prover can obtain proofs for hard theorems that other powerful provers (for example, Lee's prover `CLIN` or `Otter` [17]) cannot. The question we are trying to answer in this paper is how much of the power of this prover is due to a good choice of semantics, and how much is due to other features of the method.

We begin with a review of background knowledge. Then we give a motivated example and intuitive overview, followed by detailed discussions. An example is presented to illustrate the method. Some results are also given to show how our method performs compared with other powerful theorem provers, and the importance of a good choice of semantics. Finally we discuss some related work and conclude the paper.

2 Background

In this paper, general knowledge of first-order logic and refutational clause-form theorem proving is assumed. In the following, we will review some important definitions.

The semantic information about a first-order formula G is specified by a *domain* which is a nonempty set of objects, and an *interpretation* which assigns values (or meaning) to the constants, functions and predicates in G. A *structure* is a pair $I=(D,T)$, where D is the domain and T gives the interpretation. In this paper, we represent a structure I by a (probably infinite) set of *all* ground literals that are true in I. A structure I is a *model* of a formula G if G is evaluated to true in I. In particular, we represent a model of a ground clause set S as a set of ground literals which, when assigned true, make S true. A *natural semantics* is a structure that is a model of the general axioms contained in the theorem. A *trivial semantics* is a structure in which the true literals are exactly the positive (or negative) ones.

We now briefly describe how hyper-linking works: A literal L of a clause is *hyper-linked* with another literal M from some other clause if $L\theta = \neg M\theta$ with a most general unifier θ. The hyper-linking operation generates an instance $C\Theta$ of a clause C by hyper-linking *every* literal L_i of C with some literal M_i in other clauses such that $L_i\Theta = \neg M_i\Theta$. Instances like $C\Theta$ are generated and then explicitly grounded (by replacing all variables by a unique constant); a Davis-Putnam like procedure then checks the grounded set for satisfiability. A proof is found if the grounded set is unsatisfiable.

3 Motivation and Definitions

In this section, we give the motivation of our method by an example. With the discussion, we also give several important definitions pertinent to our method.

Let S be the set of clauses $\{P(a), \neg P(x) \vee P(f(x)), \neg P(f(f(a)))\}$. The Herbrand base is then $\{P(a), P(f(a)), P(f(f(a))), P(f(f(f(a)))), ...\}$. A structure I can then be viewed as a mapping from the Herbrand base to $\{true, false\}$. For each literal L, either $I \models L$ or $I \not\models L$. In the former case we say that I *satisfies* L (L is true in I) and in the latter case, I *does not satisfy* L (L is false in I). We can represent the set of all such (Herbrand) structures by an infinite semantic tree, as shown in Fig. 1.

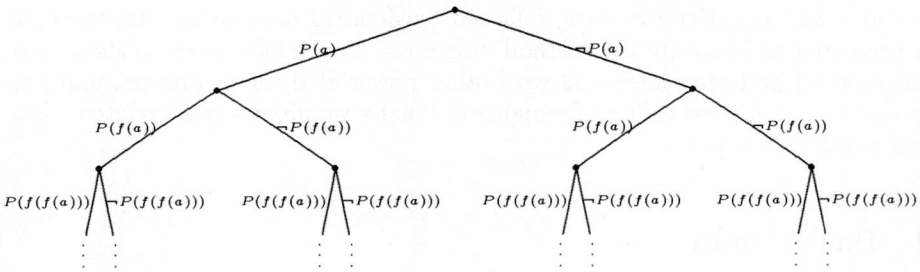

Fig. 1. A semantic tree for the clause set $\{P(a), \neg P(x) \vee P(f(x)), \neg P(f(f(a)))\}$.

In this binary tree, infinite paths correspond to structures. For example, the leftmost path corresponds to the structure $\{P(a), P(f(a)), P(f(f(a))), ...\}$.

Semantic hyper-linking can be viewed as a systematic search for a model in this tree. If this search fails, then we have shown that S is unsatisfiable and a proof has been found. Supposing that we always take the leftmost choice first, and backtrack when a contradiction is found, we obtain the tree in Fig. 2.

The circled nodes are *contradiction nodes* and for each such node we have an associated ground instance of an input clause that is not satisfied by the path to the node. The set of ground instances obtained in this way is then $\{\neg P(f(f(a))), \neg P(f(a)) \vee P(f(f(a))), \neg P(a) \vee P(f(a)), P(a)\}$. This set of clauses is unsatisfiable, and this can be shown by a Davis-and-Putnam like procedure [11].

We can view this search process in the following way: We generate a sequence I_1, I_2, I_3, I_4 of four structures:

$$I_1 = \{P(a), P(f(a)), P(f(f(a))), P(f(f(f(a)))), \ldots\}$$
$$I_2 = \{P(a), P(f(a)), \neg P(f(f(a))), P(f(f(f(a)))), \ldots\}$$

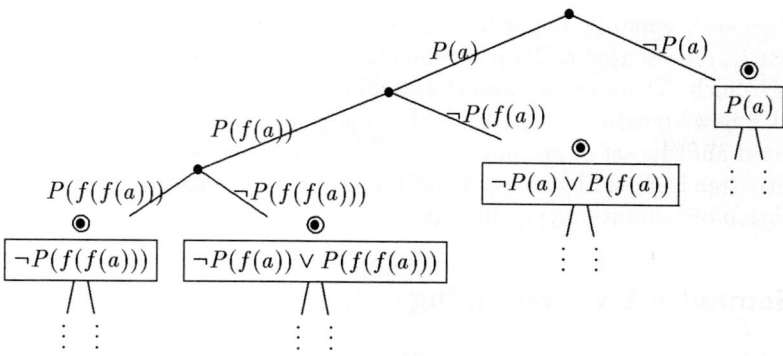

Fig. 2. The contradiction nodes and associated ground instances in Fig. 1 after model finding.

$$I_3 = \{\, P(a), \neg P(f(a)), P(f(f(a))), P(f(f(f(a)))), \ldots \,\}$$
$$I_4 = \{\, \neg P(a), P(f(a)), P(f(f(a))), P(f(f(f(a)))), \ldots \,\}$$

We note that $I_1 \not\models \neg P(f(f(a)))$, $I_2 \not\models \neg P(f(a)) \vee P(f(f(a)))$, $I_3 \not\models \neg P(a) \vee P(f(a))$, $I_4 \not\models P(a)$, each of which explains why the search backtracks on individual contradiction nodes.

We represent these (infinite) structures I_i by a finite *explicit part* and an infinite *implicit part*, as follows: Let I be the initial structure I_1. Then I_2 is $I[\{\,\neg P(f(f(a)))\,\}]$, I_3 is $I[\{\,\neg P(f(a))\,\}]$, and I_4 is $I[\{\,\neg P(a)\,\}]$. We now give a formal definition for $I[\mathcal{L}]$.

Definition 1. Let $\mathcal{L}$ be a set of ground literals, and $\overline{\mathcal{L}}$ be the set of complements of literals in $\mathcal{L}$. $\mathcal{L}^* = \mathcal{L} \cup \overline{\mathcal{L}}$. □

Definition 2. For a structure I and a set $\mathcal{L}$ of ground literals, a new structure $I[\mathcal{L}]$ is I except that the literals L in $\mathcal{L}$ are satisfied by $I[\mathcal{L}]$. Formally, $I[\mathcal{L}] \models L$ is defined as follows:

$$I[\mathcal{L}] \models L \equiv (L \in \mathcal{L} \text{ or } (L \notin \mathcal{L}^* \text{ and } I \models L))$$

□

The explicit part of $I[\mathcal{L}]$ is $\mathcal{L}$, which is represented declaratively by a set of literals. The implicit part is I which is represented procedurally, by procedures for computing truth values in I and, if possible, testing satisfiability in I. Thus each structure I_i (except for the initial one) generated during the search for models, needs not be explicitly constructed. And throughout the whole process, only the specification of the initial structure, I, is needed as the implicit part of any $I[\mathcal{L}]$. We call the literals in each $\mathcal{L}$ the *eligible literals*. In the following discussion, it will become clear that eligible literals are those eligible for generating one or more ground instances C such that $I[\mathcal{L}] \not\models C$.

In general, semantic hyper-linking generates a sequence I_j of structures, and for each I_j, represented as $I[\mathcal{L}]$, we find one (or more) ground clauses C such that $I_j \not\models C$. Such C causes the model searching in the semantic tree to backtrack to find a new structure I_{j+1}. A proof is found after no more structures can be generated and the set of ground clauses is unsatisfiable (namely, as in Fig. 2, the semantic tree is closed by contradiction nodes in all branches). This is the major motivation of semantic hyper-linking.

4 Semantic Hyper-Linking

With a structure I_i represented as $I[\mathcal{L}]$, the key idea of semantic hyper-linking is to find ground instances that are not satisfied by $I[\mathcal{L}]$. The fundamental problems are: Given I and finite set $\mathcal{L}$ of literals and clause C, does $I[\mathcal{L}] \models C$, and if not, find a ground instance $C\Theta$ of C such that $I[\mathcal{L}] \not\models C\Theta$. The following is an important observation.

Remark. If $I[\mathcal{L}] \not\models D$ and D is a ground clause then for all L in D, $I[\mathcal{L}] \not\models L$. We can express D as $D_1 \cup D_2$ where the D_i are disjoint and $D_1 \subseteq \mathcal{L}^*$ and $D_2 \cap \mathcal{L}^* = \phi$. Then $D_1 \subseteq \overline{\mathcal{L}}$ and $I \not\models D_2$. □

If C is a non-ground clause, then we are looking for such ground instances D of C. So the idea is to express C as $C_1 \cup C_2$ where $C_1 \cap C_2 = \phi$, and then find a substitution α such that $C_1\alpha \subseteq \overline{\mathcal{L}}$ and find a substitution β such that $C_2\alpha\beta$ is ground and $I \not\models C_2\alpha\beta$. Also we need to have that $C_2\alpha\beta \cap \mathcal{L}^* = \phi$. Now, α can be found just by a subsumption test, since we are essentially testing if C_1 subsumes $\overline{\mathcal{L}}$; such tests are well-known in the theorem proving community. We look for α by hyper-linking C_1 literals with those in $\mathcal{L}$. Then we search for β by a process of exhaustive enumeration.

If I is *decidable*, given a clause C possibly containing variables, it is decidable whether $I \models C$. With such a decidable I, we can then test if $I \models C_2\alpha$. If so, we know that no β exists because every instance of $C_2\alpha$ is satisfied by I. Otherwise, we proceed to do exhaustive enumeration to find β. Since there might be many possible α and C_2 for a clause C and $I[\mathcal{L}]$, such a test greatly reduces the number of $C_2\alpha$ and makes the exhaustive enumeration much faster. Some examples of using a decidable I are discussed in [10].

Definition 3. The process of generating $C_1\alpha$ is called *partial hyper-linking*, the instance $C\alpha$ is a *partial hyper-link*. The *ground instance generation* process on partial hyper-links generates $C_2\alpha\beta$. After partial hyper-linking and ground instance generation, a ground instance $C\alpha\beta$ is obtained and $I[\mathcal{L}] \not\models C\alpha\beta$. □

Example 1. Here is an example of the method of finding α and β. Let I be $\{P(a), P(f(a)), P(f(f(a))), P(f(f(f(a)))), ...\}$ as before. Let $\mathcal{L}$ be $\{\neg P(f(f(a)))\}$. Let the clause C be $\{\neg P(x), P(f(x))\}$. Suppose we choose C_1 as $\{P(f(x))\}$ and C_2 as $\{\neg P(x)\}$. Then we seek α such that $C_1\alpha \subseteq \overline{\mathcal{L}}$. For this we take α to be $\{x \leftarrow f(a)\}$. In this case $C_2\alpha$ is ground and $I \not\models C_2\alpha$. So we choose β as the identity substitution, and obtain $C\alpha\beta$ as $\{\neg P(f(a)), P(f(f(a)))\}$. The reader can check that $I[\mathcal{L}] \not\models C\alpha\beta$. We note that in this case, it was not necessary to use exhaustive enumeration since $C_2\alpha$ is already a ground clause. This is an interesting special case that is much faster than the general case, and we make more use of it (to be discussed in Sec. 5). □

For a structure I_j, after one (or more) ground clauses C are generated such that $I_j \not\models C$, a new structure I_{j+1} is constructed by building a model for all the ground clauses already generated. Since the ground clause set is finite, a finite model M will be built up. This is equivalent to finding a finite path in the semantic tree (as shown in Sec. 3) from the root to a *model node*. More important, none of the structures $I_1, \ldots, I_j$ satisfies the ground clause set. In other words, those paths closed by contradiction nodes need not be considered again. This allows the model searching in the semantic tree to be carried out in an *incremental* manner. Thus we call it *incremental model finding* [7, 8]. From M, a finite subset $\mathcal{L}$ of ground literals can be constructed such that $I_{j+1} = I[\mathcal{L}]$. Note that $\mathcal{L} \subseteq M \subseteq I_{j+1}$ and I_{j+1} satisfies all ground clauses so far. Each loop in which a new structure is found, is called a *round*. The overall semantic hyper-linking algorithm is shown in Fig. 3. Note that in Fig. 3, T is the ground clause set.

5 Finding Better Models

Recall that a process of exhaustive enumeration is needed to obtain β. This is expensive, of course, so we try to avoid it whenever possible. Thus we make this test for contradicting I_j more powerful, to include not only *direct contradictions* (that is, such ground instances C) but also *indirect contradictions* (that is, logical consequences of S that are not satisfied by I_j). This is done to avoid the enumeration involved in generating β. Let I_j be $I[\mathcal{L}_j]$ and suppose the model for the ground clause set is M_j. Note that $\mathcal{L}_j \subseteq M_j$. For each I_j we generate a set S_j of ground instances of clauses in S without using exhaustive enumeration, and we test whether $M_j \cup S_j$ is satisfiable. (Here M_j is considered as a conjunction of literals.) If not, we know that we can backtrack and go on to the next M_j (and thus I_j) without performing the exhaustive enumeration involved in finding β. This advances the I_j in larger steps without using semantics and helps to find proofs faster. This process is especially useful in areas in which defined concepts are common, such as in set theory where set union and set intersection and other operations can be defined in terms of set membership. It turns out that such definitions correspond to clauses in which the enumeration of β is not necessary. This helps a lot in set theory and temporal logic and similar areas.

Example 2. We give a simple example. Let the ground model M be $\{a = b, b = c, c = d, a \neq d\}$. Suppose we have the axiom $\{x \neq y, y \neq z, x = z\}$. Then we note that this axiom implicitly contradicts M; from $a = b$ and $b = c$ we can derive $a = c$. From this and $c = d$ we can derive $a = d$. But this contradicts

```
Algorithm Semantic Hyper-Linking(S,I)
input
  S: input clause set
  I: input semantics
output
  satisfiable: S is satisfiable
  unsatisfiable: S is unsatisfiable
begin
  T = { } % the ground clause set
  I_now = I % current semantics
  loop % each loop corresponds to a round
    if I_now |= S
    then
      S is satisfiable
      return
    else
      D= a ground instance of a clause in S such that I_now |≠ D
      T = T ∪ { D }
      if T is unsatisfiable
      then
        S is unsatisfiable % Found a proof
        return
      else
        % Find a model M for T
        % Find the eligible literal set L from M
        I_now =a new semantics I[L] such that I[L] |= T
  forever
end
```

Fig. 3. Semantic Hyper-Linking Algorithm

$a \neq d$ contained in M. Our prover searches for such implicit contradictions and has several specialized methods for finding them. □

In [7, 8], several specialized methods are discussed in detail to find "better" models by checking for direct (using *model filtering*) or indirect contradictions (using *model literal replacement* and *UR resolution*). They make the model finding backtrack by generating clauses containing negations of the model literals. It is also shown that they are quite powerful and help to find proofs much faster.

6 Soundness and Completeness

Semantic hyper-linking is obviously sound since it generates instances or logic consequences of the input clauses.

On the issue of completeness, a complexity ordering $<$ has to be defined on ground clauses (for example, comparison of largest literals when written as character strings), and for a clause C there are only finitely many clauses D such that $D < C$. Recall the discussion in Sec. 4. When generating instances C such that $I[\mathcal{L}] \not\models C$, if we always choose the ground clauses C such that C is minimal in the complexity ordering, semantic hyper-linking is complete. A sketch of the completeness proof is given in [22], a more detailed proof can be found in [6]; in [7] it is also proved that completeness is retained with those procedures to find better ground models.

It is important to note that the completeness proof is independent of the input structure I. So in practice any structure, even a trivial one (for example, a structure in which the true literals are exactly the positive ones) can be used. This is especially useful since a natural semantics is not always available.

However, the restriction to enumerate smallest literals first imposes a difficulty on enumerating large literals if they are needed to find the proof. To cope with this problem, *rough resolution* [9] can be used to remove large literals so they do not have to be generated at all.

7 An Example

In this section, we will walk through an example of moderate size. It can help to illustrate the major ideas of semantic hyper-linking.

This theorem is to prove that, in a subgroup H of a group G, for any element $a \in H$, its inverse a^{-1} is still in H. Two predicates are used: $s(x)$ means x is in the subgroup H; $p(x, y, z)$ means $xy = z$. Function $i(x)$ denotes x^{-1}; constant e is the identity of G.

The list of axioms is as follows:
Identity: $\forall x \in G$, $ex = x$ and $xe = x$.
$C1 : \{\, p(e, x, x) \,\}$
$C2 : \{\, p(x, e, x) \,\}$
Inverses: $\forall x \in G$, $x^{-1}x = e$ and $xx^{-1} = e$.
$C3 : \{\, p(i(x), x, e) \,\}$
$C4 : \{\, p(x, i(x), e) \,\}$
Associativity: $\forall x, y, z \in G$, $(xy)z = x(yz)$ and $x(yz) = (xy)z$.
$C5 : \{\, \neg p(x, y, v), \neg p(y, z, w), \neg p(x, w, u), p(v, z, u) \,\}$
$C6 : \{\, \neg p(x, y, v), \neg p(y, z, w), \neg p(w, z, u), p(x, w, u) \,\}$
H is a subgroup: $\forall x, y \in H$, $xy^{-1} \in H$
$C7 : \{\, \neg s(x), \neg s(y), \neg p(x, i(y), z), s(z) \,\}$
The theorem is $\forall x(s(x) \supset s(i(x)))$, the negation of the theorem introduces a Skolem constant b:
$C8 : \{\, s(b) \,\}$
$C9 : \{\, \neg s(i(b)) \,\}$
Suppose we choose the input structure I to be the additive group Z_6 of integer modulo 6, and a cyclic group <2> as the subgroup. So $G = \{\, 0, 1, 2, 3, 4, 5 \,\}$, and $H = \{\, 0, 2, 4 \,\}$. Then e^I is 0 and we choose b^I to be 2. Also, $s^I(x)$ means that $x \in \{\, 0, 2, 4 \,\}$ and $i^I(x)$ is $6 - x$. It can be checked that $s(i(b))$ is satisfied by I since $b^I = 2$ and $i^I(2) = 4$ and $s^I(4)$ is true. On the other hand, $p(b, i(b), b)$ is not satisfied by I.

We list in Fig. 4–8 the search of the semantic tree by the model finding in each round. Intermediate clause sets (indicated by •) are omitted. Each leaf is boxed and is either a model node, denoted by a model, or a contradiction node, denoted by a ⊙ and the contradicted ground clause. The complexity ordering is on the maximal literal size, when written as a string, of a clause.

Note that in this example, the first (left) cases are always positive literals. This is just a coincident—the first cases should always be *true* in the input structure I (explained in [7, 8]). In general, true literals could also be negative.

1. At the beginning, $I_1 = I$. Note that $I_1 \not\models C9$ ($\{\neg s(i(b))\}$). After a model $M_2 = \{s(b), \neg s(i(b))\}$ is found for the ground set $\{C8, C9\}$, a new structure $I_2 = I[\{\neg s(i(b))\}]$ is constructed. And $\neg s(i(b))$ is the only eligible literal. Figure 4 shows the search for M_2.

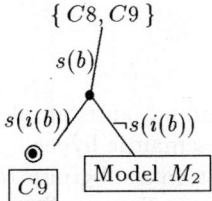

Fig. 4. Model finding in round 1

2. To find ground clauses false in I_2, we split $C7$ into $C7_1 = \{s(z)\}$ and $C7_2 = \{\neg s(x), \neg s(y), \neg p(x, i(y), z)\}$. Recall the remark from Sec. 4 that $C7_1\alpha_1$ is D_1 and $C7_2\alpha_1$ has ground instances D_2. Then we generate partial hyper-link $C7\alpha_1$, for $\alpha_1 = \{z \leftarrow i(b)\}$, by linking the literal $s(z)$ of $C7$ with $\neg s(i(b))$. Then $C7_1\alpha_1 = \{s(i(b))\}$ and $C7_2\alpha_1$ is $\{\neg s(x), \neg s(y), \neg p(x, i(y), i(b))\}$. From this partial hyper-link we generate the (smallest) ground instance of $C7\alpha_1$ by finding $\beta_1 = \{x \leftarrow e, y \leftarrow b\}$. We then obtain $C10 = C7\alpha_1\beta_1$:

$$C10 : \{s(i(b)), \neg s(e), \neg s(b), \neg p(e, i(b), i(b))\}$$

Note that $I \not\models C7_2\alpha_1\beta_1$ and $I_2 \not\models C10$. Now the ground clause set is $\{C8, C9, C10\}$. From this ground clause set, incremental model finding finds a model $M_3 = \{s(b), \neg s(i(b)), s(e), \neg p(e, i(b), i(b))\}$. Thus a new structure $I_3 = I[\{\neg s(i(b)), \neg p(e, i(b), i(b))\}]$ ($\neg s(i(b))$ and $\neg p(e, i(b), i(b))$ are eligible literals) is constructed. The semantic tree is shown in Fig. 5.

3. In the third round, we pick $C1_1 = C1$ and $C1_2 = \{\}$ and hyper-link $C1_1$ with $\neg p(e, i(b), i(b))$ to generate the ground instance $C11 = C1\alpha_2$, where $\alpha_2 = \{x \leftarrow b\}$:

$$C11 : \{p(e, i(b), i(b))\}$$

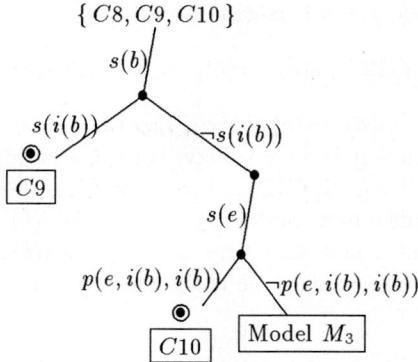

Fig. 5. Model finding in round 2

Note that partial hyper-link $C7\alpha_1$ is still generated in this round[3], but $C11$ is smaller than any of $C7\alpha_1$'s ground instances (except $C10$ which has been generated in round 2).

Now the ground set is $\{C8, C9, C10, C11\}$ and $I_3 \not\models C11$, so the model finding process backtracks and finds a new model $M_4 = \{s(b), \neg s(i(b)), \neg s(e), p(e, i(b), i(b))\}$. From M_4, a new structure $I_4 = I[\{\neg s(i(b)), \neg s(e)\}]$ is constructed. The semantic tree is shown in Fig. 6.

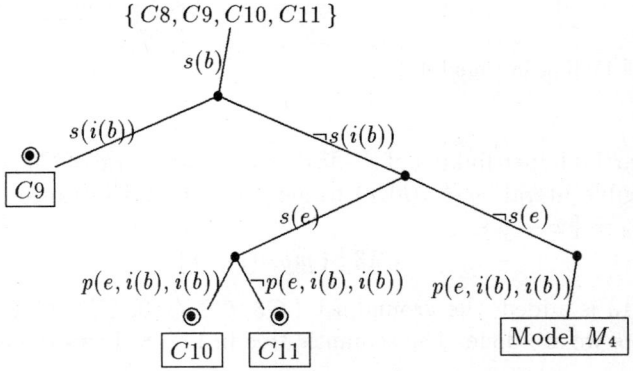

Fig. 6. Model finding in round 3

4. The next partial hyper-linking generates two partial hyper-links from $C7$: $C7\alpha_1$ (as shown before) and $C7\alpha_3$, obtained by linking the literal $s(z)$ of $C7$ with $\neg s(e)$, where $alpha_3$ is $\{z \leftarrow e\}$. Thus $C7_1\alpha_3 = \{s(e)\}$ and $C7_2\alpha_3 = \{\neg s(x), \neg s(y), \neg p(x, i(y), e)\}$. One smallest ground instance $C12 = C7\alpha_3\beta_3$,

---

[3] We are not showing all possible hyper-links; we only show those that are useful for a proof.

where $\beta_3 = \{\,x \leftarrow b, y \leftarrow b\,\}$, is generated from $C7\alpha_3$:

$$C12: \{\,s(e), \neg s(b), \neg s(b), \neg p(b, i(b), e)\,\}$$

The clause $\{\,s(e), \neg s(e), \neg s(e), \neg p(e, i(e), e)\,\}$ is also generated from $C7\alpha_3$ but we delete it since it is a tautology. With $C12$ added, the ground set becomes $\{\,C8, C9, C10, C11, C12\,\}$. Since $I_4 \not\models C12$, the model finding process backtracks and finds a new model $M_5 = \{\,s(b), \neg s(i(b)), \neg s(e), p(e, i(b), i(b)), \neg p(b, i(b), e)\,\}$. And a new structure $I_5 = I[\{\,\neg s(i(b)), \neg s(e), \neg p(b, i(b), e)\,\}]$ is constructed. The semantic tree is shown is Fig. 7.

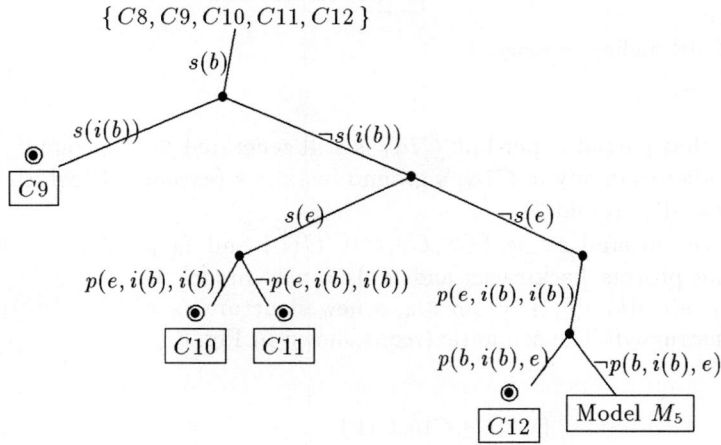

Fig. 7. Model finding in round 4

5. In the partial hyper-linking of the last round, one literal of $C4$ is hyper-linked with eligible literal $\neg p(b, i(b), e)$ to generate ground instance $C13 = C4\alpha_4$, where $\alpha_4 = \{\,x \leftarrow b\,\}$.

$$C13: \{\,p(b, i(b), e)\,\}$$

After $C13$ is added, the ground set $\{\,C8, C9, C10, C11, C12, C13\,\}$ is unsatisfiable we are done. The semantic tree in Fig. 8 shows the contradiction.

8 Implementation and Some Results

We have implemented a prover, called CLIN-S, in Prolog based on our method. There are three major components in CLIN-S: semantic hyper-linking, rough resolution and UR resolution. Rough resolution eliminates big literals, UR resolution eliminates Horn parts of proofs, and what is left is non Horn and with small literals, where semantic hyper linking performs well. Work among different components is well balanced by the time spent on them.

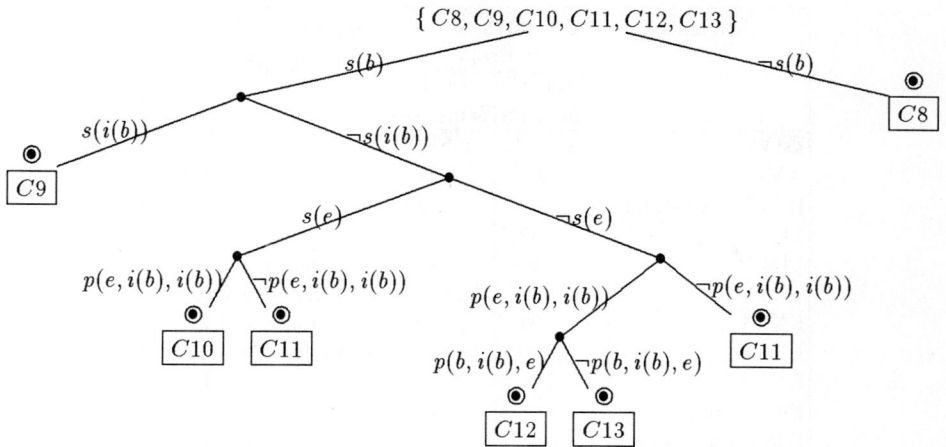

Fig. 8. Proof found in round 5

With no guidance, CLIN-S has proved several moderately difficult theorems including problems in set theory, temporal logic, modal logic, geometry, standard toy problems, and propositional calculus. Semantics sometimes helps to solve those problems. Table 1 contains results on some theorems using CLIN-S compared with Otter and CLIN. All the results, measured in seconds except when noted, are obtained on a DEC 5500. Note that those results mainly show that semantic hyper-linking is powerful and could prove those theorems without special guidance.

IMV [1] (intermediate value theorem) and AM8 [2] (attaining maximum theorem) are two theorems in real analysis. LIM1-3 are the first three LIM+ problems proposed by Bledsoe [3].

EXQ problems, EXQ1-3, are three examples from quantification theory [24] in the order of increasing difficulty. They are difficult for most theorem provers probably due to the non-Horn property. We proved all three EXQ problems on the first try fairly quickly. We think EXQ3 has never been proven before fully automatically.

HALT is the undecidability of the halting problem [4]. We use the version from non-standard clause translation [20]. Since it is difficult to devise reasonable semantics for the halting problem, we used a trivial semantics in which the true literals are exactly those positive ones. Even with such a bad choice for semantics, our prover still can prove it easily. P38 is the 38th problem in [19]. Again, a natural semantics is not available and we tried CLIN-S on both versions from standard and non-standard clause translation.

9 Related Work

Wang [25] used semantics in a hierarchical deduction theorem prover. With a heuristic search strategy called *partial set of support* (PSS) based on syntactic

Problems	CLIN-S (in Prolog)		Otter* (in C)	CLIN** (in Prolog)
	Natural Semantics	Trivial Semantics		
IMV	4,073§	> 6,000	> 3 hours	> 3 hours‡
AM8	2,626¶	1,443	> 3 hours	> 3 hours‡
HALT (nonstand)	—†	129	> 3 hours	10,476
LIM1	52	329	> 3 hours	> 3 hours
LIM2	60	353	> 3 hours	> 3 hours
LIM3	697	> 5,000	> 3 hours	> 3 hours
EXQ1	94	99	> 3 hours	> 3 hours
EXQ2	100	66	> 3 hours	> 3 hours
EXQ3	111	125	> 3 hours	> 3 hours
P38 (stand)	—†	1,119	> 3 hours	> 3 hours
P38 (nonstand)	—†	58	> 3 hours	> 3 hours

* Use binary, UR and hyper-resolution with the negated theorem in the set of support.
** Use default setting.
§ Can be proved in 243 sec with only semantic hyper-linking and model finding strategies.
‡ Can be proved in < 100 sec using predicate replacement.
¶ Can be proved in 650 sec with only semantic hyper-linking, rough resolution and model finding strategies.
† Not attempted.

Table 1. Performance of semantic hyper-linking on some hard theorems.

features, his prover performed pretty well even without semantics [26].

Nie and Plaisted [18] implemented a complete semantically backward chaining theorem prover, which is a semantic refinement of the modified problem reduction format by Plaisted [21]. It is a natural goal-subgoal system and, similar to most previous methods, discards unreachable subgoals.

Model construction has also become an important topic recently. Instead of searching for refutation, the prover constructs models until none can be found. Winker and Wos [27, 28] generated finite models and solved several open problems in ternary Boolean algebra.

Manthey and Bry [16] proposed a model generation method to build up partial models from the non-Horn part and check the models by backward chaining on the Horn part.

Slaney and Lusk [23] use an approach called *dynamic semantic resolution*, which combines model finding and the resolution procedure. They search for a finite model that satisfies the retained clauses during the proof process. They use a procedure to actually find a model dynamically. Their technique only applies to finite models.

Caferra and Zabel [5] proposed a method to simultaneously search for refutation and models in resolution style prover. The model construction is done by using equations as constraints on application of resolution.

10 Conclusion

We have presented semantic hyper-linking to use semantics in an instance-based theorem prover. Semantic hyper-linking can be seen as a combination of Gilmore's procedure and hyper-linking, with the assistance of semantics to prune the search space.

A prover CLIN-S has been implemented and has obtained fast proofs, with no other human guidance, on some problems that are difficult or impossible for other provers. Some fast proofs with trivial semantics also show that semantic hyper-linking is a powerful syntactic proving technique as well. However, we found that a natural semantics is often more helpful—just like diagrams are helpful for mathematicians to prove theorems. For example, suppose we have an axiom $x = x$. Under a semantics in which all true literals are exactly the negative ones, ground instances of $x = x$ will be generated first and obviously most of them are not useful for finding the proof. With a natural semantics, $x = x$ should be satisfied and semantic hyper-linking can focus on ground instances related to the negated theorems. A good natural semantics should satisfy the general axioms in the input and only falsify the clauses directly related to the negated theorem. However, it is still not clear how to measure the quality of different natural semantics for a given problem.

References

1. A. M. Ballantyne and W. W. Bledsoe. On generating and using examples in proof discovery. *Machine Intelligence*, 10:3–39, 1982.
2. W. W. Bledsoe. Using examples to generate instantiations of set variables. In *Proc. of the 8th IJCAI*, pages 892–901, Karlsruhe, FRG, 1983.
3. W. W. Bledsoe. Challenge problems in elementary calculus. *J. Automated Reasoning*, 6:341–359, 1990.
4. M. Bruschi. The halting problem. AAR Newsletter, March 1991.
5. Ricardo Caferra and Nicolas Zabel. A method for simultaneous search for refutations and models by equational constraint solving. *J. Symbolic Computation*, 13:613–641, 1992.
6. Heng Chu. *Semantically Guided First-Order Theorem Proving Using Hyper-Linking*. PhD thesis, University of North Carolina at Chapel Hill, 1994. Expected.
7. Heng Chu and David A. Plaisted. Model finding in semantically guided instane-based theorem proving. *Foundamenta Informaticae Journal*, 1993. To appear.
8. Heng Chu and David A. Plaisted. Model finding strategies in semantically guided instance-based theorem proving. In Jan Komorowski and Zbigniew W. Raś, editors, *Proceedings of the 7th International Symposium on Methodologies for Intelligent Systems (ISMIS-93)*, pages 19–28, 15–18 June 1993.
9. Heng Chu and David A. Plaisted. Rough resolution: A refinement of resolution to remove large literals. In *Proceedings of the Eleventh National Conference on Artificial Intelligence (AAAI-93)*, pages 15–20, 11–15 July 1993.

10. Heng Chu and David A. Plaisted. The use of presburger formulas in semantically guided theorem proving. Presented in *The Third International Symposium on Artificial Intelligence and Mathematics*, January 1994.
11. M. Davis and H. Putnam. A computing procedure for quantification theory. *J. ACM*, 7(3):201–215, 1960.
12. P. C. Gilmore. A proof method for quantification theory: its justification and realization. *IBM J. Res. Dev.*, pages 28–35, 1960.
13. J. Herbrand. Researches in the theory of demonstration. In J. van Heijenoort, editor, *From Frege to Gödel: a source book in Mathematical Logic, 1879–1931*, pages 525–581. Harvard Univ. Press, 1974.
14. Shie-Jue Lee. *CLIN: An Automated Reasoning System Using Clause Linking*. PhD thesis, University of North Carolina at Chapel Hill, 1990.
15. Shie-Jue Lee and David. A. Plaisted. Eliminating duplication with the hyper-linking strategy. *J. Automated Reasoning*, 9:25–42, 1992.
16. Rainer Manthey and François Bry. SATCHMO: a theorem prover implemented in Prolog. In E. Lusk and R. Overbeek, editors, *Proc. of CADE-9*, pages 415–434, Argonne, IL, 1988.
17. William W. McCune. *OTTER 2.0 Users Guide*. Argonne National Laboratory, Argonne, Illinois, March 1990.
18. Xumin Nie and David A. Plaisted. A complete semantic back chaining proof system. In Mark E. Stickel, editor, *Proc. of CADE-10*, pages 16–27, Kaiserslautern, Germany, 1990.
19. F. J. Pelletier. Seventy-five problems for testing automatic theorem provers. *J. Automated Reasoning*, 2:919–216, 1986.
20. D. Plaisted and S. Greenbaum. A structure-preserving clause form translation. *J. Symbolic Computation*, 2:293–304, 1986.
21. David A. Plaisted. Non-Horn clause logic programming without contrapositives. *J. Automated Reasoning*, 4:287–325, 1988.
22. David. A. Plaisted, Geoffrey D. Alexander, Heng Chu, and Shie-Jue Lee. Conditional term rewriting and first-order theorem proving. In *Proceedings of the Third International Workshop on Conditional Term-Rewriting Systems*, Pont-à-Mousson, France, 8–10 July 1992. Invited Talk.
23. John K. Slaney and Ewing L. Lusk. Finding models: Techniques and applications. Tutorial in CADE-11, 1992.
24. H. Wang. Formalization and automatic theorem-proving. In *Proc. of IFIP Congress 65*, pages 51–58, Washington, D.C., 1965.
25. Tie Cheng Wang. Designing examples for semantically guided hierarchical deduction. In *Proc. of the 9th IJCAI*, pages 1201–1207, Los Angeles, CA, 1985.
26. Tie-Cheng Wang and W. W. Bledsoe. Hierarchical deduction. *J. Automated Reasoning*, 3:35–77, 1987.
27. S. Winker. Generation and verification of finite models and counterexamples using an automated theorem prover answering two open questions. *J. ACM*, 29:273–284, 1982.
28. L. Wos and S. Winker. Open questions solved with the assistance of AURA. In W. Bledsoe and D. Loveland, editors, *Automated Theorem Proving: After 25 Years*, pages 5–48. American Mathematical Society, Providence, RI, 1984.
29. Larry Wos. *Automated Reasoning: 33 Basic Research Problems*. Prentice Hall, Englewood Cliffs, NJ, 1988.

The Applicability of Logic Program Analysis and Transformation to Theorem Proving *

D.A. de Waal and J.P. Gallagher

Department of Computer Science, University of Bristol, Queen's Building,
University Walk, Bristol BS8 1TR, U.K, +44 (0272) 303307
e-mail: andre@uk.ac.bristol.compsci, john@uk.ac.bristol.compsci.

Abstract. Analysis and transformation techniques developed for logic programming can be successfully applied to automatic theorem proving. In this paper we demonstrate how these techniques can prune the search space of the theorem prover, by detecting inference rules and clauses that cannot contribute to proofs. The specialisation techniques developed in this paper are applied to first order clausal theorem provers, but are independent of the logic and the proof system and can therefore be applied to all theorem provers written as logic programs.

1 Introduction

Analysis and transformation techniques developed for logic programming can be successfully applied to automatic theorem proving. In this paper we demonstrate how these techniques can be used to obtain useful information. We show how failure branches (possibly non-terminating) can be eliminated from the proof search tree and how a non-terminating deduction in a proof system can be turned into failure. Good speedups can result.

Our motivation for this work is the following: we want to speed up theorem provers by avoiding inference rules not needed for proving a given theorem and sentences in the theory that do not contribute to the proof of a given theorem.

Our method is based on representing the prover, object theory and theorem to be proved as a logic meta-program, object program and a goal. Techniques for optimising logic program computations are then applied. The first Futamura projection [5], defining compilation, is a well-known transformation technique based on partial evaluation [21]. It provides an adequate framework in which this problem can be expressed elegantly and can be briefly explained as follows.

An interpreter for some language L is a meta-program taking a program in L as data. The result of restricting the interpreter to a particular object program P_0 is a specialised interpreter that can only interpret one program P_0. The specialised interpreter may be regarded as a compiled version of P_0.

More specifically, when the interpreter is a theorem prover for some language (logic) L and the object program P_0 is an object theory in L, the result of restricting the theorem prover to a particular object theory is a specialised theorem

---
* Work supported by ESPRIT Project PRINCE (5246)

prover that can only prove theorems in the given object theory. The specialised theorem prover is the result of compiling the theory P_0 into the language in which the theorem prover is written. Restricting the theorem prover further to certain formulas (theorems) or classes of formulas will in general give even better specialisation results (this information is used in the analysis phase, see Section 2.4).

If the meta-language is expressive enough such that any theorem prover (or more conservatively, a large number of theorem provers) can be written using this language, we may regard it as a specification language and state different theorem provers as sets of inference rules in it. Combining partial evaluation with state of the art abstract interpretation techniques may now give us a way of detecting inference rules and object clauses that are unnecessary for proving a given theorem given some theorem prover and object theory. This information may be used to prune *useless* derivations or attempted derivation and enables the specialised prover to prove theorems considerably faster than is possible with the original theorem prover and just partial evaluation. The following diagrammatically illustrates the process: In the above figure, the "real" theorem prover

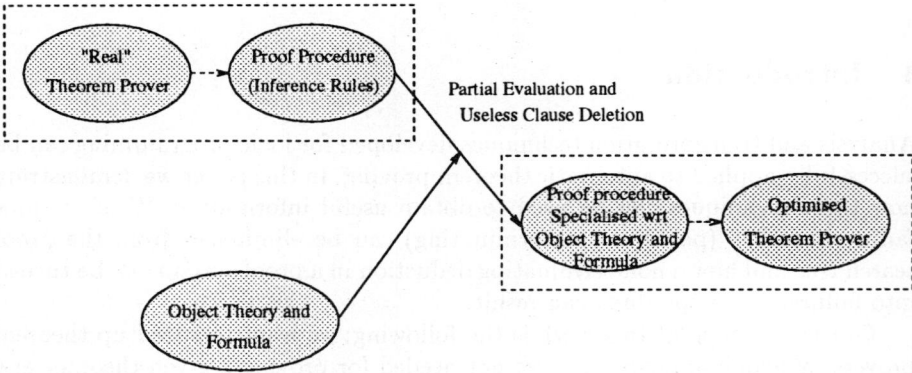

Fig. 1. Specialisation of Theorem Provers

and the proof procedure stated as inference rules together with the procedure for generating proofs may be the same. Alternatively, the "real" theorem prover might be implemented in some fast procedural language and the proof procedure may be regarded as a specification of the theorem prover. In both cases, the results of applying the method described in this paper may then be used to optimise the "specifications" and/or the original theorem prover.

As the technique described in this paper is not dependent at all on the proof system, this technique can be applied to any proof procedure and an analysis may indicate the following:

1. areas of the search space that may be avoided as they will never contribute to a proof,
2. failure for infinitely failed derivations or

3. failure to find a proof before a proof is even attempted.

Two first order clausal theorem provers are analysed. The first is a clausal theorem prover by Poole and Goebel [23]. This prover is based on the model elimination proof procedure by Loveland [17]. The second theorem prover is the nH-Prolog proof system by Loveland [18], developed to solve efficiently near-Horn or almost-Horn problems [18] (although it handles any clausal theory).

In the next section some preliminaries needed to understand the rest of the paper are given. In Section 3 the Poole-Goebel clausal theorem prover is described and we show how this prover can be specialised with respect to some object theory. The Naive nH-Prolog proof system is then described and non-termination of this proof system turned into failure. In Section 4 we give some performance results. The paper ends with a discussion and conclusions.

2 Preliminaries

We now introduce some concepts from logic program analysis and transformation. The terminology is consistent with [15].

Definition 1. useless clause
   Let P be a normal program and let $C \in P$ be a clause. Then C is useless if for all goals G, C is not used in any SLDNF-refutation of $P \cup \{G\}$.

Normally a programmer would try not to write useless clauses and so their detection has previously been regarded as debugging. Such a clause might be considered "badly-typed". Related topics are discussed in [19] and [30]. However useless clauses arise more frequently in programs that are generated by transforming some other program. In such cases removal of useless clauses is an important transformation. A stronger notion of uselessness is obtained by considering particular computations and a fixed computation rule.

Definition 2. useless clause with respect to a computation
   Let P be a normal program, G a normal goal, R a safe computation rule and let $C \in P$ be a clause. Let T be the SLDNF-tree of $P \cup \{G\}$ via R. Then C is useless with respect to T if C is not used in refutations in T or any sub-refutation associated with T.

Obviously a clause that is useless by Definition 1 is also useless with respect to any goal by Definition 2. It is undecidable whether a given clause is useless. Therefore we introduce the notion of *safe approximation* of a program in order to establish sufficient conditions for uselessness.

Definition 3. safe approximation
   Let P and P' be normal programs. Then P' is a safe approximation of P if for all definite goals G,

- if $P' \cup \{G\}$ has a finitely failed SLDNF-tree then $P \cup \{G\}$ has no SLDNF-refutation.

This definition is equivalent to saying that if a definite goal G succeeds in P then it succeeds, loops or flounders in P'. So a stronger and more intuitive condition is as follows:

- if $P \cup \{G\}$ has an SLDNF-refutation then $P' \cup \{G\}$ has an SLDNF-refutation.

For our purposes we are interested in using approximations to detect useless clauses in P, and for this the weaker condition of Definition 3 is adequate.

Definition 4. Let G be a normal goal. Then G^+ denotes the definite goal obtained by deleting all negative literals from G.

Lemma 5. *safe approximation for detecting useless clauses*
  Let P and P' be normal programs, where P' is a safe approximation of P. Let $A \leftarrow B$ be a clause in P. Then $A \leftarrow B$ is useless with respect to P if $P' \cup \{\leftarrow B^+\}$ has a finitely failed SLDNF-tree.

Proof. Assume $P' \cup \{\leftarrow B^+\}$ has a finitely failed SLDNF-tree. From Definition 3 $P \cup \{\leftarrow B^+\}$ has no SLDNF-refutation, hence $P \cup \{\leftarrow B\}$ has no SLDNF-refutation, hence $A \leftarrow B$ is not used in any refutation, hence by definition it is useless in P.

This lemma shows that useless clauses are detectable if failure in the approximation is decidable.

2.1 Regular Approximations

We consider a class of safe approximations defined by *regular program*.

Definition 6. A *regular unary clause* is of the form $p(f(X_1, \ldots, X_n)) \leftarrow t_1(X_1), \ldots, t_n(X_n)$ where $X_1, \ldots, X_n$ are distinct variables.

Definition 7. A *regular definition* of a predicate $p(X_1, \ldots, X_n)$ is a clause of the form $p(X_1, \ldots, X_n) \leftarrow t_1(X_1), \ldots, t_n(X_n)$ where $X_1, \ldots, X_n$ are distinct variables, and the predicates $t_1, \ldots, t_n$ are defined by regular unary clauses (where $t_1, \ldots, t_n$ are different from p).

Definition 8. A *regular program* is a program consisting of regular unary clauses and regular definitions of predicates.

Regular programs correspond to deterministic finite state automata [30], and therefore computations with them are decidable.

Lemma 9. *Let P be a regular program and G a goal. Then $P \cup \{G\}$ is decidable; it has a finite failure tree iff it has no refutation.*

$reverse([\,],[\,]) \leftarrow true$
$reverse([X|Xs],Ys) \leftarrow reverse(Xs,Zs), append(Zs,[X],Ys)$

$append([\,],Ys,Ys) \leftarrow true$
$append([X|Xs],Ys,[X|Zs]) \leftarrow append(Xs,Ys,Zs)$

Fig. 2. Naive Reverse

$reverse(X1,X2) \leftarrow t1(X1), t2(X2)$

$t1([\,]) \leftarrow true$
$t1([X1|X2]) \leftarrow any(X1), t1(X2)$
$t2([\,]) \leftarrow true$
$t2([X1|X2]) \leftarrow any(X1), t2(X2)$

$append(X1,X2,X3) \leftarrow t3(X1), any(X2), any(X3)$

$t3([\,]) \leftarrow true$
$t3([X1|X2]) \leftarrow any(X1), t3(X2)$

Fig. 3. Regular Approximation of Naive Reverse

We give an example in Figure 2 for the reader unfamiliar with approximation using regular programs. A regular approximation of the program in Figure 2 is shown in Figure 3, where $any(t)$ is true for any term t in the Herbrand Universe of the program. This program was produced automatically by the logic program specialisation system SP [7] at the University of Bristol.

There are other ways of generating regular programs that are more expressive (and hence could give more precise approximations) but we chose this form because it is easy to manipulate and it is straightforward to check success or failure of derivations. Another rather direct method for finding a regular approximation of a logic program is given in [4].

2.2 Computation of Regular Approximations

A regular approximation of a definite program can be computed by *abstract interpretation*. Space does not permit a detailed explanation, which can be found in [8], and the method is also summarised in [9]. A fast and precise algorithm for computing regular approximations is given in [10]. For any program P, we construct a regular program that is a safe approximation of P.

Briefly, the method is based on abstract interpretation of the standard fixpoint semantics of a definite program P, given by its T_P operator. An abstract version of the T_P operator is defined, more specifically, a monotonic operator that maps one regular program to another. Its least fixed point is computed as the limit of an ascending sequence, and it can easily be shown that this represents a safe approximation of the least fixed point of T_P. The efficient computation of fixed points has been the subject of much study in the literature of abstract interpretation. Useless clauses with respect to a given computation can also be computed by an extension of this technique using so-called "magic-set" style transformation [24], [8].

2.3 Program Specialisation

Partial evaluation of a normal program P with respect to a goal G is a well known method of program specialisation [5], [16], [6], [21]. The correctness requirement [16] is that if P' is a partial evaluation of P with respect to a goal G, then

1. $P \cup \{G\}$ has an SLDNF answer substitution θ iff $P' \cup \{G\}$ does.
2. $P \cup \{G\}$ has a finitely-failed SLDNF tree iff $P' \cup \{G\}$ does.

Partial evaluation algorithms that satisfy this requirement are well-established. Thus partial evaluation preserves the results of computations of $P \cup \{G\}$, including looping.

In [2] another specialisation technique was defined, the detection and removal of useless clauses from a program. The following result was proved.

Lemma 10. *Let P be a normal program and G a normal goal. Let $C \in P$ be a clause and let $P' = P - \{C\}$. If C is useless with respect to the computation of $P \cup \{G\}$, then*

1. *if $P \cup \{G\}$ has an SLDNF-refutation with computed answer θ then $P' \cup \{G\}$ has an SLDNF-refutation with computed answer θ; and*
2. *if $P \cup \{G\}$ has a finitely failed SLDNF-tree then $P' \cup \{G\}$ has a finitely failed SLDNF-tree.*

Thus the terminating computations of $P \cup \{G\}$ are preserved by deleting useless clauses. However, it is important to realise that the transformation may turn a nonterminating computation into a terminating computation (an infinitely failed computation may be turned into finite failure and the specialised program may yield more answers than the original program).

2.4 A Two-Stage Specialisation Procedure

The two specialisation methods can be combined. Given a normal program P, and a normal goal G, perform the following two steps.

1. Partially evaluate P wrt G (or some G' such that G is an instance of G'). Call the partially evaluated program P_1.
2. Compute a regular approximation of clauses in P_1 with respect to $P_1 \cup \{G\}$. This yields P_2 which is used to detect and delete useless clauses from P_1, giving P_3.

By the correctness of partial evaluation [16], and by Lemma 10, $P_3 \cup \{G\}$ preserves the results of all terminating computations of $P \cup \{G\}$.

3 Analysis of Proof Procedures

In this section we review two first order predicate calculus proof procedures. The first procedure is a clausal theorem prover by Poole and Goebel [23] and the second a Naive nH-Prolog proof system by Loveland [18].

The procedures are presented as meta-programs for three reasons. First, it is not the aim of this paper to develop efficient theorem provers based on the above proof procedures, but to demonstrate the applicability of analysis and transformation techniques, and the procedures as presented here suffice for this aim. Second, as the theorem prover you wish to analyse might not be implemented in a logic programming language, this approach demonstrates how an analysable version may be created without reimplementing all the intricacies present in the original prover. Third, the theory of meta-programming has received a lot of attention during the last few years and notable improvements in meta-programming style and specialisation techniques have cleared the way for this approach to be taken seriously [11], [7], [1] .

The analysis performed therefore provides us with information about the inference rules needed as well as the representation needed. This information may now be used in different ways:

1. to construct a specialised prover that takes advantage of this information,
2. to switch to a different and possibly much simpler proof system that will improve the proof process or
3. to change the representation.

3.1 Poole-Goebel Theorem Prover

Consider the clausal theorem prover in Figure 4 described by Poole and Goebel in [23]. This procedure is based on the model elimination proof procedure [17].

The program given in Figure 4 is similar to the procedure in [23]. A depth first iterative deepening mechanism has been added and the program has been rewritten in a style that makes the two inference rules explicit. Different sets of inference rules can be substituted for the above two to get proof procedures for different logics.

$neg(X, Y)$ is true if X is the negation of Y with X and Y both in their simplest form. *clause* is not the usual Prolog *clause*. $clause((X : -Y))$ is true if $X \leftarrow Y$ is a contrapositive of an arbitrary clause of the form $a_1 \vee \ldots \vee a_n \leftarrow b_1 \wedge \ldots \wedge b_m$ and X is a literal. $literal(G)$ is true if G is an atom or the negation of an atom in the object language. $member(X, Y)$ is true if X is a member of the list Y.

The program given in Figure 4 can be regarded as a sound and complete specification of the model elimination theorem prover (assuming the underlying Prolog system is sound - includes the occur check). As the occur check relates to the underlying system on which this program will be executed, the analysis method is independent of whether the occur check is implemented or not. Obviously, other optimisations could be made [26], but we are interested only in the

$solve(Goal, Anc) \leftarrow literal(Goal),$
    $depth\_bound(Depth),$
    $prove(Goal, Anc, Depth).$

$prove(G, A, D) \leftarrow D > 0,$
    $infer(G, A, A1, G1), D1 \text{ is } D - 1,$
    $proveall(G1, A1, D1).$

$proveall([\ ], \_, \_).$
$proveall([G|R], A, D) \leftarrow prove(G, A, D),$
    $proveall(R, A, D).$

$infer(G, A, \_, [\ ]) \leftarrow member(G, A).$ % Ancestor Resolution
$infer(G, A, [G\_neg|A], Body) \leftarrow$ % Input Resolution
    $clause((G : -Body)),$
    $neg(G, G\_neg).$

Fig. 4. Model Elimination Proof Procedure

success set of this program, not the search space. Any results inferred for this program will therefore also be valid for the "real" theorem prover.

3.2 Obtaining a Specialised Prover

The proof procedure may be specialised with respect to a specific class of theorems (queries of the form $solve(T, [\ ])$) or with respect to a given theorem (e.g. $solve(c(\_), [\ ])$). The more limited the queries, the more precise the results are likely to be for a given approximation method. As the theorems that we want to prove are known in the examples that follow, we generally approximate with respect to a given theorem. In general, this does not preserve the completeness of the proof system, but this can easily be restored by defining a set of queries which have to be tried [22] and doing the above analysis for each query.

We illustrate the first analysis by considering the following object theory from [20], which is a version of Popplestone's "Blind Hand Problem", together with the proof procedure in Section 3.1. This problem is also used in [26] as a benchmark.

Our aim is to detect necessary inference rules and clauses for proving the above theorem. Since this is not a definite clause theory (and cannot be turned into one by reversing signs on literals) it is not clear which ancestor resolutions are needed. We now apply our analysis method from Section 2.4 to the model elimination proof procedure and the DBA BHP object theory. First, we partially evaluate the program in Fig. 4 with respect to the object theory in Fig. 5 (clauses 1 to 13) and general goal $solve(\_, \_)$ (any theorem). The result is a specialised program with 14 different *prove* procedures and 14 different *member* procedures, one procedure for each positive or negative literal occurring in the object theory. Second, we compute a regular approximation of the partially evaluated program

1. $\neg a(X,Z,Y) \vee \neg h(Z,Y) \vee i(X,p(Y))$
2. $\neg h(W,Y) \vee h(Z,g(Z,Y))$
3. $a(X,Z,g(Z,Y)) \vee \neg h(W,Y) \vee \neg i(X,Y)$
4. $\neg h(Z,Y) \vee h(Z,l(Y))$
5. $a(s,e,n)$
6. $\neg i(X,l(Y))$
7. $\neg a(X,e,Y) \vee r(X)$
8. $\neg a(X,Z,Y) \vee a(X,Z,l(Y))$
9. $\neg a(X,Z,Y) \vee a(X,Z,p(Y))$
10. $\neg a(X,Z,Y) \vee b(X,p(g(Z,l(Y))))$
11. $c(Y) \vee \neg q(X,t,Y) \vee \neg r(X)$
12. $\neg a(X,W,Y) \vee \neg b(X,Y) \vee q(X,Z,g(Z,Y))$
13. $\neg a(X,W,Y) \vee a(X,W,g(Z,Y)) \vee i(X,Y)$
14. $\neg c(Y)$

Fig. 5. A Version of Popplestone's Blind Hand Problem (DBA BHP)

with respect to the goal $solve(c(\_),[\,])$ (where $c(\_)$ is the theorem to be proved and $[\,]$ indicates the start of the proof).

An empty approximation for all the different *member* procedures results which indicates that ancestor resolution always fails when trying to prove $c(\_)$. Furthermore, empty approximations for the *prove* procedures for the following predicates results, namely $h(\_,\_), i(\_,\_), \neg a(\_,\_,\_), \neg h(\_,\_), \neg c(\_), \neg q(\_,\_,\_)$, $\neg b(\_,\_)$ and $\neg r(\_)$. This indicates that all branches in the object level search tree, that correspond to the proof of one of these literals, are failure branches and can therefore be deleted.

We therefore have the following information available that can be fed back into the "real" theorem prover (see Introduction).

- Ancestor resolution is never needed to prove $c(\_)$.
- Clauses 1 to 4 can be deleted from the object theory (at least one *prove* procedure corresponding to a literal in the body of each contrapositive generated by the four clauses can never succeed).
- The only literals that can contribute to a proof are $a(\_,\_,\_), c(\_), q(\_,\_,\_)$, $b(\_,\_), r(\_)$ and $\neg i(\_,\_)$.

Partial evaluation and approximation made it possible to detect that the Blind Hand problem is an instance of a class of theories where input resolution is the only inference rule needed to prove the given theorem. It also shows that a limiting factor in proving the above theorem in current logic programming systems is not its inference rule, but the representation chosen for the object theory (its inability to represent negative literals in the head of a clause).

3.3 Naive nH-Prolog

As a second proof procedure consider the naive nH-Prolog proof system as described in [18]. The intuitive idea of this prover is to try to solve the Horn part of a theorem first and to postpone the non-Horn part (which should be small for near-Horn problems) until the end. Each non-Horn problem that has to be solved corresponds to one application of the splitting rule. Naive nH-Prolog is an incomplete proof procedure.

Due to space limitations, the inference rules for nH-Prolog are not given, but can be found in the extended version of this paper [3].

3.4 Failure Detection

We illustrate the second and third analyses described in Section 3.2 by analysing the following object theory from [18] in conjunction with the naive nH-Prolog proof system. The following object theory was used there to demonstrate the incompleteness of the naive nH-Prolog proof system.

$$\begin{array}{ll} q \leftarrow a,b & b \leftarrow e \\ a \leftarrow c & b \leftarrow f \\ a \leftarrow d & e \vee f \\ c \vee d & \end{array}$$

The following deduction was also given that showed that naive nH-Prolog would never terminate when trying to prove the query q (which is a logical consequence of the theory).

$$\begin{array}{lll} (0) & ?-q. & \\ (1) & :-a,b & \\ & \vdots & \\ (6) & :-q \ \#f[d] & \{restart\} \\ & \vdots & \\ (17) & :- \ \#d \ [f,d] & \\ (18) & :-q \ \#f[d] & \{restart\} \\ & \vdots & \end{array}$$

In general it is not so easy to establish nontermination of a proof system, because it is not known how many deductions are needed to detect a loop. Hundreds of deductions may be necessary to identify such a property. Even worse, a proof system may not terminate and there may be no reoccurring pattern of deductions. In this case, it is impossible to detect nontermination with a scheme as presented in the example. It is however possible to detect failure of this proof system to prove the given query by applying the method described in Section 2.4.

We again apply our analysis method to the naive nH-Prolog proof system and the object theory given above. An empty approximation for *solve* results and indicates that the above proof system is unable to prove the above theorem starting from query q. We still do not know if it is because there does not exist a proof or because of the incompleteness of the proof system. However, we have removed some indecision with respect to the above proof system and can now move on to another proof system (that may be complete) and try to prove the above query.

It is worth emphasising that this method may indicate failure for more formulas (non-theorems) than is possible with the original proof system. This is because the analysis method may approximate an infinitely failed tree by a finitely failed tree (the finite failure set of the original prover may not be preserved) [2]. This increase in power of the prover provided by the analysis method may be very important.

4 Performance Results

In this section we give some performance results for the "Blind Hand Problem" as well as some other problems from the TPTP Problem library [28]. The timings for finding all proofs to the "Blind Hand Problem" one hundred times on a SPARC station IPC using a depth bound of eight to eleven are given in the following table. All timings are in seconds. ME indicates the proof procedure given in Figure 4, PE the partially evaluated proof procedure, and PE+D the partially evaluated proof procedure with useless clauses deleted.

Blind Hand Problem					
Depth Bound	ME	Time PE	PE+D	Speedup (ME/PE)	Speedup (ME/PE+D)
8	4.55	2.69	0.53	1.7	8.5
9	25.73	14.81	2.63	1.7	9.8
10	157.00	99.41	14.12	1.6	11.1
11	1030.00	617.79	83.60	1.7	12.3

After specialisation, the "Blind Hand Problem" is solved very efficiently. All the redundant ancestor resolutions have been eliminated and a large number of contrapositives have been deleted.

The approximation time for this example was 5.8 seconds. This might seem a long time compared to the speedup achieved, but as we have only used a prototype implementation to construct the approximation, we point out that this analysis time can be greatly reduced. A recent implementation of a regular approximation procedure in C gives at least an order of magnitude improvement over our prototype implementation [13].

The speedups in the above table suggest that when larger depth bounds are required to prove difficult theorems, the possible speedup achievable will be much greater. We also feel that as the depth bounds needed to prove large theorems increase, there will come a point where the analysis time will become insignificant compared to the time taken to prove the theorem and the time saved by the deletion of useless clauses.

The next table contains some results from applying the specialisation technique developed in this paper to problems from Computing Theory, Number Theory and Algebra in the TPTP Library. Results from one puzzle and one miscellaneous problem are also included.

TPTP Number	Problem	Approx Time	Time ME	Time PE	PE+D	Speedup (ME/PE+D)
MSC002-1	Blind Hand	5.8	4.5	2.7	0.53	8.6
COM002-2	Correctness	35.9	9.5	9.5	0.47	20.0
PUZ033-1	Winds and Widows	89.7	18.0	18.0	0.42	42.9
NUM014-1	Prime Number	4.8	4.2	2.3	0.90	4.7
GRP001-5	Group Commutativity	1.7	2.7	2.6	1.23	2.2

Analysis and Transformation Results

All times are for finding all solutions 100 times (to the smallest depth where a proof can be found). The results of specialisation are very efficient programs. However, the analysis times are still relatively large and we do not show an overall gain. Nevertheless, as explained above, we feel that when tackling more difficult problems with bigger search spaces and given a more efficient approximation implementation, the tradeoff will be worthwhile.

Most interesting first-order theorems lie somewhere between being definite and extensively requiring the full power of the first-order theorem prover. For the model elimination proof procedure, the speedups obtainable therefore depend on the extent to which ancestor resolution is required, the depth bound required for the proof, the branching factor of the search tree and the number of contrapositives in the object theory that may be deleted. Further experiments will indicate the influence of each factor on the speedup achieved.

5 Discussion

In [23] the model elimination proof procedure was analysed with respect to the following set of propositional clauses.

$$a \leftarrow b \wedge c$$
$$a \vee b \leftarrow d$$
$$c \vee e \leftarrow f$$
$$\neg g \leftarrow e$$
$$g \leftarrow c$$
$$g$$
$$f \leftarrow h$$
$$h$$
$$d$$

Their analysis indicated that the negative ancestor check may be necessary for a, $\neg a$, b and $\neg b$. When applying the method in this paper to the above object theory and the program given in Figure 4, empty approximations are derived for all the specialised *member* procedures in the above theory, except for the *member* procedures for literals $\neg a$, f and h. This shows that our method is at least as precise as the method developed in [23], for this example (and that both are approximations).

In [27] Sutcliffe describes syntactically identifiable situations in which reduction does not occur in chain format linear deduction systems. He develops three analysis methods to detect such linear-input subdeductions. The most powerful of the tree methods is the Linear-Input Subset for Literals (LISL) analysis. The following object theory was analysed using this method.

$$r \vee \neg p(a) \vee \neg q \qquad\qquad t \vee u$$
$$\neg p(a) \vee q \qquad\qquad \neg u$$
$$p(a) \vee \neg q \qquad\qquad s \neg p(b)$$
$$p(a) \vee q \qquad\qquad p(b)$$
$$\neg r \vee \neg t \vee \neg s$$

His analysis method indicates that no reductions are needed for the subset $\{r, \neg t, u, \neg s, \neg p(b)\}$ and query $r \wedge \neg p(a) \wedge q$. Our analysis method infers exactly the same results when applied to the program given in Figure 4 and the above object theory. However, it is easy to construct an example where our analysis will indicate failure and LISL analysis will not infer any useful information. For example, consider the following object theory which is a subset of clauses given above.

$$\neg p(a) \vee q \qquad\qquad p(a) \vee \neg q$$

In this case LISL analysis infers no useful information, but our analysis applied to the program given in Figure 4 and the above two clauses gives an empty approximation for *solve* which indicates failure.

Wakayama [29] defines a class of Case-Free Programs that requires no factoring and ancestor resolution. A characterisation of Case-Free programs is given, but no procedure is given for detecting instances of Case-Free programs. The method in this paper can be used to detect instances of Case-Free programs. However, we do not claim that our method will detect all Case-Free programs. Our method is comparable when considering all possible derivations from a given theory, but also allows Case-Freeness of particular proofs to be inferred, as in the DBA BHP example.

In [25] and [14] Kautz and Selman use the idea of approximation for propositional theorem proving. A "Horn Approximation" of any propositional theory can be constructed. Since the complexity of proving theorems in Horn theories is less than the general case it may be worth testing formulas for theoremhood in the approximating theory before applying a general propositional theorem prover. Some non-theorems can be thrown out fast. This is an example of the same principle of reasoning by approximation.

In that work the main benefit of using an approximation is efficiency, whereas in the approximations we propose the main benefits are decidability and efficiency, since our approximating theories are regular theories for which theoremhood is decidable.

It is worth emphasising that this method is independent of the proof procedure which is not the case for the methods developed by Poole and Goebel, Sutcliffe and Wakayama. Any theorem prover that can be written as a logic program can be analysed and optimised using this method. Furthermore, this method can be significantly improved by making use of other classes of decidable theories in the approximation with a lesser information loss (better approximation) than for regular unary logic programs. We also would like to develop a general approach using meta-programming to specify theorem provers. Such an approach might be adapted from work such as [12].

6 Conclusions

We presented a novel way of analysing theorem provers with general analysis and transformation techniques developed for logic programming. Two first order clausal theorem provers were analysed with these techniques and nontrivial results inferred. We also indicated how the analysis information may be used to guide choices regarding representation and inference rules. In future work we intend to refine the analysis presented in this paper and to develop better approximations based on other decidable theories.

Acknowledgements

We would like to thank Gerd Neugebauer for his constructive remarks and many valuable suggestions that contributed greatly to the overall appearance of this paper and David Warren and John Lloyd for their comments.

References

1. D.A. de Waal. The power of partial evaluation. In Y. Deville, editor, *Logic Program Synthesis and Transformation,* Louvain-la-Neuve 1993, Workshops in Computing, pages 113–123. Springer-Verlag, 1994.
2. D.A. de Waal and J. Gallagher. Logic program specialisation with deletion of useless clauses. In D. Miller, editor, *Proceedings of the 1993 International Symposium, Vancouver,* 1993.
3. D.A. de Waal and J.P. Gallagher. The applicability of logic program analysis and transformation to theorem proving. Technical Report CSTR-93-15, University of Bristol, September 1993.
4. T. Frühwirth, E. Shapiro, M. Vardi, and E. Yardeni. Logic programs as types for logic programs. In *6th IEEE Symposium on Logic in Computer Science, Amsterdam,* 1991.
5. Y. Futamura. Partial evaluation of computation process - an approach to a compiler-compiler. *Systems, Computers, Controls,* 2(5):45–50, 1971.
6. J. Gallagher. Tutorial on specialisation of logic programs. In *The ACM-SIGPLAN Symposium on Partial Evaluation and Semantics-based Program Manipulation (PEPM'93), Copenhagen, June 1993.* ACM Press.
7. J. Gallagher. A system for specialising logic programs. Technical Report TR-91-32, University of Bristol, November 1991.
8. J. Gallagher and D.A. de Waal. Regular approximations of logic programs and their uses. Technical Report CSTR-92-06, University of Bristol, March 1992.
9. J. Gallagher and D.A. de Waal. Deletion of redundant unary type predicates from logic programs. In K.K. Lau and T. Clement, editors, *Logic Program Synthesis and Transformation,* Workshops in Computing, pages 151–167. Springer-Verlag, 1993.
10. J. Gallagher and D.A. de Waal. Fast and precise regular approximation of logic programs. Technical Report TR-93-19, Dept. of Computer Science, University of Bristol, 1993. (accepted for presentation at the Eleventh International Conference on Logic Programming).

11. C.A. Gurr. Specialising the ground representation in the logic programming language Gödel. To be published in the Proceedings of the Third International Workshop on Logic Program Synthesis and Transformation, Louvain-la-Neuve, 1993.
12. R. Harper, F. Honsell, and G. Plotkin. A framework for defining logics. *Journal of the ACM*, 40(1):143-184, 1993.
13. P. Van Hentenryck, A. Cortesi, and B. Le Charlier. Type analysis of prolog using type graphs. Technical report, Brown University, Department of Computer Science, December 1993.
14. H. Kautz and B. Selman. Forming concepts for fast inference. In *Proceedings of the Tenth National Conference on Artificial Intelligence, San Jose, California*. AAAI/MIT Press, 1992.
15. J.W. Lloyd. *Foundations of Logic Programming: 2nd Edition*. Springer-Verlag, 1987.
16. J.W. Lloyd and J.C. Shepherdson. Partial Evaluation in Logic Programming. *Journal of Logic Programming*, 11(3 & 4):217-242, 1991.
17. D.W. Loveland. *Automated theorem proving: a logical basis*. North-Holland, 1978.
18. D.W. Loveland. Near-Horn Prolog and beyond. *Journal of Automated Reasoning*, 7(1):1-26, 1991.
19. K. Marriott, L. Naish, and J-L. Lassez. Most specific logic programs. In *Proceedings of the Fifth International Conference and Symposium on Logic Programming*, Washington, August 1988.
20. D. Michie, R. Ross, and G.J. Shannan. G-deduction. *Machine Intelligence*, 7:141-165, 1972.
21. C.K. Gomard N.D. Jones and P. Sestoft. *Partial Evaluation and Automatic Program generation*. Prentice Hall, 1993.
22. G. Neugebauer. Reachability analysis. To be published in the Proceedings of the Third International Workshop on Logic Program Synthesis and Transformation, Louvain-la-Neuve, 1993.
23. D.L. Poole and R. Goebel. Gracefully adding negation and disjunction to Prolog. In E. Shapiro, editor, *Third International Conference on Logic Programming*, pages 635-641. Lecture Notes in Computer Science, Springer-Verlag, 1986.
24. H. Seki. On the power of Alexander templates. In *Proceedings of the 8th ACM SIGACT-SIGMOD-SIGART Symposium on Principles of Database Systems*, Philadelphia, Pennsylvania, 1989.
25. B. Selman and H. Kautz. Knowledge compilation using Horn approximation. In *Proceedings of the Ninth National Conference on Artificial Intelligence*. AAAI/MIT Press, 1991.
26. M.E. Stickel. A Prolog Technology Theorem Prover. In *International Symposium on Logic Programming*, Atlantic City, NJ, pages 211-217, Feb. 6-9 1984.
27. G. Sutcliffe. Linear-input subset analysis. In D. Kapur, editor, *11th International Conference on Automated Deduction*, pages 268-280. Lecture Notes in Computer Science, Springer-Verlag, 1992.
28. C. Sutter, G. Sutcliffe, and T. Yemenis. The TPTP problem library. Technical Report TPTP v1.0.0 - TR 12.11.93, James Cook University, Australia, November 1993.
29. T. Wakayama. Case-free programs: An abstraction of definite Horn programs. In M.E. Stickel, editor, *10th International Conference on Automated Deduction*, pages 87-101. Lecture Notes in Computer Science, Springer-Verlag, 1990.
30. E. Yardeni and E.Y. Shapiro. A type system for logic programs. *Journal of Logic Programming*, 10(2):125-154, 1990.

Detecting Non-Provable Goals[*]

Stefan Brüning

FG Intellektik, FB Informatik, Technische Hochschule Darmstadt
Alexanderstraße 10, D–64283 Darmstadt (Germany)
Phone: [49](6151)16-6632, FAX: [49](6151)16-5326
E-mail: stebr@intellektik.informatik.th-darmstadt.de

Abstract. In this paper we present a method to detect non-provable goals. The general idea, adopted from cycle unification, is to determine in advance how terms may be modified during a derivation. Since a complete predetermination is obviously not possible, we analyze how terms may be changed by, roughly speaking, adding and deleting function symbols. Such changes of a term are encoded by an efficiently decidable clause set. The satisfiability of such a set ensures that the goal containing the term under consideration cannot contribute to a successful derivation.

1 Introduction

To develop proof methods which are as general *and* adequate[2] as possible is one of the main goals of research in the field of automated theorem proving. General methods, like the resolution principle [16] and the connection method [1] are well known. However, calculi developed from these methods are only adequate if they are augmented by techniques which control the proof process and reduce the search space.

In this paper we address this problem by proposing a new technique to augment top-down backward-chaining calculi — like model elimination [14] or SLD-resolution [13] — such that goals which cannot contribute to a successful derivation (i.e. a refutation) are identified. This is clearly of importance since such goals are a source of infinite looping.

In the field of automated theorem proving, a lot of work deals with the avoidance of redundancy. Well-known techniques (for example cf. [12]) include clause subsumption, the tautology principle, and the identical ancestor pruning rule. In the field of logic programming, various techniques to augment SLD-resolution based interpreters were examined. Most of them (for example cf. [5, 18]) are based on subsumption tests between (sequences of) goals at runtime.

However, all these mechanisms eliminate *logical* redundancy, ie. a derivation is pruned if its success would imply the existence of a hopefully smaller successful derivation. Non-provable goals which are not redundant in this sense cannot be

---

[*] This research was supported by the Deutsche Forschungsgemeinschaft (DFG) within project KONNEKTIONSBEWEISER under grant no. Bi 228/6-2.
[2] We will consider a technique as being adequate if, roughly speaking, it solves simpler problems faster than more difficult ones [2].

handled with these techniques. For instance, techniques based on subsumption are not applicable — except for the detection of identical goals — if the goals under consideration are ground. To cope with this problem mostly only the purity principle [1] is employed. Pure goals, however, are trivially non-provable because they cannot be used to perform any inference step. Thus, the scope of the purity principle is obviously limited.

The approach we have in mind has its roots in cycle unification [4]. Cycle unification deals with the problem to compute a complete set of answer substitutions for a goal $\leftarrow P(s_1, \ldots, s_n)$ wrt a logic program consisting of one fact $P(t_1, \ldots, t_n)$ and a recursive two-literal clause $P(l_1, \ldots, l_n) \leftarrow P(r_1, \ldots, r_n)$. For that, the maximal number of required iterations through the cycle is determined by studying variable dependencies. But unfortunately even for this small class of problems the question is undecidable in general if the cycle is unrestricted, i.e. $l_1, \ldots, l_n, r_1, \ldots, r_n$ are arbitrary terms [10].

However, the general idea underlying cycle unification is appealing: Given a clause set and a goal G, try to predetermine what may happen to G during a derivation. Since a complete predetermination is not feasible one has to restrict the attention to certain aspects of a derivation.

In this paper we adopt this idea and present a technique which, roughly speaking, analyzes how a term may be changed during a derivation by "adding" and "deleting" function symbols. Using this information, many derivation chains that cannot contribute to a refutation can be pruned.

Example 1. Consider the following clause set where "..." stands for a number of additional arguments. The clauses are numbered for reference:

(1) $\{\neg P(a, \ldots)\}$
(2) $\{P(x, \ldots), \neg Q(g(x), \ldots)\}$
(3) $\{Q(x, \ldots), \neg P(g(x), \ldots)\}$
(4) $\{P(x, \ldots), R_1(x, \ldots)\}$
(5) $\{\neg R_1(g(a), \ldots)\}$
(6) $\{R_1(x, \ldots), \neg R_2(g(x), \ldots)\}$
(7) $\{R_2(x, \ldots), \neg R_1(g(x), \ldots)\}$

Suppose a derivation starts by selecting clause (1). Assuming a top-down backward-chaining proof procedure, like model elimination, clauses (2) and (3) can be used to derive a goal of the form $\neg P(g^{2n}(a), \ldots)$. Applying clause (4) once yields a goal G of the form $R_1(g^{2n}(a), \ldots)$. In order to derive the empty clause we may now apply clauses (5), (6), and (7).

Considering the term structure of clauses (6) and (7), one can recognize that each subgoal of G containing the predicate symbol R_1 is of the form $R_1(g^{2m}(a), \ldots)$ with $m < n$. Therefore clause (5) can never be applied to derive the empty clause.

Thus, just by this simple analysis it is possible to decide that a successful derivation starting with clause (1) does not exist. Theorem provers like SETHEO [12] or PTTP [19] do not apply such a kind of analysis and will loop infinitely.

A first method to analyze possible changes of terms in a corresponding way was presented in [7]. The idea is to represent such changes as a formal language which is encoded by a corresponding acceptor. The proposed technique works on arbitrary clause sets and enables to prune derivations where other techniques are not applicable. But besides the fact that it employs complex operations on languages and acceptors, it is not possible to include dependencies caused by multiple occurrences of variables. This is a serious drawback since it prevents a successful pruning in many cases.

The approach proposed in this paper is conceptually easier and, furthermore, more powerful. Given a clause set S and a goal G, possible changes of a term occurring in G are encoded by a decidable clause set. As the main result we show that the problem to prove the unsatisfiability of such a set can be used as a complete approximation[3] of the original problem to find a refutation for $S \cup \{G\}$. The resulting technique only requires a conceptually simple decision procedure based on ordered resolution and can take dependencies into account which are caused by multiple occurrences of variables.

The paper is structured as follows. In Section 2 position graphs and further basic concepts of our approach are introduced. In Section 3 we show that the information given by a position graph can be encoded by a decidable clause set. An efficient decision procedure for such clause sets based on ordered resolution is presented in Section 4 and the complexity of our technique is studied briefly.

In Section 3 only Horn-clause sets are considered. Since our main motivation for this work is to develop proof techniques for general first-order calculi, a corresponding extension to arbitrary clause sets is presented in Section 5.

Because our approach has some common aspects with Plaisted's technique of abstraction, this relationship is investigated in Section 6. Aspects of a future implementation are discussed in Section 7. In Section 8 we conclude. Due to lack of space the proofs of lemmas and theorems stated in this work are omitted. They can be found in [6].

2 Checking Possible Changes of Terms

In this section we introduce basic concepts and sketch a first approach to check possible changes of terms. We assume the reader to be familiar with the basic terminology of first-order logic, SLD-resolution, and model elimination (ME).

A clause is *Horn* if it contains at most one positive literal. As usual a Horn-clause is called *fact* if it contains only one positive and no negative literal, *query* if it contains no positive literal, and *rule* otherwise. A *program* is a finite clause set consisting of rules and facts. Speaking of a *goal* we mean in the context of SLD-resolution a literal contained in a query. In the context of ME, a goal is a B-Literal occurring in an ME-chain. Predicate symbols are denoted by uppercase letters whereas lowercase letters are used for function (f, g, h), variable (x, y, z, v, w), and constant symbols (a, b, c). Let L^d denote the negation of a literal L. For a set of literals $C = \{L_1, \ldots, L_n\}$ we define $C^d = \{L_1^d, \ldots, L_n^d\}$.

A usual we can associate a labeled tree with each term (eg. see [22]). In this paper we call the leaf nodes of such trees, i.e. nodes that are labelled with variable or constant symbols, *positions*. If p is a position in the associated tree of a term t, we sometimes simply say that p *occurs* in t.

---

[3] We call a problem P a *complete approximation* of another problem P' if the impossibility to solve P implies the impossibility to solve P' (cf. for instance [17]).

Since we cannot control possible changes of arbitrary terms we now have to precise which parts of a term are subject to the control.

Definition 1. Let t be a term. Let $t = f_1(\ldots, t_1, \ldots)$ with $t_1 = f_2(\ldots, t_2, \ldots)$, $\ldots, t_{n-1} = f_n(\ldots, t_n, \ldots)$ where t_i is the k_i-th argument of f_i and t_n is either a variable or a constant symbol. Let p denote the position corresponding to t_n.

We call $f_1^{k_1} \ldots f_n^{k_n}$ the *prefix* of p (in t) and the term $f_1^{k_1} \ldots f_n^{k_n}(t_n)$ the *prefix-term* of p (in t) where $f_1^{k_1}, \ldots, f_n^{k_n}$ are new monadic function symbols.

Example 2. Let $t = f(g(a,b), x)$. The set $\{f^1g^1, f^1g^2, f^2\}$ contains the prefixes of all positions occurring in t. The set $\{f^1g^1(a), f^1g^2(b), f^2(x)\}$ contains the corresponding prefix-terms.

Note that the exponents are needed to store argument positions. For instance, $g^1(a)$ is a prefix-term of $g(a,b)$ but not of $g(b,a)$. Neglecting this information makes it impossible to distinguish the terms $g(a,b)$ and $g(b,a)$ by their prefix-terms.

The usefulness of checking possible changes of prefix-terms during SLD-derivations has already been emphasized in [22] but no method to perform such checks automatically was given. For this purpose we use so-called position graphs. Position graphs encode possible changes of prefix-terms.

Definition 2. Let S be a clause set. A *position graph* PG_S of S is a directed graph whose nodes are positions of terms of literals in S. Let p and q be two positions occurring in different literals P and Q, respectively. (p, q) is an edge of PG_S if one of the following conditions holds:

c-edge: (P, Q) is a weakly unifiable connection[4]. For some i, t_p is the i-th argument of P, t_q is the i-th argument of Q, p occurs in t_p, q occurs in t_q, and the prefix-terms of p (in t_p) and q (in t_q) are unifiable.
nc-edge: P and Q are different non-pure literals (wrt S)[5] occurring in the same clause of S. The variables corresponding to p and q are identical.

For a position p of PG_S we define the sets $in(p)$ and $out(p)$ as:
$$in(p) = \{p' \mid (p', p) \text{ is a c-edge of } PG_S\}$$
$$out(p) = \{p' \mid (p, p') \text{ is an nc-edge of } PG_S\}$$

For illustration the reader may already consider Example 3. To some extent, c-edges reflect the connection structure of a given clause set whereas nc-edges reflect variable dependencies within a clause.

The edges of a position graph can be labelled in order to represent possible changes of a prefix-term. The *label* $l(p, q)$ of an nc-edge (p, q) is $t_1 t_2^{-1}$ where t_1 and t_2 denote the prefixes of p and q, respectively. A c-edge is labelled with the

---

[4] A connection is an unordered pair of literals with the same predicate symbol and different sign. A connection (P, Q) is *weakly unifiable* iff P^d and Q' are unifiable where $Q' = Q\sigma$ for some renaming substitution σ [8].
[5] A literal L is *pure* wrt S if there is no literal L' in S such that (L, L') is a weakly unifiable connection.

empty word ϵ. If all edges of a position graph are labelled in this way we speak of a *labelled position graph*.

The idea underlying these labels is the following. Consider the label $t_1(t_2)^{-1}$ of an nc-edge of a rule R. If R is applied to a goal L, then (roughly speaking) a prefix-term t of a position occurring in L has to be unified with $t_1(x)$. Thus, t_1 represents the prefix which is "eliminated" from t. Correspondingly, t_2 represents the prefix which is "added" afterwards.

It is useful to restrict a position graph to contain only those positions and edges which may become relevant for changing a certain prefix-term during a derivation. To this end we define a corresponding reachability-relation and introduce position graphs which are reduced wrt a position.

Definition 3. Let PG_S be a position graph of a clause set S. Let $(p_1, p_2), \ldots, (p_{n-1}, p_n)$ denote a sequence of edges in PG_S. We call the sequence *allowed* iff (p_i, p_{i+1}) is a c-edge iff i is odd and an nc-edge otherwise.

A position p_n in PG_S is p_1-*reachable* iff either $n = 1$ or there exists an allowed sequence of edges $(p_1, p_2), \ldots, (p_{n-1}, p_n)$ in PG_S.

Definition 4. A position graph of a clause set S *wrt a position* p, written PG_S^p, contains exactly those positions of PG_S which are p-reachable and exactly those edges that occur in an allowed sequence of edges starting in p.

Example 3. Consider the following program S

$$P(a, b).$$
$$Q(g(x, y), y) \leftarrow P(x, y).$$
$$P(f(x), y) \leftarrow Q(g(x, y), y).$$

together with the query $G = \leftarrow P(f(g(a, b)), b)$. Let q denote the position in G corresponding to the constant symbol a. The position graph $PG_{S \cup \{G\}}^q$ is depicted in Figure 1.

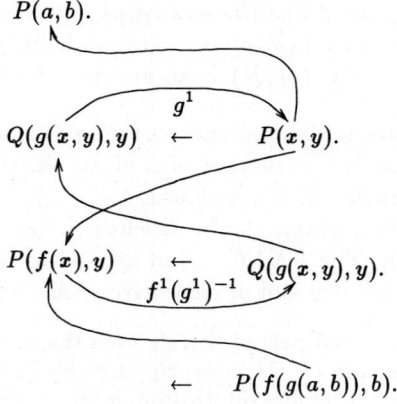

Fig. 1. Labelled Position Graph for Example 3. Labels of c-edges are omitted.

The main question is how the information encoded by position graphs can be exploited automatically. A first answer to this question was given in [7]. It is motivated by the fact that, **given** a labelled position graph PG_S^q, if we mark the position q as start-position **and every** position p with $in(p) \neq \emptyset$ and $out(p) = \emptyset$ as end-position, the resulting graph defines a finite automaton and therefore a regular language $\mathcal{L}$. A word included in $\mathcal{L}$ represents the change of the prefix-term of q during a (possibly successful) derivation of the query containing q.

Example 4. Reconsider the position graph depicted in Figure 1. Let q be the position corresponding to the constant symbol a in the query and let p be the position corresponding to a in the fact $P(a,b)$. Marking q as start-position and p as end-position, the resulting graph represents a finite automaton which accepts the regular language $\{(f^1(g^1)^{-1}g^1)^i \mid i > 0\}$. Taking into account that the successive generation and elimination of the same function symbol yields the identity (this corresponds to rules of the form $f^{-1}f \mapsto \epsilon^6$), we get the language $\mathcal{L} = \{(f^1)^i \mid i > 0\}$. Since $f^1 g^1 \notin \mathcal{L}$, it is impossible to eliminate the whole prefix $f^1 g^1$ during a derivation. Thus, the sole fact can never be applied and we can conclude that no refutation of $S \cup \{G\}$ exists.[7]

Although this technique detects non-provable goals in many situations, some serious problems remain. One major drawback is the inability to exploit dependencies caused by multiple occurrences of variables in the body of a clause. If a derivation fails because some goal cannot be solved due to such dependencies, the pruning technique is not capable to detect this.

Another deficiency is that the decision procedure based on language acceptors is in fact complex (for instance one has to intersect acceptors [7]). This is important not only theoretically but also in view of an implementation.

3 A Translation to Monadic Clause Sets

In what follows we present an alternative approach to exploit position graphs. Given a position graph $PG_{S \cup \{G\}}^q$ of a program S and a query G, we show that it is possible to generate a clause set S' such that S' encodes exactly the possible changes of the prefix-term of q, where q is a position in G. As the main result we prove that S' is unsatisfiable if there is an SLD-refutation of $S \cup \{G\}$.

This approach is conceptually easier than the one sketched in the previous section. Instead of using language acceptors and performing difficult intersections, we simply have to test the satisfiability of a clause set employing purely deductive methods. More important, it enables to check dependencies caused by multiple occurrences of variables.

The following definition shows how a position graph is encoded by a so-called monadic clause set.

---

[6] To apply such rules push-down acceptors instead of finite automata have to be used.
[7] We kept the example as simple as possible. If the program is extended by a rule of the form $P(x,y) \leftarrow P(f(x),y)$ an attempt to find a refutation leads to infinite looping. The analysis which detects that the query cannot be solved follows exactly the same line.

Definition 5. Let PG_S^q be a position graph of a clause set S wrt a position q. The corresponding *monadic clause set* M_S^q is defined as follows.

Let p be a q-reachable position which occurs in the argument t of a literal L and has the prefix-term t_p (in t). p is associated with a literal L_p which contains a unique monadic predicate symbol, the same sign as L, and argument t_p. L_p' emerges from L_p^d by substituting t_p by a variable.

1. Let p be a q-reachable position such that either $p = q$ or $out(p) = \emptyset$. Such a position is called *start-position* or *end-position*, respectively. Then M_S^q contains the clause $\{L_p\}$.
2. Let p be a q-reachable position and $(p, p_1), \ldots, (p, p_n)$ be all nc-edges starting in p. Then M_S^q contains the clause $\{L_p, L_{p_1}, \ldots, L_{p_n}\}$.
3. Let (p_1, p_2) be a c-edge of PG_S^q. Then M_S^q contains the clause $\{L_{p_1}', L_{p_2}'\}$ where the variables occurring in L_{p_1}' and L_{p_2}' are the same.

This definition needs explanation. As mentioned above, the set of clauses M_S^q encodes the position graph PG_S^q. Each position p of the graph is associated with a literal L_p. Clauses containing only one literal represent start- or end-positions. Clauses containing several literals encode edges of PG_S^q. The nc-edges $(p, p_1), \ldots, (p, p_n)$ starting in a position p are represented together in a single clause $\{L_p, L_{p_1}, \ldots, L_{p_n}\}$. This clause can be seen as an abstraction of the original clause which just includes the positions relevant for p. A c-edge (p_1, p_2) is represented by the clause $\{L_{p_1}', L_{p_2}'\}$ whose literals are unifiable with the literals representing p_1 and p_2. Thus, it connects literals contained in clauses which represent nc-edges, start-positions, or end-positions.

Example 5. Reconsider the position graph of Example 3 given in Figure 1. According to Definition 5, $M_{S \cup \{G\}}^q$ consists of the following clauses:

(1) $\{\neg P_1(f^1(g^1(a)))\}$ (4) $\{Q_2(g^1(x)), \neg P_3(x)\}$ (7) $\{P_3(x), \neg P_4(x)\}$
(2) $\{P_4(a)\}$ (5) $\{P_1(x), \neg P_2(x)\}$ (8) $\{P_3(x), \neg P_2(x)\}$
(3) $\{P_2(f^1(x)), \neg Q_1(g^1(x))\}$ (6) $\{Q_1(x), \neg Q_2(x)\}$

Clauses (1) and (2) represent the start and end-position contained in the query and the fact which have the prefix-terms $f^1(g^1(a))$ and a, respectively. Clauses (3) and (4) encode the nc-edges of $PG_{S \cup \{G\}}^q$ whereas clauses (5) to (8) represent the c-edges of $PG_{S \cup \{G\}}^q$.

Let S be a program and G a query containing a position q. It is easy to recognize the correspondence between the original problem to find an SLD-refutation of $S \cup \{G\}$ and the task to check the unsatisfiability of $M_{S \cup \{G\}}^q$. Whenever there is an SLD-derivation of $S \cup \{G\}$ which modifies the prefix-term of q, there is a corresponding SLD-derivation of $M_{S \cup \{G\}}^q$ (with query $\{L_q\}$) which modifies the prefix-term exactly in the same way. Thus, if there exists an SLD-refutation for $S \cup \{G\}$, there exists an SLD-refutation for $M_{S \cup \{G\}}^q$, too. Hence, the problem to show the unsatisfiability of $M_{S \cup \{G\}}^q$ is a complete approximation of the problem to find an SLD-refutation for $S \cup \{G\}$.

Theorem 6. *Let $PG^q_{S\cup\{G\}}$ be a position graph of a program S and a query G wrt a position q occurring in G.*
If $M^q_{S\cup\{G\}}$ is satisfiable there exists no SLD-refutation of $S \cup \{G\}$.

In order to guarantee that the prefix-term of q can be modified in an appropriate way, the unsatisfiability of a monadic clause set $M^q_{S\cup\{G\}}$ has to depend on the clause $\{L_q\}$. But in case S is a program, this obviously holds since $\{L_q\}$ is the sole query in $M^q_{S\cup\{G\}}$.

One main difference to the approach sketched in Section 2 is the ability to take dependencies caused by multiple occurrences of a variable into account. This is because the nc-edges starting in a position are encoded by *one* clause.[8]

Example 6. Let S be the following clause set.

$$Q(a,b).$$
$$Q(f(v),f(w)) \leftarrow Q(v,w).$$
$$P(y) \leftarrow Q(y,y).$$
$$\leftarrow P(x).$$

Consider the position graph PG^q_S where q denotes the position corresponding to the variable x in the query. A pruning technique neglecting the dependency caused by the occurrences of the variable y in the literal $Q(y,y)$ fails to recognize that S is satisfiable. The monadic clause set M^q_S, however, includes a clause of the form $\{P_2(y), \neg Q_{1,1}(y), \neg Q_{1,2}(y)\}$ which contains this dependency. This causes M^q_S to be satisfiable and therefore, according to Theorem 6, S is satisfiable.

4 Using Ordered Resolution as Decision Procedure

Many variants of ordered resolution have been proposed as efficient decision procedures for special restricted kinds of clause sets (for a summary see [9]). In particular it was shown that clause sets restricted to monadic predicate and function symbols can be decided by ordered resolution. Since the class of monadic clause sets, as given in Definition 5, is even more restricted, we can show that it is sufficient to use a conceptually simple calculus for our purposes.

The following definitions are taken (with slight modifications) from [9].

Definition 7. Let $<_A$ be any A-ordering[9]. C and D are variable disjoint clauses and M and N are subsets of C and D, respectively, such that $N^d \cup M$ are unifiable by the m.g.u. σ. The atom B of $(N^d \cup M)\sigma$ is called the resolved atom. If for no atom L in $(C-M)\sigma \cup (D-N)\sigma$ $B <_A L$ holds, $(C-M)\sigma \cup (D-N)\sigma$ is called the *OR-resolvent* of C and D.

---

[8] One might achieve the same results using tree-automata instead of push-down acceptors. But even if this is true (what we have not investigated in), a corresponding technique would be conceptually much more complicated.

[9] An *A-ordering* $<_A$ is a binary relation on atoms such that (1) $<_A$ is irreflexive, (2) $<_A$ is transitive, and (3) for all atoms C, D and all substitutions σ: $C <_A D$ implies $C\sigma <_A D\sigma$.

Definition 8. Let $<_A$ be any A-ordering. For a clause set S we define $\mathrm{OR}(S)$ as the set of all OR-resolvents (wrt $<_A$) of S. Additionally we define: (1) $R^0(S) = S$, (2) $R^{i+1}(S) = R^i(S) \cup \mathrm{OR}(R^i(S))$, and (3) $R^*(S) = \bigcup_i R^i(S)$. We say that a clause C is *derivable* from a clause set S iff $C \in R^*(S)$.

Let $<_d$ be the A-ordering defined as follows: Let P and Q be two monadic literals. $P <_d Q$ holds iff (i) $\mathrm{depth}(P) < \mathrm{depth}(Q)$[10] and (ii) P and Q either contain the same variable or are both ground. OR-resolution wrt this A-ordering (and optionally augmented by subsumption reduction) is denoted as OR_d-resolution.

From [9] (Theorem 5.3) we know that ordered resolution is complete for any A-ordering. In combination with the following theorem this implies that monadic clause sets can be decided by OR_d-resolution. Thus, we can use a conceptionally simple and efficient decision procedure. For instance, it does not need splitting (eg. see [9]) as other decision procedures for more general classes of clause sets.

Theorem 9. *Let S be a monadic clause set. Using OR_d-resolution only finitely many clauses are derivable from S.*

Tractability. A common way to show tractability of a resolution based calculus applied to a certain class of clause sets is to ensure that the maximal number s of literals in a derived clause and the maximal depth d of these literals are bounded by a constant. This implies that only polynomially many (in s and d) clauses are derivable. OR_d-resolution applied to monadic clause sets fulfills the second criterion.

However, the first criterion is not met in general. This is because a clause of a monadic clause set may contain more than two literals. But if we restrict ourselves to consider for each position p at most one nc-edge starting in p, we can ensure that each derivable clause contains not more than two literals. Applying this restriction therefore ensures tractability. The price is that dependencies caused by multiple variable occurrences cannot be controlled any longer.

In case we do not want to restrict position graphs, the number of derivable clauses may be exponential. However, the task is clearly easier to solve than the original problem and we have a powerful decision procedure at hand. We conjecture that the worst case only becomes relevant in few examples.

Generalizations. It is clearly of interest whether our results can be enhanced. Until now we considered changes of only one prefix-term. But one can show that our results are sharp in the sense that slight generalizations cannot be achieved. In [6] it is proven that (1) a complete predetermination of possible changes of two prefix-terms in parallel is impossible, and (2) that it is impossible to predetermine completely possible changes of terms which contain besides monadic function symbols one binary function symbol. However, a control of several prefix-terms is feasible if only the possibility to eliminate (generate) prefixes is considered. Such a technique was sketched in [7].

---

[10] $\mathrm{depth}(L)$ denotes the maximal term depth of a term occurring in L.

5 Application to Arbitrary Clause Sets

In the previous sections we restricted our attention to Horn-clause sets. In what follows we provide an extension to arbitrary clause sets. Consequently, it is not sufficient to consider SLD-derivations. Instead we shall study ME-derivations. Due to lack of space these extensions are only motivated superficially. A thorough treatment can be found in [6].

The results of Section 3 cannot be taken over in an easy fashion since Theorem 6 does not hold for non-Horn clause sets in general. This is because during a derivation — speaking in terms of ME — we not only have to take extension steps into account. Additionally we have to consider reduction steps.

This brings up a problem which can be sketched as follows. Let S be a clause set, Q a query in S, and q a position occurring in Q. In case S is Horn it is easy to see that given an ME-derivation D (containing no reduction steps) of S with top-clause Q, there is an ME-derivation D' of M_S^q with top-clause $\{L_q\}$ such that two successive extension steps in D' can be related directly to an extension step in D. Hence, the structure of derivations is preserved to some extent.

In case S is non-Horn and D contains reduction steps, there is no such correspondence in general. This may cause M_S^q to be satisfiable although an ME-refutation of S with top-clause Q exists. This can be illustrated as follows.

Example 7. Consider the following unsatisfiable (non-Horn) clause set S:

$$\{\{Q(v,w,v), Q(w,v,w)\}, \{P(z), \neg Q(z,y,x), \neg Q(a,x,y)\}, \{\neg P(a)\}\}$$

Let q be the position corresponding to the constant symbol a in the rightmost clause. It is easy to find an ME-refutation of S with top-clause $\{\neg P(a)\}$ which includes two reduction steps. But consider a possible ME-derivation D' of M_S^q with top-clause $\{L_q\}$. For each allowed cyclic path $(p, p_1), (p_2, p_3), \ldots, (p_n, p)$ in P_S^q, (p, p_1) is a c-edge and (p_n, p) an nc-edge (or vice versa). Considering Definition 5 this implies that D' cannot contain any reduction step. On the other hand P_S^q contains no end-position, and therefore no ME-refutation of M_S^q with top-clause $\{L_q\}$ can exist. Since $M_S^q \setminus \{\{L_q\}\}$ is satisfiable one can conclude that M_S^q is satisfiable.

Obvious ways to overcome this problem are to simplify the original clause set to a Horn-clause set before applying our technique[11] or to restrict the application to goals which can be proven using exclusively extension steps [20]. Employing such solutions, however, one either may have to eliminate possibly big and important parts of a clause set or to restrict the applicability of the technique only to some possibly small parts of a derivation. Therefore, we propose an alternative way to cope with this problem. Roughly speaking, the idea is to restrict position graphs in such a way that they cannot contain cycles which, on the one hand, can be put down to the non-Horn structure of the current clause set and, on the other hand, may be the reason for a monadic clause set to be

---

[11] Simplifying general clause sets in order to apply efficient procedures afterwards is a common approach in the field of approximative reasoning [17].

falsely satisfiable. The following definition provides a corresponding syntactic criterion.

Definition 10. We call a position graph PG *acceptable* if it fulfills the following criterion: For $1 \leq i \leq n$ let p_i and q_i be different positions which occur in the same literal. If there exist allowed sequences of edges

$$(p_i, r_{i,1}), (r_{i,1}, r_{i,2}), \ldots, (r_{i,m_i-1}, r_{i,m_i}), (r_{i,m_i}, q_i)$$

in PG such that (i) $(p_i, r_{i,1})$ and (r_{i,m_i}, q_i) are both c-edges for $1 \leq i \leq n$, and (ii) (q_i, p_{i+1}) is an nc-edge for $1 \leq i < n$[12], then (q_n, p_1) is no nc-edge of PG.

To verify that the position graph PG_S^q of the clause set S given in Example 7 is in fact not acceptable, just set $n = 2$, let p_1 and q_1 correspond to the variables x and y in the literal $\neg Q(z, y, x)$, and let p_2 and q_2 correspond to the variables y and x in the literal $\neg Q(a, x, y)$. Obviously, there are sequences of edges $(p_1, r), (r, r'), (r', q_1)$ and $(p_2, r), (r, r'), (r', q_2)$ in PG_S^q such that the first and the last element of these sequences are c-edges. (let r be the position corresponding to the second occurrence of the variable v in the literal $Q(v, w, v)$ and let r' be the position corresponding to the variable v in the literal $Q(w, v, w)$). Since (q_2, p_1) is an nc-edge, PG_S^q is not acceptable.

The criterion of acceptability is not too restrictive. As proven in [6], all position graphs M_S^q, where S is a Horn-clause set and q a position in a query, are acceptable. On the other hand, many position graphs for non-Horn-clause sets are acceptable, too. For instance, a position graph for the clause set in Example 1 is acceptable. In those cases where the criterion is violated, this can be overcome without requiring too much simplifications of the respective position graph. One simply has to remove the nc-edge (q_n, p_1) and (for reasons of symmetry) the nc-edge (p_1, q_n). In Example 7 it is therefore sufficient to remove the nc-edges connecting the occurrences of the variable x in the second clause.

Taking acceptable position graphs into account, it is possible to prove the following theorem which is similar to Theorem 6. Note that in order to apply this theorem, the results of Section 4 can be used without any modifications.

Theorem 11. *Let S be a clause set and G a goal occurring in an ME-derivation D of S. Let $PG_{S \cup \{\{G\}\}}^q$ be an acceptable position graph where q is a position occurring in $\{G\}$,*

If $M_{S \cup \{\{G\}\}}^q$ is satisfiable then D cannot be completed to an ME-refutation.[13]

In case we want to guarantee that the unsatisfiability of $M_{S \cup \{\{G\}\}}^q$ depends on $\{L_q\}$ we have to add a corresponding condition which states that the derivation of the empty clause from $M_{S \cup \{\{G\}\}}^q$ has to involve $\{L_q\}$. This causes neither theoretical nor practical problems (see [6] for the technical details).

---

[12] Hence, the positions $p_1, \ldots, p_n, q_1, \ldots, q_n$ occur in the same clause.
[13] The satisfiability of $M_{S \cup \{\{G\}\}}^q$ does not imply that it is impossible to find a subrefutation for G. Such a subrefutation, however, cannot contribute to an overall solution.

6 Relation to Abstraction

Our approach is related to the technique of abstraction introduced in [15]. In this section we point out common aspects[14] and differences.

Roughly speaking, abstraction can be described as the process of mapping a clause set S onto a simpler clause set T. It is important that also resolution proofs from S can be mapped on a simpler proof from T. Having found a proof from T it may be possible to find a proof for S by inverting the mapping.

Definition 12. An *abstraction* is an association of a set $f(C)$ of clauses with each clause C such that f has the following properties:

1. If clause C_3 is a resolvent of C_1 and C_2 and $D_3 \in f(C_3)$, then there exist $D_1 \in f(C_1)$ and $D_2 \in f(C_2)$ such that some resolvent of D_1 and D_2 subsumes D_3.
2. $f(\Box) = \{\Box\}$.
3. If C_1 subsumes C_2, then for every abstraction D_2 of C_2 there is an abstraction D_1 of C_1 such that D_1 subsumes D_2.

If f is a mapping with these properties, then we call f an *abstraction mapping* of clauses. Also, if $D \in f(C)$ we call D an abstraction of C.

Examples for abstractions given in [15] are (beside others) the ground abstraction which maps a clause to all its ground instances and an abstraction $f(C) = \{C'\}$ where C' is C with certain arguments of certain function or predicate symbols deleted.

Theorem 13 [15]. *Let S be a set of clauses and f an abstraction mapping for S. If S is unsatisfiable so is $f(S)$.*

This result is similar to Theorem 6. It expresses that an abstraction of a clause set can be used in the same way as a monadic clause set. The main difference to our approach is that abstraction focusses on global structural features of literals and clauses — such as the arity of predicates — whereas our approach captures an aspect of a derivation — the change of a prefix-term — which in many cases cannot be easily described by such global structural features.

The following two examples illustrate this relationship. Whereas the first one shows that a clause set which is equivalent to the corresponding monadic clause set may be obtained by abstraction, the second one shows that for some monadic clause sets no corresponding abstraction can be found easily.

Example 8. Reconsider Example 3. Suppose a mapping is applied which deletes from each predicate and function symbol the second argument. The resulting clause set is equivalent to $M^q_{S \cup \{G\}}$ since, roughly speaking, those argument positions are preserved which are included in the position graph $PG^q_{S \cup \{G\}}$.

---

[14] We only take so-called syntactic abstractions into account. Semantic abstractions, also introduced in [15], are hard to compare with our approach.

Example 9. Consider the Horn-clause set S consisting of the fact $\{P(f(b), f(a))\}$, the rule $\{P(x, f(x)), \neg P(y, f(x))\}$, and the query $\{\neg P(f(a), z)\}$. Let q denote the position corresponding to the constant symbol a in the query.

It is easy to verify that M_S^q is satisfiable. However, an abstraction which is defined by just dropping arguments is unsatisfiable. The definition of an abstraction mapping yielding a satisfiable clause set would be in fact complicated (and even if it is possible the same concepts as in Section 2 and 3 have to be used).

Concerning the extension of our approach to arbitrary clause sets, matters are even worse. This is because a correspondence between derivations of "original" clause sets and derivations of respective monadic clause sets is only given in the Horn case (see Section 5). In the non-Horn case this does not hold any longer whereas abstractions preserve the proof structure in any case. It is furthermore not clear what might correspond to acceptability in the context of abstraction.

7 Future Work

To test our technique adequately, the main task for the future is to implement it in the ME based theorem prover KoMeT which is currently developed in our group. Obviously, this has to be done with care.

As much work as possible has to be shifted into a preprocessing step. For instance, most computations needed to construct acceptable position graphs and the corresponding monadic clause sets do not have to be performed during a deduction. This can be achieved basically in three steps: (1) a graph PG_S is built which contains all c-edges and nc-edges possibly occurring in a position graph of a clause set S, (2) nc-edges that violate acceptability are removed from PG_S, and (3) for each edge and for each end-position a corresponding monadic clause is generated. At runtime, given a concrete goal G, one only has to generate clauses corresponding to the start-position, say p, and for the c-edges connecting p with another position. Finally, the remaining clauses of $M_{S \cup \{G\}}^p$ are selected employing essentially the concept of p-reachability.[15]

To perform step (2) efficiently one can build up a table that contains the information which position q is reachable from another position p by an allowed sequence of edges $e_1, \ldots, e_n$ such that e_1 and e_n are c-edges. This can be done in polynomial time using classical graph algorithms. Acceptability can then be checked by inspecting this table.

We are currently also investigating the possibility to reduce the amount of OR-resolution steps (which is required for an application of our technique during a deduction) by partially evaluating the clause set yielded after step (3). This seems to be particularly useful if S is Horn.

---

[15] Note that unless we neglect variable dependencies (see Section 4), what in many cases avoids the detection of useless goals, there is only one position graph for each goal. Otherwise we have to select between different graphs which requires heuristics taking the prefix of p and the labels of the respective nc-edges into account.

Such implementation refinements, however, will not prevent that an application of our technique to each goal produces an unacceptable computational overhead. To show that our approach can be employed successfully, we have to investigate in strategies for invoking the application of our technique.

There are obviously many such strategies whose usefulness has to be validated by testing them on different classes of clause sets. For example, a strategy can depend on the depth of terms occurring in goals. Invoking our technique if this depth exceeds a user defined limit, useless derivation chains that steadily build up terms can be detected. In order to identify useless loops it is furthermore useful to consider regularities in the sequence of clause instances used to derive a goal. To reduce the number of applications of our technique it is also reasonable to take only such goals into account whose predicates are defined recursively. Another possibility is to mark some literals of a given clause set by the user to indicate which kinds of goals are likely to produce infinite loops.

Besides using our technique during a deduction it is also worth considering to employ it only in a preprocessing step to detect useless clauses, i.e. clauses that cannot contribute to a successful derivation since they contain non-provable literals (see also [21]).

8 Conclusion

We proposed a new approach to augment top-down backward-chaining calculi such that goals that cannot contribute to a successful derivation are identified.[16] The general idea, which was adopted from cycle unification, is to analyze possible changes of a term during a derivation. Due to undecidability results we restricted ourselves to consider possible changes of prefix-terms.

This kind of term manipulation during a derivation can be modelled using (acceptable) position graphs. Position graphs themselves are represented by so-called monadic clause sets. It was proven that such sets can be used as complete approximations and that they can be decided by using a conceptually simple and efficient procedure based on ordered resolution. Furthermore we discussed the complexity of our technique and emphasized that our results are sharp in the sense that it is not possible to control more complex forms of term-manipulation such as the changes of two prefix-terms in parallel.

We investigated the relation of our approach to the technique of abstraction. We showed that for some examples equivalent results can be achieved, whereas for other examples our approach is more suitable for our purposes. Finally we discussed aspects of a future implementation.

There exist other related approaches in the field of logic programming not mentioned so far. Most of them are based on abstract interpretation. There, the general idea is to approximate the fixpoints of logic programs (eg. see [11, 21]). For some classes of logic programs fixpoints can be approximated accurately. In

---

[16] Although we only considered SLD-resolution and ME throughout this paper, our results should be equally applicable to other calculi like most connection calculi presented in [3].

most cases, however, a depth restriction has to be imposed during the interpretation process such that precision is lost. This is different from our approach.

References

1. W. Bibel. *Automated Theorem Proving*. Vieweg Verlag, 1987. Second edition.
2. W. Bibel. Perspectives on automated deduction. In *Automated Reasoning: Essays in Honor of Woody Bledsoe*, pages 77–104. Kluwer Academic, Utrecht, 1991.
3. W. Bibel. *Deduction: Automated Logic*. Academic Press, London, 1993.
4. W. Bibel, S. Hölldobler, and J. Würtz. Cycle unification. *Proceedings of the Conference on Automated Deduction*, pages 94–108. Springer, Berlin, 1992.
5. R. N. Bol, K. R. Apt, and J. W. Klop. An analysis of loop checking mechanisms for logic programming. *Theoretical Computer Science*, 86:35–79, 1991.
6. S. Brüning. Detecting Non-Provable Goals. Technical report, FG Intellektik, FB Informatik, TH Darmstadt, 1993.
7. S. Brüning. Search Space Pruning by Checking Dynamic Term Growth. *Proceedings of the International Conference on Logic Programming and Automated Reasoning*, pages 52–63. Springer, 1993.
8. E. Eder. Properties of substitutions and unifications. *Journal of Symbolic Computation*, 1:31–46, 1985.
9. C. Fermüller, A. Leitsch, T. Tammet, and N. Zamov. *Resolution Methods for the Decision Problem*. LNAI 679. Springer, 1993.
10. P. Hanschke and J. Würtz. Satisfiability of the smallest binary program. *Information Processing Letters*, 45(5):237–241, April 1993.
11. G. Janssens and M. Bruynooghe. Deriving Descriptions of Possible Values of Program Variables by Means of Abstract Interpretation. *Journal of Logic Programming*, 13:205–258, 1992.
12. R. Letz, J. Schumann, S. Bayerl, and W. Bibel. SETHEO — A High–Performance Theorem Prover for First–Order Logic. *Journal of Automated Reasoning*, 8:183–212, 1992.
13. J. W. Lloyd. *Foundations of Logic Programming*. Springer, second edition, 1987.
14. D. W. Loveland. Mechanical theorem proving by model elimination. *Journal of the ACM*, 15:236–251, 1986.
15. D. A. Plaisted. Theorem Proving with Abstraction. *Artificial Intelligence*, 16:47–108, 1981.
16. J. A. Robinson. A machine-oriented logic based on the resolution principle. *Journal of the ACM*, 12(1):23–41, 1965.
17. B. Selman and H. Kautz. Knowlede Compilation Using Horn Approximations. In *Proceedings of the AAAI National Conference on Artificial Intelligence*, 1991.
18. D. E. Smith, M. R. Genesereth, and M. L. Ginsberg. Controlling recursive inference. *Artificial Intelligence*, 30:343–389, 1986.
19. M. E. Stickel. A Prolog technology theorem prover. *10th International Conference on Automated Deduction*, pages 673–674, Springer, 1990.
20. G. Sutcliffe. Linear-Input Subset Analysis. *Proceedings of the Conference on Automated Deduction*, pages 268–280. Springer, 1992.
21. D. A. de Waal and J. Gallagher. Logic program specialisation with deletion of useless clauses (poster abstract). *Proceedings of the 1993 Logic programming Syposium*, page 632, MIT Press,1993.
22. E. Yardeni and E. Shapiro. A Type System for Logic Programs. *Journal of Logic Programming*, 10:125–153,1991.

Panel Discussion: A Mechanically Proof-Checked Encyclopedia of Mathematics: Should We Build One? Can We?

Robert S. Boyer

University of Texas at Austin and Computational Logic, Inc.

At the suggestion of the CADE–12 Program Committee Chairman, Professor Alan Bundy, a panel discussion on the topic "A Mechanically Proof-Checked Encyclopedia of Mathematics: Should We Build One? Can We?" has been organized, with participants Robert S. Boyer, N. G. de Bruijn, Gérard Huet, and Andrzej Trybulec. This discussion will consider an idea, a vision, that has been shared and pursued by a number of groups for at least several decades, an idea that is also described in the following, anonymously authored document *The QED Manifesto*, which has been influenced by contributions and criticisms from numerous quarters.

Robert S. Boyer has worked on Nqthm, also known as the Boyer–Moore Prover. See, for example *A Computational Logic Handbook*, Robert S. Boyer and J Strother Moore, Academic Press, 1988. Nqthm may be obtained by anonymous ftp from ftp.cli.com in /pub/nqthm/nqthm-1992/. Boyer may be reached at the Department of Computer Sciences, University of Texas at Austin, Austin, Texas, U.S.A., or as boyer@cli.com.

N. G. de Bruijn has worked on the Automath project. See, for example, "Checking Mathematics with Computer Assistance," N. G. de Bruijn, *Notices of the American Mathematical Society*, January 1991, Volume 38, Number 1, pp. 8–15. De Bruijn may be reached at the Department of Mathematics and Computing Science, Eindhoven University of Technology, PO Box 513, 56000MB Eindhoven, The Netherlands, or as wsdwnb@win.tue.nl.

Gérard Huet has worked on the Coq project. See, for example, "The Calculus of Constructions," Th. Coquand and G. Huet, *Information and Computation*, Volume 76, 1988, pp. 95–120. Coq may be obtained by anonymous ftp from ftp.inria.fr as the file /INRIA/Projects/formel/coq/V5.8.3/coq.tar.Z. Huet may be reached at INRIA, B.P. 105, 78153 Le Chesnay Cedex, France, or as Gerard.Huet@inria.fr.

Andrzej Trybulec has worked on the Mizar project. See, for example, the entire contents of the proof-checked journal *Formalized Mathematics (*a computer assisted approach*)*, which may be ordered from Fondation Philippe le Hody, Mizar Users Group, Av. F. Roosevelt 134 (Bte 7), 1050 Brussels, Belgium. Mizar may be obtained by anonymous ftp from menaik.cs.ualberta.ca in pub/Mizar. Trybulec may be reached at Institute of Mathematics, Warsaw University in Białystok, 15-267 Białystok, Akademicka 2, Poland or as TRYBULEC@PLBIAL11.BITNET.

The QED Manifesto

Authorship and copyright information for the QED Manifesto may be found at the end.

1 What Is the QED Project and Why Is It Important?

> The development of mathematics toward greater precision has led, as is well known, to the formalization of large tracts of it, so that one can prove any theorem using nothing but a few mechanical rules.
> GÖDEL

> If civilization continues to advance, in the next two thousand years the overwhelming novelty in human thought will be the dominance of mathematical understanding.
> WHITEHEAD

QED is the very tentative title of a project to build a computer system that effectively represents all important mathematical knowledge and techniques. The QED system will conform to the highest standards of mathematical rigor, including the use of strict formality in the internal representation of knowledge and the use of mechanical methods to check proofs of the correctness of all entries in the system.

The QED project will be a major scientific undertaking requiring the cooperation and effort of hundreds of deep mathematical minds, considerable ingenuity by many computer scientists, and broad support and leadership from research agencies. In the interest of enlisting a wide community of collaborators and supporters, we now mention five reasons that the QED project should be undertaken.

First, the increase of mathematical knowledge during the last two hundred years has made the knowledge, let alone understanding, of all, or even of the most important, mathematical results something beyond the capacity of any human. For example, few mathematicians, if any, will ever understand the entirety of the recently settled structure of simple finite groups or the proof of the four color theorem. Remarkably, however, the creation of mathematical logic and the advance of computing technology have also provided the means for building a computing system that represents all important mathematical knowledge in an entirely rigorous and mechanically usable fashion. The QED system we imagine will provide a means by which mathematicians and scientists can scan the entirety of mathematical knowledge for relevant results and, using tools of the QED system, build upon such results with reliability and confidence but without the need for minute comprehension of the details or even the ultimate foundations of the parts of the system upon which they build. Note that the approach will almost surely be an incremental one: the most important and applicable results will likely become available before the more obscure and purely theoretical ones are tackled, thus leading to a useful system in the relatively near term.

Second, the development of high technology is an endeavor of fabulously increasing mathematical complexity. The internal documentation of the next generation of microprocessor chips may run, we have heard, to thousands of pages. The specification of a major new industrial system, such as a fly-by-wire airliner or an autonomous undersea mining operation, is likely to be even an order of magnitude greater in complexity, not the least reason being that such a system would perhaps include dozens of microprocessors. We believe that an industrial designer will be able to take parts of the QED system and use them to build reliable formal mathematical models of not only a new industrial system but even the interaction of that system with a formalization of the external world. We believe that such large mathematical models will provide a key principle for the construction of systems substantially more complex than those of today, with no loss but rather an increase in reliability. As such models become increasingly complex, it will be a major benefit to have them available in stable, rigorous, public form for use by many. The QED system will be a key component of systems for verifying and even synthesizing computing systems, both hardware and software.

The third motivation for the QED project is education. Nothing is more important than mathematics education to the creation of infrastructure for technology-based economic growth. The development of mathematical ability is notoriously dependent upon *doing* rather than upon *being told* or *remembering*. The QED system will provide, via such techniques as interactive proof checking algorithms and an endless variety of mathematical results at all levels, an opportunity for the one-on-one presenting, checking, and debugging of mathematical technique, which it is so expensive to provide by the method of one trained mathematician in dialogue with one student. QED can provide an engaging and non-threatening framework for the carrying out of proofs by students, in the same spirit as a long-standing program of Suppes at Stanford for example. Students will be able to get a deeper understanding of mathematics by seeing better the role that lemmas play in proofs and by seeing which kinds of manipulations are valid in which kinds of structures. Today few students get a grasp of mathematics at a detailed level, but via experimentation with a computerized laboratory, that number will increase. In fact, students can be used (eagerly, we think) to contribute to the development of the body of definitions and proved theorems in QED. Let us also make the observation that the relationship of QED to education may be seen in the following broad context: with increasing technology available, governments will look not only to cut costs of education but will increasingly turn to make education and its delivery more cost-effective and beneficial for the state and the individual.

Fourth, although it is not a practical motivation, nevertheless perhaps the foremost motivation for the QED project is cultural. Mathematics is arguably the foremost creation of the human mind. The QED system will be an object of significant cultural character, demonstrably and physically expressing the staggering depth and power of mathematics. Like the great pyramids, the effort required (especially early on) may be great; but the rewards can be even more staggering than this effort. Mathematics is one of the most basic things that unites all people, and helps illuminate some of the most fundamental truths of nature, even of being itself. In the last one hundred years, many traditional cultural values of our civilization have taken a severe beating, and the advance of science has received no small blame for this beating. The QED system will provide a beautiful and compelling monument to the fundamental reality of truth. It will thus provide some antidote to the degenerative effects of cultural relativism and nihilism. In providing motivations for things, one runs the danger of an infinite regression. In the end, we take some things as inherently valuable in themselves. We believe that the construction, use, and even contemplation of the QED system will be one of these, over and above the practical values of such a system. In support of this line of thought, let us cite Aristotle, the Philosopher, the Father of Logic,

> That which is proper to each thing is by nature best and most pleasant for each thing; for man, therefore, the life according to reason is best and pleasantest, since reason more than anything else is man.

We speculate that this cultural motivation may be the foremost motivation for the QED project. Sheer aesthetic beauty is a major, perhaps the major, force in the motivation of mathematicians, so it may be that such a cultural, aesthetic motivation will be the key motivation inciting mathematicians to participate.

Fifth, the QED system may help preserve mathematics from corruption. We must remember that mathematics essentially disappeared from Western civilization once, during the dark ages. Could it happen again? We must also remember how unprecedented in the history of mathematics is the clarity, even perfection, that developed in this century in regard to the idea of formal proof, and the foundation of essentially the entirety of known mathematics upon set theory. One can easily imagine corrupting forces that could undermine these achievements. For example, one might suspect that there is already a trend towards believing some recent "theorems" in physics because they offer some predictive power rather than that they have any meaning, much less rigorous proof, with a possible erosion in established standards of rigor. The QED system could offer an antidote to any such tendency. The standard, impartial answer to the question "Has it been proved?" could become "Has it been checked by the QED system?" Such a mechanical proof checker could provide answers immune to pressures of emotion, fashion, and politics.

2 Some Objections to the Idea of the QED Project and Some Responses

> The peculiarity of the evidence of mathematical truths is that all the argument is on one side. There are no objections, and no answer to objections.   MILL

Objection 1. Paradoxes, Incompatible Logics, etc. Anyone familiar with the variety of mathematical paradoxes, controversies, and incompatible logics of the last hundred years will realize that it is a myth that there is certainty in mathematics. There is no fundamentally justifiable view of mathematics which has wide support, and no widely agreeable logic upon which such an edifice as QED could be founded.

Reply to Objection 1. Although there are a variety of logics, there is little doubt that one can describe all important logics within an elementary logic, such as primitive recursive arithmetic, about which there is no doubt, and within which one can reliably check proofs presented in the more controversial logics. We plan to build the QED system upon such a *root logic*. But the QED system is to be fundamentally unbiased as to the logics used in proofs. Or if there is to be a bias, it is to be a bias towards universal agreement. Proofs in all varieties of classical, constructive, and intuitionist logic will be found rigorously presented in the QED system—with sharing of proofs between logics where justified by metatheorems. For example, Gödel showed how to map theorems in classical number theory into intuitionist number theory, and E. Bishop showed how to develop much of modern mathematics in a way that is simultaneously constructive and classical. A mathematical logic may be regarded as being very much like a model of the world—one can often profit from using a model even if one ultimately chooses an alternative model because it is more suited to one's purposes. Furthermore, merely because some logic is so overly strong as to be ultimately found inconsistent or so weak as to ultimately fail to be able to express all that one hopes, one can nevertheless often transfer almost all of the technique developed in one logic to a subsequent, better logic.

Objection 2. Intellectual property problems. Such an enterprise as QED is doomed because as soon as it is even slightly successful, it will be so swamped by lawyers with issues of ownership, copyright, trade secrecy, and patent law that the necessary wide cooperation of hundreds of mathematicians, computer scientists, research agencies, and institutions will become impossible.

Reply to Objection 2. In full cognizance of the dangers of this objection, we put forward as a fundamental and initial principle that the entirety of the QED system is to be in the international public domain, so that all can freely benefit from it, and thus be inspired to contribute to its further development.

Objection 3. Too much mathematics. Mathematics is now so large that the hope of incorporating all of mathematics into a system is utterly humanly impossible, especially since new mathematics is generated faster than it can be entered into any system.

Reply to Objection 3. While it is certainly the case that we imagine anyone being free to add, in a mechanically checked, rigorous fashion, any sort of new mathematics to the QED system, it seems that as a first good objective, we should pursue checking *named* theorems and algorithms, the sort of things that are commonly taught in universities, or cited as important in current mathematics and applications of mathematics.

Objection 4. Mechanically checked formality is impossible. There is no evidence that extremely hard proofs can be put into formal form in less than some utterly ridiculous amount of work.

Reply to Objection 4. Based upon discussions with numerous workers in automated reasoning, it is our view that using current proof-checking technology, we can, using a variety of systems and expert users of those systems, check mathematics at within a factor of ten, often much better, of the time it takes a skilled mathematician to write down a proof at the level of an advanced undergraduate textbook. QED will support proof checking at the speeds and efficiencies of contemporary proof-checking systems. In fact, we see one of the benefits of the QED project as being a demonstration of the viability of mechanically-assisted (-enforced) proof-checking.

Objection 5. If QED were feasible, it would have already been underway several decades ago.

Reply to Objection 5. Many of the most well-known projects related to QED were commenced in an era in which computing was exorbitantly expensive and computer communication between geographically remote groups was not possible. Now most secretaries have more computing power than was available to most entire QED-related projects at their inception, and rapid communication between most mathematics and computer science departments through email, telnet, and ftp has become almost universal. It also now seems unlikely that any one small research group can, alone, make a major dent in the goal of incorporating all of mathematics into a single system, but at the same time technology has made widespread collaboration entirely feasible, and the time seems ripe for a larger scale, collaborative effort. It is also worth adding that research agencies may now be in a better position to recognize the Babel of incompatible reasoning systems and symbolic computation systems that have evolved from a plethora of small projects without much attention to collaboration. Then perhaps they can work towards encouraging collaboration, to minimize the lack of interoperability due to diversity of theorem-statement languages, proof languages, programming languages, computing platforms, quality, and so on.

Objection 6. QED is too expensive.

Reply to Objection 6. While this objection requires careful study at some point, we note that simply concentrating the efforts of some currently-funded projects could go a long way towards getting QED off the ground. Moreover, as noted above, students could contribute to the project as an integrated part of their studies once the framework is established, presumably at little or no cost. We can imagine a number of professionals contributing as well. In particular, there is currently a large body of tenured or retired mathematicians who have little inclination for advanced research, and we believe that some of these could be inspired to contribute to this project. It may be a good idea to have a QED governing board to recognize contributions.

Objection 7. Good mathematicians will never agree to work with formal systems because they are syntactically so constricting as to be inconsistent with creativity.

Reply to Objection 7. The written body of formal logic rightly repulses most mathematical readers. Whitehead and Russell's *Principia Mathematica* did not establish mathematics in a notation that others happily adopted. The traditional definition of formal logics is in a form that no one can stand to use in practice, e.g., with function symbols named f1, f2, f3, ... The absence of definitional principles for almost all formal logics is an indication that from the beginning, formal logics became something to be studied (for properties such as completeness) rather than to be used by humans, the practical visions of Leibniz and Frege notwithstanding. The developers of proof checking and theorem-proving systems have done little towards making their syntax tolerable to mathematicians. Yet, on this matter of syntax, there is room for the greatest hope. Although the subject of mechanical theorem-proving in general is beset with intractable or unsolvable problems, a vastly improved computer-human interface for mathematics is something easily within the grasp of current computer theory and technology. The work of Knuth on TEX and the widespread adoption of TEX by mathematicians and mathematics journals demonstrates that it is no problem for computers to deal with any known mathematical notation. Certainly, there is hard work to be done on this problem, but it is also certainly within the capacity of computer science to arrange for any rigorously definable syntax to be something that can be conveniently entered into computers, translated automatically into a suitable internal notation for formal purposes, and later reproduced in a form pleasant to humans. It is certainly feasible to arrange for the users of the QED system to be able to shift their syntax as often as they please to any new syntax, provided only that it is clear and unambiguous. Perhaps the major obstacle here is simply the current scientific reward system: precisely because new syntaxes, new parsers, and new formatters are so easy to design, little or no credit (research, academic, or financial) is currently available for working on this topic. Let us add that we need take no position on the question whether mathematicians can or should profit from the use of formal notations in the discovery of serious, deep mathematics. The QED system will be mainly useful in the final stages of proof reporting, similar to writing proofs up in journals, and perhaps possibly never in the discovery of new insights associated with deep results.

Objection 8. The QED system will be so large that it is inevitable that there will be mistakes in its structure, and the QED system will, therefore, be unreliable.

Reply to Objection 8. There is no doubt considerable room for error in the construction of the QED system, as in any human enterprise. A key motivation in Babbage's development of the computer was his objective of producing mathematical tables that had fewer errors than those produced by hand methods, an objective that has certainly been achieved. It is our experience that even with the primitive proof checking systems of today, errors made by humans are frequently found by the use of such tools, errors that would perhaps not otherwise be caught. The standard of success or failure of the QED project will not be whether it helps us to reach the kingdom of perfection, an unobtainable goal, but whether it permits us to construct proofs substantially more accurately than we can with current hand methods. In defense of the QED vision, let us assert that we believe that room for error can be radically reduced by (a) expressing the full foundation of the QED system in a few pages of mathematics and (b) supporting the development of essentially independent implementations for the basic checker. It goes without saying that in the development of any particular subfield of mathematics, errors in the statements of definitions and other axioms are possible. Agreement by experts in each mathematical subfield that the definitions are *right* will be a necessary part of establishing confidence that mechanically checked theorems establish what is intended. There is no mechanical method for guaranteeing that a logical formula says what a user intuitively means.

Objection 9. The cooperation of mathematicians is essential to building the QED edifice of proofs. However, because it is likely to remain very tedious to prove theorems formally with mechanical proof checkers for the foreeable future, mathematicians will have no incentive to help.

Reply to Objection 9. To be developed, QED does not need to attract the support of all or most mathematicians. If only a tenth of one percent of mathematicians could be attracted, that will probably be sufficient. And in compensation for the extra work currently associated with entering formal mathematics in proof checking systems, we can point out that some mathematicians may find the following benefit sufficiently compensatory: in formally expressing mathematics, one's own thoughts are often sharply clarified. One often achieves an appreciation for subtle points in proofs that one might otherwise skim over or skip. And the sheer joy of getting all the details of a hard theorem *exactly right*, because formalized and machine checked, is great for many individuals. So we conjecture that enough mathematicians will be attracted to the endeavor provided it can be sufficiently organized to have a real chance of success.

Objection 10. The QED project represents an unreasonable diversion of resources to the pursuit of the checking of ordinary mathematics when there is so much profitably to be done in support of the verification of hardware and software.

Reply to Objection 10. Current efforts in formal, mechanical hardware and software verification are exceptionally introspective, focusing upon internal matters such as compilers, operating systems, networks, multipliers, and busses. From a mathematical point of view, essentially all these verifications fall into a tiny, minor corner of elementary number theory. But eventually, verification must reach out to consider the intended effect of computing systems upon the external, continuous world with which they interact. If one attempts to try to verify the use of a DSP chip for such potentially safety critical applications as telecommunications, robot vision, speech synthesis, or cat scanning, one immediately sees the need for such basic engineering mathematics as Fourier transforms, not something at which existing verification systems are yet much good. By including the rigorous development of the mathematics used in engineering, the QED project will make a crucial contribution to the advance of the verification of computing systems.

Objection 11. The notion that interesting mathematics can ever, in practice, be formally checked is a fantasy. Whitehead and Russell spent hundreds of pages to prove something as trivial as that 0 is not 1. The notion that computing systems can be verified is another fantasy, based upon the misconception that mathematical proof can guarantee properties of physical devices.

Reply to Objection 11. That many interesting, well-known results in mathematics can be checked by machine is manifest to those who take the trouble to read the literature. One can mention merely as examples of mathematics mechanically checked from first principles: Landau's book on the foundations of analysis, Girard's paradox, Rolle's theorem, both Banach's and Knaster's fixed point theorems, the mean value theorem for derivatives and integrals over Banach-space valued functions, the fundamental counting theorem for groups, the Schroeder-Bernstein theorem, the Picard-Lindelof theorem for the existence of ODEs, Wilson's theorem, Fermat's little theorem, the law of quadratic reciprocity, Ramsey's theorem, Gödel's incompleteness theorem, and the Church-Rosser theorem. That it is possible to verify mechanically a simple, general purpose microprocessor from the level of gates and registers up through an application, via a verified compiler, has been demonstrated. So there is no argument against proof-checking or mechanical verification in principle, only an ongoing and important engineering debate about cost-effectiveness. The noisy verification debate is largely a comedy of misunderstanding. In reaction to a perceived sanctimony of some verification enthusiasts, some opponents impute to all enthusiasts grandiose claims that complete satisfaction with a computing product can be established by mathematical means. But any verification enthusiast ought to admit that, at best, verification establishes a consistency between one mathematical theory and another, e.g., between a formal specification of intended behavior of a system and a formal representation of an implementation, say in terms of gates and memory. Mathematical proof can establish neither that a specification is what any user *really wants* nor that a description of gates and memory corresponds to physical reality. So whether the results of a computation will be pleasing to or good for humans is something that cannot be formally stated, much less proved.

3 Some Background, Being a Critique of Current Related Efforts

> Although the root of logic is the same for all, the *hoi polloi* live as though they have a private understanding. HERACLITUS

In some sense project QED is already underway, via a very diverse collection of projects. Unfortunately, progress seems greatly slowed by duplication of effort and by incompatibilities. If the many people already involved in work related to QED had begun cooperation twenty-five years ago in pursuing the construction of a single system (or federation of subsystems) incorporating the work of hundreds of scientists, a substantial part of the system, including at least all of undergraduate and much of first year graduate mathematics and computer science, could already have been incorporated into the QED system by now. We offer as evidence the nontrivial fragments of that body of theorems that has been successfully completed by existing proof-checking systems.

The idea of QED is perhaps 300 years old, but one can imagine tracing it back even 2500 years. We can agree that many groups and individuals have made substantial progress on parts of this project, yet we can ask the question, is there today any project underway which can be reasonably expected to serve as the basis for QED? We believe not, we are afraid not, though we would be delighted to join any such project already underway. One of the reasons that we do not believe there is any such project underway is that we think that there exist a few basic, unsolved technical problems, which we discuss below. A second reason is that few researchers are interested in doing the hard work of checking proofs—probably due to an absence of belief that much of the entire QED edifice will ever be constructed. Another reason is that we are familiar with many automated reasoning projects but see very serious problems in many of them. Here are some of these problems.

1. Too much code to be trusted. There have been a number of automated reasoning systems that have checked many theorems of interest, but the amount of code in some of these impressive systems that must be correct if we are to have confidence in the proofs produced by these systems is vastly greater than the few pages of text that we wish to have as the foundation of QED.

2. Too strong a logic. There have been many good automated reasoning systems that *wired in* such powerful rules of inference or such powerful axioms that their work is suspect to many of those who might be tempted to contribute to QED—those of an intuitionistic or constructivist bent.

3. Too limited a logic. Some projects have been developed upon intuitionistic or constructive lines, but seem unlikely, so far anyway, to support also the effective checking of theorems in classical mathematics. We regard this *boot-strapping problem*—how to get, rigorously, from checking theorems in a weak logic to theorems in a powerful classical logic, in an effective way—to be a key unsolved technical obstacle to QED.

4. Too unintelligible a logic. Some people have attempted to start projects on a basis that is extremely obscure, at least when observed by most of the community. We believe that if the initial, base, root logic is not widely known, understood, and accepted, there will never be much enthusiasm for QED, and hence it will never get off the ground. It will take the cooperation of many, many people to build the QED system.

5. Too unnatural a syntax. Just as QED must support a variety of logics, so too must it support a variety of syntaxes, enough to make most groups of mathematicians happy when they read theorems they are looking for. It is unreasonable to expect mathematicians to have to use some computer oriented or otherwise extremely simplified syntax when concentrating on deep mathematical thoughts. Of course, a rigorous development of the syntaxes will be essential, and it will be a burden on human readers using the QED proof tree to *know* not only the logical theory in which any theorem or procedure they are reading is written but also to know the syntax being used.

6. Parochialism. There are many projects that have started over from scratch rather than building upon the work of others, for reasons of remoteness, ignorance of previous work, personalities, unavailability of code due to intellectual property problems, and issues of grants and publications. We are extremely sensitive to the fact that the issue of credit for scientific work in a large scale project such as this can be a main reason for the failure of the QED project. But we can be hopeful that if a sufficient number of scientists unite in supporting the QED project, then partial contributions to QED's advancement will be seen in a very positive light in comparison to efforts to start all over from scratch.

7. Too little extensibility. In 20 years there have been perhaps a dozen major proof-checking projects, each representing an enormous amount of activity, but which have *plateaued out* or even evaporated. It seems that when the original authors of these systems cease actively working on their systems, the systems tend to die. Perhaps this problem stems from the fact that insufficient analysis was given to the basic problems of the root logic. Without a sufficient amount of extensibility, everyone so far seems to have reached a point in which checking new proofs is too much work to do by machine, even though one knows that it is relatively easy for mathematicians to keep making progress by hand. The reason, we suspect, is that mathematicians are using some reflection principles or layers of logics in ways not yet fully understood, or at least not implemented. Mathematicians great contribution has been the continual re-evaluating, re-conceptualizing, connecting, extending and, in cases, discarding of theorems and areas. So each generation stands on the shoulders of the giants before, as if they had always been there. We are far from being able to represent mechanically such evolutionary mathematical processes. Existing mathematical logics are typically as *static* as possible, often not even permitting the addition of new definitions! Important work in logic needs to be done to design logics more adaptable to extension and evolution.

8. Too little heuristic search support. While it is in principle possible to generate entries in the QED system entirely by hand, it seems extremely likely that some sort of automated tools will be necessary, including tools that do lots of search and use lots of heuristics or strategies to control search. Some systems which have completely eschewed such search and heuristic techniques might have gotten much further in checking interesting theorems through such techniques.

9. Too little care for rigor. It is notoriously easy to find "bugs" in algorithms for symbolic computation. To make matters worse, these errors are often regarded as of no significance by their authors, who plead that the result returned is true "except on a set of measure zero," without explicitly naming the set involved. The careful determination, nay, even proof, of precisely which conditions under which a result is true is essential for building the structure of mathematics so that one can depend on it. The QED system will support the development of symbolic algebra programs in which formal proofs of correctness of derivations are provided, along with the precise statement of conditions under which the results are true.

10. Complete absence of inter-operability. One safe generalization about current automated reasoning or symbolic computation systems is that it is always somewhere between impossible and extremely difficult to use any two of them together reliably and mechanically. It seems almost essential to the inception of any major project in this area to choose a logic and a syntax that is original, i.e., incompatible with other tools. One major exception to this generalization is the base syntax and logic for resolution systems. Here, standard problem sets have been circulated for years. But even for such resolution systems there is no standard syntax for entering problems involving such fundamental mathematical constructs as induction schemas or set-builder notation.

11. Too little attention paid to ease of use. The ease of use of automated reasoning systems is perhaps lower than for any other type of computing system available! In general, while anyone can use a word processor, almost no one but an expert can use a proof checker to check a difficult theorem. Perhaps this can be explained by the fact that the designers of such systems have had to put so much of their energies and attention into rigor, that they simply did not have enough energy left for good interface design.

4 What Is To Be Done?

> The idea is to make a language such that everything we write in it is interpretable as correct mathematics ... This may include the writing of a vast mathematical encyclopedia, to which everybody (either a human or a machine) may contribute what he likes. The idea of a kind of formalized encyclopedia was already conceived and partly carried out by Peano around 1900, but that was still far from what we might call automatically readable.
> DE BRUIJN

Leadership. It seems certain that inviting deliberation by many interested parties at the planning stage is important not only to get the QED project off on a correct footing but also to encourage many to participate in the project. Until we can establish general agreement within a large, critical mass of scientists (including many distinguished mathematicians) that the QED project is probably worth doing, and until a basic manifesto agreeable to them can be drafted, possibly using parts of this document as a starting point, it is not clear whether there will be any further progress on this project. Given the extraordinary scope of this project, it is also essential that research agency leadership be obtained. It is perhaps unlikely that any one agency would be willing to undertake the funding of the entirety of such a large project. So an agreement by many agencies to cooperate will probably be essential. The requirements for leadership, both by scientists and by research agencies, are so major that it is perhaps premature to speculate about what other things should be done, in what order. Nevertheless, we will speculate about a few issues.

What planning steps should be taken to start the QED project? An obvious first concern is to enumerate and describe in some detail the kinds of things that would be found in the QED system, including

> logics
> axioms
> definitions
> theorems (including an analysis of the major parts of mathematics)
> proofs
> proof-checkers
> decision procedures
> theorem-proving programs
> symbolic computation procedures
> modeling software
> simulation software
> tools for experimentation
> numerical analysis software
> graphical tools for viewing mathematics
> interface tools for using the QED system

Crucial to this initial high level organization effort is deciding what parts of mathematics will be represented, how that mathematics will be organized, and how it will be presented. It is conceivable that years of consideration of these points should precede implementation efforts. One can imagine that a reorganization of mathematics on the order of the scope of the Bourbaki project is necessary. One can imagine major projects in the development of formal higher-level languages in which mathematics can be formally discussed and major projects devoted simply to writing the most important theorems, definitions, and proof sketches in such languages. Because different proofs of the same theorem can differ substantially in complexity, and because entering formal proofs into a proof checking system is very expensive, it is highly cost effective to consider many proofs of a theorem before setting out to verify one of them. It has been suggested by several people that a useful and relatively easy early step would be to assemble, in ftp-able form, a comprehensive survey of the parts of mathematics that have been checked by various automated reasoning systems.

A second planning step would be to establish some milestones or some priority list of objectives. For example, one could attempt to outline which parts of mathematics should be added to the system in what order. Simultaneously, an analysis of what sorts of cooperation and resources would be necessary to achieve the earlier goals should be performed.

A third planning step would be to accumulate the basic mathematical texts that are to be formalized. It is entirely possible that the QED project will greatly overlap with an effort to build an electronic library of mathematical information. It is not part of the idea of a library that the documents should be in any particular language or subjected to any sort of rigor check. But it would of great inherent value, and great value to the QED project, to have the important works of mathematics available in machine readable form and organized for ease of access.

A fourth planning step would be to attempt to achieve consensus about the statement of the most important definitions and theorems in mathematics. Until there is agreement on the formalization of the basic concepts and theorems of the important parts of mathematics, it will be hardly appropriate to begin the difficult task of building formal proofs of theorems. The formalization of statements is an extremely difficult and error-prone activity.

Although the scientific obstacles to building QED are formidable, the social, psychological, political, and economic obstacles seem much greater. In principle, we can imagine a vast collection of people successfully collaborating on such an effort. But the problems of actually getting such a collaboration to occur are possibly insurmountable. "Why," an individual researcher could well ask, "should I risk my future by working on what will be but a small part of a vast undertaking? What sort of recognition will I receive for contributing to yet one more computing system?" These are good questions, and it is not clear what the answer is. To a major extent, status in mathematics and computing is a function of publications in major journals—status for research funding, status for tenure decisions, status for promotion. It is far from clear how contributing pieces to the QED system could provide a substitute for such signs of status. Perhaps here research agencies or even university faculties and administrators could be of assistance in establishing a new societal framework in which such cooperation was encouraged.

Even given the cooperation of all the necessary people and assuming good luck in overcoming scientific hurdles, there are many issues of a very difficult but somewhat mundane character involving: version control, distribution, and support. A system with hundreds of contributors will create management difficulties perhaps not even imaginable to the small groups of researchers who have worked in the past on parts of the QED idea.

It has been suggested about the low-level QED data files that they should be humanly readable and permit comments, and that the character set should be email-able.

Non-Copyright: This document is in the public domain and so unlimited alteration, reproduction, and distribution by anyone are permitted.

Authorship: This preliminary discussion of project QED (very tentative name) is an amalgam of many ideas that many people have had and for which perhaps no one alive today deserves much credit. We are deliberately avoiding any authorship or institutional affiliation at this early stage in the project (and may decide to do so forever) in the hope that many will want to join in the project as principals, even as originators (to the extent that anyone alive today could be thought to be an originator of this project). Some of those involved in the project would much rather that QED be completed than that they, as individuals, be lucky enough to partake significantly in the project, much less get any public credit for its completion. It may seem paranoid to avoid personalities, but we are inspired by the extraordinary cooperation achieved in the Bourbaki series in an atmosphere of anonymity.

To join an Internet electronic discussion group devoted to the QED project, send a message with the single line

`subscribe qed`

to `majordomo@mcs.anl.gov`. The line above should be the content of the message, not the subject line. The subject line is ignored. An archive of this discussion group is in the directory `/pub/qed/archive/` available by anonymous ftp from `info.mcs.anl.gov`. A current version of this manifesto may be found there as the file `/pub/qed/manifesto`.

The TPTP Problem Library

Geoff Sutcliffe[1], Christian Suttner[2], Theodor Yemenis[2]

[1] Dep't of Computer Science, James Cook University, Townsville, Australia
geoff@cs.jcu.edu.au, +61 77 815085
[2] Institut für Informatik, Technische Universität München, Germany
{suttner,gemenis}@informatik.tu-muenchen.de, +49 89 521098

Abstract. This paper provides a description of the TPTP library of problems for automated theorem provers. The library is available via Internet, and is intended to form a common basis for the development of and experimentation with automated theorem provers. To support this goal, this paper provides:
- the motivations for building the library;
- a description of the library structure including overview information;
- a description of the tptp2X utility program;
- guidelines for obtaining and using the library.

1 Introduction

This paper describes the TPTP (Thousands of Problems for Theorem Provers) Problem Library. The TPTP is a library of problems for automated theorem proving (ATP) systems, for 1st order logic. The TPTP is comprehensive, and thus provides an overview, and a simple, unambiguous reference mechanism for ATP problems. All problems in the TPTP are presented in an unambiguous format, and automatic conversion to other known ATP formats is provided.

The principal motivation for this project is to move the testing and evaluation of ATP systems from the present ad hoc situation onto a firm footing. This is necessary, as results currently being published seldom provide an accurate reflection of the intrinsic power of the ATP system being considered. A common library of problems is necessary for meaningful system evaluations, meaningful system comparisons, repeatability of testing, and the production of statistically significant results. Guidelines for using TPTP problems and presenting results are given in Section 4.2.

1.1 State of the Art

A large number of interesting problems have accumulated over the years in the ATP community. Besides publishing particularly interesting individual problems, from early on researchers have collected problems in order to obtain a basis for experimentation. The first major publication[3] in this regard was [MOW76],

---
[3] The first circulation of problems for testing theorem provers to our knowledge is due to L. Wos in the late sixties.

which provides an explicit listing of clauses for 63 problems, many of which are still relevant today. In the same year Wilson and Minker [WM76] tested six resolution strategies on a collection of 86 named problems. The problem clauses are not supplied in [WM76], however. A second major thrust was provided by [Pel86], which lists 75 problems. Other more recent collections are [BLM+86], [Qua92a], [MW92], and [McC93], to name a few. Also, the Journal of Automated Reasoning's Problem Corner regularly provides interesting challenge problems. However, problems published in hardcopy form are often not suitable for testing ATP systems, because they have to be transcribed to electronic form. This is a cumbersome, error-prone process, and is feasible for only very small numbers of problems. A problem library in electronic form was made publicly available by Argonne National Laboratories (Otter format, [McC90]) in 1988 [ANL]. This library has been a major source of problems for ATP researchers. Other electronic collections of problems are available, but have not been announced officially (e.g., that distributed with the SPRFN ATP system [SPR]). Although some of these collections provide significant support to researchers, and formed the early core of the TPTP library, none (with the possible exception of the ANL library) was specifically designed to serve as a common basis for ATP research. Rather, these collections typically were built in the course of research into a particular ATP system. As a result, there are several factors that limit their usefulness as a common basis for research. In particular, existing problem collections are often:

- hard to discover and obtain,
- limited in scope and size,
- outdated,
- formatted and tuned for a particular ATP system,
- inconsistent in their presentation of equally named problems,
- undocumented,
- provided without a regular update service for new problems,
- provided without a reliable error correction service.

As a consequence, several problems arise. Firstly, system development and system evaluations typically rely on a limited range of problems, depending on the collections of problems available to the researcher. Secondly, the presentation format used in a collection may not be appropriate for the desired purpose, and a comparatively large effort is required just to make the problems locally usable (which in practice often means that such a collection of problems is simply ignored). Thirdly, using a particular collection may lead to biases in results, because the problems have been designed and tuned specifically for a particular ATP system. Fourthly, the significance and difficulty of a problem, with respect to the current state-of-the-art in ATP systems, is hard to assess by newcomers to the field. Existing test problems are often not adequate anymore (e.g., Schubert's Steamroller [Sti86]), while others may be solvable only with specialized techniques (e.g., LIM+ [Ble90]) and therefore are much too hard to start with. Finally, many copies and variants of the same "original" problem may exist in different libraries. Coupled with a lack of documentation, this means that unambiguous identification of problems, and therefore a clear interpretation of performance figures for given problems, has become difficult.

The problem of meaningfully interpreting results is even worse than indicated. Commonly a few problems are selected and hand-tuned (clauses and literals are arranged in a special order, irrelevant clauses are omitted, lemmas are added in, etc) specifically for the ATP system being tested. However, the presentation of a problem can significantly affect the nature of the problem, and changing the clauses clearly makes a different problem altogether. Nevertheless the problem may be referenced under the same name as it was presented elsewhere. As a consequence the experimental results reveal little. Some researchers avoid this ambiguity by listing the clause sets explicitly. But obviously this usually cannot be done for a large number of problems or for large individual problems. The only satisfactory solution to these issues is a common and stable library of problems, which the TPTP provides.

1.2 What is Required?

The goal for building the TPTP has been to overcome the current drawbacks, and to centralize the burden of problem collection and maintenance to one place. The TPTP tries to address all relevant issues. In particular, the TPTP

- is available by anonymous ftp.
  The TPTP is thus easily available to the research community. Awareness of the TPTP is assured by extensive formal and informal announcements.
- spans a diversity of subject matters.
  This reduces biases in the development and testing of theorem provers, which arise from the use of a limited range of problems. It also provides an overview of the domains that ATP is used in. The significance of each domain may be measured by the number of problems available.
- is large enough for statistically significant testing.
  In contrast to common practise, an ATP system should be evaluated over a large number of problems, rather than a small set of judiciously selected examples. The large size of the TPTP makes this possible.
- is comprehensive and up-to-date.
  The TPTP contains all problems known to the community. There is no longer a need to look elsewhere.
- is presented in a well structured and documented format.
  All problems in the TPTP are presented in an unambiguous format. Automatic conversion to other known formats is provided, thus eliminating the necessity for transcription. The design and arrangement of the problems is independent of any particular ATP system.
- provides a common, independent, source for problems.
  This provides unambiguous problem reference, and makes the comparison of results meaningful.
- contains problems varying in difficulty, from very simple problems through to open problems.

This allows all interested researchers, from newcomers to experts, to rely on the same problem library.
- provides statistics for each problem and for the library as a whole.
This gives information about the syntactic nature of the problems.
- will provide a rating for the difficulty of each problem.
This is important for several reasons: (1) It simplifies problem selection according to the user's intention. (2) It allows the quality of an ATP system to be judged. (3) Over the years, changes in the problem ratings will provide an indicator of the advancement in ATP. The problem ratings are currently being worked on, and will be part of a future TPTP release.
- documents each problem.
This contributes to the unambiguous identification of each problem.
- provides standard axiomatizations that can be used in new problems.
This simplifies the construction of new problems.
- is a channel for making new problems available to the community, in a simple and effective way.
- removes the necessity for the duplicated effort of maintaining many libraries.
- specifies guidelines for its use for evaluating ATP systems.
As the TPTP is a standard library of problems, it is possible to set ATP evaluation guidelines which make reported results meaningful. This will in turn simplify and improve system comparisons, and allow ATP researchers to accurately gauge their progress.

The TPTP problem library is an ongoing project, with the aim to provide all of the desired properties.

Current Limitations of the TPTP. The current version of the TPTP library is limited to problems expressed in 1st order logic, presented in clause normal form. In particular, there are no problems for induction and nonclassical theorem proving. However, see Section 5 for upcoming and planned extensions.

2 Inside the TPTP

Scope. Release v1.0.0 of the TPTP contains 1577 abstract problems, which result in 2295 ATP problems, due to alternative presentations (see Section 2.2). Tables 1, 2, and 3 provide some statistics about release v1.0.0 of the TPTP.

The problems have been collected from various sources. The two principal sources have been existing problem collections and the ATP literature. Other sources include logic programming, mathematics, puzzles, and correspondence with ATP researchers. Many people and organizations have contributed towards the TPTP. In particular, the foundations of the TPTP were laid with David Plaisted's SPRFN collection [SPR]; many problems have been taken from Argonne National Laboratory's ATP problem library [ANL] (special thanks to Bill McCune here); Art Quaife has provided several hundred problems in set theory and algebra [Qua92b]; the Journal of Automated Reasoning, CADE Proceedings,

Table 1. Statistics on the TPTP.

Number of problem domains	23
Number of abstract problems	1577
Number of problems	2295
Number of non-Horn problems	1449 (63%)
Number of problems with equality	1805 (79%)
Number of propositional problems	111 (5%)
...being non-Horn	56 (50%)
Total number of clauses	322156
Total number of literals	970456

Table 2. Statistics for non-propositional TPTP problems.

Measure	Minimum	Maximum	Average	Median
Number of clauses	2	6404	124	68
Percentage of non-Horn clauses	0%	99%	6%	4%
...in non-Horn problems	0%	99%	9%	4%
Percentage of unit clauses	0%	100%	32%	22%
Number of literals	2	18966	276	162
Percentage of equality literals	0%	100%	47%	47%
...in equality problems	4%	100%	57%	47%
Number of predicate symbols	1	4042	12	4
Number of function symbols	0	94	27	13
Percentage of constants	0%	100%	46%	33%
Number of variables	0	32000	311	271
Percentage of singletons	0%	100%	6%	7%
Maximal clause size	1	25	4	5
Maximal term depth	0	51	4	6

Table 3. Statistics for propositional TPTP problems.

Measure	Minimum	Maximum	Average	Median
Number of clauses	2	8192	444	36
Percentage of non-Horn clauses	0%	99%	20%	1%
...in non-Horn problems	1%	99%	40%	45%
Percentage of unit clauses	0%	100%	16%	1%
Number of literals	2	106496	3303	132
Number of predicate symbols	1	840	69	13
Maximal clause size	1	21	5	3

and Association for Automated Reasoning Newsletters have provided a wealth of material; smaller numbers of problems have been provided by a number of further contributors (see also the Acknowledgements at the end of Section 5).

The problems in the TPTP are syntactically diverse, as is indicated by the ranges of the values in Tables 2 and 3. The problems in the TPTP are also semantically diverse, as is indicated by the range of domains that are covered. The problems are grouped into 23 domains, covering topics in the fields of logic, mathematics, computer science, and more. The domains are presented and discussed in Section 2.1.

Releases. The TPTP is managed in the manner of a software product, in the sense that fixed releases are made. Each release of the TPTP is identified by a release number, in the form v<Version>.<Edition>.<Patchlevel>. The Version number enumerates major new releases of the TPTP, in which important new features have been added. The Edition number is incremented each time new problems are added to the current version. The Patch level is incremented each time errors, found in the current edition, are corrected. All changes are recorded in a history file, as well as in the file for an affected problem.

2.1 The TPTP Domain Structure

An attempt has been made to classify the totality of the TPTP problems in a systematic and natural way. The resulting domain scheme reflects the natural hierarchy of scientific domains, as presented in standard subject classification literature. The current classification is based mainly on the Dewey Decimal Classification [Dew89] and the Mathematics Subject Classification used for the Mathematical Reviews by the American Mathematical Society [MSC92]. We define four main fields: logic, mathematics, computer science, and other. Each field contains further subdivisions, called *domains*. These TPTP domains constitute the basic units of our classification. The full classification scheme is shown in Figure 1.

2.2 Problem Versions and Standard Axiomatizations.

There are often many ways to formulate a problem for presentation to an ATP system. Thus, in the TPTP, there are often alternative presentations of a problem. The alternative presentations are called *versions* of the underlying *abstract problem*. As the problem versions are the objects that ATP systems must deal with, they are referred to simply as problems, and the abstract problems are referred to explicitly. Each problem is stored in a separate physical file. The primary reason for different versions of an abstract problem is the use of different axiomatizations. This issue is discussed below. A secondary reason is the use of different formulations of the theorem to be proven.

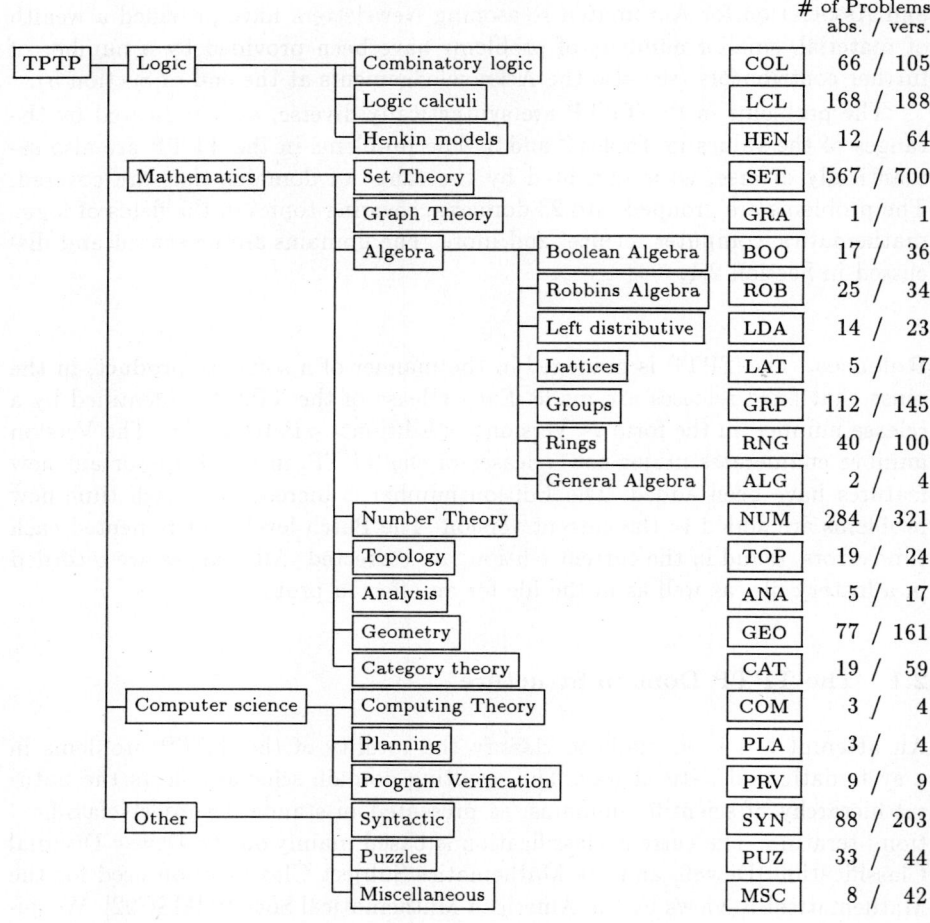

Fig. 1. The domain structure of the TPTP.

Different axiomatizations. Commonly, different axiomatizations of a theory exist, and, in general, most are equally acceptable. In the TPTP an axiomatization is acceptable if it is complete (in the sense that it captures some closed theory) and it has not had any lemmas added. Such axiomatizations are called *standard* axiomatizations. Within the ATP community some problems have been created with *non-standard* axiomatizations. An axiomatization is non-standard if it has been *reduced* (i.e., axioms have been removed) and the result is an *incomplete* axiomatization, or if it has been *augmented* (i.e., lemmas have been added) and the result is a *redundant* axiomatization. Non-standard axiomatizations are typically used to find a proof of a theorem (based on the axiomatization) using a particular ATP system. In any 'real' application of an ATP system, a standard axiomatization of the application domain would typically have to be used, at least initially. Thus the use of standard axiomatizations is desirable, because it reflects such 'real' usage. In the TPTP, for each collected problem that uses a non-standard axiomatization, a new version of the problem is created with a standard axiomatization. The standard axiomatization is created by reducing or augmenting the non-standard axiomatization appropriately.

The standard axiomatizations used in the TPTP are kept in separate axiom files, and are included in problems as appropriate. If an axiomatization uses equality, the required axioms of substitution are kept separate from the theory specific axioms. The equality axioms of reflexivity, symmetry, and transitivity, which are also required when equality is present, are also kept separately, as they are often used independently of the substitution axioms.

2.3 Problem and Axiomatization Naming

Providing unambiguous names for all problems is necessary in a problem library. A naming scheme has been developed for the TPTP, to provide unique, stable names for abstract problems, problem versions, and axiomatizations. File names are in the form DDDNNN-V[MMM].p for problem files, and DDDNNN-E.TT for axiom files. DDD is the domain name abbreviation, NNN is the index within the domain, V is the version number, MMM is an optional generation parameter, E is the axiomatization extension number, .p indicates a problem file, and TT is either .ax for theory specific axioms or .eq for equality substitution axioms. The complete file names are unique within the TPTP.

Abstract problems and axiomatizations have also been allocated semantic names. The semantic names can be used to augment file names, so as to give an indication of the contents. While these names are provided for users who like to work with mnemonic names, only the standard syntactic names are guaranteed to provide unambiguous reference. The semantic names are formed from a set of specified abbreviations. The semantic names can be added to the syntactic file names using a script that is provided.

2.4 Problem Presentation

The physical presentation of the TPTP problem library is such that ATP researchers can easily use the problems. The TPTP file format, for both problem files and axiom files, has three main sections. The first section is a header section that contains information for the user. The second section contains **include** instructions for axiom files. The last section contains the clauses that are specific to the problem.

The syntax of the problem and axiom files is that of Prolog. This conformance makes it trivial to manipulate the files using Prolog. All information in the files that is not for use by ATP systems is formatted as Prolog comments, with a leading %. All the information for ATP systems is formatted as Prolog facts. A utility is provided for converting TPTP files to other known ATP system formats (see Section 3). A description of the information contained in TPTP files is given below. Full details are given in [SSY93].

The Header Section

The header contains information about the problem, for the user. It is divided into four parts. The first part identifies and describes the problem. The second part provides information about occurrences of the problem. The third part gives the problem's ATP status and a table of syntactic measurements made on the problem. The last part contains general information about the problem. An example of a TPTP header, extracted from the problem file GRP039-7.p, is shown in Figure 2.

```
%--------------------------------------------------------------------
% File     : GRP039=SubGI2Norm-7 : Released v0.0.0, Updated v0.11.5.
% Domain   : Group Theory (Subgroups)
% Problem  : Subgroups of index 2 are normal
% Version  : [McCharen, et al., 1976] (equality) axioms : Augmented.
% English  : If O is a subgroup of G and there are exactly 2 cosets in
%            G/O, then O is normal [that is, for all x in G and y in O,
%            x*y*inverse(x) is back in O].

% Refs     : McCharen J.D., Overbeek R.A., Wos L.A. (1976), Problems and
%            Experiments for and with Automated Theorem Proving Programs
%            IEEE Transactions on Computers C-25(8), 773-782.
% Source   : [McCharen, et al., 1976]
% Names    : GP2 [McCharen, et al., 1976]

% Status   : unsatisfiable
% Syntax   : Number of clauses         :   29 (   2 non-Horn)(17 units)
%            Number of literals        :   47 (  32 equality)
%            Number of predicate symbols :  2 (   0 propositions)
%            Number of function symbols :   8 (   5 constants)
%            Number of variables       :   42 (   0 singletons)
%            Maximal clause size       :    4
%            Maximal term depth        :    3

% Comments : element_in_O2(A,B) is A^-1.B. The axioms with element_in_O2
%            force index 2.
% Bugfixes : v0.8.1 - Clauses multiply_inverse_* and multiply_identity_*
%            fixed.
%--------------------------------------------------------------------
```

Fig. 2. Example of a problem file header (GRP039-7.p).

The header fields contain the following information:

- The % File field contains the problem's syntactic and semantic names, the TPTP release in which the problem first appeared, and the TPTP release in which the problem was last updated.

- The % Domain field identifies the domain from which the problem is drawn.
- The % Problem field provides a one line, high-level description of the abstract problem.
- The % Version field gives information that differentiates this version of the problem from other versions of the problem.
- The % English field provides a full description of the problem if the one line description is too terse.
- The % Refs field provides a list of references to items in which the problem has been presented.
- The % Source field acknowledges, with a citation, where the problem was (physically) obtained from.
- The % Names field lists existing known names for the problem (as it has been named in other problem collections or the literature).
- The % Status field gives the problem's ATP status [4], one of satisfiable, unsatisfiable, open, or broken.
- The % Syntax field lists various syntactic measures of the problem's clauses.
- The % Comments field contains free format comments about the problem.
- The % Bugfixes field documents any changes which have been made to the clauses of the problem.

The Include Section

The include section contains include instructions for TPTP axiom files. An example of an include section, extracted from the problem file GRP039-7.p, is shown in Figure 3.

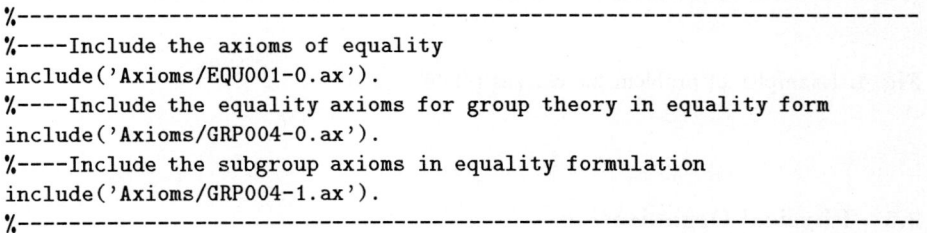

Fig. 3. Example of a problem file include section (GRP039-7.p).

Each of the include instructions indicates that the clauses in the named axiom file should be included at that point. Axiom files are presented in the same format as problem files, and include instructions may also appear in axiom files. Full versions of TPTP problems (without include instructions) can be created by using the tptp2X utility (see Section 3).

---

[4] This field has not yet been filled in all problems of the TPTP.

The Clauses Section

TPTP problems are presented in clausal normal form. The literals that make up a clause are presented as a Prolog list of terms. Each literal is a term whose functor is either ++ or --, indicating a positive or negative literal respectively. The single argument of the sign, i.e., the atom of the literal, is a Prolog term. The signs ++ and -- are assumed to be defined as prefix operators in Prolog.

Each clause has a name, in the form of a Prolog atom. Each clause also has a status, one of axiom, hypothesis, or theorem. The hypothesis and theorem clauses are those that are derived from the negation of the theorem to be proved. The status hypothesis is used only if the clauses can clearly be determined as such; otherwise their status is theorem. The name, status, and literal list of each clause are bundled as the three arguments of a Prolog fact, whose predicate symbol is input_clause. These facts are in the clauses section of the problem file.

Two examples of clauses, extracted from the problem file GRP039-1.p, are shown in Figure 4.

```
%---------------------------------------------------------------
%----Definition of subgroup of index 2
input_clause(an_element_in_O2,axiom,
    [++subgroup_member(element_in_O2(A,B)),
     ++subgroup_member(B),
     ++subgroup_member(A)]).

input_clause(prove_d_is_in_subgroup,theorem,
    [--subgroup_member(d)]).
%---------------------------------------------------------------
```

Fig. 4. Examples of problem file clauses (GRP039-1.p).

2.5 Physical Organization

The TPTP is physically organized into five subdirectories:

- The **Problems** directory contains a subdirectory for each domain, as shown in Figure 1. Each subdirectory contains the problem files for that domain.
- The **Axioms** directory contains the axiom files.
- The **TPTP2X** directory contains the tptp2X utility, described in Section 3.
- The **Scripts** directory contains some useful C shell scripts.
- The **Documents** directory contains comprehensive online information about the TPTP, in specific files. This provides quick access to relevant overview data. In particular, it provides a simple means for selecting problems with specific properties, by using standard system tools (e.g., **grep**, **awk**).

3 The tptp2X Utility

The tptp2X utility converts TPTP problems from the TPTP format to formats used by existing ATP systems. Currently, the tptp2X utility supports the following output formats:

- the MGTP format [FHKF92];
- the Otter .in format [McC90] (the set of support and inference rules to be used are also specified, and are included in the output file);
- the PTTP format [Sti84];
- the SETHEO .lop format [STvdK90];
- the SPRFN format [Pla88];
- the TPTP format, substituting include instructions with the actual clauses.

It is simple to add new formatting capabilities to the tptp2X utility. The tptp2X utility can also be used to rearrange the clauses and literals in a problem. This facilitates testing the sensitivity of an ATP system to the order in which the clauses and literals are presented. Full details of the tptp2X utiltity are given in [SSY93].

4 You and the TPTP

4.1 How to FTP the TPTP

The TPTP can be obtained by anonymous ftp from the Department of Computer Science, James Cook University, or the Institut für Informatik, Technische Universität München. The ftp hosts are coral.cs.jcu.edu.au (137.219.17.4) and flop.informatik.tu-muenchen.de (131.159.8.35), respectively. At both sites the TPTP is in the directory pub/tptp-library. There are three files. Information about the TPTP is in the ReadMe file (9.3 KByte), a technical report [SSY93] is in TR-v1.0.0.ps.Z (357 KByte), and the library itself is packaged in TPTP-v1.0.0.tar.Z (2.4 MByte). Please read the ReadMe file, as it contains up-to-date information about the TPTP.

4.2 Important: Using the TPTP

By providing this library of ATP problems, and a specification of how these problems should be presented to ATP systems, it is our intention to place the testing, evaluation, and comparison of ATP systems on a firm footing. For this reason, you should abide by the following conditions when using TPTP problems and presenting your results:

- The specific version, edition, and patch level of the TPTP, used as the problem source, must be stated (see the introduction of Section 2).
- Each problem must be referenced by its unambiguous syntactic name.

- No clauses/literals may be added/removed without explicit notice. (This holds also for removing equality axioms when built-in equality is provided by the prover.)
- The clauses/literals may not be rearranged without explicit notice. If clause or literal reversing is done by the tptp2X utility, the reversals must be explicitly noted.
- The header information in each problem may not be used by the ATP system without explicit notice. Any information that is given to the ATP system, other than that in the input_clauses, must be explicitly noted (including any system switches or default settings).

Abiding by these rules will allow unambigous identification of the problem, the arrangement of clauses, and further input to the ATP system which has been used.

5 Present and Future Activity

5.1 The Present

Users' experiences with the TPTP will tell if we have achieved the design goals described in Section 1, except for the problem ratings. The problem ratings are currently being worked on. Note that these ratings should decrease as advances in automated theorem proving are made. Therefore the long term maintenance of the individual problem ratings will provide an objective measure of progress in the field.

An upcoming addition to the TPTP is a set of Prolog programs for generating arbitrary sizes of generic problems (e.g., the "pigeon holes" problem). This will implicitly provide any size of such problems, and also reduce the physical size of the TPTP. These programs are being worked on. In order to ensure the usability of the programs, they will be tested on a public domain Prolog.

The tptp2X utility will soon be modified to be capabile of applying arbitrary sequences of transformations to problems. The transformations will include those currently in tptp2X, and others that become available (e.g., Mark Stickel has provided a transformation for his upside-down meta-interpretation approach [Sti94]). Automatic removal of equality axioms will also be provided, to ease the life of researchers whose ATP systems have built in mechanisms for equality reasoning.

5.2 The Future

We have several short and long term plans for further development of the TPTP. The main ideas are listed here.

- Performance tables for popular ATP systems.
  Extensive experimental results, for publically available ATP systems, are currently being collected (based on all the TPTP problems). These will be published in some form.

- ATP system evaluation guidelines.
  General guidelines outlining the requirements for ATP system evaluation will be produced.
- The BSTP Benchmark Suite.
  A benchmark suite (the BSTP) will be selected from the TPTP. The BSTP will be a small collection of problems, and will provide a minimal set of problems on which an ATP system evaluation can be based. The BSTP will be accompanied by specific guidelines for computing a performance index for an ATP system.
- Full 1st order logic.
  The TPTP will be extended to include problems in full 1st order logic notation. ATP systems with automatic conversion to clausal form then can derive additional information regarding the problem, such as which functors are Skolem functors.
- Various translators.
  Translators between various logical forms will be provided.
  - from full 1st order logic to clausal normal form
  - from non-Horn to Horn form
  - from 1st order to propositional form
- Non-classical and higher order logics.
  In the longer term, the TPTP may be extended to include problems expressed in non-classical and higher order logics.

5.3 Acknowledgements

We are indebted to the following people and organizations who have contributed problems and helped with the construction of the TPTP:

Geoff Alexander, the ANL group (especially Bill McCune), the Automated Reasoning group in Munich, Dan Benanav, Woody Bledsoe, Maria Poala Bonacina, Heng Chu, Tom Jech, Reinhold Letz, Thomas Ludwig, Xumin Nie, Jeff Pelletier, David Plaisted, Joachim Posegga, Art Quaife, Alberto Segre, John Slaney, Mark Stickel, Bob Veroff, and TC Wang.

References

[ANL] ANL Problem Library. Available by anonymous ftp from info.msc.anl.gov, Maths and Computer Science Division, Argonne National Laboratory, Argonne, IL.

[Ble90] W.W. Bledsoe. Challenge Problems in Elementary Calculus. *Journal of Automated Reasoning*, 6:341 – 359, 1990.

[BLM+86] R. Boyer, E. Lusk, W. McCune, R. Overbeek, M. Stickel, and L. Wos. Set Theory in First-Order Logic: Clauses for Gödel's Axioms. *Journal of Automated Reasoning*, 2:287 – 327, 1986.

[Dew89] M. Dewey. *Dewey Decimal Classification and Relative Index*. Forest Press, 20 edition, 1989.

[FHKF92] M. Fujita, R. Hasegawa, M. Koshimura, and H. Fujita. Model Generation Theorem Provers on a Parallel Inference Machine. In *Proceedings of the International Conference on Fifth Generation Computer Systems*, pages 357–375, 1992.

[McC90] W.W. McCune. Otter 2.0 Users Guide. Technical Report ANL-90/9, Argonne National Laboratory, 1990.

[McC93] W.W. McCune. Single Axioms for Groups and Abelian Groups with Various Operations. *Journal of Automated Reasoning*, 10(1):1–13, 1993.

[MOW76] J. McCharen, A. Overbeek, and L. Wos. Problems and Experiments for and with Automated Theorem-Proving Programs. *IEEE Transactions on Computers*, C-25(8):773 – 782, August 1976.

[MSC92] *Mathematical Subject Classification*. American Mathematical Society, Provedence, RI, 1992. Mathematical Reviews, Subject Index.

[MW92] W. McCune and L. Wos. Experiments in Automated Deduction with Condensed Detachment. In *Proceedings of CADE-11*, pages 209–223, Saratoga Springs, USA, 1992. Springer LNAI 607.

[Pel86] F.J. Pelletier. Seventy-five Problems for Testing Automated Theorem Provers. *Journal of Automated Reasoning*, 2:191–216, 1986.

[Pla88] D.A. Plaisted. Non-Horn Clause Logic Programming without Contrapositives. *Journal of Automated Reasoning*, 4(3):287–325, 1988.

[Qua92a] A. Quaife. Automated Deduction in von Neumann-Bernays-Godel Set Theory. *Journal of Automated Reasoning*, 8(1):91–147, 1992.

[Qua92b] A. Quaife. Problems based on NBG Set Theory. Emailed from A. Quaife to G. Sutcliffe, 1992.

[SPR] SPRFN. The problem collection distributed with the SPRFN ATP system [Pla88].

[SSY93] C.B. Suttner, G. Sutcliffe, and T. Yemenis. The TPTP Problem Library (TPTP v1.0.0). Technical Report FKI-184-93, Institut für Informatik, Technische Universität München, Munich, Germany; Technical Report 93/11, Department of Computer Science, James Cook University, Townsville, Australia, 1993.

[Sti84] M.E. Stickel. A Prolog Technology Theorem Prover. *New Generation Computing*, 2(4):371–383, 1984.

[Sti86] M.E. Stickel. Schubert's Steamroller Problem: Formulations and Solutions. *Journal of Automated Reasoning*, 2:89 – 101, 1986.

[Sti94] M.E. Stickel. Upside-Down Meta-Interpretation of the Model Elimination Theorem-Proving Procedure for Deduction and Abduction. *Journal of Automated Reasoning*, to appear, 1994.

[STvdK90] J. Schumann, N. Trapp, and M. van der Koelen. SETHEO/PARTHEO Users Manual. Technical Report SFB Bericht 342/7/90 A, Institut für Informatik, TU München, 1990.

[WM76] G.A. Wilson and J. Minker. Resolution, Refinements, and Search Strategies: A Comparative Study. *IEEE Transactions on Computers*, C-25(8):782–801, 1976.

Combination Techniques for Non-Disjoint Equational Theories*

Eric Domenjoud, Francis Klay, Christophe Ringeissen

CRIN-CNRS & INRIA-Lorraine
BP 239, 54506 Vandœuvre-lès-Nancy Cedex (France)
e-mail: {Eric.Domenjoud,Francis.Klay,Christophe.Ringeissen}@loria.fr

Introduction

Modularity is a central problem in automated deduction in equational theories. The problem may be stated as follows: given two equational theories E_1 and E_2 and algorithms for solving a problem P in each theory, can one build automatically an algorithm for solving the problem P in $E_1 \cup E_2$. Almost all results known up to now are restricted to the case where the signatures of E_1 and E_2 are disjoint, in which case, we speak of disjoint theories. Many authors studied the problem of combining unification algorithms for disjoint theories [6, 14, 13] and the best result is due to F. Baader and K. Schulz [1] who described a general technique for combining decision procedures for unifiability and, by the way, solved the problem of combining unification algorithms for non-finitary theories. This result was extended in two different ways: F. Baader and K. Schulz themselves [2] extended it to take into account some restricted forms of disunification problems, and Ch. Ringeissen [10] described an extension where theories may share constants. This last result was, up to now, the only modularity result for non-disjoint equational theories. Some authors also considered the problem of combining matching algorithms for disjoint equational theories [8, 12] which is again more complicated than just putting together two matching algorithms.

We consider here the combination of non-disjoint theories, and give sufficient conditions under which algorithms for solving the word problem, the matching problem and the unification problem in the union may be built automatically from algorithms for each theory. The problem is obviously unsolvable in general, and we restrict our attention to theories sharing constructors which have to be defined in a suitable way. This restriction is however still not sufficient in general. Indeed, some properties like simplicity, which turns out to be very useful for unification, are lost when combining theories sharing constructors. A theory is simple if a term is never equivalent to one of its strict subterms. If we consider the theories presented respectively by $fgx = hx$ and $khx = gx$, which share the constructors h and g according to our definition, we have $kfgx = gx$ modulo their union which is thus not simple. Each of these theories is however obviously simple.

Due to lack of space, almost all proofs are omitted and may be found in the extended version.

---

* This work has been partially supported by the Esprit Basic Research Working Group 6028 CCL and the GDR Programmation of CNRS.

1 Preliminaries

We first recall some basic definitions and notations about terms, substitutions and equational theories.

$T(F, X)$ denotes the set of terms built over the countable set of function symbols F and the infinite countable set of variables X. A term is ground if it contains no variable. $T(F)$ denotes the set of ground terms built over F. Given a term t and a position p in t, $t(p)$ denotes the symbol at position p in t. The root position is denoted by ϵ so that $t(\epsilon)$ is the top-symbol (or the root) of t. We write $t[s_1, \ldots, s_n]_{p_1, \ldots, p_n}$ to indicate that s_i's ($i = 1, \ldots, n$) are the subterms of t at positions p_i's and if the positions are irrelevant or clear from the context, we simply write $t[s_1, \ldots, s_n]$ to indicate that s_i's are subterms of t. $t[s_1, \ldots, s_n]_{p_1, \ldots, p_n}$ also denotes the term t whose subterm at positions $p_1, \ldots, p_n$ have been replaced with s_i's. $\mathcal{V}(t)$ is the set of variables occurring in t.

A substitution is a mapping from X to $T(F, X)$ which is the identity almost everywhere. It extends to a unique morphism form $T(F, X)$ to $T(F, X)$. If σ is a substitution, the set of variables for which $\sigma(x) \neq x$ is called the domain of σ and is denoted by $\text{Dom}(\sigma)$. The range of σ is defined as $\text{Ran}(\sigma) = \{\sigma(x) \mid x \in \text{Dom}(x)\}$. If the domain of σ is $\{x_1, \ldots, x_n\}$, σ is also written as $\{x_1 \mapsto \sigma(x_1), \ldots, x_n \mapsto \sigma(x_n)\}$. The application of σ to a term t is written in postfix notation as $t\sigma$. If $t = s\sigma$, t is an instance of s. Similarly, the composition of substitutions is written from left to right. If $\mu = \rho\sigma$ for some substitution σ, μ is an instance of ρ and σ is idempotent if $\sigma\sigma = \sigma$.

An axiom is an unordered pair of terms written $l = r$. An equational presentation is a pair (F, A) where F is a finite countable set of function symbols and A is a finite set of axioms $l = r$ with $l, r \in T(F, X)$. For any equational presentation $E = (F, A)$, $=_E$ denotes the equational theory generated by A on $T(F, X)$, that is the smallest congruence containing all instances of axioms in A. F is called the signature of E. By abuse of terminology, we do not distinguish between a theory and its presentation so that we speak of the theory E. Two terms t and s such that $t =_E s$ are said E-equal or equal modulo E. A theory E is consistent if there exists at least one model of E with more than one element. E is collapsing if there exists a non-variable term t and a variable x with $t =_E x$. E is regular if for any terms t and s, $t =_E s \Rightarrow \mathcal{V}(t) = \mathcal{V}(s)$. The union of two equational theories (F_1, A_1) and (F_2, A_2) is the theory $(F_1 \cup F_2, A_1 \cup A_2)$.

A strict ordering $>$ is well founded (or noetherian) if there exists no infinite decreasing sequence $x_1 > x_2 > \cdots$. An ordering $>$ on $T(F, X)$ is monotonic if for any function symbol f and any terms t and s, $t > s \Rightarrow f(\ldots, t, \ldots) > f(\ldots, s, \ldots)$.

Definition 1 (E-**normal forms and constructors**). Let $E = (F, A)$ be a consistent theory, X be a infinite countable set of variables and $>$ be a monotonic well founded ordering on $T(F \cup X)$ (where variables are treated as constants) such that any congruence class of E in $T(F \cup X)$ contains a unique minimal element with respect to $>$. The minimal element of the class of t modulo E will be denoted by $t\!\downarrow_E$ and will be called the E-normal form of t. A symbol $h \in F$ is a constructor of E if and only if

$$\forall t_1, \ldots, t_n \in T(F \cup X) : h(t_1, \ldots, t_n)\!\downarrow_E = h(t_1\!\downarrow_E, \ldots, t_n\!\downarrow_E)$$

A constructor h of E occurs at the top in E if there exist terms $t, s_1, \ldots, s_n \in T(F, X)$ with $t(\epsilon) \neq h$ such that $t =_E h(s_1, \ldots, s_n)$.

Two equational theories $E_1 = (F_1, A_1)$ and $E_2 = (F_2, A_2)$ share constructors if all symbols in $F_1 \cap F_2$ are constructors of $(F_1 \cup F_2, A_1)$ and $(F_1 \cup F_2, A_2)$.

A constructor h is shared at the top if it occurs at the top in (F_1, A_1) or (F_2, A_2).

Note: The E-normal forms and the constructors depend highly on the ordering $>$. For instance, if we consider the theory $fx = gx$ then depending on the ordering, either f or g is a constructor. To be completely formal, we should speak of $(E, >)$-normal form and $>$-constructors. Since we always consider the same ordering $>$, we omit to mention it.

Example 1. Let us consider the theories $E_i = (\{+_i, h\}, AC(+_i) \cup \{h(x) +_i h(y) = h(x +_i y)\})$ where $AC(+_i)$ denotes the axioms of associativity and commutativity for $+_i$. We consider the rpo ordering induced by the precedence $+_2 > +_1 > h > \cdots > x_2 > x_1$ with left to right status for $+_i$. h is then a constructor of E_1, E_2 and $E_1 \cup E_2$. Since h occurs at the top in E_1 and E_2, it is shared at the top.

One may note that if E is presented by a convergent rewriting system, our definition of constructors meets the standard one.

The fundamental property of constructors which interests us is the following:

Proposition 2. *Let h be a constructor of a theory E then for any terms t_i's and s_j's,*
$$h(t_1, \ldots, t_n) =_E h(s_1, \ldots, s_n) \iff \forall i : t_i =_E s_i$$

Now let g be another constructor of E (with respect to the same ordering) distinct from h, then for any terms t_i's and s_j's,
$$h(t_1, \ldots, t_n) \neq_E g(s_1, \ldots, s_n)$$

In all the rest of the paper, we only consider consistent theories sharing constructors. The set of shared constructors is denoted by SF. The following theorem justifies our interest in theories sharing constructors.

Theorem 3. *If theories E_1 and E_2 share constructors then $E_1 \cup E_2$ is a conservative extension of E_1 and E_2 and any constructor of E_1 and E_2 is a constructor of $E_1 \cup E_2$.*

Definition 4 (Pure terms and aliens). A term $t \in T(F_i, X)$ is said i-pure (i is 1 or 2). A term $t \in T(SF, X)$ is called a shared term. A strict subterm s of a term t is an *alien* if $s(\epsilon) \in F_i \backslash F_j$ ($i \neq j$) and all symbols on the path form the root of t to s belong to F_j.

Note that from this definition, a term which is pure may however have aliens if its top-symbol is shared. This seems odd, but turns out to be useful.

Definition 5. A variable-abstraction π is a mapping from $T(F, X)$ to X such that $\pi(t) = \pi(s) \iff t =_E s$. The i-abstraction of a term t, denoted by t^{π_i}, is defined inductively by:
- If $t = f(t_1, \ldots, t_n)$ and $f \in F_i$ then $t^{\pi_i} = f(t_1^{\pi_i}, \ldots, t_n^{\pi_i})$
- else $t^{\pi_i} = \pi(t)$

2 Word problem

The first problem we are faced with when considering the union of theories sharing constructors, is to decide equality in the union. The problem of deciding whether two terms are equal modulo a theory is also called the word problem. It would be nice to have a modularity result stating that the word problem is decidable in the union if it is decidable in each theory. We were not able to establish such a result in general, but we have nevertheless a modularity result for a slightly more general problem which, in some cases, reduces to a word problem. Since two terms are equal modulo E if and only if their E-normal forms are identical, the idea for deciding whether two terms t and s are equal is to compute for each of them some reduced form which *looks like* its E-normal form.

Definition 6 (Layers reduced forms). A term $t \in T(F_1 \cup F_2, X)$ is in layers reduced form if and only if all its aliens are in layers reduced form, t is not equal modulo $E_1 \cup E_2$ to one of its aliens, and either $t(\epsilon) \in SF \cup X$, or t is not equal modulo $E_1 \cup E_2$ to a variable or a term whose top-symbol is a shared constructor.

Layers reduced forms enjoy the following properties:

Proposition 7. *If t is a term in layers reduced form, then $t^{\pi_i} =_{E_i} (t{\downarrow}_E)^{\pi_i}$*
If t and s are in layers reduced form then $s =_{E_1 \cup E_2} t \iff s^{\pi_i} =_{E_i} t^{\pi_i}$

As a corollary, we get

Proposition 8. *$E_1 \cup E_2$-equality of terms in layers reduced form is decidable if E_i-equality is decidable for $i = 1, 2$.*

However, the decidability of equality in each theory might not be sufficient to compute a layers reduced form for any term. Therefore we introduce a new kind of problems.

Definition 9 (Symbol matching). Let $E = (F, A)$ be a consistent theory and h be a symbol in F. The symbol matching problem on h modulo E consists in deciding for any term $t \in T(F, X)$ whether there exist terms t_i's such that $t =_E h(t_1, \ldots, t_n)$.

Remarks:

1. Since matching modulo an arbitrary theory E is semi-decidable, t_i's may be effectively computed as soon as we know that they exist. Some general unification procedure [5, 4] may be used for this purpose.
2. the symbol matching problem reduces to a word problem if h is a constant.
3. If the symbol h does not occur at the top in E, then the symbol matching problem on h is trivially unsatisfiable.

Definition 10. Let t be a i-pure term. The term $t{\Downarrow}_{E_i}$ is recursively defined by:

- if $t =_{E_i} x$ for some variable x then $t{\Downarrow}_{E_i} = x$
- else if for some shared constructor h, and i-pure terms $t_1, \ldots, t_n$, $t =_{E_i} h(t_1, \ldots, t_n)$ then $t{\Downarrow}_{E_i} = h(t_1{\Downarrow}_{E_i}, \ldots, t_n{\Downarrow}_{E_i})$.
- else $t{\Downarrow}_{E_i} = t$.

Proposition 11. *For any term $t \in T(F_i, X)$, $t\Downarrow_{E_i}$ is in layers reduced form. Moreover, if E_i-equality is decidable and for any shared constructor h, the symbol matching problem on h is decidable modulo E_i, then $t\Downarrow_{E_i}$ may be computed in finite time.*

We may now give an operational definition of layers reduced forms.

Definition 12. For any term t, the term $t\Downarrow$ is defined recursively by:

- if t is i-pure then $t\Downarrow = t\Downarrow_{E_i}$
- else $t = C[t_1, \ldots, t_n]$ where t_k's are aliens of t. Let $t' = C[t_1\Downarrow, \ldots, t_n\Downarrow] = C'[s_1, \ldots, s_m]$ where s_k's are aliens of t' and variables of t' that do not occur in an alien. Now let $\bar{t}$ be the i-pure term (i may be 1 or 2) obtained by replacing each s_k by a variable, where $E_1 \cup E_2$-equal s_k's are replaced by the same variable, and let ρ be a substitution assigning to each of these variables one of the s_k's it replaces. Now $t\Downarrow = (\bar{t}\Downarrow_{E_i})\rho$.

Proposition 13. *For any term t, $t\Downarrow$ is a term in layers reduced form equal to t modulo $E_1 \cup E_2$. Moreover, if E_1-equality and E_2-equality is decidable and for any shared symbol h, the symbol matching problem on h is decidable modulo E_1 and E_2, then $t\Downarrow$ may be computed in finite time.*

We get then the modularity theorem:

Theorem 14. *Let E_1 and E_2 be two theories sharing constructors. Assume that the word problem is decidable modulo E_1 and E_2 and for any shared constructor h, the symbol matching problem on h is decidable modulo E_1 and E_2. Then the word problem is decidable modulo $E_1 \cup E_2$ and for any shared constructor h, the symbol matching problem on h is decidable modulo $E_1 \cup E_2$*

Corollary 15. *If all constructors shared at the top are constants, and equality is decidable modulo E_1 and E_2 then equality modulo $E_1 \cup E_2$ is decidable.*

3 Matching

The second problem we are interested in is the combination of matching algorithms for theories sharing constructors. The idea for solving this problem is to perform an abstraction of terms, i.e. replace aliens with fresh variables, and then solve pure match-equations in each theory. Unfortunately, purification of match-equations may introduce new variables x in right hand sides and the related solved equations $x \stackrel{?}{=} s$. The specificity of matching problems is then lost since we have to deal with unification. However, this unification can be turned into matching if purification is only performed in left hand sides of match-equations and solutions of match-equations are ground (when variables of right hand sides are considered as constants) i.e. theories are regular. In all the remaining of this section, we only consider regular theories and we assume that a complete matching algorithm is known for E_1 and E_2.

The combination technique for matching algorithms relies on the computation of a layers reduced form (see section 2) of right hand sides of match-equations. Since we assume that matching is decidable in each theory, from proposition 13, these layers reduced forms may be computed. By replacing then aliens with free constants (two $E_1 \cup E_2$-equal aliens are replaced with the same constant), a matching algorithm modulo E_i may be used to solve the problem $s \leq^? t$ if s is i-pure and t is in layers reduced form.

The transformation rules for matching problems modulo $E_1 \cup E_2$ are given in figure 1. In rule MatchSolve, $CSS_{E_i}(s \leq^? t\Downarrow)$ denotes a complete set of solutions modulo E_i of $s \leq^? t\Downarrow$.

LeftPurif
$$\frac{\Gamma \wedge s[u] \leq^? t}{\Gamma \wedge s[x] \leq^? t \wedge x =^? u} \quad \text{if } u \text{ is an alien of } s \text{ and } x \text{ is a fresh variable}$$

Merge
$$\frac{\Gamma \wedge x \leq^? t \wedge x =^? s}{\Gamma \wedge x \leq^? t \wedge s \leq^? t}$$

MatchSolve
$$\frac{\Gamma \wedge s \leq^? t}{\Gamma \wedge \bigwedge_{k \in K} x_k \leq^? t_k} \quad \text{if } s \in T(F_i, X) \backslash X, \{x_k \mapsto t_k\}_{k \in K} \in CSS_{E_i}(s \leq^? t\Downarrow)$$

Delete
$$\frac{\Gamma \wedge x \leq^? t \wedge x \leq^? t'}{\Gamma \wedge x \leq^? t} \quad \text{if } t =_E t'$$

Fail
$$\frac{\Gamma \wedge x \leq^? t \wedge x \leq^? t'}{\bot} \quad \text{if } t \neq_E t'$$

Fig. 1. Set of rules $\mathcal{RM}$ for matching in the union of regular theories

---

3.1 Soundness

The following lemma states how to solve a match-equation with an i-pure left hand side, thanks to the E_i-matching algorithm.

Lemma 16. *If s is i-pure, σ is a substitution in E-normal form and t a term in layers reduced form, then $s\sigma =_E t \iff s\sigma^{\pi_i} =_{E_i} t^{\pi_i}$.*

Therefore, we can conclude that a complete set of solutions modulo $E_1 \cup E_2$ of $s \leq^? t\Downarrow$ is obtained from $CSS_{E_i}(s \leq^? (t\Downarrow)^{\pi_i})$. Since equational theories are assumed regular, solutions are ground and so they may be written out as match-equations. Note that MatchSolve must be applied in a non-deterministic way in order to preserve all solutions.

3.2 Completeness

It is easy to check that the normal forms w.r.t $\mathcal{RM}$ are the conjunctions of solved match-equations $\wedge_{k \in K} x_k \leq^? t_k$ or $\bot$ or $\top$. Assume Γ is not a conjunction of solved match-equations. If there exists an equation $x =^? s$, then either $x \leq^? t$ is a match-equation in Γ and Merge applies or $s[x] \leq^? t$ with $s \notin X$ and either LeftPurif or Matchsolve applies. Otherwise, if there exists a match-equation $s \leq^? t$ with $s \notin X$, then either LeftPurif or MatchSolve applies. Otherwise, there exists two solved match-equations $x \leq^? t$ and $x \leq^? t'$ in Γ and either Delete or Fail applies. Thus, a transformation rule in $\mathcal{RM}$ can always be applied to Γ.

3.3 Termination

Proposition 17. *$\mathcal{RM}$ terminates for any control.*

Sketch of proof. For any problem Γ, we consider the following complexity measures:

- TS_{mul} is the multiset of theory sizes of non-variable left hand sides of match-equations and non-variable right hand sides of equations in Γ. The theory size of $t = C[t_1, \ldots, t_n]$ where t_i's are all aliens of t is defined by $TS(t) = 1 + \sum_{i=1}^{n} TS(t_i)$.
- NEQ is the number of equations in Γ.
- $NMEQ$ is the number of match-equations in Γ.

These measures are combined lexicographically. The situation is summarized in the table below:

rules	TS_{mul}	NEQ	$NMEQ$
LeftPurif	↓		
Merge	=	↓	
MatchSolve	↓		
Delete	=	=	↓

□

We get then the modularity theorem:

Theorem 18. *If E_1 and E_2 are regular theories sharing constructors and a complete and finite algorithm is known for matching modulo E_1 and E_2 then a complete and finite algorithm may be built for matching modulo $E_1 \cup E_2$.*

4 Unification

The last problem we are interested in is the unification problem. As we shall see, this problem is more difficult in the case of theories sharing constructors than in the case of disjoint theories. In order to be able to establish a modularity theorem, we had to restrict our attention to finitary theories in which no non-constant constructor is shared at the top. One should notice that from definition 1, this forbids collapsing theories but not non-regular ones. For the sake of simplicity, we present the algorithm only for regular theories because the non-regular case requires many additional definitions. However, the algorithm remains almost the same since most of the treatment is encoded in the unification algorithm of each theory. We just have to assume the existence of a more powerful unification algorithm in each theory: namely an algorithm for unification with free function symbols. The formalism we take for designing our combination algorithm is mostly borrowed from A. Boudet [3] and F. Baader & K. Schulz [1] with some modifications due to the special nature of the problem we address.

4.1 Preprocessing and data structure

The algorithm we describe is devoted to the transformation of a conjunction Γ of equations into a finite set of solved forms. As in the case of disjoint equational theories, the first step of the algorithm consists in purifying the problem. We first abstract all aliens of terms in Γ by fresh variables and add the corresponding equations to the problem. Repeated application of this operation obviously terminates and yields equations between pure terms which are either shared terms or terms with an unshared symbol at the top. Remind that from definition 4, a pure term may have aliens. We could however have equations $t =^? s$ where t is a 1-pure term and s is 2-pure term and none of them is a shared term. Such equations are said impure. In our case, since the theories are not collapsing, if such an equation has a solution σ, there must exist a term u with a shared constructor at the top such that $t\sigma =_E u$ and $s\sigma =_E u$. This means that u is a constant which occurs at the top in E_1 and E_2. The set of such constants will be denoted by SF_0. The equation may thus be split into $t =^? c \wedge s =^? c$ where c is chosen non-deterministically in SF_0. In order to remain complete, we have to collect all problems generated this way. A very similar situation is the case where the problem contains two equations $x =^? t$ and $x =^? s$ where t is a 1-pure term, s is a 2-pure term and none of them is a shared term. Again, a constant c is chosen non-deterministically in SF_0. In order to make the process terminate, we have to replace x with c everywhere in the problem and add the equation $x =^? c$. Once the purification process terminates, in each generated problem Γ', we may distinguish three kinds of equations: equations between shared terms, and equations $s =^? t$ between i-pure terms ($i = 1$ or 2) where s and t are either shared terms or terms with an unshared symbol at the top, and s or t is not a shared term. Since shared terms are built only from constructors, the subproblem consisting of the conjunction of all equations between shared terms may be solved in the empty theory, yielding either failure or the minimal idempotent solution σ_0. In the former case, Γ' has no solution. In the latter one, we apply σ_0 to the remaining equations and keep it apart from the problem.

4.2 The combination algorithm

Once a problem has been preprocessed it may be written as $\langle \sigma_0; \Gamma_1 \wedge \Gamma_2 \rangle$ where

- σ_0 is an idempotent substitution from X to $T(SF, X)$
- Γ_1 (resp. Γ_2) contains the equations $s =^? t$ between 1-pure (resp. 2-pure) terms where s or t is not a shared term.
- No variable in the domain of σ_0 occurs in an equation in $\Gamma_1 \wedge \Gamma_2$.

— For each equation $x =^? t \in \Gamma_1$ (resp. Γ_2), Γ_2 (resp. Γ_1) does not contain an equation $x =^? s$.

Each Γ_i will be solved using a unification algorithm for E_i. The following proposition states that doing so we do not loose any solution.

Proposition 19. *Let $s =^? t$ be a i-pure equation and σ a substitution in E-normal form. Then $s\sigma =_E t\sigma \iff s\sigma^{\pi_i} =_{E_i} t\sigma^{\pi_i}$.*

At this point, the main difference between the disjoint and the non-disjoint case is that a variable may get instantiated in both theories without generating a conflict. We avoid this situation by keeping in Γ_i only equations of the same form as the ones generated by preprocessing. Solving in one theory could instantiate a variable x by a term of the form $C[t_1, \ldots, t_n]$ where $C[\ ]$ contains only constructors and variables and t_i's have unshared symbols at the top. In this case, we abstract t_i's by fresh variables x_i's, replace x everywhere by $C[x_1, \ldots, x_n]$ and add the equations $x_i =^? t_i$ to the problem.

Applying repeatedly resolution modulo E_1 and E_2 will eventually produce problems Γ_1 and Γ_2 which are both solved. It could however happen that $\Gamma_1 \wedge \Gamma_2$ is not solved if a compound cycle occurs. A compound cycle is a subproblem of the form $x_1 =^? t_1[x_2] \wedge \cdots \wedge x_{2n-1} =^? t_{2n-1}[x_{2n}] \wedge x_{2n} =^? t_{2n}[x_1]$ where $x_{2i-1} =^? t_{2i-1}[x_{2i}] \in \Gamma_1$, $x_{2i} =^? t_{2i}[x_{2i+1}] \in \Gamma_2$ and no $t_i[x_{i+1}]$ is reduced to x_{i+1} (by convention, $x_{2n+1} = x_1$). Since we are actually only interested in x_i's, we write such a cycle as $C[x_1, \ldots, x_{2n}]$.

In the disjoint case, if theories are regular, such cycles have no solution. In our case, a compound cycle could be broken by instantiating some x_i with a constant in SF_0. This leads to the rule Cycle in figure 2.

We assume that we know for each theory E_i, a complete and finite unification algorithm which takes as input a problem of the form (P, V) where P is a conjunction of equations and V is a finite set of variables. Solutions of (P, V) are solutions of P which instantiate all variables in V either by a variable or by a constant in SF_0. Such an algorithm may be built if a complete and finite algorithm is known for unification with free constants (For the case of non-regular theories, we would require a unification algorithm with free function symbols). $CSS_{E_i}(P, V)$ denotes the complete set of solutions returned by the algorithm.

At any time during the unification algorithm, we may distinguish three classes of variables in the problem:

1. *Initial variables* which are the variables occurring in the problem before preprocessing.
2. *Abstraction variables* which are variables coming from an abstraction, either during preprocessing or during the algorithm itself.
3. *Introduced variables* which are variables introduced by the unification algorithms for each theory.

We make the very natural assumption that the unification algorithm for each theory may recognize initial, abstraction and introduced variables and never assigns an introduced variable to a non-introduced one or an abstraction variable to an initial one. With this assumption, our combination algorithm will always make an introduced variable appear in at most one Γ_i. We may thus also suppose that the domain of each solution does not contain an introduced variable. This does not compromise the soundness of our algorithm.

The combination algorithm is described by the two rules given in figure 2. In the rule UnifSolve$_i$, ρ_{SF} is obtained by abstracting aliens in the range of ρ by fresh variables. ρ_{F_i} is the substitution such that $x\rho = x\rho_{SF}\rho_{F_i}$ for all $x \in \text{Dom}(\rho)$. $\hat{\rho}_{F_i}$ denotes the conjunction $\wedge_{x \in \text{Dom}(\rho_{F_i})} x =^? x\rho_{F_i}$.

UnifSolve$_i$
$$\frac{\langle \sigma_0; \Gamma_i \wedge \Gamma_j \rangle}{\langle \sigma_0 \rho_{SF}; \hat{\rho}_{F_i} \wedge \Gamma_j \rho_{SF} \rangle}$$

if

- Γ_i is unsolved
- V_j ($j \neq i$) is the set of variables instantiated in Γ_j.
- $\rho \in CSS_{E_i}(\Gamma_i, V_j)$

Cycle
$$\frac{\langle \sigma_0; \Gamma \wedge \mathcal{C}[x_1, \ldots, x_{2n}] \rangle}{\langle \sigma_0; \Gamma \wedge \mathcal{C}[x_1, \ldots, x_{2n}] \rangle \{x_i \mapsto c\}}$$

if

- Γ_1 and Γ_2 are solved
- $\mathcal{C}[x_1, \ldots, x_{2n}]$ is a compound cycle
- $c \in SF_0$

Note: Both these rules are non-deterministic. In UnifSolve$_i$ we must consider each solution in $CSS_{E_i}(\Gamma_i, V_j)$. In Cycle, we must consider all possible choices of the index i and the constant c.

Fig. 2. Set of rules $\mathcal{RU}$ for unification

4.3 Completeness

We have to check that the normal forms w.r.t. $\mathcal{RU}$ are solved forms.

If either Γ_1 or Γ_2 is not solved then UnifSolve$_i$ applies. Otherwise, if $\Gamma_1 \wedge \Gamma_2$ is not solved then it contains a compound cycle $\mathcal{C}[x_1, \ldots, x_{2n}]$ and Cycle applies.

Consequently, a transformation rule in $\mathcal{RU}$ can always be applied to a problem Γ if Γ is not solved.

4.4 Termination

Proposition 20. $\mathcal{RU}$ *terminates for any control.*

Sketch of proof. For each problem $\langle \sigma_0; \Gamma_1 \wedge \Gamma_2 \rangle$, we consider the following complexity measures which are combined lexicographically.

- NIV is the number of initial variables occurring in $\Gamma_1 \wedge \Gamma_2$
- UIV is the number of initial variables which occur not instantiated in $\Gamma_1 \wedge \Gamma_2$
- NAV is the number of abstraction variables occurring in $\Gamma_1 \wedge \Gamma_2$
- USP is the number of unsolved subproblems Γ_i

Three cases must be distinguished for the rule $\mathsf{UnifSolve}_i$:

1. If some initial variable is identified with another initial variable or instantiated by a constant in SF_0 or some variable is instantiated by a term whose top-symbol belongs to $SF \backslash SF_0$ then NIV decreases. Indeed, in the last case, this variable is necessarily an initial variable which was not previously instantiated. It is then replaced everywhere by an abstraction of its value. Note also that it is the first case in which NAV may increase.
2. else, if some initial variable is newly instantiated then UIV decreases.
3. else, if an abstraction variable is identified with an initial or abstraction variable or instantiated by a constant in SF_0 then NAV decreases.
4. else, the substitution returned by the unification algorithm for E_i is of the form $\{x_1 \mapsto t_1, \ldots, x_n \mapsto t_n\}$ where x_i's are either initial or abstraction variables and each t_i has an unshared symbol at the top. In this case, the substitution ρ_{SF} is the identity. Thus if Γ_j ($j \neq i$) was previously solved, it remains solved so that USP decreases. Furthermore no new abstraction variable is introduced.

For Cycle, since the variable which gets instantiated by a shared constant was previously instantiated, it is either an initial variable in which case NIV decreases, or an abstraction variable, in which case NAV decreases.

The situation is summarized in the table below.

rules	NIV	UIV	NAV	USP
$\mathsf{UnifSolve}_i(1)$	↓			
$\mathsf{UnifSolve}_i(2)$	=	↓		
$\mathsf{UnifSolve}_i(3)$	=	=	↓	
$\mathsf{UnifSolve}_i(4)$	=	=	=	↓
Cycle(1)	↓			
Cycle(2)	=	=	↓	

□

Theorem 21. *If E_1 and E_2 are regular theories sharing constructors, all constructors shared at the top are constants and a finite and complete algorithm is known for unification with free function symbols modulo E_1 and E_2, then a complete and finite algorithm may be built for unification modulo $E_1 \cup E_2$.*

5 Undecidability results

We exhibit now two families of theories for which no uniform algorithm exists for deciding unifiability. Each theory in these families is the union of two theories sharing constructors in which unification with free function symbols is decidable. However, in each case, one of the conditions of theorem 21 is not satisfied. This shows that weakening these conditions is a difficult problem. The first family shows that there exists no uniform combination technique for combining unification for simple linear finitary theories sharing non-constant constructors at the top. The second one shows that there exists no general technique for combining unification algorithms for simple linear infinitary theories sharing no constructor at the top. We recall that a theory is simple if no term is equivalent to one of its strict subterms. For each case, undecidability is proved by showing that unification allows to solve a Post correspondence problem:

Definition 22 (Post correspondence problem). Let $\mathcal{A}$ and $\mathcal{C}$ be finite disjoint alphabets, and φ and φ' be two morphisms from $\mathcal{A}^*$ to $\mathcal{C}^*$. The Post correspondence problem for φ and φ' consists in finding a non-empty sequence $\alpha \in \mathcal{A}^+$ such that $\varphi(\alpha) = \varphi'(\alpha)$.

Theorem 23 (Post 1947 [9]). *There exists no uniform algorithm for solving the Post correspondence problem. The problem remains undecidable when φ and φ' are injective.*

In the following, we shall consider families of theories built from $\mathcal{A}$, $\mathcal{C}$, φ and φ'.

5.1 Undecidability of unification in the union of finitary theories sharing non-constant constructors at the top

We consider the theory $E_{\varphi,\varphi'}$ presented by the convergent rewriting system

$$\{f(a_i x, y) \to \omega_i f(x, a_i y) \mid i = 1, \ldots, n\} \cup \{f(\bot, y) \to h(y)\}$$
$$\cup \{f'(a_i x, y) \to \omega'_i f'(x, a_i y) \mid i = 1, \ldots, n\} \cup \{f'(\bot, y) \to h(y)\}$$

where φ and φ' are injective and ω_i and ω'_i denote respectively $\varphi(a_i)$ and $\varphi'(a_i)$. Due to injectivity, $\varphi(a_i)$ and $\varphi'(a_i)$ are non-empty words.

The following proposition in conjunction with theorem 23 shows that there exists no uniform decision algorithm for unifiability modulo $E_{\varphi,\varphi'}$.

Proposition 24. *If unifiability is decidable modulo $E_{\varphi,\varphi'}$ then the Post correspondence problem for φ and φ' is decidable.*

Sketch of proof. Any solution of the unification problem

$$P \equiv \bigvee_i f(a_i x, \bot) =^? f'(a_i x, \bot)$$

is of the form $x \mapsto \alpha\bot$ where $\alpha \in \mathcal{A}^*$ and for some a_i, $\varphi(a_i\alpha) = \varphi'(a_i\alpha)$. Thus P has a solution if and only if the Post correspondence problem for φ and φ' has a solution. $\square$

$E_{\varphi,\varphi'}$ is the union of E and E' presented by the convergent rewriting systems

$$\{f(a_i x, y) \to \omega_i f(x, a_i y) \mid i = 1, \ldots, n\} \cup \{f(\bot, y) \to h(y)\}$$

and

$$\{f'(a_i x, y) \to \omega'_i f'(x, a_i y) \mid i = 1, \ldots, n\} \cup \{f'(\bot, y) \to h(y)\}$$

Moreover, all shared symbols are clearly constructors of both E and E'.

Proposition 25. *E (resp. E') is simple and there exists a complete and finite algorithm for unification modulo E (resp. E') with free function symbols.*

As a corollary, we get

Corollary 26. *There exists no uniform technique for combining unification algorithms for simple linear finitary theories sharing constructors at the top.*

5.2 Undecidability of unification in the union of infinitary theories sharing no symbol at the top

We consider now the theory $E_{\varphi,\varphi'}$ presented by the convergent rewriting system

$$\{f(a_i x, y) \to f(x, \overline{\omega_i} y) \mid i = 1, \ldots, n\} \cup \{f'(a_i x, y) \to f'(x, \overline{\omega'_i} y) \mid i = 1, \ldots, n\}$$

where for any word λ, $\overline{\lambda}$ denotes the word obtained by reversing λ. We do not need anymore to suppose that φ and φ' are injective.

The following proposition in conjunction with theorem 23 shows that there exists no uniform decision algorithm for unifiability modulo $E_{\varphi,\varphi'}$.

Proposition 27. *If unifiability is decidable modulo $E_{\varphi,\varphi'}$ then the Post correspondence problem for φ and φ' is decidable.*

Sketch of proof. Any solution of the unification problem

$$P \equiv \bigvee_i f(a_i x, \bot) =^? f(\bot, y) \wedge f'(a_i x, \bot) =^? f'(\bot, y)$$

is of the form $\{x \mapsto \alpha \bot, y \mapsto \overline{\varphi(a_i \alpha)} \bot\}$ where $\alpha \in \mathcal{A}^*$ and $\varphi(a_i \alpha) = \varphi'(a_i \alpha)$. Thus, P has a solution if and only if the Post correspondence problem for φ and φ' has a solution. □

$E_{\varphi,\varphi'}$ is the union of E and E' presented by the convergent rewriting systems

$$\{f(a_i x, y) \to f(x, \overline{\omega_i} y) \mid i = 1, \ldots, n\} \text{ and } \{f'(a_i x, y) \to f'(x, \overline{\omega'_i} y) \mid i = 1, \ldots, n\}$$

Moreover, all shared symbols are clearly constructors of both E and E'.

Proposition 28. *E (resp. E') is simple, and unification modulo E (resp. E') with free function symbols is infinitary and decidable.*

As a corollary, we get

Corollary 29. *There exists no uniform technique for combining unification algorithms for simple linear infinitary theories sharing constructors, even if no constructor is shared at the top.*

Conclusion and perspectives

We have established modularity results for the word problem, the matching problem and the unification problem in theories sharing constructors. For the case of unification, the result seems rather weak but the undecidability results given in section 5 show that the problem is hard. However it seems possible to extend our results in various directions. As already mentioned, we are actually able to handle non-regular theories sharing constructors, provided that only constants are shared at the top. Even this restriction may be somehow relaxed if finitely many contexts built from constructors are shared at the top [11]. In this case, theory conflicts that occur in the combination algorithm may be solved by introducing explicitely a shared term taken from a finite set. Another extension, which in practice would be very useful, is to combine collapsing theories sharing constructors. This seems very difficult since we are then unable to bound the terms allowing to solve a theory conflict.

Another problem consists in combining decision algorithms for unification. This means that we do not assume that we know for each theory a unification algorithm but only a decision algorithm for unifiability. The combination becomes possible with the very strong assumption that in some sense, an equality step in each theory looks only at finitely many shared symbols [11].

Concerning modularity, one may note that our definition of constructors is actually not really modular in the sense that they are defined by the mean of an ordering on the whole combined theory. It would be nice to be able to define for each theory complete sets of constructors so that theories sharing constructors are defined in a syntactic way. That's to say: (F_1, A_1) and (F_2, A_2) share constructors if $F_1 \cap F_2 \subseteq C_1 \cap C_2$ where C_i is a complete set of constructors of (F_i, A_i). One possible idea would be to define constructors of (F, A) by the mean of a simplification ordering on $T(F)$ such that any congruence class of (F, A) contains a minimal element. If for any F' containing F, we are able to extend such an ordering to a simplification ordering on $T(F')$ such that any congruence class of (F', A) still contains a minimal element, then using a result of Kurihara & Ohuchi [7] we get a simplification ordering for the union of theories such that constructors are preserved.

At last, we could imagine to share non-free constructors that might for instance be defined as constructors modulo another theory. For example, this would allow to share a symbol which is commutative in both theories.

Acknowledgements

We would like to thank the PROTHEO group at Nancy for many fruitful discussions, especially Hélène Kirchner, Claude Kirchner and Michaël Rusinowitch.

References

1. Franz Baader and Klaus Schulz. Unification in the union of disjoint equational theories: Combining decision procedures. In *Proceedings 11th International Conference on Automated Deduction, Saratoga Springs (N.Y., USA)*, pages 50–65, 1992.

2. Franz Baader and Klaus U. Schulz. Combination techniques and decision problems for disunification. In Claude Kirchner, editor, *Rewriting Techniques and Applications, 5th International Conference, RTA-93*, LNCS 690, pages 301–315, Montreal, Canada, June 16–18, 1993. Springer-Verlag.
3. A. Boudet. *Unification dans les mélanges de théories équationelles*. Thèse de Doctorat d'Université, Université de Paris-Sud, Orsay (France), February 1990.
4. D. Dougherty and P. Johann. An improved general E-unification method. In M. E. Stickel, editor, *Proceedings 10th International Conference on Automated Deduction, Kaiserslautern (Germany)*, volume 449 of *Lecture Notes in Computer Science*, pages 261–275. Springer-Verlag, July 1990.
5. J. Gallier and W. Snyder. Complete sets of transformations for general E-unification. *Theoretical Computer Science*, 67(2-3):203–260, October 1989.
6. Claude Kirchner. *Méthodes et outils de conception systématique d'algorithmes d'unification dans les théories équationnelles*. Thèse de Doctorat d'Etat, Université de Nancy I, 1985.
7. M. Kurihara and A. Ohuchi. Modularity of simple termination of term rewriting systems with shared constructors. *Theoretical Computer Science*, 103(2):273–282, 1992.
8. T. Nipkow. Combining matching algorithms: The regular case. In N. Dershowitz, editor, *Proceedings 3rd Conference on Rewriting Techniques and Applications, Chapel Hill (N.C., USA)*, volume 355 of *Lecture Notes in Computer Science*, pages 343–358. Springer-Verlag, April 1989.
9. E. Post. Recursive unsolvability of a problem of thue. *The Journal of Symbolic Logic*, 12:1–11, 1947.
10. Ch. Ringeissen. Unification in a combination of equational theories with shared constants and its application to primal algebras. In *Proceedings of the 1st International Conference on Logic Programming and Automated Reasoning, St. Petersburg (Russia)*, volume 624 of *Lecture Notes in Artificial Intelligence*, pages 261–272. Springer-Verlag, 1992.
11. Ch. Ringeissen. *Combinaison de Résolutions de Contraintes*. Thèse de Doctorat d'Université, Université de Nancy I, December 1993.
12. Ch. Ringeissen. Combination of matching algorithms. In P. Enjalbert, E. W. Mayr, and K. W. Wagner, editors, *Proceedings 11th Annual Symposium on Theoretical Aspects of Computer Science, Caen (France)*, volume 775 of *Lecture Notes in Computer Science*, pages 187–198. Springer-Verlag, February 1994.
13. M. Schmidt-Schauß. Combination of unification algorithms. *Journal of Symbolic Computation*, 8(1 & 2):51–100, 1989. Special issue on unification. Part two.
14. K. Yelick. Unification in combinations of collapse-free regular theories. *Journal of Symbolic Computation*, 3(1 & 2):153–182, April 1987.

Primal Grammars
and
Unification Modulo a Binary Clause*

Gernot Salzer**

Technische Universität Wien, Austria

Abstract. In resolution theorem proving as well as in its descendant, logic programming, we are frequently confronted with binary clauses causing lengthy or infinite computations because of their self-resolvents. In this paper we investigate *unification modulo a binary clause*, which can be used as a short cut through loops of the form $L \rightarrow R$. For a certain class of binary clauses we show that (i) its unification problem is decidable and (ii) the unifiers can be finitely schematized. This is done by first reducing the binary clauses to a simpler form and then employing primal grammars [12]. Our work extends results obtained in the context of cycle unification.

1 Introduction

In spite of their seeming simplicity, binary clauses have been studied extensively in the past and at present. The reasons are at least twofold. On the one hand the expressive power of a single binary clause (in conjunction with an appropriate goal and fact) is already equivalent to that one of Turing machines, as recently has been shown [9]. On the other hand binary clauses are quite common in automated theorem proving and logic programming: permutation clauses like symmetry as well as many recursions in logic programs take the form of binary clauses. Hence considerable effort has been spent on controlling the behavior of such clauses in order to detect non-terminating computations and to speed up terminating ones (see for instance [6, 7, 21, 22]).

Example 1. [3] Consider the following three programs differing from each other on their last facts and on the associated queries.

$$
\begin{array}{lll}
(1) & P(x) \leftarrow Q(x,y), R(y) & \\
(2) & Q(a,a) \leftarrow & \\
(3) & Q(f(x), g(y)) \leftarrow Q(x,y) & \\
(4) \ R(g^{20}(a)) \leftarrow & (4') \ R(b) \leftarrow & (4'') \ R(x) \leftarrow \\
(G) \ \leftarrow P(x) & (G') \ \leftarrow P(x) & (G''') \ \leftarrow Setof(x, P(x), y)
\end{array}
$$

---

* The final version of this article was completed while visiting CRIN/INRIA Lorraine (Nancy).
** Technische Universität Wien, Karlsplatz 13/E185-2, A-1040 Wien (Austria); Email: salzer@logic.tuwien.ac.at

In the first program the query G requires 20 steps of backtracking through clause 2 in order to yield $P(f^{20}(a))$. The second program does not terminate on goal G', since none of the infinitely many solutions for $Q(x,y)$ yields a binding for y satisfying fact $4'$. Finally, the third program does not terminate on G'' since y would have to be the infinite set $\{a, f(a), f(f(a)), \ldots\}$ of all values for x computed by the query $\leftarrow P(x)$. Apparently the reason for the long computation in the first case and the loops in the other ones is the binary clause (3) with its infinitely many self-resolvents.

In this paper we isolate this problem by investigating *unification modulo a binary clause*. We identify a class of binary clauses, for which unification is decidable and for which the set of unifiers can be finitely schematized by means of primal grammars, a formalism first presented in [12]. In the example above the atom $Q(a,a)$ is unifiable with $Q(x,y)$ modulo $Q(f(x), g(y)) \leftarrow Q(x,y)$; the infinitely many unifiers of the form $\{x \leftarrow f^n(a), y \leftarrow g^n(a)\}$ can be finitely schematized using primal grammars and thus used in further computations.

The paper is organized as follows. After introducing basic notations in Section 2, we formally define unification modulo a binary clause in Section 3; a way of enumerating all unifiers is described in Section 4. Section 5 introduces the class of mono-cyclic clauses and states the main result of this paper, viz. the decidability of unification modulo this kind of clauses. In Section 6 mono-cyclic clauses are reduced to simply mono-cyclic ones. After presenting primal grammars informally in Section 7 we sketch the unification algorithm for mono-cyclic clauses in Section 8. Finally, Section 9 discusses the improvements gained by the use of unification modulo a binary clause.

2 Basic Notations and Definitions

We assume the reader to be familiar with the basic terminology of clause-oriented theorem proving and logic programming. To fix notation we review some definitions.

A binary clause is a clause of the form $\{\neg L, R\}$ and is written in the following either as $R \leftarrow L$ or as $L \rightarrow R$. By $\forall C$ we denote the universal closure of the clause C. The set of variables occurring in a clause or an atom A is denoted by $\text{var}(A)$.

As usual, substitutions are defined by sets of the form $\{x_1 \leftarrow t_1, \ldots, x_n \leftarrow t_n\}$, with the understanding that an application of the substitution replaces x_i by t_i. Substitutions are denoted by lowercase Greek letters and written in postfix notation; thus, for a term, atom or clause t and substitutions λ, μ the expressions $t\lambda$ and $\lambda\mu$ are to be interpreted as $\lambda(t)$ and $\mu \circ \lambda$, respectively. The n-fold composition of λ is written λ^n, defined by $\lambda^0 = \{\}$ and $\lambda^{n+1} = \lambda^n \lambda$. The domain of λ is the set $\text{dom}(\lambda) = \{x \mid x \neq x\lambda\}$, its codomain the set $\text{cod}(\lambda) = \{x\lambda \mid x \in \text{dom}(\lambda)\}$; the set of variables occurring in $\text{cod}(\lambda)$ is denoted by $\text{vcod}(\lambda)$.

3 Unification Modulo a Binary Clause

The basic intention of cycle unification [2] as well as of unification modulo a binary clause is to handle situations like the one sketched in example 1 more efficiently. In general terms the problem can be stated as follows. Given two arbitrary clauses $C_G = \{\ldots, \neg G, \ldots\}$ and $C_F = \{\ldots, F, \ldots\}$—in the context of logic programming a goal and a rule with head F—one would like to characterize all the possible ways C_G might interact with C_F in the presence of a binary clause $C = \{R, \neg L\}$; furthermore one would like to derive a finite description of the substitutions applied to G and F. Such a description in hand one could for instance eliminate the infinite loop in the second program of example 1, provided the description has the property that fact $4'$ can be effectively checked against it.

Cycle unification

A cycle unification problem consists of two atoms G and F and a binary clause C, written as $\langle G \xrightarrow{\circ} F \rangle_C$. A substitution θ is a solution of $\langle G \xrightarrow{\circ} F \rangle_C$ iff $\forall G\theta$ is a logical consequence of $\forall F$ and $\forall C$. As in unification theory, the natural questions to ask are: (i) Is cycle unification decidable? (ii) How many independent[3] solutions exist for a given problem? and (iii) Is there an algorithm enumerating a complete and minimal set of solutions?

[2] and [23] describe classes of cycle unification problems with finite sets of independent solutions; [17] deals with a class possessing an infinite number of solutions, but admitting a finite schematization. The general case of unrestricted cycle unification, however, is shown to be undecidable in [11] and [8].

Due to this undecidability result we are forced to impose restrictions on the general problem to be able to answer the above questions positively. Before introducing such restrictions we modify our framework, since cycle unification turns out to be inadequate in the context of logic programming and clausal theorem proving.

Example 2. Consider the logic program

$$\leftarrow P(a, a)$$
$$P(x, y) \leftarrow P(f(x), y)$$
$$P(f(a), z) \leftarrow Q(z)$$
$$Q(b) \leftarrow$$

containing the cycle unification problem $CU = \langle P(a, a) \xrightarrow{\circ} P(f(a), z) \rangle_C$, where $C = P(x, y) \leftarrow P(f(x), y)$. The minimal set of independent solutions for CU consists of a single substitution, viz. the empty substitution. However, this solution does not help in deriving that the goal fails. The reason obviously is that cycle unification does not record substitutions concerning variables in F.

---

[3] Two substitutions θ_1 and θ_2 are independent, if there is no substitution λ such that $\theta_1 = \theta_2\lambda$ or $\theta_1\lambda = \theta_2$

From Cycle unification to Unification Modulo a Binary Clause

A substitution θ is a solution of $\langle G \xrightarrow{\circ} F \rangle_C$ iff $\forall G\theta$ is a logical consequence of $\forall F$ and $\forall C$, i.e., iff $\forall F \rightarrow \forall G\theta$ follows from $\forall C$. In order to take the replacements for the variables in F into account we modify this definition by accepting θ as solution only if $\forall (F\theta \rightarrow G\theta)$ is a logical consequence of $\forall C$.

Example 2 (continued). $P(f(a), z)\theta \rightarrow P(a, a)\theta$ is a logical consequence of the binary clause $P(f(x), y) \rightarrow P(x, y)$ only if $\theta = \{z \leftarrow a\}$. Thus the only possibility to continue the program after exiting the loop is the goal $Q(z)\theta \rightarrow = Q(a) \rightarrow$, which correctly fails.

Definition 1. Two atomic formulas F and G are *equal modulo the binary clause* $C = L \rightarrow R$, denoted as $F \Rightarrow_C G$, iff $\forall F \rightarrow G$ is a logical consequence of $\forall L \rightarrow R$.

Note that we slightly misuse the term 'equal' to maintain the analogy to equational unification: due to the directional character of implication, 'equality' modulo a binary clause is not symmetric. Rather $L \rightarrow R$ has to be viewed as a rewrite rule: F and G are equal iff F can be rewritten to G.

Example 3. $P(a)$ and $P(f(f(a)))$ are equal modulo $C = P(x) \rightarrow P(f(x))$, i.e., $P(a) \Rightarrow_C P(f(f(a)))$. However, $P(f(f(a))) \not\Rightarrow_C P(a)$.

The word problem modulo a binary clause is decidable, i.e., the proposition $F \Rightarrow_C G$ can be tested effectively. $F \Rightarrow_C G$ holds iff the clause set $\{C, F\eta, \neg G\eta\}$ is unsatisfiable, where η is a substitution replacing every variable in F and G by a unique Skolem constant. In [19] it is shown that the satisfiability of clause sets consisting of one binary clause and two ground unit clauses is a decidable property.

It is worth noting that our word problem is nothing but the Horn clause implication problem for binary clauses. In [14] it is shown that Horn clause implication is undecidable for ternary clauses. In our context this result means that equality modulo a ternary clause, i.e., modulo a Horn clause with two literals in its body, is most probably[4] undecidable.

Equality modulo a binary clause can also be viewed as a specific instance of theory resolution [20]. Let T be the set $\{C\}$, and let $C_1 = D_1 \vee F$ and $C_2 = D_2 \vee \neg G$ be two clauses. Then $D_1 \vee D_2$ is a T-resolvent iff $F \Rightarrow_C G$.

Definition 2. Two atomic formulas F and G are *unifiable modulo a binary clause C*, iff there is a substitution θ such that $F\theta \Rightarrow_C G\theta$; θ is called a *unifier* of F and G.

Complete sets of most general unifiers are defined as usual (see for instance [1]).

---

[4] Marcinkowski and Pacholski did not exactly show that the satisfiability of one ternary clause with *two* ground unit clauses is undecidable, but only with *some* ground unit clauses.

Example 4. $P(a)$ and $P(y)$ are unifiable modulo $P(x){\rightarrow}P(f(x))$. The complete set of most general unifiers is given by the infinite set

$$\{\{y{\leftarrow}a\},\{y{\leftarrow}f(a)\},\{y{\leftarrow}f(f(a))\},\ldots\} \ .$$

Unifiability modulo a binary clause is an undecidable property, i.e., there is no algorithm determining for arbitrary F, G and C whether F and G are unifiable modulo C. This follows from a result by Hanschke and Würtz [11], who showed that the satisfiability of clause sets consisting of one binary clause and two arbitrary unit clauses is undecidable.

Each cycle unification problem $\langle G \xrightarrow{o} F \rangle_C$ can be rephrased as unification of F' and G modulo C, where F' is a renamed copy of F; the unifiers of the latter problem, when restricted to the variables in G, are exactly the solutions of the former one.

4 Enumerating the Unifiers

Let $C_n = L_n{\rightarrow}R_n$ denote the n-fold self-resolvent of a binary clause $C = L{\rightarrow}R$. More formally, let $C_1 = C$ and let C_{n+1} be the result of resolving C_n and C upon R_n and L, respectively, i.e., $C_{n+1} = L_n\mu{\rightarrow}R\eta\mu$ where η is a renaming substitution such that $\text{var}(R_n) \cap \text{var}(L\eta) = \emptyset$ and μ is a most general unifier of R_n and $L\eta$. The self-resolvents are unique up to renaming [19], hence it does not matter whether the clauses C_n and C are resolved upon L_n and R or upon R_n and L. Throughout this paper equality of clauses is always understood as equality up to renaming.

Using the notion of self-resolvents the unifiers of two atoms F and G modulo a binary clause C can be characterized in the following way.

Lemma 3. *θ is unifier of F and G modulo C iff*

(a) θ is a unifier of F and G, or
(b) there is a self-resolvent C_n and a substitution θ' such that $C_n\theta' = (F{\rightarrow}G)\theta$.

Hence a complete set of unifiers of F and G modulo C is given by

$$\{\theta_n \mid n \geq 0,\ \theta_n \text{ defined}\} \ ,$$

where θ_0 is a most general unifier of F and G, and θ_n is a most general unifier of $C_n\eta = L_n\eta{\rightarrow}R_n\eta$ and $F{\rightarrow}G$ restricted to the variables occurring in F and G.[5]

Example 4 (continued). The self-resolvents of $C = P(x){\rightarrow}P(f(x))$ are given by $C_n = P(x){\rightarrow}P(f^n(x))$. Unifying these clauses with $P(a){\rightarrow}P(y)$ we obtain $\theta_n = \{y{\leftarrow}f^n(a)\}$ for $n \geq 0$.

---

[5] Here we view binary clauses as terms headed by the binary function symbol $\rightarrow$. η is again an appropriate substitution renaming the variables of C_n and $F{\rightarrow}G$ apart.

As unification modulo a binary clause is undecidable, our aim is to identify subclasses of the general problem, for which unification becomes decidable. Additionally, since even for simple problems like the one in Example 4 a minimal complete set of unifiers is of infinite cardinality, we look for finite representations of the solutions.

5 Mono-Cyclic Clauses

One necessary precondition for a unification problem to have infinitely many most general unifiers is the existence of an infinite number of self-resolvents. In [17] it is shown that *variable-preserving* clauses have this property.

Definition 4. A binary clause C is called *variable-preserving* iff it can be written as $A\lambda_1 \to A\lambda_2$, where the atom A and the substitutions λ_1, λ_2 satisfy the following conditions:

(C1) $\text{dom}(\lambda_1) \cup \text{dom}(\lambda_2) \subseteq \text{var}(A)$ and $\text{dom}(\lambda_1) \cap \text{dom}(\lambda_2) = \emptyset$.
(C2) $\text{vcod}(\lambda_1) \subseteq \text{var}(A) - \text{dom}(\lambda_2)$ and $\text{vcod}(\lambda_2) \subseteq \text{var}(A) - \text{dom}(\lambda_1)$.

Condition C1 requires the minimality of λ_1 and λ_2. As an example, the clause $P(f(x)) \to P(f(f(x)))$ could be written as $A = P(x)$, $\lambda_1 = \{x \leftarrow f(x), y \leftarrow a\}$ and $\lambda_2 = \{x \leftarrow f(f(x))\}$. However, $y \notin \text{var}(A)$ and $\text{dom}(\lambda_1) \cap \text{dom}(\lambda_2) = \{x\}$. But obviously there is another decomposition with 'smaller' substitutions satisfying C1: $A' = P(f(x))$, $\lambda'_1 = \{\}$ and $\lambda'_2 = \{x \leftarrow f(x)\}$.

Condition C2 guarantees that the number of variables occurring in C does not grow upon self-resolution. It excludes clauses like $P(y,z) \to P(s[x,y], t[x,z])$, which are used in [11] to encode Post's correspondence problem.

Variable-preserving clauses admit the following characterization of their self-resolvents.

Lemma 5. *For all $n \geq 1$ the n-fold self-resolvent C_n of the variable-preserving clause C exists and is given by $C_n = A\lambda_1^n \to A\lambda_2^n$, where A, λ_1 and λ_2 are defined as above.*

Example 5. Let $C = P(f(x), y, z) \to P(x, f(y), g(y, z))$. Choosing $A = P(x, y, z)$, $\lambda_1 = \{x \leftarrow f(x)\}$ and $\lambda_2 = \{y \leftarrow f(y), z \leftarrow g(y, z)\}$ we obtain

$$C_n = P(x,y)\{x \leftarrow f(x)\}^n \to P(x,y)\{y \leftarrow f(y), z \leftarrow g(y,z)\}^n \ .$$

So far we know of no general way to decide unification modulo a variable-preserving clause. Therefore we impose one further restriction on λ_1 and λ_2, leading us to a decidable unification problem.

Definition 6. The *variable dependency relation* corresponding to a substitution λ, denoted by $\leadsto_\lambda$, is defined as

$$\leadsto_\lambda = \{(x,y) \mid x \in \text{dom}(\lambda), y \in \text{var}(\lambda(x))\} \ .$$

$\leadsto_\lambda^*$ denotes the reflexive and transitive closure of $\leadsto_\lambda$. A variable x is *cyclic* in λ, if there is a sequence $c = \langle x_1, \ldots, x_k \rangle$ such that $x_1 = x$, $x_i \neq x$, $x_{i-1} \leadsto_\lambda x_i$ and $x_k \leadsto_\lambda x_1$ ($2 \leq i \leq k$); c is called a cycle of length k starting with x. A variable starting no cycle is called *acyclic*. A cyclic variable x is *mono-cyclic* with period k, if all cycles starting with x have length k. A substitution λ is *mono-cyclic*, if all variables in dom(λ) are acyclic or mono-cyclic.

Example 6. $\lambda_1 = \{x \leftarrow f(x,y), y \leftarrow z, z \leftarrow g(z)\}$ is mono-cyclic, since x and z are mono-cyclic and y is acyclic.

$\lambda_2 = \{x \leftarrow f(y,z), y \leftarrow g(x), z \leftarrow g(x)\}$ is not mono-cyclic. x is mono-cyclic, whereas y and z are not: y for instance starts the cycles $\langle y, x \rangle$ and $\langle y, x, z, x \rangle$, which obviously are of different length.

Definition 7. A binary clause is *mono-cyclic* iff it is variable-preserving and additionally

(C3) λ_1 and λ_2 are mono-cyclic. (λ_1 and λ_2 are the same as in Definition 4.)

Theorem 8. *Unification modulo a mono-cyclic clause is decidable. Furthermore, the unifiers are always schematizable by a finite number of primal substitutions.*[6]

The class of mono-cyclic clauses is a generalization of directly recursive clauses [17] and also covers classes defined in [2]. Unifying cycles [23], however, are not properly contained in our class: the clause $P(x,x) \to P(f(y), f(a))$ is a unifying cycle, but no mono-cyclic clause.

In the remaining sections we sketch the proof of Theorem 8 as well as the unification algorithm.

6 Reducing the Problem

In this section we show how mono-cyclic substitutions can be transformed to simply mono-cyclic ones. This will allow us to reduce unification modulo mono-cyclic clauses to the unification of primal grammars, described in the next section.

Definition 9. A substitution λ is *simply mono-cyclic* if every variable of its domain is mono-cyclic with period 1.

Simply mono-cyclic substitutions differ from mono-cyclic ones insofar as they contain neither acyclic variables nor cycles of lengths greater than 1. The following two lemmas show that for every mono-cyclic substitution λ there are numbers k, l and substitutions ρ, σ, τ, such that

$$\{\lambda^n \mid n \geq 1\} = \{\lambda^1, \ldots, \lambda^k\} \cup \bigcup_{i=0}^{l-1} \{\rho \sigma^i \mu^n \tau \mid n \geq 0\} ,$$

where $\mu = \sigma^l$ is simply mono-cyclic.

---
[6] Primal substitutions map variables to primal terms and thus are able to schematize an infinite number of first order substitutions. Primal terms are discussed in Section 7.

Lemma 10. *Let λ be a substitution containing a component $x \leftarrow s$ such that x is acyclic. Furthermore, let $\rho = \{y \leftarrow t \in \lambda \mid x \stackrel{*}{\leadsto}_\lambda y\}$, $\tau = \lambda - \rho$ and $\sigma = (\tau\rho) - \{x \leftarrow s\}$.*

(a) $\lambda^n = \rho\sigma^{n-1}\tau$ for all $n \geq 1$.
(b) If a variable $v \neq x$ is acyclic (mono-cyclic with period k) in λ, then it is also acyclic (mono-cyclic with period k) in σ.

By repeated applications of this lemma it is possible to remove all acyclic variables, i.e., the set $\{\lambda^n \mid n \geq 1\}$ can be rewritten to $\{\lambda^1, \ldots, \lambda^k\} \cup \{\rho\sigma^n\tau \mid n \geq 0\}$, where $\text{dom}(\sigma)$ contains just the cyclic variables of λ.

Example 7. $\lambda = \{x \leftarrow g(x, u), u \leftarrow f(y), y \leftarrow g(y, v), v \leftarrow a\}$ contains the acyclic variables u and v. Applying Lemma 10 twice we obtain $\{\lambda^n \mid n \geq 1\} = \{\lambda\} \cup \{\rho\sigma^n\tau \mid n \geq 0\}$, where $\rho = \{u \leftarrow f(y), y \leftarrow g(y, a), v \leftarrow a\}$, $\sigma = \{x \leftarrow g(x, f(y)), y \leftarrow g(y, a)\}$ and $\tau = \{x \leftarrow g(g(x, u), f(y)), y \leftarrow g(y, v)\}$.

The next lemma transforms the cycles of a mono-cyclic substitution to cycles of length 1.

Lemma 11. *Let σ be a substitution containing a mono-cyclic variable x with period l, and let $\mu = \sigma^l$.*

(a) x is mono-cyclic with period 1 in μ.
(b) If a variable v is acyclic (mono-cyclic) in σ, then it is also acyclic (mono-cyclic) in μ. Furthermore, if v is mono-cyclic with period 1 in σ, then it also has period 1 in μ.

Hence, if σ is mono-cyclic and its cycles are of length $l_1, \ldots, l_m$, respectively, then $\{\sigma^n \mid n \geq 0\}$ is equivalent to $\bigcup_{i=0}^{l-1}\{\sigma^i\mu^n \mid n \geq 0\}$, where l is the least common multiple of $l_1, \ldots, l_m$ and $\mu = \sigma^l$ contains only cycles of length 1.

Example 8. Let $\sigma = \{x \leftarrow f(y, y), y \leftarrow g(x)\}$. Both variables x and y are mono-cyclic with period 2. Hence $\{\sigma^n \mid n \geq 0\}$ is equivalent to $\{\mu^n \mid n \geq 0\} \cup \{\sigma\mu^n \mid n \geq 0\}$, where $\mu = \sigma^2 = \{x \leftarrow f(g(x), g(x)), y \leftarrow g(f(y, y))\}$. Obviously all cycles in μ have length 1.

Note that for this transformation to work it is essential that the original substitution is mono-cyclic. $\sigma' = \{x \leftarrow f(x, y), y \leftarrow g(x)\}$, for instance, is cyclic but not mono-cyclic; there is no l such that all cycles in σ'^l have unit length.

7 Primal Grammars

Within the last years a zoo of different species of schematizations appeared in literature. Up to now there are recurrence terms [4], ω-terms [3], I-terms [5], R-strings [15], R-forms [16], R-terms [18] and last but not least primal grammars [12, 13]. The aim of these schematizations is to finitely describe infinite sequences of structurally similar terms arising in clausal theorem proving and during completion in term rewriting, with the hope to avoid infinite loops and divergence.

Some schematizations are better behaved than others: well-behaved ones like R-strings, ω-, I- and R-terms as well as a restricted class of primal grammars have the pleasant property that the unification of these schematizations is decidable and furthermore the number of (meta-)unifiers is always finite. For other schematizations like recurrence terms and full primal grammars unification is undecidable. Finally, for R-forms the decidability of unification is unknown.

The various schematizations can be partially ordered according to their expressiveness: R-strings and ω-terms are incomparable, but both are subsumed by I-terms, which are subsumed by R-terms, which again are subsumed by primal grammars. The latter formalism also covers the recurrence-terms. R-forms are more expressive than R-terms and incomparable to primal grammars.

In this paper we use primal grammars, being currently the most expressive of the well-behaved formalisms. Instead of repeating the rather involved definitions we give an example illustrating the main ideas. Suppose we have given the infinite set $\{P(x,y)\mu^n \mid n \geq 0\}$ of atoms, where $\mu = \{x \leftarrow f(x,y), y \leftarrow g(y)\}$. Using primal grammars this set can be schematized by the primal term $t = P(\hat{x}(u;x,y), \hat{y}(u;x,y))$ and the prime rewrite system

$$\hat{x}(0;x,y) \rightarrow x, \quad \hat{x}(s(u);x,y) \rightarrow f(\hat{x}(u;x,y), \hat{y}(u;x,y)),$$
$$\hat{y}(0;x,y) \rightarrow y, \quad \hat{y}(s(u);x,y) \rightarrow g(\hat{y}(u;x,y)) \ .$$

$\hat{x}$ and $\hat{y}$ are auxiliary function symbols possessing two kinds of arguments: so-called counter expressions (u, $s(u)$ and 0 above) preceding the semicolon, and primal terms (e.g., first order variables) following the semicolon. Counter expressions are built from counter variables like u, the constant 0 and the unary function symbol s. The terms schematized by t are all those first order terms which can be obtained by first substituting a ground counter expression of the form $s(\cdots s(0) \cdots)$ for the counter variable u and then rewriting the resulting term to normal form. The form of the rewrite rules guarantees that all normal forms are first order terms containing no auxiliary function symbols. As an example, $t\{u \leftarrow s(s(0))\} = P(\hat{x}(s(s(0));x,y), \hat{y}(s(s(0));y))$ rewrites to the normal form $P(f(f(x,y),g(y)), g(g(y)))$.

Unifying two primal terms t_1 and t_2 means to check every first order term schematized by t_1 against every one schematized by t_2 for unifiability, possibly restricted by counter variables common to both primal terms. In general there will be an infinite number of unifiable pairs of terms and thus an infinite number of first order unifiers. There are two things one could expect from a unification algorithm for primal terms. First, it should be able to decide whether t_1 and t_2 are unifiable at all, i.e., whether there is any pair of terms schematized by t_1 and t_2 that can be first order unified. Secondly, the unification algorithm should compute some finite representation of the infinitely many first order unifiers, using again primal terms for schematization.

As an example, let $t_1 = \hat{g}(u;x)$ and $t_2 = \hat{h}(v)$ with the following rewrite rules

for $\hat{g}$ and $\hat{h}$:
$$\hat{g}(0; x) \to x, \quad \hat{g}(s(u); x) \to f(\hat{g}(u; x)),$$
$$\hat{h}(0) \to a, \quad \hat{h}(s(v)) \to f(\hat{h}(v)) \ .$$

t_1 schematizes the set $T_1 = \{x, f(x), f(f(x)), \ldots\}$, t_2 the set $T_2 = \{a, f(a), \ldots\}$. Every term in T_1 unifies with an infinite number of terms in T_2, with the unifiers looking like $\{x \leftarrow a\}, \{x \leftarrow f(a)\}$ etc. These unifiers can be schematized by the single meta-unifier $\{x \leftarrow \hat{h}(v'), v \leftarrow u+v'\}$. Note the difference between this unification problem and the unification of $t_1 = \hat{g}(u; x)$ with $t'_2 = \hat{h}(u)$: t'_2 schematizes the same set of terms as t_2, but since the two primal terms share the counter variable u, we may only unify terms obtained from substituting the same value for u. Hence all of the unifiers as well as the meta-unifier coincide with the substitution $\{x \leftarrow a\}$.

It turned out that primal grammars as defined in [12] are too powerful: their unification problem is undecidable. This is mainly due to their ability to schematize sets of terms with an infinite number of different variables by means of *marked variables*. However, recently it has been shown that the unification problem for *flat* primal grammars, i.e., for primal grammars without marked variables, is decidable and that the unifiers can always be described by a finite number of meta-unifiers [10, 13]. Fortunately flat primal grammars are sufficient for our purpose.

Definition 12. Let μ be a simply mono-cyclic substitution, u a counter variable, $\hat{x}$ an auxiliary function symbol for each $x \in \mathrm{dom}(\mu)$ and $\mathrm{vcod}(\mu) = \{x_1, \ldots, x_k\}$.[7] The primal substitution corresponding to μ is defined as

$$\hat{\mu}_u = \{x \leftarrow \hat{x}(u; x_1, \ldots, x_k) \mid x \in \mathrm{dom}(\mu)\} \ ,$$

where the semantics of each auxiliary symbol $\hat{x}$ is defined by the rewrite rules

$$\hat{x}(0; x_1, \ldots, x_n) \to x, \quad \hat{x}(s(u); x_1, \ldots, x_n) \to x\mu\hat{\mu}_u$$

Lemma 13. *Let t be an arbitrary first order term, τ an arbitrary first order substitution and μ a simply mono-cyclic substitution. Then the set $\{t\mu^n\tau \mid n \geq 0\}$ is schematized by the primal term $t\hat{\mu}_u\tau$, where u is a counter variable.*

8 The Unification Algorithm

Now we are in the position to outline the unification algorithm for mono-cyclic clauses. Suppose we want to unify F and G modulo $C = A\lambda_1 \to A\lambda_2$, where C is mono-cyclic. Transforming $\{\lambda_1^n \mid n \geq 1\}$ and $\{\lambda_2^n \mid n \geq 1\}$ simultaneously according to section 6, the set $\{C_n \mid n \geq 1\}$ of self-resolvents is equivalent to

$$\{C_1, \ldots, C_k\} \cup \bigcup_{i=0}^{l-1} \{A\rho_1\sigma_1^i\mu_1^n\tau_1 \to A\rho_2\sigma_2^i\mu_2^n\tau_2 \mid n \geq 0\} \ ,$$

---
[7] Note that $\mathrm{dom}(\mu) \subseteq \mathrm{vcod}(\mu)$ since all variables in the domain are cyclic.

where $\mu_1 = \sigma_1^l$ and $\mu_2 = \sigma_2^l$ are simply mono-cyclic. By Lemma 13 each of the sets $\{A\rho_1\sigma_1^i\mu_1^n\tau_1 \rightarrow A\rho_2\sigma_2^i\mu_2^n\tau_2 \mid n \geq 0\}$ can be represented by a single primal term. Hence, the infinite set of self-resolvents collapses to the finite set of primal clauses

$$\{C_1, \ldots, C_k\} \cup \{A\rho_1\sigma_1^i\hat{\mu}_{1u}\tau_1 \rightarrow A\rho_2\sigma_2^i\hat{\mu}_{2u}\tau_2 \mid 0 \leq i \leq l-1\} \ .$$

A complete and finite description $\{\theta_0, \theta_1, \ldots, \theta_k\} \cup \Theta_0 \cup \cdots \cup \Theta_{l-1}$ of the unifiers of F and G can now be obtained in the following way:

1. θ_0 is the most general unifier of F and G.
2. $\theta_1, \ldots, \theta_k$ are obtained by unifying $F \rightarrow G$ with renamed copies of $C_1, \ldots, C_k$.
3. For $0 \leq i \leq l-1$, Θ_i is the finite set of meta-unifiers obtained by unifying $F \rightarrow G$ with a renamed copy of $A\rho_1\sigma_1^i\hat{\mu}_{1u}\tau_1 \rightarrow A\rho_2\sigma_2^i\hat{\mu}_{2u}\tau_2$, using the unification algorithm for primal grammars.

Of course some of the θ_i may not exist. If two primal terms are non-unifiable, the corresponding set Θ of meta-unifiers is empty.

Example 9. Let $F = P(a, a, a)$, $G = P(x_1, x_2, a)$ and $C = P(x, y, z) \rightarrow P(f(y), g(x, z), a)$. The binary clause C is mono-cyclic, we have $A = P(x, y, z)$, $\lambda_1 = \{\}$ and $\lambda_2 = \{x \leftarrow f(y), y \leftarrow g(x, z), z \leftarrow a\}$. The set $\{\lambda_2^n \mid n \geq 1\}$ transforms to $\{\rho\mu^n\tau \mid n \geq 0\} \cup \{\rho\sigma\mu^n\tau \mid n \geq 0\}$, where $\rho = \{z \leftarrow a\}$, $\sigma = \{x \leftarrow f(y), y \leftarrow g(x, a)\}$, $\mu = \sigma^2$ and $\tau = \{x \leftarrow f(y), y \leftarrow g(x, z)\}$. Hence the self-resolvents of C are represented by the two primal clauses

$$C_1 = P(x, y, z) \rightarrow P(\hat{x}(u; f(y), g(x, z)), \hat{y}(u; f(y), g(x, z)), a) \text{ and}$$
$$C_2 = P(x, y, z) \rightarrow P(f(\hat{y}(u; f(y), g(x, z))), g(\hat{x}(u; f(y), g(x, z)), a), a)$$

and the rewrite rules

$$\hat{x}(0; x, y) \rightarrow x, \quad \hat{x}(s(u); x, y) \rightarrow f(g(\hat{x}(u; x, y), a)) \ ,$$
$$\hat{y}(0; x, y) \rightarrow y, \quad \hat{y}(s(u); x, y) \rightarrow g(f(\hat{y}(u; x, y)), a) \ .$$

Unifying $F \rightarrow G$ with C_1 and restricting the result to the variables x_1 and x_2 we obtain the single meta-unifier $\{x_1 \leftarrow \hat{x}(u; f(a), g(a, a)), x_2 \leftarrow \hat{y}(u; f(a), g(a, a))\}$, which schematizes the following infinite set of first order unifiers:

$$\{\{x_1 \leftarrow f(a), x_2 \leftarrow g(a, a)\}, \{x_1 \leftarrow f(g(f(a), a)), x_2 \leftarrow g(f(g(a, a)), a)\}, \ldots\} \ .$$

Similarly we obtain a second meta-unifier from $F \rightarrow G$ and C_2. Hence F and G are unifiable modulo C, the unifiers can be represented by two meta-substitutions.

9 Conclusion

Controlling the self-application of binary clauses is important for theorem proving as well as for logic programming. This paper is a step further towards controlling cycles having a more complex behavior than previously investigated ones. The most important aspect of the unification algorithm seems to be the use of schematizations; on the one hand they enable us to finitely represent the infinite sets of unifiers, on the other hand they have the property to be finitely unifiable themselves. In this way it becomes possible to detect infinite loops and speed up computations.

Example 1 (continued). The fact $Q(a,a)$ and the subgoal $Q(x,y)$ are unifiable modulo the binary clause $Q(x,y) \to Q(f(x),g(y))$. The single meta-unifier is given by $\{x \leftarrow \hat{x}(u), y \leftarrow \hat{y}(u))\}$, where $\hat{x}$ and $\hat{y}$ are defined by the rewrite rules

$$\hat{x}(0) \to a, \quad \hat{x}(s(u)) \to f(\hat{x}(u)),$$
$$\hat{y}(0) \to a, \quad \hat{y}(s(u)) \to g(\hat{y}(u)),$$

Hence the second subgoal becomes instantiated to $R(\hat{y}(u))$. In the first program this meta-literal is unified with the fact $\to R(g^{20}(a))$ yielding $\{u \leftarrow s^{20}\}$. Goal (G) contains only the variable x, hence the final answer is $\{x \leftarrow \hat{x}(s^{20})\} = \{x \leftarrow f^{20}(a)\}$.

In the second program the subgoal $R(\hat{y}(u))$ and the fact $\to R(b)$ are non-unifiable, the programs stops with failure.

Fact (4″) imposes no restrictions on the bindings for u, hence the original goal $Setof(x, P(x), y)) \to$ succeeds with the substitution $\{y \leftarrow \{\hat{x}(u)\}\}$.

Summarizing, the unification modulo a binary clause eliminates in this example both infinite loops. The gain in the first program depends on the implementation of the meta-unification algorithm; the overhead introduced by the use of more complicated data-structures might well slow down the computation in this case.

Future work

By extending the expressiveness of schematizations while preserving their unifiability we hope to close the gap between mono-cyclic clauses and variable-preserving clauses. For instance, the binary clause

$$P(x,y) \to P(f(x,y), g(x))$$

is variable-preserving, but not mono-cyclic; thus its self-resolvents cannot be schematized by one of the known formalisms.

To really evaluate the usefulness of unification modulo a binary clause good implementations of the meta-unification algorithms are needed. So far these are still in an experimental state and not yet incorporated into a theorem prover.

Acknowledgements

I would like to thank the referees for their valuable comments and suggestions. To one of the referees the first two sentences of the conclusion will sound familiar; thank you for donating these words.

References

1. F. Baader and J. H. Siekmann. Unification theory. In D. M. Gabbay, C. J. Hogger, and J. A. Robinson, editors, *Handbook of Logic in Artificial Intelligence and Logic Programming*. Oxford University Press, Oxford, UK, 1993. To appear.
2. W. Bibel, S. Hölldobler, and J. Würtz. Cycle unification. In D. Kapur, editor, *11th International Conference on Automated Deduction*, LNAI 607, pages 94–108, Saratoga Springs, New York, USA, June 15–18, 1992. Springer-Verlag.
3. H. Chen and J. Hsiang. Logic programming with recurrence domains. In J. Leach Albert, B. Monien, and M. Rodríguez, editors, *Automata, Languages and Programming (ICALP'91)*, LNCS 510, pages 20–34. Springer-Verlag, 1991.
4. H. Chen, J. Hsiang, and H.-C. Kong. On finite representations of infinite sequences of terms. In S. Kaplan and M. Okada, editors, *Conditional and Typed Rewriting Systems, 2nd International Workshop*, LNCS 516, pages 100–114, Montreal, Canada, June 11–14, 1990. Springer-Verlag.
5. H. Comon. On unification of terms with integer exponents. Technical Report 770, LRI, Orsay, France, 1992. To appear in: Mathematical System Theory.
6. D. de Schreye, M. Bruynooghe, and K. Verschaetse. On the existence of nonterminating queries for a restricted class of prolog-clauses. *Artificial Intelligence*, 41:237–248, 1989.
7. P. Devienne. Weighted graphs: a tool for studying the halting problem and time complexity in term rewriting systems and logic programming. *TCS*, 75:157–215, 1990.
8. P. Devienne, P. Lebègue, and J.-C. Routier. The emptiness problem of one binary recursive Horn clause is undecidable. In *International Logic Programming Symposium '93*, Vancouver, Oct. 1993. MIT Press.
9. P. Devienne, P. Lebègue, J.-C. Routier, and J. Würtz. One binary binary Horn clause is enough. In *STACS '94*, LNCS, Caen, Mar. 1994. Springer-Verlag.
10. R. Galbavý and M. Hermann. Unification of infinite sets of terms schematized by primal grammars. Technical Report 92-R-220, CRIN, Nancy, France, 1992.
11. P. Hanschke and J. Würtz. Satisfiability of the smallest binary program. *IPL*, 45(5):237–241, Apr. 1993.
12. M. Hermann. On the relation between primitive recursion, schematization, and divergence. In H. Kirchner and G. Levi, editors, *Proceedings 3rd Conference on Algebraic and Logic Programming*, LNCS 632, pages 115–127, Volterra (Italy), 1992. Springer-Verlag.
13. M. Hermann. Divergence des systèmes de réécriture et schématisation des ensembles infinis de termes. Habilitation, Université de Nancy I and CRIN-CNRS Inria Lorraine, Nancy (France), Mar. 1994.
14. J. Marcinkowski and L. Pacholski. Undecidability of the horn-clause implication problem. In *33rd Annual IEEE Symposium on Foundations of Computer Science*, pages 354–362, Los Alamitos, 1992.

15. G. Salzer. Deductive generalization and meta-reasoning, or how to formalize Genesis. In *Proc. 7. Österreichische Artificial Intelligence Tagung*, Informatik-Berichte 287, pages 103–115. Springer-Verlag, 1991.
16. G. Salzer. *Unification of Meta-Terms*. PhD thesis, Technische Universität Wien, 1991.
17. G. Salzer. Solvable classes of cycle unification problems. In J. Dassow, editor, *International Meeting of Young Computer Scientists*, Topics in Computer Science, Smolenice, Slovakia, 1992. Gordon&Breach. To appear in 1994[8].
18. G. Salzer. The unification of infinite sets of terms and its applications. In A. Voronkov, editor, *Logic Programming and Automated Reasoning (LPAR'92)*, LNAI 624, pages 409–420, St. Petersburg, Russia, July 1992. Springer-Verlag.
19. M. Schmidt-Schauß. Implication of clauses is undecidable. *TCS*, 59:287–296, 1988.
20. M. E. Stickel. Automated deduction by theory resolution. *JAR*, 1:333–355, 1985.
21. J. D. Ullman and A. van Gelder. Efficient tests for top-down termination of logical rules. *JACM*, 35(2):345–373, 1988.
22. A. van Gelder. Efficient loop detection in prolog using the tortoise-and-hare technique. *J. Logic Programming*, 4:23–31, 1987.
23. J. Würtz. Unifying cycles. In *European Conference on Artificial Intelligence (ECAI'92)*, pages 60–64, Aug. 1992.

---

[8] This article is also available via anonymous ftp, host `logic.tuwien.ac.at`, file `pub/salzer/papers/imycs92.dvi.Z`

Conservative Query Normalization on Parallel Circumscription

Kouji Iwanuma

Department of Electrical Engineering and Computer Science
Yamanashi University
Takeda 4-3-11, Kofu, Yamanashi, 400, Japan.
E-mail: iwanuma@esi.yamanashi.ac.jp
Tel: +81 552 52 1111

Abstract. In this paper, we study a deductive computation for parallel circumscription based on *query normalization*. At first, we give two fundamental transformation rules *M-resolution* and *V-resolution*. M-resolution is an *equivalent* transformation rule for computing negative information upon circumscribed predicates occurring in queries. V-resolution is for computing variable predicates, and *nearly conserves* the satisfiability of queries. Next, we give *Conservative Query* (CQ) transformation rule by integrating M-resolution and V-resolution. CQ-transformation takes a general form of Negation as Failure rule in logic programming. It is applicable to parallel circumscription over an arbitrary first-order clausal theory. After we extend CQ-transformation by incorporating it with Robinson's resolution procedure, we discuss fundamental properties for high-speed execution based on *compilation* of CQ-transformation.

1 Introduction

Circumscription, proposed by McCarthy [12], is a very influential formalism for commonsense reasoning. Circumscription, denoted by $CIRC[T; \Gamma; \Delta]$, is formalized as a second-order sentence, and its deductive computation is very difficult. So far, in order to make it easy, several equivalent transformation rules of $CIRC[T; \Gamma; \Delta]$ itself into a first-order sentence [4, 8, 9, 10] or into a logic program [3, 13] have been studied. On the other hand, there exist a few researches [3, 13] on query transformation rules for $CIRC[T; \Gamma; \Delta]$ over a ground theory T. In this paper, we study a recursive query transformation based on *normalization*, for parallel circumscription $CIRC[T; \Gamma; \Delta]$ over an arbitrary first-order clausal theory T.

Etherington [1] showed $CIRC[T; \Gamma; \Delta] \models A$ is equivalent to $T \models A$, if the query A has neither negative occurrences of any *circumscribed* predicates $p \in \Gamma$ nor occurrences of any *variable* predicates $q \in \Delta$ (we say, such a desirable formula A is *normal*). This suggests a possibility that the second-order computation for circumscription can be reduced into the first-order one for $T \models A$. Of course, a query A is not in the normal form generally. However, several non-normal queries may be transformed into a normal form. The problem is how to mechanically normalize such a query.

In this paper, at first, we give two fundamental transformation rules *M-resolution* and *V-resolution*. The former is an equivalent transformation rule for eliminating negative occurrences of circumscribed predicate $p \in \Gamma$ in a query. The latter is for removing occurrences of variable predicates $q \in \Delta$, and *nearly* conserves the satisfiability of queries. V-resolution is motivated by the excellent algorithm SCAN proposed by Gabbay and Ohlbach [4], and takes a more appropriate form for computing parallel circumscription. Next, we give *Conservative Query* (CQ) transformation consisting of M-resolution and V-resolution. CQ-transformation takes a general form of Negation as Failure rule, which has been well studied in logic programming [11], and is applicable to parallel circumscription over an arbitrary first-order clausal theory. However CQ-transformation is possible only for queries initially involving no variable predicates, so we moreover extend CQ-transformation by incorporating it with Robinson's resolution procedure. The extended one, called CIRC-resolution, can be considered as a generalization of SLDNF-resolution. Finally, we investigate fundamental possibilities of high-speed execution based on *compilation* of CQ-transformation, and especially study a *directional theory* proposed by Hou et al. [6].

This paper is organized as follows: Section 2 is for preliminaries. In Section 3, we give M-resolution and V-resolution. Section 4 is for CQ-transformation and CIRC-resolution. Section 5 is a discussion on a restricted CQ-transformation toward compilation. Section 6 is a brief comparison with related literature.

2 Preliminaries

In this paper, we consider a *second-order language* $\mathcal{L}$ *with the first-order equality*, which is defined just as a first-order language [9]. We suppose $\mathcal{L}$ does not contain any function variables. For simplicity, we consider only 1-*ary* function or predicate constants, except for the equality $=$, and similarly for predicate variables (but we will use 2-ary predicate symbols for examples). We define *terms, atoms, literals, formulas, clauses* and *sentences*, as usual. We sometimes identify a set of formulas with the conjunction of its all elements, and similarly, a set of literals with a clause. $\exists T$ and $\forall T$ indicate the existential closure and the universal closure of a formula T, respectively. Let $\alpha \leq \beta$ denote the sentence $\forall x[\alpha(x) \supset \beta(x)]$ for predicate constants (or variables) α and β.

Definition 1. Let $\Gamma = \{p_1, \ldots, p_m\}$ and $\Delta = \{q_1, \ldots, q_n\}$ be disjoint sets of predicate constants and $T[p_1, \ldots, p_m, q_1, \ldots, q_n]$ be a first-order sentence. Then, the *parallel circumscription of* Γ *over* T *w.r.t. variable* Δ, denoted by $CIRC[T; \Gamma; \Delta]$ (or, by $CIRC[T; p_1, \ldots, p_m; q_1, \ldots, q_n]$), is the second-order sentence

$$T \wedge \forall P_1 \ldots P_m Q_1 \ldots Q_n \left[\left(T[P_1, \ldots, P_m, Q_1, \ldots, Q_n] \wedge \bigwedge_{i=1}^{m}(P_i \leq p_i) \right) \supset \bigwedge_{i=1}^{m}(p_i \leq P_i) \right],$$

where $P_1, \ldots, P_m, Q_1, \ldots, Q_n$ are appropriate predicate variables. $p \in \Gamma$ and $q \in \Delta$ are called *a circumscribed predicate* and *a variable predicate*, respectively. If $\Delta = \emptyset$, $CIRC[T; \Gamma; \Delta]$ is called *predicate circumscription*.

Throughout this paper, we assume that T of $CIRC[T; \Gamma; \Delta]$ to be a set of clauses not involving the equality, and similarly for queries A.

A *structure* M for $\mathcal{L}$ consists of a non-empty set $|M|$, called the *domain of individuals*, functions from $|M|$ to $|M|$ representing function constants, and subsets of $|M|$ representing predicate constants. Let $|\mathcal{K}|_M$ denote the extension of a (function or predicate) constant $\mathcal{K}$ in M. The equality is interpreted as the identity relation on $|M|$. That is, we consider only *normal* models.

An *assignment* σ *into* M is a function defined on the individual (or predicate) variables $\mathcal{V}$ of $\mathcal{L}$ such that $\sigma(\mathcal{V})$ is a member of $|M|$ (or respectively, a subset of $|M|$). An assignment is used to evaluate free variables occurring in an expression.

Let $|t|_M^\sigma$ denote the extension of a term t in a structure M w.r.t. an assignment σ. Let $M \models_\sigma A$ indicate a formula A is true on M w.r.t. σ. Similarly, $M \models A$ and $\models A$ express $M \models_\sigma A$ for every assignment σ, and $M \models A$ for every structure M, respectively. A structure M is called a *model* of A if $M \models A$. We write $T \models A$ if A is true in every model of a sentence T.

Definition 2. Let Γ and Δ be disjoint sets of predicate constants. For any two models M and N of T, we write $M \leq_{\Gamma;\Delta} N$ if

1. $|M| = |N|$.
2. $|\mathcal{K}|_M = |\mathcal{K}|_N$ for every (function or predicate) constant symbol $\mathcal{K}$ not in $\Gamma \cup \Delta$.
3. $|p|_M \subset |p|_N$ for every p in Γ.

The relation $\leq_{\Gamma;\Delta}$ is a pre-order. A model M of T is $\langle \Gamma, \Delta \rangle$-*minimal* if there is no model N such that $N \leq_{\Gamma;\Delta} M$ and not $M \leq_{\Gamma;\Delta} N$. The models of circumscription can be characterized as follows:

Proposition 3 (Lifschitz [9]). *A structure M is a model of $CIRC[T; \Gamma; \Delta]$ iff M is a $\langle \Gamma, \Delta \rangle$-minimal model of T.*

As is well known, *Unique Name Assumption* (UNA) plays an important role in commonsense reasoning. For example, if not assume UNA as a premise, $CIRC[block(a); block]$ by itself can not entail even $\neg block(b)$. In this paper, we adopt the well-known Clark's equality theory (see Lloyd [11]) as a formalization of UNA. His theory makes the equality to denote the unifiability between terms, and characterizes the phenomenon "failure of unification", which is quite important to terminate a recursive query computation. From now on, we shall always assume UNA (i.e., Clark's equality theory) as a premise, together with circumscription.

Definition 4. Let Γ be a set of predicate constants. We write Γ^+ (or Γ^-) to indicate the set of all positive (or respectively, negative) literals of any $p \in \Gamma$.

$\Gamma^{\pm}$ denotes $\Gamma^{+} \cup \Gamma^{-}$. C/Γ^{+} is a set of literals $L \in \Gamma^{+}$ occurring in a clause C. If $\Gamma = \{p\}$, then we abbreviate C/Γ^{+} as C/p^{+}. C/Γ^{-} and C/p^{-} are defined in a similar way.

A formula A is said to be *normal* w.r.t. $CIRC[T; \Gamma; \Delta]$, if A has no literals of $\Gamma^{-} \cup \Delta^{\pm}$. The following is a variant of the result established by Etherington [1].

Lemma 5. *If A is normal w.r.t. $CIRC[T; \Gamma; \Delta]$, then*

$$UNA, CIRC[T; \Gamma; \Delta] \models A \quad \text{iff} \quad T \models A.$$

Therefore, if A is normal, then the second-order computation for whether $UNA, CIRC[T; \Gamma; \Delta] \models A$ can be reduced into the first-order one. This is a very important property, because a non-normal formula sometimes can be transformed into the normal form. The problem is how to mechanically normalize such a formula.

3 Basic Transformation Rules

3.1 M-resolution

In Iwanuma et al. [7], we investigate an equivalent query transformation, called here *M-resolution*, for predicate circumscription. It can eliminate negative occurrences of circumscribed predicates in a query. We reformulate and adjust it to parallel circumscription. A *most general unifier* (mgu) is defined as usual [11].

Definition 6. Let C be a clause, L be a negative literal $\neg p(t)$ in C,

1. Let D be a clause not involving common variables of C and Ms be a non-empty subset $\{M_1, \ldots, M_n\}$ of D/p^{+}. If there exists a mgu θ such that $L\theta = \neg M_1 \theta = \ldots = \neg M_n \theta$, then the clause $(C - \{L\})\theta \cup (D - Ms)\theta$ is called the *M-resolvent* of C against D upon L using Ms, and is denoted by $M\text{-}Res_{C,L}[D; Ms]$. [1]
2. Let T be a set of clauses. $M\text{-}Res_{C,L}(T)$ is the set of M-resolvents

$$\{ M\text{-}Res_{C,L}[D; Ms] \mid D \in T, Ms \subset D/p^{+} \text{ and } Ms \neq \emptyset\}$$

If L is a positive, we define $M\text{-}Res_{C,L}[D; Ms]$ and $M\text{-}Res_{C,L}(T)$ as the dual of ones for the above negative case, respectively.

Theorem 7. *Let A be a set of clauses, C be a clause in A, and L be a negative literal in C/Γ^{-}.*

$$UNA, CIRC[T; \Gamma; \Delta] \models A \equiv [(A - \{C\}) \cup M\text{-}Res_{C,L}(T)].$$

Proof. Theorem 4.5 in Iwanuma et al. [7] showed $UNA, CIRC[T; \Gamma; \emptyset] \models A \equiv [(A - \{C\}) \cup M\text{-}Res_{C,L}(T)]$. Since $\models CIRC[T; \Gamma; \Delta] \supset CIRC[T; \Gamma; \emptyset]$, this is obvious. □

---

[1] Iwanuma et at. [7] uses the alternative notations $Res_{C,L,D}(Ms)$ and $Res_{C,L}(T)$ instead of $M\text{-}Res_{C,L}[D; Ms]$ and $M\text{-}Res_{C,L}(T)$, respectively.

Theorem 7 states, if $M\text{-}Res_{C,L}(T) = \emptyset$, then A is equivalent to $A - \{C\}$. $M\text{-}Res_{C,L}(T) = \emptyset$ means that any computations of the positive literal $\neg L$ are finitely failed. On the other hand, the elimination of C from A means that the negative literal L can be identified with true. Thus the replacement C in A with $M\text{-}Res_{C,L}(T)$ can be regarded as a generalization of Negation as Failure rule in logic programming [11].

Remark. $M\text{-}Res_{C,L}(T)$ must be the set of all possible resolvents of C against any factors of each $D \in T$. Considering all factors is inevitable to guarantee the soundness of M-resolution. See the following example:

Example 1. Consider $CIRC[T; block, on]$, where T consists of

$$\forall xy(on(x,y) \supset block(x) \lor block(y)), \quad (D_1)$$
$$on(a,a),$$
$$\neg rectan(a).$$

Suppose a query A is $\forall z(block(z) \supset rectan(z))$. Clearly, $CIRC[T; block, on] \not\models A$. In this case, $M\text{-}Res_{C, \neg block(z)}(T)$ is the set consisting of the clauses

$$\forall z \, (on(z,z) \supset rectan(z)), \quad (D_2)$$
$$\forall yz \, (on(z,y) \supset rectan(z) \lor block(y)), \quad (D_3)$$
$$\forall xz \, (on(x,z) \supset block(x) \lor rectan(z)). \quad (D_4)$$

Hence, UNA, $CIRC[T; block, on] \models A \equiv D_2 \land D_3 \land D_4$. D_2 is a M-resolvent of A against a factor of D_1, and can not be omitted, because $CIRC[A; block, on] \not\models D_2$ and $CIRC[A; block, on] \models D_3 \land D_4$.

3.2 V-resolution

V-resolution is a rule for removing variable predicates occurring in a query for circumscription. At first, we show two examples to briefly explain intuition hidden behind V-resolution.

Example 2. Consider $CIRC[T; block; hard]$, where T is simply

$$\forall x(hard(x) \supset block(x)) \land \forall x(soft(x) \lor hard(x)).$$

Let a query A be $\neg block(a)$. A is equivalently transformed into $\neg hard(a)$. Now, the problem is to compute $CIRC[T; block; hard] \models \neg hard(a)$. Notice $hard$ is a variable predicate, and T indicates if $M \models soft(a)$ for a model M of T, then the extension $|hard|_M$ need not involve the object $|a|$. Hence, if $CIRC[T; block; hard] \models soft(a)$, then $CIRC[T; block; hard] \models \neg hard(a)$ must hold. These suggest the clause $soft(a)$ is appropriate as a V-resolvent of $\neg hard(a)$.

Example 3. Consider $CIRC[T; block; hard]$, where T is

$$\forall x(hard(x) \supset block(x)) \land \forall x(hard(a) \lor hard(b)).$$

Let a query A be $\forall x(block(x) \supset heavy(x))$. A is transformed into $A' = \forall x(\neg hard(x) \lor heavy(x))$ by M-resolution. Consider a sufficient condition for $M \models A'$ for a minimal model M. We have to consider three cases. The first is for the truth value of $hard(a)$. $M \models heavy(a)$ leads to $M \models \neg hard(a) \lor heavy(a)$, regardless of the value of $hard(a)$. Also, $M \models hard(b)$ allows $M \models \neg hard(a)$. That is, the formula $heavy(a) \lor hard(b)$ is a sufficient condition, and is appropriate as a V-resolvent The second case is on $hard(b)$. we obtain the formula $heavy(b) \lor hard(a)$ similarly. The last is for $M \models hard(a) \land hard(b)$, and $heavy(a) \lor heavy(b)$ is obtained in the same way (Notice, the last is really for the case of $M \models a = b$, so can be ignored, because UNA have been assumed as a premise).

V-resolution is a rule integrating them suggested in Example 2 and 3. A *variant* of an expression E is usual. Let $E_{[1]}, \ldots, E_{[n]}$ denote n variants of E.

Definition 8. Let C be a clause and L be a negative literal $\neg p(t)$ in C.

1. Let D be a clause not involving common variables of C and Ms be a nonempty subset $\{M_1, \ldots, M_n\}$ of D/p^+. Let $C_{[1]}, \ldots, C_{[n]}$ be n variants of C, $L_{[1]}, \ldots, L_{[n]}$ be n variants of L and each $L_{[i]}$ occurs in $C_{[i]}$, respectively. If there exists a mgu θ such that $\neg M_1 \theta = L_{[1]}\theta, \neg M_2 \theta = L_{[2]}\theta, \ldots$, and $\neg M_n \theta = L_{[n]}\theta$, then the clause

$$(C_{[1]} - \{L_{[1]}\})\theta \cup \ldots \cup (C_{[n]} - \{L_{[n]}\})\theta \cup (D - Ms)\theta,$$

is called the *V-resolvent* of C against D upon L using Ms, and is denoted by $V\text{-}Res_{C,L}[D; Ms]$.

2. Let T be a set of clauses. $V\text{-}Res_{C,L}(T)$ is the set of V-resolvents

$$\{ V\text{-}Res_{C,L}[D; Ms] \mid D \in T, Ms \subset D/p^+ \text{ and } Ms \neq \emptyset \}$$

If L is positive, we define $V\text{-}Res_{C,L}[D; Ms]$ and $V\text{-}Res_{C,L}(T)$ as the dual of ones for the above negative case, respectively.

See Section 4 for examples of V-resolution. V-resolution is motivated by an excellent algorithm SCAN, established by Gabbay and Ohlbach [4]. SCAN is for eliminating second-order quantifier $\exists P_i$ in a formula of the form $\exists P_1 \ldots P_n(T)$, where T is an arbitrary first-order formula. The resulting formula is equivalent to the original formula if SCAN *terminates*. They showed an example for an equivalent transformation of predicate circumscription *itself* into a first-order sentence by SCAN.

V-resolution is a modification of SCAN for deleting variable predicates occurring in a *query*, not in a parallel circumscription. V-resolution does not preserve the logical equivalence of queries, but, as will be shown in Theorem 11, *nearly conserves* satisfiability of queries upon parallel circumscription.

As compared with SCAN, V-resolution is much simpler. SCAN needs to generate all factors of each clause constituting a first-order theory, whereas V-resolution never produce such a factor. Also, SCAN does not assume UNA as its setting, and uses a rule corresponding to the below CV-resolution (see Definition 13). This means SCAN uses a more general framework than the one used in V-resolution, but on the other hand, means that SCAN can not use the phenomenon "failure of unification". This causes SCAN produces extremely more resolvents to an original first-order theory T than V-resolution produces for a query A upon T. Moreover, SCAN must often fail to terminate the transformation of $\exists P_1 \ldots P_n(T)$ even in the cases where V-resolution (more precisely, CQ-transformation shown later) successfully terminates for a query A upon T.

Definition 9. A literal L is *safe* in a set A of clauses if there exist no literal M such that M is unifiable with $\neg L$. [2]

Clearly, $CIRC[T; \Gamma; \Delta] \models C \supset V\text{-}Res_{C,L}(T)$. Theorem 11 states the converse implication *nearly* holds if a selected literal L is *safe* in a clause C.

Definition 10 (Przymusinski [14]). Let Δ be a set of predicate constants. Two models M and N of T are Δ-*similar* if they differ only on the interpretation of some $q \in \Delta$. If $\Delta = \{q\}$, we abbreviate the word "Δ-similar" as "q-similar".

Theorem 11. *Let A be a set of clauses, C be a clause in A, L be a literal of a predicate q in C, and T be an arbitrary set of clauses. If L is safe in A, and there is a model M of $UNA \wedge T$ such that*

$$M \models V\text{-}Res_{C,L}(T) \wedge (A - \{C\}),$$

then there is a model N of $UNA \wedge T$ such that N is q-similar to M and $N \models A$.

We need some preparations to prove Theorem 11.

Definition 12. Let C be a clause $\forall(\bigvee_{i=1}^{m} p(s_i) \vee \bigvee_{j=1}^{n} \neg p(t_j) \vee C')$, where C' involves no occurrences of p. The *constrained clause* (C-clause) of C w.r.t. p, denoted by C^p, is the clause

$$\forall \left[\left(\bigvee_{i=1}^{m} (p(x_i) \vee x_i \neq s_i) \right) \vee \left(\bigvee_{j=1}^{n} (\neg p(y_j) \vee y_j \neq t_j) \right) \vee C' \right],$$

where x_i and y_j are fresh variables. The above $p(x_i)$ and $\neg p(y_j)$ are called *constrained literals* (C-literals) in C^p.

Notice a clause C and its C-clause C^p are clearly logically equivalent.

Definition 13. Let C^p be the C-clause of a clause C w.r.t. a predicate p, and L^p be a negative C-literal $\neg p(x)$ in C^p.

---

[2] See *Remark* in Section 4 about the need of the safety condition.

1. Let D be a clause not involving common variables of C, and Ms be a nonempty subset $\{p(t_1), \ldots, p(t_n)\}$ of D/p^+. Let $C^p_{[1]}, \ldots, C^p_{[n]}$ be n variants of C^p, $\neg p(x_1), \ldots, \neg p(x_n)$ be n variants of L^p, where each $\neg p(x_i)$ occurs in $C^p_{[i]}$ respectively. Let θ be a substitution $\{x_1/t_1, \ldots, x_n/t_n\}$. The clause

$$(C^p_{[1]} - \{\neg p(x_1)\})\theta \cup \ldots \cup (C^p_{[n]} - \{\neg p(x_n)\})\theta \cup (D - Ms),$$

is called the *constrained V-resolvent* (CV-resolvent) of C^p against D upon L^p using Ms, and is denoted by $CV\text{-}Res_{C^p,L^p}[D;Ms]$.

2. Let T be a set of clauses. $CV\text{-}Res_{C^p,L^p}(T)$ is the set of CV-resolvents

$$\{\ CV\text{-}Res_{C^p,L^p}[D;Ms]\ \mid\ D \in T,\ Ms \subset D/p^+ \text{ and } Ms \neq \emptyset\}$$

If L^p is positive, we define $CV\text{-}Res_{C^p,L^p}[D;Ms]$ and $CV\text{-}Res_{C^p,L^p}(T)$ as the dual of ones for the negative case, respectively.

Note $CV\text{-}Res_{C^p,L^p}[D;Ms]$ is always defined, but $V\text{-}Res_{C,L}[D;Ms]$ is not so.

Lemma 14. *Let C be a clause, L be a literal of a predicate p occurring in C, and T be a set of clauses. Let C^p be the C-clause of C w.r.t. p, and L^p be the C-literal in C^p corresponding to L. We have*

$$UNA \models V\text{-}Res_{C,L}(T) \equiv CV\text{-}Res_{C^p,L^p}(T).$$

Proof. Since we assume here UNA as a premise, we can prove this lemma in a similar way as for Lemma 15.3 (b) in Lloyd [11]. □

Proof of Theorem 11. We only consider the case L is a negative literal $\neg q(s)$. The positive case can be proved similarly. We assume C is in the form of $\forall(\neg q(s) \lor E)$. At first, we shall construct a structure N according to M, as follows:

- $|N| = |M|$.
- $|\mathcal{K}|_N = |\mathcal{K}|_M$ for every (function or predicate) constant symbol $\mathcal{K}$ which is not q.
- $|q|_N$ is the set $(|q|_M - \mathcal{D})$, where $\mathcal{D}$ is $\{\ |x|^\sigma_M \in |M|\ \mid$ there is an assignment σ such that $M \models_\sigma q(x) \land x = s \land \neg E^q$, where x is a fresh variable$\}$.

Notice, the above $\mathcal{D}$ is identical with the set of all elements $|s|^\sigma_M$ such that $M \models_\sigma q(s) \land \neg E$ for an assignment σ. Obviously, N is q-similar to M, and satisfies UNA. Therefore, it is enough to prove (A) $N \models A$ and (B) $N \models T$.

Firstly, we shall prove (A) by independently showing two facts: (A-1) $N \models (A - \{C\})$ and (A-2) $N \models C$.

Case A-1: Since $L = \neg q(s)$ is safe in A, A has no positive literal unifiable with $\neg L(= q(s))$. Since M is a model of UNA, the well-known Clark's theorem (see Lemma 15.2 in [11]) guarantees that any positive literals $q(t)$ of q appearing in A satisfy $M \models \forall(s \neq t)$, i.e., $|t|^\sigma_M$ never belong to the above $\mathcal{D}$ for any σ. Consequently, from the assumption $M \models (A - \{C\})$, $N \models (A - \{C\})$ is obvious.

Case A-2: Given an assignment σ, if $M \models_\sigma \neg q(s) \vee E$, then $N \models_\sigma C$ is obvious, because E has no positive literals unifiable with $\neg L$ either. Otherwise, $|s|_M^\sigma$ must be a member of the above $\mathcal{D}$, so $|s|_M^\sigma \notin |q|_N$. Clearly, $N \models_\sigma \neg q(s) \vee E$.

Next let us prove (B). We have to show $N \models D$ for every $D \in T$. If $D \in T$ has no positive occurrences of q, then $N \models D$ is obvious. Otherwise, D must be of the form $\forall[q(t_1) \vee \ldots \vee q(t_k) \vee F]$, where F is a formula not containing positive occurrences of q. Therefore, it is enough to prove, for any assignment σ,

$$N \models_\sigma q(t_1) \vee \ldots \vee q(t_k) \vee F. \tag{5}$$

From the assumption, we have $M \models_\sigma q(t_1) \vee \ldots \vee q(t_k) \vee F$ for an arbitrary given assignment σ. We consider two cases:

Case B-1: All terms t_i ($i = 1, \ldots, k$) do not satisfy $M \models_\sigma q(t_i) \wedge t_i = s \wedge \neg E^q$. In this case, $M \models_\sigma q(t_i) \supset \neg(t_i = s \wedge \neg E^q)$ holds for each t_i. Therefore, if there exists a term t_i such that $M \models_\sigma q(t_i)$, then clearly $N \models_\sigma q(t_i)$ because of $|t_i|_M^\sigma \notin \mathcal{D}$. Otherwise, $M \models_\sigma F$ must be true. F has no positive occurrences of q, hence $N \models_\sigma F$ holds. (5) is obvious.

Case B-2: Otherwise. We assume for simplicity that all terms t_i satisfying $M \models_\sigma q(t_i) \wedge t_i = s \wedge \neg E^q$ are $t_1, \ldots, t_j$ ($1 \leq j \leq k$). If there is a term t_h ($j+1 \leq h \leq k$) such that $M \models_\sigma q(t_h)$, then (5) holds obviously. If not, we pay our attention on the assumption $M \models V\text{-}Res_{C,L}(T)$. Since M is a model of UNA, we have, according to Lemma 14,

$$M \models CV\text{-}Res_{C^q, L^q}(T),$$

where C^q is the C-clause $\forall(\neg q(x) \vee x \neq s \vee E^q)$, and L^q is the C-literal $\neg q(x)$ in C^q. Therefore, we clearly have $M \models CV\text{-}Res_{C^q, L^q}[D; Ms]$ for the set $Ms = \{t_1, \ldots, t_j\} \subset D/p^+$. That is,

$$M \models \forall \left[\bigvee_{i=1}^{j} (t_i \neq s_{[i]} \vee E^q_{[i]}) \vee \bigvee_{h=j+1}^{k} q(t_h) \vee F \right], \tag{6}$$

where $s_{[i]}$ and $E^q_{[i]}$ ($1 \leq i \leq j$) are variants of s and E^q, respectively. Notice again, all t_i ($1 \leq i \leq j$) satisfy $M \models_\sigma q(t_i) \wedge t_i = s \wedge \neg E^q$, and have no common variables in any $s_{[1]}, \ldots, s_{[j]}$ and $E^q_{[1]}, \ldots, E^q_{[j]}$. Therefore, we obviously have, for an appropriate assignment ψ such that $\psi(z) = \sigma(z)$ for all variables z occurring in D,

$$M \models_\psi \bigwedge_{i=1}^{j} (t_i = s_{[i]} \wedge \neg E^q_{[i]}).$$

From (6), we have for the above ψ,

$$M \models_\psi \left[\bigvee_{i=1}^{j} (t_i \neq s_{[i]} \vee E^q_{[i]}) \vee \bigvee_{h=j+1}^{k} q(t_h) \vee F \right].$$

Therefore, $M \models_\psi q(t_{j+1}) \vee \ldots \vee q(t_k) \vee F$ holds. From the definition of ψ, $M \models_\sigma q(t_{j+1}) \vee \ldots \vee q(t_k) \vee F$ also holds. Consequently, we have $M \models_\sigma F$, because we have assumed $M \models_\sigma \neg q(t_{j+1}) \wedge \ldots \wedge \neg q(t_k)$. Since F has no positive occurrences of q, $N \models_\sigma F$ is clear. (5) is obvious. □

4 CQ-transformation and CIRC-resolution

In this section, we give an integrated transformation rule, called *Conservative Query* (CQ) transformation, consisting of M-resolution and V-resolution. However, CQ-transformation applies only to a query initially not involving variable predicates. We shall moreover extend CQ-transformation by incorporating it with Robinson's resolution procedure. The extended one, called *CIRC-resolution*, is able to compute existential queries possibly involving variable predicates.

Definition 15. A *CQ-selection rule* R w.r.t. $CIRC[T; \Gamma; \Delta]$ is a function such that, given a set A of clauses, R returns a pair $\langle L, C \rangle$ where L is in $\Gamma^- \cup \Delta^\pm$ appearing in A, and C is a clause in A involving L. CQ-selection rule R is *safe w.r.t. $CIRC[T; \Gamma; \Delta]$* when, if a selected L is in $\Delta^\pm$, then L is always safe in A.

Definition 16. Let A be a set of clauses and R be a CQ-selection rule. Then a *CQ-transformation of A w.r.t. $CIRC[T; \Gamma; \Delta]$ via R* is a (finite or infinite) sequence $B_0, B_1, B_2, \ldots$ of sets of clauses, such that

1. $B_0 = A$.
2. Let $\langle L, C \rangle$ be a selected pair from B_i by R.
   (a) If L is in Γ^-, then B_{i+1} is $(B_i - \{C\}) \cup M\text{-}Res_{C,L}(T)$.
   (b) If L is in $\Delta^\pm$, then B_{i+1} is $(B_i - \{C\}) \cup V\text{-}Res_{C,L}(T)$.

A CQ-transformation G via R is *safe*, if R is safe. Also, G is *normalizing* if G is a finite sequence $B_0, B_1, \ldots, B_n$, and B_n is normal.

Theorem 17 (Conservativeness of CQ-transformation). *Let $B_0, B_1, B_2, \ldots$ is a safe CQ-transformation of a set A of clauses w.r.t. $CIRC[T; \Gamma; \Delta]$. If A involves no literals in $\Delta^\pm$, then we have, for each $i = 0, 1, 2, \ldots$,*

$$UNA, CIRC[T; \Gamma; \Delta] \models A \quad \textit{iff} \quad UNA, CIRC[T; \Gamma; \Delta] \models B_i.$$

Proof. Each B_{i+1} is a logical consequence of T and B_i. Therefore, "only if" is trivial. For "if" part, we need the following proposition:

Proposition 18 (Przymusinski [14]). *Let M and N be Δ-similar models. If a sentence A has no elements in $\Delta^\pm$, then $M \models A$ iff $N \models A$.*

Now, we begin a proof. Consider two cases: If the literal L selected from B_i is in Γ^-, then Theorem 7 guarantees that any $\langle \Gamma, \Delta \rangle$-minimal model M of $UNA \wedge T$ satisfying $B_{i+1} = (B_i - \{C\}) \cup M\text{-}Res_{C,L}(T)$ is also a model of B_i. If L is in $\Delta^\pm$, then Theorem 11 entails that, if a model M of $UNA \wedge T$ satisfies $B_{i+1} = (B_i - \{C\}) \cup V\text{-}Res_{C,L}(T)$, then there is a Δ-similar model N of $UNA \wedge T$

such that $N \models B_i$. As a whole, for each $i = 0, 1, 2, \ldots$, if a $\langle \Gamma, \Delta \rangle$-minimal model M of $UNA \wedge T$ satisfies B_i, then there is a model N such that N is Δ-similar with M, and $N \models A$. Remember A involves no elements of $\Delta^{\pm}$. Therefore, by Proposition 18, we also have $M \models A$. □

Example 4. Consider $CIRC[T; on; ball, soft]$, where T is

$$\forall xy(\neg on(x,y) \supset ball(x) \vee ball(y)),$$
$$\forall x\ (ball(x) \supset soft(x)),$$
$$\neg soft(c).$$

Let a query A be $\neg on(c, a)$. The following CQ-transformation from A is possible:

$$\{\underline{\neg on(c,a)}\} \stackrel{M}{\Longrightarrow} \{ball(c) \vee \underline{ball(a)}\} \stackrel{V}{\Longrightarrow} \{ball(c) \vee \underline{soft(a)}\} \stackrel{V}{\Longrightarrow} \emptyset$$

The underlined literals are selected at each stage. All of them are safe. Eventually, we obtain the empty set of clauses, i.e., a tautology. By Theorem 17, $UNA, CIRC[T; on; ball, soft] \models A$.

Example 5. Consider $CIRC[T; hard; block, on]$, where T is

$$\forall x\ (block(x) \supset hard(x)),$$
$$\forall xy(on(x,y) \supset block(x) \vee block(x)),$$
$$on(a,b),$$
$$heavy(a) \wedge heavy(b).$$

Let a query A be $\forall(hard(x) \supset heavy(x))$. A safe normalizing CQ-transformation can be constructed as follows:

$$\{\forall x(\underline{\neg hard(x)} \vee heavy(x))\} \stackrel{M}{\Longrightarrow} \{\forall x(\underline{\neg block(x)} \vee heavy(x))\} \stackrel{V}{\Longrightarrow}$$

$$\left\{ \begin{array}{l} \forall xy(\neg on(x,y) \vee heavy(x) \vee block(y)), \\ \forall xy(\underline{\neg on(x,y)} \vee block(x) \vee heavy(y)), \\ \forall xy(\underline{\neg on(x,y)} \vee heavy(x) \vee heavy(y)) \end{array} \right\} \stackrel{V}{\Rightarrow} \cdots \Rightarrow$$

$$\left\{ \begin{array}{l} heavy(a) \vee \underline{block(b)}, \\ \underline{block(a)} \vee \underline{heavy(b)}, \\ \underline{heavy(a)} \vee heavy(b) \end{array} \right\} \stackrel{V}{\Rightarrow} \cdots \Rightarrow \left\{ \begin{array}{l} heavy(a) \vee hard(b), \\ hard(a) \vee heavy(b), \\ heavy(a) \vee heavy(b) \end{array} \right\}$$

Consequently, we obtain a normal formula, which is a logical consequence of T. Theorem 17 entails $UNA, CIRC[T; hard; block, on] \models A$.

Remark on Safety Condition. The safety condition for CQ-transformation is inevitable as shown in Example 6 below. A check on the safety condition needs some computation. In Section 5, we will show, if T is *directional*, then the safety condition is automatically satisfied during CQ-transformation.

Example 6. Consider $CIRC[T; ab; fly]$, where T consists of

$$\forall x\,(bird(x) \land \neg ab(x) \supset fly(x)),$$
$$\forall x\,(ostrich(x) \land \neg ab(x) \supset \neg fly(x)),$$
$$bird(a),$$
$$ostrich(a).$$

Since $T \models ab(a)$, we have UNA, $CIRC[T; ab; fly] \models ab(a)$. However, the following inconsistent deduction for the query $\neg ab(a)$ is possible if we ignore the safety condition for CQ-transformation:

$$\{\underline{\neg ab(a)}\} \stackrel{M}{\Longrightarrow} \left\{ \begin{array}{c} bird(a) \supset \underline{fly(a)}, \\ ostrich(a) \supset \underline{\neg fly(a)} \end{array} \right\} \Rightarrow \stackrel{V}{\cdots} \Rightarrow \{bird(a) \land ostrich(a) \supset ab(a)\}.$$

The obtained normal formula is a logical consequence of T, so this deduction entails UNA, $CIRC[T; ab; fly] \models \neg ab(a)$, which is a contradiction. V-resolutions performed for $fly(a)$ and $\neg fly(a)$ are inconsistent of each other.

CQ-transformation is applicable only for a query initially involving no variable predicates. Since CQ-transformation is a natural extension of Negation as Failure rule, there is a simple and natural way to integrate resolution and CQ-transformation. Next, we give such an integrated rule, called CIRC-resolution, which can be applied to queries possibly involving variable predicates. For simplifying our discussion, we only consider existential queries. CIRC-resolution is considered as a natural extension of SLDNF-resolution [11].

Definition 19. A *CIRC-resolution* from a set A of clauses w.r.t. $CIRC[T; \Gamma; \Delta]$ consists of a (finite or infinite) sequence $C_0, C_1, \ldots$ of clauses, and a sequence $H_0, H_1, \ldots$ of sets of clauses such that each pair of C_i and H_i (for $i = 0, 1, \ldots$) satisfies one of the following conditions:

1. C_i is a clause belonging to $T \cup A$, and $H_i = \emptyset$.
2. C_i is a resolvent of clauses C_j, C_k $(0 \le j, k < i)$, and $H_i = \emptyset$.
3. Suppose C_{i-1} $(i \ge 1)$ is $L_1 \lor \ldots \lor L_j \lor \ldots \lor L_n$, and L_j is a positive ground literal in Γ^+. Suppose there is a safe normalizing CQ-transformation of $\neg L_j$, and S is a resulting normalized sentence of $\neg L_j$. Then C_i is the clause $L_1 \lor \ldots \lor L_{j-1} \lor L_{j+1} \lor \ldots \lor L_n$, and $H_i = S$.

A CIRC-resolution G *generates a hypothesis* H if G and H satisfy the followings:

1. G consists of finite sequences $C_0, \ldots, C_n$ and $H_0, \ldots, H_n$.
2. C_n is the empty clause NIL.
3. $H = \bigcup_{i=0}^{n} H_i$.

Theorem 20. *Let A be a disjunction of existential sentences $\exists (B_1 \land \ldots \land B_n)$. If there is a CIRC-resolution G from $\neg A$ w.r.t. $CIRC[T; \Gamma; \Delta]$, and G generates a hypothesis H, then UNA, $CIRC[T; \Gamma; \Delta] \models A$ iff $T \models H$.*

Proof. This is obvious. □

Example 7. Consider $CIRC[T; ab; bird, fly, peng]$, where T is

$$\forall x\,(bird(x) \land \neg ab(x) \supset fly(x)), \qquad (D_7)$$
$$\forall x\,(peng(x) \supset bird(x)), \qquad (D_8)$$
$$\forall x\,(peng(x) \supset \neg fly(x)), \qquad (D_9)$$
$$bird(a). \qquad (D_{10})$$

Suppose a query A is $\exists x. fly(x)$. We try to perform CIRC-resolution to $\neg A$. First, we perform resolution for A with D_7, and next with D_{10}. The resulting formula is the clause $ab(a)$. $ab(a)$ is a positive and ground literal of a circumscribed predicate. Thus, the following CQ-transformation is possible:

$$\{\neg ab(a)\} \stackrel{M}{\Longrightarrow} \{\neg bird(a) \lor fly(a)\} \stackrel{V}{\Longrightarrow} \{\neg peng(a) \lor fly(a)\} \stackrel{V}{\Longrightarrow} \emptyset$$

The generated hypothesis is the tautology. UNA, $CIRC[T; ab; bird, fly, peng] \models A$ clearly holds.

A modified answer extraction technique by Helft et al. [5] is also applicable to CIRC-resolution.

5 A Restricted CQ-transformation

The main aim of this section is to establish a fundamental principle (not detailed techniques) for a high-speed execution of CQ-transformation based on *compilation*. Compilation is one of well-known techniques for speeding up theorem-proving. Stickel [15] developed PTTP, which is a compiler to produce a Prolog code exactly simulating Loveland's ME-deduction over a given first-order theory. The produced Prolog code is furthermore translated into a Warren Abstract Machine's code, and is executed quite efficiently. His compilation technology is extremely excellent, and have been followed by several researchers.

The rules (1) and (2) for CIRC-resolution in Definition 19 are only for a simple first-order computation, and a more sophisticated rule, such as ME-deduction, is applicable instead of them. Therefore, if ME-deduction is adopted as a basic first-order rule invoked by CIRC-resolution, then we can use the technology established by Stickel. Remaining problems is how to generate a simulation code of the rule (3) in CIRC-resolution, especially, how to simulate CQ-transformation itself in a Prolog-based computation.

Main difficulty in simulating CQ-transformation by a logic program arises from V-resolution. A V-resolvent $V\text{-}Res_{C,L}[D; Ms]$ must be computed by using several variants of a query C, whereas M-resolution uses only C itself. As shown in Iwanuma et al. [7], the skeleton of M-resolution can essentially be simulated [3] by derivation of a finitely failed SLD-tree of a logic program. This suggests that a Prolog-based execution of CQ-transformation becomes possible if V-resolution can be ignored. From now on, we study some cases where V-resolution must be identical with M-resolution. Throughout these investigations, we shall establish

---

[3] This derives that a *fair* transformation strategy becomes a normalizing strategy for a restricted CQ-transformation consisting only of M-resolution. See [7].

a fundamental principle to compile circumscription. The detail of compile techniques is omitted here, and will be reported in a forthcoming paper. Simple, but useful cases are as follows:

Definition 21. Let L be a positive (or, negative) literal of a predicate p. The *complement of L is definitely defined in T* if each clause $C \in T$ has at most one negative (or respectively, positive) occurrences of p.

Proposition 22. *Let C be a clause, L be a literal and T be a set of clauses. If L is ground, or its complement is definitely defined in T, then $V\text{-}Res_{C,L}(T)$ is identical with $M\text{-}Res_{C,L}(T)$.*

Of course, a dynamic check of conditions like the above at run-time of CQ-transformation is possible. However it is obviously better to achieve a check statically at compile-time, not at running-time of CQ-transformation. Thus, we give a class of formulas, called *directional theories*, which is a slightly modified one of Hou et al. [6]. V-resolution over directional theories is forced to be identical with M-resolution, and the direction in an arbitrary formula can be verified at compile-time quite easily.

Let $C[p^+]$ (or $C[p^-]$) indicate that a clause C contains a positive (or respectively, negative) occurrence of p.

Definition 23. The *enlarging predicate set* $\Delta_E \subset \Delta$ and the *reducing predicate set* $\Delta_R \subset \Delta$ w.r.t. $CIRC[T; \Gamma; \Delta]$ are respectively defined as follows:

1. If $q \in \Delta$ occurs positively (or negatively) in a clause $C[p^+]$ of T for some $p \in \Gamma$, then q is in Δ_E (or respectively, in Δ_R).
2. If $q \in \Delta$ occurs positively in $C[r_1^+]$ of T for some $r_1 (\neq q)$ in Δ_R or in $C[r_2^-]$ of T for some r_2 in Δ_E, then q is in Δ_E.
3. If $q \in \Delta$ occurs negatively in $C[r_1^+]$ of T for some r_1 in Δ_R or in $C[r_2^-]$ of T for some $r_2 (\neq q)$ in Δ_E, then q is in Δ_R.
4. If $q \in \Delta_E$ (or $\in \Delta_R$) has more than one negative (or respectively, positive) occurrences in a clause of T, then q is also in Δ_R (or respectively, in Δ_E).

A sentence T is *directional* w.r.t. $CIRC[T; \Gamma; \Delta]$ if $\Delta_E \cap \Delta_R = \emptyset$.

For example, the theories T in Example 1, 4 and 7 are directional, but the ones in Example 5 and 6 are not. Intuitively, the direction represents that it is possible to uniquely determine whether the extension of each variable predicate in Δ must be enlarged, or be reduced, for minimizing extensions of $p \in \Gamma$. These directional theories are quite natural, and very often appear in practical knowledge representation. The following is immediate from Proposition 22: Hence, if T is directional, every CQ-transformation over T can be simply constructed only with M-resolution. Moreover, it is unnecessary to check on the safety of selected literals.

Lemma 24. *Let $G = B_0, B_1, \ldots$ be a CQ-transformation w.r.t. $CIRC[T; \Gamma; \Delta]$. If T is directional, then the complement of every literal L_i selected at each B_i is definitely defined in T. Additionally, if the initial query B_0 contains no occurrences of $\Delta^\pm$, then every L_i is always safe in B_i.*

6 Comparison

Roughly speaking, transformation methods of circumscription $CIRC[T; \Gamma; \Delta]$ into a first-order sentence, as shown in [8, 9, 10], can not transform it if T is recursive w.r.t. some $p \in \Gamma \cup \Delta$. Although SCAN [4] may transform such a recursive theory, it needs a huge amount of computations, and often falls into loop (in more cases than CQ-transformation does). On the other hand, the methods into logic programs, as shown in [3, 13], can not transform it if T is non-definite w.r.t. some $p \in \Gamma \cup \Delta$. CQ-transformation is applicable to both of them uniformly. There are a few researches on query transformation for circumscription. Notable results are MILO-resolution by Przymusinski [14] and Ginsberg's prover [2]. Both of them are essentially for *ground* clausal theories. Extension of them for first-order theories is obscure, because it needs careful consideration on several problems being inherent to first-order computations, such as factoring operation and computation strategies, etc. CQ-transformation (and CIRC-resolution) is applicable to an arbitrary first-order clausal theory, and its normalizing strategy has been clarified for a restricted version mentioned above.

References

1. D. Etherington, *Reasoning with Incomplete Information*, (Pitman Pub., 1988).
2. M.L. Ginsberg, A circumscriptive theorem prover, *Artif. Intell.* **39** (1989) 209-230.
3. M. Gelfond and V. Lifschitz, Compiling Circumscriptive Theories into Logic Programs, *Lect. Notes in Artif. Intell.* **346** (1988) 74-99.
4. D. Gabbay and H.J. Ohlbach, Quantifier elimination in second-order predicate logic, *KR '92*, (1992) 425-435
5. N. Helft, K. Inoue and D. Poole, Query Answering in Circumscription, *IJCAI '91* (1991) 426-431
6. B. H. Hou, A. Togashi and S. Noguchi, A Partial Translation of Default Logic to Circumscription, *J. Infor. Proc.* **12** (4) (1990) 343-353.
7. K. Iwanuma, M. Harao and S. Noguchi, A query transformation for computing predicate circumscription, *Pacific Rim Inter. Conf. on Artificial Intelligence '90*, Nagoya, Japan (1990) 351-356.
8. P.G. Kolaitis and C.H. Papadimitriou, Some computational aspects of circumscription, *J. ACM* **37** (1) (1990) 1-14.
9. V. Lifschitz, Computing Circumscription, *IJCAI '85* (1985) 121-127.
10. V. Lifschitz, Pointwise circumscription, in: M.L. Ginsberg ed., *Readings in Nonmonotonic Reasoning* (Morgan Kaufmann Pub., 1988), 179-193.
11. J.W. Lloyd, *Foundations of Logic Programming* (Springer-Verlag, 2nd Ed., 1987).
12. J. McCarthy, Circumscription: A Form of Nonmonotonic Reasoning, *Artif. Intell.* **13** (1980) 27-39.
13. T. C. Przymusinski, On the Relationship Between Logic Programming and Nonmonotonic Reasoning, *AAAI-88* (1988) 444-448.
14. T. C. Przymusinski, An Algorithm to Compute Circumscription, *Artif. Intell.*, **38** (1989) 49-73.
15. M.E. Stickel, A prolog technology theorem prover: a new exposition and implementation in prolog, *Theoret. Comput. Sci.* **104** (1992) 109-128.

Bottom-up Evaluation of Datalog Programs with Arithmetic Constraints

Laurent Fribourg   Marcos Veloso Peixoto

L.I.E.N.S (URA 1327 CNRS)
45, rue d'Ulm, 75005 Paris - France
email: fribourg@dmi.ens.fr, veloso@dmi.ens.fr

Abstract. We consider a class of recursive logic programs over integers that are often met in Temporal Deductive Databases, or when proving termination of Prolog programs. We show that the least fixpoint associated with a program of this class is a linear arithmetic formula that can be generated by a refined process of bottom-up evaluation.

1 Motivations

There are two basic approaches for reasoning with logic programs. The first approach is "top-down": it tries to solve a given query by backward chaining reasoning. The second approach is "bottom-up": it infers atomic consequences of the program by forward chaining reasoning. The first approach is rather used in Logic Programming (SLD-resolution) [22], while the second one is rather used in Deductive Databases for querying Datalog programs (ie., logic programs with constants and variables, but no function symbol) [5].

We describe in this paper a bottom-up evaluation process for Datalog programs (or logic programs) whose variables are interpreted as integers. Our initial motivation was to use such a process in order to prove formulas combining lists and linear arithmetic [11]. Let us illustrate this point with an example taken from [3], p. 101.

Let $delta(L, x, y, z)$ be an atom where L is a list of (numeric) characters, x is the length of L, y is a character and z is the distance between the last occurrence of y in L and the end of L. The predicate $delta$ can be recursively defined by the logic program:

$delta([\,], 0, y, 0)$.
$delta([u|L], x+1, y, z)$ $:- z < x, delta(L, x, y, z)$.
$delta([u|L], x+1, y, z)$ $:- z = x, u = y, delta(L, x, y, z)$.
$delta([u|L], x+1, y, z+1) :- z = x, u \neq y, delta(L, x, y, z)$.

Suppose now that we want to prove the formula:

$\forall L, t, w, x, y, z \ (\ delta(L, x, y, z) \Longrightarrow (x + v \leq w \wedge t \leq v \Longrightarrow t + z \leq w)\ )$

It is difficult to prove this formula directly (e.g., by induction), but it becomes easy if we are provided with a lemma saying that $delta(L, x, y, z)$ implies $z \leq x$. A priori such a lemma is not known, but a bottom-up evaluation process can automatically generate such a lemma. The bottom-up evaluation process is not performed directly on the *delta* program, but instead over a simplified form (that is merely obtained by deletion of the list argument L and elimination of the constraints $u = y$ and $u \neq y$):

$delta'(0, y, 0)$.
$delta'(x + 1, y, z) \;\; :- z \leq x, delta'(x, y, z)$.
$delta'(x + 1, y, z + 1) :- z = x, delta'(x, y, z)$.

A refined form of bottom-up evaluation over *delta'* that we will explain in this paper produces the relation $0 \leq z \leq x$ as a generic output. The lemma $delta(L, x, y, z) \Longrightarrow z \leq x$ then follows directly from the fact that, by construction, $delta(L, x, y, z)$ implies $delta'(x, y, z)$.

Apart from the automatic generation of lemmas, an interesting application of bottom-up evaluation of Datalog programs over integers is the proof of termination of logic programs [6, 14]. From a more general point of view, bottom-up evaluation procedures for Datalog programs over integers constitute an interesting subject of research *per se* and have been extensively studied in recent years. One reason for this interest is the emergence of Constraint Logic Programming and Constraint Query Languages (e.g. [16, 17, 19, 20, 23, 24, 28, 30]). Another reason lies in the use of integers for representing temporal data and the use of bottom-up evaluation for computing temporal queries [2, 4, 9].

2 Comparison with related work

Bottom-up evaluation procedures for Datalog programs with integers have been developed by researchers of the Deductive Database community [4, 9, 28] on one hand, and by researchers interested by the proof of termination of Prolog programs [6, 14] on the other hand.

Revesz has elaborated a procedure that always terminates for a class of Datalog program with integers, but this class does not allow for the incrementation of the recursion arguments [28]. Chomicki and Imielinski have considered a class of programs that is useful for modeling time in Temporal Databases: time is represented by a numeric argument ("temporal" argument) which is decremented at each recursive call. The class of Chomicki and Imielinski allows for only one temporal argument [9]. Baudinet, Niézette and Wolper consider a class of Datalog programs that allows for several temporal arguments, but their procedure does not terminate in general; besides, the tuples of the extensive database of their programs (i.e., the tuples satisfying the non recursive rules) correspond to linear arithmetic formulas of a restricted form ("linear repeating points") [4].

As in Baudinet, Niézette and Wolper [4], our programs allow for incrementation over several recursion arguments. In contrast to [4], our programs can take any linear arithmetic formula as for an extensive database. We have besides identified a subclass of programs for which our bottom-up evaluation procedure is guaranteed to terminate (class of programs with two recursive rules and one inequality constraint per rule).

The idea of automatically generating linear inequalities among variables of a program was exploited in an imperative framework by [8], using linear programming techniques. Researchers interested by the mechanization of termination proofs have designed bottom-up evaluation procedures for Datalog programs with integers: given a Prolog procedure, the aim is to infer inequalities among the sizes of the arguments of the auxiliary predicates called by this procedure, and to use these inequalities for proving the procedure termination [6, 14]. The method of [6] is able to infer disjunctions of sets of inequalities, but each inequality is necessarily of the form $arg(i) + c > arg(j)$, thus involving at most two arguments. The method of [14] can infer sets of more general inequalities (e.g., $arg(i) + arg(j) > arg(k) + c$), but cannot infer disjunctions of such formulas. This is also the case in the work of [8] and in methods that use top-down strategies instead of bottom-up [27, 31]. In contrast, our method has no such limitation, and is able to infer (when it terminates) the arithmetic formula that is characteristic of the program.

The main limitation of our method comes from the syntactical restrictions attached to the class of programs that we consider: the recursion scheme is necessarily direct and linear (no mutual recursion, no multiple recursive calls), and within each recursive rule, the recursion arguments are necessarily increased by a constant. We believe that such a class of programs is still important (as already stressed out by the work on temporal database [2, 4, 9]), and claim that our method behaves better for this class than the other methods.

Another difficulty with our approach is that the linear arithmetic formulas generated by our method are sometimes quite intricated. These formulas can however be simplified using techniques of variable elimination (see, e.g., [21, 29]).

3 Preliminaries

Throughout the paper, we will adopt the following conventions: the letters u, v, w, x, y, z denote variables of type integer; the letters a, b, c, d, e, f, g (with possible subscripts) denote constants of type integer and the letter p is reserved to denote a symbol of predicate. The letters i, j, k, m, n, q, r, s will be used to denote variables of type natural number in the metalanguage.

Let π be a program defined by a non-recursive base rule R_0 and q linear recursive rules R_k ($1 \leq k \leq q$) of the following form:

$p(x, y, z)$ $:- \xi(x, y, z).$ rule R_0
$p(x + a_k, y + b_k, z + c_k) :- \phi_k(x, y, z), p(x, y, z)$ rule R_k

where ξ denotes a linear arithmetic formula, $\phi_k(x, y, z)$ denotes an inequality of the form $d_k x + e_k y + f_k z + g_k \odot_k 0$, with $\odot_k \in \{<, >, \leq, \geq\}$.

For the sake of notation simplicity, we have assumed that the predicate p has arity 3. (Our results remain valid for any other arity.) We use arithmetic formulas to extend the notion of Herbrand atom. We define a *generalized Herbrand atom* as a pair composed of an atom and an arithmetic formula. Note that an Herbrand atom of the form $p(6, 5, 7)$ can be considered as a generalized Herbrand atom composed of $p(x, y, z)$ together with $x = 6 \wedge y = 5 \wedge z = 7$. The immediate consequence operator for the program π (see [1, 10]) is adapted in order to be applied to generalized Herbrand atoms (*cf* [15]). Given a set I of generalized Herbrand atoms, let, for $1 \leq k \leq q$:

$$T_k(I) = \{ p(x+a_k, y+b_k, z+c_k) / \phi_k(x,y,z) \text{ holds and } p(x,y,z) \in I \}$$

The operator T_k can be seen as the immediate consequence operator associated with the rule R_k ($1 \leq k \leq q$). The immediate consequence operator T associated with π can be defined by:

$$T(I) = T_1(I) \bigcup \ldots \bigcup T_q(I)$$

Let now T^s be defined inductively as:

$$T^1(I) = T(I)$$
$$T^{s+1}(I) = T(T^s(I)), \text{ for } s \geq 1$$

The output S of the bottom-up evaluation of π is $\bigcup_{s>0} T^s(I_\xi)$, where $I_\xi = \{ p(x,y,z) / \xi(x,y,z) \text{ holds } \}$. This output is the least fixpoint of the operator T; it is also the least generalized Herbrand model of π [15, 10]. Our goal is to show that the infinite union S can be expressed as a finite arithmetic formula. More precisely, we want to show that an atom $p(u, v, w)$ belongs to S iff u, v, w satisfy a finite *linear* arithmetic formula (i.e., an arithmetic formula with $<, =, +$ but no $\times$).

4 General Presentation of the Method

In this section we present the basic notions on which our method for constructing a finite representation of the output of the bottom-up evaluation process is based. We consider first programs π containing two recursive rules[1] R_1 and R_2. A program π is of the form:

$p(x,y,z)$	$:- \xi(x,y,z).$	rule R_0
$p(x+a_1, y+b_1, z+c_1)$	$:- \phi_1(x,y,z), p(x,y,z).$	rule R_1
$p(x+a_2, y+b_2, z+c_2)$	$:- \phi_2(x,y,z), p(x,y,z).$	rule R_2

We will use a geometrical representation for computing the output of the bottom-up evaluation of π. This output can be expressed as:

$$\bigcup_{r_1, r_2, \ldots, r_s \in \{1,2\}} T_{r_1} T_{r_2} \ldots T_{r_s}(I_\xi)$$

We will associate a geometrical path to each operator composition $T_{r_1} T_{r_2} \ldots T_{r_s}$: each number 1 in the sequence $r_1 r_2 \ldots r_s$ will be represented by a horizontal movement to the right, and each number 2 a vertical downward movement. This is schematized in figure 1.

More formally, a path P, denoted $\ll (0,0), \ldots, (m,n) \gg$, is a finite sequence of points of $\mathbb{N} \times \mathbb{N}$, starting at point $(0,0)$ and ending at point (m,n), such that if a point (i,j) distinct from (m,n) is in P, then either $(i+1, j)$ or $(i, j+1)$ is in P. A horizontal path H (resp. vertical path V) is a path where all points are of form $(i, 0)$ (resp. $(0, j)$).

---
[1] Programs containing a single recursive rule are degenerated cases of such programs.

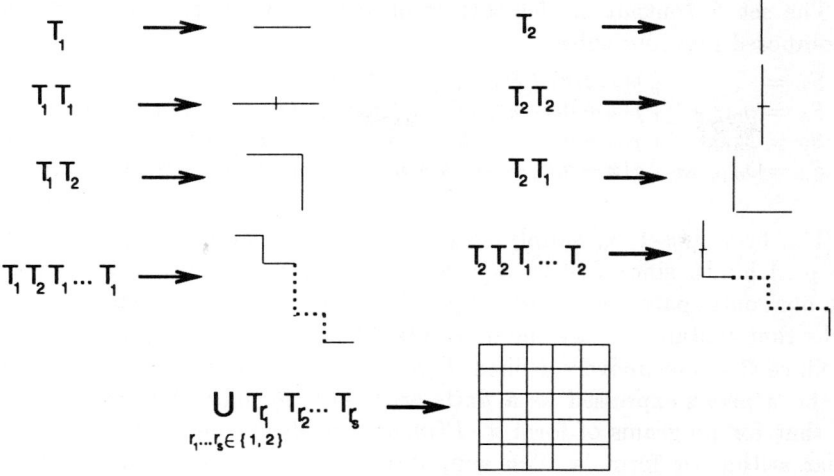

Figure 1

Each point (i,j) represents a tuple $(x+ia_1+ja_2, y+ib_1+jb_2, z+ic_1+jc_2)$. There is a horizontal constraint $\Phi_1(i,j)$ associated with each segment (i,j)-$(i+1,j)$. This constraint is an abbreviation for $\phi_1(x+ia_1+ja_2, y+ib_1+jb_2, z+ic_1+jc_2)$ and represents the atomic formula that must be satisfied for applying rule R_1 at point (i,j). Likewise, there is a vertical constraint $\Phi_2(i,j)$ associated with each segment (i,j)-$(i,j+1)$, representing the atomic formula that must be satisfied for applying rule R_2. This expression is an abbreviation for $\phi_2(x+ia_1+ja_2, y+ib_1+jb_2, z+ic_1+jc_2)$.

The *global constraint* $\mathbf{C}_P$ associated with a path P from $(0,0)$ to (m,n) is the conjunction of the constraints of all the segments of P. It is defined as:

$$\mathbf{C}_P \stackrel{\text{def}}{=} C_P(m,n)$$

where:

$$C_P(0,0) \stackrel{\text{def}}{=} \text{true}.$$
$$C_P(i+1,j) \stackrel{\text{def}}{=} C_P(i,j) \wedge \Phi_1(i,j), \qquad \text{if } (i+1,j) \in P.$$
$$C_P(i,j+1) \stackrel{\text{def}}{=} C_P(i,j) \wedge \Phi_2(i,j), \qquad \text{if } (i,j+1) \in P.$$

We denote the set of all the paths P from $(0,0)$ to (m,n) by $\Delta(m,n)$, and the disjunction of the global constraints of all the paths from $(0,0)$ to (m,n) by $\Gamma(m,n)$, i.e.:

$$\Gamma(m,n) \stackrel{\text{def}}{=} \bigvee_{P \in \Delta(m,n)} \mathbf{C}_P$$

An atom $p(x+ma_1+na_2, y+mb_1+nb_2, z+mc_1+nc_2)$ is generated by bottom-up evaluation iff x, y, z satisfy ξ and there exits a path P from $(0,0)$ to (m,n) of global constraint $\mathbf{C}_P$. The output S of the bottom-up evaluation of a program π is thus given by:

$$S = \bigcup_{m,n \in \mathbb{N}} \{ p(x+ma_1+na_2, y+mb_1+nb_2, z+mc_1+nc_2) \ / \ \xi(x,y,z) \wedge \Gamma(m,n) \}$$

The set S (output of the bottom-up evaluation of a program π) can be decomposed into four subsets:

$S_0 = \quad \{ p(x,y,z) \ / \ \xi(x,y,z) \wedge \Gamma(0,0) \}$
$S_H = \cup_{m \in \mathbb{N}^*} \ \{ p(x + ma_1, y + mb_1, z + mc_1) \ / \ \xi(x,y,z) \wedge \Gamma(m,0) \}$
$S_V = \cup_{n \in \mathbb{N}^*} \ \{ p(x + na_2, y + nb_2, z + nc_2) \ / \ \xi(x,y,z) \wedge \Gamma(0,n) \}$
$S_Z = \cup_{m,n \in \mathbb{N}^*} \ \{ p(x + ma_1 + na_2, y + mb_1 + nb_2, z + mc_1 + nc_2) \ / \ \xi(x,y,z) \wedge \Gamma(m,n) \}$

The first subset S_0 simply corresponds to the atoms $p(x,y,z)$ for which $\xi(x,y,z)$ holds, since $\Gamma(0,0) = \text{true}$. The subset S_H (resp. S_V) corresponds to a horizontal path (resp. vertical path). The last set S_Z corresponds to all the paths that contain at least one horizontal and one vertical segments.

Since $\mathbf{C}_P$ is recursively defined, $\Gamma(m,n)$ is itself recursively defined and cannot be *a priori* expressed as a *finite* arithmetic formula. Nevertheless we will see that for programs of form π, $\Gamma(m,n)$ can always be simplified as a finite linear arithmetic formula. This simplification will be made possible due to the existence of certain ordering relations over the constraints Φ_1 and Φ_2. These ordering relations are formally defined below and schematized in figure 2.

Proposition 1. For any program π, either 1 or 2 are satisfied.

1. $\Phi_1(i,j)$ implies $\Phi_1(i+1,j)$,   for all i,j
2. $\Phi_1(i+1,j)$ implies $\Phi_1(i,j)$,   for all i,j

We say that π is of class $\rightarrow$ (resp. class $\leftarrow$) if it satisfies 1 (resp. 2).

Proposition 2. For any program π, either 1 or 2 are satisfied.

1. $\Phi_1(i,j)$ implies $\Phi_1(i,j+1)$,   for all i,j
2. $\Phi_1(i,j+1)$ implies $\Phi_1(i,j)$,   for all i,j

We say that π is of class $\Downarrow$ (resp. class $\Uparrow$) if it satisfies 1 (resp. 2).

Proposition 3. For any program π, either 1 or 2 are satisfied.

1. $\Phi_2(i,j)$ implies $\Phi_2(i+1,j)$,   for all i,j
2. $\Phi_2(i+1,j)$ implies $\Phi_2(i,j)$,   for all i,j

We say that π is of class $\Rightarrow$ (resp. class $\Leftarrow$) if it satisfies 1 (resp. 2).

Proposition 4. For any program π, either 1 or 2 are satisfied.

1. $\Phi_2(i,j)$ implies $\Phi_2(i,j+1)$,   for all i,j
2. $\Phi_2(i,j+1)$ implies $\Phi_2(i,j)$,   for all i,j

We say that π is of class $\downarrow$ (resp. class $\uparrow$) if it satisfies 1 (resp. 2).

Combining the 4 propositions above, we can partition the programs π into 16 disjoint (sub)classes of programs. For example, a program is in class $\rightarrow\downarrow\Rightarrow\Downarrow$ iff it belongs simultaneously to classes $\rightarrow$, $\downarrow$, $\Rightarrow$ and $\Downarrow$.

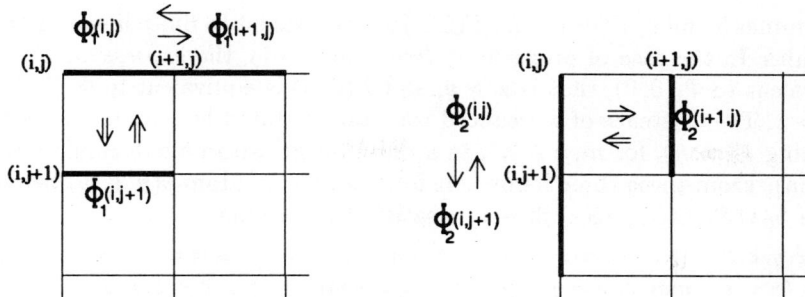

Figure 2: Constraint implications $\Rightarrow \Leftarrow \rightarrow \leftarrow \Uparrow \Downarrow \uparrow \downarrow$

Example 1

Consider the following program π_1:

$p(x, y, z) \quad :- \xi(x, y, z).$
$p(x + 2, y + 1, z) :- x > y, \ p(x, y, z).$
$p(x + 3, y + 2, z) :- y > z, \ p(x, y, z).$

where $\xi(x, y, z)$ is a linear arithmetic formula. The expression $\Phi_1(i, j)$ stands here for $x + 2i + 3j > y + i + 2j$ (i.e., $x + i + j > y$) and $\Phi_2(i, j)$ stands for $y + i + 2j > z$ (see figure 3). By comparing the constraints of the points (i, j) and $(i + 1, j)$ and those of (i, j) and $(i, j + 1)$, we determine that π_1 is in class $\rightarrow \downarrow \Rightarrow \Downarrow$. □

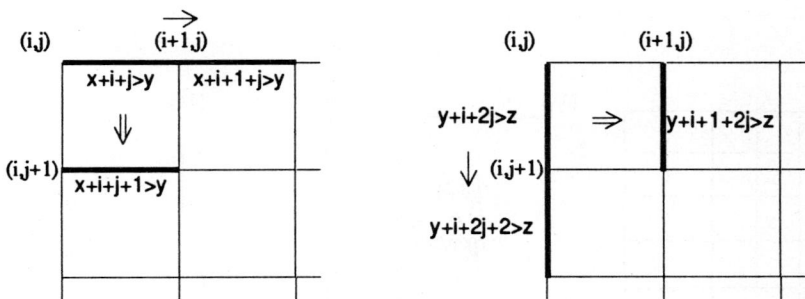

Figure 3: Constraint implications for π_1

We can easily prove that:

Lemma 5. *Given a program π, $\Gamma(m, 0)$ is either equivalent to $\Phi_1(0, 0)$ (if π is in class $\rightarrow$) or equivalent to $\Phi_1(m - 1, 0)$ (if π is in class $\leftarrow$).*

Lemma 6. *Given a program π, $\Gamma(0, n)$ is either equivalent to $\Phi_2(0, 0)$ (if π is in class $\downarrow$) or equivalent to $\Phi_2(0, n - 1)$ (if π is in class $\uparrow$).*

Lemma 7. *Given a program π and a path P from $(0, 0)$ to (m, n), for $n, m \in \mathbb{N}^*$, the global constraint C_P is equivalent to the expression $\Phi_1(i, j) \wedge \Phi_2(i', j')$, for some $(i, j), (i', j') \in P$. This expression is called the* reduced constraint *of P.*

By lemmas 5 and 6, $\Gamma(m,0)$ and $\Gamma(0,n)$ are equivalent to finite linear arithmetic formulas. In the case of program π_1 (see example 1), the expression $\Gamma(m,0)$ is equivalent to $\Phi_1(0,0)$, that is $x > y$, and $\Gamma(0,n)$ is equivalent to $\Phi_2(0,0)$, that is $y > z$. The existence of a "reduced constraint" stated by lemma 7 is useful for reducing $\Gamma(m,n)$, for $m,n \in \mathbb{N}^*$, to a finite linear arithmetic formula (see next section). From these three lemmas it follows that an atom $p(u,v,w)$ belongs to S (i.e $S_0 \cup S_H \cup S_V \cup S_Z$) iff u,v,w satisfy the formula:

$$\exists x,y,z \quad (\quad (u=x \quad \wedge \quad v=y \quad \wedge \quad w=z \quad \wedge \quad \xi(x,y,z))$$
$$\vee \exists m \ (u = x + ma_1 \wedge v = y + mb_1 \wedge w = z + mc_1 \wedge m > 0 \wedge \xi(x,y,z) \wedge \Gamma'(m,0))$$
$$\vee \exists n \ (u = x + na_2 \wedge v = y + nb_2 \wedge w = z + nc_2 \wedge n > 0 \wedge \xi(x,y,z) \wedge \Gamma'(0,n))$$
$$\vee \exists m,n \ (u = x + ma_1 + na_2 \wedge v = y + mb_1 + nb_2 \wedge w = z + mc_1 + nc_2 \wedge m > 0 \wedge n > 0 \wedge$$
$$\xi(x,y,z) \wedge \Gamma'(m,n) \ ) \qquad)$$

where Γ' denotes the finite arithmetic formula equivalent to Γ.

5 Simplification of $\Gamma(m,n)$

In this section, we explain how the expression $\Gamma(m,n)$ can be simplified as a linear arithmetic formula $\Gamma'(m,n)$. This simplification is essentially due to the orientations of the constraint implications of the various classes. We focus hereafter on one typical class.

Proposition 8. *For any programs π of class $\rightarrow\downarrow\Rightarrow\Downarrow$ and for all $m,n \in \mathbb{N}^*$, the simplified form $\Gamma'(m,n)$ of $\Gamma(m,n)$ is $(\Phi_1(0,0)\wedge\Phi_2(m,0))\vee(\Phi_2(0,0)\wedge\Phi_1(0,n))$.*

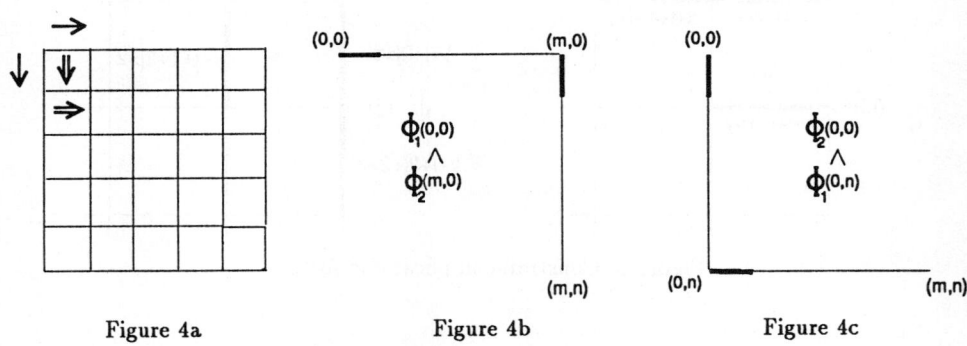

Figure 4a   Figure 4b   Figure 4c

Proof. Let P be a path from $(0,0)$ to (m,n), with $m,n \in \mathbb{N}^*$. Because of the directions of $\downarrow\Rightarrow$, the first vertical constraint of P implies any other vertical constraint of P. Similarily, because of the directions of $\rightarrow\Downarrow$, the first horizontal constraint of P implies any other horizontal constraint of P. Hence, the reduced constraint of P is equal to the conjunction of its first vertical and its first horizontal constraints.

If the path P begins with a horizontal constraint, its reduced constraint implies the reduced constraint $\Phi_1(0,0) \wedge \Phi_2(m,0)$ of the path of figure 4b. If the path

P begins with a vertical constraint, its reduced constraint implies the reduced constraint $\Phi_2(0,0) \wedge \Phi_1(0,n)$ of the path of figure 4c. So the disjunction $\Gamma(m,n)$ of the reduced constraints of all the paths from $(0,0)$ to (m,n) is equivalent to the formula $(\Phi_1(0,0) \wedge \Phi_2(m,0)) \vee (\Phi_2(0,0) \wedge \Phi_1(0,n))$. □

For the program π_1 of example 1, the expression $\Gamma'(m,n)$ is $(x > y \wedge y + m > z) \vee (y > z \wedge x + n > y)$, for $m, n \in \mathbb{N}^*$.

Similar results can be established for the following 11 other classes:

Class	$\Gamma'(m,n)$, for $m,n \in \mathbb{N}^*$.
→↓⇐⇓	$\Phi_1(0,n) \wedge \Phi_2(0,0)$
→↓⇒⇑	$\Phi_1(0,0) \wedge \Phi_2(m,0)$
←↑⇒⇑	$\Phi_1(m-1,0) \wedge \Phi_2(m,n-1)$
←↑⇐⇓	$\Phi_1(m-1,n) \wedge \Phi_1(0,n-1)$
←↓⇒⇑	$\Phi_1(m-1,0) \wedge \Phi_2(m,0)$
→↑⇐⇓	$\Phi_1(0,n) \wedge \Phi_2(0,n-1)$
←↑⇐⇑	$(\Phi_1(m-1,0) \wedge \Phi_2(m,n-1)) \vee (\Phi_1(m-1,n) \wedge \Phi_2(0,n-1))$
←↓⇒⇓	$(\Phi_1(m-1,n) \wedge \Phi_2(0,0)) \vee (\Phi_1(m-1,0) \wedge \Phi_2(m,0)) \vee (\Phi_1(k-1,0) \wedge \Phi_1(m-1,n) \wedge \Phi_2(k,0))$
→↑⇐⇑	$(\Phi_1(0,n-1) \wedge \Phi_2(0,n-1)) \vee (\Phi_1(0,0) \wedge \Phi_2(m,n-1)) \vee (\Phi_1(0,0) \wedge \Phi_1(k,n) \wedge \Phi_2(k,n-1))$
→↑⇒⇓	$(\Phi_1(0,0) \wedge \Phi_2(m,n-1)) \vee (\Phi_1(0,n) \wedge \Phi_2(0,n-1)) \vee (\Phi_1(0,k) \wedge \Phi_2(0,k-1) \wedge \Phi_2(m,n-1))$
←↓⇐⇑	$(\Phi_1(m-1,0) \wedge \Phi_2(m,0)) \vee (\Phi_1(m-1,n) \wedge \Phi_2(0,0)) \vee (\Phi_1(m-1,k) \wedge \Phi_2(0,0) \wedge \Phi_2(m,k))$

The proof of these equivalences is analogous to the proof of proposition 1. For programs in the four remaining classes (→↓⇐⇑, ←↑⇒⇓, ←↓⇐⇓ and →↑⇒⇑), the expressions $\Gamma(m,n)$ can still be simplified but the corresponding equivalence proofs are much more complicated than before (see [12] for full details).

6 Some extensions

In this section, we consider several extensions of the form of program π.

6.1 Constraints with equality

We now consider constraints with the equality symbol = (rather than <, ≤). Suppose that our program is thus of the form:

$p(x,y,z)$:− $\xi(x,y,z)$. rule R_0
$p(x+a_1, y+b_1, z+c_1)$:− $\phi_1(x,y,z), p(x,y,z)$. rule R_1
$p(x+a_2, y+b_2, z+c_2)$:− $\phi_2(x,y,z), p(x,y,z)$. rule R_2

where $\phi_1(x,y,z)$ is still of the form $d_1 x + e_1 y + f_1 z + g_1 \odot 0$, with $\odot \in \{<,>,\leq,\geq\}$, but $\phi_2(x,y,z)$ is now of the form $d_2 x + e_2 y + f_2 z + g_2 = 0$.

As in section 4, the reduction of S_0, S_H and S_V to an arithmetic formula is easy. In order to reduce S_Z (i.e., in order to compute $\Gamma'(m,n)$ for $m,n \in \mathbb{N}^*$), we will distinguish four cases, according to certain properties of invariance of the constraint ϕ_2. We say that ϕ_2 is *invariant* with respect to rule R_1 (resp. R_2) iff $d_2 a_1 + e_2 b_1 + \ldots + f_2 c_1 = 0$ (resp. $d_2 a_2 + e_2 b_2 + \ldots + f_2 c_2 = 0$).

The four cases are the following:

1. The constraint ϕ_2 is invariant w.r.t. R_2, but not w.r.t. R_1.
2. The constraint ϕ_2 is invariant w.r.t. R_1, but not w.r.t. R_2.

3. The constraint ϕ_2 is invariant w.r.t. both R_1 and R_2.
4. The constraint ϕ_2 is neither invariant w.r.t. R_1 nor w.r.t. R_2.

We illustrate how to adapt the method described in sections 4 and 5 on a typical example of case 1.

Example 2

Let π_2 be the program:

$p(x,y,z)$:– $\xi(x,y,z)$.
$p(x+1,y,z)$:– $x \geq z$, $p(x,y,z)$.
$p(x+1,y,z+1)$:– $x = z$, $p(x,y,z)$.

The constraint $x = z$ is invariant w.r.t. R_2, but not w.r.t. R_1. This means that once we apply rule R_1 (after a sequence of application of R_2), the constraint $x = z$ will never be satisfied again and no further application of R_2 can occur. Hence, all sequences of rule application are of the form $R_1^i R_2^n R_1^{m-i}$ (see figure 6).

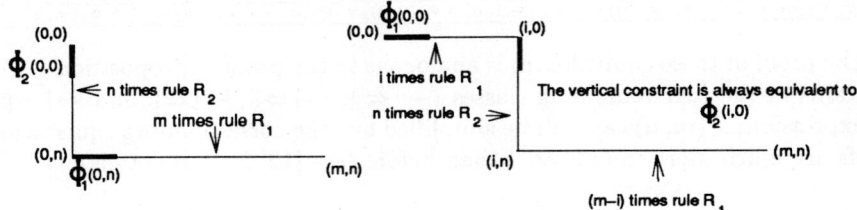

Figure 6

For applying rule R_2 after i applications of R_1, the constraint $\phi_2(i,0)$ (i.e., $x + i = z$) must be satisfied. Futhermore, since the program is of class $\rightarrow \Downarrow$, the first horizontal constraint of a path always implies any other horizontal constraint of this path. So the reduced constraint of a path is either $\phi_1(0,n) \wedge \phi_2(0,0)$ (i.e., $x \geq z \wedge x = z$) if $i = 0$, or $\phi_1(0,0) \wedge \phi_2(i,0)$ (i.e., $x \geq z \wedge x+i = z$) otherwise.

It follows that $\Gamma'(m,n)$ is equivalent to $(\phi_1(0,0) \wedge \phi_2(i,0)) \vee (\phi_1(0,n) \wedge \phi_2(0,0))$, for some $0 < i \leq m$. This expression reduces here to $x \geq z$.

We have:

$S_0 = \{ p(x,y,z) \ / \ \xi(x,y,z) \}$
$S_H = \cup_{m \in \mathbb{N}^*} \{ p(x+m,y,z) \ / \ \xi(x,y,z) \wedge x \geq z \}$
$S_V = \cup_{n \in \mathbb{N}^*} \{ p(x+n,y,z+n) \ / \ \xi(x,y,z) \wedge x = z \}$
$S_Z = \cup_{m,n \in \mathbb{N}^*} \{ p(x+m+n,y,z+n) \ / \ \xi(x,y,z) \wedge x \geq z \}$

Note that the program *delta'* displayed in section 1 coincides with π_2 when $(x = 0 \wedge z = 0)$ is taken as for $\xi(x,y,z)$. We have then:

$S = S_0 \cup S_H \cup S_V \cup S_Z = \{ delta'(0,y,0) \} \cup \{ delta'(m,y,0) \ / \ m > 0 \} \cup \{ delta'(n,y,n) \ / \ n > 0 \} \cup \{ delta'(m+n,y,n) \ / \ m > 0 \wedge n > 0 \}$.

This set can be easily put under the form $\{ delta'(x,y,z) \ / \ 0 \leq z \leq x \}$. □

The 3 other cases, as well as the case where the symbol of both arithmetic constraints Φ_1 and Φ_2 (and not only Φ_2) are $=$, can be treated similarly [12].

6.2 Several Recursive Rules

The method described in the previous sections applies to programs having (at most) two recursive rules. This method can however be used to treat programs made of q recursive rules ($q > 2$). Roughly speaking, one partitions the set of q rules into several subsets of (at most) two rules, and applies iteratively the previous procedure to each of these subsets. At each application, one starts from the set of atoms generated at the preceding step. The process terminates when no new atoms are generated, but it can possibly loop forever (enlarging at each step the current set of generated atoms). However in practice, there is often an appropriate partitioning of the rules set and an appropriate ordering of the rules application that make the process terminate.

Example 3

Let π_3 be the program:

$p(x,y,z)$	$:- x = 0, y = 0, z = 0$	R_0
$p(x+1, y+1, z) :-$	$p(x,y,z)$	R_1
$p(x, y+1, z+1) :-$	$2x > y,\ p(x,y,z)$	R_2
$p(x, y+1, z)$	$:- 2x > y,\ p(x,y,z)$	R_3

We proceed as follows:
1. The iterated application of R_1 and R_2 (starting from the basic tuple $(0,0,0)$ and using the method of section 5) generates:
$S_1 = \{p(m_1, m_1 + n_1, n_1)\ /\ m_1 + 1 > n_1, n_1 \geq 0,\ m_1 \geq 0\}$
2. The iterated application of R_3 to the set S_1 generates:
$S_2 = \{p(m_1, m_1 + n_1 + n'_1, n_1)\ /\ m_1 + 1 > n'_1 + n_1,\ n'_1 \geq 0, n_1 \geq 0,$
$m_1 \geq 0\}$
3. The iterated application of R_1 and R_2 to the set S_2 (using the method of section 5) generates:
$S'_1 = \{p(m_1 + m_2, m_1 + n_1 + n'_1 + m_2 + n_2, n_1 + n_2)\ /\ m_1 + 1 > n'_1 + n_1,$
$m_1 + m_2 + 1 > n'_1 + n_1 + n_2, n'_1 \geq 0, n_1 \geq 0, m_1 \geq 0, m_2 \geq 0, n_2 \geq 0\}$

The iterated application of R_3 to the set S'_1 generates a set which is included in S_2. Our method thus terminates and the output S of the bottom-up evaluation process of π_3 is the union $S_1 \cup S_2 \cup S'_1$, which corresponds to the disjunction of three conjunctions of inequalities. This disjunction can be actually further simplified, and S can be more concisely expressed as:
$\{p(x,y,z)\ /\ 2x \geq y\ \wedge\ y \geq z + x\ \wedge\ y \geq 0\ \wedge\ z \geq 0\}$ □

6.3 Several Constraints

We show now how certain programs with more than one constraint per rule can be transformed into a program of class π. The basic idea consists to eliminate the constraints which belong to the base rule and are invariant w.r.t to the recursive rules, then to compute the output S' of the bottom-up evaluation of the resulting program, finally to reintroduce the invariant constraints into S'. (This is similar to the idea of "pushing" constraints, see e.g. [23, 30]; cf [26, 18]). This technique is illustrated below.

Example 4

Consider the following program π_4:

$p(v,w,x,y,z)$	$:- v \neq w, x = 0, y = 0, z = 0$	R_0
$p(v,w,x+1,y+1,z)$	$:- v \neq w, \ p(v,w,x,y,z)$	R_1
$p(v,w,x,y+1,z+1)$	$:- v \neq w, 2x > y, \ p(v,w,x,y,z)$	R_2
$p(v,w,x,y+1,z)$	$:- 2x > y, \ p(v,w,x,y,z)$	R_3

We first observe that $v \neq w$ is a constraint of the basic rule R_0 which is invariant w.r.t. all the recursive rules of the program. So we eliminate this constraint from π_4, thus obtaining a program π'_4. A bottom-up evaluation process on π'_4 (similar to that performed on program π_3 of example 3) yields:
$S' = \{\ p(v,w,x,y,z) \ / \ 2x \geq y \ \wedge \ y \geq z+x \ \wedge \ y \geq 0 \ \wedge \ z \geq 0\ \}$
The output for π_4 is therefore:
$S = \{\ p(v,w,x,y,z) \ / \ v \neq w \ \wedge \ 2x \geq y \ \wedge \ y \geq z+x \ \wedge \ y \geq 0 \ \wedge \ z \geq 0\ \}$
The program π_4 corresponds to a simplified version of the Boyer-Moore Majority Algorithm [25], and the output S can be used as a lemma to prove the correctness of this algorithm. □

Others extensions with several constraints per rule are studied in [12].

7 Applications

7.1 Generation of lemmas

As explained at the beginning of this paper, our initial motivation for studying bottom-up evaluation of Datalog programs with integers, is its applicability to the automatic generation of lemmas. More precisely, given a logic program defining a predicate of the form $p(L,x,y,z)$ where L is a variable denoting a list of integers, and x,y,z are variables denoting integers, it is often possible to transform the atom $p(L,x,y,z)$ into an atom of the form $p'(x,y,z)$ defined only over integer arguments (see [11]). The bottom-up evaluation of $p'(x,y,z)$ then yields an arithmetic relation $\psi(x,y,z)$ that also holds for $p(L,x,y,z)$, generating the lemma: $p(L,x,y,z) \Rightarrow \psi(x,y,z)$.

In the motivation section (see also example 2), we have thus sketched out how to generate the lemma $delta(L,x,y,z) \Rightarrow z \leq x$. The method applies also if the hypothesis contains more than one predicate. For example, one can generate lemmas of the form $p_1(L,x_1,y_1,z_1) \wedge p_2(L,x_2,y_2,z_2) \Rightarrow \psi(x_1,y_1,z_1,x_2,y_2,z_2)$; it suffices to replace the conjunction $p_1(L,x_1,y_1,z_1) \wedge p_2(L,x_2,y_2,z_2)$ by an equivalent atom $p_3(L,x_1,y_1,z_1,x_2,y_2,z_2)$ and to apply the previous method to this new atom. This application to the automatic generation of lemmas is useful for proving safety properties of communicating automata and protocols [13].

7.2 Termination of Prolog programs

As seen in section 2, bottom-up evaluation of Datalog programs with integers is useful for proving the termination of logic programs. The Datalog programs correspond to logic programs in which the original arguments have been replaced by their sizes. (This can be viewed as a form of "abstract interpretation" [7].) As an illustration consider the following Prolog program (borrowed from [27]).

$split([\,], y, [\,], [\,])$.
$split([u|L], y, [u|M], N) :- u \leq y, split(L, y, M, N)$.
$split([u|L], y, M, [u|N]) :- u > y, split(L, y, M, N)$.

By replacing the list arguments by their sizes (numbers of elements) and removing the auxiliary constraints $u \leq y$ and $u > y$, one transforms the above program into the Datalog program:

$split'(x, y, z, v) \qquad :- x = 0, z = 0, v = 0.$
$split'(x+1, y, z+1, v) :- split'(x, y, z, v).$
$split'(x+1, y, z, v+1) :- split'(x, y, z, v).$

Our method applied to *split'* generates an arithmetical formula equivalent to $x = z + v$. This means that the size of the fourth argument of *split* is the sum of the sizes of the first and third arguments. This relation among the sizes of the arguments of *split* can be used for showing the termination of a *quicksort* program (see [27]). Note that Plümer's (top-down) procedure generates the weaker relation $x \geq z + v$.

7.3 Temporal Deductive Databases

A Datalog program is a logic program without function symbols. A Temporal Datalog program is a logic program where function symbols (e.g. *succ*, +) are allowed for the "temporal" arguments. Given a Temporal Datalog program and a query $p(x, y, z)$, the query processing problem consists to determine whether or not there exists a ground substitution σ such that $\sigma(p(x, y, z))$ logically follows from the program (and to determine all of these substitutions, if any). This problem is undecidable for general temporal Datalog programs, but has been proven decidable for temporal Datalog programs which have a *unique* temporal argument over which incrementation is done systematically (no arithmetic constraint) [9]. Our process of bottom-up evaluation makes the query processing problem decidable for a new class of temporal Datalog programs: these programs have an *arbitrary number* of temporal arguments over which incrementation is done conditionally (according to the satisfaction of an arithmetic constraint); on the other hand, our programs contain *at most two* recursive rules.

8 Conclusion

We have developed a bottom-up evaluation procedure for a special class of recursive Datalog programs over integers that are often met in Temporal Deductive Databases, or when proving termination of logic programs.

When the procedure terminates, it computes a linear arithmetic formula which characterizes the least fixpoint associated with the program. The procedure always terminates (unlike the other same-purpose procedures found in the literature) for programs made of at most two recursive rules with at most one arithmetic constraint per rule. Moreover, using additional techniques such as rule partitioning and constraint "pushing", our procedure also terminates in many cases where the programs have more than two recursive rules and more than one constraint per rule.

We have assumed in this paper that the domain of the variables and constants were the domain of integer numbers, but our method can be applied if the domain of the variables and constants is the domain of *rational* or *real* numbers. We have also assumed that ξ was a linear arithmetic formula, but the method applies as well for *general arithmetic* formulas. Further generalizations on the form of program π are given in [12].

References

1. K. R. Apt, M. H. Van Emden. "Contributions to the Theory of Logic Programming". *J.ACM* 29, 1982, pp. 841-862.
2. M. Baudinet. "Temporal Logic Programming is Complete and Expressive". *Proc. 16th ACM Symp. on Principles of Programming Languages*, Austin, 1989, pp. 267-280.
3. R. S. Boyer and J. S. Moore. "Integrating Decision Procedures into Heuristic Theorem Provers: A Case of Linear Arithmetic", *Machine Intelligence 11*, J. E. Hayes, D. Michie and J. Richards (eds.), Clarendon Press, Oxford, 1988, pp. 83-124.
4. M. Baudinet, M. Niezette and P. Wolper. "On the Representation of Infinite Temporal Data Queries". *Proc. 10th ACM Symp. on Principles of Database Systems*, Denver, 1991, pp. 280-290.
5. F. Bancillon and R. Ramakrishnan. "An Amateur's Introduction to Recursive Query Processing Strategies", *Proc. ACM Conf. on Management of Data*, Washington, 1986, pp. 16-52.
6. A. Brodsky and Y. Sagiv. "Inference of Inequality Constraints in Logic Programs". *Proc. 10th ACM Symp. on Principles of Database Systems*, Denver, 1991, pp. 227-240.
7. P. Cousot and R. Cousot. "Abstract Interpretation: A Unified Lattice Model for Static Analysis of Programs by Construction or Approximation of Fixpoints", *Conference Record of the 4th ACM Symposium on Principles of Programming Languages*, Los Angeles, 1977, pp. 238-252.
8. P. Cousot and N. Halbwachs. "Automatic Discovery of Linear Restraints among Variables of a Program". *Conference Record 5th ACM Symp. on Principles of Programming Languages*, Tucson, 1978, pp. 84-96.
9. J. Chomicki and T. Imielinski. "Temporal Deductive Databases and Infinite Objects". *Proc. 7th ACM Symp. on Principles of Database Systems*, Austin, 1988, pp. 61-81.
10. M. H. Van Emden and R. A. Kowalski. "The Semantics of Predicate Logic as a Programming Language", *J.ACM* 23:4, 1976, pp. 733-742.
11. L. Fribourg. "Mixing List Recursion and Arithmetic". *Proc. 7th IEEE Symp. on Logic in Computer Science*, Santa Cruz, 1992, pp. 419-429.

12. L. Fribourg and M. Veloso Peixoto. "Bottom-up Evaluation of Datalog Programs with Arithmetic Constraints". Technical report, LIENS 92-13, June 1992.
13. L. Fribourg and M. Veloso Peixoto. "Concurrent Constraint Automata". Technical report, LIENS 93-10, May 1993.
14. A. Van Gelder. "Deriving Constraints among Argument Sizes in Logic Programs", *Proc. 9th ACM Symp. on Principles of Database Systems*, Nashville, 1990, pp. 47-60.
15. J. Jaffar and J.L. Lassez. "Constraint Logic Programming", *Proc. 14th ACM Symp. on Principles of Programming Languages*, 1987, pp. 111-119.
16. G. Kuper. "On the Expressive Power of the Relational Calculus with Arithmetic Constraints". *Proc. 3rd International Conference on Database Theory*, Paris, 1990, pp. 203-313.
17. P. Kanellakis, G. Kuper and P. Revesz. "Constraint Query Languages". *Proc. 9th ACM Symp. on Principles of Database Systems*, Nashville, 1990, pp. 299-313.
18. D.B. Kemp and P.J. Stuckey. "Analysis Based Constraint Query Optimization". *Proc. 10th Intl. Conf. on Logic Programming*, Budapest, The MIT Press, 1993, pp. 666-682.
19. F. Kabanza, J.M. Stevenne and P. Wolper. "Handling Infinite Temporal Data". *Proc. 9th ACM Symp. on Principles of Database Systems*, Nashville, 1990, pp. 392-403.
20. J-L Lassez. "Querying Constraints". *Proc. 9th ACM Symp. on Principles of Database Systems*, Nashville, 1990, pp. 288-298.
21. J-L Lassez. "Parametric Queries, linear constraints and variable elimination". *Proc. Conference on Design and Implementation of Symbolic Computer*, LNCS 429, 1990, pp. 164-173.
22. J. Lloyd. "Foundations of Logic Programming". Second Edition, Springer Verlag, Berlin, 1987.
23. A. Levy and Y. Sagiv. "Constraints and Redundancy in Datalog". *Proc. 11th ACM Symp. on Principles of Database Systems*, San Diego, 1992, pp. 67-80.
24. R. Van der Meyden. "Reasoning with Recursive Relations: Negation, Inequality and Linear Order (Extended Abstract)". *Proc. ILPS Workshop on Deductive Databases*, San Diego, 1991.
25. J. Misra and D. Gries. "Finding Repeated Elements", *Science of Computer Programming 2*, 1982, pp. 143-152.
26. K. Marriot and P.J. Stuckey. "The 3 R's of Optimizing Constraint Logic Programs: Refinement, Removal and Reordering". *Proc. 20th ACM Symp. on Principles of Programming Languages*, Charleston, 1993, pp. 334-344.
27. L. Plümer. "Termination Proofs for Logic Programs based on Predicate Inequalities".*Proc. 7th International Conference on Logic Programming*, Jerusalem, 1990, pp. 634-648.
28. P. Revesz . "A Closed Form for Datalog Queries with Integer Order". *Proc. 3rd International Conference on Database Theory*, Paris, 1990, pp. 187-201.
29. D. Srivastava. "Subsumption in Constraint Query Languages with Linear Arithmetic Constraints". *Proc. 2nd International Symp. on Artificial Intelligence and Mathematics*, Fort Lauderdale, 1992.
30. D. Srivastava and R. Ramakrishnan. "Pushing Constraint Selections". *Proc. 11th ACM Symp. on Principles of Database Systems*, San Diego, 1992, pp. 301-315.
31. J. D. Ullman and A. Van Gelder. "Efficient Test for Top-Down Termination", *J.ACM* 35:2, 1988, pp. 345-373.

On Intuitionistic Query Answering in Description Bases

Véronique Royer[1] and J. Joachim Quantz[2]

[1] ONERA DES/SIA, avenue de la division Leclerc, B.P. 72, F-92322 Chatillon CEDEX
royer@onera.fr
[2] Technische Universität Berlin, KIT-VM11, FR 5–12, Franklinstr. 28/29, D-10587 Berlin
jjq@cs.tu-berlin.de

Abstract. In this paper we present a weak and a strong intuitionistic calculus for query answering in Description Logics (DL). Given the standard model-theoretic semantics for DL, a complete query-answering calculus has to perform complex case analyses to cope with implicit disjunctions stemming from some of the concept-forming operators in DL. To avoid this complexity we propose an intuitionistic approach to query answering based on the Sequent-Calculus-style axiomatization of DL we have developed in [20] and [21]. By taking into account only the intuitionistic inference schemata of this axiomatization, we obtain a strong intuitionistic query-answering calculus. An additional restriction to reasoning about explicit objects allows a further simplification of the proof theory and yields a weak intuitionistic calculus.

We prove completeness of these calculi wrt axiomatic semantics based on the Intuitionistic Sequent Calculus. For the weak calculus we also give a least fixed point semantics as known from Deductive Databases and Logic Programming.

Keywords: Description Logics, Intuitionistic Sequent Calculus,
Least Fixed Point Semantics, Query Answering

1 Introduction

Description Logics (DL) can be seen as a formal elaboration of the ideas underlying *Semantic Networks* and *Frames*. Following the debate on the lacking formal foundation of these representation formats in the mid-70's, Brachman proposed the representation language KL-ONE, which is described in an overview, which was circulated in the beginning of the 1980's and was finally published in 1985 [4]. In the last decade, substantial theoretical research concerning complexity and decidability of different dialects of Description Logics has been conducted (e.g. [6]). In parallel, several DL systems have been implemented and used in various applications (e.g. [5, 11, 18]).

In the following, we will use a simple DL, whose syntax is given below. The basic alphabet for DL formulae is given by a set $\mathcal{C}_n$ of concept names c_n, a set $\mathcal{R}_n$ of role names r_n, and a set $\mathcal{O}$ of object names o. Complex concept and role terms are formed using boolean constructors like conjunction, disjunction, or negation or so-called *role restriction* constructors, for defining concepts as sets of objects satisfying particular restrictions wrt role terms. The DL formulae of the form $t_1 \sqsubseteq t_2$ or $t_1 \doteq t_2$ are usually called *terminological formulae* whereas formulae of the form $c(o)$ and $r(o_1, o_2)$ are called *assertional formulae*.

$$t \rightarrow c, r$$
$$c \rightarrow \top, \bot, c_n, c_1 \sqcap c_2, c_1 \sqcup c_2, \neg c, \forall r{:}c, \exists r{:}c, \geq nr{:}c, \leq nr{:}c$$
$$r \rightarrow r_n, r_1 \sqcap r_2$$
$$\gamma \rightarrow t_1 \sqsubseteq t_2, t_1 \doteq t_2, c(o), r(o_1, o_2)$$

For this language a model-theoretic semantics can be given where a model M of a set of DL formulae Γ is a pair $\langle D, [\![\cdot]\!]^\mathcal{I}\rangle$. D is a set, called the domain, and $[\![\cdot]\!]^\mathcal{I}$ is an interpretation function mapping concepts into subsets of D, roles into subsets of $D \times D$, and object names injectively (Unique Name Assumption) into D, in accordance with the following equations (we use $r(d)$ to denote $\{e : \langle d, e\rangle \in r\}$):

$$[\![\top]\!]^\mathcal{I} = D$$
$$[\![\bot]\!]^\mathcal{I} = \emptyset$$
$$[\![c_1 \sqcap c_2]\!]^\mathcal{I} = [\![c_1]\!]^\mathcal{I} \cap [\![c_2]\!]^\mathcal{I}$$
$$[\![c_1 \sqcup c_2]\!]^\mathcal{I} = [\![c_1]\!]^\mathcal{I} \cup [\![c_2]\!]^\mathcal{I}$$
$$[\![\neg c]\!]^\mathcal{I} = D \setminus [\![c]\!]^\mathcal{I}$$
$$[\![\forall r : c]\!]^\mathcal{I} = \{d \in D : [\![r]\!]^\mathcal{I}(d) \subseteq [\![c]\!]^\mathcal{I}\}$$
$$[\![\exists r : c]\!]^\mathcal{I} = \{d \in D : [\![r]\!]^\mathcal{I}(d) \cap [\![c]\!]^\mathcal{I} \neq \emptyset\}$$
$$[\![\geq nr{:}c]\!]^\mathcal{I} = \{d \in D : |[\![r]\!]^\mathcal{I}(d) \cap [\![c]\!]^\mathcal{I}| \geq n\}$$
$$[\![\leq nr{:}c]\!]^\mathcal{I} = \{d \in D : |[\![r]\!]^\mathcal{I}(d) \cap [\![c]\!]^\mathcal{I}| \leq n\}$$
$$[\![r_1 \sqcap r_2]\!]^\mathcal{I} = [\![r_1]\!]^\mathcal{I} \cap [\![r_2]\!]^\mathcal{I}$$

Satisfaction of formulae is then defined as follows:

$$M \models t_2 \sqsubseteq t_1 \text{ iff } [\![t_2]\!]^\mathcal{I} \subseteq [\![t_1]\!]^\mathcal{I}$$
$$M \models t_1 \doteq t_2 \text{ iff } [\![t_1]\!]^\mathcal{I} = [\![t_2]\!]^\mathcal{I}$$
$$M \models c(o) \text{ iff } [\![o]\!]^\mathcal{I} \in [\![c]\!]^\mathcal{I}$$
$$M \models r(o_1, o_2) \text{ iff } \langle [\![o_1]\!]^\mathcal{I}, [\![o_2]\!]^\mathcal{I}\rangle \in [\![r]\!]^\mathcal{I}$$

M is a model of a formula γ iff $M \models \gamma$; it is a model of a set of formulae Γ iff it is a model of every formula in Γ. A formula γ is entailed by a set of formulae Γ (written $\Gamma \models \gamma$) iff every model of Γ is also a model of γ.

Note that Description Logics in general are subsets of First-Order Logic with Equality (FOL)—concepts are translated into unary predicate, roles into binary predicates, and object names into constants. Due to lack of space we here give only the FOL translations of **all**, **some**, **atleast**, and **atmost** (for a complete translation see [21]):

$$\overline{\forall r : c} \stackrel{\text{def}}{=} \lambda x (\forall y \bar{r}(x)(y) \rightarrow \bar{c}(y))$$
$$\overline{\exists r : c} \stackrel{\text{def}}{=} \lambda x (\exists y \bar{r}(x)(y) \wedge \bar{c}(y))$$
$$\overline{\geq nr{:}c} \stackrel{\text{def}}{=} \lambda x (\exists y_1 \ldots y_n \bar{r}(x)(y_1) \wedge \bar{c}(y_1) \ldots \bar{r}(x)(y_n) \wedge \bar{c}(y_n)$$
$$\wedge dif(y_1, \ldots, y_n))$$
$$\overline{\leq nr{:}c} \stackrel{\text{def}}{=} \lambda x (\exists y_1 \ldots y_{n+1} \bar{r}(x)(y_1) \wedge \bar{c}(y_1) \ldots \bar{r}(x)(y_{n+1}) \wedge \bar{c}(y_{n+1})$$
$$\rightarrow equ(y_1, \ldots, y_{n+1}))$$

$$equ(y_1, ..., y_n) \stackrel{\text{def}}{=} \vee_{i \neq j}(y_i = y_j)$$
$$dif(y_1, ..., y_n) \stackrel{\text{def}}{=} \neg equ(y_1, ..., y_n)$$

When using DL systems for knowledge representation the modeled knowledge is usually separated into a TBox and an ABox. A TBox is a list of definitions $t_n \doteq t$ or $t_n \sqsubseteq t$ containing exactly one definition for all names t_n in $c_n \cup r_n$. Furthermore each c_n (r_n) occurring on the right-hand side of a definition must have been defined itself in an earlier definition (there are thus no cyclic definitions). The ABox is a list of assertions $c(o)$ and $r(o_1, o_2)$.

Most DL systems also allow the introduction of so-called rules or implication links $c_1 \Rightarrow c_2$. These rules are either interpreted as material implications, being thus equivalent to the formula $c_1 \sqsubseteq c_2$, or as forward chaining trigger rules applied whenever an object o is known to be an instance of c_1 [2, 7]. In this paper we will not address the issue of rules and restrict our investigation to DL modelings containing only a TBox and an ABox.

Note that the distinction between TBox and ABox corresponds to a distinction in Deductive Databases between the *intensional database*, which contains rule-based knowledge usually in the form of horn clauses, and the *extensional database*, which contains contigent knowledge modeled as facts. The process of automated reasoning then consists basically in matching facts with antecedents of rules and computing the corresponding succedents. Semantically, the informational content of the deductive database consists of all the knowledge entailed by the extensional and intensional databases. More precisely, as "knowledge" usually means "data", i.e. ground atomic facts, from the database perspective, the computational issue for Deductive Databases is to compute the set of atomic ground facts which derive from facts and rules.

To make this analogy explicit we will in the following speak of *description bases* DB consisting of a terminological base TB and an assertional base AB. We thereby indicate a *procedural interpretation* of terminological knowledge, which has to play the role of deduction rules for assertions. This is the common idea of turning definitions into computation rules, which is quite classical in Logic Programming, Rewrite Systems etc. To underline this procedural interpretation we will therefore use *terminological clauses* of the form $t_1 \rightarrow t_2$ in the TB, which correspond to the subsumption formulae $t_1 \sqsubseteq t_2$ in the TBox.

There are two major differences between description bases and Deductive Databases, however:

1. There are logical entailment relations between subsumption formulae in description bases, but there are none between deductive rules in Deductive Databases.
2. To answer queries involving role-restriction operators complex assertional reasoning can be necessary in description bases, whereas there is no reasoning about facts in Deductive Databases.

The first difference can be illustrated by considering a description base containing the terminological clause $c_1 \rightarrow c_2$ and the assertion $\forall r{:}c_1(o)$. To derive the assertion $\forall r{:}c_2(o)$ we have to add the terminological clause $\forall r{:}c_1 \rightarrow \forall r{:}c_2$, which is logically entailed by $c_1 \rightarrow c_2$.

The complexity of assertional reasoning is illustrated by the following description base taken from [7]:

friend(*john,susan*) loves(*susan,peter*) graduate(*susan*)
friend(*john,peter*) loves(*peter,mary*) ¬ graduate(*mary*)

Given the model-theoretic semantics, one can show by case reasoning that this AB entails the fact ∃friend:(graduate ⊓ ∃loves:¬ graduate)(*john*). In any model of AB either *peter* is graduate or *peter* is not. In both cases the assertion is true: in the first case, *peter* is the graduate friend loving a non-graduate (*mary*); in the second case *susan* is the graduate friend loving a non-graduate (*peter*).

Such reasoning translates into complex syntactical proofs in Sequent Calculus. The corresponding sequent proof shows that right contraction rules, hence right disjunctions, are necessary to simulate the particular reasoning by case in this example. The lesson is that trying to cope with complete assertional inferences involves being able to reason with disjunctions and contraction rules which duplicate arbitrarily the contexts of proofs, at the expense of computational costs. Therefore, any query-answering procedure based on linear goal-subgoal strategies as applied in Logic Programming will fail on this example. In the following we will therefore pursue an intuitionistic approach which avoids reasoning with right contractions and right disjunctions.

2 Intuitionistic Proof Theory

Above we have pointed out the analogy between DL and Deductive Databases. Definite databases, where intensional rules are definite Horn clauses (i.e. Horn clauses with exactly one positive literal), can be given a proof-theoretical account—the *least fixed point semantics* (e.g. [16]). Indeed, the informational content of any definite deductive database can be computed as the least fixed point of some *consequence operator*. The consequence operator thus formalizes the intuition that the rules of the intensional databases can be processed as procedural rules to generate new knowledge from knowledge matching their antecedents. The differences between DL and Deductive Databases sketched above indicate that additional inference mechanisms have to be provided if we want to apply the deductive paradigm to DL. Thus a terminological base has to contain the logical closure, i.e. closure under terminological entailment, of its corresponding TBox. Similarly, assertion bases have by themselves a great inferential power, contrarily to the extensional facts of intensional databases. The deductive paradigm must account for both intrinsic entailment relations, assertional and terminological.

In [20] we have presented a systematic method towards axiomatizing DL, namely by deriving inference schemata via rewriting Sequent Calculus proofs of terminological formulae. In [21] we have applied this method to the expressive terminological fragment underlying BACK V5 [11]. We did not address the axiomatization of assertional inferences, however, which we will deal with in this paper.

The following is a list of intuitionistic inference schemata for the terminological fragment investigated in this paper. Most inference rules are presented in inconsistency form (variants of them are obtained by using negation rules, intuitionistically). The propositional rules are basically the rules for the propositional fragment of the Intuitionistic Sequent Calculus and are therefore omitted here. More details about the complete axiomatization can be found in [21].

$$\top \sqsubseteq c \rightleftharpoons \top \sqsubseteq \forall r{:}c$$

$$c \sqsubseteq \bot \Rightarrow \exists r{:}c \sqsubseteq \bot$$
$$\rightarrow \forall r : c \doteq \neg \exists r{:}\neg c$$
$$c_2 \sqcap c_1 \sqsubseteq \bot,\, r_1 \sqsubseteq r_2 \rightarrow \forall r_2{:}c_2 \sqcap \exists r_1{:}c_1 \sqsubseteq \bot$$
$$c_1 \sqcap c_2 \sqsubseteq c_3,\, r_2 \sqsubseteq r_1,\, r_2 \sqsubseteq r_3 \rightarrow \forall r_1{:}c_1 \sqcap \forall r_2{:}c_2 \sqsubseteq \exists r_3{:}c_3$$
$$c_1 \sqcap c_2 \sqsubseteq c_3,\, r_2 \sqsubseteq r_1,\, r_2 \sqsubseteq r_3 \rightarrow \forall r_1{:}c_1 \sqcap \exists r_2{:}c_2 \sqsubseteq \exists r_3{:}c_3$$
$$\rightarrow \exists r : c \doteq {\geq}1r{:}c$$
$$q < p,\, c_1 \sqsubseteq c_2,\, r_1 \sqsubseteq r_2 \rightarrow {\leq}qr_2{:}c_2 \sqcap {\geq}pr_1{:}c_1 \sqsubseteq \bot$$
$$c_2 \sqcap c_1 \sqsubseteq \bot,\, r_1 \sqsubseteq r_2 \rightarrow \forall r_2{:}c_2 \sqcap {\geq}pr_1{:}c_1 \sqsubseteq \bot$$
$$\rightarrow {\geq}nr{:}c \doteq \neg {\leq}n-1r{:}c$$
$$q < p,\, r_1 \sqsubseteq r_2,\, r_1 \sqsubseteq r_3,$$
$$c_1 \sqcap c_3 \sqsubseteq c_2 \rightarrow {\leq}qr_2{:}c_2 \sqcap {\geq}pr_1{:}c_1 \sqcap \forall r_3{:}c_3 \sqsubseteq \bot$$
$$r_1 \sqcap r_2 \sqsubseteq r_3,\, r_1 \sqcup r_2 \sqsubseteq r_4,$$
$$c_1 \sqcap c_2 \sqsubseteq c_3,\, c_1 \sqcup c_2 \sqsubseteq c_4,$$
$$k + n < p + q \rightarrow {\geq}pr_1{:}c_1 \sqcap {\geq}qr_2{:}c_2 \sqcap {\leq}kr_3{:}c_3 \sqsubseteq {\leq}nr_4{:}c_4$$

The rules used for defining a least fixed point semantics can then be obtained by taking the logical closure of a TB under such inference schemata.

Definition 1. Let TB be a terminological base, Σ a set of intuitionistic inference schemata, and θ a terminological clause. **Intutitionistic terminological entailment** is defined as follows: TB $\models_{IT} \theta$ iff TB $\vdash_\Sigma \theta$.

Suppose we have defined a Consequence Operator Φ and let Φ^ω be its least fixed point. We can then define a semantics DB $\models \alpha$ iff $\alpha \in \Phi^\omega(\text{DB})$. (We will give a formal definition below.) We are now interested in specifying a proof theory $\vdash$ such that DB $\vdash \alpha$ iff DB $\models \alpha$. A couple of remarks seem in order:

1. A purely forward chaining proof theory is not feasible since the set of TB clauses obtained by closing the original TB under the inference schemata is usually infinite (consider, for example, the schema $c_1 \sqsubseteq c_2 \rightarrow \exists r{:}c_1 \sqsubseteq \exists r{:}c_2$). We will therefore need a backward chaining proof strategy which takes the query into account.
2. As can be seen from the "graduate friends" example, the existence of implicit disjunctions would make a backward chaining approach relying on Classical Logic inefficient. If the Right Contraction rule is applicable, an arbitrary number of duplications is possible, which means that no control of the proof process is possible.

Of course the second problem has been the motivation for restricted inferences in Linear Logic [9], Logic Programming [14], and Deductive Databases [1]. In the following, we will use the standard approaches developed in these fields. We will therefore define alternative, intuitionistic semantics for DL. We will therefore define alternative, intuitionistic semantics for DL.

There seem to be two possibilities to define an intuitionistic semantics for DL: we can either exploit the standard translation of DL into FOL and specify an *axiomatic semantics* based on derivability wrt the Intuitionistic Sequent Calculus; or we can employ techniques from Deductive Databases and specify a *least fixed point semantics* based on intuitionistic inference schemata for DL. Before addressing these semantic issues we will sketch in the next section the basic ideas underlying our query-answering calculus by considering the "modern team" example. In Section 4 we will then present a weak and a strong intuitionistic query-answering calculus. Finally, in Section 5 we prove completeness of these calculi wrt axiomatic semantics. The weak calculus is also shown to be complete wrt a least fixed point semantics.

3 The Modern Team Example

Before presenting our proof theories let us consider the "modern team" example. The modern team TBox consists of the following definitions:[3]

$$
\begin{aligned}
&\text{man} \sqsubseteq \text{human} \\
&\text{woman} \sqsubseteq \text{human} \sqcap \neg\, \text{man} \\
&\text{leader} \sqsubseteq \text{member} \\
&\text{team} \doteq \forall \text{member:human} \sqcap\, \leq 4\, \text{member} \sqcap\, \geq 1\, \text{leader} \\
&\text{modern-team} \doteq \text{team} \sqcap \forall \text{leader:woman}
\end{aligned}
$$

The modern team Abox contains the following descriptions:

modern-team(mt)   member(mt,$dick$)   man($dick$)
≤ 3 member(mt)   member(mt,tom)   man($dick$)
                  member(mt,$mary$)

It can be proved model-theoretically but also intuitionistically that *mary* is the leader of *mt*. (*tom, dick*, and *mary* are all members of team *mt* which has at most three members. Hence, due to the *Unique Name Assumption*, these are all members of *mt*. As there must be at least one leader for *mt*, necessarily a female one, this cannot be *tom* or *dick*, who are men. Consequently, *mary* is necessarily leader of *mt* and must be a woman.) Note that this is a "moderate" form of case reasoning, namely the elimination of possibilities by Modus Tollens: man(tom), leader(mt,x) → ¬ man(x), therefore ¬ leader(mt,tom).

Now let us try to formalize how one could proceed by syntactical reasoning in answering the query "who is the leader of *mt*?", which we will write as leader?(mt,x). Note that this query is *open* and contains the variable x. As an answer we expect to retrieve all the objects o in $\mathcal{O}$ such that DB $\models$ leader(mt,o).

The definitions in the TBox are rewritten into *terminological clauses*[4] in order to stress their procedural meaning. Typically, any terminological definition of the form $c \doteq c_1 \sqcap \ldots \sqcap c_n$ gives rise to:

1. one clause stating a sufficient condition for c, namely $c_1 \sqcap \ldots \sqcap c_n \to c$
2. n clauses stating necessary conditions for c, namely $c \to c_i$

---

[3] We use $\geq n\, r$ and $\leq n\, r$ to abbreviate $\geq nr{:}\top$ and $\leq nr{:}\top$ respectively.
[4] Terminological clauses are like Horn clauses except that quantified concept terms may occur in them instead of just atomic ones.

Processing the necessary conditions 'team $\rightarrow \geq 1$ leader' and 'modern-team $\rightarrow$ team' in a backward chaining manner, leads to new queries team?(mt) and modern-team?(mt). The computation of queries and corresponding answers is formalized by the following *query-rules* and *answer-rules*:[5]

queries, answers	derivation rules
≥ 1 leader?(mt)	r?(x,y) $\Longrightarrow \geq n$ r?(x)
team?(mt)	c_2?(x) , $c_1 \rightarrow c_2 \Longrightarrow c_1$?(x) (*TB clause triggering*)
modern-team?(mt)	c_2?(x) , $c_1 \rightarrow c_2 \Longrightarrow c_1$?(x)
modern-team!(mt)	c?(x) , c(o) $\Longrightarrow$ c!(o) (*Lookup in AB*)
team!(mt) , ≥ 1 leader!(mt)	Lookup in AB
$\forall$leader:woman!(mt)	c_1!(x) , $c_1 \rightarrow c_2 \Longrightarrow c_2$!(x) (*TB clause triggering*)

The third rule formalizes the basic answering process of just looking up in the ABox. Therefore it can only derive answers with respect to *known* objects, hence could not answer the open query leader?(mt,x). Answer-rules performing some kind of *Skolemization* are necessary for that, therefore introducing new names to denote implicit objects:

$$\text{leader!}(mt,s) | r?(o,x) , \geq n \text{ r!}(o) \Longrightarrow \wedge_{i=1}^{i=n} r!(o,s_i) \quad \text{(for new names } s_i\text{)}.$$

The query-answering process goes on collecting more evidence or *constraints* about the still undefined answer "s", in particular querying about cardinality upper bounds:

$\leq n$ leader?(mt)	r?(x,y) $\Longrightarrow \leq n$ r?(x)
$\leq n$ member?(mt)	$\leq n$ r_1?(mt) , $r_1 \rightarrow r_2 \Longrightarrow \leq n$ r_2?(mt)
member?(mt,y)	$\leq n$ r?(x) $\Longrightarrow$ r?(x,y)
member!(mt,s) , member!(mt,tom)	Lookup in AB
member!(mt,dick) , member!(mt,mary)	"
≤ 3 member!(mt)	"
s=tom $\vee$ s= mary $\vee$ s=dick	$\wedge_{i=1}^{i=n+1} r!(x,y_i)$, $\leq n$ r!(x) $\Longrightarrow \vee_{i\neq j}(y_i = y_j)$ (+ UNA)
woman!(s)	$\forall$r:c!(x), r!(x,y) , $\Longrightarrow$ c!(y)
$\neg$man!(s)	TB clause triggering

Note that the second rule takes into account TBox entailment, namely the antimonotonicity of the **atmost** constructor. Now, one has obtained equality constraints which can be solved by exploiting the information about the male/female status of objects, according to the classical equality (substitution) axioms:

man?(y)	$\neg$man!(x) $\Longrightarrow$ man?(y) (if x$\notin \mathcal{O}$)
man!(tom) , man!(dick)	Lookup in AB
s $\neq$ tom , s $\neq$ dick	$\neg$c!(x) $\wedge$ c!(y) $\Longrightarrow$ x $\neq$ y
s = mary	$x \neq x_1, x = x_1 \vee \ldots \vee x = x_n \Longrightarrow x = x_2 \vee \ldots \vee x = x_n$
leader!(mt,mary)	y = z , r!(x,y) $\Longrightarrow$ r!(x,z)

Finally, it should be noted that quite complex inferences have been performed, involving both the elimination of an existential quantification (≥ 1 leader(mt)) and case reasoning wrt the derived disjunctive constraint: $s = tom \vee s = mary \vee s = dick$. As opposed

---

[5] They are "hybrid" rules operating both on query-answer assertions, ABox assertions and TBox clauses.

to the "married friends" example this is still *intuitionistic reasoning* because only *left* disjunctions are necessary for the corresponding sequent proof. Note also that when only man(*tom*) is known in the ABox, there is no way to determine the identity of "s" without ambiguity.

4 Intuitionistic Query Answering

The previous example has given an informal idea of how one could proceed to specify and automatize a query-answering procedure in description bases. The present section introduces formally two *query-answering calculi*. They look much like the Alexander Method for Deductive Databases [12]. However, as the structure of subsumption formulae and assertions are far more complex than in definite deductive databases, the adequate analogue is rather the Sequent Calculus (SC).

Intuitively, all the logical inference rules of SC (except from Weakening and Cut) are reversible rules, i.e. they can be processed backwards or forwards, depending on whether one is doing deductive reasoning (going from premisses to conclusions) or abductive reasoning (going from conclusions to premisses). The query rules correspond to the latter interpretation—they correspond to goal decomposition (as in logic programs) or formula decomposition (as in sequent proofs); the answer rules come from the former interpretation—they correspond to ordinary deduction. Actually, the query-answering calculus for description bases copes with that dual abduction/deduction interpretation, not for ordinary SC, but for the axiomatic calculus of description logics we have proposed in Section 2.

In order to define rigorously the notion of queries and answers wrt a description base, we have to extend the language of assertions.

Definition 2. Let DB= (TB, AB) be a DL description base, $\mathcal{O}$ the set of object names, and $\mathcal{V}$ a set of variable symbols. A **query assertion** in DB has the form c?(x) or r?(x,y), where x and y are members of $\mathcal{O} \cup \mathcal{V}$. An **answer assertion** in DB has the form c!(x) or r!(x,y), where x and y are members of $\mathcal{O} \cup \mathcal{V}$.

Note that new variable parameters, different from the object names originally asserted in DB, are introduced for querying as well as answering. In fact, one distinction between the weak and the strong calculus presented below is that the strong calculus will also use variables in answer assertions, whereas the weak one does not.

Also, we do not allow for assertions stating explicit equality or inequality assumptions between *object names* of $\mathcal{O}$. Instead, in conformity with the tradition of databases and DL, we adopt the Unique Name Assumption (UNA)—any two different objects names stand for two distinct individuals in every model. Equality or inequality assertions about *variables parameters* of $\mathcal{V}$ will be dealt with *internally* in the strong calculus, but there will be no explicit query about (in)equalities.

4.1 The Weak Intuitionistic Query-Answering Calculus

We will first specify a weak intuitionistic calculus dealing only with explicit objects, but not with implicit ones.

Triggering of TB clauses

infer:	$c_i(o)$, $1 \leq i \leq n$, $c_1 \sqcap ... \sqcap c_n \to c$	$\to c(o)$
query:	$c?(x)$, $c_1 \sqcap ... \sqcap c_n \to c$	$\Longrightarrow c_i?(x)$, $1 \leq i \leq n$
answer:	$c_i!(u)$, $1 \leq i \leq n$, $c_1 \sqcap ... \sqcap c_n \to c$	$\Longrightarrow c!(u)$
infer:	$c \sqcup c_i(o)$, $1 \leq i \leq n$, $c_1 \sqcap ... \sqcap c_n \to c$	$\to c(o)$
query:	$c?(x)$, $c_1 \sqcap ... \sqcap c_n \to c$	$\Longrightarrow (c_i \sqcup c)?(x)$, $1 \leq i \leq n$
answer:	$(c_i \sqcup c)!(u)$, $1 \leq i \leq n$, $c_1 \sqcap ... \sqcap c_n \to c$	$\Longrightarrow c!(u)$

plus the same rules for clauses with roles terms

Lookup in AB

infer:	$c(o)$	$\to c(o)$
answer:	$c?(x)$, $c(o)$, $o \in \mathcal{O}$	$\Longrightarrow c!(o)$
answer:	$r?(x,y)$, $r(o,o')$, $o, o' \in \mathcal{O}$	$\Longrightarrow r!(o,o')$

Conjunction elimination

infer:	$c_1 \sqcap c_2(o)$	$\rightleftharpoons c_1(o) \wedge c_2(o)$
query:	$(c_1 \sqcap c_2)?(x)$	$\Longrightarrow c_1?(x) \wedge c_2?(x)$
answer:	$c_1!(u)$, $c_1!(u)$	$\Longrightarrow (c_1 \sqcap c_2)!(u)$

Disjunction elimination (intuitionistic)

query:	$(c_1 \sqcup c_2)?(x)$	$\Longrightarrow c_1?(x)$
query:	$(c_1 \sqcup c_2)?(x)$	$\Longrightarrow c_2?(x)$
answer:	$c_1!(u)$	$\Longrightarrow (c_1 \sqcup c_2)!(u)$
answer:	$c_2!(u)$	$\Longrightarrow (c_1 \sqcup c_2)!(u)$

De-Skolemization: atleast introduction

infer:	$r(o,o_i)$, $c(o_i)$, $o_i \in \mathcal{O}, 1 \leq i \leq n$	$\to \geq nr{:}c(o)$
query:	$\geq nr{:}c?(x)$	$\Longrightarrow r?(x,y) \wedge c?(y)$
answer:	$\wedge_{i=1}^{i=n} r!(u,v_i) \wedge c!(v_i) \wedge dif(v_1,...,v_n)$	$\Longrightarrow \geq nr{:}c!(u)$

Propagation: all elimination

infer:	$\forall r{:}c(o)$, $r(o,o')$	$\to c(o')$
query:	$c?(x)$	$\Longrightarrow \forall r{:}c?(y) \wedge r?(y,x)$
answer:	$\forall r{:}c!(u)$, $r!(u,v)$	$\Longrightarrow c!(v)$

Fig. 1. The Weak Intuitionistic Query-Answering Calculus.

Definition 3. The **weak intuitionistic query-answering calculus** is defined by the rules listed in Fig. 1, where all the parameters are taken from $\mathcal{O}$.

We will write DB $\vdash_{WQA} \alpha$ if the query or answer assertion α is derivable in the weak calculus from a description base DB.

The limitation to reasoning with explicit objects in the weak calculus can be illustrated by considering a description base containing only the terminological clause $\exists r{:}c_1 \to \exists r{:}c_2$ and the assertion $\exists r{:}c_1(o)$. The query $\exists r{:}c_2?(o)$ will succeed, whereas the query $c_2?(x)$

will fail. This is due to the fact that there is no assertion about an explicit object belonging to c_2. The weak calculus thus lacks some *skolemization* rule for creating *skolem objects*, like the creative rule ∃-right in SC or the skolemization procedure in Resolution. For the same reason, the answer leader!(*mt,mary*) is not derivable in the weak calculus—there is no known answer to the query leader?(*mt,s*). The constraint leader!(*mt,s*) cannot be derived and hence the constraint $s = tom \lor s = dick \lor s = mary$ cannot be used in the proof.

Note that the avoidance of reasoning with implicit objects (i.e. object names not occurring within assertions of AB), even if only temporarily in the course of proofs, distinguishes the weak calculus from model theoretic *epistemic* calculi as proposed, for example, in [13, 19]. In these calculi the answer set of an epistemic query is defined as the intersection of the answer sets wrt arbitrary models. leader!(*mt,mary*) is an epistemic answer to leader?(*mt,x*) in the model theoretical sense, because in all the models of AB respecting TB *mary* satisfies invariantly the query. The derivation of this answer fails in our rather syntactic calculus, because the answer cannot be found using an "epistemic" procedure appealing only to explicitly known individuals.

4.2 The Strong Intuitionistic Query-Answering Caculus

In order to deal with implicit objects during answer derivations we need a stronger calculus:

Definition 4. The **strong intuitionistic query-answering calculus** is obtained from the weak intuitionistic query-answering calculus by:

1. extending the weak intuitionistic rules to apply to answer predicates having variable parameters (e.g. c!(x), where $x \in \mathcal{O} \cup \mathcal{V}$);
2. adding the rules shown in Fig. 2.

We will write DB $\vdash_{SQA} \alpha$ if the query or answer assertion α is derivable in the strong calculus from a description base DB.

Let us outline the subtle difference between the two kinds of answers dealt with by these new rules. c!(o), where o is a "known" object of $\mathcal{O}$, is a valid answer, whereas c!(x), where x is a fresh parameter created internally during the query-answering process, only is if x can be proved to be equal to some known object using equality reasoning.

Indeed, whenever the parameter introduction rules are allowed, equality reasoning comes into play, in order to manage possible equalities between skolem objects (which do not need to satisfy UNA). Let us consider the example, TB = $\{c_1 \rightarrow c_3, c_2 \rightarrow c_3\}$ and AB = $\{\exists r{:}c_1(o_1), c_2(o_2), r(o_1,o_2), \leq 1r{:}c_3(o_1)\}$. Then we will derive $c_1!(o_2)$. We could not obtain the same answer with the weak calculus, because no assertion about an explicit object being a c_1 is available for exploiting the constraint introduced by the **atmost** term.

Let us finally notice that the strong calculus has no introduction rules for **all** or **atmost** queries. The reason is because assertional bases lack implication ; implicative reasoning is thus performed only at the level of terminological inferences, by TB clauses triggering.

Skolemization: **atleast** *elimination*

infer:	$\geq nr{:}c(o)$	$\rightarrow$	$\wedge_{i=1}^{i=n}(r(o,x_i) \wedge c(x_i))$
			$\wedge\, dif(x_1,...,x_n)$, ($x_i$ new)
query:	$c?(y)$	$\Longrightarrow$	$\geq nr{:}c?(x)$
query:	$r?(x,y)$	$\Longrightarrow$	$\geq nr{:}c?(x)$
answer:	$\geq nr{:}c!(u)$	$\Longrightarrow$	$\wedge_{i=1}^{i=n}(r!(u,v_i) \wedge c!(v_i))$
			$\wedge\, dif(v_1,...,v_n)$, ($v_i$ new)

Cardinality reasoning : **atmost** *elimination*

infer: $\leq nr_1{:}c_1(u), c_1(v_i), r_1(u,v_i), 1 \leq i \leq n$,
$r_1(u,v_{n+1}), c_1 \sqcup c(v_{n+1})\, dif(v_1,...,v_{n+1})$ $\rightarrow c(v_{n+1})$

query: $c?(x)$ $\Longrightarrow \leq nr_1{:}c_1?(y)$
$\wedge\, (c_1 \sqcup c)?(x) \wedge r_1?(y,x)$

answer: $\leq nr_1{:}c_1!(u)), r_1!(u,v_i), c_1!(v_i), 1 \leq i \leq n$
$r_1!(u,v_{n+1}), (c_1 \sqcup c)!(v_{n+1}), dif(v_1,...,v_{n+1}) \Longrightarrow c!(v_{n+1})$

Negation reasoning

infer:	$\forall r{:}c(o), \neg c(o')$	$\rightarrow$	$\neg r(o,o')$
query:	$\neg r?(x,y)$	$\Longrightarrow$	$\forall r{:}c?(x) \wedge \neg c?(y)$
answer:	$\forall r{:}c!(u), \neg c!(v)$	$\Longrightarrow$	$\neg r!(u,v)$

infer: $\leq nr{:}c(u), c(v_i), r(u,v_i), 1 \leq i \leq n$,
$r(u,v_{n+1}), dif(v_1,...,v_{n+1})$ $\rightarrow \neg c(v_{n+1})$

query: $\neg c?(x)$ $\Longrightarrow \leq nr{:}c?(y) \wedge r?(y,x)$

answer: $\leq nr{:}c!(u), c!(v_i), r!(u,v_i), 1 \leq i \leq n$,
$r!(u,v_{n+1}), dif(v_1,...,v_{n+1})$ $\Longrightarrow \neg c!(v_{n+1})$

infer: $\leq nr{:}c(u), c(v_i), r(u,v_i), 1 \leq i \leq n$,
$c(v_{n+1}), dif(v_1,...,v_{n+1})$ $\rightarrow \neg r(u,v_{n+1})$

query: $\neg r?(x,y)$ $\Longrightarrow \leq nr{:}c?(y) \wedge c?(y)$

answer: $\leq nr{:}c!(u), c!(v_i), r!(u,v_i), 1 \leq i \leq n$,
$c!(v_{n+1}), dif(v_1,...,v_{n+1})$ $\Longrightarrow \neg r!(u,v_{n+1})$

Equality reasoning

	$o, o' \in \mathcal{O}$	$\Longrightarrow$	$o \neq o'$ (*Unique Name Axioms*)
	$x \neq x_1, x = x_1 \vee ... \vee x = x_n$	$\Longrightarrow$	$x = x_2 \vee ... \vee x = x_n$ (*identity axioms*)
infer:	*the SC substitution axioms*		
answer:	$c!(u), u = v$	$\Longrightarrow$	$c!(v)$
answer:	$r!(o,u), u = v$	$\Longrightarrow$	$r!(o,v)$
query:	$r!(u,b), b \notin \mathcal{O}$	$\Longrightarrow$	$\neg r?(u,x)$
answer:	$r!(u,b), \neg r!(u,v)$	$\Longrightarrow$	$b \neq v$
query:	$c!(b), b \notin \mathcal{O}$	$\Longrightarrow$	$\neg c?(x)$
answer:	$c!(v), \neg c!(u)$	$\Longrightarrow$	$u \neq v$
infer:	$\leq nr_1{:}c_1(u), c(v_i), r(u,v_i), 1 \leq i \leq n+1$	$\rightarrow$	$equ(v_1,...,v_{n+1})$
query:	$b \notin \mathcal{O}, c!(b), r!(u,b)$	$\Longrightarrow$	$\leq nr{:}c?(u)$
answer:	$\leq nr{:}c!(u), c!(v_i), r!(u,v_i), 1 \leq i \leq n+1$	$\Longrightarrow$	$equ(v_1,...,v_{n+1})$

Fig. 2. Additional rules of the Strong Intuitionistic Query-Answering Caculus.

Operator	left-rule	right-rule
Axiom/Cut-Rule	$\Gamma, \alpha \Rightarrow \Delta, \alpha$	$\dfrac{\Gamma_1, \alpha \Rightarrow \Delta_1 \; ; \; \Gamma_2 \Rightarrow \Delta_2, \alpha}{\Gamma_1, \Gamma_2 \Rightarrow \Delta_1, \Delta_2}$
Equality	$\Rightarrow x = x$	$x = y, \alpha(x) \Rightarrow \alpha(y)$
Intuitionistic Weakening	$\dfrac{\Gamma \Rightarrow \Delta}{\Gamma, \alpha \Rightarrow \Delta}$	$\dfrac{\Gamma \Rightarrow}{\Gamma \Rightarrow \alpha}$
Intuitionistic Contraction	$\dfrac{\Gamma, \alpha, \alpha \Rightarrow \Delta}{\Gamma, \alpha \Rightarrow \Delta}$	
Intuitionistic $\vee$	$\dfrac{\Gamma, \alpha \Rightarrow \Delta \; ; \; \Gamma, \beta \Rightarrow \Delta}{\Gamma, \alpha \vee \beta \Rightarrow \Delta}$	$\dfrac{\Gamma \Rightarrow \Delta, \alpha}{\Gamma \Rightarrow \Delta, \alpha \vee \beta}$
$\wedge$	$\dfrac{\Gamma, \alpha, \beta \Rightarrow \Delta}{\Gamma, \alpha \wedge \beta \Rightarrow \Delta}$	$\dfrac{\Gamma \Rightarrow \Delta, \alpha \; , \; \Gamma \Rightarrow \Delta, \beta}{\Gamma \Rightarrow \Delta, \alpha \wedge \beta}$
$\rightarrow$	$\dfrac{\Gamma, \beta \Rightarrow \Delta \; ; \; \Gamma \Rightarrow \alpha, \Delta}{\Gamma, \alpha \rightarrow \beta \Rightarrow \Delta}$	$\dfrac{\Gamma, \alpha \Rightarrow \Delta, \beta}{\Gamma \Rightarrow \Delta, \alpha \rightarrow \beta}$
Intuitionistic $\neg$	$\dfrac{\Gamma, \alpha \Rightarrow}{\Gamma \Rightarrow \neg \alpha}$	$\dfrac{\Gamma \Rightarrow \alpha}{\Gamma, \neg \alpha \Rightarrow \Delta}$
$\forall$	$\dfrac{\Gamma, A(t) \Rightarrow \Delta}{\Gamma, \forall x A(x) \Rightarrow \Delta}$	$\dfrac{\Gamma \Rightarrow \Delta, A(a)}{\Gamma \Rightarrow \Delta, \forall x A(x)}$
$\exists$	$\dfrac{\Gamma, A(a) \Rightarrow \Delta}{\Gamma, \exists x A(x) \Rightarrow \Delta}$	$\dfrac{\Gamma \Rightarrow \Delta, \exists x A(x)}{\Gamma \Rightarrow \Delta, A(t)}$

Fig. 3. Intuitionistic Sequent Calculus [10]. (a stands for an *eigen-parameter*, i.e. a parameter not occurring in the lower sequent; t stands for a term occurring in the lower sequent.)

5 Completeness

We will now give an axiomatic and a least fixed point semantics and show completeness of our intuitionistic calculi wrt these semantics. For the axiomatic semantics consider the Intuitionistic Sequent Calculus [10, 23] shown in Fig. 3. It is obtained from the classical Sequent Calculus [8] by replacing the classical $\neg$, $\vee$, weakening, and contraction rules by their intuitionistic versions.

Definition 5. We will use $\vdash_{SISC}$ to refer to derivability in the Intuitionistic Sequent Calculus shown in Fig. 3; we will use $\vdash_{WISC}$ to refer to derivability in the Calculus obtained by dropping the rules $\exists$-left and $\forall$-right.

Proposition 6. *The weak intuitionistic query-answering calculus is sound and complete wrt derivability in the Intuitionistic Sequent Calculus without $\exists$-left and $\forall$-right:*

$$DB, c?(x) \vdash_{WQA} c!(o) \text{ iff } \vdash_{WISC} \overline{TB}, \overline{AB}, UNA \Rightarrow \overline{c}(\overline{o})$$

$$DB, r?(x,y) \vdash_{WQA} r!(o_1, o_2) \text{ iff } \vdash_{WISC} \overline{TB}, \overline{AB}, UNA \Rightarrow \overline{r}(\overline{o_1}, \overline{o_2})$$

The strong intuitionistic query-answering calculus is sound and complete wrt derivability in the Intuitionistic Sequent Calculus:

$$DB, c?(x) \vdash_{SQA} c!(o) \text{ iff } \vdash_{SISC} \overline{TB}, \overline{AB}, UNA \Rightarrow \overline{c}(\overline{o})$$

$DB, r?(x,y) \vdash_{SQA} r!(o_1,o_2)$ iff $\vdash_{SISC} \overline{TB}, \overline{AB}, UNA \Rightarrow \overline{r}(\overline{o_1},\overline{o_2})$

Proof (Sketch). The proof exploits the translation into sequents according to the FOL translation of DL. Actually it consists in analyzing the SC proofs of the sequents of the form:

$$(*) \quad \overline{TB}, \overline{AB}, UNA \Rightarrow \alpha$$

Typically all the rules are obtained by examining the different possibilities of eliminating the logical connectors and quantifiers from the FOL translations of concepts or roles occurring in these sequents.

The only exception is when $\alpha = c(a)$ is a universally quantified assertion like $\leq nr{:}c(a)$ or $\forall r{:}c(a)$. The key part of the proof is to show that no query rule is necessary in this case to eliminate the universal quantifiers in $c(a)$ and that answering just proceeds by triggering one appropriate terminological clause. Formally, one shows by structural induction about the antecedent terms of (*) that whenever there is an intuitionistic sequent proof of (*), there must be concept assertions $\{c_i(a), 1 \leq i \leq n\}$ such that the following sequents are provable.

$$\overline{TB}, \overline{AB}, UNA \Rightarrow c_i(a), \ 1 \leq i \leq n$$
$$\overline{TB}, c_1(a), ..., c_n(a) \Rightarrow c(a)$$

Proving the last sequent is exactly proving the terminological inference $TB \vdash c_1 \sqcap ... \sqcap c_n \sqsubseteq c$.

For the weak calculus we can also give a least fixed-point semantics:

Definition 7. Let $DB = (TB,AB)$ be a DL description base and α a DL assertion. The weak intuitionistic consequence operator Φ_{TB} is defined as follows. For any set of assertions S, $\Phi_{TB}(S)$ is defined as the smallest set such that:

1. $S \subseteq \Phi_{TB}(S)$
2. If $\vdash_{WISC} \overline{S}, UNA \Rightarrow \overline{\alpha}$ then $\alpha \in \Phi_{TB}(S)$
3. If $TB \models_{IT} c_1 \sqcap ... \sqcap c_n \to c$ and $c_1(o), ..., c_n(o) \in S$, then $c(o) \in \Phi_{TB}(S)$

As usual we rely on Tarski's Theorem and the fact that Φ_{TB} is a monotonic continuous operator on the power set of assertions ordered by set inclusion [22]. Therefore, $\Phi_{TB}^{\omega}(AB) = \bigcup_{n=1}^{n=\infty} \Phi_{TB}^n(AB)$.

Proposition 8. *Let $DB = (TB,AB)$ be a DL description base.*

$$DB, c?(x) \vdash_{WQA} c!(o) \text{ iff } c(o) \in \Phi_{TB}^{\omega}(AB)$$

Proof (sketch). Condition (2) in the definition of the consequence operator is taken care of by the structural rules of the weak calculus. Condition (3) is taken care of by the triggering rules.

In order to define a similar least fixed point semantics for the strong calculus we would have to include skolemization rules in the definition of the consequence operator and to allow skolem objects in the triggering conditions of terminological clauses. The resulting semantics, however, hardly fits the usual standards of least fixed point semantics in Deductive Databases or Logic Programming, because it is far less driven by the syntax of rules and facts than the weak version. The weak calculus is thus closer related to the general paradigm of Deductive Databases and Logic Programming than the strong calculus.

6 Conclusion

In this paper, we have given a proof-theoretical characterization of Query Answering in description bases. We have here basically axiomatized assertional reasoning, by means of explicit query and answer rules. The link with terminological reasoning is performed by the triggering of terminological clauses. This is in some analogy with the least-fixed point semantics of Deductive Databases.

On the other hand, the restriction to intuitionistic inferences relies on some analogy with Logic Programming. Indeed restricting to intuitionistic disjunctions is necessary to get some linear control of the query derivation process as in Logic Programming.

The intuitionistic calculi presented in this paper allow an efficient answering of open queries due to the *explicit control of the proof strategy* given by the existence of query and answer rules. This contrasts to the tableaux-based technique used for example in [7], which is mainly intended for answering closed queries and can only be inefficiently applied to open queries[6].

Since we have followed the standard strategy of Deductive Databases, an extension of our approach to recursive description bases seems possible by integrating methods similar to the ones developed in these fields (e.g. [3, 12]).

Let us now conclude by some comments about complexity. Even with intuitionistic semantics, some complexity still arises from left-implication elimination as performed in the secong triggering rule for TB clauses in Fig. 1, because arbitrarily long disjunctive queries potentially arise. This source of complexity can be eliminated by restricting to languages without explicit disjunctions and without full negation[7], as is usually done in implemented systems like BACK V5 [11]. Then the rules involving disjunctions in Fig. 1 and Fig. 2 are no more needed.

The Strong Calculus remains intrinsically complex because of Skolemization and Equality reasoning. Finally, in the Weak Intuitionistic Calculus, some unavoidable source of complexity remains because of the need to perform pure terminological inferences, which presupposes some specific automated system. To increase efficiency, at the expense of completeness, one should take the approach of systems based on *flexible inference strategies*, where some inferences only are performed [17].

---

[6] The query $c?(x)$ has to be mapped to queries $c?(o_i)$ for all $o_i \in \mathcal{O}$.

[7] Negation is restricted to so-called "primitive terms" [15].

References

1. K. Apt, H. Blair, A. Walker, "Towards a Theory of Declarative Knowledge", in *Workshop on Foundations of Deductive Databases and Logic Programming*, Washington, 1986
2. F. Baader, B. Hollunder, "Embedding Defaults into Terminological Knowledge Representation Formalisms", in *KR-92*, 306–317
3. F. Bancilhon, D. Maier, Y. Sagic, J. Ullman, "Magic Sets and other Strange Ways of Implementing Logic Programs", in *Principles of Database Systems (PODS-86)*, 1–15, 1986
4. R.J. Brachman, J.G. Schmolze, "An Overview of the KL-ONE Knowledge Representation System" *Cognitive Science* 9, 171–216, 1985
5. R. Brachman, D.L. McGuiness, P.F. Patel-Schneider, L. Alperin Resnick, A. Borgida, "Living with CLASSIC: When and How to Use a KL-ONE-like Language", in ,J. Sowa (Ed.), *Principles of Semantic Networks: Explorations in the Representation of Knowledge*, San Mateo: Morgan Kaufmann, 401–456, 1991
6. F.M. Donini, M. Lenzerini, D. Nardi, W. Nutt, "Tractable Concept Languages" *IJCAI-91*, 458–463
7. F.M. Donini, M. Lenzerini, D. Nardi, A. Schaerf, W. Nutt, "Adding Epistemic Operators to Concept Languages", in *KR-92*, 342–353
8. J. Gallier, *Logic for Computer Science; Foundations of Automatic Theorem Proving*, New York: Harper and Row, 1986
9. J.Y. Girard, "Linear Logic", *Theoretical Computer Science*, 50, 1987
10. J.Y. Girard, Y. Lafont, P. Taylor, *Proofs and Types*, Cambridge: Cambridge University Press, 1989
11. T. Hoppe, C. Kindermann, J.J. Quantz, A. Schmiedel, M. Fischer, BACK V5 *Tutorial & Manual*, KIT Report 100, Technische Universität Berlin, 1993
12. J.M. Kerisit, R. Lescoeur, J. Rohmer, "The Alexander Method: a Technique for the Processing of Recursive Axioms in Deductive Databases", *New Generation Computing* 3(4), 1986
13. H.J. Levesque, "Foundations of a Functional Approach to Knowledge Representation", *Artificial Intelligence* 23, 155–212, 1984
14. J.W. Lloyd, *Foundations of Logic Programming*, Berlin: Springer, 1987
15. B. Nebel, *Reasoning and Revision in Hybrid Representation Systems*, Lecture Notes in Artificial Intelligence 422, Berlin: Springer, 1990
16. T. Przyymusinski, "On the Declarative and Procedural Semantics of Logic Programs", *Journal of Automated Reasoning* 5, 167–205, 1989
17. J.J. Quantz, G. Dunker, V. Royer, "Flexible Inference Strategies for DL Systems", to appear in *International Workshop on Description Logics*, Bonn, 1994
18. J.J. Quantz, B. Schmitz, "Knowledge-Based Disambiguation for Machine Translation", *Minds and Machines* 4, 39–57, 1994
19. R. Reiter, "On Asking What a Database Knows", in J.W. LLoyd (Ed.), *Computational Logic*, Berlin: Springer, 96–113, 1990
20. V. Royer, J.J. Quantz, "Deriving Inference Rules for Terminological Logics", in , D. Pearce, G. Wagner (eds), *Logics in AI, Proceedings of JELIA'92*, Berlin: Springer, 84–105, 1992
21. V. Royer, J.J. Quantz, *Deriving Inference Rules for Description Logics: a Rewriting Approach into Sequent Calculi*, KIT Report 112, Technische Universität Berlin, 1993
22. J.E. Stoy, *Denotational Semantics: The Scott–Strachey Approach to Programming Language Theory*, Cambridge: MIT Press, 1977
23. G. Sundholm, "Systems of Deduction", in D. Gabbay, F. Guenthner (eds), *Handbook of Philosophical Logic, Vol. I*, Dordrecht: Reidel, 133–188, 1983

Deductive Composition of Astronomical Software from Subroutine Libraries

Mark Stickel[1], Richard Waldinger[1],
Michael Lowry[2], Thomas Pressburger[2], Ian Underwood[2]

[1] Artificial Intelligence Center
SRI International
Menlo Park, CA 94025
{stickel,waldinger}@ai.sri.com

[2] Artificial Intelligence Research Branch
Recom Technologies
NASA Ames Research Center
Moffett Field, CA 94035
{lowry,pressburger,underwood}@ptolemy.arc.nasa.gov

Abstract. Automated deduction techniques are being used in a system called Amphion to derive, from graphical specifications, programs composed from a subroutine library. The system has been applied to construct software for the planning and analysis of interplanetary missions. The library for that application is a collection of subroutines written in FORTRAN-77 at JPL to perform computations in solar-system kinematics. An application domain theory has been developed that describes the procedures in a portion of the library, as well as some basic properties of solar-system astronomy, in the form of first-order axioms.

Specifications are elicited from the user through a menu-driven graphical user interface; space scientists have found the graphical notation congenial. The specification is translated into a theorem, which is proved constructively in the astronomical domain theory by an automated theorem prover, SNARK. An applicative program is extracted from the proof and converted to FORTRAN-77. By the method of its construction, the program is guaranteed to meet the given specification and requires no further verification, provided, of course, that the specification, domain theory, and system itself are correct.

Amphion has successfully constructed more than a hundred programs to solve problems, formulated at NASA Ames, JPL, and Stanford, which involve typical computations involving the sun, planets, moons, and spacecraft. The system is currently being alpha tested at JPL.

1 Introduction

Automatic deductive program synthesis has been studied for many years but has never been used in practice. By restricting our attention to the construction of programs composed from subroutine libraries, rather than the primitive instructions of a programming language, and by adapting domain-specific control strategies, we have applied deductive methods to construct useful software.

Subroutine Libraries

Subroutine libraries are one of the most prevalent forms of software reuse, particularly within the scientific programming community. However, end users often do not make effective use of libraries. Sometimes this happens because the subroutines are not adequately documented. But even when excellent documentation is provided, users often have neither the time nor the inclination to familiarize themselves with it. In either case, the result is that most users lack the expertise to properly identify and compose the routines appropriate to their application. In domains with mature subroutine libraries, one can greatly improve the productivity and quality of software engineering by automating the effective use of those libraries.

Subroutines are commonly accessed by indexing key words in their documentation, a very approximate method. In the work of Rollins and Wing [RW 91], logic programming techniques are invoked to retrieve appropriate subroutines, according to their specifications, but composing them is left up to the user. In this work, deductive methods—that is, methods of automated reasoning or theorem proving—are applied to the composition of subroutines into software. In that sense it most closely resembles the work of Tyugu and his associates [Tyu 88], in which software is also composed from subroutine libraries, to meet specifications expressed in intuitionistic propositional logic.

Although deductive methods are independent of the application domain, we discuss their application to the construction of software for performing computations in solar-system astronomy. Such computations are necessary in the planning and data analysis for interplanetary scientific missions. For example, observing the location of a moon of a nearby planet is often the best way of determining the position of the observing spacecraft.

Amphion

The Intelligent Software Project of the Artificial Intelligence Research Branch at NASA Ames, led by Michael Lowry, has been developing a system called Amphion[3] to automate the composition of software from subroutine libraries. Software requirements are specified in a graphical notation. An interactive interface, which is domain-independent but employs the vocabulary of the domain,

---

[3] Amphion built a wall around Thebes by charming the stones into place with a magic lyre.

presents the user with a menu of alternatives and elicits the specification gradually. The user need not know the contents of the library, the syntax of the specification language, or the target programming language.

The graphical specification is translated automatically into a first-order-logic theorem, and a program is developed from the logical form of the specification using SRI's automated deduction system SNARK, which has been implemented by Mark Stickel. The resulting program is subjected to common-subexpression elimination and translated into FORTRAN-77. The translation package invokes Refine™ transformations from Kestrel's KIDS system [Smi 90]. It would be a relatively small change to produce a final program in a language other than FORTRAN.

SPICE

Amphion's first application domain is software for planning and interpreting space-science observations. The software is based on SPICELIB, a library of procedures for solar-system geometry. These routines, written in FORTRAN-77 at the Navigation Ancillary Information Facility (NAIF) at JPL, perform basic computations involving the sun, planets, moons, and spacecraft. Various systems of time measurement (e.g., ephemeris time, which is used in astronomical tables, and spacecraft clock time) and multiple frames of reference come into play. Light is not assumed to travel instantaneously across astronomical distances.

The NAIF library procedures are used by astronomers and researchers as primitives to build more complex software. The subroutines embody considerable expertise and cannot easily be recreated. Although the routines are well documented, users seem reluctant to invest the time and effort to learn about them. They frequently attempt to reimplement routines that already exist in SPICELIB because they did not find them in the documentation, or if they are sufficiently influential, they prevail on the authors of the library to retrieve the appropriate routines and compose them into the required software.

2 Deductive Component

The emphasis of this paper is on the role of SNARK, the deductive subsystem, in Amphion. Other papers will focus on the astronomical aspects of the system and on the graphical interface.

Deductive Approach

A program is developed from the logical form of the specification by a deductive approach, which is based on work of Zohar Manna, of Stanford University, and Richard Waldinger [MW 92]. We prove a mathematical theorem that expresses the existence of an output that meets the specified conditions. The graphical specification language corresponds to only a subset of predicate logic, but in principle knowledgeable users can introduce logical specifications directly.

The proof is conducted in a classical logic but is restricted to be constructive—in other words, in proving the existence of the required output, we are forced to indicate a method for finding it. That method becomes the basis for a program to compute the output, which may be extracted from the proof. This program is guaranteed, by the way it was constructed, to meet the specification—it requires no further verification.

The structure of the program reflects the proof from which it was extracted. If the proof relies on reasoning by cases (e.g., by application of the resolution rule [Rob 65]), the resulting program may contain a conditional expression. If the proof depends on the mathematical induction principle, the program may invoke recursion or other repetitive constructs.

The theorem is proved valid in an *application domain theory* that provides the knowledge on which the software depends. The specifications of the available subroutines, the constructs of the specification language, and properties of the application domain are expressed by axioms in the domain theory. The application domain theory also determines the options offered to the user by the graphical interface.

Program synthesis differs in its technical emphasis when its output is expressed in terms of subroutine calls rather than the primitives of a programming language. When most of the recursive and iterative constructs are embedded in subroutines, the major technical challenge is to effectively decompose the problem and glue together subroutines. While general program synthesis imposes severe demands on a deductive system, the theorems that arise when software is composed from a subroutine library appear to be within the range of existing deductive technology.

SNARK

To automate a deductive approach requires an automated deduction system, or theorem prover. SNARK is especially suitable for program synthesis and other applications in artificial intelligence and software engineering. SNARK is invoked as a subsystem of Amphion, but it can also be used independently or as a component of other systems.

The current implementation of SNARK, in COMMON LISP, includes the resolution [Rob 65] and paramodulation [WR 69] rules for handling the constructs of first-order logic with equality, like McCune's OTTER [McC 90]. It also will employ the principle of mathematical induction, like Boyer and Moore's NQTHM [BM 88]. Proofs are developed within Manna and Waldinger's *deductive tableau* framework [MW 93] and can be restricted to be constructive so that programs can be extracted. Clause form is optional—if the user prefers, formulas may employ a full set of logical connectives in arbitrary form.

It is intended that the SNARK user will be able to introduce new inference rules, but in the current implementation the user chooses among a fixed set of rules. An indexing mechanism allows the system to retrieve from its memory only those formulas that are syntactically relevant.

SNARK (like OTTER) is agenda-driven—it draws conclusions from a formula when that formula reaches the top of its agenda. The user does have the ability to influence the strategy adopted by the system, for example, by providing the function used to order the agenda. Although interactive handles are being attached to SNARK, the system is fundamentally automatic.

3 The Astronomical Domain

For the astronomical application, the specifications of a portion of the subroutines of the SPICELIB library are represented by axioms in the application domain theory. Other axioms describe properties of the specification constructs and the geometry and space kinematics on which the construction of the software depends. At this moment the domain theory consists of more than 200 axioms, all of which are available when we attempt to prove the specification theorem. It is beside the point of this paper to describe the domain theory (largely the work of Lowry and Pressburger) in any detail. But let us present enough of the theory to suggest its contents.

The Astronomical Domain Theory

A fundamental entity in the domain theory is a space-time location (sometimes called an event), a position in space at a certain time; for two events to be identical, they must correspond to the same position and time.

The relation *lightlike?*(e_1, e_2) holds if a photon could leave the position corresponding to event e_1, at the time corresponding to that event, and arrive at the position and time corresponding to event e_2; the symbol *lightlike?* is a specification construct, not a subroutine in the library.

The function *ephemeris-object-and-time-to-event* yields an event corresponding to the position of a given astronomical object (e.g., a planet or spacecraft) at a given time; this is also a specification concept. Objects and times are abstract entities, independent of any representation system for designating astronomical objects or units for measuring time.

The specification function *a-sent*(o,d,ta) computes the time a photon must leave the origin object o in order to arrive at the destination object d at time ta. This function is defined in part by the axiom *lightlike?-of-a-sent*:

```
(all (o d ta)
     (lightlike? (ephemeris-object-and-time-to-event o
                    (a-sent o d ta))
                 (ephemeris-object-and-time-to-event d ta)))
```

(The axioms and theorems are written in LISP notation, e.g., (a-sent o d ta) instead of *a-sent*(o,d,ta).) In other words, a photon could leave object o at time (a-sent o d ta) and arrive at object d at time ta.

The specification constructs deal mainly with abstract entities. But each abstract entity corresponds to one or more concrete entities, which depend on a particular representation scheme or system of units. In particular, an abstract astronomical body such as Jupiter is assigned a NAIF library symbol, called its NAIF id; the NAIF id of Jupiter is 599. Each abstract time corresponds to a concrete ephemeris time and to a concrete spacecraft clock time. The function $abs(fn, c)$ is used to denote the abstract entity corresponding to the concrete entity c; here, fn is the abstraction function that maps concrete entities into abstract ones. (For technical reasons, abstraction functions are reified; that is, they are denoted by constants and terms rather than by function symbols.) For example, (abs ephemeris-time-to-time et) stands for the abstract time corresponding to the ephemeris time et, and (abs naif-id-to-body 599) stands for Jupiter.

The subroutines in the library apply to concrete entities, not abstractions. For example, the subroutine (sent onid dnid eta) is analogous to the abstract function (a-sent o d t) but applies to concrete NAIF ids for origin and destination bodies, onid and dnid, respectively, and a concrete arrival time eta in ephemeris-time units, rather than their abstract counterparts. The precise relationship between the specification function a-sent and the subroutine sent is expressed by the following *a-sent-to-sent* axiom:

```
(all (onid dnid eta)
     (= (a-sent (abs naif-id-to-body onid)
                (abs naif-id-to-body dnid)
                (abs ephemeris-time-to-time eta))
        (abs ephemeris-time-to-time (sent onid dnid eta)))).
```

In other words, the result of first translating the concrete entities into abstractions and then computing a-sent is the same as the result of computing sent on the concrete entities and then translating to abstract time.

The Sample Problems

With the assistance of astronomers at JPL and Stanford, and based on his own experience, Underwood assembled a collection of fifteen sample problems representative of what might be requested of a NAIF consultant. The problems require solar-system computations typical of those required for scientific missions; some were from software that had been developed for the Hubble Space Telescope Science Institute. Although a NAIF expert would be able to construct programs to solve these problems in less than half an hour, NAIF experts are in short supply; a programmer unfamiliar with the NAIF library might require several days to learn its contents before composing the software.

Amphion was able to construct programs for all fifteen sample problems completely automatically, without user interaction. Once the specifications were elicited from the user, the system required less than three minutes to construct each program.

Let us look at one of the sample problems.

Shadow of Io

The first problem we considered involved determining the location of the shadow cast on Jupiter by its moon Io, as observed at a given time on Voyager 2 (Figure 1). The point *pi* indicates the shadow.

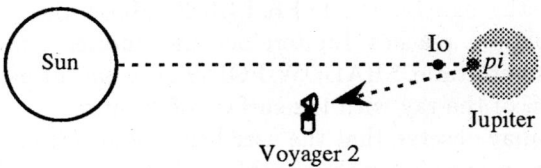

Fig. 1. Where is the shadow of Io on Jupiter?

The corresponding graphical specification (Figure 2) may appear confusing, but would be much clearer if we could show the step-by-step interaction between user and system. After a one-hour tutorial, a novice user may require a half hour to construct such a specification; an experienced user can do it in a few minutes.

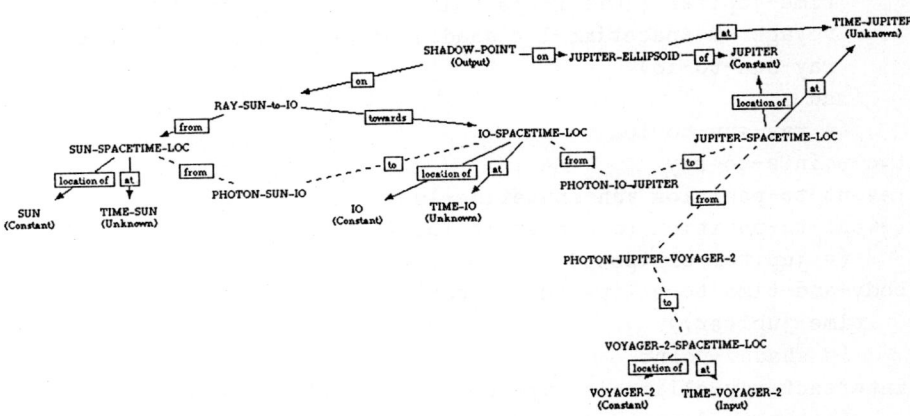

Fig. 2. Shadow of Io Graphical Specification

In Figure 2, PHOTON-SUN-IO designates a photon that passes from the sun at a certain time and reaches Io at another; the purpose of speaking about photons is to specify times. Similarly PHOTON-IO-JUPITER is perhaps the same photon as it leaves Io and reaches Jupiter, and PHOTON-JUPITER-VOYAGER-2 is the photon as it leaves Jupiter and arrives at Voyager 2. The input to the program (indicated at the lower right) is the time that photon reaches Voyager 2.

RAY-SUN-to-IO is a ray (that is, a half-infinite line) that originates at the sun and passes through Io. JUPITER-ELLIPSOID is the surface of Jupiter at the time the photon reaches Jupiter; because Jupiter rotates and moves, its surface changes with time. SHADOW-POINT, the output of the program, is the first intersection of the ray with the surface of Jupiter.

The reader may observe that the user has chosen certain simplifications and approximations in specifying this problem. For instance, the sun and Io are regarded as points, and though Jupiter is sometimes regarded as a spheroid, PHOTON-IO-JUPITER arrives at the center of Jupiter, not its surface. The decision as to which simplifications may be made is left up to the user.

The theorem obtained from this specification is given in Figure 3.

```
(all (time-voyager-2-c)
     (find (shadow-point-c)
    (exists
     (time-sun sun-spacetime-loc time-io io-spacetime-loc
        time-jupiter jupiter-spacetime-loc time-voyager-2
        voyager-2-spacetime-loc shadow-point jupiter-ellipsoid
        ray-sun-to-io)
     (and
      (= ray-sun-to-io
(two-points-to-ray
 (event-to-position sun-spacetime-loc)
 (event-to-position io-spacetime-loc)))
      (= jupiter-ellipsoid
(body-and-time-to-ellipsoid jupiter
     time-jupiter))
      (= shadow-point
(intersect-ray-ellipsoid ray-sun-to-io jupiter-ellipsoid))
      (lightlike? jupiter-spacetime-loc voyager-2-spacetime-loc)
      (lightlike? io-spacetime-loc jupiter-spacetime-loc)
      (lightlike? sun-spacetime-loc io-spacetime-loc)
      (= voyager-2-spacetime-loc
(ephemeris-object-and-time-to-event voyager-2 time-voyager-2))
      (= jupiter-spacetime-loc
(ephemeris-object-and-time-to-event jupiter time-jupiter))
      (= io-spacetime-loc
(ephemeris-object-and-time-to-event io time-io))
      (= sun-spacetime-loc
(ephemeris-object-and-time-to-event sun time-sun))
```

```
        (= shadow-point (abs (coords-to-point j2000) shadow-point-c))
        (= time-voyager-2
(abs ephemeris-time-to-time time-voyager-2-c))))))
```

Figure 3: Shadow of Io Theorem

The quantifier **find** is a constructive version of the existential quantifier **exists**; in proving the existence of **shadow-point-c** (which corresponds to SHADOW-POINT in the graphical specification), the system is forced to indicate a method for finding it.

Although this and the other theorems required for the astronomical application are not mathematically deep, some of the authors of this paper will confess to being unable to prove them from the axioms by hand. SNARK required about 40 seconds (on a Sun 670MP) to prove this one. The program extracted from the proof, as translated by Amphion into FORTRAN-77, is given in Figure 4.

```
        SUBROUTINE SHADOW ( TIMEVO, SHADOW )

C       Input Parameters
        DOUBLE PRECISION TIMEVO

C       Output Parameters
        DOUBLE PRECISION SHADOW ( 3 )

C       Function Declarations
        DOUBLE PRECISION SENT

C       Parameter Declarations
        INTEGER JUPITE
        PARAMETER (JUPITE = 599)
        INTEGER VOYGR2
        PARAMETER (VOYGR2 = -32)
        INTEGER SUN
        PARAMETER (SUN = 10)
        INTEGER IO
        PARAMETER (IO = 501)

C       Variable Declarations
        DOUBLE PRECISION RADJUP ( 3 )
        DOUBLE PRECISION TJUPIT
        DOUBLE PRECISION PJUPIT ( 3 )
        DOUBLE PRECISION TIO
        DOUBLE PRECISION MJUPIT ( 3, 3 )
```

```
      DOUBLE PRECISION PIO ( 3 )
      DOUBLE PRECISION TSUN
      DOUBLE PRECISION PSUN ( 3 )
      DOUBLE PRECISION DPSPI ( 3 )
      DOUBLE PRECISION DPJPS ( 3 )
      DOUBLE PRECISION XDPSPI ( 3 )
      DOUBLE PRECISION XDPJPS ( 3 )
      DOUBLE PRECISION P ( 3 )
      DOUBLE PRECISION DPJUPP ( 3 )

C     Dummy Variable Declarations
      INTEGER DMY0
      DOUBLE PRECISION DMY20 ( 3 )
      DOUBLE PRECISION DMY30 ( 3 )
      DOUBLE PRECISION DMY40 ( 3 )
      LOGICAL DMY90

      CALL BODVAR ( JUPITE, 'RADII', DMY0, RADJUP )
      TJUPIT = SENT ( JUPITE, VOYGR2, TIMEV0 )
      CALL FINDPV ( JUPITE, TJUPIT, PJUPIT, DMY20 )
      CALL BODMAT ( JUPITE, TJUPIT, MJUPIT )
      TIO = SENT ( IO, JUPITE, TJUPIT )
      CALL FINDPV ( IO, TIO, PIO, DMY30 )
      TSUN = SENT ( SUN, IO, TIO )
      CALL FINDPV ( SUN, TSUN, PSUN, DMY40 )
      CALL VSUB ( PIO, PSUN, DPSPI )
      CALL VSUB ( PSUN, PJUPIT, DPJPS )
      CALL MXV ( MJUPIT, DPSPI, XDPSPI )
      CALL MXV ( MJUPIT, DPJPS, XDPJPS )
      CALL SURFPT ( XDPJPS, XDPSPI, RADJUP ( 1 ), RADJUP ( 2 ),
     .RADJUP ( 3 ), P, DMY90 )
      CALL VSUB ( P, PJUPIT, DPJUPP )
      CALL MTXV ( MJUPIT, DPJUPP, SHADOW )

      END
```

Figure 4: Shadow of Io Program

Again there is little point to reading the entire program; we printed it to emphasize the difference between the program and its specification. After the long sequence of declarations and initializations, the program invokes the SPICELIB procedure **bodvar**, which computes the radii of Jupiter; because the surface of Jupiter is an ellipsoid, it has three radii, which are stored in an array. Then the library function **sent** computes the time **tjupit** a photon must have left Jupiter to reach Voyager 2 at input time **timev0**. The procedures **findpv** and **bodmat** then compute the position and the orientation of Jupiter at time **tjupit**. The orientation of Jupiter is represented by a three-by-three matrix of double-precision numbers. And so on.

In short, the specification deals with abstract entities, such as planets, times, and ellipsoids; the program deals with integers, double-precision numbers, and matrices.

4 Strategic Considerations

We do not provide a systematic description of SNARK here, but we do describe some of the heuristic features that SNARK employed to solve the astronomical problems.

Recursive Path Ordering

SNARK employs term rewriting and the paramodulation rule [WR 69] for reasoning about equality. It has been found possible to avoid replacing one term with another if the second term is greater than the first with respect to a certain kind of ordering, a *recursive-path ordering* [Der 82]. The recursive-path orderings are syntactic relations defined on the terms of our language. SNARK allows the user to declare a recursive-path ordering before beginning a proof. The user provides an ordering on the constants and function symbols of the language, and that determines a corresponding ordering on the terms, which is used to control the paramodulation rule. It has been established [HR 91] that the recursive-path-ordering strategy is complete for first-order logic with equality. If a sentence has a proof, it can be proved with the strategy, regardless of the choice of ordering.

SNARK's success in this domain depends on its use of the recursive-path-ordering strategy and on the choice of a particular ordering. Indeed, there are examples in which SNARK found a proof in less than a minute with a plausible ordering, but failed to find a proof in a reasonable time if that ordering was reversed or if ordering information was omitted altogether.

We found that a good heuristic for ordering the terms was, roughly speaking, to direct SNARK to replace abstract, noncomputable symbols with concrete, computable ones, which could appear in the final program extracted from the proof. With little effort, it was possible to declare an ordering that would enable SNARK to construct a program. A single ordering sufficed for all the problems in the astronomical domain. In general, we do not expect Amphion users to have to supply a recursive path ordering—that is done when the application domain theory is formulated.

The SPICE Agenda-Ordering Function

We have remarked that SNARK is an agenda-driven theorem prover. When it infers a new formula, it places it on an agenda, a list of formulas, to wait its turn to be processed. A formula is not processed until it reaches the head of the agenda; then it is removed from the agenda and all its immediate consequences are added.

The place at which a new formula is added to the agenda is determined by the *agenda-ordering function*. Although a default agenda-ordering function is provided with the system, the SNARK user may choose another or provide a new one, written in COMMON LISP. One of the ways SNARK has been specialized to the astronomical domain is with a new SPICE agenda-ordering function, written by Pressburger. This strategy gives special attention to goals with literals containing the predicate symbol `lightlike?` for which one of the arguments is ground (variable-free) and the other contains a variable; there are axioms in the domain theory, such as the axiom *lightlike?-of-a-sent* given previously, that are capable of solving any such literals. To a lesser extent, the strategy favors goals with fewer abstract function symbols. The effect of this strategy is to first determine the space-time locations of all the bodies in the problem, and then to replace all the abstract function symbols with concrete ones, which correspond to SPICELIB routines.

The choice of agenda-ordering function can be critical. One problem we have encountered requires less than three minutes with the SPICE agenda ordering but more than an hour with the SNARK default agenda ordering. All the astronomical problems were solved with this same SPICE agenda ordering; we do not expect Amphion users to have to change this ordering.

The Set-of-Support Strategy

The mathematical applications on which theorem provers are commonly tested require relatively deep proofs in theories with few axioms. In contrast, the astronomical domain, like most software-engineering applications, requires us to find mathematically less sophisticated proofs in theories with a large number of axioms, which represent the subject knowledge of the domain. For such a problem, it is appropriate to invoke the *set-of-support* strategy [WRC 65] to focus attention on the goal—the theorem to be proved—at the expense of the axioms. This strategy requires that every formula we infer be descended from the goal. Otherwise, with so many axioms, it is hard to decide in advance which of them are relevant to the proof. In fact, the set-of-support strategy turned out to be crucial in the astronomical domain—theorems that are proved in under a minute with set of support cannot be proved within the available space without it.

When we employ the set-of-support and the recursive-path-ordering strategies at the same time, however, we lose completeness—there may be some valid theorems we will be unable to prove without violating the restrictions of one of the strategies. (In fact, once we combine the recursive-path-ordering strategy with the constructiveness restriction, which guarantees that we can extract programs from proofs, we may already have lost completeness.) The domain theory contains some logically redundant axioms to circumvent this incompleteness, but this is something of a stopgap measure. In the future, a hybrid strategy that allows some reasoning forward from axioms and some reasoning backward from the goal may be employed in combination with the recursive-path-ordering strategy.

5 Performance

Since the test cases were run, demonstrations of Amphion have been given by Lowry at NASA Ames, JPL, and other sites. Members of the audience unfamiliar with the system were invited to specify their own programs. In almost all cases, the graphical notation was adequate to specify the new program and SNARK was capable of proving the corresponding theorem and constructing the specified program.

In all our test cases, including those proposed by participants in NASA and JPL demonstrations, the specification has been formulated in less than half an hour; an experienced Amphion user needs just a few minutes. It is often more convenient to revise the stored specification of a similar problem than to construct a new specification from scratch. The theorems have required less than ten minutes—usually less than three minutes—for SNARK to prove, and the translation into FORTRAN is completed in seconds.

For one problem, the desired program relied on properties of subroutines in SPICELIB that had not yet been axiomatized. It required less than half an hour to introduce the new axioms; the system was then able to construct the new program.

Once SNARK has found one proof and extracted the corresponding program, we can restart it to find other proofs and perhaps other programs. This ability is not used by Amphion, because we have not found that the various programs differed in any significant way; they were doing more or less the same things in different orders.

The system has recently been installed at NAIF so that JPL astronomers can use the system regularly, on an experimental basis.

6 What Next?

The problems solved so far have been relatively simple, none requiring more than two of three pages of FORTRAN code and none including if-statements or loops, except implicitly at the subroutine level. While SNARK does regularly introduce conditionals, for example by application of the resolution rule, its ability to introduce iterative or recursive constructs is rudimentary. It currently contains no induction rule, so we must provide the appropriate well-founded relation, on which the induction is based, and enter the induction hypothesis as an axiom. Although none were encountered in the sample problem set or in demonstrations, there are problems in the domain that do require iterations that cannot be relegated to subroutines. When we do employ induction, it may be advisable to use a nonclausal representation of formulas; so far, all formula have been kept in clausal form.

For more complex problems, it will be necessary to decompose the specification into subspecifications of more manageable modules. Simple decompositions might be achieved automatically, with the help of tactics that could be built into SNARK itself. Other decompositions will be performed interactively through the graphical interface. In this way, the user would specify the original problem and its decomposition into modules with the same mechanism.

Once SNARK has successfully constructed a module, its specification can be added to the theory as a new axiom. If that axiom is used in a proof, the module will be invoked by the corresponding program. Thus, if the decomposition is done appropriately, SNARK will be able to compose the modules to solve the main problem. Whether the decomposition is accomplished automatically or with user assistance, the correctness of the resulting program and its modules is guaranteed by the method of their construction, provided that the domain theory is correct.

Nothing in the techniques we are using restricts us to SPICELIB or to the astronomical domain. We are currently considering other application domains in which the same technology would be valuable. Characteristic of a potentially fruitful domain are the existence of a mature subroutine library, many of whose users are imperfectly acquainted with its contents. Deductive methods are particularly attractive when the correctness of the derived software is critical. In such a domain, it is plausible that existing deductive technology will suffice to give computationally naive users access to a large library of subroutines and enable them to compose software of practical power and high reliability.

7 Acknowledgements

We would like to thank the National Science Foundation for support of some of this research, under Grant CCR-8922330.

References

[BM 88] R. S. Boyer and J S. Moore, *A Computational Logic Handbook*, Academic Press, Boston, MA (1988).

[Der 82] N. Dershowitz, Orderings for Term-Rewriting Systems, *Journal of Theoretical Computer Science*, 17,3 (1982), 279-301.

[HR 91] J. Hsiang and M. Rusinowitch, Proving Refutation Completeness of Theorem-Proving Strategies: The Transfinite Semantic Tree Method, *Journal of the ACM*, 38,3 (1991), 559-587.

[McC 90] W. McCune, *Otter 2.0 User's Guide*, Technical Report ANL-90/9, Argonne National Laboratory, Argonne, IL (1990).

[MW 92] Z. Manna and R. Waldinger, Fundamentals of Deductive Program Synthesis, *IEEE Transactions on Software Engineering*, 18,8 (1992), 674-704.

[MW 93] Z. Manna and R. Waldinger, *Deductive Foundations of Computer Programming*, Addison-Wesley, Reading, MA (1993).

[Rob 65] J. A. Robinson, A Machine-Oriented Logic Based on the Resolution Principle. *Journal of the ACM* 12 (1965) 23-41.

[RW 91]   E. J. Rollins and J. M. Wing, Specifications as Search Keys for Software Libraries, *Eighth International Conference on Logic Programming*, Paris, June 1991.

[Smi 90] D. R. Smith, KIDS: A Semiautomatic Program Development System. *IEEE Transactions on Software Engineering* 16,9 (1990) 1024–1043.

[Tyu 88] E. H. Tyugu, *Knowledge-Based Programming*, Turing Institute Press, Glasgow, Scotland, 1988.

[WR 69]   L. Wos and G. Robinson, Paramodulation and Theorem Proving in First-Order Theories with Equality. In B. Meltzer and D. Michie (editors), *Machine Intelligence* 4, American Elsevier, New York, NY (1969) 135–150.

[WRC 65] L. Wos, G. A. Robinson, and D. F. Carson, Efficiency and Completeness of the Set-of-Support Strategy in Theorem Proving. *Journal of the ACM*, 12,4 (1965), 536–541.

Proof Script Pragmatics in IMPS*

William M. Farmer, Joshua D. Guttman, Mark E. Nadel, F. Javier Thayer

The MITRE Corporation
202 Burlington Road
Bedford, MA 01730-1420, USA

Telephone: 617-271-2749; Fax: 617-271-3816

E-mail: {farmer,guttman,men,jt}@mitre.org

Abstract. This paper introduces the IMPS proof script mechanism and some practical methods for exploiting it.

1 Introduction

IMPS, an Interactive Mathematical Proof System [4, 2], is intended to serve three ultimate purposes:

- To provide mathematics education with a mathematics laboratory for students to develop axiomatic theories, proofs, and rigorous methods of symbolic computation.
- To provide mathematical research with mechanized support covering a range of concrete and abstract mathematics, eventually with the help of a large theory library of formal mathematics.
- To allow applied formal methods to use flexible approaches to formalizing problem domains and proof techniques, in showing software or hardware correctness.

Thus, the goal of IMPS is to provide mechanical support for traditional methods and activities of mathematics, and for traditional styles of mathematical proof. Other automated theorem provers may be intended for quite different sorts of problems, and they can therefore be designed on quite different principles. For instance, some are meant to act as the back ends for AI systems that need to prove simple theorems about simplified worlds. However, theorem provers of this latter kind will not be able to serve the purposes we have mentioned, for which a wide range of traditional mathematical techniques must be supported.

In this paper we will focus on the IMPS *proof script* mechanism; we intend to illustrate why it aids in developing the large bodies of mathematical theory that are needed for these goals. As such, the paper is intended as a pragmatic one rather than a theoretical one. We aim to emphasize concrete, practical ways of getting proofs done. We believe that the IMPS proof script mechanism aids the

---

* Supported by the MITRE-Sponsored Research program.

user in carrying out proofs, in tailoring and reusing previously developed proof ideas, and in conveying the essential content and structure of proofs.

Our pragmatic approach derives from our view that, in mechanized theorem proving, what happens on the outside may be more important than what happens on the inside. For example, a user may be more sensitive to the time it takes him to do some simple data entry tasks, or to find a clever encoding of a mathematical idea, than he is to the time it takes the machine to explore deduction steps. Frequently, one should be less concerned with what it is possible to do and more concerned with what can be done conveniently.

In particular, our aim is to provide flexible and convenient ways of manipulating and reusing proofs. The operations that will serve this purpose cannot be determined from proof theory, but primarily from experience. Given our aim, it is not necessary to adopt an object logic in which proofs themselves are first class objects, because we do not aim to prove things about proofs. On the contrary, it is more important to make it very easy for a user to get his hands on his proofs and (especially) partial proof attempts; to make it very cheap for him to restart proofs and reexecute portions; and to encourage an experimental attitude to proof construction.

Mechanizing mathematics is widely acknowledged to be hard work. It forces us to a more formal level, and at the same time to a more concrete representational level, than we would normally adopt in standard mathematical practice. Thus, offsetting the benefit that we gain confidence in the correctness of our proofs, there are the extra burdens that most everyone who has dealt with a mechanized theorem prover has surely experienced. Effective styles of usage are needed to mitigate these burdens and to provide new methods exploiting the more concrete structure we have at our disposal. Below, we will describe some techniques we have found so far.

IMPS is self-consciously an interactive system; however, we believe that current mechanized theorem provers intended for mathematics are all interactive in one sense or another. The kinds of interaction can vary from the crafting of an appropriate sequence of lemmas to reach the theorem, to the setting of various switches before control is passed to the machine, to the user supplying most of the proof by hand. Perhaps a more interesting distinction than the familiar but flawed one of "degree of autonomy" is the distinction between systems in which the user interacts only between attempts to construct a proof, and those in which he interacts during the process of constructing a proof. IMPS is emphatically of the second kind. We find that, in the first kind of system, the human helps the machine to prove the theorem, while in the second kind, the machine can help the human. Partly for this reason, IMPS provides relatively large proof steps in many of its proof commands, which aids the user in correlating the steps in constructing an IMPS proof with the successive portions of an intuitive proof sketch.

In order to encourage a substantial community of users to adopt IMPS, we want to describe a range of styles of effective usage. These styles of usage are best refined only after substantial experience has been gained in using the system for a particular type of problem. Over time, these styles of usage evolve in tandem with improvements or adaptations in the theorem prover itself. In this paper we convey some aspects of a successful style for using the IMPS facilities for interactive and script-based proof.

The paper is organized as follows. Section 2 introduces special procedures for applying theorems called *macetes*, which play a fundamental role in the IMPS proof system. Sections 3–5 describe the IMPS proof script mechanism and ways it can be put to use. Section 6 briefly compares IMPS proof scripts and macetes with traditional tactics. And Section 7 contains a conclusion.

2 Macetes

Macetes supplement the IMPS simplifier [4, 5] in order to provide more flexibility to the user. The simplifier applies universally quantified equalities as rewrite rules in a manner which is usually beyond the user's control. In particular, it is not possible for the user to direct the simplifier to apply only those theorems that belong to a specified set of theorems. This rigidity of rewrite rule application clearly clashes with normal mathematical practice, where theorems are usually applied, individually or in groups, in a way which is dependent on content and ultimately determined by the mathematics practitioner.

In IMPS the macete mechanism is designed to provide a simple facility to extend the simplifier in straightforward ways (or build simple simplifiers from scratch) so that the user has more control over what theorems get applied. Macetes are of two basic kinds: atomic and compound. For instance, when a theorem is installed (after it has been proved), a corresponding atomic macete (called a theorem macete) is automatically created. These theorem macetes do a variety of kinds of conditional rewriting depending on the syntactic form of the underlying theorem. Theorem macetes are created for all theorems, even those that are not conditional equalities [5]. A theorem may be applied as a macete using ordinary matching or using *translation matching*, an inter-theory form of expression matching which allows a theorem to be applied outside of its home theory [3]. Atomic macetes also include simplification and beta-reduction.

Compound macetes are specified using an extremely simple language for determining control of the process of applying atomic macetes. This language provides a few simple constructors for sequencing and iteration of arbitrary macetes.

Macetes are applied via special proof commands that add at most one inference to the deduction graph. Since the number of macetes that are loaded into the system may be large (1000 macetes is typical), the facility would not be too practical if the user had to guess the right one out of the blue. To deal with this IMPS provides a special menu—the "macete menu"—to tell the user which macetes may be applicable to a given subgoal. In situations where over 1000 macetes are available, there are rarely more than 10 macetes presented to the user. Consequently, this menu can be used to provide guidance on what to do next. It provides information to the novice about what is available in the theory library, and it provides feedback to the expert about the essential content of his subgoal.

3 Proof Scripts in IMPS

A logical deduction is represented in IMPS as a kind of directed graph called a *deduction graph* [4]. An IMPS user initially sees proof as an interactive process—frequently with much trial and error—acting on a deduction graph. In this view, developing the proof means issuing a sequence of commands, with the IMPS user interface supplying a good deal of information about what steps may be useful. Each command is applied to a specific node in the deduction graph and produces (zero or more) additional nodes. Roughly speaking, the graph structure represents the relation of entailment between the nodes. When an inference supports a node with subgoal nodes, and all of the subgoals are recognized to be true ("grounded"), then the node is also grounded. A proof is complete when its original goal node is grounded. Throughout the proof, the user has a current node, which may be freely changed between commands. After each command, the system selects a current node. The next command will apply to this newly selected node unless the user explicitly changes the current node. When a command has added new unsupported subgoals, the new current node is generally the leftmost; when a command has grounded its node, the system generally chooses the leftmost unsupported descendent of the nearest ungrounded ancestor.[1] IMPS enforces a goal directed style of reasoning, in which the proof is constructed from the conclusion backwards using a sequent-based system of rules.

A *proof script* (or *script* for short) is a sequence of certain s-expressions that, when executed, applies a sequence of commands to a deduction graph. An example of a proof script is shown in Figure 1. The structure of the deduction graph and the default way of selecting a new current node determine how these commands are applied to the deduction graph. An intimate knowledge of the syntax for proof scripts is unimportant: no one types scripts. Instead, the user interacts with the system through its interface, usually selecting from menus generated on the fly, to carry out the commands. At any point in the process, the user can ask IMPS to create a proof script that is a transcript of the proof or partial proof. In practice, all proof scripts are created from these basic transcripts of interactive execution, by editing them to introduce control structures.

The s-expressions in a proof script are of several kinds:

- Command forms do the ultimate work of adding new nodes to the deduction graph. In the example shown in Figure 1, applying macetes such as `tr%subseteq-antisymmetry` and `indicator-facts-macete` and doing the direct inferences are command forms.
- Node motion forms cause script execution to continue at a node other than the natural continuation node; for instance, (`jump-to-node top`).

---

[1] A more elaborate scheme is used when the nearby portion of the deduction graph is not tree-like.

- Assignment forms may define local macetes or invocable scripts (see below) which can be referenced at other places within the script. This is especially useful in certain kinds of proofs, such as proofs by symmetry, which involve two or more instances of the same argument. (`label-node top`) is a different kind of assignment form.
- Conditionals subordinate the execution of a portion of a script to the validity of a certain condition. Typical conditions are that the assertion of the goal sequent matches a given expression or that a particular subscript succeeds in adding inferences to the proof.
- Iteration forms provide for execution of a subscript while a specified condition holds, or over a specified range of nodes. In Figure 1, `for-nodes` begins an iteration over the set of unsupported descendents of the target node.
- Block and comment forms provide structure and documentation.

```
((label-node top)
 (apply-macete-with-minor-premises tr%subseteq-antisymmetry)
 (script-comment
   "Replace equation with two inclusions.")
 direct-inference
 (jump-to-node top)
 (for-nodes
   (unsupported-descendents)
   (block
     insistent-direct-inference
     (apply-macete-with-minor-premises indicator-facts-macete)
     beta-reduce-repeatedly))))
```

Fig. 1. Script for proving sets are equal

While a script is executing, it maintains two distinguished nodes in the deduction graph. First, there is the *head* node, which remains fixed. Second, there is the *current* node, which starts off as the head node, and as execution progresses, evolves according to the default selection rules described above or the dictates of explicit node motion forms included in the script.

We regard a script execution as an attempt to provide a proof, or part of a proof, for a particular subgoal node, namely the one selected as the head node. For this reason if, part way through execution, the head node should become grounded, the remainder of the script is discarded. There is nothing more for it usefully to do. We have found that this principle greatly improves the robustness and predictability of the script mechanism. Without it, there is the risk that "overachieving" scripts will carry out meaningless proof steps in some adjacent portion of the deduction, with the consequence that later proof commands may by default apply to the wrong subgoals. As a special case, a *block* is an anonymous script procedure with no parameters. In effect, it merely introduces a head node and discards any of the nested commands that may remain after the head node is grounded.

These remarks describe the possible "moves" in interacting with IMPS. Later we will we describe some of the "strategies" we employ in playing the game.

3.1 Invocable Scripts

The most basic user level inference steps are given by built-in IMPS *proof commands*; these are Lisp procedures[2] which call *primitive inferences* in useful patterns. Assuming that the primitive inferences, several of which carry out sophisticated reasoning steps, are correctly implemented, proof commands are guaranteed only to make sound inferences, because they modify deduction graphs only by calling primitive inferences. There are approximately 60 built-in proof commands in IMPS. A single command will, in general, add several nodes, and often several levels of nested subgoals, to the deduction graph. Moreover, the same command, issued in different contexts, may add different numbers of nodes or levels.

In theory, new proof commands can be added to IMPS by directly writing new Lisp procedures, although writing such procedures is usually difficult. An *invocable script* is a new proof command created by the user from a proof script. It can be invoked—either interactively or in other scripts—just like the basic IMPS proof commands implemented in the underlying Lisp. When requested, IMPS tells the user which proof commands, whether built-in or user-defined, are possibly applicable to a given subgoal.

The parameters to an invocable script are untyped, and referenced positionally by positive integers. Their actual values may be:

- integers;
- strings, used to represent expressions; or
- symbols, used to represent theorems, macetes, and commands primarily.

An invocable script may be defined at the top level, as a globally available command, using the **def-script** form as in Figure 2. Alternatively, it may be purely local to an encompassing proof script. For instance, the local invocable script in Figure 3 contraposes against the assumption matching the pattern given as argument, after which it uses a group of theorems about the algebra of fractions as a macete, before finally calling the simplifier. This local invocable script is then used three times in the proof fragment that follows.

Invocable scripts may take other invocable scripts or other commands as arguments, as was illustrated by the example in Figure 2. It repeatedly applies direct inferences (sequent calculus right introduction rules) and antecedent inferences (sequent calculus left introduction rules) backwards to generate a set of subgoals, before finally applying the command given by its actual parameter to every leaf node introduced in this process. This invocable script is frequently called with the parameter **simplify**, although the example of Figure 1 could be rewritten by passing the contents of the block as its argument.

---

[2] Yale's T dialect of Scheme [10] is in fact the implementation language.

```
(def-script command-on-direct-descendents 1
  ((label-node compound)
   direct-and-antecedent-inference-strategy
   (jump-to-node compound)
   (for-nodes
    (unsupported-descendents)
    $1)))
```

Fig. 2. Invocable script with a command parameter

```
(let-script
 contrapose-denom-remove 1
 ;; The arg is the pattern to contrapose on
 ;;
 ((contrapose $1)
  (apply-macete-with-minor-premises
   fractional-expression-manipulation)
  simplify))
($contrapose-denom-remove "with(r:rr,r<0);")
($contrapose-denom-remove "with(r:rr,r=0);")
($contrapose-denom-remove "with(r:rr,r=1);")
```

Fig. 3. Local invocable script and its application

4 Proof by Emacs

A crucial advantage of the IMPS script language is that simple textual manipulations of the scripts allow a user to reuse proofs or portions of proofs in a highly predictable way. In many cases, a very superficial understanding of many portions of a proof is enough to enable a user to transform it into a proof of another theorem.

As an example, consider Figure 4, which was created while developing parts of freshman calculus using nonstandard analysis. It establishes the theorem that the limit of a sum equals the sum of the individual limits. The proof script contained within it (below the word **proof**) is given almost as it appears in our files; we have added only the marker **&** appearing at the right of some of the lines. It is not necessary to understand the script completely. That is part of the point.

Suppose that we now want to prove the analogous theorem about the limit of a product. This theorem about products does not follow from the theorem about sums. Moreover, in most standard treatments the proofs are considerably different. However, in our nonstandard treatment there is a proof of the theorem for products that is very much like the theorem for sums. The extra work that one would expect to have to do is encapsulated in a lemma corresponding to the

```
(def-theorem sum-of-limits
  "forall(f,g:[rr,rr], c:rr, #(lim(f,c)) and #(lim(g,c)) implies
    lim(lambda(x:rr,f(x)+g(x)),c)=lim(f,c)+lim(g,c))"         &
  (theory nsa-theory)
  (proof
    (direct-and-antecedent-inference-strategy
      (apply-macete-with-minor-premises
        iota-free-characterization-of-lim)
      direct-and-antecedent-inference-strategy
      (apply-macete-with-minor-premises ast-composition-binary)
      beta-reduce-repeatedly
      (force-substitution "ast(+)" "++" (0))                  &
      (move-to-sibling 1)
      simplify
      extensionality
      (unfold-single-defined-constant (0) ++)                 &
      (apply-macete-with-minor-premises additivity-of-st)     &
      (apply-macete-with-minor-premises
        iota-free-characterization-of-lim-existence)
      (unfold-single-defined-constant-globally ++)            &
      (apply-macete-with-minor-premises ast-extends-compound)
      (apply-macete-with-minor-premises
        lim-existence-implies-finite-on-monad)
      direct-and-antecedent-inference-strategy
      (apply-macete-with-minor-premises
        lim-existence-implies-finite-on-monad)
      direct-and-antecedent-inference-strategy)))
```

Fig. 4. Limit of a sum

lemma additivity-of-st.

A user more or less familiar with what was going on, but who did not necessarily follow the proof completely, can recognize that only the lines with a & to the right seem to have anything to do with addition in particular. The other lines are more or less generic with respect to the issue here. We now change just those lines to the corresponding ones for multiplication, generally using global-replace or query-replace in Emacs; hence the description "proof by Emacs." In this case, + is replaced with * and additivity is replaced by multiplicativity, which suffices to produce both the theorem to be proved and also the proof script. When we run this new script on the new theorem, we see that a complete proof is obtained. One might ask how the user is to know the "corresponding" lines. Although ultimately this is a matter of mathematical understanding, IMPS can provide some assistance, as our next example will illustrate.

Continuing our excursion through freshman calculus, consider the analogous theorem on the limit of a quotient. Suppose we try exactly the same approach. Imagine we have changed the addition to division, as we changed the addition to multiplication above. We might mistakenly assume that the analogue of `multiplicativity-of-st` is called `divisibility-of-st`, and make the change accordingly. We now run the script, but when it tries to execute

(`apply-macete-with-minor-premises divisibility-of-st`)

it returns the error message that there is no such macete. IMPS will help us find the correct name. We then jump back into the script at the point this was attempted, or rerun the portion of the script up to that point. Now, the correct choice `st-of-quotient` will pop up in the menu of applicable macetes, and it can be inserted. One might argue that our naming convention was not very consistent, but how much consistency should you expect, especially when theory libraries are developed by different users?

We are not done yet, as the observant reader will have noticed, because this "theorem" is not true. When we run the new script, we no longer get an error message, but after all the commands have been executed, the goal node is not grounded. The user now examines the ungrounded nodes and considers what IMPS was unable to prove. The user might examine, for example, the default current node. A quick look, and perhaps an IMPS simplification, will make it quite clear that the problem is that the hypothesis saying that the limit of the denominator is nonzero is missing. Without this hypothesis, there is no way to discharge the proof obligation to show that the denominator is nonzero or that the quotient is defined. Amending the hypothesis, rerunning the script, and doing a few additional steps will ground the corrected theorem. Naturally, the process will not always proceed as smoothly, but it would be unrealistic to expect that it would.

In extreme cases of proof by Emacs, no editing whatever is needed to reuse a script. In proving that a finite set S with n elements has 2^n subsets, a key ingredient for the induction is that if you remove any element x from S, then every subset of S is either a subset of $S \setminus \{x\}$ or of the form $A \cup \{x\}$, where A is such a subset. Figure 5 contains a proof script for the lemma that asserts this. There is an analogous theorem that states that if a set S has n elements, then for $k \leq n$, S has $\binom{n}{k}$ subsets of cardinality k. The inductive proof of that theorem depends on an analogue of the above lemma for decomposing the k element subsets. It turns out that the script displayed in Figure 5 will also ground this theorem, even though the proof in the strict sense (for instance, represented as a deduction graph) is entirely different.

Of course, not all IMPS proofs are, or could be, "done by Emacs" in any significant way. On the other hand, however, it is often the case in mathematics that a claim is made that the proof of theorem B is similar to the proof of theorem A, with the implicit or sometimes explicit suggestion that the details are left to the reader. Often the reader does not check the details, and more significantly, neither does the author. Not infrequently, what looks quite similar on a superficial level is quite different when all the details need to be supplied. This sometimes leads to errors.

In such cases, "proof by Emacs" can be extremely useful, particularly because it is cheap and easy. A user can try a number of scripts quickly without taking

```
(def-theorem power-decomposition
  "forall(s:sets[ind_1], x:ind_1, n:nn, x in s implies
    power(s)=
    power(s diff singleton{x}) union power(s) inters filter(x))"
  (theory generic-theory-1)
  (proof
   (direct-and-antecedent-inference-strategy
    unfold-defined-constants
    set-equality-script
    direct-and-antecedent-inference-strategy
    (contrapose "with(p:prop,not(p));")
    simplify-insistently
    (contrapose "with(p:prop,not(p));")
    simplify
    direct-and-antecedent-inference-strategy
    set-containment-script
    direct-and-antecedent-inference-strategy
    (incorporate-antecedent
     "with(f,x:sets[ind_1],x subseteq f)")
    simplify-insistently )))
```

Fig. 5. Power set decomposition lemma

much time or effort, and especially, without requiring much hard thinking, which is especially important at the end of a long day. When a successful proof is found, one can then reflect upon it at one's convenience, to absorb the mathematical content. Although one might waste one's time in vain attempts, the underlying soundness of IMPS insures that if one does ground the theorem, then it has really been proved, even if the user does not yet understand exactly why it worked.

The relatively high level of IMPS scripts is needed: it preserves the similarity on the superficial level, which often breaks down if one must descend to greater detail. Recall that a single command in a script will often generate many steps in the underlying proof, and the same command may generate different proof steps in different contexts. Consequently, by using scripts we can take advantage of similarity on the more schematic level. IMPS automatically supplies the necessary differences in detail.

"Proof by Emacs" may prove a useful technique in cooperative mathematics research and mathematics education. It may be possible to borrow proofs which might involve techniques or concepts outside the expertise of the borrower, but which he could then adapt for his purposes. This often happens informally today, but with far less assurance that the adaptation is legitimate. Similarly, instructors can give students worked examples to adapt, allowing the students to verify their adaptations for themselves. Of course, this sort of thing is also done now, except that either details are omitted or students do not get timely reinforcement. This would also allow the assignment of much larger, more realistic problems, in place of the much shorter ones that are customary today, and yet would no doubt save a great deal of wear and tear on the instructor.

"Proof by Emacs" is one among a range of techniques that IMPS supports to allow the mathematician to do what he would like to do anyway, but in a more convenient and reliable way.

5 Proof by Symmetry

Various essentially different kinds of reasoning are lumped together under the term "proof by symmetry." In some cases, a proof by symmetry may be formalized by constructing a theory interpretation from an axiomatic theory $\mathcal{T}$ into itself; for instance, the right cancellation law in groups follows from the left cancellation law, using the "symmetry" (interpretation) that maps the group operation $\cdot$ to $\lambda x, y \, . \, y \cdot x$. The IMPS mechanisms supporting this form of reasoning are discussed in [4, 3]. In this section, we will instead focus on cases which do not easily fit that pattern, but in which portions of a proof are symmetrical with each other. The formula shown in Figure 6 involving the floor function[3] supplies a very simple example: In proving the right hand side from the left hand side, the two halves of the nested biconditional are symmetrical. In fact, in proving the left hand side from the right hand side, we also create two symmetrical subgoals, namely to prove that $\text{floor}(x) \leq \text{floor}(y)$ and that $\text{floor}(y) \leq \text{floor}(x)$, but these are handled trivially by the IMPS simplifier.

$$\forall x, y : \mathbf{R} \quad \text{floor}(x) = \text{floor}(y) \iff (\forall m : \mathbf{Z} \quad m \leq x \iff m \leq y)$$

Fig. 6. Floor equality criterion

The user starts the proof by breaking apart the logical connectives, after which he confronts, as his current subgoal, the task of showing $m \leq y$ assuming

$$m \leq x \quad \text{and} \quad \text{floor}(x) = \text{floor}(y).$$

To do so, he instantiates the first theorem shown in Figure 7 with the arguments x and m, and instantiates the second theorem with y. Simplification completes this case. Since the second case is obviously true for the same reason, he can request that IMPS print the text of his proof so far, as shown in Figure 8. The user may then edit this text to construct a locally defined script, shown in Figure 9. Finally, the user types and executes the proof command ($do-case "y" "x"), which carries out the same inferences with the roles of x and y interchanged. The full proof, as it is recorded in an IMPS theory file, is given in Figure 10. This final

---

[3] The symbol "floor" is a constant defined in the IMPS theory of real arithmetic as that function which for every real x returns the unique integer n such that $n \leq x < n+1$.

$$\forall x : \mathbf{R}, n : \mathbf{Z} \quad n \leq x \supset n \leq \text{floor}(x)$$

$$\forall x : \mathbf{R} \quad \text{floor}(x) \leq x$$

Fig. 7. Lemmas used in the proof

```
direct-and-antecedent-inference-strategy
(instantiate-theorem floor-not-much-below-arg ("x" "m"))
(instantiate-theorem floor-below-arg ("y"))
simplify
```

Fig. 8. Transcript of first case

presentation has the advantage that the symmetry between the two subgoals is made explicit for a later reader by the two $do-case forms. Thus, our approach to proof by symmetry eases the user's burden in the course of developing the proof, and makes the structure of the proof easier to read off its final form.

We have used this technique in many examples, where the individual subgoals may be far more demanding. Another frequent source of symmetrical subgoals—apart from biconditionals in the goal—is the instances of the anti-symmetry of $\leq$. Instances of the trichotomy of $<$ also frequently furnish two symmetrical cases (the strict inequalities) as well as the nonsymmetrical, and frequently quite easy, case with the equality.

6 Comparison with Tactics

Tactic-based theorem proving [8] has been a major area of research in automated reasoning since the development of Edinburgh LCF [6]. In fact, the ML programming language was invented for writing LCF tactics. Today tactics are used in several major theorem proving systems, including HOL [7], Isabelle [9], and Nuprl [1].

Tactics are procedures that automate part of the proof process. They come in many flavors. For example, a *refinement tactic* is a procedure which generates a list of subgoals from a given goal in a deduction, and a *transformational tactic* is a procedure that constructs a new deduction from an old deduction [1]. The former notion of a tactic is fairly restrictive, while the latter notion is quite broad. Although the notion of a tactic varies widely, all tactics are constructed or applied in a special way so they are guaranteed to be sound with respect to the proof system being used.

An IMPS proof script is a kind of tactic which serves the same purposes as other kinds of tactics:

```
(let-script
 do-case 2
 ((instantiate-theorem floor-not-much-below-arg ((% " ~a " $1) "m"))
  (instantiate-theorem floor-below-arg ((% " ~a " $2)))
  simplify))
```

Fig. 9. Script for the symmetrical cases

```
direct-and-antecedent-inference-strategy
(let-script
 do-case 2
 ((instantiate-theorem floor-not-much-below-arg ((% " ~a " $1) "m"))
  (instantiate-theorem floor-below-arg ((% " ~a " $2)))
  simplify))
($do-case "x" "y")
($do-case "y" "x")
(apply-macete-with-minor-premises <=-anti-symmetry)
(apply-macete-with-minor-premises floor-not-much-below-arg)
(command-on-direct-descendents simplify)
```

Fig. 10. Final proof script

– To create new proof commands (rules of inference).
– To represent executable proof sketches.
– To store proofs in a compact, replayable form.

A proof script is more general than a refinement tactic since the behavior of a proof script is dependent on the structure, contents, and current node of the deduction graph to which it is applied. On the other hand, a proof script is more restrictive than a transformational tactic since the script programming language is restricted and since scripts can change the structure of a deduction graph only by adding new nodes. From our point of view, proof scripts have just about the right level of generality: they are powerful enough to do useful things, but controlled enough to be easily manipulated as text. Proof scripts also have several idiosyncrasies that set them apart from other kinds of tactics:

– The systematic use of simplification.
– The application of theorems via macetes.
– The use of the "current node" idea to linearize the execution of scripts.
– The use of "blocks" to make scripts more robust.

A macete is also a kind of tactic, but it has a very limited purpose: to apply and retrieve theorems (or organized collections of theorems). Macetes play an extremely important role in the IMPS proof process. Like proof scripts, macetes are idiosyncratic:

- Macetes apply a theorem or collection of theorems at any location in (the assertion of) a goal, even deeply within it.
- As we mentioned in Section 2, macetes use both ordinary expression matching and translation matching so that theorems can be applied both inside and outside of their home theories.
- Theorem macetes (see Section 2) are automatically created by IMPS when theorems are installed.

Even though the means to program macetes is quite restricted in comparison to tactics and IMPS proof scripts, in practice it is very easy for the IMPS user to build useful macetes.

7 Conclusion

For any proof system, we may distinguish the knowledge explicitly formalized in it from the body of more procedural knowledge which is not formalized, but which a person must grasp to use it effectively. Proving interesting theorems, whether with pencil and paper or with mechanized theorem provers, requires a great deal of each.

In IMPS, the explicitly formalized knowledge required is primarily codified in the theory library, a large and open-ended collection of axiomatic theories and theorems. Theory interpretations serve as links to interconnect the theories in the library. The current theory library contains about 50 named theories and over 1100 replayable proofs. The theories include formalizations of the real number system and objects like sets and sequences; theories of abstract mathematical structures such as groups and metric spaces; and theories to support specific applications of IMPS in computer science. Several of the theorems proved reach about the level of the fundamental theorem of calculus. The most developed area of mathematics in the theory library is abstract mathematical analysis.

The more implicit, procedural knowledge is encoded in several ways. Compound macetes provide one way, as many of them amount to rudimentary procedures for (logically sound) symbolic computation. Proof scripts are intended as another repository of this sort of practical knowledge. User-defined commands introduced by **def-script** encapsulate knowledge of how to perform some conceptually unified, higher-level manipulation on a goal. In "proof by Emacs," the proof of one theorem serves as a template indicating how to prove related theorems. One form of proof by symmetry uses the procedural knowledge accumulated in proving one case to codify what is common between the cases. More generally, we consider the proof scripts in the theory library as a mine of ideas and techniques for getting a wide variety of substantial theorems rigorously proved with IMPS.

In this paper we have tried to convey a portion of this procedural knowledge for the case of IMPS. We think that the developers and users of various proof systems can make valuable contributions to our common goals by formulating and exchanging this kind of information, in addition to the more usual information about explicit logical techniques.

Acknowledgments

We are grateful for the suggestions received from the referees.

References

1. R. L. Constable, S. F. Allen, H. M. Bromley, W. R. Cleaveland, J. F. Cremer, R. W. Harper, D. J. Howe, T. B. Knoblock, N. P. Mendler, P. Panangaden, J. T. Sasaki, and S. F. Smith. *Implementing Mathematics with the Nuprl Proof Development System*. Prentice-Hall, Englewood Cliffs, New Jersey, 1986.
2. W. M. Farmer, J. D. Guttman, and F. J. Thayer. IMPS: System description. In D. Kapur, editor, *Automated Deduction—CADE-11*, volume 607 of *Lecture Notes in Computer Science*, pages 701–705. Springer-Verlag, 1992.
3. W. M. Farmer, J. D. Guttman, and F. J. Thayer. Little theories. In D. Kapur, editor, *Automated Deduction—CADE-11*, volume 607 of *Lecture Notes in Computer Science*, pages 567–581. Springer-Verlag, 1992.
4. W. M. Farmer, J. D. Guttman, and F. J. Thayer. IMPS: an Interactive Mathematical Proof System. *Journal of Automated Reasoning*, 11:213–248, 1993.
5. W. M. Farmer, J. D. Guttman, and F. J. Thayer. The IMPS user's manual. Technical Report M93B-138, The MITRE Corporation, Bedford, MA, November 1993.
6. M. Gordon, R. Milner, and C. P. Wadsworth. *Edinburgh LCF: A Mechanised Logic of Computation*, volume 78 of *Lecture Notes in Computer Science*. Springer-Verlag, 1979.
7. M. J. C. Gordon. HOL: A proof generating system for higher-order logic. In G. Birtwistle and P. A. Surahmanyam, editors, *VLSI Specification, Verification, and Synthesis*, pages 73–128. Kluwer, 1987.
8. R. Milner. The use of machines to assist in rigorous proof. In C. A. R. Hoare and J. C. Shepherdson, editors, *Mathematical Logic and Programming Languages*, pages 77–88. Prentice/Hall International, 1985.
9. L. C. Paulson. Isabelle: The next 700 theorem provers. In P. Odifreddi, editor, *Logic and Computer Science*, pages 361–368. Academic Press, 1990.
10. J. A. Rees, N. I. Adams, and J. R. Meehan. *The T Manual*. Computer Science Department, Yale University, fifth edition, 1988.

A Mechanization of Strong Kleene Logic for Partial Functions*

Manfred Kerber   Michael Kohlhase

Fachbereich Informatik, Universität des Saarlandes
66041 Saarbrücken, Germany
+49-681-302-{4628|4627}
{kerber|kohlhase}@cs.uni-sb.de

Abstract. Even though it is not very often admitted, partial functions do play a significant role in many practical applications of deduction systems. Kleene has already given a semantic account of partial functions using three-valued logic decades ago, but there has not been a satisfactory mechanization. Recent years have seen a thorough investigation of the framework of many-valued truth-functional logics. However, strong Kleene logic, where quantification is restricted and therefore not truth-functional, does not fit the framework directly. We solve this problem by applying recent methods from sorted logics. This paper presents a resolution calculus that combines the proper treatment of partial functions with the efficiency of sorted calculi.

1 Introduction

Many practical applications of deduction systems in mathematics and computer science rely on the proper treatment of partial functions. Although there are work-arounds for most concrete situations, there has been a considerable interest in the community for clean formalizations of partial functions.

One of the key problems to be solved when formalizing partial functions is to decide what happens if partial functions are applied to arguments not in their domain. In mathematical practice expressions like $\frac{0}{0} = 1$ or $odd(predecessor(0))$ are thought to be neither true nor false. This phenomenon can be handled in the well-known systems for intuitionistic logic, where the law of the excluded middle does not hold, hence $\frac{0}{0} = 1$ can be (and in fact is) neither true nor false, since neither the truth nor the falsehood of this expression can be shown. However, most mathematicians do not want to give up the law of the excluded middle, because it is basic for a strong proof technique, the indirect proof. The standard way to deal with this situation in classical mathematics is to reject expressions like $\frac{0}{0}$ as "meaningless". This phenomenon of "truth value gaps" is studied in various systems of free logic [22, 10, 15]. A related approach that seems more available for mechanization has first been advocated by Kleene in [14]. He introduces an additional individual $\bot$ denoting meaningless individuals and a third truth value u, standing for the "undefined" truth value. At first glance this seems to be a great deviation from mathematical practice, which only acknowledges two truth values, but the third truth value simply labels situations that would be rejected in mathematical practice anyway.

In recent years, methods for the operationalization of many-valued logics have been developed by Carnielli [6], Hähnle [12], Baaz and Fermüller [2]. All of these lo-

---
* This work was supported by the Deutsche Forschungsgemeinschaft (SFB 314)

gics have in common that they are *truth-functional*, that is, composed formulae obtain their truth values from their components and (for quantifiers) from *all* instances of the scope. Therefore a direct utilization of these methods is impossible for Kleene logic, since his quantifiers only range over defined values, that is, not over ⊥. Kleene's approach has been utilized by Tichy [21], Lucio-Carrasco and Gavilanes-Franco [16] to give logical systems for partial functions. Both approaches offer unsorted operationalizations of the systems in sequent calculi by providing special subcalculi for reasoning about definedness.

Other authors (cf. [5, 8, 19, 24]) have avoided the problems that accompany treating a third truth value, and simply consider all atomic expressions containing a meaningless term as false. This has the advantage that partial functions can be handled within the classical two-valued framework. However, the serious drawback is that the results of these logic systems can be unintuitive to the working mathematician. For instance it is mathematical consensus that the following equation should only hold provided that y is not 0:

$$\forall x_{\mathbb{R}}, y_{\mathbb{R}}, z_{\mathbb{R}} \cdot z = \frac{x}{y} \Rightarrow x = y * z$$

However, in the abovementioned systems this is a theorem, since for $y = 0$ the atom $z = \frac{x}{0}$ obtains the truth value f which in turn makes the implication true. We formalize Kleene's ideas for partial functions in an order-sorted three-valued logic, called $\mathcal{SKL}$, that uses the Kleene's strong interpretation of connectives and quantifiers and adapts techniques from Weidenbach's logic [24] to handle definedness information. It will turn out (cf. example 3.9) that the formula above is not a theorem in our formalization, since the case $y = 0$ is a counterexample.

2 Strong Order-Sorted Kleene Logic ($\mathcal{SKL}$)

In [14] Kleene presents a logic, which he calls *strong three-valued logic* for reasoning about partial recursive predicates on the set of natural numbers. He argues that the intuitive meaning of the third truth value should be "undefined" or "unknown" and introduces the truth tables shown in definition 2.6. Similarly Kleene enlarges the universe of discourse by an element ⊥ denoting the undefined number. In his exposition the quantifiers only range over natural numbers, in particular he does not quantify over the undefined individual (number).

The approach of this paper is to make Kleene's meta-level discussion of defined and undefined individuals explicit by structuring the universe of discourse with the sort $\mathfrak{D}$ for all defined individuals. Furthermore all functions and predicates are strict, that is, if one of the arguments of a compound term or an atom evaluates to ⊥, then the term evaluates to ⊥ or the truth value of the atom is u. Just as in Kleene's system, our quantifiers only range over individuals in $\mathfrak{D}$, that is, individuals that are not undefined. This is in contrast to the well-understood framework for truth-functional many-valued logics, where the concept of definedness and defined quantification cannot be easily introduced, since quantification is truth-functional and depends on the truth values for all (even the undefined) instantiations of the scope. Kleene's concept of bounded quantification is essential for our program of representing partial functions, since in a truth-functional approach no proper universally quantified expression can evaluate to the truth value t (dually for the existential quantifier), since all functions and predicates are assumed strict.

In the following we present the logic system $\mathcal{SKL}$, which is a sorted version of what we believe to be a faithful formalization of Kleene's ideas from [14]. We treat the sorted version here, since we need the machinery for dynamic sorts in the calculus to be able to treat the sort $\mathfrak{D}$ (sort techniques as that from [24, 25] give us the bounded quantification). We will call formulations of $\mathcal{SKL}$ where $\mathfrak{D}$ is the only sort in the signature *strong unsorted Kleene logic*. The further use of sorts gives the well-known advantages of sorted logics for the conciseness of representation and reduction of search spaces.

Syntax

Definition 2.1 (Signature) A *signature* $\Sigma := (\mathcal{S}, \mathcal{V}, \mathcal{F}, \mathcal{P})$ consists of the following disjoint sets

- $\mathcal{S}$ is a finite set of *sorts* including the sort $\mathfrak{D}$. We define $\mathcal{S}^* := \mathcal{S} \setminus \{\mathfrak{D}\}$
- $\mathcal{V}$ is a set of *variable symbols*. Each variable x is associated with a unique sort S, which we write in the index, i.e. x_S. We assume that for each sort $S \in \mathcal{S}$ there is a countably infinite supply of variables of sort S in $\mathcal{V}$.
- $\mathcal{F}$ is a set of *function symbols*.
- $\mathcal{P}$ is the set of *predicate symbols*.

The sets $\mathcal{F}$ and $\mathcal{P}$ are subdivided into the sets $\mathcal{F}^k$ of *function symbols of arity k* and $\mathcal{P}^k$ of *predicate symbols of arity k*. Note that individual constants are just nullary functions. We call a signature *unsorted* if $\mathcal{S}^*$ is empty, that is, if $\mathfrak{D}$ is the only sort.

Definition 2.2 (Terms and Formulae) We define the set of *terms* to be the set of variables together with *compound terms* $f(t^1, \ldots, t^k)$ for terms $t^1, \ldots, t^k$ and $f \in \mathcal{F}^k$.

If $P \in \mathcal{P}^k$, then $P(t^1, \ldots, t^k)$ is a *proper atom*. If t is a term and S a sort then $t \leqslant S$ is a *sort atom*. The set of *formulae* contains all atoms and with formulae A and B the formulae $A \wedge B$, $A \vee B$, $A \Rightarrow B$, $\neg A$, $!A$, $\forall x_S. A$ and $\exists x_S. A$. Here the intended meaning of $!A$ is that A is defined.

Semantics

In this section we will define the three valued semantics for $\mathcal{SKL}$ by postulating an "undefined individual" $\bot$ in the universe of discourse. Note that this is similar to the classical flat CPO construction [20], but Kleene's interpretation of truth values does not make u minimal. Since we are not interested in least fix-points, monotonicity does not play a role in this paper.

Definition 2.3 (Strict Σ-Algebra) Let Σ be a signature, then a pair $(\mathcal{A}, \mathcal{I})$ is called a *strict Σ-algebra*, iff
1. the *carrier set* $\mathcal{A}$ is a set of at least two elements that contains $\bot$,
2. the *interpretation function* $\mathcal{I}$ obeys the following restrictions:
   (a) For all function symbols f, the function $\mathcal{I}(f): \mathcal{A}^k \longrightarrow \mathcal{A}$ is strict for $\bot$, that is, $\mathcal{I}(f)(a_1, \ldots, a_k) = \bot$, if $a_i = \bot$ for (at least) one i.
   (b) If P is a predicate symbol, then the relation $\mathcal{I}(P) \subseteq \mathcal{A}^k$ is strict for $\bot$, that is, $\mathcal{I}(P)(a_1, \ldots, a_k) = $ u, if $a_i = \bot$ for (at least) one i.
   (c) If $S \neq \mathfrak{D}$ is a sort, then $\mathcal{I}(S)$ is a total, unary, and strict relation, that is, $\mathcal{I}(S)(a) \in \{\text{f}, \text{t}\}$, if $a \neq \bot$ and $\mathcal{I}(S)(\bot) = $ u.
   (d) $\mathcal{I}(\mathfrak{D})(\bot) = $ f and $\mathcal{I}(\mathfrak{D})(a) = $ t, if $a \neq \bot$. Note that in contrast to all other sorts and predicates, the denotation of $\mathfrak{D}$ is not a strict relation.

We define the *carrier* $\mathcal{A}_S$ of sort S as $\mathcal{A}_S := \{a \in \mathcal{A} \mid \mathcal{I}(S)(a) = \mathsf{t}\}$. Note that in contrast to other sorted logics, it is not assumed that the $\mathcal{A}_S$ are non-empty. This fact will require special treatments in the transformation to clause normal form and for instantiations in the resolution calculus. Furthermore $\perp \notin \mathcal{A}_S$ for any $S \in \mathcal{S}$.

By systematically deleting $\perp$ and u from the carrier and the truth values we can canonically transform strict Σ-algebras into algebras of partial functions. These are an algebraic account of the standard interpretation in mathematics, where partiality of functions is directly modeled by right-unique relations. Obviously these notions of algebras have a one-to-one correspondence, so both approaches are equivalent.

Definition 2.4 (Σ-assignment) Let $(\mathcal{A}, \mathcal{I})$ be a strict Σ-algebra, then we call a total mapping $\varphi: \mathcal{V} \longrightarrow \mathcal{A}$ a Σ-*assignment*, iff $\varphi(x_S) \in \mathcal{A}_S$, provided $\mathcal{A}_S$ is non-empty and $\varphi(x_S) = \perp$ if $\mathcal{A}_S = \emptyset$. We denote the Σ-assignment that coincides with φ away from x and maps x to a with $\varphi, [a/x]$.

Definition 2.5 Let φ be a Σ-assignment into a strict Σ-algebra $(\mathcal{A}, \mathcal{I})$ then we define the *value function* $\mathcal{I}_\varphi$ *from formulae to* $\mathcal{A}$ inductively to be

1. $\mathcal{I}_\varphi(f) := \mathcal{I}(f)$, if f is a function or a predicate.
2. $\mathcal{I}_\varphi(x) := \varphi(x)$, if x is a variable.
3. $\mathcal{I}_\varphi(f(t^1, \ldots, t^k)) := \mathcal{I}(f)(\mathcal{I}_\varphi(t^1), \ldots, \mathcal{I}_\varphi(t^k))$, if f is a function or predicate.
4. $\mathcal{I}_\varphi(t \in S) = \mathcal{I}(S)(\mathcal{I}_\varphi(t))$.

Note that this definition applies to $\mathcal{P}$ and $\mathcal{F}$ alike, thus we have given the semantics of all atomic formulae. The semantic status of sorts is that of total unary predicates; in particular in $\mathcal{A}$ we have $\mathcal{I}_\varphi(t \in S) = \mathsf{u}$, iff $\mathcal{I}_\varphi(t) = \perp$.

Definition 2.6 The value of a formula dominated by a connective is obtained from the value(s) of the subformula(e) in a truth-functional way. Therefore it suffices to define the truth tables for the connectives:

$\wedge$	f	u	t
f	f	f	f
u	f	u	u
t	f	u	t

$\vee$	f	u	t
f	f	u	t
u	u	u	t
t	t	t	t

$\Rightarrow$	f	u	t
f	t	t	t
u	u	u	t
t	f	u	t

$\neg$	
f	t
u	u
t	f

!	
f	t
u	f
t	t

Kleene does not use the ! operator as a connective but treats it on the meta-level. Note while it is useful it is not necessary for the treatment. Furthermore, even this connective does not render $\mathcal{SKL}$ truth-functionally complete, since, just like negation and conjunction, ! is *normal*, that is, when restricted to $\{\mathsf{f}, \mathsf{t}\}$ all connectives yield values in $\{\mathsf{f}, \mathsf{t}\}$.

The semantics of the quantifiers is defined with the help of function $\tilde{\forall}$ and $\tilde{\exists}$ from the non-empty subsets of the truth values in the truth values. We define

$$\mathcal{I}_\varphi(\mathbf{Q}x_S. A) := \tilde{\mathsf{Q}}(\{\mathcal{I}_{\varphi,[a/x]}(A) \mid a \in \mathcal{A}_S\})$$

where $\mathsf{Q} \in \{\forall, \exists\}$ and furthermore

$$\tilde{\forall}(T) := \begin{cases} \mathsf{t} & \text{for } T = \{\mathsf{t}\} \\ \mathsf{u} & \text{for } T = \{\mathsf{t}, \mathsf{u}\} \text{ or } \{\mathsf{u}\} \\ \mathsf{f} & \text{for } \mathsf{f} \in T \end{cases} \quad \text{and} \quad \tilde{\exists}(T) := \begin{cases} \mathsf{t} & \text{for } \mathsf{t} \in T \\ \mathsf{u} & \text{for } T = \{\mathsf{f}, \mathsf{u}\} \text{ or } \{\mathsf{u}\} \\ \mathsf{f} & \text{for } T = \{\mathsf{f}\} \end{cases}$$

Note that with this definition quantification is separated into a truth-functional part $\tilde{\forall}$ and an instantiation part that only considers members of $\mathcal{A}_S$. Since $\bot$ is not a member of any $\mathcal{A}_S$, quantification never considers it and therefore cannot be truth-functional even for the unsorted case.

For lack of space we will in the following often only treat the (sufficient) subset $\{\wedge, \neg, !, \forall\}$ of logical symbols, since all others can be defined from these.

Definition 2.7 (Σ-Model) Let A be a formula, then we call a strict Σ-algebra $\mathcal{M} := (\mathcal{A}, \mathcal{I})$ a Σ-model for A (written $\mathcal{M} \models A$), iff $\mathcal{I}_\varphi(A) = \mathsf{t}$ for all Σ-assignments φ. With this notion we can define the notions of *validity, (un)-satisfiability,* and *entailment* in the usual way.

Remark 2.8 The "tertium non datur" principle of classical logic is no longer valid, since formulae can be undefined, in which case they are neither true nor false. We do however have a "quartum non datur" principle, that is, formulae are either true, false, or undefined, which allows us to derive the validity of a formula by refuting that it is false or undefined. We will use this observation in our resolution calculus.

Extended Example
We will formalize an extended example from elementary algebra that shows the basic features of $\mathcal{SKL}$. Here the sort $\mathbb{R}^*$ denotes the real numbers without zero. Note that we use the sort information to encode definedness information for inversion: $\frac{1}{x}$ is defined for all $x \in \mathbb{R}^*$, since $\mathbb{R}^*$ is subsort of $\mathfrak{D}$ by definition. Naturally, we give only a reduced formalization of real number arithmetic that is sufficient for our example. (For instance, we could add expressions like $\frac{1}{0} \notin \mathfrak{D}$.) Consider the formula $A := (A1 \wedge A2 \wedge A3 \wedge A4 \wedge A5) \Rightarrow T$ with

A1 $\forall x_{\mathbb{R}}\ x \neq 0 \Rightarrow x \in \mathbb{R}^*$
A2 $\forall x_{\mathbb{R}^*}\ \frac{1}{x} \in \mathbb{R}^*$
A3 $\forall x_{\mathbb{R}^*}\ x^2 > 0$

A4 $\forall x_{\mathbb{R}}\ \forall y_{\mathbb{R}}\ x - y \in \mathbb{R}$
A5 $\forall x_{\mathbb{R}}\ \forall y_{\mathbb{R}}\ x - y = 0 \Rightarrow x = y$
T $\forall x_{\mathbb{R}}\ \forall y_{\mathbb{R}}\ x \neq y \Rightarrow \left(\frac{1}{x-y}\right)^2 > 0$

An informal mathematical argumentation why T is entailed by $A1 \wedge \ldots \wedge A5$ can be as follows:

Let x and y be arbitrary elements of $\mathbb{R}$. If $x = y$, the premise of T is false, hence the whole expression true (in this case the conclusion evaluates to u). If $x \neq y$, then the premise is true and the truth value of the whole expression is equal to that of the conclusion $\left(\frac{1}{x-y}\right)^2 > 0$. Since $x \neq y$ we get by A5 that $x - y \neq 0$ and by A4 that $x - y \in \mathbb{R}$, hence by A1 $x - y \in \mathbb{R}^*$ and by A2 $\frac{1}{x-y} \in \mathbb{R}^*$, which finally gives $\left(\frac{1}{x-y}\right)^2 > 0$ together with A3.

However, if we analyze the justification of this argumentation, we see that there is a hidden assumption, namely the totality of the binary predicate $>$ on $\mathbb{R} \times \mathbb{R}$. In fact the formula A is not a tautology, since it is possible to interpret the $>$ predicate as undefined for the second argument being zero, so that A3 as well as T evaluate to u, while the other Ai evaluate to t, hence the whole expression evaluates to u. There are two solutions of this problem, namely adding further formulae Ai, in which the definiteness of the predicates are specified, or – what is normally done in mathematics – to start with a formula where the Ai are assumed to be true, that is, neither false nor undefined. We will discuss the alternatives in remark 3.10 and modify the example accordingly.

Relativization into Truth-Functional Logic

In this section we show that we can always systematically transform $\mathcal{SKL}$ formulae to formulae in an unsorted truth-functional three-valued logic $\mathbf{K}^3$ in a way that respects the semantics. However, we will see that this formulation will lose much of the conciseness of the presentation and enlarge the search spaces involved with automatic theorem proving.

At first glance it may seem that $\mathcal{SKL}$ is only an order-sorted variant of a three-valued instance of the truth-functional many-valued logics that were very thoroughly investigated by Carnielli, Hähnle, Baaz and Fermüller [2, 6, 12]. However, since all instances of this framework are truth-functional, even unsorted $\mathcal{SKL}$ does not fit into this paradigm, since quantification excludes the undefined element. In $\mathcal{SKL}$ quantification is bounded by sorts, which are all subsorts of the sort $\mathfrak{D}$ of defined objects. Therefore relativization not only considers sort information, it also has to care about definedness aspects in quantification.

Informally, $\mathbf{K}^3$-formulae are just first-order formulae (with the additional unary connective !). While the three-valued semantics of the connectives is just that given in definition 2.6, the semantics of quantifications uses unrestricted instantiation:

$$\mathcal{I}_\varphi(\forall x. \ A) := \widetilde{\forall}(\{\mathcal{I}_{\varphi,[a/x]}(A) \mid a \in \mathcal{A}\})$$

Definition 2.9 We define transformations $\mathbf{Rel}^S$ and $\mathbf{Rel}^\mathfrak{D}$, that map $\mathcal{SKL}$-sentences to unsorted $\mathcal{SKL}$-sentences and further to $\mathbf{K}^3$-sentences. $\mathbf{Rel}^S$ is the identity on terms and atoms, homomorphic on connectives and

$$\mathbf{Rel}^S(\forall x_S. \ \Phi) := \forall x_\mathfrak{D}. \ S(x) \Rightarrow \mathbf{Rel}^S(\Phi)$$

Note that in order for these sentences to make sense in unsorted $\mathcal{SKL}$ we have to extend the set of predicate symbols by unary predicates S for all sorts $S \in \mathcal{S}^*$. Furthermore, for any of these new predicates we need the axiom: $\forall x_\mathfrak{D}. \ !S(x)$. The set of all these axioms is denoted by $\mathbf{Rel}^S(\Sigma)$.

We define $\mathbf{Rel}^\mathfrak{D}$ to be the identity (only dropping the sort references from the variables) on terms and proper atoms and

$$\mathbf{Rel}^\mathfrak{D}(t \triangleleft \mathfrak{D}) := \mathfrak{D}(t) \quad \text{and} \quad \mathbf{Rel}^\mathfrak{D}(\forall x_\mathfrak{D}. \ A) := \forall x. \ \mathfrak{D}(x) \Rightarrow \mathbf{Rel}^\mathfrak{D}(A)$$

Just as above we have to extend the set of predicate symbols by a unary predicate $\mathfrak{D}$ and need a set $\mathbf{Rel}^\mathfrak{D}(\Sigma)$ of signature axioms, which contains the axioms

$$\forall x_1, \ldots, x_n. \ P^n(x_1, \ldots, x_n) \vee \neg P^n(x_1, \ldots, x_n) \Rightarrow (\mathfrak{D}(x_1) \wedge \ldots \wedge \mathfrak{D}(x_n))$$
$$\forall x_1, \ldots, x_n. \ \mathfrak{D}(f(x_1, \ldots, x_n)) \Rightarrow (\mathfrak{D}(x_1) \wedge \ldots \wedge \mathfrak{D}(x_n))$$

for any predicate symbol $P \in \mathcal{P}^n$, such that $P \neq \mathfrak{D}$ and for any function symbol $f \in \mathcal{F}^n$, together with the axioms

$$\forall x. \ !\mathfrak{D}(x)^2 \quad \text{and} \quad \exists x. \ \mathfrak{D}(x)$$

These axioms axiomatize the $\mathcal{SKL}$ notion of definedness in $\mathbf{K}^3$. In particular the last axioms state that the predicate $\mathfrak{D}$ is two-valued and non-empty, in contrast to all other sort predicates which are strict, thus three-valued, and may be empty. The

---

[2] Since this is an axiom (see remark 3.10), we could also have used $\forall x. \ \mathfrak{D}(x) \vee \neg \mathfrak{D}(x)$ here.

other axioms force all functions and predicates to be interpreted strictly with respect to the $\mathfrak{D}$ predicate. Note that in the case of nullary function symbols the signature axioms have the form $\mathfrak{D}(c^0)$, which state that individual constants are defined.

Theorem 2.10 (Sort Theorem) *Let Φ be a set of sentences, then the following statements are equivalent*

1. Φ has a Σ-model.
2. $\mathbf{Rel}^S(\Phi)$ has a $\Sigma \cup S^*$-model that satisfies $\mathbf{Rel}^S(\Sigma)$.
3. $\mathbf{Rel}^{\mathfrak{D}} \circ \mathbf{Rel}^S(\Phi)$ has a $\mathbf{K}^3$-model that satisfies $\mathbf{Rel}^{\mathfrak{D}}(\Sigma \cup S^*) \cup \mathbf{Rel}^{\mathfrak{D}}(\mathbf{Rel}^S(\Sigma))$.

Proof sketch: Using standard sort techniques we can use the signature axioms of Σ to identify sorted models in the class of unsorted models, for details see [13]. □

As a consequence of the sort theorem, the standard operationalization for many-valued logics [2, 6, 12] can be utilized to mechanize strong order-sorted Kleene logic and in fact the system of Lucio-Carrasco and Gavilanes-Franco [16] can be seen as a standard many-valued tableau operationalization [12, 3] of the relativization of $\mathcal{SKL}$. However, as the extended example shows, we can do better by using sorted methods, since relativization expands the size and number of input formulae and furthermore expands the search spaces involved in automatic theorem proving by building up many meaningless branches. Note that already the formulation of unsorted $\mathcal{SKL}$ where we only have the required sort $\mathfrak{D}$ is more concise than the relativized version and, as we will see, the theory of definedness is treated by the sorts in the $\mathcal{RPF}$ calculus (cf. section 3). Thus the $\mathcal{RPF}$ calculus is closer to informal practice than the relativization in this respect.

Extended Example (continued)

The relativization $\mathbf{Rel}^S(\mathbf{Rel}^{\mathfrak{D}}(A))$ of the formula A in the extended example is the $\mathbf{K}^3$-formula (R1 $\wedge$ R2 $\wedge$ R3 $\wedge$ R4 $\wedge$ R5) $\Rightarrow$ RT.

R1 $\forall x.\, \mathfrak{D}(x) \Rightarrow (\mathbb{R}(x) \Rightarrow (x \neq 0 \Rightarrow \mathbb{R}^*(x)))$
R2 $\forall x.\, \mathfrak{D}(x) \Rightarrow (\mathbb{R}^*(x) \Rightarrow \mathbb{R}^*(\frac{1}{x}))$
R3 $\forall x.\, \mathfrak{D}(x) \Rightarrow (\mathbb{R}^*(x) \Rightarrow x^2 > 0)$
R4 $\forall x.\, \mathfrak{D}(x) \Rightarrow (\mathbb{R}(x) \Rightarrow (\forall y.\, \mathfrak{D}(y) \wedge \mathbb{R}(y) \Rightarrow \mathbb{R}(x - y)))$
R5 $\forall x.\, \mathfrak{D}(x) \Rightarrow (\mathbb{R}(x) \Rightarrow (\forall y.\, \mathfrak{D}(y) \Rightarrow (\mathbb{R}(y) \Rightarrow (x - y = 0 \Rightarrow x = y))))$
RT $\forall x.\, \mathfrak{D}(x) \Rightarrow (\mathbb{R}(x) \Rightarrow (\forall y.\, \mathfrak{D}(y) \Rightarrow (\mathbb{R}(y) \Rightarrow (x \neq 0 \Rightarrow \left(\frac{1}{x}\right)^2 > 0))))$

The set of signature axioms $\mathbf{Rel}^{\mathfrak{D}}(\Sigma \cup S^*) \cup \mathbf{Rel}^{\mathfrak{D}}(\mathbf{Rel}^S(\Sigma))$ is the following set of $\mathbf{K}^3$-formulae:

R$^=$ $\forall x, y.\, (x = y \vee x \neq y) \Rightarrow \mathfrak{D}(x) \wedge \mathfrak{D}(y)$ $\quad$ R^2 $\forall x.\, \mathfrak{D}(x^2) \Rightarrow \mathfrak{D}(x)$
R$^>$ $\forall x, y.\, (x > y \vee x \not> y) \Rightarrow \mathfrak{D}(x) \wedge \mathfrak{D}(y)$ $\quad$ $\mathfrak{D}^!$ $\forall x.\, \mathfrak{D}(x) \vee \neg \mathfrak{D}(x)$
R$^-$ $\forall x, y.\, \mathfrak{D}(x - y) \Rightarrow \mathfrak{D}(x) \wedge \mathfrak{D}(y)$ $\quad$ $\mathfrak{D}^\emptyset$ $\exists x.\, \mathfrak{D}(x)$
R$^/$ $\forall x.\, \mathfrak{D}(\frac{1}{x}) \Rightarrow \mathfrak{D}(x)$ $\quad$ R$^{\mathbb{R}}$ $\forall x.\, \mathfrak{D}(x) \Rightarrow\, !\mathbb{R}(x)$
R^0 $\mathfrak{D}(0)$ $\quad$ R$^{\mathbb{R}^*}$ $\forall x.\, \mathfrak{D}(x) \Rightarrow\, !\mathbb{R}^*(x)$

3 Resolution

In this section we present a resolution calculus $\mathcal{RPF}$ with dynamic sorts that is a generalization of Weidenbach's work [24, 25] with ideas from [2, 12]. In the literature [23, 7, 18, 11] there are various calculi for sorted logics that vary in deductive power but have in common that the sort information available to sort reasoning remains static over the length of a proof. These methods are not sufficient for our purposes, since definedness[3] cannot in general be decided by syntactic means only, but is usually given in the form of logical axioms that have to be reasoned about in the calculus itself. In contrast to these Weidenbach's logics allows the declaration of conditional sort (and thus definedness) information. When these conditions are proved in the course of the proof, additional sort information becomes available for the sort reasoning mechanism. There are two variants of this calculus (unsorted unification [24] and sorted unification [25]), we have generalized both for our purposes, but in this paper we only present the first (simpler) version due to the lack of space. We refer the reader to the full version of this paper [13] for the other variant.

Clause Normal Form

Definition 3.1 Let A be a formula, then we call A^α (the formula A indexed with the intended truth value $\alpha \in \{f, u, t\}$), a *labeled formula*. We will call a labeled atom L^α a *literal* and a set of literals $\{L_1^{\alpha_1}, \ldots, L_n^{\alpha_n}\}$ a *clause*. We say that a Σ-model $\mathcal{M}$ *satisfies* a clause C, iff it satisfies one of its literals $L^\alpha \in C$, that is, $\mathcal{I}_\varphi(L^\alpha) = \alpha$. $\mathcal{M}$ satisfies a set of clauses iff it satisfies each clause. In order to conserve space, we employ the "," as the operator for the disjoint union of sets, so that C, L^α means $C \cup \{L^\alpha\}$ and L^α is not a member of C. Furthermore we adopt Hähnle's notion of multi-labels in the form $C, A^{\alpha\beta}$ to mean C, A^α, A^β.

Definition 3.2 (Transformations to Clause Normal Form)

$$\frac{C,(A \wedge B)^t}{C,A^t \quad C,B^t} \qquad \frac{C,(A \wedge B)^u}{C,A^{ut} \quad C,B^{ut} \quad C,A^u,B^u} \qquad \frac{C,(A \wedge B)^f}{C,A^f,B^f}$$

$$\frac{C,(\neg A)^t}{C,A^f} \qquad \frac{C,(\neg A)^u}{C,A^u} \qquad \frac{C,(\neg A)^f}{C,A^t}$$

$$\frac{C,(\forall x_S.\, A[x_S])^t}{C,A[x_S]^t} \qquad \frac{C,(\forall x_S.\, A[x_S])^u}{C,A[f(y^1,\ldots,y^n)]^u \quad C,A[x_S]^{ut} \quad C,(f(y^1,\ldots,y^n) \triangleleft S)^t}$$

$$\frac{C,(\forall x_S.\, A[x_S])^f}{C,A[f(y^1,\ldots,y^n)]^f \quad C,(f(y^1,\ldots,y^n) \triangleleft S)^t}$$

$$\frac{C,(!A)^t}{C,A^{tf}} \qquad \frac{C,(!A)^u}{C} \qquad \frac{C,(!A)^f}{C,A^u} \qquad \frac{C,(t \triangleleft \mathfrak{D})^u}{C} \qquad \frac{C,(t \triangleleft S)^u}{C,(t \triangleleft \mathfrak{D})^f}$$

where $\{x_S, y^1, \ldots, y^n\} = \mathbf{Free}(A)$ and f is a new function symbol of arity n. Here $\mathbf{Free}(A)$ denotes the set of free variables of A.

---

[3] Cohn's Boolean lattice of sorts [7] has $\bot$ elements for ill-sorted terms and formulae. While the notation is similar to ours, this concept should not be confused with undefinedness. In contrast to Kleene's interpretation, all of Cohn's connectives are strict and no expression containing ill-sorted elements can be a tautology, making his calculus and his notion of ill-sortedness inappropriate for the treatment of undefinedness.

Note that this set of transformations is confluent, therefore any total reduction of a set Φ of labelled sentences results in a unique set of clauses. We will denote this set with $\mathbf{CNF}(\Phi)$.

General Assumption 3.3 The clause normal form transformations as presented above are not complete, that is, they do not transform every given labelled formula into clause form, since the rules for quantified formulae insist that the bound variable occurs in the scope. In fact the handling of degenerate quantifications poses some problems in the presence of possibly empty sorts, as quantification over empty sets are vacuously true. In this situation we have three possibilities, either to forbid degenerate quantifications, or empty sorts, or treat degenerate quantifications in the clause normal form transformations. For this paper we chose the first, since degenerate quantifications do not make much sense mathematically and do not appear in informal mathematics. See [13] for the other possibilities. Thus we will asssume that in all formulae in this paper the bound variables of quantifications occur in the scopes.

As usual the reduction to clause normal form conserves satisfiability.

Theorem 3.4 *Let Φ be a set of labelled sentences, then the clause normal form $\mathbf{CNF}(\Phi)$ is satisfiable, iff Φ is.*

Resolution Calculus ($\mathcal{RPF}$)

Now we proceed to give a simple resolution calculus, which utilizes unsorted unification. However, despite its name the calculus still utilizes the sort information present in the clause set and therefore gives considerably improved search behavior over unsorted methods as in [16]. In [13], we have further improved the calculus by using a sorted unification algorithm, which delegates parts of the search into the unification algorithm. For unsorted substitutions a naive resolution rule is unsound. Therefore we have to add a residual (the sort constraint) that ensures the well-sortedness of the unifier.

Definition 3.5 (Sort Constraints) Let $\sigma = [t^1/x^1_{S_1}], \ldots, [t^n/x^n_{S_n}]$ be a substitution, then we define the *sort constraint for σ* to be the clause

$$\mathcal{SC}(\sigma) := \{(t^1 {\in} S_1)^{\mathsf{fu}}, \ldots, (t^n {\in} S_n)^{\mathsf{fu}}\}$$

Definition 3.6 (Resolution Inference Rules ($\mathcal{RPF}$))

$$\frac{L^\alpha, C \qquad M^\beta, D}{\sigma(C), \sigma(D), \mathcal{SC}(\sigma)} \, Res \qquad \frac{L^\alpha, M^\alpha, C}{\sigma(L^\alpha), \sigma(C), \mathcal{SC}(\sigma)} \, Fac$$

$$\frac{(t{\in}\mathfrak{D})^{\mathsf{f}}, C \qquad L^\gamma, D}{\rho(C), \rho(D), \mathcal{SC}(\rho)} \, Strict$$

where $\alpha \neq \beta$ and $\gamma \in \{\mathsf{t}, \mathsf{f}\}$. For *Res* and *Fac* the substitution σ is the most general (unsorted) unifier of L and M and for *Strict* there exists a subterm s of L, such that ρ is a most general unifier of t and s. Here we have assumed α, β and γ to be single truth values, naturally the rules can be easily extended to sets of truth values.

Remark 3.7 Note that clauses containing A^{fut} are tautologous and can therefore be deleted in the generation of the clause normal form as well as in the deduction process. The calculus can be extended by the usual subsumption rule, allowing to delete clauses that are subsumed (super-sets).

Definition 3.8 Let A be a sentence and Φ be the clause normal form of the set $\{\{A^f\},\{A^u\}\}$ then we say that A can be *proven in* $\mathcal{RPF}$ ($\vdash A$), iff there is a derivation of the empty clause $\square$ from Φ with the inference rules above.

Example 3.9 Now we can come back to the example from the exposition. The assertion is not a theorem of $\mathcal{SKL}$, since the clause normal form of the instance $\{\{(1=\frac{1}{0} \Rightarrow 1=0*1)^f\},\{(1=\frac{1}{0} \Rightarrow 1=0*1)^u\}\}$, namely:

$$(1=\tfrac{1}{0})^u, (1=\tfrac{1}{0})^t$$
$$(1=0*1)^u, (1=0*1)^f$$

is satisfiable. In fact in any reasonable formalization of elementary algebra $1=\frac{1}{0}$ is undefined, whereas $1=0*1$ is false. Thus, since $\mathcal{RPF}$ is sound (cf. 3.11), the example cannot be a theorem.

Remark 3.10 In practical applications most problems will be of the form $A := (A_1 \wedge \ldots \wedge A_n \Rightarrow C)$ where the A_i are the assumptions and C is the intended conclusion. In contrast to classical first-order predicate logic where it suffices to take the clause normal form of $\{\{A_1^t\},\ldots,\{A_n^t\},\{C^f\}\}$ the situation here is more complex, since in $\mathcal{SKL}$ we also have to refute the case that A gets the value u. It is however easy to see, that we can start the calculation of the clause normal form with the set

$$\{\{A_1^{ut}\},\ldots,\{A_n^{ut}\},\{C^{fu}\}\} \qquad \text{or with the sets}$$

$$\{\{A_1^{ut}\},\ldots,\{A_n^{ut}\},\{A_1^u\},\ldots,\{A_n^u\},\{C^u\}\} \qquad (*)$$
$$\{\{A_1^t\},\ldots,\{A_n^t\},\{C^{fu}\}\} \qquad (**)$$

which have to be refuted by the resolution calculus independently. In the second case the refutation can be split in two independent proofs, thus reducing the search space considerably. Nevertheless, the refutation of the set (*) is impractical except for trivial examples. Fortunately in mathematical practice the assumptions A_i often have the status of axioms, which are assumed to be true independently of the theorem[4]. Then the problem is really of the form

$$A' := (A_1 \wedge !A_1 \wedge \ldots \wedge A_n \wedge !A_n \Rightarrow C)$$

The clause normal form of A' is just that of (**), which is close to the classical case in derivational complexity. In particular the background theory formalized by the A_i results in exactly the same clauses as in the classical case.

---

[4] This is also the very idea of the set of support strategy in resolution theorem proving.

Extended Example (continued)

Following the discussion above we will continue our extended example with the calculation of the clause normal form (∗∗) of A1∧!A1 ∧ ... ∧ A5∧!A5 ⇒ T. Since $\mathbb{R}$ and $\mathbb{R}^*$ are not empty, we can use slightly simplified quantification rules in the clause normal form transformations (see [13]).

A1 $(x_\mathbb{R} = 0)^t, (x_\mathbb{R} \in \mathbb{R}^*)^t$
A2 $(\frac{1}{x_{\mathbb{R}^*}} \in \mathbb{R}^*)^t$
A3 $(x_{\mathbb{R}^*}^2 > 0)^t$
A4 $(x_\mathbb{R} - y_\mathbb{R} \in \mathbb{R})^t$
A5 $(x_\mathbb{R} - y_\mathbb{R} = 0)^f, (x_\mathbb{R} = y_\mathbb{R})^t$

The price for the formal treatment of three-valued partiality has to be paid in the complicated clause normal form of the formula T with the label fu.

T1 $(c \in \mathbb{R})^t$
T2 $(d \in \mathbb{R})^t$
T3 $(e \in \mathbb{R})^t$
T4 $(f \in \mathbb{R})^t$
T5 $(g(y_\mathbb{R}) \in \mathbb{R})^t$
T6 $(c = d)^f, (e = f)^{fu}$

T7 $(c = d)^f, \left(\left(\frac{1}{e-f}\right)^2 > 0\right)^{fu}$
T8 $\left(\left(\frac{1}{c-d}\right)^2 > 0\right)^f, (e = f)^{fu}$
T9 $\left(\left(\frac{1}{c-d}\right)^2 > 0\right)^f, \left(\left(\frac{1}{e-f}\right)^2 > 0\right)^{fu}$

Eight further clauses resulting from the theorem are not shown here, four are tautologies, four others not needed for the derivation below.

T6 & A5→R1 $(c - d = 0)^f, (e = f)^{fu}, (c \in \mathbb{R})^{fu}, (d \in \mathbb{R})^{fu}$
R1 & A1→R2 $(c - d \in \mathbb{R}^*)^t, (e = f)^{fu}, (c - d \in \mathbb{R})^{fu}, (c \in \mathbb{R})^{fu}, (d \in \mathbb{R})^{fu}$
R2 & A4→R3 $(c - d \in \mathbb{R}^*)^t, (e = f)^{fu}, (c \in \mathbb{R})^{fu}, (d \in \mathbb{R})^{fu}$
R3 & T1→R4 $(c - d \in \mathbb{R}^*)^t, (e = f)^{fu}, (d \in \mathbb{R})^{fu}$
R4 & T2→R5 $(c - d \in \mathbb{R}^*)^t, (e = f)^{fu}$
T8 & A3→R6 $(e = f)^{fu}, \left(\left(\frac{1}{c-d}\right) \in \mathbb{R}^*\right)^{fu}$
R5 & A2→R7 $(e = f)^{fu}, (c - d \in \mathbb{R}^*)^{fu}$
R7 & R5→R8 $(e = f)^{fu}$

Analogously, clause T7 can be reduced with T9 to R16.

... & ... →R16 $\left(\left(\frac{1}{e-f}\right)^2 > 0\right)^{fu}$
R16 & A3 →R17 $\left(\frac{1}{e-f} \in \mathbb{R}^*\right)^{fu}$
R17 & A2 →R18 $(e - f \in \mathbb{R}^*)^{fu}$
R18 & A1 →R19 $(e - f = 0)^t, (e - f \in \mathbb{R})^{fu}$
R19 & A4 →R20 $(e - f = 0)^t, (e \in \mathbb{R})^{fu}, (f \in \mathbb{R})^{fu}$
R20 & A5 →R21 $(e = f)^t, (e \in \mathbb{R})^{fu}, (f \in \mathbb{R})^{fu}$
R21 & T3 →R22 $(e = f)^t, (f \in \mathbb{R})^{fu}$
R22 & T4 →R23 $(e = f)^t$
R8 & R23→R24 □

Soundness and Completeness

Theorem 3.11 (Soundness) $\mathcal{RPF}$ *is sound.*

Proof sketch: The soundness of the resolution and factoring rules is established in the usual way taking into account that the sort constraints make the substitutions "well-sorted" and thus compatible with the semantics: The sort constraints add two sort literals $(t{\in}S)^{\mathsf{f}}, (t{\in}S)^{\mathsf{u}}$ per component of the substitution, which only can be refuted if indeed $(t{\in}S)^{\mathsf{t}}$.

The *Strict* rule is sound, because functions and predicates in $\mathcal{SKL}$ are strict and thus undefined subterms of a literal make the literal undefined. □

Definition 3.12 Let $C := \{L_1^{\alpha_1}, \ldots, L_n^{\alpha_n}\}$ be a clause, then the *conditional instantiation* $\sigma{\downarrow}(C)$ *of* σ *to* C is defined by

$$\sigma{\downarrow}(C) := \{\sigma(L_1^{\alpha_1}), \ldots, \sigma(L_n^{\alpha_n})\} \cup \mathcal{SC}(\sigma|_{\mathbf{Free}(C)})$$

The following result from [24] is independent of the number of truth values.

Lemma 3.13 *Conditional instantiation is sound: for any clause C, substitution σ and Σ-model $\mathcal{M}$ we have that $\mathcal{M} \models \sigma{\downarrow}(C)$, whenever $\mathcal{M} \models C$.*

Definition 3.14 Let A be a sentence and $\mathbf{CNF}(A)$ be the clause normal form of A, then we define the *Herbrand set of clauses* $\mathbf{CNF}_H(A)$ *for* A as
$\mathbf{CNF}_H(A) := \{\sigma{\downarrow}(C) \mid C \in \mathbf{CNF}(A), \sigma \text{ ground substitution}, \mathbf{Dom}(\sigma) = \mathrm{Free}(C)\}$

Definition 3.15 We will call two literals L^α and L^β *complementary*, if $\alpha \neq \beta$ and literals L^γ and $(t{\in}\mathfrak{D})^{\mathsf{f}}$ ⊥-*complementary*, if t is a subterm of L and $\gamma \in \{\mathsf{t},\mathsf{f}\}$.

Definition 3.16 (Herbrand Model) Let Φ be a set of clauses, then the *Herbrand base* $\mathcal{H}(\Phi)$ *of* Φ is defined to be the set of all ground atoms containing only function symbols that appear in the clauses of Φ. If there is no individual constant in Φ, we add a new constant c. A *valuation* ν is a function $\mathcal{H}(\Phi) \longrightarrow \{\mathsf{f}, \mathsf{u}, \mathsf{t}\}$, such that for all atoms $L, M \in \mathcal{H}(\Phi)$ the literals $L^{\nu(L)}$ and $M^{\nu(M)}$ are not ⊥-complementary. Note that these literals are not complementary since ν is a function. The Σ-*Herbrand model* $\mathcal{H}$ *for* Φ *and* ν is the set $\mathcal{H} := \{L^\alpha \mid \alpha = \nu(L), L \in \mathcal{H}(\Phi)\}$.

We say that a Σ-Herbrand model $\mathcal{H}$ *satisfies a clause set* Φ iff for all ground substitutions σ and clauses $C \in \Phi$ we have $\sigma{\downarrow}(C) \cap \mathcal{H} \neq \emptyset$. A clause set is called Σ-*Herbrand-unsatisfiable* iff there is no Σ-Herbrand-model for Φ.

Theorem 3.17 (Herbrand Theorem) *Let A be a formula, then the clause normal form $\mathbf{CNF}(A)$ has a Σ-model iff $\mathbf{CNF}_H(A)$ has a Σ-Herbrand-model.*

Proof: Let $\mathcal{M} = (\mathcal{A}, \mathcal{I})$ be a Σ-model for $\Phi := \mathbf{CNF}(A)$. The set

$$\mathcal{H} := \{L^\alpha \mid L \in \mathcal{H}(\Phi), \alpha = \mathcal{I}_\varphi(L)\}$$

is a Σ-Herbrand model for $\Psi := \mathbf{CNF}_H(A)$ if φ is an arbitrary Σ-assignment, since obviously $\mathcal{I}_\varphi$ is a valuation. To show that indeed $\mathcal{H}$ is a Σ-Herbrand model for Ψ, we assume the opposite, that is, there is a clause $C \in \Psi$, such that $\mathcal{H} \cap C = \emptyset$. Since $C \in \Psi$ there is a substitution $\sigma = [t^i/x^i_{S_i}]$ and a clause $D \in \Phi$, such that $C = \sigma{\downarrow}(D) = \sigma(D) \cup \mathcal{SC}(\sigma)$.

Without loss of generality we can assume that $\mathcal{I}(S_i)(\mathcal{I}_\varphi(t^i)) = \mathsf{t}$, since otherwise $\mathcal{I}_\varphi(t^i \mathbin{\in} S^i) \in \{\mathsf{f}, \mathsf{u}\}$, and therefore $(t^i \mathbin{\in} S^i)^\gamma \in \mathcal{H}$ for $\gamma \in \{\mathsf{f}, \mathsf{u}\}$, which contradicts the assumption. Thus the mapping $\psi := \varphi, [\mathcal{I}_\varphi(t^i)/x^i]$ is a Σ-assignment.

Note that since $\mathcal{M}$ is a model of Φ, we have that $\mathcal{M} \models D$ and therefore there is a literal $L^\alpha \in D$, such that $\alpha = \mathcal{I}_\psi(L) = \mathcal{I}_\varphi(\sigma(L))$, hence $\sigma(L) \in \mathcal{H}$, which contradicts the assumption.

For the converse direction let $\mathcal{H}$ be a Σ-Herbrand model for Ψ. To construct a Σ-model $\mathcal{M}$ for Φ we first construct an algebra of partial functions $(\mathcal{A}, \mathcal{I})$ (and then pass to the corresponding strict Σ-algebra). Let

$$\mathcal{A} := \{t \mid \exists L^\alpha \in \mathcal{H} \text{ where } \alpha \in \{\mathsf{f}, \mathsf{t}\} \text{ and } t \text{ subterm of } L\}$$

and let $\mathcal{I}(S)$, $\mathcal{I}(f^n)$ and $\mathcal{I}(P^n)$ be partial functions, such that

$$\mathcal{I}(S)(t) = \mathsf{t} \text{ iff } (t \mathbin{\in} S)^\mathsf{t} \in \mathcal{H}$$
$$\mathcal{I}(f^n)(t^1, \ldots, t^n) := f^n(t^1, \ldots, t^n) \text{ iff } f^n(t^1, \ldots, t^n) \in \mathcal{A}$$
$$\mathcal{I}(P^n)(t^1, \ldots, t^n) := \alpha \text{ iff } (P^n(t^1, \ldots, t^n))^\alpha \in \mathcal{H}$$

Now let $\mathcal{M}$ be the strict Σ-algebra corresponding to the partial Σ-algebra $(\mathcal{A}, \mathcal{I})$. We proceed by convincing ourselves that $\mathcal{M} \models \Phi$. Let $C \in \Phi$ and $\varphi := [t^i/x^i_{S_i}]$ be an arbitrary Σ-assignment. Since $\mathcal{A}$ is a set of ground terms φ is also a ground substitution and moreover $(t^i \mathbin{\in} S_i)^\mathsf{t} \in \mathcal{H}$ by construction of $\mathcal{I}$.

$\mathcal{H}$ is a Σ-Herbrand model for Ψ and thus $\varphi\!\downarrow\!(C) \cap \mathcal{H} = (\varphi(C) \cup \mathcal{SC}(\varphi)) \cap \mathcal{H} \neq \emptyset$. Because $\mathcal{H}$ cannot contain complementary literals we must already have a literal $\varphi(L^\alpha) \in \varphi(C) \cap \mathcal{H}$. Now let ν be the valuation associated with $\mathcal{H}$. Since $\varphi(L^\alpha) \in \mathcal{H}$ we have $\alpha = \nu(\varphi(L)) = \mathcal{I}_\varphi(L)$, which implies $\mathcal{M} \models_\varphi L^\alpha$. We have taken C and φ arbitrary, so we get the assertion. □

Corollary 3.18 *A set Φ of ground unit clauses is unsatisfiable iff it contains two complementary or $\bot$-complementary literals.*

Theorem 3.19 (Ground Completeness) *Let Φ be an unsatisfiable set of ground clauses, then there exists a $\mathcal{RPF}$ derivation of the empty clause from Φ.*

Proof: The proof is analogous to the standard k-parameter proof of Anderson and Bledsoe [1]. We show be induction on $k := \sum_{C \in \Phi}(\text{card}(C) - 1)$ that there exists a refutation for Φ.

If $k = 0$ then Φ is a set of ground unit clauses. Therefore by Corollary 3.18 and the assumed unsatisfiability there has to be a pair of complementary or $\bot$-complementary literals in Φ. Thus a single application of the rule *Res* or *Strict* yields the empty clause.

If $k > 0$, then there is a non-unit clause $C =: C_1 \cup C_2 \in \Phi$. If $\Phi = \Phi' \cup \{C\}$ then the k parameters for $\Phi_1 := \Phi' \cup \{C_1\}$ and $\Phi_2 := \Phi' \cup \{C_2\}$ are smaller than k and therefore by inductive hypothesis there are refutations for Φ_1 and Φ_2 which can be combined to a refutation for Φ, since Φ is ground. □

Theorem 3.20 (Completeness) $\mathcal{RPF}$ *is complete.*

Proof sketch: For the proof of this assertion we combine the completeness result from the ground case with a lifting argument. It turns out that the lifting property can be established by methods from [24], since they are independent of the number of truth values. □

4 Conclusion

We have developed an order sorted three-valued logic for the formalization of informal mathematical reasoning with partial functions. This system formalizes and generalizes the system proposed by Kleene in [14] for the treatment of partial functions over natural numbers to general first-order logic. In fact we believe that the unsorted version of our system without the ! operator is a faithful formalization of Kleene's ideas. Furthermore we have presented a sound and complete resolution calculus with dynamic sorts for our system, which uses the sort mechanism to capture the fact that in Kleene's logic quantification only ranges over defined individuals.

Our calculus can be seen as an extension of classical logic that combines methods from many-valued logics (cf. [2, 12]) for a correct treatment of the undefined and order-sorted logics (see [24, 25]) for an adequate treatment of the defined. It differs from the sequent calculus in [16] in that the use of dynamic sort techniques greatly simplifies the calculus. In particular in the version with sorted unification as described in [13] most definedness preconditions can be taken care of in the unification. Thus we believe that our system is not only more faithful to Kleene's ideas (definedness inference is handled in the unification at a level below the calculus) but also more efficient for the sort techniques involved.

Of course further extensions of the system described here have to be considered in order to be feasible for practical mathematics. In particular this calculus does not address the question of the efficient mechanization of equality, here paramodulation (cf. [17]) or even superposition ([4]) methods would be interesting to study. However, we believe that this endeavor will mainly involve the development of the sort aspects for these calculi, because we think that the aspects of three-valuedness will not be critical.

On the other hand, the mechanization of higher-order features is essential for the formalization of mathematical practice. Higher-order logics are especially suitable for formalizing partial functions, since functions are first class objects of the systems, that can even be quantified over. In this direction the work of Farmer et al. [8, 9] has shown that partial functions are a very natural and powerful tool for formalizing mathematics. We expect that our three-valued approach, which remedies some problems of their simpler two-valued approach (see the discussion in the introduction and in example 3.9) can be generalized in much the same manner and will be a useful tool for formalizing mathematics.

Acknowledgments: We would like to thank Christian Fermüller and Ortwin Scheja for stimulating discussions and the anonymous referees for useful comments.

References

1. R. Anderson and W.W. Bledsoe. A linear format for resolution with merging and a new technique for establishing completeness. *Journal of the ACM*, 17:525–534, 1970.
2. Matthias Baaz and Christian G. Fermüller. Resolution for many-valued logics. In A. Voronkov, editor, *Proceedings of International Conference on Logic Programming and Automated Reasoning*, pages 107–118, St. Petersburg, Russia, 1992. Springer Verlag. LNAI 624.
3. Matthias Baaz, Christian G. Fermüller, and Richard Zach. Dual systems of sequents and tableaux for many-valued logics. Technical Report TUW-E185.2BFZ.2-92, Technische Universität Wien, 1993. Short version in *Proceedings of the 23rd International Symposium on Multiple Valued Logic*, Sacramento, California, USA, 1993. IEEE Press.

4. Leo Bachmair and Harald Ganzinger. Non-clausal resolution and superposition with selection and redundancy criteria. In A. Voronkov, editor, *Proceedings of International Conference on Logic Programming and Automated Reasoning*, pages 273–284, St. Petersburg, Russia, 1992. Springer Verlag. LNAI 624.
5. Michael J. Beeson. *Foundations of Constructive Mathematics.* Springer Verlag, 1985.
6. Walter A. Carnielli. On sequents and tableaux for many-valued logics. *Journal of Non-Classical Logic*, 8(1):59–76, 1991.
7. Anthony G. Cohn. A more expressive formulation of many sorted logics. *Journal of Automated Reasoning*, 3:113–200, 1987.
8. William M. Farmer. A partial functions version of Church's simple theory of types. Technical Report M88-52, The MITRE Corporation, Bedford, Massachusetts, USA, February 1990.
9. William M. Farmer, Joshua D. Guttman, and F. Javier Thayer. IMPS: An Interactive Mathematical Proof System. *Journal of Automated Reasoning*, 11(2):213–248, October 1993.
10. Bas C. van Fraassen. Singular terms, truth-value gaps, and free logic. *The Journal of Philosophy*, LXIII(17):481–495, 1966.
11. Alan M. Frisch. The substitutional framework for sorted deduction: Fundamental results on hybrid reasoning. *Artificial Intelligence*, 49:161–198, 1991.
12. Reiner Hähnle. *Automated Deduction in Multiple-Valued Logics*, Oxford University Press, 1994.
13. Manfred Kerber and Michael Kohlhase. A mechanization of strong Kleene logic for partial functions. SEKI-Report SR-93-20 (SFB), Universität des Saarlandes, Saarbrücken, Germany, 1993.
14. Stephen Cole Kleene. *Introduction to Metamathematics.* Van Nostrand, 1952.
15. H. Leblanc and R. Thomason. Completeness theorems for some presupposition-free logics. *Fundamenta Mathematicae*, 62:125–164, 1968.
16. Francisca Lucio-Carrrasco and Antonio Gavilanes-Franco. A first order logic for partial functions. In *Proceedings STACS'89*, pages 47–58. Springer Verlag, 1989. LNCS 349.
17. Arthur Robinson and Larry Wos. Paramodulation and TP in first order theories with equality. *Machine Intelligence*, 4:135–150, 1969.
18. Manfred Schmidt-Schauß. *Computational Aspects of an Order-Sorted Logic with Term Declarations.* Springer Verlag, 1989. LNAI 395.
19. R. Schock. *Logics without Existence Assumptions.* Almquist & Wisell, 1968.
20. Dana S. Scott. Outline of a mathematical theory of computation. Technical Monograph PRG-2, Oxford University Computing Laboratory, November 1970.
21. Pawel Tichy. Foundations of partial type theory. *Reports on Mathematical Logic*, 14:59–72, 1982.
22. Bertrand Russell. On denoting. *Mind (New Series)*, 14:479–493, 1905.
23. Christoph Walther. *A Many-Sorted Calculus Based on Resolution and Paramodulation.* Research Notes in Artificial Intelligence. Pitman and Morgan Kaufmann, 1987.
24. Christoph Weidenbach. A resolution calculus with dynamic sort structures and partial functions. SEKI-Report SR-89-23, Fachbereich Informatik, Universität Kaiserslautern, Kaiserslautern, Germany, 1989. Short version in ECAI'90, p.668–693.
25. Christoph Weidenbach. A sorted logic using dynamic sorts. Technical Report MPI-I-91-218, Max-Planck-Institut für Informatik, Saarbrücken, Germany, 1991. Short version in IJCAI'93, p.60–65.

Algebraic Factoring and Geometry Theorem Proving

Dongming Wang

Laboratoire d'Informatique Fondamentale et d'Intelligence Artificielle
Institut IMAG, 46, avenue Félix Viallet, 38031 Grenoble Cédex, France

Abstract. Two methods for polynomial factorization over algebraic extension fields are reviewed. It is explained how geometric theorems may be proved by using irreducible zero decomposition for which algebraic factoring is necessary. A set of selected geometric theorems are taken as examples to illustrate how algebraic factoring can help understand the ambiguity of a theorem and prove it even if its algebraic formulation does not precisely correspond to the geometric statement. Among the polynomials occurring in our examples which need to be algebraically factorized, 12 are presented. Experiments with the two factoring methods for these polynomials are reported in comparison with the Maple's built-in factorizer. Our methods are always faster and any of the 12 polynomials can be factorized within 40 CPU seconds on a SUN SparcServer 690/51. Timings for proving the example theorems are also provided.

1 Introduction

To prove a geometric theorem $\mathfrak{T}$ by using Wu type methods [2-5, 13, 20-25], one first expresses the hypothesis and the conclusion of $\mathfrak{T}$ respectively as sets of polynomial equations $\mathbb{H} = 0$ and $\mathbb{C} = 0$, and then determines whether every $C = 0$ ($C \in \mathbb{C}$) is a consequence of $\mathbb{H} = 0$. The algebraic formulation may not precisely correspond to the original theorem and its geometric statement. This is mainly due to the occurrence of geometric degeneracies and ambiguities which one might be unaware of or has trouble to deal with when formulating the theorem. To make the algebraic formulation precise, one has to state the theorem tediously and use polynomial inequalities whose treatment is computationally much more difficult. Therefore, we sometimes even prefer to use "imprecise" algebraic formulations in order to deal with geometric degeneracies and ambiguities more effectively and to avoid using inequalities. In such cases, the algebraic form of a geometric theorem may no longer be a theorem in the logical sense. We need to decide whether or not the conclusion $\mathbb{C} = 0$ holds true on the entire configuration defined by $\mathbb{H} = 0$, and if not as it is often the case, on which part of the configuration it holds true. To do so, it may be necessary to decompose the set $\mathbb{H}$ of hypothesis polynomials into irreducible components. Such a decomposition often requires heavy computation due to various time-consuming substeps. One of the substeps is factorization of polynomials over successive algebraic extension fields which has been considered as the bottle-neck and the most difficult part of Wu's method [20, 21] and has been left aside without much investigation in the context of geometry theorem proving. There are a few papers in which algebraic factorization and the reducibility problem are considered for special cases. For example, a factoring method based on the determination of perfect squares for the discriminants of quadratic polynomials was proposed in [2]. Wu [22] described a method for dealing with the reducibility problem due to certain geometric ambiguities by introducing oriented lines and circles. See also a recent paper by Yang, Zhang and Hou [25]. Some other techniques were used to avoid the reducibility of the involved polynomials (see [5, 13] for example).

On the other hand, several general methods for polynomial factorization over algebraic extension fields have been developed in the research community of computer algebra (see [1, 7-9, 11, 18, 19] for example). Some of these methods with practical improvements have been implemented in the available computer algebraic systems with reasonable efficiency. However, we are not aware of any work on the application of these methods to geometry theorem proving. Meanwhile, Hu and the author proposed a simple method for algebraic factoring [6] based on equations solving. More recently the author presented another method [16] which has some similarity to Trager's [11]. Our second method is quite efficient, whereas so is the first only for polynomials of lower degree.

In this paper algebraic factoring is considered together with geometry theorem proving. This can be of interest because algebraic approaches to geometry theorem proving require algebraic factoring on the one hand, while polynomials from geometry theorem proving are good (not artificially constructed) examples for testing the efficiency and applicability of factoring methods to practical problems on the other hand. In the following sections we shall give a brief review of our two methods of factorization over algebraic extension fields and explain how geometric theorems can be proved by using irreducible zero decomposition for which algebraic factoring is necessary. Our attention here is paid to algebraic factorization rather than zero decomposition and others. As test examples we take a set of popularized, famous geometric theorems whose algebraic proofs may require algebraic factorization. We report on the computing times with the two methods for these examples in comparison with the latest release of the Maple V's built-in factorizer. It is shown that our methods are faster for all the cases. In fact, any of the 12 polynomials can be algebraically factorized by using our methods within 40 CPU seconds on a SUN SparcServer 690/51.

2 Methods for Algebraic Factoring

Let $u_1, \ldots, u_d$ be d transcendental elements (indeterminates), abbreviated sometimes to u, and Q the rational number field. Let $K_0 = Q(u_1, \ldots, u_d)$ be the extension field obtained from Q by adjoining $u_1, \ldots, u_d$ and $K_i = K_0(\eta_1, \ldots, \eta_i)$ ($1 \leq i \leq r$) the algebraic extension field obtained from K_0 by adjoining successively the algebraic elements $\eta_1, \ldots, \eta_i$, where η_i has minimal polynomial

$$A_i = a_{i0} y_i^{m_i} + a_{i1} y_i^{m_i - 1} + \cdots + a_{im_i}, \quad a_{ij} \in K_{i-1}, \; j = 0, \ldots, m_i$$

for each i. As the minimal polynomials A_i are explicitly given, we shall simply write $K_0(y_1, \ldots, y_i)$ for K_i without introducing the η's. Without loss of generality, we assume that $A_i(u, y_1, \ldots, y_i) \in K_0[y_1, \ldots, y_i]$ for each i. Then in the terminology of characteristic sets $\mathcal{A} = \{A_1, \ldots, A_r\}$ forms an irreducible ascending set. For simplicity of wording, we call $A_1, \ldots, A_r$ the adjoining polynomials and $\{A_1, \ldots, A_r\}$ the adjoining ascending set of the field K_r for $y_1, \ldots, y_r$, and denote by $\text{prem}(P, \mathcal{A})$ the pseudo-remainder of any polynomial P w.r.t. (with respect to) $\mathcal{A}$.

Our problem amounts to factorizing any polynomial

$$F(y) = f_0 y^n + f_1 y^{n-1} + \cdots + f_n \in K_0[y_1, \ldots, y_r, y], \quad n > 1,$$

where

$$f_i = \sum_{\substack{0 \leq k_j \leq m_j - 1 \\ 1 \leq j \leq r}} f_{ik_1 \cdots k_r} y_1^{k_1} \cdots y_r^{k_r}, \quad f_{ik_1 \cdots k_r} \in K_0, \quad i = 0, \ldots, n,$$

over the extension field K_r. We assume that one knows how to factorize polynomials over K_0 and $F(y)$ is already irreducible over K_0.

2.1 Method A

Suppose that $F(y)$ can be factorized over K_r as

$$F(y) \doteq f_0 \cdot G(y) \cdot H(y), \qquad (1)$$

where

$$\begin{aligned} G(y) &= y^s + g_1 y^{s-1} + \cdots + g_s, \\ H(y) &= y^t + h_1 y^{t-1} + \cdots + h_t, \end{aligned} \quad s + t = n, \ 1 \leq s, t \leq n - 1$$

and the dot equality means that $\text{prem}(F - f_0 GH, \mathcal{A}) = 0$. The above g_i and h_j can be written as

$$\begin{aligned} g_i &= \sum_{\substack{0 \leq k_l \leq m_l - 1 \\ 1 \leq l \leq r}} g_{ik_1 \cdots k_r} y_1^{k_1} \cdots y_r^{k_r}, \\ h_j &= \sum_{\substack{0 \leq k_l \leq m_l - 1 \\ 1 \leq l \leq r}} h_{jk_1 \cdots k_r} y_1^{k_1} \cdots y_r^{k_r}, \end{aligned} \quad \begin{aligned} & g_{ik_1 \cdots k_r}, h_{jk_1 \cdots k_r} \in K_0, \\ & i = 1, \ldots, s, \ j = 1, \ldots, t. \end{aligned}$$

Here, the number of $g_{ik_1 \cdots k_r}$'s and $h_{jk_1 \cdots k_r}$'s is $(s+t)m_1 \cdots m_r = M$. We rename these indeterminate coefficients with a fixed order as $x_1 \prec x_2 \prec \cdots \prec x_M$. Now expand $F - f_0 GH$, compute its pseudo-remainder R w.r.t. $\mathcal{A}$ and equate the coefficients of all the power products of R in $y_1, \ldots, y_r, y$ to 0. We shall obtain a system of M polynomial equations

$$V_1(u, x_1, \ldots, x_M) = 0, V_2(u, x_1, \ldots, x_M) = 0, \ldots, V_M(u, x_1, \ldots, x_M) = 0 \qquad (2)$$

in the indeterminates $x_1, \ldots, x_M$ with coefficients in K_0. Thus, whether the polynomial F can be factorized over K_r is equivalent to whether the above system of polynomial equations has a solution for $x_1, \ldots, x_M$ in the field K_0. So the problem of algebraic factoring is reduced to the problem of solving polynomial equations which can be done by any available method. In [6, 12], we explained how the solvability and solutions can be determined by using the characteristic set method of Ritt-Wu with Gauss' lemma. Here as well as in the following subsection the reader is assumed to be familiar with Ritt-Wu's method [3, 10, 14, 20, 21].

2.2 Method B

Let $\mathcal{A}^* = \mathcal{A} \cup \{F\}$ and compute a characteristic set $\mathcal{C}$ of $\mathcal{A}^*$ w.r.t. the variable ordering $y \prec y_1 \prec \cdots \prec y_r$. Denote the first polynomial in $\mathcal{C}$ by C_0. It happens *in general* that all polynomials other than C_0 in $\mathcal{C}$ are linear in their leading variables, while C_0 involves the variables u and y only. If this is the case, $\mathcal{C}$ is said to be *quasilinear*. If $\mathcal{C}$ is not quasilinear, we make a linear transformation by substituting $y - c_1 y_1 - \cdots - c_r y_r$ for y, where $c_1, \ldots, c_r$ are randomly chosen integers, and recover the transformation later by substituting $y + c_1 y_1 + \cdots + c_r y_r$ back for y. The probability of obtaining a quasilinear characteristic set with such a linear transformation is 1.

In any case, if C_0 is reducible over $\boldsymbol{K}_0$, we try to determine possible factors of F over $\boldsymbol{K}_r$ by computing the greatest common divisor (g.c.d. over $\boldsymbol{K}_r$) of F with each $\boldsymbol{K}_0$-factor of C_0. Practically, we can start this determination as soon as $\mathcal{A}^*$ is triangularized, without need to arrive at an exact characteristic set. Note that during the computation one should try to remove some factors (over $\boldsymbol{K}_0$) if possible. Some factors of F over $\boldsymbol{K}_r$ may also be determined from those $\boldsymbol{K}_0$-factors and the initials of other polynomials in $\mathcal{C}$ which involve the variables u and y only.

If $\mathcal{C}$ is quasilinear and C_0 is irreducible over $\boldsymbol{K}_0$, then $\mathcal{C}$ as an ascending set is clearly irreducible. It is important that in this case (when $\mathcal{C}$ is a characteristic set of $\mathcal{A}^*$), the g.c.d. of F and C_0 over $\boldsymbol{K}_r$ must be a true irreducible factor of F over $\boldsymbol{K}_r$. One can see this by noting that each irreducible factor of F over $\boldsymbol{K}_r$ corresponds to an ascending set in the irreducible zero decomposition of $\mathcal{A}^*$ over $\boldsymbol{K}_0$, independent of the variable ordering. Therefore, all the irreducible factors of F over $\boldsymbol{K}_r$ can be determined by producing such quasilinear characteristic sets and with g.c.d. computations. The reader may refer to [16] for details and implementation strategies. This factoring algorithm is rather efficient and can be used to factorize a number of polynomials of reasonable size.

3 Geometry Theorem Proving via Irreducible Zero Decomposition

Let $\mathfrak{T}_{\mathrm{alge}}(\mathbb{H} \Rightarrow \mathbb{C})$ be an algebraic formulation of a geometric theorem $\mathfrak{T}$, where $\mathbb{H}$ is a set of polynomials corresponding to the hypothesis and $\mathbb{C}$ contains, without loss of generality, a single polynomial C corresponding to the conclusion of $\mathfrak{T}$. Proving the theorem $\mathfrak{T}$ is equivalent to deciding whether any zero of $\mathbb{H}$ is a zero of C, and if not, which parts of the zeros of $\mathbb{H}$ are zeros of C. An elementary version of Wu's method [20, 21] proceeds roughly as follows: first compute a characteristic set $\mathcal{C}$ of $\mathbb{H}$ and let J be the product of initials of the polynomials in $\mathcal{C}$ (maybe plus some removed factors), and then compute the pseudo-remainder R of C w.r.t. $\mathcal{C}$. If R is identically equal to 0, then $\mathfrak{T}$ is proved to be true under the subsidiary condition $J \neq 0$. This simple procedure works pretty well for confirming a large number of geometric theorems and is practically efficient. However, if R happens to be non-zero, one cannot immediately tell whether $\mathfrak{T}$ is false or not and thus has to examine the reducibility of $\mathcal{C}$ and to perform decomposition further. $\mathcal{C}$ is reducible often when some geometric ambiguities such as bisection of angles and contact of circles are involved in $\mathfrak{T}$. Moreover, if one wishes to know whether $\mathfrak{T}$ is true as well when $J = 0$ which in general corresponds to some degenerate cases of $\mathfrak{T}$, one has to carry out a further decomposition too.

For a complete version of Wu's method, we compute, by using either Ritt-Wu's algorithm [10, 20, 21] or a more efficient one described in [17], a sequence of irreducible ascending sets (or triangular forms) $\mathcal{C}_1, \ldots, \mathcal{C}_e$ such that

$$\text{Zero}(\mathbb{H}) = \bigcup_{i=1}^{e} \text{Zero}(\mathcal{C}_i/J_i), \tag{3}$$

where J_i is the product of initials of the polynomials in $\mathcal{C}_i$, $\text{Zero}(\mathbb{H})$ denotes the set of common zeros of the polynomials in $\mathbb{H}$ and $\text{Zero}(\mathcal{C}_i/J_i) = \text{Zero}(\mathcal{C}_i) \setminus \text{Zero}(J_i)$ for each i. Thus, determining whether C vanishes on $\text{Zero}(\mathbb{H})$ or on which part of $\text{Zero}(\mathbb{H})$ C vanishes is equivalent to determining whether G vanishes on $\text{Zero}(\mathcal{C}_i/J_i)$ for all i or only for some i. The latter can be verified by direct computation of the pseudo-remainder R_i of C w.r.t. each $\mathcal{C}_i$. The algebraic form $\mathfrak{T}_{\text{alge}}(\mathbb{H} \Rightarrow C)$ of the theorem $\mathfrak{T}$ is universally true if and only if all R_i are identically 0. Usually, $\mathfrak{T}_{\text{alge}}(\mathbb{H} \Rightarrow C)$ is not universally true, that is, some R_i are identically 0 but the others are not. This is because either $\mathfrak{T}$ itself is only generically true and the components for which $R_i \not\equiv 0$ correspond to the non-degeneracy conditions, or $\mathfrak{T}_{\text{alge}}(\mathbb{H} \Rightarrow C)$ is true only on some significant components. In the latter case, a component for which $R_i \not\equiv 0$ corresponds either to a non-degeneracy condition of $\mathfrak{T}$ or to a configuration which is not what the theorem $\mathfrak{T}$ means and is added to $\mathbb{H}$ due to the imprecise algebraic formulation of $\mathfrak{T}$.

The above procedure is theoretically more complete yet practically less efficient. It can be used not only to prove theorems but also to help understand the geometric ambiguity of a theorem which one might not be aware of at first sight. In fact, it is also a non-trivial requirement for anyone to formulate a theorem with 100% accuracy either geometrically or algebraically. The examples given in the following section will illustrate this point. Nevertheless, to compute irreducible zero decompositions one needs to examine the reducibility of ascending sets for which polynomial factorization over successive algebraic extension fields is required. This was considered as the bottle-neck of Wu's method because of the previous lack of efficient procedures for algebraic factoring. Applying the methods reviewed in Section 2, we observed that algebraic factoring is not the most time-consuming step in Wu's method as well as methods based one our elimination procedures [17]. There are other steps such as verifications of the 0 remainder and control of the branch expansion which are even computationally harder and for which there is a need of much work, though a number of useful techniques have already been developed (see [4, 15] for example). In the following section, we shall consider some typical yet thorny examples of geometry theorem proving where algebraic factoring may be necessary.

In the zero decomposition (3) some components may be considered redundant. Such redundant components can be removed completely if we proceed further to arrive at an irreducible decomposition of the algebraic variety defined by $\mathbb{H}$ (cf. [15]) or partially if we adopt an alternative decomposition

$$\text{Zero}(\mathbb{H}) = \bigcup_{i=1}^{s} \text{Zero}(\text{PB}(\mathcal{C}_i)), \tag{4}$$

where PB($\mathcal{C}_i$) denotes the prime basis of the ideal having characteristic set $\mathcal{C}_i$ (cf. [3, 4, 14, 15]). Then, we need to compute the pseudo-remainder R_i only for those i for which PB($\mathcal{C}_i$) is not removed but actually appears in (4) or the corresponding variety decomposition. We do not discuss this aspect in more detail.

4 Examples from Geometry Theorem Proving

For each of the following examples we try to use an algebraic formulation as "natural" as possible in which no effort is made to avoid the reducibility and thus the algebraic factoring problem. We are aware that most of the reducibility cases can be avoided by some tricky formulation which takes into account of geometric information. However, we insist not to utilize any of such tricks in this paper where the factoring problem is concerned. We also try to figure out the proof of a theorem even if its algebraic formulation does not precisely correspond to the geometric statement and thus is not a theorem in the logical sense. This will help us understand the geometric ambiguity reflected in the algebraic form of the theorem.

Example 1 (Incenter and Excenters [21]) The bisectors of the three angles of an arbitrary triangle ABC intersect three-to-three at four points.

We take coordinates for the three vertices of ABC as $A(-u_1, 0), B(u_1, 0), C(u_2, u_3)$. Let the three bisectors of the angles A, B, C meet the y-axis at $A'(0, y_1), B'(0, y_2), C'(0, y_3)$ respectively. The hypothesis of the theorem consists of $\angle CAA' = \angle A'AB$, $\angle ABB' = \angle B'BC, \angle BCC' = \angle C'CA$ and the conclusion to be proved is: the three lines AA', BB', CC' are concurrent. By taking tangent[1] for the equalities of the angles the hypothesis conditions correspond to the following three polynomial equations

$H_1 = u_3 y_1^2 + 2 u_1 u_2 y_1 + 2 u_1^2 y_1 - u_1^2 u_3 = 0,$
$H_2 = u_3 y_2^2 - 2 u_1 u_2 y_2 + 2 u_1^2 y_2 - u_1^2 u_3 = 0,$
$H_3 = u_3 y_3^2 - u_3^2 y_3 - u_2^2 y_3 + u_1^2 y_3 - u_1^2 u_3 = 0.$

With the variable ordering $y_1 \prec y_2 \prec y_3$, the above polynomials already form a characteristic set $\mathcal{C} = \{H_1, H_2, H_3\}$. Direct verification shows that the pseudo-remainder of the conclusion polynomial C w.r.t. $\mathcal{C}$ is non-zero. In order to prove the theorem, we need to examine the reducibility of $\mathcal{C}$. This involves first checking the reducibility of H_2 over $\boldsymbol{K}_1 = \boldsymbol{Q}(u_1, u_2, u_3, y_1)$, where y_1 is an algebraic element having adjoining polynomial H_1. It is verified that

$\qquad H_2$ is irreducible $\qquad\qquad\qquad\qquad\qquad\qquad\qquad\qquad\qquad\qquad$ (F1)

over $\boldsymbol{K}_1$. Next, we check whether H_3 is reducible over $\boldsymbol{K}_2 = \boldsymbol{K}_1(y_2)$, where y_2 is an algebraic element having adjoining polynomial H_2. It is found that H_3 can be

---
[1] In order to avoid trigonometric functions and radicals, we use some alternatives, for example, the equality of tangent of angles instead of the equality of angles and the square of distance instead of distance, in expressing geometric statements. In the hypothesis relations below no differentiation is made between internal and external bisectors.

factorized as
$$H_3 \doteq \frac{1}{4u_1^4}(2u_1^2y_3 - u_1^2u_3 - u_1^2y_1 + u_1u_2y_1 - u_1u_2y_2 - u_1^2y_2 - u_3y_1y_2) \cdot$$
$$(2u_1^2u_3y_3 + 2u_1^4 - 2u_1^2u_2^2 - u_1^2u_3^2 + u_1^2u_3y_1 - u_1u_2u_3y_1 \quad \text{(F2)}$$
$$+ u_1u_2u_3y_2 + u_1^2u_3y_2 + u_3^2y_1y_2).$$

Using this factorization, $\mathcal{C}$ is immediately decomposed over[2] $Q(u_1, u_2, u_3)$ into two irreducible components. The algebraic form of the theorem is true on one component and false on the other. This corresponds to the geometric fact that among the 8 sets of three (internal or external) bisectors of the three respective angles, the bisectors in 4 sets are concurrent at four points and those in the other sets are not. □

Example 2 (Steiner's Theorem [24, 25]). Let ABC', BCA' and CAB' be three equilateral triangles drawn all outward or all inward on the three sides of an arbitrary triangle ABC. Then the three lines AA', BB' and CC' are concurrent.

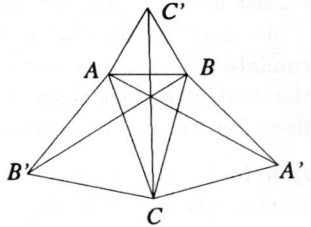

Take the coordinates of the points as $A(0,0)$, $B(1,0)$, $C(u_1, u_2)$, $C'(y_1, y_2)$, $A'(y_3, y_4)$ and $B'(y_5, y_6)$. Then the hypothesis of the theorem can be expressed as

$$\begin{aligned}
H_1 &= 2y_1 - 1 = 0, & (AC' = BC') \\
H_2 &= y_1^2 + y_2^2 - 1 = 0, & (AC' = AB) \\
H_3 &= y_3^2 + y_4^2 - u_1^2 - u_2^2 = 0, & (AB' = AC) \\
H_4 &= y_3^2 + y_4^2 - (y_3 - u_1)^2 - (y_4 - u_2)^2 = 0, & (AB' = CB') \\
H_5 &= (y_5 - 1)^2 + y_6^2 - (u_1 - 1)^2 - u_2^2 = 0, & (BA' = BC) \\
H_6 &= (y_5 - 1)^2 + y_6^2 - (y_5 - u_1)^2 - (y_6 - u_2)^2 = 0, & (BA' = CA') \\
D &= u_2 \neq 0, & (C \text{ is not on } AB)
\end{aligned}$$

where $D \neq 0$ is added to exclude the degenerate case in which A, B, C are collinear and whether the equilateral triangles are outward or inward is not distinguished.

To prove the theorem, we compute an irreducible decomposition for $\text{Zero}(\mathbb{H}/D)$ over Q, where $\mathbb{H} = \{H_1, \ldots, H_6\}$. Then 9 triangular forms $\mathcal{T}_1, \ldots, \mathcal{T}_9$ can be obtained by using our method such that

$$\text{Zero}(\mathbb{H}/u_2) = \bigcup_{i=1}^{9} \text{Zero}(\mathcal{T}_i/u_2), \quad (5)$$

where

$$\begin{aligned}
\mathcal{T}_1 &= [T_1, T_2, T_3, T_4, T_5, T_6], & \mathcal{T}_5 &= [u_2^2 + u_1^2, T_1, T_2, T_4, T_5, T_6], \\
\mathcal{T}_2 &= [T_1, T_2, T_3', T_4, T_5', T_6], & \mathcal{T}_6 &= [u_2^2 + u_1^2, T_1, T_2, T_4, T_5', T_6], \\
\mathcal{T}_3 &= [T_1, T_2, T_3', T_4, T_5, T_6], & \mathcal{T}_7 &= [u_2^2 + u_1^2 - 2u_1 + 1, T_1, T_2, T_3, T_4, T_6], \\
\mathcal{T}_4 &= [T_1, T_2, T_3, T_4, T_5', T_6], & \mathcal{T}_8 &= [u_2^2 + u_1^2 - 2u_1 + 1, T_1, T_2, T_3', T_4, T_6], \\
& & \mathcal{T}_9 &= [2u_1 - 1, 4u_2^2 + 1, T_1, T_2, T_4, T_6],
\end{aligned}$$

---

[2] When using zero decomposition over $Q(u_1, \ldots, u_d)$ rather than over Q for simplicity, we assume that the parameters $u_1, \ldots, u_d$ are appropriately chosen and consider any inequation in $u_1, \ldots, u_d$ as a non-degeneracy condition of the theorem under discussion.

$$T_1 = 2y_1 - 1,$$
$$T_2 = 4y_2^2 - 3,$$
$$T_3 = -2y_3 + 2u_2y_2 + u_1,$$
$$T_4 = -2u_2y_4 - 2u_1y_3 + u_2^2 + u_1^2,$$
$$T_5 = -2y_5 - 2u_2y_2 + u_1 + 1,$$
$$T_6 = -2u_2y_6 + 2y_5 - 2u_1y_5 + u_2^2 + u_1^2 - 1,$$
$$T_3' = -2y_3 - 2u_2y_2 + u_1,$$
$$T_5' = -2y_5 + 2u_2y_2 + u_1 + 1.$$

By computing $\text{prem}(C, T_i)$ (where C is the conclusion-polynomial), we find that the algebraic form of the theorem is true only on T_1 and false on all the other components. Therefore, it is true under some subsidiary conditions including $T_3' \neq 0$ and $T_5' \neq 0$. It is not very difficult to figure out that $T_3' = 0$ if and only if one of the two triangles ABC' and CAB' is drawn inward and the other outward, and $T_5' = 0$ if and only if one of ABC' and BCA' is drawn inward and the other outward. The theorem is true if and only if the three triangles ABC', CAB', BCA' are drawn all inward or all outward.

During the computation of (5), several polynomials have to be factorized over algebraic extension fields. Two of them are given as follows:

$$4y_3^2 - 4u_1y_3 - 3u_2^2 + u_1^2 \doteq T_3 \cdot T_3', \tag{F3}$$

$$4y_5^2 - 4u_1y_5 - 4y_5 - 3u_2^2 + 2u_1 + u_1^2 + 1 \doteq T_5 \cdot T_5' \tag{F4}$$

over $Q(u_1, u_2, y_2)$ with y_2 having adjoining polynomial $4y_2^2 - 3$. □

Example 3 (Morley's Theorem [3-5, 20]). The neighbouring trisectors of the three angles of any triangle intersect to form 27 triangles in all, of which 18 are equilateral. The hypothesis and the conclusion of this theorem consist of $\angle ABC = 3\angle PBC$, $\angle ACB = 3\angle PCB, \angle CAB = 3\angle RAB, \angle ABR = \angle PBC, \angle ACQ = \angle PCB$, $\angle BAR = \angle QAC$ and $PQ = PR, PQ = QR$ respectively. Let the coordinates of the points be taken as $A(y_2, y_1), B(u_1, 0), C(u_2, 0), P(0, 1), Q(y_6, y_5), R(y_4, y_3)$. Then both the hypothesis and the conclusion can be expressed as polynomial equations with index triples[3] $\mathbb{H} = \{[6\ y_2\ 1], [6\ y_2\ 1], [90\ y_4\ 3], [9\ y_4\ 1], [9\ y_6\ 1], [41\ y_6\ 1]\}$ and $\mathbb{C} = \{[6\ y_6\ 2], [6\ y_6\ 2]\}$ w.r.t. the ordering $u_1 \prec u_2 \prec y_1 \prec \cdots \prec y_6$. $\mathbb{H}$ can be decomposed over $Q(u_1, u_2)$ into two irreducible triangular forms

$$\mathcal{T} = [T_1, T_2, T_3, T_4, T_5, T_6],$$
$$\mathcal{T}^* = [T_1, T_2, T_3^*, T_4, T_5, T_6],$$

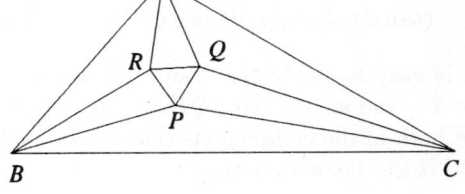

where

$T_1 = 3u_1^2 - 9u_1^2u_2^2 + 3u_1^2y_1u_2^2 - u_1^2y_1 + 8u_1y_1u_2 + 3u_2^2 + 3y_1 - y_1u_2^2 - 1,$

$T_2 = y_1u_2^3 - 3y_1u_2 + 3u_2^2y_2 - 3u_2^3 - y_2 + u_2,$

$T_3 = -4u_1^2 - 16y_3u_1u_2 - y_3^2u_1^2 + 4y_3u_1^2 + 3y_3^2 + 12u_1^2u_2^2 - y_3^2u_2^2 + 8u_2u_1y_3^2$
$\quad\quad + 3y_3^2u_1^2u_2^2 - 12y_3u_1^2u_2^2,$

$T_4 = y_3u_1^2 - 2u_1^2 + 2y_4u_1 - y_3,$

$T_5 = 3u_2^2y_3u_1^3y_5 - 6u_1^3u_2^2y_5 + 2u_1^3y_5 - u_1^3y_3y_5 - 6u_1^2u_2^3y_3 + 3u_1^2u_2^3y_5y_3 - 6u_2u_1^2y_3$
$\quad\quad + 7u_1^2y_5y_3u_2 - 6u_2^2u_1y_5 + 7u_2^2u_1y_5y_3 + 3y_3u_1y_5 + 2u_1y_5 + 2u_2^3y_3 - u_2^3y_3y_5$
$\quad\quad + 3y_3y_5u_2 + 2y_3u_2,$

---
[3] Associated with a non-constant polynomial P there is an index triple $[t\ v\ d]$ with t the number of terms in P, v the leading variable of P and d the degree of P in v.

$$T_6 = y_1 u_2^2 - u_2 y_1 y_6 + u_2 y_2 y_5 - u_2^2 y_5 - y_1 y_5 - y_2 y_6 + y_2 u_2 + u_2 y_6 - u_2^2,$$
$$T_3^* = 3y_3 - y_3 u_1^2 + 8y_3 u_1 u_2 + 3y_3 u_1^2 u_2^2 - u_2^2 y_3 - 2u_2 u_1 + 6u_2^3 u_1 + 2u_1^2 - 6u_1^2 u_2^2.$$

In computing the zero decomposition, no algebraic factorization is needed. The pseudo-remainders of the conclusion-polynomials are both 0 w.r.t. T, but not 0 w.r.t. T^*. Therefore, under some non-degeneracy conditions the algebraic form of the theorem is true on one component and false on the other.

Let us recall the tricky formulation of Wu [20], in which the constraint $\angle CBP + \angle PCB + \angle BAR \equiv \theta \mod 2\pi$ with $\tan^2 \theta = 3$ was imposed. After the addition of this to $\mathbb{H}$ the component T^* is then excluded, so that only T remains. Therefore, we may arrive at the same conclusion as Wu without using his trick in the formulation.

We note that T_3 is of degree 2 and T_3^* of degree 1 in y_3. This can be explained roughly as follows. After the trisectors are fixed for two angles of the triangle, the trisectors for the third angle would have three possibilities in forming the triangle PQR. T_3 corresponds to two of these possibilities for which PQR is equilateral, and T_3' corresponds to the third possibility for which PQR is not equilateral in general. To see the former more clearly, let us introduce a new variable y_0 and add $T_0 = y_0^2 - 3$ to $\mathbb{H}$. Then T_3 can be factorized, over $Q(u_1, u_2, y_0)$ with y_0 having adjoining polynomial T_0, as

$$T_3 \doteq \frac{T_3' T_3''}{D} = \frac{(H + 2u_1 y_0 + 2u_1 y_0 u_2^2)(H - 2u_1 y_0 - 2u_1 y_0 u_2^2)}{3u_1^2 u_2^2 - u_1^2 + 8u_2 u_1 - u_2^2 + 3}, \tag{F5}$$

where $H = 3y_3 - y_3 u_1^2 - u_2^2 y_3 + 8y_3 u_1 u_2 + 3y_3 u_1^2 u_2^2 + 2u_1^2 - 8u_2 u_1 - 6u_1^2 u_2^2$, so $\{T_0\} \cup T_1$ can be further decomposed into two irreducible triangular forms

$$T' = [T_0, T_1, T_2, T_3', T_4, T_5, T_5], \quad T'' = [T_0, T_1, T_2, T_3'', T_4, T_5, T_5].$$

We may prove instead of $PQ = PR, PQ = QR$ the conclusions $\tan^2 \angle QPR = 3, \tan^2 \angle PQR = 3$ which can be written as

$$(\tan \angle QPR + y_0)(\tan \angle QPR - y_0) = 0, \quad (\tan \angle PQR + y_0)(\tan \angle PQR - y_0) = 0.$$

It is easy to verify that $\tan \angle QPR + y_0 = 0$ and $\tan \angle PQR - y_0 = 0$ are true on T', and so are $\tan \angle QPR - y_0 = 0$ and $\tan \angle PQR + y_0 = 0$ on T''. That is, on both of the components that correspond to the two possibilities of T_3 in forming PQR the theorem is true. □

Example 4 (Thébault-Taylor's Theorem [3, 22, 24, 25]). Given a triangle ABC and a point D on the side BC, let C_2 be any Thébault circle with center T tangent to the circumscribed circle C_0 of the triangle and the lines AD and BC. Then among the inscribed and escribed circles of ABC there is just one C_1 with center I such that TI passes through the center of another Thébault circle C_3 tangent to C_0 and AD, BC.

We use the algebraic formulation given in [24], in which the hypothesis set $\mathbb{H}$ consists of 7 polynomials with index triples $[11\ x_1\ 2], [35\ x_2\ 2], [35\ x_3\ 2], [3\ x_4\ 1], [3\ x_5\ 1], [12\ x_6\ 1], [13\ x_7\ 1]$ and the conclusion consists of a single polynomial C with index triple $[11\ x_7\ 1]$ in the variables $u_1 \prec u_2 \prec u_3 \prec x_1 \prec \cdots \prec x_7$. $\mathbb{H}$ can be decomposed over $Q(u_1, u_2, u_3)$ into four irreducible triangular forms $T_1, \ldots, T_4$ with

$$\mathcal{T}_1 = [T_1, T_2, T_3, T_4, \ldots, T_7], \quad \mathcal{T}_3 = [T_1, T_2', T_3, T_4, \ldots, T_7],$$
$$\mathcal{T}_2 = [T_1, T_2, T_3', T_4, \ldots, T_7], \quad \mathcal{T}_4 = [T_1, T_2', T_3', T_4, \ldots, T_7],$$
$$T_1 = 4x_1^2 u_1^2 u_2^4 + 6u_1^2 u_2^4 - 2u_2^2 - 2u_2^6 - 2u_2^2 u_1^4 - 2u_2^6 u_1^4 + u_1^2 + u_1^2 u_2^8 + 4u_2^4 u_3$$
$$-4u_2^4 u_1^4 u_3 - 4u_2^4 u_2^2 u_1^2,$$
$$T_2 = 2x_2 u_1^2 u_2^2 - 2x_2 u_2^3 u_1 + 2x_2 u_2 u_1 - 2x_2 u_2^2 + u_1^2 u_2^4 - 2u_1^2 u_2^2 + u_1^2 + 2u_2^3 u_3 u_1$$
$$-2u_1 u_3 u_2 + u_2^4 - 2u_2^2 - 2u_1 x_1 u_2 + 2u_1 x_1 u_2^3 + 1,$$
$$T_3 = 2x_3 u_2^3 u_1 - 2x_3 u_1^2 u_2^2 - 2x_3 u_2 u_1 + 2x_3 u_2^2 - u_1^2 u_2^4 + 2u_1^2 u_2^2 - u_1^2 - 2u_2^3 u_3 u_1$$
$$+2u_1 u_3 u_2 - u_2^4 + 2u_2^2 - 2u_1 x_1 u_2 + 2u_1 x_1 u_2^3 - 1,$$
$$T_4 = -x_4 u_1 u_2 + u_1^2 u_2^2 - 1, \quad T_5 = u_1^2 - x_5 u_1 u_2 - u_2^2,$$
$$T_6 = 2x_6 u_1^2 - 2x_6 u_1^2 u_2^4 + 2x_6 x_4 u_1^2 u_2^2 + 2x_6 x_5 u_1^2 u_2^2 - x_5 u_1^2 - x_5 u_1^2 u_2^4 - x_4 u_1^4$$
$$-x_4 u_1^2 u_2^4 - u_1^4 + u_2^4 + u_1^4 u_2^4 - 1,$$
$$T_7 = x_7 u_2^2 - x_7 u_1^4 u_2^4 + x_7 u_2^2 u_1^4 - x_7 u_1^2 - x_6 u_2^2 + x_6 u_2^2 u_1^4 + x_6 u_1^2 - x_6 u_1^2 u_2^4$$
$$+2x_6 x_5 u_1^2 u_2^2 - x_5 u_1^2 u_2^4 - x_5 u_1^2 + u_2^4 - u_1^4,$$
$$T_2' = -2x_2 u_1^2 u_2^2 - 2x_2 u_2^3 u_1 + 2x_2 u_2 u_1 + 2x_2 u_2^2 - u_1^2 u_2^4 + 2u_1^2 u_2^2 - u_1^2 + 2u_2^3 u_3 u_1$$
$$-2u_1 u_3 u_2 - u_2^4 + 2u_2^2 - 2u_1 x_1 u_2 + 2u_1 x_1 u_2^3 - 1,$$
$$T_3' = 2x_3 u_1^2 u_2^2 + 2x_3 u_2^3 u_1 - 2x_3 u_2 u_1 - 2x_3 u_2^2 + u_1^2 u_2^4 - 2u_1^2 u_2^2 + u_1^2 - 2u_2^3 u_3 u_1$$
$$+2u_1 u_3 u_2 + u_2^4 - 2u_2^2 - 2u_1 x_1 u_2 + 2u_1 x_1 u_2^3 + 1.$$

The pseudo-remainder of C is 0 w.r.t. $\mathcal{T}_1$, but not 0 w.r.t. $\mathcal{T}_2, \mathcal{T}_3$ and $\mathcal{T}_4$. Hence, the algebraic form of the theorem is true on one component and false on all the others. The largest polynomial occurring in the reduction of the proof contains 168 terms. More than half of the computing time was spent for the following algebraic factorizations:

$$8x_2 u_2^4 + 4x_2 u_1^2 - 2u_3 u_1^2 + u_1^4 u_2^8 + 4u_3 u_2^2 + 4u_2^6 u_3 - 8u_2^4 u_3 + 2u_2^4 + u_1^4$$
$$-8x_2^2 u_2^4 u_3 u_1^2 - 2u_1^4 u_2^4 - 8u_2^4 u_1^4 u_3 - 12u_2^4 u_3 u_1^2 + 4x_1 u_1^2 u_2^4 + 8u_2^3 u_3 u_1^2$$
$$+4u_2^6 u_1^4 u_3 + 4u_1^4 u_3 u_2^2 - 2u_2^8 u_3 u_1^2 + 8u_2^6 u_3 u_1^2 - 4x_2^2 u_1^4 u_2^4 - 4x_2 u_2^6 u_1^4$$
$$+8x_2 u_1^4 u_2^4 + 4x_2 u_1^2 u_2^8 + 8x_2 u_1^2 u_2^4 - 8x_2 u_1^2 u_2^2 - 8x_2 u_2^6 u_1^2 - 2x_1 u_1^4 u_2^8 \quad \text{(F6)}$$
$$-u_2^8 + 4x_2^2 u_2^4 + 8x_2^2 x_1 u_1^2 u_2^4 - 4x_2 u_2^6 - 4x_2 u_2^2 - 2x_1 u_1^2 - 4x_2 u_2^2 u_1^4 - 1$$
$$\doteq -\frac{u_2^2 (2\, u_1^2 x_1 - 2\, u_1^2 u_3 - u_1^4 + 1)}{u_1^4 u_2^2 - u_1^2 + u_2^2 - u_1^2 u_2^4} \cdot T_2 \cdot T_2',$$

$$8x_3 u_1^2 u_2^4 - 8x_3^2 x_1 u_1^2 u_2^4 + 8x_3 u_1^4 u_2^4 - 8u_2^4 u_1^4 u_3 - 12u_2^4 u_3 u_1^2 - 4x_1 u_1^2 u_2^4$$
$$+8u_2^3 u_3 u_1^2 + 4u_2^6 u_1^4 u_3 + 4u_1^4 u_3 u_2^2 - 2u_2^8 u_3 u_1^2 + 8u_2^6 u_3 u_1^2 + 2x_1 u_1^2 u_2^8$$
$$-4x_3^2 u_1^4 u_2^4 - 4x_3 u_2^2 u_1^4 - 4x_3 u_2^6 u_1^4 + 8x_3 u_2^4 - 2u_3 u_1^2 + u_1^4 u_2^8 + 4u_3 u_2^2$$
$$+4u_2^6 u_3 - 8u_2^4 u_3 + 2u_2^4 + u_1^4 - u_2^8 - 8x_3^2 u_2^4 u_3 u_1^2 + 4x_3 u_1^2 u_2^8 + 2x_1 u_1^2 \quad \text{(F7)}$$
$$+4x_3^2 u_2^4 + 4x_3 u_1^2 - 4x_3 u_2^2 - 4x_3 u_2^2 - 8x_3 u_1^2 u_2^2 - 8x_3 u_2^6 u_1^2 - 2u_1^4 u_2^4 - 1$$
$$\doteq \frac{u_2^2 (2\, u_1^2 x_1 + 2\, u_1^2 u_3 + u_1^4 - 1)}{u_1^4 u_2^2 - u_1^2 + u_2^2 - u_1^2 u_2^4} \cdot T_3 \cdot T_3'$$

over $\mathbf{Q}(u_1, u_2, u_3, x_1)$ with x_1 having adjoining polynomial T_1. □

Example 5 (Steiner-Lehmus' Theorem [23]). Any triangle ABC whose two internal bisectors AA' and BB' are equal is an isosceles triangle.

We take the coordinates of the points as $A(-1,0), B(1,0), C(x_1, x_2), A'(x_3, x_4)$, $B'(x_5, x_6)$. Then the hypothesis of the theorem consists of

$$H_1 = -x_2 x_3^2 - 2\, x_2 x_3 - x_2 + 2\, x_1 x_4 x_3 + 2\, x_1 x_4$$
$$+2\, x_4 x_3 + 2\, x_4 + x_2 x_4^2 = 0, \qquad (\angle CAA' = \angle A'AB)$$

$$H_2 = x_6^2 x_2 + 2\,x_6 x_5 x_1 - 2\,x_6 x_5 - 2\,x_6 x_1 + 2\,x_6$$
$$-x_5^2 x_2 + 2\,x_5 x_2 - x_2 = 0, \qquad (\angle ABB'' = \angle B'BC)$$
$$H_3 = x_6 x_1 - x_2 + x_6 - x_5 x_2 = 0, \qquad (B' \text{ is on } AC)$$
$$H_4 = x_4 x_1 + x_2 - x_4 - x_3 x_2 = 0, \qquad (A' \text{ is on } BC)$$
$$H_5 = (x_3 + 1)^2 + x_4^2 - (x_5 - 1)^2 - x_6^2 = 0. \qquad (AA' = BB')$$

The problem is to decide when $C = x_1 = 0$ ($AC = BC$). With the ordering $x_1 \prec \cdots \prec x_6$, $\{H_1, \ldots, H_5\}$ can be decomposed over $\boldsymbol{Q}$ by Ritt-Wu's method into 15 irreducible ascending sets and by ours into 21 irreducible triangular forms. There are 6 ascending sets in which x_2 is contained. These ascending sets correspond to the degenerate case in which A, B, C are collinear. Among the remaining 9 ascending sets, four contain x_1 as their first polynomials, so the algebraic form of the theorem is true on these components and false on the others. During the zero decomposition, several algebraic factorizations have to be computed. Two of them are given as follows:

$$2\,x_3^2 + 2\,x_3 - 1 \doteq \frac{1}{2}(2\,x_3 - 3\,x_2 + 1)(2\,x_3 + 3\,x_2 + 1), \tag{F8}$$

$$-x_5^2 + x_1 x_5 + x_5 - 4\,x_1 - 5$$
$$\doteq \frac{1}{16}(-4\,x_5 + x_1 x_2 - 5\,x_2 + 2\,x_1 + 2)(4\,x_5 + x_1 x_2 - 5\,x_2 - 2\,x_1 - 2) \tag{F9}$$

over $\boldsymbol{Q}(x_2)$ with adjoining polynomial $3\,x_2^2 - 1$ for x_2 and $\boldsymbol{Q}(x_1, x_2)$ with adjoining ascending set $\{x_1^2 - 6\,x_1 - 11, x_1 x_2^2 + 3\,x_2^2 + 52\,x_1 + 76\}$ for x_1, x_2 respectively. □

For geometry theorem proving, polynomials needed to be factorized as well as the adjoining polynomials are usually quadratic. This is mainly because in the current state the considered geometric theorems only involve figures like triangles and circles whose algebraic character is no more than quadratic and the algebraic formulations are often made carefully and as simple as possible to avoid polynomials of high degree. However, if we do not take good care of algebraic formulation or we consider geometric figures with algebraic character of high order, we will have to factorize polynomials over algebraic extension fields with adjoining polynomials of degree greater than 2. Two of such examples are presented as follows.

Example 1'. The same theorem stated in Example 1. Now we express the hypothesis of the theorem as $\angle CAD = \angle DAB, \angle ABD = \angle DBC, \angle BCH = \angle HCA$ and take the coordinates of the points as $A(x_1, 0), B(x_2, 0), C(0, x_3), D(x_4, x_5), H(x_6, 0)$ (cf. [21], p. 194). The conclusion to be proved is: the point D lies on the line CH. The hypothesis conditions correspond to the following three polynomial equations

$$H_1 = -x_4^2 x_3 + 2 x_4 x_3 x_1 - x_1^2 x_3 - 2 x_1 x_5 x_4 + 2 x_1^2 x_5 + x_3 x_5^2 = 0,$$
$$H_2 = 2 x_2^2 x_5 - x_3 x_2^2 - 2 x_5 x_4 x_2 + 2 x_4 x_3 x_2 + x_3 x_5^2 - x_4^2 x_3 = 0,$$
$$H_3 = x_1 x_6^2 + x_2 x_6^2 + 2 x_6 x_3^2 - x_3^2 x_2 - 2 x_6 x_1 x_2 - x_3^2 x_1 = 0.$$

The characteristic set $\mathcal{C}$ of $\mathbb{H} = \{H_1, H_2, H_3\}$ consists of three polynomials

$$C_1 = 4x_4^2 x_2^2 - x_2^2 x_3^2 - 4x_4 x_1 x_2^2 + 12x_4^2 x_2 x_1 - 8x_4^3 x_2 + 4x_3^2 x_4 x_2 - 2x_2 x_3^2 x_1$$
$$\quad - 4x_1^2 x_4 x_2 + 4x_4^4 + 4x_4^2 x_1^2 - 4x_4^2 x_3^2 - x_3^2 x_1^2 - 8x_4^3 x_1 + 4x_3^2 x_4 x_1,$$
$$C_2 = -2x_2 x_5 + x_3 x_2 + 2x_4 x_5 - 2x_4 x_3 + x_3 x_1 - 2x_1 x_5$$

and $C_3 = H_3$, where the degrees of C_1, C_2, C_3 in their leading variables x_4, x_5, x_6 are $4, 1, 2$ respectively. The pseudo-remainder of the conclusion polynomial w.r.t. $\mathcal{C}$

is not identically 0, so we need to examine the reducibility of $\mathcal{C}$. In fact, C_3 can be factorized into two polynomials as

$$C_3 \doteq \frac{1}{x_1 + x_2}(x_1x_6 + x_2x_6 - 2x_1x_4 - 2x_2x_4 + 2x_4^2) \cdot \quad \text{(F10)}$$
$$(x_1x_6 + x_2x_6 + 2x_1x_4 + 2x_2x_4 - 2x_4^2 + 2x_3^2 - 2x_1x_2)$$

over $\boldsymbol{Q}(x_1, x_2, x_3, x_4)$ with x_4 having adjoining polynomial C_1, so $\mathbb{H}$ can be decomposed over $\boldsymbol{Q}(x_1, x_2, x_3)$ into two irreducible components. The algebraic form of the theorem is true on one of them and false on the other. □

Example 6 (Feuerbach's Theorem [21]). The nine-point circle of any triangle is tangent to the inscribed and escribed circles of the triangle.

Referring to the algebraic formulation given in [21], we can easily verify that the conclusion polynomial G there can be factorized over $\boldsymbol{Q}$ and the set of hypothesis polynomials can be decomposed over $\boldsymbol{Q}(x_1, x_2, x_3)$ with no need of algebraic factorization into four irreducible ascending sets. With respect to each ascending set, there is one and only one of the pseudo-remainders of the four factors of G that is identically equal to 0. This phenomenon can be easily explained from a geometric point of view. We have tried a more natural algebraic formulation different from Wu's. In our case, the set of hypothesis polynomials can be decomposed into four irreducible ascending sets, too, over the corresponding rational function field and a similar phenomenon appears. However, with our formulation we have to perform the following algebraic factorizations for the irreducible zero decomposition:

$$16\,u_2^2y_9^2 - 2\,u_2^2u_1^2 - u_2^4 - 2\,u_2^2 + 2\,u_1^2 - u_1^4 - 1$$
$$\doteq (4\,u_2y_9 + 2\,y_1^2 - u_2^2 - u_1^2 - 1)(4\,u_2y_9 - 2\,y_1^2 + u_2^2 + u_1^2 + 1), \quad \text{(F11)}$$

$$16\,y_1u_2^2y_{10}^2 + 16\,u_1u_2^2y_{10}^2 + 16\,u_2^2u_1y_1^2 - 18\,u_2^2u_1 + 6\,u_2^2u_1^2y_1 - u_1^4y_1 - y_1$$
$$-32u_2^2y_1^3 + 22u_2^2y_1 + 7y_1u_2^4 - 2u_2^2u_1^3 - u_2^4u_1 - u_1 + 2u_1^3 + 2u_1^2y_1 - u_1^5$$
$$\doteq \frac{y_1 + u_1}{u_1^2} \cdot (4u_1u_2y_{10} - 4u_2^2y_1 - 6u_1y_1^2 + 4y_1^3 - 4y_1 + u_1^3 + u_1u_2^2 + 5u_1) \cdot \quad \text{(F12)}$$
$$(4u_1u_2y_{10} + 4u_2^2y_1 + 6u_1y_1^2 - 4y_1^3 + 4y_1 - u_1^3 - u_1u_2^2 - 5u_1)$$

over $\boldsymbol{Q}(u_1, u_2, y_1)$ with adjoining polynomial $y_1^4 - u_2^2y_1^2 - u_1^2y_1^2 - y_1^2 + u_1^2$ for y_1. □

5 Experiments and Comparison

We felt that our first algebraic factoring method based on solving polynomial equations is rather efficient according to our preliminary hand-calculations made in [6, 12]. It is indeed so for polynomials of low degree. However, when coming to its implementation and trying some more complex examples on computer, we were soon aware that the method is inefficient for polynomials of high degree. Our second method appears to be superior to the first for polynomials of high degree. However, the adjoining polynomials and the polynomials factorized in the previous examples are all of low degree yet with parametric variables. They are somewhat special in the sense that they arise from the algebraic formulations of geometric theorems.

Our factoring methods have both been implemented in the author's **charsets** package whose first version was included in the Maple share library for distribution in January 1991. We use the Maple built-in factorizer for multivariate polynomial factorization over Q and the characteristic set method for solving the polynomial equations (2). The performance of an improved implementation of the second method by us in Maple 4.3 for 40 test examples (which were partly taken from available publications and partly randomly generated) was reported in [16]. For all those examples to which the Maple 4.3 built-in algorithm is applicable, our algorithm is always faster.

Below we report on the computing times for methods A and B with our current implementation in Maple V.2 and the Maple V.2 built-in function **factor** on a SUN SparcServer 690/51 for the previous examples of algebraic factoring. The timings are given in CPU seconds and include the time for garbage collection. The computation in the cases indicated with * was rejected by the Maple system for the reason "object too large." We note that for most of the examples the built-in **factor** in Maple V and its predecessors is not applicable. The latest release Maple V.2 has considerably extended **factor** for polynomials over multiple algebraic extension fields by implementing the well-known methods of Trager [11], Lenstra [8, 9] and others, and has much improved its efficiency.

The meanings of the heading entries in Table 1 are explained as follows: No – the example number; d – the number of transcendental elements for the extension field; $\deg(\mathcal{A})$ – the degrees of the adjoining polynomials in $\mathcal{A}$ in their leading variables, separated by the slant (noting that all the 12 polynomials factorized are quadratic); Maple V.2 – time for Maple V.2's built-in **factor**; Method A – time for our first algorithm and Method B – time for our second algorithm in Maple V.2.

Table 1. Timings for factorizing 12 example polynomials

No	d	$\deg(\mathcal{A})$	Method A	Method B	Maple V.2
(F1)	3	2	.417	.317	1.484
(F2)	3	2/2	1.916	30.183	138.55
(F3)	2	2	.333	.584	1.5
(F4)	2	2	.333	.566	1.333
(F5)	2	2	1.583	2.016	3.233
(F6)	3	2	6.867	20.4	41.516
(F7)	3	2	6.65	21.083	40.583
(F8)	0	2	.283	.266	.517
(F9)	0	2/2	1.45	.883	2.5
(F10)	3	4	8.05	40.583	142.317
(F11)	2	4	4.283	11.967	29.45
(F12)	2	4	*	27.317	57.916

The timings in Table 1 show that both of our methods are faster than Maple V.2's built-in factorizer for all the 12 polynomials with one exception in which the factorization was not completed for Maple's system error "object to large." Also, our Method A is faster than Method B. The timings reported here only give one an impression on the computational cost of algebraic factorization for polynomials from geometry theorem proving. They do not reflect the overall performance of the methods under discussion. In fact, we have tried the methods for the 40 examples listed in [16]. Method A is obviously slower than Method B for polynomials of high degree, while a combination of methods A and B is faster than Maple V.2's built-in factorizer no longer for all the examples.

The total times for proving the example theorems and for the involved algebraic factorizations (using Method A) are provided in Table 2 for reference and comparison, where our method is based on the decomposition algorithms described in [17]. K indicates the field over which the irreducible zero decomposition is computed. For the unsuccessful cases, the times are 42.8, 48.217 and 39.284 seconds respectively when Method B is used for algebraic factoring.

Table 2. Timings for proving 6 example theorems

Ex	Our Method	Wu's Method	Algebraic Factoring	K
1	2.565	3.132	2.333	$Q(u_1, u_2, u_3)$
2	7.21	25.866	3.984	Q
3	18.166	54.133	0	$Q(u_1, u_2)$
4	16.15	19.45	13.517	$Q(u_1, u_2, u_3)$
5	33.483	95	4.884	Q
1'	8.75	13.567	8.05	$Q(x_1, x_2, x_3)$
6	*	*	*	$Q(u_1, u_2)$

Finally, we remark that Method A can also be used to factorize polynomials algebraically when the extension field is not given [12]. This is because over which field a polynomial F is to be factorized depends on in which field the algebraic equations (2) are to be solved. If (2) has a solution in some algebraic extension field, then the polynomial can be factorized over this field. For example, let us suppose that the polynomial

$$F = 2u_3X^2Y^2 - u_2^2u_3X^2 + u_2^2u_3Y^2 - 2u_1u_3X^3 + 2u_2^3YX - u_3^3X^2 - 2u_1u_2^2YX$$
$$-2u_3^2Y^3 + u_3Y^4 + u_3X^4 + 2u_2u_3^2YX + 2u_1u_3^2YX - 2u_1u_3Y^2X + 2u_1u_2Y^3$$
$$+2u_1u_2u_3X^2 + u_3^3Y^2 + 2u_1u_2YX^2 - 2u_3^2X^2Y - 2u_2^2Y^3 - 2u_2^2YX^2 - 2u_1u_2u_3Y^2$$

can be factorized as the product of two polynomials F_1 and F_2 of degree 2 in X with indeterminate coefficients. Then by writing down the polynomial equations we found that they have a solution for the indeterminate coefficients so that F_1 and F_2 can be written as

$$F_1 = (X - \frac{u_1 - a}{2})^2 + (Y + \frac{u_1^2 - au_2 - u_1u_2 - a^2}{2u_3})^2 + \frac{a(u_1^2 - 2u_1u_2 - a^2)}{2(u_1 - u_2 + a)},$$
$$F_2 = (X - \frac{u_1 + a}{2})^2 + (Y + \frac{u_1^2 + au_2 - u_1u_2 - a^2}{2u_3})^2 - \frac{a(u_1^2 - 2u_1u_2 - a^2)}{2(u_1 - u_2 - a)},$$

where $a = (u_3^2 + u_1^2 - 2u_1u_2 + u_2^2)^{1/2}$. That is, F has been factorized over the extension field $Q(u_1, u_2, u_3, a)$.

Acknowledgments. This work is supported by FWF and CEC under ESPRIT BRP 6471 (MEDLAR II) and CNRS-MRE under project inter-PRC "Mécanisation de la Déduction." Part of this paper was written in October/November 1993 when I was visiting the Department of Mathematics, University of Saarland. I wish to thank Wolfram Decker for his kind invitation and also Sorin Popescu and Michael Messollen for interesting discussions.

References

[1] Abbott, J. A.: On the factorization of polynomials over algebraic fields. Ph.D thesis. School of Math. Scis., Univ. of Bath, England (1989).
[2] Chou, S. C.: Proving elementary geometry theorems using Wu's algorithm. In: Automated theorem proving: after 25 years, Contemp. Math. **29** (1984) 243–286.
[3] Chou, S. C.: Mechanical geometry theorem proving. D. Reidel Publ. Co., Dordrecht-Boston-Lancaster-Tokyo (1988).
[4] Chou, S. C., Gao, X. S.: Ritt-Wu's decomposition algorithm and geometry theorem proving. In: Proc. CADE-10, Lecture Notes in Comput. Sci. **449** (1990) 207–220.
[5] Gao, X. S.: Transcendental functions and mechanical theorem proving in elementary geometries. J. Automat. Reason. **6** (1990) 403–417.
[6] Hu, S., Wang, D. M.: Fast factorization of polynomials over rational number field or its extension fields. Kexue Tongbao **31** (1986) 150–156.
[7] Landau, S.: Factoring polynomials over algebraic number fields. SIAM J. Comput. **14** (1985) 184–195.
[8] Lenstra, A. K.: Lattices and factorization of polynomials over algebraic number fields. In: Proc. EUROCAM'82, Marseille (1982) 32–39.
[9] Lenstra, A. K.: Factoring multivariate polynomials over algebraic number fields. SIAM J. Comput. **16** (1987) 591–598.
[10] Ritt, J. F.: Differential algebra. New York: Amer. Math. Soc. (1950).
[11] Trager, B. M.: Algebraic factoring and rational function integration: In: Proc. ACM Symp. Symb. Algebraic Comput., Yorktown Heights (1976) 219–226.
[12] Wang, D. M.: Mechanical approach for polynomial set and its related fields (in Chinese). Ph.D thesis. Academia Sinica, China (1987).
[13] Wang, D. M.: On Wu's method for proving constructive geometric theorems. In: Proc. IJCAI-89, Detroit (1989) 419–424.
[14] Wang, D. M.: Characteristic sets and zero structure of polynomial sets. Lecture notes. RISC-LINZ, Joh Kepler Univ., Austria (1989).
[15] Wang, D. M.: Irreducible decomposition of algebraic varieties via characteristic sets and Gröbner bases. Comput. Aided Geom. Design **9** (1992) 471–484.
[16] Wang, D. M.: A method for factorizing multivariate polynomials over successive algebraic extension fields. Preprint. RISC-LINZ, Joh Kepler Univ., Austria (1992).
[17] Wang, D. M.: An elimination method for polynomial systems. J. Symb. Comput. **16** (1993) 83–114.
[18] Wang, P. S.: Factoring multivariate polynomials over algebraic number fields. Math. Comput. **30** (1976) 324–336.
[19] Weinberger, P. J., Rothschild, L. P.: Factoring polynomials over algebraic number fields. ACM Trans. Math. Software **2** (1976) 335–350.
[20] Wu, W. T.: Basic principles of mechanical theorem proving in elementary geometries. J. Syst. Sci. Math. Sci. **4** (1984) 207–235.
[21] Wu, W. T.: Basic principles of mechanical theorem proving in geometries (Part on elementary geometries, in Chinese). Beijing: Science Press (1984).
[22] Wu, W. T.: On reducibility problem in mechanical theorem proving of elementary geometries. Chinese Quart. J. Math. **2** (1987) 1–20.
[23] Wu, W. T., Lu, X. L.: Triangles with equal bisectors (in Chinese). Beijing: People's Education Press (1985).
[24] Yang, L., Zhang, J. Z.: Searching dependency between algebraic equations: an algorithm applied to automated reasoning. MM Res. Preprints 7 (1992) 105–114.
[25] Yang, L., Zhang, J. Z., Hou, X. R.: An efficient decomposition algorithm for geometry theorem proving without factorization. MM Res. Preprints 9 (1993) 115–131.

Mechanically proving geometry theorems using a combination of Wu's method and Collins' method

Nicholas Freitag McPhee[1], Shang-Ching Chou[2], and Xiao-Shan Gao[2]

[1] University of Minnesota, Morris; Morris, MN, 56267, USA, mcphee@cda.mrs.umn.edu, (612) 589-6321
[2] Department of Computer Science, The Wichita State University; Wichita, KS, 67208, USA, {chou, gao}@cs.twsu.edu, (316) 689-3918***

Abstract. Wu's method has been shown to be extremely successful in quickly proving large numbers of geometry theorems. However, it is not generally complete for real geometry and is unable to handle inequality problems. Collins' method is complete for real geometry and is able to handle inequality problems, but is not, at the moment, able to prove some of the more challenging theorems in a practical amount of time and space. This paper presents a combination that is capable of proving theorems beyond the theoretical reach of Wu's method and the (current) practical reach of Collins' method. A proof of Pompeiu's theorem using this combination is given, as well as a list of several other challenging theorems proved using this combination.

1 Background

In this paper we present a combination of two very powerful algebraic methods for proving geometry theorems: Wu's method and Collins' method. This combination is capable of proving theorems that are beyond either the theoretical or practical reach of either method alone. Due to limitations of space we will assume that the reader is already somewhat familiar with *what* these two methods do, if not *how* they do it; we will treat each method as a black box in this paper, so there is no need for any particular understanding of the internal workings of either method. For more complete discussions of Collins' method, as well as several examples of impressive uses of it, see any of [9, 1, 8]. For more complete discussions of Wu's method see any of [3, 10, 11]; [3] catalogues 512 geometry theorems automatically proved using Wu's method. The work presented here is based on that reported in [10], which contains many of the details omitted here.

Before we begin, it is necessary to lay out some assumptions. In particular, throughout this paper we will take K to be a fixed computable ordered field of characteristic zero, such as $\mathbf{Q}$ (the field of rational numbers), and $K[y_0, y_1, \ldots, y_{m-1}] = K[\mathbf{y}]$ to be the polynomial ring with indeterminates $y_0, y_1, \ldots, y_{m-1}$ (e.g., polynomials in $y_0, \ldots, y_{m-1}$ with rational coefficients). $\mathcal{F}$ will typically be a fixed extension field of K, such as $\mathbf{R}$, the field of real numbers, or $\mathbf{C}$, the field of complex numbers.

*** Chou and Gao were supported in part by NSF Grant CCR-9117870.

Given these assumptions, both Wu's method and Collins' method can address geometry problems whose translation into algebra takes the form

$$(\forall \mathbf{a} : \mathbf{a} \in \mathcal{F}^m : G(\mathbf{a})) \ . \tag{1}$$

Here G is a logical proposition consisting of the standard propositional connectives[4] and polynomial equations, inequations, and inequalities,[5] with the polynomials in G coming from $K[\mathbf{y}]$.

Wu's method is very efficient, but cannot, in general, handle inequality problems, i.e., G cannot contain inequalities ($>$, $\geq$). Also, while it is complete when $\mathcal{F}$ is algebraically closed, it is incomplete if the field $\mathcal{F}$ associated with the geometry is not algebraically closed. Thus it is unable to answer certain questions in real geometry.

Collins' method, on the other hand, can handle inequality problems, and can in fact handle mixed universal and existential quantifiers instead of just universals. Collins' method is also complete when the associated field is real closed, and is thus able to address many problems in real geometry not in the scope of Wu's method. Currently, its main drawback is one of efficiency; Wu's method is able to prove a large body of geometry theorems not presently within the practical reach of Collins' method.

The earliest work on combining Wu's method and Collins' method to prove inequality theorems is apparently [6], where the authors consolidated and extended a variety of ideas on proving inequality theorems based on the Rabinowitsch/Seidenberg device for converting polynomial inequalities to equalities through the introduction of new variables. They were able to prove a significant collection of inequality problems, but their proofs required human intervention at the end to determine whether certain rational functions were, for example, positive definite. It was pointed out that Collins' method could theoretically be used for this step, but at the moment many of the more complex problems lie beyond the practical reach of the best implementations.

In [7], Wu's method was extended through the addition of what could be considered a special case of Collins' method. The resulting combination was used to mechanically prove several problems in real geometry, the most impressive being the proof of the non-existence of the 8_3 configuration in the real plane. Here again, all inequalities were converted to equalities through the Rabinowitsch/Seidenberg device, at the expense of introducing more new variables.

Here we present a combination of Wu's method and Collins' method that is complete for the class of geometry problems whose algebraic translation is of the form (1), where G can contain equations, inequations, and inequalities, and where the field $\mathcal{F}$ is real closed. Thus this combination is able to address a class of geometry Wu's method cannot address alone, and while it is considerably less general than Collins' method (since it cannot handle mixed quantifiers), it has proved considerably faster than current implementations of Collins' method alone. Unlike most of its predecessors, this combination is fully automatic.

---

[4] Throughout we will use the symbols $\wedge, \vee, \neg, \equiv, \Rightarrow, \Leftarrow$ for, respectively, *and, or, not, logically equivalent, implies, follows from*.

[5] We will use "equation" to refer to a statement of the form $p = 0$, "inequation" to refer to a statement of the form $p \neq 0$, and "inequality" to refer to statements of the form $p > 0$, $p \geq 0$, $p < 0$, or $p \leq 0$.

2 The combination algorithm

2.1 Definitions

Throughout we will refer to four-tuples of the form

$$(EQS, NEQS, GTS, GEQS) ,$$

the intended meaning of which is captured in the following definition:

Definition 1 $Sols_\mathcal{F}$. Let EQS, $NEQS$, GTS, and $GEQS$ be finite subsets of $K[\mathbf{y}]$, and define $Sols_\mathcal{F}$ by[6]

$$Sols_\mathcal{F}(EQS, NEQS, GTS, GEQS) =$$
$$\{\, \mathbf{a} \in \mathcal{F}^m \mid$$
$$\quad (\forall e : e \in EQS : e(\mathbf{a}) = 0)$$
$$\quad \wedge (\forall n : n \in NEQS : n(\mathbf{a}) \neq 0)$$
$$\quad \wedge (\forall s : s \in GTS : s(\mathbf{a}) > 0)$$
$$\quad \wedge (\forall g : g \in GEQS : g(\mathbf{a}) \geq 0) \,\} \ .$$

Also, given a predicate P defined on $\mathcal{F}^m$, we'll extend the definition of $Sols_\mathcal{F}$ by

$$Sols_\mathcal{F}(P) = \{\mathbf{a} \in \mathcal{F}^m \mid P(\mathbf{a})\} \ .$$

□

Thus $Sols(EQS, NEQS, GTS, GEQS)$ is the sub-space of $\mathcal{F}^m$ over which all the polynomials in EQS are zero, those in $NEQS$ are non-zero, those in GTS are strictly positive, and those in $GEQS$ are non-negative.

Assume then that the geometric conditions of a problem have been converted to algebraic conditions in the form of (1). Then our goal is to show that

$$(\forall \mathbf{a} : \mathbf{a} \in \mathcal{F}^m : G(\mathbf{a})) ,$$

which is equivalent to

$$\neg(\exists \mathbf{a} : \mathbf{a} \in \mathcal{F}^m : \neg G(\mathbf{a})) \ . \tag{2}$$

We can write $\neg G(\mathbf{a})$ in disjunctive normal form, and then convert (2) to an equivalent statement about $Sols$:

$$\left(\bigcup_i Sols_\mathcal{F}(EQS_i, NEQS_i, GTS_i, GEQS_i) \right) = \emptyset \ . \tag{3}$$

Then proving geometry theorems whose algebraic statement take the form of (1) reduces to showing the validity of equations of form (3), which in turn reduces to verifying statements of the form

$$Sols_\mathcal{F}(EQS, NEQS, GTS, GEQS) = \emptyset \ . \tag{4}$$

---

[6] The function $Sols_\mathcal{F}$ is a generalization of the function $Zero_\mathcal{F}$ from, e.g., [5].

Thus our goal in this paper is to present an algorithm to answer questions of form (4), where $\mathcal{F}$ is a real closed field. Since the results presented here are valid for any real closed field $\mathcal{F}$, $Sols_\mathcal{F}$ will typically be shortened to just $Sols$.

The algorithm will be presented as a series of simplifications, or rewrite rules, that take a tuple

$$(EQS, NEQS, GTS, GEQS)$$

and return a set of 'simpler' tuples

$$\{(EQS_i, NEQS_i, GTS_i, GEQS_i)\} \tag{5}$$

such that

$$Sols(EQS, NEQS, GTS, GEQS) = \emptyset$$
$$\equiv \tag{6}$$
$$(\forall i :: Sols(EQS_i, NEQS_i, GTS_i, GEQS_i) = \emptyset).$$

Therefore the problem of deciding (4) reduces to deciding the emptiness of

$$Sols(EQS_i, NEQS_i, GTS_i, GEQS_i)$$

for each new tuple. Note that if the set (5) is empty, then (6) implies that (4) is trivially true. One then continues to repeatedly apply these rules to the tuples resulting from prior simplifications, until either all the tuples have been shown to be empty, thereby confirming (4), or until some tuple is found that clearly has a solution, thus denying (4).

Each of the simplification rules relies on one of three basic mechanisms: Ritt-Wu decomposition, using Collins' method to eliminate a variable, or using an ascending chain to simplify an inequality. The use of Ritt-Wu decomposition will be similar to that discussed in any of [3, 10, 5, 4, 12], so little more will be said about it here.[7] Two techniques (a weak form and a strong form) for choosing and eliminating a variable via Collins' method are discussed in Sect. 2.4. The techniques for using an ascending chain to simplify inequalities are discussed in [10], and will not be covered here.[8]

2.2 The simplification steps

We assume we are given a tuple

$$(EQS, NEQS, GTS, GEQS),$$

and want to find a set of 'simpler' tuples

$$\{(EQS_i, NEQS_i, GTS_i, GEQS_i)\}$$

---

[7] We will assume a familiarity with terms such as "(irreducible) ascending chain", "reduced (with respect to a chain)", "initial (of a polynomial)", "class (of a variable or polynomial)", "leading variable", and "leading degree"; definitions can be found in the listed sources. [10], for example, defines these terms and presents a new, more deterministic, version of the Ritt-Wu decomposition algorithm.

[8] These simplifications are not theoretically necessary, as Simplification 2 could be removed from the algorithm without affecting either its soundness or completeness. In practice, though, this step is very valuable in speeding up proofs.

satisfying (6). This is accomplished by considering each of the following simplifications in turn, and applying the first one whose conditions are met.

Throughout, assume that none of the polynomial sets EQS, $NEQS$, GTS, and $GEQS$ contain a constant. If at any point a constant is generated, then it is either irrelevant (e.g., the constant 0 in EQS), or causes (4) to be trivially true (e.g., the constant -4 in GTS). Also note that since we are to use the first applicable simplification, when analyzing a particular simplification we can in each case assume the negation of the conditions for the previous cases.

Simplification 0: *EQS is empty.* Here there are two cases. One is that each of $NEQS$, GTS, and $GEQS$ is empty, in which case (4) is trivially false, and our original statement is not a theorem. In the other case we use the methods in Sect. 2.4 to choose a variable from $NEQS \cup GTS \cup GEQS$, and then use Collins' method to eliminate this variable as described in Sect. 2.4. □

Simplification 1: *$EQS \neq \emptyset$, and EQS is not an ascending chain.* Here we use Ritt-Wu decomposition to find a set of pairs of polynomial sets $\{(ASC_i, IS_i)\}$ such that

1. each ASC_i is a non-trivial (irreducible)[9] ascending chain;
2. each IS_i contains at least the factors of the initials of ASC_i;
3. $\quad Zero(EQS/(NEQS \cup GTS)) = \bigcup_i Zero(ASC_i/IS_i)$.

Then we have a set of new tuples $\{(ASC_i, IS_i, GTS, GEQS)\}$ such that

$$Sols(EQS, NEQS, GTS, GEQS) = \bigcup_i Sols(ASC_i, IS_i, GTS, GEQS) \ .$$

□

Simplification 2: *Some $f \in GTS \cup GEQS \cup NEQS$ can be simplified.* In this and all following cases we can assume that EQS is in fact an (irreducible) ascending chain not consisting of a constant. If there is a polynomial f in GTS, $GEQS$, or $NEQS$ that is not reduced with respect to the ascending chain EQS, then we can use EQS to simplify f using the methods described in [10]. □

Simplification 3: *There is a polynomial in EQS of even leading degree.* Here we use Collins' method (as described in Sect. 2.4) to eliminate the leading variable of f, where f is the polynomial in EQS of highest class having even leading degree. □

Simplification 4: *Either GTS or $GEQS$ is non-empty.* Here we choose a variable from $GTS \cup GEQS$ as described in Sect. 2.4, and then use Collins' method to eliminate this variable as described in Sect. 2.4. □

Simplification 5: *Both GTS and $GEQS$ are empty.* Remember that in this case we know that EQS is in fact a non-trivial irreducible ascending chain whose members are all of *odd leading degree*. We also know that $NEQS$ includes the factors of the

---

[9] Only Simplification 5 requires that the ascending chain is in fact irreducible, and our experience suggests that this simplification rule is rarely needed. Thus it is quite practical to use a coarse form of Ritt-Wu decomposition that doesn't guarantee irreducibility, such as that presented in either [5, 4].

initials of each $f \in EQS$. In this case

$$Sols_\mathcal{R}(EQS, NEQS, \emptyset, \emptyset)$$
$$=$$
$$Zero_\mathcal{R}(EQS/NEQS)$$
$$\neq$$
$$\emptyset$$

for any real closed field $\mathcal{R}$, and so (4) must be false. If one is using a coarse form of Ritt-Wu decomposition (see, e.g., [5, 4]) that does not guarantee that the ascending chain EQS is irreducible, then $Zero_\mathcal{R}(EQS/NEQS)$ might still be empty, even though all the polynomials in EQS are of odd degree.[10] In this instance one can either apply a refined form of Ritt-Wu decomposition to produce irreducible chains, or continue the simplification process by using Collins' method (as described in Sect. 2.4) to eliminate the leading variable of the polynomial of highest class in EQS. □

2.3 Termination, Soundness, and Completeness

To argue that this algorithm terminates is more tedious than instructive, so we'll refer the interested reader to [10] for the details. It should be noted, however, that we are currently only able to prove that the algorithm terminates when using the strong form of variable elimination (see below). In our experience, however, computations using the weak form have always terminated, usually generating faster, shorter proofs. Thus it is our opinion that the weak form is still valuable even without a termination proof.

To show that this combination algorithm is sound we need to show that in each of the simplifications from the previous section the set of tuples

$$\{(EQS_j, NEQS_j, GTS_j, GEQS_j)\}$$

generated has the property that

$$Sols(EQS, NEQS, GTS, GEQS) = \emptyset$$
$$\Leftarrow \quad\quad\quad\quad\quad\quad\quad\quad (7)$$
$$(\forall j :: Sols(EQS_j, NEQS_j, GTS_j, GEQS_j) = \emptyset) \ .$$

Completeness follows if in each case (7) can be shown, except with ($\Leftarrow$) replaced by ($\equiv$).

The soundness and completeness of each of Simplifications 0, 3, 4, and 5 reduces to showing the soundness and completeness of the variable elimination methods in Sect. 2.4. In Sect. 2.4, the strong form of these methods are shown to be both sound and complete, and the weak form is shown to be sound but not necessarily complete. The soundness and completeness of Simplification 1 is a direct result of the fact that Ritt-Wu decomposition preserves zeros; see, e.g., [12, 11, 3, 10] for details. The soundness and completeness of Simplification 2 is shown in [10].

---

[10] As a simple example, consider the reducible chain $\{x^3 - x^2 + x - 1, (x-1)*y + 1, (x^2 + 1)*z + 1\}$. Since the first polynomial factors to $(x-1)*(x^2+1)$, it is clear that this chain has no common solutions in any field.

2.4 Variable elimination

This section will cover two forms of variable elimination, a *strong form* and a *weak form*. In each case we will discuss the manner in which variables are chosen for elimination, the details of how the elimination is accomplished, and the soundness and completeness of this elimination process.

Choosing a variable to eliminate.

The strong form. Here we are given a set of polynomials S and asked to choose some variable, appearing in at least one of these polynomials, for elimination via quantifier elimination. Technically each variable is a candidate but, not surprisingly, the choice made can significantly affect performance. Experience suggests that the performance of quantifier elimination depends primarily on the total number of variables in the expression from which the quantifier is being eliminated, so the current approach is to choose a variable that forces the fewest *other* variables to be included. Thus for each variable y, the union is formed of the set of variables in each polynomial containing y. The variable y that generates the smallest such set is chosen, since it will require quantifier elimination to handle the fewest variables.

A few definitions will help make this precise:

Definition 2 (vars; necessary_vars). Assume $p \in K[y_0, \ldots, y_{m-1}]$, and define the function $vars$ by

$$vars(p) = \{ y_i \mid y_i \text{ in } p \} .$$

Also, for a set of polynomials S and a variable y, define $necessary\_vars$ by

$$necessary\_vars(S, y) = (\cup p : p \in S \wedge (y \text{ in } p) : vars(p)) .$$

□

Choosing a variable then reduces to finding the variable y satisfying

$$y = (\min y_i : 0 \leq i < m : |necessary\_vars(S, y_i)|).$$

We could further refine this process in the case of ties by taking polynomial degrees or variable ordering into account, but this is not currently done.

The weak form. In the strong form of variable elimination, every polynomial containing y must be considered when forming the union of forced variables. In the weak form, however, at most one polynomial from EQS is considered, namely the one with y as its leading variable; if no such polynomial exists then no equality polynomials need to be considered. The reasons for this modification are discussed further below.

Elimination of a variable.

Strong form. Here assume we have a variable y and a set of conditions encoded in a tuple

$$T = (ASC, NEQS, GTS, GEQS) ,$$

and we want to use quantifier elimination to eliminate y from every polynomial in T.

In general, to eliminate y from the tuple, one could use Collins' method to find a quantifier free formula Θ (free of y) that is logically equivalent to

$$(\exists y :: (\forall f : f \in ASC : f = 0)$$
$$\wedge \ (\forall f : f \in NEQS : f \neq 0)$$
$$\wedge \ (\forall f : f \in GTS : f > 0) \tag{8}$$
$$\wedge \ (\forall f : f \in GEQS : f \geq 0))) .$$

In practice, however, there are several ways in which (8) can be modified so that the quantifier elimination process is faster.

In particular, we will partition the polynomial sets ASC, $NEQS$, GTS, and $GEQS$ such that

$$ASC = ASC' \cup ASC_y$$
$$NEQS = NEQS' \cup NEQS_y$$
$$GTS = GTS' \cup GTS_y$$
$$GEQS = GEQS' \cup GEQS_y ,$$

where ASC_y is the set of polynomials $p \in ASC$ such that either

- p contains y, or
- p is of total degree 1,

and $ASC' = ASC - ASC_y$, where $-$ is set difference. $NEQS_y$, $NEQS'$, etc., are defined similarly. We will then use Collins' method to eliminate y from the conditions encoded by ASC_y, $NEQS_y$, GTS_y, and $GEQS_y$,[11] yielding a logically equivalent formula Θ that is free of y. We can then convert Θ to disjunctive normal form, and convert the resulting set of disjuncts to a set of tuples

$$\{(EQS_i, NEQS_i, GTS_i GEQS_i)\} .$$

For a real closed field $\mathcal{R}$, we then have (see [10] for details),

$$Sols_\mathcal{R}(ASC, NEQS, GTS, GEQS) = \emptyset$$
$$\equiv$$
$$\bigwedge_i (Sols(ASC' \cup EQS_i, NEQS' \cup NEQS_i,$$
$$GTS' \cup GTS_i, GEQS' \cup GEQS_i) = \emptyset) .$$

Thus simplifying $(ASC, NEQS, GTS, GEQS)$ to

$$\{(ASC' \cup EQS_i, NEQS' \cup NEQS_i, GTS' \cup GTS_i, GEQS' \cup GEQS_i)\}$$

is both sound and complete.

---

[11] While it is not strictly necessary to include in these sets the polynomials of total degree 1 not containing y, doing so helps reduce the number of cases generated (see [10] for details).

Weak form. In the strong form of elimination, the variable y is eliminated from *every* polynomial in T containing y; in the weak form we do not. While we still eliminate y from every polynomial in $NEQS$, GTS, and $GEQS$, we only eliminate y from the (at most) one polynomial $f_i \in EQS$ such that y is the leading variable of f_i. This has the advantage of removing several polynomials, and occasionally a few variables, from the elimination process, thus speeding things up.

That this is sound is a consequence of the simple fact that

$$(P \Rightarrow P')$$
$$\Rightarrow$$
$$Sols(P) \subset Sols(P')$$
$$\equiv$$
$$Sols(P) = \emptyset \Leftarrow Sols(P') = \emptyset \ .$$

Thus if applying the weak form of variable elimination to a tuple

$$(EQS, NEQS, GTS, GEQS)$$

produces a set of new tuples

$$\{(EQS_i, NEQS_i, GTS_i, GEQS_i)\} \ ,$$

then we have

$$Sols(EQS, NEQS, GTS, GEQS) = \emptyset$$
$$\Leftarrow$$
$$\left(\bigcup_i Sols(EQS_i, NEQS_i, GTS_i, GEQS_i)\right) = \emptyset \ .$$

Therefore if we can show that all the 'child' tuples have no solutions, then the original tuple has no solutions as well. If, on the other hand, one of the 'child' tuples does indeed have solutions, it is not guaranteed that the original tuple has solutions as well, so the weak form is not complete. To date, however, few proofs have failed on this account, and the weak form usually generates faster, shorter proofs. If completeness is a great concern, however, one could attempt a proof with the weak form first, and, if that fails, try again with the strong form.

3 An example: Pompeiu's theorem

Having presented the combination of Wu's method and Collins' method in detail in the previous section, here we will show its use on a challenging problem that can not currently be practically proved using either Wu's method or Collins' method alone: Pompeiu's theorem. The proof is entirely automatic and required no human intervention, and the text given is an only slightly edited version of the proof generated automatically.

In this section we will present first the algebraic statement of the theorem, and then the actual proof. The details of the one call to Hong's implementation of Collins' method are listed after the proof.

3.1 The algebraic form

Example 1 (Pompeiu's theorem). If ABC is an equilateral triangle, and P is some point not on the circumcircle of ABC, then the segments AP, BP, CP can be used to form a triangle (i.e., the sum of the lengths of any two exceeds the length of the third) (see Figure 1).[12] □

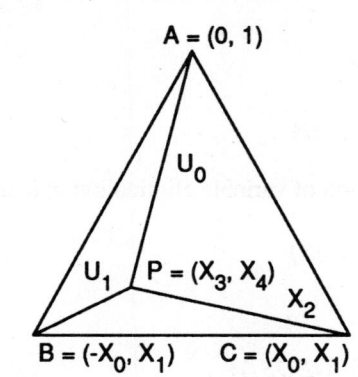

Fig. 1. Pompeiu's theorem: If ABC is equilateral, then for any P not on the circumcircle the segments AP, BP, CP can be used to form a triangle.

In our conversion to algebraic form, we'll restrict the theorem to a specific equilateral triangle: one whose circumcenter is the origin having $(0, 1)$ as a vertex.[13] Thus, we'll use the following assignment of coordinates:

$$A = (0, 1)$$
$$B = (-X_0, X_1)$$
$$C = (X_0, X_1)$$
$$P = (X_3, X_4) \ ,$$

and also let the variables U_0, U_1, and X_2 represent respectively the (positive) lengths of the segments AP, BP, and CP. We'll also take the variables to be ordered as

$$U_0 \prec U_1 \prec X_0 \prec X_1 \prec X_2 \prec X_3 \prec X_4 \ .$$

Given these coordinates, converting the various geometric conditions to algebraic ones yields:

---

[12] This theorem is often presented with no mention of the necessary non-degenerate condition that P not be on the circumcircle of ABC. Also, the theorem can be generalized by allowing P to be off the plane of ABC. In this case the non-degenerate condition is that P is not on the sphere through A, B, and C centered at the circumcenter of ABC.

[13] The combination succeeds in proving the theorem with a more general choice of coordinates, but the text of the resulting proof is much longer than that presented here.

- ABC is an equilateral triangle: Given these coordinates, this is equivalent to $X_0 = \pm\sqrt{3}/2$ and $X_1 = -1/2$, or

  $$4 * X_0^2 - 3 = 0 \wedge 2 * X_1 + 1 = 0 .$$

- U_0 is the (positive) length of AP:

  $$X_4^2 - 2 * X_4 + X_3^2 - U_0^2 + 1 = 0 \wedge U_0 > 0$$

- U_1 is the (positive) length of BP:

  $$X_4^2 - 2 * X_1 * X_4 + X_3^2 + 2 * X_0 * X_3 + X_1^2 + X_0^2 - U_1^2 = 0 \wedge U_1 > 0$$

- X_2 is the (positive) length of CP:

  $$X_4^2 - 2 * X_1 * X_4 + X_3^2 - 2 * X_0 * X_3 - X_2^2 + X_1^2 + X_0^2 = 0 \wedge X_2 > 0$$

- P is not on the circumcircle of ABC:

  $$X_4^2 + X_3^2 - 1 \neq 0$$

- **Conclusion** $AP + BP > CP$

  $$X_2 - U_0 - U_1 < 0$$

Now, to simplify the discussion somewhat, we introduce a convention for selecting the components of a tuple

$$S = (eqs, neqs, gts, geqs)$$

by defining

$$S.EQS = eqs$$
$$S.NEQS = neqs$$
$$S.GTS = gts$$
$$S.GEQS = geqs .$$

Using this convention, and remembering that our goal is to show that the conjunction of the hypotheses and the negation of the conclusion has no solutions, we can express our goal as showing that

$$Sols(SS(1)) = \emptyset ,$$

where

$$SS(1).EQS = \{$$
$$4 * X_0^2 - 3,\ 2 * X_1 + 1,\ X_4^2 - 2 * X_4 + X_3^2 - U_0^2 + 1,$$
$$X_4^2 - 2 * X_1 * X_4 + X_3^2 + 2 * X_0 * X_3 + X_1^2 + X_0^2 - U_1^2,$$
$$X_4^2 - 2 * X_1 * X_4 + X_3^2 - 2 * X_0 * X_3 - X_2^2 + X_1^2 + X_0^2$$
$$\}$$
$$SS(1).NEQS = \{X_4^2 + X_3^2 - 1\}$$
$$SS(1).GTS = \{U_0, U_1, X_2\}$$
$$SS(1).GEQS = \{X_2 - U_0 - U_1\} .$$

3.2 The proof

The proof presented here took just under a minute. The total CPU time was just over 44 seconds (47+ seconds real time), with roughly 42 seconds of the CPU time being in the one call to Hong's implementation of Collins' method.[14] The details of this one call to Hong's program will be presented after the proof.

We start by factoring the polynomials in $SS(1).NEQS$, and applying the Ritt-Wu decomposition algorithm to $SS(1).EQS$. This yields a single new tuple $SS(1.0)$ such that

$$(Sols(SS(1)) = \emptyset) \equiv (Sols(SS(1.0)) = \emptyset) ,$$

where

$SS(1.0).EQS = \{$
  $4 * X_0^2 - 3, \ 2 * X_1 + 1,$
  $X_2^4 + (-U_0^2 - U_1^2 - 3) * X_2^2 + U_0^4 + (-U_1^2 - 3) * U_0^2 + U_1^4 - 3 * U_1^2 + 9,$
  $4 * X_0 * X_3 + X_2^2 - U_1^2, \ 6 * X_4 - X_2^2 + 2 * U_0^2 - U_1^2$
$\}$
$SS(1.0).NEQS = \{X_4^2 + X_3^2 - 1, X_0, X_2, U_0, U_1\}$
$SS(1.0).GTS = \{U_0, U_1, X_2\}$
$SS(1.0).GEQS = \{X_2 - U_0 - U_1\}$

The problem then reduces to deciding whether $Sols(SS(1.0)) = \emptyset$. The polynomial

$$X_4^2 + X_3^2 - 1$$

from $SS(1.0).NEQS$ is reducible with respect to $SS(1.0).EQS$, and can thus be simplified using the methods described in [10]. This simplification yields a single new tuple, $SS(1.0.0)$, that is identical to $SS(1.0)$ except that

$$X_4^2 + X_3^2 - 1$$

from $SS(1.0).NEQS$ is replaced in $SS(1.0.0).NEQS$ by

$$X_2^2 + U_0^2 + U_1^2 - 6 .$$

Since

$$(Sols(SS(1.0)) = \emptyset) \equiv (Sols(SS(1.0.0)) = \emptyset) ,$$

the problem now reduces to deciding whether $Sols(SS(1.0.0)) = \emptyset$. All the polynomials in $SS(1.0.0).GTS$, $SS(1.0.0).GEQS$, and $SS(1.0.0).NEQS$ are reduced with respect to $SS(1.0.0).EQS$. However, $SS(1.0.0).EQS$ contains two polynomials of even leading degree, and of these

$$X_2^4 + (-U_0^2 - U_1^2 - 3) * X_2^2 + U_0^4 + (-U_1^2 - 3) * U_0^2 + U_1^4 - 3 * U_1^2 + 9$$

has the highest leading variable. We can then use Collins' method to eliminate the leading variable, X_2, as discussed in Sect. (2.4). This generates no new tuples, thereby proving that $Sols(SS(1.0.0)) = \emptyset$ and completing the proof.

---

[14] All timings were produced on a NeXT Cube with a Motorola 68040 processor and 24Mb of RAM.

3.3 Use of Collins' method

Our implementation of this combination uses Hoon Hong's excellent implementation of Collins' method, based on the improvements in [9]. The proof of Pompeiu's theorem just presented required one use of this implementation. The variable being eliminated was X_2, and so, using the weak form of variable elimination, almost all of the conditions in $Sols(SS(1.0.0))$ involving X_2 needed to be included in this call to Collins' method. The exceptions to this were the last two equality conditions, since they had leading variables above X_2 in the variable ordering.

Thus this one use of Collins' method was to eliminate the lone quantifier (and thus the variable X_2) from the following:

$$(\exists X_2 ::$$
$$X_2^4 + (-U_0^2 - U_1^2 - 3) * X_2^2 + U_0^4 + (-U_1^2 - 3) * U_0^2 + U_1^4 - 3U_1^2 + 9 = 0$$
$$\land U_0 \neq 0 \land U_1 \neq 0 \land X_2 \neq 0 \land X_2^2 + U_0^2 + U_1^2 - 6 \neq 0$$
$$\land U_0 > 0 \land U_1 > 0 \land X_2 > 0 \land X_2 - U_0 - U_1 \geq 0) \ .$$

Hong's program determined that this is logically equivalent to *False*, thereby proving that for *all* values of U_0 and U_1 there is *no* X_2 satisfying this condition. This computation took just under 42 seconds, constituting the bulk of the time required for the entire proof.

4 Conclusions

4.1 Advantages of the combination

After going over an example in some detail, the obvious question is "What advantages did the combination have over either Wu's method or Collins' method alone?"

Pompeiu's theorem depends fundamentally on inequalities, so the combination had a clear advantage over Wu's method due to Wu's method's inability to address inequality problems. On the other hand, one could theoretically use Collins' method alone to prove Pompeiu's theorem. At the moment, however, this theorem appears to lie outside the practical grasp of even the best implementations of Collins' method; see [10] for details. Thus while it is theoretically possible to use only Collins' method to prove this theorem, at the moment the combination appears to have a clear practical advantage.

4.2 Other theorems proved

As well as Pompeiu's theorem, numerous other theorems have been proved using this approach; we list several of the more interesting ones here.

- **Steiner-Lehmus theorem:** Any triangle with two equal internal bisectors is isosceles (listed in [6] as a challenge problem for geometry theorem provers capable of handling inequalities). This problem was solved in [11] (also see [2]), but this solution required a substantial amount of human intervention. It requires the use of inequalities, putting it outside the theoretical reach of Wu's method, and at the moment this problem seems beyond (or at least at the fringes of) the practical reach of Collins' method, as we have let Hong's implementation of Collins' method run for over 10 days with no success. The combination presented here has been used to settle this question fully automatically, with the proof taking just under 3 minutes. The full details can be found in [10].

- **Variants of the Steiner-Lehmus theorem:** The Steiner-Lehmus theorem is a consequent of the following simple variant. Assume ABC is a triangle such that $AB > AC$. Then the angle bisector from B to AC is longer than the angle bisector from C to AB (i.e., the longer bisector goes to the shorter side). This variant can be easily proved using this combination, as can other variants obtained by replacing "angle bisectors" with either "medians" or "altitudes".
- **Nonexistence of an 8_3 configuration in the real plane:** One of the most challenging theorems proved with this combination is the theorem that there is no 8_3 configuration in the real plane (see [7]). This problem is particularly interesting because the statement that no 8_3 configuration exists is valid in real geometries, but not in complex geometry. Thus while the theorem has no obvious inequality component, it is outside of the theoretical scope of Wu's method alone, but is soluble using the combination.
- **Euler's theorem:** If R and r are the radii of, respectively, the circumscribed and inscribed circles of some triangle, then $R \geq 2 * r$, with equality holding only if the triangle is equilateral.
- **Pasch's theorem:** Let ABC be a triangle, and let l be a line intersecting BC at D, AC at E, and AB at F. If D is between B and C, and E is between A and C, then we have that F is *outside* the segment $\overline{AB}$.
- **Heron's formula for the area of a triangle:** Assume that a, b, and c are the lengths of the sides of a triangle, and let $s = (a + b + c)/2$. Then the area of the triangle is

$$\sqrt{s * (s-a) * (s-b) * (s-c)} \ .$$

Like Pompeiu's theorem, this also requires working with the (positive) lengths of segments rather than their squares, and thus requires inequalities.

4.3 Ideas for future research

While we have had considerable success with this combination, there are certainly improvements that can be made. For example, there are elements of the algorithm that, while they 'seem reasonable', are somewhat ad hoc and deserve further investigation; some of these are discussed further in [10].

One interesting research direction would be to examine the possibility of combining Wu's method with weaker methods than full Collins' method. The combination presented in the previous chapter uses much less than the full generality of Collins' method, since we only use it to eliminate a single universal quantifier at a time. In [7] the non-existence of an 8_3 configure in the real plane was proved with just a simple special case of Collins' method, and it seems plausible that further research into such special cases could prove fruitful.

5 Acknowledgments

This work could not have proceeded without Hoon Hong's excellent implementation of Collins' method graciously provided by Hong, George Collins, and Jeremy Johnson. We also gratefully acknowledge the helpful suggestions of the referees.

References

1. D. S. Arnon. Geometric reasoning with logic and algebra. *Artificial Intelligence Journal*, 37:37–60, 1988.
2. S. C. Chou. *Proving and discovering theorems in elementary geometries using Wu's method*. PhD thesis, Department of Mathematics, University of Texas, Austin, 1985.
3. S. C. Chou. *Mechanical geometry theorem proving*. D. Reidel Publishing Company, Dordrecht, Netherlands, 1988.
4. S. C. Chou and X. S. Gao. Ritt-Wu's decomposition algorithm and geometry theorem proving. In M. E. Stickel, editor, *Proceedings of the 10th international conference on automated deduction*. Springer-Verlag, 1990.
5. S. C. Chou and X. S. Gao. Techniques for Ritt-Wu's decomposition algorithm. In *Proceedings of the IWMM*. International Academic Publishing, 1992.
6. S. C. Chou, X. S. Gao, and D. S. Arnon. On the mechanical proof of geometry theorems involving inequalities. Technical Report TR–89–31, Department of Computer Sciences, University of Texas at Austin, October 1989.
7. S. C. Chou, X. S. Gao, and N. F. McPhee. A combination of Ritt-Wu's method and Collins' method. Technical Report TR–89–28, Department of Computer Sciences, University of Texas at Austin, October 1989.
8. G. E. Collins. Quantifier elimination for real closed fields by cylindrical algebraic decomposition. In *Proceeding Second GI Conference on Automata Theory and Formal Languages*, volume 33 of *Lecture Notes In Computer Science*, pages 134–183. Springer-Verlag, 1975.
9. Hoon Hong. *Improvements in CAD-based quantifier elimination*. PhD thesis, Computer and Information Science Research Center, The Ohio State University, 1990. Technical Report OSU-CISRC-10/90-TR29.
10. Nicholas Freitag McPhee. *Mechanically proving geometry theorems using Wu's method and Collins' method*. PhD thesis, University of Texas at Austin, 1993.
11. Wu Wen-tsün. Basic principles of mechanical theorem proving in geometries. *J. of Sys. Sci. and Math. Sci.*, 4(3):207–235, 1984. Republished in *Journal of Automated Reasoning*, 2(4):221–252, 1986.
12. Wu Wen-tsün. On zeros of algebraic equations — an application of Ritt's principle. *Kexue Tongbao*, 31(1):1–5, 1986.

Str+ve and Integers*

Larry M. Hines

Computer Science, University of Texas
Austin, Texas 78712
hines@cs.utexas.edu

Abstract. The STR+VE theorem prover is designed to prove theorems about dense linear orders. In this paper, we discuss two extensions of STR+VE to theorems about the integers. We describe a version of STR+VE which proves theorems solely about integers. Also, we describe a version for theorems over two separate linear orders, one dense and one the integers.

1 Introduction

Theorems involving integer inequalities arise across a variety of theorem proving applications. Several automated provers [1, 4, 6, 13, 14, 16, 17] are specifically designed to prove these theorems.

These provers are decision procedures for Presburger Arithmetic or its extensions. They play a substantial role in proof and program verification, but alone are insufficient for proof discovery. These provers either prohibit uninterpreted function symbols or prohibit existential quantification. Hence, only when given a correct instantiation of x, can these propositional provers verify formulas such as $\exists x(f(x) = 0)$. In proof and program verification, instantiations are known. Conversely in proof discovery and program synthesis, appropriate instantiations are unknown. These application domains require a first order logic prover.

Here we present an integer inequality prover to discover proofs. The prover is first order which allows both arbitrary quantification and uninterpreted function symbols. The prover derives the correct instantiations in tandem with the proof derivation.

This integer prover is based upon STR+VE, a theorem prover designed primarily to prove theorems about dense linear orders. STR+VE will prove (in theory) any theorem in first order logic [10]. Nonetheless, it is most useful in application domains containing dense linear orders, whereby STR+VE's effectiveness results from the restrictive strategy controlling the dense linear order axioms' usage.

Dense linears orders do not occur in some relevant applications, namely, program verification, induction-based theorem proving, and temporal reasoning systems. In these domains, discrete linear orders, such as the integers, are prevalent. As in dense orders, unrestricted inference from discrete order axioms such

---

* This work is supported in part by National Science Foundation Grants 26-1099-26 and CCR-9101980.

as transitivity leads to an unmanagable number of resolvents. Our integer prover controls inference by the same strategy by STR+VE.

The integer version of STR+VE, STR+VE-I, contains new integer versions of STR+VE's standard inference rules (chaining, variable elimination, self-chaining), a new interpreted function symbol (floor) and new inference rules for floor.

The new integer *chaining* inference rule reflects differences between standard transitivity and integral transitivity. The *variable elimination* rule is modified as well. Both the new chaining and variable elimination rules employ a new interpreted function symbol, *floor*. Floor of x is the largest integer less than or equal to x. Additional inference rules, a new version of chaining, and a new version of variable elimination enforce this interpretation of floor.

New inference rules are constructed following the "Austin approach" to automated theorem proving, which incorporates knowledge of a proof sequence and, thereby, deletes intermediate steps or results. For instance, the new integer chaining rule can be regarded as a sequence applying scalar multiplication, the definition of floor, scalar multiplication again, addition and transitivity. STR+VE-I neither computes the intermediate results nor adds them to the prover's clause set. This is because the above sequence is packaged as a single rule. Moreover, STR+VE-I's restrictions on transitivity implicitly applies to each sequence step.

STR+VE-I represents a first order logic theorem prover for integers with these new inference rules. STR+VE-I's control strategy is identical to STR+VE's. Briefly, STR+VE-I selects central variables, then removes their *shielding terms* with a series of chaining rule applications. Variable elimination occurs once the shielding terms have been removed. Next a ground decision procedure[1] completes the refutation as the clauses become ground (all of their variables are removed).[2]

Prohibition against applying chaining to variables is crucial to STR+VE's control strategy.[3] Consequently, STR+VE-I proves integer inequality theorems with comparable effectiveness with which STR+VE proves dense inequality theorems.

Integer inequalities also arise in formulas for dense linear orders, whereby, the integers are typically employed as sequence indices. Neither the standard version of STR+VE nor STR+VE-I alone can address these theorems. The proof methods of both provers are needed. Combining separate proof procedures can be difficult in general [3]; however, STR+VE and STR+VE-I differ only with respect to their respective inference rules. The provers use the same control strategy.

The "two order" version of STR+VE, STR+VE-2, employs integer rules for integer terms and dense rules for dense terms by typing each term as either integer or dense. STR+VE's control strategy and, consequently, its effectiveness remains intact. (See Section 4).

Our ultimate goal is automating proof discovery for all linearly ordered groups. STR+VE-2 advances toward that goal.

---

[1] Such as SUP-INF [1, 16].
[2] See [12] for details on STR+VE's central variable strategy.
[3] This prohibition is weakened in STR+VE$\subseteq$ [11].

2 Terminology and Normalization

A *term* is a variable, a non-numeric constant or a function other than +, s∗ and floor applied to an appropriate number of arguments. A non-variable term containing a variable is a *shielding term*.

An *addend* is either r or (s∗ r X) where r is a rational number and X is a term or an expression. In the later, r is called X's coefficient. An *expression* is either an addend, (+ $X_1...X_n$), or (floor $X_1...X_n$) where each X_i is an expression.

Expressions are normalized by "pushing in" scalar multiplication. A normalized addend is (s∗ r X) where X is a term or is a normalized floor expression. An expression, (+ $X_1...X_n$) or (floor $X_1...X_n$), is normalized if each X_i is a normalized addend. The X_i's are called the expression's addends.

An inequality literal, (< X Y) or (≤ X Y), is normalized if X and Y are normalized expressions such that the addends of X and Y have positive integer coefficients.

Throughout this paper, J is used as an integer-typed term. Lowercase j's are integer-typed variables. I, K and L are used as normalized expressions containing only integer-typed addends. B is used as a densely-typed term. Lowercase b's are densely-typed variables. Y is a term of any type and lowercase y's are variables of any type. W, X and Z are normalized expressions.

r is a positive rational number and s a negative rational number. n and m are positive integers.

◁ is an inequality sign, either < or ≤.

Finally, we will use $\lfloor X_1 + ... + X_n \rfloor$ for (floor $X_1...X_n$).

3 Integers Only

The axioms of dense linear orders are encoded within STR+VE's four inference procedures.[4] The chaining inference rule[5] invokes the transitivity axioms:

$$X < Y \wedge Y < Z \to X < Z \tag{1}$$

$$X \leq Y \wedge Y < Z \to X < Z \tag{2}$$

$$X < Y \wedge Y \leq Z \to X < Z \tag{3}$$

$$X \leq Y \wedge Y \leq Z \to X \leq Z. \tag{4}$$

Variable elimination invokes the interpolation axiom of dense orders

$$\exists Y (X < Y \wedge Y < Z) \iff X < Z. \tag{5}$$

Normalization and self-chaining invoke the irreflexivity of <.

---

[4] STR+VE's ground decision procedure is essentially a combination of chaining, self-chaining and normalization. It is not included here to save space.

[5] Another aspect of chaining in STR+VE is factoring. There is no separate factoring rule. It is combined with chaining [10, 15], but this feature need not be changed to incorporate the integers. So, factoring is not shown here to simplify the presentation.

STR+VE incorporates group axioms of addition [7], as well as scalar multiplication. Group axioms are encoded for each of the aforementioned procedures. The inclusion of these axioms, as seen below, complicates the otherwise simple task of extending STR+VE to the integers.

Nevertheless, incorporating axioms for the integers into STR+VE-I requires modifying only these four inference procedures. STR+VE's strategy for controlling these procedures' application remains unchanged.

3.1 Integer Chaining

The conclusion of (1) for the integers can be strengthened to $X + 1 < Z$ because of the discrete nature of the integers. If the group axioms of addition are omitted included in STR+VE, then this strengthening create the only necessary change for chaining.

However, STR+VE's chaining does incorporate the group axioms of addition and the scalar multiplication order axiom,

$$0 < n \wedge X < Y \rightarrow n \cdot X < n \cdot Y. \tag{6}$$

For example, from $1 \leq 3 \cdot J$ and $2 \cdot J \leq K$, the standard chaining rules derives $2 \leq 3 \cdot K$. For the integers, a stronger conclusion is possible. $1 \leq 3 \cdot J$ is equivalent to $1 \leq J$ because J's type is integer. Now, chaining derives $2 \leq K$.

In general, dense chaining derives $I \leq n \cdot K$ from $I \leq n \cdot J$ and $m \cdot J \leq K$. When $m > 1$, we may derive the stronger conclusion that K is greater than or equal to m times the smallest integer larger than I/n.

To formalize this, we introduce *floor* into STR+VE-I. $\lfloor X \rfloor$ is interpreted as the greatest integer less than or equal to X. The integer transitivity axioms are

$$I < n \cdot J \wedge m \cdot J \triangleleft K \Longrightarrow m \left\lfloor \frac{I}{n} \right\rfloor + m \triangleleft K \text{ and} \tag{7}$$

$$I \leq n \cdot J \wedge m \cdot J \triangleleft K \Longrightarrow m \left\lfloor \frac{I-1}{n} \right\rfloor + m \triangleleft K. \tag{8}$$

These axioms are encoded into STR+VE-I's integer chaining rule. An *integer chain resolvent* of clauses $C_1 \vee I \triangleleft_1 n \cdot J'$ and $C_2 \vee m \cdot J'' \triangleleft_2 K$ is

$$(C_1 \vee C_2 \vee m \left\lfloor \frac{I'}{n} \right\rfloor + m \triangleleft_2 K)\sigma$$

where J' and J'' are unifable with mgu σ, either J' or J'' is a shielding term, neither J' nor J'' is a variable and I' is I if $\triangleleft_1$ is $<$ and is $I - 1$ otherwise.

3.2 Integer Variable Elimination

STR+VE's variable elimination rule incorporates the interpolation axiom of dense linear orders. Although, interpolation (i.e., between any two numbers is another number) does not apply to the integers, variable elimination is possible. An integer exists strictly between two integers if and only if one of the integers is two greater than the other.

As with chaining, scalar multiplication complicates variable elimination rule and floor is used to formalize the resulting axioms.

$$\forall j(I < n \cdot j \vee m \cdot j \lhd K) \Longrightarrow m \cdot \lfloor \tfrac{I}{n} \rfloor \lhd K \text{ and} \qquad (9)$$

$$\forall j(I \leq n \cdot j \vee m \cdot j \lhd K) \Longrightarrow m \cdot \lfloor \tfrac{I-1}{n} \rfloor \lhd K. \qquad (10)$$

An *integer variable elimination* resolvent of a clause

$$C \vee \bigvee_{i=1}^{k} I_i \lhd_i n_i \cdot j \vee \bigvee_{l=1}^{k'} m_l \cdot j \lhd'_l K_l$$

is

$$C \vee \bigvee_{i=1}^{k} \bigvee_{l=1}^{k'} m_l \cdot \lfloor \tfrac{I'_i}{n_i} \rfloor \lhd'_l K_l$$

where j does not occur in any I_i or K_l, or in C and I'_i is I_i if $\lhd_i$ is $<$ and is $I_i - 1$ otherwise.

3.3 Self-Chaining

At its simplest, STR+VE's self-chaining inference rule encodes the irreflexivity of less than, namely that $(Y \not< Y)$. This applies whether Y is dense or integer.

Consequently, STR+VE-I employs STR+VE's standard self-chaining rule unchanged.

3.4 Integer Normalization

The normalization procedure applies to each newly created literal in STR+VE insuring the definitions given in Section 2. Normalization uses the associativity and commutativity of + to sort and flatten expressions and to combine addends with common terms.

The group axioms of + apply to the integers, so no part of this procedure need be deleted. However, the integers permit an additional reduction.

If the greatest common divisor of m and n is 1 and $1 < m$, the literal $n \lhd m \cdot I$ reduces to $\lfloor \tfrac{n}{m} \rfloor + 1 \leq I$.

This reduction insures that the example cited in Section 3.1 is appropriately solved because $1 \leq 3 \cdot J$ reduces to $1 \leq J$. Hence, chaining (dense or integer) derives $2 \leq K$. Nonetheless, the general case with $I \leq n \cdot J$ remains unsolved because the greatest common divisor of I and n may not be known.

3.5 Inference Rules for Floor

Having introduced floor into STR+VE-I, we must provide mechanisms to remove it.

A STR+VE-I user may not invoke the floor function directly because it is restricted to be *an internal function only*. The floor function arises only as a result of chaining on integer-typed terms or variable eliminating integer-typed variables. Examination of the integer chaining and integer variable elimination rules demonstrates that every floor expression appears on an inequality's left side with a positive integer coefficient.

We need not provide rules that clash a left side floor expression against a right side floor expression. The left sided terms, however, must be removed. Directly applying $X - 1 < \lfloor X \rfloor$ is incomplete, acts against the actions of integer chaining and applies unrestricted to every floor expression in the system.

Consider $I \leq 3 \cdot J$ and $2 \cdot J \leq K$. Integer chaining derives $2 \lfloor \frac{I-1}{3} \rfloor + 2 \leq K$. Applying $X - 1 < \lfloor X \rfloor$ to this conclusion derives $2 \cdot I + 1 < 3 \cdot K$. From this and $2 \leq I$, integer chaining derives $2 \lfloor \frac{2}{2} \rfloor + 2 \leq 3 \cdot K + 1$ which reduces to $1 \leq K$.

A stronger conclusion is warranted. From $2 \leq I$ and $I \leq 3 \cdot J$, integer chaining derives $2 \leq 3 \cdot J$. Then integer chaining derives $2 \leq K$ from $2 \cdot J \leq K$.

STR+VE-I does not apply any of its inference rules directly to the floor expressions. Instead STR+VE-I removes floor expressions by removing their addends as it does with + expressions. STR+VE-I applies the appropriate inference rules to these addends assuming that they are not themselves floor expressions. Therefore, unifying floor expressions by associative-communicative unification is avoided.

3.5.1 Floor Chaining

STR+VE-I does not directly apply $X - 1 < \lfloor X \rfloor$ to remove floor expressions for the reasons given above. Instead, STR+VE-I removes the floor expression's addends. To remove these addends, we chose a rule based upon transforming a second inequality to have a matching right sided floor expression. For example, given $m \cdot \lfloor r \cdot J + X \rfloor \leq K$, an inequality $I < n \cdot J$ can be transformed into $\lfloor r \cdot \lfloor \frac{I}{n} \rfloor + r + X \rfloor \leq \lfloor r \cdot J + X \rfloor$. Integer transitivity axiom (8) is then applicable, deriving $m \cdot \lfloor r \cdot \lfloor \frac{I}{n} \rfloor + r + X \rfloor \leq K$.

Formally, we have

$$I < n \cdot J \wedge m \cdot \lfloor r \cdot J + X \rfloor \triangleleft K \Longrightarrow m \lfloor r \cdot \lfloor \tfrac{I}{n} \rfloor + r + X \rfloor \triangleleft K \text{ and} \quad (11)$$

$$I \leq n \cdot J \wedge m \cdot \lfloor r \cdot J + X \rfloor \triangleleft K \Longrightarrow m \lfloor r \cdot \lfloor \tfrac{I-1}{n} \rfloor + r + X \rfloor \triangleleft K, . \quad (12)$$

To apply floor chaining, STR+VE-I searches, as usual, for a term containing the central variable. If such a term is found and it is a floor expression, STR+VE-I searches the addends within it. If an addend is a shielding term, then floor chaining is applied. (This is the motivation for the re-definition of shielding term given in Section 2.)

Floor chaining introduces another location in which a floor expression can appear. Floor expressions can appear with a positive rational coefficient as an addend within another floor expression as seen in the conclusions of (11) and (12). For example, $\lfloor \frac{I}{n} \rfloor$ appears within $\lfloor r \cdot \lfloor \frac{I}{n} \rfloor + r + X \rfloor$ with coefficient r in the first floor chaining rule.

Consequently, the axioms on which the floor chaining rules are based must be more general. To do so, we first define "subterm paths".

Let $X[u]$ indicate the subterm selected by the path u. A path is a list of positive integers. $X[()]$ selects X and $(f\ x_1...x_n)[(l.u)]$ selects x_l where $l \leq n$.

Let $X[u/v]$ be the result of replacing the selected subterm with v. For example, (floor X (s* 3 (floor X Y)))$[(1\ 2\ 2)/Z]$ replaces X in (floor X Y) with Z. The result is (floor X (s* 3 (floor Z Y))).

Using paths, the floor transitivity axioms are

$$I < n \cdot J \wedge m \cdot L \triangleleft K \Longrightarrow m \cdot L[(2\ l.u)/ \lfloor \tfrac{I}{n} \rfloor + 1] \triangleleft K \text{ and} \quad (13)$$

$$I \leq n \cdot J \wedge m \cdot L \triangleleft K \Longrightarrow m \cdot L[(2\ l.u)/ \lfloor \tfrac{I-1}{n} \rfloor + 1] \triangleleft K \quad (14)$$

where u is a path such that $L[u]$ is a floor expression, $L[(l.u)]$ is addend and $L[(2\ l.u)]$ is J.

A *floor chain resolvent* of clauses $C_1 \vee I \triangleleft_1 n \cdot J'$ and $C_2 \vee m \cdot L \triangleleft_2 K$ is

$$(C_1 \vee C_2 \vee m \cdot L[(2\ l.u)/ \lfloor \tfrac{I'}{n} \rfloor + 1] \triangleleft_2 K)\sigma$$

where $L[u]$ is a floor expression,
J' and $L[(2\ l.u)]$ are unifable with mgu σ,
J' or $L[(2\ l.u)]$ is a shielding term,
J' nor $L[(2\ l.u)]$ is a variable and
I' is I if $\triangleleft_1$ is $<$ and is $I-1$ otherwise.

3.5.2 Variable Elimination

STR+VE-I seeks to remove a variable from a clause by removing the variable's shielding terms. These terms are of the form $(f\ x_1...x_n)$ where f is not $+$, s* or floor. Once all the shields are removed, the variable appears only within $+$, s* or floor and the variable is ready to be eliminated.

The normalization procedure, as amended in Section 3.4, insures that when such a variable occurs within a floor expression, its coefficients are positive. The coefficients of the surrounding floor expressions are also positive. Hence, given a normalized literal, $L(j) \triangleleft K$, where j is unshielded in $L(j)$, then $L(j)$ must be a non-decreasing function with respect to j. Therefore,

$$\forall j(n \cdot j \triangleleft_1 I \rightarrow L(j) \triangleleft_2 K) \iff L(\lfloor \tfrac{I'}{n} \rfloor) \triangleleft_2 K \quad (15)$$

where I' is I if $\triangleleft_1$ is $\leq$ and is $I-1$ otherwise.

We also have that

$$\forall j(n \cdot j \triangleleft_1 I \rightarrow L(j) \triangleleft_2 K + m \cdot j) \Rightarrow L(\lfloor \tfrac{I'}{n} \rfloor) \triangleleft_2 K + m \cdot \lfloor \tfrac{I'}{n} \rfloor) \quad (16)$$

where I' is similarly defined.

Floor variable elimination is defined based upon these axioms.
Let C be a clause $C' \vee \bigvee_{i=1}^{k} I_i \vartriangleleft_i n_i \cdot j \vee \bigvee_{l=1}^{k_1} L_l(j) \vartriangleleft'_l K_l$
$$\vee \bigvee_{l=k_1+1}^{k_2} L_l(j) \vartriangleleft'_l m_l \cdot j + K_l$$
where j does not occur in any I_i or K_l, or in C and j appears unshielded within L_l.

Let $D(X)$ be $\bigvee_{l=1}^{k_1} L_l(\lfloor X \rfloor) \vartriangleleft'_l K_l \vee \bigvee_{l=k_1+1}^{k_2} L_l(\lfloor X \rfloor) \vartriangleleft'_l m_l \cdot \lfloor X \rfloor + K_l$.

Then $C' \vee \bigvee_{i=1}^{k} D(\frac{I'_i}{n_i})$ is a *floor variable elimination resolvent* of a C where I'_i is I_i if $\vartriangleleft_i$ is $<$ and is $I_i - 1$ otherwise.

Because Axiom (15) is "if and only if", applying it to remove a variable j is sound and preserves unsatisfiability. Applying Axiom 16 is sound, but does not preserve unsatisfiability.

Nonetheless, STR+VE-I applies the floor variable elimination rule as a rewrite rule. The parent clause is replaced by the resulting clause. When $k_1 \neq k_2$ unsatisfiability may not be preserved. Hence, STR+VE-I is not complete for the integers.

3.5.3 Self-Chaining

As stated above, self-chaining cancels two unifiable terms within the same literal. Because the order (dense or integral) is a group, STR+VE's self-chaining inference rule implicitly encodes $Y - Y = 0$ as well. Applying this cancellation of inverses within floor yields $\lfloor J + (-J) + X \rfloor = \lfloor X \rfloor$. But negative integral terms are pushed outside of floor by normalization (Section 3.5.4).

Thus, positive integral terms within floor must be self-chained against integral terms appearing on the right side of the literal or combined with other positive integral terms within the floor.

For example, consider $\lfloor \frac{1}{2}(f\,j) \rfloor < (f\,5)$ where f is an integral function. By applying $X < \lfloor X \rfloor + 1$ and unifying $(f\,j)$ and $(f\,5)$, we get $\frac{1}{2}(f\,5) < (f\,5) + 1$ which reduces to $0 < (f\,5) + 2$.

A *floor self-chain resolvent* of a clause, $C \vee L \vartriangleleft K + m \cdot J$, is defined as
$$(C \vee L[(u/L[(1.u)] + ... + L[(k.u)] - 1] \vartriangleleft K + m \cdot J)\sigma$$
where $L[u]$ is a floor expression, $L[u]$ has k addends, J and $L[(2\,l.u)]$ are unifable with mgu σ, J or $L[(2\,l.u)]$ is a shielding term and J nor $L[(2\,l.u)]$ is a variable.

A *floor factor resolvent* of a clause, $C \vee L \vartriangleleft K + m \cdot J$, is defined as
$$(C \vee L \vartriangleleft K + m \cdot J)\sigma$$
where $L[u]$ is a floor expression and there is a subset of two or more of $L[u]$'s addends such that no addend's term is a variable, such that the addends' terms are unifable with mgu σ, and such that sum of the addends' coefficients is an integer.

For example, let C be the unit clause
$$\lfloor \tfrac{1}{2}(f(g\,j)) + \tfrac{1}{2}(f\,j') + \tfrac{1}{2}(f(g\,2)) + \tfrac{1}{2} \rfloor < (k\,j\,j').$$

Then the floor factor resolvents of C are

$$(f\,(g\,j)) + \lfloor \tfrac{1}{2}(f\,(g\,2)) + \tfrac{1}{2} \rfloor < (k\,j\,(g\,j)),$$
$$(f\,(g\,2)) + \lfloor \tfrac{1}{2}(f\,j') + \tfrac{1}{2} \rfloor < (k\,2\,j') \text{ and}$$
$$(f\,(g\,2)) + \lfloor \tfrac{1}{2}(f\,(g\,j)) + \tfrac{1}{2} \rfloor < (k\,j\,(g\,2)).$$

3.5.4 Floor Normalization

In addition to the normalizations given in Sections 2 and 3.4, STR+VE-I includes the following.

1. When $1 \leq r$, $\lfloor r \cdot J + X \rfloor$ reduces to $\lfloor (r - \lfloor r \rfloor) \cdot J + X \rfloor + \lfloor r \rfloor \cdot J$. So, positive coefficients of a floor's integer terms are strictly between 0 and 1. Floor addends with common terms are combined before this normalization is applied.
2. $\lfloor s \cdot J + X \rfloor$ reduces to $\lfloor (s - \lfloor s \rfloor) \cdot J + X \rfloor + \lfloor s \rfloor \cdot J$. As a consequence, the coefficients of a floor's integer terms are positive. Floor addends with common terms are combined before this normalization is applied.
3. If $n \cdot X < K + n$ reduces to false then $n \cdot \lfloor X \rfloor \vartriangleleft K$ is false.
4. If $n \cdot X < K$ reduces to true then $n \cdot \lfloor X \rfloor \vartriangleleft K$ is true.
5. $\lfloor \tfrac{1}{m} \cdot \lfloor X \rfloor \rfloor$ reduces to $\lfloor \tfrac{1}{m} X \rfloor$.

3.6 STR+VE-I

STR+VE-I employs six inference rules in place of STR+VE's three (chaining, self-chaining and variable elimination). STR+VE-I's six rules are integer chaining, self-chaining, floor chaining, floor variable elimination, floor self-chaining and floor factoring. Floor variable elimination subsumes integer variable elimination, so that rule is not used. Floor variable elimination is a rewrite rule in STR+VE-I just as variable elimination is in STR+VE.

Also, STR+VE-I employs STR+VE's normalizations along with six addition normalizations listed in Sections 2 and 3.5.4.

STR+VE-I applies the same strategy as STR+VE. Central variables are selected. Integer chaining, self-chaining, floor chaining, floor self-chaining or floor factoring remove each central variable's shielding term. The last three rules (floor chaining, floor self-chaining and floor factoring) are applicable if the shielding term is within a floor expression. Otherwise, integer chaining or self-chaining is applicable.

Floor variable elimination removes the variable once all the shielding terms have been removed. Clauses are entered into the ground prover once all their variables have been eliminated.

The ground prover consists of integer chaining, self-chaining, floor chaining, floor self-chaining and STR+VE-I's normalization in a matrix format.

4 Two Separate Linear Orders: One Dense and One Integer

Integers appear in many formulas as sequence indices. The sequence can be the rational numbers or some other order. Moreover, this second order may constitute the primary concern of the formula whereas the integers may be secondary. "Every Cauchy sequence is bounded" is one example. Proving these formulas requires reasoning about the integers as well as the primary order.

In this section, we combine STR+VE-I and STR+VE to obtain a prover, STR+VE-2, for these "two order" formulas.

Terms interpreted as integers must be distinguished in stating formulas from terms interpreted as dense. The two orders do not overlap, terms of one type are never added to terms of the other type. Similarly, the order of two differently typed terms is never asserted. Thus, each literal is distinguishable as integer or dense.

For each functional type, $((t_1, ..., t_n) \mapsto t_{n+1})$, the t_i's are restricted to be either integer or dense, but not both. So, unifying two terms of the same type does not lead to the unification of subterms of different types.

Using the terminology of theory resolution [18], we can now say that the *key* literals of integer chaining, $I \triangleleft_1 n \cdot J'$ and $m \cdot J'' \triangleleft_2 K$ have integer typed. The residue, $m \cdot \left\lfloor \frac{I'}{n} \right\rfloor + m \triangleleft_2 K$, also has integer typed. In fact, the key literals of all of the integer rules (chaining, self-chaining, normalization, floor chaining, floor variable elimination, floor self-chaining, floor factoring and floor normalization) have integer type. Likewise, the residues are integer typed.

The dense rules have densely typed key literals and densely typed residues. So, no key literal of a dense rule is a key literal of an integer rule. Also, residue literals of dense rules are not keys of integer rules. These two statement are also true when dense and integer are reverse.

The rule sets are independent. Combining the dense and integer versions is straight-forward because the versions employ the same strategy. Each selects a central variable, removes its shielding terms and then eliminates the variable. A ground prover finishes the refutation when enough variables are removed to have a ground unsatisfiable clause set.

Only two modifications are required to enforce the rules' type restrictions. First, the shielding term's type determines which rules, dense or integer, removes it. Second, the variable's type determines which type of variable elimination is employed.

4.1 Quantifiers Specifying Type

As stated above, input formulas must identify which terms are integer and which terms are dense. STR+VE-2's input formulas are in a LISP-style prefix notation. ALL and SOME indicate universal and existential quantification.

(all i P) indicates P holds for every i in the dense linear order. (some i P) means P holds for some i in the dense linear order.

This notation has been generalized slightly to allow the type of i to be specified. (all (i . integers) P) indicates P holds for every integer i. (some (i . integers) P) indicates P holds for some integer i.

Lastly, the type of a function's range must be specified. (function f P) indicates P holds for every densely valued functional interpretation of f. (function (f . integers) P) indicates P holds for every integral valued functional interpretation of f.

4.2 STR+VE-2

STR+VE-2 retains STR+VE's central variable strategy. The standard (or dense) chaining or self-chaining rules remove the densely typed shielding terms. Integer chaining, integer self-chaining, floor chaining, floor self-chaining or floor factoring remove the integer typed shielding terms.

The standard (or dense) variable elimination rule removes densely typed variables and the floor variable elimination rule removes integer typed variables. Both versions of variable elimination function as rewrite rules, replacing the parent clause with the resolvent.

Normalization and two ground provers finish the refutation. STR+VE-2's normalization reduces integer literals as it does dense literals. STR+VE-2 also applies the normalizations given in Sections 2 and 3.5.4 to integer literals.

The integer literals of ground clauses enter the integer ground prover and the dense literals enter the dense ground prover. Literals refuted by the respective ground provers are deleted.

This "two order" version of STR+VE subsumes both STR+VE-I and STR+VE.

If a formula contains only integral terms, integral inference rules alone apply. These rules' key literals are integral and their residual literals are likewise integral. Thus, the literals in each resolvent generated are integral and STR+VE-2's dense rules are never used. Because the strategies of STR+VE-I and STR+VE-2 are identical, STR+VE-2 functions on entirely integer typed formulas exactly as STR+VE-I.

Similarly, STR+VE-2 functions on entirely densely typed formulas exactly as STR+VE.

5 Conclusion

The STR+VE versions described here advances us toward the larger goal of automating proof discovery for all linearly ordered groups.

STR+VE-I, the integer version of STR+VE, is designed to prove integer inequality theorems. All infinite cyclic groups are isomorphic to the integers, so that theorems for infinite cyclic groups convert to the integers and, in turn, may be proved by STR+VE-I.

Not all infinite discrete linearly ordered groups are isomorphic to the integers. For example, the lexicograchically ordered group of integer pairs is not isomorphic to the integers. Research is currently underway generalizing STR+VE-I to apply to these groups.

STR+VE-2 proves theorems for the integers and separate dense linear orders by combining the techniques of STR+VE and STR+VE-I. STR+VE-2 subsumes both STR+VE and STR+VE-I.

Although application domains for STR+VE-2 are extensive, we are currently investigating enlarging its application area in three ways. The first is the aforementioned generalization of STR+VE-I. The direct goal is to extend STR+VE-I to all discrete linearly ordered groups.

The second line of investigation seeks to extend STR+VE-2 to dense linear orders containing integers. Rational and real numbers represent two well-known dense orders containing integers. These dense orders arise in many applications, especially when integral subgroups are discernable. Fix point numbers are then representable by

$$\forall x (\exists i \in \mathbb{Z}(i \leq n \cdot x < i + 1 \wedge (\text{Fix } x) = \tfrac{1}{n} i)$$

where n is the chosen power of 10.

The third extension incorporates full multiplication and division into STR+VE. Extending STR+VE in this manner enlarges it's application area to include linearly ordered fields.

6   Acknowledgements

Laura Buss edited portions of this paper and her efforts greatly improved the presentation.

References

1. Bledsoe, W. W., "The Sup-Inf Method in Presburger Arithmetic". The University of Texas at Austin, Math Department Memo ATP-18. December 1974. Essentially the same as: A new method for proving certain Presburger formulas. Fourth IJCAI, Tbilisi, USSR, September 3-8, 1975.
2. Bledsoe, W. W., and L. M. Hines, "Variable Elimination and Chaining in a Resolution-Base Prover for Inequalities", *Proc. 5th Conference on Automated Deduction*, Les Arcs, France, Springer-Verlag, (July 1980) 70-87.
3. Boyer, R. S. and J S. Moore, "Integrating Decision Procedures into Heuristics Theorem Provers: A Case Study of Linear Arithmetic", Technical Report ICSCA-CMP-44, University of Texas, Austin, TX, January 1985.
4. Brooks, R. A., *Model-Based Computer Vision*, UMI Research Press, Ann Arbor, MI.
5. Davis, M. and H. Putman, "A Computing Procedure for Quantification Theory", *J. Association of Computing Machinery* 7, (1960) 201-215.
6. Käufl, T., "Reasoning about Systems of Linear Inequalities", *Proc. 9th International Conference on Automated Deduction*, Argonne, Illinois, Springer-Verlag, (May 1988) 469-486.
7. Hines, L. and W. W. Bledsoe, "The STR+VE Prover", Technical Report ATP-94, University of Texas at Austin, 1990.

8. Hines, L., "Building In Axioms and Lemmas", Ph.D. Dissertation, University of Texas at Austin, (May 1988).
9. Hines, L., "Hyper-Chaining and Knowledge-Based Theorem Proving", *Proc. 9th International Conference on Automated Deduction*, Argonne, Illinois, Springer-Verlag, (May 1988) 469-486.
10. Hines, L., "Completeness of a Prover for Dense Linear Orders", *Journal of Automated Reasoning*, Vol. 8, 45-75, 1992, Netherlands, Kluwer Academic Pulishers.
11. Hines, L., "STR+VE⊆: The STR+VE-based Subset Prover", *10th International Conference on Automated Deduction*, July 1990, Kaiserslautern, Germany, Springer-Verlag, 193-206.
12. Hines, L., "The Central Variable Strategy of STR+VE", *11th International Conference on Automated Deduction*, June 1992, Saratoga Springs, NY, Springer-Verlag, 35-49.
13. King, J. C., *A Program Verifier*, Phd Dissertation, Carnegie-Mellon University, 1969.
14. Nelson, G., and D. Oppen, "A Simplifier Based on Efficient Decision Algorithms", *Proc. Fifth ACM Symp. on Principles of Programming Languages*, 1978.
15. Rabinov, A., "A Restriction of Factoring in Binary Resolution." *Proc. 9th International Conference of Automated Deduction*, Argonne, Illinois, USA, Springer-Verlag, (May 1988) 582-591.
16. Shostak, R., "A Practical Decision Procedure for Arithmetic with Function Symbols", *JACM*, April 1979.
17. Slagle, J. R., and L. Norton, "Experiments with an Automatic Theorem Prover Having Partial Ordering Rules", *CACM 16*, 1973, pp 683-688.
18. Stickel, M. E., "Automated Deduction by Theory Resolution, *IJCAI-85*, Los Angeles, California 1985, pp 1181-1186.

A Example

The sum of the limit of two sequences is the limit of the sum.

```
(FUNCTION (F . DENSE)
  (FUNCTION (G . DENSE)
    (-> (& [ALL (EF . DENSE)
            (-> (< 0 EF)
                (SOME (N . INTEGERS)
                  (ALL (I . INTEGERS)
                    (-> (< N I)
                        (ALL (J . INTEGERS)
                          (-> (< N J)
                              (& (<= (F I) (+ (F J) EF))
                                 (<= (F J) (+ (F I) EF)))]
           [ALL (EG .DENSE)
            (-> (< 0 EG)
                (SOME (N . INTEGERS)
                  (ALL (I . INTEGERS)
                    (-> (< N I)
```

```
                    (ALL (J . INTEGERS)
                       (-> (< N J)
                           (& (<= (G I) (+ (G J) EG))
                              (<= (G J) (+ (G I) EG)))]
        [ALL (E . DENSE)
          (-> (< 0 E)
              (SOME (N . INTEGERS)
                 (ALL (I . INTEGERS)
                   (-> (< N I)
                       (ALL (J . INTEGERS)
                          (-> (< N J)
                              (& (<= (+ (F I) (G I))
                                     (+ (F J) (G J) E))
                                 (<= (+ (F J) (G J))
                                     (+ (F I) (G I) E)))]
```

6. $(0 < E)$
4. $(0 < (J\ x0)\ -x0)$
5. $(0 < (I\ x0)\ -x0)$
2. $(0 <= (N1\ x2)\ -x0)\ (0 <= (N1\ x2)\ -x1)\ (0 <= (G\ x0)\ -(G\ x1)\ x2)$
   $(0 <= -x2)$
3. $(0 <= (N\ x2)\ -x0)\ (0 <= (N\ x2)\ -x1)\ (0 <= (F\ x0)\ -(F\ x1)\ x2)$
   $(0 <= -x2)$
1. $(0 < -E\ -(F\ (J\ x0))\ (F\ (I\ x0))\ -(G\ (J\ x0))\ (G\ (I\ x0)))$
   $(0 < -E\ (F\ (J\ x0))\ -(F\ (I\ x0))\ (G\ (J\ x0))\ -(G\ (I\ x0)))$
   Derived by (CHAIN 3 1).
7. $(0 <= -(J\ x1)\ (N\ x2))\ (0 <= (N\ x2)\ -x0)\ (0 <= -x2)$
   $(0 < -E\ (F\ (J\ x1))\ -(F\ (I\ x1))\ (G\ (J\ x1))\ -(G\ (I\ x1)))$
   $(0 < -E\ (F\ (I\ x1))\ -(F\ x0)\ -(G\ (J\ x1))\ (G\ (I\ x1))\ x2)$
   Derived by (CHAIN 4 7).
8. $(0 <= (N\ x1)\ -x0)\ (0 < (N\ x1)\ -x2)\ (0 <= -x1)$
   $(0 < -E\ (F\ (J\ x2))\ -(F\ (I\ x2))\ (G\ (J\ x2))\ -(G\ (I\ x2)))$
   $(0 < -E\ (F\ (I\ x2))\ -(F\ x0)\ -(G\ (J\ x2))\ (G\ (I\ x2))\ x1)$
   Derived by (CHAIN 8 3).
9. $(0 <= -(J\ x3)\ (N\ x4))\ (0 <= (N\ x1)\ -x0)\ (0 < (N\ x1)\ -x3)$
   $(0 < -E\ -(F\ (I\ x3))\ (F\ x2)\ (G\ (J\ x3))\ -(G\ (I\ x3))\ x4)$
   $(0 < -E\ (F\ (I\ x3))\ -(F\ x0)\ -(G\ (J\ x3))\ (G\ (I\ x3))\ x1)$
   $(0 <= (N\ x4)\ -x2)\ (0 <= -x1)\ (0 <= -x4)$
   Derived by (CHAIN 4 9).
10. $(0 <= (N\ x2)\ -x1)\ (0 < (N\ x2)\ -x4)\ (0 <= (N\ x3)\ -x0)$
    $(0 < -E\ -(F\ (I\ x4))\ (F\ x0)\ (G\ (J\ x4))\ -(G\ (I\ x4))\ x3)$
    $(0 < -E\ (F\ (I\ x4))\ -(F\ x1)\ -(G\ (J\ x4))\ (G\ (I\ x4))\ x2)$
    $(0 < (N\ x3)\ -x4)\ (0 <= -x2)\ (0 <= -x3)$
    Derived by (SELF-CHAIN 10).
12. $(0 <= -(I\ x1)\ (N\ x3))\ (0 <= (N\ x2)\ -x0)\ (0 < (N\ x2)\ -x1)$
    $(0 < -E\ (F\ (I\ x1))\ -(F\ x0)\ -(G\ (J\ x1))\ (G\ (I\ x1))\ x2)$
    $(0 < -E\ (G\ (J\ x1))\ -(G\ (I\ x1))\ x3)\ (0 < (N\ x3)\ -x1)\ (0 <= -x2)\ (0 <= -x3)$
    Derived by (CHAIN 5 12).

16. $(0 <= (N\ x1)\ -x0)\ (0 < (N\ x1)\ -x3)\ (0 < (N\ x2)\ -x3)$
    $(0 < -E\ (F\ (I\ x3))\ -(F\ x0)\ -(G\ (J\ x3))\ (G\ (I\ x3))\ x1)$
    $(0 < -E\ (G\ (J\ x3))\ -(G\ (I\ x3))\ x2)\ (0 <= -x1)\ (0 <= -x2)$
    Derived by (SELF-CHAIN 16).
19. $(0 <= -(I\ x0)\ (N\ x1))\ (0 < (N\ x1)\ -x0)\ (0 < (N\ x2)\ -x0)$
    $(0 < -E\ -(G\ (J\ x0))\ (G\ (I\ x0))\ x1)$
    $(0 < -E\ (G\ (J\ x0))\ -(G\ (I\ x0))\ x2)\ (0 <= -x1)\ (0 <= -x2)$
    Derived by (CHAIN 5 19).
20. $(0 < (N\ x0)\ -x2)\ (0 < (N\ x1)\ -x2)\ (0 < -E\ (G\ (J\ x2))\ -(G\ (I\ x2))\ x0)$
    $(0 < -E\ -(G\ (J\ x2))\ (G\ (I\ x2))\ x1)\ (0 <= -x0)\ (0 <= -x1)$
    Derived by (CHAIN 20 2).
21. $(0 <= -(J\ x1)\ (N1\ x4))\ (0 < (N\ x0)\ -x1)\ (0 < (N\ x3)\ -x1)$
    $(0 <= (N1\ x4)\ -x2)\ (0 < -E\ -(G\ (J\ x1))\ (G\ (I\ x1))\ x0)$
    $(0 < -E\ -(G\ (I\ x1))\ (G\ x2)\ x3\ x4)\ (0 <= -x0)\ (0 <= -x3)\ (0 <= -x4)$
    Derived by (CHAIN 4 21).
22. $(0 < (N\ x1)\ -x4)\ (0 < (N\ x2)\ -x4)\ (0 <= (N1\ x3)\ -x0)$
    $(0 < (N1\ x3)\ -x4)\ (0 < -E\ -(G\ (J\ x4))\ (G\ (I\ x4))\ x1)$
    $(0 < -E\ -(G\ (I\ x4))\ (G\ x0)\ x2\ x3)\ (0 <= -x1)\ (0 <= -x2)\ (0 <= -x3)$
    Derived by (CHAIN 2 22).
24. $(0 <= -(J\ x3)\ (N1\ x6))\ (0 < (N\ x1)\ -x3)\ (0 < (N\ x5)\ -x3)$
    $(0 <= (N1\ x2)\ -x0)\ (0 < (N1\ x2)\ -x3)\ (0 <= (N1\ x6)\ -x4)$
    $(0 < -E\ -(G\ (I\ x3))\ (G\ x0)\ x1\ x2)\ (0 <= -x1)\ (0 <= -x2)$
    $(0 < -E\ (G\ (I\ x3))\ -(G\ x4)\ x5\ x6)\ (0 <= -x5)\ (0 <= -x6)$
    Derived by (CHAIN 4 24).
25. $(0 < (N\ x2)\ -x6)\ (0 < (N\ x4)\ -x6)\ (0 <= (N1\ x3)\ -x0)$
    $(0 < (N1\ x3)\ -x6)\ (0 <= (N1\ x5)\ -x1)\ (0 < (N1\ x5)\ -x6)$
    $(0 < -E\ -(G\ (I\ x6))\ (G\ x0)\ x2\ x3)\ (0 <= -x2)\ (0 <= -x3)$
    $(0 < -E\ (G\ (I\ x6))\ -(G\ x1)\ x4\ x5)\ (0 <= -x4)\ (0 <= -x5)$
    Derived by (SELF-CHAIN 25).
27. $(0 <= -(I\ x0)\ (N1\ x3))\ (0 < (N\ x2)\ -x0)\ (0 < (N\ x4)\ -x0)$
    $(0 < (N1\ x3)\ -x0)\ (0 <= (N1\ x5)\ -x1)\ (0 < (N1\ x5)\ -x0)$
    $(0 < -E\ (G\ (I\ x0))\ -(G\ x1)\ x4\ x5)\ (0 < -E\ x2\ x3)$
    $(0 <= -x2)\ (0 <= -x3)\ (0 <= -x4)\ (0 <= -x5)$
    Derived by (SELF-CHAIN 27).
30. $(0 <= -(I\ x0)\ (N1\ x2))\ (0 <= -(I\ x0)\ (N1\ x4))$
    $(0 < (N\ x1)\ -x0)\ (0 < (N\ x3)\ -x0)\ (0 < (N1\ x2)\ -x0)$
    $(0 < (N1\ x4)\ -x0)\ (0 < -E\ x1\ x2)\ (0 < -E\ x3\ x4)$
    $(0 <= -x1)\ (0 <= -x2)\ (0 <= -x3)\ (0 <= -x4)$
    Derived by (VE CHAIN 5 30).
32. Box.

What is a Proof?

Richard Platek

richard@com.oracorp

Odyssey Research Associates
and
Cornell University, Ithaca, NY 14853, USA

As in all areas of Philosophy, those of us who claim we have no need or interest in the Philosphy of Mathematics are usually in the grip of a rather primitive form of some standard philosophical position. Many of us 'practical' people could benefit from hearing the various philosophical positions which underlie the vulgar attitudes we may be familiar with stated and debated in their classical purity. The famous twentieth century controversies over mathematical foundations and the nature of proof (e.g., Hilbert vs. Brouwer) actually have a long pedigree and reflect fundamental oppositions as to the nature of reality, thought, the role of mind, etc. The aim of this talk is help the listener place contemporary discussions of the meaning of mathematics within the proper historical context.

Termination, geometry and invariants

Ursula Martin

um@cs.st-and.ac.uk

School of Mathematical and Computational Sciences, University of St. Andrews,
St. Andrews, Scotland

Many different techniques have been developed for proving the termination of programs or functions. Some are purely ad-hoc arguments developed solely to solve an immediate problem. However a typical termination proof involves finding a quantity which decreases at each step of a computation, or in other words finding a well-founded ordering which is respected by the process we are considering. An extraordinary diversity of well-founded division orderings on structures such as vectors, strings, partitions or terms have been devised to prove termination, particularly of rewrite systems. A full survey is given in [4, 14].

Such orderings can also be used to constrain the search space in the search for solutions or normal forms in a wide variety of algorithms such as Knuth Bendix completion, for computation in algebras or groups, Gröbner basis, for computation in polynomial rings, or resolution. It turns out that the choice of ordering may have a surprising effect on efficiency [2, 3, 9, 11].

The association of ordinals to proofs that computations terminate goes back at least as far as Turing, who in 1949 [15] outlined what is now generally known as Floyd's method of analysing program correctness, and used an ordinal technique to prove termination of an example which calculates factorial by repeated addition.

Ordinals enable us to link termination proofs with classical proof and recursion theory. Following Gallier [5] one might regard the link as being Kruskal's tree theorem, which lies behind both the proofs of well-foundedness of many orderings on terms and the structure of various ordinal hierarchies. The links have been drawn more closely: for example Cichon[1] calculates the order types of term orderings and use these results to link termination proofs with the recursion theoretic hierarchy.

Order types are an example of an invariant: although there are many different orderings there may be relatively few order types which occur, and Cichon's result links these to other apparently unconnected mathematical notions.

A geometrical approach gives rise to numeric invariants of orderings. Martin [7] classified division orderings on multisets in terms of certain cones in a real vector space, extending results obtained for Gröbner basis algorithms [2].

More recently Martin, Scott and others [7, 13, 12] have developed a classification of division orderings on strings in terms of a family of numerical invariants which characterise an ordering by a sequence of real functions. The characterisation can be refined by counting occurrences of certain sub-strings, essentially using the theory of Lie elements in a free algebra. Certain orderings, the recursive path orderings, seem ubiquitous and have a somewhat mysterious alternative characterisation as the (essentially) unique ordering of highest order-type[10]. Work has begun on a similar techniques for classifying orderings on terms [8].

This classification has practical implications, since we may attempt to decide if a certain system is terminating in such an ordering by solving certain linear or polynomial constraints. Indeed rather than work with a particular ordering we may work instead with such a set of constraints, which may change through the execution of an algorithm, and termination may be assured provided that at each stage we know that the constraints have a solution. The ordering associates numeric and logical invariants to the underlying algorithms and data structures. The link between these complex invariants, the mathematical structure of the search space, and the underlying algorithms is as yet largely unexplored.

References

1. E A Cichon, Bounds on Derivation Lengths from Termination Proofs, International Journal of Foundations of Computer Science, to appear.
2. J Davenport, Y Siret and E Tournier, Computer Algebra, Academic Press 1988
3. D De Schreye and Kristof Verschaetse, Tutorial on termination of logic programs, *in* Springer Lecture Notes in Computer Science 649, Meta-Programming in Logic, 1992
4. N Dershowitz, Termination of Rewriting, Journal of Symbolic Computation 3 (1987) 69-116
5. J H Gallier, What's so special about Kruskal's theorem and the ordinal Γ_0? A survey of some results on Proof theory, Annals of Pure and Applied Logic 53 (1991)
6. D. Knuth and P. Bendix, Simple Word Problems in Universal Algebras, *in* Computational Problems in Abstract Algebra, Pergamon Press 1970, ed J. Leech.
7. U. Martin, A geometrical approach to multiset orderings, Theoretical computer Science 67 (1989) 37-54
8. U Martin, Linear interpretations by counting patterns, *in* Springer Lecture Notes in Computer Science 690, 5th International Conference on Rewriting Techniques and Applications, Montreal, June 1993
9. U Martin and M F K Lai, Some experiments with a completion theorem prover, Journal of Symbolic Computation (1992) 13, 81-100
10. Ursula Martin and Elizabeth Scott, The order types of termination orderings on terms, strings and multisets, *in* proceedings of the Eighth IEEE Conference on Logic in Computer Science, Montreal, 1993
11. W W McCune, OTTER 2.0, *in* Proceedings of the 10th International conference on Computer Aided Deduction, Lecture Notes in Computer Science , Springer Verlag, 1990
12. T Saito, M Katsura, Y Kobayashi and K Kajitori, On Totally Ordered Free Monoids, *in* Words, Languages and Combinatorics, World Scientific Publishing Co, Singapore 1992
13. E A Scott, Weights for total division orderings on strings, to appear, Theoretical Computer Science 1994.
14. J Steinbach, Simplification orderings- history of results, to appear Fundamentae Informaticae, 1994.

15. A M Turing, Checking a large routine, in: Report of a Conference on High Speed Automatic Calculating Machines, Univ Math Lab Cambridge, 1949, 67–69

Ordered Chaining for
Total Orderings*

Leo Bachmair[1] and Harald Ganzinger[2]

[1] Department of Computer Science, SUNY at Stony Brook,
Stony Brook, NY 11794, U.S.A,
leo@sbcs.sunysb.edu
[2] Max-Planck-Institut für Informatik,
Im Stadtwald, D-66123 Saarbrücken, Germany,
hg@mpi-sb.mpg.de

Abstract. We design new inference systems for total orderings by applying rewrite techniques to chaining calculi. Equality relations may either be specified axiomatically or built into the deductive calculus via paramodulation or superposition. We demonstrate that our inference systems are compatible with a concept of (global) redundancy for clauses and inferences that covers such widely used simplification techniques as tautology deletion, subsumption, and demodulation. A key to the practicality of chaining techniques is the extent to which so-called variable chainings can be restricted. Syntactic ordering restrictions on terms and the rewrite techniques which account for their completeness considerably restrict variable chaining. We show that variable elimination is an admissible simplification technique within our redundancy framework, and that consequently for dense total orderings without endpoints no variable chaining is needed at all.

1 Introduction

The axioms of the theories of partial and total orderings are extremely prolific in the context of resolution-based theorem proving. Many theorem provers build in transitivity by so-called chaining rules, which allow one to derive $(C \vee D \vee u < v)\sigma$ from premises $C \vee u < s$ and $D \vee t < v$ after unifying s and t by σ. Even though chaining is a considerable improvement over naive resolution, in this general form it still generates a huge search space. First, chaining inferences are always possible if s or t is a variable. For instance, the totality axiom $x < y \vee y \leq x$ can always be applied in four different ways, leading to the derivation of many equivalent variants of each clause. The situation is even worse for total orderings with additional structure, and in particular dense orderings with no endpoints. The clauses which express density and the absence of endpoints

---

* The research described in this paper was supported in part by the German Science Foundation (Deutsche Forschungsgemeinschaft) under grant Ga 261/4-1, by the German Ministry for Research and Technology (Bundesministerium für Forschung und Technologie) under grant ITS 9102/ITS 9103 and by the ESPRIT Basic Research Working Group 6028 (Construction of Computational Logics). The first author was also supported by the Alexander von Humboldt Foundation.

can also be chained with any other clause. For example, the clause $x < rx$ expressing the non-existence of a right endpoint, generates infinitely many inequalities $x < r^n x$ just by chaining with itself. Secondly, the same clause $C \vee u < s$ can be used as the first or as the second premise of a chaining inference, so that one needs to chain with both terms, u and s. Third, and related to the first two points, chaining is designed to generate the full (possibly infinite) transitive closure of a given set of inequalities, even though for any particular refutation a finite subset will be sufficient. This lack of goal-orientedness forms another obstacle to any efficient proof search. These problems apply to transitive relations in general; solutions have been proposed for certain theories.

Bledsoe and Hines (1980) have investigated dense total orderings without endpoints and have developed techniques for eliminating certain occurrences of variables from formulas. In their inference system, no chaining through variables is performed and no explicit inferences with transitivity, density, totality and the "no endpoints" axioms are computed. Completeness results for particular such systems of restricted chaining are proved by Bledsoe, Kunen, and Shostak (1985) and Hines (1992). Theorem provers developed from these theoretical investigations have performed successfully in proving theorems such as the continuity of the sum of two continuous functions or the intermediate value theorem; see Bledsoe and Hines (1980), Hines (1988) or Hines (1990).

Ordered paramodulation (Hsiang and Rusinowitch 1991) and superposition (Bachmair and Ganzinger 1990) are chaining-based inference systems for equality relations (the congruence properties of which also require chaining through subterms). Paramodulation into or below variables was first shown to be unnecessary by Brand (1975). Various syntactic restrictions avoid that equalities are always applied in both directions. In many cases even an infinite transitive closure of a given set of equalities can be represented by a finite convergent rewrite system. Only that system, and not the complete transitive closure of the given set of equalities (or clauses), is computed by the ordered variants of paramodulation and superposition.

Our aim is to combine the two approaches, keeping their best features, but avoiding their drawbacks. For that purpose we adapt the term rewriting techniques for arbitrary transitive relations described in Bachmair and Ganzinger (1993b) to total orderings.

1.1 Results

We present refutationally complete inference systems for total orderings, in which chaining is restricted by syntactic ordering constraints in much the same way as we know it from superposition calculi. That is, chaining inferences as above are performed only if the term $s\sigma$ is maximal in both (instances by σ of the) parent clauses. These ordering constraints immediately rule out many forms of chaining through variables. For instance, a term $x\sigma$ cannot be maximal if the variable x is shielded in a clause, that is, occurs as an argument of a function symbol. We also go beyond simple chaining in that we eliminate all occurrences of the maximal term in one single inference that combines several chaining steps. The effect is similar to hyper-resolution, in that the results of intermediate chaining steps need not be explicitly generated. Explicit inferences with transitivity and totality are shown to be redundant. This inference system applies to arbitrary total orderings and avoids most, though not all, variable chainings. The ordering constraints, like in the equational case, derive from a presentation of the full transitive closure of the given binary relation by an appropriate rewrite closure.

We prove refutational completeness of our inference systems in the presence of a general notion of redundancy for clauses and inferences by which most of the commonly applied simplification and elimination techniques (e.g., tautology elimination or subsumption) can be justified. For total orderings in particular we show that variable elimination, as proposed by Bledsoe and Hines (1980), is a simplification rule in our sense: a clause becomes redundant once variable elimination has been applied to it. In other words, variable elimination can be made mandatory. Ordering constraints for inferences and mandatory elimination of unshielded variables together achieve that chaining through variables can be completely avoided for dense total orderings with no endpoints. Variable elimination is also excluded in the inference systems proposed by Bledsoe, Kunen, and Shostak (1985) and Hines (1992), though these calculi, unlike ours, are not compatible with tautology deletion. Our proofs are comparatively simple and in particular profit from our ability of treating simplification techniques such as variable elimination as a separate issue.

The first of the two calculi which we present in this paper lacks an efficient treatment of equality. Like in previous approaches (Bledsoe and Hines 1980, Bledsoe, Kunen, and Shostak 1985, Hines 1992) an equality $s \approx t$ is represented by two inequalities $s \leq t$ and $t \leq s$. The disadvantage of this approach is that the implicitly specified equality relation requires additional congruence axioms for each function symbol, while such powerful simplification mechanisms as demodulation cannot be used. Therefore we introduce another system, in which we combine chaining with superposition (Bachmair and Ganzinger 1990). Equality is thus built into the inference rules; explicit inferences with congruence axioms or functional reflexive axioms and superposition into or below variables are not needed. The extended inference system also considerably improves earlier chaining systems with paramodulation (Slagle 1972), and for dense orderings without endpoints superposition from variables is not needed either. In addition, when superposing into inequalities only the maximal term of the inequality needs to be replaced. As before, the completeness result allows one to discard redundant clauses and inferences and admits simplifications such as demodulation or condensement. Certain technical definitions and proofs that had to be omitted due to lack of space, can be found in (Bachmair and Ganzinger 1993a).

We have implemented most the techniques investigated here as an extension of the Saturate system (Nivela and Nieuwenhuis 1993), and have obtained promising experimental results.

2 Preliminaries

2.1 Orderings

A (strict) *partial ordering* is a transitive and irreflexive binary relation; a *quasi-ordering* a reflexive and transitive binary relation. The reflexive closure of a strict ordering is a quasi-ordering. On the other hand, if $\leq$ is a quasi-ordering, then its *strict part* $<$, defined by: $x < y$ if and only if $x \leq y$ but not $y \leq x$, is a strict ordering. An ordering $<$ is said to be *total* if $x < y$ or $y < x$, whenever x and y are distinct. The ordering is *dense* if for all x and y with $x < y$, there exists an element z, such that $x < z$ and $z < y$. An ordering $<$ has no *left* (resp. *right*) *endpoint* if for every x there exists a y such that $y < x$ (resp. $x < y$). By an ordering *without endpoints* we mean one that has neither a left nor a right endpoint. For instance, the usual less-than relation on the natural numbers is a total ordering with a left, but no right, endpoint. The less-than relation on the real numbers is a dense total ordering without endpoints.

2.2 Predicate logic

We consider first-order predicate logic with equality; more specifically, first-order languages with (uninterpreted) function symbols, variables, and the (interpreted) predicate symbols $<$, $\leq$, and $\approx$.[3] A *term* is an expression $f(t_1, \ldots, t_n)$ or x, where f is a function symbol of arity n, x is a variable, and $t_1, \ldots, t_n$ are terms. An *atomic formula* (or simply *atom*) is an expression $s \approx t$, called an *equality*, or an expression $s < t$ or $s \leq t$, called a strict or non-strict *inequality*, respectively, where s and t are terms. A *literal* is an expression A (a *positive* literal) or $\neg A$ (a *negative* literal), where A is an atomic formula. We also write $s \not\approx t$ instead of $\neg(s \approx t)$ and use a similar notation for inequalities. When we speak of the *priority* of a predicate symbol, we take $<$ to have the highest and $\approx$ the lowest priority. The symbol $\triangleleft$ is used to denote (strict or non-strict) inequalities; the symbol $\underline{\triangleleft}$ to denote equalities and inequalities. A *clause* is a finite multiset of literals. We write a clause by listing its literals, $\neg A_1, \ldots, \neg A_m, B_1, \ldots, B_n$; or as a disjunction $\neg A_1 \vee \cdots \vee \neg A_m \vee B_1 \cdots \vee B_n$; or as a sequent $A_1, \ldots, A_m \to B_1, \ldots, B_n$. An expression is said to be *ground* if it contains no variables.

By a (*Herbrand*) *interpretation* we mean a set I of ground atomic formulas. We say that an atom A is *true* (and $\neg A$, *false*) in I if $A \in I$; and that A is *false* (and $\neg A$, *true*) in I if $A \notin I$. A ground clause is *true* in an interpretation I if at least one of its literals is true in I; and is *false* otherwise. In general, a clause is said to be true in I if all its ground instances are true. The *empty clause* is false in every interpretation. We say that I is a *model* of a set of clauses N (or that N is *satisfied* by I) if all elements of N are true in I. Occasionally, a model of N will also be called an *N-interpretation*. A set N is *satisfiable* if it has a model, and *unsatisfiable* otherwise.

We intend $\approx$ to be interpreted as an equality relation[4] and $<$ as a strict ordering. Also, $x \leq y$ is meant to be interpreted as $x < y \vee x \approx y$. (From a logical point of view, the symbol $\leq$ is therefore superfluous. But for theorem proving purposes shorter formulas are generally preferable, and it is better to avoid replacing a non-strict inequality by a disjunction.) The most problematic properties of these relations—for an automated theorem prover—are the transitivity properties. By a *transitivity interpretation* we mean a model of the set

---

[3] Uninterpreted predicate symbols can be encoded in a many-sorted language with function symbols and equality, by representing a formula $P(t_1, \ldots, t_n)$ as an equality $f_P(t_1, \ldots, t_n) \approx true_P$.

[4] An equality relation is a congruence on ground terms.

TR of all *transitivity axioms*

$$x \trianglelefteq_1 y, y \trianglelefteq_2 z \to x \trianglelefteq z$$

where $\trianglelefteq$ is the highest-priority symbol of $\trianglelefteq_1$ and $\trianglelefteq_2$. For example, $x \not< y \lor y \not\leq z \lor x < z$ is a transitivity axiom. The set PO, consisting of the axioms in TR and the axioms (i) $x \not< x$, (ii) $x \leq x$, and (iii) $x \not< y \lor x \leq y$, encodes that $<$ is a strict ordering and $\leq$ a quasi-ordering (with $<$ contained in its strict part). Interpretations satisfying PO are called partial orderings for short. If in addition the totality axiom (iv) $x < y \lor y \leq x$ is satisfied, the interpretation is called a total ordering. In that case, $<$ coincides with the strict part of $\leq$.

2.3 Chaining

Many theorem provers build in transitivity by so-called *chaining rules*,

$$\frac{C, u < s \quad D, t < v}{C\sigma, D\sigma, u\sigma < v\sigma}$$

where σ is a most general unifier of s and t; see Slagle (1972). For instance, with chaining we may deduce $u < v$ from $u < s$ and $s < v$. Specific variants of this inference rule can be found in calculi for dense total orderings (Bledsoe and Hines 1980, Bledsoe, Kunen, and Shostak 1985) and in paramodulation calculi for equality (Robinson and Wos 1969). In its full generality, the chaining rule is not practical, as the search space spanned by the inferences may be huge. *Variable chainings*, that is, chainings with a premise $C \lor u < x$ or $D \lor y < v$, are particularly prolific. Fortunately, they can be completely excluded in the case of dense total orderings without endpoints (Bledsoe, Kunen, and Shostak 1985) and considerably restrained in the case of paramodulation (Brand 1975 was the first to prove that paramodulation into a variable is unnecessary).

Chaining essentially generates the transitive closure of a given binary relation. For example, from $a < b$ and $b < c$ and $c < d$ we may deduce $a < c$ and $b < d$ and $a < d$. In other words, whenever there is a (finite) chain of equalities and/or inequalities

$$x_1 \trianglelefteq_1 x_2 \trianglelefteq_2 \cdots \trianglelefteq_{n-1} x_n$$

the equality or inequality $x_1 \trianglelefteq x_n$ can be deduced, where $\trianglelefteq$ is the highest-priority predicate among all $\trianglelefteq_i$. The basic idea of the term rewriting approach is to consider not arbitrary chains, but only those in which the intermediate terms $x_2, \ldots, x_{n-1}$ are in a certain sense simpler than the endpoints x_1 and x_n. Moreover, the equality or inequality $x_1 \trianglelefteq x_n$ is not necessarily deduced, but instead the corresponding chain may be implicitly represented by a rewrite system. We employ these ideas for the design of refutationally complete chaining systems in which various constraints are imposed on inference rules. Terminating rewrite systems are important in this context.

2.4 Rewrite systems

Let I be a set of ground atomic formulas and $\succ$ be a simplification ordering. We use the equalities and inequalities in I as *rewrite rules*. More precisely, we write $u \triangleleft_I v$ if $u \triangleleft v$ is an inequality in I and write $u \approx_I v$ if $u = w[s]$ and $v = w[t]$, for some term w and equality $s \approx t$ or $t \approx s$ in I. Furthermore, given $u \trianglelefteq_I v$, we write $u \trianglelefteq_I^L v$, if $u \succ v$, and $u \trianglelefteq_I^R v$, if $v \succ u$. The subscripts are dropped if I is clear from the context.

By a *proof* (in I) we mean a finite sequence $u_0 \triangleleft_1 u_1 \triangleleft_2 u_2 \ldots u_{n-1} \triangleleft_n u_n$ (in I), where $n \geq 0$. More specifically, we speak of a proof of $u_0 \leq u_n$. We speak of a proof of $u_0 < u_n$ if in addition at least one of the symbols $\triangleleft_i$ is $<$ (and hence $n \geq 1$); and of a proof of $u_0 \approx u_n$ if *all* symbols $\triangleleft_i$ are $\approx$. (If $n = 0$, we have a proof of $u_0 \approx u_0$.) By the *transitive closure* of I we mean the set of all atoms $u \trianglelefteq v$ provable in I. A *rewrite proof* is a sequence

$$u_0 < u_1$$

or

$$u_0 \trianglelefteq_1^L \cdots \trianglelefteq_{m-1}^L u_m \trianglelefteq_m^R \cdots \trianglelefteq_n^R u_n.$$

By the *rewrite closure* of I we mean the set of all atoms $u \trianglelefteq v$ that are provable by a rewrite proof in I.

In proving the completeness of chaining systems we construct Herbrand models for certain clause sets, describing interpretations by rewrite closures. A key question is under which circumstances the rewrite closure of a set I is a transitivity interpretation; a question, it turns out, that is related to commutation properties of the rewrite relations $\trianglelefteq_I$.

If a sequence $u_0 \triangleleft_1 \cdots \triangleleft_n u_n$ is not a rewrite proof then either (i) there is a subsequence $u_{i-1} \trianglelefteq_i^R u_i \trianglelefteq_{i+1}^L u_{i+1}$, or (ii) u_{i-1} and u_i are identical or incomparable (i.e., $u_{i-1} \not\succ u_i$ and $u_i \not\succ u_{i-1}$), for some i. A subsequence $u \trianglelefteq_1^R t \trianglelefteq_2^L v$, called a *peak*, is said to *commute* if $u \trianglelefteq v$ is provable by a rewrite proof, where $\trianglelefteq$ is the highest-priority symbol of $\trianglelefteq_1$ and $\trianglelefteq_2$. Commutation allows us to deal with case (i). If $\succ$ is a complete simplification ordering, then case (ii) applies only if $u_{i-1} = u_i$.

For example, if I contains the atoms $a < b$ and $b < c$ and $c < d$, and if $a \succ b \succ c \succ d$, then $a <^L b$ and $b <^L c$ and $c <^L d$. The relation $<^L$ evidently commutes with the (empty) relation $<^R$. If we choose the ordering differently, $b \succ c \succ d \succ a$, there is a peak, $a <^R b <^L c$, that does not commute. There are rewrite proofs of $a < b$ and $b < c$, but not of $a < c$.

The following lemma relates commutation and transitivity.

Lemma 1. *Let $\succ$ be a complete simplification ordering and I be a Herbrand interpretation that contains no strict inequality $t < t$. The rewrite closure of I is a transitivity interpretation if and only if all peaks in I commute.*

The lemma provides the starting point for our investigation of chaining techniques for total orderings, as peaks can be made to commute by applying suitable chaining inferences. (In the above example the peak $a <^R b <^L c$ commutes once the inequality $a < c$ has been deduced from $a < b$ and $b < c$ by chaining.) Similar, so-called "critical pair lemmas" form the basis of all completion procedures;

Levy and Agustí (1993) appear to have been the first to apply these techniques to non-symmetric rewrite relations. We go beyond usual completion procedures in that we consider general clauses, and thus have to deal with negative literals and disjunctions of literals. Particular emphasis will be given to the question of (the necessity of) variable chainings. For a discussion of the general aspects of these questions we refer to our work on rewrite techniques for transitive relations (Bachmair and Ganzinger 1993b). In this paper we look at total orderings, primarily dense total orderings without endpoints.

3 Maximal chaining

Our inference system is parameterized by complete reduction orderings $\succ$.[5] Let us assume an arbitrary such ordering to be given.

3.1 Inference rules

The predicate $\leq$ for non-strict inequality can be defined in terms of strict inequality and equality. It is also possible, though, to express equality in terms of inequality, representing $u \approx v$ by (the conjunction of) two inequalities $u \leq v$ and $v \leq u$. This, indeed, is the framework in which previous chaining inference systems have been formalized, the advantage being that no specific inference mechanisms for equality are needed. The necessary properties of the implicit equality relation are specified by the set $EE_{\mathcal{F}}$ of clauses

$$\bigvee_{1 \leq i \leq n} x_i < y_i \vee \bigvee_{1 \leq i \leq n} y_i < x_i \vee f(x_1, \ldots, x_n) \leq f(y_1, \ldots, y_n)$$

where f ranges over all function symbols in $\mathcal{F}$; see Bledsoe, Kunen, and Shostak (1985). The necessity of including these axioms is a disadvantage. Later on, we will build equality into the inference mechanism via paramodulation, thereby obviating any explicit equality axioms. (Another advantage of equational inference systems is that further simplification techniques such as demodulation become available, which are indispensable in actual implementations.)

In this section all atomic formulas are assumed to be (strict or non-strict) inequalities. Totality can be expressed as a clause $x < y \vee y \leq x$ or $x \leq y \vee y \leq x$. By *TA* we denote the set of these two clauses. They can be used to transform negative inequalities into positive ones: replace $x \not< y$ by $y \leq x$ and $x \not\leq y$ by $y < x$. Thus, we only need to consider disjunctions of positive inequalities.

We build the irreflexivity and transitivity properties directly into the inference mechanism.

Irreflexivity Resolution:
$$\frac{C, s < t}{C\sigma}$$
where σ is the most general unifier of s and t and $s\sigma$ is a maximal term in $C\sigma$.

---
[5] A reduction ordering is *complete* if it is total on ground terms. Complete reduction orderings on ground terms are simplification orderings.

Maximal Chaining:

$$\frac{C\ ,\ u_1 \triangleleft_1 s_1, \ldots, u_m \triangleleft_m s_m \quad D\ ,\ t_1 \triangleleft'_1 v_1, \ldots, t_n \triangleleft'_n v_n}{C\sigma\ ,\ D\sigma\ ,\ \bigvee_{1 \leq i \leq m, 1 \leq j \leq n} u_i \sigma \triangleleft_{i,j} v_j \sigma}$$

where (i) σ is the most general unifier of $s_1, \ldots, s_m, t_1, \ldots, t_n$ and $\triangleleft_{i,j}$ is the highest-priority symbol of $\triangleleft_i$ and $\triangleleft'_j$, (ii) $u_i \sigma \not\succeq s_1 \sigma$, for all $1 \leq i \leq m$, and $u\sigma \not\succeq s_1 \sigma$, for all terms u in C, and (iii) $v_i \sigma \not\succeq t_1 \sigma$, for all $1 \leq i \leq n$, and $v\sigma \not\succeq t_1 \sigma$, for all terms v in D, and $t_1 \sigma$ occurs in $D\sigma$ only in inequalities $v \triangleleft t_1 \sigma$.

For example, from $a < f(c)$ and $f(x) < a \lor f(x) < b$ we may deduce $a < a \lor a < b$ (assuming $f(c) \succ a \succ b$ in the given ordering). Maximal chaining is designed to eliminate the maximal terms in a clause. The conclusion of a maximal chaining inference can also be obtained by a sequence of $m \times n$ ordinary chaining inferences, interspersed with factoring inferences; but an important difference is that with maximal chaining the intermediate clauses need not be deduced.

Thus, from $a < f(c)$ and $f(x) < a \lor f(x) < b$ we may obtain, first $a < a \lor f(c) < b$, and then $a < a \lor a < b$ by ordinary chaining. The intermediate clause $a < a \lor f(c) < b$ cannot be obtained by maximal chaining. In this regard, the effect of maximal chaining is similar to that of hyper-resolution. This, incidentally, is also a significant difference between maximal chaining and "equivalence factored chaining," as defined by Hines (1992). A single equivalence factored chaining step may also combine several ordinary chaining steps, but intermediate clauses can also be deduced. For instance, both $a < a \lor f(c) < b$, and $a < a \lor a < b$ are equivalence factored chain resolvents.[6]

The inference system $\mathsf{MC}^{\succ}$ consisting of irreflexivity resolution and maximal chaining represents our basic inference mechanism for total orderings. (We also assume that the premises of an inference have no variables in common; if necessary, the variables in one premise are renamed.) Below we shall outline further improvements to the calculus for richer structures, such as dense total orderings without endpoints. These improvements ultimately derive from the concept of redundancy (of clauses and inferences) discussed in the next section.

3.2 Redundancy and Refutational Completeness

Simplification and deletion techniques, such as subsumption, tautology deletion, condensement, demodulation, contextual rewriting, etc., have proved to be indispensable for actual implementations of theorem provers. As in our previous work on superposition calculi, we provide a general framework for formalizing and reasoning about such techniques, based on an appropriate notion of redundancy.

Redundancy of inferences and clauses depends on a well-founded ordering on clauses that is obtained by lifting the given ordering term ordering $\succ$ first to literals and then to clauses. The ordering is the key to our refutational completeness proofs and therefore unshielded variables are also taken into account

---

[6] Another difference is that maximal chaining employs an ordering restriction on terms, whereas equivalence factored chaining imposes a "shielding term restriction."

in its definition. Details on the clause ordering, which we denote by the symbol $\succ$ also, can be found in Bachmair and Ganzinger (1993a).

Given sets of clauses N and S, we say that a ground instance $C\sigma$ of some clause C (which need not be an element of N) is S-*redundant* with respect to N if there exist ground instances $C_1\sigma_1, \ldots, C_k\sigma_k$ of N such that (i) $C\sigma$ is true in every model of $S \cup \{C_1\sigma_1, \ldots, C_k\sigma_k\}$ and (ii) $C\sigma \succ C_j\sigma_j$, for all j with $1 \leq j \leq k$. The clause C is S-redundant if all its ground instances are.

Tautologies are S-redundant in this sense, for any set S, and most cases of proper subsumption are also covered by this notion of redundancy. According to the definition, the axioms in S are all S-redundant. We are particularly interested in PO-redundancy. (Since these axioms are built into the chaining mechanism, we expect them to be redundant.)

A ground inference with conclusion B and maximal premise C is called S-*redundant* with respect to N if either some premise is S-redundant, or else there exist ground instances $C_1, \ldots, C_k$ of N such that B is true in every model of $S \cup \{C_1, \ldots, C_k\}$ and $C \succ^c C_j$, for all j with $1 \leq j \leq k$. A non-ground inference is called S-redundant if all its ground instances are S-redundant.

We say that a set of clauses N is *saturated up to S-redundancy* if all inferences from N are S-redundant. Since the ground versions of the inferences in $\mathsf{MC}^\succ$ are simplifying in that their conclusion is smaller than their maximal premise, they can be rendered redundant by adding the conclusion to the given set of clauses. Thus, computing the closure of a clause set under these inferences yields a saturated set.

The following theorem establishes the refutational completeness of MC in the presence of redundancy.

Theorem 2. *Let N be a set of clauses with subset $EE_\mathcal{F}$, where $\mathcal{F}$ is the given set of function symbols. If N is saturated up to PO-redundancy with respect to MC, then it has a total ordering as model if and only if it does not contain the empty clause.*

This theorem applies to arbitrary total orderings. The chaining mechanism can be improved for more specific theories, such as dense total orderings without endpoints.

3.3 Variable elimination

A *variable chaining* is a maximal chaining in which one of the terms s_i or t_j is a variable. Variable chaining can be quite prolific, as the unification of terms required for an inference always succeeds if the terms are variables. Fortunately, the ordering constraints in conditions (ii) and (iii) considerably cut down on the number of variable chainings. More specifically, the ordering constraints can only be satisfied if each term s_i or t_j is either a non-variable or an *unshielded variable*; that is, a variable that does not occur in a subterm $f(\ldots, x, \ldots)$. For example, the variable x is unshielded in $a < x \lor x < b$. Certain unshielded variables can be eliminated in any total ordering. More occurrences of unshielded variables can be eliminated in total orderings without endpoints. If the ordering is also dense *all* unshielded variables can be eliminated.

Variable Elimination:

$$\frac{C\ ,\ u_1 \triangleleft_1 x, \ldots, u_m \triangleleft_m x\ ,\ x \triangleleft'_1 v_1, \ldots, x \triangleleft'_n v_n}{C\ ,\ \bigvee_{1 \leq i \leq m, 1 \leq j \leq n} u_i \triangleleft_{i,j} v_j}$$

where x is an unshielded variable not occurring in C, u_i and v_i, and where $\triangleleft_{i,j}$ is the lowest-priority symbol of $\triangleleft_i$ and $\triangleleft'_j$.

This inference rule is sound for total dense orderings without endpoints. It has been used by Bledsoe, Kunen, and Shostak (1985) in their chaining calculus. Weaker variable elimination rules can be applied to non-dense orderings. For example, in any total ordering $a < b$ is equivalent to $a < x \vee x < b$. In an ordering without left endpoint a disjunction $C \vee \bigvee_i t_i < x$, where x does not occur in C or any term t_i, is equivalent to C. In a total, "discrete" ordering, in which $s(x)$ denotes the "successor" of x, the disjunction $a \leq x \vee x \leq b$ is equivalent to $a \leq s(b)$. This equivalence may also be used as a variable elimination rule.

Lemma 3. *Let C' be a clause $C \vee \bigvee_i u_i \triangleleft_i x \vee \bigvee_j x \triangleleft_j v_j$ with an unshielded variable x not occurring in C, u_i or v_i; and let C'' be $C \vee \bigvee_{i,j} u_i \triangleleft_{i,j} v_j$, where $\triangleleft_{i,j}$ is the lowest-priority symbol of $\triangleleft_i$ and $\triangleleft'_j$. Then C' is either a totality axiom or else is PO-redundant with respect to any set N that contains C'' and the totality axioms.*

The lemma not only indicates that variable elimination is simplifying, but that in the presence of the totality axioms the premise in any of the elimination rules is rendered redundant by its conclusion. The variable elimination rule may therefore be called *simplification* rules: their premises can be *replaced* by the respective conclusion. In other words, variable elimination can be made *mandatory*, in that no other inference rule needs to be applied to a clause with an unshielded variable, which also means that variable chainings are not needed. The compatibility of variable elimination with other chaining systems has been shown by Richter (1984) and Hines (1992).

By $\mathsf{CV}^\succ$ we denote the calculus consisting of irreflexivity resolution and maximal chaining plus variable elimination. Also, let DO denote the set PO plus certain appropriately chosen clauses to encode the properties of a dense ordering without endpoints.

Lemma 4. *Let N be a set of clauses that is saturated up to PO-redundancy with respect to MC and contains $EE_\mathcal{F}$ as a subset, where $\mathcal{F}$ is the given set of function symbols. If N contains no unshielded variables, then $N \cup DO$ is also saturated up to PO-redundancy.*

The significance of this lemma rests on the fact that through simplification it is possible to eliminate all unshielded variables from clauses. Clauses $C \vee x \leq x$ are PO-tautologies and, hence, redundant. A clause $C \vee x < x$ becomes redundant once C has been deduced by irreflexivity resolution. (This is the only situation in which irreflexivity resolution needs to be applied to a clause $C \vee s < t$, where s or t is a variable.) All other unshielded variables can be eliminated by variable elimination.

4 Chaining with superposition

The chaining calculus described above requires explicit equality axioms. Equality can also be built in via paramodulation (originally introduced by Robinson and Wos 1969), which in fact may be seen as a form of "subterm chaining" (as discussed in Bachmair and Ganzinger 1993b). We now allow all three symbols $\approx$, $<$ and $\leq$ to occur in clauses. This is logically redundant as $\leq$ can be expressed as a disjunction in $\approx$ and $<$, but from a practical point of view, and as was confirmed by our experimentation with the Saturate system, it is more appropriate to handle the three relations simultaneously and specifically so as to avoid duplication of terms as much as possible. For the same reason we allow negative equalities $s \not\approx t$ that would otherwise have to be transformed into disjunctions $s < t \lor t < s$. Thus, clauses may contain positive and negative equalities in addition to positive inequalities.

Totality TE can be expressed as a clause $x < y \lor y < x \lor x \approx y$ and the anti-symmetry axiom AS as $x \leq y$, $y \leq x \rightarrow x \approx y$.

The calculus CS consists of three parts: a chaining calculus for inequalities, a superposition calculus for equalities, and further chaining rules that connect equalities with inequalities. Superposition is a form of paramodulation in which ordering constraints are imposed on the inference rules. We use the superposition calculus of Bachmair and Ganzinger (1994). Symmetry of equality can be built in by ordered resolution with the clause $u \not\approx v \lor v \approx u$. (In practice, equations are usually considered as unordered pairs to avoid unnecessary duplication of formulas.)

Reflexivity Resolution:

$$\frac{u \not\approx v \, , \, C}{C\sigma}$$

where σ is a most general unifier of u and v and $u\sigma \not\approx v\sigma$ is a maximal literal in $C\sigma$.

Superposition:

$$\frac{C \, , \, t \approx s \quad D \, , \, L'[s']}{C\sigma \, , \, D\sigma \, , \, L[t]\sigma}$$

where (i) σ is a most general unifier of s and s'; (ii) $t\sigma \approx s\sigma$ is strictly maximal in $C\sigma$ and $t\sigma \not\preceq s\sigma$; (iii) L' is either an equality $u[s] \approx v$ or $v \approx u[s]$, or the negation thereof, such that $v\sigma \not\preceq u\sigma$ and $L'\sigma$ is strictly maximal in $D\sigma$, if it is a positive equality, and just maximal in $D\sigma$ if it is negative; and (v) s' is not a variable.

Equality Factoring:

$$\frac{C \, , \, t \approx s \, , \, t' \approx s'}{C\sigma \, , \, t\sigma \not\approx t'\sigma \, , \, t'\sigma \approx s'\sigma}$$

where (i) σ is a most general unifier of s and s', (ii) $t\sigma \not\preceq s\sigma$, and (iii) $s\sigma \approx t\sigma$ is maximal in $(C \lor s \approx t \lor s' \approx t')\sigma$.

Superposition is a restricted form of subterm chaining applied to equations. We need a similar form of "maximal subterm chaining" with equations into inequations. These inference rules are designed to reduce maximal terms.

Equality Chaining Left:

$$\frac{C\,,\,u \approx s \quad D\,,\,t_1[s_1]_p \triangleleft_1 v_1,\ldots,t_n[s_n]_p \triangleleft_n v_n}{C\sigma\,,\,D\sigma\,,\,t_1[u]_p \triangleleft_1 v_1,\ldots,t_n[u]_p \triangleleft_n v_n}$$

where (i) $\sigma = \rho\tau$ with ρ the most general unifier of $s, s_1, \ldots, s_n$, and τ the most general unifier of $t_1[s_1]\rho, \ldots, t_n[s_n]\rho$, (ii) $u\sigma \not\preceq s\sigma$, and $u\sigma \approx s\sigma$ is strictly maximal in $C\sigma$, (iii) $v_i\sigma \not\preceq t_1\sigma$, for all $1 \leq i \leq n$, and $v\sigma \not\succ t_1\sigma$, for all terms v in D, and $t_1\sigma$ occurs in $D\sigma$ only in equalities or inequalities $v \triangleleft t_1\sigma$, and (iv) none of the s_i is a variable.

Equality Chaining Right:

$$\frac{C\,,\,u \approx s \quad D\,,\,v_1 \triangleleft_1 t_1[s_1]_p,\ldots,v_n \triangleleft_n t_n[s_n]_p}{C\sigma\,,\,D\sigma\,,\,v_1 \triangleleft_1 t_1[u]_p,\ldots,v_n \triangleleft_n t_n[u]_p}$$

where (i) $\sigma = \rho\tau$ with ρ is the most general unifier of $s, s_1, \ldots, s_n$, τ the most general unifier of $t_1[s_1]\rho, \ldots, t_n[s_n]\rho$, (ii) $u\sigma \not\preceq s\sigma$, and $u\sigma \approx s\sigma$ is strictly maximal in $C\sigma$, (iii) $v_i\sigma \not\preceq t_1\sigma$, for all $1 \leq i \leq n$, and $v\sigma \not\succ t_1\sigma$, for all terms v in D, and $t_1\sigma$ occurs in $D\sigma$ only in equalities, and (iv) none of the s_i is a variable.

Finally, we also need to be able to deduce equalities implicitly specified by inequalities, so that the anti-symmetry axiom is satisfied. For that purpose we modify the maximal chaining rule as follows.

Inequality Chaining:

$$\frac{C\,,\,u_1 \triangleleft_1 s_1,\ldots,u_m \triangleleft_m s_m \quad D\,,\,t_1 \triangleleft'_1 v_1,\ldots,t_n \triangleleft'_n v_n}{C\sigma\,,\,D\sigma\,,\,\bigvee_{1 \leq i \leq m, 1 \leq j \leq n} ineq(i,j)}$$

where (i) σ is the most general unifier of $s_1, \ldots, s_m, t_1, \ldots, t_n$, (ii) $ineq(i,j)$ is the strict inequality $u_i\sigma < v_j\sigma$ if $\triangleleft_i$ or $\triangleleft'_j$ is $<$, and the disjunction $u_i\sigma < v_j\sigma \vee u_i\sigma \approx t_1\sigma$ otherwise,[7] (iii) $u_i\sigma \not\preceq s_1\sigma$, for all $1 \leq i \leq m$, and $u\sigma \not\preceq s_1\sigma$, for all terms u in C, and (iv) $v_i\sigma \not\preceq t_1\sigma$, for all $1 \leq i \leq n$, and $v\sigma \not\succ t_1\sigma$, for all terms v in D, and $t_1\sigma$ occurs in $D\sigma$ only in inequalities $v \triangleleft t_1\sigma$.

For example, from $a \leq f(x)$ and $f(a) \leq b \vee f(y) < c$ we may deduce $a < b \vee f(a) \approx a \vee a < c$ (assuming $f(a)$ is the maximal term in the clause). Inequality chaining is like maximal chaining, except that when two non-strict inequalities are chained we split the resulting non-strict inequality into a strict inequality and an equality.

The calculus CS consists of the above inference rules plus irreflexivity resolution. (Maximal chaining is replaced by inequality chaining.)

Theorem 5. *Let N be a set of clauses that is saturated up to PO-redundancy (with respect to* CS*). Then N has a total ordering with equality as a model if and only if it does not contain the empty clause.*

---

[7] Multiple copies of an equation $u_i\sigma \approx t_1\sigma$ resulting from different inequalities $t_j < v_j$ and $t_k < v_k$ can of course be merged into one.

The variable elimination rule applies as previously, but the presence of equality complicates matters somewhat. To sum up what is explained in detail in our full paper, variable elimination is not necessarily a simplification rule, as some instances of a premise may not be rendered redundant by the conclusion. Fortunately, those instances of a clause, in which an unshielded variable x is instantiated by a maximal term, do become redundant as a result of variable elimination. Therefore, inferences involving the unshielded variable (chaining through x, superposition from x, [ir]reflexivity resolution applied to [in]equalities with x) are unnecessary. (The ordering constraints for such inferences require x to be instantiated with the maximal term, in which case the corresponding instances of the premises are redundant.) Hence variable chaining is not needed.

5 Experimental Results

We have implemented a theorem prover based on the above inferences systems (with explicit equality) which can handle arbitrary transitive relations, not just orderings and equality, cf. Bachmair and Ganzinger (1993b). The implemented inference systems also allow one to define selection functions on negative literals, so that negative chaining inferences need to be computed only with maximal or selected literals. The completeness of selection can be proved in a similar way as for superposition (Bachmair and Ganzinger 1994).

Our implementation is based on the Saturate system by Nivela and Nieuwenhuis (1993) which implements superposition together with a variety of simplification and elimination techniques. The system attempts to detect those properties of a given clause set (transitivity, reflexivity, irreflexivity, symmetry, totality, density, no endpoints, monotonicity) that are central to the selection of the appropriate instance of our family of chaining-based inference systems. Apart from implementing the inference rules we have also extended some of the mechanisms for redundancy proofs to a specific handling of inequations. For instance, we have extended the equational tautology checker to inequations, applying ground versions of the chaining inference systems as decision procedures to that end. We also reduce newly generated formulas containing inequational atoms by unit equations as well as inequations. (Reduction by negative chaining with unit inequations is sound if a tautology can be obtained that way.)

The system, which is implemented in Prolog, performs quite well on many non-trivial examples, but is a prototype in many respects. It is not at all tuned to fast proof search and there are no sophisticated inference selection schemes, clause weightings or indexing data structures. Its major purpose is to serve as a test-bed for trying out various combinations of chaining-based inference systems with ordering constraints and checks for redundancy. The table below shows some benchmarks: ivt refers to the intermediate value theorem; scf to the the proof of continuity of the sum of two continuous functions (challenge problems 3 and 5, respectively, of Bledsoe 1990).; and perm to one of the main lemmas about the composition of permutations of subarrays that one needs for the verification of quicksort. The last problem is interesting as it combines equality (to represent permutations as bijective functions) with inequality (for reasoning about the index ranges of subarrays).

problem	length	inferences	non-red	time	simpl	constraints
ivt	65	636	152	76	46	11
scf3	47	343	166	93	44	10*
scf5	56	858	268	240	150	21*
perm	67	450	210	143	40	79
perm	70	805	252	96	55	6*

length: proof length (including axioms)
inferences: number of computed inferences
non-red: number of computed inferences that were not proved to be redundant
time: total runtime in seconds on a Sparc10 under SICStus
simpl: time spent on simplification and redundancy proofs
constraints: time spent on solving ordering constraints

A central component of the system is a constraint solver for the lexicographic path ordering. Constraint solving can at times be very slow (e.g., in perm), as the underlying problem is NP-complete (Nieuwenhuis 1993). In fact this part can be prohibitively expensive, so that our system provides a choice between a full and a partial constraint solving mode. In the latter mode the ordering constraints on premises of inferences are checked individually, *before* a unifier is applied ("a priori constraints"). A (*) indicates runs with a priori constraint checking. These usually generate substantially more inferences, but the simpler constraint solving process may still result in a better overall runtime, as can be seen from the two runs of the perm example.

Ordering restrictions considerably cut down the search space of chaining, but ordered chaining by itself would still be too prolific and not enough goal-oriented to get any of these examples in reasonable time. The power of the prover stems from its combination of order-constrained inferences together with simplification and elimination of redundant clauses and inferences. From the benchmarks one observes that among the inferences that satisfy the ordering constraints, only 25%–45% are non-redundant ("non-red") with respect to the applied criteria. (But simplification and checking for redundancy usually is much more expensive than computing inferences.) Our abstract notion of redundancy is also of great value in checking whether certain simplification techniques are compatible with the inference system at hand. For instance for proving ivt it turned out that a simplification called resolution subsumption was extremely useful. To show that it was compatible and to add it to the system was very easy.

Our prover performs especially well in cases where transitive relations do not interfere too much with the algebraic structure of function symbols. This is the case for most of the lemmas that are needed for verifying sorting programs. At present, our system does not perform well on examples, such as scf5, with associative-commutative operators, such as addition, that are monotonic with respect to the ordering. Associativity and commutativity are dealt with by ordered rewriting, which is not a very practical method, despite the fact the our system applies specific checks for ground confluence of derived equations in such a case. We got the proof for scf5 only by using AC-unification instead of syntactic unification. But with this modification our implementation is incomplete as the lexicographic path ordering is not AC-compatible and as we do not generate extended clauses. In fact, we even had to slightly modify the problem formulation and run the problem with a priori constraints to make up for this incompleteness. An open problem which we are currently working on is how to combine decision methods for theories such as linear arithmetic with general chaining inferences. Once that has been worked out, there should be no need to resort to ad hoc techniques and problem massaging in examples such as scf5.

6 Summary

We have presented chaining-based inference systems for total orderings that extend previous work in several respects: we impose ordering constraints on chaining inferences; we build equality directly into the inference mechanism; and we establish refutational completeness in the presence of a notion of redundancy that covers such important simplification techniques as tautology deletion, subsumption, condensement, and demodulation. The completeness proofs of our results are comparatively simple. We deal with variable elimination as a simplification rule, separately from the chaining rules; an approach that better clarifies the connection between these essential components of chaining systems.

The improvements are not only of theoretical significance, as experimental evidence indicates that equational inference mechanisms, such as superposition and demodulation, are preferable to explicit axiomatizations of equality. Some form of equational reasoning appears to be used in the provers described by Bledsoe and Hines (1980), Hines (1988) and Hines (1990), and inference systems with equational reasoning capabilities may provide a better approximation to actual implementation practice than other chaining systems.

The superposition calculus, as described above, contains an explicit factoring rule for equalities, whereas maximal chaining implicitly encodes factoring for inequalities. An alternative to maximal chaining would be ordered chaining (as in Bachmair and Ganzinger 1993b) with inequality factoring.

In algebraic structures such as ordered rings one may have monotonicity or anti-monotonicity of functions (such as + and −) with respect to the ordering. The techniques we have discussed should be extended to such cases. An important question in this context is to what extent chaining through or below variables is necessary. Our techniques may also be useful in the context of chaining and variable elimination for set theory, an application that has been studied by Hines (1990).

Acknowledgements. We would like to thank the anonymous referees for their comments.

References

L. BACHMAIR AND H. GANZINGER, 1990. On Restrictions of Ordered Paramodulation with Simplification. In M. Stickel, editor, *Proc. 10th Int. Conf. on Automated Deduction, Kaiserslautern*, Lecture Notes in Computer Science, vol. 449, pp. 427–441, Berlin, Springer-Verlag.

L. BACHMAIR AND H. GANZINGER, 1993. **Ordered Chaining for Total Orderings.** Technical Report MPI-I-93-250, Max-Planck-Institut für Informatik, Saarbrücken. Full version of this paper.

L. BACHMAIR AND H. GANZINGER, 1993. Rewrite Techniques for Transitive Relations. Technical Report MPI-I-93-249, Max-Planck-Institut für Informatik, Saarbrücken. To appear in Proc. LICS'94.

L. BACHMAIR AND H. GANZINGER, 1994. Rewrite-based equational theorem proving with selection and simplification. *Journal of Logic and Computation*, Vol. 4, No. 3, pp. 1–31. Revised version of Technical Report MPI-I-91-208. To appear.

W. BLEDSOE, K. KUNEN AND R. SHOSTAK, 1985. Completeness results for inequality provers. *Artificial Intelligence*, Vol. 27, pp. 255–288.

W. W. BLEDSOE AND L. M. HINES, 1980. Variable Elimination and Chaining in a Resolution-based Prover for Inequalities. In Wolfgang Bibel, Robert Kowalski, editors, *5th Conference on Automated Deduction*, LNCS 87, pp. 70–87, Les Arcs, France, Springer-Verlag.

W.W. BLEDSOE, 1990. Challenge problems in elementary calculus. *Journal of Automated Reasoning*, Vol. 6, pp. 341–359.

D. BRAND, 1975. Proving theorems with the modification method. *SIAM Journal of Computing*, Vol. 4, pp. 412–430.

L. HINES, 1990. Str+ve⊆: The Str+ve-based Subset Prover. In Mark E. Stickel, editor, *10th International Conference on Automated Deduction*, LNAI 449, pp. 193–206, Kaiserslautern, FRG, Springer-Verlag.

L. M. HINES, 1992. Completeness of a Prover for Dense Linear Logics. *Journal of Automated Reasoning*, Vol. 8, pp. 45–75.

L.M. HINES, 1988. Hyper-chaining and knowledge-based theorem proving. In *Proc. 9th Int. Conf. on Automated Deduction*, Lecture Notes in Computer Science, pp. 469–486. Springer-Verlag.

J. HSIANG AND M. RUSINOWITCH, 1991. Proving refutational completeness of theorem proving strategies: The transfinite semantic Tree method. *Journal of the ACM*, Vol. 38, No. 3, pp. 559–587.

J. LEVY AND J. AGUSTÍ, 1993. Bi-rewriting, a Term Rewriting Technique for Monotonic Order Relations. In C. Kirchner, editor, *Rewriting Techniques and Applications*, Lecture Notes in Computer Science, vol. 690, pp. 17–31, Berlin, Springer.

R. NIEUWENHUIS, 1993. Simple LPO constraint solving methods. *Information Processing Letters*, Vol. 47, No. 2, pp. 65–69.

P. NIVELA AND R. NIEUWENHUIS, 1993. Saturation of first-order (constrained) clauses with the *Saturate* system. In Claude Kirchner, editor, *Rewriting Techniques and Applications, 5th International Conference, RTA-93*, LNCS 690, pp. 436–440, Montreal, Canada, Springer-Verlag.

M.M. RICHTER, 1984. Some reordering properties for inequality proof trees. Lecture Notes in Computer Science, vol. 171, pp. 183–197. Springer-Verlag.

G.A. ROBINSON AND L. T. WOS, 1969. Paramodulation and theorem proving in first order theories with equality. In B. Meltzer, D. Michie, editors, *Machine Intelligence 4*, pp. 133–150. American Elsevier, New York.

J. R. SLAGLE, 1972. Automatic theorem proving for theories with Built-in Theories Including Equality, Partial Orderings, and Sets. *Journal of the ACM*, Vol. 19, pp. 120–135.

Simple Termination Revisited

Aart Middeldorp

Institute of Information Sciences and Electronics
University of Tsukuba, Tsukuba 305, Japan
e-mail: ami@softlab.is.tsukuba.ac.jp
tel: +81-298-535538

Hans Zantema

Department of Computer Science, Utrecht University
P.O. Box 80.089, 3508 TB Utrecht, The Netherlands
e-mail: hansz@cs.ruu.nl
tel: +31-30-534116

ABSTRACT

In this paper we investigate the concept of simple termination. A term rewriting system is called simply terminating if its termination can be proved by means of a simplification order. The basic ingredient of a simplification order is the subterm property, but in the literature two different definitions are given: one based on (strict) partial orders and another one based on preorders (or quasi-orders). In the first part of the paper we argue that there is no reason to choose the second one, while the first one has certain advantages.

Simplification orders are known to be well-founded orders on terms over a finite signature. This important result no longer holds if we consider infinite signatures. Nevertheless, well-known simplification orders like the recursive path order are also well-founded on terms over infinite signatures, provided the underlying precedence is well-founded. We propose a new definition of simplification order, which coincides with the old one (based on partial orders) in case of finite signatures, but which is also well-founded over infinite signatures and covers orders like the recursive path order.

1. Introduction

One of the main problems in the theory of term rewriting is the detection of termination: for a fixed system of rewrite rules, determine whether there exist infinite reduction sequences or not. Huet and Lankford [8] showed that this problem is undecidable in general. However, there are several methods for proving termination that are successful for many special cases. A well-known method for proving termination is the recursive path order (Dershowitz [2]). The basic idea of such a path order is that, starting from a given order (the so-called *precedence*) on the operation symbols, in a recursive way a well-founded order on terms is defined. If every reduction step in a term rewriting system corresponds to a decrease according to this order, one can conclude that the system is terminating. If the order is closed under contexts and substitutions then the decrease only has to be checked for the rewrite rules instead of all reduction steps. The bottleneck of this kind of method is how to prove that a relation defined recursively on terms is indeed a well-founded order. Proving irreflexivity and transitivity often turns out to be feasible, using some induction and case analysis. However, when stating an arbitrary recursive definition of such an order, well-foundedness is very hard to prove directly. Fortunately, the powerful *Tree Theorem* of Kruskal implies that if the order satisfies some simplification property, well-foundedness is obtained for free. An order satisfying this property is called a *simplification order*. This notion of simplification comprises two ingredients:
- a term decreases by removing parts of it, and
- a term decreases by replacing an operation symbol with a smaller (according to the precedence) one.

If the signature is infinite, both of these ingredients are essential for the applicability of Kruskal's Tree Theorem. It is amazing, however, that in the term rewriting literature the notion of simplification order is motivated by the applicability of Kruskal's Tree Theorem but only covers the first ingredient. For infinite signatures one easily defines non-well-founded orders that are simplification orders according to that definition. Therefore, the usual definition of simplification order is only helpful for proving termination of systems over finite signatures. Nevertheless, it is well-known that simplification orders like the recursive path order are also well-founded on terms over infinite signatures (provided the precedence on the signature is well-founded).

In this paper we propose a definition of a simplification order that matches exactly the requirements of Kruskal's Tree Theorem, since that is the basic motivation for the notion of simplification order. According to this new definition all simplification orders are well-founded, both over finite and infinite signatures. For finite signatures the new and the old notion of simplification order coincide. A term rewriting system is called *simply terminating* if there is a simplification order that orients the rewrite rules from left to right. It is immediate from the definition that every recursive path order over a well-founded precedence can be extended to a simplification order, and hence it is well-founded. Even if one is only interested in finite term rewriting systems this is of interest: *semantic labelling* ([15]) often succeeds in proving termination of a finite but "difficult" (non-simply terminating) system by transforming it into an infinite system over an infinite signature to which the recursive path order readily applies.

In the literature simplification orders are sometimes based on preorders (or quasi-orders) instead of (strict) partial orders. A main result of this paper is that there are no compelling reasons for doing so. We prove (constructively) that every term rewriting system that can be shown to be terminating by means of a simplification order based on preorders, can be shown to terminating by means of a simplification order (based on partial orders). Since basing the notion of simplification order on preorders is more susceptible to mistakes and results in stronger proof obligations, simplification orders should be based on partial orders. (As explained in Section 3 these remarks already apply to finite signatures.) As a consequence, we prefer the partial order variant of *well-quasi-orders*, the so-called *partial well-orders*, in case of infinite signatures. By choosing partial well-orders instead of well-quasi-orders a great part of the theory is not affected, but another part becomes cleaner. For instance, in Section 5 we prove a useful result stating that a term rewriting system is simply terminating if and only if the union of the system and a particular system that captures simplification is terminating. Based on well-quasi-orders a similar result does not hold.

A useful notion of termination for term rewriting systems is *total termination* (see [6, 14]). For finite signature one easily shows that total termination implies simple termination. In Section 6 we show that for infinite signatures this does not hold any more: we construct an infinite term rewriting system whose terminating can be proved by a polynomial interpretation, but which is not simply terminating.

2. Termination

In order to fix our notations and terminology, we start with a very brief introduction to term rewriting. Term rewriting is surveyed in Dershowitz and Jouannaud [4] and Klop [9].

A *signature* is a set $\mathcal{F}$ of *function symbols*. Associated with every $f \in \mathcal{F}$ is a natural number denoting its arity. Function symbols of arity 0 are called *constants*. Let $\mathcal{T}(\mathcal{F}, \mathcal{V})$ be the set of all terms built from $\mathcal{F}$ and a countably infinite set $\mathcal{V}$ of *variables*, disjoint from $\mathcal{F}$. The set of variables occurring in a term t is denoted by $\mathcal{V}ar(t)$. A term t is called *ground* if $\mathcal{V}ar(t) = \varnothing$. The set of all ground terms is denoted by $\mathcal{T}(\mathcal{F})$.

We introduce a fresh constant symbol $\square$, named *hole*. A *context* C is a term in $\mathcal{T}(\mathcal{F} \cup \{\square\}, \mathcal{V})$ containing precisely one hole. The designation *term* is restricted to members of $\mathcal{T}(\mathcal{F}, \mathcal{V})$. If C is a context and t a term then $C[t]$ denotes the result of replacing the hole in C by t. A term s is a *subterm* of a term t if there exists a context C such that $t = C[s]$. A subterm s of t is *proper* if $s \neq t$. We assume familiarity with the *position* formalism for describing subterm occurrences. A *substitution* is a map σ from $\mathcal{V}$ to $\mathcal{T}(\mathcal{F}, \mathcal{V})$ with the property that the set $\{x \in \mathcal{V} \mid \sigma(x) \neq x\}$ is finite. If σ is a substitution and t a term then $t\sigma$ denotes the result of applying σ to t. We call $t\sigma$ an *instance* of t. A binary relation R on terms is *closed under contexts* if $C[s]\ R\ C[t]$ whenever $s\ R\ t$, for all contexts C. A binary relation R on terms is *closed under substitutions* if $s\sigma\ R\ t\sigma$ whenever $s\ R\ t$, for all substitutions σ. A *rewrite relation* is a binary relation on terms that is closed under contexts and substitutions.

A *rewrite rule* is a pair (l, r) of terms such that the left-hand side l is not a variable and variables which occur in the right-hand side r occur also in l, i.e., $\mathcal{V}ar(r) \subseteq \mathcal{V}ar(l)$. Since we are interested in (simple) termination in this paper, these two restrictions rule out only trivial cases. Rewrite rules (l, r) will henceforth be written as $l \to r$.

A *term rewriting system* (TRS for short) is a pair $(\mathcal{F}, \mathcal{R})$ consisting of a signature $\mathcal{F}$ and a set $\mathcal{R}$ of rewrite rules between terms in $\mathcal{T}(\mathcal{F}, \mathcal{V})$. We often present a TRS as a set of rewrite rules, without making explicit its signature, assuming that the signature consists of the function symbols occurring in the rewrite rules. The smallest rewrite relation on $\mathcal{T}(\mathcal{F}, \mathcal{V})$ that contains $\mathcal{R}$ is denoted by $\to_\mathcal{R}$. So $s \to_\mathcal{R} t$ if there exists a rewrite rule $l \to r$ in $\mathcal{R}$, a substitution σ, and a context C such that $s = C[l\sigma]$ and $t = C[r\sigma]$. The subterm $l\sigma$ of s is called a *redex* and we say that s rewrites to t by *contracting* redex $l\sigma$. We call $s \to_\mathcal{R} t$ a *rewrite* or *reduction step*. The transitive closure of $\to_\mathcal{R}$ is denoted by $\to_\mathcal{R}^+$ and $\to_\mathcal{R}^*$ denotes the transitive and reflexive closure of $\to_\mathcal{R}$. If $s \to_\mathcal{R}^* t$ we say that s *reduces* to t. The converse of $\to_\mathcal{R}^*$ is denoted by $\leftarrow_\mathcal{R}^*$.

A TRS $(\mathcal{F}, \mathcal{R})$ is called *terminating* if there are no infinite reduction sequences $t_1 \to_\mathcal{R} t_2 \to_\mathcal{R} t_3 \to_\mathcal{R} \cdots$ of terms in $\mathcal{T}(\mathcal{F}, \mathcal{V})$. In order to simplify matters, we assume throughout this paper that the signature $\mathcal{F}$ contains a constant symbol. Hence a TRS is terminating if and only if there do not exist infinite reduction sequence involving only ground terms.

A (strict) *partial order* $\succ$ is a transitive and irreflexive relation. The reflexive closure of $\succ$ is denoted by $\succeq$. The converse of $\succeq$ is denoted by $\preccurlyeq$. A partial order $\succ$ on a set A is *well-founded* if there are no infinite descending sequences $a_1 \succ a_2 \succ \cdots$ of elements of A. A partial order $\succ$ on A is *total* if for all different elements $a, b \in A$ either $a \succ b$ or $b \succ a$. A *preorder* (or *quasi-order*) $\succsim$ is a transitive and reflexive relation. The converse of $\succsim$ is denoted by $\precsim$. The *strict part* of a preorder $\succsim$ is the partial order $\succ$ defined as $\succsim\backslash\precsim$. Every preorder $\succsim$ induces an equivalence relation $\sim$ defined as $\succsim \cap \precsim$. It is easy to see that $\succ\ =\ \succsim\backslash\sim$. A preorder is said to be well-founded if its strict part is a well-founded partial order.

A rewrite relation that is also a partial order is called a *rewrite order*. A well-founded rewrite order is called a *reduction order*. We say that a TRS $(\mathcal{F}, \mathcal{R})$ and a partial order $\succ$ on $\mathcal{T}(\mathcal{F}, \mathcal{V})$ are *compatible* if $\mathcal{R}$ is contained in $\succ$, i.e., $l \succ r$ for every rewrite rule $l \to r$ of $\mathcal{R}$. It is easy to show that a TRS is terminating if and only if it is compatible with a reduction order.

DEFINITION 2.1. We say that a binary relation R on terms has the *subterm property* if $C[t]\ R\ t$ for all contexts $C \neq \Box$ and terms t.

DEFINITION 2.2. Let $\mathcal{F}$ be a signature. The TRS $\mathcal{E}mb\,(\mathcal{F})$ consists of all rewrite rules

$$f(x_1, \ldots, x_n) \to x_i$$

with $f \in \mathcal{F}$ a function symbol of arity $n \geq 1$ and $i \in \{1,\ldots,n\}$. Here $x_1,\ldots,x_n$ are pairwise different variables. We abbreviate $\to^+_{\mathcal{E}mb(\mathcal{F})}$ to $\rhd_{emb}$ and $\leftarrow^*_{\mathcal{E}mb(\mathcal{F})}$ to $\unlhd_{emb}$. The latter relation is called *embedding*.

The following easy result relates the subterm property to embedding.

LEMMA 2.3. *A rewrite order $\succ$ on $\mathcal{T}(\mathcal{F},\mathcal{V})$ has the subterm property if and only if it is compatible with the TRS $\mathcal{E}mb(\mathcal{F})$.* □

3. Simple Termination — Finite Signatures

Throughout this section we are dealing with *finite* signatures only.

DEFINITION 3.1. A *simplification order* is a rewrite order with the subterm property. A TRS $(\mathcal{F},\mathcal{R})$ is *simply terminating* if it is compatible with a simplification order on $\mathcal{T}(\mathcal{F},\mathcal{V})$.

Since we are only interested in signatures consisting of function symbols with fixed arity, we have no need for the *deletion property* (cf. [2]). Dershowitz [1, 2] showed that every simply terminating TRS is terminating. The proof is based on the beautiful Tree Theorem of Kruskal [10].

DEFINITION 3.2. An infinite sequence $t_1, t_2, t_3, \ldots$ of terms in $\mathcal{T}(\mathcal{F},\mathcal{V})$ is *self-embedding* if there exist $1 \leq i < j$ such that $t_i \unlhd_{emb} t_j$.

THEOREM 3.3 (KRUSKAL'S TREE THEOREM—FINITE VERSION). *Every infinite sequence of ground terms is self-embedding.* □

THEOREM 3.4. *Every simply terminating TRS is terminating.*
PROOF. Easy consequence of Theorem 3.3 and Lemma 2.3. □

The following well-known result is especially useful for showing that a given TRS is *not* simply terminating, see [14].

LEMMA 3.5. *A TRS $(\mathcal{F},\mathcal{R})$ is simply terminating if and only if $(\mathcal{F}, \mathcal{R} \cup \mathcal{E}mb(\mathcal{F}))$ is terminating.* □

In the term rewriting literature the notion of simplification order is sometimes based on preorders instead of partial orders. Dershowitz [2] obtained the following result.

THEOREM 3.6. *Let $(\mathcal{F},\mathcal{R})$ be a TRS. Let $\succsim$ be a preorder on $\mathcal{T}(\mathcal{F},\mathcal{V})$ which is closed under contexts and has the subterm property. If $l\sigma \succ r\sigma$ for every rewrite rule $l \to r \in \mathcal{R}$ and substitution σ then $(\mathcal{F},\mathcal{R})$ is terminating.* □

A preorder that is closed under contexts and has the subterm property is sometimes called a *quasi-simplification order*. Observe that we require $l\sigma \succ r\sigma$ *for all substitutions* σ in Theorem 3.6. It should be stressed that this requirement cannot be weakened to the compatibility of $(\mathcal{F}, \mathcal{R})$ and $\succ$ (i.e., $l \succ r$ for all rules $l \to r \in \mathcal{R}$) if we additionally require that $\succsim$ is closed under substitutions, as is incorrectly done in Dershowitz and Jouannaud [4]. For instance, the relation $\to^*_\mathcal{R}$ associated with the TRS

$$\mathcal{R} = \begin{cases} f(g(x)) & \to & f(f(x)) \\ f(g(x)) & \to & g(g(x)) \\ f(x) & \to & x \\ g(x) & \to & x \end{cases}$$

is a rewrite relation with the subterm property (because $\mathcal{R}$ contains $\mathcal{E}mb(\{f, g\})$). Moreover, $l \to^*_\mathcal{R} r$ but not $r \to^*_\mathcal{R} l$, for every rewrite rule $l \to r \in \mathcal{R}$. So $\mathcal{R}$ is included in the strict part of $\to^*_\mathcal{R}$. Nevertheless, $\mathcal{R}$ is not terminating:

$$f(g(g(x))) \to_\mathcal{R} f(f(g(x))) \to_\mathcal{R} f(g(g(x))) \to_\mathcal{R} \cdots$$

The point is that the strict part of $\to^*_\mathcal{R}$ is not closed under substitutions. Hence to conclude termination from compatibility with $\succsim$ it is essential that both $\succ$ and $\succsim$ are closed under substitutions.

Dershowitz [2] writes that Theorem 3.6 generalizes Theorem 3.4. We have the following result.

THEOREM 3.7. *A TRS $(\mathcal{F}, \mathcal{R})$ is simply terminating if and only if there exists a preorder $\succsim$ on $\mathcal{T}(\mathcal{F}, \mathcal{V})$ that is closed under contexts, has the subterm property, and satisfies $l\sigma \succ r\sigma$ for every rewrite rule $l \to r \in \mathcal{R}$ and substitution σ.* □

The proof is given in Section 5, where the above theorem is generalized to TRSs over arbitrary, not necessarily finite, signatures.

So every TRS whose termination can be shown by means of Theorem 3.6 is simply terminating, i.e., its termination can be shown by a simplification order. Since it is easier to check $l \succ r$ for finitely many rewrite rules $l \to r$ than $l\sigma \succsim r\sigma$ but not $r\sigma \succsim l\sigma$ for finitely many rewrite rules $l \to r$ and infinitely many substitutions σ, there is no reason to base the definition of simplification order on preorders.

4. Partial Well-Orders

Theorem 3.4 does not hold if we allow infinite signatures. Consider for instance the TRS $(\mathcal{F}, \mathcal{R})$ consisting of infinitely many constants a_i and rewrite rules $a_i \to a_{i+1}$ for all $i \geq 1$. The rewrite order $\to^+_\mathcal{R}$ vacuously satisfies the subterm property, but $(\mathcal{F}, \mathcal{R})$ is not terminating:

$$a_1 \to_\mathcal{R} a_2 \to_\mathcal{R} a_3 \to_\mathcal{R} \cdots$$

So in case $\mathcal{F}$ is infinite, compatibility with $\mathcal{E}mb\,(\mathcal{F})$ does not ensure termination. In the next section we will see that the results of the previous section can be recovered by suitably extending the TRS $\mathcal{E}mb\,(\mathcal{F})$.

DEFINITION 4.1. Let $\succ$ be a partial order on a signature $\mathcal{F}$. The TRS $\mathcal{E}mb\,(\mathcal{F},\succ)$ consists of all rewrite rules of $\mathcal{E}mb\,(\mathcal{F})$ together with all rewrite rules

$$f(x_1,\ldots,x_n) \;\;\to\;\; g(x_{i_1},\ldots,x_{i_m})$$

with f an n-ary function symbol in $\mathcal{F}$, g an m-ary function symbol in $\mathcal{F}$, $n \geqslant m \geqslant 0$, $f \succ g$, and $1 \leqslant i_1 < \cdots < i_m \leqslant n$ whenever $m \geqslant 1$. Here $x_1,\ldots,x_n$ are pairwise different variables. We abbreviate $\to^+_{\mathcal{E}mb\,(\mathcal{F},\succ)}$ to $\succ_{emb}$ and $\leftarrow^*_{\mathcal{E}mb\,(\mathcal{F},\succ)}$ to $\preccurlyeq_{emb}$. The latter relation is called *homeomorphic embedding*.

Since $\mathcal{E}mb\,(\mathcal{F},\varnothing) = \mathcal{E}mb\,(\mathcal{F})$, homeomorphic embedding generalizes embedding. In the next section we show that all results of the previous section carry over to infinite signatures if we require compatibility with $\mathcal{E}mb\,(\mathcal{F},\succ)$, provided the partial order $\succ$ satisfies a stronger property than well-foundedness. This property is explained below.

DEFINITION 4.2. Let $\succ$ be a partial order on a set A.
- An infinite sequence $(a_i)_{i \geqslant 1}$ over A is called *good* if there exist indices $1 \leqslant i < j$ with $a_i \preccurlyeq a_j$, otherwise it is called *bad*.
- An infinite sequence $(a_i)_{i \geqslant 1}$ over A is called a *chain* if $a_i \preccurlyeq a_{i+1}$ for all $i \geqslant 1$. We say that $(a_i)_{i \geqslant 1}$ contains a chain if it has a subsequence that is a chain.
- An infinite sequence $(a_i)_{i \geqslant 1}$ over A is called an *antichain* if neither $a_i \preccurlyeq a_j$ nor $a_j \preccurlyeq a_i$, for all $1 \leqslant i < j$.

LEMMA 4.3. *Let $\succ$ be a partial order on a set A. The following statements are equivalent.*
- *Every partial order that extends $\succ$ (including $\succ$ itself) is well-founded.*
- *Every infinite sequence over A is good.*
- *Every infinite sequence over A contains a chain.*
- *The partial order $\succ$ is well-founded and does not admit antichains.*

PROOF. Similar as done in [7] for well-quasi-orders. □

DEFINITION 4.4. A partial order $\succ$ on a set A is called a *partial well-order* (PWO for short) if it satisfies one of the four equivalent assertions of Lemma 4.3.

Using the terminology of PWOs, Theorem 3.3 can now be read as follows: if $\mathcal{F}$ is a finite signature then $\rhd_{emb}$ is a PWO on $\mathcal{T}(\mathcal{F})$.

By definition every PWO is a well-founded order, but the reverse does not hold. For instance, the empty relation on an infinite set is a well-founded order but not a PWO. Clearly every total well-founded order (or well-order) is a PWO. Any partial order extending a PWO is a PWO. The following lemma states how new PWOs can be obtained by restricting existing PWOs.

LEMMA 4.5. *Let $\succ$ be a PWO on a set A and let $\sqsupset$ be a PWO on a set B. Let $\varphi\colon A \to B$ be any function. The partial order $\succ'$ on A defined by $a \succ' b$ if and only if $a \succ b$ and $\varphi(a) \sqsupseteq \varphi(b)$ is a PWO.*

PROOF. Let $(a_i)_{i \geqslant 1}$ be any infinite sequence over A. Since $\succ$ is a PWO this sequence admits a chain

$$a_{\phi(1)} \preccurlyeq a_{\phi(2)} \preccurlyeq a_{\phi(3)} \preccurlyeq \cdots.$$

Since $\sqsupset$ is a PWO on B there exist $1 \leqslant i < j$ with $\varphi(a_{\phi(i)}) \sqsubseteq \varphi(a_{\phi(j)})$. Transitivity of $\preccurlyeq$ yields $a_{\phi(i)} \preccurlyeq a_{\phi(j)}$. Hence $a_{\phi(i)} \preccurlyeq' a_{\phi(j)}$, while $\phi(i) < \phi(j)$. We conclude that $(a_i)_{i \geqslant 1}$ is a good sequence with respect to $\succ'$, so $\succ'$ is a PWO. □

COROLLARY 4.6. *The intersection of two PWOs on a set A is a PWO on A.*

PROOF. Choose the function φ in Lemma 4.5 to be the identity on A. □

THEOREM 4.7 (KRUSKAL'S TREE THEOREM—GENERAL VERSION). *If $\succ$ is a PWO on a signature $\mathcal{F}$ then $\succ_{emb}$ is a PWO on $\mathcal{T}(\mathcal{F})$.* □

PWOs are closely related to the more familiar concept of *well-quasi-order*.

DEFINITION 4.8. A *well-quasi-order* (WQO for short) is a preorder that contains a PWO.

The above definition is equivalent to all other definitions of WQO found in the literature. Kruskal's Tree Theorem is usually presented in terms of WQOs. This is not more powerful than the PWO version: notwithstanding the fact that the strict part of a WQO is not necessarily a PWO, it is very easy to show that the WQO version of Kruskal's Tree Theorem is a corollary of Theorem 4.7, and vice-versa.

Let $\succ$ be a PWO on a signature $\mathcal{F}$. A natural question is whether we can restrict $\succ_{emb}$ while retaining the property of being a PWO on $\mathcal{T}(\mathcal{F})$. In particular, do we really need all rewrite rules in $\mathcal{E}mb(\mathcal{F}, \succ)$? In case there is a uniform bound on the arities of the function symbols in $\mathcal{F}$, we can greatly reduce the set $\mathcal{E}mb(\mathcal{F}, \succ)$. That is, suppose there exists an $N \geqslant 0$ such that all function symbols in $\mathcal{F}$ have arity less than or equal to N. Now we can apply Lemma 4.5: choose φ to be the function that assigns to every function symbol its arity and take $\sqsupset$ to be the empty relation on $\{1, \ldots, N\}$. Hence the partial order $\succ'$ on $\mathcal{F}$ defined by $f \succ' g$ if and only if f and g have the same arity and $f \succ g$ is a PWO. The corresponding set $\mathcal{E}mb(\mathcal{F}, \succ')$ consists, besides all rewrite rules of the form $f(x_1, \ldots, x_n) \to x_i$, of all rewrite rules $f(x_1, \ldots, x_n) \to g(x_1, \ldots, x_n)$ with f and g n-ary function symbols such that $f \succ g$. This construction does not work if the arities of function symbols in $\mathcal{F}$ are not uniformly bounded. Consider for instance a signature $\mathcal{F}$ consisting of a constant a and n-ary function symbols f_n for every $n \geqslant 1$ (and let $\succ$ be any PWO on $\mathcal{F}$). The sequence

$$f_1(a),\ f_2(a, a),\ f_3(a, a, a),\ \ldots$$

is bad with respect to $\succ'_{emb}$. Finally, one may wonder whether the restriction to all rewrite rules $f(x_1, \ldots, x_n) \to g(x_{i+1}, \ldots, x_{i+m})$ with f an n-ary function symbol, g an m-ary function symbol, $n \geqslant m \geqslant 0$, $n - m \geqslant i \geqslant 0$, and $f \succ g$ is sufficient. This is also not the case, as can be seen by extending the previous signature with a constant b and considering the sequence

$$f_2(b,b), \; f_3(b,a,b), \; f_4(b,a,a,b), \; \ldots .$$

Of course, if the signature $\mathcal{F}$ is finite then the rules of $\mathcal{E}mb(\mathcal{F})$ are sufficient since the empty relation is a PWO on any finite set.

5. Simple Termination — Infinite Signatures

Kurihara and Ohuchi [11] were the first to use the terminology simple termination. They call a TRS $(\mathcal{F}, \mathcal{R})$ simply terminating if it is compatible with a simplification order on $\mathcal{T}(\mathcal{F}, \mathcal{V})$. Since compatibility with a simplification order doesn't ensure the termination of TRSs over infinite signatures, see the example at the beginning of the previous section, this definition of simple termination is clearly not the right one. Ohlebusch [12] and others call a TRS $(\mathcal{F}, \mathcal{R})$ simply terminating if it is compatible with a *well-founded* simplification order on $\mathcal{T}(\mathcal{F}, \mathcal{V})$. This is a very artificial way to ensure that every simply terminating is terminating, more precisely, termination of simply terminating TRSs has nothing to do with Kruskal's Tree Theorem; simply terminating TRSs are terminating by definition. We propose instead to bring the definition of simple termination in accordance with (the general version of) Kruskal's Tree Theorem.

DEFINITION 5.1. A *simplification order* is a rewrite order on $\mathcal{T}(\mathcal{F}, \mathcal{V})$ that contains $\succ_{emb}$ for some PWO $\succ$ on $\mathcal{F}$. A TRS $(\mathcal{F}, \mathcal{R})$ is *simply terminating* if it is compatible with a simplification order on $\mathcal{T}(\mathcal{F}, \mathcal{V})$.

Because the empty relation is a PWO on any finite set, this definition coincides with the one in Section 3 in case of finite signatures.

THEOREM 5.2. *Every simply terminating TRS is terminating.*
PROOF. Let $(\mathcal{F}, \mathcal{R})$ be compatible with a simplification order $\sqsupset$ on $\mathcal{T}(\mathcal{F}, \mathcal{V})$. Let $\succ$ be any PWO such that $\succ_{emb}$ is included in $\sqsupset$. Theorem 4.7 shows that the restriction of $\succ_{emb}$ to ground terms is a PWO. Hence the extension $\sqsupset$ of $\succ_{emb}$ is well-founded on ground terms. Therefore $(\mathcal{F}, \mathcal{R})$ is terminating. □

The following result extends the very useful Lemma 3.5 to arbitrary TRSs.

LEMMA 5.3. *A TRS $(\mathcal{F}, \mathcal{R})$ is simply terminating if and only if the TRS $(\mathcal{F}, \mathcal{R} \cup \mathcal{E}mb(\mathcal{F}, \succ))$ is terminating for some PWO $\succ$ on $\mathcal{F}$.*
PROOF.

⇒ Let $(\mathcal{F}, \mathcal{R})$ be compatible with the simplification order $\sqsupset$ on $\mathcal{T}(\mathcal{F}, \mathcal{V})$. By definition there exists a PWO $\succ$ on $\mathcal{F}$ such that $\succ_{emb} \subseteq \sqsupset$. If $l \to r \in \mathcal{E}mb(\mathcal{F}, \succ)$ then $l \succ_{emb} r$ and therefore $l \sqsupset r$. Hence $\mathcal{E}mb(\mathcal{F}, \succ)$ is also compatible with $\sqsupset$. So $(\mathcal{F}, \mathcal{R} \cup \mathcal{E}mb(\mathcal{F}, \succ))$ is simply terminating. Theorem 5.2 shows that $(\mathcal{F}, \mathcal{R} \cup \mathcal{E}mb(\mathcal{F}, \succ))$ is terminating.

⇐ Suppose $(\mathcal{F}, \mathcal{R} \cup \mathcal{E}mb(\mathcal{F}, \succ))$ is terminating for some PWO $\succ$ on $\mathcal{F}$. Let $\sqsupset$ be the rewrite order associated with the TRS $(\mathcal{F}, \mathcal{R} \cup \mathcal{E}mb(\mathcal{F}, \succ))$. Clearly $\succ_{emb} \subseteq \sqsupset$. Hence $\sqsupset$ is a simplification order. Since $(\mathcal{F}, \mathcal{R})$ is compatible with $\sqsupset$, we conclude that it is simply terminating.

□

It should be stressed that there is no equivalent to the above lemma if we base the definition of simplification order on WQOs. This is one of the reasons why we favor PWOs.

In the remainder of this section we generalize Theorem 3.7 (and hence Theorem 3.6) to arbitrary TRSs. Our proof is based on the elegant proof sketch of Theorem 3.6 given by Plaisted [13]. The proof employs *multiset extensions* of preorders. A *multiset* is a collection in which elements are allowed to occur more than once. If A is a set then the set of all finite multisets over A is denoted by $\mathcal{M}(A)$. The *multiset extension* of a partial order $\succ$ on A is the partial order $\succ_{mul}$ defined on $\mathcal{M}(A)$ defined as follows: $M_1 \succ_{mul} M_2$ if $M_2 = (M_1 - X) \uplus Y$ for some multisets $X, Y \in \mathcal{M}(A)$ that satisfy $\varnothing \neq X \subseteq M_1$ and for all $y \in Y$ there exists an $x \in X$ such that $x \succ y$. Dershowitz and Manna [5] showed that the multiset extension of a well-founded partial order is again well-founded.

DEFINITION 5.4. Let $\succsim$ be a preorder on a set A. For every $a \in A$, let $[a]$ denote the equivalence class with respect to the equivalence relation $\sim$ containing a. Let $A/\sim = \{[a] \mid a \in A\}$ be the set of all equivalence classes of A. The preorder $\succsim$ on A induces a partial order $\succ$ on $A/\sim$ as follows: $[a] \succ [b]$ if and only if $a \succ b$. (The latter $\succ$ denotes the strict part of the preorder $\succsim$.) For every multiset $M \in \mathcal{M}(A)$, let $[M] \in \mathcal{M}(A/\sim)$ denote the multiset obtained from M by replacing every element a by $[a]$. We now define the *multiset extension* $\succsim_{mul}$ of the preorder $\succsim$ as follows: $M_1 \succsim_{mul} M_2$ if and only if $[M_1] \succ^=_{mul} [M_2]$ where $\succ^=_{mul}$ denotes the reflexive closure of the multiset extension of the partial order $\succ$ on $A/\sim$.

It is easy to show that $\succsim_{mul}$ is a preorder on $\mathcal{M}(A)$. The associated equivalence relation $\sim_{mul} = \succsim_{mul} \cap \precsim_{mul}$ can be characterized in the following simple way: $M_1 \sim_{mul} M_2$ if and only if $[M_1] = [M_2]$. Likewise, its strict part $\succ\!\!\!\sim_{mul}$ has the following simple characterization: $M_1 \succ\!\!\!\sim_{mul} M_2$ if and only if $[M_1] \succ_{mul} [M_2]$. Observe that we denote the strict part of $\succsim_{mul}$ by $\succ\!\!\!\sim_{mul}$ in order to avoid confusion with the multiset extension $\succ_{mul}$ of the strict part $\succ$ of $\succsim$, which is a smaller relation.

The above definition of multiset extension of a preorder can be shown to be equivalent to the more operational ones in Dershowitz [3] and Gallier [7], but since we define the multiset extension of a preorder in terms of the well-known multiset extension of a partial order, we get all desired properties basically for free. In particular, using the fact that multiset extension preserves well-founded partial orders, it is very easy to show that the multiset extension of a well-founded preorder is well-founded.

DEFINITION 5.5. If $t \in \mathcal{T}(\mathcal{F}, \mathcal{V})$ then $S(t) \in \mathcal{M}(\mathcal{T}(\mathcal{F}, \mathcal{V}))$ denotes the finite multiset of all subterm occurrences in t and $F(t) \in \mathcal{M}(\mathcal{F})$ denotes the finite multiset of all function symbol occurrences in t.

LEMMA 5.6. Let $\succsim$ be a preorder on $\mathcal{T}(\mathcal{F}, \mathcal{V})$ with the subterm property. If $s \succ t$ then $S(s) \succ_{mul} S(t)$.

PROOF. We show that $s \succ t'$ for all $t' \in S(t)$. This implies $\{s\} \succ_{mul} S(t)$ and hence also $S(s) \succ_{mul} S(t)$. If $t' = t$ then $s \succ t'$ by assumption. Otherwise t' is a proper subterm of t and hence $t \succsim t'$ by the subterm property. Combining this with $s \succ t$ yields $s \succ t'$. □

LEMMA 5.7. Let $\succsim$ be a preorder on $\mathcal{T}(\mathcal{F}, \mathcal{V})$ which is closed under contexts. Suppose $s \succsim t$ and let C be an arbitrary context.
- If $S(s) \succ_{mul} S(t)$ then $S(C[s]) \succ_{mul} S(C[t])$.
- If $S(s) \succsim_{mul} S(t)$ then $S(C[s]) \succsim_{mul} S(C[t])$.

PROOF. Let $S_1 = S(C[s]) - S(s)$ and $S_2 = S(C[t]) - S(t)$. For both statements it suffices to prove that $S_1 \succsim_{mul} S_2$. Let $p \in \mathcal{P}os(C[s])$ be the position of the displayed s in $C[s]$. There is a one-to-one correspondence between terms in S_1 (S_2) and positions in $\mathcal{P}os(C) - \{p\}$. Hence it suffices to show that $s' \succsim t'$ where $s' = C[s]_{|q}$ and $t' = C[t]_{|q}$ are the terms in S_1 and S_2 corresponding to position q, for all $q \in \mathcal{P}os(C) - \{p\}$. If p and q are disjoint positions then $s' = t'$. Otherwise $q < p$ and there exists a context C' such that $s' = C'[s]$ and $t' = C'[t]$. By assumption $s \succsim t$. Closure under contexts yields $s' \succsim t'$. We conclude that $S_1 \succsim_{mul} S_2$. □

After these two preliminary results we are ready for the generalization of Theorem 3.7 to arbitrary TRSs.

THEOREM 5.8. A TRS $(\mathcal{F}, \mathcal{R})$ is simply terminating if and only if there exists a preorder $\succsim$ on $\mathcal{T}(\mathcal{F}, \mathcal{V})$ that is closed under contexts, contains the relation $\sqsupset_{emb}$ for some PWO $\sqsupset$ on $\mathcal{F}$, and satisfies $l\sigma \succ r\sigma$ for every rewrite rule $l \to r \in \mathcal{R}$ and substitution σ. □

PROOF. The "only if" direction is obvious since the reflexive closure $\succeq$ of the simplification order $\succ$ used to prove simple termination is a preorder with the desired properties. For the "if" direction it suffices to show that $(\mathcal{F}, \mathcal{R} \cup \mathcal{E}mb(\mathcal{F}, \sqsupset))$ is a terminating TRS, according to Theorem 5.3. First we show that either $S(s) \succ_{mul} S(t)$ or $S(s) \sim_{mul} S(t)$ and $F(s) \sqsupset_{mul} F(t)$ whenever $s \to t$ is a reduction step in the TRS $(\mathcal{F}, \mathcal{R} \cup \mathcal{E}mb(\mathcal{F}, \sqsupset))$. So let $s = C[l\sigma]$ and $t = C[r\sigma]$ with $l \to r \in \mathcal{R} \cup \mathcal{E}mb(\mathcal{F}, \sqsupset)$. We distinguish three cases.
- If $l \to r \in \mathcal{R}$ then $l\sigma \succ r\sigma$ by assumption and $S(l\sigma) \succ_{mul} S(r\sigma)$ according to Lemma 5.6. The first part of Lemma 5.7 yields $S(s) \succ_{mul} S(t)$.

- If $l \to r \in \mathcal{E}mb(\mathcal{F})$ then $l\sigma = f(t_1,\ldots,t_n)$ and $r\sigma = t_i$ for some $i \in \{1,\ldots,n\}$. Therefore $S(l\sigma) \succsim_{mul} S(r\sigma)$ since $S(t_i)$ is properly contained in $S(f(t_1,\ldots,t_n))$. Clearly $l\sigma \sqsupset_{emb} r\sigma$ and thus also $l\sigma \succsim r\sigma$. An application of the first part of Lemma 5.7 yields $S(s) \succsim_{mul} S(t)$.
- If $l \to r \in \mathcal{E}mb(\mathcal{F},\sqsupset) - \mathcal{E}mb(\mathcal{F})$ then $l\sigma = f(t_1,\ldots,t_n)$ and $r\sigma = g(t_{i_1},\ldots,t_{i_m})$ with $f \sqsupset g$, $n \geqslant m \geqslant 0$, and $1 \leqslant i_1 < \cdots < i_m \leqslant n$ whenever $m \geqslant 1$. We have of course $l\sigma \sqsupset_{emb} r\sigma$ and thus also $l\sigma \succsim r\sigma$. Since the multiset $\{t_{i_1},\ldots,t_{i_m}\}$ is contained in the multiset $\{t_1,\ldots,t_n\}$, we obtain $S(l\sigma) \succsim_{mul} S(r\sigma)$ and $F(l\sigma) \sqsupset_{mul} F(r\sigma)$. The second part of Lemma 5.7 yields $S(s) \succsim_{mul} S(t)$. We obtain $F(s) \sqsupset_{mul} F(t)$ from $F(l\sigma) \sqsupset_{mul} F(r\sigma)$.

Kruskal's Tree Theorem shows that $\sqsupset_{emb}$ is a PWO on $\mathcal{T}(\mathcal{F})$. Hence $\succsim$ is a well-founded preorder on $\mathcal{T}(\mathcal{F})$. Since multiset extension preserves well-founded preorders, $\succsim_{mul}$ is a well-founded preorder on $\mathcal{M}(\mathcal{T}(\mathcal{F}))$. Because $\sqsupset$ is a PWO on the signature $\mathcal{F}$ it is a well-founded partial order. Hence its multiset extension $\sqsupset_{mul}$ is a well-founded partial order on $\mathcal{M}(\mathcal{F})$. We conclude that $(\mathcal{F}, \mathcal{R} \cup \mathcal{E}mb(\mathcal{F},\sqsupset))$ is a terminating TRS. $\square$

6. Other Notions of Termination

In this final section we investigate the relationship between simple termination and other restricted kinds of termination as introduced in [14]. First we recall some terminology. Let $\mathcal{F}$ be a signature. A *monotone* $\mathcal{F}$-algebra $(\mathcal{A}, \succ)$ consists of a non-empty $\mathcal{F}$-algebra $\mathcal{A}$ and a partial order $\succ$ on the carrier A of $\mathcal{A}$ such that every algebra operation is strictly monotone in all its coordinates, i.e., if $f \in \mathcal{F}$ has arity n then

$$f_\mathcal{A}(a_1,\ldots,a_i,\ldots,a_n) \succ f_\mathcal{A}(a_1,\ldots,b_i,\ldots,a_n)$$

for all $a_1,\ldots,a_n, b_i \in A$ with $a_i \succ b_i$ ($i \in \{1,\ldots,n\}$). We call a monotone $\mathcal{F}$-algebra $(\mathcal{A}, \succ)$ *well-founded* if $\succ$ is well-founded. We define a partial order $\succ_\mathcal{A}$ on $\mathcal{T}(\mathcal{F},\mathcal{V})$ as follows: $s \succ_\mathcal{A} t$ if $[\alpha](s) \succ [\alpha](t)$ for all assignments $\alpha\colon \mathcal{V} \to A$. Here $[\alpha]$ denotes the homomorphic extension of α. Finally, a TRS $(\mathcal{F}, \mathcal{R})$ is said to be *compatible* with $(\mathcal{A}, \succ)$ if $(\mathcal{F}, \mathcal{R})$ and $\succ_\mathcal{A}$ are compatible.

It is not difficult to show that the relation $\succ_\mathcal{A}$ is a rewrite order on $\mathcal{T}(\mathcal{F}, \mathcal{V})$, for every monotone $\mathcal{F}$-algebra $(\mathcal{A}, \succ)$. If $(\mathcal{A}, \succ)$ is well-founded then $\succ_\mathcal{A}$ is a reduction order. It is also straightforward to show that a TRS $(\mathcal{F}, \mathcal{R})$ is terminating if and only if it is compatible with a well-founded monotone $\mathcal{F}$-algebra. Simple termination can be characterized semantically as follows.

DEFINITION 6.1. A monotone $\mathcal{F}$-algebra is called *simple* if it is compatible with the TRS $\mathcal{E}mb(\mathcal{F}, \succ)$ for some partial well-order $\succ$ on $\mathcal{F}$.

It is straightforward to show that a TRS $(\mathcal{F}, \mathcal{R})$ is simply terminating if and only if it is compatible with a simple monotone $\mathcal{F}$-algebra.

DEFINITION 6.2. A TRS $(\mathcal{F}, \mathcal{R})$ is called *totally terminating* if it is compatible with a well-founded monotone $\mathcal{F}$-algebra $(\mathcal{A}, \succ)$ such that $\succ$ is a total order on the carrier set of $\mathcal{A}$. If the carrier set of $\mathcal{A}$ is the set of natural numbers and $\succ$ is the standard order then the TRS is called *ω-terminating*. If in addition the operation $f_\mathcal{A}$ is a polynomial for every $f \in \mathcal{F}$, the TRS is called *polynomially terminating*.

Total termination has been extensively studied in [6]. Clearly every polynomially terminating TRS is ω-terminating and every ω-terminating is totally terminating. For both assertions the converse does not hold, as can be shown by the counterexamples $\mathcal{R}_1 = \{f(g(h(x))) \to g(f(h(g(x))))\}$ and $\mathcal{R}_2 = \{f(g(x)) \to g(f(f(x)))\}$ respectively. An easy observation ([14]) shows that every totally terminating TRS over a finite signature is simply terminating. Again the converse does not hold as is shown by the well-known example $\mathcal{R}_3 = \{f(a) \to f(b), g(b) \to g(a)\}$.

Somewhat surprisingly, for infinite signatures total termination does not imply simple termination any more: we prove that the non-simply terminating TRS $(\mathcal{F}, \mathcal{R}_4)$ is even polynomially terminating. Here $\mathcal{F}$ is the signature $\{f_i, g_i \mid i \in \mathbb{N}\}$ and $\mathcal{R}_4$ consists of all rewrite rules

$$f_i(g_j(x)) \to f_j(g_j(x))$$

where $i, j \in \mathbb{N}$ with $i < j$. First we prove that $(\mathcal{F}, \mathcal{R}_4)$ is not simply terminating. Let $\succ$ be any PWO on $\mathcal{F}$. Consider the infinite sequence $(f_i)_{i \geqslant 1}$. Since every infinite sequence is good, we have $f_j \succ f_i$ for some $i < j$. Hence $\mathcal{E}mb(\mathcal{F}, \succ)$ contains the rewrite rule $f_j(x) \to f_i(x)$, yielding the infinite reduction sequence

$$f_i(g_j(x)) \to f_j(g_j(x)) \to f_i(g_j(x)) \to \cdots$$

in the TRS $(\mathcal{F}, \mathcal{R}_4 \cup \mathcal{E}mb(\mathcal{F}, \succ))$. Lemma 5.3 shows that $(\mathcal{F}, \mathcal{R}_4)$ is not simply terminating.

For proving polynomial termination of $(\mathcal{F}, \mathcal{R}_4)$, interpret the function symbols as the following polynomials over $\mathbb{N}$:

$$f_{i\mathcal{A}}(x) = x^3 - ix^2 + i^2 x \quad \text{and} \quad g_{i\mathcal{A}}(x) = x + 2i$$

for all $i, x \in \mathbb{N}$. Let $i \in \mathbb{N}$. The interpretation $g_{i\mathcal{A}}$ of g_i is clearly strictly monotone in its single argument. The same holds for the interpretation of f_i since

$$f_{i\mathcal{A}}(x+1) - f_{i\mathcal{A}}(x) = (x+1-i)^2 + 2x^2 + x + i > 0$$

for all $x \in \mathbb{N}$. It remains to show that $f_{i\mathcal{A}}(g_{j\mathcal{A}}(x)) > f_{j\mathcal{A}}(g_{j\mathcal{A}}(x))$ for all $i, j, x \in \mathbb{N}$ with $i < j$. Fix i, j, x and let $y = g_{j\mathcal{A}}(x) = x + 2j$. Then

$$f_{i\mathcal{A}}(g_{j\mathcal{A}}(x)) - f_{j\mathcal{A}}(g_{j\mathcal{A}}(x)) = f_{i\mathcal{A}}(y) - f_{j\mathcal{A}}(y) = y(j-i)(y-j-i) > 0$$

since $j > i$ and $y \geqslant 2j > j + i > 0$. We conclude that $(\mathcal{F}, \mathcal{R}_4)$ is polynomially terminating.

Summarizing the relationship between the various kinds of termination we obtain Figure 1; for $\mathcal{R}_5$ and $\mathcal{R}_6$ we simply take the union of $\mathcal{R}_4$ with $\mathcal{R}_1$ and $\mathcal{R}_2$

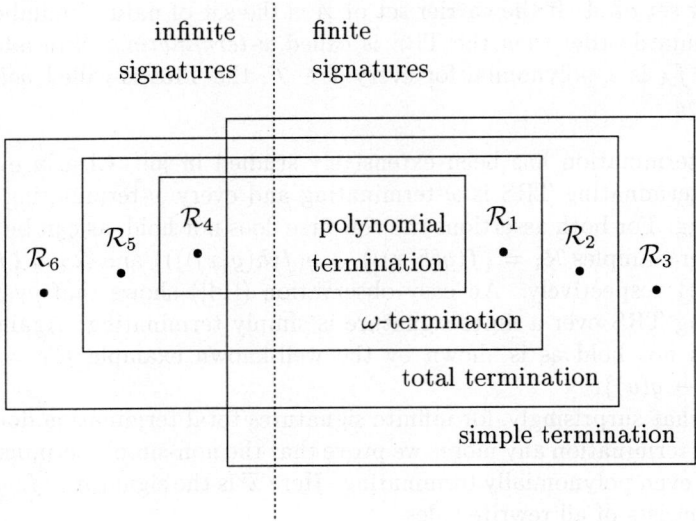

FIGURE 1.

respectively. Uwe Waldmann (personal communication) was the first to prove total termination of a non-simply terminating system similar to $\mathcal{R}_4$, using a much more complicated total well-founded order.

The class of simply terminating TRSs is properly included in the class of all TRSs that are compatible with a well-founded rewrite order having the subterm property. Nevertheless, it's quite big. For instance, it includes all TRSs whose termination can be shown by means of the recursive path order (Dershowitz [2]) and its variants. This can be seen as follows. It is known that $\succ_{rpo}$ is a rewrite order on $\mathcal{T}(\mathcal{F},\mathcal{V})$ with the subterm property (cf. [2]). It is not difficult to show that $\succ_{rpo}$ extends $\succ_{emb}$, for any precedence $\succ$ on the signature $\mathcal{F}$. Hence $\succ_{rpo}$ is a simplification order whenever the precedence $\succ$ is a PWO. In particular, if the signature is finite then every $\succ_{rpo}$ is a simplification order. If $\succ$ is a well-founded precedence on an arbitrary signature then $\succ_{rpo}$ is included in a simplification order (and hence well-founded). This follows from the *incrementality* of the recursive path order (i.e., if $\succ \subseteq \sqsupset$ then $\succ_{rpo} \subseteq \sqsupset_{rpo}$) and the well-known fact that every well-founded partial order can be extended to a total well-founded partial order. Hence every TRS $(\mathcal{F},\mathcal{R})$ that is compatible with $\succ_{rpo}$ for some well-founded precedence $\succ$ on $\mathcal{F}$ is simply terminating.

References

1. N. Dershowitz, *A Note on Simplification Orderings*, Information Processing Letters **9**(5), pp. 212–215, 1979.

2. N. Dershowitz, *Orderings for Term-Rewriting Systems*, Theoretical Computer Science **17**(3), pp. 279–301, 1982.

3. N. Dershowitz, *Termination of Rewriting*, Journal of Symbolic Computation **3**(1), pp. 69–116, 1987.

4. N. Dershowitz and J.-P. Jouannaud, *Rewrite Systems*, in: Handbook of Theoretical Computer Science, Vol. B (ed. J. van Leeuwen), North-Holland, pp. 243–320, 1990.

5. N. Dershowitz and Z. Manna, *Proving Termination with Multiset Orderings*, Communications of the ACM **22**(8), pp. 465–476, 1979.

6. M.C.F. Ferreira and H. Zantema, *Total Termination of Term Rewriting*, Proceedings of the 5th International Conference on Rewriting Techniques and Applications, Montreal, Lecture Notes in Computer Science **690**, pp. 213–227, 1993.

7. J. Gallier, *What's so Special about Kruskal's Theorem and the Ordinal Γ_0? A Survey of Some Results in Proof Theory*, Annals of Pure and Applied Logic **53**, pp. 199–260, 1991.

8. G. Huet and D. Lankford, *On the Uniform Halting Problem for Term Rewriting Systems*, report 283, INRIA, 1978.

9. J.W. Klop, *Term Rewriting Systems*, in: Handbook of Logic in Computer Science, Vol. II (eds. S. Abramsky, D. Gabbay and T. Maibaum), Oxford University Press, pp. 1–112, 1992.

10. J.B. Kruskal, *Well-Quasi-Ordering, the Tree Theorem, and Vazsonyi's Conjecture*, Transactions of the American Mathematical Society **95**, pp. 210–225, 1960.

11. M. Kurihara and A. Ohuchi, *Modularity of Simple Termination of Term Rewriting Systems*, Journal of the Information Processing Society Japan **31**(5), pp. 633–642, 1990.

12. E. Ohlebusch, *A Note on Simple Termination of Infinite Term Rewriting Systems*, report nr. 7, Universität Bielefeld, 1992.

13. D.A. Plaisted, *Equational Reasoning and Term Rewriting Systems*, To appear in: Handbook of Logic in Artificial Intelligence and Logic Programming, Vol. I (eds. D. Gabbay and J. Siekmann), Oxford University Press, 1993.

14. H. Zantema, *Termination of Term Rewriting by Interpretation*, Proceedings of the 3rd International Workshop on Conditional Term Rewriting Systems, Pont-à-Mousson, Lecture Notes in Computer Science **656**, pp. 155–167, 1993. Full version to appear as *Termination of Term Rewriting: Interpretation and Type Elimination* in Journal of Symbolic Computation, 1994.

15. H. Zantema, *Termination of Term Rewriting by Semantic Labelling*, report RUU-CS-93-24, Utrecht University, 1993. Submitted.

Termination Orderings for Rippling*

David A. Basin[1] and Toby Walsh[2]

[1] Max-Planck-Institut für Informatik
Saarbrücken, Germany
Email: basin@mpi-sb.mpg.de   Phone: (49) (681) 302-5435

[2] INRIA-Lorraine, 615, rue du Jardin Botanique,
54602 Villers-les-Nancy, France
Email: walsh@loria.fr   Phone: (33) 83 59 30 15

Abstract. Rippling is a special type of rewriting developed for inductive theorem proving. Bundy et. al. have shown that rippling terminates by providing a well-founded order for the annotated rewrite rules used by rippling. Here, we simplify and generalize this order, thereby enlarging the class of rewrite rules that can be used. In addition, we extend the power of rippling by proposing new domain dependent orders. These extensions elegantly combine rippling with more conventional term rewriting. Such combinations offer the flexibility and uniformity of conventional rewriting with the highly goal directed nature of rippling. Finally, we show how our orders simplify implementation of provers based on rippling.

1 Introduction

Rippling is a form of goal directed rewriting developed at Edinburgh [5, 3] and in parallel in Karlsruhe [11, 12] for inductive theorem proving. In inductive proof, the induction conclusion typically differs from the induction hypothesis by the addition of some constructors or destructors. Rippling uses special annotations, called *wavefronts*, to mark these differences. They are then removed by annotated rewrite rules, called *wave-rules*. Rippling has several attractive properties. First, it is highly goal directed, attempting to remove just the differences between the conclusion and hypothesis, leaving the common structure preserved. And second, it terminates yet allows rules like associativity to be used both ways.

The contributions of this paper are to simplify, improve, and generalize the specification of wave-rules and their associated termination orderings. Wave-rules have previously been presented via complex schematic definitions that intertwine the properties of structure preservation and the reduction of a well-founded measure (see [3] and §7). As these properties may be established independently, our definition of wave-rules separates these two concerns. Our main focus is on new measures. We present a family of measures that, despite their simplicity, admit strictly more wave-rules than the considerably more complex specification given in [3].

This work has several practical applications. By allowing rippling to be combined with new termination orderings, the power of rippling can be greatly extended. Al-

---

* The second author's current address is: IRST, Location Panté di Povo, I38100 Trento, Italy. toby@irst.it

though rippling has been designed primarily to prove inductive theorems it has recently been applied to other problem domains. We show that in rippling, as in conventional rewriting, the ordering used should be domain dependent. We provide several new orderings for applying rippling to new domains within induction (e.g. domains involving mutually recursive functions) and outside of induction (e.g. equational problem solving). In doing so, we show for the first time how rippling can be combined with conventional rewriting.

Another practical contribution is that our work greatly simplifies the implementation of systems based on rippling. Systems like Clam [4] require a procedure, called a *wave-rule parser*, to annotate rewrite rules. Clam's parser is based upon the complex definition of wave-rules in [3] and as a result is itself extremely complex and faulty. We show how, given a simple modular order, we can build simple modular wave-rule parsers. We have implemented such parsers and they have pleasant properties that current implementations lack (e.g. notions of correctness and completeness); our work hence leads to a simpler and more flexible mechanization of rippling.

The paper is organized as follows. In §2 we give a brief overview of rippling. In §3 we define an order on a simple kind of annotated term and use this in §4 to build orderings on general annotated terms. Based on this we show in §5 how rewrite rules may be automatically annotated. In §6 we describe how new orders increase the power and applicability of rippling. In §7 we compare this work to previous work in this area and discuss some practical experience. Finally we draw conclusions.

2 Background

We provide a brief overview of rippling. For a complete account please see [3].

Rippling arose out of an analysis of inductive proofs. For example, if we wish to prove $P(x)$ for all natural numbers, we assume $P(n)$ and attempt to show $P(s(n))$. The hypothesis and the conclusion are identical except for the successor function $s(.)$ applied to the induction variable n. Rippling marks this difference by the annotation, $P(\boxed{s(\underline{n})})$. Deleting everything in the box that is not underlined gives the skeleton, which is preserved during rewriting. The boxed but not underlined term parts are wavefronts, which are removed by rippling.

Formally, a *wavefront* is a term with at least one proper subterm deleted. We represent this by marking a term with *annotation* where wavefronts are enclosed in boxes and the deleted subterms, called *waveholes*, are underlined. Schematically, a wavefront looks like $\boxed{\xi(\underline{\mu_1}, ..., \underline{\mu_n})}$, where $n > 0$ and μ_i may be similarly annotated. The part of the term not in the wavefront is called the *skeleton*. Formally, the skeleton is a non-empty set of terms defined as follows.

Definition 1 (Skeleton)

1. $skel(t) = \{t\}$ for t a constant or variable
2. $skel(f(t_1, ..., t_n)) = \{f(s_1, ..., s_n) | \forall i.\, s_i \in skel(t_i)\}$
3. $skel(\boxed{f(\underline{t_1}, ..., \underline{t_n})}) = skel(t_1) \cup ... \cup skel(t_n)$ for the t_i in waveholes.

We call a term *simply annotated* when all its wavefronts contain only a single wave-hole and *generally annotated* otherwise. In the simply annotated case, the skeleton function returns a singleton set whose member we call the skeleton. E.g. the skeleton of $f(\boxed{s(\underline{a})}, \boxed{s(\underline{b})})$ is $f(a,b)$.

We define *wave-rules* to be rewrite rules between annotated terms that meet two requirements: they are skeleton preserving and measure decreasing. This is a simpler and more general approach to defining wave-rules than that given in [3] where these requirements were intertwined into the syntactic specification of a wave-rule.[3] *Skeleton preservation* in the simply-annotated case means that both the LHS (left-hand side) and RHS (right-hand side) of the wave-rule have an identical skeleton. In the multi-hole case we demand that *some* of the skeletons on the LHS are preserved on the RHS and no new skeletons are introduced, i.e. $skel(LHS) \supseteq skel(RHS)$.

Wavefronts in wave-rules are also *oriented*. This is achieved by marking the wavefront with an arrow indicating if the wavefront should move up through the skeleton term tree or down towards the leaves. Oriented wavefronts dictate a measure on terms that rippling decreases. The focus of this paper is on these measures.

Below are some examples of wave-rules (s is successor and $<>$ is infix append).

$$\boxed{s(\underline{U})}^{\uparrow} \times V \Rightarrow \boxed{(\underline{U \times V}) + V}^{\uparrow} \qquad (1)$$

$$\boxed{s(\underline{U})}^{\uparrow} \geq \boxed{s(\underline{V})}^{\uparrow} \Rightarrow U \geq V \qquad (2)$$

$$\boxed{\underline{U} + \underline{V}}^{\uparrow} \times W \Rightarrow \boxed{\underline{U \times W} + \underline{V \times W}}^{\uparrow} \qquad (3)$$

$$(\boxed{\underline{U} <> \underline{V}}^{\uparrow}) <> W \Rightarrow U <> (\boxed{\underline{V} <> \underline{W}}^{\uparrow}) \qquad (4)$$

$$U <> (\boxed{\underline{V} <> \underline{W}}^{\uparrow}) \Rightarrow \boxed{(\underline{U <> V}) <> \underline{W}}^{\uparrow} \qquad (5)$$

$$\boxed{\underline{U} + \underline{V}}^{\uparrow} = \boxed{\underline{W} + \underline{Z}}^{\uparrow} \Rightarrow \boxed{\underline{U = W} \wedge \underline{V = Z}}^{\uparrow} \qquad (6)$$

(1) and (2) are typical of wave-rules based on a recursive definitions. The remainder come from lemmas. Methods for turning definitions and lemmas into wave-rules is the subject of §5. Note that annotation in the wave-rules must match annotation in the term being rewritten. This allows use of rules like associativity of append, (4) and (5), in both directions; these would loop in conventional rewriting. Note also that in (6) the skeletons of the RHS are a strict subset of those of the LHS.

As a simple example of rippling, consider proving the associativity of multiplication using structural induction. In the step-case, the induction hypothesis is

$$(x \times y) \times z = x \times (y \times z)$$

and the induction conclusion is

$$(\boxed{s(\underline{x})}^{\uparrow} \times y) \times z = \boxed{s(\underline{x})}^{\uparrow} \times (y \times z).$$

---

[3] This generalization is, however, briefly discussed in their further work section.

The wavefronts in the induction conclusion mark the differences with the induction hypothesis. Rippling on both sides of the induction conclusion using (1) yields (7) and then with (3) on the LHS gives (8).

$$(\boxed{x \times y + y}^\uparrow) \times z = \boxed{(x \times (y \times z)) + y \times z}^\uparrow \quad (7)$$

$$\boxed{((x \times y) \times z) + y \times z}^\uparrow = \boxed{(x \times (y \times z)) + y \times z}^\uparrow \quad (8)$$

As the wavefronts are now at the top of each term, we have successfully rippled-out both sides of the equality. We can complete the proof by simplifying with the induction hypothesis.

The example illustrates how rippling preserves skeletons during rewriting. Provided rippling does not get *blocked* (no wave-rule applies yet we are not completely rippled-out), we are guaranteed to be able to simplify with the induction hypothesis (called *fertilization* in [2]). This explains the highly goal directed nature of rippling.

We can also ripple wavefronts towards the position of universally quantified variables in the induction hypothesis. Such positions are called *sinks* because wavefronts can be absorbed there; when we appeal to the induction hypothesis, universally quantified variables will be matched with the content of the sinks. Rippling towards sinks at the leaves of terms is called *rippling-in*. Wavefronts are oriented with arrows pointing out (upwards) or in (downwards) indicating if they are moving towards the root or leaves. *Transverse* wave-rules like (4) are used to turn outward directed wavefronts inwards.

3 Ordering Simple Wave-Rules

In this section we consider only simply annotated terms (whose wavefronts have a single wavehole). In the next section we generalize to orders for generally annotated terms with multiple waveholes. We begin with motivation, explaining generally the kinds of orders we wish to define. Afterwards, we propose several concrete measures that are similar, though simpler, to those given by Bundy *et. al.* in [3]. They are able to order all the wave-rules given in [3] and in addition allow rule orientations not possible using the measure given there (see §7).

We consider annotated terms as decorated trees where the tree is the skeleton and the wavefronts are boxes decorating the nodes. See, for example, the first tree in Fig. 1 which represents the term $\boxed{s(\underline{U})}^\uparrow \geq \boxed{s(\underline{V})}^\uparrow$. Our orders are based on assigning measures to annotation in these trees. We can define progressively simpler orders by simplifying these annotated trees to capture the notion of progress during rippling that we wish to measure.

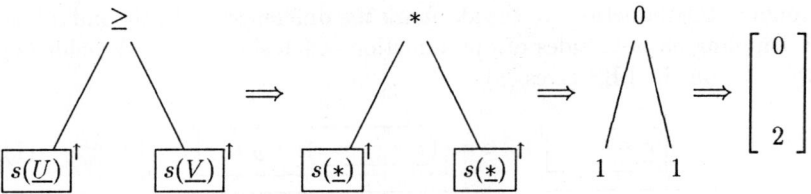

Fig. 1. Defining a measure on annotated terms.

To begin with, since rippling is skeleton preserving, we needn't account for the contents of the skeleton in our orderings. That is, we can abstract away function symbols in the skeleton, for example, mapping each function to a variadic function constant "*". This gives, for example, the second tree in Fig. 1. In §6.2, we return to this abstraction and examine termination orderings that do allow the skeleton to be changed during rewriting.

A further abstraction is to ignore the names of function symbols within wave-fronts and assign some kind of numeric weight to wave-fronts. For example, we may tally up the values associated with each function symbol as in a Knuth-Bendix ordering. The simplest kinds of weights that we may assign to wave-fronts measure their *width* and their *size*. Width is the number of nested function symbols between the root of the wavefront and the wavehole. Size is the number of function symbols and constants in a wavefront. For simplicity, we will consider just the width unless otherwise stated. This gives, for example, the third tree in Fig. 1. Of course, there are problem domains where we want our measure to reflect more of the structure of wave-fronts. §6.1 contains an example of this showing how the actual contents may be compared using a conventional term ordering.

Finally, a very simple notion of progress during rippling is simply that wave-fronts move up or down through the skeleton tree. Under this view, the tree structure may be ignored: it is not important which branch a wave-front is on, only its height in the skeleton tree. Under this notion of progress, we can apply an abstraction that maps the tree onto a list, level by level. For instance, we can use the sum of the weights at a given depth. Applying this abstraction gives the final tree in Fig. 1. Again, note that depths are *relative* to the skeleton and not depth in the erased term tree.

To make this more formal and concrete, we introduce some definitions. A *position* is simply a path address (written "Dewey decimal style") in the term tree of the skeleton and the subterm of t at position p is denoted by t/p. If s is a subterm of t at position p, its *depth* is the length of p. The *height* of t, written $|t|$, is the maximal depth of any subterm in t. For an annotated term t, the *out-weight of a position* p is the sum of the weights of the (possibly nested) outwards oriented wavefronts at p. The *in-weight* is defined identically except for inward directed wavefronts. We now define a measure on terms corresponding to the final tree in Fig. 1 based on weights of annotation relative to their depths.

Definition 2 (Out/In Measure) *The* out-measure, $MO(t)$ *(*in-measure, $MI(t)$*)* *of an annotated term t is a list whose i-th element is the sum of out-weights (in-weights) for all term positions in t at depth i.*

For example, in the following palindrome function over lists ("::" is infix cons)

$$palin(\boxed{H :: \underline{T}}^\uparrow, Acc) \Rightarrow \boxed{H :: palin(T, \boxed{H :: \underline{Acc}}^\downarrow)}^\uparrow \qquad (9)$$

and the skeleton of both sides is $palin(T, Acc)$ and the out-measure of the LHS is [0,1] and the RHS is [1,0]. The in-measures are [0,0] and [0,1] respectively.

We now define a well-founded ordering on these measures which reflects the progress that we want rippling to make during rewriting. Consider, a simple wave-rule like (1),

$$\boxed{s(\underline{U})}^\uparrow \times V \Rightarrow \boxed{(\underline{U \times V}) + V}^\uparrow.$$

The LHS out-measure is $[0,1]$, and the RHS is $[1,0]$. Rippling has progressed here as the one out-oriented wavefront has moved up the term. In general, rippling progresses if one out-oriented wavefront moves up or disappears, while nothing deeper moves downwards. If the out-measure on a term before rippling is $[l_1, ..., l_k]$ and after $[r_1, ..., r_k]$ then there must be some depth j where $l_j > r_j$ and for all $i > j$ we have $l_i = r_i$. This is simply the lexicographic order on the reverse of the two lists (compared with $>$ on the natural numbers).[4] Progress for in-oriented wavefronts is similar and reflects that these wavefronts should move towards leaves; that is, we use the lexicographic order on the in-measures. Of course, both outward and inward oriented wavefronts may occur in the same rule. For example, consider (9). As in [3], we define a composite ordering on the out and in measures. We order the out measure before the in measure since this enables us to ripple wavefronts out and either to reach the top of the term, or at some point to turn the wavefront down and to ripple it in towards the leaves.

Definition 3 (Composite Ordering) $t \succ s$ *iff* $\langle MO(t), MI(t) \rangle >_o \langle MO(s), MI(s) \rangle$ *where $>_o$ is the lexicographic order on pairs whose first components are compared with $>_{revlex}$ and the second with $>_{lex}$, the reversed and unreversed lexicographic order on lists of equal length.*

Given the well-foundedness of $>$ on the natural numbers and that lexicographic combinations of well-founded orders are well-founded we can conclude the following.

Lemma 1 *The composite ordering is well-founded.*

We lack space here to discuss implementations of rippling. Two different implementations are considered in [3] and [12]. For both calculi, $\succ$ (and $\succ^*$ of the next

---

[4] Note that these lists are the same length as the skeletons of both sides are identical; however, when we generalize the measure to multi-holed waves, the skeletons may have different depths and we pad with trailing zeros where necessary.

section) is monotonic and stable over the substitutions produced during rippling. It follows from standard techniques that if all wave-rules are oriented so that $l \succ r$ then rippling terminates [8].

4 Ordering Multi-Wave-Rules

We now generalize our order for simply annotated terms to those with generalized annotation, that is, multiple waveholes in a single wavefront. Wave-rules involving such terms are called *multi-wave-rules* in [3] and we have already seen an example of this in (6). The binomial equation is another example.

$$binom(\boxed{s(\underline{X})}^{\uparrow}, \boxed{s(\underline{Y})}^{\uparrow}) = \boxed{binom(X, \boxed{s(\underline{Y})}^{\uparrow}) + \underline{binom(X,Y)}}^{\uparrow} \qquad (10)$$

We define orders for generally annotated terms in a uniform way from the previous ordering by reducing generally annotated terms to sets of simply annotated terms and extending $\succ$ to such sets. This reduction is accomplished by considering ways that general annotation can be *weakened* to simple annotation by "absorbing" waveholes. Weakening a multi-wave term like (10) erases some of the waveholes (underlining) though always leaving at least one wavehole. A wavefront is *maximally weak* when it has exactly one wavehole. A term is *maximally weak* when all its wavefronts are maximally weak. Maximally weak terms are simply annotated and this allows us to use the previously defined measure $\succ$ on these terms.

Returning to the binomial example, (10) has only the following two weakenings.

$$binom(\boxed{s(\underline{X})}^{\uparrow}, \boxed{s(\underline{Y})}^{\uparrow}) = \boxed{binom(X, \boxed{s(\underline{Y})}^{\uparrow}) + \underline{binom(X,Y)}}^{\uparrow} \qquad (11)$$

$$binom(\boxed{s(\underline{X})}^{\uparrow}, \boxed{s(\underline{Y})}^{\uparrow}) = \boxed{binom(X, s(Y)) + \underline{binom(X,Y)}}^{\uparrow} \qquad (12)$$

Both of these are maximally weak as each wavefront has a single hole.

Let $weakenings(s)$ be the set of maximal weakenings of a term s. We now define an ordering on generally annotated terms l and r.

Definition 4 (General ordering) $l \succ^* r$ *iff* $weakenings(s) \succ\!\!\!\succ weakenings(t)$ *where* $\succ\!\!\!\succ$ *is the multiset ordering over the order* $\succ$ *on simply annotated terms.*

This order is sensible as all the elements of the weakening sets are simply annotated and can be compared with $\succ$. Also observe that if l and r are simply annotated then their weakenings are $\{l\}$ and $\{r\}$ and $l \succ^* r$ agrees with $l \succ r$. In general, we will drop the superscript on $\succ^*$ and use context (e.g., at least one argument has multiple holes) to disambiguate.

As the multi-set extension of a well-founded ordering is well-founded [10] we immediately have the following lemma.

Lemma 2 $\succ^*$ *is well-founded.*

As an example, consider (10). The LHS weakenings are

$$\{binom(\boxed{s(\underline{X})}^\uparrow, \boxed{s(\underline{Y})}^\uparrow)\}.$$

The RHS weakenings are

$$\{\boxed{binom(X, \boxed{s(\underline{Y})}^\uparrow) + binom(X,Y)}^\uparrow, \boxed{binom(X + s(Y)) + \underline{binom(X,Y)}}^\uparrow\}.$$

The sole member of the first set is $\succ$-greater than both members of the second set. This equation is measure decreasing and hence a wave-rule when used left to right.

5 Parsing

These orders are simple and admit simple mechanization. We begin with simply annotated terms and then sketch the generalization to multi-waves. We have implemented the routines we describe and in §7 we report on practical experience.

A wave-rule $l \to r$ must satisfy two properties: the preservation of the skeleton, and a reduction of the measure. We achieve these separately. An *annotation phase* first annotates l and r with unoriented wavefronts so their skeletons are identical; this guarantees that rippling is skeleton preserving. An *orientation phase* then orients the wavefronts so that $l \succ r$. We sum this up by the slogan

$$WAVE\text{-}RULE = ANNOTATION + ORIENTATION. \tag{13}$$

5.1 Annotation

To annotate terms we can use the *ground difference unification* algorithm given in [1]. Since parsing is an off-line computation (performed once before theorem proving), it is also reasonable to find skeleton preserving annotation via generate-and-test: generate candidate annotations and test if the resulting terms have the same skeleton. Consider, for example, annotating the recursive definition of the palindrome function. There are four possible skeletons: $palin(T, Acc)$, T, Acc, and H. The first of these corresponds to the annotation

$$palin(\boxed{H :: \underline{T}}, Acc) \Rightarrow \boxed{H :: palin(T, \boxed{H :: \underline{Acc}})}. \tag{14}$$

The remaining annotations are *trivial* in that both sides are completely within wavefronts except for some subterm at the leaves. For example,

$$\boxed{palin(H :: T, \underline{Acc})} \Rightarrow \boxed{H :: palin(T, H :: \underline{Acc})}.$$

Such trivial wave-rules can usually be ignored as they they make no progress moving wavefronts (although they can be used for wavefront normalization, see §6.1).

5.2 Orientation

Given annotated, but unoriented rules, we must now orient them by placing arrows on the wavefronts. We do this by picking an orientation for wavefronts on the LHS of the wave-rule and finding an orientation on the RHS such that $l \succ r$. In Clam the wave-rules used are oriented with wavefronts on the LHS exclusively out or in. Other combinations are, of course, possible. In general the number of wavefronts, n in the LHS is very small, typically one or two in [3]; hence, it is not much extra effort to consider all 2^n orientations and for each of these generate an orientation for the RHS.[5] In practice this is manageable; see §7.

For each orientation of l we must orient r. If l contains at least one outward oriented wavefront there will always be a measure decreasing orientation of r, namely with all wavefronts oriented in. However, orienting wavefronts inwards prohibits later rippling out whilst orienting outwards does not. If rippling-out blocks, we can always redirect wave-rules inwards with the rewrite rule. $\boxed{F(\underline{X})}^{\uparrow} \Rightarrow \boxed{F(\underline{X})}^{\downarrow}$. This rule is structure preserving and measure decreasing. Hence, we orient r's annotation so that it is measure decreasing and $\succ$-*maximal*; that is, for all orientations r_o, if $l \succ r_o$ then $r \succeq r_o$ ($\succeq$ is the union of the identity relation with $\succ$).

One can find a maximal orientation using generate and test, but it is possible to do much better. Below we sketch an algorithm, linear in $|r|$. Its input is two annotated terms l and r where l is oriented and r unoriented. The output is r oriented and $\succ$-maximal. In what follows, suppose $|l|$ (and hence $|r|$) equals k. Let $t_i^{\uparrow}$ be the sum of out-weights at depth i, $t_i^{\downarrow}$ be the sum of in-weights at depth i, and $flip(t, d, n)$ be the operation that non-deterministically flips down n arrows in t at depth d (there may be multiple choices corresponding to different branches or multiple wavefronts at the same position). We assume below that l has at least one wavefront oriented up. If this is not the case then all of r's wavefronts must be oriented down and this is a maximal orientation iff $l \succ r$. Otherwise orientation proceeds as follows. We first orient all the wavefronts in r upwards and then execute the first of the following statements that succeeds.

1. choose the maximum i such that $l_i^{\uparrow} > r_i^{\uparrow}$ and $\forall j \in \{i+1..k\}.flip(r, j, r_j^{\uparrow} - l_j^{\uparrow})$
2. $\forall i \in \{0..k\}.flip(r, i, r_i^{\uparrow} - l_i^{\uparrow})$ and succeed if $MI(l) >_{lex} MI(r)$
3. choose the minimum i such that $l_i^{\uparrow} \neq 0$, $flip(r, i, r_i^{\uparrow} - l_i^{\uparrow} - 1)$ and $\forall j \in \{i+1..k\}.flip(r, j, r_j^{\uparrow} - l_j^{\uparrow})$

Each of the three statements can be executed in linear time. Note that the first two may fail (there does not exist a maximum i in the first case, or in the second the test $MI(l) >_{lex} MI(r)$ fails) but the third case will always succeed.

Lemma 3 *The orientation algorithm computes all $\succ$-maximal r where $l \succ r$.*

Proof (sketch): If the first statement succeeds then $\forall j \in \{i+1..k\}.l_j^{\uparrow} = r_j^{\uparrow}$ and $l_i^{\uparrow} > r_i^{\uparrow}$ so $MO(l) >_{revlex} MO(r)$ and r is maximal. Otherwise, $\forall i.l_i^{\uparrow} \leq r_i^{\uparrow}$ so we

---

[5] This requires of course an implementation that efficiently indexes wave-rules so that extra wave-rules do not degrade the performance of rippling.

flip arrows down to equate out-orders and test $MI(l) >_{lex} MI(r)$. If this succeeds, we have a maximal r. Otherwise we still have $\forall i.l_i^! \leq r_i^!$ but flipping arrows in r to equate out-orders is insufficient as r then has a larger in-order. However, by assumption, l has at least one outward wavefront with a least depth i, so we can flip enough arrows at this depth so $r_i = l_i - 1$. Thus $l \succ r$ and r is maximal. □

This parser for simply annotated terms is correct (it only returns wave-rules) and complete (it returns all maximal wave-rules under the orderings we define). As an example, consider (9) with the LHS oriented all out. We begin by orienting both wavefronts in the RHS out. The two sides thus have the measures $\langle [0,1], [0,0] \rangle$ and $\langle [1,1], [0,0] \rangle$ respectively. Hence step 1 fails. Moreover, if we equate the out-measures by turning down the annotation at depth 0, this gives the RHS a measure of $\langle [0,1], [1,0] \rangle$ so step 2 fails. Finally we succeed in step 3 by turning down the arrow at depth 1 giving the RHS a measure of $\langle [1,0], [0,1] \rangle$. The resulting oriented annotation is given in (9).

5.3 Multi-waves and sinks

The above ideas generalize easily to multi-wave-rules. For reasons of space we only sketch this. We generate skeleton preserving annotations analogous to the single-hole case but allow multi-holed wavefronts. Usually both sides are simply annotated and we may use the above orientation algorithm. Alternatively, after fixing an orientation for the LHS of the wave-rule we may orient the RHS by cycling through possible orientations. For each orientation we compare the weakenings of the two sides under the multi-set ordering over our measure and we pick the RHS orientation with the greatest measure. There are various ways the efficiency of this can be enhanced. E.g. we need only compute weakenings of each side once; with "orientation variables" we may propagate the different orientations we select for the RHS to orientations on the weakening set before comparison under the multi-set measure.

One kind of annotation we haven't yet discussed in our measures is *sinks* (see §2). This is deliberate as we can safely ignore sinks in both the measure and the parser. Sinks only serve to decrease the applicability of wave-rules by creating additional preconditions; that is, we only ripple inwards if there is a sink underneath the wavefront. But if rippling terminates without such a precondition, it terminates with it as well. Sinks (and also recent additions to rippling such as colours [15]) can be seen as not effecting the termination of rippling but rather the *utility* of rippling. That is, they increase the chance that we will be able to fertilize with the hypothesis successfully.

6 Extensions to Rippling

By introducing new termination orders for rippling, we can combine rippling with conventional term rewriting. Such extensions greatly extend the power and applicability of rippling both within and outwith induction. In addition, by design, our orderings are not dependent upon rippling preserving skeletons. This allows us to use rippling in new domains involving, for example, mutual recursion or definition unfolding where the skeleton needs to be modified; such applications were previously outside the scope of rippling. We feel that these extensions offer the promise of the "best of both worlds": that is, the highly goal directed nature of rippling combined with the flexibility and uniformity of conventional rewriting. To test these ideas, we have implemented an *Annotated Rewrite System*, a simple PROLOG program which manipulates annotated terms, and in which we can mix conventional term rewriting and rippling. All the examples below have been proven by this system.

6.1 Unblocking

Rippling can sometimes become blocked. Usually the blockage occurs due to the lack of a wave-rule to move the differences out of the way; in such a situation the wave-rule may be speculated automatically using techniques presented in [13]. However, sometimes the proof becomes blocked because a wavefront needs to be rewritten so that it matches either a wave-rule (to allow further rippling) or a sink (to allow fertilization). This is best illustrated by an example.

Consider the following theorem, where rev is naive reverse, $qrev$ is tail-recursive reverse using an accumulator, $<>$ is infix append, and $::$ infix cons.

$$\forall L, M.\ qrev(L, M) = rev(L) <> M \qquad (15)$$

To prove this theorem, we perform an induction on L. The induction hypothesis is

$$qrev(l, M) = rev(l) <> M$$

and the induction conclusion is

$$qrev(\boxed{h :: \underline{l}}^{\uparrow}, \lfloor m \rfloor) = rev(\boxed{h :: \underline{l}}^{\uparrow}) <> \lfloor m \rfloor . \qquad (16)$$

where m is a skolem constant which sits in a sink, annotated with "$\lfloor \ \rfloor$".

We will use wave-rules taken from the recursive definition of $qrev$, and rev.

$$rev(\boxed{H :: \underline{T}}^{\uparrow}) \Rightarrow \boxed{rev(T) <> (H :: nil)}^{\uparrow} \qquad (17)$$

$$qrev(\boxed{H :: \underline{T}}^{\uparrow}, L) \Rightarrow qrev(T, \boxed{H :: \underline{L}}^{\uparrow}) \qquad (18)$$

On the LHS, we ripple with (18) to give

$$qrev(l, \boxed{\boxed{h :: \underline{m}}^{\downarrow}}) = rev(\boxed{h :: \underline{l}}^{\uparrow}) <> \lfloor m \rfloor .$$

On the RHS, we ripple with (17) and then (4), the associativity of $<>$ to get

$$qrev(l, \boxed{\boxed{h :: \underline{m}}^{\downarrow}}) = rev(l) <> (\boxed{\boxed{(h :: nil) <> \underline{m}}^{\uparrow}}) . \qquad (19)$$

Unfortunately, the proof is now blocked. We can neither further ripple nor fertilize with the induction hypothesis. The problem is that we need to simplify the wavefront on the righthand side. Clam currently uses an ad-hoc method to try to perform wavefront simplification when rippling becomes blocked. In this case (19) is rewritten to

$$qrev(l, \lfloor \boxed{h :: \underline{m}}^{\downarrow} \rfloor) = rev(l) <> (\lfloor \boxed{h :: \underline{m}}^{\downarrow} \rfloor).$$

Fertilization with the induction hypothesis can now occur.

In general, unblocking steps are not sanctioned under the measure proposed earlier, or that given in [3]; their uncontrolled application during rippling can lead to non-termination. But we can easily create new orders where unblocking steps are measure decreasing. These new orders allows us to combine rippling with conventional rewriting of wavefronts in an elegant and powerful way. Namely, unblocking rules will be measure decreasing wave-rules accepted by the parser and applied like other wave-rules.

We define an unblocking ordering by giving (as before) an ordering on simply annotated terms, which can then be lifted to an order on multi-wave terms. To order simply annotated terms, we take the lexicographic order of the simple wave-rule measure proposed above (using size of the wavefront as the notion of weight) paired with $>_{wf}$, an order on the *contents* of wavefronts. As a simply annotated term may still contain multiple wavefronts, this second order is lifted to a measure on sets of wavefronts by taking its multi-set extension. The first part of the lexicographic ordering will ensure that anything which is normally measure decreasing remains measure decreasing and the second part will allow us to orient rules that only manipulate wavefronts. This combination provides a termination ordering that allows us to use rippling to move wavefronts about the skeleton and conventional rewriting to manipulate the contents of these wavefronts.

For the reverse example, the normalization ordering is very simple; we use the following wave-rules.

$$\boxed{nil <> \underline{L}}^{\downarrow} \Rightarrow L \tag{20}$$

$$\boxed{(H :: T) <> \underline{L}}^{\downarrow} \Rightarrow \boxed{H :: (T <> \underline{L})}^{\downarrow} \tag{21}$$

The first is already parsed as a wave-rule using our standard measures, but we need to add the second. This rule doesn't change the size of the wavefront or its position but only its form. Hence we want this to be decreasing under some normalization ordering. There are many such orderings; here we take $>_{wf}$ to be the recursive path ordering [7] on the terms in the wavefront where <> has a higher precedence than :: and all other function symbols have an equivalent but lower priority. The measure of the LHS of (21) is now greater than that of the RHS as its wavefront is $(H :: T) <> *$ which is greater than $H :: (T <> *)$ in the recursive path ordering (to convert wavefronts into well formed terms, waveholes are marked with the new symbol *).

Unblocking steps which simplify wavefronts are useful in many proofs enabling both immediate fertilization (as in this example) and continued rippling. Wavefronts can even be unblocked using a different set of rules to that used for rippling.

6.2 Mutual Recursion and Skeleton Simplification

Rippling can also become blocked because the skeleton (and not a wavefront) needs to be rewritten. This happens in proofs involving mutually recursive functions, definition unfolding, and other kinds of rewriting of the skeleton. Consider

$$\forall x.\, even(s(s(0)) \times x)$$

where $even$ has the following wave-rules.

$$even(\boxed{s(\underline{U})}^{\uparrow}) \Rightarrow odd(U) \qquad (22)$$

$$odd(\boxed{s(\underline{U})}^{\uparrow}) \Rightarrow even(U) \qquad (23)$$

Note that (22) and (23) are not wave-rules in the conventional sense since they are not skeleton preserving. However, they do decrease the annotation measure. Rules (22) and (23) can be viewed as a more general type of wave-rule of the form $LHS \Rightarrow RHS$ which satisfy the constraint $skeleton(LHS) \equiv skeleton(RHS)$ where $\equiv$ is some equivalence relation. In this case, the equivalence relation includes the granularity relation in which $even(x)$ and $odd(x)$ are in the same equivalence class. Rippling with this more general class of wave-rules still gives us a guarantee of termination. However weakening the structure preservation requirement can reduce the utility of rippling since now we are only guaranteed to rewrite the conclusion into a member of the equivalence class of the hypothesis.

To prove the theorem, we will also need the following wave-rules.

$$\boxed{s(\underline{U})}^{\uparrow} + V \Rightarrow \boxed{s(\underline{U+V})}^{\uparrow} \qquad (24)$$

$$U + \boxed{s(\underline{V})}^{\uparrow} \Rightarrow \boxed{s(\underline{U+V})}^{\uparrow} \qquad (25)$$

The theorem can be proved without (25) but this requires a nested induction and generalization, complications which need not concern us here.

The proof begins with induction on x. The induction hypothesis is

$$even(s(s(0)) \times n)$$

and the induction conclusion is

$$even(s(s(0)) \times \boxed{s(\underline{n})}^{\uparrow}). \qquad (26)$$

Unfortunately rippling is immediately blocked. To continue the proof, we simplify the skeleton of the induction conclusion by exhaustively rewriting (26) using the unannotated version of (1) and the following rules.

$$0 \times V \Rightarrow 0 \qquad (27)$$
$$0 + V \Rightarrow V \qquad (28)$$

This gives

$$even(\boxed{s(\underline{n})}^\uparrow + \boxed{s(\underline{n})}^\uparrow). \qquad (29)$$

Note that the skeleton was changed by this rewriting. The induction hypothesis can, however, be rewritten using the same rules so that it matches the skeleton of (29). Of course, arbitrary rewriting of the skeleton may not preserve the termination of rippling. To justify these unblocking steps we therefore introduce a new termination order which combines lexicographically a measure on the skeleton with the measure on annotations.[6] We then admit rewrite rules provided their application decreases this combined measure. This new order allows us to combine rippling with conventional rewriting of the skeleton in an elegant and powerful way. In this case, the recursive path order on skeletons (with precedence $\times > + > s > 0$) is again adequate. Note that though termination is guaranteed, again skeleton preservation has been weakened. Since the skeleton can be changed during rippling, we are no longer able to guarantee that we can fertilize at the end of rippling. However, provided the skeleton is rewritten identically in both the hypotheses and the conclusion, we will still be able to fertilize.

To return to the proof, rippling (29) with (24) gives

$$even(\boxed{s(n + \boxed{s(\underline{n})}^\uparrow)}^\uparrow).$$

Then with (25) gives

$$even(\boxed{s(s(\underline{n+n}))}^\uparrow).$$

We now ripple with the mutually recursive definition of even, (22),

$$odd(\boxed{s(\underline{n+n})}^\uparrow).$$

Note that this step also changes the skeleton. However, as the measure decreases and as the skeleton stays in the same equivalence class, such rewriting is permitted. Finally rippling with (23) gives

$$even(n + n).$$

This matches the (rewritten) induction hypothesis and so completes the proof.

The power of rippling is greatly enhanced by its combination with traditional rewriting. For example, proofs involving mutually recursive functions, or other kinds of skeleton simplification (e.g., definition unfolding) were not previously possible with rippling. The use of conventional term rewriting to simplify the skeleton is a natural dual to the use of conventional rewriting to simplify wavefronts; indeed they are orthogonal and can be combined to allow even more sophisticated rewriting.

---

[6] With more complex theorems, the height of the skeleton may increase; the addition of the height of the skeleton to the order ensures termination in such situations.

6.3 Other Applications

Rippling has found several novel uses of outside of induction. For example, it has been used to sum series [14], to prove limit theorems [15], and guide equational reasoning [11]. However, new domains, especially non-inductive ones, require new orderings to guide proof. For example, consider the PRESS system [6].[7] To solve algebraic equations, PRESS uses a set of methods which apply rewrite rules. The three main methods are: *isolation*, *collection*, and *attraction*. Below are examples of rewrite rules used by each of these methods.

$$\text{ATTRACTION}: \quad \boxed{\log(\underline{U}) + \log(\underline{V})}^{\uparrow} \Rightarrow \boxed{\log(\underline{U} \times \underline{V})}^{\uparrow}$$

$$\text{COLLECTION}: \quad \boxed{\underline{U} \times \underline{U}}^{\uparrow} \Rightarrow \boxed{\underline{U}^2}^{\uparrow}$$

$$\text{ISOLATION}: \quad \boxed{\underline{U}^2}^{\uparrow} = V \Rightarrow U = \boxed{\pm\sqrt{\underline{V}}}^{\uparrow}$$

PRESS uses preconditions and not annotation to determine rewrite rule applicability. Attraction must bring occurrences of unknowns closer together. Collection must reduce the number of occurrences of unknowns. Finally, isolation must make progress towards isolating unknowns on the LHS of the equation. These requirements can easily be captured by annotation and PRESS can thus be implemented by rippling. The above wave-rules suggest how this would work. PRESS wave-rules are structure preserving, where the preserved structure is the unknowns. The ordering defined on these rules reflects the well-founded progress achieved by the PRESS methods. Namely, we lexicographically combine orderings on the number of waveholes for collection, their distance (shortest path between waveholes in term tree) for attraction, and our width measure on annotation weight for isolation.

7 Related Work and Experience

The measures and orders we give are considerably simpler than those in [3]. There, the properties of structure preservation and the reduction of a measure are intertwined. Bundy *et al.* describe wave-rules schematically and show that any instance of these schemata is skeleton preserving and measure decreasing under an appropriately defined measure. Mixing these two properties makes the definition of wave-rules very complex. For example, the simplest kind of wave-rule proposed are so-called *longitudinal* wave-rules (which ripple-out) defined as rules of the form,

$$\eta(\boxed{\xi_1(\underline{\mu_1^1},..,\underline{\mu_1^{p_1}})}^{\uparrow},..,\boxed{\xi_n(\underline{\mu_n^1},..,\underline{\mu_n^{p_n}})}^{\uparrow}) \Rightarrow \boxed{\zeta(\eta(\underline{\varpi_1^1},..,\underline{\varpi_n^1}),..,\eta(\underline{\varpi_1^k},..,\underline{\varpi_n^k}))}^{\uparrow}$$

---

[7] Due to space constraints, we only sketch this application. The idea of reconstructing PRESS with rippling was first suggested by Nick Free and Alan Bundy.

that satisfy a number of side conditions. These include: each ϖ_i^j is either an unrippled wavefront, $\boxed{\xi_i(\underline{\mu_i^1},\ldots,\underline{\mu_i^{p_i}})}$, or is one of the waveholes, $\underline{\mu_i^l}$; for each j, at least one ϖ_i^j must be a wavehole. η, the ξ_is, and ζ are terms with distinguished arguments; ζ may be empty, but the ξ_is and η must not be. There are other schemata for *traverse* wave-rules and *creational* wave-rules[8]. These schemata are combined in a general format, so complex that in [3] it takes four lines to print. It is notationally involved although not conceptually difficult to demonstrate that any instance of these schemata is a wave-rule under our size and width measures.

Consider the longitudinal schema given above. It is clear that evey skeleton on the RHS is a skeleton of the LHS because of the constraint on the ϖ_j^i. What is trickier to see is that it is measure decreasing. Under our order this is the case if LHS $\succ^*$ RHS. We can show something stronger, namely, for every $r \in$ *weakenings*$(RHS). \exists l \in$ *weakenings*$(LHS). l \succ r$. To see this observe that any such r must be a maximal weakening of

$$r' = \boxed{\zeta(\eta(\varpi_1^1,\ldots,\varpi_n^1),\ldots,\eta(\varpi_1^j,\ldots,\varpi_n^j),\ldots\eta(\varpi_1^k,\ldots,\varpi_n^k))}^\uparrow$$

for some $j \in \{1..k\}$. Corresponding to r' is an l' which is a weakening of the LHS where $l' = \eta(t_1,\ldots,t_n)$ and the t_i correspond to the ith subterm of $\eta(\varpi_1^j,\ldots,\varpi_n^j)$ in r': if ϖ_i^j is an unrippled wavefront then $t_i = \varpi_i^j = \boxed{\xi_i(\underline{\mu_i^1},\ldots,\underline{\mu_i^{p_i}})}$, and alternatively if ϖ_i^j a wavehole μ_i^l then $t_i = \boxed{\xi_i(\underline{\mu_i^1},\ldots,\mu_i^l,\ldots,\underline{\mu_i^{p_i}})}$. Now r is a maximal weakening of r' so there is a series of weakening steps from r to r'. Each of these weakenings occurs in a ϖ_i^j and we can perform the identical weakening steps in the corresponding t_i in l' leading to a maximal weakening l. As l and r are maximally weak they may be compared under $\succ$. Their only differences are that r has an additional wavefront at its root and is missing a wavefront at each ϖ_i^j corresponding to a wavehole. The depth of ϖ_i^j is greater than the root and at this depth the outmeasure of l is greater than r (under any of the weights defined in §3) and at all greater depths they are identical. Hence $l \succ r$.

Similar arguments hold for the other schemata given in [3] and from this we can conclude that wave-rules acceptable under their definition are acceptable under ours. Moreover it is easy to construct simple examples that are wave-rules under our formalism but not theirs; for example, the following two rules are measure decreasing but are not instances of their schema.

$$rot(\boxed{s(\underline{X})}^\uparrow,\boxed{H::\underline{T}}^\uparrow,Acc) \Rightarrow rot(X,T,\boxed{H::\underline{Acc}}^\uparrow)$$

$$\boxed{0+\underline{X}}^\uparrow \Rightarrow X$$

---

[8] Creational wave-rules are used to move wavefronts between terms during induction proofs by destructor induction. They complicate rippling in a rather specialized and uninteresting way. Our measures could be easily generalized to include such creational rules.

Aside from being more powerful, there are additional advantages to the approach taken here. Our notion of wave-rules and measures are significantly simpler and therefore easier to understand. As a result, they are easier to implement. The definition of wave-rules given in [3] is not what is recognized by the Clam waverule parser as it returns invalid wave-rules under either our definition or that of [3] and misses many valid ones. For example, Clam's current parser fails to find even wave-rules as simple as the following.

$$divides(\boxed{\underline{X+Y}}^{\uparrow}, Y) \Rightarrow \boxed{s(\underline{divides(X,Y)})}^{\uparrow}$$

We have therefore implemented the parser described in §5. The parser is simple, just a couple of pages of Prolog, yet allows new orderings based on different annotation measures to be easily incorporated. Although parsing is in the worst case exponential in the size of the rewrite rule, the parser typically takes under 5 seconds to return a complete set of maximal wave-rules (which seems reasonable for an off-line procedure). The parser is part of our annotated rewrite system and will be shortly integrated into the Clam theorem prover.

8 Conclusions

An ordering for proving the termination of rippling along with a schematic description of wave-rules was first given in [3]. We have simplified, generalized and improved both this termination ordering, and the description of wave-rules. In addition, we have shown that different termination orderings for rippling can be profitably used within and outwith induction. Such new orderings can combine the highly goal directed features of rippling with the flexibility and uniformity of more conventional term rewriting. We have, for example, given two new orderings which allow unblocking, definition unfolding, and mutual recursion to be added to rippling in a principled (and terminating) fashion; such extensions greatly extend the power of the rippling heuristic. To support these extensions, we have implemented a simple *Annotated Rewrite System* which annotates and orients rewrite rules, and with which we can rewrite annotated terms. We have used this system to perform experiments combining rippling and conventional term rewriting. We confidently expect that this combination of rippling and term rewriting has an important rôle to play in many areas of theorem proving and automated reasoning.

Acknowledgments

Many of the ideas described here stem from conversations with members of the Edinburgh MRG group, in particular with Alan Bundy, Ian Green, and Andrew Ireland. We also wish to thank Sean Matthews, David Plaisted, Michael Rusinowitch, and Andrew Stevens for comments on earlier drafts. The first author was funded by the German Ministry for Research and Technology (BMFT) under grant ITS 9102. The second author was supported by a SERC and a HCM Postdoctoral Fellowship.

References

1. D. Basin and T. Walsh. **Difference unification**. In *Proceedings of the 13th IJCAI*. International Joint Conference on Artificial Intelligence, 1993.
2. R.S. Boyer and J S. Moore. *A Computational Logic*. Academic Press, 1979. ACM monograph series.
3. A. Bundy, A. Stevens, F. van Harmelen, A. Ireland, and A. Smaill. Rippling: A heuristic for guiding inductive proofs. *Artificial Intelligence*, 62:185–253, 1993.
4. A. Bundy, F. van Harmelen, C. Horn, and A. Smaill. The Oyster-Clam system. In M.E. Stickel, editor, *10th International Conference on Automated Deduction*. 1990.
5. A. Bundy, F. van Harmelen, A. Smaill, and A. Ireland. Extensions to the rippling-out tactic for guiding inductive proofs. In M.E. Stickel, editor, *10th International Conference on Automated Deduction*, pages 132–146. Springer-Verlag, 1990.
6. A. Bundy and B. Welham. Using meta-level inference for selective application of multiple rewrite rules in algebraic manipulation. *Artificial Intelligence*, 16(2):189–212, 1981.
7. N. Dershowitz. Orderings for term-rewriting systems. *Theoretical Computer Science*, 17(3):279–301, March 1982.
8. N. Dershowitz. Termination of Rewriting. In J.-P. Jouannaud, editor, *Rewriting Techniques and Applications*. Academic Press, 1987.
9. N. Dershowitz and J.-P. Jouannaud. Rewrite systems. In J. van Leeuwen, editor, *Handbook of Theoretical Computer Science*, volume B. North-Holland, 1990.
10. N. Dershowitz and Z. Manna. Proving termination with multiset orderings. *Comms. ACM*, 22(8):465–476, 1979.
11. D. Hutter. Guiding inductive proofs. In M.E. Stickel, editor, *10th International Conference on Automated Deduction*. 1990.
12. D. Hutter. Colouring terms to control equational reasoning. An Expanded Version of PhD Thesis: Mustergesteuerte Strategien für Beweisen von Gleichheiten (Universität Karlsruhe, 1991), in preparation.
13. A. Ireland and A. Bundy. Using failure to guide inductive proof. Technical report, Dept. of Artificial Intelligence, University of Edinburgh, 1992.
14. T. Walsh, A. Nunes, and A. Bundy. The use of proof plans to sum series. In D. Kapur, editor, *11th Conference on Automated Deduction*. 1992.
15. T. Yoshida, A. Bundy, I. Green, T. Walsh, and D. Basin. Coloured rippling: the extension of a theorem proving heuristic. Technical Report, Dept. of Artificial Intelligence, University of Edinburgh, 1993. Under review for ECAI-94.

A Novel Asynchronous Parallelism Scheme for First-Order Logic*

David B. Sturgill and Alberto Maria Segre

Cornell University, Ithaca NY 14853-7501, USA

Abstract. We present a new parallelization technique designed to speed the solution of first-order inference problems by detecting and pruning unnecessary search. Our technique, called *nagging*, is naturally fault tolerant and relatively robust in the presence of high communication latency. As a result, it is well suited to execution on a network of loosely-coupled processors. Our implementation has been tested on as many as 100 workstations and exhibits encouraging scaling properties. In this paper, we introduce nagging in the context of existing work on parallel search for logical inference, present an informal analysis of its operation, and present some empirical results for our PTTP-style prototype.

1 Introduction

We are interested in developing automated systems capable of solving large logistics problems, such as those which arise naturally in manufacturing and transportation scheduling. More generally, we are interested in developing techniques that permit the user to formally describe some domain of interest (e.g., airlift/logistics problems) and then to pose questions which may be answered according to this domain description. First-order logic provides a powerful framework for formalizing these and other similar domains, and many powerful inference techniques exist for reasoning within this framework.

As Stickel's work on PTTP demonstrates[16], it is possible to construct high-performance, search-based systems to solve general first-order logic problems via model elimination. PTTP exploits highly efficient inference mechanisms originally developed within the Prolog community. Unfortunately, these mechanisms are often inappropriate or insufficient for large first-order search problems. Various techniques have been identified to alleviate these disadvantages by augmenting the basic model elimination search procedure. The principal contribution of this paper is to describe and evaluate a new technique for parallelizing search for PTTP-style first-order logical inference systems. By effectively bringing multiple processing elements to bear, we can increase the maximum size of those problems which can realistically be solved.

---

* Support for this research was provided by the Office of Naval Research through grant N00014-90-J-1542, by the Advanced Research Project Agency through Rome Laboratory Contract Number F30602-93-C-0018 through Odyssey Research Associates, Incorporated, and by the Air Force Office for Scientific Research through a Graduate Student Fellowship.

In the following sections, we first describe related work in parallelizing search in the context of theorem proving and logic programming. We then propose a new, asynchronous, parallelization technique designed to speed the effective solution time for a large class of deduction problems. We describe an implementation of our technique in a PTTP-style system based on an extended version of the Warren Abstract Machine (WAM) architecture[1, 18], and present experimental results that illustrate the effectiveness of our approach. Finally, we consider some open issues that remain to be explored and sketch some directions for future research.

2 Related Work

Most work in parallelizing logical inference falls into one of two broad classes. *OR parallelism*[2, 12] employs multiple processors to examine alternative solutions concurrently, while *AND parallelism*[8, 7] uses more than one processor to construct different portions of a single solution in parallel.[2]

OR-parallel approaches break the search space into smaller, independent, subspaces which are then explored in parallel by separate processes. In partitioning the search space OR parallelism exploits the natural divisions prescribed by the nondeterministic choice points. Since a sequential theorem prover cannot know *a priori* which OR choices lead to success and which lead to failure, OR parallelism may improve performance by not waiting for one choice to fail before starting work on the others. As soon as one process finds a solution in its slice of the search space, the solution is reported and work done by the remaining processes may be discarded. Consequently, OR-parallel processes may execute in complete independence. Inter-process communication is necessary only when a solution is found or when one process exhausts its search space and must consult its neighbors for additional work (i.e., load balancing).

In contrast, much of the work in parallel logic programming has focused on AND-parallel techniques. The idea behind AND parallelism is that, if there are several different subproblems to satisfy (goals to resolve away), it may be be possible to speed execution by working on each of them in parallel. Unlike OR parallelism, AND-parallel evaluation is complicated by the interaction of logic variables. AND-parallel processes must ensure that the solutions to their associated subgoals agree in how they bind shared logic variables. The need to resolve these variable binding conflicts typically either necessitates some additional communication among AND-parallel processes or restricts when AND parallelism may be productively applied. For this reason, various approaches to AND parallelism have concentrated on developing policies for how and when shared variables may be bound and on techniques for determining when AND parallelism may be applied without risk of binding conflicts. While it may be the case that more speedup is potentially available with AND-parallel approaches,

---

[2] Some systems combine both OR-parallel and AND-parallel techniques[3, 5] or exploit additional sources of parallelism, such as parallel unification.

these techniques are often not as straightforward in full first-order logic as they are in Prolog's restrictive definite clause framework.

3 A New Approach to Parallelism

We present here a new parallel search strategy that exhibits many desirable features of both AND- and OR-parallel techniques. Our approach, called *nagging*, is geared toward a loosely-coupled network of heterogeneous processors and is naturally fault tolerant.

3.1 Nagging

Nagging is a search pruning technique that exploits the dependencies between conjunctive goals. Consider the conjunctive goal $G_1 \wedge \ldots \wedge G_n$. If some G_i is unsatisfiable then $G_1 \wedge \ldots \wedge G_n$ is certainly unsatisfiable. If, while attempting to satisfy $G_1 \wedge \ldots \wedge G_n$, it is determined that G_i is unprovable irrespective of the proofs for $G_{j \neq i}$, work on all remaining $G_{j \neq i}$ may be immediately abandoned. Of particular interest here is the fact that determining the unsatisfiability of G_i alone may entail substantially less search than determining the unsatisfiability of $G_1 \wedge \ldots \wedge G_n$ under a sequential left-to-right search discipline.

In a conventional backward-chaining search, the goal stack always represents a conjunct that must be satisfied if the current search path is to succeed. As a result of this stack-order maintenance of the goals in the current resolvent, a theorem prover may commit to satisfying some goals early in the search but not actually examine them until much later. In Fig. 1, for example, goals B, C and D are pushed onto the goal stack as soon as goal A is expanded. However, goal D will not be examined until proofs for B and C have been completed. Naturally this goal ordering policy can result in phenomenally bad search behavior. This is evident when goals such as D are intrinsically unsatisfiable or when variable bindings made early in the search render them unprovable. In this situation, not only is the inevitable failure undetected until all intermediate goals are proven, but, even when the failure is detected, naive backtracking may generate a great deal of search among intermediate goals before correcting the root cause of the failure.

Nagging is designed to avoid this undesirable search behavior by detecting guaranteed failures early and preempting the search to avoid them. Parallelism via nagging employs two types of processes, a single master proving process and one or more nagging processes. The master process searches for a solution using some serial inference procedure.[3] Nagging processes operate asynchronously and in parallel with the master process. When a nagging process becomes idle, it examines the state of the master process as captured by its stack of open goals

---

[3] For our purposes, this inference procedure is model elimination. Since we rely on iterative deepening for search completeness, any individual search for a proof represents only a single iteration of the overriding iterative deepening search and is therefore resource bounded.

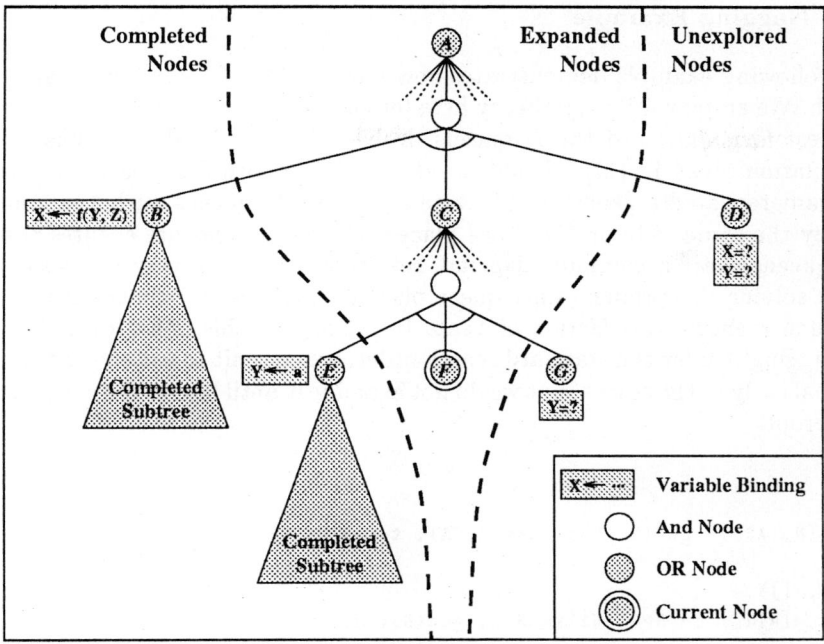

Fig. 1. Partial proof constructed during search to satisfy goal A

$G_1, \ldots, G_n$. The nagging process selects some G_i from this set and attempts to prove it using the same resource bound employed by the master process. If the nagging process is able to prove G_i then no search pruning will occur; the master process may yet be able to find a proof for G_i that is consistent with proofs for the remaining $G_1, \ldots G_{i-1}, G_{i+1}, \ldots G_n$. If, however, the nagging process fails to find a proof for G_i, then we are guaranteed that the master process will be unable to satisfy all of the goals in its goal stack, $G_1 \wedge \ldots G_i \ldots \wedge G_n$. When the nagging process exhibits such a failure, it can force the master to backtrack far enough to either remove G_i from the goal stack or retract a variable binding that was imposed on G_i. If, on the other hand, G_i comes off the master's goal stack while the nagging process is still underway, the nagging process' search is known to be irrelevant and can be interrupted, allowing the nagging process to select a new subproblem.

Like OR parallelism, nagging centers around the idea that OR choices made by a sequential theorem prover may eventually result in failure. While OR parallelism seeks to alleviate this problem by considering alternative OR choices concurrently, nagging employs parallel processes that attempt to demonstrate the infeasibility of the current OR choices more quickly than the sequential search.

3.2 Nagging Example

The following example demonstrates how nagging may be effective at speeding search. We employ a Prolog theory here for clarity. Consider the naive generate-and-test formulation of the N-queens problem in Fig. 2, which is based on a formulation given in [17].[4] Under a left-to-right, depth-first search order this program repeatedly generates placements for all N queens such that no two occupy the same rank or file. Once placed, the test predicate ensures that no two queens share a common diagonal. Intuitively, this is a very bad way to go about solving this problem since queen placements made early in the search may preclude eventual satisfaction of test. In particular, this is an ideal situation for nagging. Under the standard search order, we commit to satisfying the test subgoal early in the search, but we do not examine it until late in the construction of a proof.

```
queen(N, A) :- gen(N, L), perm(L, A), test(A).

gen(0, []).
gen(N, [N|T]) :- N>0, N1 is N-1, gen(N1, T).

perm([], []).
perm([H|T], [A|P]) :- del(A, [H|T], L), perm(L, P).

del(X, [X|Y], Y).
del(X, [Y|Z], [Y|W]) :- del(X, Z, W).

test([]).
test([H|T]) :- safe(T, H, 1), test(T).

safe([], _, _).
safe([H|T], U, D) :- H-U =\= D, U-H =\= D, D1 is D+1, safe(T, U, D1).
```

Fig. 2. A naive generate-and-test formulation of N-queens

Figure 3 presents a snapshot of a partial proof constructed during the search for a solution to queen(6, A). At this point, the first two queens have been placed in a manner that guarantees failure of test. However, this failure will not be detected until the remaining four queens are placed. Even then, the search

---
[4] Although there are obvious opportunities for a skilled programmer to improve performance by reformulating this particular domain, other domains are not as easily repaired. As this example indicates, nagging does not require that the particular defects in a domain theory be identified. It can therefore help to avoid pathological search behavior that is either not obvious to the programmer or not easily repaired by domain reformulation.

will backtrack through all (4! − 1) variant placements of the last four queens before it reevaluates the placement of the original two.

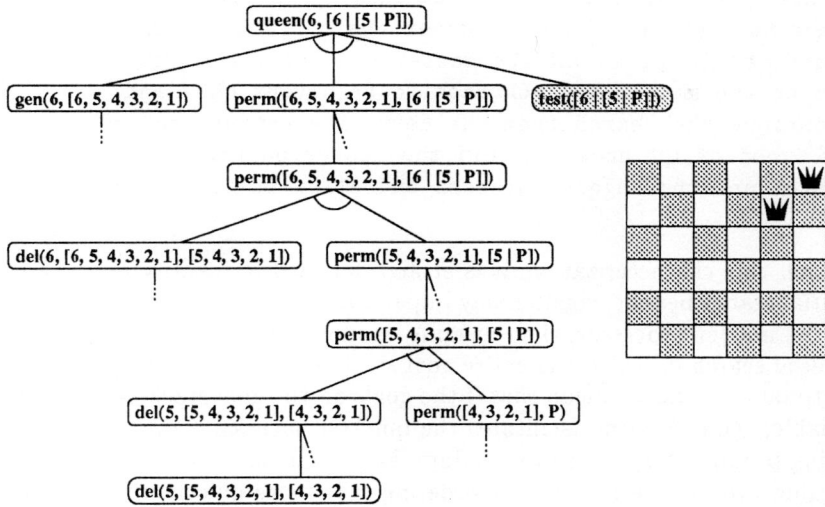

Fig. 3. Snapshot of a partial proof tree during an attempt to solve queen(6, A)

If nagging is employed in this situation, it may be possible to avoid much of this unnecessary search. If a nagging process elects to examine the goal test([6|[5|P]]), then the inevitable failure can be quickly detected. The master process will then be forced to immediately backtrack and reconsider the placement of the first two queens.

3.3 Preliminary Analysis

As presented here, nagging may be seen as a form of parallel, speculative reordering of the goal stack; a nagging process selects goals that the master process may not examine until much later and effectively promotes them to an earlier position in the search. The nagging process then attempts to exhibit a failure under this permuted order before the master process discovers the failure under the standard order. In general, the nagging process need not be restricted to examining individual goals on the master's goal stack. Unsatisfiability of any subset of the goal stack is sufficient to render the current search path infeasible. In fact unsatisfiability of any ancestor goal of any element of the goal stack is sufficient to justify pruning. Accordingly, nagging may be generalized as follows:

```
let Q₁...Qₙ stand for the sequence of resolvents
    generated along the master's current search path
Select one such resolvent Qᵢ
Permute the order of goals in Qᵢ yielding Qᵢ'
Truncate Qᵢ' ≡ Gᵢ₁,...Gᵢₖᵢ to produce some Qᵢ'' ≡ Gᵢ₁,...Gᵢⱼ, 1 ≤ j ≤ kᵢ
Attempt to prove Gᵢ₁ ∧ ... ∧ Gᵢⱼ under the current resource bound
if proof of Qᵢ'' fails and the master has not yet backtracked past Qᵢ
    force the master to backtrack to resolvent Qᵢ₋₁ and
    continue the search from the next alternative choice.
elseif proof of Qᵢ'' underway and the master backtracks past Qᵢ
    interrupt the nagging process and allow it to nag again.
```

Given this characterization, it is straightforward to identify bounds on the potential search benefit nagging may impart. If nagging processors are no faster than the master processor, then nagging cannot speed the search beyond that of a serial search in which the entire goal stack is always optimally ordered. By an optimal ordering we mean that if the goals on the goal stack are consistently satisfiable, their ordering facilitates the quickest success; if they are not, their ordering produces the most rapid failure. Nagging results in speedup only when a nagging process' permuted goal ordering fails more quickly than the master process' ordering. Since such a permuted ordering can be no better than optimal, a nagging process will take at least as long to exhibit this failure as an optimally ordered search. In contrast, nagging can never cause the master process to explore more nodes than it would without nagging. Since nagging only serves to prune branches from the search that are guaranteed to eventually fail, it can only reduce the number of nodes expanded by the master process. Aside from implementation overhead, nagging will not adversely impact performance.[5]

We can expect nagging to be most effective in domains where there is a large degree of variation in the search spaces entailed by different permutations of the goal stack and where the default ordering is seldom globally optimal. Since different goal stack orderings may entail arbitrarily differing search space sizes, nagging may result in superlinear speedup in poorly ordered domain formulations. In practice, the sensitivity of search space size to permutation of the goal stack is largely determined by the interaction of AND-shared variables. Consequently, nagging is likely to do well in domains where conjunctive goals share unbound variables and where the satisfiability of each goal is highly dependent on the bindings of these shared variables. This is contrary to many conventional AND-parallelism techniques where AND-shared variables are problematic.

Although the operation of nagging may be compared with intelligent backtracking, it differs from conventional intelligent backtracking schemes in several respects. Since nagging is not constrained by the standard search order, it may

---

[5] This claim assumes that the master process is non-learning; information gained in failed branches is not used to subsequently reorder or prune the search. Naturally, if a prover learns from failed search branches, backtracking out of these branches early may preclude learning and adversely impact performance.

detect failures in goals that the master process has not yet examined. Nagging may also detect when some goal G is unsatisfiable regardless of how its free variables are bound, while a sequential search may have to repeatedly attempt to prove different instantiations of G. Furthermore, while intelligent backtracking induces a fair amount of overhead, the bookkeeping necessitated by nagging is minimal and relatively infrequent. Finally, as our prototype implementation demonstrates, nagging and intelligent backtracking work well in combination.

4 Prototype Implementation

To empirically examine the effectiveness of nagging, we have developed a prototype implementation which we call the *Distributed, Adaptive, Logical Inference* (DALI) engine. In the spirit of Stickel's PTTP [16], DALI extends the architecture of the Warren Abstract Machine [1] to first-order completeness. In addition, DALI employs further extensions such as subgoal caching [14] and a form of intelligent backtracking to enhance search efficiency.

DALI is currently implemented in C and nagging is realized over a local network of Unix workstations using sockets. When work is transferred between processors it is done by reconstructing the state of computation as described in [2, 12, 4] rather than by copying it. This helps to reduce communication overhead by only requiring transmission of the sequence of nondeterministic choices made along the master process' current search path.

5 Experimental Results

We present here some experimental results obtained using a version of the DALI implementation described above. While the N-queens problem is suitable for expository purposes, the effectiveness of nagging in such contrived domains is unlikely to be indicative of its performance on realistic problem formulations. To evaluate nagging in more representative first-order domains, we make use of a collection of 85 theorem proving problems provided by Mark Stickel[16].[6]

Our intent is to demonstrate the effectiveness of nagging in its most basic manifestation. Accordingly, our experiment is designed to compare the performance of a particularly naive configuration of the DALI system with an equivalent serial system. For the experiments reported here, nagging processes select subproblems according to the following simple criteria:

---

[6] This set includes the following individual problem instances: burstall, shortburst, prim, has_parts1, has_parts2, ances, num1, group1, group2, ew1, ew2, ew3, rob1, rob2, dm, qw, mqw, dbabhp, apabhp, ex4t1, ex4t2, ex5, ex6t1, ex6t2, wos1, wos2, wos3, wos4, wos5, wos6, wos7, wos8, wos9, wos10, wos11, wos12, wos13, wos14, wos15, wos16, wos17, wos18, wos19, wos20, wos21, wos22, wos23, wos24, wos25, wos26, wos27, wos28, wos29, wos30, wos31, wos32, wos33, ls5, ls17, ls23, ls26, ls28, ls29, ls35, ls36, ls37, ls41, ls55, ls65, ls68, ls75, ls76t1, ls76t2, ls87, ls100, ls103, ls105, ls106, ls108, ls111, ls112, ls115, ls116, ls118 and ls121

- Some resolvent Q_i is selected randomly from among the set of all resolvents along the current search path, $Q_1,..Q_n$. Selection is weighted to favor Q_j over Q_k for $1 \leq j < k \leq n$. This corresponds to a bias in favor of higher subproblem granularity. Unsatisfiability of Q_j will result in greater search pruning than will the unsatisfiability of Q_k. However, goals extracted from Q_j are likely to represent larger search problems than those from Q_k.
- Resolvents that were created too recently are not considered for nagging. This also represents a bias in favor of subproblems that have been under investigation for a longer period of time. This policy represents an effort to limit the overhead of transmitting a subproblem to a nagging process to situations where there is reason to believe the subproblem is sufficiently large.
- Once some Q_i is selected, it is permuted randomly subject to the following constraint: goals in Q_i that are either created in or specialized by variable bindings made in the inference from Q_{i-1} to Q_i are ordered before other goals in Q_i. This represents a policy of forcing the nagging process to concentrate on goals that are more likely to be unsatisfiable in Q_i than their counterparts were in Q_{i-1}.
- After permuting Q_i to form Q'_i, the nagging process attempts to prove Q'_i using the same resource bound as the master process. This corresponds to having the nagging process consider a permuted copy of the master's entire goal stack.

Finally, in order to avoid confusing the speedup due to nagging with the search reduction that might be obtained via the use of intelligent backtracking, both serial and parallel systems tested here employ identical forms of intelligent backtracking, similar to that described in [10].

All 85 test problems are attempted both by this configuration of DALI using 5 nagging processors as well as by an identical serial version of the system. A resource limit of 5,000,000 inferences is imposed on each problem; any problem as yet unsolved within this resource bound is abandoned and marked unsolved. For this experiment, the master process is operating on a dedicated Sun Sparc 670MP "Cypress" system with 128MB or real memory. Nagging processors are drawn from a pool of 38 Sun workstations of equal or lesser power and memory. Note that the nagging processors represent a shared resource, and are therefore also supporting other users in a variety of tasks during the experiment.

Of the 85 problems in the test suite, both serial and parallel systems left 12 problems unsolved. Figure 4 plots the parallel solution times (sum of system and user times as measured by the operating system) for each of the 73 solved problems against the corresponding serial solution times. Note that both axes are presented on a log scale.

Data points above the solid $y = x$ line in Fig. 4 correspond to trials in which the parallel search actually took longer to solve a particular problem than the serial search. Points below the solid line denote trials that experienced some amount of speedup. The dotted $y = \frac{1}{6}x$ line represents linear speedup with respect to the number of processors employed. Note that it may be somewhat

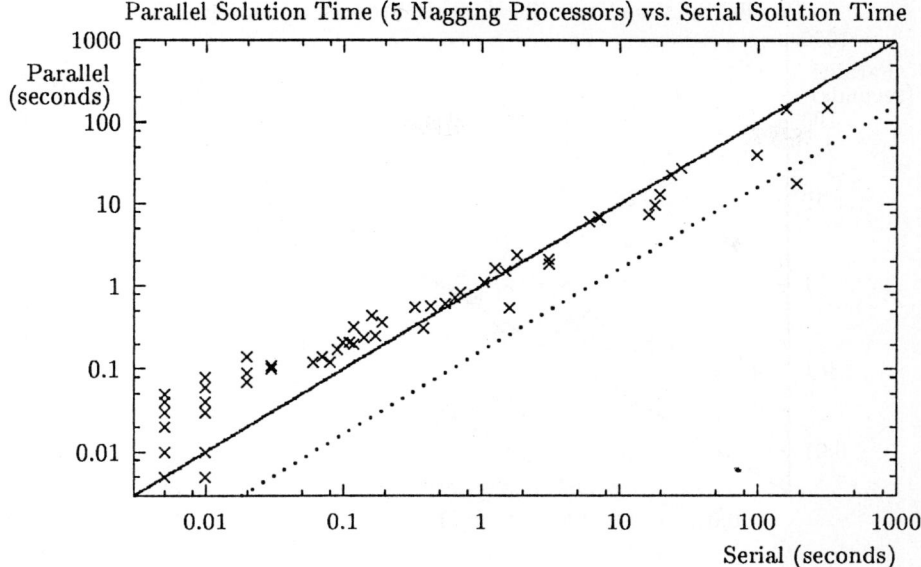

Fig. 4. Parallel execution times with 5 nagging processors vs. sequential execution times for 73 solved test problems as measured by the master process. Points below the solid line represent problems solved more quickly in parallel. Points below the dotted line correspond to instances of superlinear speedup. The vertical and horizontal alignment of points in the lower left portion of the graph is due to the granularity of the system timer.

unfair to expect near-linear speedups in our experiments given that the nagging processors are generally less powerful than the master processor and are also being used by other users during the experiment. Thus the use of N processors actually corresponds to rather less than using N times the memory and CPU power of the master processor. Nevertheless, the existance of data points below the dotted line correspond to instances of superlinear speedup, where the speedup is even greater than the number of processors. Observe that, for problems that are solved quickly by a serial system (i.e., those data points on the left side of Fig. 4), the added overhead of nagging dominates any beneficial search pruning it provides. In contrast, as problems become larger, nagging seems to be more effective, exhibiting superlinear speedup for some of the largest problems.

Fig. 5 shows the results obtained for a similar experiment using 30 nagging processors. Since this system employs more processors, the threshold for superlinear speedup is correspondingly higher. Consequently, the dotted line in Fig. 5 appears lower in the graph, at $y = \frac{1}{31}x$. As with 5 nagging processors, we

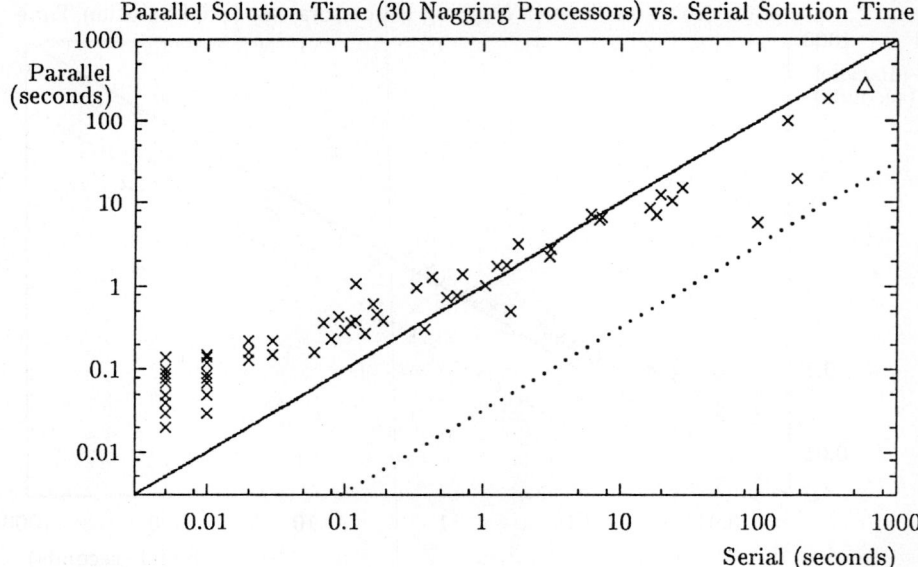

Fig. 5. Parallel execution time with 30 nagging processors vs. sequential execution times. The lone triangular data point denotes the problem that was solved by the parallel system but unsolved by the serial system within the allotted resource bound. The abscissa represents the serial system's time to failure (a lower bound on the actual serial solution time) while the ordinate represents the parallel system's solution time as usual. Thus this data point's real position lies shifted to the right by some unspecified distance.

see that nagging is generally more effective on the large problems, where the overhead due to nagging can be amortized over a larger search space and where the pruning due to nagging is more effective. As might be expected, the preformance penalty on smaller problems is larger than with 5 nagging processors. On the other hand, while the use of 30 nagging processors does not demonstrate superlinear speedup on any one problem, it does on average find solutions to the larger problems faster than with only 5 nagging processors. In addition, the use of 30 nagging processors leads to the solution of one additional problem within the allotted resource bound. Since we do not have a valid value for the abscissa, we plot the extra problem's parallel solution time against the serial system's time to failure, a valid lower bound on actual serial time to solution. Thus the actual position of the extra data point lies shifted to the right by some unknown amount, which would improve the apparent performance of the parallel system in the plot.

6 Discussion

In this paper we have described an asynchronous parallelization scheme for theorem proving and have presented some preliminary experimental results supporting its effectiveness in practice. Our approach shares several desirable properties associated with conventional OR- and AND-parallel search techniques:

- As with many OR-parallel models, assignment of work is initiated by idle processors; busy processors don't have to constantly stop and consider how they might delegate some of their workload.
- As in OR parallelism, subproblems investigated via nagging are largely independent, and communication is comparatively infrequent. Furthermore, since nagging affords some control over granularity of a nagged subproblem, there are opportunities to regulate the frequency of communication.
- Like some of the techniques for parallelizing Prolog [6], nagging need not affect the order in which solutions are generated. As described here, nagging does not change the order in which the master process explores the search space; it only serves to prune those search branches that are doomed to failure.
- As illustrated in Conery's AND/OR process tree [5], AND- and OR-parallel techniques may be cleanly applied both recursively and in conjunction to facilitate greater speedup and scalability. Nagging admits a similar recursive interpretation and may, itself, be combined with other sources of parallelism.

Nagging also displays some additional desirable properties that are largely unique to this approach. Parallelism via nagging automatically features a limited degree of fault tolerance. Since nagging as just described only serves to force the master process to backtrack out of search branches that are guaranteed to fail, any search space pruned by nagging would otherwise be exhausted by the master eventually. Consequently, the master process has no real dependence on the nagging processes and never has to wait for messages from them. If messages are lost or delayed, the master may perform unnecessary search, but it will eventually generate an identical solution.

Note also that conventional OR- and AND-parallel search strategies exhibit a high degree of dependence on the local syntactic structure of the domain specification. For example, OR-parallel strategies typically partition the search space where multiple clause definitions are applicable; AND-parallel approaches usually parallelize a conjunctive query or the antecedents of a rule. Nagging is less restricted by local syntactic features of the domain formulation. A nagging process is free to examine sets of goals from distant regions in the AND/OR search tree, thereby exploiting solution constraints that are not inherently local.

The experimental results reported here are particularly encouraging given that the parameters of the nagging implementation are somewhat arbitrary and intended to simplify presentation rather than to exhibit maximal speedup. Continuing investigation has shown that nagging may be rendered even more effective by a variety of techniques. For example, preventing nagging processors from working on redundant subproblems typically requires a small amount of additional overhead that is greatly outweighed by the concomitant improvement in performance. More sophisticated heuristic selection criteria for picking nagged subproblems have also been shown to result in improved performance.

In addition, there are many opportunities to make more effective use of the information obtained during a nagging process' search. Although nagging is most effective when nagged subproblems are found to be unsatisfiable, nagged subproblems that are successfully proven can also be used to some advantage. In the spirit of AND parallelism, the proof may be sent back to the master process and grafted into its partial proof tree. Of course, in the event that the nagged subproblem is a reordering of the master process' entire goal stack, the proof represents a complete proof of the original problem. In this case, the nagging process' proof constitutes a faster solution, albeit perhaps not the first solution in the Prolog sense. Thus, unlike in the nagging configuration used for the experiments reported here, nagging processes can lead to faster solutions both by succeeding (and providing a partial proof) or by failing (and pruning the master's search space). Similarly, there are cases in which intermediate information obtained during a nagging process' search may be used to eagerly prune search in the master process as well as in sibling nagging processes.

Although using a larger number of nagging processes can lead to greater speedup, it also imposes greater overhead on the master process. The master process must divide its attention between performing search and keeping up with requests from the nagging processes. As an alternative to a flat arrangement of nagging processes we have also investigated hierarchical topologies for nagging, or *meta-nagging*. In this scheme, each nagging process may itself serve as a master to other nagging processes. While a nagging process attempts to asynchronously prune the search of its master process, it may be assisted by other nagging processors that attempt to prune its search. Meta-nagging reduces the overhead on the top-level master process by not requiring it to service requests from all nagging processes directly. In addition, appropriately configured meta-nagging may be more effective in the presence of non-uniform message latencies.

Finally, we are investigating the interaction between parallelism and existing search speedup techniques. A large class of serial speedup techniques, including success and failure caching [15, 14] and explanation-based learning [13], provide a means for detecting and exploiting search order bias (the *adaptive* component of DALI). Solutions that are consistent with this learned bias are found more quickly, at the expense of longer search for solutions inconsistent with the bias. Of course, in order to be effective, these techniques all rely on the assumption that future queries will be consistent with the learned bias. Since a parallel nagging strategy offers substantial latitude in deciding how work should be distributed, it may be possible to select nagged subproblems that are consistent with a particular nagging processor's learned bias, thereby improving the effectiveness of these techniques.

We are actively pursuing all of these research directions. A different configuration of the DALI system incorporating more sophisticated nagged-problem selection techniques, meta-nagging, subgoal caching, and the exploitation of sucessfully-proven nagged subproblems has already been applied to a broad range of first-order logic problems using 100 workstations[15]. In time, we hope our efforts will produce an effective distributed inference mechanism ideally suited to solving large-scale practical problems.

References

1. H. Aït-Kaci, *Warren's Abstract Machine*, The MIT Press, Cambridge, Massachusetts, 1991.
2. O.L. Astrachan and D.W. Loveland, "METEORs: High Performance Theorem Provers using Model Elimination," in *Automated Reasoning, Essays in Honor of Woody Bledsoe*, Boyer, R. S.(ed.),Kluwer Academic Publishers, Boston, 1991, pp. 31-59.
3. M. Avvenuti, P. Corsini and G. Frosini, "A Distributed Interpreter for Inherent AND/OR Parallelism," in *Implementations of Distributed Prolog*, Kacsuk, P., and Wise, M.J.(eds.), John Wiley & Sons, Chichester, England, 1992, pp. 65-87.
4. W.F. Clocksin, "Principles of the DelPhi Parallel Inference Machine," *Computer Journal*, **32**:1 , (1989), pp. 386-392.
5. J.S. Conery, *Parallel Execution of Logic Programs*, Kluwer Academic Publishers, Boston, 1987.
6. P. Corsini, G. Frosini and G. Speranza, "The Parallel Interpretation of Logic Programs in Distributed Architectures," *Computer Journal*, **32**:1 , (1990), pp. 29-35.
7. D. DeGroot, "Restricted AND-Parallelism," *Proceedings of the International Conference on Fifth Generation Computer Systems*, November, 1984, pp. 471-478.
8. S.A. Delgado-Rannauro, "OR-Parallel Logic Computational Models," in *Implementations of Distributed Prolog*, Kacsuk, P., and Wise, M.J.(eds.), John Wiley & Sons, Chichester, England, 1992, pp. 3-26.
9. C. Elkan, "Conspiracy Numbers and Caching for Searching And/Or Trees and Theorem-Proving," *Proceedings of the Eleventh International Joint Conference on Artificial Intelligence*, Detroit, MI, August 1989, pp. 341-346.
10. V. Kumar and Y-J. Lin, "An Intelligent Backtracking Scheme for Prolog," *Proceedings of the IEEE Symposium on Logic Programming*, August 1987, pp. 406-414.
11. F. Kurfeß, "Potentiality of Parallelism in Logic," *Parallelization in Inference Systems*, Dagstuhl Castle, Germany, December 1990, pp. 3-25.
12. J. Schumann and R. Letz, "PARTHEO: A High-Performance Parallel Theorem Prover," *10th International Conference on Automated Deduction*, Kaiserslautern, FRG, July, 1990, pp. 40-56.
13. A.M. Segre and C. Elkan, "A High Performance Explanation-Based Learning Algorithm," *Artificial Intelligence*, (To appear, 1994).
14. A.M. Segre and D. Scharstein, "Bounded-Overhead Caching for Definite-Clause Theorem Proving," *Journal of Automated Reasoning*, 11, (1993), pp. 83-113.
15. A.M. Segre and D.B. Sturgill, "Using Hundreds of Workstations to Solve First-Order Logic Problems," *Proceedings of the Twelfth National Conference on Artificial Intelligence*, To appear, July 1994.

16. M.E. Stickel, "A Prolog Technology Theorem Prover: Implementation by an Extended Prolog Compiler," *Journal of Automated Reasoning*, 4, (1988), pp. 353-380.
17. E. Tick, *Parallel Logic Programming*, The MIT Press, Cambridge, Massachusetts, 1991.
18. D.H.D. Warren, *An Abstract Prolog Instruction Set*, Technical Note 309, SRI International, Menlo Park, California, October 1983.

Proving with BDDs and Control of Information

Jean Goubault

Bull Corporate Research Center, rue Jean Jaurès, Les Clayes sous Bois, France
Email: Jean.Goubault@frcl.bull.fr Tel: (33 1) 30 80 69 28

Abstract. We present a new automated proof method for first-order classical logic, aimed at limiting the combinatorial explosion of the search. It is non-clausal, based on BDDs (binary decision diagrams) and on new strategies that control the size and traversal of the search space by controlling the amount of information, in Shannon's sense, gained at each step of the proof. Our prover does not search blindly for a proof, but thinks a lot to decide of a course of action. Practical results show that this pays off.

1 Introduction

We present a complete refutation method for first-order classical logic that aims at controlling the growth and at guiding the traversal of the search space intelligently. Our starting point is [10], which proves that finding whether a given proposition in this logic is obvious, for several different reasonable definitions of non-obviousness, is Σ_2^p-complete. This not only means that proving is hard, but also that any complete proof method is built on, or hides, two nested levels of combinatorial search: typically, an outer level searching for a refutation, and an inner level that does propositional unsatisfiability checks. These two levels are incompressible, i.e. we need good strategies for both levels to get a good prover. We propose to use BDDs (binary decision diagrams), a very efficient tool for propositional logic, as the propositional substrate; we then control the search for a refutation by information-theoretic tools, in the sense of Shannon [18].

The plan of the paper is as follows: we introduce BDDs in Section 2. Then, representing formulas as BDDs, we develop in Section 3 a proof method for first-order logic; Section 4 defines strategies to guide this method, by maximizing the gain of information during the proof search process. Section 5 illustrates it on a small example, and reports on the behavior of our prover on classical test problems. Section 6 is the conclusion.

2 BDDs

Recall that the language of first-order logic consists in formulas built on atomic formulas with $\wedge$, $\vee$, $\neg$, and quantifications $\forall$ and $\exists$. Atomic formulas, in turn, are applications of predicate symbols to lists of terms, and terms are either variables or applications of function symbols to lists of terms. We shall deal only with skolemized formulas $\forall x_1, \ldots, x_n \cdot \Phi$, where Φ is quantifier-free.

Let T and F be two constants outside the language of first-order logic, T representing truth, and F representing falsehood. Shannon's decomposition principle states that for every quantifier-free formula Φ, and every atom A:

$$\Phi = (A \Rightarrow \Phi[\mathrm{T}/A]) \wedge (\neg A \Rightarrow \Phi[\mathrm{F}/A]) \tag{1}$$

$\Phi[\Phi'/A]$ being the result of replacing A by Φ' in Φ. We write this as $\Phi = A \to \Phi[\mathrm{T}/A]; \Phi[\mathrm{F}/A]$.

Note that A does not appear in either $\Phi[\mathrm{T}/A]$ or $\Phi[\mathrm{F}/A]$. Repetitively choosing an atom A appearing in Φ, decomposing it w.r.t. A, and continuing the process on $\Phi[\mathrm{T}/A]$ and $\Phi[\mathrm{F}/A]$ until no formula can be decomposed further builds a binary tree whose nodes are labeled by atoms and whose leaves are either T or F. Such a tree is called a *Shannon tree* for the original formula.

BDDs are reduced and shared versions of Shannon trees, what Bryant [4] calls ROBDDs. *Sharing* is an implementation concern, and means that any two isomorphic subtrees of the Shannon tree are stored at the same address in memory, making it a DAG (directed acyclic graph) and saving memory. *Reducing* means converting every subDAG of the form $A \to \Phi; \Phi$ into Φ. This is a logical consequence of the law of the excluded middle.

If we decompose a formula w.r.t. an enumeration of the atoms appearing in it in a given total order $<$, then BDDs are compact canonical forms for quantifier-free formulas. Moreover, the usual propositional logical operations are easily defined by recursive descent of the BDDs to combine. Negation is defined by $\neg \mathrm{T} = \mathrm{F}$, $\neg \mathrm{F} = \mathrm{T}$, $\neg(A \to \Phi_+; \Phi_-) = A \to \neg\Phi_+; \neg\Phi_-$, and all other boolean operations $\oplus$ ($\vee, \wedge, \Rightarrow, \Leftrightarrow$, etc.) are defined by $(A \to \Phi_+^1; \Phi_-^1) \oplus (A \to \Phi_+^2; \Phi_-^2) = A \to (\Phi_+^1 \oplus \Phi_+^2); (\Phi_-^1 \oplus \Phi_-^2)$, with appropriate definitions on T and F (this is called *orthogonality* of $\oplus$). By remembering in a cache the results of previous computations, negation can be computed in linear time, and binary operations can be computed in quadratic time w.r.t. the sizes of the input BDDs.

Other operations on BDDs are equally easily definable. Replacing an atom A by Φ' in Φ is defined by $\mathrm{T}[\Phi'/A] = \mathrm{T}$, $\mathrm{F}[\Phi'/A] = \mathrm{F}$, and if $\Phi = A' \to \Phi_+; \Phi_-$, then $\Phi[\Phi'/A]$ is $(\Phi' \Rightarrow \Phi_+) \wedge (\neg\Phi' \Rightarrow \Phi_-)$ if $A = A'$, otherwise $(A' \Rightarrow \Phi_+[\Phi'/A]) \wedge (\neg A' \Rightarrow \Phi_-[\Phi'/A])$. This can be simplified in a number of particular cases (when Φ' is T, F, or an atom, notably). If A is an atom, we let $\vec{\exists} A \cdot \Phi$ be $\Phi[\mathrm{T}/A] \vee \Phi[\mathrm{F}/A]$ and $\vec{\forall} A \cdot \Phi$ be $\Phi[\mathrm{T}/A] \wedge \Phi[\mathrm{F}/A]$. We also use the notion of *reduction* of a BDD w.r.t. an equivalence relation E between atoms: Φ/E is Φ where every atom A is replaced by the least (w.r.t. $<$) A' equivalent to A modulo E.

As BDDs are canonical forms for propositional formulas, building a BDD from a propositional formula is NP-hard: it should not be surprising that BDDs can grow exponentially large w.r.t. the boolean formula they represent. The simplicity of the structure, together with the fact that their explosion in size rarely happens in practical applications until now have made BDDs a structure of choice notably in hardware verification [4, 3], in propositional logic and in first-order logic [6, 17]. From now on, we consider BDDs to be reduced ordered BDDs.

3 Proof Search

A skolemized formula $F = \forall x_1, \ldots, x_n \cdot M$, with M quantifier-free, is unsatisfiable iff for some $k \in \mathbb{N}$ and some substitutions σ_i, $1 \leq i \leq k$, $M\sigma_1 \wedge \ldots \wedge M\sigma_k$ is propositionally unsatisfiable. Alternatively, let ρ_i be renamings mapping each variable x to a new variable x_i, such that $x_i = y_j$ iff $x = y$ and $i = j$; call $M\rho_i$ the ith *copy* of M. Let M^k be $M\rho_1 \wedge \ldots \wedge M\rho_k$, then F is unsatisfiable iff for some k and σ, the BDD for $M^k \sigma$ is reduced to F. We first concentrate on finding σ for k fixed (Section 3.1), then constrain the search for k (Section 3.2).

3.1 Quasi-linear Strategy with Set of Support

Let Φ be a BDD (typically representing M^k). Call F-path a path from the root of Φ to an F leaf; this is a sequence of *literals*, a literal being an atom A and a sign, $-$ (resp. $+$) if the path chooses the branch on which A is false (resp. true). Viewing T-paths as conjunctions of literals, Φ is a disjunction of T-paths: this is a compact representation of a set of vertical paths, as used in the connection, or the matings method [2, 1]. Dually, considering F-paths as disjunctions of literals (with the signs reversed), Φ is a conjunction of F-paths, we get a compact representation of a set of clauses. We shall prefer the latter point of view, for reasons we explain in Section 4.1.

Now, Φ represents a set of clauses for M^k through its F-paths. In general, M has the form $Ax \wedge \neg T$, where Ax is a conjunction of axioms, assumed consistent, and we wish to take this information into account. So, as in the set of support strategy in resolution [19], we identify a subconjunction Φ'' of the F-paths of Φ to represent Ax^k. This determines the conjunction Φ' of the remaining F-paths in Φ, which we call the *witness* and plays the rôle of the set of support. This is expressed in purely logical terms by:

$$\Phi = \Phi' \wedge \Phi'', \quad \Phi' \vee \Phi'' = \mathrm{T}, \quad \Phi'' \sigma \neq \mathrm{F} \qquad (2)$$

for any substitution σ. (2) entails $\Phi'' = \Phi \vee \neg \Phi'$ and $\Phi' = \Phi \vee \neg \Phi''$.

Now, we wish to find σ such that $\Phi\sigma = \mathrm{F}$. We would also like to reason on Φ' instead of on the larger Φ; to do this, we first need the following:

Lemma 1. *If Φ, Φ' and Φ'' verify conditions (2):*

1. *If $\Phi' = \mathrm{F}$, then Φ is refuted;*
2. *If $\Phi' = \mathrm{T}$, then Φ is irrefutable (i.e, there is no σ such that $\Phi\sigma = \mathrm{F}$);*
3. *If A is any atom, let $\Phi^0 = \dot{\exists}A \cdot \Phi$, $\Phi'^0 = \Phi^0 \vee \dot{\forall}A \cdot \Phi'$, $\Phi''^0 = \dot{\exists}A \cdot \Phi''$. Then, if for some σ, $\Phi\sigma$ is F, but $A\sigma \neq A'\sigma$ for all atoms A' in Φ, Φ', Φ'' other than A, then $\Phi^0 \sigma$ is F. Besides, Φ^0, Φ'^0 and Φ''^0 verify conditions (2).*

Proof: Assume that Φ, Φ' and Φ'' verify conditions (2). If $\Phi' = \mathrm{F}$, then $\Phi = \Phi' \wedge \Phi'' = \mathrm{F}$, proving item 1. If $\Phi' = \mathrm{T}$, then $\Phi'' = \Phi \vee \neg \Phi' = \Phi$; but then $\Phi\sigma = \mathrm{F}$ iff $\Phi''\sigma = \mathrm{F}$, which was assumed impossible; this proves item 2. If $\Phi\sigma = \mathrm{F}$ and A is an atom such that $A\sigma \neq A'\sigma$ for all atoms A' in Φ, Φ', Φ'' other than A, notice that for all $\varphi \in \{\Phi, \Phi', \Phi''\}$, and $\tau \in \{T, F\}$, $\varphi[\tau/A]\sigma = \varphi\sigma[\tau/A\sigma]$, so that $(\dot\exists A \cdot \varphi)\sigma = \dot\exists A\sigma \cdot \varphi\sigma$. In particular, $\Phi^0\sigma = \dot\exists A\sigma \cdot \Phi\sigma = \dot\exists A\sigma \cdot F = F$.

Now check conditions (2) for Φ^0, Φ'^0 and Φ''^0. First, $\Phi'^0 = \Phi^0 \vee \neg\Phi''^0$ since the latter is $\Phi^0 \vee \neg\dot\exists A \cdot (\Phi \vee \neg\Phi') = \Phi^0 \vee \neg((\dot\exists A \cdot \Phi) \vee \neg(\dot\forall A \cdot \Phi')) = \Phi^0 \vee \neg(\Phi^0 \vee \neg(\dot\forall A \cdot \Phi')) = (\Phi^0 \vee \neg\Phi^0) \wedge (\Phi^0 \vee \dot\forall A \cdot \Phi') = \Phi'^0$. So $\Phi'^0 \vee \Phi''^0 = T$, and $\Phi'^0 \wedge \Phi''^0 = \Phi^0 \wedge \Phi''^0 = \dot\exists A \cdot (\Phi \wedge \Phi'') = \dot\exists A \cdot \Phi = \Phi^0$. Finally, if $\Phi''^0\sigma = F$, then $\dot\exists A\sigma \cdot \Phi''\sigma = F$, so $\Phi''\sigma = F$, contradicting (2). □

W.r.t. the search for a σ such that $\Phi\sigma = F$, the only atoms A' of interest in item 3 are those occurring in Φ, Φ' or Φ''. But if A' occurred in Φ' and not in Φ, the only way to instantiate Φ to F would be to instantiate Φ^0 to F, since $\Phi = \Phi^0$. By first replacing Φ' by Φ'^0 and Φ'' by Φ''^0, we can get rid of this case; from now on, we won't mention this normalization explicitly. Then, as $\Phi'' = \Phi \vee \neg\Phi'$, all atoms occurring in Φ'' will also occur in Φ. This entails that the only atoms A' of interest in item 3 are those occurring in Φ alone.

Our proof procedure operates by generating new values for Φ and its witness Φ' in a non-deterministic way, so that conditions (2) are maintained. We also carry along a finite set Σ of substitutions such that no instance of any $\sigma \in \Sigma$ is allowed to be used for refuting Φ. Let q be a set of triples (Φ, Φ', Σ), initialized to $\{(\Phi, \Phi', \emptyset)\}$, where Φ is the initial M^k and Φ' its associated witness: if M is of the form $Ax \wedge \neg T$, where Ax is a set of axioms, assumed satisfiable, and T is the goal to prove, we let Φ' be $(\neg Ax \vee \neg T)^k$, so that Φ'' is implicitly initialized to Ax^k. The procedure does:

1. If $q = \emptyset$, fail (no refutation for the initial formula).
2. Extract some (Φ, Φ', Σ) from q, and remove it from q.
3. If $\Phi' = F$, then Φ is refuted; return on success.
4. If $\Phi' = T$, go back to step 1.
5. Otherwise, choose an atom A in Φ' (hence in Φ). Compute $\Phi^0 = \dot\exists A \cdot \Phi$, $\Phi'^0 = \Phi^0 \vee \dot\forall A \cdot \Phi'$.
6. Compute the set $\Delta\Sigma$ of mgus (*most general unifiers*) unifying A with a complementary atom in Φ (i.e. an atom having a sign opposite to A on some T-path of Φ), except those that are instances of substitutions in Σ.
7. Add to q the triple $(\Phi^0, \Phi'^0, \Sigma \cup \Delta\Sigma)$, and all the $(\Phi\sigma, \Phi'\sigma, \Sigma)$, for $\sigma \in \Delta\Sigma$. Go back to step 1.

(Notice how this encodes a non-deterministic search for σ, with the help of BDDs to check for propositional unsatisfiability as an oracle, thus mirroring the Σ_2^p structure of the search [10]. As the problem is complete for Σ_2^p, we cannot escape such a two-tiered structure, provided $P \neq NP$ and $NP \neq \Sigma_2^p$.)

Items 5 and 6 look for all possible ways of "eliminating A from Φ'", for all A's: if Φ is to be refuted by σ, Lemma 1 says that $\Phi'\sigma$ must reduce to F. So A must disappear from Φ', either by being rewritten to some atom A' with which it is unified (then σ must be an instance of some element of $\Delta\Sigma$), or by generating Φ^0, Φ'^0, assuming σ won't unify A to any A' (use Lemma 1, item 3).

The purpose of Σ is to prevent redundant searches, i.e. testing the same σ on both Φ and Φ^0 after step 7; indeed, $\Phi^0\sigma$ might be produced in later steps either by instantiating Φ^0 by σ, or by constructing $(\Phi\sigma)^0$: Σ forbids the first production. This is a first use of *instance subtraction*, a technique invented and developed by J.-P. Billon in the framework of resolution; here, this has the effect of a subsumption test on substitutions.

Notice that if Φ' consists of only one clause (F-path), this is a linear strategy (see [12] for the method, and definitions of phantom literals $[A]$, top clauses, etc.) guided by top clause Φ', A (resp. $\neg A$) being the literal resolved upon. If A is chosen as high as possible in the BDD, this is similar to ordered linear resolution. Adding $(\Phi^0, \Phi'^0, \Sigma \cup \Delta\Sigma)$ means deleting a phantom literal (A or $\neg A$). The $(\Phi\sigma, \Phi'\sigma, \Sigma)$'s, on the other hand, correspond to all resolutions between the top clause and the others on literal A (or $\neg A$). But these resolvents still contain A in the form of $A\sigma$. To prevent redundant unifications with $A\sigma$, the linear strategy makes A a phantom $[A]$. We eliminate these redundancies by using Σ.

Contrarily to the linear strategy, Φ' is a set of clauses (F-paths) rather than a unique clause. This allows us to deal with all possible top clauses in parallel, instead of inducing some backtracking if the chosen top clause was not the right one. All treatments in common to all possible top clauses are indeed factorized by the way we combine F-paths when we build Φ^0 and Φ'^0.

We call this procedure *quasi-linear refutation with set of support*. It is complete, as soon as we allow for an arbitrary number k of copies (see Section 3.2). The choices of A, and of the triple to extract from q in step 2 are arbitrary: we show in Section 4 how to get (hopefully) the most favorable choices.

To implement this, we maintain and use a *connection graph* that links all unifiable atoms in Φ, where links are labeled with their mgu. (This is an abuse of language, as usual connection graphs [11] do not link atoms, but literals.) We code this graph by two maps (finite-domain set-theoretic functions):

- `links` maps atoms to sets of mgus: for each atom A in Φ, `links`(A) is the set of mgus σ between A and some other atom A' in Φ, with $\sigma \notin \Sigma$;
- `vertices` maps mgus to equivalence relations among atoms. For each $\sigma \in$ `links`(A), with A in Φ, `vertices`(σ) is the relation $\cong_\sigma$ defined as $A \cong_\sigma A'$ iff $A\sigma = A'\sigma$; we encode $\cong_\sigma$ as a map from atoms with non-trivial equivalence classes (i.e, of cardinal ≥ 2) to their equivalence classes.

The splitting of the graph in two maps is justified by the fact that the objects of interest in our procedure are not the atoms, but the substitutions. So, if σ is an mgu, and $E = $ `vertices`(σ), then $\Phi\sigma = (\Phi/E)\sigma$, and $\Phi\sigma = $ F iff $\Phi/E = $ F. This allows us to never actually *instantiate* Φ (or Φ') by σ, by merely *reducing* Φ (and Φ') by E. We need to carry σ around by converting our triples into 4-tuples $(\Phi, \Phi', \Sigma, \sigma)$; we call σ the *context*. We also need to instantiate links by σ, i.e. to unify their labels with σ, cutting the link if this failed; we use `links` to maintain the coherence of the connection graph efficiently.

Maps and sets are best implemented by shared splay-trees, as in the HimML set-based language [5]. Binary operations on them (overwriting, union, intersection, ...) are then doable in linear time in their cardinals on average; adding, removing elements from the domain of maps or from sets, and applying a map to an element is logarithmic on average; and these times do not depend on the structure of elements, in particular not on the level of nesting of sets.

The set Σ is harder to implement efficiently. What we really need is to represent the possibly infinite set of all *instances* of substitutions in Σ. To do this, we first normalize substitutions σ w.r.t. a total order $\prec$ on terms such that $x \prec t$ for every variable x and non-variable term t: a substitution can be written as a set of multiequations $x_1^j = x_2^j = \ldots = x_{n_j}^j = t^j$, $1 \leq j \leq m$, having disjoint sets of x_i^j's and t^j, and where t^j is the greatest term for $\prec$ in the jth multiequation; then, a normal form for σ is the list of all couples (x_i^j, t^j), $1 \leq j \leq m$, $1 \leq i \leq n_j$, sorted by increasing x_i^j's. A modification of Martelli and Montanari's algorithm [14] that works in time $O(n \log n)$ to do this is given in [6]. Then, we factorize all these normal forms by sharing their common prefixes (with some interleaving) into a non-deterministic finite automaton-like structure representing Σ, where states are variables, and transitions are empty or are labeled with terms to match (instead of letters to compare to for equality); details can be found in [6].

Finally, it is never necessary to consider twice the same unifier in step 7. Indeed, replacing Φ by Φ^0 or the $\Phi\sigma$'s in step 7 partitions the set of possibly refuting substitutions between those which unify A with no other atom, and those which unify A with A', for each other A'. So, we additionally maintain a global variable *iset*, initially $\emptyset$; in step 7, we replace *iset* by *iset* $\cup \Delta\Sigma$; then we exclude all unifiers in *iset* from consideration in step 6. This is another example of Billon's instance subtraction principle. Since substitutions are in normal form, HimML sets can be used to represent *iset*, with $O(\log \operatorname{card} \mathit{iset})$ average access time; alternatively, since our normal forms are words on an alphabet of bindings, we could represent *iset* as a deterministic finite state automaton.

3.2 Amplifications

In general, we have to refute a conjunction of skolemized first-order formulas $\forall x_1, \ldots, x_{n_i} \cdot N_i$, $1 \leq i \leq m$. (axioms, assumed consistent, and negated goals.) To do this, we have to copy each N_i a certain number u_i of times (what we call the *logical multiplicity* of N_i), and find a refuting substitution for their conjunction. As this is inefficient, we initialize u_i to 0, and increase it as we need to; we call this *amplification*, in the spirit of [1].

First, we need to build copies of the N_i's. Assume the free variables in N_i are x_1, x_2, ... (for simplicity, w.l.o.g. we assume they are the same for all N_i.) Then, we represent the kth copy $N_{i,k}$ of N_i as N_i where each x_j is replaced by a new variable $x_j^{i,k}$, uniquely determined by j, i and k. Since we never actually instantiate any BDD, much less any atom, for each atom A, every variable in A has the same i and k values. So, instead of actually performing the replacement, we represent the kth copy of A coming from N_i by the triple (A, i, k); we let $i = k = 0$ by convention if A has no variables.

Conceptually, imagine that N_i has been copied an infinite number of times and apply the procedure of the last section. Fortunately, we only need a finite number, u_i^{max}, of these copies to refute the conjunction of the N_i's, and because conjunction is commutative, we can impose these to be copies 1 through u_i^{max}. This can be used to determine the number of copies we need to actually perform in memory; we call this number c_i the *physical number of copies*. Both u_i and c_i evolve through the proof search process. In our proof procedure, Φ is always of the form $(\exists A_1, \ldots, A_p \cdot \Phi_0)\sigma$, where Φ_0 is the conjunction of all $N_{i,k}$'s, the A_i's are atoms and σ is the context. We define u_i as the highest k such that $x_j^{i,k}$ appears in σ (in its domain, or free in a term of its range) or such that one of the A_l's is a kth copy of an atom. ($u_i = 0$ when $p = 0$, $\sigma = []$.) That is, u_i is the highest copy of N_i that we cannot swap freely for other copies, because it has been fixed by some binding in the context, or by some elimination of an atom A while transforming Φ into $\Phi^0 = \exists A \cdot \Phi$. Then:

Lemma 2. *The only atoms (A, i, k) and (A', i', k') we need to unify are when:*

- $i = 0$, $i' \neq 0$ *and* $k' \leq u_{i'} + 1$;
- *or symmetrically,* $i \neq 0$, $i' = 0$ *and* $k \leq u_i + 1$;
- *or* $i \neq 0$, $i = i'$, *and either* $u_i = 0$, $k = 1$, $k' \leq 2$, *or* $u_i \neq 0$, $k \leq u_i$, $k' \leq u_i + 1$;
- *or* $i \neq 0$, $i' \neq 0$, $i \neq i'$, *and:*
  - $u_i = 0$, $u_{i'} = 0$, $k = 1$, *and* $k' = 1$;
  - *or* $u_i = 0$, $u_{i'} \neq 0$, $k = 1$, *and* $k' \leq u_{i'} + 1$;
  - *or symmetrically* $u_i \neq 0$, $u_{i'} = 0$, $k \leq u_i + 1$ *and* $k' = 1$;
  - *or* $u_i \neq 0$, $u_{i'} \neq 0$, $k \leq u_i + 1$, $k' \leq u_{i'} + 1$ *and either* $k \neq u_i + 1$ *or* $k' \neq u_{i'} + 1$.

Proof: Note that Φ and its witness Φ' are invariant by permuting the mth and m'th copies of N_i, as soon as $m \geq u_i$ and $m' \geq u_i$, for every i. Also, note that $k = 0$ iff $i = 0$ iff A is closed.

If $i = 0$, the only atoms other than A and unifiable with A are not closed, so $i' \neq 0$. If $i = 0$, then $i' \neq 0$, and by permuting copies if necessary, we can impose $k' \leq u_{i'} + 1$. If $i' = 0$, $i \neq 0$, symmetrically we impose $k \leq u_i + 1$. So now assume that $i \neq 0$ and $i' \neq 0$.

If $i = i'$, we cannot permute copies of N_i and $N_{i'}$ independently. But if $u_i = 0$, by permuting k with 1, we can impose $k = 1$; and if $k' = k$, this imposes $k' = 1$, otherwise we can impose $k' = 2$. If $u_i \neq 0$, we can first impose $k \leq u_i$ or $k' \leq u_i$. Indeed, assume that Φ_0 is not refutable by any substitution σ unifying an atom with copy number $k \leq u_i$ and another with copy number $k' > u_i$. Consider the equivalence relation E on atoms induced by σ, and look at it as a partition; it is the union of two disjoint partitions E_1 et E_2, E_1 applying only on atoms with copy numbers less than or equal to u_i, E_2 on atoms with copy numbers greater than u_i. Writing Φ_0 as the conjunction of Φ_1 (copies $\leq u_i$) and of Φ_2 (copies $> u_i$), $\Phi_0/(E_1 \cup E_2) = $ F iff $(\Phi_1/E_1) \wedge (\Phi_2/E_2) = $ F. But as Φ_1 and Φ_2 have disjoint sets of atoms, either $\Phi/E_1 = $ F or $\Phi/E_2 = $ F; as card $E_1 < $ card E and card $E_2 < $ card E, in both cases, E cannot be of minimal cardinal: contradiction. Now, $k \leq u_i$ or $k' \leq u_i$, but as $i = i'$, we can assume w.l.o.g. that $k \leq u_i$, swapping the roles of A and A' if needed. Then, if $k' > u_i + 1$, we can permute k' with $u_i + 1$ to impose $k' \leq u_i + 1$.

If $i \neq i'$, then copies of N_i and $N_{i'}$ can be permuted independently. If $u_i = 0$ (resp. $u_{i'} = 0$), we can impose $k = 1$; and if $u_i \neq 0$ (resp. $u_{i'} \neq 0$), we can impose $k \leq u_i + 1$ (resp. $k' \leq u_{i'} + 1$) by permuting copies if needed. Moreover, the case $u_i \neq 0$, $u_{i'} \neq 0$, $k = u_i + 1$ and $k' = u_{i'} + 1$ is useless, by a similar argument on equivalence relations as above. □

This provides a bound on the copy number k of each atom that we need to consider in step 6 of our proof procedure. In practice, it allows us to let the physical number of copies c_i be $\max(u_i, 1) + 1$. In particular, all N_i's will be copied twice initially ($u_i = 0$, hence $c_i = 2$).

To implement this, we maintain counters u_i and c_i, and distinguish two kinds of amplification: *physical amplification* adds N_{i,c_i} conjunctively to Φ and increments c_i, while *logical amplification* increments u_i and then calls the physical amplification subroutine until $c_i = \max(u_i, 1) + 1$. To maintain the witness Φ', either N_i is the negated goal, and we add N_{i,c_i} conjunctively to Φ'; or N_i is one of the axioms, so to maintain the irrefutability condition of Φ'' (last condition in 2), we replace Φ' by $\Phi' \wedge (N_{i,c_i} \vee \neg \Phi)$ (Φ being the unamplified Φ; this means we leave Φ'' unchanged). We do this on each triple generated at step 7.

Besides, conjunction is also idempotent, so we also forbid unifiers that would make two copies of the same N_i identical. (for simplicity, we limit ourselves to textual identity, as tested by a comparison of the terms bound to their respective free variables.) Indeed, w.l.o.g. we can restrict the search to refuting substitutions that leave different copies different. This is the last example of Billon's instance subtraction principle.

4 Control of Information

It remains to decide of strategies, first for choosing A in step 5, second for drawing a triple from the queue q in step 2. q is just a way to account for the non-determinism in the search for a proof: in general, we get a *search tree*, usually infinite, whose nodes are the triples (Φ, Φ', Σ); their sons are the $(\Phi^0, \Phi'^0, \Sigma \cup \Delta \Sigma)$ and the $(\Phi \sigma, \Phi' \sigma, \Sigma)$ as defined in step 7. The first strategy will determine the shape of the search tree, and the second the way we explore it.

4.1 Controlling the Size of the Search Space

One of the natural measures of non-obviousness of a proposition identified in [10] is the *step count*, i.e. the least number of non-trivial unifications we need to effect to build a refuting substitution for the initial formula M. This is the least depth level in the search tree where a refuting substitution can be found.

Our idea is the following: for a given depth level (non-obviousness level), we get there faster if the search tree is narrower (the narrower, the smaller). To keep it narrow, we minimize the branching factor at each choice point. For each A we choose in step 5, if the number of unifiers of A with another atom A' is p, the search tree branches $p + 1$-fold: p possible instantiations $\Phi \sigma$, and one atom elimination possibility Φ^0. We therefore minimize p, by drawing A such that card links(A) is minimal. When there are several atoms minimizing this, we choose the highest in the BDD; this usually minimizes the size of Φ^0.

Actually, we only count unifiers between A and some *complementary* atom. This complicates the minimization procedure, but it remains polynomial-time computable ($O(m^2(n+m))$) with m atoms and n nodes in Φ [6]): we attach to each node a set pos (resp. neg) of atoms occurring positively (resp. negatively) on T-paths underneath; it is given by pos(T) =pos(F) = $\emptyset$ (resp. neg(T) =neg(F) = $\emptyset$), and pos($A' \to \Phi_+;\Phi_-$) =pos(Φ_+)$\cup$pos(Φ_-) $\cup$ $f(\Phi_+, A')$ (resp. neg($A' \to \Phi_+;\Phi_-$) =neg(Φ_+)$\cup$neg(Φ_-) $\cup$ $f(\Phi_-, A')$), where $f(\Phi, A')$ is $\emptyset$ if $\Phi = $ F and $\{A'\}$ otherwise. Using bit vectors for sets of atoms, we compute the set pos(Φ[F/A])$\cup$neg(Φ[T/A]) of atoms complementary to A in time $O(mn)$, filter out links(A) by it, and repeat for all m atoms.

The approach of Section 3.1 is justified by the fact that we can minimize the branching factor $p+1$ in polynomial time. A more natural approach, more akin to the connection method, would strive to eliminate T-paths instead [8]; we found no polynomial-time minimizing procedure in this case, though.

An interesting case is $p = 0$. Then there is no choice: we *must* replace (Φ, Φ', Σ) by $(\Phi^0, \Phi'^0, \Sigma)$. We call an atom A for which $p = 0$ a *pure atom*, and this replacement *elimination* of A. To speed up the procedure, we detect this case just after extracting the triple from q in step 2, eliminating all its pure atoms right away. Pure atoms subsume the notion of pure literal: A (resp. $\neg A$) is a pure literal when Φ is the conjunction of some BDD ϕ and of $A \Rightarrow \psi$ (resp. $\neg A \Rightarrow \psi$), with A not occurring in ϕ or ψ; then refuting Φ reduces to refuting ϕ. But pure atoms offer more: in propositional logic for instance, all atoms are pure, so any BDD reduces to T or F right away, whereas eliminating pure literals cannot reduce further, say, $A \to (B \to $ F; T); $(B \to $ T; F).

The significance of the branching factor has first been noticed in [8]: $\log(p+1)$ is the average *loss of information*, in the sense of Shannon [18] (a change of *entropy*), incurred by the choice between one among $p+1$ ways of refuting Φ.

4.2 Maximal Information Gain First Search

When we explore the search tree, going from one node $(\Phi_1, \Phi'_1, \Sigma_1)$ to one of its sons $(\Phi_2, \Phi'_2, \Sigma_2)$, we also gain some information. By quantifying this gain, we shall guide the search for a refutation in the search tree.

The information gained between Φ'_1 and Φ'_2 must be a difference between amounts of information stored in Φ'_1 and Φ'_2. Now, the amount of information stored in a BDD Φ' should be an entropy. It should be 0 when we can conclude: if Φ' is F (we have found a refutation), and if Φ' is T (we cannot find a refutation). Otherwise, it should be negative. Moreover, as we are interested in getting a BDD without any validating truth valuation, i.e. without T-paths, this entropy should measure the obstacle that the T-paths in Φ' represent. So, it is natural to define this information $I(\Phi')$ as the sum over all paths of the entropies associated with the probability that it is a T-path in Φ'. Let b_f (resp. b_v) be the number of F-paths (resp. T-paths) in Φ'. Then $I(\Phi') = \sum_{b \text{ path in } \Phi'} P(b \text{ T-path}) \log P(b \text{ T-path}) = \sum \frac{b_v}{b_f + b_v} \log \frac{b_v}{b_f + b_v} =$

$b_v \log \frac{b_v}{b_f+b_v}$. Thermodynamically speaking, paths would be independent particles in a gas, which can be in one of two states, T or F, and we compute the partial entropy of the T particles.

b_v and b_f can be computed on a node-per-node basis, so that $I(\Phi')$ is computed in time polynomial in the size of Φ'. Then, the change in information ΔI between Φ'_1 and its son Φ'_2 equals $I(\Phi'_2) - I(\Phi'_1)$, minus $\log(p+1)$. We define the *information gain* $G(\Phi')$ at node (Φ, Φ', Σ) as the sum of the ΔI's associated with the edges forming the path from the root $(\Phi_0, \Phi'_0, \emptyset)$ to (Φ, Φ', Σ) in the search tree.

The idea now, is to choose the triple (Φ, Φ', Σ) in step 2 of our proof procedure that maximizes $G(\Phi')$ among all triples in q, i.e. q is a priority queue sorted by $G(\Phi')$. We call this *maximal information gain first search*. Intuitively, we always examine first the triple that has progressed most towards a refutation, aiming at finding a refutation as soon as possible. Then:

Theorem 1 *Maximal information gain first search is complete.*

Proof: We prove that on any infinite branch of the search tree, $G(\Phi')$ is not bounded from below; then, if there is a refutation node, it is obtained with finite information gain, so it will eventually be preferred to all infinite branches.

Consider an infinite branch of the search tree. At depth d, $G(\Phi') = I(\Phi') - I_0 - \sum_{i=1}^{d} \log(p_i + 1)$, where I_0 is the initial stored information, and the $p_i + 1$'s are the successive branching factors. If $G(\Phi')$ was bounded from below, all but a finite number of p_i's would be zero, as $I(\Phi') \leq 0$. So, below some depth level, we could only eliminate pure atoms and perform amplifications. But if an atom is pure after amplifying, it is also before, so eliminating pure atoms (which the procedure does systematically) would reduce Φ' to T or to F right away. So the branch cannot be infinite. □

This theorem is weak, and does not tell anything about the speed of convergence of the strategy; it does not even take $I(\Phi')$ into account. However, the main reason in practice why $G(\Phi')$ decreases on infinite branches is due to the $I(\Phi')$ term. Notice that amplifications increase the number of F-paths and of T-paths (except when $\Phi' = F$), thus $I(\Phi')$ decreases since $\frac{\partial I(\Phi')}{\partial b_v} = \log \frac{b_v}{b_f+b_v} + 1 - \frac{b_v}{b_f+b_v} < 0$, and $\frac{\partial I(\Phi')}{\partial b_f} = -\frac{b_v}{b_f+b_v} < 0$ (unless $b_v = 0$, i.e. $\Phi' = F$). This allows the prover to limit the number of amplifications efficiently.

We shall see in Section 5 that this strategy is not only complete, but efficient, and usually leads straight to a refutation. Because of this, q never expands much, contrarily to breadth-first search. This strategy is also able to find a refutation rather deep in the search tree without exploring the full subtree above this depth level, avoiding the main handicap of iterative deepening depth-first search.

5 Experimental Results

We have implemented a prover based on the previous techniques in an interpreter [7] for the HimML language, an extension of Standard ML with fast sets and maps. Skolemization and ordering of the BDDs are decided by the smart method of [9]. As our proof procedure is rather intricate, we first describe a small example, then give experimental results based on standard benchmark problems.

5.1 Illustration

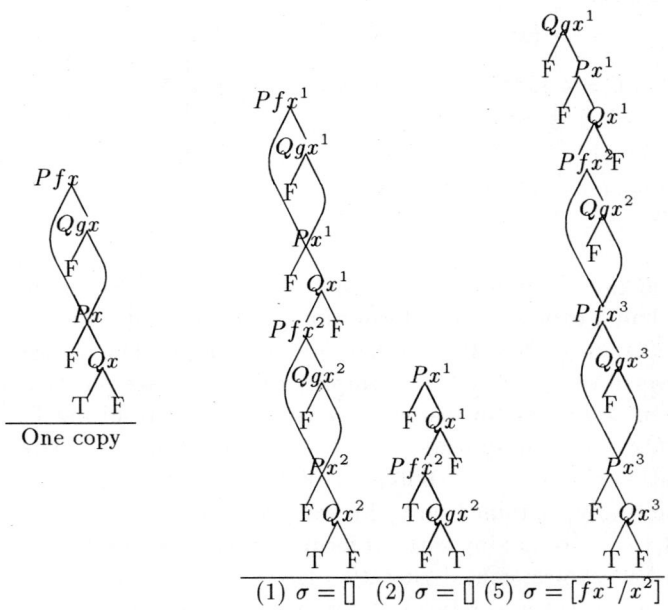

Fig. 1. BDDs for Problem 19

As an example, we choose Pelletier's problem 19 [15], which is simple, yet needs 3 copies: we have to refute $(Pfx \Rightarrow Qgx) \wedge Px \wedge \neg Qx$. (we drop parentheses since all symbols are unary.) Figure 1 (left) shows a BDD for it; we draw $A \rightarrow \Phi_+; \Phi_-$ by putting Φ_- as left son of A, and Φ_+ as its right son.

Figure 2 shows the corresponding search tree. At the root (node (1)), we have amplified the formula twice physically, yielding the BDD on Figure 1, column (1). Among the atoms there, the one that minimizes the branching factor $p+1$ is Pfx^1, with $p=1$ (unifying Pfx^1 with Px^2; Pfx^2 is not considered because it would identify copies 1 and 2). Transforming Φ into Φ^0 means going to node (2) in Figure 2; applying the substitution $[fx^1/x^2]$ leads to node (5).

At node (2), we eliminate the pure atoms Px^2 and Qx^2 (they are pure since their only mgus would make different copies equal), and get the BDD in Figure 1, column (2). Logical amplification bumps u_1 from 0 to 1, then no physical amplification is needed since $c_1 = 2$ already. The information stored in (2) is

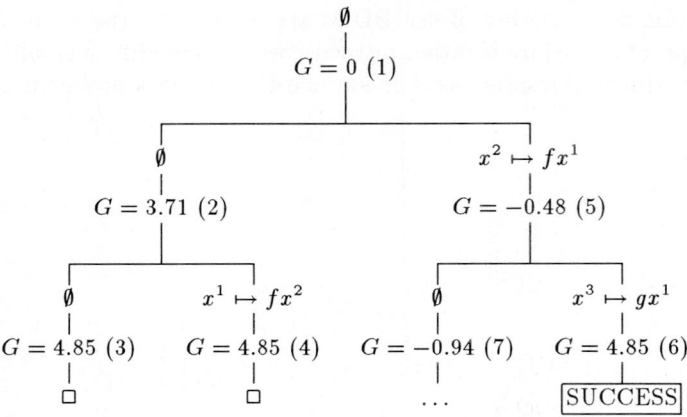

Fig. 2. The search tree for Problem 19

-1.83, vs. -6.23 at the root; so we gain 4.40 from (1) to (2), minus $\log 2 = 0.69$ because we had 2 choices. The information gain G at node (2) is thus 3.71.

At node (5), we get the BDD in Figure 1, column (5). The context is $[fx^1/x^2]$, so u_1 is bumped to 2, and we must amplify the BDD once physically, to change c_1 to 3. Then, Pfx^1 is eliminated by purity, and the resulting BDD does not depend on Qx_2, so it disappears, too. The information stored here is -6.02, so the information gain is $G = (-6.02) - (-6.23) - 0.69 = -0.48$.

Now, q initially contained only (1); we have now replaced it by (2) and (5), so $q = \{(2),(5)\}$. To maximize the information gain, we extract node (2) from it; this is a bad choice: considering Px^1 leads to nodes (3) and (4), which are both failure nodes (cannot refute). After examining and dismissing (3) and (4), which have the highest information gain (4.85), (5) is examined, leading to nodes (7) and (6) by considering Qgx^1 as minimizing atom. Then, (6) has the highest information gain, and is a success node (refutation found).

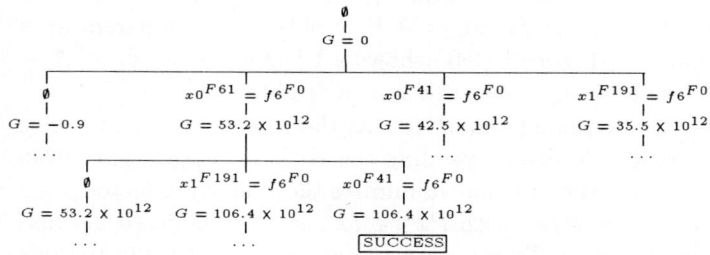

Fig. 3. The search tree for Problem 47

The strategy didn't guess the correct branch in the search tree; this enabled us to illustrate the way backtracking is handled through q. This is an exception, however. On most benchmark problems we tested, the prover went to the solution almost without backtracking. One of the most surprising problems is Schubert's Steamroller (#47), whose associated search tree is shown in Figure 3. (Only non-deterministic steps are shown on the search tree, since the purity test handles the others; this explains why the solution is found at only depth 2.) This problem is reputed unsolvable in backward chaining, even with set of support strategies. However, our prover does just that: it first guesses that the animal we look for is the wolf, and then correctly appeals to the most complicated axiom in the set (the one we would use last, normally) to justify it. Because it is a good guess, the proof is found without backtracking. (The ... on the figure means that the subtree rooted there was not explored.) This is no magic: these guesses are the ones that maximize the information gain when we look at atoms that minimize the branching factor. Nonetheless, because the BDDs are quite large, the information stored there is astronomical. This also explains why the proof took so long (see Figure 4).

5.2 Benchmarks

#	t	G	↔	s
18	0.0	0	−	1
19	3.0	9.0	3	13
20	3.5	65.	2	13
21	0.6	1.0	2.5	6
22	0.8	1.0	3	7
23	0.0	0	−	1
24	4.3	3.4×10^2	2	5
25	4.0	9.1×10^2	2	5
26	54.	4.5×10^5	3.5	15
27	2.1	1.7×10^3	2	5
28	2.0	31.	2	15
29	32.	4.0×10^3	3.5	8
30	1.1	59.	2	5
31	1.3	99.	2	5
32	2.2	2.2×10^2	2	7
33	0.0	0	−	1
34	4.3	36.	3	10
59	1.2	13.	4	9

#	t	G	↔	s
35	0.2	0.7	2	3
36	2.3	34.	2	5
37	79.	8.9×10^4	3	7
38	0.0	0	−	1
39	0.2	3.0	2	3
40	1.1	15.	2.5	6
41	1.3	11.	2.5	6
42	2.3	4.6	4.5	10
43	2.3	2.1×10^2	2	7
44	15.	9.0×10^2	1.8	45
45	11.	8.1×10^3	2	7
46	11.	1.4×10^3	2	7
47	1037.	6.3×10^{13}	3.5	8
50	1.2	2.4	2.4	13
55b	10.	1.8×10^4	2	5
57	1.1	4.1	3	4
59	1.2	13.0	4	9
60	2.5	34.	2	11
62	0.9	11.	2.3	8
66	3.1	12.	3	10
67	2.3	14.	3	7
68	2.5	14.	3	7
69	5.7	68.	3	7
86	12.	5.6×10^2	3.5	8

#	t	G	↔	s
48	37.	2.9×10^3	8	9
49	1.5×10^2	2.6×10^4	3	7
51	—			
52	—			
53	—			
54	—			
55	—			
56	5.3×10^2	1.1×10^6	7	15
58	3.0×10^2	6.2×10^3	9.5	20
61	3.4×10^2	3.4×10^3	9.5	20
63	1.1×10^2	1.3×10^4	10	11
64	1.1×10^2	1.3×10^4	10	11
65	1.3×10^2	3.3×10^3	10	21
73	—			

Fig. 4. Results of the prover

We ran the HimML interpreter [7] on a MIPS R3000 machine, which is about 20% to 30% faster than a SparcStation 2. We tested all 75 Pelletier's problems [15], except #74 and #75, which are descriptions of whole classes of problems, and #70, which is open. We added a version without equality of "Aunt Agatha and the Butler" (#55), due to Manthey and Bry [13], which we refer to as #55b, and Pelletier and Rudnicki's "non-obviousness problem" [16], which we name #86. The results are shown in Figure 4; all figures are given with a precision of 2 digits. All propositional problems were omitted; they were solved too fast to be measured, except #71 (Urquhart's U-problems, which are exponential in

n for resolution, were tested up to $n = 165$; computation time remained under
$1s.$, although garbage collection approached $40s.$) and #72 (pigeonhole formulas,
which run in $0.17 \times 1.7^n s.$ for n pigeons and $n-1$ holes).

The left-hand side array shows results on monadic problems, the middle
one on non-monadic problems, and the right-hand side one on problems with
equality. t is the time, in seconds; G is the information gain of the first found
refutation; $\leftrightarrow$ is the average branching factor during the search; and s is the size
of the part of the search tree the prover explored. Equality is dealt with naively,
by merely adding the usual axioms for the equality symbol.

Problems #18, #23, #33 and #38 are already reduced to F by the smart
skolemizer we use: once false dependencies have been removed, they can be
solved on the propositional level, using BDDs. Apart from them, our prover is
quite regular in its achievements: only 6 problems, all with equality, could not
be solved in less than 20 minutes. Then, only #47 (the Steamroller) and most
problems with equality take a long time — on the order of minutes. What is
striking is the low branching factor, and the small size of the search tree we
explore, in all cases. Pelletier's problems are simple, but it may be hard to make
a prover understand why; ours manages to detect this simplicity.

In particular, #66, #67, #68, #69 are encodings of propositional logic theorems, which are not hard as such; but the encodings hide their simplicity, to
the point that, to our knowledge, no other prover has been reported in the literature to solve them, until now. #86, known as a hard problem, is solved in 12.5
seconds: only Satchmo does better, as far as we know; quite a feat, considering
that Satchmo exploits the full power of a Prolog engine, whereas our prover
runs on an interpreter for a set-based language. Finally, our prover can cope
with equality in more than half the cases, although it does not know anything
about equality but axioms: not that we consider this approach feasible, but it
demonstrates the robustness of our approach for general first-order proof search.

6 Conclusion

The first-order proof procedure we developed is remarkable in that it is able to
recognize most simple theorems as simple, even if their simplicity is well hidden.
This procedure is based on BDDs on the propositional level, on a sophisticated
proof search mechanism having some similarities with techniques known previously in the framework of resolution, and finally on strategies for controlling the
growth of the search space and traversing this search space as best possible.

Under the practical assumption that BDDs do not grow too much (i.e, superpolynomially), which seems supported by most examples of the benchmark
suite, our procedure takes a long but polynomial time to think about how it is
going to refute the formula, instead of blindly exploring all possibilities at the
risk of spending an exponential amount of time to find the same results. We
think that although the prover takes its time, it is time well spent.

Future plans include recoding the prover in a compiled language instead of
an interpreter, to gain several orders of magnitude in efficiency, and specializing

this proof procedure to particular theories, notably the theory of equality. It would also be interesting to find a way of decreasing the risks of explosion of the sizes of BDDs, either by using improved implementations of BDDs, or by dispensing with them partially or in whole; the latter would probably be hard to do, as our procedure depends crucially on having rich canonical forms for propositional formulas, notably for controlling the amount of non-determinism and computing information gains.

References

1. P. Andrews. Theorem proving via general matings. *JACM*, 28(2):193–214, 1981.
2. W. Bibel. *Automated Theorem Proving*. Vieweg, 2nd, revised edition, 1987.
3. J.-P. Billon. Perfect normal forms for discrete functions. Technical Report 87019, Bull, 1987.
4. R. E. Bryant. Graph-based algorithms for boolean functions manipulation. *IEEE Trans. Computers*, C35(8):677–692, 1986.
5. J. Goubault. Implementing functional languages with fast equality, sets and maps: an exercise in hash consing. Technical report, Bull, 1992.
6. J. Goubault. *Démonstration automatique en logique classique : complexité et méthodes*. PhD thesis, École Polytechnique, Palaiseau, France, 1993.
7. J. Goubault. The HimML reference manual. Technical report, Bull, 1993.
8. J. Goubault. Syntax independent connections. In D. et al.. Basin, editor, *Workshop on Theorem Proving with Analytic Tableaux and Related Methods*, number MPI-I-93-213. Max Planck Institut für Informatik, 1993.
9. J. Goubault. A BDD-based skolemization procedure. In M. D'Agostino, editor, *Workshop on Theorem Proving with Analytic Tableaux and Related Methods*, 1994.
10. J. Goubault. The complexity of resource-bounded first-order classical logic. In *11th STACS*, 1994.
11. R. Kowalski. A proof procedure using connection graphs. *JACM*, 22(4):572–595, 1975.
12. R. Kowalski and D. Kuehner. Linear resolution with selection function. *Artificial Intelligence*, 2:227–260, 1971.
13. R. Manthey and F. Bry. SATCHMO: a theorem prover implemented in Prolog. In *9th CADE*, pages 415–434, 1988.
14. A. Martelli and U. Montanari. An efficient unification algorithm. *ACM TOPLAS*, 4(2):258–282, 1982.
15. F. J. Pelletier. Seventy-five problems for testing automatic theorem provers. *JAR*, 2:191–216, 1986. Errata in JAR, 4:235–236, 1988.
16. F. J. Pelletier and P. Rudnicki. Non-obviousness. *AAR Newsl.*, 6:4–5, 1986.
17. J. Posegga. *Deduction with Shannon Graphs or: How to Lift BDDs to First-order Logic*. PhD thesis, Institut für Logik, Komplexität und Deduktionssystem, Universität Karlsruhe, Karlsruhe, FRG, 1993.
18. C. E. Shannon and W. Weaver. *The Mathematical Theory of Communication*. Illinois University Press, 1949. Reprinted by Illini Books, 1963.
19. L. Wos, G. A. Robinson, and D. F. Carson. Efficiency and completeness of the set of support strategy in theorem proving. *JACM*, 12(4):536–541, 1965.

Extended Path–Indexing

Peter Graf*

Max-Planck-Institut für Informatik
Im Stadtwald
66123 Saarbrücken, Germany
email: graf@mpi-sb.mpg.de
voice: (+49) 681 302 5433

Abstract. The performance of a theorem prover crucially depends on the speed of the basic retrieval operations, such as finding terms that are unifiable with (instances of, or more general than) some query term. Among the known indexing methods for term retrieval in deduction systems, Path–Indexing exhibits a good performance in general. However, as Path–Indexing is not a perfect filter, the candidates found by this method still have to be subjected to a unification algorithm in order to detect failures resulting from occur–checks or indirect clashes. As perfect filters, discrimination trees and abstraction trees thus outperform Path–Indexing in some cases. We present an improved version of Path–Indexing that provides both the query trees and the Path–Index with indirect clash and occur–check information. Thus compared to the standard method we dismiss much more terms as possible candidates.

1 Introduction

The performance of a theorem prover can be increased by speeding up the retrieval and maintenance of data referred to by the deduction system [5]. In this context we use a tool which provides fast access to a set of terms, a so-called *index*. We distinguish between the following tasks: Find terms in the index which are unifiable with some *query term*, find terms which are instances of the query term, and find terms which are more general. Because of these abilities, indexing is exploited in order to support different tasks in automated reasoning. In order to find resolution partners for a given literal, for example, a theorem prover has to find unifiable literals. Subsumption of clauses can be detected by the retrieval of generalizations or instances of literals of clauses. Even the retrieval of rewrite rules, demodulators, and paramodulants can be accelerated by indexing if the indexing mechanism also supports retrieval in the subterms of the indexed term set.

The importance of indexing has been shown by the OTTER theorem prover [6]. Due to the consistent use of Path–Indexing [10, 7] and Discrimination Tree Indexing [7, 2, 1], this prover became one of the most powerful and fastest

---

* This work was supported by the "Schwerpunkt Deduktion (Projekt: EDDS)" of the German Science Foundation (DFG).

deduction systems. Additionally, further techniques such as Abstraction Tree Indexing [8, 9], Coordinate Indexing [10], and Codeword Indexing [4, 11] have been introduced in the literature.

Standard Path–Indexing. The retrieval process is determined by the query term, i.e. the term we are searching partners for, and the set of entries of the index. In the simplest case these entries are terms too. Path–Indexing is based on two different data structures. Each query results in the construction of a new *query tree* which describes complex restrictions on the entries of the index. Query trees depend on the query term. Entries which fulfill these restrictions are candidates for being unifiable with the query term. In order to find entries with special path properties we use *Path–Lists*. Path–Lists depend on the terms which were inserted into the index. Strictly spoken, our index consists of a set of Path–Lists and a retrieval mechanism using query trees. However, we sometimes refer to a set of Path–Lists as an index.

In order to find terms which are unifiable with the query term $g(a)$, for example, we create a query tree which contains the following information: Terms unifiable with $g(a)$ are either variables or they have g as a top symbol and the argument of g is either a variable or the constant a. Obviously, query trees may contain three different types of nodes: LEAF-nodes, AND-nodes, and OR-nodes. For each of the properties "term is a variable", "top symbol is g and argument is a variable", and "term is $g(a)$" we maintain a Path–List which is a list of pointers to entries of the index. Eventually, we evaluate the query tree by computing unions (at OR-nodes) and intersections (at AND-nodes) of the Path–Lists referred to in the LEAF-nodes.

Note that Path–Indexing does not identify variables during the search, because we store properties of the form "term is a variable" only. We do not detect, for instance, that the term $f(x,x)$ is not unifiable with $f(y, f(y,z))$, because it treats this unification problem as if it had to unify the terms $f(x_1, x_2)$ and $f(y_1, f(y_2, z))$. As a consequence we have to apply an ordinary unification algorithm to the entries found and the query term in order to detect failures resulting from occur–checks and indirect clashes in the case of non–linear terms.

However, Path–Indexing provides some features which other indexing techniques do not have at all or are at least very expensive: A Path–Index may be scanned by several query trees in parallel, which allows parallel or recursive processes to work on the same index. Using query trees we get the results one by one and do not need to evaluate the whole tree. Finally, it is possible to insert entries to and delete entries from a Path–Index even when the retrieval is still in progress.

Extended Path–Indexing. Extending Path–Indexing with term constraints, however, can drastically reduce the number of candidates found in a retrieval and as a consequence accelerate retrieval, because fewer unifications have to be performed. The detection of unification failures resulting from occur–checks and indirect clashes in a unification problem does not always require the exact knowledge of the structure and the variables of the terms. Let us consider the unification of $f(x,x)$ and $f(y,f(y,z))$ again. Obviously, the two terms cannot be unified, because the two arguments of $f(x,x)$ are identical while the two arguments of $f(y,f(y,z))$ cannot be unified since y occurs in $f(y,z)$.

We extend Path–Indexing by creating new lists for terms which have identical subterms at specific positions and for terms which have non–unifiable subterms at these positions. Depending on the properties of the query term, we can reduce the number of entries found by omitting those entries which have non–unifiable subterms at a position where the query term has identical subterms and vice versa.

The advantage of this approach is, that the decision in which set to put a term, depends only on the term itself and needs to be done just once for each entry of our Path–Index, no matter how many retrievals will follow.

2 Preliminaries

We use the standard notions for first order logic. Let V and F be two disjoint sets of symbols. V denotes the set of *variable* symbols. The set of n-ary *function* symbols is F_n and $F = \cup F_i$. Furthermore, T is the set of *terms* with $V \subseteq T$ and $f(t_1, \ldots, t_n) \in T$ if $f \in F_n$ and $t_i \in T$. Function symbols with arity 0 are called *constants*. For two terms s and t which are identical we write $s = t$. In our examples we use the symbols x, y, and z as variables and the symbols f, g, and h as function symbols. Constants are denoted by a, b, and c.

A *substitution* σ is a mapping from variables to terms represented by the set of pairs $\{x_1 \mapsto t_1, \ldots, x_n \mapsto t_n\}$ with $\sigma(x_i) = t_i$ for $1 \leq i \leq n$. A *unifier* for two terms s and t is a substitution σ such that $\sigma(s) = \sigma(t)$. If such a unifier exists s and t are called *unifiable*. Terms may be *non–unifiable* for different reasons. *Clashes* occur when two non-variable symbols are not identical. In contrast to *indirect* clashes and failures resulting from *occur–checks*, a *direct* clash can be detected without considering partial substitutions. For example, the terms $f(a,y)$ and $f(x,b)$ are unifiable and the unifier is $\{x \mapsto a, y \mapsto b\}$. Furthermore, we get a direct clash when unifying $f(a,x)$ and $f(b,y)$, an indirect clash when unifying $f(x,x)$ and $f(a,b)$, and the occur–check detects the failure when unifying $f(x,x)$ and $f(y,g(y))$.

A *position* in a term is a finite sequence of natural numbers and the empty position ε. The subterm of a term t at position p is denoted by t/p and $t/\varepsilon = t$. We define the set of all positions $O(t)$ of a term t with the help of the function $\circ$ which represents the concatenation of positions.

Definition 1. Let $t = f(t_1, \ldots, t_n)$ be a term. The set of *positions of term t* is recursively defined by $O(t) = \{\varepsilon\} \cup \{i \circ p \mid p \in O(t_i), t \notin F_0, t \notin V\}$.

Definition 2. Let p be a position. The set of *terms that contain the position p* is defined by $\mathcal{T}_p = \{t \mid t \in T, p \in O(t)\}$.

For example, $O(h(a,g(b),x)) = \{\varepsilon, \varepsilon \circ 1, \varepsilon \circ 2, \varepsilon \circ 2 \circ 1, \varepsilon \circ 3\}$. Constants and variables have the position $\{\varepsilon\}$ only.

A vector of pairs (p_i, s_i) where p_i is a natural number or ε and s_i is a function symbol or the special symbol $*$ is called a *path*. Paths are denoted by $[p_1, s_1, \ldots, p_n, s_n]$. Paths are an extension of positions. That means, that we simply add a symbol to each of the natural numbers of a position. For example, we can extend the position $\varepsilon \circ 1$ to the path $[\varepsilon, f, 1, a]$. This path represents a set of terms having f as the top symbol and the first argument a. In Path–Indexing paths do not contain variables. All variables are replaced by the same special symbol $*$. For reasons of simplicity we often do not write the first position p_1; it is always the empty position ε. We will also omit commas if the result is unambiguous. The following definition of sets of terms with special path properties uses the function *top* which returns the top symbol of a term.

Definition 3. Let $p = p_1 \circ \ldots \circ p_n$ be a position and t a term. The set of *terms that contain a path* $[p_1, s_1, \ldots, p_n, s_n]$ is recursively defined: We have $t \in \mathcal{T}_{[p_1, s_1, \ldots, p_n, s_n]}$ iff all of the three conditions hold.
(1) $t \in \mathcal{T}_{[p_1, s_1, \ldots, p_{n-1}, s_{n-1}]}$ if $n > 1$
(2) $t \in \mathcal{T}_p$
(3) $s_n = \begin{cases} * & \text{if } t/p \in V \\ \text{top}(t/p) & \text{otherwise} \end{cases}$ if $n > 0$

The terms $h(g(a), b, a)$, $h(x, b, c)$, $h(f(a, c), b, x)$, for example, are all members of the set $\mathcal{T}_{[h2b]}$. The term $h(a, g(b), x)$ is a member of the sets $\mathcal{T}_{[h]}$, $\mathcal{T}_{[h1a]}$, $\mathcal{T}_{[h2g]}$, $\mathcal{T}_{[h2g1b]}$, and $\mathcal{T}_{[h3*]}$.

3 Standard Path–Indexing

Although we will introduce Path–Indexing for the retrieval of terms which are unifiable with some query term, this indexing technique is also able to support the search for instances and more general terms. This paper, however, will not focus on these features.

Path–Indexing is based on two different data structures. Each query results in the construction of a new *query tree* and in order to find entries with special path properties we use *Path–Lists*.

Query Trees. Query trees describe complex restrictions on the entries of the index. They depend on the query term and are built for non-variable[1] query terms only. A term unifiable with $t = f(t_1, \ldots, t_n)$ is either a variable *or* a complex term $s = f(s_1, \ldots, s_n)$ *and* the arguments t_i and s_i are pairwise unifiable. We recursively apply this idea to the arguments of t and get a complex tree whose leaves are marked with paths. Consider, for example, the first (complete) query tree in Fig. 1 which represents the restrictions that terms unifiable with the non-variable query term $f(a, g(b))$ have to fulfill.

Obviously, our complete query tree contains redundant information. For example, it is not necessary to check the path $[f]$, because all paths contained in

---
[1] A variable x is unifiable with a term t if $x \notin V(t)$. Therefore, finding unifiable entries does not result in a restriction on the terms of the index, because variables are not identified in Path–Indexing; all entries of the index have to be found.

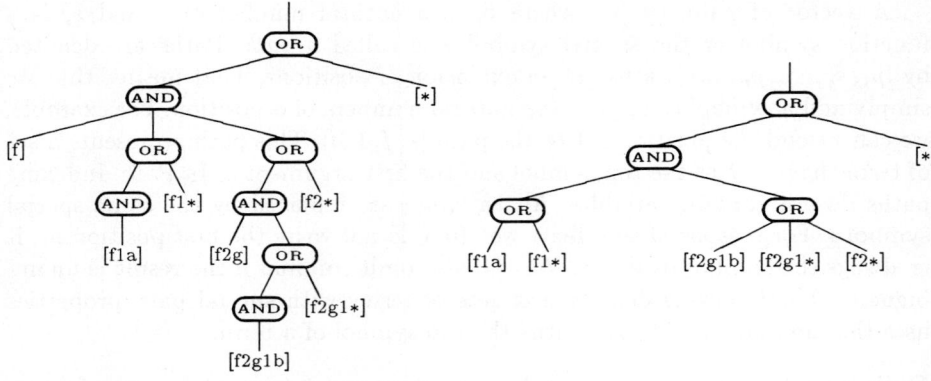

Fig. 1. Complete and simplified query tree for the query term $f(a, g(b))$

the other subtrees of the corresponding AND-node start with the function symbol f. Additionally, we do not need AND- or OR-nodes if there is only a single subtree below. Therefore, the complete query tree can be simplified. Note that a simplified query tree does not depend on the data in the index. However, we might still be able to reduce the size of the tree if we take the current entries of the index into consideration. Assume that there were no entries with paths [*], [f2g1b], and [f2g1*]. The minimal query tree one can get under this assumption is depicted in Fig. 2.

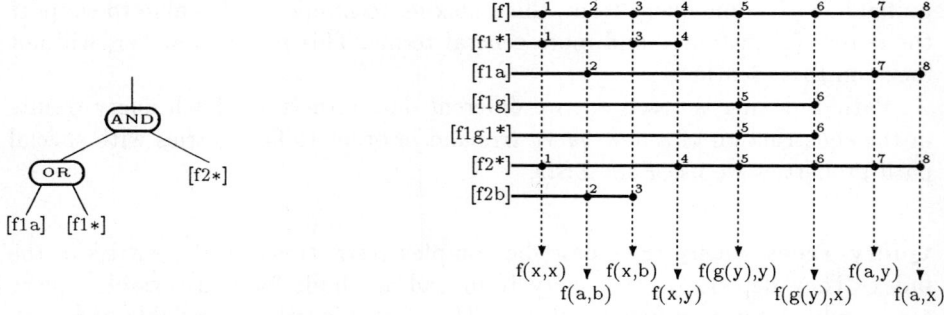

Fig. 2. Minimal query tree for $f(a, g(b))$ and Path–Index

Path–Lists. The index itself consists of a list of secondary indexes, so–called *Path–Lists*. A Path–List is a list of pointers to terms that share a special path property. To find a specific Path–List we use a hashing mechanism. Figure 2 shows an example of an index. Path–Lists are sorted, which is necessary in order to compute AND-nodes in the query tree efficiently. In our implementation we sort the terms by address.

Retrieval. Retrieval brings together query trees and Path–Lists. Each leaf of the query tree maintains a cursor on the appropriate Path–List which is initialized by a pointer to the first element of the list. The query tree is evaluated by a recursive algorithm. Application of the algorithm to the root node of the query tree yields either an entry of the index or NULL, i.e. there are no more entries left. This paragraph explains how unions (at OR-nodes) and intersections (at AND-nodes) of ordered sets are computed efficiently. A more detailed description can be found in [3].

Each time a leaf is evaluated, the pointer to the current entry is returned and the cursor is moved one position further. An OR-node returns the minimum of the results of the recursive calls on its subtrees. The computations in AND-nodes are most complex. We have to find the smallest address of a term that has all the properties described by the subtrees of the AND-node, i.e. the address has to be found in all subtrees of the AND-node.

We start the evaluation of the query tree of Fig. 2 at the AND-node, which yields two calls. One for the OR-node and one for [f2∗]. The OR-node causes the evaluation of [f1a] and [f1∗]. The LEAF-nodes are evaluated, their cursors are moved, and the terms 2 and 1 are returned. The OR-node computes the minimum and returns 1 to the AND-node. The evaluation of the LEAF-node [f2∗] also results in 1 and therefore $f(x,x)$ is the result of all subtrees of the AND-node and the first result of our retrieval. The next evaluation of the query tree is less simple, because the OR-node returns 3 as a result while the evaluation of [f2∗] results in 4. The OR-node is evaluated again (it returned the smaller value) and 4 is the next answer, which corresponds to $f(x,y)$. Similarly, we find $f(a,y)$ and $f(a,x)$.

4 Non–Unifiability

The notion of non–unifiability is most important when extending Path–Indexing in order to make it a better filter. In this context we have to answer the question of how we can derive two sets of information from two different terms such that the union of the information sets results in a test for non–unifiability of the terms. As we talk about indexing terms it is of fundamental importance that the information set for each term which is inserted into the index can be computed at insertion time. During retrieval, we compute the information set for the query term once and compare it with the information sets of the terms in the index. In order to define non–unifiability, we need some new notions.

Definition 4. Let p and q be positions. The set of terms that contain the *two different positions p and q* is defined by $\mathcal{T}_{p,q} := \{\, t \mid t \in \mathcal{T}_p,\, t \in \mathcal{T}_q,\, p \neq q \,\}$.

We have three disjoint subsets of $\mathcal{T}_{p,q}$.

Definition 5. The set of terms that contain *identical subterms* at two different positions p and q is defined by $\mathcal{T}^{=}_{p,q} := \{\, t \mid t \in \mathcal{T}_{p,q},\, t/p = t/q \,\}$. The set of terms that contain *non–unifiable subterms* at two different positions is defined by $\mathcal{T}_{p,q}^{\neq} := \{t \mid t \in \mathcal{T}_{p,q}, \neg \exists \sigma . \sigma(t/p) = \sigma(t/q)\}$. The set of *remaining* terms is $\mathcal{T}_{p,q}^{*} := \mathcal{T}_{p,q} \setminus (\mathcal{T}_{p,q}^{=} \cup \mathcal{T}_{p,q}^{\neq})$.

Lemma 6. $\mathcal{T}_{p,q}^{=}$, $\mathcal{T}_{p,q}^{\neq}$, and $\mathcal{T}_{p,q}^{*}$ are disjoint and $\mathcal{T}_{p,q}^{=} \cup \mathcal{T}_{p,q}^{\neq} \cup \mathcal{T}_{p,q}^{*} = \mathcal{T}_{p,q}$.

Additionally, we define three disjoint subsets of $\mathcal{T}_{p,q}^{\neq}$.

Definition 7. The set of terms with *non–unifiable variable–term pairs* at positions p and q is defined by $\mathcal{T}_{p,q}^{\neq v} := \{t \mid t \in \mathcal{T}_{p,q}^{\neq}, t/p \in V\}$. The set of terms with *non–unifiable, non–variable subterms* at positions p and q is defined by $\mathcal{T}_{p,q}^{\neq t} := \{t \mid t \in \mathcal{T}_{p,q}^{\neq}, t/p \notin V, t/q \notin V\}$.

Lemma 8. $\mathcal{T}_{p,q}^{\neq v}$, $\mathcal{T}_{q,p}^{\neq v}$, and $\mathcal{T}_{p,q}^{\neq t}$ are disjoint and $\mathcal{T}_{p,q}^{\neq v} \cup \mathcal{T}_{q,p}^{\neq v} \cup \mathcal{T}_{p,q}^{\neq t} = \mathcal{T}_{p,q}^{\neq}$.

The set $\mathcal{T}_{p,q}^{=}$ consists of two disjoint subsets.

Definition 9. The set of terms with *equal variables* at positions p and q is defined by $\mathcal{T}_{p,q}^{=v} := \{t \mid t \in \mathcal{T}_{p,q}^{=}, t/p \in V\}$. The set of terms with *equal non-variable subterms* at positions p and q is defined by $\mathcal{T}_{p,q}^{=t} := \{t \mid t \in \mathcal{T}_{p,q}^{=}, t/p \notin V\}$.

Lemma 10. $\mathcal{T}_{p,q}^{=v}$ and $\mathcal{T}_{p,q}^{=t}$ are disjoint and $\mathcal{T}_{p,q}^{=v} \cup \mathcal{T}_{p,q}^{=t} = \mathcal{T}_{p,q}^{=}$.

Only for the sets $\mathcal{T}_{p,q}^{\neq v}$ and $\mathcal{T}_{q,p}^{\neq v}$ the order of the subscripts p and q is important. For all other notations the order of the subscripts does not matter.

Two terms are non–unifiable if they have two different top symbols or if there are two positions which occur in both terms such that for one term the subterms at these positions are identical and the subterms of the other term at these positions are non–unifiable. In order not only to detect indirect clashes we also consider situations in which the two subterms consist of a variable and a term which contains this variable.

Consider the term $t = f(f(g(a), g(a)), f(g(x), x))$, for example. We have $t \in \mathcal{T}_{1 \circ 1, 1 \circ 2}^{=t}$, because $g(a)$ occurs at the positions $1\circ1$ and $1\circ2$ in t. Additionally, $t \in \mathcal{T}_{2 \circ 2, 2 \circ 1}^{\neq v}$, because x occurs in $g(x)$, and finally, $t \in \mathcal{T}_{1,2}^{\neq t}$, because $f(g(a), g(a))$ and $f(g(x), x)$ cannot be unified.

Definition 11. Let s and t be two non-variable terms. The predicate of *non-unifiability* is defined by

$$\text{NU}(s,t) \text{ iff } \text{top}(s) \neq \text{top}(t) \vee \exists k, l \in O(s) \cap O(t) \; ((s \in \mathcal{T}_{k,l}^{=} \wedge t \in \mathcal{T}_{k,l}^{\neq}) \vee$$
$$(s \in \mathcal{T}_{k,l}^{\neq} \wedge t \in \mathcal{T}_{k,l}^{=}) \vee (s \in \mathcal{T}_{k,l}^{\neq v} \wedge t \in \mathcal{T}_{l,k}^{\neq v})).$$

Note that the predicate serves as a filter only; NU is sufficient for non-unifiability, but not necessary, i.e. NU does not detect all failures. However, $\text{NU}(s,t)$ will never be **TRUE** if the terms s and t are unifiable. Additionally, NU is defined for non-variable terms only. In Table 1 we see some examples.

Obviously, the concept of non–unifiability cannot generally be used in Path–Indexing, because there are far too many combinations of positions to provide a Path–List for each one. In the next section we will restrict the definition of NU in order to get a concept applicable to Path–Indexing.

Table 1. Some examples of NU

s	t	NU(s,t)	s	t	NU(s,t)
$f(a,b)$	$g(y)$	TRUE	$f(x,g(x))$	$f(g(y),y)$	TRUE
$f(x,x)$	$f(g(y),y)$	TRUE	$f(x,g(x))$	$f(y,g(y))$	FALSE
$f(x,x)$	$f(g(y),g(y))$	FALSE	$f(x,f(x,y))$	$f(f(u,v),f(v,u))$	TRUE
$f(x,x)$	$f(f(y,y),f(a,b))$	TRUE	$h(a,x,x)$	$h(y,y,b)$	FALSE[2]
$f(x,x)$	$f(a,b)$	TRUE	$f(x,b)$	$f(g(x),b)$	FALSE[3]

5 Extended Path–Indexing

Standard Path–Indexing uses query trees as complex descriptions of restrictions on terms. We will now extend both the query trees and the index itself by not only checking direct clashes as in ordinary Path–Indexing but also considering non–unifiability. More precisely, we add non–unifiability restrictions to the query trees which consider indirect clashes and occur–checks. The index, however, remains a filter only, although there will be more lists in the index. We will still have to apply an ordinary unification algorithm in order to test unifiability and to extract the unifying substitutions.

5.1 Adapting Non–Unifiability

We have learned from Standard Path–Indexing that the unification partners for terms with special properties can only be found in special sets. Table 2 sums up the connections between the sets when non–unifiability is taken into account. The column "where to search" can directly be derived from the definition of NU. From the Lemmata 6, 8 and 10 we get the third column of the table.

Table 2. Connections of the term sets

query term	where to search	where *not* to search
$t \in \mathcal{T}_{p,q}^=$	$s \in \mathcal{T}_{p,q}^* \cup \mathcal{T}_{p,q}^{=v} \cup \mathcal{T}_{p,q}^{=t}$	$s \notin \mathcal{T}_{p,q}^{\neq v} \cup \mathcal{T}_{q,p}^{\neq v} \cup \mathcal{T}_{p,q}^{\neq t}$
$t \in \mathcal{T}_{p,q}^{\neq v}$	$s \in \mathcal{T}_{p,q}^* \cup \mathcal{T}_{p,q}^{\neq v} \cup \mathcal{T}_{p,q}^{\neq t}$	$s \notin \mathcal{T}_{p,q}^{=v} \cup \mathcal{T}_{p,q}^{=t} \cup \mathcal{T}_{q,p}^{\neq v}$
$t \in \mathcal{T}_{p,q}^{\neq t}$	$s \in \mathcal{T}_{p,q}^* \cup \mathcal{T}_{p,q}^{\neq}$	$s \notin \mathcal{T}_{p,q}^{=v} \cup \mathcal{T}_{p,q}^{=t}$
$t \in \mathcal{T}_{p,q}^*$	$s \in \mathcal{T}_{p,q}$	$s \notin \emptyset$

The reason for considering the sets in which a partner for the query term t *cannot* occur is simple: Query trees represent restrictions on the term sets.

---

[2] NU fails, because the indirect clash can only be detected using the information $x = y$, which can be derived when unifying the terms only.

[3] Here the result of NU is not correct, because the sets of variables of s and t are not disjoint.

On the one hand, we will extend the query trees with additional restrictions resulting from non–unifiability which, on the other hand, will be tested with the help of the disjoint sets $\mathcal{T}_{p,q}^{=v}$, $\mathcal{T}_{p,q}^{=t}$, $\mathcal{T}_{p,q}^{\neq v}$, $\mathcal{T}_{q,p}^{\neq v}$, and $\mathcal{T}_{p,q}^{\neq t}$. We avoid the usage of $\mathcal{T}_{p,q}^{\neq}$, because $\mathcal{T}_{p,q}^{\neq}$ contains the subsets $\mathcal{T}_{p,q}^{\neq v}$ and $\mathcal{T}_{q,p}^{\neq v}$ which have to be created for the case $t \in \mathcal{T}_{p,q}^{\neq v}$ anyway. Using $\mathcal{T}_{p,q}^{=v}$ and $\mathcal{T}_{p,q}^{=t}$ instead of $\mathcal{T}_{p,q}^{=}$ results in better constraints in the query tree.

We have also stated that it is impossible to provide a Path–List for each possible combination of positions in a term. In order to restrict the number of combinations we introduce the notion of *distance* of positions.

Definition 12. Let $p = p_1 \circ \ldots \circ p_n$ and $q = q_1 \circ \ldots \circ q_m$ be two different positions. Let $k = \max\{i \mid \forall j.\, 1 \leq j \leq i \leq \min\{n,m\} : p_j = q_j\}$ be the length of the longest common prefix of p and q. The *distance* of p and q is defined by
$$\mathrm{dist}(p,q) := \begin{cases} \infty & \text{if } k = \min\{n,m\} \\ \max\{n,m\} - k & \text{otherwise} \end{cases}.$$

For example, $\mathrm{dist}(1 \circ 2 \circ 1, 1 \circ 1 \circ 2 \circ 2) = 3$, $\mathrm{dist}(1,2) = 1$, and $\mathrm{dist}(1 \circ 2 \circ 1, 1 \circ 2 \circ 2) = 1$. Additionally, we have $\mathrm{dist}(1 \circ 2, 1) = \infty$, because 1 is a prefix of $1 \circ 2$. Identical positions also have distance ∞.

Definition 13. Let p and q be two different positions. The set of terms that contain two positions with a *maximum distance* of $n > 0$ is defined by
$\mathcal{T}_{p,q,n} := \{\, t \mid t \in \mathcal{T}_{p,q},\, \mathrm{dist}(p,q) \leq n \,\}$.
The maximum distance of p and q is also called *NU–depth*.

In case the NU–depth $n > 0$ is constant, we may drop the subscript and write $\mathcal{T}_{p,q}$ instead of $\mathcal{T}_{p,q,n}$. In the following we regard the definitions of Section 4 as if they had been made using $\mathcal{T}_{p,q,n}$ for a constant NU–depth. Note, that if we had restricted the NU–depth to 1, for instance, the non–unifiability of $f(x, f(x, y))$ and $f(f(u, v), f(v, u))$ would not have been detected.

5.2 Extending the Data Structures

Extended Query Trees. We apply our new definition of non–unifiability to query trees. Actually, there is no big difference between standard and extended query trees, except for some constraints which are added to specific nodes. As an example we present the extended query tree for terms s_i which are unifiable with the term $t = f(x, g(x))$ in Fig. 3. The NU–depth is set to 1. The three constraints $s_i \notin \mathcal{T}_{1,2}^{=v}$, $s_i \notin \mathcal{T}_{1,2}^{=t}$ and $s_i \notin \mathcal{T}_{2,1}^{\neq v}$ are due to the fact that $t \in \mathcal{T}_{1,2}^{\neq v}$ (see Table 2). Constrained nodes affect the evaluation of the query tree in a very simple and efficient way: When addresses of terms are bubbled up from the leaves to the root of the tree, only those addresses pass a constrained node which do not occur in the sets described by the constraint. Note that a constraint may consist of more than one set.

We now focus on the positions of the constraints in the query tree. The retrieved set of terms would not change even if we attached all constraints to

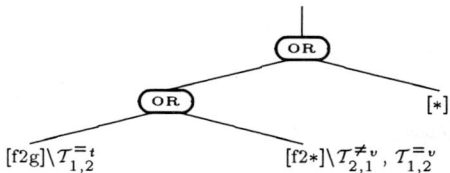

Fig. 3. Extended query tree for the query term $f(x, g(x))$, NU–depth=1

the root node of the query tree. However, each constraint has its distinct position in the tree in order to reduce evaluation costs. In Fig. 3 the constraint $s_i \notin \mathcal{T}_{2,1}^{\neq v}$ is attached to the leaf [f2*], because $\mathcal{T}_{2,1}^{\neq v}$ contains terms with variables at position 2 only. In parallel, $\mathcal{T}_{1,2}^{=t}$ which consists of non-variable terms at position 2 only will not be searched for variables resulting from leaf [f2*] if the constraint is attached to the leaf [f2g]. As extended query trees can also be simplified and minimized, we can omit some constraints in case the appropriate node in the query tree is deleted.

Let us take a closer look at where in the query tree the constraints have to be added. Consider Fig. 4, a fragment of the query tree which corresponds to the query term $t \in \mathcal{T}_{p,q}$ at position p. Depending on the properties of t, we have to add sets to constraints at specific positions of the extended query tree. The table in Fig. 4 shows the insertion rules.

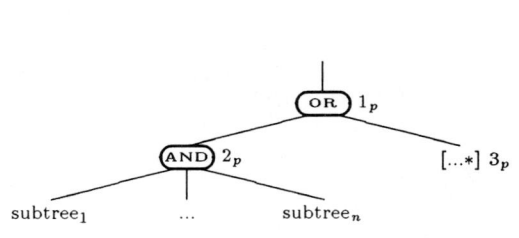

	add set	node
$t \in \mathcal{T}_{p,q}^=$	$\mathcal{T}_{p,q}^{\neq t}$	2_p
	$\mathcal{T}_{p,q}^{\neq v}$	3_p
	$\mathcal{T}_{q,p}^{\neq v}$	3_q
$t \in \mathcal{T}_{p,q}^{\neq v}$	$\mathcal{T}_{p,q}^{=t}$	2_p
	$\mathcal{T}_{p,q}^{=v}$	3_p
	$\mathcal{T}_{q,p}^{\neq v}$	3_q
$t \in \mathcal{T}_{p,q}^{\neq t}$	$\mathcal{T}_{p,q}^{=t}$	2_p
	$\mathcal{T}_{p,q}^{=v}$	3_p

Fig. 4. Add sets to constraints

Note that the set $\mathcal{T}_{q,p}^{\neq v}$ is added to the constraint of node 3_q of the query tree fragment which represents the query term at position q. Additionally, we will result in less recursive calls when evaluating the extended query tree if we do not add the set $\mathcal{T}_{p,q}^{\neq t}$ to node 2 but to node 1 of subtree$_1$. In Fig. 5 the extended query tree for terms which are unifiable with the query term $f(g(x), g(h(x)))$ is depicted. The NU–depth is set to 2.

Just like in Standard Path–Indexing the query tree is simplified and minimized before the retrieval is started. Again we may delete those LEAF-nodes

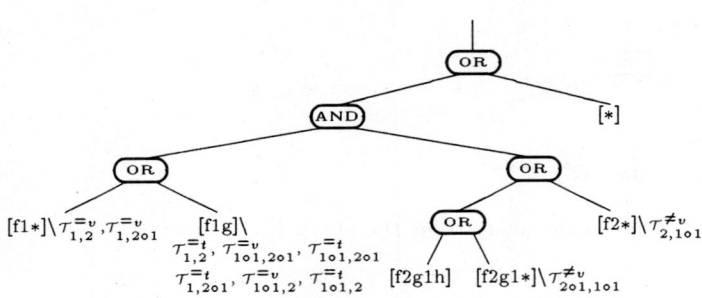

Fig. 5. Extended query tree for the query term $f(g(x), g(h(x)))$, NU-depth=2

which correspond to empty Path–Lists. Additionally, we may delete all empty sets in constraints. In case an AND-node or an OR-node is deleted because it has just one subtree, the constraint is inherited to the root of the subtree. In this way we get a minimal extended query tree which is based on the extended Path–Index of Fig. 6.

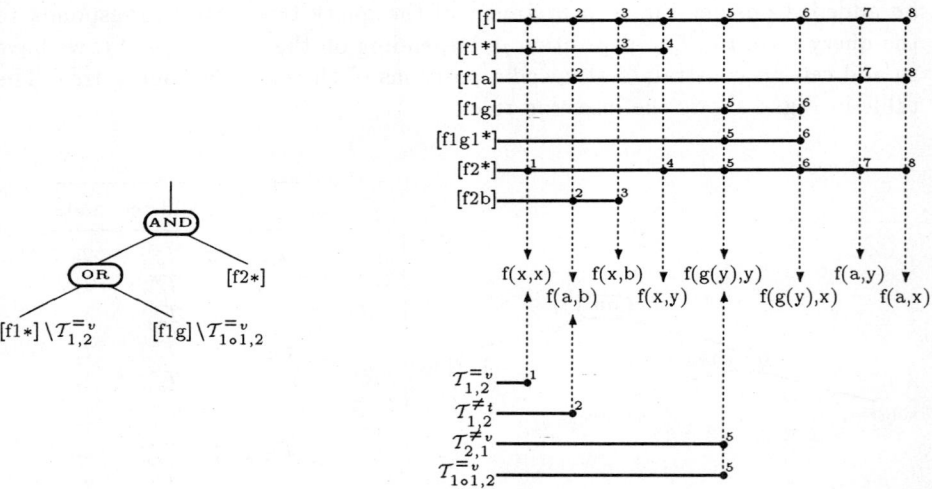

Fig. 6. Minimal extended query tree for $f(g(z), g(h(z)))$ and extended Path–Index

Extended Path–Lists. Path–Lists are created when terms are inserted into the index. In order to provide the Path–Lists for the tests of the constraints, one has to detect for every possible combination of positions p and q of a term s with dist$(p,q) \leq$ NU-search depth, whether s is in one of the sets $\mathcal{T}_{p,q}^{=v}$, $\mathcal{T}_{p,q}^{=t}$, $\mathcal{T}_{p,q}^{\neq v}$, $\mathcal{T}_{q,p}^{\neq v}$, or $\mathcal{T}_{p,q}^{\neq t}$. In parallel to ordinary Path–Lists extended Path–Lists are represented as sorted lists. In this context it is important that these lists are sorted according to the same sorting criterion as the ordinary Path–Lists. Note that the insertion of a term does not need to result in an entry in an extended Path–List, see $f(x,b)$ or $f(x,y)$ for example.

Extended Retrieval. The retrieval algorithm which allows for extended query trees is just a slight modification of the ordinary retrieval algorithm. The only difference is the fact that the entry which has been found in a node of an extended query tree is tested against the constraints, i.e. we search for the entry in the appropriate extended Path–Lists. Obviously, the search takes advantage of the fact that the lists are sorted and may start at the position in the list where the last comparison stopped. In case the entry is found in one of the lists, a new retrieval on the current node of the tree is started, which will yield another entry. Otherwise the entry is moved towards the root of the extended query tree.

In the example depicted in Fig. 6 we would find the terms $f(x,x)$, $f(x,y)$, $f(g(y),y)$, and $f(g(y),x)$ if we used Standard Path–Indexing. However, Extended Path–Indexing exploits the constraints and the found terms are $f(x,y)$ and $f(g(y),x)$ only, because $f(x,x)$ is rejected at the OR-node by constraint $\mathcal{T}_{1,2}^{=v}$ and $f(g(y),y)$ is rejected at the LEAF-node [f1g] by constraint $\mathcal{T}_{1\circ 1,2}^{=v}$.

6 Experiments

Term Sets. For our experiments we used the term sets which were introduced in [7]. The term sets were taken from typical OTTER applications. The sets are paired. There is a set of positive literals and a set of negative literals in each pair. Unifiable terms are searched in order to find resolution partners and to detect unit conflicts. The sets EC–pos and EC–neg consist of 500 terms each and are derived from a theorem in equivalential calculus. Two representative members of EC–pos and EC–neg are $P(e(e(x, e(y, e(z, e(e(u, e(v, z)), e(v, u)))), e(y, x)))$ and $\neg P(e(e(x, e(e(y, e(z, x)), e(z, y))), e(e(u, e(e(v, e(w, u)), e(w, v))), e(e(v6, e(e(v7, e(v8, v6)), e(v8, v7))), e(e(v9, e(e(e(b, a), e(e(e(a, e(b, c)), c), v9)), v10)), v10)))))$. The sets CL–pos and CL–neg have 1000 members and are derived from a theorem in combinatory logic. Two representative members of CL–pos and CL–neg are $g(x, g(g(g(g(g(B, B), y), z), u), v)) = g(g(g(B, x), g(y, z)), g(u, v))$ and $g(f(g(g(N, x), y)), g(g(g(N, x), y), f(g(g(N, x), y)))) \neq g(g(g(x, f(g(g(N, x), y))), y), f(g(g(N, x), y)))$. Finally, the sets BOOL–pos and BOOL–neg are derived from a theorem in the relational formulation of Boolean algebra and consist of 6000 terms each. Two representative members of BOOL–pos and BOOL–neg are $Sum(x, p(x, y), p(x, s(x, y)))$ and $\neg Sum(p(c2, n(x)), p(c2, x), c4)$.

Memory Requirements. In Table 3 we compare the memory requirements of Standard Path–Indexing and extended Path–Indexing. Extended Path–Indexing occupies the more memory the larger the NU–depth is. The NU–depth should be selected carefully, as the memory requirements may drastically increase. In most cases a NU–depth of 1 or 2 is sufficient, as our retrieval times in Table 5 will show.

Table 3. Memory Requirement [KBytes]

indexed set	Standard Path–Indexing	Extended Path–Indexing			
		1	2	3	4
EC–pos	115	184	281	345	371
EC–neg	639	951	1759	2754	3822
CL–pos	437	570	887	1341	1885
CL–neg	1400	1908	3395	6223	10974
BOOL–pos	576	821	1550	1985	1985
BOOL–neg	1125	1498	2710	4047	5151

Filter Properties. Table 4 gives a summary of how many terms pass the indexing filter. The column "no filter" shows how many term pairs would have to be unified without using a filter. In our examples we created an index for the "indexed set" and started a query for each term of the "query set". Obviously, Standard Path–Indexing is a bad filter for the term sets EC–pos and EC–neg. The filter properties for CL–pos and CL–neg are average. Standard Path–Indexing is a very good filter for the term sets BOOL–pos and BOOL–neg. A "perfect filter" would not find any unifiable term pairs and therefore our examples are rather extreme. For all indexing problems Extended Path–Indexing significantly improves the filter properties. The larger the NU–depth the better the filter.

Table 4. Filter Properties [Terms Found]

indexed set	query set	no filter	Standard Path–Indexing	Extended Path–Indexing				perfect filter
				1	2	3	4	
EC–pos	EC–neg	250000	249104	88791	52208	50434	50421	0
EC–neg	EC–pos	250000	249104	88791	52208	50434	50421	0
CL–pos	CL–neg	1000000	154399	118319	13684	3206	385	0
CL–neg	CL–pos	1000000	154399	118319	13684	3206	385	0
BOOL–pos	BOOL–neg	36000000	40773	20322	408	0	0	0
BOOL–neg	BOOL–pos	36000000	40773	20322	408	0	0	0

Retrieval Times. The experiments were run on a Sun SPARCstation SLC computer with 16 MBytes of RAM. The size of the hash table was limited to 500, the index depth (introduced in [7]) was limited to 15. We give a survey on the retrieval times in Table 5. The times include the construction of the query trees as well as the time spent for test unifications.

In the EC and CL examples Extended Path–Indexing is faster than Standard Path–Indexing. However, Standard Path–Indexing is already a very good filter for the BOOL examples. The performance cannot be improved, because we have to cope with a very large number of rather small terms for which the the test unifications are easy.

Finally, Extended Path–Indexing should be used in the cases of non–linear terms, which are not unifiable because of indirect clashes and occur–checks. In these cases Standard Path–Indexing does not provide good filter properties. The best NU-depth depends on the terms themselves. Taking the memory requirements into consideration a small NU–depth of 1 or 2 should be preferred.

Although they have no application in theorem proving, unifications of the form X–pos X–pos and X–neg X–neg have also been tested. They produced similar results.

Table 5. Retrieval Times [Seconds]

indexed set	query set	no filter	Standard Path–Indexing	Extended Path–Indexing			
				1	2	3	4
EC–pos	EC–neg	41.5	67.8	41.6	36.0	38.2	42.1
EC–neg	EC–pos	44.7	57.5	34.6	29.0	28.6	28.9
CL–pos	CL–neg	107.5	43.5	38.8	11.9	13.7	20.7
CL–neg	CL–pos	111.2	46.1	41.4	17.2	19.7	24.3
BOOL–pos	BOOL–neg	301.5	22.8	26.3	35.0	39.1	40.2
BOOL–neg	BOOL–pos	299.8	5.9	6.6	7.6	8.5	8.9

Implementation. Standard and Extended Path–Indexing are implemented in C and are available via anonymous ftp. They are as well as implementations of discrimination and abstraction trees part of "A Collection of Indexing Data Structures (ACID)" developed at MPI. In the future we will try to further improve ACID which is a library for efficient data structures and algorithms for theorem provers. Our implementations do not depend on term data structures and can very easily be embedded into other software. Our Path–Indexing software allows parallel access to the same index and is able to delete and insert entries into the index when retrieval is interrupted. For more information send e-mail to *acid@mpi-sb.mpg.de*.

7 Conclusion

We presented an extension of Standard Path–Indexing which is able to drastically reduce retrieval time in the cases in which Standard Path–Indexing itself is not a good filter. However, the advantages of Standard Path–Indexing are preserved. In Extended Path–Indexing indirect clashes and failures resulting from occur-checks are detected with the help of extended Path–Lists at the time the query tree is evaluated. Therefore, less test unifications have to be performed. The memory requirements can be reduced by an appropriate choice of the NU–depth.

Acknowledgements

Thanks to Leo Bachmair, Hans Jürgen Ohlbach, and David Plaisted for their comments on earlier versions of this paper. I also like to thank William McCune for providing the term sets. Eventually, I am grateful to Christoph Meyer for his contributions to the implementation of Extended Path–Indexing.

References

1. L. Bachmair, T. Chen, and I.V. Ramakrishnan. Associative-commutative discrimination nets. In *Proceedings TAPSOFT '93, LNCS 668*, pages 61–74. Springer Verlag, 1993.
2. J. Christian. Flatterms, discrimination nets, and fast term rewriting. *Journal of Automated Reasoning*, 10(1):95–113, February 1993.
3. P. Graf. Path indexing for term retrieval. Technical Report MPI-I-92-237, Max-Planck-Institut für Informatik, Saarbrücken, Germany, December 1992.
4. L.J. Henschen and S.A. Naqvi. An improved filter for literal indexing in resolution systems. In *Proceedings of the 6th International Joint Conference on Artificial Intelligence*, pages 528–529, 1981.
5. E. Lusk and R. Overbeek. Data structures and control architectures for the implementation of theorem proving programs. In *5th International Conference on Automated Deduction*, pages 232–249. Springer Verlag, 1980.
6. W. McCune. Otter 2.0. In *10th International Conference on Automated Deduction*, pages 663–664. Springer Verlag, 1990.
7. W. McCune. Experiments with discrimination-tree indexing and path-indexing for term retrieval. *Journal of Automated Reasoning*, 9(2):147–167, October 1992.
8. H.J. Ohlbach. Abstraction tree indexing for terms. In *Proceedings of the 9th European Conference on Artificial Intelligence*, pages 479–484. Pitman Publishing, London, August 1990.
9. H.J. Ohlbach. Compilation of recursive two-literal clauses into unification algorithms. In *Proc. of AIMSA 1990*. Bulgaria, 1990.
10. M. Stickel. The path–indexing method for indexing terms. Technical Note 473, Artificial Intelligence Center, SRI International, Menlo Park, CA, October 1989.
11. M. Wise and D. Powers. Indexing prolog clauses via superimposed codewords and field encoded words. In *Proceedings of the IEEE Conference on Logic Programming*, pages 203–210, 1984.

Exporting and Reflecting Abstract Metamathematics

Robert L. Constable

rc@cs.cornell.edu

Department of Computer Science, Cornell University, Ithaca, NY 14853, USA

The automated reasoning community has achieved remarkable results over the past 50 years, and we see clearly how to carry on substantially further. On the way, we will circumvent various technical obstacles, and we will discover new truths. I want to suggest a way of dealing with two important technical obstacles. The suggested approach will also lead us to a deeper understanding of a fundamental mechanism of language — *reflection*. The first problem is how to safely extend the automatic capabilities of a theorem prover, and the second is how to guarantee the correctness of the prover in the first place.

In tactic-oriented provers, the programming of new refinement tactics is a safe way to extend the prover. It is safe because refinement tactics reduce new inference steps to primitive rules. While this is safe, it is increasingly expressive as tactics do more work. Moreover, not all reasoning steps proceed by such reductions, e. g. decision procedures work differently as do transformation tactics.

I propose that we prove formal metamathematical results which are useful in building and extending provers, and then invoke these results via reflection rules. Moreover, we should develop the results abstractly so that they can be exported and specialized to a number of logics. In this way the community can collectively assist in the accumulation and use of relevant metaknowledge.

In this paper, I illustrate this method by considering the simple example of a propositional calculus *decision procedure*. I sketch its development for abstract logic, definable in Nuprl; then I show how to apply this to Nuprl via its reflection rule. A similar idea is developed in the paper *Using Reflected Decision Procedures* co-authored with William Aitken and Judith Underwood.

Associative-Commutative Deduction with Constraints*

Laurent Vigneron
CRIN-CNRS & INRIA-Lorraine
BP 239, 54506 Vandœuvre-lès-Nancy Cedex, France
E-mail : Laurent.Vigneron@loria.fr

Abstract

Associative-commutative equational reasoning is known to be highly complex for theorem proving. Hence, it is very important to focus deduction by adding constraints, such as unification and ordering, and to define efficient strategies, such as the basic requirements à la Hullot. Constraints are formulas used for pruning the set of ground instances of clauses deduced by a theorem prover. We propose here an extension of AC-paramodulation and AC-superposition with these constraint mechanisms ; we do not need to compute AC-unifiers anymore. The method is proved to be refutationally complete, even with simplification. The power of this approach is exemplified by a very short proof of the equational version of SAM's Lemma using DATAC, our implementation of the strategy.

1 Introduction

Automated deduction with equality and associative-commutative (AC) operators, i.e. binary operators f satisfying the axioms $f(f(x,y),z) = f(x,f(y,z))$ and $f(x,y) = f(y,x)$, has been a long standing open problem. Our solution for dealing with AC axioms is to work in the AC-congruence classes and to use associative and commutative identity equality, matching and unification. This idea of replacing axioms by ad-hoc mechanisms such as unification algorithms and inference rules was first proposed by Plotkin in [Plo72], and has been extended to term rewriting systems including associativity and commutativity in [PS81].

The paramodulation rule has been extended to AC theories in [RV91, Pau92, Wer92] and in [RV93], where we have proposed an inference system based on the ordered paramodulation strategy. In this paper, we add important restrictions in the line of [NR92a, BGLS92] for obtaining a basic strategy for AC-paramodulation and AC-superposition. The principle of the basic strategy, extension of the basic narrowing of [Hul80], is to forbid paramodulation in terms introduced by a substitution in a previous deduction. For instance, considering the paramodulation step (with the unifier $\{x \mapsto a\}$)

$$\frac{(f(a) = b) \quad P(f(x)) \vee Q(x,b)}{P(b) \vee Q(\boxed{a},b)}$$

$f(x)$ is replaced by b, assuming that $f(a) \not\preceq b$ and $P(f(a)) \not\preceq Q(a,b)$. The subterm a of the deduced clause is blocked (framed) to show that no more paramodulation step is allowed in it. A refinement of this strategy is to block in addition the replacing term in the deduced clause (i.e. b in $P(b)$). This refinement is not valid for the superposition strategy [Rus91].

In the following, we refine and extend this strategy by associating to each clause a set of constraints as in [KKR90]. Then, when a new clause is generated, it inherits the constraints of

---
*This work is partially supported by the GDR Programmation and the Esprit Basic Research working group 6028, CCL.

its ancestors and also the AC-unification constraints (not computed) and ordering constraints produced by the last inference step. Applied to the previous example, we get :

$$\frac{(f(a)=b) \; [\![T_1]\!] \qquad P(f(x)) \vee Q(x,b) \; [\![T_2]\!]}{P(y) \vee Q(x,b) \; [\![T_1 \wedge T_2 \wedge T_3]\!]}$$

where T_3 consists in the AC-unification constraints $f(x)=_{AC} f(a)$ and $y=_{AC} b$ (the last one corresponds to the basic strategy), and the ordering constraints $f(a) \succ b$ and $P(f(x)) \succ Q(x,b)$ for simulating an ordered strategy.

In a first version of this work[1], we had to compute AC-unifiers in a very restrictive case. Later Nieuwenhuis and Rubio [NR94] have proposed a very close result where no AC-unifier was needed (but with other drawbacks). In this paper, we have been able to get rid of the last few AC-unifiers needed by a very simple trick, inspired by the proof technique of Nieuwenhuis and Rubio. After discussion with them, this trick is compatible with their method.

Related works on the basic strategy and constraints are detailed in Section 2 ; we describe there our motivations for applying such techniques to deduction in AC theories. Our notations are described in the third section. Section 4 presents inference rules and the lifting lemma. Section 5 introduces several important simplification rules that are compatible with our inference rules. The method for proving the refutational completeness is explained in Section 6 and the detailed proof can be found in [Vig93]. The last Section 7 details the proof of SAM's Lemma, which was produced by our implementation of the method described in this paper.

2 Related Work and Motivations

The notion of basic strategy was first introduced by Slagle [Sla74] with the definition of *blocked inferences*, i.e. inferences using substitutions which do not contain a reducible term, in the sense of term rewriting. This concept was extended to theories with permutative axioms (including AC axioms) by Ballantyne and Lankford [BL79]. In another setting but with close ideas, Kapur, Musser and Narendran [KMN85] have defined the *prime superposition*, which takes into account that many superpositions are useless in the Knuth-Bendix completion procedure. But, the first strategy including the main ideas of the basic strategy is the *modification method* of Brand [Bra75].

In his thesis, Hullot [Hul80] has introduced a related concept called *basic narrowing*, which consists in forbidding narrowing steps in subterms introduced by previous inference steps. It was extended to *basic superposition* and *basic paramodulation* by Nieuwenhuis and Rubio [NR92a], and Bachmair, Ganzinger, Lynch and Snyder [BGLS92]. Both these works apply the model construction technique of [BG90] to derive the refutational completeness of the strategy. In [NR92a], subterms where superpositions are forbidden never appear explicitly, since they are handled by unification constraints, while in [BGLS92], unifiers are computed and the blocked subterms are replaced by variables and appear in a substitution associated to a clause. To compare the basic superposition and paramodulation strategies, the first one permits to restrict to left-hand sides of equations, while the second one permits to block the right-hand sides when they replace subterms of a clause.

In this paper, we extend the basic paramodulation and superposition to AC theories. We use explicit constraints as in [NR92b] to get an homogeneous framework with ordering and unification constraints at the same level.

Such ordering constraints were used by Peterson [Pet90] and Martin and Nipkow [MN90] to orient equations like the commutativity axiom : $x \cdot y \to y \cdot x$ if $x > y$. But, constraints are also fundamental to trace the chosen orientations. Indeed, when an equation $(s=t)$ is oriented as $s \to t$, it has to be coherent with further inference steps. This coherence is imposed by keeping

---
[1] Presented at the 2nd Workshop CCL, La Escala, Spain, September 1993.

the constraint $s > t$ which is added to the deduced clause.
A second kind of constraints we consider are AC-unification constraints. They permit to limit the number of generated clauses. Consider for instance the paramodulation from the equation $(x * x * x * x = x)$ into the clause $P(y_1 * y_2 * y_3 * y_4)$, where $*$ is AC ; a clause will be generated for each most general AC-unifier, but there are 34 359 607 481 of them. Using an AC-unification constraint, only one clause is deduced : $P(x) [\![x * x * x * x =_{AC} y_1 * y_2 * y_3 * y_4]\!]$. The advantage of such constraints was already pointed out in [KKR90].

The main result of this paper is the definition of a refutationally complete set of inference rules, compatible with simplification, for deduction with constraints in AC theories. These inference rules are non-trivial refinements of rules defined in [RV93], by including constraints of

- AC-unification, to express the basic strategy and limit the number of deduced clauses,
- Ordering, to express the orientation of equations and the ordered strategy, which consists in always applying inference steps between maximal literals of clauses.

The additional refinements defined in [RV93] remain complete, such as the maximal positions requirement, which consists in forbidding paramodulations at non maximal positions of AC-operators. Moreover, we never use the AC axioms explicitly ; they are handled though AC-unification constraints and the inference rules application.

Nieuwenhuis and Rubio [NR92b] have defined a complete set of inference rules based on superposition, with ordering and unification constraints. Their work deals with the empty theory. They have recently extended this result to AC theories [NR94]. However, we have important refinements on the rules named contextual paramodulation and extended paramodulation (Definitions 5 and 7) they do not have.

Both systems had the same problem : to keep the completeness property, the basic strategy has to be weakened by keeping the possibility of applying indirectly a paramodulation (or superposition) in the constraint part (Example 1). First, we wanted to avoid such inference steps, and we defined an inference system which had to compute some AC-unifiers. But, a nicer solution is to never compute AC-unifiers and also to permit some paramodulations in constraints as proposed by Nieuwenhuis and Rubio in [NR94]. So, the inference system presented in this paper is a light transformation of our first system : no AC-unifiers are needed, and we show that paramodulations in constraints have to be allowed only in very restrictive cases (Definition 5) which can be tested with a very simple control. For these special cases, we used a trick inspired by the proof technique of Nieuwenhuis and Rubio. We conjecture that we have obtained an optimal version of basicness.

Last, our method is implemented and we report experiments on non-trivial problems.

3 Notations and Definitions

This section introduces the basic notions relevant to our work, based on the standard notations and definitions for term rewriting and unification given in [DJ90, JK91].

We assume that the operators from a given subset $\mathcal{F}_{AC}$ of $\mathcal{F}$ are associative and commutative, which means that for $f \in \mathcal{F}_{AC}$ the axioms $f(f(x,y), z) = f(x, f(y,z))$ and $f(x,y) = f(y,x)$ are implicit in the theory to be considered. The congruence on $\mathcal{T}(\mathcal{F}, \mathcal{X})$, generated by these AC axioms, is called AC-equality and written $=_{AC}$. We have to mention that, in this paper, we do not use the flatten representation for terms. The set of AC-matchers from a term s into a term t is denoted by $AC\_match\{s,t\}$.

The relation $\sqsubset_{AC}$ represents the subterm property modulo AC, i.e. $u \sqsubset_{AC} v$ if there is a term v', AC-equal to v, such that u is a strict subterm of v'. $u \sqsubseteq_{AC} v$ if $u \sqsubset_{AC} v$ or $u =_{AC} v$. For instance, $a * b \sqsubset_{AC} b * (a * c)$, where $*$ is AC. This relation is extended to sets of terms : $\{s_1, \ldots, s_m\} \sqsubseteq_{AC} \{t_1, \ldots, t_n\}$ if $\forall i, \exists j,\ s_i \sqsubseteq_{AC} t_j$.

We denote by $\cap_{AC}$ (resp. $\cup_{AC}$ and $\subseteq_{AC}$) the intersection (resp. union and inclusion) of sets, using the AC-equality for comparing objects, i.e. terms or atoms. For example, $\{a*b, b, c\} \cap_{AC} \{b*a, c, d\} = \{a*b, c\}$, $\{a*b\} \cup_{AC} \{b*a, c\} = \{a*b, c\}$ and $\{a*b, c\} \subseteq_{AC} \{b*a, b, c\}$, where $*$ is AC. The extension of $\subseteq_{AC}$ to multisets is denoted by $\subseteq\!\!\subseteq_{AC}$. For instance, $\{a*b, c, c\} \subseteq\!\!\subseteq_{AC} \{b*a, c, b, c\}$, but $\{a*b, c, c\} \not\subseteq\!\!\subseteq_{AC} \{b*a, b, c\}$.

$\mathcal{MP}os(t)$ denotes the set of all non variable positions p in t such that either $\mathcal{H}ead(t_{|p}) \notin \mathcal{F}_{AC}$, or, if $p = p' \cdot i$, $\mathcal{H}ead(t_{|p})$ and $\mathcal{H}ead(t_{|p'})$ are not the same AC-operator.

Given a term t and an AC-operator f, $\mathcal{H}terms(t, f)$ is the multiset of the subterms below f at the root of t; For instance, $\mathcal{H}terms(a*((a*x)*g(b)), *)$ is $\{a, a, x, g(b)\}$.

Our ordered paramodulation strategy is based on a *complete simplification ordering* $\succ$ on terms and atoms, total on ground AC-congruence classes (see definition in [RV93]) and AC-compatible [NR91, RN93]. By overloading of notation, given a literal L and a clause D, $L \succ D$ (resp. $L \prec D$) means that the atom corresponding to L is greater than each (resp. less than one) atom of D.

In the following, we use *constrained clauses*, denoted by $C\,[\![T]\!]$ as in [NR92b], where C is a clause in the first-order logic and T is a conjunction of constraints, atomic formulas of type $s =_{AC} t$ or $s \succ t$, for terms or atoms s and t. The solutions of a set of constraints T are the ground substitutions defined by:

$$Sol(c \wedge c') = Sol(c) \cap Sol(c')$$
$$Sol(s =_{AC} t) = \{\sigma \mid s\sigma \text{ and } t\sigma \text{ are AC-equal}\}$$
$$Sol(s \succ t) = \{\sigma \mid s\sigma \succ t\sigma\}$$

Some other constraints are used to test emptiness of the intersection of sets of terms, and to test inclusion of multisets of terms. $C\,[\![\sigma]\!]$ will denote a shorthand of $C\,[\![\bigwedge_{x \in \mathcal{D}om(\sigma)} x =_{AC} x\sigma]\!]$. A set of constraints T is *satisfiable* if it admits at least one solution.

4 Inference Rules

Before defining our inference rules, let us recall the main ideas of our theorem proving strategies. First, the ordered strategy forces the application of inference rules between maximal literals of clauses. Second, the AC-unification constraints, which are never evaluated, simulate the basic strategy. When a clause is deduced by an inference step, we do not specify that its set of constraints has to be satisfiable. Indeed, there are two strategies : to check the satisfiability of the constraints either at each step, or only when the empty clause is generated. The first solution is costly, but it avoids further useless deductions. So, the choice of the strategy is left to the user.

4.1 Paramodulation Strategy

We define in the following 6 inference rules, presented as fractions with the initial clauses on top and the deduced clause below. In the given examples, $*$ is an AC-operator.

Definition 1 (Constrained AC-Factoring)

$$\frac{L_1 \vee \ldots \vee L_n \vee D \,[\![T_1]\!]}{L_1 \vee D \,[\![T_1 \wedge T_2]\!]} \quad \text{where } T_2 \text{ is } L_1 =_{AC} \ldots =_{AC} L_n \wedge L_1 \succ D$$

The next rule simulates a resolution step with the reflexivity axiom ($x = x$).

Definition 2 (Constrained AC-Reflexion)

$$\frac{\neg(s = t) \vee D \,[\![T_1]\!]}{D \,[\![T_1 \wedge T_2]\!]} \quad \text{where } T_2 \text{ is } s =_{AC} t \wedge (s = t) \succ D$$

Definition 3 (Constrained AC-Resolution)

$$\frac{A_1 \vee D_1 \; [\![T_1]\!] \qquad \neg A_2 \vee D_2 \; [\![T_2]\!]}{D_1 \vee D_2 \; [\![T_1 \wedge T_2 \wedge T_3]\!]} \quad \text{where } T_3 \text{ is } A_1 =_{AC} A_2 \; \wedge \; A_1 \succ D_1 \vee D_2$$

The previous two inference rules are the only ones for generating the *empty clause* (denoted $\square$), i.e. a clause with no literals and a satisfiable set of constraints.

The next two inference rules are the paramodulation and the contextual paramodulation. The first one is well known, but the second one is specific to AC theories ; this rule and the extended paramodulation, defined later, simulate the associativity axiom. The experienced reader will remark that these rules simulate the use of extensions of clauses [PS81]. An example of contextual paramodulation (without constraints) is : $\frac{(a*b=c) \quad P((a*d)*b)}{P(c*d)}$, where the contextual paramodulation is applied in the subterm $(a * d) * b$, which is AC-equal to $(a * b) * d$, and the term d is what we call the context of the paramodulation step.

In the next two rules, the position p, where the paramodulation is applied, has to be non-variable and maximal for AC-operators (see definition of $\mathcal{MPos}$ in Section 3). Indeed, if a (contextual) paramodulation can be applied at a non-maximal position of an AC-operator f, a contextual paramodulation at the corresponding maximal position of f is equivalent. For instance, applying a paramodulation (or contextual paramodulation) in the subterm $f(t_1, t_2)$ of the clause $P(f(f(t_1, t_2), t_3))$ is equivalent to a contextual paramodulation in the subterm $f(f(t_1, t_2), t_3)$ where t_3 is in the context.

Definition 4 (Constrained AC-Paramodulation)

$$\frac{(l=r) \vee D_1 \; [\![T_1]\!] \qquad L \vee D_2 \; [\![T_2]\!]}{L[x_r]_p \vee D_1 \vee D_2 \; [\![T_1 \wedge T_2 \wedge T_3]\!]}$$

where x_r is a new variable, $p \in \mathcal{MPos}(L)$
and T_3 is $L_{|_p} =_{AC} l \; \wedge \; x_r =_{AC} r \; \wedge \; l \succ r \; \wedge \; (l=r) \succ D_1 \; \wedge \; L \succ D_2$

Note that in the deduced clause, the term r is replaced by a new variable (x_r) to avoid further paramodulation steps inside.

In the next rule, we use a new kind of variables, called *extension variables* and indexed by an AC-operator, for representing the context of contextual paramodulation steps. These special variables implement the same notion of irreducibility subterms used by Nieuwenhuis and Rubio in [NR94], and we detail their purpose in the following.

Definition 5 (Constrained AC-Contextual Paramodulation)

$$\frac{(l=r) \vee D_1 \; [\![T_1]\!] \qquad L \vee D_2 \; [\![T_2]\!]}{L[f(x_r, z_f)]_p \vee D_1 \vee D_2 \; [\![T_1 \wedge T_2 \wedge T_3 \wedge T'_3]\!]}$$

where x_r is a new variable, $p \in \mathcal{MPos}(L)$,
   z_f is a new extension variable for the AC-operator f ,
and T_3 is $L_{|_p} =_{AC} f(l, z_f) \; \wedge \; x_r =_{AC} r \; \wedge \; l \succ r \; \wedge \; (l=r) \succ D_1 \; \wedge \; L \succ D_2$
and T'_3 is the conjunction of the constraints $\mathcal{H}terms(l, f) \not\subseteq_{AC} \mathcal{H}terms(t, f)$, for
   each term t in $\mathcal{H}terms(L_{|_p}, f)$ which is not an extension variable for f.

This rule allows to replace only parts of the subterm at position p of the literal L, provided that the operator at p is AC. So, the idea is that L is divided in two parts : the first one has to be AC-unifiable with l, and the second one (referred by z_f) is the context. The constraints on $\mathcal{H}terms$ mean that we did not apply a paramodulation into a non-extension variable of L, assuming we forget the use of maximal positions of AC-operators ; moreover, they guarantee that the root operator of l is f.

The role of extension variables z_f is important but does not need a strong control. Indeed, in other inference rules, they are considered as other variables. But, in a contextual paramodulation, we do not put back the context, i.e. the non-used subterms, in the deduced clause and it produces the loss of completeness as shown in the next example. To correct this, we allow further contextual paramodulations finding the term to be replaced in this context, marked by an extension variable. This is the reason why the condition on $\mathcal{H}terms$ is not applied to extension variables. As mentioned earlier, extension variables implement the same notion of irreducibility as in [NR94]. Indeed, in the proof of completeness, each term appearing in the constraints is assumed to be irreducible ; but, when a ground term t is associated to an extension variable z_f, the irreducibility is required only for the terms of $\mathcal{H}terms(t,f)$.

Example 1 Given three clauses $(a*b=c)$ (1) , $P((a*a)*(b*b))$ (2) and $\neg P(c*c)$ (3) , we can prove that they form an AC-inconsistent system in the following way :
A contextual paramodulation from (1) into (2) at the subterm $(a*a)*(b*b)$ produces the clause $P(x*z_*)\,[\![\,x=_{AC}c \land z_*=_{AC}a*b\,]\!]$ (4) ; then, another contextual paramodulation from (1) into (4) at the subterm $x*z_*$ permits to deduce $P(y*z'_*)\,[\![\,z=_{AC}a*b \land z'_*=_{AC}c \land y=_{AC}c\,]\!]$ (5) . Then, a resolution between (3) and (5) produces the empty clause $\square\,[\![\,z=_{AC}a*b \land z'=_{AC}c\,]\!]$. If the condition on $\mathcal{H}terms$ was generalized to variables of extension, we could not deduce the clause (5) and find a contradiction since the left-hand side $a*b$ of the equation would be found in a variable. ♦

But, this is the only role of such extension variables and the next rule has the highest priority.

Definition 6 (Deletion of Extension Variables) *If there is an extension variable z_f in the constraints T of a clause $C\,[\![\,T\,]\!]$ which does not appear in C, this clause has to be replaced by $C\,[\![\,T\sigma\,]\!]$, where σ is the substitution $\{z_f \mapsto z\}$ and z is a new variable.*

This definition is motivated by the following properties :

1. If an extension variable z_f appears only in the constraints of a clause, it cannot be used by a contextual paramodulation step and also loses its special interest.

2. Any extension variable z_f has a single occurrence in a clause : it is created by the contextual paramodulation rule, and, to be duplicated, it has to instantiate another variable. But, since unifiers are not computed, variables are never instantiated and also an extension variable cannot appear twice in a clause.

3. As a consequence, an extension variable may only appear in a clause under the AC-operator for which it is defined.

In the last two inference rules, *a further refinement can be defined.* If the literal L is a negative equation $\neg(s=t)$, and if the position of the (contextual) paramodulation is in the term s, we can add the ordering constraint $s \succ t$, specifying that this inference step has to be applied in the maximal side of the equation.

The following extended paramodulation rule can be viewed as a mutual contextual paramodulation between two clauses, at the head of the left-hand side of their maximal equation.

Definition 7 (Constrained AC-Extended Paramodulation)

$$\frac{(l_1=r_1)\vee D_1\ [\![\,T_1\,]\!] \qquad (l_2=r_2)\vee D_2\ [\![\,T_2\,]\!]}{(f(x_{r_1},z_1)=f(x_{r_2},z_2))\vee D_1\vee D_2\ [\![\,T_1\wedge T_2\wedge T_3\,]\!]}$$

where x_{r_1}, x_{r_2}, z_1 and z_2 are new variables, $f\in\mathcal{F}_{AC}$
and T_3 is $f(l_1,z_1)=_{AC}f(l_2,z_2) \land x_{r_1}=_{AC}r_1 \land x_{r_2}=_{AC}r_2$
$\land\ l_1\succ r_1 \land l_2\succ r_2 \land (l_1=r_1)\succ D_1 \land (l_2=r_2)\succ D_2$
$\land\ \mathcal{H}terms(l_1,f)\cap_{AC}\mathcal{H}terms(l_2,f)\neq\emptyset$
$\land\ \mathcal{H}terms(z_1,f)\cap_{AC}\mathcal{H}terms(z_2,f)=\emptyset$

In the deduced clause, the terms r_1 and r_2 and the contexts z_1 and z_2 are replaced by variables ; it is different from the contextual paramodulation where the context is marked by an extension variable ; indeed, our proof technique for completeness (transfinite semantic trees) has shown that paramodulations in these contexts were useless. The last constraints force the existence of a maximal overlap between l_1 and l_2. In other words, we have to consider the maximal set of shared subterms between l_1 and l_2.

So, we have defined a set of inference rules that drastically limits the number of possible deductions, thanks to the ordering and AC-unification constraints, but particularly because we never compute any AC-unifier ; we need only to check the AC-unifiability. In the following, we show that our results also apply to the superposition strategy.

4.2 Superposition Strategy

The previous inference rules can be refined for the superposition strategy. The main difference between superposition and paramodulation is that superposition is only applied into maximal sides of equational literals. So, the three paramodulation rules are replaced by rules of *Constrained AC-Superposition*, *Constrained AC-Contextual Superposition* and *Constrained AC-Extended Superposition*. Their definitions differ from the paramodulation ones by :

1. If a (contextual) superposition is applied from an equation $(l = r)$ into the term s of a positive equational literal $(s = t)$, the constraints $s \succ t$ and $(s = t) \succ (l = r)$ are added.

2. In inference rules, the constraints representing the maximality of an equational literal $(l = r)$ or $\neg(l = r)$ (where l is assumed to be greater than r) in a clause have to be modified :
   - $(l = r)$ is maximal in a clause $(l = r) \vee D$ if $(l = r) \succ L$ for any positive literal L of D and, $l \succ s$ and $l \succ t$ for any negative literal $\neg(s = t)$ of D.
   - $\neg(l = r)$ is maximal in a clause $\neg(l = r) \vee D$ if $(l = r) \succ \neg L$ for any negative literal L of D and, $(l = l) \succ L$ for any positive literal L of D.

3. The right-hand sides are no more replaced by new variables ; they appear explicitly in the deduced clause.

For completeness, a new rule has to be added :

Definition 8 (Constrained AC-Equational Factoring)

$$\frac{(u_1 = v_1) \vee (u_2 = v_2) \vee D \ [\![T_1]\!]}{(u_1 = v_1) \vee \neg(v_1 = v_2) \vee D \ [\![T_1 \wedge T_2]\!]}$$

where T_2 is $u_1 =_{AC} u_2 \wedge u_1 \succ v_1 \wedge (u_1 = v_1) \succ (u_2 = v_2) \vee D$

This inference rule transforms a clause so that it does no more contain two positive equational literals with the same left-hand side.

Nieuwenhuis and Rubio have defined a similar system of inference rules for superposition [NR94]. Both inference systems do not compute any AC-unifier and permit l to be in a variable of $L_{|_p}$ for the contextual superposition rule. But, they do not restrict it to extension variables. So, they allow useless inference steps which consist in applying a superposition step in a variable, as exemplified by the following :

Example 2 Given three clauses $((a*a)*x = a*x)\,(1)$, $(b*c = c)\,(2)$ and $P((a*b)*(a*c))\,(3)$, we can deduce the clause $P(a*x) \ [\![x =_{AC} b*c]\!] \ (4)$ by a superposition from (1) into (3). Then they allow a contextual superposition from (2) into (4) which produces $P(c*y) \ [\![y =_{AC} a]\!] \ (5)$.

If we forget that we are working at maximal positions of AC-operators, this last step corresponds to a superposition from (2) into the variable x, which is well-known to be useless. With our constraints, we forbid it and a similar clause can be derived by a contextual superposition from (2) into (3) which produces $P(c * z_*)[\![z_* =_{AC} a*a]\!]$ (4') and a superposition from (1) into (4') which produces $P(a*x)[\![x =_{AC} c]\!]$ (5').
The clauses (5) and (5') are apparently different, but if we use the propagation of constrained subterms in both sequences of deduction, the terms a and c appeared in the constraints. So, the deduced clauses should be $P(y*x)[\![x =_{AC} c \wedge y =_{AC} a]\!]$ (5''). ♦

Moreover, Nieuwenhuis and Rubio do not have our restrictions on $\mathcal{H}terms$ for the contextual and extended superpositions.

Now, we have to introduce the lifting lemma for our inference rules, since the proof of completeness, based on transfinite semantic trees, is first obtained in the ground case (see Section 6).

4.3 Lifting Lemma

The purpose of a lifting lemma is to show that inferences on ground clauses are instances of inferences on the corresponding general clauses. But, lifting a paramodulation step may be impossible, when it replaces a subterm introduced by the substitution which makes the clause ground. The solution of [Pet83] is to restrict paramodulation to ground clauses where variables are substituted by irreducible terms w.r.t. a term rewriting system. The technique has been applied to prove the AC-completeness of inference rules described in [RV93], and we are going to generalize it for the constrained paramodulation and superposition strategies.

The next proposition is a consequence of the stability property of the ordering $\succ$.

Proposition 9 *Let A and B be two objects. If $A\sigma \succ B\sigma$ for a substitution σ, then $A \not\prec B$.*

A consequence of this proposition is :

Corollary 10 *Let σ be a ground AC-unifier of objects $A_1, \ldots, A_m$, and B_1 and B_2 be two objects such that : $B_1\sigma \succ B_2\sigma$. There exists an AC-mgu τ of $A_1, \ldots, A_m$ such that : $B_1\tau \not\prec B_2\tau$.*

This corollary permits to lift ordering constraints, and the substitution τ, defined by $\sigma = \tau\rho$ (for a ground substitution ρ), allows us to lift the AC-unification constraints.

We have to check that whenever the conditions for applying inference rules to the ground level are satisfied, they are also valid at the general level. These conditions are (for paramodulation and superposition rules) :

1. p is a non-variable position in L

2. The constraints on $\mathcal{H}terms$ are satisfiable

The first condition remains valid at the general level since it concerns an occurrence in C (for a constrained clause $C[\![T]\!]$). For the second ones, using the substitutions σ, τ and ρ defined above, and given two terms s and t, if the constraint $\mathcal{H}terms(s\sigma, f) \not\subseteq_{AC} \mathcal{H}terms(t\sigma, f)$ is satisfied, the constraint $\mathcal{H}terms(s\tau, f) \not\subseteq_{AC} \mathcal{H}terms(t\tau, f)$ is satisfiable by ρ, since $\sigma = \tau\rho$. The reasoning is the same if we test emptiness or non-emptiness of the intersection between $\mathcal{H}terms(s\sigma, f)$ and $\mathcal{H}terms(t\sigma, f)$. So, all these constraints are satisfiable at the general level and we can state the main lemma :

Lemma 11 (Lifting Lemma) *Let $C_1[\![T_1]\!], \ldots, C_n[\![T_n]\!]$ be clauses and σ a ground solution of all the T_i, $i = 1 \ldots n$. If an inference rule R applies to $C_1[\![\sigma]\!], \ldots, C_n[\![\sigma]\!]$ and produces a clause $C[\![\sigma_c]\!]$, the same inference rule R applies to clauses $C_1[\![T_1]\!], \ldots, C_n[\![T_n]\!]$ and generates a clause $D[\![T]\!]$ such that C and D are AC-equal, and σ_c is a solution of T.*

5 Simplification Rules

In this section, we define subsumption and simplification rules compatible with the constrained inference rules described in the previous section. But, these rules have strong conditions to be applied ; we also propose new ones which are more often applicable. These rules are fundamental for the efficiency of our theorem prover, since they allow us to delete redundant clauses.

5.1 Constrained Subsumption

The purpose of subsumption is to eliminate clauses which are redundant. In the classical definition, *a clause C_1 subsumes a clause C_2 if there is a substitution ρ such that $C_1\rho \subseteq C_2$*. Since clauses are considered as multisets, C_2 has at least as many literals as C_1. It is extended to :

Definition 12 (Constrained AC-Subsumption) *A constrained clause $C_1 \llbracket T_1 \rrbracket$ subsumes a constrained clause $C_2 \llbracket T_2 \rrbracket$ if there is a solution σ_1 of T_1 such that : for each solution σ_2 of T_2, there is a substitution ρ such that $C_1\sigma_1\rho$ is a subset of $C_2\sigma_2$, i.e.* $\exists \sigma_1 \in Sol(T_1)$, $\forall \sigma_2 \in Sol(T_2)$, $\exists \rho$, $C_1\sigma_1\rho \subseteq_{AC} C_2\sigma_2$

The standard way for applying this rule consists in the deletion of the *strictly subsumed* clauses, i.e. of clauses which are subsumed by a clause, but which do not subsume this clause. However, in our case, this causes the loss of the completeness. Indeed, as shown in [BGLS92], it may happen when the terms of the co-domain of $\sigma_1\rho$, $CoDom(\sigma_1\rho)$, are not included in those of σ_2, $CoDom(\sigma_2)$. So, first, we check that this condition is satisfied.

Definition 13 (Application of the Strict Constrained AC-Subsumption) *If a clause $C_1 \llbracket T_1 \rrbracket$ subsumes strictly a clause $C_2 \llbracket T_2 \rrbracket$ and if $CoDom(\sigma_1\rho) \subseteq_{AC} CoDom(\sigma_2)$, the clause $C_2 \llbracket T_2 \rrbracket$ is deleted.*

This definition has to be modified for taking into account the extension variables introduced by the contextual paramodulation and superposition rules (Definition 5). The only modification is in the calculus of $CoDom(\sigma_i)$ in previous definition : if z_f is an extension variable appearing under the AC-operator f in C_i, the set of terms $\mathcal{H}terms(z_f\sigma_i, f)$ will replace $z_f\sigma_i$ in $CoDom(\sigma_i)$.

From now on, whenever we will talk about the co-domain of a substitution, it will refer to this new definition.

5.2 Constrained Simplification

Usually, a clause $L \vee D$ is simplified by an equality $(l = r)$ at a position p of the literal L if there is a substitution ρ such that : $L_{|p} = l\rho$ and $l\rho \succ r\rho$. This definition becomes :

Definition 14 (Constrained AC-Simplification) *A clause $(l=r) \vee D_1 \llbracket T_1 \rrbracket$ simplifies a clause $L \vee D_2 \llbracket T_2 \rrbracket$ at a position $p \in \mathcal{MP}os(L)$ if :*

$\exists \sigma_1 \in Sol(T_1)$, $\forall \sigma_2 \in Sol(T_2)$, $\exists \rho \in AC\_match\{l\sigma_1, L_{|p}\sigma_2\}$ solution of $T_1\sigma_1$,
$$\begin{cases} D_2\sigma_2 =_{AC} D_2'\sigma_2 \vee D_2''\sigma_2 \quad \text{and} \quad D_1\sigma_1\rho \subseteq_{AC} D_2'\sigma_2 \\ l\sigma_1\rho \succ r\sigma_1\rho \\ (l=r)\sigma_1\rho \prec L\sigma_2 \vee D_2''\sigma_2 \end{cases}$$

This defines a conditional simplification, since the simplifying clause may contain other literals than $(l = r)$, provided that they are in the instances of the clause D_2 too. Notice that set inclusion is used rather than multiset inclusion in the definition. It means that there may be a literal which appears twice in $D_1\sigma_1\rho$ and only once in $D_2'\sigma_2$. Moreover, our proof technique imposes that a literal of $L \vee D_2$, which is not in D_1, has to be greater than $(l = r)$.

Definition 15 (Constrained AC-Contextual Simplification) $(l = r) \vee D_1 \, [\![T_1]\!]$ *simplifies contextually* $L \vee D_2 \, [\![T_2]\!]$ *at a position* $p \in \mathcal{MP}os(L)$ *if* :

$\exists \sigma_1 \in \mathcal{S}ol(T_1)$, $\exists \nu \in AC\_mgus\{f(y,z), L_{|_p}\}$,
$\forall \sigma_2 \in \mathcal{S}ol(T_2)$, $\exists \mu$, $\sigma_2 = \nu\mu$, $\exists \rho \in AC\_match\{l\sigma_1, y\nu\mu\}$ *solution of* $T_1\sigma_1$,

$$\begin{cases} \text{either } \mathcal{H}ead(y\nu) = \mathcal{H}ead(l\sigma_1\rho) = f \in \mathcal{F}_{AC} \\ \text{or } y\nu \text{ is an extension variable for } f \text{ and } \mathcal{H}ead(y\nu\mu) = f \\ D_1\sigma_1\rho \subseteq_{AC} D_2\sigma_2 \\ l\sigma_1\rho \succ r\sigma_1\rho \end{cases}$$

This rule is the counterpart of the contextual paramodulation. But, here we cannot use the notion of extension variable to express the context ; we have to split $L_{|_p}$ explicitly into two parts by calculating AC-unifiers ($AC\_mgus\{f(y,z), L_{|_p}\}$) : one is matched by $l\sigma_1$ ($y\nu$), and the other (the context) has to be reinserted in the deduced clause ($z\nu$). The substitution μ is needed to relate ν and σ_2. We do not have to check that there is an atom in $L\sigma_2 \vee D_2\sigma_2$ which is greater than $(l = r)\sigma_1\rho$, since $L\sigma_2$ obviously satisfies this condition.
The next example shows the loss of completeness if we used the technique of extension variables, like in the contextual paramodulation.

Example 3 Given the AC-inconsistent system, where f is AC and $c \succ d$,

$$P(f(f(a,d), f(b,c))) \quad (1) \qquad (f(a,b) = b) \quad (2)$$
$$\neg P(f(f(d,d), b)) \quad (3) \qquad (c = d) \quad (4)$$

The clause (1) can be simplified by (2) into $P(f(b, z_f)) \, [\![z_f =_{AC} f(c,d)]\!]$ (1) . But then, there is no way to prove the AC-inconsistency. With the previous definition, the simplified clause is $P(f(b, f(c,d)))$ (1) , which can be simplified by (4) into $P(f(b, f(d,d)))$ (1) . A resolution step between (1) and (3) yields the contradiction. ♦

The classical application of such simplification rules is to replace the clause $L \vee D$ by $L[r]_p \vee D$. But, we encounter the same problems as for the subsumption ; so a condition on the co-domains of substitutions is again necessary :

Definition 16 (Application of the Constrained AC-Simplifications) *If a clause* $C_1 \, [\![T_1]\!]$ *simplifies a clause* $C_2 \, [\![T_2]\!]$ *at a position* p *and if* $CoDom(\sigma_1\rho) \sqsubseteq_{AC} CoDom(\sigma_2)$, *the clause* $C_2 \, [\![T_2]\!]$ *is replaced by* $L[r]_p \vee D_2 \, [\![T_2 \wedge T'_1\sigma_1 \wedge L_{|_p} =_{AC} l\sigma_1 \wedge \sigma_{1|_{Var(r)}}]\!]$ *for the simplification rule, and* $L[f(r, z\nu)]_p \vee D_2 \, [\![T_2 \wedge T'_1\sigma_1 \wedge y\nu =_{AC} l\sigma_1 \wedge \sigma_{1|_{Var(r)}} \wedge \nu_{|_{Var(L_{|_p})}}]\!]$ *for the contextual simplification rule* ; T'_1 *is the subset of non AC-unification constraints of* T_1,

5.3 Special Subsumption and Simplification

We saw that conditions on co-domains of substitutions have to be respected to keep the completeness of our inference rules. But, they limit the number of applicable simplifications. To balance with this, we modify the definitions of subsumption and simplification rules so that, even if conditions on co-domains are not satisfied, the rules can be applied provided that we partially instantiate the skeleton of the simplifying clause C_1 [BGLS92].
However, this is not sufficient since a constraint may have several solutions. Hence, to partially instantiate C_1 may produce the loss of some of these substitutions. The solution we propose is to create a new clause, the instantiation of $C_1 \, [\![\sigma_1]\!]$, and to transform $C_1 \, [\![T_1]\!]$ into $C_1 \, [\![T_1 - \sigma_1]\!]$, i.e. specifying that σ_1 is no more a solution of the constraints. To denote this, we can either use AC-disunification, i.e. $C_1 \, [\![T_1 \wedge \{\bigvee_{x \in \mathcal{D}om(\sigma_1)} x \neq_{AC} x\sigma_1\}]\!]$, or ordering constraints, i.e. $C_1 \, [\![T_1 \wedge \{\bigvee_{x \in \mathcal{D}om(\sigma_1)} x \prec x\sigma_1 \vee x \succ x\sigma_1\}]\!]$. But, the second technique raises problems with the new variables introduced by the AC-unification algorithm.

So, for the subsumption and the simplification rules, if $CoDom(\sigma_1\rho) \not\sqsubseteq_{AC} CoDom(\sigma_2)$, the actions are :

- to create the new clause $C_1\tau \llbracket \sigma_1' \wedge T_1'\sigma_1' \rrbracket$, where T_1' is the subset of non AC-unification constraints of T_1, $\sigma_1 = \tau\sigma_1'$ and $CoDom(\sigma_1'\rho) \sqsubseteq_{AC} CoDom(\sigma_2)$,
- to transform the clause $C_1 \llbracket T_1 \rrbracket$ into $C_1 \llbracket T_1 - \sigma_1 \rrbracket$,
- for simplification and contextual simplification rules, in the deduced clause, to instantiated the right-hand side of the equation r by τ and to use σ_1' instead of σ_1 in the constraints.

5.4 Some Particular Simplification Rules

The simplification rules defined previously are applied only when all the instances of a clause can be simplified by the same clause, and at the same position. This is very restrictive, and we can imagine that, if only one instance of a clause is simplifiable, it may be interesting to simplify it. This is why we define *local simplification* rules, by specifying that we consider only one solution ($\exists\sigma_2 \in Sol(T_2)$). Applications of local simplifications consist first in creating a new clause, the result of the simplification step, second in transforming the clause $L \vee D_2 \llbracket T_2 \rrbracket$, by adding a constraint to specify that σ_2 is discarded from the set of solutions of T_2.

Example 4 If we apply a local simplification from the clause $(a*b=c) \llbracket \rrbracket$ (1) into the clause $P(a*y,z) \llbracket y*z =_{AC} d*b \rrbracket$ (2), using the substitution $\sigma_2 = \{y \mapsto b, z \mapsto d\}$, the generated clause is $P(c,z) \llbracket y =_{AC} b \wedge z =_{AC} d \rrbracket$ (3), and the clause (2) is replaced by $P(a*y,z) \llbracket T \rrbracket$, where T is the set of constraints $y*z =_{AC} d*b \wedge \{y \neq_{AC} b \vee z \neq_{AC} d\}$, transformed to $y =_{AC} d \wedge z =_{AC} b$. ♦

With the same principle, we define the *local strict subsumption* rule, which is applied when an instance of a clause is strictly subsumed.

An important simplification is that, given a constrained clause $C \llbracket T \rrbracket$ and a solution σ of T, *if the clause $C \llbracket \sigma \rrbracket$ is simplifiable in $CoDom(\sigma)$, it is redundant.* Directly from the proof, we see that we never need ground instances of clauses that are reducible in their substitution part. This deletion is also *locally* applied, and we notice that, if every instance of a clause is simplifiable in its substitution (even by different clauses) this clause is redundant and can be deleted.

Finally, a number of other classical simplification rules can be adapted (locally or not) to the constrained deduction. For instance, we can delete clauses which contain a literal AC-equal to an identity ($s = s$) and clauses which contain literals L and $\neg L'$, where L and L' are AC-equal. Another one is the *clausal simplification* rule, which deletes all instances of a literal L in every clause of S, if $\neg L$ is a clause of S. We can also delete every instance of $\neg(x = x)$ (it is a kind of *trivial reflexion*). A clausal simplification step may be viewed as a resolution step, followed by the deletion of one of the parent clauses.

6 Refutational Completeness

Let INF be the set of inference rules and simplification rules described in previous sections. A *derivation* from a set of clauses S_0 is a sequence $S_0, S_1, \ldots$ of sets of constrained clauses obtained by successive applications of rules of INF. Let S^* denote $\bigcup_{i \geq 0} S_i$. A derivation $S_0, S_1, \ldots$ is *fair* if each clause $C \llbracket T \rrbracket$ of $\bigcap_{i \geq j} INF(S_i)$ is subsumed by a clause $C' \llbracket T' \rrbracket$ of S^*. A clause $C \llbracket T \rrbracket$ is *persistent* in a fair derivation $S_0, S_1, \ldots$ if there is an index k such that $C \llbracket T \rrbracket$ belongs to $\bigcap_{i \geq k} S_i$. A set of clauses S is *AC-unsatisfiable* if it has no *AC-consistent* model, i.e. no model coherent with the AC axioms.

The following theorem states that INF is refutationally complete in AC theories.

Theorem 17 *Let S be an AC-unsatisfiable set of clauses with empty constraints. Then, every fair derivation from S generates the empty clause.*

The method we use for proving this theorem is based on the notion of transfinite semantic trees introduced in [HR91], which was adapted to AC theories in [RV93]. We present here only a sketch of the proof, but it is detailed in [Vig93].
Sketch of Proof : *In the empty theory, the proof is done by defining the transfinite tree of all possible interpretations of the Herbrand base $\mathcal{A}(\mathcal{P}, \mathcal{F})$, and considering its maximal consistent semantic tree (MCT), i.e. the maximal subtree of interpretations which do not falsify a ground instance of a clause of S^*. Then, the rightmost path of that subtree is proved empty, which implies that MCT is empty too, and also that the empty clause belongs to S^*.*

But, this rightmost path of MCT may not be a model for AC theories. So, we define by induction a sequence Π of nodes in MCT, which is the rightmost AC-consistent path. Then, we prove that Π is empty, by showing that its last node falsifies a clause deduced by an inference step between clauses labeling successors of nodes of Π. MCT has to be empty too, and also S^* contains the empty clause.

The problem with the basic strategy is that no paramodulation is allowed in the constrained part of a clause. Indeed, when a ground instance $C[\![\sigma]\!]$ of a constrained clause labels a node I, it means that the interpretation I falsifies the ground clause $C\sigma$; but, to show that MCT is empty, a paramodulation in a subterm introduced by the substitution σ in $C\sigma$ may be needed. So, to avoid this problem, we impose that a ground clause $C[\![\sigma]\!]$ labels a node I of a semantic tree if $C\sigma$ is falsified by this interpretation and if each term of $\mathcal{CoDom}(\sigma)$ is irreducible by the ground equations of I.

Another problem is raised by the simplification rules : the clauses used by an inference step may never appear in the same S_i. But, fortunately, this never keeps from generating the empty clause, since we show that every node, successor of a node of Π, can be labeled by a persistent clause, i.e. a clause which is never simplified. □
The case of non-empty initial constraints can be treated as in [NR92b].

7 Example of Resolution : SAM's Lemma

We present a fully detailed example of a proof by refutation, using inference rules described in Sections 4 and 5. However, these rules are not yet completely implemented in our prover DATAC ; indeed, AC-unifications are eagerly solved. Note that, in our implementation, clauses $\neg A_1 \vee \ldots \vee \neg A_m \vee B_1 \vee \ldots \vee B_n$ are represented as sequents $A_1, \ldots, A_m \Rightarrow B_1, \ldots, B_n$. So, superposition and simplification rules are decomposed in left- and right- rules, whether they are applied to the left or to the right of the sign $\Rightarrow$.
DATAC is written in CAML Light, a functional language of the ML family. It runs on SUN, HP and IBM PC workstations. The AC-unification and AC-matching procedures are based on Stickel's and Hullot's algorithms respectively.

The example below is the well known SAM's Lemma, posed by Bumcroft in 1965 as an open problem in lattice theory, and taken from the paper *Semi-Automated Mathematics* of Guard et al. [GOBS69]. We use its equational formulation from McCune [McC88]. To our knowledge, the only completely automatic proof of an equational version of this lemma was done by RRL [ZK88]. The lemma states that :

if x and $meet(u, v)$ are complements, and y and $join(u, v)$ are complements,
then $meet(join(y, meet(x, v)), join(y, meet(x, u)))$ is y.

Operators *meet* and *join*, written "." and "+" in the proof, are associative and commutative. The operator $COMP$, which denotes the complement, is commutative. The greatest element is 1. The least element is 0. Lattices are characterized by clauses 1 to 8, modularity by 9 ;

the complement is defined by clauses 10 to 12, and the skolemized negation of the theorem is expressed by clauses 13 to 15. The ordering used is a variant of the APO, with the precedence $COMP >_p = \text{ and } . >_p + >_p B >_p A >_p R_2 >_p R_1 >_p 1 >_p 0$.

>> Initial set of clauses :

Clause 1 : $\Rightarrow x_1.x_1 = x_1$ Clause 2 : $\Rightarrow x_1 + x_1 = x_1$
Clause 3 : $\Rightarrow x_1.(x_1 + x_2) = x_1$ Clause 4 : $\Rightarrow x_1 + (x_1.x_2) = x_1$
Clause 5 : $\Rightarrow x_1.0 = 0$ Clause 6 : $\Rightarrow x_1 + 0 = x_1$
Clause 7 : $\Rightarrow x_1.1 = x_1$ Clause 8 : $\Rightarrow x_1 + 1 = 1$
Clause 9 : $x_1.x_2 = x_1 \Rightarrow x_2.(x_1 + x_3) = x_1 + (x_3.x_2)$
Clause 10 : $COMP(x_1, x_2) \Rightarrow x_1.x_2 = 0$ Clause 11 : $COMP(x_1, x_2) \Rightarrow x_1 + x_2 = 1$
Clause 12 : $x_1 + x_2 = 1, x_1.x_2 = 0 \Rightarrow COMP(x_1, x_2)$
Clause 13 : $\Rightarrow COMP(R_1, A + B)$ Clause 14 : $\Rightarrow COMP(R_2, A.B)$
Clause 15 : $(R_1 + (R_2.A)).(R_1 + (R_2.B)) = R_1 \Rightarrow$

>> AC-Resolution between 10 and 13

Clause 16 : $\Rightarrow z_1.z_2 = 0$ [$z_1 =_{AC} R_1 \wedge z_2 =_{AC} B + A$]

>> AC-Resolution between 10 and 14

Clause 18 : $\Rightarrow z_1.z_2 = 0$ [$z_1 =_{AC} R_2 \wedge z_2 =_{AC} B.A$]

>> Left AC-Superposition from 9 into 15

Clause 99 : $z_1.z_2 = R_1 , z_1 + (z_3.z_2) = R_1 \Rightarrow$
     [$z_1 =_{AC} R_1 \wedge z_2 =_{AC} R_1 + (R_2.B) \wedge z_3 =_{AC} R_2.A$]

>> Left AC-Clausal Simplification in 99 thanks to 3

Clause 99 : $z_1 + (z_3.z_2) = R_1 \Rightarrow$ [$z_1 =_{AC} R_1 \wedge z_2 =_{AC} R_1 + (R_2.B) \wedge z_3 =_{AC} R_2.A$]

>> AC-Extended Superposition between 3 and 9

Clause 123 : $x_1.z_1 = x_1 \Rightarrow (x_1 + (x_4.z_1)).x_3 = x_3.z_2$
     [$z_1 =_{AC} x_2 + x_3 \wedge z_2 =_{AC} x_4 + x_1 \wedge \mathcal{H}terms(x_3,.) \cap_{AC} \{x_4 + x_1\} = \emptyset$]

>> Left AC-Contextual Superposition from 3 into 123

Clause 170 : $x_1.z. = z_3 \Rightarrow (z_3 + (x_4.z_1)).x_3 = x_3.z_2$
     [$z_1 =_{AC} x_2 + x_3 \wedge z_2 =_{AC} x_4 + (x_1.x_2) \wedge z_3 =_{AC} x_1.x_2 \wedge z. =_{AC} x_2$
     $\wedge \mathcal{H}terms(x_3,.) \cap_{AC} \{x_4 + (x_1.x_2)\} = \emptyset \wedge \{x_2 + x_3\} \not\subseteq_{AC} \mathcal{H}terms(x_1,.)$
     $\wedge ((x_1.x_2).(x_2 + x_3) = x_1.x_2) \succ (((x_1.x_2) + (x_4.(x_2 + x_3))).x_3 = x_3.(x_4 + (x_1.x_2)))$]

>> Trivial AC-Reflexion in 170

Clause 170 : $\Rightarrow (z_3 + (x_4.z_1)).x_3 = x_3.z_2$
     [$z_1 =_{AC} x_2 + x_3 \wedge z_2 =_{AC} x_4 + (x_1.x_2) \wedge z_3 =_{AC} x_1.x_2$
     $\wedge \mathcal{H}terms(x_3,.) \cap_{AC} \{x_4 + (x_1.x_2)\} = \emptyset \wedge \{x_2 + x_3\} \not\subseteq_{AC} \mathcal{H}terms(x_1,.)$
     $\wedge ((x_1.x_2).(x_2 + x_3) = x_1.x_2) \succ (((x_1.x_2) + (x_4.(x_2 + x_3))).x_3 = x_3.(x_4 + (x_1.x_2)))$]

>> Right AC-Superposition from 16 into 170

Clause 322 : $\Rightarrow z_1.z_2 = (z_3 + 0).z_1$
     [$z_1 =_{AC} A \wedge z_2 =_{AC} R_1 + (x_1.B) \wedge z_3 =_{AC} x_1.B \wedge \{A + B\} \not\subseteq_{AC} \mathcal{H}terms(x_1,.)$]

>> Right AC-Simplification from 6 into 322

Clause 322 : $\Rightarrow z_1.z_2 = z_3.z_1$
     [$z_1 =_{AC} A \wedge z_2 =_{AC} R_1 + (x_1.B) \wedge z_3 =_{AC} x_1.B \wedge \{A + B\} \not\subseteq_{AC} \mathcal{H}terms(x_1,.)$]

>> Left AC-Contextual Simplification from 322 into 99

Clause 99 : $z_1 + ((z_3.z_2).z_4) = R_1 \Rightarrow$
     [$z_1 =_{AC} R_1 \wedge z_2 =_{AC} A \wedge z_3 =_{AC} R_2.B \wedge z_4 =_{AC} R_2$]

>> Left AC-Contextual Simplification from 1 into 99

Clause 99 : $z_1 + (z_4.(z_2.z_3)) = R_1 \Rightarrow$ [$z_1 =_{AC} R_1 \wedge z_4 =_{AC} R_2 \wedge z_3 =_{AC} B \wedge z_2 =_{AC} A$]

>> Left AC-Simplification from 18 into 99

Clause 99 : $z_1 + 0 = R_1 \Rightarrow$ [$z_1 =_{AC} R_1$]

>> Left AC-Clausal Simplification in 99 thanks to 6

Clause 99 : $\square$

>> The initial system is AC-inconsistent <<

To give some statistics on this proof, from 15 initial clauses, 187 applications of inference rules and 474 simplifications, 307 clauses were generated ; it shows that our prover is not enough optimized for the choice of inferences to try. But, the more interesting point is that only 103 clauses were retained. These deletions are caused by simplifications in constraints (71), forward subsumptions (16) and eliminations of trivial clauses (132). Moreover, forward subsumption and the detection of trivial clauses have avoided the creation of 146 redundant clauses.

8 Conclusion and Further Works

We have designed a refutationally complete inference system for AC theories which never compute AC-unifiers and which incorporates basicness and ordering requirements. These refinements are expressed with constraints in a very natural way, and we have shown that Constrained AC-paramodulation is compatible with simplification rules. Such rules are fundamental for efficiency of theorem provers. Another way to use our inference rules would be to build the precedence on operators while resolving a system, such as some implementations of Knuth-Bendix. Some further optimizations are possible, as the propagation of blocked subterms in a clause [BGLS92]. For instance, in the introductory example, the subterm b may be blocked in the atom $Q(\boxed{a}, b)$ to obtain the clause $P(\boxed{b}) \vee Q(\boxed{a}, \boxed{b})$, since it is already blocked in $P(\boxed{b})$. Using constraints, it would be denoted by $P(y) \vee Q(x,y) \, [\![x =_{AC} a \, \wedge \, y =_{AC} b]\!]$.

These strategies have been implemented in our theorem prover **DATAC**, and experiments with non-trivial examples are successful. The method used here to prove completeness of inference systems in AC theories has been adapted to other strategies such as superposition [Rus91]. But, an essential point to be studied and implemented is the propagation of ordering constraints.

Acknowledgments : I wish to thank Michaël Rusinowitch and Claude and Hélène Kirchner for their pertinent remarks and their support for this work. Especially, I would like to thank Robert Nieuwenhuis and Albert Rubio for discussions on topics related to this paper.

References

[BG90]   L. Bachmair and H. Ganzinger. On restrictions of ordered paramodulation with simplification. In M. E. Stickel, editor, *Proc. 10th CADE Conf., Kaiserslautern (Germany)*, volume 449 of *LNCS*, pages 427–441. Springer-Verlag, July 1990.

[BGLS92] L. Bachmair, H. Ganzinger, C. Lynch, and W. Snyder. Basic paramodulation and superposition. In *Proc. 11th CADE Conf., Saratoga Springs (N.Y., USA)*, pages 462–476, 1992.

[BL79]   A. M. Ballantyne and D. S. Lankford. The refutation completeness of blocked permutative narrowing and resolution. In *Proc. of Fourth Workshop on Automated Deduction, Austin, Texas*, 1979.

[Bra75]   D. Brand. Proving theorems with the modification method. *SIAM J. of Computing*, 4:412–430, 1975.

[DJ90]   N. Dershowitz and J.-P. Jouannaud. Rewrite systems. In J. van Leeuwen, editor, *Handbook of Theoretical Computer Science*. Elsevier Science Publishers B. V. (North-Holland), 1990.

[GOBS69] J. R. Guard, F. C. Oglesby, J. H. Bennett, and Settle. Semi-automated mathematics. *JACM*, 16:49–62, 1969.

[HR91]   J. Hsiang and M. Rusinowitch. Proving refutational completeness of theorem-proving strategies : the transfinite semantic tree method. *JACM*, 38(3):559–587, July 1991.

[Hul80]   J.-M. Hullot. *Compilation de Formes Canoniques dans les Théories équationelles*. Th. 3e cycle, Université de Paris Sud, Orsay (France), 1980.

[JK91]   J.-P. Jouannaud and C. Kirchner. Solving equations in abstract algebras: a rule-based survey of unification. In J.-L. Lassez and G. Plotkin, editors, *Computational Logic. Essays in honor of Alan Robinson*, chapter 8, pages 257–321. MIT Press, Cambridge (MA, USA), 1991.

[KKR90]   C. Kirchner, H. Kirchner, and M. Rusinowitch. Deduction with symbolic constraints. *Revue d'Intelligence Artificielle*, 4(3):9–52, 1990. Special issue on Automatic Deduction.

[KMN85]   D. Kapur, D. R. Musser, and P. Narendran. Only prime superpositions need be considered in the Knuth-Bendix procedure, 1985. Computer Science Branch, Corporate Research and Development, General Electric, Schenectady, New York.

[McC88]   W. McCune. Challenge equality problems in lattice theory. In E. Lusk and R. Overbeek, editors, *Proc. 9th CADE Conf., Argonne (Ill., USA)*, volume 310 of *LNCS*, pages 704–709. Springer-Verlag, 1988.

[MN90]   U. Martin and T. Nipkow. Ordered rewriting and confluence. In M. E. Stickel, editor, *Proc. 10th CADE Conf., Kaiserslautern (Germany)*, volume 449 of *LNCS*. Springer-Verlag, 1990.

[NR91]   P. Narendran and M. Rusinowitch. Any ground associative-commutative theory has a finite canonical system. In R. V. Book, editor, *Proc. 4th RTA Conf., Como (Italy)*. Springer-Verlag, 1991.

[NR92a]   R. Nieuwenhuis and A. Rubio. Basic superposition is complete. In B. Krieg-Brückner, editor, *Proceedings of ESOP'92*, volume 582 of *LNCS*, pages 371–389. Springer-Verlag, 1992.

[NR92b]   R. Nieuwenhuis and A. Rubio. Theorem proving with ordering constrained clauses. In D. Kapur, editor, *Proceedings of CADE-11*, volume 607 of *LNCS*, pages 477–491. Springer-Verlag, 1992.

[NR94]   R. Nieuwenhuis and A. Rubio. AC-superposition with constraints: no AC-unifiers needed. In A. Bundy, editor, *Proceedings of the 12th International Conference on Automated Deduction, Nancy (France)*, LNCS. Springer-Verlag, 1994.

[Pau92]   E. Paul. A general refutational completeness result for an inference procedure based on associative-commutative unification. *JSC*, 14(6):577–618, 1992.

[Pet83]   G. Peterson. A technique for establishing completeness results in theorem proving with equality. *SIAM J. of Computing*, 12(1):82–100, 1983.

[Pet90]   G. E. Peterson. Complete sets of reductions with constraints. In M. E. Stickel, editor, *Proc. 10th CADE Conf., Kaiserslautern (Germany)*, volume 449 of *LNCS*, pages 381–395. Springer-Verlag, 1990.

[Plo72]   G. Plotkin. Building-in equational theories. *Machine Intelligence*, 7:73–90, 1972.

[PS81]   G. Peterson and M. E. Stickel. Complete sets of reductions for some equational theories. *JACM*, 28:233–264, 1981.

[RN93]   A. Rubio and R. Nieuwenhuis. A precedence-based total ac-compatible ordering. In C. Kirchner, editor, *Proc. 5th RTA Conf., Montreal (Canada)*, volume 690 of *LNCS*, pages 374–388. Springer-Verlag, 1993.

[Rus91]   M. Rusinowitch. Theorem-proving with resolution and superposition. *JSC*, 11:21–49, 1991.

[RV91]   M. Rusinowitch and L. Vigneron. Automated deduction with associative commutative operators. In P. Jorrand and J. Kelemen, editors, *Fundamental of Artificial Intelligence Research*, volume 535 of *LNCS*, pages 185–199. Springer-Verlag, 1991.

[RV93]   M. Rusinowitch and L. Vigneron. Automated deduction with associative commutative operators. Internal rep. 1896, INRIA, May 1993. To appear in *J. of Applicable Algebra in Engineering, Communication and Computation*.

[Sla74]   J. R. Slagle. Automated theorem-proving for theories with simplifiers, commutativity and associativity. *JACM*, 21(4):622–642, 1974.

[Vig93]   L. Vigneron. Basic AC-paramodulation. In F. Orejas, editor, *Proc. 2nd CCL Workshop, La Escala (Spain)*, September 1993.

[Wer92]   U. Wertz. First-order theorem proving modulo equations. Technical Report MPI-I-92-216, MPI Informatik, April 1992.

[ZK88]   H. Zhang and D. Kapur. First-order theorem proving using conditional rewrite rules. In E. Lusk and R. Overbeek, editors, *Proc. 9th CADE Conf., Argonne (Ill., USA)*, volume 310 of *LNCS*, pages 1–20. Springer-Verlag, 1988.

AC-superposition with constraints: no AC-unifiers needed

Robert Nieuwenhuis and Albert Rubio*

Technical University of Catalonia, Pau Gargallo 5, 08028 Barcelona, Spain
E-mail: {roberto,rubio}@lsi.upc.es.

Abstract. We prove the completeness of (basic) deduction strategies with constrained clauses modulo associativity and commutativity (AC). Here each inference generates *one single conclusion* with an additional equality $s =_{AC} t$ in its constraint (instead of one conclusion for each minimal AC-unifier, i.e. exponentially many). Furthermore, *computing AC-unifiers is not needed at all*. A clause $C \; [\![\, T \,]\!]$ is redundant if the constraint T is not AC-unifiable. If C is the empty clause this has to be decided to know whether $C \; [\![\, T \,]\!]$ denotes an inconsistency. In all other cases any sound method to detect unsatisfiable constraints can be used.

1 Introduction

Some fundamental ideas on applying symbolic constraints to theorem proving were given in [KKR90], where a constrained clause is a shorthand for its (infinite) set of ground instances satisfying the constraint. In a constrained equation $f(x) \simeq a \; [\![\, x = g(y) \,]\!]$, the equality '=' of the constraint is usually interpreted in $\mathcal{T}(\mathcal{F})$ (syntactic equality), or in some quotient algebra $\mathcal{T}(\mathcal{F})/\equiv_E$ where E is an equational theory. The methods in [KKR90] require to *propagate* parts of the constraints to the clause. For example, if '=' is syntactic equality, by such a propagation step the previous equation can be replaced by the logically equivalent one $f(g(y)) \simeq a \; [\![\, true \,]\!]$.

Avoiding propagation is essential for exploiting the constraints. By doing so for (syntactic) equality constrained clauses, in [NR92a] the completeness of *basic superposition* is proved (independently also in [BGLS92]), and by adding *ordering constraints* the search space is further reduced by *inheriting* the ordering restrictions of the inference rules ([NR92b]). Constrained (equational) superposition can e.g. be expressed like:

$$\frac{s' \simeq t' \; [\![\, T' \,]\!] \quad s \simeq t \; [\![\, T \,]\!]}{s[t']_p \simeq t \; [\![\, T' \wedge T \wedge s' \succ t' \wedge s \succ t \wedge s|_p = s' \,]\!]} \quad \text{where } s|_p \notin \mathcal{V}ars(s)$$

In this (elegant and powerful) representation for ordered inference rules, information from the meta-level (the ordering and unifiability restrictions) is kept and inherited to restrict future inferences: clauses with unsatisfiable constraints are tautologies, hence redundant. These inference rules are *basic* (i.e. no inferences are needed on subterms introduced by the unifiers of previous inferences) because the accumulated

---
* Both authors upported by the Esprit Working Group CCL, ref. 6028

unification problems are kept in the equality constraints (no propagation is needed) and future superpositions can take place only on non-variable subterms of the clause part.

Here we use these techniques for deduction modulo the theory of associativity and commutativity (AC), applying in our completeness proofs an essential ingredient which we recently gave for this purpose in [RN93]: an AC-compatible simplification ordering that is total on AC-distinct ground terms. This ordering is also defined on terms with variables, which makes it applicable in practice for checking the — non-ground— ordering restrictions (the first and only —as far as we know— other such ordering of [NR91] is defined only on ground terms). Our completeness proofs are based on the *model generation* framework with its *abstract redundancy notions* for detecting redundant clauses and inferences during the theorem proving process, defined by Bachmair and Ganzinger in [BG91]. Here we have adapted these techniques to AC-deduction, similarly to Wertz's work ([Wer92]), although he deals with explicit *extended* clauses, while we simulate them by means of specific AC-inference rules, like in [RV93]. We believe that our treatment of the AC-case is interesting in itself (apart from the constraints) because of its simplicity. Albert Rubio's PhD. thesis [Rub94] contains a very complete analysis of all these constrained deduction methods (also covering the AC case).

We prove the refutation completeness of a superposition-based inference system, where each inference has *one single conclusion* with an additional equality $s =_{AC} t$ in its constraint, instead of one conclusion for each minimal AC-unifier (double exponentially many: e.g. $x + x + x$ and $y_1 + y_2 + y_3 + y_4$ have more than a million minimal unifiers [Dom92]). This also eliminates the need of computing AC-unifiers in AC-deduction (or completion) methods [PS81, JK86, BD89] which have motivated a huge amount of research on computing complete sets of AC-unifiers (see [BS93] for a recent survey). A clause C with an AC-equality constraint T of the form $s_1 =_{AC} t_1 \land \ldots \land s_n =_{AC} t_n$ can be proved redundant by means of efficient incomplete methods detecting cases of unsatisfiability of T. If C is the empty clause one can decide the AC-unifiability of T (which is NP-complete, cf. [KN92]) to know whether an inconsistency has been derived or not.

The first results on (almost basic) constrained deduction modulo AC were reported by Laurent Vigneron. In a recent version of his work [Vig94] he also avoids the computation of AC-unifiers (by applying our notion of irreducibility, defin. 4.4) and defines several additional restrictions. His proofs are completely different from ours and based on transfinite semantic trees as in [RV93]. He also reports on an implementation of these methods. Advantages of our application of the model generation method wrt. semantic trees are simplicity: our full proofs can be given in a very reasonable amount of space (this paper), and also that the known extensions to the model generation method for constrained paramodulation like *redex orderings* and *variable abstraction* [BGLS92] can be smoothly incorporated here. Similarly, the *abstract* redundancy notions express very sharp bounds on *concrete* redundancy methods (like the ones given by Vigneron which indeed fit into our abstract ones).

In section 3 of this paper we give a superposition-based inference system, which is proved refutationally complete in section 4. Section 5 is on the compatibility with simplification and deletion methods, and section 6 is on Knuth-Bendix-like completion of (conditional) equations modulo AC.

2 Basic notions and terminology

Here an *equation* is a multiset of terms $\{s,t\}$, which will be written in the form $s \simeq t$. A first-order clause is a pair of (finite) multisets of equations Γ (the *antecedent*) and Δ (the *succedent*), denoted by $\Gamma \rightarrow \Delta$. By *(ordering and equality) constraints* we mean quantifier-free first-order formulae built over the binary predicate symbols $\succ$ and $=_{AC}$ relating terms in $T(\mathcal{F},\mathcal{X})$, where $=_{AC}$ denotes AC-equality, and $\succ$ denotes the AC-compatible simplification ordering of [RN93] on ground terms. We say that a ground substitution σ *satisfies* a constraint T if $T\sigma$ is (or evaluates to) true in this sense.

We extend $\succ$ to an ordering on ground equations (in fact, to their *occurrences* in clauses) and to clauses, s.t. terms in the antecendent get a slightly higher complexity than in the succedent. An occurrence of an equation $t \simeq t'$ in an antecendent is (the two-fold multiset) $\{\{t, t'\}\}$, and in a succedent it is $\{\{t\}, \{t'\}\}$. Now the two-fold multiset extension of $\succ$ is total on (AC-distinct occurrences of) ground equations, and the three-fold multiset extension of $\succ$ is a total ordering on (AC-distinct) ground clauses. We will ambiguously use $\succ$ to denote all these orderings on terms, equations and clauses. An equation e is called *maximal* in a ground clause C if there is no equation e' in C such that $e' \succ e$ and *strictly maximal* if there is no e' with $e' \succeq e$ (i.e. $e' \succ e$ or $e' =_{AC} e$).

An interpretation I is a congruence on ground terms. It satisfies a ground clause $\Gamma \rightarrow \Delta$, denoted $I \models \Gamma \rightarrow \Delta$, if $I \not\supseteq \Gamma$ or else $I \cap \Delta \neq \emptyset$. An interpretation I satisfies (is a model of) a constrained clause $C \llbracket T \rrbracket$, denoted $I \models C \llbracket T \rrbracket$, if it satisfies every ground instance of $C \llbracket T \rrbracket$, i.e. every $C\sigma$ such that σ is ground and $T\sigma$ is true. Therefore, clauses with unsatisfiable constraints are tautologies, and $C \llbracket T \rrbracket$ is the *empty clause* only if C is empty and T is satisfiable. I satisfies a set of clauses S, denoted by $I \models S$, if it satisfies every clause in S. A clause C follows from a set of clauses S (denoted by $S \models C$), if C is satisfied by every model of S. For dealing with non-equality predicates, atoms A can be expressed by equations $A \simeq true$ where *true* is a special symbol (minimal in $\succ$).

We use the definitions of [DJ90] for rewriting-related notions. However, to avoid confusion with the arrow $\rightarrow$ of clauses, we denote ground rewrite rules (ground equations $s \simeq t$ with $s \succ t$) by $s \Rightarrow t$. The congruence generated by a set of equations (or rewrite rules) E (which is an interpretation) will be denoted by E^*.

It is well-known that a term s can be *flattened* by removing all AC-operators f that are immediately below another f. For example, if f and g are AC-operators, then $h(f(f(a,a), f(b, g(c, g(d, e)))))$ is flattened into $h(f(a, a, b, g(c, d, e)))$. We do not use flattening in this paper, except for illustrating the following. The symbols that are not removed under flattening are in a *maximal position*: if p is a position in a ground term s, we define $maxpos(s, p)$ to be p if $top(s|_p)$ is not an AC-operator and else $maxpos(s, p)$ is the maximal prefix p' of p such that $p' = \lambda$ or $p' = p'' \cdot n$ with $top(s|_{p''}) \neq top(s|_p)$. Let s and t be two AC-equal terms, i.e. $s =_{AC} t$. Then their flattened forms are equal up to permutation of arguments of AC-operators: a one-to-one correspondance can be established in this way between maximal positions in s and in t. We will sometimes speak about the *corresponding* position in t of some maximal position in s. Note that if u and v are subterms at corresponding maximal positions of resp. s and t, then $u =_{AC} v$.

3 Inference rules

Definition 3.1 *The following inference system is called* BOAC *(for basic, ordering constrained and AC). It is* strict *superposition-based, i.e. the ordering constraints OC added in all rules always encode that the equations $s \simeq t$ and $s' \simeq t'$ are maximal wrt. $\succ$ in the premise to which they belong (strictly maximal if they belong to the succedent, except the factoring rule, where only $s \simeq t$ is maximal), with $s \succ t$ and $s' \succ t'$ (except in equality resolution).*

In the AC rules, where x and x' are new variables, the term s' can be restricted to be headed by the AC-symbol f. This can be expressed in the constraint language and added to the constraint. As usual, AC-top superposition is only needed if s and s' are headed by f and share some top-level subterm (not headed with f) but x and x' do not (some restrictions implied by this condition can be formulated in the constraint language).

Of course, the superposition inferences are needed only if $s|_p$ is non-variable, and AC-superposition is needed only if moreover $s|_p$ (which has an f as top symbol) is not immediately below another f. Some examples are given at the end of next section.

1. *strict superposition right:*

$$\frac{\Gamma' \to \Delta', s' \simeq t'\ [\,T'\,] \qquad \Gamma \to \Delta, s \simeq t\ [\,T\,]}{\Gamma', \Gamma \to \Delta', \Delta, s[t']_p \simeq t\ [\,T' \wedge T \wedge OC \wedge\ s|_p =_{AC} s'\,]}$$

2. *strict superposition left:*

$$\frac{\Gamma' \to \Delta', s' \simeq t'\ [\,T'\,] \qquad \Gamma, s \simeq t \to \Delta\ [\,T\,]}{\Gamma', \Gamma, s[t']_p \simeq t \to \Delta', \Delta\ [\,T' \wedge T \wedge OC \wedge\ s|_p =_{AC} s'\,]}$$

3. *equality resolution:*

$$\frac{\Gamma, s \simeq t \to \Delta\ [\,T\,]}{\Gamma \to \Delta\ [\,T \wedge OC \wedge\ s =_{AC} t\,]}$$

4. *factoring:*

$$\frac{\Gamma \to \Delta, s' \simeq t', s \simeq t\ [\,T\,]}{\Gamma, t \simeq t' \to \Delta, s \simeq t\ [\,T \wedge OC \wedge\ s =_{AC} s'\,]}$$

5. *AC-strict superposition right:*

$$\frac{\Gamma' \to \Delta', s' \simeq t'\ [\,T'\,] \qquad \Gamma \to \Delta, s \simeq t\ [\,T\,]}{\Gamma', \Gamma \to \Delta', \Delta, s[f(t', x)]_p \simeq t\ [\,T' \wedge T \wedge OC \wedge\ s|_p =_{AC} f(s', x)\,]}$$

6. *AC-strict superposition left:*

$$\frac{\Gamma' \to \Delta', s' \simeq t'\ [\,T'\,] \qquad \Gamma, s \simeq t \to \Delta\ [\,T\,]}{\Gamma', \Gamma, s[f(t', x)]_p \simeq t \to \Delta', \Delta\ [\,T' \wedge T \wedge OC \wedge\ s|_p =_{AC} f(s', x)\,]}$$

7. *AC-top-superposition:*

$$\frac{\Gamma' \to \Delta', s' \simeq t'\ [\,T'\,] \qquad \Gamma \to \Delta, s \simeq t\ [\,T\,]}{\Gamma', \Gamma \to \Delta', \Delta, f(t', x') \simeq f(t, x)\ [\,T' \wedge T \wedge OC \wedge\ f(s', x') =_{AC} f(s, x)\,]}$$

4 Refutation completeness

Definition 4.2 *Let $C \, [\![\, T \,]\!]$ be a constrained clause and let σ be a ground substitution such that $T\sigma$ is true. Then $C\sigma$ is called a ground* instance *with σ of $C \, [\![\, T \,]\!]$.*

If moreover $C\sigma$ is of the form $\Gamma \to \Delta, t \simeq t'$ where $top(t)$ is an AC symbol f, and $t \simeq t'$ is the strictly maximal equation in $C\sigma$ with $t \succ t'$, then for every ground term v, the clause $\Gamma \to \Delta, f(t,v) \simeq f(t',v)$ is also a ground instance of $C \, [\![\, T \,]\!]$, called an extended *instance (with the* context v*) of $C\sigma$.*

Definition 4.3 *If, for a given ground instance $C\sigma$ of a clause C, a variable x only appears in equations $x \simeq t$ of the succedent of C with $x\sigma \succ t\sigma$ then x is called a succedent-top variable of $C\sigma$, denoted $x \in stvars(C\sigma)$.*

Definition 4.4 *Let R be a set of ground rewrite rules, let $C\sigma$ be a ground instance of a clause C and let x be a variable in $Vars(C)$. Then x is said to be variable irreducible in $C\sigma$ wrt. R if,*

1. *$x\sigma$ is irreducible wrt. R, or*
2. *$x \in stvars(C\sigma)$ and $x\sigma$ is irreducible wrt. all rules $l \to r \in R$ s.t. $x\sigma \simeq t\sigma \succ l \simeq r$ for all $x \simeq t$ in C, or*
3. *x only occurs in C immediately below some AC-symbol f and all subterms t of $x\sigma$ with $top(t) \neq f$ are irreducible wrt. R.*

If this property holds for all x in $Vars(C)$ then $C\sigma$ is variable irreducible[2] wrt. R.

Definition 4.5 *Let S be a set of constrained clauses. Now for each ground instance C of a constrained clause in S, we inductively define the cases in which C generates certain ground rewrite rules, in terms of the set R_C of rules generated by instances smaller (wrt. $\succ$) than C.*

Let AC_C denote the set of ground instances $s \simeq s'$ of equations in AC with $C \succ s \simeq s'$ and s and s' irreducible wrt. R_C, and let I_C denote the interpretation $(R_C \cup AC_C)^*$.

If C is a ground instance of a clause in S of the form $\Gamma \to \Delta, s \simeq t$ where $s \simeq t$ is strictly maximal (wrt. $\succ$) in C with $s \succ t$ and $I_C \not\models \Gamma \to \Delta$ then

1. *if C is a non-extended instance that is variable irreducible wrt. R_C and all u with $u =_{AC} s$ are irreducible by R_C, then C generates all rules $u \Rightarrow t$ with $u =_{AC} s$.*
2. *if C is an extended instance of some D that has generated some rule, then for each u with $u =_{AC} s$ and u irreducible by R_C, C generates a rule $u \Rightarrow t$.*

Finally, we define $R_S = \bigcup R_C$, $AC_S = \bigcup AC_C$, and $I_S = (R_S \cup AC_S)^*$.

Lemma 1. *Let S be a set of constrained clauses, and C and D instances of clauses in S with $C \succ D$. Then $R_C \supseteq R_D$, $AC_C \supseteq AC_D$, and $I_C \supseteq I_D$.*

---

[2] Here point 2. is based on the irreducibility notion of [BGLS92], and point 3. is the crucial trick in our proof for lifting in the AC-case and thus avoiding the computation of AC-unifiers, cf. example 4.8. Vigneron applies our idea in [Vig94] and extends it by allowing point 3 only if x is a new variable of AC-superposition, which permits to impose some more restrictions on the inferences (this also works for the proofs given here).

Proof. The first point, $R_C \supseteq R_D$, holds by definition. For $AC_C \supseteq AC_D$: suppose some $s \simeq s' \in AC_D$ and $s \simeq s' \notin AC_C$. Then there is some instance D' producing a rule $l \Rightarrow r$ that reduces s or s', with $s \simeq s' \in AC_{D'}$. But the equation $l \simeq r$ is strictly maximal in D' with $l \succ r$ and since $l \Rightarrow r$ reduces $s \simeq s'$, we have $s \succeq l$ and $s \succ r$, i.e. $s \simeq s' \succ l \simeq r$ and therefore $s \simeq s' \succ D'$, which contradicts $s \simeq s' \in AC_{D'}$. From the two previous points follows $I_C \supseteq I_D$. □

Lemma 2. *Let S be a set of constrained clauses.*

1. *If instances C and D with $C \succ D$ generate resp. $l \Rightarrow r$ and $l' \Rightarrow r'$ then $l \succ l'$.*
2. *For all $s_1 \simeq s_2$ in AC_S, the terms s_1 and s_2 are irreducible wrt. R_S.*
3. *For all $l \Rightarrow r$ and $l' \Rightarrow r'$ in R_S, if $l =_{AC} l'$ then r and r' are the same term.*
4. *There are no overlaps between left hand sides of rules of R_S.*
5. *If $u =_{AC} v$ for ground terms u and v, then u is reducible by R_S iff v is.*

Lemma 3. *Let S be a set of constrained clauses, and let s and t be ground terms. Then $I_S \models s \simeq t$ implies $s \to^*_{R_S} s' =_{AC_S} t' \leftarrow^*_{R_S} t$ for some s' and t'.*

Proof. If $I_S \models s \simeq t$ then there is a proof $s \leftrightarrow^*_{R_S \cup AC_S} t$, whose critical peaks can be overlaps of the form $t_1 \leftarrow_{R_S} t_2 \to_{R_S} t_3$, or of the form $t_1 \leftrightarrow_{AC_S} t_2 \to_{R_S} t_3$. The first situation cannot happen (by point 4. of the previous lemma). The other kind of overlap must be of the form $u[l[s_1]] \leftrightarrow_{AC_S} u[l[s_2]] \to_{R_S} u[r]$ (by point 2.). Since there is a rule $l[s_2] \Rightarrow r$, and $l[s_1] =_{AC} l[s_2]$, the term $l[s_1]$ is also reducible by R_S (point 5.), and moreover, it must be reducible at topmost position, because otherwise the corresponding AC-equal subterm in $l[s_2]$ would also be reducible. Now if $l[s_2]$ is reducible by some rule $l[s_2] \Rightarrow r'$, then $r' = r$ (point 3.). The fact that all overlaps $t_1 \leftrightarrow_{AC_S} t_2 \to_{R_S} t_3$ can be directly replaced by $t_1 \to_{R_S} t_3$ implies (by induction on the number of such AC-steps in $s \leftrightarrow^*_{R_S \cup AC_S} t$) that in fact $s \to^*_{R_S} s' =_{AC_S} t' \leftarrow^*_{R_S} t$. □

Lemma 4. *If $\Gamma\sigma \to \Delta\sigma, s\sigma \simeq t\sigma$ is an instance C of S that generates rules $u \Rightarrow t\sigma$ for all $u =_{AC} s\sigma$ then $I_S \not\models \Gamma\sigma \to \Delta\sigma$.*

Proof. Since C generates rules, $I_C \not\models \Gamma\sigma \to \Delta\sigma$. If $I_C \supseteq \Gamma\sigma$ then also $I_S \supseteq \Gamma\sigma$. It remains to be shown that $I_S \cap \Delta\sigma = \emptyset$.

We know $I_C \cap \Delta\sigma = \emptyset$, and the rules generated by instances bigger than C have left hand sides bigger than u, so they cannot contribute to rewrite proofs of $\Delta\sigma$, and the only rules in $R_S \setminus R_C$ that could be used are the $u \Rightarrow t\sigma$, if there is some equation $s'\sigma \simeq t'\sigma$ in $\Delta\sigma$ such that $s'\sigma$ is one of the u's and $I_C \models t\sigma \simeq t'\sigma$. But then the following inference by factoring can be made:

$$\frac{\Gamma \to \Delta, s' \simeq t', s \simeq t \, [\![T]\!]}{\Gamma, t' \simeq t \to \Delta, s \simeq t \, [\![T \wedge OC \wedge s =_{AC} s']\!]}$$

Its conclusion has a ground instance D smaller than C of the form $\Gamma\sigma, t'\sigma \simeq t\sigma \to \Delta\sigma, s\sigma \simeq t\sigma$ where $I_D \not\models D$ and which generates the rules $u \Rightarrow t\sigma$ for all $u =_{AC} s\sigma$, contradicting the fact that they have been generated by C. □

Lemma 5. *Let S be a set of constrained clauses. If a ground instance $C\sigma$ of a clause $C \, [\![T]\!]$ in S, with C of the form $\Gamma \to \Delta, s \simeq t$ generates rules $u \Rightarrow t\sigma$ with $u =_{AC} s\sigma$ then $C\sigma$ is variable irreducible wrt. R_S.*

Proof. We know $C\sigma$ is variable irreducible wrt. $R_{C\sigma}$, since it has generated a rule, so we only have to show the variable irreducibility of x in $Vars(C)$ wrt. rules $l \Rightarrow r$ in $R_S \setminus R_{C\sigma}$. Now if $s\sigma \succ x\sigma$ then no such rule $l \Rightarrow r$ reduces $x\sigma$, since $l \succeq s\sigma \succ x\sigma$. Otherwise, since $s\sigma$ is the maximal term in $C\sigma$, we have $x\sigma =_{AC} s\sigma$ and only some rule $u \Rightarrow t\sigma$ can reduce $x\sigma$. But then $x \in stvars(C\sigma)$ and $u \simeq t\sigma \succeq x\sigma \simeq t'\sigma$ for every $x \simeq t'$ in C, i.e. x is variable irreducible in $C\sigma$ wrt. R_S. □

Definition 4.6 *Let S be a set of constrained clauses. We denote by $irred(S)$ the set of non-extended ground instances of clauses in S that are variable irreducible wrt. R_S.*

Lemma 6. *Let S be a set of constrained clauses not containing the empty clause and closed under BOAC. Then $I_S \models irred(S) \cup AC$.*

Proof. We will derive a contradiction from the existence of a minimal (wrt. $\succ$) instance C in $irred(S) \cup AC$ such that $I_S \not\models C$.

If C is an instance $f(u_1, u_2) \simeq f(u_2, u_1)$ of the commutativity axiom $f(x, y) \simeq f(y, x)$, then since $C \notin AC_S$, one of $f(u_1, u_2)$ and $f(u_2, u_1)$ is reducible by R_S and therefore both of them, since they are AC-equal. If u_1 or u_2 is reducible, e.g. u_1 rewrites into some u, then $I_S \not\models f(u, u_2) \simeq f(u_2, u)$ which is a smaller instance than C of AC, contradicting the minimality of C. (Note that this smaller instance really exists since the AC-axioms do not have constraints). If u_1 and u_2 are irreducible, then let wlog. $f(u_1, u_2)$ be reducible by a rule $f(u_1, u_2) \Rightarrow r$ in R_S. But then, since u_1 and u_2 are irreducible, the same instance that generates $f(u_1, u_2) \Rightarrow r$ also produces $f(u_2, u_1) \Rightarrow r$, contradicting $I_S \not\models f(u_1, u_2) \simeq f(u_2, u_1)$.

If C is an instance $f(f(u_1, u_2), u_3) \simeq f(u_1, f(u_2, u_3))$ of the associativity axiom, then since $C \notin AC_S$, one of $f(f(u_1, u_2), u_3)$ and $f(u_1, f(u_2, u_3))$ is reducible by R_S and therefore both of them, since they are AC-equal. If u_1, u_2 or u_3 is reducible, then a contradiction is obtained as in the previous case for commutativity. If one of $f(f(u_1, u_2), u_3)$ and $f(u_1, f(u_2, u_3))$ is reducible at topmost position by a rule $l \Rightarrow r$ generated by the non-extended instance then also the other one is reducible at the topmost position by a rule with the same right hand side, contradicting $I_S \not\models f(f(u_1, u_2), u_3) \simeq f(u_1, f(u_2, u_3))$.

Otherwise, let the rule reducing $f(f(u_1, u_2), u_3)$ be $l \Rightarrow r$ with $l =_{AC} s\sigma$ and $r = t\sigma$ or $l =_{AC} f(s\sigma, v)$ and $r = f(t\sigma, v)$ generated by an instance with σ (possibly extended with the context v) of a clause D of the form $\Gamma \rightarrow \Delta, s \simeq t$. Then $f(f(u_1, u_2), u_3) =_{AC} f(s\sigma, v_1)$ for some v_1, where v_1 is, respectively, v (topmost reduction with extended rule), or u_3 (reduction of $f(u_1, u_2)$ with non-extended rule), or $f(u_3, v)$ (reduction of $f(u_1, u_2)$ with extended rule), and $f(f(u_1, u_2), u_3)$ is rewritten into some s_1 with $s_1 =_{AC_{\prec C}} f(t\sigma, v_1)$ (we use $AC_{\prec C}$ to denote the instances of AC that are smaller wrt. $\succ$ than C; note that $I_S \models AC_{\prec C}$ by minimality of C).

Similarly, let the rule reducing $f(u_1, f(u_2, u_3))$ be $l' \Rightarrow r'$ with $l' =_{AC} s'\sigma$ and $r' = t'\sigma$ or $l' =_{AC} f(s'\sigma, v')$ and $r' = f(t'\sigma, v')$ generated by an instance with σ (possibly extended with the context v') of a clause D' of the form $\Gamma' \rightarrow \Delta', s' \simeq t'$ (we can use the same σ since D and D' do not share variables). Then $f(u_1, f(u_2, u_3)) =_{AC} f(s'\sigma, v_2)$ for some v_2, where v_2 is, respectively, v' (topmost reduction with extended rule), or u_1 (non-topmost reduction with non-extended rule), or $f(u_1, v')$

(non-topmost reduction with extended rule), and $f(u_1, f(u_2, u_3))$ is rewritten into some s_2 with $s_2 =_{AC_{<C}} f(t'\sigma, v_2)$.

Now there is an inference by AC-top-superposition between D' and D:

$$\frac{\Gamma' \rightarrow \Delta', s' \simeq t' \ [\![T']\!] \qquad \Gamma \rightarrow \Delta, s \simeq t \ [\![T]\!]}{\Gamma', \Gamma \rightarrow \Delta', \Delta, f(t', x') \simeq f(t, x) \ [\![T' \wedge T \wedge OC \wedge \ f(s', x') =_{AC} f(s, x)]\!]}$$

whose conclusion D_1 has an instance with a ground substitution θ defined by $\theta = \sigma \cup \{x \mapsto v_1, x' \mapsto v_2\}$.

$D_1\theta$ is variable irreducible wrt. R_S since $D'\theta$ and $D\theta$ are variable irreducible wrt. R_S (by lemma 5) and for every variable $y \in Vars(D_1)$, either $y\theta$ is irreducible wrt. R_S or we have:

1. $y \in stvars(D\theta)$ or $y \in stvars(D'\theta)$ in some $y \simeq t_y$ Then also $y \in stvars(D_1\theta)$ in the same $y \simeq t_y$, and therefore y is variable irreducible in $D_1\theta$ wrt. R_S.
2. y only occurs in D or D' immediately below some AC-symbol. Then y also only occurs in D_1 immediately below the same AC-symbol and therefore it is variable irreducible in $D_1\theta$ wrt. R_S.
3. y is one of the new variables, x or x' (wlog. x). Since $f(f(u_1, u_2), u_3) =_{AC} f(s\sigma, v_1)$, for every subterm u at a maximal position of v_1 with $top(u) \neq f$ there is an u' with $u =_{AC} u'$ in u_1, u_2 or u_3, which implies that such u are irreducible wrt. R_S, and since x occurs only in D_1 immediately below the AC-symbol f, it follows that x is variable irreducible in $D_1\theta$ wrt. R_S.

$D_1\theta$ is an existing instance smaller than C, and moreover variable irreducible, so from the minimality of C we have $I_S \models D_1\theta$. We know $I_S \not\models \Gamma'\sigma, \Gamma\sigma \rightarrow \Delta'\sigma, \Delta\sigma$, so it must be the case that $I_S \models f(t', x')\theta \simeq f(t, x)\theta$, i.e. $I_S \models f(t'\sigma, v_2) \simeq f(t\sigma, v_1)$, but then $f(f(u_1, u_2), u_3) \rightarrow_{R_S} s_1 =_{AC_{<C}} f(t\sigma, v_1)$ and $f(u_1, f(u_2, u_3)) \rightarrow_{R_S} s_2 =_{AC_{<C}} f(t'\sigma, v_2)$ contradict $I_S \not\models C$, which completes the proof for the case where C is an instance of associativity.

If C is an instance with σ of a clause $D \ [\![T]\!]$ in S, then there are several cases to be analyzed, depending on the maximal equation in C:

1. C has two maximal equations in its succedent.
2. C has one strictly maximal equation in its succedent.
3. C has a maximal equation $u \simeq u'$ in its antecedent, with $u =_{AC} u'$.
4. C has a maximal equation $u \simeq u'$ in its antecedent, with $u \succ u'$.

1. C has two maximal equations in its succedent, i.e. D is $\Gamma \rightarrow \Delta, s' \simeq t', s \simeq t$, and there are two AC-equal maximal equations $s'\sigma \simeq t'\sigma$ and $s\sigma \simeq t\sigma$, that is, we have $s\sigma =_{AC} s'\sigma \succ t\sigma =_{AC} t'\sigma$, (note that $s\sigma \neq_{AC} t\sigma$, since otherwise in fact $s\sigma =_{AC_{<C}} t\sigma$, which, since $I_S \models AC_{<C}$, would contradict $I_S \not\models C$). But then the following inference by factoring can be made:

$$\frac{\Gamma \rightarrow \Delta, s' \simeq t', s \simeq t \ [\![T]\!]}{\Gamma, t' \simeq t \rightarrow \Delta, s \simeq t \ [\![T \wedge OC \wedge \ s =_{AC} s']\!]}$$

Its conclusion D_1 has a ground instance $D_1\sigma$ of the form $\Gamma\sigma, t'\sigma \simeq t\sigma \rightarrow \Delta\sigma, s\sigma \simeq t\sigma$ which is not satisfied by I_S, is smaller than C, and variable irreducible (as above) wrt. R_S. This contradicts the minimality of C.

2. Let $D\sigma$ be a clause $\Gamma\sigma \to \Delta\sigma, s\sigma \simeq t\sigma$, with a strictly maximal equation $s\sigma \simeq t\sigma$, and $s\sigma \succ t\sigma$ (we have $s\sigma \neq_{AC} t\sigma$ as in case 1, except if C consists only of the equation $s\sigma \simeq t\sigma$, but then $s\sigma =_{AC} t\sigma$ would follow from $AC_{\prec C}$ plus instances of AC that are AC-equal to C, which are shown true in I_S by the first part of this proof).

Since $I_S \not\models C$, the instance C has not generated the rule $s\sigma \Rightarrow t\sigma$. This must be because some u with $s\sigma =_{AC} u$ is reducible by R_C with a rule generated by an instance C' smaller than C.

Let this rule be the one that reduces u at a position p with $top(u|_p) = f$ (for some AC or non-AC symbol f) where p is innermost in the following sense: no other rule reduces u in a position p' below $maxpos(u,p)$ with $top(u|_{p'}) \neq f$. Note that such a rule always exists. Now there are two main subcases **2.1** and **2.2**, depending on whether the rule has been generated by an extended instance or by a non-extended instance.

2.1 Let C' be a non-extended instance $D'\sigma$ of some clause $D' \llbracket T' \rrbracket$ (we can use the same σ since D and D' do not share variables) with D' of the form $\Gamma' \to \Delta', s' \simeq t'$, where the rule is $u' \Rightarrow t'\sigma$ for some u' with $s'\sigma =_{AC} u' = u|_p$.

2.1.1 If $maxpos(u, p) = p$ then $u' =_{AC} s\sigma|_{p'}$ for the corresponding maximal position p' in $s\sigma$ and therefore $s'\sigma =_{AC} s\sigma|_{p'}$. Then $s\sigma|_{p'}$ cannot be below a variable, as $D\sigma$ is variable irreducible wrt. R_S: if s were in $stvars(D\sigma)$, then it would be irreducible wrt. rules smaller than $s\sigma \simeq t\sigma$, which is not the case; case 3. of definition 4.4 does not apply either, since $maxpos(s\sigma, p') = p'$. Then the inference by strict superposition right

$$\frac{\Gamma' \to \Delta', s' \simeq t' \llbracket T' \rrbracket \quad \Gamma \to \Delta, s \simeq t \llbracket T \rrbracket}{\Gamma', \Gamma \to \Delta', \Delta, s[t']_{p'} \simeq t \llbracket T' \wedge T \wedge OC \wedge s|_{p'} =_{AC} s' \rrbracket}$$

can be made, and, since all the conditions for its application hold, its conclusion D_1 has a ground instance $D_1\sigma$ of the form $\Gamma'\sigma, \Gamma\sigma \to \Delta'\sigma, \Delta\sigma, s\sigma[t'\sigma]_{p'} \simeq t\sigma$ that is not satisfied by I_S, is smaller than C, and (as above) variable irreducible wrt. R_S. This contradicts the minimality of C.

2.1.2 If $p_1 = maxpos(u, p) \neq p$ then $u|_{p_1} =_{AC} s\sigma|_{p'}$ for the corresponding maximal position p' in $s\sigma$ and $u|_{p_1} =_{AC} f(u', v) =_{AC} f(s'\sigma, v)$ for some v, and therefore $s\sigma|_{p'} =_{AC} f(s'\sigma, v)$. Then $s\sigma|_{p'}$ cannot be a variable, as before, since again $maxpos(s\sigma, p') = p'$. Then the inference by AC-strict superposition right

$$\frac{\Gamma' \to \Delta', s' \simeq t' \llbracket T' \rrbracket \quad \Gamma \to \Delta, s \simeq t \llbracket T \rrbracket}{\Gamma', \Gamma \to \Delta', \Delta, s[f(t', x)]_{p'} \simeq t \llbracket T' \wedge T \wedge OC \wedge s|_{p'} =_{AC} f(s', x) \rrbracket}$$

can be made, and, since all the conditions for its application hold, its conclusion a ground instance with σ of the form $\Gamma'\sigma, \Gamma\sigma \to \Delta'\sigma, \Delta\sigma, s\sigma[f(t'\sigma, v)]_{p'} \simeq t\sigma$ that is not satisfied by I_S, is smaller than C, and variable irreducible wrt. R_S (as above, and taking into account that all subterms u of $s\sigma|_{p'}$ with $top(u) \neq f$ are irreducible. wrt. R_S, since we have considered an innermost reduction in this sense). This again contradicts the minimality of C.

2.2 The other subcase is that u is reducible by R_C with a rule generated by an extended instance C' smaller than C. Let C' be an extended instance of some clause $D' \llbracket T' \rrbracket$ of the form $\Gamma' \to \Delta', s' \simeq t'$, where $top(s'\sigma)$ is the AC-symbol f and the rule is $u' \Rightarrow f(t'\sigma, v)$ for some u' with $u' =_{AC} f(s'\sigma, v)$ and $u|_p = u'$.

Let p_1 be $maxpos(u, p)$. Then $u|_{p_1} =_{AC} s\sigma|_{p'}$ for the corresponding position p' in $s\sigma$ and $u|_{p_1} =_{AC} f(u', v') =_{AC} f(s'\sigma, v')$ for some v', which is v if $p_1 = p$, and $f(v, v'')$ otherwise, for some v''. Therefore $s\sigma|_{p'} =_{AC} f(s'\sigma, v')$ for some p' and v'. Then $s\sigma|_{p'}$ cannot be a variable, as before, since again $maxpos(s\sigma, p') = p'$, another inference by AC-strict superposition right can be made, whose conclusion, as above, has an instance that contradicts the minimality of C.

3. C has a maximal equation in its antecedent whose members are AC-equal, i.e. D is $\Gamma, s \simeq t \to \Delta$, and $s\sigma =_{AC} t\sigma$. Then the following inference by equality resolution can be made:
$$\frac{\Gamma, s \simeq t \to \Delta \, [\![T]\!]}{\Gamma \to \Delta \, [\![T \wedge OC \wedge s =_{AC} t]\!]}$$

Its conclusion has a ground instance smaller than C of the form $\Gamma\sigma \to \Delta\sigma$ which is not satisfied by I_S, is smaller than C, and variable irreducible wrt. R_S, which contradicts the minimality of C.

4. $D\sigma$ is $\Gamma\sigma, s\sigma \simeq t\sigma \to \Delta\sigma$, with a maximal equation $s\sigma \simeq t\sigma$, and $s\sigma \succ t\sigma$.

Since $I_S \not\models C$, we have $I_S \models s\sigma \simeq t\sigma$, and by lemma 3 $s\sigma$ must be reducible by R_C (with a rule generated by an instance C' smaller than C).

Let this rule be the one that reduces $s\sigma$ in an innermost position p with $top(s\sigma|_p) = f$ as in case 2: no other rule reduces $s\sigma$ in a position p' below $maxpos(s\sigma, p)$ with $top(u|_{p'}) \neq f$. Now a contradiction is obtained exactly as it is done in case 2, but always inferences by (AC-) superposition left are considered instead of superposition right. □

The following theorem states the refutational completeness of our inference system, provided the initial set of clauses has no constraints. It is well-known that with arbitrary initial clauses these methods are incomplete[3].

Theorem 4.7 *(Refutation Completeness of BOAC)*
Let S be the closure under BOAC of a set S_0 of clauses with empty constraints. Then the empty clause is in S iff $S_0 \cup AC$ is inconsistent.

Proof. The left-to-right implication is trivial, since the inference rules of BOAC are sound, i.e. S_0 and S are logically equivalent modulo AC. For the right-to-left implication, suppose the empty clause is not in S. We prove that then $S_0 \cup AC$ is consistent, as it has an AC-model, namely I_S. From the previous lemma, we have that $I_S \models irred(S)$ and $I_S \models AC$. Since S_0 is contained in S, in particular $I_S \models C\sigma$ for each $C \, [\![true]\!]$ in S_0 with $x\sigma$ irreducible wrt. R_S for each x in $Vars(C)$ (which are existing instances of clauses in S_0 since these clauses have no constraints). But since I_S is a congruence containing R_S, it also satisfies the instances of S_0 that are reducible by R_S, since these instances follow from the irreducible ones and R_S. This means $I_S \models S_0$. □

Recall that $C \, [\![T]\!]$ is the empty constrained clause if C is empty and T is satisfiable, and that our inference rules are sound also if no ordering constraints are

---
[3] cf. [NR92b] for counter examples, and for a discussion about what kind of initial clauses can be allowed, like arbitrary constrained clauses without equality literals.

added, but that the AC-equality constraints are essential for soundness. Therefore we only have to decide the satisfiability of the AC-equality part of T to know whether $C \llbracket T \rrbracket$ implies the inconsistency of S_0 or not. It is an open problem whether the satisfiability of our kind of ordering constraints is decidable or not, but this is not needed here. We only need sufficiently powerful methods for detecting as many unsatisfiable ordering constraints as possible, which can be done applying the extension of our ordering to terms with variables.

Example 4.8 *If f is an AC symbol and e.g. $a \succ c \succ d$ then the set*

1. $\quad \rightarrow f(a,b) \simeq c$
2. $\quad \rightarrow f(f(a,a), f(b,b)) \simeq d$
3. $f(c,c) \simeq d \rightarrow$

is inconsistent[4]. *Now an AC-strict superposition right inference by 1. on 2. with conclusion 4.* $\rightarrow f(c,x) \simeq d \llbracket x =_{AC} f(a,b) \rrbracket$ *can be made (like normal superposition on the topmost f with a so-called extension $f(f(a,b),x) \simeq f(c,x)$). This seems insufficient to obtain a refutation: an inference with 1. seems to be needed below x in 4. But this can also be done by another AC-inference on 4. with 1. on the maximal f, producing 5.* $\rightarrow f(c,y) \simeq d \llbracket x =_{AC} f(a,b) \wedge y =_{AC} c \rrbracket$. *This is why def. 4.4-3. considers in clause 4. the variable x instantiated with $f(a,b)$ to be variable irreducible (our results imply that blocking, i.e. forbidding inferences with reducible unifiers, is complete for AC iff "reducible" is replaced by "variable reducible"). Finally, 5. produces the empty clause with 3.*

Example 4.9 *The clause $\rightarrow f(x,0) \simeq x$ would usually be considered self-extending, i.e. no AC-superposition with this clause would be needed. But this is incompatible with basicness: $f(0, f(0,0)) \simeq 0 \rightarrow$ cannot be refuted by basic superposition on maximal positions ([Wer92]), since only $x \simeq 0 \rightarrow \llbracket x =_{AC} f(0,0) \rrbracket$ could be obtained.*

By our basic AC-superposition rule, one gets $f(y,x) \simeq 0 \rightarrow \llbracket x =_{AC} 0 \wedge y =_{AC} 0 \rrbracket$, and then $z \simeq 0 \rightarrow \llbracket x =_{AC} 0 \wedge y =_{AC} 0 \wedge z =_{AC} 0 \rrbracket$, and finally, by equality resolution, the empty clause $\rightarrow \llbracket x =_{AC} 0 \wedge y =_{AC} 0 \wedge z =_{AC} 0 \rrbracket$.

5 Compatibility with simplification and deletion

In the previous section, for simplicity reasons, we have only proved the refutational completeness of the inference system, i.e. that inconsistent sets closed under the inference rules will contain the empty clause. But in fact theorem 4.7 also holds for sets that are closed *up to redundant inferences*, the so-called saturated sets. Here we will define theorem proving derivations that systematically compute saturated sets of clauses, even if *redundant clauses* are deleted during the theorem proving process and redundant inferences are not considered. The empty clause will be deduced in such derivations starting from an inconsistent initial set of clauses without constraints. We will first consider abstract redundancy notions and give practical methods for simplification and deletion later on. In the following we consider only inferences by $BOAC$:

---

[4] This example is adapted from[BD89]. We thank Leo Bachmair for pointing this example to us and for his comments.

Definition 5.10 *(cf. [BG91, NR92b] for more details).*
Let S be a set of constrained clauses.

1. *A clause $C \llbracket T \rrbracket$ is redundant in S if for every set of rules R s.t. $\Rightarrow_R \subseteq \succ$ and for every ground instance $C\sigma$ that is variable irreducible wrt. R, there exist ground instances $e_1 \ldots e_m$ of AC and variable irreducible wrt. R ground instances $D_1 \ldots D_n$ of clauses in S with $C\sigma \succ e_i$ and $C\sigma \succ D_i$ and $R \cup \{D_1, \ldots, D_m, e_1, \ldots, e_m\} \models C\sigma$.*
2. *An inference π with premise $C \llbracket T \rrbracket$, or premises $C' \llbracket T' \rrbracket$ and $C \llbracket T \rrbracket$, and conclusion $D \llbracket T''' \rrbracket$ is redundant in S if for every set of rules R s.t. $\Rightarrow_R \subseteq \succ$ and for every ground inference of the form $\pi\sigma$ whose premises and conclusion are variable irreducible instances wrt. R, there exist ground instances $e_1 \ldots e_m$ of AC and variable irreducible wrt. R ground instances $D_1 \ldots D_n$ of clauses in S such that $C\sigma \succ e_i$ and $C\sigma \succ D_i$ and $R \cup \{D_1, \ldots, D_m, e_1, \ldots, e_m\} \models D\sigma$.*
3. *S is saturated if every inference with premises in S is redundant in S.*
4. *A theorem proving derivation is a sequence of sets of (constrained) clauses $S_0, S_1, \ldots$ where each S_{i+1} is obtained from S_i by adding a clause following from $S_i \cup AC$ or by removing a clause that is redundant in S_i.*
5. *The set S_∞ of persistent clauses in $S_0, S_1, \ldots$ is defined as $\cup_j (\cap_{k \geq j} S_k)$.*
6. *A theorem proving derivation $S_0, S_1, \ldots$ is fair if, for every inference π with persistent premises, there is some S_j s.t. π is redundant in S_j.*

Lemma 7. *If $S_0, S_1, \ldots$ is a fair theorem proving derivation then S_∞ is saturated.*

Lemma 8. *Let S be a saturated set of constrained clauses not containing the empty clause. Then $I_S \models irred(S) \cup AC$.*

Proof. By adapting lemma 6 as done in [NR92b]: the conclusions of the inferences now need not belong to S, but instead the inferences are redundant, i.e. the conclusion is deducible from certain instances smaller than the maximal premise wich leads to a contradiction as in lemma 6. □

Theorem 5.11 *Let $S_0, S_1, \ldots$ be a fair theorem proving derivation, where S_0 is a set of clauses with empty constraints. Then $S_0 \cup AC$ is inconsistent iff the empty clause is in some S_j. Moreover, if some S_j is saturated and does not contain the empty clause, then $I_{S_j} \models S_j \cup AC$.*

5.1 Practical simplification and deletion methods

Definition 5.12 *A variable x is not lower bounded by a constraint T if $T\sigma \equiv true$ implies $T\sigma' \equiv true$, for every pair of ground substitutions σ and σ', s.t. $x\sigma \succ x\sigma'$ and $y\sigma' = y\sigma$ for every other variable y in the domain of σ.*

Lemma 9. *Let S be a set of constrained clauses. A clause $C \llbracket T \rrbracket$ is redundant in S if for every ground instance $C\sigma$ of it, there exist ground instances $e_1 \ldots e_m$ of AC and ground instances $D_i \sigma_i$ of clauses $D_1 \llbracket T_1 \rrbracket \ldots D_n \llbracket T_n \rrbracket$ in S, with $C\sigma \succ e_i$ and $C\sigma \succ D_i \sigma_i$, such that $\{e_1, \ldots, e_m, D_1 \sigma_1, \ldots, D_n \sigma_n\} \models C\sigma$ and for every variable x in each D_i:*

1. x is not lower bounded by T_i, or else
2. there is a variable $y \in Vars(C)$ s.t. $x\sigma_i = y\sigma$, with $y \notin stvars(C, \sigma)$ and there is no AC-symbol f such that y only occurs in C immediately below f.

A similar lemma as the previous one can be given for the redundancy of inferences. The lemma provides the conditions that have to be fulfilled by any *practical* method we want to use in this framework for proving that during the theorem proving process a certain clause can be deleted (e.g. because it is a tautology, or because it is subsumed by another one) or replaced by a simpler one (by demodulation or other simplification techniques). Indeed, we can construct counter examples for simplification methods that do not fulfil these conditions (cf. [NR92b] for the non-AC case).

Because of lack of space we do not discuss here which adapted versions for subsumption, simplification, etc. fall into the conditions of the previous lemma (this is a simple exercise; we are not aware of any such sound practical method that cannot be reformulated to follow from the previous lemma). Let us only mention that it is also possible to propagate (called *to weaken* in [NR92b]) parts of the constraints of the clauses used in redundancy proofs (which may require computing AC-unifers) to make conditions 1. and 2. of the previous lemma to hold.

6 Saturated sets and Knuth-Bendix completion

Suppose S is a finite saturated set without the empty clause, obtained from an initial set without constraints (in the following, such a set will be simply called saturated). Since $I_S \models S$, obtaining such an S proves the consistency of the theory (one can normally only prove inconsistencies), and on the other hand it is an efficient tool for theorem proving in this theory, since no inferences have to be computed between clauses of S. In fact, in some cases, depending on the syntactic properties of S, decision procedures for the theory are obtained. This is the case e.g. for saturated sets of equations E, which are convergent for both *rewriting modulo AC*, denoted $\rightarrow_{E/AC}$, and for *extended AC-rewriting* denoted $\rightarrow_{E\backslash AC}$. Let $l\sigma \simeq r\sigma$ be an instance of an equation of E with $l\sigma \succ r\sigma$. Then $s \rightarrow_{E/AC} u[r\sigma]$ if $s =_{AC} u[l\sigma]$. Furthermore, extended AC-rewriting is defined as $s[u] \rightarrow_{E\backslash AC} s[r\sigma]$ if $u =_{AC} l\sigma$ and $s[u] \rightarrow_{E\backslash AC} s[f(r, x)\sigma]$ if $u =_{AC} f(l, x)\sigma$.

Theorem 6.13 *Let E be a saturated set of constrained equations. For all terms s and t*
$$E \cup AC \models s \simeq t \quad \text{iff} \quad s \rightarrow^*_{E/AC} \circ \leftrightarrow^*_{AC} \circ \leftarrow^*_{E/AC} t \quad \text{iff} \quad s \rightarrow^*_{E\backslash AC} \circ \leftrightarrow^*_{AC} \circ \leftarrow^*_{E\backslash AC} t.$$

Proof. We prove that every ground term s (possibly with new Skolem constants to which our ordering $\succ$ can be extended) can be rewritten into some minimal (wrt. $\succ$) representative of its ($E \cup AC$)-congruence class (note that all minimal representatives of the same class are AC-equal since $\succ$ is total). We proceed by induction wrt. $\succ$, i.e. it suffices to prove the reducibility wrt. $\rightarrow_{E/AC}$ and $\rightarrow_{E\backslash AC}$ of non-minimal s.

Let t be such a minimal representative of the congruence class of s. If $s \succ t$, the only inference rules that can be applied in a refutation of $s \simeq t \rightarrow$ are (AC and non-AC) strict superposition left steps on s with some equation $l \simeq r \,[\![\, T \,]\!]$ of E. Then the conclusion is a clause C of the form $s[r]_p \simeq t \to [\![T \wedge s \succ t \wedge l \succ r \wedge s|_p =_{AC} l]\!]$ (in the case of a non-AC step) or $s[f(r,x)]_p \simeq t \to [\![T \wedge s \succ t \wedge l \succ r \wedge s|_p =_{AC} f(l,x)]\!]$ (in an AC step). Such a step must exist, since $E \cup AC \models s \simeq t$, and a ground instance $C\sigma$ must exist such that the constraint is true, which implies that $l\sigma \succ r\sigma$. But then s is reducible by extended AC-rewriting, as $s|_p =_{AC} l\sigma$ or $s|_p =_{AC} f(l,x)\sigma$ for some p and, for rewriting modulo AC, we have $s =_{AC} u[l\sigma]_p$. □

Let us remark that if r has no "extra variables" (variables that are not in l), then AC-matching provides such a ground σ and we can check whether σ fulfils the equality constraint part T' of T (the ordering part can be ignored). Otherwise, it suffices to instantiate the extra variables with the adequate mgu of T', and the remaining variables with the smallest constant (this provides a smaller *ground* term of the same $(E \cup AC)$-congruence class iff such a term exists). The mgu's can also be computed once and for all for E before rewriting (for this particular purpose, computing AC-unifiers may even be unnecessary, but this has to be studied in detail).

A similar decision result (by refutation or by conditional rewriting) holds for ground queries $S \cup AC \models s_1 \simeq t_1 \wedge \ldots \wedge s_n \simeq t_n$ if S contains only Horn clauses $\Gamma \to [\![T]\!]$ (which need not to be used) and Horn clauses $\Gamma \to s \simeq t [\![T]\!]$ where for each mgu θ of the equality part of T and for all ground instances with $\theta\sigma$ the positive equation $s\theta\sigma \simeq t\theta\sigma$ is strictly maximal and if $s\theta\sigma \succ t\theta\sigma$ then $Vars(s\theta) \supseteq Vars(\Gamma\theta)$.

References

[BD89] Leo Bachmair and Nachum Dershowitz. Completion for rewriting modulo a congruence. *Theoretical Computer Science*, 2 and 3(67):173–201, 1989.

[BG91] Leo Bachmair and Harald Ganzinger. Rewrite-based equational theorem proving with selection and simplification. Technical Report MPI-I-91-208, Max-Planck-Institut für Informatik, Saarbrücken, August 1991. To appear in Journal of Logic and Computation.

[BGLS92] Leo Bachmair, Harald Ganzinger, Christopher Lynch, and Wayne Snyder. Basic paramodulation and superposition. In Deepak Kapur, editor, *11th International Conference on Automated Deduction*, LNAI 607, pages 462–476, Saratoga Springs, New York, USA, June 15–18, 1992. Springer-Verlag.

[BS93] Franz Baader and Jörg Siekmann. Unification theory. In D.M. Gabbay, C.J. Hogger, and J.A. Robinson, editors, *Handbook of Logic in Artificial Intelligence and Logic Programming*. Oxford University Press (to appear), 1993.

[DJ90] Nachum Dershowitz and Jean-Pierre Jouannaud. Rewrite systems. In Jan van Leeuwen, editor, *Handbook of Theoretical Computer Science*, volume B: Formal Models and Semantics, chapter 6, pages 244–320. Elsevier Science Publishers B.V., Amsterdam, New York, Oxford, Tokyo, 1990.

[Dom92] Eric Domenjoud. A technical note on AC-unification. the number of minimal unifiers of the equation $\alpha x_1 + \ldots + \alpha x_p = \beta y_1 + \ldots + \beta y_q$. *Journal of Automated Reasoning*, 8(1):39–44, 1992.

[JK86] Jean-Pierre Jouannaud and Hélène Kirchner. Completion of a set of rules modulo a set of equations. *SIAM Journal of Computing*, 15:1155–1194, 1986.

[KKR90] Claude Kirchner, Hélène Kirchner, and Michaël Rusinowitch. Deduction with symbolic constraints. *Revue Française d'Intelligence Artificielle*, 4(3):9–52, 1990.

[KN92] Deepak Kapur and Paliath Narendran. Complexity of unification problems with associative commutative operators. *Journal of Automated Reasoning*, 9:261–288, 1992.

[NR91] Paliath Narendran and Michael Rusinowitch. Any ground associative commutative theory has a finite canonical system. In *Fourth int. conf. on Rewriting Techniques and Applications*, LNCS 488, pages 423–434, Como, Italy, April 1991. Springer-Verlag.

[NR92a] Robert Nieuwenhuis and Albert Rubio. Basic superposition is complete. In B. Krieg-Brückner, editor, *European Symposium on Programming*, LNCS 582, pages 371–390, Rennes, France, February 26–28, 1992. Springer-Verlag.

[NR92b] Robert Nieuwenhuis and Albert Rubio. Theorem proving with ordering constrained clauses. In Deepak Kapur, editor, *11th International Conference on Automated Deduction*, LNAI 607, pages 477–491, Saratoga Springs, New York, USA, June 15–18, 1992. Springer-Verlag.

[PS81] G.E. Peterson and M.E. Stickel. Complete sets of reductions for some equational theories. *Journal Assoc. Comput. Mach.*, 28(2):233–264, 1981.

[RN93] Albert Rubio and Robert Nieuwenhuis. A precedence-based total AC-compatible ordering. In C. Kirchner, editor, *5th International Conference on Rewriting Techniques and Applications*, LNCS 690, pages 374–388, Montreal, Canada, June 16–18, 1993. Springer-Verlag.

[Rub94] Albert Rubio. Automated deduction with ordering and equality constrained clauses. PhD. Thesis, Technical University of Catalonia, Barcelona, Spain, 1994.

[RV93] Michael Rusinowitch and Laurent Vigneron. Automated deduction with associative commutative operators. Rapport de Recherche 1896, Institut National de Recherche en Informatique et en Automatique, INRIA-Lorraine, May 1993.

[Vig94] Laurent Vigneron. Basic AC-Paramodulation. In Allan Bundy, editor, *12th International Conference on Automated Deduction*, LNAI, pages –, Nancy, France, June 1994. Springer-Verlag.

[Wer92] Ulrich Wertz. First-order theorem proving modulo equations. Technical Report MPI-I-92-216, Max-Planck-Institut für Informatik, Saarbrücken, April 1992.

The Complexity of Counting Problems in Equational Matching

Miki Hermann[1,*]   Phokion G. Kolaitis[2,†]

[1] CRIN (CNRS) and INRIA-Lorraine, BP 239, 54506 Vandœuvre-lès-Nancy, France. (e-mail: hermann@loria.fr)

[2] Computer and Information Sciences, University of California, Santa Cruz, CA 95064, U.S.A. (e-mail: kolaitis@cse.ucsc.edu)

Abstract

We introduce a class of counting problems that arise naturally in equational matching and study their computational complexity. If E is an equational theory, then #E-Matching is the problem of counting the number of complete minimal E-matchers of two given terms. #E-Matching is a well-defined algorithmic problem for every finitary equational theory. Moreover, it captures more accurately the computational difficulties associated with finding complete sets of minimal E-matchers than the corresponding decision problem for E-matching does.

In 1979, L. Valiant developed a computational model for measuring the complexity of counting problems and demonstrated the existence of #P-*complete* problems, i.e., counting problems that are complete for counting non-deterministic Turing machines of polynomial-time complexity. Using the theory of #P-completeness, we analyze the computational complexity of #E-matching for several important equational theories E. We establish that if E is one of the equational theories A, C, AC, I, U, ACI, Set, ACU, or ACIU, then #E-Matching is a #P-complete problem. We also show that there are equational theories, such as the restriction of AC-matching to linear terms, for which the underlying decision matching problem is solvable in polynomial time, while the associated counting matching problem is #P-complete.

1   Introduction and Summary of Results

Since the pioneering work of Plotkin [Plo72] over twenty years ago, the study of matching and unification modulo a fixed equational theory E has occupied a central place in automated deduction and has found numerous applications to several other branches of computer science, including logic programming, program verification, and database query languages. Researchers in this area have investigated a variety of equational theories and have examined in depth certain algorithmic aspects of matching and unification modulo an equational theory E.

---

*Partially supported by *Institut National Polytechnique de Lorraine* grant 910 0146 R1.

†Part of the research reported here was carried out while this author was visiting CRIN & INRIA-Lorraine supported by the University of Nancy 1 and INRIA-Lorraine. Research of this author is also supported by a 1993 John Simon Guggenheim Fellowship and by NSF Grant CCR-9108631.

There are two main algorithmic problems arising in the study of E-matching and E-unification. The first is a decision problem, namely, given two terms s and t, decide whether or not there is an E-matcher (or an E-unifier) of s and t. The second problem is to design matching and unification algorithms such that, given two terms s and t, the algorithm terminates and returns a set which is empty, if s and t are not E-matchable (respectively, not E-unifiable), or, otherwise, is a complete set of E-matchers of s and t (respectively, a complete set of E-unifiers). The second problem is, of course, meaningful only for theories for which the first problem is solvable and which, moreover, are *finitary*, i.e., for every term s and t there is a finite set of complete E-matchers (E-unifiers). For such theories, algorithms for the second problem should preferably return a complete set of *minimal* E-matchers (*minimal* E-unifiers).

Benanav, Kapur, Narendran [BKN87] and Kapur, Narendran [KN86] established that the decision problem for E-matching is NP-complete for many important equational theories E, including associativity A, commutativity C, associativity-commutativity AC, and extensions of AC with idempotency I or existence of unit U. Benanav, Kapur, and Narendran [BKN87] discovered also one exception to these NP-completeness phenomena, namely they proved that the decision problem for AC1-matching is solvable in polynomial time, where AC1 is AC restricted to *linear terms*, i.e., every variable occurs at most once in a term being matched. Concerning E-unification, Kapur and Narendran [KN92a] showed that the decision problem for AC-unification is NP-complete, which came as a surprise, since the prevailing intuition is that AC-unification is harder than AC-matching and, thus, this decision problem ought to have complexity higher than NP.

Although it is undoubtedly useful to pinpoint the computational complexity of the underlying decision problem, in practice it is far more important to analyze the complexity of E-matching and E-unification algorithms that return complete sets of (minimal) E-matchers or (minimal) E-unifiers. So far, relatively little progress has been made in deriving tight upper and lower bounds for the complexity of such algorithms. A notable exception is the case of AC-unification for which Kapur and Narendran [KN92b] found an algorithm that runs in doubly exponential time and returns a complete set of AC-unifiers, albeit not necessarily a minimal one. This upper bound is quite tight, since Domenjoud [Dom92] produced a set of AC-unification problems with n variables whose complete set of minimal AC-unifiers has $O(2^{2^n})$ elements.

Assume that E is some finitary equational theory and $\mathcal{A}$ is an algorithm such that, given two terms s and t as input, it returns a complete set of minimal E-matchers of s and t, if s and t can be matched. In this case, the algorithm $\mathcal{A}$ can also be used to compute the cardinality of a complete set of minimal E-matchers. Thus, we are able to solve at the same time a *counting problem* associated with E-matching, namely the problem of counting the number of complete minimal E-matchers. Notice that this problem is always well defined, since it is known (cf. [FH86]) that for every two terms s and t all sets of complete minimal E-matchers of s and t are of the same cardinality. Our goal in this paper is to initiate a systematic study of the computational complexity of *counting problems* in equational matching. We believe that these counting problems are quite natural and that they deserve to be studied in their own right. Moreover, we feel that counting problems reflect more accurately the computational difficulties of equational matching than the corresponding decision problems do.

Counting problems arise naturally in many areas of computer science and combinatorial mathematics. In 1979, Valiant [Val79a] developed a computational model for classifying the complexity of counting problems and introduced the class #P of functions that are computed by a *counting* Turing machine in polynomial time, i.e., a non-deterministic Turing machine that runs in polynomial time and has an auxiliary output device on which it prints in binary notation the number of its accepting computations on a given input. Valiant [Val79a] showed that the class #P has *complete* problems under certain restricted type of reductions that either preserve the number of solutions (*parsimonious* reductions) or, at least, make it possible to compute the number of solutions of one problem from the number of solutions of another problem (*counting* reductions). Quite often, NP-completeness proofs for decision problems can be translated to #P-completeness proofs for the corresponding counting problems by observing that the polynomial transformation in the proof of NP-hardness preserves the number of solutions. In particular, this is the case for #3-SAT, the prototypical #P-complete problem, which asks for the number of satisfying assignments of a 3CNF Boolean formula. On the other hand, Valiant [Val79a] demonstrated the existence of polynomial-time decision problems, such as perfect matching in bipartite graphs, whose associated counting problem is #P-complete. Several other problems were subsequently shown to exhibit this behavior in Valiant [Val79b].

In this paper, we apply the theory of #P-completeness to the study of counting problems in equational matching. If E is a finitary equational theory, then the #E-*Matching Problem* is the problem of computing the cardinality of a complete set of minimal E-matchers of two given terms s and t. We examine several important equational theories E and first show that their #E-Matching problem is a member of the class #P. Usually, membership of a counting problem in #P follows more or less directly from the definition of the problem. This, however, turns out not to be the case with counting problems in equational matching. In fact, proving that a particular #E-Matching problem is in #P often requires extensive use of different syntactic and structural properties of the underlying equational theory E. After deriving upper bounds for the complexity of counting problems in equational matching, we obtain tight lower bounds and, thus, establish that #E-Matching is a #P-complete problem for several equational theories E. In particular, we show the #P-completeness of #A-Matching, #C-Matching, and #AC-Matching. Similar #P-completeness results are obtained for the equational theories of idempotency I, existence of unit U, their AC extensions, and the restriction of ACI to Set matching. We also examine AC1-matching, the restriction of AC-matching to linear terms and establish that #AC1-Matching is #P-complete. This is achieved by showing that the problem of counting the number of perfect matchings in bipartite graphs can be reduced in a parsimonious way to #AC1-Matching. Since AC1-matching has a polynomial-time decision problem (cf. [BKN87]), we have a new manifestation of the phenomenon that a counting problem can be harder than its associated decision problem.

The results reported here on the one hand give a rather complete picture of the complexity of counting problems in equational matching and on the other yield a new family of #P-complete problems of different character than the counting problems studied thus far by researchers in computational complexity.

2 Counting Problems in Equational Matching

In this section, we will define the basic concepts and introduce the family of counting problem arising in equational matching. We also present here a minimum amount of the necessary background material from computational complexity and unification.

2.1 Counting Problems and the Class #P

A *counting Turing machine* is a non-deterministic Turing machine equipped with an auxiliary output device on which it prints in binary notation the number of its accepting computations on a given input. A counting Turing machine has *time complexity* $t(n)$ if the longest accepting computation of the machine over all inputs of size n is at most $t(n)$. By varying the functions $t(n)$, we obtain different complexity classes of counting functions. Thus, #P is the class of functions that are computable by counting Turing machines of polynomial-time complexity. Similarly, #EXPTIME is the class of functions that are computable by counting Turing machines of exponential-time complexity. Counting Turing machines and the complexity class #P were introduced and studied in depth by Valiant [Val79a, Val79b]. Here, we will work with a slightly different, but essentially equivalent, description of the class #P that appears in Kozen [Koz92].

Assume that Σ and Γ are nonempty alphabets and let $w: \Sigma^* \to \mathcal{P}(\Gamma^*)$ be a function from the set Σ^* of strings over Σ to the power set $\mathcal{P}(\Gamma^*)$ of Γ^*. If x is a string in Σ^*, then we refer to $w(x)$ as the *witness set* for x and to the elements of $w(x)$ as *witnesses* for x. Every such *witness function* w can be identified with the following *counting problem* w: given a string x in Σ^*, find the number of witnesses for x in the set $w(x)$. In what follows, $|x|$ is the length of a string x, and $|S|$ is the cardinality of a set S.

Definition 2.1 ([Koz92]) The class #P is the class of counting problems w such that:

(1) There is a polynomial-time algorithm to determine, for given strings x and y, if $y \in w(x)$;

(2) There exists a natural number $k \geq 1$ (which can depend on the counting problem w) such that $|y| \leq |x|^k$ for all $y \in w(x)$.

A typical member of #P is the counting problem #SAT: given a string x encoding a Boolean formula, find the number of truth assignment satisfying x. In this case, the witness set $w(x)$ consists of all truth assignments (encoded as strings) satisfying x.

Counting problems relate to each other via *counting reductions* and *parsimonious reductions*, which are stronger than the polynomial-time reductions between NP-problems.

Definition 2.2 Let $w: \Sigma^* \to \mathcal{P}(\Gamma^*)$ and $v: \Pi^* \to \mathcal{P}(\Delta^*)$ be two counting problems.

A *polynomial many-one counting* (or *weakly parsimonious*) *reduction* from w to v consists of a pair of polynomial-time computable functions $\sigma: \Sigma^* \to \Pi^*$ and $\tau: N \to N$ such that $|w(x)| = \tau(|v(\sigma(x))|)$. A *parsimonious reduction* from w to v is a counting reduction σ, τ from w to v such that τ is the identity function.

A counting problem w is *#P-hard* if there are counting reductions from it to all problems in #P. If in addition w is a member of #P, then we say that w is *#P-complete*.

Proving that a counting problem is #P-hard is viewed as evidence that this problem is truly intractable. Actually, in complexity theory it is generally believed that #P-hard problems are not members of the class FPH, the functional analog of the polynomial hierarchy PH. In particular, no #P-hard problem is known to belong to the class FP^{NP} of all functions that are computable in polynomial time using NP oracles (cf. Johnson [Joh90, section 4.1]). Thus, to the extent of course that one can compare decision problems with counting problems, a #P-completeness result suggests a higher level of intractability than an NP-completeness result.

The following #P-complete problems will be of particular use to us in the sequel. Notice that the underlying decision problem for #3-SAT is NP-complete, while for the other two it is solvable in polynomial time. In fact, the decision problem for Positive 2-SAT is trivial, since every positive 2-SAT formula is satisfiable.

#3-SAT [Val79b]
Input: Set V of variables and Boolean formula F over V in conjunctive normal form with exactly three literals in each clause.
Output: Number of truth assignments for the variables in V that satisfy F.

#POSITIVE 2-SAT (appeared in [Val79b] as #MONOTONE 2-SAT)
Input: Set V of Boolean variables and Boolean formula F over V in conjunctive normal form such that each clause of F consists of exactly two positive literals.
Output: Number of truth assignments for the variables in V that satisfy F.

#PERFECT MATCHINGS [Val79a]
Input: Bipartite graph G with $2n$ nodes.
Output: Number of *perfect matchings* in G, i.e., sets of n edges such that no pair of edges shares a common node.

2.2 Equational Theories

A *signature* $\mathcal{F}$ is a set of function symbols of designated arities. If $\mathcal{F}$ is a signature and $\mathcal{X}$ is a set of variables, we let $T(\mathcal{F}, \mathcal{X})$ denote the set of all terms over the signature $\mathcal{F}$ and the variables in $\mathcal{X}$. We also write $V(t)$ for the set of variables occurring in a term t. As usual, a *ground term* is a term without variables. A *substitution* is a mapping $\rho: \mathcal{X} \longrightarrow T(\mathcal{F}, \mathcal{X})$ such that $x\rho = x$ for all but finitely many variables x. Thus, a substitution ρ can be identified with its restriction on the finite set $\mathcal{D}om(\rho) = \{x \in \mathcal{X} \mid x\rho \neq x\}$, which is called the *domain* of ρ. A substitution ρ is *ground* if $x\rho$ is a ground term for all $x \in \mathcal{D}om(\rho)$. Every substitution can be extended to an endomorphism on the algebra of terms.

An *equation* is a pair of terms $l = r$. Each equation is viewed as an *equational axiom*, namely as the first-order sentence $(\forall x_1)\ldots(\forall x_m)(l = r)$ obtained from the equation by universal quantification over all variables occurring in the terms l and r. If E is a set of equational axioms, then the *equational theory* Th(E) *presented by* E is the smallest congruence relation over $T(\mathcal{F}, \mathcal{X})$ containing E and closed under substitutions, i.e., Th(E) is the smallest congruence containing all pairs $l\rho = r\rho$, where $l = r$ is in E and ρ is a substitution. By an abuse of terminology, we will often say "the equational theory E" instead of the correct "the equational theory Th(E) presented by E". We write $s =_E t$ to denote that the pair (s, t) of terms is a member of Th(E).

E-equality on terms can be extended to substitutions by setting $\rho =_E \sigma$ if and only if $(\forall x \in \mathcal{X})(x\rho =_E x\sigma)$. If V is a set of variables and ρ, σ are substitutions, we put $\rho =_E^V \sigma$ if and only if $(\forall x \in V)(x\rho =_E^V x\sigma)$. We also consider the preorder $\leq_E^V$ on substitutions defined by the condition: $\sigma \leq_E^V \rho$ if and only if $(\exists \eta)(\sigma\eta =_E^V \rho)$. In turn, this preorder gives rise to the following equivalence relation $\equiv_E^V$ on substitutions:

$$\rho \equiv_E^V \sigma \iff \rho \leq_E^V \sigma \text{ and } \sigma \leq_E^V \rho.$$

Notice that, in general, $\rho \leq_E^V \sigma$ does not imply that $\rho \equiv_E^V \sigma$, and, by the same token, $\rho \equiv_E^V \sigma$ does not imply that $\rho =_E \sigma$. It is easy to see, however, that these three relations coincide on ground substitutions with the same domain.

In the sequel, we will be concerned with equational theories presented by finite sets E whose axioms are among the following:

Associativity	A(f):	$f(x, f(y, z)) = f(f(x, y), z)$
Commutativity	C(g):	$g(x, y) = g(y, x)$
Idempotency	I(f):	$f(x, x) = x$
Existence of Unit	U(f):	$f(x, 1) = x$
Homomorphism	H(f, g, h):	$f(g(x, y)) = h(f(x), f(y))$
Endomorphism	End(f, g):	$f(g(x, y)) = g(f(x), f(y))$

We will also consider AC1-Matching, which is the restriction of AC-matching to linear terms, and *Set Matching*, i.e., the special case of ACI-matching in which there is only one ACI-symbol and it occurs on the top of the matched terms.

2.3 Counting Matching Problems in Equational Theories

In what follows, let s be a term, let $V = V(s)$ be the set of variables of s, and let t be a ground term. An E-*matcher of s and t* is a substitution ρ such that $s\rho =_E t$. If such an E-matcher exists, then we say that the term s E-*matches* the ground term t. The E-*matching problem* is the decision problem to determine, given a term s and a ground term t, whether s E-matches t.

A *complete set of E-matchers of s and t* is a set S of substitutions such that the following hold: (1) each substitution $\rho \in S$ is an E-matcher of s and t, and, moreover, $\mathcal{D}om(\rho) \subseteq V$; (2) for every E-matcher σ of s and t, there is a substitution $\rho \in S$ such that $\rho \leq_E^V s$. We say that S is a *complete set of minimal E-matchers of s and t* if, in addition, every two distinct members of S are $\leq_E^V$-incomparable.

In general, it may be the case that s E-matches t, but there is no complete set of minimal E-matchers of s and t. On the other hand, it is well known that if a complete set of minimal E-unifiers of s and t exists, then it is unique up to $\equiv_E^V$.

Proposition 2.3 ([FH86]) *Let s be a term, let t be a ground term, and assume that S_1 and S_2 are two complete sets of minimal E-unifiers of s and t. Then there is a one-to-one and onto mapping $f: U_1 \to U_2$ such that $f(\rho) \equiv_E^V \rho$ for every substitution ρ in U_1, where V is the set of variables of s. As a result, all complete sets of minimal E-matchers of s and t are of the same cardinality.*

From now on, we assume that E is a set of equational axioms such that if s E-matches t, then there exists a complete set of minimal E-matchers of s and t. We let $\mu\mathrm{CSM}_E(s, t)$ denote the (unique up to $\equiv_E^V$) complete set of minimal E-matchers of s and t, if s E-matches t, or the empty set, otherwise.

Siekmann and Szabó [SS82] initiated a study of equational theories based on the cardinalities of the sets $\mu\mathrm{CSM}_E(s, t)$. In particular, E-matching is said to be *unitary* if for every term s and every ground term t we have that $|\mu\mathrm{CSM}_E(s, t)| \leq 1$. E-matching is said to be *finitary* if for every term s and every ground term t the set $\mu\mathrm{CSM}_E(s, t)$ is finite. It is well known that AC-matching is finitary, but not unitary.

One might contemplate studying finitary equational theories by classifying them into a hierarchy as follows: the first level of the hierarchy consists of all unitary theories, while the k-th level of it, $k \geq 2$, consists of all theories E such that for every term s and every ground term t we have that $|\mu\text{CSM}_E(s,t)| \leq k$. This, however, turns out not to be meaningful, because Book and Siekmann [BS86] showed that if E is a set of equational axioms such that E-matching is not unitary, then E-matching is *unbounded*, which means that for every natural number $k \geq 2$ there is a term s and a ground term t such that $|\mu\text{CSM}_E(s,t)| > k$.

If E is a set of equational axioms such that E-matching is finitary, then we associate with E the following *counting problem*:

#E-MATCHING
Input: A term s and a ground term t
Output: Cardinality of the set $\mu\text{CSM}_E(s,t)$.

Our goal in this paper is to study the computational complexity of #E-Matching problems for finitary equational theories E. In view of the aforementioned result by Book and Siekmann's [BS86], this appears to be the only reasonable approach to analyzing finitary theories according to the cardinalities of the sets $\mu\text{CSM}_E(s,t)$. Notice that if E-matching is unitary, then the complexity of #E-Matching is essentially the same as the complexity of the decision problem for E-matching. In particular, if BR is the set of equational axioms for Boolean rings, then #BR-Matching can be computed in polynomial time using NP oracles (i.e., it belongs to the class FP^{NP}), since BR-matching is unitary and its decision problem is in NP (cf. Martin and Nipkow [MN89]).

We investigate the computational complexity of the #E-Matching problem for several finitary, non-unitary equational theories E and in each case we establish that #E-Matching is a #P-complete problem. A proof of #P-completeness has two distinct parts, namely one must first show that the problem at hand is indeed a member of the class #P and then establish that every problem in #P has a counting reduction to it. We undertake each of these tasks separately in the next two sections.

3 Membership of #E-Matching Problems in #P

Quite often, membership of a counting problem in #P follows immediately from the definition of the problem. This is, for example, the case with most #P-complete problems considered by Valiant [Val79a, Val79b], including #SAT, #3-SAT, #Positive 2-SAT, and #Perfect Matchings. In contrast, if E is an arbitrary finitary equational theory, then it is not at all obvious that the associated #E-Matching problem is a member of #P. Actually, as we will soon see, in order to prove that such problems are in #P we must reflect on particular properties of the underlying equational theory E.

In the sequel, we will make systematic use of the description of #P given in Definition 2.1. Thus, for each #E-Matching problem considered here we must find a function w defined on pairs (s,t), where s is a term and t is a ground term, such that the two conditions of Definition 2.1 are fulfilled. At first sight, a natural unambiguous choice for the witness set $w(s,t)$ appears to be the set

$$w(s,t) = \{[\rho]_{\equiv_E^V} \mid \rho \text{ is a member of a complete set of minimal E-matchers for } s \text{ and } t\},$$

where $[\rho]_{\equiv_E^V}$ is the equivalence class of the substitution ρ with respect to the congruence $\equiv_E^V$. However, since the members of a witness set must be strings over some alphabet, this raises the question of how to represent $\equiv_E^V$-equivalence classes by strings. It is clear that if we take a string consisting of the entire equivalence class (by concatenating all its members), then we may not be able to fulfill the second condition, because for many equational theories E, including AC, some $\equiv_E^V$-equivalence classes of matchers of s and t may have exponentially many members (in the size of s and t). Thus, our only alternative is to represent each $\equiv_E^V$-equivalence class by a unique *canonical representative* of it and then take as witness set the complete set of minimal E-matchers of s and t that are the canonical representatives of their $\equiv_E^V$-equivalence class. In what follows, we show that for many important equational theories it is possible to find canonical representatives such that both conditions of Definition 2.1 are satisfied. These canonical representatives will be defined in a uniform way for the class of *regular theories*. For each specific theory E, however, we will have to use particular properties of E in order to establish that the canonical representatives satisfy the desired conditions.

An equational theory E is *regular* if for every axiom $(l = r) \in E$ we have $V(l) = V(r)$. As Fages and Huet [FH86] put it, "in regular theories variables cannot disappear". All equational theories considered here are regular. In regular theories each matcher must be a ground substitution. Thus, as explained in 2.2, if ρ and σ are E-matchers of the terms s and t, then ($\rho \leq_E^V \sigma \iff \rho \equiv_E^V \sigma \iff \rho =_E \sigma$). Using this, one can easily derive the following result, which appeared first in Fages and Huet [FH86, Proposition 4.1].

Proposition 3.4 *Let E be a regular equational theory, let s be a term, and let t be a ground term such that s E-matches t. Then the following are true:*

(1) There exists a complete sets of minimal E-matchers of s and t.

(2) A set S is a complete set of minimal E-matchers of s and t if and only if S is a complete set of E-matchers of s and t such that no two distinct members of S are equal with respect to $=_E^V$, where V is the set of variables of s.

(3) A substitution ρ is a member of a complete set of minimal E-matchers of s and t if and only if ρ is an E-matcher of s and t.

Let E be a regular equational theory and assume that for every pair of terms s and t such that s E-matches t we have selected a unique representative from each $\equiv_E^V$-equivalence class of E-matchers of s and t, where V is the set of variables of s. If $[\rho]_{\equiv_E^V} = [\rho]_{=_E^V}$ is such an equivalence class, then we let ρ^* denote its selected representative and we say that ρ^* is the *canonical representative* of $[\rho]_{\equiv_E^V}$. Once the canonical representatives have been selected, we can define a *canonical witness function* w^* for #E-Matching such that the witness set $w^*(s,t)$ consists of the canonical representatives of all equivalence classes of E-matchers of s and t. By combining Definition 2.1 with Proposition 3.4, we obtain the following useful sufficient criterion for membership in #P.

Proposition 3.5 *Let E be a regular equational theory and let w^* be a canonical witness function for #E-Matching such that the following conditions hold:*

Condition (1a). *There is a polynomial-time algorithm to determine, given a term s, a ground term t, and a ground substitution ρ, whether ρ is an E-matcher of s and t.*

Condition (1b). *There is a polynomial time algorithm to determine, given a term s, a ground term t, and an E-matcher ρ of s and t, whether ρ is the canonical representative of its $\equiv_E^V$-equivalence class $[\rho]_{\equiv_E^V}$.*

Condition (2). *There is a natural number $k \geq 1$ such that $|\rho^*| \leq (|s| + |t|)^k$ for all canonical representatives ρ^* in $w^*(s,t)$.*

Then the #E-Matching problem is a member of the class #P.

We now examine specific equational theories and for each of them we show that it is possible to find canonical witness functions such that the above conditions are satisfied.

3.1 Verifying Conditions (1a) and (1b)

Notice that Condition (1a) amounts to having a polynomial-time algorithm to determine, given a term s, a ground term t, and a ground substitution ρ, whether $s\rho =_E^V t$. Thus, Condition (1a) is automatically satisfied by every equational theory E for which E-equality of terms (i.e., $t_1 \stackrel{?}{=}_E t_2$) can be tested in polynomial time.

Benanav, Kapur, and Narendran [BKN87] showed that AC-equality can be tested in polynomial time by reducing this problem to the existence of a matching of given size in bipartite graphs. As pointed out in Kapur and Narendran [KN86], ACI-equality can also be tested in polynomial time by using the same algorithm and removing identical arguments of an idempotent function after flattening terms. It follows that Set-equality can be checked in polynomial time as well. By removing units instead of identical arguments in flattened terms, we can check ACU-equality in polynomial time. ACIU-equality can be checked in polynomial time by combining ACI-equality testing with ACU-equality testing. Another modification of the AC-equality method can be used to test for C-equality in polynomial time. For this, instead of testing for the existence of a given size matching, we test for the existence of perfect matchings in a graph.

A-equality can be tested in polynomial time by first reducing the terms to the right-associative form and then checking for syntactic equivalence. I-equality can be tested in polynomial time by first reducing each term to its I-normal form using the leftmost innermost strategy and then checking the reduced terms in I-normal form for syntactic equivalence. The same approach works also for U-equality and, hence, can be extended to IU-equality. H-equality (assuming $g \neq h$) can be tested in polynomial time by a similar method with the leftmost outermost strategy. This extends also to ACH.

We now focus on Condition (1b). We will first define the canonical representatives of the equivalence classes $[\rho]_{\equiv_E^V}$ for an arbitrary regular theory E and then show that for each specific equational theory studied here there is a polynomial-time algorithm to find the canonical representative ρ^* of the class $[\rho]_{\equiv_E^V}$ from a given member ρ of it.

Recall that if E is a regular theory, s is a term, and t is a ground term, then every E-matcher of s and t is a ground substitution. If we have a *total precedence* $\succ$ on the signature $\mathcal{F}$, then the *lexicographic path ordering* $\succ_{lpo}$ induced by it is a well-founded ordering that is total on ground terms [Der87]. Thus, we can compare two E-matchers $\rho = [x_1 \mapsto t_1, \ldots, x_n \mapsto t_n]$ and $\rho' = [x_1 \mapsto t'_1, \ldots, x_n \mapsto t'_n]$ by comparing them lexicographically through the terms t_i and t'_i, provided the sequence of variables x_1, ..., x_n is fixed. Hence, there exists a well-ordering relation $\succ_{lpo}^{lex}$ on E-matchers and so we can choose the $\succ_{lpo}^{lex}$-smallest E-matcher in each equivalence class as its *canonical representative*. The above definition of canonical representatives is uniform for all regular theories. What turns out, however, to be different for each specific theory considered here is the algorithm for computing the canonical representatives in polynomial time.

For certain equational theories E it is possible to orient appropriately their axioms and obtain a *convergent rewrite system*, i.e., a rewrite system that is both *confluent* and *terminating*. More specifically, if we introduce the rewrite rules

$$A: \quad f(f(x,y),z) \rightarrow f(x,f(y,z)) \qquad I: \quad f(x,x) \rightarrow x$$
$$H: \quad f(g(x,y)) \rightarrow h(f(x),f(y)) \qquad U: \quad f(x,1) \rightarrow x,$$

then the corresponding rewrite systems for A, I, U, IU, and H are convergent. Thus, if E is one of A, I, U, IU, or H, and the substitution $\rho = [x_1 \mapsto t_1, \ldots, x_n \mapsto t_n]$ is an E-matcher of the terms s and t, then for each term t_i we can find its *normal form* $t_i\downarrow_R$, where R is the corresponding convergent rewrite system. This normal form exists, because R is terminating, and is unique, because R is confluent. Thus, the substitution $[x_1 \mapsto t_1\downarrow_R, \ldots, x_n \mapsto t_n\downarrow_R]$ coincides with the canonical representative of $[\rho]_{\equiv_E^V}$.

The algorithms for determining E-equality for A, I, U, IU, and H can be also used to find the canonical representatives of ground substitutions in polynomial time. The leftmost innermost strategy is complete for I, U, and IU, since we cannot create new redexes below a position at which we apply a rewrite rule and each application of a rewrite rule eliminates one occurrence of an E-symbol. The leftmost outermost strategy is complete for H, provided $g \neq h$, since each application of the rewrite rule eliminates one occurrence of the symbol g. Moreover, in each case the number of steps in the derivation is linear in the size of the input substitution.

The above method does not work for AC or for C. For the case of AC, we use the following polynomial-time algorithm to find canonical representatives. Given an AC-matcher $\rho = [x_1 \mapsto t_1, \ldots, x_n \mapsto t_n]$, transform each term t_i to its right-associative form and flatten it, obtaining this way the flattened term $\bar{t}_i$. The flattening is done from left to right, i.e., if f is an AC-symbol, then a subterm $f(s_1, f(s_2, f(\ldots f(s_{k-1}, s_k))))$ in the right-associative form is flattened to $f(s_1, \ldots, s_k)$. After this, for each subterm of $\bar{t}_i$ headed by an AC-symbol we sort the immediate subterms in a bottom-up way. This means that a flattened subterm $f(s_1, \ldots, s_k)$ is permuted to $f(s_{\pi(1)}, \ldots, s_{\pi(k)})$, where $s_{\pi(k)} \succ_{lpo} s_{\pi(k-1)} \succ_{lpo} \cdots \succ_{lpo} s_{\pi(2)} \succ_{lpo} s_{\pi(1)}$. This way, we obtain the flattened and sorted term t_i^*, $1 \leq i \leq n$, and, hence, the representative ρ^*. The same method, without transformation to the right-associative form and without flattening, works for C. Transformation to the right-associative form is polynomial, flattening is polynomial (cf. [BKN87, KN92a]), sorting of terms is polynomial, the lexicographic path ordering is computed in polynomial time [Sny93], and the number of AC-symbols (C-symbols) is limited by the size of the input ρ. Thus, the algorithm runs in polynomial time.

The AC extensions of the rewrite systems I, U, IU, and H are confluent and terminating modulo AC (although the termination must be proved by an AC-compatible ordering). Thus, we can proceed hierarchically. First, we rewrite an E-matcher ρ to the AC-normal form ρ^+ with respect to the confluent and terminating rewrite system R modulo AC. The system R corresponds to one for I, U, IU, or H. Then the E-matcher ρ^+ is transformed to ρ^* using the previously described method for AC. This hierarchical method is correct, since rewriting modulo AC means rewriting AC-equivalence classes. Thus, we have R-equivalence classes and within them AC-equivalence classes.

3.2 Verifying Condition (2)

An equational theory E is *permutative* if for each axiom $(l = r) \in E$ the multisets of symbols in l and r are equal. The theories A, C, AC are permutative, whereas if a theory contains idempotency I, unit U, homomorphism H, or endomorphism End as one of its axioms, then it is not permutative. If E is a permutative theory and s, t are terms such that $s =_E t$, then $|s| = |t|$. It follows that if E is a permutative theory, s is a term, and t is a ground term such that s E-matches t, then for every E-matcher ρ of s and t we have that $|\rho| \leq |t|$, since $s\rho =_E t$. Thus, Condition (2) holds for every permutative theory. By combining the above remarks with our findings in the preceding Section 3.1, we obtain our first result establishing membership of several #E-Matching problems in #P.

Theorem 3.6 *Let E be one of the equational theories* A, C, AC. *Then the #E-Matching problem is in the class #P.*

In order to analyze non-permutative theories, we turn again to the rewrite systems used in the previous Section 3.1. We say that an equational theory E is *simplifying* if for every axiom $(l = r) \in$ E, the term r is a proper subterm of the term l. By the same token, we say that a rewrite system R is *simplifying* if for every rewrite rule $(l \to r) \in R$, the term r is a proper subterm of the term l. If E and Q are two equational theories, then E is Q-*simplifying* if for every axiom $(l = r) \in$ E − Q there exists a proper subterm l' of l, such that $l' =_Q r$. A rewrite system R is Q-*simplifying* if for every rewrite rule $(l \to r) \in R$ there exists a proper subterm l' of l such that $l' =_Q r$.

Clearly, the theories I, U, and IU are simplifying, while the theories ACI, ACU, and ACIU are AC-simplifying. It is also clear that if a rewrite step $l \longrightarrow_R r$ is carried out in a simplifying rewrite system R, then $|r| < |l|$. Using these facts, it is not hard to verify that Condition (2) holds for each of the equational theories I, U, IU, and their AC extensions. This completes the proof of our second result about membership in #P.

Theorem 3.7 *Let E be one of the equational theories* I, U, IU, Set, ACI, ACU, *and* ACIU. *Then the #E-Matching problem is in the class #P.*

Finally, we consider the homomorphism axiom H and its AC extension. Although the homomorphism rewrite rule $f(g(x,y)) \to h(f(x), f(y))$ increases the size of terms, each application of it eliminates one occurrence of the symbol g. This makes it possible to derive a bound on the size of canonical representatives, namely one can easily show that if ρ is an H-matcher of s and t, then $|\rho^*| \leq 2|t|$. Moreover, a similar bound can be obtained for ACH-matchers. Thus, we have established the following result.

Theorem 3.8 *#H-Matching and #ACH-Matching are both in the class #P.*

The rewrite rule $f(g(x,y)) \to g(f(x), f(y))$ for Endomorphism does not eliminate the occurrence of g. It is an open problem whether #End-Matching is in the class #P.

4 #P-Hardness of #E-Matching Problems

In this section, we derive lower bounds for the complexity of #E-Matching problems.

Theorem 4.9 *Let E be one of the equational theories* A, C, AC. *Then #E-Matching is a #P-complete problem.*

Proof: (*Sketch*) As shown in Theorem 3.6, each of these counting problems is a member of #P. Benanav, Kapur, and Narendran [BKN87] showed that A-matching and C-matching are NP-hard decision problems by reducing 3-SAT to these problems. These reductions turn out to be parsimonious and, thus, #A-Matching and #C-Matching are #P-hard problems. Benanav, Kapur, and Narendran [BKN87] also showed that AC-matching is an NP-hard decision problem by reducing Monotone 3-SAT to AC-matching, where Monotone 3-SAT is the satisfiability decision problem for 3CNF Boolean formulas in which each clause consists either of only positive literals or of only negative literals. The reduction given in [BKN87] is not parsimonious. We show that #AC-Matching is a #P-hard problem by producing a parsimonious reduction of #Positive 2-SAT to it.

Consider an instance of #Positive 2-SAT with variables $X = \{x_{11}, x_{12}, \ldots, x_{n1}, x_{n2}\}$ and clauses $C = \{c_1, \ldots, c_n\}$, where $c_i = x_{i1} \vee x_{i2}$. Let f be an AC-symbol, let g be a n-ary function symbol that is neither associative nor commutative, and let 1 and 0 be two constant symbols that will be used to simulate the truth or falsity of a Boolean variable. We also let $Y = \{y_1, \ldots, y_n\}$ be a set of new variables, one for each clause, such that $X \cap Y = \emptyset$. With each clause $c_i = x_{i1} \vee x_{i2}$, we associate the term $f(x_{i1}, x_{i2}, y_i)$ and the ground term $f(1, 1, 0)$. We let s denote the term $g(f(x_{11}, x_{12}, y_1), \ldots, f(x_{n1}, x_{n2}, y_n))$ and let t denote the ground term $g(f(1,1,0), \ldots, f(1,1,0))$. In the full paper we show that the number of truth assignments satisfying $c_1 \wedge \cdots \wedge c_n$ is equal to the cardinality of the set $\mu\text{CSM}_{AC}(s, t)$. □

Recall that AC1-matching is the restriction of AC-matching to terms in which each variable occurs at most once. Benanav, Kapur, and Narendran [BKN87] showed that AC1-matching can be decided in polynomial time. In contrast, we show next that the counting problem associated with AC1-matching is harder than its decision problem

Theorem 4.10 *#AC1-Matching is a #P-complete problem.*

Proof: (*Sketch*) Membership of #AC1-matching in #P is a direct consequence of membership in #P of #AC-matching. We will show that #AC1-matching is #P-hard by producing a parsimonious reduction from #Perfect Matchings to #AC1-Matching.

Suppose that we are given a bipartite graph $G = (S, T, E)$ with $2n$ nodes, where $S = \{s_1, \ldots, s_n\}$ and $T = \{t_1, \ldots, t_n\}$ is the partition of the nodes. Let a be a constant symbol, f a unary function symbol, g a (n+1)-ary function symbol that is neither associative nor commutative, and h an AC-symbol. We also consider the sets of variables $X = \{x_{ij} \mid i, j = 1, \ldots, n\}$ and $Y = \{y_1, \ldots, y_n\}$, where $X \cap Y = \emptyset$.

With each node s_i in the set S we associate the term $s_i^* = g(s_i^1, \ldots, s_i^n, s_i^{n+1})$, where

$$s_i^j = \begin{cases} f(x_{ii}) & \text{if } 1 \leq i, j \leq n \text{ and } i = j \\ x_{ij} & \text{if } 1 \leq i, j \leq n \text{ and } i \neq j \\ y_i & \text{if } 1 \leq i \leq n \text{ and } j = n+1 \end{cases}$$

Intuitively, we view the nodes $s_1, \ldots, s_n$ in S as vectors of a "matrix":

$$\begin{aligned} s_1^* &= g(f(x_{11}), x_{12}, x_{13}, \ldots, x_{1n}, y_1) \\ s_2^* &= g(x_{21}, f(x_{22}), x_{23}, \ldots, x_{2n}, y_2) \\ &\vdots \\ s_n^* &= g(x_{n1}, x_{n2}, \ldots, x_{n,n-1}, f(x_{nn}), y_n) \end{aligned}$$

in which the subterms $f(x_{ii})$ occupy the main diagonal, while the variables $y_1, \ldots, y_n$ are along the last column. With each node t_i in T we associate the ground term $t_i^* = g(t_i^1, \ldots, t_i^n, t_i^{n+1})$, where

$$t_i^j = \begin{cases} f(a) & \text{if } (s_j, t_i) \in E \\ a & \text{otherwise} \end{cases}$$

Thus, we view the nodes $t_1,\ldots,t_n$ in T as vectors of another "matrix"

$$\begin{aligned} t_1^* &= g(t_1^1,\ldots,t_1^n,f(a)) \\ t_2^* &= g(t_2^1,\ldots,t_2^n,f^2(a)) \\ &\vdots \\ t_n^* &= g(t_n^1,\ldots,t_n^n,f^n(a)). \end{aligned}$$

The intuition behind the second matrix is that it represents the adjacency matrix (extended by the column of $f^k(a)$'s) of the edge relation E of the graph G, where the terms $f(a)$ and a are used to simulate the values 1 and 0, respectively.

In the full paper we show that for each i and j, $1 \le i,j \le n$, there is an edge $(s_i, t_j) \in E$ if and only if the term s_i^* AC1-matches the ground term t_j^*. As a result, the above is a parsimonious reduction of #Perfect Matchings to #AC1-Matching. □

We obtain next lower bounds for #E-Matching, where E is idempotency I, or unit U, or one of the AC extensions of I and U. For this, we need to consider yet another counting problem, whose #P-completeness was proved in Creignou and Hermann [CH93].

#1-in-3-SAT [CH93]
Input: Set V of variables and Boolean formula F over V in conjunctive normal form with exactly three literals in each clause.
Output: Number of truth assignments for the variables in V such that they make true exactly one literal in each clause.

A parsimonious reduction from #3-SAT to #1-in-3-SAT is obtained by replacing each 3-clause $c_i = l_{i1} \vee l_{i2} \vee l_{i3}$ by the clauses $c_{i1} = l_{i1} \vee x_{i2} \vee x_{i3}$, $c_{i2} = \bar{l}_{i2} \vee x_{i2} \vee y_{i2}$, $c_{i3} = \bar{l}_{i3} \vee x_{i3} \vee y_{i3}$, and $c_{i4} = x_{i3} \vee y_{i2} \vee z_i$.

Theorem 4.11 *Let E be one of the equational theories I, U, IU, Set, ACI, ACU, and ACIU. Then #E-Matching is a #P-complete problem.*

Proof: (*Hint*) By Theorem 3.7, each of these counting problems is a member of #P. The #P-hardness of #I-Matching, #U-Matching, and #ACU-Matching is proved by producing parsimonious reductions from #1-in-3-SAT. The reduction to #U-Matching and to #ACU-Matching is an adaptation of the NP-hardness reduction of U-unification and ACU-unification, given in [TA87]. The #P-hardness of IU-Matching and #ACIU-Matching is established by exhibiting parsimonious reductions from Positive 2-SAT to each of these problems. Finally, the reduction from 3-SAT to Set-matching in [KN86] is linear and, hence, weakly parsimonious. Thus, #ACI-Matching is #P-hard as well. □

Finally, we obtain #P-hardness results for the counting problems associated with the AC extensions of homomorphism H and Endomorphism End. For this, we produce parsimonious reductions from #Positive 2-SAT. Details are given in the full appear.

Theorem 4.12 *#ACH-Matching is a #P-complete problem and #ACEnd-Matching is a #P-hard problem.*

5 Concluding Remarks

In this paper we introduced and studied the class of #E-Matching problems, which are the counting problems that ask for the cardinalities of complete sets of minimal E-matchers in some finitary equational theory E. Using the theory of #P-completeness, we identified the complexity of #E-Matching problems for several equational theories E. The table below summarizes our findings and compares the complexity of counting problems in equational matching with the complexity of the corresponding decision problems.

Theory	Decision	Counting	Theory	Decision	Counting
A	NP-complete	#P-complete	I	NP-complete	#P-complete
C	NP-complete	#P-complete	U	NP-complete	#P-complete
AC	NP-complete	#P-complete	IU		#P-complete
AC1	P	#P-complete	ACI	NP-complete	#P-complete
ACH		#P-complete	Set	NP-complete	#P-complete
ACEnd		#P-hard	ACU	NP-complete	#P-complete
BR	NP-complete	FP^{NP}	ACIU		#P-complete

Although in most cases the NP-completeness of the decision problem is accompanied by the #P-completeness of the associated counting problem, it should be emphasized that in general there is no relation between the complexities of these two problems, as manifested by the results about AC1 and BR.

The work presented here suggests that a similar investigation should be carried out for #E-Unification problems, i.e., counting problems that ask for the cardinalities of the complete sets of minimal E-unifiers in some finitary equational theory E. Although our results imply that several #E-unification problems are #P-hard, we already know that there are equational theories E, such as AC, for which #E-Unification is not a member of #P. Indeed, Domenjoud [Dom92] found AC-unification problems with n variables whose complete set of minimal AC-unifiers has $O(2^{2^n})$ elements. Since a counting problem in #P takes values that are bounded by a single exponential in the size of the input, it follows that #AC-Unification is not in #P. It is an interesting open problem to analyze the computational complexity of #AC-Unification and to determine whether it is complete for some higher counting complexity class, such as #EXPTIME or #PSPACE. Results along these lines will delineate the computational difference between matching and unification in a precise manner and will confirm the intuition that unification is harder than matching.

References

[BKN87] D. Benanav, D. Kapur, and P. Narendran. Complexity of matching problems. *Journal of Symbolic Computation*, 3:203–216, 1987.

[BS86] R. Book and J. Siekmann. On unification: equational theories are not bounded. *Journal of Symbolic Computation*, 2:317–324, 1986.

[CH93] N. Creignou and M. Hermann. On #P-completeness of some counting problems. Research report 93-R-188, Centre de Recherche en Informatique de Nancy, 1993.

[Der87]   N. Dershowitz. Termination of rewriting. *Journal of Symbolic Computation*, 3(1 & 2):69–116, 1987. Special issue on Rewriting Techniques and Applications.

[Dom92]   E. Domenjoud. Number of minimal unifiers of the equation $\alpha x_1 + \cdots + \alpha x_b =_{AC} \beta y_1 + \cdots + \beta y_q$. *Journal of Automated Reasoning*, 8:39–44, 1992.

[FH86]   F. Fages and G. Huet. Complete sets of unifiers and matchers in equational theories. *Theoretical Computer Science*, 43(1):189–200, 1986.

[Joh90]   D.S. Johnson. A catalog of complexity classes. In J. van Leeuwen, editor, *Handbook of Theoretical Computer Science, Volume A: Algorithms and Complexity*, chapter 2, pages 67–161. North-Holland, Amsterdam, 1990.

[KN86]   D. Kapur and P. Narendran. NP-completeness of the set unification and matching problems. In J.H. Siekmann, editor, *Proceedings 8th International Conference on Automated Deduction, Oxford (England)*, volume 230 of *Lecture Notes in Computer Science*, pages 489–495. Springer-Verlag, July 1986.

[KN92a]   D. Kapur and P. Narendran. Complexity of unification problems with associative-commutative operators. *Journal of Automated Reasoning*, 9:261–288, 1992.

[KN92b]   D. Kapur and P. Narendran. Double-exponential complexity of computing a complete set of AC-unifiers. In *Proceedings 7th IEEE Symposium on Logic in Computer Science, Santa Cruz (California, USA)*, pages 11–21, 1992.

[Koz92]   D.C. Kozen. *The design and analysis of algorithms*, chapter 26: Counting problems and #P, pages 138–143. Springer-Verlag, 1992.

[MN89]   U. Martin and T. Nipkow. Boolean unification — the story so far. *Journal of Symbolic Computation*, 7(3 & 4):275–294, 1989.

[Plo72]   G. Plotkin. Building-in equational theories. *Machine Intelligence*, 7:73–90, 1972.

[Sny93]   W. Snyder. On the complexity of recursive path orderings. *Information Processing Letters*, 46:257–262, 1993.

[SS82]   J. Siekmann and P. Szabó. Universal unification and classification of equational theories. In D.W. Loveland, editor, *Proceedings 6th International Conference on Automated Deduction, New York (NY, USA)*, volume 138 of *Lecture Notes in Computer Science*, pages 369–389. Springer-Verlag, June 1982.

[TA87]   E. Tidén and S. Arnborg. Unification problems with one-sided distributivity. *Journal of Symbolic Computation*, 3(1 & 2):183–202, 1987.

[Val79a]   L.G. Valiant. The complexity of computing the permanent. *Theoretical Computer Science*, 8:189–201, 1979.

[Val79b]   L.G. Valiant. The complexity of enumeration and reliability problems. *SIAM Journal on Computing*, 8(3):410–421, 1979.

Representing Proof Transformations for Program Optimization

Penny Anderson

INRIA, Unité de Recherche de Sophia-Antipolis
2004 Route des Lucioles, BP 93
06902 Sophia-Antipolis CEDEX, France
tel. 93.65.78.16
anderson@sophia.inria.fr

Abstract. In the *proofs as programs* methodology a program is derived from a formal constructive proof. Because of the close relationship between proof and program structure, transformations can be applied to proofs rather than to programs in order to improve performance. We describe a method for implementing transformations of formal proofs and show that it is applicable to the optimization of extracted programs. The method is based on the representation of derived logical rules in Elf, a logic programming language that gives an operational interpretation to the Edinburgh Logical Framework. It results in declarative implementations with a general correctness property that is verified automatically by the Elf type checking algorithm. We illustrate the technique by applying it to the problem of transforming a recursive function definition to obtain a tail-recursive form.

1 Introduction

Research in constructive logics and type theories has resulted in a good understanding of the relation of formal constructive proof to computation, and has produced a number of systems (e.g., Nuprl [6], Coq [7]) capable of supporting the development of programs by constructive theorem-proving (the *proofs as programs* strategy). However, proof development presents some of the same difficulties that program development does: notably problems of adaptation and reuse. This observation has stimulated work on extending the strategy with proof transformations; the idea is to attempt to optimize the program by transforming the proof, rather than extracting the program from the proof and transforming it afterwards. Research in this area suggests that this approach can address problems in program development that are not easily solved either by syntactic program transformation techniques or through theorem-proving support alone. Such problems include the specialization of a program to its input [10], [17] and the optimization of the call structure of a recursive function definition [26], [17], [18].

One standard approach to metaprogramming tasks for formal logic like proof transformation has been to use a separate programming language, such as ML, as a metalanguage for a type theory considered as an object logic. A more recently developed strategy, which has been applied to programming language semantics [13], [12], [19], theorem provers [8], [9], and the metatheory of deductive systems [29], is the use of a higher-order logic programming language as a logical framework. This is the basis of the approach taken here. The Elf programming language [27] supports this strategy by providing an operational interpretation to the Edinburgh Logical Framework (LF) [14]. LF is a dependently-typed λ-calculus designed to serve as a framework for the encoding of formal deductive systems. Its type system is expressive enough to support straightforward encodings of many logics used in reasoning about programming languages, formal logic, and the like. The decidability of the LF type system gives practical force to the *judgments as types* and the corresponding *proof-checking as type-checking* principles. Since a deduction represented in LF is valid exactly when its representation is well-typed, the Elf implementation of LF type-checking provides a proof checker for encoded deductive systems.

In this paper we describe a methodology for implementing proof transformations in Elf that yields a declarative formulation with easily verified correctness properties. It is closely related to the deduction transformations of [29]. The method exploits the higher-order unification capabilities of Elf to express a transformation by means of higher-order patterns in the spirit of Huet and Lang [16]. Although the technique is restricted to transformations based on derived rules of the encoded logic, it has been applied successfully [1] to program optimizations such as the introduction of tail recursive form, finite differencing [23], and modification of the domain of a recursive function definition. Experiments to date have been limited to a simple system of natural deduction for first-order logic, but the approach is applicable to any logic that can be encoded in LF. In combination with support for theorem proving and tools for analysis and heuristics like those of [17] it could provide an easily verifiable and extensible basis for interactive or automated systems for the development and maintenance of verified programs.

The paper is organized as follows: Sect. 2 gives some necessary background on an Elf encoding of natural deduction proofs and implementation of program extraction. In Sect. 3 we describe the basics of the transformation encoding methodology by means of a small example. Sect. 4 shows how the method can be used to implement a transformation to obtain a tail-recursive function definition. We apply the transformation to an example program and demonstrate that transforming the proof can result in more optimization than is obtained by syntactic program transformation. We conclude with a summary and assessment in Sect. 5.

2 Representation of Proofs and Program Extraction

We implemented basic support for the proofs as programs strategy in Elf. Program extraction, program execution, proof transformations, and other forms of term manipulation were implemented as Elf programs. The use of LF and Elf permits concise encodings that directly reflect the inference systems they represent. This is often sufficient for proof search – for instance, it yields implementations of a programming language interpreter and type assignment system, and of program extraction. Although the representation of natural deduction does not give a theorem-prover, the equivalence of proof checking and Elf type checking is the basis for the verification of our encodings of proof transformations.

Elf gives an operational semantics to LF type declarations: an Elf program is a collection of type declarations and an Elf query is a type. The query may contain free variables, which are treated like Prolog logic variables; it succeeds if the system can construct an inhabitant of the query type from the declarations that constitute the program.

We sketch an encoding in the style of Harper et al. [14] of natural-deduction style proofs for intuitionistic many-sorted first-order predicate calculus. Individual terms, propositions and proofs are represented as object-level Elf terms.

Terms are either natural numbers or lists of natural numbers:

```
i : type.
zero : i.
succ : i -> i.
ilist : type.
nl : ilist.                 %empty list
cns : i -> ilist -> ilist.  %cons
```

There is no declaration corresponding to variables; they are represented as Elf bound variables through the use of higher-order abstract syntax.

Logical constants and connectives are encoded as follows (we omit the declarations that permit the use of the binary connectives as infix operators):

```
o : type.                       %formulas
true : o.                       %truth
false : o.                      %absurdity
~ : o -> o.                     %negation
& : o -> o -> o.                %conjunction
\/ : o -> o -> o.               %disjunction
=> : o -> o -> o.               %implication
forall : (i -> o) -> o.         %universal quantification
exists : (i -> o) -> o.         %existential quantification
lforall : (ilist -> o) -> o.    %list quantifiers
lexists : (ilist -> o) -> o.
```

Elf permits declared constants to begin with a non-letter; thus => is the constant representing implication. LF non-dependent function types are notated using ->.

Higher-order abstract syntax is used to represent the binding properties of the quantifiers by meta-level abstraction.

```
|- : o -> type.

truei   : |- true.
falsee  : {C:o} |- false -> |- C.
andi    : |- A -> |- B -> |- (A & B).
andel   : |- (A & B) -> |- A.
ander   : |- (A & B) -> |- B.
oril    : {B:o} |- A -> |- (A \/ B).
orir    : {A:o} |- B -> |- (A \/ B).
ore     : (|- A -> |- C) -> (|- B -> |- C) -> |- (A \/ B) -> |- C.
impliesi : (|- A -> |- B) -> |- (A => B).
impliese : |- (A => B) -> |- A -> |- B.
noti    : (|- A -> |- false) -> |- (~ A).
note    : |- (~ A) -> |- A -> |- false.
foralli : ({x:i} |- (A x)) -> |- (forall A).
foralle : {T:i} |- (forall A) -> |- (A T).
existsi : {A:i -> o} {T:i} |- (A T) -> |- (exists A).
existse : ({x:i} |- (A x) -> |- C) -> |- (exists A) -> |- C.
lforalli : ({x:ilist} |- (A x)) -> |- (lforall A).
lforalle : {L:ilist} |- (lforall A) -> |- (A L).
lexistsi : {A:ilist -> o} {L:ilist} |- (A L) -> |- (lexists A).
lexistse : ({x:ilist} |- (A x) -> |- C) -> |- (lexists A) -> |- C.
lind    :  {A:ilist -> o}
           |- (A nl)
           -> ({l:ilist} |- (A l) -> {x:i} |- (A (cns x l)))
           -> |- (lforall A).
```

Fig. 1. Elf encoding of inference rules

The encoding of proofs (Fig. 1) uses the dependent types of LF to obtain a proof checker. The dependent type constructor `|- : o -> type` provides an association between a proof and its end-formula, which permits the encoding of inference rules to enforce the constraint that any well-typed proof term is a valid proof. The Elf syntax `{x:A} B` represents the LF dependent type construction $\Pi x : A. B$. In the figure many Π-quantifiers are omitted, since Elf takes free variables in clauses to be implicitly Π-quantified, and types for them can often be inferred automatically[25]. Higher-order abstract syntax supports the representation of the introduction and discharge of assumptions, as well as the side conditions restricting the free occurrences of variables in assumptions.

We defined a simply typed λ-calculus with primitive recursion for representing extracted programs. A typing discipline and operational semantics for the language were given in the form of deductive systems encoded as Elf programs. The programming language syntax, evaluation and type inference were implemented in the style of Michaylov and Pfenning [19], which extends the higher-order representations developed in λProlog by Hannan and Miller in [13], [12]. We omit details of the language description and give example programs in the syntax of Standard ML [22].

Our implementations of type and program extraction are an adaptation to the Elf setting of *modified realizability* [24], [15], [30], [31], [32], in which extracted types and programs are simplified to remove uninformative subterms while retaining computationally useful information. The simplification is performed during extraction, based on the fact that program values of interest are generated only by proofs of disjunctions and existentially quantified formulas. In particular, a proof of $\neg A$ is always uninformative; we exploit this fact in the example of Sect. 4.

Type and program extraction can be expressed as deductive systems inductively defined over formulas and proof objects respectively; these are straightforward to transcribe into executable Elf programs. The interested reader is referred to [2] for details. For an understanding of the rest of this paper, it is sufficient to keep in mind that the use of induction corresponds to primitive recursion in the extracted program, and that a proof of $\exists x\,.\,A(x)$ with x a list and $A(x)$ uninformative corresponds to a list expression in the program.

3 Proof Transformations as Derived Rules

What do we mean by a transformation based on a derived rule? A small example is the following transformation to swap the members of a pair. For any proof $\mathcal{P}$ of $A \wedge B$ there is a proof of the following form:

$$\dfrac{\dfrac{\mathcal{P}}{A \wedge B}\wedge\mathrm{E}_R \quad \dfrac{\mathcal{P}}{A \wedge B}\wedge\mathrm{E}_L}{B \wedge A}\wedge\mathrm{I}$$

Thus we may conclude $B \wedge A$ from $A \wedge B$ by a derived rule of the object logic. A derived rule involving metavariables for formulas (as A and B above) can be directly encoded in Elf by putting Π-quantified variables for the metavariables, as in the following encoding of the derivation.

```
%%% Rule statement
flipand : |- A & B -> |- B & A -> type.
%%% Rule derivation
fliptrans : {A:o} {B:o} {P: |- A & B}
            flipand P (andi (ander P) (andel P)).
```

This also implements a proof transformation: given to the Elf interpreter, it can be used to transform any proof of a formula $A \wedge B$ to one of $B \wedge A$. Here is an example query, giving as input a proof of $0 = 0 \wedge 1 = 1$:

```
?- flipand (andi (eq_refl zero) (eq_refl (succ zero)))
           Output_proof.
```

In response to the query Elf succeeds with the substitution:

```
Output_proof =
  andi (ander (andi (eq_refl zero)
                    (eq_refl (succ zero))))
       (andel (andi (eq_refl zero)
                    (eq_refl (succ zero)))).
```

The transformation acts indirectly as a program transformation: from a proof of $A \wedge B$ the system extracts a program expression e with a pair type $\alpha \times \beta$; from the transformed proof of $B \wedge A$ it extracts the expression $\langle \mathbf{snd}(e), \mathbf{fst}(e) \rangle$ with the type $\beta \times \alpha$.

The type-correctness of the clause fliptrans provides strong guarantees of some properties of the transformation. If the transformation succeeds then Output_proof is bound to a term $\mathcal{P}'$ that encodes a valid proof of $B \wedge A$. This property is distinct from the guarantee given by program extraction, which ensures that the program satisfies its specification regardless of how the transformation is encoded. Properties specific to this transformation can also be easily proved. From properties of extraction proved in [1] it follows that if the program expression extracted from $\mathcal{P}$ has type $\alpha \times \beta$, where α is the type extracted from A and β the type extracted from B, then the program expression extracted from $\mathcal{P}'$ has type $\beta \times \alpha$. It is not hard to show that in fact the values are exchanged. It is also easy to see that the transformation is total.

4 A Proof Transformation for the Introduction of Tail-recursion

A well-known programming strategy for the improvement of a recursive function definition is the introduction of an accumulator argument to achieve a tail-recursive form, which can be compiled to a loop. This strategy has long been studied by researchers in transformational programming, for example in [5], [4], and many others. Here we describe a transformation that performs the analogous optimization at the level of proofs; its application to a small example program shows that in some cases the proof transformation can perform more optimization than techniques restricted to the program level.

A Simple Proof and Program. We present the problem in terms of a program to select all elements of a list that are at least 2. Following is a specification for the program, with a simple proof.

Specification 1

$$\forall l . \exists r . [\forall y . y \in r \Leftrightarrow (y \in l \land y \geq 2)]$$

Proof 2 The proof is by induction on the list l. If l is empty, choose r to be empty as well. Otherwise $l = x :: l'$ and, by the induction hypothesis, there is an r' such that $\forall y . y \in r' \Leftrightarrow (y \in l' \land y \geq 2)$. Then if $x \geq 2$ let r be $x :: r'$; otherwise let r be r'. □

The following program is extracted from a formalized version of the proof:

Program 3
```
fun select [] = []
  | select (x::l) =
     let val r = (select l) in if x >= 2 then x::r else r end
```

This can easily be transformed at the program level to tail-recursive form using the fold-unfold system [5].

Program 4
```
fun sel [] k = (reverse k)
  | sel (x::l) k = if x >= 2 then (sel l (x::k)) else (sel l k)
fun select l = sel l []
```

The transformed definition is tail-recursive, but contains a call to **reverse** which could be eliminated since the specification says nothing about the order of the elements in the list to be computed. But since this change does not preserve functional equivalence, it is not straightforward to accomplish via purely syntactic program transformation. One could avoid the difficulty by observing that the input is treated here as a set, not a list, and working in a more appropriate algebra. Depending on the context in which the function is used, this may or may not be easy to do. The use of a proof transformation yields a similar tail-recursive program that computes with lists but does not use the reverse function.

A Proof Structure that Guarantees Tail-recursive Form. The view of proofs as programs rests on the correspondence between proof structure and program structure. Recursion in the program corresponds to induction in the proof. In a constructive proof of $\forall l . \exists r . \Phi(l, r)$ by list induction, the inductive case constructs a witness: some term t such that for arbitrary x, $\Phi(x :: l, t)$. The witness is usually some function of a parameter r' obtained by eliminations from the induction hypothesis $\exists r'. \Phi(l, r')$. It corresponds to the value computed in the body of the extracted recursive function definition.

A program is tail-recursive when it does no computation with the result of its recursive call. Correspondingly, an inductive proof is realized by a tail-recursive program when the witness t is identical to the parameter r' of the induction hypothesis. That is, if the proof appeals to the induction hypothesis to obtain a parameter r' satisfying $\Phi(l, r')$, and shows that *the same value* r' also satisfies $\Phi(x :: l, r')$ for any x, the extracted function is tail-recursive.

To obtain a proof of this form, it is necessary to modify the induction hypothesis of the original proof. This corresponds to the well-known technique of generalizing a loop invariant [11]. Generalization to a particular form is the basis for our proof transformation.

A Derived Rule for Tail-recursion Introduction. Given an initial specification of the form $\forall l . \exists r . \Phi(l, r)$, a useful generalization is a formula

$$\forall l . \forall k . \exists s . \exists r . \Phi(l, r) \land \Psi(k, r, s),$$

where k corresponds to the accumulator argument in the extracted program. The transformation produces an inductive subproof of this formula, which corresponds to the auxiliary function definition **sel** of the fold-unfold derivation. The existentially quantified variable s corresponds to the result computed by the auxiliary tail-recursive function. However, the extracted program will compute a pair of lists corresponding to $\langle s, r \rangle$. To eliminate the unwanted computation of r, we insert a double negation in the new formula, obtaining

$$\forall l . \forall k . \exists s . \neg\neg \exists r . \Phi(l, r) \land \Psi(k, r, s).$$

This technique exploits the use of modified realizability: negations are uninformative and do not induce computation in the extracted program.

The derived rule that represents the transformation expresses what additional proof obligations must be met for the construction of the output proof. The base and inductive cases of the input supply the first two premises of the rule. Premises (3), (4), and (5) represent proof obligations. Their proofs and the formula Ψ must be supplied to the transformation in addition to the input proof.

$$\frac{\begin{array}{l}(1)\ \Phi([], \mathbf{r_0}) \\ (2)\ \forall x, l, r . \Phi(l, r) \supset \Phi(x :: l, \mathbf{f}(x, r)) \\ (3)\ \forall k . \Psi(k, \mathbf{r_0}, \mathbf{s_0}(k)) \\ (4)\ \forall x, k, r, s . \Psi(\mathbf{h}(x, k), r, s) \supset \Psi(k, \mathbf{f}(x, r), s) \\ (5)\ \forall l, r, s . \Phi(l, r) \land \Psi(\mathbf{k_0}, r, s) \supset \Phi(l, s)\end{array}}{\forall l . \exists s . \neg\neg \Phi(l, s)} \text{TR}$$

The premises of the rule are presented on separate lines and numbered for ease of reference. Bold-face identifiers, e.g. $\mathbf{r_0}$, are parameters of the inference rule.

When applied, e.g. $\mathbf{f}(x,r)$, they represent expressions in which the terms to which they are applied may occur free.

The structure of the output proof is specified by the derivation of the rule. For simplicity's sake we ignore the double negation in the conclusion, which is trivial to derive, and informally give the derivation of an intermediate output proof $\mathcal{P}'$ of $\forall l \,.\, \exists s \,.\, \Phi(l, s)$.

Proof 5 We first show $\forall l \,.\, \forall k \,.\, \exists s \,.\, \exists r \,.\, \Phi(l, r) \land \Psi(k, r, s)$ by induction on the list l. If $l = [\,]$ then for arbitrary k by premises (1) and (3) we can choose $\mathbf{r_0}$ for r and $\mathbf{s}_0(k)$ for s. Otherwise $l = x :: l'$. Let k' be an arbitrary list. We instantiate the induction hypothesis with $\mathbf{h}(x, k')$ for k to obtain s', r' such that $\Phi(l', r') \land \Psi(\mathbf{h}(x, k'), r', s')$. By premise (2) it follows that $\Phi(x :: l', \mathbf{f}(x, r'))$, and by premise (4) we have $\Psi(k', \mathbf{f}(x, r'), s')$. Then we can choose $\mathbf{f}(x, r')$ for r and s' for s. Since k' is arbitrary we have $\forall k \,.\, \exists s \,.\, \exists r \,.\, \Phi(l, r) \land \Psi(k, r, s)$.

Then for any list l we have s, r such that $\Phi(l, r) \land \Psi(\mathbf{k_0}, r, s)$, and from premise (5) $\Phi(l, s)$. □

Unlike the example of Sect. 3 the rule TR does not have the conclusion $\forall l \,.\, \exists r \,.\, \Phi(l, r)$ of the input proof as a premise. Thus it cannot be used directly to transform Proof 2; first the input proof must be analyzed to obtain premises (1) and (2) of the rule. The next section shows how higher-order unification can be used to do the analysis.

Representation in Elf. We describe a representation that produces the intermediate proof $\mathcal{P}'$; [1] gives details of how the double negation is obtained. The proof transformation is encoded in two stages: the first analyzes the input to obtain the premises (1) and (2) of the derived rule; the second encodes the rule itself.

To express the transformation in terms of the derived rule we define a judgment (Fig. 2) relating the input proof, the proof obligations, and the output proof. The judgment is implemented by a single clause (Fig. 3) that specifies how to obtain premises (1) and (2) of the derived rule TR by unification with the input. Higher-order unification is essential here; it enables us to pick out premise (2) as a (metalevel) function from terms and proofs to proofs. The instantiations for `Prem1` and `Prem2` are passed to the derived rule, which is invoked by the subgoal `<- tr_drule R0 ....` It is in solving this subgoal that the Elf interpreter constructs the output proof.

The encoding (Fig. 4) of the rule TR follows the same methodology as the example of Sect. 3. Meta-level Π-quantification represents the metavariables for formulas and individual expressions. Comparison of the encodings of the premises with the statement of the rule TR shows the use of a technique called lifting: object-level universal quantifiers and implications have been replaced by meta-level Π-quantification. The lifted representation is faithful to the object logic; this is a consequence of the fact that constructive minimal many-sorted predicate logic is representable in LF via the propositions-as-types interpretation, as

```
tail_rec :
{R0} {S0} {K0} {F} {H}
{Phi: ilist -> ilist -> o} {Psi: ilist -> ilist -> ilist -> o}
%% Input proof:
|- (lforall ([l] lexists ([r] (Phi l r))))
-> ({k:ilist} |- (Psi k R0 (S0 k)))                        %% Premise 3
-> ({x:i} {k:ilist} {r:ilist} {s:ilist}                    %% Premise 4
       |- (Psi (H x k) r s) -> |- (Psi k (F x r) s))
-> ({l} {r} {s}                                            %% Premise 5
       |- ((Phi l r) & (Psi K0 r s)) -> |- (Phi l s))
%% Output proof:
-> (|- (lforall ([l] (lexists ([r] Phi l r)))))
-> type.
```

Fig. 2. Transformation judgment

```
tl_rec_clause :
{R0} {S0} {K0} {F} {H} {Phi} {Psi}
{Prem1} {Prem2} {Prem3} {Prem4} {Prem5}
{Output_proof}
tail_rec R0 S0 K0 F H Phi Psi
(lind ([l] lexists ([r] Phi l r))
      (lexistsi ([r] Phi nl r) R0 Prem1)                   %% Premise 1
      ([l] [p] [x]
         (lexistse ([r] [q]
           lexistsi _ (F x r) (Prem2 x l r q))             %% Premise 2
         p)))
Prem3 Prem4 Prem5 Output_proof
<- tr_drule R0 S0 K0 F H Phi Psi Prem1 Prem2 Prem3 Prem4 Prem5
            Output_proof.
```

Fig. 3. Transformation clause

described in, e.g., [3]. (For details of the correspondence between the lifted representation and a representation using object-level quantification the interested reader is referred to [1].) When the lemma proofs end in introductions (the most likely case) lifting has the benefit of avoiding the generation of detours in the output proof, which correspond to unwanted redices in the extracted program.

As in the example of Sect. 3, the clause (Fig. 5) implementing this judgment is a straightforward formalization of the rule derivation (Proof 5) and specifies the structure of the output proof.

Some care must be taken when coding transformations this way to avoid encountering unification problems that are not solved by the Elf implementation. If an encoding restricts its use of logic variables for pattern matching to *higher-*

```
tr_drule :
 {R0} {S0} {K0} {F} {H}
 {Phi: ilist -> ilist -> o} {Psi: ilist -> ilist -> ilist -> o}
 %% Input base case (premise 1):
 |- Phi nl R0
 %% Input inductive case (premise 2):
 -> ({x:i} {l} {r} |- Phi l r -> |- Phi (cns x l) (F x r))
 %% Premise 3
 -> ({k:ilist} |- Psi k R0 (S0 k))
 %% Premise 4
 -> ({x:i} {k:ilist} {r:ilist} {s:ilist}
       |- Psi (H x k) r s -> |- Psi k (F x r) s)
 %% Premise 5
 -> ({l} {r} {s} |- (Phi l r) & (Psi K0 r s) -> |- Phi l s)
 %% Output proof:
 -> (|- lforall [l] (lexists [s] (Phi l s)))
 -> type.
```

Fig. 4. The derived rule TR

```
tr_rule_derive :
 tr_drule R0 S0 K0 F H Phi Psi Prem1 Prem2 Prem3 Prem4 Prem5
 (lforalli [l]
   %% Recovery of original specification:
   lexistse ([s] [p']
    lexistse ([r] [p]
     (lexistsi _ s (Prem5 l r s p)))
    p')
   (lforalle K0 (lforalle l
     %% New inductive sub-proof:
     (lind ([l] lforall [k] lexists [s] lexists [r]
              (Phi l r) & (Psi k r s))
       %% Base case:
       (lforalli [k]
         lexistsi _ (S0 k) (lexistsi _ R0 (andi Prem1 (Prem3 k))))
       %% Step case:
       ([l] [p] [x] lforalli [k]
         (lexistse ([s] [q'] (lexistse ([r] [q]
            lexistsi _ s
              (lexistsi _ (F x r)
                (andi (Prem2 x l r (andel q)) (Prem4 x k r s (ander q)))))
           q'))
         (lforalle (H x k) p)))))))).
```

Fig. 5. Derivation of TR

order patterns in the sense of [28], only deterministic unification problems will arise during execution. These problems fall within a decidable subcase of higher-order unification discovered by Miller [20] for the simply typed lambda calculus and extended to LF by Pfenning. Unification problems that cannot be solved by the Elf implementation can still be solved by an Elf program for unification, using a technique adapted from Miller [21]. This complicates the encoding but presents no essential difficulty.

With this clause and formalizations Psi, P_3, P_4, P_5 of Ψ and the proof obligations, we can transform a formalization P of Proof 2 by submitting the following query to Elf:

?- tail_rec R0 S0 F H K0 Phi Psi P P_3 P_4 P_5 Q.

Here the italic variables (e.g., P) stand for closed LF object terms provided by the user. The text in typewriter font (e.g., R0) is input as-is by the user; thus the variables R0...Phi and Q are Elf logic variables. If the query succeeds the Elf interpreter displays the terms bound to them in the course of the search. The term bound to Q is a representation of the transformed proof.

The resulting proof can be further transformed by the same methods (which impose no further proof obligations) to obtain a proof of $\forall l . \exists s . \neg\neg \Phi(l, s)$ in which unwanted computation is suppressed.

The following is an informal statement of the new proof of Specification 1 obtained by applying this transformation to the `select` example:

Proof 6 We first show

$$\forall l . \forall k . \exists s . \exists r . [\forall y . y \in r \Leftrightarrow (y \in l \wedge y \geq 2)] \wedge [\forall y . y \in s \Leftrightarrow (y \in r \vee y \in k)]$$

by induction on l. If $l = []$ then choose $s = k$ and $r = []$. Otherwise $l = x :: l'$. If $x \geq 2$, then by instantiating the induction hypothesis with l' and $x :: k$, we obtain a list r' containing all elements ≥ 2 of l' and a list s' containing all the elements of r' and $x :: k$. Then choose $r = x :: r'$ and $s = s'$. Otherwise $\neg x \geq 2$; again by the induction hypothesis there is a list r' as before, and a list s' containing all the elements of r' and k. Let $r = r'$ and $s = s'$.

Now let k be the empty list, so that s contains exactly the elements of r. Then $\forall y . y \in s \Leftrightarrow (y \in l \wedge y \geq 2)$ and $\forall l . \exists r . [\forall y . y \in r \Leftrightarrow (y \in l \wedge y \geq 2)]$ as required. □

Applied to the example the implementation yields the following extracted program:

Program 7

```
fun sel l =
  let fun sel' nil = (fn k => k)
      | sel' (x::l') = let val p = (sel' l') in
          fn k => (p (if x >= 2 then (x::k) else k)) end
  in (sel' l nil) end;
```

This program is tail-recursive, and is an improvement over Program 4, as it does not use the **reverse** function. The difference is a consequence of working with the specification and its proof throughout the development rather than restricting the reasoning to the properties of the program alone. It is an example of the effect noted by Goad [10]: using proof transformation a program can be optimized by exploiting the fact that there is no need to preserve functional equivalence of the successive stages of program development.

Correctness Properties. The encoding provides two sorts of correctness guarantees for the transformation. Because type-checking is proof-checking, the type-correctness of the transformation ensures that the transformed proof is a valid proof of a particular known end-formula. This is a general property that holds for any proof transformation encoded this way. The technique also supports reasoning about properties specific to this transformation. For example, in [1] we give an Elf program to recognize tail-recursive object programs. Using the techniques of [29] we give a partial representation in Elf of a proof that the tail-recursion transformation yields a tail-recursive program if it succeeds.

5 Conclusion

We have a demonstrated a methodology for encoding proof transformations that is applicable to program optimization problems. By limiting the scope of the technique to transformations based on derived rules of the encoded logic, we obtain encodings with correctness properties that are easy to establish. Although this is a strong limitation it appears that many techniques from the field of program transformation may be amenable to this kind of encoding. We should point out that the choice of Elf as a basis for the encoding does not limit the implementable transformations to those based on derived rules. Elf programs can be written to implement more complex transformations, for example, proof normalization. Such transformations share the general correctness property that the resulting proof is valid and its end-formula is known, but other properties are typically more difficult to establish.

This style of representation is largely independent of the choice of theorem-proving techniques and heuristic issues. Thus it could perhaps be usefully combined with some of the work of Madden [17] [18] on the use of automatic theorem proving techniques and analysis tools tailored for fully automated program development by proof transformation.

Acknowledgement

I would like to thank Frank Pfenning for providing the Elf system and for supervising my research. Thanks also to the referees for their helpful comments.

References

1. Penny Anderson. *Program Derivation by Proof Transformation.* PhD thesis, Department of Computer Science, Carnegie Mellon University, October 1993. Available as Technical Report CMU-CS-93-206.
2. Penny Anderson. Program extraction in a logical framework setting. In *Proceedings of the 5th International Conference on Logic Programming and Automated Reasoning*, July 1994. To appear.
3. Henk Barendregt. Introduction to generalized type systems. *Journal of Functional Programming*, 1(2):125–154, April 1991.
4. R.S. Bird. The promotion and accumulation strategies in transformational programming. *ACM Transactions on Programming Languages and Systems*, 6(4):487–504, October 1984.
5. R. M. Burstall and John Darlington. A transformation system for developing recursive programs. *Journal of the Association for Computing Machinery*, 24(1):44–67, January 1977.
6. Robert L. Constable et al. *Implementing Mathematics with the Nuprl Proof Development System.* Prentice-Hall, Englewood Cliffs, New Jersey, 1986.
7. Gilles Dowek, Amy Felty, Hugo Herbelin, Gérard Huet, Christine Paulin-Mohring, and Benjamin Werner. The Coq proof assistant user's guide. Rapport Technique 134, INRIA, Rocquencourt, France, December 1991. Version 5.6.
8. Amy Felty. *Specifying and Implementing Theorem Provers in a Higher-Order Logic Programming Language.* PhD thesis, Department of Computer and Information Science, University of Pennsylvania, July 1989. Available as Technical Report MS-CIS-89-53.
9. Amy Felty. Implementing tactics and tacticals in a higher-order logic programming language. *Journal of Automated Reasoning*, 11:43–81, 1993.
10. Christopher A. Goad. Computational uses of the manipulation of formal proofs. Technical Report Stan-CS-80-819, Stanford University, August 1980.
11. David Gries. *The Science of Programming.* Springer-Verlag, 1981.
12. John Hannan. *Investigating a Proof-Theoretic Meta-Language for Functional Programs.* PhD thesis, University of Pennsylvania, January 1991. Available as MS-CIS-91-09.
13. John Hannan and Dale Miller. A meta-logic for functional programming. In H. Abramson and M. Rogers, editors, *Meta-Programming in Logic Programming*, pages 453–476. MIT Press, 1989.
14. Robert Harper, Furio Honsell, and Gordon Plotkin. A framework for defining logics. *Journal of the Association for Computing Machinery*, 40(1):143–184, January 1993.
15. Susumu Hayashi. An introduction to PX. In Gerard Huet, editor, *Logical Foundations of Functional Programming*. Addison-Wesley, 1990.
16. Gérard Huet and Bernard Lang. Proving and applying program transformations expressed with second-order patterns. *Acta Informatica*, 11:31–55, 1978.
17. Peter Madden. *Automated Program Transformation Through Proof Transformation.* PhD thesis, University of Edinburgh, 1991.
18. Peter Madden. Automatic program optimization through proof transformation. In D. Kapur, editor, *Proceedings of the 11th International Conference on Automated Deduction*, pages 446–460, Saratoga Springs, New York, June 1992. Springer-Verlag LNAI 607.

19. Spiro Michaylov and Frank Pfenning. Natural semantics and some of its metatheory in Elf. In L.-H. Eriksson, L. Hallnäs, and P. Schroeder-Heister, editors, *Proceedings of the Second International Workshop on Extensions of Logic Programming*, pages 299–344, Stockholm, Sweden, January 1991. Springer-Verlag LNAI 596.
20. Dale Miller. A logic programming language with lambda-abstraction, function variables, and simple unification. In Peter Schroeder-Heister, editor, *Extensions of Logic Programming: International Workshop, Tübingen FRG, December 1989*, pages 253–281. Springer-Verlag LNCS 475, 1991.
21. Dale Miller. Unification of simply typed lambda-terms as logic programming. In K. Furukawa, editor, *Proceedings of the Eighth International Conference on Logic Programming*, pages 255–269. MIT Press, July 1991.
22. Robin Milner, Mads Tofte, and Robert Harper. *The Definition of Standard ML*. MIT Press, Cambridge, Massachusetts, 1990.
23. Robert Paige and Shaye Koening. Finite differencing of computable expressions. Technical Report LCSR-TR-8, Laboratory for Computer Science Research, Rutgers University, August 1980.
24. Christine Paulin-Mohring. Extracting F_ω programs from proofs in the calculus of constructions. In *Sixteenth Annual Symposium on Principles of Programming Languages*, pages 89–104. ACM Press, January 1989.
25. Frank Pfenning. On the undecidability of partial polymorphic type reconstruction. *Fundamenta Informaticae*, 199? To appear. Preliminary version available as Technical Report CMU–CS–92–105, School of Computer Science, Carnegie Mellon University, Pittsburgh, Pennsylvania, January 1992.
26. Frank Pfenning. Program development through proof transformation. *Contemporary Mathematics*, 106:251–262, 1990.
27. Frank Pfenning. Logic programming in the LF logical framework. In Gérard Huet and Gordon Plotkin, editors, *Logical Frameworks*, pages 149–181. Cambridge University Press, 1991.
28. Frank Pfenning. Unification and anti-unification in the Calculus of Constructions. In *Sixth Annual IEEE Symposium on Logic in Computer Science*, pages 74–85, Amsterdam, The Netherlands, July 1991.
29. Frank Pfenning and Ekkehard Rohwedder. Implementing the meta-theory of deductive systems. In D. Kapur, editor, *Proceedings of the 11th International Conference on Automated Deduction*, pages 537–551, Saratoga Springs, New York, June 1992. Springer-Verlag LNAI 607.
30. James T. Sasaki. *Extracting Efficient Code from Constructive Proofs*. PhD thesis, Cornell University, May 1986. Available as Technical Report TR 86–757.
31. Helmut Schwichtenberg. On Martin-Löf's theory of types. In *Atti Degli Incontri di Logica Mathematica*, pages 299–325. Dipartimento di Matematica, Università di Siena, 1982.
32. Helmut Schwichtenberg. A normal form for natural deductions in a type theory with realizing terms. In Ettore Casari et al., editors, *Atti del Congresso Logica e Filosfia della Scienza, oggi. San Gimignano, December 7–11, 1983*, pages 95–138, Bologna, Italy, 1985. CLUEB.

Exploring Abstract Algebra in Constructive Type Theory

Paul Jackson*

Department of Computer Science
Cornell University
Ithaca NY 14853
USA
Tel: [+1] 607-255-1372
E-mail: jackson@cs.cornell.edu

Abstract. I describe my implementation of computational abstract algebra in the Nuprl system. I focus on my development of multivariate polynomials. I show how I use Nuprl's expressive type theory to define classes of free abelian monoids and free monoid algebras. These classes are combined to create a class of all implementations of polynomials. I discuss the issues of subtyping and computational content that came up in designing the class definitions. I give examples of relevant theory developments, tactics and proofs. I consider how Nuprl could act as an algebraic 'oracle' for a computer algebra system and the relevance of this work for abstract functional programming.

1 Introduction

1.1 Aims and Motivation

One aim of the Nuprl project is to explore the use of constructive type theory for specifying, verifying and synthesizing functional programs. To date, effort has been concentrated on developing programs over concrete data-types. I would like to see how well Nuprl could support the development of programs using abstract data types (ADT's); ADT's are widely advocated as an effective structuring principle and support for them is provided in a variety of modern programming languages. Type theory can provide an effective framework for formally reasoning about ADT's. Type theory is particularly convenient in that both ADT specifications and implementations are described in a single language [21, 3].

In previous work on using Nuprl to verify hardware designs for floating-point arithmetic, I noticed that algebraically-similar kinds of reasoning were coming up frequently over different concrete datatypes such as the integers, the rationals and bit-vectors. For example, a significant part of the reasoning concerned iterated sums over monoids. It was obvious that work could be saved

---

* Supported by NASA GSRP Fellowship NGT-50786 and ONR Grant N00014-92J-1764

by first making definitions and proving theorems in an appropriately abstract setting.

A newer motivation comes from observing the development of computer algebra systems such as Mathematica, Maple and Axiom. Such systems provide a wealth of functionality, but not the rigor or the abstractness of a theorem proving environment. I want to explore possibilities for symbiotic interactions between computer algebra systems and theorem provers, and speculate on how in the long term they might be integrated. Already promising work has been done in the area [7, 12]. Here at Cornell I am currently investigating links between Nuprl and the Weyl system [23]. In particular I am exploring the use of Nuprl as an algebraic oracle to Weyl.

Nuprl's type theory provides a certain range of options for making explicit various computational aspects of algebra. I would like to understand which options are worth formalizing and which ones I am missing, both from the point of view of providing assistance to computer algebra systems and of developing abstract programs. For example, most computer algebra systems are interested in algebraic structures with decidable equality relations, so I am reasoning primarily with structures with boolean-valued equality functions.

1.2 Background

Nuprl [8, 15] is an interactive tactic-based theorem prover in the LCF tradition. It uses a constructive type theory similar to that of Martin-Löf [19]. The type theory is briefly reviewed in Sect. 2.

Nuprl has been developed in the last 10 years by over a dozen people. It has a well developed user interface and a large collection of tactics for such tasks as forward and backward chaining, rewriting and arithmetic reasoning. Libraries have been built up in number theory [13], analysis [6], hardware verification [4], hardware synthesis [1] and meta-reasoning [14]. Many of the tactics and theory developments mentioned in this paper are covered in greater detail in my thesis [16].

1.3 Organization of this Paper

To illustrate my work, I have chosen to show a sequence of definitions leading up to the class of multivariate polynomial algebras. Section 2 contains an introduction to Nuprl's type theory that should be adequate for understanding the definitions, and the definitions themselves are presented in Sect. 3. Section 4 gives some practical details on my work to date and Sect. 5 describes applications of the work that we are actively pursuing. Finally, Sect. 6 summarizes my accomplishments.

1.4 Related Work

I know of several other efforts to develop abstract algebra in a theorem proving environment. Gunter[10] working with HOL has proven group isomorphism theorems and shown the integers mod n to be an implementation of abstract groups. Harrison and Thèry [12] have looked at the interaction between HOL and Maple, where Maple performs algebraic manipulations and integrations that HOL then verifies. The IMPS people have a notion of *little theories* [9] which they use for proving theorems about groups and rings. Anthony Bailey has developed a concrete theory of polynomials in one variable in the LEGO system and has proven the correctness of Euclid's algorithm over these polynomials [2]. The Mizar project seems to have done a fair amount in algebra, but hasn't yet tackled polynomials.

Clarke and Xudong have been adding theorem proving capabilities to Mathematica to create their Analytica system [7]. They have impressive results in proving equivalences of sums of series, but their work has been hindered by the lack of rigor inherent in the Mathematica environment.

The Larch [11] group has worked on program specification and verification using ADT's in a first-order-logic setting.

2 Type Theory Preliminaries

I give here an informal overview of the types in Nuprl's type theory I have been working with:

- The *booleans* $\mathbb{B}$ and the *integers* $\mathbb{Z}$.
- A *dependent-function* (Π) type constructor $\rightarrow$. If A is a type and B_x is a family of types, indexed by $x \in A$, then $x{:}A \rightarrow B_x$ is the type of functions f, such that $f(a) \in B_a$ for all $a \in A$. If B_x is the same for all $x \in A$, I write the type as simply $A \rightarrow B$. I assume that $\rightarrow$ associates to the right. Each type $A \rightarrow B$ is considered as containing only the computable functions from A to B rather than all set theoretic functions.
- A *dependent-product* (Σ) type constructor $\times$. If A is a type and B_x is a family of types, indexed by $x \in A$, then $x{:}A \times B_x$ is the type of pairs $\langle a, b \rangle$, such that $a \in A$ and $b \in B_a$. If B_x is the same for all $x \in A$, I write the type as simply $A \times B$. Sometimes I write $A \times A$ as A^2. I assume that $\times$ associates to the right.
- A *set* type constructor $\{\,:\,|\,\}$. If A is a type and P_x is a proposition in which x of type A occurs free, then $\{x{:}A|P_x\}$ is the type of those element x of A for which P_x is true. The type of non-negative integers $\mathbb{N}$ and positive integers $\mathbb{N}^+$ are constructed using the subset type constructor.
- *Universes* of types $\mathbb{U}_i$ for $i = 1, 2, 3 \ldots$. $\mathbb{U}_i$ includes $\mathbb{U}_j$ for all $j < i$ as base types and is closed under the type constructors listed above. In Sect. 3 I drop these universe level indices without risk of confusion, although the indices must be stated when proving lemmas in the Nuprl system.

Every type has a natural equality relation on it. I write $x =_T y$ or $x = y \in T$ when I want to say that x and y are members of T and are equal by the equality relation associated with T. I write $x = y$ if T is obvious from context. Functions always respect type equalities, so if function f has type $S \to T$, then $fx =_T fx'$ whenever $x =_S x'$.

The equality associated with a type can be weakened using Nuprl's *quotient type* constructor; if R is an equivalence relation on type T, then the quotient type constructed from T and R is written $x,y{:}T//xRy$. The inhabitants of $x,y{:}T//xRy$ are the *same* as the inhabitants of T; the quotient type does not group elements of T into equivalence classes. Inhabitants are considered equal when they are related by R.

I use fairly usual notation for programming language constructs. Function application is designated by juxtaposition. For example, I write $f\ a$. Often I use infix notation for binary function application. For example, if $* \in (T \times T) \to T$, then for $*\langle a, b\rangle$ I write $a * b$.

Logic is injected into type theory using the propositions-as-types correspondence. Each predicate-logic expression corresponds to a type with the type being inhabited iff the predicate-logic expression is provable. The proof of a logical expression specifies exactly how to construct the term that inhabits the corresponding type. Sometimes the inhabitant is interesting; for example it might be a function that computes something useful. In this case, we can view the logical expression corresponding to the type it inhabits as a kind of program specification. When I talk about the *computational content* of a logical expression, I am referring to the possible inhabitants of the corresponding type. Nuprl's logic is well-suited to constructive mathematics, but it also can support classical styles of reasoning.

3 Polynomial Algebra Development

3.1 The Approach

In mathematics, one talks about *the* ring of polynomials $A[S]$, given some ring A and some set of indeterminates S. However, a computer algebra system might have several *different* implementations of polynomials. Mathematically they are all isomorphic, but computationally they have distinct characteristics.

I am interested here in using type theory to characterize this practice, so I define a type PolyAlg(S, A) that is the class of *all* implementations of polynomials over indeterminates S and ring A. This definition gets at the general properties of polynomials and abstracts away from the features of particular implementations. At the same time, I craft the definition such that a wide variety of useful computable functions could be constructed from an arbitrary inhabitant of the class.

My approach is based on the standard abstract approach found in textbooks such as Lang [18] or Bourbaki [5]:

1. Monomials in S are elements of a free abelian monoid over S.
2. Polynomials over S with coefficients from A are elements of a free monoid algebra over the monoid of monomials and the ring A.

From the above two constructions I derive the ring of polynomials and such functions as the injections from A and S, and 'substitution' function for instantiating indeterminates with elements of any ring B given an homomorphism from A to B.

I require implementations of classes to provide boolean-valued functions for deciding equality. Without these functions, one is rather limited in the kinds of computable functions that can be defined. This is standard practice in most computer algebra systems. However, the class definitions I give don't rely in any essential way on these functions.

3.2 The Free Abelian Monoid of Monomials

I start by defining EqType, a class for types with associated equality functions:

$$\mathsf{EqTypeSig} \doteq S{:}\mathbb{U} \times (S^2 \to \mathbb{B})$$

$$\mathsf{EqType} \doteq \{\langle S, eq\rangle{:}\mathsf{EqTypeSig} \mid \forall a, b{:}S.\ a\ eq\ b \iff a =_S b\}\ .$$

The class EqType is the set of those elements of the signature EqTypeSig for which the equality function agrees with the equality relation of the type. I refer to elements of this class as *equality types*. I assume in what follows that an element $\mathcal{S}$ of EqType has form $\langle S, eq\rangle$.

The class signature for monoids, MonSig is:

$$\mathsf{MonSig} \doteq M{:}\mathbb{U} \times (M^2 \to \mathbb{B}) \times (M^2 \to M) \times M$$

and the class for abelian monoids AbMon is:

$$\begin{aligned}
\mathsf{AbMon} \doteq \{&\langle M, eq, *, e\rangle{:}\mathsf{MonSig}| \\
&\forall a, b{:}M.\ a\ eq\ b \iff a =_M b \\
&\wedge\ \forall a, b, c{:}M.\ (a * b) * c = a * (b * c) \\
&\wedge\ \forall a, b, c{:}M.\ a * b = b * a \\
&\wedge\ \forall a{:}M.\ (a * e) = a \\
\}\ .&
\end{aligned}$$

I assume in what follows that an element $\mathcal{M}$ of AbMon has form $\langle M, eq, *, e\rangle$.

The definition of FAbMon(S), the class of free abelian monoids (with computable equalities) over the equality type $\mathcal{S}$ is:

$$\begin{aligned}
\mathsf{FAbMon}(\mathcal{S}) \doteq\ &\mathcal{M}{:}\mathsf{AbMon} \\
&\times\ \iota{:}S \to M \\
&\times\ \mathcal{M}'{:}\mathsf{AbMon} \to \phi{:}(S \to M') \\
&\to \{!\ \hat{\phi}{:}\mathsf{MonHom}(\mathcal{M}, \mathcal{M}') \mid \phi = \hat{\phi} \circ \iota\}\ .
\end{aligned}$$

Here, $\mathsf{MonHom}(\mathcal{M}, \mathcal{M}')$, the set of monoid homomorphisms from $\mathcal{M}$ to $\mathcal{M}'$, is defined as:

$$\begin{aligned}
\mathsf{MonHom}(\mathcal{M}, \mathcal{M}') \doteq \{\phi{:}M \to M' \mid\ &\phi\ e = e' \\
\wedge\ &\forall a, b{:}M.\ \phi\ (a * b) = \phi\ a *'\ \phi\ b\}
\end{aligned}$$

and I use the abbreviation:

$$\{!x{:}T|P_x\} \doteq \{x{:}T|P_x \wedge \forall x'{:}T.\ P_{x'} \Rightarrow x = x'\}\ .$$

$\{!x{:}T|P_x\}$ should be be read as 'the type containing the unique x of type T such that P_x holds'.

The definition of $\mathsf{FAbMon}(\mathcal{S})$ is based on the characterization of a free abelian monoid over some set S as being an abelian monoid $\mathcal{M}$ and an injection ι of S into M, such that for any abelian monoid $\mathcal{M}'$ and mapping ϕ of S into M', there is a unique abelian monoid homomorphism $\hat{\phi}$ from M to M' which satisfies the equation $\phi = \hat{\phi} \circ \iota$. This equation can be stated pictorially by saying for each $\mathcal{M}'$ and ϕ there is a unique $\hat{\phi}$ such that the following diagram commutes:

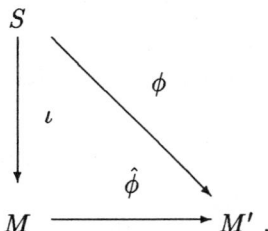

The definition of $\mathsf{FAbMon}(S)$ captures the requirement that there is an appropriate mapping into any abelian monoid by insisting that a function be supplied that generates this mapping. Specifically, let the triple $\langle \mathcal{M}, \iota, \Psi \rangle$ be some element of $\mathsf{FAbMon}(S)$. If Ψ is given as arguments an abelian monoid $\mathcal{M}'$ and a function ϕ, it must return the unique monoid homomorphism $\hat{\phi}$ from $\mathcal{M}$ to $\mathcal{M}'$ that satisfies the equation $\phi = \hat{\phi} \circ \iota$.

I could have written the definition of $\mathsf{FAbMon}(S)$ as:

$\mathsf{FAbMonSig}(\mathcal{S}) \doteq \mathcal{M}{:}\mathsf{AbMon} \times S \to M$

$\mathsf{FAbMon}(\mathcal{S}) \doteq \{\langle \mathcal{M}, \iota \rangle{:}\mathsf{FAbMonSig}(\mathcal{S})\ |$
$\quad\quad \forall \mathcal{M}'{:}\mathsf{AbMon}\ \forall \phi{:}(S \to M')$
$\quad\quad \exists!\hat{\phi}{:}\mathsf{MonHom}(\mathcal{M}, \mathcal{M}').\ \phi = \hat{\phi} \circ \iota\ \},$

not requiring Ψ to be explicitly supplied, but this would not have been nearly as computationally interesting.

An example of a useful function that can easily be generated from Ψ is as follows: if $\mathcal{S}$ is an equality type of indeterminates, then $\Psi\ \langle \mathbb{Z}, eq_\mathbb{Z}, +, 0 \rangle\ (\lambda x.1)$, is a function for calculating the total degree of a monomial (the sum of the powers of the indeterminates).

If one applies Ψ to the monoid of monomials and the injection function of some other implementation of $\mathsf{FAbMon}(\mathcal{S})$, one gets a function that translates representations of monomials into that used by the other implementation. Similarly, one could use the Ψ of that other implementation to translate back. These translation functions are inverses of one another and hence set up a constructive isomorphism between the two implementations. (This is just an instance of the fact that in category theory, initial objects are unique up to isomorphism.) A computable function defined over one implementation can be lifted to any other implementation using these translation functions. Therefore, all implementations are equally expressive, and every implementation is in a sense 'computationally complete'. Without the Ψ functions being explicitly required of implementations I would not have this computational completeness.

The standard mathematical representation for the carrier of a free abelian monoid over S is the set of functions of finite support of type $S \to \mathbb{N}$. (A function has *finite support* if it returns 0 for all but a finite number of elements of its domain.) A naive translation of this into type theory gives the type

$$\{f{:}(S \to \mathbb{N}) \mid f \text{ has finite support}\} \ .$$

This type is not good constructively since there is no way of *computing* a finite support of elements of this type and hence giving a computable definition for Ψ. To make the finite support explicit, one could use instead the type:

$$\{\langle s, f\rangle{:}\text{FinSet}(S) \times (S \to \mathbb{N}) \mid f \text{ has finite support } s\}$$

or the type $x, y{:}\text{A-List}(S, \mathbb{N}^+)//x$ permutation of y, where

A-List$(T, T') \doteq \{z{:}(T \times T')\text{List} \mid$ all elements of z have distinct left sides$\}$.

The quotient type is needed here because my class definition for free abelian monoids expects the equality associated with the carrier type of implementations to be the monoid equality; for example, it requires that all functions taking elements of the monoid carrier as arguments to respect this equality. However, one can have two a-lists that represent the same monomial but that are distinct according to the standard equality on lists. If type FinSet(S) of finite sets of elements of S is implemented using lists, one would similarly have to use a quotient type to hide the list order.

Nuprl's quotient type is an important tool for abstraction because it hides structural detail. Consider verifying the correctness of functions in a class implementation that use quotiented alists as described above. Nuprl's type theory forces one to show that the functions' behaviors are independent of the order of pairs in elements of the carrier, and so in this case the order is hidden.

3.3 The Polynomial Algebra

The definition of the class FMonAlg$(\mathcal{G}, \mathcal{A})$ of free monoid algebras over the abelian monoid $\mathcal{G}$ and the ring $\mathcal{A}$ is similar in structure to that of the free abelian monoid:

FMonAlg$(\mathcal{G}, \mathcal{A}) \doteq \mathcal{B}{:}\mathcal{A}$-Algebra
$\qquad \times \ \iota{:}\text{MonHom}(\mathcal{G}, \mathcal{B} \downarrow_{mmon})$
$\qquad \times \ \mathcal{B}'{:}\mathcal{A}$-Algebra
$\qquad \quad \to \phi{:}\text{MonHom}(\mathcal{G}, \mathcal{B}' \downarrow_{mmon})$
$\qquad \quad \to \{!\ \hat{\phi}{:}\mathcal{A}\text{-AlgHom}(\mathcal{B}, \mathcal{B}') \mid \phi = \hat{\phi} \circ \iota\ \}$

where $\mathcal{A}$-Algebra is the class of algebras over the ring $\mathcal{A}$, $\downarrow_{mmon}$ is a forgetful class morphism that projects out the multiplicative monoid of an algebra and $\mathcal{A}$-AlgHom$(\mathcal{B}, \mathcal{B}')$ is the type of $\mathcal{A}$-Algebra homomorphisms from $\mathcal{B}$ to $\mathcal{B}'$. As with the free monoid algebra, I require the universal projection function to be an explicit part of implementations.

I use this class to form polynomial algebras by supplying a free abelian monoid of monomials for $\mathcal{G}$ and considering $\mathcal{A}$ as the ring of coefficients. The addition and multiplication operations of the free monoid algebra are then the standard corresponding polynomial operations. A standard mathematical implementation of the carrier of a free monoid algebra is as the set of functions of finite support of type $G \to A$. One constructive implementation is:

x, y:A-List$(G, A^{-0}) // x$ permutation of y .

where A^{-0} is the type of the non-zero elements of the ring carrier A. If $\mathcal{G}$ is a monoid of monomials, then this implementation is considering polynomials as a-lists of monomials and their coefficients with the order of the a-lists quotiented out.

Finally, by defining a polynomial algebra implementation to be a pair of the implementation of monomials and the implementation of polynomials over the monomials, I arrive at a definition of an class for polynomial algebras:

$$\mathsf{PolyAlg}(\mathcal{S}, \mathcal{A}) \doteq \mathcal{M}\mathsf{:FAbMon}(\mathcal{S})$$
$$\times \mathsf{FMonAlg}(\mathsf{mon}(\mathcal{M}), \mathcal{A})$$

where $\mathcal{S}$ is an equality type for the indeterminates, $\mathcal{A}$ is the coefficient ring and mon$(\mathcal{M})$ is a function that selects out the monoid part of the triple $\mathcal{M}$.

I can define functions that project out from implementations of PolyAlg$(\mathcal{S}, \mathcal{A})$ the ring of polynomials and functions related to this ring. For example, injections from A and S, and a universal summation/evaluation function over polynomials that given a homomorphism from coefficient ring $\mathcal{A}$ to some other ring $\mathcal{B}$ and an assignment of values in B to the polynomial indeterminates, creates a function that maps polynomials over $\mathcal{A}$ into B.

3.4 Choices in Making Definitions

In algebra, subtyping relationships between algebraic classes are ubiquitous. With the simplest form of subtyping, *set subtyping*, the underlying signatures are the same; every member of the class of gaussian monoids is a member of the class of monoids. A richer form of subtyping, *forgetful subtyping*, involves forgetting components of signatures; every member of the class of groups can be considered a member of the class of monoids if one forgets the inverse operation. Algebraic notation without forgetful subtyping quickly becomes extremely cluttered with trivial forgetful class morphisms. I can implement set subtyping with Nuprl's set type, but unfortunately, the rigid nature of Σ types prevents me from implementing forgetful subtyping. An ad-hoc partial solution that I've adopted for now is to minimize the number of class signatures; in my implementation of the abelian monoid class, I actually use the group signature class rather than a monoid signature class. When specifying an inhabitant of the abelian-monoid class I then have to supply a dummy inverse function.

There are numerous proposals [22] for forms of dependent record types that support forgetful subtyping. I am very interested in trying to adapt one of them to Nuprl's type theory.

Many algebraic definitions have computational content when considered constructively. The free class definitions in Sect. 3 have these universal projection functions. I have experimented with defining the permutation relation on lists such that the computational content of the proposition 'list x is a permutation of list y' when the proposition is true is a permutation function that permutes list x into list y. The content of a relation expressing membership of an element x of a commutative ring A in a finitely-generated ideal $\langle a_1 \ldots a_n \rangle$ could be a list of elements $c_1 \ldots c_n$ of A such such that $x = c_1 a_1 + \ldots c_n a_n$.

One challenge in writing class definitions is in deciding what computational content to explicitly bring out and what to leave implicit. Consider the discussion in Sect. 3.2 of alternative definitions for FAbMon($\mathcal{S}$). Definitions become very tedious if one takes a conservative approach and makes explicit all computational content. If the equality relation on the carrier type of members of algebraic classes has computational content, then every predicate involving equality such as $\forall x, y.\ x + y \Rightarrow y + x$ or $\forall x, y.\ x = y \Rightarrow y = x$ also has computational content. When defining classes in Nuprl, I would have to include such predicates in my class signatures rather than the right-hand-side of set types, and I would lose all set subtyping properties.

In Nuprl's type theory, the equality relation naturally associated with a type corresponds to a unit type when true and a void type when false; its computational content is always trivial. I have used these natural equality relations in all my class definitions; I get set subtyping properties and get the benefit of the built-in support in Nuprl's type theory for these relations. An example of the built-in support is that elements of function types always respect these equality relations. This design decision seems in accord with current practice in constructive algebra [20] and in the computer algebra system Axiom [17] where much attention has been paid to constructivity.

4 Practical Work

4.1 Theory Development

All the class definitions described in Sect. 3 have been entered into the Nuprl system, but as yet I haven't proved much using them. I would like eventually to tackle reasoning about algorithms in computational algebra and these algorithms assume much more detailed structure than my definitions provide. For example, it is fundamental that every polynomial can be uniquely expressed as a sum of monomials. The standard way of proving this involves first constructing a canonical implementation of the polynomial algebra class involving a representation of functions of finite support. I also need to reason about orderings on monomials.

My theory development efforts to date have been mostly directed at building up a set of foundational theories that are sufficient to support reasoning about implementations of polynomials. Relevant theories I have developed include:

1. Permutations: I have shown that the set of permutation functions on a type is a group and that every permutation on a finite type can be expressed as a composition of swaps. I have defined a permutation relation on lists and shown that iterated abelian operations over lists respect this relation. This theory of permutations supports the definition of finite sets and multisets.
2. Gaussian Monoids: I have proven that every non-unit can be uniquely factorized into primes up to associates and permutations.
3. Quotient algebras: I have developed the basic theory of normal subgroups and ideals, have defined quotienting operations on groups and rings and have started on proving isomorphism theorems.

One ongoing part of my work is developing a discipline for using Nuprl's quotient types. Quotient types are needed for the carrier type in implementations of my algebraic classes, as explained in Sect. 3. I am also relying on them to form quotient algebras.

4.2 Tactic Development

Tactics are functions in the programming language ML for executing proof development strategies. Some examples are given in the next section. I identify three kinds of tactics that I have developed that have been particularly useful in my algebraic proofs to date.

1. Tactics for arithmetic reasoning. I have implemented a procedure for solving linear arithmetic problems and have augmented it to automatically take advantage of the linear properties of non-linear arithmetic expressions (for example the list length function). These tactics proved themselves invaluable when reasoning about permutation functions (bijections on $\mathbb{N}n \to \mathbb{N}n$) and the 'select' function for picking the ith element from a list.
2. Tactics for rewriting with respect to equivalence relations. These tactics automatically index into Nuprl's library to look up lemmas about which relations functions respect. Example equivalence relations that I have come across in algebra are the permutation relation on lists, the 'associated' relation that comes up in divisibility theory, and the 'equal mod an ideal' relation that comes up in ring theory.
3. Tactics for automating the proof of type inclusion relations. Sometimes these tactics call on other tactics such as the arithmetic reasoning tactics when integer subrange types were involved.

4.3 An Example Proof

To illustrate the level and style of reasoning in Nuprl I show in Fig. 1 a proof that free abelian monoids over a fixed set are unique up to isomorphism. Figure 2 gives the two lemmas that are explicitly referenced in the proof.

One proves a theorem in Nuprl by first entering it as a goal to be proved and then repeatedly invoking tactics to break goals into simpler subgoals. In Fig. 1, the initial goal is indicated by ① and the tactics immediately follow each BY.

The logical formulas are shown exactly as they appeared when the theorem was interactively proved. Goals and tactics are entered using a structured editor, so there is no need for the notation to be completely unambiguous. For example, the $|\cdot|$ is the carrier projection function for both elements of EqType and AbMonoid. Most of the notation should be self-explanatory. The .inj and .umap postfix operators project out the second and third components of elements of the FAbMon{i}(S) class. The display of the projection operator for the first component has been suppressed to improve readability since it is obvious where it is used. (Visual notation can be changed in Nuprl in a matter of seconds, so it is easy if desired to make it visible.) The {i} parameters specify the level of the carrier universe terms in the class definitions and are implicitly universally-quantified over.

At ②, univeral quantifiers were stripped. The ... and ...a indicate applications of the Auto tactic that does certain obvious steps of inference and solves *well-formedness* subgoals. Nuprl's type theory is sufficiently rich that the well-formedness of expressions is in general undecidable and so well-formedness is checked by proof. At ③, the existential quantifiers were instantiated by the given terms and the definition Invfuns was split open into its two component parts. At ④, lemma free_abmon_endomorph_is_id was backchained through. In order to verify the instantiation of f in this lemma, the Auto tactic automatically picked up on a lemma stating that the composition of two homomorphisms is a homomorphism.

At ⑤, RW CompIdNormC 0 normalized the function compositions (the o's) in the conclusion, making them right-associated. CompIdNormC is a rewriting function for normalizing expressions over the monoid $\langle T \to T, \circ, \text{Id}(T) \rangle$ for any type T. Its ML definition is:

```
let CompIdNormC = MkMonoidNormC 'comp_id_monoid_wf' ;;
```

where comp_id_monoid_wf is the name of a lemma that states that for all types T, $\langle T \to T, \circ, \text{Id}(T) \rangle$ is a monoid . I have written functions for creating rewriting function similar to MkMonoidNormC for algebras including groups, abelian groups and rings. The RewriteWith tactic repeatedly rewrites with the indicated lemmas and hypotheses. The tactics at ⑤ completed this branch of the proof. The other branch of the proof, ⑥, was proven with tactics identical to those at ④ and ⑤.

```
free_abmon_unique:

① ⊢ ∀S:EqType{i}
   |   ∀M:FAbMon{i}(S)
   |    ∀N:FAbMon{i}(S)
   |     ∃f:MonHom(M,N). ∃g:MonHom(N,M). InvFuns(|M|,|N|,f,g)
   |
② BY (UnivCD...a)
   |
   1. S: EqType{i}
   2. M: FAbMon{i}(S)
   3. N: FAbMon{i}(S)
   ⊢ ∃f:MonHom(M,N). ∃g:MonHom(N,M). InvFuns(|M|,|N|,f,g)
   |
③ BY (InstConcl ['M.umap N N.inj';'N.umap M M.inj']
   |     THENM D 0...a)
   |\
   | ⊢ (N.umap M M.inj o M.umap N N.inj) = Id{|M|} ∈ (|M| → |M|)
   | |
④ | BY (BLemma 'free_abmon_endomorph_is_id'...a)
   | |
   | ⊢ ((N.umap M M.inj o M.umap N N.inj) o M.inj) = M.inj ∈ (|S| → |M|)
   | |
⑤ | BY (RW CompIdNormC 0
   |       THENM RewriteWith [] ''free_abmon_umap_properties_1'' 0 ...)
   |
   ⊢ (M.umap N N.inj o N.umap M M.inj) = Id{|N|} ∈ (|N| → |N|)
   |
⑥ BY ...
```

Fig. 1. Proof of uniqueness of free abelian monoid up to isomorphism

```
free_abmon_umap_properties_1:

  ∀S:EqType{i}
  ∀M:FAbMon{i}(S)
   ∀N:AbMonoid{i}
    ∀p:|S| → |N|. ((M.umap N p o M.inj) = p ∈ (|S| → |N|))

free_abmon_endomorph_is_id:

  ∀S:EqType{i}
   ∀M:FAbMon{i}(S)
    ∀f:MonHom(M,M)
     ((f o M.inj) = M.inj ∈ (|S| → |M|))
     ⇒ (f = Id{|M|} ∈ (|M| → |M|))
```

Fig. 2. Supporting lemmas for free-abelian-monoid uniqueness proof

5 Applications

5.1 Interaction with Computer Algebra Systems

I am exploring ways in which Nuprl could usefully interact with the Weyl computer algebra system being developed here at Cornell [23]. One scenario is for Nuprl to behave as an *algebraic oracle* for the computer algebra system. In the course of calculations, Weyl creates new instances of algebraic structures and sometimes would like to decide which algorithm to use based on properties of these structures. However the properties might require some theorem proving work to make them apparent. This is where Nuprl comes in.

A simple trial example I am looking at concerns reasoning about rings over integers, rationals and polynomials that have been quotiented by finitely generated ideals. A query might proceed as follows:

1. Weyl asks Nuprl Q1: "is the quotiented ring $\mathbb{Z}/(1009)$ an integral domain?"
2. Nuprl looks up a lemma about quotienting rings by ideals and reduces Q1 to Q2: "is the principle ideal (1009) a prime ideal?"
3. Nuprl looks up a second lemma about when principle ideals over the integers generated by positive numbers are prime ideals and reduces Q2 to Q3: "is the natural number 1009 prime?"
4. Weyl happens to have an efficient primality testing algorithm implemented, so Nuprl asks Weyl Q3.
5. Weyl answers "yes" to Q3, and so Nuprl replies "yes" to Weyl's Q1.

Currently, Nuprl answers such queries by exhaustively forward and backward chaining through a small database of lemmas. As I increase the size of the database, I will be interested in exploring more efficient inference strategies for answering queries.

5.2 Algebraic Programming

I can use Nuprl's Σ type and set type to construct ADT class definitions in a way similar to that which I used for the abelian monoid class in Sect. 3.2. I can treat classes as an abstract specifications of submodules of a functional program and verify the correctness of the program given an arbitrary implementations of these classes. I also can verify the correctness of particular implementations of the classes. Nuprl's type theory is therefore a good uniform framework for formally developing structured functional programs using ADT's.

Moreover, I have shown in this paper that Nuprl's type theory is adequate for defining free or initial classes. Final classes should be no more difficult to construct. The importance of this is that in the ADT community there has been a debate going on about the relative merits of loose, initial and final algebraic specifications [22]. In Nuprl, all three paradigms can be explored.

6 Conclusions

I have presented here some of the first steps I have taken in implementing abstract algebra in an interactive theorem proving setting. To my knowledge, this is the first attempt at developing a general theory of polynomial algebras in either a classical or a constructive theorem proving environment. The definitions used in developing the polynomial algebras illustrate the expressiveness of Nuprl's type theory, making full use of its Π, Σ, set and quotient types.

As I described in Sect. 5, the work has promising applications in the formal development of software, as well as in the development of new more powerful and more rigorous computer algebra systems.

7 Acknowledgements

My formulation of the class definitions for the polynomial algebra benefited from many discussions with Robert Constable. The interface work between Nuprl and Weyl is being done in close collaboration with Weyl's architect, Richard Zippel. I thank the anonymous referees for their comments on an original draft of this paper.

References

1. Mark Aagaard and Miriam Leeser. The implementation and proof of a boolean simplification system. Technical Report EE–CEG–90–2, Cornell School of Electrical Engineering, March 1990. In the *Oxford Workshop on Designing Correct Circuits*, September, 1990.
2. Anthony Bailey. Representing algebra in LEGO. Master's thesis, University of Edinburgh, November 1993.
3. David A. Basin and Robert L. Constable. Metalogical frameworks. In Gérard Huet and Gordon Plotkin, editors, *Logical Environments*. Cambridge University Press, 1993.
4. David A. Basin and Peter Del Vecchio. Verification of combinational logic in Nuprl. In M. E. Leeser and G. M. Brown, editors, *Hardware Specification, Verification, and Synthesis: Mathematical Aspects*, pages 333–357. Springer Verlag, 1990. LNCS 408.
5. Nicolas Bourbaki. *Algebra, Part I*. Elements of Mathematics. Addison-Wesley, 1974.
6. Jawahar Chirimar and Douglas J. Howe. Implementing constructive real analysis: Preliminary report. In *Constructivity in Computer Science*, volume 613 of *Lecture Notes in Computer Science*, pages 165–178. Springer-Verlag, 1992.
7. Edmund Clarke and Zhao Xudong. Analytica - a theorem prover in mathematica. In D. Kapur, editor, *11th Conference on Automated Deduction*, volume 607 of *Lecture Notes in Artifical Intelligence*, pages 761–765. Springer-Verlag, 1992.
8. Robert Constable et al. *Implementing Mathematics with The Nuprl Development System*. Prentice-Hall, NJ, 1986.

9. William M. Farmer, Joshua D. Guttman, and F. Javier Thayer. Little theories. In D. Kapur, editor, *11th Conference on Automated Deduction*, volume 607 of *Lecture Notes in Artifical Intelligence*, pages 567–581. Springer-Verlag, 1992.
10. Elsa L. Gunter. Doing algebra in simple type theory. Technical Report MS-CIS-89-38, Department of computer and Information Science, University of Pennsylvania, 1989.
11. John V. Guttag and James J. Horning. *Larch: Languages and Tools for Formal Specification*. Texts and Monographs in Computer Science. Springer-Verlag, 1993.
12. John Harrison and Laurent Thèry. Extending the HOL theorem prover with a computer algebra system to reason about the reals. In *Proceedings of the HOL '93 Workshop on Higher Order Logic Theorem Proving and its Applications*, 1993.
13. Douglas J. Howe. Implementing number theory: An experiment with Nuprl. In *Eighth Conference on Automated Deduction*, volume 230 of *Lecture Notes in Computer Science*, pages 404–415. Springer-Verlag, July 1987.
14. Douglas J. Howe. *Automating Reasoning in an Implementation of Constructive Type Theory*. PhD thesis, Cornell University, 1988.
15. Paul B. Jackson. Nuprl and its use in circuit design. In R.T. Boute V. Stavridou, T.F.Melham, editor, *Proceedings of the 1992 International Conference on Theorem Provers in Circuit Design*, IFIP Transactions A-10. North-Holland, 1992.
16. Paul B. Jackson. *Enhancing the Nuprl Theorem Prover and Applying it to Constructive Algebra*. PhD thesis, Cornell University, 1994. Forthcoming.
17. Richard D. Jenks and Robert S. Sutor. *AXIOM: the Scientific Computation System*. Springer-Verlag, 1992.
18. Serge Lang. *Algebra*. Addison-Wesley, 2nd edition, 1984.
19. Per Martin-Löf. Constructive mathematics and computer programming. In *Sixth International Congress for Logic, Methodology, and Philosophy of Science*, pages 153–175, Amsterdam, 1982. North Holland.
20. Ray Mines, Fred Richman, and Wim Ruitenburg. *A Course in constructive Algebra*. Universitext. Springer-Verlag, 1988.
21. John C. Mitchell and Gordon Plotkin. Abstract types have existential type. In *Conference Record of the Twelfth Annual ACM Symposium on Principles of Programming Languages*. Association for Computing Machinery, SIGACT, SIGPLAN, 1985.
22. Martin Wirsing. Algebraic specification. In J. van Leeuwen, editor, *Handbook of Theoretical Computer Science*, volume B: Formal Models and Semantics, chapter 13. Elsevier, 1990.
23. Richard Zippel. The Weyl computer algebra substrate. In Alfonso Miola, editor, *Design and Implementation of Symbolic Computation Systems*, volume 722 of *Lecture Notes in Computer Science*, pages 303–318. Springer Verlag, 1993.

Tactic Theorem Proving with Refinement-Tree Proofs and Metavariables

Amy Felty and Douglas Howe

AT&T Bell Laboratories, 600 Mountain Ave., Murray Hill, NJ 07974, USA

Abstract. This paper describes a prototype of a programmable interactive theorem-proving system. The main new feature of this system is that it supports the construction and manipulation of tree-structured proofs that can contain both metavariables and derived rules that are computed by tactic programs. The proof structure encapsulates the top-down refinement process of proof construction typical of most interactive theorem provers. Our prototype has been implemented in the logic programming language λProlog, from which we inherit a general kind of higher-order metavariable. Backing up, or undoing, of proof construction steps is supported by solving unification and matching constraints.

1 Introduction

Interactive proof construction typically proceeds top-down, starting with the statement of the theorem to be proven, and then successively refining goals into subgoals. This pattern is characteristic of most interactive systems, although there are large differences in the kinds of refinement that can be performed and in the underlying ideas of proof and state of the system. Our focus is on refinement by *tactics*. Loosely speaking, a tactic is a program that reduces a goal (typically a sequent) to a sequence of subgoals such that the goal can be inferred from the subgoals.

The set of refinements used to construct a proof has a natural tree structure. A system that directly supports this structure can provide a number of useful operations for building, manipulating and reading proofs. For example, subproofs of interest, or the steps leading up to a particular subgoal, can be quickly located via user-directed navigation of the tree. Bad proof strategies can be corrected by locating the first bad step in the tree, then removing just the part of the argument that depended on the step. The inferences in a proof can be replayed when, for example, a definition is changed, allowing one to recover all the portions not affected by the change. Proofs can be cut-and-pasted, and used for reasoning by analogy. [2] contains more examples of such operations.

Metavariables are supported in a number of existing systems [8, 3, 13, 10, 9, 4]. There are many compelling examples of their usefulness in interactive proofs. A common example is "existential introduction". Removing the existential quantifier in the goal $\vdash \exists x.\ p(x)$ requires a witness. We can use a metavariable X to stand for the witness term, refining the goal to the subgoal $\vdash p(X)$. We can instantiate X to a specific term later in the proof, at which time it will be replaced wherever it occurs. The point of postponing choice of the witness is that the details of the proof of $p(X)$ may make it possible to automatically find an instantiation of X, or at least make it easier for the user to determine it. For example, $p(X)$ may be an equation that can be manipulated to present an explicit solution $X = t$ for X.

There are many similar examples. Metavariables can be used to stand for programs to be synthesized from constructive proofs [11, 13], for induction hypotheses (or program invariants, as in [8]) that can be strengthened as proof progresses, and for "don't care" arguments to functions, such as type arguments or domain-membership evidence for constructive partial functions, that can be automatically deduced in a uniform way. Also, various procedures for automating reasoning make essential use of metavariables, and some of these procedures can be more usefully integrated with an interactive system if the system supports metavariables.

We have designed and implemented a theorem-proving system for constructing and manipulating refinement-tree proofs with metavariables. Proofs are trees of goals where each node g has associated with it a "justification" which specifies how the goal g can be inferred from its children. This justification can be a rule (name), in which case g and the children of g must be the conclusion and premises, respectively, of an instance of the rule, or it can be a representation T of a tactic.

A fundamental problem is to explain how T relates to g and its children $g_1, \ldots, g_n$. If we ignore metavariables, we can take T to be a program which, when applied to g, produces $g_1, \ldots, g_n$. If we allow metavariables, then this explanation no longer suffices. Suppose that after this tactic refinement was done, some metavariable instantiations were made, resulting in instances g', $g'_1, \ldots, g'_n$ of g and $g_1, \ldots, g_n$. What should be the relationship of T to g', $g'_1, \ldots, g'_n$?

One possibility is to restrict the tactic language so that we are guaranteed that the relationship above continues to hold: T applied to g' produces subgoals $g'_1, \ldots, g'_n$. We reject this possibility for two main reasons. First, although a *pure* logic programming language would preserve this relationship, we want to allow the use of a more practical programming language for tactic programming in order to enable users to achieve a high degree of automatic support for proof construction. For example, we would like to be able to accommodate the full λProlog language including its non-logical operators, and also the functional language ML, which has been proven well-suited to tactic programming.

Second, this property to be guaranteed is not respected by many commonplace and useful tactics. For example, suppose that we have a goal $X + 0 > 0 \vdash \phi(X)$, where X is a metavariable we intend to instantiate in another part of the proof, and the refinement step we wish to make is to simplify the goal using a tactic Simplify, say, that repeatedly applies common arithmetic rewrite rules. Assume the subgoal produced by Simplify is $X > 0 \vdash \phi(X)$. If, later during the proof, X gets instantiated to $1+0$, then the subgoal becomes $1+0 > 0 \vdash \phi(1+0)$, yet if Simplify were re-executed on the new goal, it would produce the subgoal $1 > 0 \vdash \phi(1)$. This example illustrates another problem with using a pure logic programming language for tactics: we do not want simplification to instantiate

X, but instead to treat it as a constant. In general, we want some tactics to instantiate metavariables, and others to treat them as constants.

One of the design goals for our system is programmability. We want to support programming of a wide range of procedures and tools for automating proof construction. We have therefore designed a simple proof structure with a small set of simple basic operations for constructing proofs. In particular, we have tried to avoid including in a proof information pertaining to the history of its construction, such as a record of metavariable instantiations. Section 2 gives a slightly simplified account of our approach to proofs. Section 5 gives some implementation details.

An important operation in interactive theorem provers is "undoing" parts of a proof attempt. Metavariables cause serious complications for this operation.

Without metavariables, subproofs P_1 and P_2 with no nodes in common are independent, and can be treated as completely separate tasks. However, metavariables can introduce dependencies between the subproofs. Suppose that a step in building P_1 involves the instantiation of a logic variable X that is mentioned in P_2. The instantiation replaces all occurrences of X in P_2, and subsequent steps in building P_2 may take advantage of this change. If we now decide that P_1 was a bad proof attempt, we could just remove it from the main proof and start a new subproof. But this would not completely undo the effect of the bad subproof P_1. We might also want to find and remove the parts of P_2 that depended on X's instantiation, and, in addition, uninstantiate any of the variables instantiated by refinements within these parts.

One kind of "undo" is directly inherited from our implementation language λProlog (see [4] for details). This is "chronological undo", which undoes proof modifications in the order they were performed. This is a useful form of undo, especially for quick local backups, and is the only form of undo in all other systems with metavariables except ALF [10]. However, it is unsatisfactory in general since it will often remove parts of the proof which are independent of the targeted parts.

We have designed a number of undo procedures that improve on chronological undo and have implemented one of them. These procedures are based on solving sequences of higher-order matching and unification problems. In section 3 we give an example session with our system that illustrates some of the problems and issues for designing undo operations for this kind of proof. Section 4 describes the undo procedure, and Section 5 gives some details of its implementation.

Our prototype has been implemented in the higher-order logic programming language λProlog [12], from which we inherit a general kind of higher-order metavariable, similar to what is found in Isabelle [13]. Our proof structure is inspired by Nuprl [2, 1]. The system is generic, in the sense that it can easily accommodate most logics that can be specified in the general style of LF [7] and Isabelle [13]. The exact style of encoding of logics in our system is very similar to [4], except that we have made a commitment to sequent-calculus presentations.

Although we have based our prototype on λProlog, the ideas are applicable to tactic systems based on ML such as LCF [6], Nuprl [2] and HOL [5]. Implementing the ideas for such systems would be straightforward, although it would be much more work because ML lacks λProlog's built-in support for metavariables, bound variables and unification.

The conclusion of the paper has a few additional comparisons with related work and a brief discussion of future work.

2 Proofs

This section gives a simplified account of our proof data-structure. Two of the principle constituents of proofs are *goals* and *tactics*. Goals are intended to be sequents in the logic being implemented (the "object logic"), and tactics are programs from some programming language. To make the presentation simple, and to keep the description here close to what has been implemented, we assume that both goals and tactics are represented as terms in the simply-typed λ-calculus. In our implementation, object logics are encoded as λ-terms and tactics are programs in λProlog, a language whose programs are all λ-terms.

There are a number of differences between the implementation and what is described here, mostly for reasons of efficiency and ease of tactic programming. Some of the more important differences are given at the end of the section.

Let Λ be the set of terms of the typed λ-calculus over some set of base types and some set of constants. We identify $\alpha\beta\eta$-equal terms. Thus, when we say that there is a substitution σ such that $\sigma(e) = \sigma(e')$, we mean that e and e' have a higher-order unifier. We distinguish a base type and define the set G of *goals* to be the set of all terms in Λ of this type.

Metavariables in this setting are simply the free variables of a goal. We will use capital letters to stand for metavariables. Ordinary variables in our representation are bound by λ-abstractions. See the implementation section for details on this representation.

A *proof* is a tree of goals where each node has associated with it a *justification*. The justification says how the goal can be inferred from its children. We will specify what the justifications are below, but for now, assume that for any justification j there is an associated pair $s(j) = (g, \overline{g})$, called a *step*, where g is a goal and $\overline{g}$ is a sequence $g_1, \ldots, g_n$ ($n \geq 0$) of goals. We place the following restriction on proofs. Let g be a node in the tree, let j be its associated justification, and let $\overline{g}$ be the sequence of children of g (in left-to-right order). Write $s(j) = (g', \overline{g}')$. We require that there be a substitution σ such that $\sigma(g') = g$ and $\sigma(\overline{g}') = \overline{g}$ (using the obvious extension of substitution to sequences of goals).

Thus $s(j)$ can be thought of as a rule schema, with premises $\overline{g}'$ and conclusion g', and the valid instances of the schema are obtained by substituting for metavariables. For example, one of the allowed justifications in our implementation of first-order logic is the constant and_i (for "and-introduction"), and

$$s(and\_i) \;=\; (H \vdash A \,\&\, B, \;(H \vdash A, \; H \vdash B)),$$

which corresponds to the rule schema

$$\frac{H \vdash A \quad H \vdash B}{H \vdash A \, \& \, B.}$$

Note that proof trees are preserved under *instantiation:* if σ is a substitution, p is a proof, and $\sigma(p)$ is obtained by replacing every goal g in p by $\sigma(g)$ (and keeping the same associated justifications), then $\sigma(p)$ is a proof.

There are three kinds of justifications. One corresponds to primitive rules of the object logic and one to tactics. There is also a justification j_{prem}, where

$$s(j_{prem}) = (G, ()).$$

This corresponds to a trivial "rule" which infers any goal from no premises. Goals in a proof tree that have j_{prem} as their justification are called *premises* of the proof. Thus proof trees represent incomplete proofs in the object logic. The root goal of a proof is called its *conclusion*.

An important operation on proofs is *refinement*, where a justification is used to extend a proof tree at a premise. In particular, let g be a premise of a proof p, let j be a justification with $s(j) = (g', (g'_1, \ldots, g'_n))$, and let σ and σ' be substitutions such that $\sigma(g) = \sigma'(g')$. Then the *refinement of p at g using j, σ and σ'* is obtained from $\sigma(p)$ by giving j to the premise $\sigma(g)$ as a new justification, and adding children $\sigma'(g'_1), \ldots, \sigma'(g'_n)$, each with j_{prem} as its justification. For example, if g is $\vdash \phi \, \& \, \psi$, σ' substitutes ϕ for A and ψ for B, and $\sigma = \emptyset$, then refining at g using and_i, σ and σ' adds children $\vdash \phi$ and $\vdash \psi$ to g.

Note that the problem of finding some σ and σ', given g and j, can be cast as a higher-order unification problem. Let τ be a substitution that renames metavariables in g' such that $\tau(g')$ and g have no metavariables in common. Then σ, σ' exist if and only if g and $\tau(g')$ are unifiable.

Justifications corresponding to inference rules of the object logic are represented as a subset R of Λ. (Typically $r \in R$ will be a constant.) For each $r \in R$ there is an associated step $s(r)$. The justifications corresponding to tactics have a number of components. These will be explained by considering the typical kind of refinement: extending a proof by applying a tactic to a premise.

Tactics are represented as a subset T of Λ. Applying a tactic $e \in T$ to a premise g of a proof p is a single operation for a user of our theorem-prover, but it actually consists of first obtaining a justification and substitution from e and g, and then using these to perform a refinement step. More precisely, first the tactic e is evaluated with argument g, producing as its value a step $s' = (g', \overline{g}')$ and a substitution σ such that $\sigma(g) = g'$. Then the *tactic justification* $j = (e, g, \sigma, s')$ is formed, where $s(j)$ is defined to be s'. g is called the *tactic argument* of j. Finally, we refine p at g using j, σ and $\emptyset$. Thus, applying e to g produces a substitution to be applied to p, new children $\overline{g}'$ for the premise being refined, and a justification for the refinement.

For example, if g is $\vdash X = 0$ and if e is a tactic that instantiates X with 0, we might have $\sigma(X) = 0$, $\vdash 0 = 0$ for g', and $(\vdash 0 = 0)$ for $\overline{g}'$. The act of refining by j and σ will replace X by 0 in the entire proof, and produce a new premise $\vdash 0 = 0$ as a child of the old premise.

Note that occurrences of tactic justifications in a proof need not have arisen from the process just described, since the constraint on a justification j is only in terms of the step $s(j)$ associated with it.

There are several reasons for including more information in a tactic justification than just its step: e and g are informative to a user, since they determine the step; e is needed for "replaying" a proof; and g and σ are required for the undo procedure. These last two points are discussed in the section on undo. The step is stored in the justification for reasons of efficiency, so that it does not have to be recomputed from e and g every time it is needed.

Some of the more important differences between the implementation and what is described above are as follows. In the account here, free variables correspond to metavariables, and free variables of the object logic are assumed to be handled by λ-abstraction in goals. In the implementation, these abstractions are handled more conveniently by distributing them throughout proofs, so there is only one λ-abstraction introduced whenever a new free variable of the object logic is needed.

In the implementation, the logic variables of λProlog are used for metavariables. This means that when, for example, we form a tactic justification, we cannot simply directly store the tactic argument in the justification, since then subsequent instantiations of metavariables in the proof might change it. So, when a tactic justification is created, the components have their logic variables "abstracted out", *i.e.* bound by λ-abstractions, to prevent them from being instantiated by further proof operations. More is said about this abstraction operation in Section 5.1.

Tactics work somewhat differently in the implementation. Since we are using a logic-programming language, and logic variables are used to implement metavariables, tactics do not need to explicitly return a substitution. To guarantee soundness, tactics produce proofs as results, from which the step is obtained as the conclusion and premises of the proof. These proofs can again contain tactic justifications. This introduces a circularity, but this is not hard to deal with (see [1] for one approach).

3 An Example Session

The following session illustrates interaction with our system, including refinement commands, navigation within a proof, and undo. We prove a simple formula from first-order logic. The tactics used here implement the basic inference rules of a sequent calculus. The lines beginning with "!:" indicate user input. All other text is output from the system (we have added some whitespace and renamed a few variables). All metavariables are printed as capital letters.

We begin the session by entering the following query to λProlog.

```
prove (exists x\ (exists y\ ( (q x imp q y) and
     (q x imp (q a or q b)) and (q a imp q b imp q x) and
```

```
            (q a imp exists z\ (q z)) and (q b imp q x)))).
```

This results in a prompt for input. The user supplies the command to run the tactic (repeat intro) which repeatedly applies some of the introduction rules for connectives occurring in formulas on the right in the sequent, resulting here in five subgoals.

```
!: tactic (repeat intro).
Address:
|- exists x\ (exists y\ ((q x imp q y) and
        (q x imp q a or q b) and (q a imp q b imp q x) and
        (q a imp exists z\ (q z)) and (q b imp q x)))
By tactic repeat intro
q X |- q Y
q X |- q a or q b
q b, q a |- q X
q a |- q Z
q b |- q X
```

After running the tactic, the node of the proof is redisplayed. The output above shows the address of the node of the proof being displayed (a list of integers, empty in this case), followed by the goal at the node, its justification following the word By, and the subgoals of the node.

The user then solves these subgoals in left-to-right order using the **next** command to go to the next premise node in the tree.

```
!: next.
Address: 1      q X |- q Y          By ?
!: tactic (hyp 1).
Address: 1      q Y |- q Y          By tactic hyp 1
!: next.
Address: 2      q Y |- q a or q b   By ?
!: tactic (then or_il_tac (hyp 1)).
Address: 2      q a |- q a or q b   By tactic then or_il_tac (hyp 1)
!: next.
Address: 3      q b, q a |- q a     By ?
!: tactic (hyp 2).
Address: 3      q b, q a |- q a     By tactic hyp 2
!: next.
Address: 4      q a |- q Z          By ?
!: tactic (hyp 1).
Address: 4      q a |- q a          By tactic hyp 1
!: next.
Address: 5      q b |- q a          By ?
```

The hyp tactic completes a proof when the formula on the right unifies with the formula on the left at the position indicated by an integer argument. In the first, second, and fourth subgoals above, this operation causes instantiation of metavariables: X to Y, Y to a, and Z to a, respectively. Note that or_il_tac in subgoal 2 chooses the first disjunct of the conclusion which, when followed by hyp 1, through the use of the tactic combinator **then**, forces the instantiation of Y to a.

Now no further progress can be made. Looking at subgoal 5 as it originally appeared above, the user realizes that X must be instantiated to b if this subgoal is to be provable. By comparing to the 5 subgoals above, we see that it was subgoal 2 that forced the instantiation of a. This can be verified by moving through the tree to the second child of the root, and then using the **show_tactic_arg** command to examine the goal which was the original argument to the tactic at that node. Entering the **uninst** command there undoes the instantiation of the variables in this sequent and any nodes that depended on them, in this case just Y. We omit the printing of the root node.

```
!: root.         !: down 2.
Address: 2       q a |- q a or q b     By tactic then or_il_tac (hyp 1)
!: show_tactic_arg.
q Y |- q a or q b
!: uninst.
Address: 2       q Y1 |- q a or q b    By ?
```

By going back to the root and printing the proof, we see the refinements at subgoals 2 and 3 have been undone, since they both depended on the instantiation of Y to a. The proof of the first subgoal which originally unified two logic variables remains, as does the proof of subgoal 4.

```
!: root.        !: show_proof.
Goal: |- exists x\ [...]    Justification: By tactic repeat intro
  Goal: q Y1 |- q Y1        Justification: By tactic hyp 1
  Goal: q Y1 |- q a or q b  Justification: By ?
  Goal: q b, q a |- q Y1    Justification: By ?
  Goal: q a |- q a          Justification: By tactic hyp 1
  Goal: q b |- q Y1         Justification: By ?
```

The remaining subgoals can be completed by choosing the second disjunct in subgoal 2 forcing the instantiation of metavariable Y1 to b.

4 Undoing Proofs

In the implemented undo procedure (**uninst**), we want to back the proof up enough so that for some node, called the *undo node*, the goal that appeared at that node before a tactic refinement was applied becomes the goal at that node in the new proof. The procedure removes any parts of the proof which would force metavariables in the undo node to be instantiated. Furthermore, it undoes any instantiations forced by the removed subproofs.

For our undo procedure, we assume that all non-premise nodes in a proof have tactic justifications. Below, we will speak of a node's associated tactic, tactic argument, substitution, and step, corresponding to the four components of its justification. In order to not have to worry about renaming of free variables, we assume (without loss of generality) that justifications from different nodes in a proof have no free variables in common. However, it is not necessary to modify a proof to meet this criteria in order to run the undo procedure. Instead, at each step, fresh variables could be introduced to rename the components of a justification.

The procedure works by finding an ordering on the nodes in the proof that corresponds to a possible method of building the tree by using alternating instantiations and tactic-refinement steps. The ordering is used to track "when" variables become instantiated in order to determine which branches to prune. To reconstruct the new tree, the appropriate branches are pruned from the original tree. Then, the same procedure used to find the ordering on the original tree is used to construct the new tree. We describe this core procedure and discuss how it is used in each of the two phases of the implemented **uninst** operation.

The main operation of the core procedure is a step-by-step reconstruction of the proof. At each step, we have the partially reconstructed tree p with a sequence of premises $g_1, \ldots, g_n$ called the *fringe*, that collects the premises that must still be expanded to obtain the complete tree. In addition, we have a mapping from fringe elements to the subproofs $p_1, \ldots, p_n$ in p rooted at the corresponding locations in the original proof. It is these subproofs that must be processed in order to complete the reconstruction. At any point, if we build a "justification tree" by taking the associated justifications of each node in $p, p_1, \ldots, p_n$ and attaching those for $p_1, \ldots, p_n$ at the appropriate nodes, we obtain the same justification tree as from the original proof.

At each step, we choose for "expansion" an element g_i of the fringe that satisfies the following two requirements. First, the tactic argument at the root node r of p_i must be an instance of g_i. Let σ be the matching substitution. Second, for every other fringe element g' mapped to subproof q, the tactic argument at the root of q must be an instance of $\sigma(g')$. Let j be the justification at node r and let τ be the substitution component of j. We refine p at g_i using j, $\sigma \circ \tau$ and the empty substitution. Thus, $\tau(\sigma(g_i))$ now replaces g_i in the new proof. We then replace g_i in the fringe by the new premises added by the refinement. The mapping is extended by mapping these elements to the corresponding subproofs rooted at the children of r.

A proof p' is obtained by *pruning* from p if it results from removing some subproofs from p (leaving new premises). p' is a *re-instantiation* of p if the two proofs have the same justification tree. Using an inductive argument, it can be shown that the set of proofs for which the above reconstruction procedure can be applied is closed under pruning, instantiation, re-instantiation, and refinement by tactics. An important invariant during reconstruction of any proof built only from these operations is that there will always be some fringe node that meets the two requirements for expansion. The ordering of nodes by relative time that they were expanded by tactic refinement, for example, can be used in reconstruction of the proof, and each successive node will meet the requirements. There may be other possible orderings, all of which correspond to possible orderings in which the tree could have been constructed. We call such an ordering a *refinement ordering*.

The proof reconstruction procedure is repeated twice in the **uninst** operation. The first time, we keep track of two kinds of information: a set of variables $\mathcal{V}$ that must remain uninstantiated and a set of addresses $\mathcal{A}$ of nodes that roots of subtrees that must be pruned. Initially, $\mathcal{V}$ is the set containing all of the free variables in the tactic argument of the undo node, and $\mathcal{A}$ is empty. At each step, after a fringe node g is chosen for expansion, $\mathcal{V}$ is updated. In particular, the match substitution σ is checked to see if it maps any of the variables in $\mathcal{V}$ to new variables. These variables must also be added to $\mathcal{V}$. After each refinement step, $\mathcal{A}$ is updated. Any nodes containing metavariables in $\mathcal{V}$ that get instantiated by this refinement step must be marked for deletion. More precisely, let τ be the substitution at the root node of the proof that g is mapped to. Let $\mathcal{V}'$ be the set of all variables X in $\mathcal{V}$ such that either $\tau(\sigma(X))$ is not a variable or there is a Y in $\mathcal{V}$ with $X \neq Y$ and $\tau(\sigma(X)) = \tau(\sigma(Y))$. If $\mathcal{V}'$ is non-empty, then the address of the node just expanded is added to $\mathcal{A}$. In addition, the address of any fringe node that contains a variable in $\mathcal{V}'$ is also added to $\mathcal{A}$. Note that no addresses will get added to $\mathcal{A}$ until the point in reconstruction after the undo node is expanded.

After the reconstruction is complete, the subproofs indicated by $\mathcal{A}$ are pruned, resulting in a new proof q. Now, the reconstruction procedure is repeated on q. This time $\mathcal{V}$ and $\mathcal{A}$ are ignored. The tree produced by this phase is a minimally instantiated version of the justification tree. It is this proof that is returned from the **uninst** operation.

A very slight modification of this procedure gives us an operation that allows the user to point at specific variables in one or more tactic arguments and ask that they remain uninstantiated in the remaining tree. This can be achieved by initializing $\mathcal{V}$ to be the selected variables only. Multiple variables from different nodes can be processed simultaneously by including them all in $\mathcal{V}$.

A slightly more complicated operation is to request that all of the variables instantiated by a particular tactic refinement are backed up along with all branches that saw these instantiations. Here the substitution at the justification of the undo node is used. Any variables that are mapped to anything other than themselves should be put into $\mathcal{V}$. This operation can be extended to include all variables instantiated in a particular subproof. Here, in addition to any variable from the substitution at the undo node, any variable from the substitution at any of its descendants that is mapped to something other than itself must be included in $\mathcal{V}$.

The operations discussed so far prune any branches that depend on instantiation of the selected variables. Another option is to attempt to replay them using the tactics in justifications. Such replay can be accommodated by introducing a new *bad* justification. Then instead of pruning, a tactic justification is changed to a bad one which retains the tactic and the structure of the tree below it. A final phase of the procedure would then attempt to replay as much of the bad proof by re-executing the tactics, only pruning when execution fails. Alternatively, the "bad" subproofs could be left to the user.

All of these undo operations work on proofs closed under tactic refinement, pruning, instantiation, and re-instantiation. We can extend the proof reconstruction procedure to handle a larger class of proofs including those pieced together through unification. However, the notion of tracking "when" a variable is instantiated no longer works in the same way. In particular there is not necessarily a refinement ordering because any two proofs pieced together may have been constructed independently. We define a heuristic for determining an order. This procedure is a simple modification of the above procedure. In the case when no node in the fringe meets the two requirements for expansion, we find a node g in the fringe maximizing a particular measure. Let g' be the conclusion of the step at the root r of the subproof that g is mapped to and let σ be a substitution such that $\sigma(g) = \sigma(g')$. That is, instead of matching with the tactic argument, we require unification with the step conclusion. This unification problem is in fact solvable for all nodes in the fringe. We now need a measure to determine when one ordering of nodes is better than another. We can use a measure that favors instantiations that are done via tactic refinement over those done simply to match up the goal of a node with the corresponding premise of a step in the parent node. The measure should have the property that any refinement ordering maximizes it. In the worst case, this heuristic reduces to reconstructing the tree by repeatedly solving unification problems in arbitrary order. In the best case, it finds an ordering corresponding to the order in which the proof could have been constructed for those trees where such an ordering exists.

5 Implementation

This section describes the implementation of our system. It starts with a brief account of the implementation language.

5.1 λProlog

λProlog is a partial implementation of higher-order hereditary Harrop (hohh) formulas [12] which extend positive Horn clauses in essentially two ways. First, they allow implication and universal quantification in the bodies of clauses, in addition to conjunctions, disjunctions, and existentially quantified formulas. In this paper, we only consider the extension to universal quantification. Second, they replace first-order terms with the more expressive simply typed λ-terms and allow quantification over predicate and function symbols. The application of λ-terms is handled by β-conversion, while the unification of λ-terms is handled by higher-order unification.

The terms of the language are the terms of Λ where the set of base types includes at least the type symbol o, which denotes the type of logic programming propositions. In this section, we adopt the syntax of λProlog. Free variables are represented by tokens with an upper case initial letter and constants are represented by tokens with a lower case initial letter. Bound variables can begin with either an upper or lower case letter. λ-abstraction is represented using backslash as an infix symbol.

Logical connectives and quantifiers are introduced into λ-terms by introducing suitable constants with their types. In particular, we introduce constants for conjunction (,), disjunctions (;), and (reverse) implication (:-) having type o -> o -> o. The constant for universal quantification (pi) is given type (A -> o) -> o for each type replacing the "type variable" A. A function symbol whose target type is o, other than a logical constant, will be considered a *predicate*. A λ-term of type o such that the head of its βη-long form is not a logical constant will be called an *atomic formula*. A *goal* is a formula that does not contain implication. A *clause* is a formula of the form (pi x_1\...(pi x_n\(A:- G))) where G is a goal formula and A is an atomic formula with a constant as its head. In presenting clauses, we leave off outermost universal quantifiers, and write (A:- G).

Search in λProlog is similar to that in Prolog. Universal quantification in goals (pi x\G) is implemented by introducing a new parameter c and trying to prove [c/x]G. Unification is restricted so that if G contains logic variables, the new constant c will not appear in the terms eventually instantiated for those logic variables.

Several non-logical features of λProlog are used in our implementation. We use the cut (!) operator to eliminate backtracking points. In addition, we have implemented a new primitive make_abs which takes any term and replaces all logic variables with λ-bindings at the top-level. It has type A -> abs A -> list mvar -> o where abs is a type constructor introduced for this purpose and the third argument is a list containing all of the logic variables in the order they occurred in the term. We use this operation to "freeze" the degree of instantiation of a term as well as to implement a match procedure. In order to correctly freeze a term, this operation must also freeze a record of any unification constraints on the logic variables occurring in the term. The current implementation does not do so. However, we have verified that our implementation does not generate constraints. We make the restriction that any programmer defined tactics also cannot generate constraints.

λProlog allows type constructors for building types. In addition to abs, we use pair and list in our implementation.

5.2 Proofs and Tactics

Below are the basic types and operations for our implementation of proofs.

```
goal            type.
agoal           (A -> goal) -> goal.
step            type.
step            goal -> list goal -> step.
prule_name      type.
prule_def       prule_name -> seq -> list seq -> o.
proof           type.
just            type.
prem_just       just -> o.
prule_just      prule_name -> just -> o.
```

```
tactic_to_just   (goal -> proof -> o) -> goal -> proof -> just -> o.
one_step_proof   step -> just -> proof -> o.
compose_proofs   proof -> list proof -> proof -> o.
aproof           (A -> proof) -> proof.
```

These are intended to form abstract data types for justifications and proofs. We have omitted several destructors for these types. All of our operations for building and modifying proofs do so via the above operations.

The type goal is the type of goals. Goals are essentially sequents. They also have some additional structure which we plan to exploit in future work. We represent hypothesis lists of sequents using function composition (as is done in Isabelle) so that higher-order unification can be used to deal with metavariables standing for subsequences of hypothesis lists. λ-abstracted sequents are also goals: the constructor agoal converts a term x\(G x) into a goal. The type step and the constructor step implement the steps of Section 2.

The object logic is assumed to be specified by a type prule_name whose members are the primitive rule names, and a predicate prule_def that associates a (sequent version of a) step with each rule name. For example, the following clauses specify the rules for and-introduction and all-introduction.

```
prule_def and_i (|- H (A and B)) ((|- H A)::(|- H B)::nil).
prule_def forall_i (|- H (forall A)) ((aseq (x\ (|- H (A x))))::nil).
```

Here |- is the constructor for basic sequents, and aseq constructs abstracted sequents.

There are three ways of building proofs. One is to use aproof to turn an abstracted proof into a proof. In a proof (aproof x\(P x)), the bound variable x represents a new object level variable whose scope is the proof (P x). The second way to build proofs is to construct a one-step proof from a step and a justification. (one_step_proof S J P) computes the step corresponding to the justification J, checks that it matches the step S=(step G Gs), then produces a proof whose root has goal G and justification J, and whose children are premises with goals from the list Gs. The premises may use the aproof constructor. This would be the case if, for example, J were the justification for the rule forall_i.

The final way to construct proofs is with compose_proofs which attaches the members of a list of proofs at the premises of another proof. This is used in the implementation of then, a combinator for sequencing tactics. Tactics are predicates of type goal -> proof -> o. Some care was taken with the composition operation in order to make tactics efficient. In particular, it produces a variant representation of a proof that delays actual computation of the composition. Usually the actual composition never needs to be performed, and when it does, it will usually be in the context of other delayed compositions, and grouped compositions can be handled much more efficiently.

The type just is an abstract type of justifications. prule_just constructs primitive rule justifications. Tactic justifications have four parts made up of two "abstracted" pairs with types (abs (pair (goal -> proof -> o) goal)) and (abs (pair (list mvar) step)), respectively. The four parts of this datatype implement the four parts of the tactic justification: the tactic, tactic argument, substitution, and step. Since metavariables are implemented directly using the logic variables of λProlog, and since we do not want variables in any of these components to be further instantiated, we use the the make_abs operation described earlier to "freeze" them. When a copy is needed, it is made by applying these abstractions to new logic variables. Instead of a set of variable/instance pairs, the substitution is represented as a list of terms such that the length of this list is the same as the length of the binder of the first pair. A substitution is applied by taking the list of new variables used to make a copy of the first abstraction and matching it against this stored list.

tactic_to_just takes a tactic T, runs it on a goal, returning the tactic's proof and the corresponding justification. Below is the main clause of its implementation. (The test that the last argument is a variable to ensure one-way behaviour is omitted here).

```
tactic_to_just T G P (trule AbsTacAp AbsSigmaStep)
:- make_abs (p T G) AbsTacAp Subs,
   T G P,
   concl P NewG,
   prems P Gs,
   make_abs (p Subs (step NewG Gs)) AbsSigmaStep Bazola.
```

trule is the hidden constructor for tactic justifications, p is a pairing constructor, and concl and prems compute the conclusion and premises of a proof.

Although it is unlikely, it is possible for a user of our system to build objects of type proof that are not proofs. For maximum security, we would need to include some further run-time checks. In a language like ML, such security could be obtained through the type system.

5.3 Undo: the uninst Command

The implementation of uninst follows the description in Section 4 with a few optimizations. One such optimization comes from using λProlog's built-in unification for our match procedure. When checking the first requirement of a fringe element and determining the match substitution σ, for example, the match procedure can directly apply the result of the match to the metavariables of the goal, automatically propagating σ to the new proof.

At the refinement step of proof reconstruction, the application of a substitution is also propagated in the new proof by logic variable instantiation. Note that, in the refinement step, instead of using the substitution in the justification, we could simply match the goal to be refined with the conclusion of the step. In fact, we need not record substitutions at all in tactic justifications. However, they serve as an optimization, allowing the propagation of the exact substitution that was originally done by executing the tactic. In addition, because matching uses λProlog unification, by avoiding matching we also avoid generating unification constraints.

6 Discussion

Isabelle [13] and Coq [3] have metavariables and support tactic-style theorem-proving, but refinement trees are implicit. Operations on these trees are limited, and, in particular, undo is chronological. This also applies to KIV [8], even though it explicitly supports a form of refinement trees. In contrast to ALF [10] and Coq, our system only supports simple types for metavariables. If the object logic has a richer type system, then types must be represented explicitly, for example as predicates in the object logic. ALF supports dependency-directed undo, but proofs are λ-terms, not refinement trees.

Plans for future work include: improving the way types are handled; designing and implementing further undo operations that handle arbitrary proofs, *e.g.* proofs that are pieced together using unification; adapting Nuprl's scheme for compact storage of proofs in files; and implementing our ideas for Nuprl.

References

1. S. F. Allen, R. L. Constable, D. J. Howe, and W. B. Aitken. The semantics of reflected proof. In *Proceedings of the Fifth Annual Symposium on Logic and Computer Science*, pages 95–107. IEEE Computer Society, June 1990.
2. R. L. Constable, et al. *Implementing Mathematics with the Nuprl Proof Development System*. Prentice-Hall, Englewood Cliffs, New Jersey, 1986.
3. G. Dowek, A. Felty, H. Herbelin, G. Huet, C. Paulin-Mohring, and B. Werner. The coq proof assistant user's guide. Technical Report 134, INRIA, December 1991.
4. A. Felty. Implementing tactics and tacticals in a higher-order logic programming language. *Journal of Automated Reasoning*, 11(1):43–81, August 1993.
5. M. Gordon. A proof generating system for higher-order logic. In *Proceedings of the Hardware Verification Workshop*, 1989.
6. M. J. Gordon, R. Milner, and C. P. Wadsworth. *Edinburgh LCF: A Mechanized Logic of Computation*, volume 78 of *Lecture Notes in Computer Science*. Springer-Verlag, 1979.
7. R. Harper, F. Honsell, and G. Plotkin. A framework for defining logics. In *The Second Annual Symposium on Logic in Computer Science*. IEEE, 1987.
8. M. Heisel, W. Reif, and W. Stephan. Tactical theorem proving in program verification. In M. Stickel, editor, *Tenth Conference on Automated Deduction*, volume 449 of *Lecture Notes in Computer Science*, pages 117–131. Springer-Verlag, 1990.
9. C. Horn. The Oyster Proof Development System. University of Edinburgh, 1988.
10. L. Magnussan. Refinement and local undo in the interactive proof editor ALF. In *Informal Proceedings of the 1993 Workshop on Types for Proofs and Programs*, 1993.
11. Z. Manna and R. Waldinger. A deductive approach to program synthesis. *Transactions on Programming Languages and Systems*, 2:90–121, 1980.
12. D. Miller, G. Nadathur, F. Pfenning, and A. Scedrov. Uniform proofs as a foundation for logic programming. *Annals of Pure and Applied Logic*, 51:125–157, 1991.
13. L. Paulson. Isabelle: The next 700 theorem provers. In P. Odifreddi, editor, *Logic and Computer Science*, pages 361–385. Academic Press, 1990.

Unification in an Extensional Lambda Calculus with Ordered Function Sorts and Constant Overloading

Patricia Johann* and Michael Kohlhase**
Fachbereich Informatik
Universität des Saarlandes
66123 Saarbrücken, Germany
{*pjohann, kohlhase*} *@cs.uni-sb.de*

Abstract. We develop an order-sorted higher-order calculus suitable for automatic theorem proving applications by extending the extensional simply typed lambda calculus with a higher-order ordered sort concept and constant overloading. Huet's well-known techniques for unifying simply typed lambda terms are generalized to arrive at a complete transformation-based unification algorithm for this sorted calculus. Consideration of an order-sorted logic with functional base sorts and arbitrary term declarations was originally proposed by the second author in a 1991 paper; we give here a corrected calculus which supports constant rather than arbitrary term declarations, as well as a corrected unification algorithm, and prove in this setting results corresponding to those claimed there.

1 Introduction

In the quest for calculi best suited for automating logic, the introduction of sorts has been one of the most promising developments. Sorts, which are intended to capture for automated deduction purposes the kinds of meta-level taxonomic distinctions that humans naturally assume structure the universe, can be employed to syntactically distinguish objects of different classes. The essential idea behind sorted logics is to assign sorts to objects and to restrict the ranges of variables to particular sorts, so that unintended inferences, which then violate the constraints imposed by this sort information, are disallowed. These techniques have been seen to dramatically reduce the search space associated with deduction in first-order systems ([Wal88], [Coh89], [Sch89]).

On the other hand, the inherently higher-order nature of many problems whose solutions one would like to deduce automatically has sparked an increasing interest in higher-order deduction. The behavior of sorted higher-order calculi, which boast both the expressiveness of higher-order logics and the efficiency of sorted calculi, is thus a natural topic of investigation. In this paper, we develop precisely such a calculus — an order-sorted lambda calculus supporting functional base sorts and constant overloading — as well as a complete unification algorithm for this calculus, which

---

* On leave from the Department of Mathematics and Computer Science, Hobart and William Smith Colleges, Geneva, NY 14456, *johann@hws.bitnet*. This material is based on work supported by the National Science Foundation, Grant No. INT-9224443.
** Supported by the Deutsche Forschungsgemeinschaft (SFB 314).

is suitable for use in an automated deduction setting. Calculi intended for actual mathematical deduction will no doubt support constant — if not arbitrary term — declarations (see Example 3.7); by incorporating constant declarations into our calculus, we treat deduction issues common to all mathematically useful extensional order-sorted higher-order logics supporting functional base sorts.

Although Huet proposed the study of a simple sorted lambda calculus in an appendix to [Hue72], the development of order-sorted higher-order calculi for use in deduction systems has only in recent years been pursued ([Koh92], [NQ92], [Pfe92]). There has, however, been considerable interest in order-sorted higher-order logic from the point of view of higher-order algebraic specifications, the theory of functional programming languages, and object-oriented programming ([Car88], [BL90], [Qia90], [CG91], [Pie91]).

In unsorted logics, the knowledge that an object is a member of a certain class of objects is expressed using unary predicates. This leads to a multitude of unit clauses in deductions, each of which carries only taxonomic information and contributes to a severe explosion of the search space. In sorted logics, predicates are replaced by sorts carrying precisely the same taxonomic information, so that their attendant unit clauses are also eliminated and the search space is correspondingly pruned. The incorporation of sort information is perhaps even more natural for higher-order than for first-order logics: type information in higher-order logics can be regarded as coding very coarse distinctions between disjoint classes of objects, so that sorts merely refine an already present structure. But more importantly, the benefits of sorts for restricting search spaces in higher-order deduction will necessarily be more pronounced than in first-order systems, since the sort hierarchy propagates into the higher-order structure of the logics.

Sorting the universe of individuals in higher-order logics gives rise to new classes of functions, namely those whose domains and codomains are (denoted by) the sorts. But in addition to sorting function universes in such a first-order manner, classes of functions defined by domains and codomains can themselves be divided into subclasses since functions are explicit objects of higher-order logics. Functional base sorts, $i.e.$, base sorts that denote classes of functions, are thus permitted. Syntactically, each sort A comes with a type, a codomain sort $\gamma(A)$, and — if of functional type — also with a domain sort $\delta(A)$. Partial orders on the set of sorts, capturing inclusion relations among the various classes of objects, are induced by covariance in the codomain sort via subsort declarations. But in the presence of functional base sorts an additional mechanism for inducing subsort information is needed: since any function of sort A is a function with domain $\delta(A)$ and codomain $\gamma(A)$, a functional sort A must always be a subsort of the sort $\delta(A) \to \gamma(A)$.

The calculus presented here supports constructs for restricting the ranges of variables to, and assigning constants membership in, certain classes of objects. Depending on the partial order induced on the sorts, certain classes of terms built from these atoms then become the objects of study — the partial order restricts the class of models for the calculus, so that terms must meet certain conditions to denote meaningful objects, $i.e.$, to be well-sorted. Notions of β- and η-reduction generalizing the corresponding reductions in the simply typed lambda calculus are defined on the class of well-sorted terms. The former is a straightforward adaptation of typed β-reduction, but the delicate interaction between extensionality and partially ordered sorts necessitates care in defining the latter. If X is a term of functional sort A, for example, and x is a variable whose range is restricted to the subsort B of $\delta(A)$, then $\lambda x.Xx$ denotes the restriction of the function (denoted by) X to the domain (specified by) B. In order to properly model extentionality by η-reduction, B must therefore be precisely the (maximal) domain of X in order for $\lambda x.Xx$ to η-reduce to X — otherwise X would be equal to a proper restriction of itself.

A similar subtle interplay between extensionality and functional base sorts renders the natural generalization of Huet's ([Hue75]) classical method for unification of simply typed lambda terms inadequate in our setting. Nevertheless, a more liberal notion of partial binding, which in particular does not require the bindings to be η-expanded, does suffice for incrementally approximating answer substitutions for arbitrary unification problems modulo $\beta\eta$-equality on well-sorted terms.

As in the simply typed lambda calculus, the need for "guessing" partial bindings for pairs so called *flex-flex pairs* gives rise to a serious explosion of the search space, but unfortunately, this cannot be avoided without sacrificing the unification completeness of our algorithm. Huet resolved this difficulty in the simply typed lambda calculus by redefining the higher-order unification problem to a form sufficient for refutation purposes: flex-flex pairs are considered to *pre-unified*, or already solved. We conjecture that it is possible to define an appropriate notion of pre-unification in our setting as well, but warn that a naive modification of the standard methods is evidently insufficient for calculi supporting functional base sorts. Specifically, pre-unification only makes sense under regular signatures, and the existence of unifiers for flex-flex pairs depends heavily on the partial order on sorts under which unification is being considered.

Unification in an extensional order-sorted lambda calculus with functional base sorts was first investigated in [Koh92]. A calculus supporting functional base sorts and arbitrary term, rather than only constant, declarations is proposed there, but its presentation is flawed in serveral places. Our calculus can be seen as a subcalculus of the one proposed in [Koh92] which has been corrected to be well-defined and to properly incorporate extensionality (see the problematic clauses 4 and 5 of Definition 2.5, and Remark 2.10, there). The notion of partial binding developed here paves the way for remedying both the ill-defined unification transformations and the flawed completeness proof of [Koh92]. For a detailed treatment of our results and the issues surrounding them, the reader is referred to the full paper [JK93].

2 The Calculus

The set of *types* $\mathcal{T}$ is obtained by inductively closing a set of *base types* $\mathcal{T}_0$ under the operation $\alpha \to \beta$; assuming right-associativity of $\to$, the *length* of a type $\alpha \equiv \alpha_1 \to \alpha_2 \to ... \to \alpha_n$, denoted $length(\alpha)$, is $n - 1$. Types are denoted by lower case Greek letters. In theorem proving applications we might have only two base types, o denoting truth-values and ι denoting the universe of individuals, with all other subdivisions of the universe being coded into sort distinctions among individuals, as described in the next subsection.

For each type $\alpha \in \mathcal{T}$, fix a countably infinite set of variables $x_\alpha, y_\alpha, z_\alpha, ...$ of type α and a countably infinite set of constants $a_\alpha, b_\alpha, c_\alpha, ...$ of type α. We assume that no two distinct variables or constants have the same type-erasure.

$\mathcal{LC}$ is the set of explicitly simply typed lambda terms over the variables and constants. We omit reference to the type of X when this will not lead to confusion. On $\mathcal{LC}$, $\beta\eta$-equality is generated by $\beta\eta$-reduction, denoted by $\xrightarrow{\beta\eta}$ and determined by the usual rules $(\lambda x.X)Y \xrightarrow{\beta} X[x := Y]$ and $\lambda x.Xx \xrightarrow{\eta} X$. $\beta\eta$-reduction is terminating and confluent (*i.e., convergent*) on $\mathcal{LC}$-terms.

The reflexive, transitive closure of a reduction relation $\xrightarrow{\nu}$ is denoted $\xrightarrow{\nu}\!\!*$, and we write $=_\nu$ for the symmetric closure of $\xrightarrow{\nu}\!\!*$. We write $X \equiv Y$ to indicate that two $\mathcal{LC}$-terms X and Y are identical up to renaming of bound variables. As is customary, we consider $\mathcal{LC}$-terms identical up to renaming of bound variables to be the same.

2.1 Order-sorted Structures

As described in the introduction, we capitalize on the fact that functions are explicit objects of higher-order logic by allowing classes of functions defined by domains and codomains to themselves be divided into subclasses. We thus postulate both functional base sorts — *i.e.*, base sorts that denote classes of functions — as well as non-functional base sorts.

Definition 2.1 A *sort system* is a quintuple $(\mathcal{S}_0, \mathcal{S}, \tau, \delta, \gamma)$ such that:

- $\mathcal{S}_0$ is a set of *base sorts* distinct from the set of type symbols. The set of *sorts* obtained by closing $\mathcal{S}_0$ under the operation $A \to B$ comprises $\mathcal{S}$.
- The *type function* τ is a mapping $\tau : \mathcal{S}_0 \to \mathcal{T}$. If $\tau(A) \in \mathcal{T}_0$, then A is said to be *non-functional*, and A is said to be *functional* otherwise; the set of non-functional (resp., functional) sorts is denoted by $\mathcal{S}^{nf}$ (resp., $\mathcal{S}^f$). For all $A \in \mathcal{S}^f$, we require that $\tau(A) = \tau(\delta(A)) \to \tau(\gamma(A))$, where the *domain sort function* δ is a map $\delta : \mathcal{S}_0^f \to \mathcal{S}$, the *codomain sort function* γ is a map $\gamma : \mathcal{S}_0 \to \mathcal{S}$ with $\gamma \mid_{\mathcal{S}^{nf}}$ the identity map, and the mappings δ and γ are extended to $\mathcal{S}$ by defining $\delta(A) = B$ and $\gamma(A) = C$ for $A \equiv B \to C \in \mathcal{S}$.

Sorts are denoted by upper case Roman letters. If the context is clear, we abbreviate by $\mathcal{S}$ the sort system $(\mathcal{S}_0, \mathcal{S}, \tau, \delta, \gamma)$. Since we are ultimately interested in sorted terms and their typed counterparts, we only consider sort systems for which τ is surjective. We further assume that for each $\alpha \in \mathcal{T}$ there exist only finitely many $A \in \mathcal{S}_0$ with $\tau(A) = \alpha$.

It will be useful to have some notational conventions for domain and codomain sorts. For any $A \in \mathcal{S}$, define the following notation: $\delta^0(A) \equiv A$, $\gamma^0(A) \equiv A$, and for $i \geq 1$, $\gamma^i(A) \equiv \gamma(\gamma^{i-1}(A))$, and $\delta^i(A) \equiv \delta(\gamma^{i-1}(A))$. Write $length(A)$ for the *length* of the sort A.

Example 2.2 Functional base sorts are useful in the study of elementary analysis, where we might postulate a non-functional base sort R denoting the reals and a functional base sort C with $\delta(C) = R$ and $\gamma(C) = R$ denoting the class of real-valued continuous functions on the reals. Since it is not possible to distinguish syntactically such continuous functions solely in terms of their domains and codomains, permitting functional base sorts indeed increases the expressiveness of a calculus.

While types represent disjoint classes of objects, certain kinds of orderings on sorts reflect permissible inclusion relations among classes of objects sorts denote. We capture a consistency condition which such orderings are required to satisfy by defining, for a sort system $\mathcal{S}$ and a pair of sorts A and B in $\mathcal{S}$ such that $\tau(A) = \tau(B)$, the set $\text{Con}(A, B)$ of *subsort declarations* (for $\mathcal{S}$) to be the set $\{[A \leq B]\}$ if $A, B \in \mathcal{S}^{nf}$, and

$$\text{Con}(\delta(A), \delta(B)) \cup \text{Con}(\delta(B), \delta(A)) \cup \text{Con}(\gamma(A), \gamma(B)) \cup \{[A \leq B]\}$$

if $A, B \in \mathcal{S}^f$. A *sort structure* (for $\mathcal{S}$) is any set of subsort declarations obtained by inductively adding sets of the form $\text{Con}(A, B)$ to the empty set. Since each set $\text{Con}(A, B)$ of subsort declarations is finite, sort structures are necessarily finite. For any sort structure Δ, we have $[A \leq B] \in \Delta$ iff $\text{Con}(A, B) \subseteq \Delta$.

Any sort structure Δ induces an *inclusion ordering* $\leq_\Delta$ (or simply "$\leq$") on $\mathcal{S}$, inductively defined by the rules of Definition 2.3.

Definition 2.3 For any sort structure Δ, the *inclusion ordering determined by* Δ contains all judgements of the form $\Delta \vdash A \leq B$ provable by the following calculus:

$$\frac{[A \leq B] \in \Delta}{\Delta \vdash A \leq B} \qquad \frac{A \in \mathcal{S}^f}{\Delta \vdash A \leq \delta(A) \to \gamma(A)}$$

$$\frac{}{\Delta \vdash A \leq A} \qquad \frac{\Delta \vdash A \leq B}{\Delta \vdash C \to A \leq C \to B}$$

$$\frac{\Delta \vdash A \leq B \qquad \Delta \vdash B \leq C}{\Delta \vdash A \leq C}$$

Clearly we cannot insist that $\Delta \vdash A \leq B$ hold for any sorts A and B with a common domain sort C and codomain sorts satisfying $\Delta \vdash \gamma(A) \leq \gamma(B)$ (assuming, for example, a standard semantics). But if Δ is a sort structure for $\mathcal{S}$, and $\sim$ is the equivalence relation induced by $\leq$, then $A, B \in \mathcal{S}^f$, $\Delta \vdash A \leq B$ implies $\Delta \vdash \delta(A) \sim \delta(B)$ and $\Delta \vdash \gamma(A) \leq \gamma(B)$. In addition, for all $A, B \in \mathcal{S}$, $\Delta \vdash A \leq B$ implies $\tau(A) = \tau(B)$, so that any sort system $\mathcal{S}$ is the disjoint union of infinitely many subsets $\mathcal{S}_\alpha = \{A \in \mathcal{S} \mid \tau(A) = \alpha\}$ of sorts such that if $A \in \mathcal{S}_\alpha$ and $B \in \mathcal{S}_\beta$ with $\alpha \not\equiv \beta$, then A and B are incomparable with respect to $\leq$. Since $\mathcal{S}$ has only finitely many base sorts per type, each subset $\mathcal{S}_\alpha$ is finite. Decidability of the inclusion ordering determined by any sort structure thus follows from the next lemma, which is proved by induction on $length(\alpha)$.

Lemma 2.4 *For any type $\alpha \in T$ and any sort structure Δ, if $\leq$ is the inclusion ordering determined by Δ, then the restriction $\leq_\alpha$ of $\leq$ to sorts of type α is effectively computable.*

Theorem 2.5 *The inclusion ordering determined by any sort structure Δ is decidable.*

It will be important that the signatures over which our well-sorted terms are built "respect function domains," i.e., that for any term X and any sorts A and B such that X has sort A and also sort B, $\delta(A) \sim \delta(B)$ holds. The proof that signatures indeed satisfy this property (see Lemma 2.11) depends in part on the consistency conditions for sort structures and in part on the fact that constant declarations meet the sort condition of the fifth clause of Definition 2.7 below, given in terms of the equivalence relation Rdom, which we now define.

Definition 2.6 Given a sort structure Δ for S and a pair of sorts A and B in S, A Rdom_Δ B holds if either $A, B \in S^{nf}$ and $\tau(A) = \tau(B)$, or if $A, B \in S^f$, $\Delta \vdash \delta(A) \sim \delta(B)$, and $\gamma(A)$ Rdom_Δ $\gamma(B)$.

We write "Rdom" for Rdom_Δ when Δ can be discerned from the context. Then A Rdom B implies $\tau(A) = \tau(B)$, and $\Delta \vdash A \leq B$ implies A Rdom B.

Definition 2.7 A *signature* Σ comprises *i*) a sort system $S = (S_0, S, \delta, \gamma, \tau)$, *ii*) a sort structure Δ (for S), *iii*) a countably infinite set $Vars_A$ of *variables* $x_A, y_A, z_A, \ldots$ for each $A \in S$, *iv*) a set $\mathcal{C}$ of typed constant symbols, and *v*) a set of *constant declarations* of the form $[c_\alpha :: A]$ for $c \in \mathcal{C}$ such that $\tau(A) = \alpha$. We assume that if $[c :: A]$ and $[c :: B]$ are constant declarations, then A Rdom B.

The requirement that $\tau(A) = \alpha$ for a constant declaration $[c_\alpha :: A]$ insures that sort assignments respect the types of constants. In a theorem proving context, any signature would have, for each $\alpha \in \mathcal{T}$, only finitely many constant declarations involving constants of type α. We will assume this restriction on signatures.

Any sorted variable can naturally be regarded as a typed variable by "forgetting" its sort information. Denoting the forgetful functor by $\overline{}$, we may regard the sorted variable x_A as the typed variable $\overline{x_A}$, *i.e.*, as $x_{\tau(A)}$. By prudently naming the variables, we can arrange that the forgetful functor is bijective on variables, thereby avoiding merely technical complications that could otherwise arise.

2.2 Term Structure

Definition 2.8 Let Σ be a signature with sort structure Δ. The set of *well-sorted* $\mathcal{LC}$-*terms* for Σ is determined inductively by the following inference rules:

$$\frac{x \in Vars_A}{\Sigma \vdash x : A} \ (var) \qquad \frac{\Sigma \vdash X : A \quad \Sigma \vdash Y : B \quad \Delta \vdash B \sim \delta(A)}{\Sigma \vdash XY : \gamma(A)} \ (app)$$

$$\frac{[c :: A] \in \Sigma}{\Sigma \vdash c : A} \ (const) \qquad \frac{x \in Vars_B \quad \Sigma \vdash X : A}{\Sigma \vdash \lambda x.X : B \to A} \ (abs)$$

$$\frac{\Sigma \vdash X : A \quad \Delta \vdash \delta(A) \sim B}{\Sigma \vdash \lambda x_B.Xx : A} \ (\eta) \qquad \frac{\Sigma \vdash X : B \quad \Delta \vdash B \leq A}{\Sigma \vdash X : A} \ (weaken)$$

Let $\mathcal{LC}_A(\Sigma) = \{X \mid \Sigma \vdash X : A\}$ and $\mathcal{LC}(\Sigma) = \bigcup_{A \in S} \mathcal{LC}_A(\Sigma)$. For any $X \in \mathcal{LC}(\Sigma)$ write $S_\Sigma(X)$ for $\{A \in S \mid X \in \mathcal{LC}_A(\Sigma)\}$. Since the inclusion ordering determined by any sort structure Δ is transitive, we need never follow one application of the rule (weaken) by another in constructing sort derivations for well-sorted $\mathcal{LC}$-terms (henceforth called $\mathcal{LC}(\Sigma)$-*terms*). We consider $\mathcal{LC}(\Sigma)$-terms which are identical up to renaming of (sorted) variables to be the same, and omit sort information whenever possible.

If Σ is a signature with sort system S and sort structure Δ, and if $\sim$ is the equivalence relation determined by Δ, then $\mathcal{LC}_A(\Sigma) = \mathcal{LC}_B(\Sigma)$ whenever $A \sim B$. Passing to the quotient signature Σ' with respect to $\sim$, *i.e.*, to the signature with sort system S' equal to $S/\sim$ obtained by replacing sorts in S by canonical $\sim$-equivalence class representatives, we arrive at a signature whose equivalence relation is trivial and such that $\mathcal{LC}_A(\Sigma') = \mathcal{LC}_A(\Sigma)$ for all sorts A. We may therefore assume that $\leq$ is a partial ordering for all signatures in the remainder of this paper. We also assume that we have ridded our sort structures of redundant subsort declarations of the form $[A \leq A]$, and that whenever $\Delta \vdash B \leq A$ for a sort structure Δ, $length(B) \leq length(A)$ holds. The latter assumption is without loss of generality under a standard semantics, and implies that $length(B) \leq length(A)$ if $\Delta \vdash B \leq A$.

A routine induction on sort derivations establishes that signatures are subterm closed, i.e., that each subterm of a well-sorted term is again well-sorted.

In any signature Σ, if $x \in Vars_A$, then x has least sort A in Σ. But because of constant overloading, not every term will necessarily have a unique least sort. For an arbitrary term X, however, if $\Sigma \vdash X : A$ and $\Sigma \vdash X : B$ then $\tau(A) = \tau(B)$. As a result, the fact that Σ has only finitely many sorts per type implies that, for $X \in \mathcal{LC}(\Sigma)$, the set of sorts $\mathcal{S}_\Sigma(X)$ is finite. It also follows that if we consider the forgetful functor to be the identity on typed constants, then it can be extended to an injection (but not necessarily a bijection) from $\mathcal{LC}(\Sigma)$ into $\mathcal{LC}$. And if Σ is a signature with empty sort structure and exactly one sort A such that $\tau(A) = \alpha$ for each $\alpha \in T_0$, then $\mathcal{LC}(\Sigma)$ is isomorphic to the fragment of $\mathcal{LC}$ containing only the finitely many constants per type appearing in constant declarations in Σ.

To prove computability of sort assignment for $\mathcal{LC}(\Sigma)$, we extend the function $\mathcal{S}_\Sigma(\cdot)$ on $\mathcal{LC}(\Sigma)$ to all of $\mathcal{LC}$. For $X \in \mathcal{LC}$ and Σ a signature, define $\mathcal{S}_\Sigma(X) = \{\mathcal{S}_\Sigma(Y) \mid Y \in \mathcal{LC}(\Sigma) \text{ and } \overline{Y} \equiv X\}$. Then $X \in \mathcal{LC} \setminus \mathcal{LC}(\Sigma)$ iff $\mathcal{S}_\Sigma(X) = \emptyset$. If there exists a $Y \in \mathcal{LC}(\Sigma)$ with $\overline{Y} \equiv X$, then it is unique; in this case, we say that $X \in \mathcal{LC}$ is *well-sorted* with respect to Σ.

Theorem 2.9 *For $X \in \mathcal{LC}$ and any signature Σ, $\mathcal{S}_\Sigma(X)$ is effectively computable.*

Proof. We will later observe that η-reduction on $\mathcal{LC}(\Sigma)$ is sort-preserving, and, assuming this, we take X to be in η-normal form. Induction on the structure of X completes the proof. □

Corollary 2.10 *For $X \in \mathcal{LC}$ and any signature Σ, it is decidable whether or not X is well-sorted with respect to Σ.*

As promised, we can prove (by induction according to the various cases for the derivations of $\Sigma \vdash X : A$ and $\Sigma \vdash X : B$) that

Lemma 2.11 *If $\Sigma \vdash X : A$ and $\Sigma \vdash X : B$, then A Rdom B. That is, any signature Σ respects function domains.*

Lemma 2.11 guarantees that for any term X and any sorts $A, B \in \mathcal{S}_\Sigma(X)$ we must have $\delta(A) = \delta(B)$. This unique domain sort for X is called its *supporting sort* and is denoted $supp(X)$. At first glance, requiring signatures to respect function domains appears to be a grave restriction on the expressiveness of a calculus, but functional extensionality itself relies heavily on the notion of implicitly specified domains of functions, which unique supporting sorts syntactically capture. Indeed, in mathematics, functions are assumed to have unique (implicitly specified) domains, and must therefore be distinguished from restrictions to subdomains: functions f and g are the same only if $fa = ga$ for all a in the common (implicitly specified) domain of f and g.

2.3 Order-sorted Reduction

As per the above discussion, η-expansion of the term X_A to $\lambda x_B.Xx$, which corresponds to restricting the function denoted by X to the sort denoted by B, should only again yield the original function if B represents the domain of the function denoted by X. This restriction is embodied in the order-sorted η-rule.

Definition 2.12 Let Σ be any signature. The following order-sorted reductions are defined for $\mathcal{LC}(\Sigma)$-terms:

- $(\lambda x.X)Y \xrightarrow{\beta} X[x := Y]$, and
- $\lambda x_B.Xx \xrightarrow{\eta} X$ if $x_B \notin FV(X)$ and $B \equiv supp(X)$.

The first rule above, assumed to happen without free variable capture, is called *(order-sorted) β-reduction*; the second is called *(order-sorted) η-reduction*. Since order-sorted $\beta\eta$-reduction generalizes ordinary typed $\beta\eta$-reduction, we write $\xrightarrow{\beta\eta}$ for order-sorted $\beta\eta$-reduction as well as for its typed version.

It is important to our program that the fundamental operations of our calculus do not allow the formation of ill-sorted terms from well-sorted ones. This ensures that our unification algorithm never has to handle ill-sorted terms. In fact, if $X \xrightarrow{\beta} Y$, then $\mathcal{S}_\Sigma(X) \subseteq \mathcal{S}_\Sigma(Y)$. A similar although slightly stronger result holds for η-reduction: if $X \xrightarrow{\eta} Y$, then $S_\Sigma(X) = S_\Sigma(Y)$.

Order-sorted $\beta\eta$-reduction is convergent. Termination is a direct consequence of the corresponding well-known result for the simply typed lambda calculus, and weak confluence — and, in light of termination, therefore confluence — follows from weak confluence of $\beta\eta$-reduction on $\mathcal{LC}$ together with the fact that $X \xrightarrow{\beta\eta} Y$ implies $supp(X) \equiv supp(Y)$. It thus makes sense to refer to *the* order-sorted $\beta\eta$-normal form of an $\mathcal{LC}(\Sigma)$-term, and *the* order-sorted long (*i.e.*, η-expanded) β-normal form of X, denoted $l\beta nf(X)$.

3 Order-sorted Higher-order Unification

When considering unification in the simply typed lambda calculus, it is customary to work modulo η-equality. We explicitly keep track of order-sorted η-equality, since the interaction between extensionality and sorts can be unexpectedly subtle. Fix an arbitrary signature Σ for use throughout the remainder of this paper.

3.1 Systems and Substitutions

We will represent unification problems by equational systems comprising the pairs of $\mathcal{LC}(\Sigma)$-terms to be simultaneously unified, and use transformations of such systems as our main tool for solving the unification problems they represent.

A *pair* is a two-element multiset of $\mathcal{LC}(\Sigma)$-terms. A *system* is a finite set Γ of pairs. A pair is η-*trivial* (or simply *trivial*) if its elements are η-equal, and Σ-*valid* if its elements are $\beta\eta$-equal; a system is Σ-*valid* if each of its pairs is Σ-valid. As usual, we write $\Gamma, \langle X, Y \rangle$ instead of $\Gamma \cup \{\langle X, Y \rangle\}$, but since Γ may or may not also contain $\langle X, Y \rangle$, such a decomposition is ambiguous. We use the notation $\Gamma; \langle X, Y \rangle$ to abbreviate $\Gamma \cup \{\langle X, Y \rangle\}$ when $\langle X, Y \rangle$ is *not* a pair in Γ. A pair $\langle X, Y \rangle$ is *solved in* Γ if it is either trivial, or for some $x \in Vars_A$, $X \xrightarrow{\eta} x$, $A \in \mathcal{S}_\Sigma(Y)$ and there are no occurrences of x in Γ other than the one indicated. In this case, x is said to be *solved in* Γ. If each pair in Γ is solved in Γ, then Γ is a *solved system*.

A *substitution* is a finitely supported map from variables to $\mathcal{LC}(\Sigma)$; a substitution θ induces a mapping on terms, which we also denote by θ. We write substitution application as juxtaposition, so that θX is the application of the substitution θ to the term X, and by $D(\theta)$ and $I(\theta)$ we denote the set of variables in the domain of θ and the set of variables introduced by θ, respectively. A substitution θ is *well-sorted* if for every $x \in Vars_A$, $A \in \mathcal{S}_\Sigma(\theta x)$. It follows that if $X \in \mathcal{LC}_A(\Sigma)$ and θ is well-sorted, then $\theta X \in \mathcal{LC}_A(\Sigma)$ as well. That the set of well-sorted substitutions is closed under composition is not hard to prove.

We can extend equalities on $\mathcal{LC}(\Sigma)$ to (well-sorted) substitutions in the usual manner: Let $=_*$ be an equational theory on $\mathcal{LC}(\Sigma)$, W be a set of variables, and θ and θ' be substitutions. Then $\theta =_* \theta'[W]$ means that for every variable in $x \in W$, $\theta x =_* \theta' x$. The subsumption relation $\theta' \leq_* \theta[W]$ holds provided there exists a substitution ρ such that $\theta =_* \rho\theta'[W]$. If W is the set of all variables, we drop the notation "$[W]$." If $=_*$ is the empty equational theory we write "$\equiv$" and "$\leq$" for the induced equality and subsumption ordering on substitutions.

We can extend substitutions on $\mathcal{LC}(\Sigma)$ to mappings on systems $\Gamma \equiv \{\langle X_i, Y_i\rangle \mid i \leq n\}$ by defining $\sigma\Gamma$ to be the system $\{\langle \sigma X_i, \sigma Y_i\rangle \mid i \leq n\}$. The normal form $l\beta n f(\Gamma)$, all of whose unsolved pairs comprise terms in long β-normal form, is defined similarly. If all terms in the unsolved pairs of Γ are in long β-normal form, we say that Γ is *in long β-normal form*. We write $FV(X)$ for the set of free variables occurring in the $\mathcal{LC}(\Sigma)$-term X and $FV(\Gamma)$ for the free variables occurring in the terms in the system Γ.

A well-sorted substitution θ is a Σ-*unifier* of a system Γ if $\theta\Gamma$ is Σ-valid. If σ is a Σ-unifier of Γ with the properties that $D(\sigma) \subseteq FV(\Gamma)$ and that for any Σ-unifier θ of Γ, $\sigma \leq_{\beta\eta} \theta$ holds, then σ is said to be a *most general Σ-unifier* of Γ. A system Γ is Σ-*unifiable* if there exists some Σ-unifier of Γ. An idempotent well-sorted substitution θ is a *normalized Σ-unifier* of a system Γ if *i*) $D(\theta) \subseteq FV(\Gamma)$, *ii*) θ is a Σ-unifier of Γ, and *iii*) for all unsolved variables x in Γ, θx is in long β-normal form. Write $U_\Sigma(\Gamma)$ for the set of all normalized Σ-unifiers of Γ. It is clear that every well-sorted substitution θ is $\beta\eta$-equal to a well-sorted substitution θ' with $D(\theta) = D(\theta')$ and $\theta' x$ in long β-normal form for each $x \in D(\theta)$. Such a substitution θ' is said to be *in long β-normal form*. Thus for any Σ-unifier θ of a system Γ, there exists a $\theta' \in U_\Sigma(\Gamma)$ such that $\theta' =_{\beta\eta} \theta[FV(\Gamma)]$. In particular, every Σ-unifiable system has a normalized Σ-unifier. For technical reasons, normalized Σ-unifiers will be important in what follows. Note that we relax the standard requirement that normalized substitutions map all variables to normal forms, and allow solved variables to be bound arbitrarily. This is justified in Lemma 3.2 below.

The remainder of this section explores the relationship between systems and their unifiers. If Γ is a solved system whose non-trivial pairs are $\langle X_1, Y_1\rangle, ..., \langle X_n, Y_n\rangle$ with $X_i \xrightarrow{\eta} x_i$ for $i = 1, ..., n$, then these pairs determine an idempotent well-sorted substitution $\sigma_\Gamma = \{x_1 \mapsto Y_1, ..., x_n \mapsto Y_n\}$, although such a pair $\langle X, Y\rangle$ with $X \xrightarrow{\eta} x \in Vars_A$ and $Y \xrightarrow{\eta} y \in Vars_A$ requires a choice as to which of x and y is to be in the domain of the substitution. We assume that a uniform way exists for making this choice, and so refer to *the* well-sorted substitution determined by a solved system. Conversely, idempotent well-sorted substitutions can be represented by solved systems without trivial pairs. If σ is such a substitution, write $[\sigma]$ for any solved system which represents it. Any system Γ can be written as $\Gamma'; [\sigma]$ where $[\sigma]$ is the set of solved pairs in Γ. We call $[\sigma]$ the *solved part* of Γ.

Transformation-based unification methods attempt to reduce systems to be unified to solved systems which represent their unifiers. The fundamental connection between solved systems and Σ-unifiers is that solved systems represent their own solutions:

Lemma 3.1 *If* $\Gamma \equiv \langle X_1, Y_1 \rangle, ..., \langle X_n, Y_n \rangle$ *is a solved system, then* σ_Γ *is a most general* Σ-*unifier for* Γ. *In fact, for any* Σ-*unifier* θ *of* Γ, $\theta =_{\beta\eta} \theta \sigma_\Gamma$.

In general, however, a system Γ will not have a single most general Σ-unifier. The next lemma shows that we need not be concerned with solved pairs when computing Σ-unifiers. This is consistent with the intuition that the solved part of a system is merely a record of an answer substitution being constructed.

Lemma 3.2 *Suppose* Γ *is a* Σ-*unifiable system with solved part* $[\sigma]$ *and unsolved part* Γ'. *If* θ *is a* Σ-*unifier of* Γ, *then for every* Σ-*unifier* ρ *of* Γ' *such that* $D(\rho) \subseteq FV(\Gamma')$ *and* $\rho \leq_{\beta\eta} \theta[FV(\Gamma')]$, $\rho\sigma$ *is a* Σ-*unifier of* Γ *and* $\rho\sigma \leq_{\beta\eta} \theta[FV(\Gamma)]$.

3.2 The Unification Algorithm

One of the key steps for sorted higher-order unification is solving the following problem: given a term $X \equiv \lambda x_1...x_k.hU_1...U_n \in \mathcal{LC}_A(\Sigma)$ in long β-normal form, find a term $G \in \mathcal{LC}_A(\Sigma)$ with head h which can be instantiated to yield X. This is a generalization of a problem in $\mathcal{LC}$ which Huet ([Hue75]) resolved by describing a set of *partial bindings* in long β-normal form capable of approximating any $\mathcal{LC}$-term by instantiation. While Huet-style partial bindings suffice for approximating arbitrary $\mathcal{LC}(\Sigma)$-terms — although not necessarily with bindings of the appropriate sorts — in our setting, we cannot require that partial bindings be η-expanded without sacrificing completeness of our Σ-unification algorithm (see Example 3.6). Below, a variable will be called *fresh* if it does not appear in any term in the current context.

Definition 3.3 If h is an atom such that either $h \in Vars_C$ or $[h :: C]$ is a constant declaration in Σ, then a *partial binding of sort* A *for head* h is any term of the form $G \equiv \lambda y_1...y_l.hV_1...V_m$, where *i)* $l = length(A)$, *ii)* $m = l + length(\tau(C)) - length(\tau(A)) \geq 0$, *iii)* $\Delta \vdash \gamma^m(C) \leq \gamma^l(A)$, *iv)* $y_j \in Vars_{\delta^j(A)}$ for $j = 1, ..., l$, and *v)* $V_i \equiv z_i y_1...y_l$ for $1 \leq i \leq m$, where $z_i \in Vars_{\delta^1(A) \to ... \to \delta^l(A) \to \delta^i(C)}$ is fresh.

For a given sort A and head h partial bindings need not exist due to conditions *ii)* and *iii)* of Definition 3.3, but because signatures respect function domains, when they do exist they are unique up to renaming of the variables z_i. If Σ is a signature without functional base sorts, then the partial bindings are η-expanded; in particular, if Σ is a signature with exactly one sort per (base) type, then the partial bindings are precisely those obtained for $\mathcal{LC}$. Writing $\mathcal{G}_A^h(\Sigma)$ for the set of partial bindings of sort A for head h, the fact that $\Sigma \vdash G : A$ for $G \in \mathcal{G}_A^h(\Sigma)$ justifies our terminology.

Call a partial binding $G \equiv \lambda y_1...y_l.hV_1...V_m$ a jth *projection binding* if $h \equiv y_j$ and an *imitation binding* if $h \in FV(G) \cup C$. The following transformations on which our algorithm is based are adapted from those of [Sny91].

Definition 3.4 The set ΣT comprises the following transformations on systems in long β-normal form (it is possible that $k = 0$ below).

- DECOMPOSE: For any atom h,

$$\Gamma; \langle \lambda x_1...x_k.hX_1...X_n, \lambda x_1...x_k.hU_1...U_n \rangle \Longrightarrow$$
$$\Gamma, \langle \lambda x_1...x_k.X_1, \lambda x_1...x_k.U_1 \rangle, ..., \langle \lambda x_1...x_k.X_n, \lambda x_1...x_k.U_n \rangle.$$

- ELIMINATE: If $x \in Vars_A$, $x \notin \{x_1,...,x_k\}$, $x \notin FV(\lambda x_1...x_k.X)$, and $\sigma = \{x \mapsto \lambda x_1...x_k.X\}$ is well-sorted, then

$$\Gamma; \langle \lambda x_1...x_k.xx_1...x_k, \lambda x_1...x_k.X \rangle \Longrightarrow \langle x, \lambda x_1...x_k.X \rangle, \sigma\Gamma.$$

- IMITATE: If $x \in Vars_A$, $h \in \mathcal{C}$ or $h \in FV(\lambda x_1...x_k.hU_1...U_m)$, $h \not\equiv x$, and $G \in \mathcal{G}_A^h(\Sigma)$ is an imitation binding, then

$$\Gamma; \langle \lambda x_1...x_k.xX_1...X_n, \lambda x_1...x_k.hU_1...U_m \rangle \Longrightarrow$$
$$\Gamma, \langle x, G \rangle, \langle \lambda x_1...x_k.xX_1...X_n, \lambda x_1...x_k.hU_1...U_m \rangle.$$

- j-PROJECT: If $x \in Vars_A$, h is a (possibly bound) atom and $G \in \mathcal{G}_A^h(\Sigma)$ is a jth projection binding for some $j \in \{1,...,n\}$ such that $head(X_j) \in \mathcal{C}$ implies $head(X_j) \equiv h$, then

$$\Gamma; \langle \lambda x_1...x_k.xX_1...X_n, \lambda x_1...x_k.hU_1...U_m \rangle \Longrightarrow$$
$$\Gamma, \langle x, G \rangle, \langle \lambda x_1...x_k.xX_1...X_n, \lambda x_1...x_k.hU_1...U_m \rangle.$$

- GUESS: If h is any atom, and x and y are free variables in $Vars_A$ and $Vars_B$, respectively, both distinct from h, and $G \in \mathcal{G}_A^h(\Sigma)$, then

$$\Gamma; \langle \lambda x_1...x_k.xX_1...X_n, \lambda x_1...x_k.yU_1...U_m \rangle \Longrightarrow$$
$$\Gamma, \langle x, G \rangle, \langle \lambda x_1...x_k.xX_1...X_n, \lambda x_1...x_k.yU_1...U_m \rangle.$$

As part of the transformations IMITATE, j-PROJECT, and GUESS, we immediately apply ELIMINATE to the new pair $\langle x, G \rangle$.

Our sort mechanism insures that applications of the transformations are such that all terms involved are well-sorted. We adopt the convention that no transformations may be done out of solved or trivial pairs, which accords with the intuition that the solved pairs in a system are merely recording an answer substitution as it is incrementally built up.

We emphasize that there is no deletion of trivial pairs in this presentation. This guarantees that if $\Gamma \Longrightarrow \Gamma'$, then $FV(\Gamma) \subseteq FV(\Gamma')$, so that when a fresh variable is chosen during a computation it is guaranteed to be new to the entire computation. This prevents us from having to manipulate the "protected sets of variables" typically found in completeness proofs in the literature, and respects the fundamental idea behind the use of transformations for describing algorithms, namely that the logic of the problem being considered can be abstracted from implementational issues.

Definition 3.5 The non-deterministic algorithm $\Sigma\mathcal{U}$ is the process of repeatedly

1. reducing all terms of the unsolved pairs in the system to long β-normal form and then applying some transformation in $\Sigma\mathcal{T}$ to an unsolved pair, and
2. returning a most general Σ-unifier if at any point in the computation the system becomes solved.

The choice of pair upon which Algorithm $\Sigma \mathcal{U}$ is to act, and the rule from $\Sigma \mathcal{T}$ to be applied, are non-deterministic. We illustrate use of Algorithm $\Sigma \mathcal{U}$:

Example 3.6 Let $[b :: \delta(A)]$ and $[c :: A]$ comprise the set of constant declarations in a signature Σ with a functional base sort A. Let $f \in Vars_A$, $x \in Vars_{\delta(A)}$, and $w \in Vars_{A \to \delta(A)}$, and consider the Σ-unifiable long β-normal form system $\Gamma \equiv \langle fx, cb \rangle, \langle wc, b \rangle$. Applying IMITATE with partial binding c to the first pair of Γ yields $\langle f, c \rangle, \langle cx, cb \rangle, \langle wc, b \rangle$. An application of DECOMPOSE results in $\langle f, c \rangle, \langle x, b \rangle, \langle wc, b \rangle$, and an application of IMITATE with binding $\lambda y.b$ for $y \in Vars_A$ to the third pair, followed by some β-reductions give the solved system $\Gamma' \equiv \langle f, c \rangle, \langle x, b \rangle, \langle w, \lambda y.b \rangle, \langle b, b \rangle$. We extract the well-sorted substitution $\sigma = \{f \mapsto c, x \mapsto b, w \mapsto \lambda y.b\}$, and anticipating Theorem 3.8, conclude that σ is a Σ-unifier of Γ' and hence of Γ. If we instead allow only η-expanded partial bindings, then the only possible IMITATE step binds f to a term of the form $\lambda y.c(zy)$ for a variable y and a fresh variable z of appropriate sorts. But then ELIMINATE cannot be performed on the pair $\langle f, \lambda y.c(zy) \rangle$ (as is required to complete the IMITATE step), since $\Sigma \not\vdash \lambda y.c(zy) : A$.

While unification in $\mathcal{LC}(\Sigma)$ is apparently more delicate than unification in $\mathcal{LC}$, the extra care pays off when sort information disallows certain undesirable unifications that would be possible in an unsorted calculus.

Example 3.7 Let Σ be a signature with base sorts D, I, and R, where the non-functional sort R denotes the real numbers, and the functional sorts D and I denote the strictly decreasing and strictly increasing functions on the reals, respectively. Suppose further that $\delta(D) = \delta(I) = R$ and $\gamma(D) = \gamma(I) = R$. Finally, let $[n :: D \to I]$ and $[4 :: R]$ comprise the set of constant declarations of Σ, where n denotes the "negation functor" mapping each function F to $-F$, and 4 denotes the real number four.

Let $x \in Vars_R$, $f \in Vars_I$, and $g \in Vars_D$, and consider the unification problem given by the pairs $\langle f4, ngx \rangle, \langle gx, 4 \rangle$. It is not hard to see that an application of IMITATE to the pair $\langle f4, ngx \rangle$ is the only possibility for computation. Letting z be fresh from $Vars_D$, we have that $nz \in \mathcal{G}_I^n(\Sigma)$, and so can apply IMITATE with this binding for f to get $\langle f, nz \rangle, \langle nz4, ngx \rangle, \langle gx, 4 \rangle$. Similarly, we conclude that only DECOMPOSE applies here, resulting in $\langle f, nz \rangle, \langle z, g \rangle, \langle x, 4 \rangle, \langle gx, 4 \rangle$. Two applications of ELIMINATE yield $\langle f, ng \rangle, \langle z, g \rangle, \langle x, 4 \rangle, \langle g4, 4 \rangle$, all of whose pairs, save the last — unsolvable — one, are solved. The only alternative to eliminating z above is applying GUESS to $\langle z, g \rangle$ in the second derived system, but this makes no progress toward a solution. Anticipating Theorem 3.13, we conclude that the original system is unsolvable, in accordance with the facts that neither the identity function nor the function which is constantly four is strictly decreasing.

Of course, if D were to denote the (not strictly) decreasing real-valued functions on the reals, then we would expect $\langle g4, 4 \rangle$ to be solvable by binding g to $\lambda y.4$. A calculus allowing arbitrary term declarations finds a middle road between the typed calculus, which permits too many bindings, and one supporting only constant declarations, which permits too few: declaring $\lambda y.4$ to be of sort D when $y \in Vars_R$, Γ yields precisely the desired solutions.

3.3 Soundness and Completeness of the Algorithm

The proof that our transformations are sound is not appreciably different from the proof for the corresponding transformations for unification in $\mathcal{LC}$.

Theorem 3.8 *(Soundness) If $\Gamma \Longrightarrow \Gamma'$, then for any well-sorted substitution θ, θ is a Σ-unifier of Γ if it is a Σ-unifier of Γ'.*

Thus if Algorithm $\Sigma\mathcal{U}$ is run on initial system Γ and returns a well-sorted substitution θ, then θ is indeed a Σ-unifier of Γ. Our main result (Theorem 3.13) is a converse. We require a few technical lemmas, the first of which is proved by induction on the derivation of $\Sigma \vdash Y : A$.

Lemma 3.9 *If $Y \equiv \lambda x_1...x_p.hU_1...U_q \in \mathcal{LC}_A(\Sigma)$ is in $\beta\eta$-normal form, then either $h \in Vars_C$ or $[h :: C]$ is a constant declaration in Σ for some sort C such that $length(A) + length(\tau(C)) - length(\tau(A)) \geq 0$ and $\Delta \vdash \gamma^q(C) \leq \gamma^p(A)$.*

Lemma 3.10 *If $X \equiv \lambda x_1...x_k.hU_1...U_n \in \mathcal{LC}_A(\Sigma)$ is in long β-normal form, then there exist a partial binding $G \in \mathcal{G}_A^h(\Sigma)$ and a well-sorted substitution ρ in long β-normal form such that $D(\rho)$ is precisely the set of fresh variables in G, ρz has smaller depth than X for each $z \in D(\rho)$, and $\rho G =_{\beta\eta} X$.*

Proof. Let $Y \equiv \lambda x_1...x_p.hU_1'...U_q'$ be the $\beta\eta$-normal form of X, where $U_i \xrightarrow{\eta} U_i'$ for $i = 1,...,q$, $p \leq k$, and $n = q + (k - p)$. Let C be the sort whose existence is guaranteed by Lemma 3.9, $m = length(A) + length(\tau(C)) - length(\tau(A)) \geq 0$, and $G \equiv \lambda x_1...x_l.hV_1...V_m \in \mathcal{G}_A^h(\Sigma)$, where $V_i = z_i x_1...x_l$ for fresh variables z_i, $i = 1,...,m$. Then $l \leq length(\tau(A)) = k$ and $n = length(\tau(C))$, so that $m = l + n - k = l + q - p$. Since $\Sigma \vdash Y : A$, we must have $p \leq l \leq k$. The substitution ρ mapping z_i to $\lambda x_1...x_l.U_i$ for $i = 1,...,q$, and z_i to $\lambda x_1...x_l.x_{p-q+i}$ for $i = q+1,...,m$ is well-sorted, has domain consisting precisely of the set of fresh variables in G, and has the property that ρz has smaller depth than X for each $z \in D(\rho)$. It is well-defined because $m - q = l - p \geq 0$, and indeed $\rho(G) =_\beta \lambda x_1...x_l.hU_1...U_q x_{p+1}...x_l =_\eta \lambda x_1...x_p.hU_1'...U_q' =_\eta X$. □

Note that with the Huet-style partial bindings, it would not necessarily be possible to find G of sort A and a substitution ρ as required:

Example 3.11 If Σ is a signature with a constant declaration $[c :: A]$ for a functional base sort A, then $\Sigma \vdash \lambda x.cx : A$ using (const) followed by an application of (η). Any Huet-style partial binding that might approximate the long β-normal form $\lambda x.cx$ must be of the form $\lambda x.c(zx)$ where z is a fresh variable of an appropriate sort, but there is no derivation of $\Sigma \vdash \lambda x.c(zx) : A$. Under our definition, however, $G \equiv c$ is itself a partial binding of sort A for head h, and ρ can be taken to be the identity substitution.

The measure μ defined by $\mu(\Gamma, \theta) = \langle \mu_1(\Gamma, \theta), \mu_2(\Gamma) \rangle$, where $\mu_1(\Gamma, \theta)$ is the multiset of the depths of the θ-bindings of unsolved variables in Γ which are also in $D(\theta)$, and $\mu_2(\Gamma)$ is the multiset of depths of terms in Γ, will provide the basis for proving termination of Algorithm $\Sigma\mathcal{U}$.

Lemma 3.12 *Let $\theta \in U_\Sigma(\Gamma)$ and let $\langle X, Y \rangle$ be an unsolved pair in a system Γ in long β-normal form. Then there exist a system Γ' and a substitution θ' such that $\Gamma \Longrightarrow \Gamma'$, $\theta \equiv \theta'[FV(\Gamma)]$, $\theta' \in U_\Sigma(\Gamma')$, and $\mu(\Gamma', \theta') < \mu(\Gamma, \theta)$.*

Proof. If $head(X) \equiv head(Y) \notin D(\theta)$, then since $\langle X, Y \rangle$ is not trivial, DECOMPOSE must apply and we must have $\theta \in U_\Sigma(\Gamma')$. Also, $\mu(\Gamma', \theta) < \mu(\Gamma, \theta)$ since $\mu_1(\Gamma', \theta) \leq \mu_1(\Gamma, \theta)$ and $\mu_2(\Gamma') < \mu_2(\Gamma)$.

Otherwise, at least one of X and Y has an unsolved variable $x \in D(\theta) \cap Vars_A$ of Γ as its head; assume X does. Then since θ is well-sorted, $\Sigma \vdash \theta x : A$, and θx is in long β-normal form since θ is normalized. Suppose $\theta x \equiv \lambda x_1...x_k.hU_1...U_n$. By Lemma 3.10, there exist $G \in \mathcal{G}_A^h(\Sigma)$ and a well-sorted substitution ρ in long β-normal form satisfying the conclusions of that lemma. Thus if $head(Y) \notin D(\theta)$ and $h \equiv head(Y)$, then IMITATE applies, if $head(Y) \notin D(\theta)$ and $h \not\equiv head(Y)$, then j-PROJECT applies for some j, and if $head(Y) \in D(\theta)$, then GUESS applies. Taking $\theta' = \theta \cup \rho$, we have that $\theta \equiv \theta'[FV(\Gamma)]$, $\theta \in U_\Sigma(\Gamma')$ since $\theta \in U_\Sigma(\Gamma)$ and ρ is in long β-normal form, and $D(\rho)$ is exactly the set of fresh variables in G. Moreover, $\mu_1(\Gamma', \theta') < \mu_1(\Gamma, \theta)$: x is removed from the set of unsolved variables in Γ which appear in $D(\theta)$, and is replaced by the set of fresh variables of G, but for each such variable z, $\theta'z \equiv \rho z$ is smaller than θx. Thus $\mu(\Gamma', \theta') < \mu(\Gamma, \theta)$.

Observe that if $head(X) \equiv head(Y) \notin D(\theta)$ does not hold, but $X \xrightarrow{\eta} x \in Vars_A$, x is not free in Y, and $\Sigma \vdash Y : A$, then ELIMINATE applies. In this case, we can take θ' to be θ by noting that $\mu_1(\Gamma', \theta) < \mu_1(\Gamma, \theta)$. □

The proof of Lemma 3.12 shows that it is possible to restrict DECOMPOSE to apply only when $head(X) \equiv head(Y) \notin D(\theta)$, although there is no way of encoding this restriction into the transformations. If we call a transformation prescribed by Lemma 3.12 a μ-*prescribed* transformation, then each application of a μ-prescribed transformation decreases the well-founded measure μ. The previous lemma guarantees that if Γ is a Σ-unifiable system in long β-normal form to which no μ-prescribed transformation in $\Sigma\mathcal{T}$ applies, then Γ is solved.

Theorem 3.13 *Let θ be a Σ-unifier of Γ. Then there exists a computation of Algorithm $\Sigma\mathcal{U}$ on Γ producing a Σ-unifier σ of Γ such that $\sigma \leq_{\beta\eta} \theta[FV(\Gamma)]$.*

Proof. Since every Σ-unifier of Γ is pointwise $\beta\eta$-equal on $FV(\Gamma)$ to some $\theta' \in U_\Sigma(\Gamma)$, we prove the theorem under the added hypothesis that $\theta \in U_\Sigma(\Gamma)$.

If Γ is not in long β-normal form, then perform reductions until a system in long β-normal form results. Note that if θ Σ-unifies Γ, then θ also Σ-unifies $l\beta nf(\Gamma)$, and that this reduction is a $\Sigma\mathcal{U}$ step. We may therefore assume without loss of generality in the remainder of this proof that Γ is in long β-normal form. We induct on the length of the longest sequence of μ-prescribed sequence of transformations available out of Γ.

If no μ-prescribed transformation from $\Sigma\mathcal{T}$ applies to Γ, then Γ is solved so we may return a most general Σ-unifier σ of Γ whose existence is guaranteed by Lemma 3.1. This action is a step of Algorithm $\Sigma\mathcal{U}$, and $\sigma \leq_{\beta\eta} \theta$. If some μ-prescribed transformation from $\Sigma\mathcal{T}$ applies to Γ yielding a system Γ' and a substitution θ' satisfying the conclusion of Lemma 3.12, then applying this transformation is a $\Sigma\mathcal{U}$ step. By the induction hypothesis, there is a computation of $\Sigma\mathcal{U}$ on Γ' producing a Σ-unifier δ of Γ' such that $\delta \leq_{\beta\eta} \theta'[FV(\Gamma')]$. It follows from Lemma 3.8 that δ is a Σ-unifier of Γ, and since $FV(\Gamma) \subseteq FV(\Gamma')$, $\delta \leq_{\beta\eta} \theta'[FV(\Gamma)]$. But $\theta' \equiv \theta[FV(\Gamma)]$, so that $\delta \leq_{\beta\eta} \theta[FV(\Gamma)]$. □

Since we have not made any assumption about the order in which transformations from $\mathcal{ET}$ are performed, and since any application of ELIMINATE to a system reduces the measure μ, we infer that the strategy of eager variable elimination is complete for unification in our calculus. It is unknown whether eager variable elimination is complete for an arbitrary calculus and equational theory, even if both are first-order.

References

[BL90] K. B. Bruce and G. Longo. A Modest Model of Records, Inheritance, and Bounded Quantification. *Information and Computation* 87, pp. 196 – 240, 1990.

[Car88] L. Cardelli. A Semantics of Multiple Inheritance. *Information and Computation* 76, pp. 138 – 164, 1988.

[CG91] P.-L. Curien and G. Ghelli. Subtyping + Extensionality: Confluence of $\beta\eta$top Reduction in $F_\leq$. In *Proc. TACS '91*, Springer-Verlag LNCS 526, pp. 731 – 749, 1991.

[Coh89] A. G. Cohn. Taxonomic Reasoning with Many-sorted Logics. *Artificial Intelligence Review* 3, pp. 89 – 128, 1989.

[Hue72] G. Huet. Constrained Resolution: A Complete Method for Higher Order Logic. Dissertation, Case Western Reserve University, 1972.

[Hue75] G. Huet. A Unification Algorithm for Typed λ-Calculus. *Theoretical Computer Science* 1, pp. 27 – 57, 1975.

[JK93] P. Johann and M. Kohlhase. Unification in an Extensional Lambda Calculus with Ordered Function Sorts and Constant Overloading. Technical Report SR-93-14, Universität des Saarlandes, 1993.

[Koh92] M. Kohlhase. An Order-sorted Version of Type Theory. In *Proc. LPAR '92*, Springer-Verlag LNAI 624, pp. 421 – 432, 1992.

[NQ92] T. Nipkow and Z. Qian. Reduction and Unification in Lambda Calculi with Subtypes. In *Proc. CADE '92*, Springer-Verlag LNAI 607, pp. 66 – 78, 1992.

[Pfe92] F. Pfenning. Intersection Types for a Logical Framework. POP-Report, Carnegie-Mellon University, 1992.

[Pie91] B. C. Pierce. Programming with Intersection Types and Bounded Polymorphism. Dissertation, Carnegie Mellon University, 1991.

[Qia90] Z. Qian. Higher-order Order-sorted Algebras. In *Proc. Algebraic & Logic Programming '90*, Springer-Verlag LNCS 463, pp. 86 – 100, 1990.

[Sch89] M. Schmidt-Schauß. Computational Aspects of an Order-sorted Logic with Term Declarations. Springer-Verlag LNAI 395, 1989.

[Sny91] W. Snyder. A Proof Theory for General Unification. Birkhäuser Boston, 1991.

[Wal88] C. Walther. Many-sorted Unification. *Journal of the ACM* 35, pp. 1 – 17, 1988.

Decidable Higher-Order Unification Problems

Christian Prehofer*

Technische Universität München**

Abstract. Second-order unification is undecidable in general. Miller showed that unification of so-called higher-order patterns is decidable and unitary. We show that the unification of a linear higher-order pattern s with an arbitrary second-order term that shares no variables with s is decidable and finitary. A few extensions of this unification problem are still decidable: unifying two second-order terms, where one term is linear, is undecidable if the terms contain bound variables but decidable if they don't.

1 Introduction

Higher-order unification is currently used in theorem provers such as Isabelle [25], TPS [1], Nuprl[1] [4] and others. The success of λ-Prolog [22] has shown the utility of higher-order constructs for programming. Other applications of higher-order unification include program synthesis [13] and machine learning [14, 6]. In this paper we consider the unification of a linear λ-term with an arbitrary second-order λ-term and develop several classes where this unification problem is decidable.

We start with an overview of the existing decidability results for higher-order unification problems in Figure 1. The column labeled Monadic refers to the unification of terms with unary function symbols only. A simply typed λ-term is a higher-order pattern, if all its free variables only have distinct bound variables as arguments. Dale Miller, as indicated in the column labeled Patterns, recently showed that unification of higher-order patterns is decidable and unitary.[2]

Section 3 reviews a set of transformation rules for full higher-order unification. Then we show in Section 4 that unification of linear higher-order patterns with an arbitrary second-order term is decidable and finitary, if the two terms share no variables. In particular, we do not have to resort to pre-unification, as equations with variables as outermost symbols on both sides (flex-flex) pairs can be finitely solved in this case. Further extensions are discussed in Section 5. The most general extension, unifying two second-order terms where one term is linear, is undecidable if the terms contain bound variables and decidable otherwise.

---

* Research supported by the DFG under grant Br 887/4-2, *Deduktive Programmentwicklung* and by ESPRIT WG 6028, *CCL*.
** Full Address: Fakultät für Informatik, Technische Universität München, 80290 München, Germany. E-mail: prehofer@informatik.tu-muenchen.de
[1] Nuprl uses only second-order pattern matching.
[2] We will adopt a relaxed notion of patterns, where unification is only finitary.

Order	Unification Problem			
	Unification	Patterns	Monadic	Matching
1	decidable			
2	undecidable		decidable	decidable
	Goldfarb '81 [12] Farmer '91 [10]	⋮	Farmer '88 [9]	Huet '73 [17, 16]
3	undecidable		undecidable	decidable
	Huet '73 [17, 16] Lucchesi '72 [20]	⋮	Narendran '90 [23]	G. Dowek '92 [7]
∞	⋮	decidable D. Miller '91[21]	⋮	? Wolfram '92 [31]

Fig. 1. Decidability of Higher-Order Unification

2 Notation and Basic Definitions

The following notation of λ-calculus are used in the sequel. For the standard theory of λ-calculus we refer to [15, 2]. We assume the following variable conventions:

- F, G, H, P, X, Y free variables,
- a, b, c, f, g (function) constants,
- x, y, z bound variables,
- α, β types.

The following grammar defines the syntax for λ**-terms**,

$$t \;=\; F \;\mid\; x \;\mid\; c \;\mid\; \lambda x.t \;\mid\; (t_1\; t_2)$$

A list of syntactic objects $s_1, \ldots, s_n$ where $n \geq 0$ is abbreviated by $\overline{s_n}$. We will use n-fold abstraction and application, i.e.

$$\lambda \overline{x_m}.f(\overline{s_n}) = \lambda x_1 \ldots \lambda x_m.((\cdots(f\; s_1)\cdots)\; s_n)$$

Substitutions are finite mappings from variables to terms and are denoted by $\{\overline{X_n \mapsto t_n}\}$. We assume the following standard conversions in λ-calculus:

α**-conversion:** $\lambda x.t =_\alpha \lambda y.(\{x \mapsto y\}t)$
β**-conversion:** $(\lambda x.s)t =_\beta \{x \mapsto t\}s$
η**-conversion:** if $x \notin \mathcal{FV}(t)$, then $\lambda x.(tx) =_\eta t$

A term in β-normal form is in long $\beta\eta$-normal form if it is η-expanded [30]. For our proofs we assume that terms are in long $\beta\eta$-normal form, for brevity we sometimes use η-normal form, which is denoted by $t\downarrow_\eta$. We assume that this transformation into long $\beta\eta$-normal form is an implicit operation, e.g. occurs when applying a substitution to a term. The **head** of a term $\lambda \overline{x_k}.v(\overline{t_n})$ is defined as $Head(\lambda \overline{x_k}.v(\overline{t_n})) = v$. **Free** and **bound variables** of a term t will be denoted as $\mathcal{FV}(t)$ and $\mathcal{BV}(t)$, respectively. We describe the subterm at a **position** p in a λ-term t by $t|_p$. A (sub-)term $t|_p$ is **ground**, if no free variables of t occur in $t|_p$. A variable is **isolated** if it occurs only once (in a term or in a system of equations). A term is **linear** if no free variable occurs repeatedly.

The set of **types** $\mathcal{T}$ for the simply typed λ-terms is generated by a set $\mathcal{T}_0$ of base types (e.g. int, bool) and the function type constructor $\rightarrow$. Notice that $\rightarrow$ is right associative, i.e. $\alpha \rightarrow \beta \rightarrow \gamma = \alpha \rightarrow (\beta \rightarrow \gamma)$. The **order** of a type $\varphi = \alpha_1 \rightarrow \ldots \rightarrow \alpha_n \rightarrow \beta, \ \beta \in \mathcal{T}_0$ is defined as

$$Ord(\varphi) = \begin{cases} 1 & \text{if } n = 0, \text{i.e. } \varphi = \beta \in \mathcal{T}_0 \\ 1 + max(Ord(\alpha_1), \ldots, Ord(\alpha_n)) & \text{otherwise} \end{cases}$$

A **language of order** n is restricted to function constants of order $\leq n+1$ and variables of order $\leq n$.

Definition 1. A simply typed λ-term s is a **relaxed higher-order pattern**, if all free variables in s only have bound variables as arguments, i.e. if $X(\overline{t_n})$ is a subterm of s, then all $t_i\downarrow_\eta$ are bound variables.

Examples are $\lambda x, y.F(x, y)$ and $\lambda x.f(G(\lambda z.x(z)))$, where the latter is at least third-order. Non-patterns are $\lambda x, y.F(a, y), \lambda x.G(H(x))$.

In most of the existing literature [21, 24], patterns are required to have distinct bound variables as arguments to a free variable. This restriction is necessary for unitary unification, but for our purpose this is not relevant and we will henceforth work with relaxed higher-order patterns and call these patterns for brevity.

We identify α-equivalent terms and assume that free and bound variables are kept disjoint [2]. Furthermore, we assume that bound variables with different binders have different names.

3 Pre-unification by Transformations

We present in the following a version of the transformation system $\mathcal{PT}$ for higher-order unification of Snyder and Gallier [30]. More precisely, we use the primed transformations for pre-unification of Section 5 in [30], where also the omitted type constraints can be found. These transformation rules in Figure 2 work on sets of pairs of terms to be unified, written as $\{u = v, \ldots\}$. Pre-unification differs from unification by the handling of so-called flex-flex pairs. These are equations of the form $\lambda \overline{x_k}.P(\ldots) = \lambda \overline{x_k}.P'(\ldots)$, which permit in general an infinite number of incomparable unifiers but are guaranteed to have at least one unifier, e.g. $\{P \mapsto \lambda \overline{x_m}.a, P' \mapsto \lambda \overline{x_n}.a\}$. The idea of pre-unification goes back to Huet [17]. It means to handle flex-flex pairs as constraints and not to attempt to solve them explicitly.

The only place where the restriction to second-order terms simplifies the system is the last rule, projection, where x_i must be a first-order object. Hence the binding to F in this case is of the simpler form $F = \lambda \overline{x_n}.x_i$, which will be important for our results.

We have restricted the application of rule (3) slightly compared to [30]: the rule of Snyder et al. can also be applied to flex-flex equations. We have excluded

> **(1) Delete**
>
> $$\{t = t\} \cup S \Rightarrow S$$
>
> **(2) Decompose**
>
> $$\{\lambda\overline{x_k}.f(\overline{t_n}) = \lambda\overline{x_k}.f(\overline{t'_n})\} \cup S \Rightarrow \bigcup_{i=1,\ldots,n}\{\lambda\overline{x_k}.t_i = \lambda\overline{x_k}.t'_i\} \cup S$$
>
> **(3) Eliminate**
>
> $$\{F = \lambda\overline{x_k}.t\} \cup S \Rightarrow \{F = \lambda\overline{x_k}.t\} \cup \{F \mapsto \lambda\overline{x_k}.t\}S$$
> $$\text{if } F \notin \mathcal{FV}(\lambda\overline{x_k}.t) \text{ and}$$
> $$Head(t) \text{ is not a free variable}$$
>
> **(4'a) Imitate**
>
> $$\{\lambda\overline{x_k}.F(\overline{t_n}) = \lambda\overline{x_k}.f(\overline{t'_m})\} \cup S \Rightarrow \{F = \lambda\overline{x_n}.f(\overline{H_m(\overline{x_n})})\} \cup$$
> $$\{\lambda\overline{x_k}.F(\overline{t_n}) = \lambda\overline{x_k}.f(\overline{t'_m})\} \cup S$$
>
> **(4'b) Project**
>
> $$\{\lambda\overline{x_k}.F(\overline{t_n}) = \lambda\overline{x_k}.v(\overline{t'_m})\} \cup S \Rightarrow \{F = \lambda\overline{x_n}.x_i(\overline{H_m(\overline{x_n})})\} \cup$$
> $$\{\lambda\overline{x_k}.F(\overline{t_n}) = \lambda\overline{x_k}.v(\overline{t'_m})\} \cup S$$
> where v is a constant or bound variable

Fig. 2. System $\mathcal{PT}$ for Higher-order Pre-unification

this case as it is not necessary for completeness (the same is done in the algorithm presented by Snyder et al.).

Theorem 2 (Snyder-Gallier). *System $\mathcal{PT}$ is a sound and complete transformation system for higher-order pre-unification.*

When applying the rules of system $\mathcal{PT}$ to a set of equations, the completeness does not depend on how the equations are selected. The only branching occurs when both immitation and projection apply to some equation. This was shown by Huet [17].

Example 1. Consider the unification problem at the root of the search tree in Figure 3, which is obtained by the transformations $\mathcal{PT}$ in Figure 2. Notice that in this example all projection substitutions are of the form $\lambda x.x$. The failure cases are caused by a clash of distinct symbols and are abbreviated. Putting the substitutions of the only successful path together gives the only solution $\{F \mapsto \lambda x.g(x,x), G_1 \mapsto G, G_2 \mapsto G\}$.

4 Unifying Linear Patterns with Second-Order Terms

In this section we show that unification of second-order λ-terms with linear patterns is decidable and finitary. Let us first use system $\mathcal{PT}$ to solve the pre-unification problem.

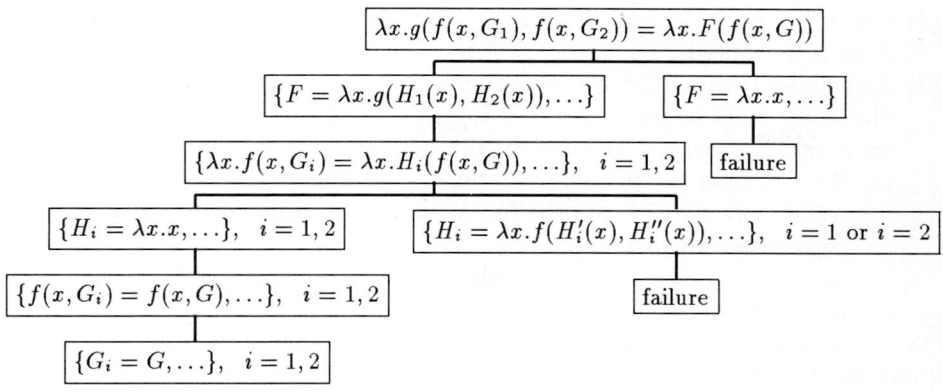

Fig. 3. Search Tree with System $\mathcal{PT}$

Lemma 3. *System $\mathcal{PT}$ terminates for two variable-disjoint terms $s = t$ if s is a linear pattern and t is second-order. Furthermore, $\mathcal{PT}$ terminates with a set of flex-flex pairs of the form $\lambda \overline{x_k}.P(\overline{y_i}) = \lambda \overline{x_k}.P'(\overline{u_i})$ where all y_i are bound variables and P is isolated.*

Proof We show that system $\mathcal{PT}$ terminates for this unification problem. We start with the equation $s = t$ and apply the transformations modulo commutativity of $=$ in Figure 2. By this we achieve that after any sequence of transformations, all free variables on the left hand sides (lhs) are isolated in the system of equations, as all newly introduced variables on the lhs are linear also. The latter can be easily seen be examining the cases $(4'a)$ or $(4'b)$, the other rules are trivial. Another important invariant is that the lhs's remain patterns, which is easy to verify.

In addition, we ignore "solved" equations of the form $F = t$ or $t = F$ which are created by transformation (3). This is necessary for the termination ordering. Since transformations $(1), (2)$ and (3) preserve the set of solutions, as shown in [30], we can assume that variable elimination (3) is applied eagerly; in particular, after a transformation $(4'a)$ or $(4'b)$, we assume that (3) is applied (with implicit β-normalization). In addition, we assume that transformation (2) is applied after $(4'a)$ and after applying $(4'b)$ to a lhs.

We use the following lexicographic termination ordering on the multiset of equations (ignoring all solved equations):

A: Compare the number of constant symbols on all lhs's, if equal
B: compare the number of occurrences of bound variables on all lhs's that are not below a free variable, if equal
C: compare the multiset of the sizes of the right-hand sides (rhs).

Now we show that the transformations reduce the above ordering:

(1) trivial
(2) A or B is reduced.
(3) Although (3) eliminates one equation, it is not trivial that it also reduces the above ordering. In particular, we do not apply (3) to flex-flex pairs, which could increase the size of some rhs if a bound variable occurs repeatedly on the lhs. Consider the possible equations (3) is applied to:
  - $\lambda \overline{x_k}.F(\overline{x_k}) = \lambda \overline{x_k}.t$: As the free variable F is isolated, A and B remain constant and C is reduced.
  - $\lambda \overline{x_k}.a(\ldots) = \lambda \overline{x_k}.F(\overline{x_k})$: The elimination of an equation with a constant a reduces A.
  - $\lambda \overline{x_k}.x_i(\ldots) = \lambda \overline{x_k}.F(\overline{x_k})$: Here B is reduced (and possibly A).

(4'a) We have two cases:
  - $\lambda \overline{x_k}.F(\overline{y_n}) = \lambda \overline{x_k}.f(\overline{t_m})$: The imitation binding for F is of the form $F = \lambda \overline{x_n}.f(\overline{H_m(\overline{x_n})})$. Now, after applying (3) and (2), we replace the above equation by a set of equations of the form $\lambda \overline{x_j}.H_i(\overline{y_n}) = \lambda \overline{x_j}.t_i$, where $i = 1, \ldots, m$. Notice that the number of constants on the lhs (A) does not increase, as all y_m are bound variables. Also, B remains unchanged. As F is isolated and hence does not occur on any right hand side, C decreases after transformation (2) is applied.
  - $\lambda \overline{x_k}.f(\overline{t_n}) = \lambda \overline{x_k}.F(\overline{u_m})$: We obtain an imitation binding as above and can apply (3) and (2). Then the number of constant symbols on the lhs's decreases, since F may not occur on the lhs's.

(4'b) We again have two cases:
  - $\lambda \overline{x_k}.F(\overline{y_n}) = \lambda \overline{x_k}.y_i(\overline{t_m})$: As $\overline{y_n}$ are bound variables, this rule applies only if the head of the rhs is a bound variables as well, say y_i. Then the case is similar to the Imitation case above, as after (4'b), transformations (3) and (2) apply.
  - $\lambda \overline{x_k}.v(\overline{t_n}) = \lambda \overline{x_k}.F(\overline{u_m})$: As we have second-order variables on the rhs, we only have projection bindings of the form $F = \lambda \overline{x_k}.x_i$. Hence the lhs's (i.e. A and B) are unchanged, whereas C decreases, as we assume terms in long $\beta\eta$-normal form.

□

So far, we have shown that pre-unification is decidable. To solve the remaining flex-flex pairs, notice that all of these are of the form

$$\lambda \overline{x_k}.P(\overline{y_m}) = \lambda \overline{x_k}.P'(\overline{u_n}),$$

where P is isolated and $\{\overline{y_m}\}$ are bound variables. Now $\lambda \overline{x_k}.P'(\overline{u_n})$ is almost an instance of the lhs, we only have to eliminate all occurrences of bound variables that are not in $\{\overline{y_m}\}$.

Example 2. Consider the pair $\lambda x, y.F(x) = \lambda x, y.F'(F''(x), F''(y))$. There are two ways to eliminate y on the rhs, i.e. $\theta_1 = \{F' \mapsto \lambda z_1, z_2.F'_1(z_1)\}$ and $\theta_2 = \{F'' \mapsto \lambda z_1.F''_1\}$, where F'_1 and F''_1 are new variables.

Eliminate
$(\theta, [\lambda\overline{x_k}.P(\overline{t_n})|R], W) \Rightarrow_{el} (\tau_{P,i}\theta, \tau_{P,i}[\lambda\overline{x_k}.P(\overline{t_n})|R], W)$
  if $\exists x \in W \cap \mathcal{BV}(\lambda\overline{x_k}.t_i)$

Proceed
$(\theta, [\lambda\overline{x_k}.v(\overline{t_n})|R], W) \Rightarrow_{el} (\theta, [\overline{\lambda\overline{x_k}.t_n}|R], W)$
  unless v is a bound variable in W

Fig. 4. System $\mathcal{EL}$ for Eliminating Bound Variables

We first define some notation to formalize this idea. We use square brackets to denote lists, i.e. appending a list R to an element t is written as $[t|R]$. The **application of a substitution to a list**, written as $\theta[\overline{t_n}]$ is defined as $[\overline{\theta t_n}]$. For a variable F of type $\overline{\alpha_n} \to \alpha_0$ we define the **i-th parameter eliminating substitution** $\tau_{F,i}$ as

$$\tau_{F,i} = \{F \mapsto \lambda\overline{x_n}.F'(x_1, \ldots, x_{i-1}, x_{i+1}, \ldots, x_n)\},$$

where F' is a new variable of appropriate type.

The transformation rules $\Rightarrow_{el}$ in Figure 4 transform triples of the form (θ, l, W), where θ is the computed substitution, l is the list of remaining terms, and W is the set of bound variables to be eliminated.

We say system $\mathcal{EL}$ succeeds if it reduces a triple to $(\theta, [], W)$. For the flex-flex pair in Example 2 system $\mathcal{EL}$ works as follows, starting with the triple $(\{\}, [\lambda x, y.F'(F''(x), F''(y))], \{y\})$. Then $\mathcal{EL}$ can either eliminate the second argument of F' or it can proceed until the triple $(\{\}, [\lambda x, y.F''(y)], \{y\})$ is reached and then eliminate y. In these two cases, $\mathcal{EL}$ succeeds with θ_1 and θ_2, respectively, as in Example 2. All other cases fail.

Observe that system $\mathcal{EL}$ is not optimal, as it can produce the same solution twice. For instance, consider the pair $\lambda x.F = \lambda x.F'(F'(x))$. There are two different transformation sequences that yield the unifier with $\{F' \mapsto \lambda s.F''\}$. More precisely, this happens only if a bound variable occurs below nested occurrences of a variable at subtrees with the same index. Let us first show the correctness of $\mathcal{EL}$.

Lemma 4 Correctness of $\mathcal{EL}$. Let $\lambda\overline{x_k}.P(\overline{y_m}) = \lambda\overline{x_k}.P'(\overline{u_n})$ be a flex-flex pair where $\{\overline{y_m}\} \subseteq \{\overline{x_k}\}$ and P does not occur in $P'(\overline{u_n})$. Assume further $W = \{\overline{x_k}\} - \{\overline{y_m}\}$. If $(\{\}, [P'(\overline{u_n})], W) \Rightarrow^*_{el} (\theta, [], W)$ then $\theta \cup \{P \mapsto \theta\lambda\overline{y_m}.P'(\overline{u_n})\}$ is a unifier of $\lambda\overline{x_k}.P(\overline{y_m}) = \lambda\overline{x_k}.P'(\overline{u_n})$.

Proof We show that $\{P \mapsto \theta\lambda\overline{y_m}.P'(\overline{u_n})\}$ is a well defined substitution, i.e. all bound variables in $\theta P'(\overline{u_n})$ are locally bound or are in $\overline{y_m}$. As any successful sequence of $\mathcal{EL}$ reductions must traverse the whole term $\lambda\overline{x_k}.P'(\overline{u_n})$ to succeed, only bound variables in $\{\overline{y_m}\}$ can remain; occurrences of $\{\overline{x_k}\} - \{\overline{y_m}\}$ are either eliminated by some substitution $\tau_{P,i}$ in rule Eliminate, or the algorithm fails as the rule Proceed does not permit these bound variables. □

The next lemma states that if θ eliminates all occurrences of variables in W from $\overline{t_n}$, then there is a sequence of $\mathcal{EL}$ reductions that approximates θ.

Lemma 5. *If* $\tau\overline{[t_n]} = \overline{[t_n]}$, $\mathcal{BV}(\overline{\theta t_n}) \cap W = \emptyset$, $\theta = \delta\tau$ *for some substitution* δ, *and* $\overline{t_n}$ *are second-order terms, then there exist a reduction* $(\tau, \overline{[t_n]}, W) \Rightarrow^*_{el} (\theta', [], W)$ *and a substitution* δ' *such that* $\theta = \delta'\theta'$.

Proof by induction on the sum of the sizes of the terms in $\overline{[t_n]}$. Clearly, each $\Rightarrow_{el}$ reduction reduces this sum. The base case, where $n = 0$, is trivial. We show that for each such problem some $\mathcal{EL}$ step applies and that the induction hypothesis can be applied. Depending on the form of t_1 and the conditions of the rules of $\mathcal{EL}$, we apply different rules. Assume t_1 is of the form $\lambda\overline{x_k}.P(\overline{u_m})$ and $\theta P = \lambda\overline{y_m}.t$. By our variable conventions, we can assume that $W \cap \mathcal{BV}(\theta P) = \emptyset$. As $\lambda\overline{x_k}.P(\overline{u_m})$ is a second-order term, some bound variable from W appears in $\theta\lambda\overline{x_k}.P(\overline{u_m})$ if and only if it appears in some $\theta\lambda\overline{x_k}.u_i$ where $y_i \in \mathcal{BV}(\lambda\overline{y_m}.t) = \mathcal{BV}(\theta P)$. Then let

$$i = Min\{j \mid \exists x \in \mathcal{BV}(\theta\lambda\overline{x_k}.u_j) \cap W\}.$$

Thus this set describes the indices of bound variables that may not occur in $\theta P = \lambda\overline{y_m}.t$ by assumption on θ, e.g. $y_i \notin \mathcal{BV}(\lambda\overline{y_m}.t)$. If the above set is empty and no j exists, we apply the second rule and can then safely apply the induction hypothesis.

In case i exists, we know that $\mathcal{BV}(\theta\overline{\lambda\overline{x_k}.u_n}) \cap W \subseteq \mathcal{BV}(\overline{\lambda\overline{x_k}.u_n}) \cap W$. Hence the Eliminate rule applies with $\tau_{P,i} = \{P \mapsto \lambda\overline{x_m}.P_0(x_1, \ldots, x_{i-1}, x_{i+1}, \ldots, x_m)\}$. Then we can apply the induction hypothesis to $(\tau_{P,i}\tau, \tau_{P,i}\overline{[t_n]}, W)$: define δ' such that $\delta'X = \delta X$ if $X \neq P$ and $\delta'P_0 = \lambda y_1, \ldots, y_{i-1}, y_{i+1}, \ldots, y_m.t$. Notice that δ' is well formed, as $y_i \notin \mathcal{BV}(\lambda\overline{y_m}.t)$. Clearly the premises for the induction hypothesis are fulfilled, as $\theta = \delta'\tau_{P,i}\tau$ follows from $\tau P = P$. Then the induction hypothesis assures that both $\mathcal{EL}$ succeeds with a substitution θ' and that a substitution δ'' exists such that $\theta = \delta''\theta'$.

The remaining cases of t_1 are trivial as the Proceed rule does not compute substitutions. $\square$

Now we can show that $\mathcal{EL}$ captures all unifiers.

Lemma 6 (Completeness of $\mathcal{EL}$). *Assume* θ *is a unifier of a flex-flex pair of the form* $\lambda\overline{x_k}.P(\overline{y_m}) = \lambda\overline{x_k}.P'(\overline{u_n})$, *where* $\{\overline{y_m}\} \subseteq \{\overline{x_k}\}$ *and all* y_m *are distinct. Assume further* $\lambda\overline{x_k}.P'(\overline{u_n})$ *is second-order and does not contain* P. *Let* $W = \{\overline{x_k}\} - \{\overline{y_m}\}$. *Then there exist a substitution* $\theta'' = \theta' \cup \{P \mapsto \theta'\lambda\overline{y_m}.P'(\overline{u_n})\}$ *and a reduction* $(\{\}, [\lambda\overline{x_k}.P'(\overline{u_n})], W) \Rightarrow^*_{el} (\theta', [], W)$ *such that* θ'' *is more general than* θ.

Proof It is clear that any unifier must eliminate all bound variables from W on the right hand side. Then the proof follows easily from Lemma 5. $\square$

To state the above lemma in a simple form, we did not allow repeated bound variables on the left hand side. In the next lemma we extend this result to repeated variables, which causes some technical overhead. Repeated variables may cause an additional number of distinct unifiers in each case, as there can be different permutations if a repeated variable occurs in the common instance. Consider for example the pair $\lambda x.F(x,x) = \lambda x.F'(x)$. There are the two solutions $\{F \mapsto \lambda y, z.F'(y)\}$ and $\{F \mapsto \lambda y, z.F'(z)\}$.

Theorem 7. *Assume t is a second-order λ-term and s is a linear pattern such that s shares no variables with t. Then the unification problem $s = t$ is decidable and finitary.*

Proof We first extend Lemma 6 to the case of repeated bound variables. Formally, consider the pair $\lambda \overline{x_k}.P(\overline{y_m}) = \lambda \overline{x_k}.v$ and assume some bound variables occur several times in $P(\overline{y_m})$. Assume $\mathcal{EL}$ succeeds with $(\theta, [], \{\overline{x_k} - \overline{y_m}\})$. Let $p(i,j)$ be the position of the j-th occurrence of x_i in θv. For this solution of $\mathcal{EL}$, all solutions for P are of the form $\{P \mapsto \lambda \overline{z_m}.v'\}$, where $Head(v'|_{p(i,j)}) = z_i$ and $y_i = x_i$ for all positions $p(i,j)$ of some x_i in θv and $Head(v'|_q) = Head(\theta v|_q)$ otherwise. Here the last equation allows for many permutations, as some x_j may occur repeatedly in $\overline{y_m}$. All these permutations are clearly independent from the remaining parts of the computed unifier, as P does not occur elsewhere, and can hence be easily computed.

From Lemma 3 we know that $\mathcal{PT}$ terminates with a set of flex-flex pairs, where the lhs is a pattern. Then by the extended Lemma 6 we can use $\mathcal{EL}$ to compute a complete and finite set of unifiers for some flex-flex pair, as $\mathcal{EL}$ terminates and is finitely branching. This unifier is applied to the remaining equations. Repeat this for all flex-flex pairs. This procedure terminates and works correctly as all lhs's are patterns and only have isolated variables. Notice that a flex-flex pair remains flex-flex when applying a unifier computed by $\mathcal{EL}$. □

It can be shown that $\mathcal{EL}$ computes at most a quadratic number of different substitutions. Let n be the number of occurrences of variables to be eliminated and let m be the maximal number of nested free variables. Then there can be at most m distinct ways to eliminate some particular variable. As m and n are both linear in the size, the maximal number is solutions, i.e. mn, is quadratic.

However, repeated bound variables on the lhs may cause an exponential number of different solutions. Consider for instance $\lambda x.F(x,x) = \lambda x.v$, where x occurs in v exactly n times. Then there are 2^n different solutions.

It would be interesting to examine whether the computed set of unifiers is minimal, in particular for $\mathcal{EL}$. However, the most concise representation of all unifiers is still a flex-flex pair. Which representation is best clearly depends on the application. For instance, flex-flex pairs may not be satisfactory for programming languages where explicit solutions are desired. For automated theorem proving, flex-flex pairs can be advantageous as they can reduce the search space in some cases.

Observe that $\mathcal{EL}$ is not complete for the third-order case. Here, if a free variable has two arguments, one can be a function. If in some solution this function is applied to the other argument, then this function could eliminate, in the above sense, the other argument. For instance, consider the third-order pair $\lambda x, y.F(x) = \lambda x, y.F'(\lambda z.F''(z), y)$. Here $\mathcal{EL}$ would not uncover the solution $\{F' \mapsto \lambda y, z.F'_0(y(z)), F'' \mapsto \lambda x.a, F \mapsto \lambda y, z.F'_0(a)\}$.

5 Extensions

In the following sections, we will examine extensions of the above decidability result. First, notice that the linearity restriction is essential; otherwise full second-order unification can easily be embedded. But even with one linear term, this embedding still works:

Example 3. Consider the unification problem

$$\lambda x.F(f(x,G)) = \lambda x.g(f(x,t_1), f(x,t_2)),$$

where t_1 and t_2 are arbitrary second-order terms. By applying the transformations $\mathcal{PT}$ it is easy to see (compare to Example 1) that in all solutions of the above problem $F = \lambda x.g(x,x)$ and $t_1 = t_2$ must be solved, which is clearly undecidable.

Motivated by this example, we consider the following two extensions. First, we assume that arguments of free variables are either bound variables or second-order ground terms. Secondly, we consider the case where an argument of a free variable contains no bound variables. These two cases can be combined in a straightforward way, as shown in Section 5.3. Thus arguments of free variables may either be ground second-order terms or terms with no bound variables. The general case where one term is linear follows easily from Example 3:

Corollary 8. *It is undecidable to determine if two second-order terms unify, even if one is linear.*

Pre-unification of two linear second-order terms is however decidable and finitary, as shown by Dowek [8].

5.1 Ground Second-Order Arguments to Free Variables

We now loosen the restriction that one term must be a linear pattern. As long as all arguments of free variables are either bound variables or ground second-order terms, we can still solve the pre-unification problem. In particular, for the second-order case, this can be rephrased as disallowing nested free variables. However, we only solve the pre-unification problem, as the resulting flex-flex pairs are more intricate than in the last section.

Similar to the above, we present a termination ordering for a particular strategy of the $\mathcal{PT}$ transformations. We will see that in essence only one new case results from these ground second-order terms. This can be handled separately by second-order matching, which is decidable and finitary. (It is also an instance of Theorem 7.) That is, whenever such a matching problem occurs, this is solved immediately (considering all its solutions). Hence we first need a lemma about matching. A **substitution is ground** if it maps variables to ground terms only.

Lemma 9. *Solving a second-order matching problem with system $\mathcal{PT}$ yields only solutions that are ground substitutions.*

This result does not hold for the higher-order case, as noted by Dowek [8]: e.g. $\{F \mapsto \lambda x.x(Y)\}$ is a solution to $F(\lambda x.a) = a$, but no complete set of ground matchers exists. Now we can show the desired theorem:

Theorem 10. *Assume s,t are λ-terms such that t is second-order, s is linear and s shares no variables with t. Furthermore, all arguments of free variables in s are either*

- *bound variables of arbitrary type or*
- *second-order ground terms of base type.*

Then the pre-unification problem $s = t$ is decidable and finitary.

Proof We give a termination ordering for system $\mathcal{PT}$ with the same additional assumptions as in the proof of Lemma 3, i.e. eager application of rules (2) and (3). In addition, we consider solving a second-order matching problem an atomic operation, with possibly many solutions. In particular, after a projection on a lhs, this step eliminates one equation and applies a (ground) substitution to the rhs. It is easy to see that the two premises, only isolated variables and no nested free variables on the lhs's, are invariant under the transformations.

We use the following (lexicographic) termination ordering on the multiset of equations (ignoring all solved equations):

A: Compare the number of occurrences of constant symbols and of bound variables that are not below a free variable on a lhs, if equal
B: compare the number of free variables in all rhs's, if equal
C: compare the multiset of the sizes of the rhs's.

The remainder of the proof is similar to Lemma 3 and is left out for lack of space. □

It might seem tempting to apply the same technique to arguments that are third-order ground terms, as third-order matching is known to be decidable. However, there can be an infinite number of matchers and without a concise representation for these the extension of the above method seems difficult.

5.2 No Bound Variables in an Argument of a Free Variable

We say a bound variable y in $\lambda \overline{x_n}.t$ is **outside bound** if $y = x_i$ for some i. The set of all outside bound variables of a term $\lambda \overline{x_n}.t$ is written as $\mathcal{OBV}(\lambda \overline{x_n}.t) = \mathcal{BV}(\lambda \overline{x_n}.t) \cap \{\overline{x_n}\}$.

We show that the remaining case, where an argument of a free variable contains no (outside-)bound variables, can be reduced to a simpler case. This method checks unifiability, but does not give a complete set of unifiers. We use the standard notation of contexts as terms with holes, written as $C[t]$.

Theorem 11. *Assume $s = \lambda \overline{x_n}.C[H(t_1, \ldots, t_i, \ldots)]$ and t are variable-disjoint λ-terms such that s is linear. Assume further t_i contains free but no bound variables, i.e. $\mathcal{OBV}(\lambda \overline{x_n}, \overline{y_m}.t_i) = \emptyset$, where $\overline{y_m}$ are all bound variables on the path to the position of t_i in $\lambda \overline{x_n}.C[H(t_1, \ldots, t_i, \ldots)]$. Then the unification problem*

$s = t$ has a solution, iff $\lambda x_0, \overline{x_n}.C[H(t_1,\ldots,x_0,\ldots)] = \lambda x_0.t$, where x_0 does not occur elsewhere, is solvable.

Proof Consider the unification problem

$$\lambda \overline{x_n}.C[H(t_1,\ldots,t_i,\ldots)] = \lambda \overline{x_n}.u$$

where H occurs only once in $\lambda \overline{x_n}.C[H(t_1,\ldots,t_i,\ldots)]$ and t_i does not contain bound variables. Assume $\{X_1,\ldots,X_m\} = \mathcal{FV}(t_i)$. Let a solution to this problem be of the form $\{H \mapsto \lambda \overline{x_n}.t_0\} \cup \{\overline{X_m \mapsto u_m}\} \cup S$. As H does not occur elsewhere, we can construct a substitution $\theta = \{H \mapsto \lambda \overline{x_n}.\{x_i \mapsto t'\}t_0\} \cup S$, where $t' = \{\overline{X_m \mapsto u_m}\}t_i$, that is a solution to

$$\lambda x_0, \overline{x_n}.C[H(t_1,\ldots,x_0,\ldots)] = \lambda x_0, \overline{x_n}.u$$

Notice that θ is well-formed, as $\lambda \overline{x_n}, \overline{y_m}.t_i$ does not contain (outside) bound variables. The other direction is simple, since x_0 does not occur elsewhere, i.e. not in an instance of $\lambda x_0, \overline{x_n}.u$. □

Notice that the above procedure only helps deciding unification problems but does not imply that pre-unification or even unification is finitary.

5.3 Putting It All Together

Now we can combine the previous results. Recall that the remaining case is undecidable in general.

Theorem 12. *Assume s,t are λ-terms such that t is second-order, s is linear and s shares no variables with t. Furthermore, all arguments of a free variable F in s are either*

- *bound variables of arbitrary type or*
- *second-order ground terms of base type or*
- *second-order terms of base type without variables bound outside of F.*

Then the unification problem $s = t$ is decidable.

Proof First apply Theorem 11 to the unification problem until s has no nested free variables. This argument can be applied repeatedly, as the lhs is linear and hence the substitutions of multiple applications do not overlap. Then Theorem 10 can be applied to decide this problem. □

A special case often considered (e.g. [12]) is terms with second-order variables, but no bound variables. Then we get the following stronger result as an instance of Theorem 12:

Proposition 13. *Assume s,t are second-order λ-terms such that s is linear and shares no variables with t. Furthermore, s contains no bound variables. Then the unification problem $s = t$ is decidable.*

The above unification problems are at least NP-hard, as they subsume second-order matching, which is NP-complete [3].

6 Applications

As mentioned in the introduction, higher-order unification is currently used in several theorem provers, programming languages, and logical frameworks. With the above results we can now develop simplified and somewhat restricted versions of the above applications that enjoy decidable unification. It should be mentioned that several systems such as Elf [27] and Isabelle[3] have already resorted to higher-order patterns, where unification behaves much like the first-order case.

There is an interesting variety of applications where linearity is a common and sometimes also useful restriction. For instance, narrowing [18] is a general method to solve equations modulo a theory given by a term rewrite system. Then we can define a second-order version of narrowing with decidable unification as long as the left-hand sides of the used rules are linear patterns. This is in fact a common restriction for constructor-based narrowing [32] and for functional logic languages [19]. Using the results of this work, different versions of higher-order narrowing are developed in [28]. Usually, the lhs's of the rewrite rules in these applications are restricted and fulfill the requirement for linear patterns. For instance, we could use rules such as

$$map(F, cons(X, Y)) \longrightarrow cons(F(X), map(F, Y)).$$

Notice that systems which work only with higher-order patterns cannot express this rule, as the right-hand side is not a pattern. Then narrowing or rewriting with this rule may yield a non-pattern term and repeated narrowing needs higher-order unification with the linear left-hand side. So far, most functional logic languages even with higher order terms only use first-order unification. Interestingly, when coding functions such as map into predicates, as for instance done in higher-order logic programming [22], the head of the literal, e.g. $mapP(F, cons(X, Y), cons(F(X), L)) :- mapP(F, Y, L)$, is not linear. However, when invoking this rule only with goals of the form $mapP(t, t', Z)$, where Z is a fresh variable,[4] then the unification problem is decidable as it is equivalent to a unification with a linear term. Thus our results also explain to some extent why unification in higher-order logic programming rarely diverges.

Furthermore we open the way for finding decidable second-order matching problems w.r.t. higher-order equational theories. First results on second-order matching modulo first-order theories can be found in [5].

Higher-order theorem provers often work with some form of a sequent calculus, where most rules have linear premises and conclusions e.g.

$$\frac{\Gamma \vdash A \quad \Gamma \vdash B}{\Gamma \vdash A \& B}$$

Furthermore, non-linear unification problems occur mostly with rewriting, e.g. with rules such as $P \& P \longrightarrow P$. For rewriting, however, only matching is required.

---
[3] Isabelle still uses full higher-order pre-unification, if the terms are not patterns.
[4] Such variables are also called "output-variables" in [29]

Another application area is type inference, which is mostly based on unification, whereby decidable static type inference for programming languages is desired. In many advanced type systems such as Girard's system F [11] variables may range over functions from types to types, i.e. second-order type variables. In particular, Pfenning [26] relates type inference in the nth-order polymorphic λ-calculus with nth-order unification. Thus progress in higher-order unification may help finding classes where type inference is decidable.

Acknowledgements. The author wishes to thank Tobias Nipkow, Gilles Dowek, and the anonymous referees for valuable comments and discussions.

References

1. Peter B. Andrews, Sunil Issar, Dan Nesmith, and Frank Pfenning. The TPS theorem proving system. In M.E. Stickel, editor, *Proc. 10th Int. Conf. Automated Deduction*, pages 641–642. LNCS 449, 1990.
2. Hendrik Pieter Barendregt. *The Lambda Calculus, its Syntax and Semantics*. North Holland, 2nd edition, 1984.
3. L. D. Baxter. *The complexity of Unification*. PhD thesis, University of Waterloo, Waterloo, Canada, 1976.
4. Robert Constable, S. Allen, H. Bromly, W. Cleaveland, J. Cremer, R. Harper, D. Howe, T. Knoblock, N. Mendler, P. Panangaden, J. Sasaki, and S. Smith. *Implementing Mathematics With the Nuprl Proof Development System*. Prentice-Hall, New Jersey, 1986.
5. Régis Curien. Second-order E-matching as a tool for automated theorem proving. In *EPIA '93*. Springer LNCS 725, 1993.
6. Michael R. Donat and Lincoln A. Wallen. Learning and applying generalised solutions using higher order resolution. In E. Lusk and R. Overbeek, editors, *9th International Conference On Automated Deduction*, pages 41–61. Springer Verlag, 1988.
7. Gilles Dowek. Third order matching is decidable. In *Proceedings, Seventh Annual IEEE Symposium on Logic in Computer Science*, pages 2–10, Santa Cruz, California, 22–25 June 1992. IEEE Computer Society Press.
8. Gilles Dowek. Personal communication, 1993.
9. W. M. Farmer. A unification algorithm for second-order monadic terms. *Annals of Pure and Applied Logic*, 39:131–174, 1988.
10. W. M. Farmer. Simple second-order languages for which unification is undecideable. *Theoretical Computer Science*, 87:25–41, 1991.
11. J.-Y. Girard, Y. Lafont, and P. Taylor. *Proofs and Types*. Cambridge Tracts in Theoretical Computer Science 7. Cambridge University Press, 1989.
12. W. D. Goldfarb. The undecidability of the second-order unification problem. *Theoretical Computer Science*, 13:225–230, 1981.
13. Masami Hagiya. Synthesis of rewrite programs by higher-order semantic unification. In S. Arikawa, S. Goto, S. Ohsuga, and T. Yokomori, editors, *Algorithmic Learning Theory*, pages 396–410. Springer, 1990.
14. Masateru Harao. Analogical reasoning based on higher-order unification. In S. Arikawa, S. Goto, S. Ohsuga, and T. Yokomori, editors, *Algorithmic Learning Theory*, pages 151–163. Springer, 1990.
15. J.R. Hindley and Jonathan P. Seldin. *Introduction to Combinators and λ-Calculus*. Cambridge University Press, 1986.

16. Gérard Huet. A unification algorithm for typed λ-calculus. *Theoretical Computer Science*, 1:27–57, 1975.
17. Gérard Huet. *Résolution d'équations dans les languages d'ordre 1,2,...ω*. PhD thesis, University Paris-7, 1976.
18. Jean-Marie Hullot. Canonical forms and unification. In W. Bibel and R. Kowalski, editors, *Proceedings of 5th Conference on Automated Deduction*, pages 318–334. Springer Verlag, LNCS, 1980.
19. M.Rodríguez-Artalejo J.C.González-Moreno, M.T.Hortalá-González. On the completeness of narrowing as the operational semantics of functional logic programming. In E.Börger, G.Jäger, H.Kleine Büning, S.Martini, and M.M.Richter, editors, *Computer Science Logic. Selected papers from CSL'92*, LNCS, pages 216–231, San Miniato, Italy, September 1992. Springer.
20. C. L. Lucchesi. The undecidability of the unification problem for third order languages. Technical Report CSRR 2059, University of Waterloo, Waterloo, Canada, 1972.
21. Dale Miller. Unification of simply typed lambda-terms as logic programming. In P.K. Furukawa, editor, *Proc. 1991 Joint Int. Conf. Logic Programming*, pages 253–281. MIT Press, 1991.
22. Gopalan Nadathur and Dale Miller. An overview of λ-Prolog. In Robert A. Kowalski and Kenneth A. Bowen, editors, *Proc. 5th Int. Logic Programming Conference*, pages 810–827. MIT Press, 1988.
23. P. Narendran. Some remarks on second order unification. Technical report, Institute of Programming and Logics, Dep. of Computer Science, State Univ. of New York at Albany, 1989.
24. Tobias Nipkow. Higher-order critical pairs. In *Proceedings, Sixth Annual IEEE Symposium on Logic in Computer Science*, pages 342–349, Amsterdam, The Netherlands, 15–18 July 1991. IEEE Computer Society Press.
25. Lawrence C. Paulson. Isabelle: The next 700 theorem provers. In P. Odifreddi, editor, *Logic and Computer Science*, pages 361–385. Academic Press, 1990.
26. F. Pfenning. Partial polymorphic type inference and higher-order unification. In *ACM Conference on Lisp and Functional Programming*, pages 153–163, Snowbird, Utah, July 1988. ACM-Press.
27. Frank Pfenning. Logic programming in the LF logical framework. In Gérard Huet and Gordon D. Plotkin, editors, *Logical Frameworks*. Cambridge University Press, 1991.
28. Christian Prehofer. Higher-order narrowing. In *Proceedings, Ninth Annual IEEE Symposium on Logic in Computer Science*. IEEE Computer Society Press, 1994. To appear.
29. U. S. Reddy. On the relationship between logic and functional languages. In D. DeGroot and G. Lindstrom, editors, *Logic Programming: Functions, Relations, and Equations*, pages 3–36. Prentice-Hall, Englewood Cliffs, NJ, 1986.
30. Wayne Snyder and Jean Gallier. Higher-order unification revisited: Complete sets of transformations. *J. Symbolic Computation*, 8:101–140, 1989.
31. D. A. Wolfram. *The Clausal Theory of Types*. Cambridge Tracts in Theoretical Computer Science 21. Cambridge University Press, 1993.
32. J.-H. You. Enumerating outer narrowing derivation for constructor-based term rewriting systems. In C. Kirchner, editor, *Unification*, pages 541–564. Academic Press, 1990. Also in J. Symbolic Computation, 1989.

Theory and Practice of Minimal Modular Higher-Order E-Unification

Olaf Müller[1] and Franz Weber[2]

[1] FZI, Haid-und-Neu-Str. 10-14, D-76131 Karlsruhe 1, Email: omueller@fzi.de
[2] Bertelsmann Distribution, An der Autobahn, D-33310 Gtersloh, (formerly FZI)

Abstract. Modular higher-order E-unification, as described in [6], produces numerous redundant solutions in many practical cases. We present a refined algorithm for finitary theories with a finite number of non-free constants that avoids most redundant solutions, analyse it theoretically, and give first experimental results. In comparison to [6], the description of the E-unification algorithm is enriched by a definition of the translation process between first and higher-order terms and an explicit handling of new variables. This explicit handling gives deeper insight into the reasons for redundant solutions and thus provides a method for their avoidance. In order to study the efficiency of this method and the performance of the modular approach in general, some benchmark examples are presented and an interpretation of their empirical evaluation is given.

1 Introduction

There is considerable practical potential in studying higher-order E-unification. First, E-unification is a key algorithm for higher-order theorem provers. Furthermore, higher-order unification is used in extensions of logic programming like λ-Prolog or ELF. Circuit design and hardware verification problems are facilitated by unification with respect to laws of boolean algebra. Finally, the inclusion of algebraic laws of programm operators into unification supports formal software development tools.

However, there is little experience with the practical implementation of higher-order E-unification algorithms. Building on top of Huet's famous algorithm [3], [2], [5] and [6] describe theoretical approaches. In [1] an approach based on combinators is developed. All of these algorithms, although correct and complete, loose one important property of the original algorithm of Huet: They do not necessarily return independent solutions. Hence, the question arises whether the possibility of redundant solutions affects the practicability of those algorithms.

In this paper we propose an algorithm that avoids redundant solutions. We show that redundant solutions are a significant practical problem and that the proposed algorithm solves this problem efficiently. Additionally, it dramatically reduces the search space of higher-order E-unification.

We start by the formal introduction of all necessary notions in section 2. In addition to the classical definition of higher-order E-unification, we define unification problems with respect to a certain set of variables of interest. This enables us to deal with redundance explicitly.

In section 3 we present the modular approach to higher-order E-unification (as described in [6]) by a set of transformation rules. The idea of modularity is to combine first-order and higher-order E-unification algorithm as black boxes. We extend the approach of [6] by handling newly introduced auxiliary variables explicitly and give a formal definition of the translation of higher-order E-unification problems into first-order E-unification problems. Furthermore we introduce a new kind of imitation rule (called K-*imitation*) which combines two rules of modular higher-order E-unification. This combined rule will be a prerequisite for the improved algorithm.

In section 4 we analyze the reasons for redundant solutions within the presented approach (see also [8]). As an improvement, we derive a search tree pruning strategy for the K-imitation rule. This strategy for finitary theories with a finite number of non-free constants avoids redundance.

Modular higher-order E-unification and the proposed search tree pruning strategy are empirically evaluated in section 5 (see also [4]), establishing the practical importance and efficiency of the pruning strategy. As criteria for efficiency, we consider the reduction of both redundance and search space. Concerning the modular approach the effectiveness of the advanced alien subterm strategy of [6] is shown. This strategy aims for pushing as much as possible into the first-order E-unification algorithm. A heuristic optimization of the strategy is described and empirically validated. Finally we conclude in section 6.

2 Preliminaries

Simply typed λ-calculus. The set of *base types* is denoted by $\mathcal{T}_o$, the set of *types* by $\mathcal{T}$, a type is written as $\overline{T} \to A$, where $A \in \mathcal{T}_o$. For every $T \in \mathcal{T}$, $\mathcal{C}_T$ and $\mathcal{V}_T$ denote pairwise disjoint denumerable sets of *function constants* and *variables*. Let $\mathcal{C} = \bigcup_{T \in \mathcal{T}} \mathcal{C}_T$ and $\mathcal{V} = \bigcup_{T \in \mathcal{T}} \mathcal{V}_T$. The set of *symbols* is defined as $\mathcal{S} = \mathcal{V} \cup \mathcal{C}$.

The constructions of *application* and *abstraction* define the set $\mathcal{L}_T$ of terms of type $T \in \mathcal{T}$ as usual. Let $\mathcal{L} = \bigcup_{T \in \mathcal{T}} \mathcal{L}_T$. If necessary, we add the type T of a term t by writing $t : T$ or t^T, T is also denoted by $\tau(t)$. Abstractions are written as $x \cdot t$, the term $x_1 \cdot \ldots \cdot x_k \cdot t$ is abbreviated by $\overline{x} \cdot t$, where t is not an abstraction, the term $(\ldots(ht_1)\ldots t_k)$ is abbreviated by $h(\overline{t})$. In addition, we use the following vector notation:

$$\overline{x \cdot r} = \overline{x} \cdot r_1, \ldots, \overline{x} \cdot r_n$$
$$r(\overline{s}) = r_1(\overline{s}), \ldots, r_n(\overline{s})$$

Free and *bound* occurrences of variables are defined as usual, the set of all bound variables in a syntactic object $\mathcal{O}$ is denoted by $\mathcal{BV}(\mathcal{O})$, and that of all free variables by $\mathcal{FV}(\mathcal{O})$. We also use $\mathcal{C}(\mathcal{O})$ in order to denote the set of all function constants in $\mathcal{O}$.

Long $\beta\eta$-Normal form, Rigid, Flexible. We assume the usual definitions of α, β, and η-conversion. In the rest of this paper, terms are only compared modulo

α-conversion. A term t is in long $\beta\eta$-normal form, denoted by $t \downarrow_{\beta\eta}$, if it is β-reduced and η-expanded. Such a term t has always the form $\overline{x} \cdot h(\overline{s})$ with $h \in \mathcal{S}$, $\overline{s}$ a vector of terms and $\tau(h(\overline{s})) \in \mathcal{T}_0$. The symbol h is called *head*, $\overline{x}$ is called *binder* and $\overline{s}$ *matrix*. A term t in long $\beta\eta$-normal form is called *flexible* if $h \in \mathcal{FV}(t)$, *rigid* if not.

Substitution. A *substitution* is a function $[x_1 \to t_1, ..., x_n \to t_n]$ with $x_i \in \mathcal{V}$, $t_i \in \mathcal{L}$ and $\tau(x_i) = \tau(t_i)$, $i = 1, ..., n$. The *domain* of a substitution σ is defined as $\mathcal{D}om(\sigma) = \{x_1, ..., x_n\}$, and the *range* as $\mathcal{R}an(\sigma) = \bigcup_{i=1,...,n} \mathcal{FV}(t_i)$. A substitution σ is called *dependent* on or *less general* than a substitution ρ, denoted by $\sigma \leq \rho$, if $\sigma = \rho\tau$ for a substitution τ.

Algebraic Theory. The set of *algebraic terms* $\alpha\mathcal{L}$ is the smallest set containing $\bigcup_{T \in \mathcal{T}_0} \mathcal{S}_T$ and satisfying that if $f \in \mathcal{C}_{\overline{B} \to A}$ with $\{B_1, ..., B_k, A\} \subseteq \mathcal{T}_0$, and $s_i \in \alpha\mathcal{L} \cap \mathcal{L}_{B_i}$, for $1 \leq i \leq n$, then $f(\overline{s}) \in \alpha\mathcal{L}$. Every such constant $f \in \mathcal{C}_{\overline{B} \to A}$ is called an *algebraic function constant*. An *algebraic theory* E is a set of unordered pairs $l \simeq r$ of terms in $\alpha\mathcal{L}$ with $\tau(l) = \tau(r)$, called *algebraic equations*. Function constants not occuring in E are said to be *free* in E. An algebraic theory E is called *finitary* if the set of all most general E-unifiers is finite for all unification problems. We use $=_{\beta\eta E}$ to denote the equivalence relation induced by E-equivalence and $\beta\eta$-reduction.

Higher-Order E-Unification. A *disagreement pair*, denoted by $s =^? t$, is an unordered pair of long $\beta\eta$-normal forms s and t with $\tau(s) = \tau(t)$. It is said to be *rigid-rigid* if both terms are rigid, *flexible-rigid* if s is rigid and t flexible, *flexible-flexible* if both terms are flexible. In the interest of clarity we make use of the following vector notation:

$$\left(\overline{\overline{x} \cdot r \stackrel{?}{=} \overline{x} \cdot s} \right) = \overline{x} \cdot r_1 \stackrel{?}{=} \overline{x} \cdot s_1, ..., \overline{x} \cdot r_n \stackrel{?}{=} \overline{x} \cdot s_n$$

A *disagreement set*, often denoted by $\mathcal{U}$, is a finite multiset of disagreement pairs. A disagreement pair $v =^? t \in \mathcal{U}$ is said to be *solved* in $\mathcal{U}$ if $v \notin \mathcal{FV}(t) \cup \mathcal{FV}(\mathcal{U} - \{v =^? t\})$. A disagreement set is called *presolved* or a *preunifier* if all of its disagreement pairs are either solved or flexible-flexible.

We fix an algebraic theory E and a disagreement set $\mathcal{U}$ from now on. A substitution σ is called an E-*unifier* of $\mathcal{U}$ if $s\sigma =_{\beta\eta E} t\sigma$ for each $s =^? t \in \mathcal{U}$. The set of all such E-unifiers is denoted by $\|\mathcal{U}\|_E$, the set of all most general E-unifiers is denoted by $mgu\|\mathcal{U}\|_E$ – in case of its existence. The prefix or index E is omitted, if we consider $E = \emptyset$. Completeness of a set of E-unifiers or preunifiers with respect to $\mathcal{U}$ is defined as usual. For a substitution σ we define $[\sigma] = \{v =^? v\sigma | v \in \mathcal{D}om(\sigma)\}$.

The algorithm for preunification introduces new auxiliary variables that are not relevant to the solution of the original unification problem. For studying the minimality of the algorithm it is necessary to project such variables i.e. to abstract from them in each step.

Definition 1 Projection of Variables. A disagreement set $\mathcal{U}$ in which only the solutions of variables of a set $\mathcal{W} \subseteq \mathcal{V}$ are of interest, is denoted by $\mathcal{U}|_{\mathcal{W}}$. Let σ be an E-unifier of $\mathcal{U}$. Then every substitution ρ with $v\rho = v\sigma$ for all $v \in \mathcal{W}$ is an E-unifier of $\mathcal{U}|_{\mathcal{W}}$. Let $\mathcal{W}^c = \{v \in \mathcal{V} | v \notin \mathcal{W}\}$. The substitution ρ with $v\rho = v\sigma$ for all $v \in \mathcal{W}$ and $v\rho = v$ for all $v \in \mathcal{W}^c$ is denoted by $\sigma|_{\mathcal{W}}$. If σ is a most general E-unifier of $\mathcal{U}$, then $\sigma|_{\mathcal{W}}$ is a most general E-unifier of $\mathcal{U}|_{\mathcal{W}}$.

In the rest of the paper, unless otherwise stated, R, S, T will stand for arbitrary types, A, B for base types, r, s, t, u for terms, c, d, f, g for function constants, x, y, z for bound variables, v, w for free variables, and σ, ρ for substitutions.

3 Modular Higher-Order E-Unification

In this section we present the algorithm for higher-order equational pre-unification as already described in [6]. Special emphasis is laid on the translation process between higher-order and first-order terms which is given precisely here for the first time. The description is short, giving a compact survey over the transformation rules of the algorithm. The rules are presented formally in a table at the end of the section and act intuitively in the following way:

1. *E-Simplification*: In this step consisting of three rules the given first-order E-unification algorithm is used as follows:
   (a) Rule *Abstraction*: Construct underlying first-order-acceptable problem by abstracting the so called alien subterms (Def. 2-5) into new variables. Alien and thus not acceptable to the E-unification algorithm are only subterms of the form $v(\overline{s})$ with free variable v. Therefore, in particular, terms like $x \cdot t$ are acceptable.
   (b) Rule *U-Select*: Choose a first-order-acceptable problem.
   (c) Rule *E-unify*: Translate this first-order-acceptable problem into a real first-order problem (Def. 6-7) and solve it by the given E-unification algorithm (Def. 8).
2. **Variable binding**: Bind free variables occurring as heads of disagreement pairs (heads of alien subterms not touched in the step above) by partial bindings as in [7]. This is done by imitating the corresponding head (rule *E-Imitation*) or by projecting on one of its arguments (rule *Projection*).

First, we define the set of all λ-abstractions that cover a given subterm, e.g. x, y cover the subterm $f(a, y)$ in $x \cdot y \cdot v(x, f(a, y))$. A subterm is given by a context.

Definition 2 Context, Substitution. For $T \in \mathcal{T}$ the set $\mathcal{P}_T$ of T-*contexts* and $\tau : \mathcal{P}_T \to \mathcal{T}$ are inductively defined as

1. $[]_T \in \mathcal{P}_T, \tau []_T = T$. Often we use $[]$ instead of $[]_T$.
2. If $k \in \mathcal{P}_T$, $t \in \mathcal{L}_R$ and $\tau(k) = R \to S$, then $k(t) \in \mathcal{P}_T, \tau(k(t)) = S$.
3. If $k \in \mathcal{P}_T$, $t \in \mathcal{L}_{R \to S}$ and $\tau(k) = R$, then $t(k) \in \mathcal{P}_T, \tau(t(k)) = S$.
4. If $k \in \mathcal{P}_T$, $x \in \mathcal{V}_R$ and $\tau(k) = S$, then $x \cdot k \in \mathcal{P}_T, \tau(x \cdot k) = R \to S$.

The set of *contexts* is defined as $\mathcal{P} = \bigcup_{T \in \mathcal{T}} \mathcal{P}_T$. Let $c \in \mathcal{S}$. Then the *substitution of r with $\tau(r) = T$ in a context* $k \in \mathcal{P}_T$ is defined as

$$c\,[r] = c$$
$$[]\,[r] = r$$
$$k\,(s)\,[r] = (k\,[r])\,(s)$$
$$s\,(k)\,[r] = s\,(k\,[r])$$
$$(x \cdot k)\,[r] = x \cdot (k\,[r])$$

Definition 3 Covering λ-abstractions. Let ε be the empty vector, "," the concatenation of vectors and $k \in \mathcal{P}$. The *covering λ-abstractions* are a function $\Lambda : \mathcal{P} \to \mathcal{V}^*$, given by the following equations:

$$\Lambda([]) = \varepsilon \qquad \Lambda(r(k)) = \Lambda(k)$$
$$\Lambda(k(r)) = \Lambda(k) \qquad \Lambda(x \cdot k) = x, \Lambda(k)$$

Definition 4 Variable Subterms, First-Order-Acceptable. Let r be a $\beta\eta$-long term, $k \in \mathcal{P}$. The term $v(\bar{s})$ with v variable, $\bar{s}$ vector of terms, is called a *variable subterm* of r in context k, if $r = k[v(\bar{s})]$. The variable subterm $v(\bar{s})$ is called *trivial*, if $\Lambda(k) = \bar{s}$. A term that includes only trivial variable subterms is called *first-order-acceptable*; the set of all these terms is denoted by $\mathcal{L}_{fo}$.

Definition 5 Maximal Alien Subterm. A variable subterm $v(\bar{s})$ in a context k is called *alien* if $\Lambda(k) \neq \bar{s}$. An alien subterm $v(\bar{s})$ of a term r in a position k is called a *maximal alien subterm* – denoted by $\mathcal{MAS}(k, v(\bar{s}))$ – if it is no subterm of any other alien subterm of r.

Thus a term is first-order-acceptable if all free variables v in it possess as arguments only the bound variables of all λ-abstractions in a top-down order. For example, $x \cdot y \cdot v(x,y)$ is first-order-acceptable, $x \cdot y \cdot v(y,x)$ and $x \cdot y \cdot v(x, f(a,y))$ are not. If a term is first-order-acceptable, it is translated into an algebraic term by transforming abstractions into applications and higher-order symbols into symbols of base type. For example, $x \cdot f(x)$ is translated into $\lambda_1(f(c_1))$ with new constants λ_1 and c_1. Otherwise, maximal alien subterms are searched and replaced by trivial variable subterms.

Definition 6 Translation. The set $\mathcal{T}_0$ of base types may be extended inductively by the set $\{\langle t \rangle | t \in \mathcal{T}\}$, where $\langle \rangle : \mathcal{T} \to \mathcal{T}_0$ is an arbitrary injective function. Let

$$\lambda_{i, S \to (T \to S)} : \langle S \rangle \to \langle T \to S \rangle,\ i \in \text{Nat}, S, T \in \mathcal{T}$$
$$c_{i, S \to (T \to S)} : \langle T \rangle,\ i \in \text{Nat}, S, T \in \mathcal{T}$$

be new and global algebraic constants. The function

$$\langle \rangle : \mathcal{C} \cup \mathcal{V} \cup \mathcal{L}_{fo} \to \alpha\mathcal{L}$$

called *translation*, is defined as an injective function given by the following conditions:

1. If $v \in \mathcal{V}$, $\tau(v) = \overline{R} \to A$, then $\langle v \rangle \in \mathcal{V}$, $\tau(\langle v \rangle) = \langle A \rangle$.
2. If $c \in \mathcal{C}$, $\tau(c) = \overline{R} \to A$, then $\langle c \rangle \in \mathcal{C}$, $\tau(\langle c \rangle) = \overline{\langle R \rangle} \to \langle A \rangle$.
3. If $r \in \mathcal{L}_{fo}$, then $\langle r \rangle = \langle r \rangle_0$ and for $\overline{s} \in \mathcal{L}_{fo}^*$

$$\langle v(\overline{x}) \rangle_n = \langle v \rangle$$

$$\langle d(\overline{s}) \rangle_n = \langle d \rangle \left(\overline{\langle s \rangle_n} \right)$$

$$\langle w^T \cdot r^S \rangle_n = \lambda_{n, S \to (T \to S)} \left(\langle r^S [w^T \to c_{n, S \to (T \to S)}] \rangle_{n+1} \right)$$

Definition 7 Backtranslation. An algebraic term is called *backtranslatable*, if all constants in it are either $\lambda_{i, S \to (T \to S)}$, $c_{i, S \to (T \to S)}$ or of the form $\langle c \rangle$. These are exactly all algebraic terms in $\mathcal{R}an(\langle \rangle)$, up to base typed variables eventually introduced during the first-order algorithm. The set of all backtranslatable terms is denoted by $\mathcal{L}_{bt}$. The function

$$\langle \rangle^{-1} : \{ \langle c \rangle | c \in \mathcal{C} \} \cup \{ c_{i, S \to (T \to S)} | \; i \in \text{Nat}, S, T \in \mathcal{T} \} \to \mathcal{S}$$

is defined by the following equations:

$$\langle \langle c \rangle \rangle^{-1} = c$$

$$\langle c_{i, S \to (T \to S)} \rangle^{-1} = x_i^T$$

with x_i^T new variables of type T. (New means here: $x_i^T = x_j^S \Leftrightarrow S = T \wedge i = j$) Let d be a constant of the form $\langle d' \rangle$ of type $\overline{\langle R \rangle} \to \langle A \rangle$ and $\forall i \cdot \forall S \cdot \forall T \cdot d \neq \lambda_{i, S \to (T \to S)}$, v a variable of type $\langle U \rangle$, $\overline{x}$ a vector of bound variables, $w_{\overline{x}}$ a new variable of type $\overline{\tau(x)} \to U$ dependent on $\overline{x}$, $\overline{s} \in \mathcal{L}_{bt}^*$ with $|\overline{s}| = \left| \overline{\langle R \rangle} \right|$ and $r \in \mathcal{L}_{bt}$. Then the function

$$\langle r \rangle_{\overline{x}}^{-1} : \mathcal{L}_{bt} \times \mathcal{V}^* \to \mathcal{L}_{fo}$$

called *backtranslation*, is given by the following equations:

$$\langle d(\overline{s}) \rangle_{\overline{x}}^{-1} = \langle d \rangle^{-1} \left(\overline{\langle s \rangle_{\overline{x}}^{-1}} \right)$$

$$\langle v \rangle_{\overline{x}}^{-1} = w_{\overline{x}}(\overline{x})$$

$$\langle \lambda_{i, S \to (T \to S)}(r) \rangle_{\overline{x}}^{-1} = x_i^T \cdot \langle r \rangle_{x_i^T, \overline{x}}^{-1}$$

Definition 8 Simplification Substitution. Let $\mathcal{U}$ be a first-order-acceptable disagreement set. For all $\sigma \in uni_E(\langle U \rangle)$ the *simplification substitution* $\theta_{\mathcal{U}}(\sigma)$ given by a first-order E-unification algorithm uni_E is defined as

$$\theta_{\mathcal{U}}(\sigma) = \left\{ H_i \to (\overline{y})_i \cdot \langle \langle H_i \rangle \sigma \rangle_{(\overline{y})_i}^{-1} : i \in 1..n \right\}$$

In this definition uni_E denotes a function which returns a complete set of E-unifiers for every first-order disagreement set, $\{H_1..H_n\}$ contains all free variables in $\mathcal{U}$, each H_i may have the type $\overline{(T)}_i \to A_i$, then $(\overline{y})_i$ is a vector of variables with $|(\overline{y})_i| = |(\overline{T})_i|$, $y_{ij} = x_j^{T_{ij}}$ and

$$x_j^{T_{ij}} = \langle c_{j, S \to (T_{ij} \to S)} \rangle^{-1}$$

for an arbitrary type S, i.e. the covering λ-abstractions $(\overline{y})_i$ have to be built using the variables of the backtranslation that are determined by the type of H_i.

Definition 9 Transformation Rule. There are two types of transformation rules. An *alternative rule*, denoted by $\mathcal{U} \stackrel{\text{condition}}{\to} \mathcal{U}'$ transforms a disagreement set $\mathcal{U}$ into a disagreement set $\mathcal{U}'$, if the condition "condition" holds. An *exclusive rule*, denoted by $\mathcal{U} \stackrel{\text{condition}}{\Rightarrow} \mathcal{U}'$ has the same meaning with the additional restriction that no further rule should be applied to $\mathcal{U}$ to build the search tree.

Definition 10 Modular Higher-Order E-Unification. Given a disagreement set $\mathcal{U}$ and a theory E the *modular higher-order E-unification* enumerates a set of preunifiers by applying a complete search strategy for search trees on the rules of Table 1. The search process starts with $|\mathcal{U}|_\emptyset$.

From now on we restrict ourselves to finitary theories E with a finite number of non-free constants. The first property guarantees a finite number of first-order unifiers and thus a finite branching with the E-unify rule, the second property guarantees a finite branching with the E-imitation rule.

Definition 11 K-Imitation. The K-imitation rule is defined in Table 2. In the following sections it will replace the E-imitation rule partially (section 4) or even completely (section 5). It combines E-imitation with the succeeding E-simplification step. However, we do not search maximal alien subterms for abstraction, but simply take the arguments $\overline{t}$ of c instead of them.

4 Minimal Modular Higher-Order E-Unification

4.1 Reasons for Dependence

Exactly two reasons are responsible for the fact that modular higher-order E-unification is not minimal in the general case. (In [8] an algorithm to eliminate redundant solutions caused by both reasons is proposed that is minimal for unification problems which have only preunifiers without flexible-flexible pairs and for which higher-order E-matching has most general unifiers.)

Dependence by branching. The first reason for dependence is the branching produced by the E-simplification step. ($\mathcal{U}$-Select) splits a unification problem into a higher-order problem $\mathcal{U}_1$ and a first-order-acceptable problem $\mathcal{U}_2$. Let us suppose that the first-order algorithm is minimal, i.e. that every two solutions σ_1 and σ_2 are independent. Then $\Theta_{\mathcal{U}_2}(\sigma_1)$ and $\Theta_{\mathcal{U}_2}(\sigma_2)$ are independent with respect to $\mathcal{U}_2$, too. But they may be dependent taking the higher-order problem $\mathcal{U}_1$ into account. This is shown by the following example.

Example 1. Consider the unification problem $v(y) \circ v(1) \stackrel{?}{=} c(1) \circ c(y)$ with $\circ$ as a C-operator written in infix notation. Applying (Abstraction) and ($\mathcal{U}$-Select) to it yields $\mathcal{U}_1 = \{H_1 =^? v(y), H_2 =^? v(1)\}$ and $\mathcal{U}_2 = \{H_1 \circ H_2 = c(1) \circ c(y)\}$

Table 1. Rules for Higher-Order E-Unification

Let $\mathcal{W} \subseteq \mathcal{V}$, $k \in \mathcal{P}$, $\overline{y} = \Lambda(k)$, H a new variable, uni_E a first-order E-unification algorithm, σ a substitution with only free $x \in \mathcal{R}an(\sigma)$, $\overline{r}$, $\overline{s}$ vectors of terms, $\overline{z}$ a vector of new variables with $|\overline{z}| = |\overline{s}|$, $\overline{w}$ a vector of new variables with $|\overline{w}| = |\overline{r}|$, $a \notin \overline{z}$ a symbol, $\overline{R_i}, i = 1..|\overline{s}|$ vectors of types and $\overline{y_i}, i = 1..|\overline{R_i}|$ vectors of new variables.

Abstraction	$\left\lvert \mathcal{U}, k\left[v\left(\overline{s}\right)\right] \stackrel{?}{=} r \right\rvert_{\mathcal{W}^c}$ $\quad \stackrel{\mathcal{MAS}(k,v(\overline{s})),\, k[v(\overline{s})]\text{ or }r\text{ rigid}}{\Rightarrow}$ $\left\lvert \mathcal{U} \cup \left\{ k\left[H\left(\overline{y}\right)\right] \stackrel{?}{=} r,\, \overline{y} \cdot H\left(\overline{y}\right) \stackrel{?}{=} \overline{y} \cdot v\left(\overline{s}\right) \right\} \right\rvert_{(\mathcal{W} \cup \{H\})^c}$	
$\mathcal{U}$-Select	$\left\lvert \mathcal{U}_1 \cup \mathcal{U}_2 \right\rvert_{\mathcal{W}^c} \stackrel{\mathcal{U}_2 \text{ first-order-acceptable}}{\Rightarrow} \left\lvert \mathcal{U}_1, \mathcal{U}_2 \right\rvert_{\mathcal{W}^c}$	
E-Unify	$\left\lvert \mathcal{U}_1, \mathcal{U}_2 \right\rvert_{\mathcal{W}^c} \stackrel{\sigma \in uni_E(\langle \mathcal{U}_2 \rangle)}{\Rightarrow} \left\lvert \mathcal{U}_1 \cup [\theta_{\mathcal{U}_2}(\sigma)] \right\rvert_{(\mathcal{W} \cup \mathcal{FV}(\sigma))^c}$	
Select	$\left\lvert \mathcal{U} \cup \left\{ r \stackrel{?}{=} s \right\} \right\rvert_{\mathcal{W}^c} \stackrel{r \stackrel{?}{=} s \text{ not solved}}{\Rightarrow} \left\lvert \mathcal{U},\, r \stackrel{?}{=} s \right\rvert_{\mathcal{W}^c}$	
Elimination	$\left\lvert \mathcal{U},\, \overline{x} \cdot v\left(\overline{x}\right) \stackrel{?}{=} \overline{x} \cdot v\left(\overline{x}\right) \right\rvert_{\mathcal{W}^c} \rightarrow \left\lvert \mathcal{U} \right\rvert_{\mathcal{W}^c}$	
Propagation	$\left\lvert \mathcal{U},\, \overline{x} \cdot v\left(\overline{x}\right) \stackrel{?}{=} r \right\rvert_{\mathcal{W}^c} \stackrel{v \notin \mathcal{FV}(r)}{\rightarrow} \left\lvert (\mathcal{U}[v \rightarrow r]) \downarrow_{\beta\eta} \cup \left\{ \overline{x} \cdot v\left(\overline{x}\right) \stackrel{?}{=} r \right\} \right\rvert_{\mathcal{W}^c}$	
E-Imitation	$\left\lvert \mathcal{U},\, \overline{x} \cdot c\left(\overline{r}\right) \stackrel{?}{=} \overline{x} \cdot v\left(\overline{s}\right) \right\rvert_{\mathcal{W}^c} \stackrel{c \notin \mathcal{C}(E) \Rightarrow a \in \{c\} \cup \mathcal{C}(E)}{\rightarrow}$ $\left\lvert \mathcal{U} \cup \left\{ \overline{x} \cdot c\left(\overline{r}\right) \stackrel{?}{=} \overline{x} \cdot v\left(\overline{s}\right),\, \overline{z} \cdot v\left(\overline{z}\right) \stackrel{?}{=} \overline{z} \cdot a\left(\overline{w\left(\overline{z}\right)} \downarrow_{\beta\eta}\right) \right\} \right\rvert_{(\mathcal{W} \cup \{\overline{w}\})^c}$	
Projection	$\left\lvert \mathcal{U},\, \overline{x} \cdot c\left(\overline{r}\right) \stackrel{?}{=} \overline{x} \cdot v\left(\overline{s}\right) \right\rvert_{\mathcal{W}^c} \stackrel{\substack{s_i : \overline{R_i} \rightarrow A \\ \overline{s} \neq \overline{x} \vee v \in \mathcal{FV}(\overline{x} \cdot c(\overline{r}))}}{\rightarrow}$ $\left\lvert \mathcal{U} \cup \left\{ \overline{x} \cdot c\left(\overline{r}\right) \stackrel{?}{=} \overline{x} \cdot v\left(\overline{s}\right),\, \left(v \stackrel{?}{=} \overline{z} \cdot z_i\left(\overline{y_i\left(\overline{z}\right)}\right) \right) \downarrow_{\beta\eta} \right\} \right\rvert_{(\mathcal{W} \cup \{\overline{y_i}\})^c}$	

Table 2. K-Imitation Rule

In addition to the notions in Table 1 let $\overline{e}$ and $\overline{f}$ be new variables with $|\overline{e}| = |\overline{r}|$, $|\overline{f}| = |\overline{s}|$ and d an arbitrary new constant.

K-Imitation	$\left\lvert \mathcal{U},\, \overline{x} \cdot c\left(\overline{r}\right) \stackrel{?}{=} \overline{x} \cdot v\left(\overline{s}\right) \right\rvert_{\mathcal{W}^c} \stackrel{\substack{\sigma \in \mathrm{mgu} \left\lVert \langle c \rangle\left(\overline{e}\right) \stackrel{?}{=} \langle d \rangle\left(\overline{f}\right) \right\rVert_E \\ \overline{s} \neq \overline{x} \vee v \in \mathcal{FV}(\overline{x} \cdot c(\overline{t}))}}{\rightarrow}$ $\left\lvert \mathcal{U} \cup \left\{ \begin{array}{c} \overline{x} \cdot \langle e\sigma \rangle_{\overline{x}}^{-1} \stackrel{?}{=} \overline{x} \cdot r \\ \hline \overline{x} \cdot \langle f\sigma \rangle_{\overline{x}}^{-1} \stackrel{?}{=} \overline{x} \cdot w\left(\overline{s}\right) \\ \hline \overline{z} \cdot v\left(\overline{z}\right) \stackrel{?}{=} \overline{z} \cdot d\left(\overline{w\left(\overline{z}\right)} \downarrow_{\beta\eta}\right) \end{array} \right\} \right\rvert_{(\mathcal{W} \cup \{\overline{w}\} \cup \mathcal{FV}(\sigma))^c}$

with new variables $H_i, i = 1, 2$. Obviously, there are two independent first-order solutions of $\mathcal{U}_2$ and we get, after propagation of H_1 and H_2, the simpler problems

$$\left\{c(y) \stackrel{?}{=} v(y), c(1) \stackrel{?}{=} v(1)\right\}, \left\{c(1) \stackrel{?}{=} v(y), c(y) \stackrel{?}{=} v(1)\right\}.$$

Not until two imitations concerning v and y do we realize that both disagreement sets lead to the same solution $[v \to z \cdot c(1), y \to 1]$. Thus the higher-order problem $\mathcal{U}_1$ has specialized the independent first-order solutions to dependent solutions of the whole problem.

Dependence by variable projection. Another source of dependence are auxiliary variables which are newly introduced during the unification process. Such variables are generated by the E-imitation, projection and E-unify rules, as the explicit handling of new variables in Table 1 shows (namely $\overline{w}$, $\overline{y_i}$ and $\mathcal{FV}(\sigma)$). Now there may be solutions whose independence is only guaranteed under consideration of these auxiliary variables. Since they are irrelevant to the original problem, they are projected in the end and the solutions become dependent. This is shown by the following example.

Example 2. Consider the unification problem $c \circ d \stackrel{?}{=} v(1)$ in long $\beta\eta$-normal form with $\circ$ as a C-operator written in infix notation. E-Imitation yields the substitution $[v \to x \cdot g(x) \circ h(x)]$, where g and h are newly introduced variables. Because of the commutativity of $\circ$ there are two solutions of the resulting problem $c \circ d =^? g(1) \circ h(1)$. The first is

$$[h \to x \cdot c, g \to x \cdot d, v \to x \cdot c \circ d]$$

which leads to $[v \to x \cdot c \circ d]$ after abstracting from the auxiliary variables. The second solution is

$$[h \to x \cdot d, g \to x \cdot c, v \to x \cdot d \circ c]$$

and leads to $[v \to x \cdot d \circ c]$ after abstracting from the auxiliary variables. Hence, both solutions are equal, since $\circ$ is commutative.

4.2 Static Searchtree Pruning

In the following, we present a strategy that avoids the kind of dependence caused by introducing new variables in the E-imitation rule. The aim is not to eliminate single redundant solutions, but to cut whole redundant branches in a preventative manner. Consider the K-imitation rule and the general imitation problem $\mathcal{U} \cup p$ with $p = \overline{x} \cdot c(\overline{t}) =^? \overline{x} \cdot v(\overline{s})$. For each pair (d, σ) with

$$\sigma \in \mathrm{mgu} \left\| \langle c \rangle \, (\overline{e}) \stackrel{?}{=} \langle d \rangle \, (\overline{f}) \right\|_E$$

a branch with unification problem

$$|\mathcal{U} \cup \mathcal{B}(\sigma, d)|_{(\{\overline{w}\} \cup \mathcal{FV}(\sigma))^c}, \quad \mathcal{B}(\sigma, d) = \left\{ \begin{array}{c} \overline{\overline{x} \cdot \langle e\sigma \rangle_{\overline{x}}^{-1} \stackrel{?}{=} \overline{x} \cdot t} \\ \overline{\overline{x} \cdot \langle f\sigma \rangle_{\overline{x}}^{-1} \stackrel{?}{=} \overline{x} \cdot w(\overline{s})} \\ \overline{z} \cdot v(\overline{z}) \stackrel{?}{=} \overline{z} \cdot d \left(\overline{w(\overline{z})} \downarrow_{\beta\eta} \right) \end{array} \right\}$$

is generated. A branch $\mathcal{A}_1 = \mathcal{U} \cup \mathcal{B}(\sigma_1, d_1)$ is dependent on a branch $\mathcal{A}_2 = \mathcal{U} \cup \mathcal{B}(\sigma_2, d_2)$ if all solutions for $\mathcal{A}_1$ are also solutions for $\mathcal{A}_2$. This property may be checked according to the following theorem found in [8]:

Theorem 12. *Let $\mathcal{V} = w_2 \cup \mathcal{FV}(\sigma_2)$ and $\mathcal{W} = w_1 \cup \mathcal{FV}(\sigma_1)$ be sets of variables obeying*

$$\mathcal{V} \cap (\mathcal{FV}(\mathcal{U}) \cup \mathcal{FV}(\mathcal{B}(\sigma_1, d_1))) = \emptyset$$
$$\mathcal{W} \cap (\mathcal{FV}(\mathcal{U}) \cup \mathcal{FV}(\mathcal{B}(\sigma_2, d_2))) = \emptyset$$

If there exists a substitution ρ with $\mathcal{B}(\sigma_2, d_2)\rho|_\mathcal{V} = \mathcal{B}(\sigma_1, d_1)$ then

$$|||\mathcal{B}(\sigma_1, d_1) \cup \mathcal{U}||_E|_{\mathcal{W}^c} \subseteq |||\mathcal{B}(\sigma_2, d_2) \cup \mathcal{U}||_E|_{\mathcal{V}^c}$$

Such dependencies may be analyzed in a static way independent of the actual matrices of the disagreement pair p at runtime of the algorithm. If there is such a static analysis for the theory E and c is not free in E, certain branches $\mathcal{B}(\sigma, d)$ can be cut every time when the K-imitation rule is applied. This optimization does not alter the completeness. We call it *static searchtree pruning* (short SSP) for K-imitation.

In the following, we calculate the dependence of branches generated by K-imitation for commutative operators:

Example 3. Let $\circ$ be a commutative operator. It cannot be imitated but through itself and we get:

$$\text{mgu} \left\| e_1 \circ e_2 \stackrel{?}{=} f_1 \circ f_2 \right\|_{C_\circ} = \{[e_1 \to f_1, e_2 \to f_2], [e_1 \to f_2, e_2 \to f_1]\}$$

Given a general imitation problem $\mathcal{U} \cup \{\overline{x} \cdot t_1 \circ t_2 =^? \overline{x} \cdot v(\overline{u})\}$ K-imitation produces the two problems

$$\left| \mathcal{U} \cup \left\{ \begin{array}{l} \overline{x} \cdot t_1 \stackrel{?}{=} \overline{x} \cdot w_1(\overline{u}) \\ \overline{x} \cdot t_2 \stackrel{?}{=} \overline{x} \cdot w_2(\overline{u}) \\ \overline{z} \cdot v(\overline{z}) \stackrel{?}{=} \overline{z} \cdot w_1(\overline{z}) \circ w_2(\overline{z}) \end{array} \right\} \right|_{\{f_1, f_2, w_1, w_2\}^c}$$

and

$$\left| \mathcal{U} \cup \left\{ \begin{array}{l} \overline{x} \cdot t_1 \stackrel{?}{=} \overline{x} \cdot w_2(\overline{u}) \\ \overline{x} \cdot t_2 \stackrel{?}{=} \overline{x} \cdot w_1(\overline{u}) \\ \overline{z} \cdot v(\overline{z}) \stackrel{?}{=} \overline{z} \cdot w_2(\overline{z}) \circ w_1(\overline{z}) \end{array} \right\} \right|_{\{f_1, f_2, w_1, w_2\}^c}$$

Applying the substitution $\rho = [w_1 \to w_2, w_2 \to w_1]$ to the second problem, we recognize that both problems are equivalent under commutativity. Therefore, only one of the two branches generated by K-imitation must be considered for the algorithm.

Analogous, it can be shown [8] that for associative and commutative operators K-imitation produces seven branches of which three are redundant.

5 Empirical Study and Analysis

In this section practical aspects concerning the minimality and performance of the algorithm are considered. Therefore, the algorithm is evaluated by several benchmarks, shown in Table 3 (from [4]). Variables are presented by capital letters, constants by small letters. In every example the equality "=" stands for a C-operator, the conjunction "∧" is considered as an AC-operator. The first four benchmarks are taken from practical proofs in the area of theory deduction and "non-clausal"-resolution. In particular, example 1 represents a proof step in [8] concerning some properties of predicate programming.

Table 3. Benchmark Table for Higher-Order E-Unification

1.	$F(u \cdot x \cdot y \cdot u \wedge x = u \wedge y) \stackrel{?}{=} \forall i \cdot p(i) \wedge q(i) = r(i) \wedge p(i)$
2.	$F(i \cdot o \cdot v \cdot e \cdot \text{read}(v,o) = \text{bagof}(e) \wedge \text{delta}(\text{abs}(i),o,v)) \stackrel{?}{=}$ $\text{ls} \sqsubseteq i \cdot o \cdot \text{read}(v,o) = \text{bagof}(\text{empty}) \wedge \text{delta}(\text{abs}(i),o,c)$
3.	$F(e \cdot b \cdot f \cdot g \cdot v \cdot (\neg b(e) \Rightarrow (v = f(e))) \wedge (b(e) \Rightarrow v = g(e))) \stackrel{?}{=}$ $\text{ls} = \forall w \cdot ((w = c) \Rightarrow \text{read}(w,o) = \text{bagof}(d)) \wedge$ $(\neg(w = c) \Rightarrow \text{read}(w,o) = \text{read}(w, \text{abs}(i)))$
4.	$F((\forall x \cdot p(x)) \wedge q) \stackrel{?}{=}$ $\text{ls} = \text{read}(c,o) = \text{bagof}(e) \wedge \forall w \cdot w \neq c \Rightarrow \text{read}(w,o) = \text{read}(w, \text{abs}(i))$
5.	$H \wedge (L = R) \wedge T(L) \stackrel{?}{=} r(b) \wedge h1 \wedge h2 \wedge (c = b)$
6.	$H \wedge (L = R) \wedge T(L) \stackrel{?}{=} r(b) \wedge h1 \wedge h2 \wedge h3 \wedge (c = b)$
7.	$c \wedge d \stackrel{?}{=} V(t)$
8.	$c \wedge d \wedge e \stackrel{?}{=} V(t)$
9.	$c \wedge d \wedge e \wedge f \stackrel{?}{=} V(t)$
10.	$V(X) \wedge V(1) \stackrel{?}{=} c(1) \wedge c(X)$
11.	$fun(a = f(a) \wedge p(S), b(X) \Rightarrow f(X) = g(X)) \stackrel{?}{=}$ $fun(F(G), H(G))$

5.1 Minimality of the Algorithm

Dependencies are a real problem. In higher-order E-unification, dependent solutions appear in a surprisingly high frequency. As Fig. 1 shows, the algorithm often returns a multiple of the ideal number of solutions (see in particular ex. 1-4). Furthermore, iterations of AC-operators in a term can even cause an explosive ascent of redundance. For instance the intentionally designed series of problems $c_1 \wedge c_2 \wedge \ldots \wedge c_n =^? V(t)$ (ex. 7-9) produces for $n = 3$ already 120 AC-equivalent solutions. The reason for such a striking phenomenon are repeated occurrences of redundant branches within already redundant branches. Since these examples (7-9) represent basic elements of higher-order problems the question of unnecessary solutions is significant in practice.

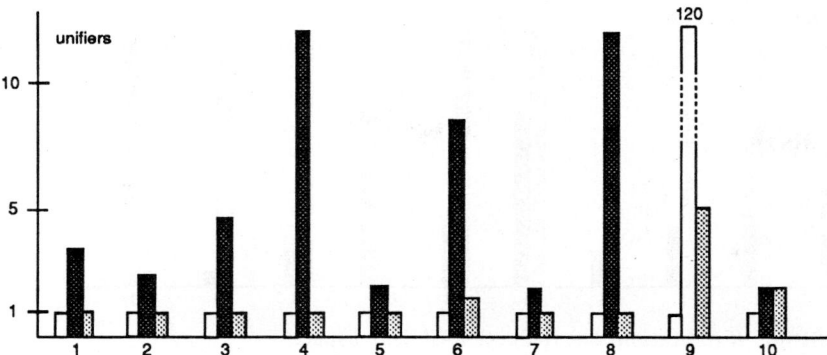

Fig. 1. SSP for K-Imitation and Dependence: The ideal number of unifiers (white) is normalized to one, the solutions returned by the algorithm without K-imitation (black) and with K-imitation (grey) are given in relation to this value.

SSP for K-imitation enables efficient reduction of dependence. SSP for K-imitation rule turns out to be a useful method to avoid dependent solutions (see Fig. 1). In almost every case all dependencies are eliminated, the only exceptions are the problems 6, 9 and 10. This result is surprising, but also reasonable: Pruning for K-imitation only cuts search paths that lead necessarily to redundant leaves. Therefore every time when the K-imitation rule is applied, a whole branch of dependent solutions is avoided. As the K-imitation rule is frequently used, the striking success of the method is explicable.

However, the method is not totally complete, not even in the case of dependence by variable projection. This is shown by examples 6 and 9. Only 20 of 265 (respectively 4 of 119) dependent solutions remain.

In addition, SSP for K-imitation cannot be used to eliminate dependence by branching. But fortunately this is not of great importance. The evaluation of the whole data material yielded not one redundant solution of this kind of dependence. The only representative may be problem 10, which was specially designed for purposes of illustration. Note also that dependencies by variable projection that arise in connection with the projection or E-unify rule did not appear either.

Therefore the elimination of dependence by SSP for K-imitation seems to be quite sufficient for most application cases.

SSP for K-imitation reduces search space. Since SSP for K-imitation cuts unnecessary branches already in a preventative manner, a lot of search space and time can be saved. Figure 2 demonstrates the reduction of the search space as a percentage of the remaining vertices of the search tree. The results are obtained by a static cutting of 3/7 (case of AC) respectively 1/2 (case of C) of the actual branches.

As a result SSP for K-imitation has proved to be an efficient and yet simply implementable optimization of higher-order E-unification.

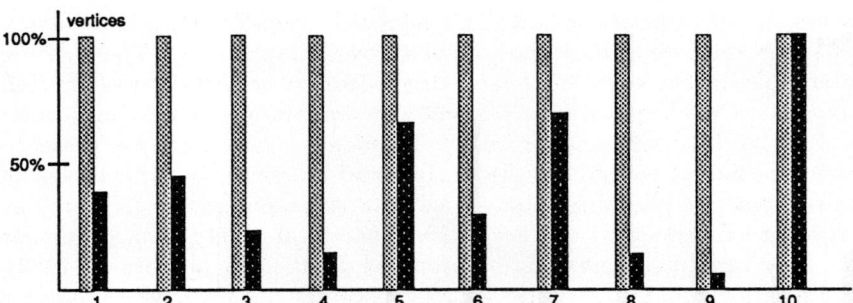

Fig. 2. SSP for K-Imitation and Reduction of Searchspace

5.2 Performance of the Algorithm

The performance of the algorithm is mainly determined by the success of the modular approach to the problem. Therefore it is interesting to study how profitably the first-order algorithms can be utilized. For this aim we use the measure $(fo > ho)$ that gives the percentage of those E-simplification problems whose translated first-order part is greater than higher-order part that cannot be translated. For the size of a term we have taken the number of symbols. The data of our experiments are displayed in Fig. 3. In addition to $(fo > ho)$, we display the number of E-imitation and projection rules as a percent of all rules applied. It is striking that $(fo > ho)$ is closely related to the number of projections. If the number of projections descends, $(fo > ho)$ descends too, and thus with it the profit of the first-order modules. At the same time the number of E-imitations increases.

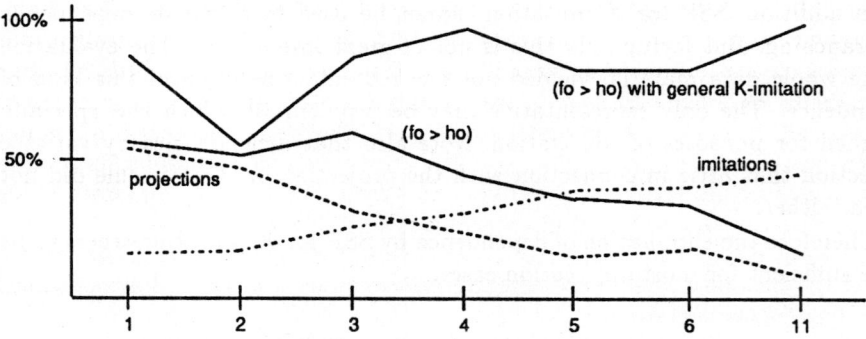

Fig. 3. Profit of First-Order Modules in relation to the kind of Rules

In order to analyse this observation, we study the application of the E-imitation and projection rules in connection with the following simplification step. Let

$$\overline{x} \cdot c\left(\overline{t}\right) \stackrel{?}{=} \overline{x} \cdot F\left(\overline{s}\right)$$

be a flexible-rigid disagreement set where F denotes a free variable.

1. By E-imitation we get – for the present without consideration of a theory E – the substitution $F \to \overline{z} \cdot c\left(\overline{G(\overline{z})}\right)$. The propagation of this substitution yields $\overline{x} \cdot c\left(\overline{t}\right) \stackrel{?}{=} \overline{x} \cdot c\left(\overline{G(\overline{s})}\right)$ with new free variables G_i. The following simplification step is reduced to a simple transition to the arguments of c, because the terms $G_i(\overline{s})$ are recognized as maximal alien subterms. We get the disagreement pairs

$$\overline{x} \cdot G(\overline{s}) \stackrel{?}{=} \overline{x} \cdot t$$

   Even under consideration of a theory E this simplification step does not exceed the elimination of the head c in the most cases. Therefore we call it a *Huet simplification step*.

2. By projection we get – if the type of an argument term s_i allows it – the substitution $F \to \overline{z} \cdot z_i \left(\overline{G_i(\overline{z})}\right)$. The propagation of this substitution yields

$$\overline{x} \cdot s_i\left(G_i(\overline{s})\right) \stackrel{?}{=} \overline{x} \cdot c\left(\overline{t}\right)$$

   Here the following simplification can possibly be very effective. The effectiveness mainly depends on the size of s_i and on the occurence of alien subterms in s_i or $\overline{t}$.

Thus the possibility of an efficient simplification step is more likely after a projection than after an E-imitation.

These considerations underpin the following model for successful search paths. It is also motivated by the observation of several examples.

1. The unification process starts with an E-simplification step which is mostly based on a large first-order problem.
2. A series of E-imitations directly succeeded by a simple Huet simplification step follows. The unchanged arguments $\overline{s}$ of the flexible head F are passed on to the new variables G_i that were introduced during the E-imitation step.
3. Now a projection is made on one of the arguments s_i. Here a unification problem arises that allows an effective E-simplification step. The search process returns to step 1.

Surely this model does not describe all successful search paths, since propagations of flexible-flexible disagreement pairs and β-reductions may destroy the regularity. Rather it presents a typical application case.

As a result, we have an obvious optimizing idea: Restrict of the application of the E-simplification rule to such cases in which the integration of a first-order algorithm leads to a striking improvement. As a first step in this direction, E-imitation should always be combined with the succeeding Huet simplification step. This means that the E-imitation rule should always be replaced by the K-imitation rule even if no algebraic theory is touched in the actual step i.e. even if no searchtree pruning is possible. We call this approach *general K-imitation*.

The success of general K-Imitation is shown in Fig. 3 using the measure ($fo > ho$). It increases tremendously up to values between 80 and 100 percent. Thus by the omission of the less effective Huet simplification steps the remaining E-simplifications become quite more effective on an average. The profit of the first-order algorithms is clearly improved. Hence we conclude that general K-imitation is highly recommandable under the concept of modularity.

6 Conclusion

We presented a higher-order E-unification algorithm, an optimization of this algorithm by static searchtree pruning, and a heuristic to steer the usage of first-order E-unification algorithms for higher-order unification. The theoretical presentation and the empirical studies, including the interpretation of our results, give a deeper understanding of the process of modular higher-order E-unification. Static search tree pruning for K-imitation and the heuristic to steer the usage of first-order E-unification seem to be fundamental methods for the construction of effective higher-order E-unification algorithms. The insight into the unification process provided in this paper could perhaps be the basis for the development of further optimizations.

Acknowledgments: The authors wish to thank Tobias Nipkow, Zhenyu Qian and Ullrich Hustadt for several instructive discussions and valuable comments on earlier versions of the paper.

References

[1] Dougherty, D. and Johann, P.: A Combinatory Logic Approach to Higher-Order E-Unification. *Proc. 11th Conf. Automated Deduction*, LNCS 607 (1992), 79-93.

[2] Gallier, J. and Snyder, W.: Complete Sets of Transformations for General E-Unification. *Theoretical Computer Science*, 67 (1988), 203-260.

[3] Huet, G.: A Unification Algorithm for Typed λ-Calculus. *Theoretical Computer Science* 1 (1975), 27-57.

[4] Müller, O.: Optimierung der modularen E-Unifikation höherer Stufe – Implementierung und Analyse, master thesis, Karlsruhe 1993.

[5] Nipkow, T. and Qian, Z.: Modular Higher-Order E-Unification. *Proc. 4th Int. Conf. Rewriting Techniques and Applications*, LNCS 488 (1991), 200-214.

[6] Qian, Z. and Wang, K.: Higher-Order E-Unification for Arbitrary Theories. *Proc. of 1992 Int. Joint Conf. and Symp. on Logic Programming*, MIT Press, 1992.

[7] Snyder, W. and Gallier, J.: Higher-Order Unification Revisited: Complete Sets of Transformations. *J. Symbolic Computation* 8 (1989), 101-140.

[8] Weber, F.: Softwareentwicklung mit Logik höherer Stufe – eine Anwendung von Theoriededuktion auf interaktives Beweisen, PhD thesis, Karlsruhe 1993.

A Refined Version of General E-Unification

Rolf Socher-Ambrosius*

*MPI Informatik, Im Stadtwald,
D-66123 Saarbrücken, Germany
Tel. +49-681-302-5367*
e-mail: socher@mpi-sb.mpg.de

Abstract Transformation-based systems for general E-unification were first investigated by Gallier and Snyder. Their system extends the well-known rules for syntactic unification by Lazy Paramodulation, thus coping with the equational theory. More recently, Dougherty and Johann improved on this method by giving a restriction of the Lazy Paramodulation inferences. In this paper, we show that their system can be further improved by a stronger restriction on the applicability of Lazy Paramodulation. It turns out that the framework of proof transformations provides an elegant and natural means for proving completeness of the inference system.

1 Introduction

This paper describes a transformation based procedure for unification in an arbitrary equational theory representable by an equational system E. Since unification is now commonly being regarded as equation solving, the transformations operate on equational systems. J. Gallier and W. Snyder [2, 3] were the first to study transformation based methods for E-unification. They devised an inference system for general E-unification consisting of the common rules for syntactic unification together with an additional *Lazy Paramodulation* rule, which takes the equational theory into account. Paramodulation steps are done lazily so that the nondeterministic algorithm induced by the transformations is complete even when paramodulation into variables is forbidden. For instance, given the equational theory

---

* This work was funded by the German Ministry for Research and Technology (BMFT) under grant ITS 9103.

$$E = \{f(a,b) \approx a, a \approx b\}$$

and the E-unification problem $\{f(x,x) \approx x\}$, no paramodulation step applies at a nonvariable position. One would thus have to paramodulate into one of the variables of the term $f(x,x)$. Instead, Gallier and Snyder do allow paramodulation into the term $f(x,x)$, trading the immediate unification of the terms $f(x,x)$ and $f(a,b)$ for an additional E-unification problem $f(x,x) \approx f(a,b)$. Their system thus allows an inference

$$\{f(x,x) \approx x\} \Rightarrow \{f(x,x) \approx f(a,b), a \approx x\} \qquad (1)$$

In [1], D. Dougherty and P. Johann improve on Gallier and Snyder's system by restricting the applicability of Lazy Paramodulation to so called *top unifiable* term-pairs. Two terms are top unifiable if they agree on those positions that are function positions in both terms. Decomposition of top unifiable term-pairs thus eventually leads to an equational system of the form $\{x_1 \approx t_1, \ldots, x_n \approx t_n\}$. The terms $f(x,x)$ and $f(a,b)$, for instance, are top unifiable, while $f(x,a)$ and $f(a,b)$ are not. Therefore, in solving the E-unification problem of the preceding paragraph, we must consider the inference (1), whereas for the problem $\{f(x,a) \approx x\}$ under the same theory E, it is not necessary to infer the equation $f(x,a) \approx f(a,b)$. Such a restricted Lazy Paramodulation rule, together with the requirement that the top unifiable term-pair is decomposed immediately, is called *Relaxed Paramodulation*. The intuitive argument for this restriction is provided by an innermost strategy applying to the subterm a of $f(x,a)$ rather than to the whole term itself.

This paper provides two additional restrictions to Gallier and Snyder's E-unification transformations. First, we show that Lazy Paramodulation can be constrained even further to apply only to so called *top left unifiable* pairs, without sacrificing completeness. Consider, for instance, the theory

$$E = \{f(x, x) \approx x, a \approx b\},$$

and the unification problem $\{f(a, b) \approx a\}$. The terms $f(x,x)$ and $f(a, b)$ are top unifiable, thus giving rise to a Relaxed Paramodulation inference

$$\{f(a, b) \approx a\} \Rightarrow \{x \approx b, x \approx a\}$$

at the root. However, applying an innermost strategy, one would preferably paramodulate into the subterm b of the unification problem, thus deriving

$$\{f(a, b) \approx a\} \Rightarrow \{f(a, a) \approx a\}.$$

The *outermost* inference step is unnecessary, because we can rely on an innermost strategy to yield the same solution $\{x \approx a\}$. This observation is generalized to restrict Lazy Paramodulation to top left unifiable pairs.

We also show that inference steps need not be applied to solved equations. This result, although both intuitive and expected, has not previously been proved.

The rest of the introduction reviews the basic notation used in the text.

Given a signature $\mathcal{F}$ of *function symbols*, each $f \in \mathcal{F}$ coming with an arity $\alpha(f) \geq 0$, and a set $\mathcal{V}$ of *variables*, the set $\mathcal{T}(\mathcal{F},\mathcal{V})$ is defined to be the set of *terms* built over $\mathcal{V}$ using the function symbols in $\mathcal{F}$. For any object o, $\mathcal{V}ar(o)$ denotes the set of variables occurring in o. A *position* in a term t is a sequence of natural numbers referring to a subterm of t, the root position is denoted by Λ. The set of positions of a term t is denoted by $\mathcal{P}os(t)$, the set of variable positions by $\mathcal{V}\mathcal{P}os(t)$, and the set of nonvariable positions by $\mathcal{F}\mathcal{P}os(t)$. If $p \in \mathcal{P}os(t)$ is a position in t, then $t|_p$ denotes the subterm of t at position p, $t[s]_p$ denotes replacement of $t|_p$ by s, and $t(p)$ denotes the function or variable symbol at position p.

A *substitution* is the unique extension of a mapping $\sigma: \mathcal{V} \to \mathcal{T}(\mathcal{F},\mathcal{V})$ with finite domain $\mathcal{D}om(\sigma)$ to the free $\mathcal{F}$-algebra $\mathcal{T}(\mathcal{F}, \mathcal{V})$ over generators $\mathcal{V}$. We write substitutions in suffix notation. A substitution σ with domain $\{x_1, \ldots, x_n\}$ will be written in the form $\{x_1 \leftarrow x_1\sigma, \ldots, x_n \leftarrow x_n\sigma\}$. The *restriction* $\sigma|_V$ of σ to a set $V \subseteq \mathcal{V}$ is the substitution $\sigma|_V$ defined by $x\sigma|_V = x\sigma$ for $x \in V$, and $y\sigma|_V = y$ for $y \notin V$.

An *equation* is an unordered pair $s \approx t$ of terms, an (equational) *system* is a finite set of equations. If $\sigma = \{x_1 \leftarrow t_1, \ldots, x_n \leftarrow t_n\}$ is a substitution, then $[\sigma]$ denotes the equational system $\{x_1 \approx t_1, \ldots, x_n \approx t_n\}$.

The relation $\leftrightarrow$ is defined by

$$s \leftrightarrow_{[p,l \approx r,\sigma]} t$$

iff $s|_p = l\sigma$ and $t = s[r\sigma]_p$. Note that because $l \approx r$ is an unordered pair, $\leftrightarrow$ is a symmetric relation. By $\leftrightarrow^*$, we denote the transitive and reflexive closure of $\leftrightarrow$. If E is an equational system, we write $s \leftrightarrow_{E,\sigma} t$ to denote that $s \leftrightarrow_{[p,l \approx r,\sigma]} t$ holds for some position $p \in \mathcal{P}os(s)$ and some equation $l \approx r \in E$. The relation $\leftrightarrow^*_E$ is more conveniently denoted by $=_E$. The relation $=_E$ can be extended to substitutions by $\sigma =_E \theta$ iff $t\sigma =_E t\theta$.

A *proof* P of $s \approx t$ in E is a sequence

$$s \leftrightarrow_{[p_1,l_1 \approx r_1,\sigma_1]} \cdots \leftrightarrow_{[p_n,l_n \approx r_n,\sigma_n]} t$$

of *proof steps*. The p_i are the positions *used by* P. A proof step at the root position Λ is called a *root step*. If $\Pi \subseteq \mathcal{P}os(s)$, then a proof P of $s \approx t$ is a *proof below* Π if any position q used by P satisfies $q \geq p$ for some $p \in \Pi$.

Let E be an equational system. The substitution σ is said to E-*unify* the equation $s \approx t$ if $s\sigma =_E t\sigma$; it E-unifies the system S if it simultaneously E-unifies each equation in S. By $u_E(S)$, we denote the set of all E-unifiers of S. Given two systems S and S', we write $S \leq_E S'$ if $u_E(S') \subseteq u_E(S)$. The corresponding *subsumption ordering* $\leq_E$ on substitutions is defined by $\sigma \leq_E \theta$ if $\sigma\theta =_E \theta$. It is not hard to see that σ is the smallest substitution that E-unifies the equational system $[\sigma]$.

A system $S = \{x_1 \approx t_1, \ldots, x_n \approx t_n\}$ is in *solved form* if each x_i occurs only once in S. An equation $s \approx t$ is called *solved* provided the system $\{s \approx t\}$ is solved. If $S = [\sigma]$ is a system in solved form and V is a set of variables, then $S\,|\,_V$ is defined by

$$S\,|\,_V = [\sigma\,|\,_V] = \{x \approx t \mid x \approx t \in S \text{ and } x \in V\}.$$

Given a system S, we say that a solved system $[\sigma]$ is an E-*solution* of S if $S \leq_E [\sigma]$. In particular, then $\sigma \in u_E(S)$. If $[\sigma]$ is an E-solution of $t \approx s$, so that $s\sigma =_E t\sigma$, then there exists a proof P of the form

$$t \stackrel{*}{\leftrightarrow}_{[\sigma],\varepsilon} t\sigma \leftrightarrow_E \ldots \leftrightarrow_E s\sigma \stackrel{*}{\leftrightarrow}_{[\sigma],\varepsilon} s \qquad (2)$$

where ε denotes the identity substitution. Such a proof P is called a *canonical* proof of $t \approx s$ in $([\sigma], E)$.

By $\mu(P)$, we denote the number of E-steps of P, and by $\mu_\sigma(P)$, the number of $[\sigma]$-steps of P. Similarly, if S is a system and P_i is a canonical proof of $t_i \approx s_i$ in $([\sigma], E)$ for each $t_i \approx s_i \in S$, then the set $\mathcal{P} = \{P_i \mid t_i \approx s_i \in S\}$ is called a canonical proof of S in $([\sigma], E)$, and $\mu(\mathcal{P}) = \Sigma_i \mu(P_i)$, and $\mu_\sigma(\mathcal{P}) = \Sigma_i \mu_\sigma(P_i)$.

Let P be a canonical proof of $t \approx s$ in $([\sigma], E)$ as in (2). We call the variable $x \in \mathcal{V}ar(t)$ *normalized* in P if there is some $p \in \mathcal{V}\mathcal{P}os(t)$ such that

(i) $t\,|\,_p = x$

(ii) if there is any step of P applying at a position above, below or at p, then leftmost such step applies above p.

As an example, the variable x is normalized in the proof

$$f(x,x,y) \leftrightarrow_{[\{x \leftarrow a, y \leftarrow b\}]} f(a,a,b) \leftrightarrow_{a \approx c} f(a,c,b) \leftrightarrow_{b \approx c} f(a,c,c)$$

because no E-step applies below the first argument position. Unlike x, the variable y is not normalized. However, it is possible to normalize y, too, by using the E-equivalent substitution $\{x \leftarrow a, y \leftarrow c\}$ in the proof

$$f(x,x,y) \leftrightarrow_{[\{x \leftarrow a, y \leftarrow c\}]} f(a,a,c) \leftrightarrow_{a \approx c} f(a,c,c).$$

In general, suppose the variable x is not normalized in P. Then if p is a position with $t|_p = x$, then the leftmost P-step applying at a position above, below or at p applies at or below p. Then P is of the form

$$t[x]_p \leftrightarrow^*_{[\sigma]} t\sigma[x\sigma]_p \leftrightarrow^*_E t'\sigma[x\sigma]_p \leftrightarrow_{(q,l\approx r,\theta)} t'\sigma[w]_p \leftrightarrow^*_E s\sigma \leftrightarrow^*_{[\sigma]} s,$$

with $q \geq p$, and in the subproof

$$t\sigma[x\sigma]_p \leftrightarrow^*_E t'\sigma[x\sigma]_p,$$

no step applies above, below or at a position r with $t|_r = x$. We define a substitution σ' by

$$y\sigma' = \begin{cases} w & \text{if } y = x \\ y\sigma & \text{otherwise} \end{cases}$$

Then $\sigma =_E \sigma'$ and the proof

$$P' = t[x]_p \leftrightarrow^*_{[\sigma']} t\sigma'[w]_p \leftrightarrow^*_E t'\sigma'[w]_p \leftrightarrow^*_E t'\sigma[w]_p \leftrightarrow^*_E s\sigma \leftrightarrow^*_E s\sigma' \leftrightarrow^*_{[\sigma']} s$$

is a proof of $t \approx s$ in $([\sigma'], E)$. Continuing this process, we eventually obtain a substitution $\overline{\sigma}$ with $\overline{\sigma} =_E \sigma$, and a proof $\overline{P}$ of $t \approx s$ in $([\overline{\sigma}], E)$ of the form

$$t[x]_p \leftrightarrow^*_{[\overline{\sigma}]} t\overline{\sigma}[w]_p \leftrightarrow^*_E s\overline{\sigma} \leftrightarrow^*_{[\overline{\sigma}]} s,$$

such that no E-step of $\overline{P}$ is below p. Then the variable x is normalized in P. Likewise, we can construct to any proof $\mathcal{P}$ of S a proof $\mathcal{P}'$ such that every variable $x \in \mathcal{V}ar(S)$ is normalized in some $P \in \mathcal{P}$. We call such a canonical proof $\mathcal{P}$ of a system S *normalized*.

2 An Inference System For E-Unification

In the following, let E be a fixed but arbitrary equational system. In order to understand the various restrictions of Lazy Paramodulation, it is useful to consider the effects they have on equational proofs. The basic idea is that if $[\sigma]$ is an E-solution of the equation $t \approx s$, then there is a canonical proof P of $t \approx s$ in $([\sigma], E)$. This proof serves as a guide to selecting an inference step. For instance, given such a proof P as in (2), we might guess the leftmost E-step $u \leftrightarrow_{[p,l\approx r,\theta]} v$, such that $p \in \mathcal{FPos}(t)$,

$$\begin{array}{cccc} t & u[l]_p & u[r]_p & s \\ \downarrow & \downarrow & \downarrow & \downarrow \\ t\sigma \leftrightarrow^*_E & u[l\theta]_p \leftrightarrow^*_E & u[r\theta]_p \leftrightarrow^*_E & s\sigma \end{array}$$

thus breaking the proof P into two parts

$$t \leftrightarrow^*_{[\sigma]} t\sigma \leftrightarrow^*_E u[l\theta]_p \leftrightarrow^*_{[\theta]} u[l]_p, \tag{3}$$

and

$$u[r]_p \leftrightarrow^*_{[\theta]} u[r\theta]_p \leftrightarrow^*_E s\sigma \leftrightarrow^*_{[\sigma]} s. \tag{4}$$

The proof (3) can be further decomposed into a subproof below the position p and another subproof using the remaining positions to yield

$$t|_p \leftrightarrow^*_{[\sigma]} (t\sigma)|_p \leftrightarrow^*_E l\theta \leftrightarrow^*_{[\theta]} l \tag{5}$$

and

$$t[*]_p \leftrightarrow_{[\sigma]} t\sigma[*]_p \leftrightarrow^*_E u[*]_p. \tag{6}$$

Here we use the symbol $*$ as a new constant, indicating a "hole" in a term. Now parts (4) and (6) yield a proof

$$t[r]_p \leftrightarrow u[r]_p \leftrightarrow_{[\theta]} u[r\theta]_p \leftrightarrow^*_E s\sigma \leftrightarrow_{[\sigma]} s \tag{7}$$

This proof transformation corresponds to the Lazy Paramodulation inference

$$\{t \approx s\} \Rightarrow \{t|_p \approx l, t[r]_p \approx s\}.$$

The derived equations have proofs (5) and (7), respectively. In this manner we can find for every E-step of the proof P that uses a function position of t a corresponding Lazy Paramodulation step.

The effect of our restriction on Lazy Paramodulation inferences now consists in a specification of the E-step to be guessed. *Almost Lazy Paramodulation* tries to guess an *innermost* E-step $u \leftrightarrow_{[p,l\approx r,\theta]} v$, with $p \in \mathcal{FP}os(t)$. For such a step, the subproof

$$(t\sigma)|_p \leftrightarrow^*_E l\theta \tag{8}$$

in (5) is below $\mathcal{VP}os(t)$. We will therefore require for an almost Lazy Paramodulation step into $t \approx s$ with $l \approx r$ at position $p \in \mathcal{FP}os(t)$ that there exist substitutions σ and θ and a proof of the form (8) below $\mathcal{VPos}(t)$. This condition will be realized by so called *top left unification*, a refinement of the notion of top unification introduced in [1].

Definition 1 (*Top Unification*) *The terms s and t are said to be* **top unifiable** *if $s(p) = t(p)$ for all $p \in \mathcal{FPos}(s) \cap \mathcal{FPos}(t)$. In this case, the set $tu(s,t)$ is defined by*

$$tu(s,t) = \{s|_p \approx t|_p \mid p \in \mathcal{Pos}(s) \cap \mathcal{Pos}(t) \cap (\mathcal{VPos}(s) \cup \mathcal{VPos}(t))\}.$$

In other words, for two terms s and t to be top unifiable means that decomposition of $s \approx t$ does not fail and eventually produces a system of the form $\{x_1 \approx t_1, \ldots, x_n \approx t_n\}$. This system is precisely the set $tu(s,t)$.

Definition 2 (*Top Left Unification*) *Given two variable disjoint terms s and t, define*

$$M(s,t) = \{(u,v) \mid u \approx x, v \approx x \in tu(s,t), x \in \mathcal{V}ar(t)\}.$$

The (ordered) pair (s, t) is said to be **top left unifiable** *if it is top unifiable and if additionally every pair $(u, v) \in M(s, t)$ is top unifiable too.*

The following example illustrates the concept of top left unification.

Example 3 *a) Let $s = f(a,b)$ and $t = f(x,x)$. Then the pair (s, t) is top unifiable, because $\mathcal{FPos}(s) \cap \mathcal{FPos}(t) = \{\Lambda\}$ and $s(\Lambda) = f = t(\Lambda)$. It is not top left unifiable, because $M(s,t) = \{(a,b)\}$ and (a,b) is not top unifiable.*

b) Let $s = f(g(u), g(v))$ and $t = f(x,x)$. Then as above, s and t are top unifiable, $M(s,t) = \{(g(u), g(v))\}$ and therefore s and t are also top left unifiable.

The following lemma is crucial for the completeness proof of our inference system for E-unification.

Lemma 4 *Let s and t be variable disjoint terms, and let σ be a substitution. If there is a canonical proof $t\sigma \leftrightarrow^*_E s\sigma$ below $\mathcal{VPos}(t)$, then the pair (t, s) is top left unifiable.*

Proof. Since there is a proof $t\sigma \leftrightarrow^*_E s\sigma$ below $\mathcal{VPos}(t)$,

$$t(p) = t\sigma(p) = s\sigma(p) = s(p)$$

holds for all $p \in \mathcal{FPos}(s) \cap \mathcal{FPos}(t)$, hence the pair (t, s) is top unifiable. Now let $(u,v) \in M(t, s)$. Then there exists an $x \in \mathcal{V}ar(s)$ such that $u \approx x$ and $v \approx x$ are both in $tu(t, s)$, and positions p and q, such that $t|_p = u$, $s|_p = x$, $t|_q = v$, and $s|_q = x$. Hence there is a proof $u\sigma \leftrightarrow^*_E x\sigma$ below $\mathcal{VPos}(u)$ and a proof $v\sigma \leftrightarrow^*_E x\sigma$ below $\mathcal{VPos}(v)$. These combine to form a proof P of the form

$$u\sigma \leftrightarrow^*_E x\sigma \leftrightarrow^*_E v\sigma$$

such that each position used by P is below $\mathcal{VPos}(u)$ or below $\mathcal{VPos}(v)$. This implies that the pair (u, v) is top unifiable. Now we have shown that (t, s) is top unifiable and that any $(u, v) \in M$ is top unifiable. The pair (t, s) thus is top left unifiable. □

The following definition introduces Almost Lazy Paramodulation.

Definition 5 *The inference system $\mathcal{T}_E$ comprises the rules shown in Figure 1.*

We write $S \Rightarrow S'$ if S' is obtained from S by application of an inference rule in $\mathcal{T}_E$.

There are two substantial differences between $\mathcal{T}_E$ and the inference systems proposed in [2] and [1]. First, $\mathcal{T}_E$ restricts Lazy Paramodulation to *unsolved equations* $s \approx t$. Second, it replaces *top unifiability* by the stronger condition of *top left unifiability*. Computation with $\mathcal{T}_E$ is illustrated by the following example.

Example 6 *Let*

$$E = \{f(x,x) \approx x, a \approx b\}$$

be an equational theory and let $S = \{f(a,b) \approx a\}$ *be an E-unification problem. The terms $f(x,x)$ and $f(a,b)$ are top unifiable with* $tu(f(x,x), f(a,b)) = \{x \approx a, x \approx b\}$. *The inference system in* [1] *thus admits the Relaxed Paramodulation inference*

$$\{f(a,b) \approx a\} \Rightarrow \{x \approx a, x \approx b, x \approx a\}.$$

This inference step is not possible in the inference system $\mathcal{T}_E$, because $f(a,b)$ and $f(x,x)$ are not top left unifiable. Instead, the unification problem S is solved via the derivation

$$\{f(a,b) \approx a\} \Rightarrow \{f(a,a) \approx a\} \qquad \text{(Almost Lazy Paramodulation with } b \approx a)$$
$$\Rightarrow \{x \approx a, a \approx a\} \qquad \text{(Almost Lazy Paramodulation with } f(x,x) \approx x)$$
$$\Rightarrow \{x \approx a\}. \qquad \text{(Trivial)}$$

Trivial	$\dfrac{\{t \approx t\} \cup S}{S}$
Decomposition	$\dfrac{\{ft_1\ldots t_n \approx fs_1\ldots s_n\} \cup S}{\{t_1 \approx s_1,\ldots,t_n \approx s_n\} \cup S}$
Variable Elimination	$\dfrac{\{x \approx t\} \cup S}{\{x \approx t\} \cup S\{x \leftarrow t\}}$

if $x \notin \mathcal{V}ar(t)$ and $x \in \mathcal{V}ar(S)$.

Almost Lazy Paramodulation	$\dfrac{\{s \approx t\} \cup S}{tu(s\vert_p \approx l) \cup \{s[r]_p \approx t\} \cup S}$

if $s \approx t$ is not solved, $p \in \mathcal{FPos}(s)$, $s\vert_p$ and l are top left unifiable, and $l \approx r \in E$.

Figure 1: The Inference System $\mathcal{T}_E$

We call an inference system T *correct* if $S \Rightarrow_T S'$ implies $S \leq_E S'$. The correctness of our inference system $\mathcal{T}_E$ follows immediately from the soundness of Gallier and Snyder's transformations.

Theorem 7 *The inference system $\mathcal{T}_E$ is correct.*

In the following, we show that the inference system $\mathcal{T}_E$ is also complete, that is, given a system S with $\mathcal{V}ar(S) = V$ and a solution $[\sigma]$ of S, there exists a derivation $S \Rightarrow^* [\sigma']$ such that $\sigma'\vert_V \leq \sigma$. Our completeness proof uses a well-founded ordering $>$ on proofs. The basic idea of the proof is as follows: Given a reducible system S, there exists a (normalized canonical) proof $\mathcal{P}$ of S. We construct a system S' with $S \Rightarrow S'$ and a proof $\mathcal{P}'$ of S' such that $\mathcal{P}' < \mathcal{P}$. The proof ordering $>$ is well-founded, so there exists a terminating derivation $S \Rightarrow^* S''$ such that S'' is irreducible by $\mathcal{T}_E$ and hence in solved form. It then remains to verify that for $S'' = [\sigma']$ the relation $\sigma'\vert_V \leq \sigma$ holds.

Definition 8 *For any proof P of $t \approx s$, we define a measure μ_1 by*

$$\mu_1(P) = |\mathcal{P}os(t)| + |\mathcal{P}os(s)|$$

Likewise, if $\mathcal{P}$ is a proof of a system S, we define $\mu_1(\mathcal{P}) = \Sigma_{P \in \mathcal{P}}\ \mu_1(P)$. We define orderings $>_1, >_2, >_3$ on proofs by

$$\mathcal{P} >_1 \mathcal{P}' \text{ if } \mu(\mathcal{P}) > \mu(\mathcal{P}')$$
$$\mathcal{P} >_2 \mathcal{P}' \text{ if } \mu_\sigma(\mathcal{P}) > \mu_\sigma(\mathcal{P}')$$
$$\mathcal{P} >_3 \mathcal{P}' \text{ if } \mu_1(\mathcal{P}) > \mu_1(\mathcal{P}').$$

Finally, the ordering $>$ on proofs is defined to be the threefold lexicographic combination $(>_1,>_2,>_3)$.

It is clear that the proof ordering $>$ is well-founded.

Lemma 9 *Let S be a system with $V = \mathcal{V}ar(S)$ and let $[\sigma]$ be an E-solution of S. Moreover, let $\mathcal{P}$ be a normalized canonical proof of S in $([\sigma], E)$.*

If S is reducible by $\mathcal{T}_E$, then there exists a system S' with $S \Rightarrow S'$ and a substitution σ' with $\sigma'|_V = \sigma$ such that S' has a normalized canonical proof $\mathcal{P}'$ in $([\sigma'], E)$ with $\mathcal{P} > \mathcal{P}'$.

Proof. We can assume without loss of generality that S contains no trivial equations.

Case 1: If $S = \{ft_1...t_n \approx fs_1...s_n\} \cup W$ and $P \in \mathcal{P}$ is a normalized canonical proof of $ft_1...t_n \approx fs_1...s_n$ in $([\sigma], E)$ that contains no root step, then for each i, $1 \leq i \leq n$, there exists a normalized canonical proof P_i of $t_i \approx s_i$ in $([\sigma], E)$. Now let $\sigma' = \sigma$,

$$S' = \{t_1 \approx s_1,..., t_n \approx s_n\} \cup W$$

and

$$\mathcal{P}' = \{P_1,..., P_n\} \cup \mathcal{P} - \{P\}.$$

Then it is clear that $\mathcal{P}'$ is a normalized canonical proof of S' in $([\sigma'], E)$ with $\mu_E(\mathcal{P}) = \mu_E(\mathcal{P}')$ and $\mu_1(\mathcal{P}) > \mu_1(\mathcal{P}')$. This implies $\mathcal{P} > \mathcal{P}'$.

Case 2: Let $S = \{t \approx s\} \cup W$, where neither s nor t is a variable, and assume $P \in \mathcal{P}$ is a normalized canonical proof of $t \approx s$ in $([\sigma], E)$ that contains a root step. Since this step uses the position $\Lambda \in \mathcal{FPos}(t)$, there is at least one proof step of P that applies at a nonvariable position of t. The proof P thus can be written in the form

$$t \overset{*}{\leftrightarrow}_{[\sigma]} t\sigma \overset{*}{\leftrightarrow}_E u \leftrightarrow_{[p,l \approx r, \eta]} u' \overset{*}{\leftrightarrow}_E s\sigma \overset{*}{\leftrightarrow}_{[\sigma]} s,$$

where the step $u \leftrightarrow u'$ is the leftmost one that applies at a nonvariable position. Hence $p \in \mathcal{FPos}(t)$, and the proof $t \overset{*}{\leftrightarrow}_{[\sigma]} t\sigma \overset{*}{\leftrightarrow}_E u$ is below $\mathcal{VPos}(t)$. The equation $l \approx r$ can be assumed to be variable disjoint from the terms s and t, and so the union $\sigma' := \sigma \cup \eta$ is well defined. From the normalized canonical proof P, we obtain normalized canonical proofs

$$P_0 = t \underset{[\sigma']}{\overset{*}{\leftrightarrow}} t\sigma' \underset{E}{\overset{*}{\leftrightarrow}} u,$$

$$P_1 = u \underset{[p, l \approx r, \sigma']}{\leftrightarrow} u',$$

and

$$P_2 = u' \underset{E}{\overset{*}{\leftrightarrow}} s\sigma' \underset{[\sigma']}{\overset{*}{\leftrightarrow}} s.$$

For $i = 0, 2$, let $m_i = \mu(P_i)$, so that

$$\mu(P) = m_0 + m_2 + 1.$$

The proof P_0 can be decomposed into the normalized canonical proofs

$$P_{01} = t|_p \underset{[\sigma']}{\overset{*}{\leftrightarrow}} (t\sigma')|_p \underset{E}{\overset{*}{\leftrightarrow}} u|_p = l\sigma' \underset{[\sigma']}{\overset{*}{\leftrightarrow}} l$$

and

$$P_{02} = t[*]_p \underset{[\sigma']}{\overset{*}{\leftrightarrow}} t\sigma'[*]_p \underset{E}{\overset{*}{\leftrightarrow}} u[*]_p.$$

Let $\mu(P_{01}) = k$, so that $\mu(P_{02}) = m_0 - k$. Since the proof P_{01} is below $\mathcal{VPos}(t)$, by Lemma 4, the pair $(t|_p, l)$ is top left unifiable. Moreover, by decomposing the proof P_{01}, we obtain a normalized canonical proof $\mathcal{P}_1$ of $tu(t|_p, l)$ in $([\sigma'], E)$ with $\mu(\mathcal{P}_1) = k$.

Likewise, from the proof P_{02} we obtain a normalized canonical proof

$$t[r]_p \underset{[\sigma']}{\overset{*}{\leftrightarrow}} t\sigma'[r\sigma']_p \underset{E}{\overset{*}{\leftrightarrow}} u[r\sigma']_p = u'$$

of length $m_0 - k$, and together with the proof P_2, we obtain a normalized canonical proof Q of $t[r]_p \approx s$ in $([\sigma'], E)$ with $\mu(Q) = m_0 - k + m_2$.

We define

$$S' = tu(t|_p, l) \cup \{t[r]_p \approx s\} \cup W$$

and

$$\mathcal{P}' = \mathcal{P}_1 \cup \{Q\} \cup (\mathcal{P} - \{P\}).$$

It is clear that $\mathcal{P}'$ is a normalized canonical proof of S' with

$$\mu(\mathcal{P}') = k + (m_0 - k + m_2) + \mu(\mathcal{P}) - (m_0 + m_2 + 1) = \mu(\mathcal{P}) - 1,$$

which implies $\mathcal{P} > \mathcal{P}'$. Finally, since $\sigma' = \sigma \cup \eta$, where $\mathcal{D}om(\eta)$ is disjoint from V, it is clear that $\sigma'|_V = \sigma$.

Case 3: Let $S = \{t \approx x\} \cup W$, with $x \in \mathcal{V}ar(t)$, and assume $P \in \mathcal{P}$ is a normalized canonical proof of $t \approx x$ in $([\sigma], E)$. Then P must contain a root step. Now the system $(S', \mathcal{P}')$ is obtained as in case 2.

Case 4: If one of cases 1 to 3 applies to S, then the assertion of the lemma is satisfied. We can thus assume any equation in S to be solved. Since S is reducible by Variable Elimination, there exists a variable $x \in V$ occurring more than once in S. Moreover, the variable x is normalized in the proof $\mathcal{P}$. $\mathcal{P}$ therefore contains a proof of the form $x \leftrightarrow_{[\sigma]} t$, and consequently S contains an equation $t \approx x$ with $t = x\sigma$. Then S is of the form $S = \{t \approx x\} \cup W$ with $x \in \mathcal{V}ar(W)$. Let $S' = \{t \approx x\} \cup W\{x \leftarrow t\}$, and let $\mathcal{P}'$ be obtained from $\mathcal{P}$ by replacing each proof step of the form $u[x] \leftrightarrow_{[\sigma]} u[t]$ by the empty proof. Then it is easy to see that $\mathcal{P}'$ is a normalized canonical proof of S' in $([\sigma], E)$. Moreover, $\mu(\mathcal{P}) = \mu(\mathcal{P}')$ and $\mu_\sigma(\mathcal{P}) > \mu_\sigma(\mathcal{P}')$, which implies $\mathcal{P} > \mathcal{P}'$.

Theorem 10 *The inference system $\mathcal{T}_E$ is complete, i.e., given an equational system S with $\mathcal{V}ar(S) = V$ and an E-solution $[\sigma]$ of S, there exists a derivation $S \Rightarrow^* [\sigma']$, such that σ' is an E-solution of S with $\sigma'|_V \leq_E \sigma$.*

Proof. If S is irreducible by $\mathcal{T}_E$, then by the proof of lemma 9, S is in solved form and the assertion of the theorem is trivially satisfied.

Now suppose S is reducible by $\mathcal{T}_E$. Then by the previous lemma, there exists a derivation

$$S = S_0 \Rightarrow S_1 \Rightarrow \ldots$$

and canonical proofs $\mathcal{P}_0, \mathcal{P}_1, \ldots$ such that $\mathcal{P}_i > \mathcal{P}_{i+1}$ for all $i \geq 0$. As the ordering $>$ on proofs is terminating, so is the derivation $S = S_0 \Rightarrow S_1 \Rightarrow \ldots$. If this derivation has length n, then $[\sigma'] = S_n$ is an E-solution of S. Moreover, by the previous lemma, $\sigma' \leq_E \sigma_n$ for some substitution σ_n with $\sigma_n|_V = \sigma$. Hence $\sigma'|_V \leq_E \sigma_n|_V = \sigma$. □

Acknowledgment

I would like to thank Patricia Johann for a thourough reading and for useful discussions on the subject of this paper.

References

1. Dougherty, D.J. and Johann, P. An Improved General E-Unification Method. *J. Symbolic Computation* **14** (1992), 303-320.

2. Gallier, J. and Snyder, W. Complete Sets of Transformations for General E-unification. *Theoretical Computer Science* **67**, 2 & 3 (1989), 203-260.

3. Snyder, W. *A Proof Theory for General E-Unification*. Birkhäuser, Boston 1991.

A Completion-Based Method for Mixed Universal and Rigid E-Unification

Bernhard Beckert

Universität Karlsruhe
Institut für Logik, Komplexität und Deduktionssysteme
76128 Karlsruhe, Germany
beckert@ira.uka.de   Tel. +49-721-608-4324

Abstract. We present a completion-based method for handling a new version of E-unification, called "mixed" E-unification, that is a combination of the classical "universal" E-unification and "rigid" E-unification. Rigid E-unification is an important method for handling equality in Gentzen-type first-order calculi, such as free-variable semantic tableaux or matings. The performance of provers using E-unification can be increased considerably, if mixed E-unification is used instead of the purely rigid version. We state soundness and completeness results, and describe experiments with an implementation of our method.

1 Introduction

We present a completion-based method for handling a new version of E-unification called "mixed" E-unification, that is a combination of the classical "universal" E-unification and "rigid" E-unification [9]. There has been a growing interest in rigid E-unification, because it is an important method for handling equality in Gentzen-type first-order calculi, such as free-variable semantic tableaux [4] or matings [9, 15]. The performance of provers using E-unification for handling equality can be increased considerably, if mixed E-unification is used instead of the purely rigid version [4].[1]

The Unfailing Knuth-Bendix-Algorithm (UKBA) [13, 1] with narrowing [14], that is generally considered to be the best algorithm for universal E-unification and has often been implemented, cannot be used to solve rigid or mixed problems. Completion-based methods for rigid E-unification have been described in [9, 10]. These, however, are non-deterministic and unsuited for implementation (they have, in fact, never been implemented). In [4] a method for solving mixed E-unification problems has been introduced that does not use completion but is based on computing equivalence classes.

---

[1] An equality has often to be applied more than once in a proof, each time with different substitutions for the variables occurring in it. In Gentzen-type calculi the mechanism to do so is to generate several instances of the equality. It is, however, often possible to recognize equalities that are "universal" w.r.t. variables they contain (e.g. equalities that occur on only one branch of a tableau). If *mixed* E-unification is used, this knowledge can be used to avoid generating additional copies of equalities.

The basic idea of our approach—and the main difference to the classical unfailing completion procedure—is that during the completion process *free* variables are never renamed, even if equalities that have variables in common are applied to each other. In addition, constraints consisting of a substitution and an *order condition* are attached to the reduction rules and terms.[2]

The paper is organized as follows: In the next section we give some basic definitions and notations; the different versions of E-unification are described in Section 3. In Sections 4 to 8 we develop our new method for mixed E-unification. Completeness and correctness results are presented in Section 9; the implementation of our method and experiments are described in Sections 10 and 11. Finally, we draw conclusions from our research in Section 12. We assume familiarity with completion-based methods [1] and (universal) E-unification [12].

2 Preliminaries

We use sequences of natural numbers to denote positions in terms; t_p is the subterm at position p in the term t (e.g. $f(a,b)_{\langle 2 \rangle} = b$). The equality predicate is denoted by $\approx$, such that no confusion with the meta-level equality $=$ can arise.

For the sake of simplicity and without any loss of generality, we use a slightly non-standard notion of substitutions: They have to be idempotent and of finite domain; **Subst** is the set of these substitutions. *id* is the empty substitution. The application of a substitution σ to a term t is denoted by $t\sigma$; if a substitution is applied to a quantified rule, equality, or term, the bound variables are never instantiated. $\leq$ denotes the specialization relation on substitutions: $\sigma \leq \tau$ iff there is a σ' such that $\sigma' \circ \sigma = \tau$.

The reduction ordering $\succ_{\text{LPO}}$ on terms is an arbitrary but fixed *lexicographic path ordering* (LPO) [7] that is total on ground terms.[3]

3 Universal, Rigid and Mixed E-Unification

To be able to mix rigid and universal E-unification, we have to use equalities (resp. reduction rules) containing two types of variables. To distinguish them syntactically, equalities $(\forall \bar{x})(l \approx r)$ and reduction rules $(\forall \bar{x})(l \rightarrow r)$ can be explicitly quantified w.r.t. variables they contain.[4]

Definition 1. A mixed E-unification problem $\langle E, s, t \rangle$ consists of a finite set E of equalities of the form $(\forall \bar{x})(l \approx r)$ and terms s and t.[5]

---

[2] In [5] a similar type of constraints is used for E-unification—but only for its purely universal version. In [4] substitutions are used to restrict the validity of terms. For a completion-based approach, however, this is not sufficient because the validity of reduction rules depends on the ordering on terms.
[3] Other reduction orderings can be used, provided the satisfiability of constraints (Def. 6) is still decidable.
[4] $(\forall \bar{x})$ is an abbreviation for $(\forall x_1) \cdots (\forall x_n)$.
[5] Without making a real restriction, we require the sets of bound and free variables in the problem to be disjoint.

A substitution σ is a **solution** to the problem, iff $E\sigma \models (s\sigma \approx t\sigma)$, where the free variables in $E\sigma$ are "held rigid", i.e. treated as constants.

The major differences between this definition and that generally given in the (extensive) literature on (universal) E-unification are: (i) the equalities in E are *explicitly* quantified (instead of considering all the variables in E to be *implicitly* universally quantified); (ii) in difference to the "normal" notion of logical consequence, free variables in $E\sigma$ are "held rigid"; (iii) the substitution σ is applied not only to the terms s und t but as well to the set E.

Definition 2. An E-unification problem $\langle E, s, t\rangle$ is **purely universal** iff there are no free variables in E; it is **purely rigid** iff there are no bound variables in E.[6]

The intention of defining different versions of E-unification is to allow the equalities in $E\sigma$ to be used differently in a proof for $E\sigma \models (s\sigma \approx t\sigma)$: in the purely universal case the equalities can be applied several times with different instantiations for the variables they contain; in the purely rigid case they can be applied more than once but with only one instantiation for each variable x (namely $\sigma(x)$); in the mixed case there are both types of variables. The following table shows some simple examples:

E	s	t	MGUs	Type
$\{f(x) \approx x\}$	$f(a)$	a	$\{x/a\}$	purely rigid
$\{f(a) \approx a\}$	$f(x)$	a	$\{x/a\}$	ground
$\{(\forall x)(f(x) \approx x)\}$	$g(f(a), f(b))$	$g(a,b)$	id	purely universal
$\{f(x) \approx x\}$	$g(f(a), f(b))$	$g(a,b)$	—	purely rigid
$\{(\forall x)(f(x,y) \approx f(y,x))\}$	$f(a,b)$	$f(b,a)$	$\{y/b\}$	mixed

The following well known [17, 9] feature of purely rigid and purely universal E-unification can be proven to be valid for mixed E-unification as well: Supposed the substitution σ is a solution to the mixed E-unification problem $\langle E, s, t\rangle$, then every specialization τ of σ is a solution to $\langle E, s, t\rangle$ as well.

Since our aim is to find *most general* unifiers (MGUs), a subsumption relation on substitutions has to be defined. One could use the specialization relation $\leq$. But, for solving mixed E-unification problems, the subsumption relation $\leq_E$ is better suited:[7]

Definition 3. Let E be a set of equalities. The subsumption relation $\leq_E$ is defined on the set of substitutions by: $\sigma \leq_E \tau$ iff there is a substitution σ' such that $E\tau \models (\sigma' \circ \sigma)(x) \approx \tau(x)$ for all variables x, where the free variables in $E\tau$ are held rigid.

The intuitive meaning of $\sigma \leq_E \tau$ is: there is a specialization of σ that can be derived from τ by applying equalities from $E\tau$. In contrary to $\leq$ the relation $\leq_E$ depends on the set E of equalities.

---

[6] If E is ground, the problem is both purely rigid and purely universal.

[7] A similar subsumption relation—for purely rigid problems—has been defined in [9].

Since purely universal E-unification is already undecidable, mixed E-unification is—in general—undecidable as well. It is, however, possible to enumerate a complete set of MGUs. Purely rigid E-unification is NP-complete and, therefore, decidable [8].[8]

Often, in particular for applications in automated theorem proving, several E-unification problems have to be solved simultaneously:

Definition 4. A finite set $\{\langle E_1, s_1, t_1\rangle, \ldots, \langle E_n, s_n, t_n\rangle\}$ of mixed E-unification problems ($n \geq 1$) is called **simultaneous** E-unification problem.

A substitution σ is a solution to the simultaneous problem iff it is a solution to every component $\langle E_k, s_k, t_k\rangle$ ($1 \leq k \leq n$).

A simultaneous E-unification problem can be solved by searching for common specializations of solutions to its components.[9]

4 Constraints

For different substitutions σ, the completion of $E\sigma$ contains different reduction rules. Nevertheless, a single completion can be computed for all $E\sigma$, if constraints are attached to the rules to restrict their validity to certain (sets of) substitutions.

The first part of the constraints we attach to reduction rules and terms is an *order condition*[10]; it expresses a restriction on the ordering of terms w.r.t. the LPO used.

Example 1. A reduction system equivalent to $E\sigma = \{x \approx y\}\sigma$ either consists of the rule $(x \to y)\sigma$ or the rule $(y \to x)\sigma$, depending on which of the terms $\sigma(x)$ and $\sigma(y)$ is greater w.r.t. the LPO used.

The expression $x \succ y$ is the natural choice for a restriction such as "the term substituted for x has to be greater than that substituted for y":

Definition 5. **Order conditions** are composed of the **atomic** order conditions $s \succ t$ (s and t are terms) using the logical connectives $\neg$, $\wedge$, $\vee$ and $\supset$, and the constants *true* and *false*.

Ground order conditions, i.e., order conditions that contain no variables, are assigned a truth value by interpreting the (predicate) symbol $\succ$ by a (fixed) LPO.

A (non-ground) order condition O is **true** iff $O\sigma$ is true for all ground substitutions σ, **false** (or inconsistent) iff its negation $\neg O$ is true, and **consistent** iff it is not false.

Since LPOs are total on ground terms, the truth value of ground order conditions is well defined; non-ground order conditions are (similar to first order formulas) either consistent or inconsistent, and may be true or false.

---

[8] In [9] a proof is given that is based on a non-deterministic algorithm for computing a *finite* set of MGUs that is complete w.r.t. $\leq_E$.
[9] Simultaneous *purely rigid* E-unification is decidable [10].
[10] These are similar to the "constraints" in [16], but there are some differences.

Example 2. The order condition $f(a) \succ a$ is true; $(x \succ y) \wedge (y \succ x)$ is false; and $x \succ y$ is consistent. The truth value of $a \succ b$ depends on the LPO used to interpret $\succ$.

In some cases, order conditions are not sufficient for describing the set of substitutions for which a reduction rule is valid:

Example 3. Suppose $E = \{f(b) \approx a,\ f(x) \approx c\}$; the reduction rule $c \to a$ is part of the completion of $E\sigma$ iff $\sigma(x) = b$ (then the equalities are a critical pair).

One could use the formula $x \approx t$ to express conditions of the form "x has to be substituted by (an instance of) t", if the predicate symbol $\approx$ were allowed in order conditions. That, however, would make the handling of conditions unnecessarily complicated. Instead the substitution $\{x/t\}$ itself becomes part of the constraint:

Definition 6. A **constraint** $c = \langle \sigma, O \rangle$ consists of a substitution σ and an order condition O such that the variables in the domain of σ do not occur in O, i.e. $O = O\sigma$.

A substitution τ **satisfies** a constraint $c = \langle \sigma, O \rangle$ iff τ is a specialization of σ and $O\tau$ is true. τ satisfies a set C of constraints iff there is a $c \in C$ satisfied by τ.

Sat(c) (resp. Sat(C)) is the set of substitutions satisfying the constraint c (resp. the set C of constraints).

Note, that sets of constraints implicitly represent *disjunctions*. To simplify the handling of constraints, we give some additional definitions and notations:

Definition 7. A constraint c_1 **subsumes** a constraint c_2 iff the substitutions satisfying c_2 satisfy as well c_1: Sat(c_2) $\subset$ Sat(c_1).

A constraint c^{-1} that is satisfied by the substitutions *not* satisfying c is called **negation** of c: Sat(c^{-1}) = **Subst** $\setminus$ Sat(c).

A constraint $c_1 \sqcap c_2$ that is satisfied by the constraints satisfying both c_1 and c_2 is called a **combination** of c_1 and c_2: Sat($c_1 \sqcap c_2$) = Sat(c_1) $\cap$ Sat(c_2).

The *empty constraint* $\epsilon = \langle id, true \rangle$ consists of the empty substitution id and the order condition $true$; it is satisfied by all substitutions.

There are efficient algorithms for computing negations and combinations of constraints. Since a constraint c_1 subsumes a constraint c_2 iff $c_1^{-1} \sqcap c_2$ is inconsistent, these are an important part of the implementation. Deciding whether a constraint is satisfiable is NP-hard [6]. The problem can however simplified considerably: The order condition $(s \succ x) \wedge (x \succ t)$ is inconsistent if there is no term between s and t (w.r.t. the LPO used). Without causing any harm, we can do without checking for such inconsistencies, that are very difficult to detect.

5 Constrained Terms and Reduction Rules

Since—syntactically—constrained reduction rules can be considered to be constrained terms,[11] it suffices to define the latter:

---

[11] Over a different signature that contains $\to$ as a function symbol.

Definition 8. A **constrained term** $\mathbf{t} = (\forall \bar{x})(t \ll c)$ is a term t with a constraint $c = \langle \sigma, O \rangle$ attached to it such that $t\sigma = t$.[12] It can be universally quantified w.r.t. some or all of the variables it contains (the quantification includes the constraint).

On first sight quantified terms may look strange, but, later on, a constrained term $\mathbf{t}$ is used to express the fact that it can be derived from another term $\mathbf{t}'$. Therefore, it is important to be able to make a distinction between rigid and non-rigid (quantified) variables.

Using constraints, for every equality an equivalent set of reduction rules can be constructed; even for those that cannot be oriented without constraints.

Example 4. The equality $f(x) \approx g(y)$ cannot be oriented without constraints, since (i) its instance $f(g(a)) \approx g(a)$ has to be oriented from left to right, while (ii) its instance $f(a) \approx g(f(a))$ has to be oriented from right to left. The constrained rules $f(x) \rightarrow g(y) \ll \langle id, f(x) \succ g(y) \rangle$ and $g(y) \rightarrow f(x) \ll \langle id, g(y) \succ f(x) \rangle$, however, define the same derivability relation as the equality $f(x) \approx g(y)$.

Other typical examples are the pair of constrained rules $x \rightarrow y \ll \langle id, x \succ y \rangle$ and $y \rightarrow x \ll \langle id, y \succ x \rangle$, that corresponds to the equality $x \approx y$; and the constrained rule $(\forall x)(\forall y)(f(x,y) \rightarrow f(y,x) \ll \langle id, f(x,y) \succ f(y,x) \rangle)$, that is equivalent to $(\forall x)(\forall y)(f(x,y) \approx f(y,x))$.

The possibility to orient every equality justifies the following definition, that assigns to each set of equalities a constrained reduction system. Since it will be the starting point of the completion process, it is called the initial system:

Definition 9. Let E be a set of equalities. Then

$$\{(\forall \bar{x})(s \rightarrow t \ll \langle id, s \succ t \rangle) \mid (\forall \bar{x})(s \approx t) \in E \text{ or } (\forall \bar{x})(t \approx s) \in E\}$$

is the **initial** constrained reduction system assigned to E.

A constrained reduction system $\mathcal{R}$ defines derivability relations $\Rightarrow_\mathcal{R}$ and $\Rrightarrow_\mathcal{R}$ on the set of constrained terms:

Definition 10. Let $\mathcal{R}$ be a constrained reduction system and $\mathbf{t} = (\forall \bar{x})(t \ll c_t)$ a constrained term. Iff there is a rule $\mathbf{r} = (\forall \bar{y})(l \rightarrow r \ll c_r)$ in $\mathcal{R}$, such that

1. $\{x_1, \ldots, x_n\} \cap \text{Var}(r) = \emptyset$ and $\{y_1, \ldots, y_m\} \cap \text{Var}(t) = \emptyset$,[13]
2. p is a position in t where $t_{|p}$ is not a variable unless $t_{|p} = l = x_i$,
3. $t_{|p}$ and l are (syntactically) unifiable with an MGU ν,
4. the combination $c_{new} = \langle \mu, O_{new} \rangle = c_t \sqcap c_r \sqcap \langle \nu, true \rangle$ is consistent,

then $\mathbf{t} \Rightarrow_\mathcal{R} \mathbf{t}'$, where $\mathbf{t}' = (\forall \bar{x})(\forall \bar{y})((t[p/r])\mu \ll c_{new})$.[14]

Iff in addition (i) $t_{|p} = l\mu$, and (ii) c_{new} subsumes c_t, then $\mathbf{t} \Rrightarrow_\mathcal{R} \mathbf{t}'$. We call the triple $\langle \mathbf{r}, p, \mu \rangle$ a justification for $\mathbf{t} \Rightarrow_\mathcal{R} \mathbf{t}'$ (resp. $\mathbf{t} \Rrightarrow_\mathcal{R} \mathbf{t}'$).

---
[12] The symbol $\ll$ means "if".
[13] This is not a real restriction, since the bound variables can be renamed.
[14] If the constraint c_{new} expresses restrictions on *bound* variables that do not occur in $t[p/r]$, these restrictions can be omitted. For example, $(\forall x)(a \rightarrow b \ll \langle id, x \succ c \rangle)$ can be reduced to $a \rightarrow b \ll \epsilon$.

The intuitive meaning of $(\forall \bar{x})(s \ll c_s) \Rightarrow_{\mathcal{R}} (\forall \bar{y})(t \ll c_t)$ is: there is a substitution σ such that $t\sigma$ can be derived from $s\sigma$ using a rule from $\mathcal{R}$, and σ satisfies the constraints c_s, c_t and that attached to the rule.

The main difference between the two derivability relations $\Rightarrow_{\mathcal{R}}$ and $\Rrightarrow_{\mathcal{R}}$ (which is a sub-relation of $\Rightarrow_{\mathcal{R}}$) is that the derivation $\mathbf{t} \Rrightarrow_{\mathcal{R}} \mathbf{t}'$ is "reversible", if the order on terms is not taken into concern. The derived term $\mathbf{t}'$ can—in combination with the rules in $\mathcal{R}$—take on the functions of $\mathbf{t}$. In contrary to that, a derivation $\mathbf{t} \Rightarrow_{\mathcal{R}} \mathbf{t}'$ is "irreversible" (provided $\mathbf{t} \not\Rrightarrow_{\mathcal{R}} \mathbf{t}'$).

Example 5. Some examples for derivations and their justification:

$(g(a,c) \ll \epsilon) \Rightarrow (g(a,b) \ll \epsilon)$ — $\langle (c \to b \ll \epsilon), \langle 2 \rangle, id \rangle$
$(f(c) \ll \epsilon) \Rightarrow (c \ll \epsilon)$ — $\langle ((\forall x)(f(x) \to x \ll \epsilon), \langle \rangle, id \rangle$
$(a \ll \epsilon) \Rightarrow (y \ll \langle \{x/a\}, a \succ y \rangle)$ — $\langle (x \to y \ll \langle id, x \succ y \rangle), \langle \rangle, \{x/a\} \rangle$
$(f(c) \ll \epsilon) \Rightarrow (c \ll \langle \{x/c\}, true \rangle)$ — $\langle (f(x) \to x \ll \epsilon), \langle \rangle, \{x/c\} \rangle$

It is useful to define a subsumption relation on constrained terms. It is similar to the relation between a term (without constraint) and its instances:

Definition 11. A constrained term $\mathbf{t}_1 = (\forall \bar{x})(t_1 \ll c_1)$ **subsumes** a constrained term $\mathbf{t}_2 = (\forall \bar{y})(t_2 \ll c_2)$, iff (i) t_2 is an instance of t_1, and (ii) the combination $c_1 \sqcap \langle \mu, true \rangle$ subsumes the constraint c_2 (Def. 7).

Example 6. The constrained term $a \ll \epsilon$ subsumes $a \ll \langle \{x/a\}, true \rangle$.

If $b \succ_{\text{LPO}} a$, then the constrained rule $x \to a \ll \langle id, x \succ a \rangle$ subsumes the rule $b \to a \ll \langle \{x/b\}, true \rangle$.

6 Completion of Constrained Reduction Systems

6.1 Goal of the Completion

The following transformation rules define a method for completing constrained reduction systems. If this rules are applied repeatedly (in a fair way) to an initial system $\mathcal{R} = \mathcal{R}^0$, a system $\mathcal{R}^\infty$ is approximated. It represents the (classical) completions of all the different instances of E.

In general, the instances of $\mathcal{R}^\infty$ will not be irreducible and, therefore, not canonical. Nevertheless, the relation $\Rightarrow_{\mathcal{R}^\infty}$ will be confluent (in a sense clarified in Lemma 19), and thus have the feature crucial for computing normal forms of constrained terms and solving E-unification problems.

The following example shows that it would not make sense to expect the instances to be canonical:

Example 7. None of the transformation rules introduced in the next section can be applied to the reduction system $\mathcal{R}^\infty = \{f(x) \to c \ll \epsilon, a \to b \ll \epsilon\}$. Nevertheless, its instance $\{f(a) \to c, a \to b\}$ is not canonical, since it can be simplified to $\{f(b) \to c, a \to b\}$.

6.2 The Transformation Rules

The rules that have to be applied to complete a reduction system are presented in form of transformation rules.[15]

Deletion A rule that has an inconsistent constraint attached to it can be removed, because it cannot be applied anyway:

(Del) $\dfrac{\mathcal{R} \cup \{(\forall \bar{x})(s \to t \ll c)\}}{\mathcal{R}}$ c inconsistent

Example 8. The rule $x \to f(x) \ll \langle id, x \succ f(x) \rangle$ can be deleted, because its constraint is inconsistent.

Subsumption A constrained rule that is subsumed by another rule (Def. 11) can be removed:

(Sub) $\dfrac{\mathcal{R} \cup \{\mathbf{r}, \mathbf{r}'\}}{\mathcal{R} \cup \{\mathbf{r}\}}$ $\mathbf{r}$ subsumes $\mathbf{r}'$

Equivalence Transformation A constraint c attached to a reduction rule can be replaced by a set $\{c_1, \ldots, c_n\}$ of constraints that—disjunctively connected—are equivalent to c (i.e. $\mathrm{Sat}(c) = \bigcup_{1 \leq i \leq n} \mathrm{Sat}(c_i)$). Since only a single constraint can be attached to a rule, n copies of the original rule are generated:

(Equ) $\dfrac{\mathcal{R} \cup \{(\forall \bar{x})(l \to r \ll c)\}}{\mathcal{R} \cup \{(\forall \bar{x})(l\sigma \to r\sigma \ll \langle \sigma, O \rangle) \mid \langle \sigma, O \rangle \in C\}}$ $\mathrm{Sat}(c) = \mathrm{Sat}(C)$, C finite

Though this equivalence rule is not necessary for the completeness of our method, it is very useful; it allows to transform constraints into a normal form, and thus simplify their handling significantly.

Example 9. The rule $f(x,y) \to f(a,b) \ll \langle id, f(x,y) \succ f(a,b) \rangle$ can be replaced by $f(x,y) \to f(a,b) \ll \langle id, x \succ a \rangle$ and $f(a,y) \to f(a,b) \ll \langle \{x/a\}, y \succ b \rangle$.

Critical Pair Rule, Combination, Simplification The transformation rules described so far allow to delete rules or to replace them by new ones without using the derivability relation $\Rightarrow_{\mathcal{R}}$. But, to complete a reduction system, $\Rightarrow_{\mathcal{R}}$ has to be taken into concern by applying one rule $\mathbf{r}_2 \in \mathcal{R}$ to another rule $\mathbf{r}_1 \in \mathcal{R}$. Suppose $\mathbf{r}_1 = (\forall \bar{x})(s \to t \ll c_1)$, $\mathbf{r}_2 = (\forall \bar{y})(l \to r \ll c_2)$, and the rule $\mathbf{r}_2$ can be applied to $\mathbf{r}_1$ to derive the rule $\mathbf{r}'_1 = (\forall \bar{x})(\forall \bar{y})(s_{new} \to t_{new} \ll c_{new})$, i.e., $\mathbf{r}_1 \Rightarrow \mathbf{r}'_1$ with a justification $\langle \mathbf{r}_2, p, \mu \rangle$. We cannot just add the new rule $\mathbf{r}'_1$ to $\mathcal{R}$: Firstly, instances of $\mathbf{r}'_1$ may be oriented differently; we therefore have to use the two symmetrical versions

$$\mathbf{r}_{new1} = (\forall \bar{x})(\forall \bar{y})(s_{new} \to t_{new} \ll c_{new} \sqcap \langle id, s_{new} \succ t_{new} \rangle)$$
$$\mathbf{r}_{new2} = (\forall \bar{x})(\forall \bar{y})(t_{new} \to s_{new} \ll c_{new} \sqcap \langle id, t_{new} \succ s_{new} \rangle) \ .$$

---
[15] The set of constrained rules below the line can be derived from the set above the line if the conditions on the right are met.

Secondly, the form of the transformation rule depends on whether (i) $\mathbf{r}_1 \Rrightarrow \mathbf{r}_1'$ (besides $\mathbf{r}_1 \Rightarrow \mathbf{r}_1'$) or not,[16] and (ii) which side of $\mathbf{r}_1$ the rule $\mathbf{r}_2$ has been applied to, i.e., whether p is a position in s or in t.

If $\mathbf{r}_1 \Rrightarrow \mathbf{r}_1'$, then $\mathbf{r}_{new1}$ and $\mathbf{r}_{new2}$ allow—together with $\mathbf{r}_2$—all the derivations possible with $\mathbf{r}_1$. If, in addition, $\mathbf{r}_2$ has been applied to the right side of $\mathbf{r}_1$, one can conclude that the constraint attached to $\mathbf{r}_{new2}$ is inconsistent. In that case the transformation is called *simplification* (Sim), since $\mathbf{r}_1$ can be replaced by the single new rule $\mathbf{r}_{new1}$:

$$\textbf{(Sim)} \quad \frac{\mathcal{R}}{(\mathcal{R} \setminus \{\mathbf{r}_1\}) \cup \{\mathbf{r}_{new1}\}} \quad p \text{ in } t,\ \mathbf{r}_1 \Rrightarrow_{\mathcal{R}} \mathbf{r}_1'$$

Else, if $\mathbf{r}_2$ has been applied to the left side of $\mathbf{r}_1$, the rule $\mathbf{r}_{new2}$ cannot be left out, because the constraint attached to it may be consistent. Such a transformation is called *composition* (Com).

$$\textbf{(Com)} \quad \frac{\mathcal{R}}{(\mathcal{R} \setminus \{\mathbf{r}_1\}) \cup \{\mathbf{r}_{new1}, \mathbf{r}_{new2}\}} \quad p \text{ in } s,\ \mathbf{r}_1 \Rrightarrow_{\mathcal{R}} \mathbf{r}_1'$$

If $\mathbf{r}_1 \not\Rrightarrow \mathbf{r}_1'$, the new rules cannot replace the old rule $\mathbf{r}_1$; it cannot be removed. Nevertheless, the transformation has to be carried out provided $\mathbf{r}_2$ has been applied to the left side of $\mathbf{r}_1$. Then $\mathbf{r}_1$ and $\mathbf{r}_2$ are a *critical pair*, and the new rules are needed to make the reduction system confluent:

$$\textbf{(CP)} \quad \frac{\mathcal{R}}{\mathcal{R} \cup \{\mathbf{r}_{new1}, \mathbf{r}_{new2}\}} \quad p \text{ in } s,\ \mathbf{r}_1 \not\Rrightarrow_{\mathcal{R}} \mathbf{r}_1'$$

In difference to the critical pair rule defined in [9] the unifier μ is only applied locally to the new rules (not to the whole system $\mathcal{R}$).

Example 10. Suppose $f \succ_{\text{LPO}} c \succ_{\text{LPO}} b \succ_{\text{LPO}} a$, and $\mathcal{R}$ contains the constrained reduction rules

$$\begin{aligned}
\mathbf{r}_1 &= f(c) \to b \ll \epsilon & \mathbf{r}_3 &= (\forall x)(f(x) \to y \ll \langle id, f(x) \succ y \rangle) \\
\mathbf{r}_2 &= b \to a \ll \epsilon & \mathbf{r}_4 &= f(x) \to y \ll \langle id, f(x) \succ y \rangle
\end{aligned}$$

The simplification rule (Sim) can be applied to $\mathbf{r}_1$ and $\mathbf{r}_2$ to replace $\mathbf{r}_1$ by the single new rule $f(c) \to a \ll \epsilon$.

The composition rule (Com) can be applied to $\mathbf{r}_1$ and $\mathbf{r}_3$ to replace $\mathbf{r}_1$ by $y \to b \ll \langle id, (f(c) \succ y \wedge y \succ b) \rangle$ and $b \to y \ll \langle id, (f(c) \succ y \wedge b \succ y) \rangle$.

The critical pair rule (CP) can be applied to $\mathbf{r}_1$ and $\mathbf{r}_4$ (note, that in $\mathbf{r}_4$ the variable x is not quantified); the new rules $y \to b \ll \langle \{x/c\}, (f(c) \succ y \wedge y \succ b) \rangle$ and $b \to y \ll \langle \{x/c\}, (f(c) \succ y \wedge b \succ y) \rangle$ have to be added.

6.3 Fair Completion Procedures

In general, an infinite number of transformation steps can be necessary to complete a reduction system. But even if the computation does not terminate, a completion $\mathcal{R}^\infty$ is approximated, consisting of the *persistent* reduction rules, that occur in all but a finite number of the resulting system. To generate a confluent reduction systems, certain fairness conditions have to be met:

---

[16] That is, $\mathbf{r}_1 \Rrightarrow \mathbf{r}_1'$ with the same justification as $\mathbf{r}_1 \Rightarrow \mathbf{r}_1'$; whether $\mathbf{r}_1 \Rrightarrow \mathbf{r}_1'$ with a different justification is not relevant.

Definition 12. $\mathcal{R} \vdash \mathcal{R}'$ means that the constrained reduction system $\mathcal{R}'$ can be derived from $\mathcal{R}$ by applying one of the transformation rules from Section 6.2.

A **transformation procedure** specifies, when supplied with an initial reduction system $\mathcal{R}^0$, in which way (in particular: in which order) the transformation rules are to be applied to generate a sequence $\mathcal{R}^0 \vdash \mathcal{R}^1 \vdash \mathcal{R}^2 \vdash \cdots$ of reduction systems. Then, the reduction system

$$\mathcal{R}^\infty = \begin{cases} \mathcal{R}^m & \text{if the sequence is of length } m \\ \bigcup_{k \geq 0} \bigcap_{m \geq k} \mathcal{R}^m & \text{if the sequence is infinite} \end{cases}$$

is called the **completion** of $\mathcal{R} = \mathcal{R}^0$, and the completion of the set E of equalities if $\mathcal{R}$ is the initial system for E.

A transformation procedure is **fair** provided:

1. There is no infinite sequence $(\mathbf{r}_i)_{i \geq 0} \subset \bigcup_{m \geq k} \mathcal{R}^m$ such that for all $i \geq 0$ the rule $\mathbf{r}_{i+1}$ has been derived from $\mathbf{r}_i$ by an equivalence transformation.
2. There is no infinite sequence $(\mathbf{r}_i)_{i \geq 0} \subset \bigcup_{m \geq k} \mathcal{R}^m$ such for all $i \geq 0$ the rule $\mathbf{r}_{i+1}$ subsumes $\mathbf{r}_i$, and $\mathbf{r}_i$ has therefore been removed.
3. For every persistent critical pair $\mathbf{r}_1, \mathbf{r}_2 \in \mathcal{R}^\infty$ there is an $i \geq 0$ such that $\mathcal{R}^{i+1}$ has been derived by applying the critical pair transformation rule to $\mathbf{r}_1, \mathbf{r}_2 \in \mathcal{R}^i$.

The first two fairness conditions are of a more technical nature: Condition 1 avoids infinite sequences of equivalence transformations. Condition 2 assures that, if there is an infinite sequence of rules subsuming each other, at least one of them is in the completion $\mathcal{R}^\infty$.

Condition 3 is the most important: it assures the application of the critical pair transformation rule to all persistent critical pairs. It is essential for achieving confluence of the completion.

Provided, the above fairness conditions are met, arbitrary heuristics can be used to choose the next transformation rule to apply.

7 Computing Normal Forms

7.1 Normalization Rules

Using constrained reduction systems and terms, a term has more than one normal form—in general an infinite number of them.

Example 11. With $\mathcal{R}^\infty = \{b \to a \ll \epsilon, d \to c \ll \epsilon\}$ the constrained term $x \ll \epsilon$ has three normal forms: $a \ll \langle\{x/b\}, true\rangle$, $c \ll \langle\{x/d\}, true\rangle$, and $x \ll \epsilon$ itself.

The above example shows that there can be redundancies in a set of normal forms: the validity of $x \ll \epsilon$ is not restricted to substitutions σ such that $\sigma(x) \neq a$ and $\sigma(x) \neq b$.

The computation of normal forms is—similar to the completion procedure—presented in form of transformation rules operating on sets of constrained terms:

Definition 13. To compute the normal forms of a set $\mathcal{T}$ of constrained terms, the rules **deletion** (Del), **equivalence** (Equ), **subsumption** (Sub), **simplification** (Sim), and **deduction** (Ded) can be applied to $\mathcal{T}$; the rules depend on a constrained reduction system $\mathcal{R}$:

(Del) $\quad \dfrac{\mathcal{T} \cup \{(\forall \bar{x})(t \ll c)\}}{\mathcal{T}} \qquad c$ inconsistent

(Equ) $\quad \dfrac{\mathcal{T} \cup \{(\forall \bar{x})(t \ll c)\}}{\mathcal{T} \cup \{(\forall \bar{x})(t\sigma \ll \langle \sigma, O \rangle) \mid \langle \sigma, O \rangle \in C\}} \qquad \begin{array}{l} \text{Sat}(c) = \text{Sat}(C), \\ C \text{ finite} \end{array}$

(Sub) $\quad \dfrac{\mathcal{T} \cup \{\mathbf{t}, \mathbf{t}'\}}{\mathcal{T} \cup \{\mathbf{t}\}} \qquad \mathbf{t}$ subsumes $\mathbf{t}'$

(Sim) $\quad \dfrac{\mathcal{T} \cup \{\mathbf{t}\}}{\mathcal{T} \cup \{\mathbf{t}'\}} \qquad \mathbf{t} \Rightarrow_\mathcal{R} \mathbf{t}'$

(Ded) $\quad \dfrac{\mathcal{T} \cup \{\mathbf{t}\}}{\mathcal{T} \cup \{\mathbf{t}, \mathbf{t}'\}} \qquad \mathbf{t} \Rightarrow_\mathcal{R} \mathbf{t}',\ \mathbf{t} \not\Rightarrow_\mathcal{R} \mathbf{t}'$

7.2 Fair Normalization Procedures

As for completion, an infinite number of normalization steps can be necessary; similar fairness conditions have to be met. A set $\mathcal{T}^\infty$ of normal forms is approximated, consisting of the *persistent* terms, that occur in all but a finite number of the sets.

Definition 14. $\mathcal{T} \vdash \mathcal{T}'$ means that the set $\mathcal{T}'$ of constrained terms can be derived from $\mathcal{T}$ by applying one of the normalization rules from Definition 13.

A **normalization procedure** specifies, when supplied with an initial set $\mathcal{T}^0$ of constrained terms and a reduction system $\mathcal{R}$, in which way the rules are to be applied to generate a sequence $\mathcal{T}^0 \vdash \mathcal{T}^1 \vdash \mathcal{T}^2 \vdash \cdots$ of sets of constrained terms. Then, the set

$$\mathcal{T}^\infty = \begin{cases} \mathcal{T}^m & \text{if the sequence is of length } m \\ \bigcup_{k \geq 0} \bigcap_{m \geq k} \mathcal{T}^m & \text{if the sequence is infinite} \end{cases}$$

is called the **set of normal forms** of $\mathcal{T} = \mathcal{T}^0$ (w.r.t. $\mathcal{R}$).

A normalization procedure is **fair** provided:

1. There is no infinite sequence $(\mathbf{t}_i)_{i \geq 0} \subset \bigcup_{m \geq k} \mathcal{T}^m$ such that for all $i \geq 0$ the term $\mathbf{t}_{i+1}$ has been derived from $\mathbf{t}_i$ by an application of equivalence (Equ).
2. There is no infinite sequence $(\mathbf{t}_i)_{i \geq 0} \subset \bigcup_{m \geq k} \mathcal{T}^m$ such that for all $i \geq 0$ the term $\mathbf{t}_{i+1}$ subsumes $\mathbf{t}_i$, and $\mathbf{t}_i$ has therefore been removed.
3. For every persistent term $\mathbf{t} \in \mathcal{T}^\infty$ that a rule $\mathbf{r} \in \mathcal{R}$ can be applied to, there is an $i \geq 0$ such that $\mathcal{T}^{i+1}$ has been derived by applying $\mathbf{r}$ to $\mathbf{t} \in \mathcal{T}^i$.

The first two fairness conditions are similar to that of fair completion procedures (Def. 12). Condition 3 assures that whenever possible deduction and simplification are applied to persistent terms.

7.3 Combining Completion and Normalization

Although a completion $\mathcal{R}^\infty$ may be infinite, one has to abandon the computation of further reduction rules at a certain point, if completion and normalization of terms are separated. It is very difficult to decide when this point is reached. Therefore, it is better to combine the completion and the normalization process:

Definition 15. A completion and normalization sequence $(\langle \mathcal{R}^i, \mathcal{T}^i \rangle)_{i \geq 0}$ consists of constrained reduction systems $\mathcal{R}^i$ and sets $\mathcal{T}^i$ of constrained terms, where (for $i \geq 0$) either (i) $\mathcal{R}^{i+1}$ has been derived from $\mathcal{R}^i$ by applying a transformation rule (Sec. 6.2) and $\mathcal{T}^i = \mathcal{T}^{i+1}$; or (ii) $\mathcal{T}^{i+1}$ has been derived from $\mathcal{T}^i$ by applying a normalization rule (Def. 13) and $\mathcal{R}^i = \mathcal{R}^{i+1}$.

Of course, when completion and normalization are combined, the fairness conditions (Def. 12 and 14) still have to be met.

8 Solving Mixed E-Unification Problems

Now we can solve an arbitrary mixed E-unification problem $\langle E, s, t \rangle$ by completing the initial reduction system $\mathcal{R}^0$ for E and computing the sets of normal forms of the constrained terms $s \ll \epsilon$ and $t \ll \epsilon$. Using these normal forms, sets $\mathcal{C}^i$ of constraints can be computed that are satisfied by solutions to the unification problem. These approximate a set $\mathcal{C}$ such that $\text{Sat}(\mathcal{C})$ is a complete set of unifiers:

Definition 16. Let $\langle E, s, t \rangle$ be a mixed E-unification problem, $\mathcal{R}^0$ the initial system for E, $\mathcal{S}^0 = \{s \ll \epsilon\}$, $\mathcal{T}^0 = \{t \ll \epsilon\}$, and $(\langle \mathcal{R}^i, \mathcal{S}^i \rangle)_{i \geq 0}$ and $(\langle \mathcal{R}^i, \mathcal{T}^i \rangle)_{i \geq 0}$ fair completion and normalization procedures. Then, for $(i = 0, 1, 2, \ldots, \infty)$ the sets $\mathcal{C}^i(\langle E, s, t \rangle)$ consist of the constraints

$$\{c_1 \sqcap c_2 \sqcap \langle \mu, \text{true} \rangle \mid (\forall \bar{x})(r_1 \ll c_1) \in \mathcal{S}^i, (\forall \bar{y})(r_2 \ll c_2) \in \mathcal{T}^i,$$
$$r_1 \text{ and } r_2 \text{ are (syntactically) unifiable with an MGU } \mu \}$$

$\mathcal{C}(\langle E, s, t \rangle)$ denotes their union $\bigcup_{i \geq 0} \mathcal{C}^i(\langle E, s, t \rangle)$.

9 Soundness, Completeness, Confluence

In this section we state soundness and completeness results for our method. Due to space restrictions the proofs are omitted; they can be found in [2].

Theorem 17 Soundness. Let $\langle E, s, t \rangle$ be a mixed E-unification problem. A substitution σ satisfying one of the constraints in $\mathcal{C}(\langle E, s, t \rangle)$ (Def. 16) is a solution to $\langle E, s, t \rangle$.

Theorem 18 Completeness. Let $\langle E, s, t \rangle$ be a mixed E-unification problem. The set $\text{Sat}(\mathcal{C}(\langle E, s, t \rangle))$ of unifiers is ground-complete w.r.t. the subsumption relation $\leq_E$ (Def. 3), i.e., for every ground unifier σ of $\langle E, s, t \rangle$ there is a substitution $\tau \in \text{Sat}(\mathcal{C}(\langle E, s, t \rangle))$ such that $\tau \leq_E \sigma$.

A ground-complete set of unifiers w.r.t. the relation $\leq$ can be computed by inverting the constrained rules in a completion $\mathcal{R}^\infty$ for E (i.e., by changing their orientation, not the validity of their constraints), and applying the inversion to the unifiers in $\mathrm{Sat}(\mathcal{C}(\langle E, s, t\rangle))$. Computing these additional solutions can be necessary—in theory—to find solutions to a simultaneous E-unification problem by combining solutions to its components. Fortunately, in practice this turns out to be very rarely the case, in particular in the semantic tableau framework.

$\stackrel{*}{\Rightarrow}_{\mathcal{R}^\infty}$ is in general not well founded. Therefore, our method is only a semi-deciding procedure for unifiability—even if the completion $\mathcal{R}^\infty$ is finite (it is an open problem, whether $\stackrel{*}{\Rightarrow}_{\mathcal{R}^\infty}$ is well founded for purely rigid E-unification problems). The following example shows that, in addition, $\stackrel{*}{\Rightarrow}_{\mathcal{R}^\infty}$ cannot be expected to be confluent:

Example 12. Supposed there are rules $f(a) \to a \ll \epsilon$ and $f(b) \to b \ll \epsilon$ in $\mathcal{R}^\infty$. Then from the constrained term $\mathbf{s} = f(x) \ll \epsilon$ terms $\mathbf{t}_1 = a \ll \langle\{x/a\}, true\rangle$ and $\mathbf{t}_2 = b \ll \langle\{x/b\}, true\rangle$ can be derived (i.e. $\mathbf{s} \Rightarrow_{\mathcal{R}^\infty} \mathbf{t}_1$ and $\mathbf{s} \Rightarrow_{\mathcal{R}^\infty} \mathbf{t}_2$).

If $\Rightarrow_{\mathcal{R}^\infty}$ were confluent, there would have to be a term derivable from both $\mathbf{t}_1$ and $\mathbf{t}_2$. That would not make any sense but contradicts soundness.

However, the derivability relation $\stackrel{*}{\Rightarrow}_{\mathcal{R}^\infty}$ can be proven to be "weak" confluent (the proof of Theorem 18 is based upon that):

Lemma 19. *If $\mathcal{R}^\infty$ is a fair completion, $\mathbf{s}$, $\mathbf{t}_1$ and $\mathbf{t}_2$ are constrained terms such that (i) $\mathbf{s} \stackrel{*}{\Rightarrow}_{\mathcal{R}^\infty} \mathbf{t}_1$ and $\mathbf{s} \stackrel{*}{\Rightarrow}_{\mathcal{R}^\infty} \mathbf{t}_2$, and (ii) the combination $c_1 \sqcap c_2$ is consistent, then there are constrained terms $\mathbf{u}_1$ and $\mathbf{u}_2$, such that (i) $\mathbf{t}_1 \stackrel{*}{\Rightarrow}_{\mathcal{R}^\infty} \mathbf{u}_1$ and $\mathbf{t}_2 \stackrel{*}{\Rightarrow}_{\mathcal{R}^\infty} \mathbf{u}_2$, and (ii) $\mathbf{u}_1$ and $\mathbf{u}_2$ have a common instance.*

10 Implementation

The completion-based method for mixed E-unification we have described, has been implemented as part of the tableau-based theorem prover $_3\mathcal{T}^A\mathcal{P}$ [3]. The implementation consists of about 2500 lines of code, written in Quintus Prolog. Besides the possibility to prove theorems from predicate logic with equality, the E-unification module can be used "stand alone" to solve simultaneous mixed E-unification problems. Complete sets of unifiers w.r.t. both $\leq$ and $\leq_E$ can be computed.[17] The experiments described in the next section have been carried out using this implementation. Upon request the source code is available from the author.

11 Experiments

If free variables occur in only one side of an equality that cannot be oriented without using constraints, a lot of different critical pairs are generated. The worst

---
[17] Because these sets are infinite in general, they can only be enumerated.

case are sets of equalities such as $E = \{x_1 \approx y_1, x_2 \approx y_2, x_3 \approx y_3\}$. Its completion consists of 126 reduction rules, that take more than 10 minutes[18] to compute. A similar example is $E = \{f(x) \approx g(y), f(f(u)) \approx g(g(v))\}$. The completion, consists of 16 rules and is computed in 6.1s; 19 critical pairs are generated. In practice, however, such equalities are only very rarely used to formulate theories; therefore, the problem does not occur too often.

The standard example for purely universal completion are the axioms from group theory. The completion generated (ten rules) is the same that is computed by an implementation of the UKBA; all rules have the empty constraint attached to them.

The following example, taken from [11], shows that our method can be superior to the UKBA for purely universal problems:

The equalities $(\forall x)(\forall y)(m(x,y) \approx m(y,x))$ and $(\forall x)(\forall y)(p(x,y) \approx p(y,x))$, expressing commutativity of m and p, combined with $(\forall x)(m(x,1) \approx x)$ and $m(a,b) \approx p(a,1)$ are difficult to complete using the UKBA, because the former cannot be oriented. Our implementation, however, computes the following completion, generating only two critical pairs (instead of 16):

$$\mathbf{r}_3 = (\forall x)(\forall y)(p(y,x) \to p(x,y) \ll \langle id, y \succ x \rangle)$$
$$\mathbf{r}_4 = (\forall x)(m(x,1) \to x \ll \epsilon)$$
$$\mathbf{r}_6 = (\forall x)(\forall y)(m(y,x) \to m(x,y) \ll \langle id, y \succ x \rangle)$$
$$\mathbf{r}_7 = m(a,b) \to p(1,a) \ll \epsilon$$
$$\mathbf{r}_8 = (\forall x)(m(1,x) \to x \ll \langle id, x \succ 1 \rangle)$$

There are other similar examples, where using constraints, the completion can be computed in a few seconds, whereas using an implementation of the UKBA, hundreds of critical pairs have to be generated, and several minutes are needed.

Experiments using the theorem prover $_3T^AP$ showed that for adding equality to semantic tableaux our method is superior to other approaches (see [4]). A major and still unsolved problem is the fact that $_3T^AP$ closes the tableau branches one after the other. In fact, finding a closing substitution for a tableau virtually never fails because a single branch could not be closed, but because the search for a single substitution closing all branches simultaneously takes too long.

12 Conclusion

The method presented is the first completion-based algorithm for mixed E-unification and the first completion-based algorithm for purely rigid E-unification that has been implemented. In addition, there are examples (Sec. 11) where it is superior to the classical UKBA for purely universal problems.

In contrary to other methods for purely rigid E-unification, the algorithm is deterministic, i.e., backtracking has never to be used to complete a system of equalities or to solve a unification problem. Moreover, the completion process does not depend on the terms to be unified. That is important for applications,

---

[18] Runtimes have been measured on a SUN SPARC 10 workstation.

where often a lot of different terms have to be unified using the same set of equalities.

Completion-based mixed E-unification is a promising way to add the handling of equality to semantic tableaux and other Gentzen-type calculi for first order logic; however, methods have to be developed for composing a substitution that closes all branches of a tableau simultaneously from a great number of substitutions closing single branches.

References

1. L. Bachmair, N. Dershowitz, and D. Plaisted. Completion without failure. In H. Aït-Kaci and M. Nivat, editors, *Resolution of Equations in Algebraic Structures, Volume 2*, chapter 1. Academic Press, 1989.
2. B. Beckert. Ein vervollständigungsbasiertes Verfahren zur Behandlung von Gleichheit im Tableaukalkül mit freien Variablen. Diploma thesis, Univ. Karlsruhe, 1993.
3. B. Beckert, S. Gerberding, R. Hähnle, and W. Kernig. The tableau-based theorem prover $_3T^Ap$ for multiple-valued logics. In *Proceedings, 11th International Conference on Automated Deduction (CADE), Albany/NY*, LNCS. Springer, 1992.
4. B. Beckert and R. Hähnle. An improved method for adding equality to free variable semantic tableaux. In *Proceedings, 11th International Conference on Automated Deduction (CADE), Albany/NY*, LNCS. Springer, 1992.
5. J. Chabin, S. Anantharaman, and P. Réty. E-unification via constrained rewriting. Unpublished, 1993.
6. H. Comon. Solving inequations in term algebras. In *Proceedings, 5th Symposium on Logic in Computer Science (LICS), Philadelphia/PA*. IEEE Press, 1990.
7. N. Dershowitz. Termination of rewriting. *J. of Symbolic Computation*, 3(1), 1987.
8. J. Gallier, P. Narendran, D. Plaisted, and W. Snyder. Rigid E-unification: NP-completeness and application to equational matings. *Information and Computation*, pages 129–195, 1990.
9. J. Gallier, P. Narendran, S. Raatz, and W. Snyder. Theorem proving using equational matings and rigid E-unification. *Journal of the ACM*, 39(2), 1992.
10. J. Goubault. Simultaneous rigid E-unifiability is NEXPTIME-complete. Technical report, Bull Corporate Research Center, 1993.
11. J. Hsiang and J. Mzali. SbREVE user's guide. Technical report, LRI, Université de Paris-Sud, 1988.
12. C. Kirchner, editor. *Unification*. Academic Press, 1990.
13. D. E. Knuth and P. B. Bendix. Simple word problems in universal algebras. In J. Leech, editor, *Computational Problems in Abstract Algebras*. Pergamon Press, Oxford, 1970.
14. W. Nutt, P. Réty, and G. Smolka. Basic narrowing revisited. *Journal of Symbolic Computation*, 7(3/4):295–318, 1989.
15. U. Petermann. A framework for integrating equality reasoning into the extension procedure. In *Proc., 2nd Workshop on Theorem Proving with Analytic Tableaux and Related Methods, Marseille*. MPI für Informatik, 92-213, Saarbrücken, 1993.
16. G. Peterson. Complete sets of reductions with constraints. In *Proceedings, 10th International Conference on Automated Deduction (CADE), Kaiserslautern*, LNCS. Springer, 1990.
17. J. Siekmann. Universal unification. *Jour. of Symbolic Computation*, 7(3/4), 1989.

On Pot, Pans and Pudding
or
How to Discover Generalised Critical Pairs

Reinhard Bündgen

Wilhelm-Schickard-Institut, Universität Tübingen
D-72076 Tübingen, Fed. Rep. of Germany
phone: x7071/295459 — fax: x7071/295958
e-mail: ⟨buendgen@informatik.uni-tuebingen.de⟩

Abstract: We develop a new critical pair criterion for term completion modulo equational theories. Our criterion relies on computing generalised critical pairs. It is compatible with most known critical pair criteria based on subconnectedness. Therefore our procedure can profit from the additional benefits of other critical pair criteria. A first test implementation has shown the practical usefulness of the new criterion for completion modulo associative and commutative theories.

1 Introduction

Term rewriting systems and term completion procedures [KB70] are powerful tools in equational theorem proving. Term completion procedures compute sets of rewrite rules which allow to decide the equivalence of two terms within an equational theory by looking at their respective normal forms. Therefore the rewrite relation must be shown to be both terminating and confluent.

For many important applications, term rewriting systems must be augmented by a built-in theory to overcome the problem of dealing with infinite reductions. In particular for the use with many algebraic domains (including automated theorem proving based on specifications of Boolean rings), some operators are known to be associative and commutative. Describing associativity and commutativity (AC) laws by rewrite rules inevitably results in infinite reductions. Therefore the AC-laws will be treated as built-in features of certain operators. As a consequence all basic operations (equality test, match, unification, reduction, ...) must be extended to operate 'modulo AC'. This idea has been worked out in [LB77, PS81, JK86].

However building in AC-laws does not come for free. In general the complexity of the completion process grows by orders of magnitude[1] when adding AC-laws. Experience has shown that most of the time needed to complete an equational specification is spent while normalising critical pairs (i. e. equations deduced from overlapping the left-hand sides of two rules) which in turn depends on the number of critical pairs computed. For plain term rewriting the number of critical pairs which can be derived from two rules is bounded by the sum of the sizes of the left-hand sides of the two rules. For rewriting modulo AC, this is no longer true, due to the fact that unification modulo AC is not unitary.

---

[1] as far as this makes sense for semi-decision procedures

The aim of this work is to reduce the number of critical pairs computed in an AC-completion process. We start with the following simple observation: Critical pairs are normally defined in terms of a *most general unifier* (or a complete set of unifiers modulo an equational theory $\mathcal{E}$). For plain rewrite systems critical pairs can also be defined using the notion of a *most general common instance* of two terms and the matching substitution which maps the terms to their common instance[2] [BL82]. Clearly the two definitions are equivalent for free terms because there the substitution matching the most general common instance *is* a most general unifier. In Buchberger's algorithm [Buc65], the key operation to compute critical pairs[3] is the computation of the least common multiple of two head monomials. These least common multiples correspond to most general instances.

Our idea is to lift this result to completion modulo equational theories. This would be extremely helpful in the case of AC-completion because the number of AC-unifiers tends to be much larger than the number of common AC-instances. And thus the number of *generalised AC-critical pairs* based on common instances would be smaller than the number of critical pairs based on common unifiers. We will show that it is in general impossible to replace $\mathcal{E}$-critical pairs by generalised $\mathcal{E}$-critical pairs. This is mainly due to the fact that for many theories (including AC) the reduction of a redex is not uniquely defined by a rule. We will however analyse under which circumstances it suffices to compute generalised critical pairs and based on this we will develop a powerful critical pair criterion. Our criterion proved to be very effective in an implementation of a Peterson-Stickel AC-completion procedure.

Critical pair criteria to speed up completion procedures have first been proposed for Buchberger's algorithm [Buc79]. This criterion has then been carried over to the Knuth-Bendix procedure [WB85, Win84, Küc86b, Küc86a]. Another line of critical pair criteria [LB79, KMN88] was motivated from the blocking technique in resolution theorem proving [Sla74]. [ZK89] exploit symmetries in rules and in [Bün91b] a criterion to 'transform' critical pairs into simpler ones is proposed. In the setting of proof transformations, Bachmair and Dershowitz [BD88] provide a uniform characterisation of critical pair criteria as proof elimination patterns. Further approaches to reduce the complexity of AC-completion can be found in [Lai89] where a different characterisation of AC-rules is defined. [BPW89, JM92] propose to use completion modulo AC1 (AC and identity) if the AC-operators allow for neutral elements. The intention behind this idea is similar to ours: The sets of complete AC1-unifiers are often (much) smaller than the respective sets of complete AC-unifiers.

The remainder of this paper is organised as follows. In the next section, we fix the notation and provide a few preliminaries. In Section 3, we analyse under which conditions a completion procedure may employ generalised critical pairs in favour of plain critical pairs. The subsequent section first shows the potential benefits of computing complete sets of common AC-instances instead of complete sets of AC-unifiers and then argues why the results of the preceding section cannot be applied to AC-completion directly. Section 5 then refines the conditions of Section 3 resulting in a new critical pair criterion. We relate the new criterion to well-known other critical pair criteria. In Section 6, we report on experimental results.

---

[2] Based on this idea, the critical pair computation procedure in [Küc82] has been designed around a call to an mgci-computation and without any explicit data structures for substitutions.

[3] i. e. s-polynomials

2 The Base Ingredients: Preliminaries

We assume the reader is familiar with the theory of term rewriting systems. For surveys on this topic see [DJ90, Klo92, Pla93].

Abstract Reduction Relations A *reduction relation* $\to_A \subseteq \mathcal{D} \times \mathcal{D}$ is an asymmetric binary relation. Then $\leftarrow_A$ is the inverse relation of $\to_A$. $\leftrightarrow_A$, $\to_A^*$ and $\leftrightarrow_A^*$ are the symmetric -, transitive and reflexive -, and the symmetric, transitive and reflexive closures of $\to_A$ respectively. The relation $\to_A$ is *terminating* if there is no infinite chain $a_1 \to_A a_2 \to_A \ldots$. The relation $\to_A$ is *confluent* if for all $a, b, c \in \mathcal{D}$ such that $b \leftarrow_A^* a \to_A^* c$ there is a $d \in \mathcal{D}$ with $b \to_A^* d \leftarrow_A^* c$. We then write $b \downarrow_A c$. To prove the confluence of a relation it often suffices to show a weaker condition. $\to_A$ is *locally confluent* if for all $a, b, c \in \mathcal{D}$ with $b \leftarrow_A a \to_A c$, $b \downarrow_A c$ follows. In [New42] Newman showed that for all terminating relations, confluence is equivalent to local confluence. Another criterion for confluence was introduced by Winkler and Buchberger in [WB85][4]. Let $\succ \supseteq \to_A$ be a strict ordering on $\mathcal{D} \times \mathcal{D}$. Then $b \leftarrow_A^* a \to_A^* c$ is *subconnected* w.r.t. $\succ$ if there are $d_1, \ldots, d_n \in \mathcal{D}$ with $a \succ d_i$ for $1 \leq i \leq n$ and $b \leftrightarrow_A d_1 \leftrightarrow_A \ldots \leftrightarrow_A d_n \leftrightarrow_A c$.

Lemma 1 (Winkler & Buchberger). *If $\succ \supseteq \to_A$ is terminating then $\to_A$ is confluent iff all $a, b, c \in \mathcal{D}$ with $b \leftarrow_A a \to_A c$ are subconnected.* □

Terms Throughout this paper, we denote by $\mathcal{F} = \bigcup_i \mathcal{F}_i$ the finite set of ranked *function symbols* (the $\mathcal{F}_n$ are the n-ary function symbols) and by $\mathcal{X}$ the set of *variables*. Then $T(\mathcal{F}, \mathcal{X})$ denotes the set of all terms freely generated by $\mathcal{F}$ and $\mathcal{X}$. For a term t, $\mathcal{X}(t)$ is the set of variables occurring in t. Let p be a *position* (or *occurrence*) in a term t. Then $t|_p$ denotes the subterm of t at position p, $t(p)$ is the symbol labelling position p and for $s \in T(\mathcal{F}, \mathcal{X})$, $t[s]_p$ is the result of replacing in t the subterm at position p by s. For two positions p and p', we say $p' < p$ if p' is a prefix of p.

A *substitution* $\sigma : \mathcal{X} \to T(\mathcal{F}, \mathcal{X})$ is a mapping from variables to terms. If we extend the application of a substitution σ to a term t, we write $t\sigma$ meaning that all variables x in t are simultaneously replaced by $\sigma(x)$. $t\sigma$ is an *instance* of t and we say that t is *more general than s* if s is an instance of t. The last two notions can be extended to substitutions in the natural way. If there is a substitution ν such that $s\nu = t\nu$ then s and t *unify*, ν is a *unifier* and $s\nu$ is a *common instance* of s and t. Two terms are *variants* (*equal modulo variable renaming*) if they are mutual instances of each other. Two substitutions σ and ρ are equal if for all $x \in \mathcal{X}$: $\sigma(x) = \rho(x)$.

Rewrite Systems A *term rewriting system* $\mathcal{R}$ is a set of *rewrite rules* $l \to r$, where l and r are terms. l is called the *left-hand side* of the rule and r is its *right-hand side*. A rule is *left-linear* if in its left-hand side every variable occurs at most once. The term t reduces in one step to s, $s \to_\mathcal{R} t$, if $s = s[l\sigma]_p$, $l \to r \in \mathcal{R}$ and $t = s[r\sigma]_p$. $s|_p$ is then called a *redex*.

In this paper, we are interested in term rewriting modulo an equational theory $\mathcal{E}$. All of the above definitions can be carried over to terms and term rewriting systems modulo $\mathcal{E}$ by setting $P(a_1, \ldots, a_n)$ modulo $\mathcal{E}$ if there are $a_i' \in [a_i]_\mathcal{E}$ such that $P(a_1', \ldots, a_n')$ for relations and properties P. We also write $\mathcal{E}$-P or $P_\mathcal{E}$ instead of P modulo $\mathcal{E}$. In particular, we write $s =_\mathcal{E} t$ if $[s]_\mathcal{E} = [t]_\mathcal{E}$. E.g. s and t *unify modulo* $\mathcal{E}$ if there is a substitution ν, $s^* \in [s\nu]_\mathcal{E}$

---

[4] For the special case of reductions in k-algebras, this criterion was already presented in [Ber78].

and $t^* \in [t\nu]_\mathcal{E}$ such that $s^* = t^*$. A set U of $\mathcal{E}$-unifiers of s and t is *complete* if for every σ with $s\sigma =_\mathcal{E} t\sigma$ there is a $\nu \in U$ and a substitution σ' such that $\sigma =_\mathcal{E} \nu\sigma'$. We write $U = CSU_\mathcal{E}(s,t)$.

In [PS81], term rewriting systems are postulated to be $\mathcal{E}$-compatible. Among others $\mathcal{E}$-*compatibility* ensures that a term s is reducible by $\mathcal{R}$ modulo $\mathcal{E}$ iff s has a redex, i.e. $s =_\mathcal{E} s[l\sigma]_p$, for some $l \to r \in \mathcal{R}$ and $p \in O(s)$. Given two rules $l_1 \to r_1$ and $l_2 \to r_2$, $p \in O(l_1)$ with $l_1(p) \notin \mathcal{X}$, $\nu \in CSU_\mathcal{E}(l_1|_p, l_2)$ then $(l_1[r_2]_p\nu, r_1\nu)$ is an $\mathcal{E}$-*critical pair* of $l_1 \to r_1$ and $l_2 \to r_2$. $l_1\nu$ is called its *superposition term*. A rewrite relation $\mathcal{R}$ is confluent modulo $\mathcal{E}$ if for all $s, t_1, t_2 \in T(\mathcal{F}, \mathcal{X})$ such that $t_1 \leftarrow_{\mathcal{R}/\mathcal{E}} s \to_{\mathcal{R}/\mathcal{E}} t_2$ there are $s_1, s_2 \in T(\mathcal{F}, \mathcal{X})$ with $s_1 =_\mathcal{E} s_2$, $t_1 \to^*_{\mathcal{R}/\mathcal{E}} s_1$ and $t_2 \to^*_{\mathcal{R}/\mathcal{E}} s_2$.

Provided that $\mathcal{E}$ allows for finite sets of complete unifiers the following *critical pair theorem* [KB70, PS81] induces an effective procedure to test for confluence of terminating term rewriting systems.

Theorem 2 (Knuth & Bendix / Peterson & Stickel). *Let $\mathcal{E}$ be an equational theory and $\mathcal{R}$ be a terminating, $\mathcal{E}$-compatible term rewriting system. Then $\mathcal{R}$ is confluent modulo $\mathcal{E}$ iff all $\mathcal{E}$-critical pairs of $\mathcal{R}$ are confluent modulo $\mathcal{E}$.* □

By Lemma 1, the $\mathcal{E}$-confluence of the critical pairs in Theorem 2 can be replaced by the requirement that each critical pair be subconnected below its superposition term.

A *term completion procedure* computes a terminating and confluent ($=$ *complete*) term rewriting system from a set of equations by (1) orienting equations to rules, (2) normalising terms in equations (and rules), (3) deleting trivial equations and (4) computing critical pairs which are considered as new equations. These steps are repeated until upon success no equations are left and no new critical pairs can be computed.

3 If ifs and ans were pots and pans ...

In this section, we want to present a *hypothetical completion procedure* based on a new critical pair theorem. As we will see, for the empty theory this new theorem coincides with the well known critical pair theorem of Knuth and Bendix [KB70] or its extensions based on subconnectedness rather than confluence [Küc86a]. We call our procedure hypothetical because we are not aware of a non-trivial theory for which this procedure does not coincide with a well-known completion procedure. However as we will see in Section 5, this procedure will lead us the way to find a new critical pair criterion.

We first state two properties of the rewrite relation and of the theory $\mathcal{E}$ modulo which we want to rewrite. These properties will be preconditions for the proof of the new theorem.

Definition 3. Let $\mathcal{E}$ be an equational theory and s and t be terms. A *complete set of common instances* of s and t modulo $\mathcal{E}$ (we write $CSCI_\mathcal{E}(s,t)$) is a set of terms such that

- for all $t^* \in CSCI_\mathcal{E}(s,t)$ there is a substitution μ such that $s\mu =_\mathcal{E} t\mu =_\mathcal{E} t^*$ and
- for all substitutions ν with $s\nu = t\nu$ there is a $t^* \in CSCI_\mathcal{E}(s,t)$ and a substitution σ such that $s\nu =_\mathcal{E} t^*\sigma$.

A set of substitutions M is a $CSCI_\mathcal{E}$-matcher of s and t modulo $\mathcal{E}$ (we write $M = CSCIM_\mathcal{E}(s,t)$) if for every $t^* \in CSCI_\mathcal{E}(s,t)$ there is a substitution $\sigma \in M$ such that $t\sigma = t^*$.

For the empty theory, $CSCI_\emptyset(s,t)$ is either empty or it consists of the most general common instance only. It is clear that normally we are interested in minimal[5] complete sets of common instances even though this is not needed for the theory to hold. Note also that in general even minimal complete sets of common instances are not uniquely defined. However we have:

Lemma 4. *Let $\mathcal{E}$ be an equational theory and s and t be terms. For all $\nu \in CSU_\mathcal{E}(s,t)$ there is a $\mu \in CSCIM_\mathcal{E}(s,t)$ and a substitution σ such that $s\nu =_\mathcal{E} s\mu\sigma$.* □

For the empty theory $CSU_\emptyset$ and $CSCIM_\emptyset$ are equal (modulo variable renaming). Unfortunately for non-empty theories, we do not always find a σ' such that $\mu =_\mathcal{E} \sigma\sigma'$. Actually, if for an arbitrary but fixed term t we have $t\sigma =_\mathcal{E} t\rho$ for two substitutions σ and ρ, we *cannot* conclude that $\sigma|_{\mathcal{X}(t)} =_\mathcal{E} \rho|_{\mathcal{X}(t)}$ where for $\mathcal{V} \subseteq \mathcal{X}$, $\sigma|_\mathcal{V}(x) = \sigma(x)$ if $x \in \mathcal{V}$ and $\sigma|_\mathcal{V}(x) = x$ otherwise.

Example 1. Let $\mathcal{E} = \{x + y \leftrightarrow y + x\}$, $a, b \in \mathcal{F}_0$ and $v, w, x, y, z, z' \in \mathcal{X}$. Then for $\sigma = \{v \mapsto a, w \mapsto b\}$, $\rho = \{v \mapsto b, w \mapsto a\}$ and $t = v + w$ we have $t\sigma =_\mathcal{E} t\rho$ but $\sigma(v) \neq_\mathcal{E} \rho(v)$.

Let $s = z + a$ then $CSU_\mathcal{E}(s,t) = \{\{v \mapsto a, w \mapsto z', z \mapsto z'\}, \{v \mapsto z', w \mapsto a, z \mapsto z'\}\}$ and we may choose $CSCIM_\mathcal{E}(s,t) = \{\{v \mapsto a, w \mapsto z', z \mapsto z'\}\}$. But there is no σ' such that $\{v \mapsto a, w \mapsto z', z \mapsto z'\}\sigma' =_\mathcal{E} \{v \mapsto z', w \mapsto a, z \mapsto z'\}$. □

Therefore we need the following characterisation of $\mathcal{E}$-unification.

Definition 5. Let $\mathcal{E}$ be an equational theory. Two terms s and t *unify nicely modulo $\mathcal{E}$* iff for any $\nu \in CSU_\mathcal{E}(s,t)$ and any $\mu \in CSCIM_\mathcal{E}(s,t)$ where $s\nu$ is an $\mathcal{E}$-instance of $s\mu$ there exists a substitution σ such that $\nu|_{\mathcal{X}(s)\cup\mathcal{X}(t)} =_\mathcal{E} \mu\sigma|_{\mathcal{X}(s)\cup\mathcal{X}(t)}$. An *equational theory $\mathcal{E}$* is called *nice* if any two terms s and t unify nicely modulo $\mathcal{E}$.

Example 2. According to the above remarks the empty theory is nice.
Let $\mathcal{E} = \{a \leftrightarrow b\}$, $a, b \in \mathcal{F}_0$, $s = f(x, a)$ and $t = f(b, y)$, then s and t unify nicely modulo $\mathcal{E}$. □

The next criterion ensures that the choice of a redex uniquely defines a rewrite of a rule. I.e. given a rule and a term with a redex at position p, there is essentially only one way to apply the rule to that redex.

Definition 6. Let $l \to r$, $l' \to r'$ be rules where $l' \to r'$ is a variant of $l \to r$. Then $l \to r$ is *top confluent (modulo $\mathcal{E}$)* if for all $\nu \in CSU_\mathcal{E}(l, l')$, $r\nu =_\mathcal{E} r'\nu$. The pair $(r\nu, r'\nu)$ is called a *top $\mathcal{E}$-critical pair*. We say $l \to r$ is *top confluent in a term rewriting system $\mathcal{R}$* if $r\nu$ and $r'\nu$ have a common $\mathcal{R}/\mathcal{E}$ normal form.

A term rewriting system $\mathcal{R}$ is *top confluent (in R) (modulo $\mathcal{E}$)* if all rules in $\mathcal{R}$ are top confluent (in R) (modulo $\mathcal{E}$).

---
[5] minimal w.r.t. the subsumption ordering

Example 3. The empty theory is top confluent. All rules whose right-hand sides are ground are top confluent. For examples of non-top confluent rules see Examples 7 and 9. □

Before we state our new critical pair theorem we introduce a generalisation of critical pairs. For the empty theory this characterisation coincides with the standard definition of critical pairs.

Definition 7. Let $l_1 \to r_1$ and $l_2 \to r_2$ be two rules, $p \in O(l_1)$ is a non-variable position in l_1 ($l_1(p) \notin \mathcal{X}$) and $\mu \in CSCIM_\mathcal{E}(l_1|_p, l_2)$. Then $(l_1[r_2]_p\mu, r_1\mu)$ is a *generalised $\mathcal{E}$-critical pair* of $l_1 \to r_1$ and $l_2 \to r_2$ (at position p).

The following Lemma can be proved by proof transformation techniques [BD88] or using the Winkler-Buchberger lemma (vis the subconnectedness criterion in [Küc86a]).

Lemma 8. *Let $\mathcal{E}$ be a nice equational theory, $\mathcal{R}$ be a $\mathcal{E}$-compatible and terminating term rewriting system, $l_1 \to r_1, l_2 \to r_2 \in \mathcal{R}$ be two (in $\mathcal{R}$) top confluent rules and all generalised $\mathcal{E}$-critical pairs of $l_1 \to r_1$ and $l_2 \to r_2$ are subconnected. Then all critical pairs of $l_1 \to r_1$ and $l_2 \to r_2$ are subconnected.*

Proof: Let $p \in O(l_1)$ where $l_1|_p \notin \mathcal{X}$ and $\nu \in CSU_\mathcal{E}(l_1|_p, l_2)$. By Lemma 4 there is a σ and a $\mu \in CSCIM_\mathcal{E}((l_1|_p, l_2)$ such that $l_1|_p\nu =_\mathcal{E} l_2\nu =_\mathcal{E} l_1|_p\mu\sigma =_\mathcal{E} l_2\mu\sigma$. Since $\mathcal{E}$ is nice, there is also a σ' with $l_1\nu = l_1\mu\sigma'$. Therefore

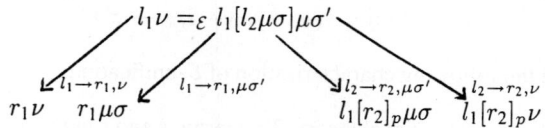

Now $(r_1\nu, r_1\mu\sigma)$ and $(l_1[r_2]_p\mu\sigma, l_1[r_2]_p\nu)$ are subconnected by top confluence of $l_1 \to r_1$ and $l_2 \to r_2$ and $(r_1\mu\sigma, l_1[r_2]_p\mu\sigma)$ is an instance of the generalised critical pair $(r_1\mu, l_1[r_2]_p\mu)$. □

Theorem 9. *Let $\mathcal{E}$ be a nice theory and $\mathcal{R}$ be a $\mathcal{E}$-compatible, terminating and (in $\mathcal{R}$) top confluent term rewriting system. Then $\mathcal{R}$ is confluent modulo $\mathcal{E}$ iff all generalised critical pairs of $\mathcal{R}$ are confluent.*

Proof: Follows from Theorem 2 and Lemma 9 □

In [ZK89] an abstract result is presented which states that from two superpositions only the most general one must be considered:

Lemma 10 (Zhang & Kapur). *Let $l_1 \to r_1, l_2 \to r_2, l_3 \to r_3 \in \mathcal{R}$, $p \in O(l_1)$, $l_1(p) \notin \mathcal{X}$, $l_1|_p\sigma = l_2\sigma$, $l_1|_p\theta = l_3\theta$. If there is a substitution τ such that $\sigma\tau = \theta$ and the critical pair $(l_1[r_2]_p\sigma, r_1\sigma)$ is confluent then the critical pair $(l_1[r_3]_p\theta, r_1\theta)$ is also confluent.* □

Note that Lemma 10 is not applicable in the proof of Lemma 8 because Lemma 10 needs a very special generalisation of critical pairs which relies on the fact that θ is an $\mathcal{E}$-instance of σ rather than $l_1\theta$ being an $\mathcal{E}$-instance of $l_1\sigma$. So our generalisation requirements are weaker.

4 Bitter Reality

In the remainder of this paper, we will try to apply our result of the previous section to theories describing associative-commutative operators. We call such theories *AC-theories*. An AC-theory is of the form

$$AC = \{f(x,y) \leftrightarrow f(y,x), f(f(x,y),z) \leftrightarrow f(x(f(y,z)) \mid f \in \mathcal{F}_{AC}\}$$

where $\mathcal{F}_{AC} \subseteq \mathcal{F}$ is a subset of the binary operators. AC-theories are very important because of their algebraic relevance. Most interesting algebraic structures (like rings, modules, polynomials, lattices, ...) contain associative-commutative operators. On the other hand AC-theories cannot be handled by pure term rewriting systems because any orientation of the AC-equations into rules violates the termination property of the resulting rewrite system. To handle AC-theories term rewriting must be generalised to term rewriting modulo AC. Term completion procedures for AC-theories have been proposed in [LB77, PS81, JK86, Lai89]. These procedures are very expensive due to the generally large number of critical pairs associated with two rules. These critical pairs must then be normalised by rather expensive AC-rewriting. The high number of critical pairs is among others due to the fact that a complete set of AC-unifiers is in general not a singleton set. AC-unification is however finitary and AC-unification algorithms have been presented in [LS76, Sti81, Fag87].

In the presence of AC-theories the application of Theorem 9 would be especially rewarding because the minimal set of complete AC-unifiers is generally much larger than the corresponding minimal complete set of common instances. Let us illustrate this by an easy example.

Example 4. Let $+ \in \mathcal{F}_{AC}, x_1, x_2, y, z, z_1, z_2, z_3 \in \mathcal{X}, s = x_1 + x_1 + x_2$ and $t = y + y + y$. Then $U = CSU_{AC}(s,t)$ has five elements:

$$U = \{ \{x_1 \mapsto z_1, \quad x_2 \mapsto z_1, \quad y \mapsto z_1\},$$
$$\{x_1 \mapsto z_1 + z_2 + z_2 + z_2, x_2 \mapsto z_1, \quad y \mapsto z_1 + z_2 + z_2\},$$
$$\{x_1 \mapsto z_1, \quad x_2 \mapsto z_1 + z_3 + z_3 + z_3, y \mapsto z_1 + z_3\},$$
$$\{x_1 \mapsto z_2 + z_2 + z_2, \quad x_2 \mapsto z_3 + z_3 + z_3, \quad y \mapsto z_2 + z_2 + z_3\},$$
$$\{x_1 \mapsto z_1 + z_2 + z_2 + z_2, x_2 \mapsto z_1 + z_3 + z_3 + z_3, y \mapsto z_1 + z_2 + z_2 + z_3\}\},$$

but $CSCI_{AC}(s,t) = \{z + z + z\}$ with $CSCIM_{AC} = \{x_1 \mapsto z, x_2 \mapsto z, y \mapsto z\}$. □

Example 5. Bürckert *et al.* published a unification benchmark for AC-unification in finitely generated Abelian semi-groups [BHK+88]. For this benchmark, we compared the orders of CSU_{AC} and $CSCI_{AC}$ for all unification problems but problem acuni-97 (computing problem acuni-97 would have cost an unreasonable amount of time and space). The average number of unifiers is 804.5 and the average number of common instances is 1.6. This number is even lower than the average number of AC1-unifiers (3.7). □

Example 6. For the examples presented in the appendix of [For87], we find 2, 5, 4, 4, 2, 2, and 35 unifiers and 2, 1, 1, 2, 2, 2 and 1 common instances. These examples also contain nested AC- and non-AC-operators. □

For further evidence of the discrepancy between $|CSU_{AC}|$ and $|CSCI_{AC}|$ see Section 6.

Unfortunately we cannot directly profit from Theorem 9 because term rewriting systems modulo AC are in general *not* top confluent as the following example illustrates.

Example 7. Let $+ \in \mathcal{F}_{AC}$ and $x, x', y, y' \in \mathcal{X}$. The rule $f(x+y) \to x$ is not top confluent because the term $f(x' + y')$ can be reduced to both x' and y'. □

This means with AC-reductions a redex and a rule do not uniquely define a rewrite. To make matters even worse, we must realize that AC is *not* a nice theory either.

Example 8. Consider the terms s and t from Example 4.

$$\nu_5 = \{x_1 \mapsto z_1+z_2+z_2+z_2, x_2 \mapsto z_1+z_3+z_3+z_3, y \mapsto z_1+z_2+z_2+z_3\} \in CSU_{AC}(s,t),$$

$$\mu = \{x_1 \mapsto z, x_2 \mapsto z, y \mapsto z\} \in CSCIM_{AC}$$

and let $s^* = f(x_1, s)$ then $s^*\mu = f(z, z + z + z)$ does not match

$$s^*\nu_5 = f(z_1 + z_2 + z_2 + z_2, z_1 + z_1 + z_1 + z_2 + z_2 + z_2 + z_2 + z_2 + z_2 + z_3 + z_3 + z_3).$$

□

So we must conclude that it is *not* sufficient to compute only generalised AC-critical pairs in order to check AC-confluence or to perform AC-completion.

5 The Tinker

As indicated by Examples 4 through 6, the potential speed ups to be gained from using generalised critical pairs instead of proper critical pairs are so promising that we want to investigate under which conditions we can exploit at least Lemma 8 in spite of the serious obstacles mentioned at the end of the previous section.

Relaxing niceness First we realize that the requirement of a nice theory is more than what is needed in the proof of Lemma 8. Let $l_1 \to r_1$ and $l_2 \to r_2$ be two top confluent rules. We can show that all critical pairs at position p can be replaced by the generalised critical pairs at position p if for every $\nu \in CSU_\mathcal{E}(l_1|_p, l_2)$ there is a $\mu \in CSCIM_\mathcal{E}(l_1|_p, l_2)$ and a substitution σ such that $l_1\nu =_\mathcal{E} l_1\mu\sigma$ while $l_2\nu$ is an $\mathcal{E}$-instance of $l_2\mu$. This requirement is clearly weaker than nice $\mathcal{E}$-unification of $l_1|_p$ and l_2. In the case that $\mathcal{E} = AC$, we can give easy criteria when this weaker requirement is fulfilled:

1. If $p = \lambda$ is the top position of l_1 the requirement follows from Lemma 4 for any $\mathcal{E}$.
2. If $l_1 \to r_1$ is left-linear, no variables at positions in $O(l_1) \setminus O(l_1|_p)$ are constrained by ν and therefore σ exists.
3. If p is such that no variable in $\mathcal{X}(l_1|_p)$ occurs in l_1 outside of $l_1|_p$ then σ exists by the same argument as above.

The above criteria are ordered w.r.t. ascending complexity of their test. The first criterion is trivial to test. Left-linearity is such an important concept in term rewriting that many implementations label the rules with left-linearity information anyway. It is also possible to preprocess each rule to be labelled by a list of (un)constrained positions as described by the third criterion.

Criteria for top confluence Now we are left with the requirement of top confluence. As experiments with AC-completion showed many rules are top confluent anyway. The following formula must be fulfilled for a rule $l \to r$ to be top $\mathcal{E}$-confluent:

$$\forall s\ \exists t\ \forall \sigma\ (l\sigma =_\mathcal{E} s \Rightarrow r\sigma =_\mathcal{E} t) \tag{1}$$

where s and t are terms and σ is a substitution. Let us develop some criteria which ensure that (1) holds.

Definition 11. Let $\mathcal{E}$ be an equational theory and $t \in T(\mathcal{F}, \mathcal{X})$. A *position* $p \in O(t)$ is $\mathcal{E}$-*constrained in* t if there is a position $p' < p$ and a term t' such that $t \leftrightarrow^*_\mathcal{E} t'$ and an equation of $\mathcal{E}$ can be applied to the top of $t'|_{p'}$.

A *variable* $x \in \mathcal{X}(t)$ is $\mathcal{E}$-*constrained in* t if there is an $\mathcal{E}$-constrained position $p \in O(t)$ with $t|_p = x$.

In the case of $\mathcal{E} = AC$ it is easy to test whether a position is AC-constrained: p is AC-constrained in t if there is a $p' < p$ with $t(p') \in \mathcal{F}_{AC}$.

Lemma 12. *Let $\mathcal{E}$ be an equational theory, $s, t \in T(\mathcal{F}, \mathcal{X})$ and $p \in O(s)$ is not $\mathcal{E}$-constrained in t. Then for all substitutions σ, σ' with $s\sigma =_\mathcal{E} s\sigma' =_\mathcal{E} t$ it follows that $s|_p \sigma =_\mathcal{E} s|_p \sigma'$.* □

In particular, for a not $\mathcal{E}$-constrained variable $x = s(p)$ we have $\sigma(x) =_\mathcal{E} \sigma'(x)$. Now we can state the following criteria for top confluence of a rule $l \to r$:

1. no variable in $\mathcal{X}(r)$ is $\mathcal{E}$-constrained or
2. there is at most one AC-constrained variable in l or
3. if r contains variables which are $\mathcal{E}$-constrained in l then for each such variable occurrence $p \in O(r)$ there are occurrences $p' < p$ and $q \in O(l)$ such that $r|_{p'} =_\mathcal{E} l|_q$ and q is not $\mathcal{E}$-constrained in l.

This list of criteria is not complete for our purposes because for Lemma 8 it is sufficient that the rules are top confluent in $\mathcal{R}$ which is weaker than pure top confluence.

Example 9. Let $\{\cdot, +\} \subseteq \mathcal{F}_{AC}$, $x, y, z \in \mathcal{X}$. The rule $R = x \cdot (y + z) \to (x \cdot y) + (x \cdot z)$ is not top confluent because it has a non-trivial top AC-critical pair

$$((x + x') \cdot y) + ((x + x') \cdot z), ((y + z) \cdot x) + ((y + z) \cdot x').$$

But R is top confluent in $\{R\}$. □

Main Result Another alternative to ensuring top confluence by the above criteria is that in a completion procedure we 'promise' to eventually make all rules top confluent by computing all top $\mathcal{E}$-critical pairs of all rules. Technically speaking this means that we require all top critical pairs to be subconnected instead of confluent.

Lemma 13. *Let $\mathcal{E}$ be an equational theory, $\mathcal{R}$ be an $\mathcal{E}$-compatible and terminating term rewriting system, $l_1 \to r_1, l_2 \to r_2 \in \mathcal{R}$ be two rules whose top critical pairs are subconnected, $p \in O(l_1)$ (with $l_1(p) \notin \mathcal{X}$) be a non-variable position in l_1. Further*

1. *for every $\nu \in CSU_\mathcal{E}(l_1|_p, l_2)$ there is a $\mu \in CSCIM_\mathcal{E}(l_1|_p, l_2)$ and a substitution σ such that $l_1\nu =_\mathcal{E} l_1\mu\sigma$ and $l_2\nu$ is an $\mathcal{E}$-instance of $l_2\mu$, and*
2. *all generalised $\mathcal{E}$-critical pairs of $l_1 \to r_1$ and $l_2 \to r_2$ at position p are subconnected.*

Then all $\mathcal{E}$-critical pairs of $l_1 \to r_1$ and $l_2 \to r_2$ at position p are subconnected.

Proof: Analogous to that of Lemma 8. Top confluence can be replaced by subconnectedness of the $\mathcal{E}$-top critical pairs and (1.) takes the part of $\mathcal{E}$ being nice. □

Example 10. Let $v, w, w', x, y, z \in \mathcal{X}$, $\cdot, + \in \mathcal{F}_{AC}$, $R_1 = x \cdot (y + z) \to (x \cdot y) + (x \cdot z)$ and $R_2 = v + -(v) \to 0$. Unifying R_2 and $y + z$ at position 2 of the left-hand side of R_1 results in a complete AC-unifier with four elements. From this unifier we can derive two critical pairs (from the four pairs which can be computed each two are equal modulo AC), but there is only one generalised AC-critcal pair $(w \cdot 0, (w \cdot w') + (w \cdot -(w')))$. □

Lemma 13 leaves us with three classes of critical pairs to be computed:

1. top critical pairs of all non-top confluent rules,
2. the generalised critical pairs covered in Lemma 13 and
3. all remaining critical pairs.

Actually classes 1 and 3 need not be distinguished and of course we may decide to compute proper critical pairs even when the computation of generalised critical pair would suffice.

Theorem 14. *Let $\mathcal{E}$ be an equational theory and $\mathcal{R}$ be an $\mathcal{E}$-compatible and terminating term rewriting system. Then $\mathcal{R}$ is confluent modulo $\mathcal{E}$ iff*

1. *all top critical pairs of $\mathcal{R}' \subseteq \mathcal{R}$ are subconnected and*
2. *all generalised $\mathcal{E}$-critical pairs are subconnected for all $l_1 \to r_1, l_2 \to r_2 \in \mathcal{R}'$ and positions $p \in O(l_1)$ (with $l_1(p) \notin \mathcal{X}$) where for every $\nu \in CSU_\mathcal{E}(l_1|_p, l_2)$ there is a $\mu \in CSCIM_\mathcal{E}(l_1|_p, l_2)$ and a substitution σ such that $l_1\nu =_\mathcal{E} l_1\mu\sigma$ and $l_2\nu$ is an $\mathcal{E}$-instance of $l_2\mu$ and*
3. *all remaining $\mathcal{E}$-critical pairs are subconnected.* □

For $\mathcal{R}' = \emptyset$, Theorem 14 reduces to the standard critical pair theorem (stated using subconnectedness). It is up to the implementor to choose an appropriate $\mathcal{R}'$, to decide whether to include only top confluent rules in $\mathcal{R}'$ or to compute the top critical pairs of $\mathcal{R}'$ and to decide according to which 'niceness criteria' he wants to apply Lemma 13. In our test implementation we chose $\mathcal{R}' = \mathcal{R}$ (i.e. we compute all top critical pairs) and use the 'niceness criteria' 1–3 presented above.

Critical Pair Criteria Critical pair criteria to reduce the number of critical pairs for which confluence must be proven have been proposed before. We want to relate our results to these criteria and show how the benefits of the different approaches can be combined.

In [Küc85, Küc86b, Küc86a], Küchlin proposed the *subconnectedness criterion*. Roughly speaking this criterion can eliminate subconnected critical pairs. It detects subconnected pairs by showing that the equality proof of the pair can be split into two smaller proofs.

Lemma 15 (Küchlin). *Let $s, u, v, w \in T(\mathcal{F}, \mathcal{X})$ and $R_0, R_1, R_2 \in \mathcal{R}$ be rules such that*

$$\begin{array}{ccc} & s & \\ R_0 \swarrow & \downarrow R_1 & \searrow R_2 \\ u & v & w. \end{array}$$

Then (u, w) is subconnected below s if both (u, v) and (v, w) are subconnected below s. □

Depending on the relative positions of the occurrences in s where the rules R_0, R_1 and R_2 apply (u, v) and (v, w) are either confluent by default or can be traced back to critical pairs between an outer rule R_0 or R_2 and the middle rule R_1. As already pointed out by Küchlin this criterion obtains its full strength if it is applied recursively. That is we impose a well founded ordering $\gg$ on the critical pairs[6] such that in order to apply Lemma 15 to a critical pair (u, w) all critical pairs which must be subconnected to ensure the subconnectedness of (u, w) must be smaller than (u, w). Several well founded orderings on critical pairs have been proposed so far. In [Küc86b], Küchlin suggests to order the pairs according to their creation time (i. e. a rule $l \to r$ may be used as middle rule only if its pairs with the outer rules have been handled before). Others restrict the position of the middle rule applications. Winkler and Buchberger allow only middle reductions on the path between the top of the outer rule and the position p of the superposition [WB85]. Kapur, Musser and Narendran allow only middle reductions inside the redex of the inner rule [KMN88].

Lemma 13 can be seen as a special case of Küchlin's subconnectedness criterion because for all $\nu \in CSU_\mathcal{E}(l_1|_p, l_2)$, the superposition term $l_1\nu$ can be reduced 'in the middle' by $l_2 \to r_2$ using a matcher $\mu\sigma$ with $\mu \in CSCIM_\mathcal{E}(l_1|_p, l_2)$. Note that for critical pairs (u_i, v_i) derived from superposing rules $l \to r$ and $l' \to r'$ at a position $p \in O(l)$ using a unifier ν_i, we can safely impose a well founded ordering $(u_i, v_i) \gg_g (u_j, v_j)$ if $\nu_j \in CSCIM_\mathcal{E}(l|_p, l')$ and $\nu_i \in CSU_\mathcal{E}(l|_p, l') \setminus CSCIM_\mathcal{E}(l|_p, l')$. Now it is clear that we can combine Lemma 13 with any implementation of the subconnectedness criterion provided the implementation relies on an critical pair ordering which is compatible with the ordering $\gg_g$ defined above. In particular, we can take advantage of the implementation proposed by Küchlin. Applying the subconnectedness criterion to a generalised critical pair is then equivalent to multiple parallel applications of the criterion to the according proper critical pairs. I. e. Lemma 13 then multiplies the effect of a single application of Lemma 15. Similarly our criterion can be combined with the critical pair transformation criterion [Bün91b] which amounts to a look ahead to a future application of a rule collapse with subsequent application of the subconnectedness criterion.

If we agree to really compute the generalised critical pairs (i. e. the generalised critical pairs are minimal w. r. t. $\gg$) then we can use any implementation of the subconnectedness criterion (including those proposed in [WB85], [KMN88] and [ZK89]) to discard critical pairs not covered by Lemma 13.

---

[6] or equivalently on the origination information of the critical pairs: the rules, the position where the rules superpose and the unifier.

6 The Proof of the Pudding is in the Eating

It remains to show that what is left over from Theorem 14 is still a powerful criterion in practice. We have implemented a first version of a Peterson-Stickel AC-completion procedure based on Theorem 14 using ReDuX [Bün93] and compared it with a plain Peterson-Stickel procedure. Our implementation sets $\mathcal{R}' = \mathcal{R}$ (i. e. we compute all top critical pairs) and uses the 'niceness criteria' 1.–3. of Section 5. Since the author is not aware of an efficient algorithm to compute complete sets of common instances modulo AC, a very naive procedure was used. It is based on computing a complete set of unifiers and then determining the $CSCIM_{AC}$ by mutually matching the respective instances. So far, criteria for top confluence and combinations with other critical pair criteria have not been realized.

Table 1 compares the behaviours of a Peterson-Stickel procedure (A) and the enhanced procedure computing generalised critical pairs (B). The experiments were run on a Sun 10-20 under SunOS 4.1.3. Columns two and three show the numbers of unifiers computed for the critical pair computation. This corresponds to the number of critical pairs which must be tested for confluence. Columns four through six show the number of matches needed. Matches are mainly needed in the reduction procedures and the extension rule computations. The number for B includes also the matches needed to compute the $CSCI_{AC}$ (column 6).

problem	# unifications A	B	# matches A	B	for mgci	total time (ms) A	B
aring1	876	610	54,974	21,774	640	11,305	6,698
acring1	671	524	24,118	14,839	377	8,313	6,341
dlattice	335	258	8,252	5,075	197	9,486	7,327
ZV2	147	105	4,272	2,611	62	1,003	680
RX1C	1,314	769	435,394	117,507	3,348	1,234,574	119,765
R235	2,757	2,039	1,315,426	811,341	5,928	794,308	541,688

Table 1. Completions with critical pairs (A) vs. generalised critical pairs (B)

aring1, *acring1* and *dlattice* stand for the completion of a ring with 1, a commutative ring with 1 and a distributive lattice. *ZV2* and *RX1C* are confluence tests for complete systems where ZV2 specifies the integers together with $\sqrt{2}$ and RX1C specifies the Gröbner base $\{xy^2 - x, x^2 - 4y^2, 4y^3 - 4y\}$ over integral polynomial rings as proposed in [Bün91a]. R235 is the completion of the complete specification of the integers together with 1/2 and 1/3 and the two equations $\frac{1}{5} + \frac{1}{5} + \frac{1}{5} + \frac{1}{5} + \frac{1}{5} \leftrightarrow 1$ and $-\frac{1}{5} \leftrightarrow -1 + \frac{1}{5} + \frac{1}{5} + \frac{1}{5} + \frac{1}{5}$.

The examples above show improvements by 20 – 40%, but also extreme improvements by a factor 10 like for RX1C are possible. The matches needed in the $CSCI_{AC}$-computation tend to be rather 'heavy' matches. Whereas the number of matches in the $CSCI_{AC}$-computation make up less than 4% of the total number of matches, the time spent in the first kind of matches amounts to 5–60% of the total time spent during the matches. Note that the total matching time may make up more than 90% of the whole completion time. Therefore a more efficient algorithm to compute the $CSCI_{AC}$ or the $CSCIM_{AC}$ can considerably affect the run time of the whole completion.

7 Conclusion

We have examined in how far the confluence of generalised critical pairs (i. e. critical pairs based on most general common instance computations) is sufficient to prove the confluence of term rewriting systems modulo an equational theory. This leads to a new critical pair criterion. A first test implementation has shown that this criterion performs well in AC-completion procedures. To profit even more from this criterion it would be desirable to have an efficient algorithm to compute the complete set of common AC-instances of two terms. Even though our criterion is compatible with many other known critical pair criteria, additional practical experience with the combination of different criteria is still needed. So far we developed our theory based on the completion procedure described in [PS81], but we are confident that using the results in [Küc86a] our criterion can be carried over to completion procedures in the setting of [JK86].

8 Acknowledgements

I thank Wolfgang Küchlin for many discussions on the subject of critical pair criteria.

References

[BD88] Leo Bachmair and Nachum Dershowitz. Critical pair criteria for completion. *Journal of Symbolic Computation*, 6:1–18, 1988.

[Ber78] George M. Bergman. The diamond lemma for ring theory. *Advances in Mathematics*, 29:178–218, 1978.

[BHK+88] H.-J. Bürckert, A. Herold, D. Kapur, J. H. Siekmann, M. E. Stickel, M. Tepp, and H. Zhang. Opening the ac-unification race. *Journal of Automated Reasoning*, 4:465–474, 1988.

[BL82] Bruno Buchberger and Rüdiger Loos. Algebraic simplification. In *Computer Algebra*, pages 14–43. Springer-Verlag, 1982.

[BPW89] Timothy B. Baird, Gerald E. Peterson, and Ralph W. Wilkerson. Complete sets of reductions modulo associativity, commutativity and identity. In Nachum Dershowitz, editor, *Rewriting Techniques and Applications (LNCS 355)*, pages 29–44. Springer-Verlag, 1989. (Proc. RTA'89, Chapel Hill, NC, USA, April 1989).

[Buc65] Bruno Buchberger. *Ein Algorithmus zum Auffinden der Basiselemente des Restklassenringes nach einem nulldimensionalen Polynomideal*. PhD thesis, Universität Innsbruck, 1965.

[Buc79] Bruno Buchberger. A criterion for detecting unnecessary reductions in the construction of Gröbner-Bases. In E. Ng, editor, *Symbolic and Algebraic Computing (LNCS 72)*, pages 3–21. Springer-Verlag, 1979. (Proc. EUROSAM'79, Marseille, France).

[Bün91a] Reinhard Bündgen. Completion of integral polynomials by AC-term completion. In Stephen M. Watt, editor, *International Symposium on Symbolic and Algebraic Computation*, pages 70 – 78, 1991. (Proc. ISSAC'91, Bonn, Germany, July 1991).

[Bün91b] Reinhard Bündgen. *Term Completion Versus Algebraic Completion*. PhD thesis, Universität Tübingen, D-7400 Tübingen, Germany, May 1991. (reprinted as report WSI 91-3).

[Bün93] Reinhard Bündgen. Reduce the redex → ReDuX. In Claude Kirchner, editor, *Rewriting Techniques and Applications (LNCS 690)*, pages 446–450. Springer-Verlag, 1993. (Proc. RTA'93, Montreal, Canada, June 1993).

[DJ90] Nachum Dershowitz and Jean-Pierre Jouannaud. Rewrite systems. In Jan van Leeuwen, editor, *Formal Models and Semantics*, volume B of *Handbook of Theoretical Computer Science*, chapter 6. Elsevier, 1990.

[Fag87] Fronçois Fages. Associative commutative unification. *Journal of Symbolic Computation*, 3:257–275, 1987.

[For87] Albrecht Fortenbacher. An algebraic approach to unification under associativity and commutativity. *Journal of Symbolic Computation*, 3:217–229, 1987.

[JK86] Jean-Pierre Jouannaud and Hélène Kirchner. Completion of a set of rules modulo a set of equations. *SIAM J. on Computing*, 14(4):1155–1194, 1986.

[JM92] Jean-Pierre Jouannaud and Claude Marché. Termination and completion modulo associativity, commutativity and identity. *Theoretical Computer Science*, 104:29–51, 1992.

[KB70] Donald E. Knuth and Peter B. Bendix. Simple word problems in universal algebra. In J. Leech, editor, *Computational Problems in Abstract Algebra*. Pergamon Press, 1970. (Proc. of a conference held in Oxford, England, 1967).

[Klo92] Jan Willem Klop. Term rewriting systems. In S. Abramsky, D. M. Gabbay, and T. S. E. Maibaum, editors, *Background: Computational Strcutures*, volume 2 of *Handbook of Logic in Computer Science*, chapter 1. Oxford University Press, 1992.

[KMN88] Deepak Kapur, David R. Musser, and Paliath Narendran. Only prime superpositions need be considered in the Knuth-Bendix completion procedure. *Journal of Symbolic Computation*, 6:19–36, 1988.

[Küc82] Wolfgang Küchlin. An implementation and investigation of the Knuth-Bendix completion algorithm. Master's thesis, Informatik I, Universität Karlsruhe, D-7500 Karlsruhe, W-Germany, 1982. (Reprinted as Report 17/82.).

[Küc85] Wolfgang Küchlin. A confluence criterion based on the generalised Knuth-Bendix algorithm. In B. F. Caviness, editor, *Eurocal'85 (LNCS 204)*, pages 390–399. Springer-Verlag, 1985. (Proc. Eurocal'85, Linz, Austria, April 1985).

[Küc86a] Wolfgang Küchlin. *Equational Completion by Proof Transformation*. PhD thesis, Swiss Federal Institute of Technology (ETH), CH-8092 Zürich, Switzerland, June 1986. (Also as *Equational Completion by Proof Simplification*, Report 86-02, Mathematics, ETH Zürich, May 1986).

[Küc86b] Wolfgang Küchlin. A generalized Knuth-Bendix algorithm. Technical Report 86-01, Mathematics, Swiss Federal Institute of Technology (ETH), CH-8092 Zürich, Switzerland, January 1986.

[Lai89] Mike Lai. On how to move mountains 'associatively and commutatively'. In Nachum Dershowitz, editor, *Rewriting Techniques and Applications (LNCS 355)*, pages 187–202. Springer-Verlag, 1989. (Proc. RTA'89, Chapel Hill, NC, USA, April 1989).

[LB77] Dallas Lankford and A. M. Ballantyne. Decision procedures for simple equational theories with commutative-associative axioms: Complete sets of commutative-associative reductions. Technical Report Report ATP-39, Department of Mathematics and Computer Sciences, University of Texas, Austin, August 1977.

[LB79] Dallas Lankford and A. M. Ballantyne. Blocked permutative narrowing and resolution. In *Proc. 4th Workshop on Automated Deduction*, pages 168–174, Austin, Texas, February 1979. (Corrigendum of June, 1979, Addendum of January, 1984).

[LS76] M. Livesey and J. Siekmann. Unification of bags and sets. Technical report, Institut für Informatik I, Universität Karlsruhe, 7500 Karlsruhe, Fed. Rep. of Germany, 1976.

[New42] M. H. A. Newman. On theories with a combinatorial definition of "equivalence". *Annals of Mathematics*, 43(2):223–243, 1942.

[Pla93]   David Plaisted. Equational reasoning and term rewriting systems. In D. M. Gabbay, C. J. Hogger, and J. A. Robinson, editors, *Logical Foundations*, volume 1 of *Handbook of Logic in Artificial Intelligence and Logic Programming*, chapter 5. Oxford University Press, 1993.

[PS81]   G. Peterson and M. Stickel. Complete sets of reductions for some equational theories. *Journal of the ACM*, 28:223–264, 1981.

[Sla74]   J. R. Slagle. Automated theorem-proving for theories with simplifiers, commutativity, and associativity. *Journal of the ACM*, 21(4):622–642, October 1974.

[Sti81]   Mark E. Stickel. A unification algorithm for associative-commutative functions. *JACM*, 28(3):423–434, July 1981.

[WB85]   Franz Winkler and Bruno Buchberger. A criterion for eliminating unnecessary reductions in the Knuth-Bendix algorithm. In *Proc. Colloquium on Algebra, Combinatorics and Logic in Computer Science*. J. Bolyai Math. Soc., J. Bolyai Math. Soc. and North-Holland, 1985. (Colloquium Mathematicum Societatis J. Bolyai, Györ, Hungary, 1983).

[Win84]   Franz Winkler. *The Church-Rosser Property in Computer Algebra and Special Theorem Proving: An Investigation of Critical-Pair/Completion Algorithms*. PhD thesis, Johannes Kepler Universität Linz, May 1984.

[ZK89]   Hanatao Zhang and Deepak Kapur. Consider only general superpositions in completion procedures. In Nachum Dershowitz, editor, *Rewriting Techniques and Applications (LNCS 355)*, pages 513–527. Springer-Verlag, 1989. (Proc. RTA'89, Chapel Hill, NC, USA, April 1989).

Semantic Tableaux with Ordering Restrictions

Stefan Klingenbeck[*] and Reiner Hähnle

University of Karlsruhe
Institute for Logic, Complexity and Deduction Systems
76128 Karlsruhe, Germany
{klingenb,reiner}@ira.uka.de   +49-721-608-{3978,4329}

Abstract. The aim of this paper is to make restriction strategies based on orderings of the Herbrand universe available for semantic tableau-like calculi as well. A marriage of tableaux and ordering restriction strategies seems to be most promising in applications where generation of counter examples is required. In this paper, starting out from semantic trees, we develop a formal tool called *refutation graphs*, which (i) serves as a basis for completeness proofs of both resolution and tableaux, and (ii) is compatible with so-called A-ordering restrictions. The main result is a first-order ground tableau procedure complete for A-ordering restrictions.

Introduction

In recent years one could observe a kind of renaissance of tableau-related methods in automated theorem proving after the field has been dominated by resolution approaches for many years[2]. Tableaux are easy to adjust to nonclassical logics, and they have already a number of advantages for classical first-order logic that have not been paid great attention to for some time, notably: (i) they have a higher ground speed than resolution; (ii) it is easy to incorporate theories [10]; (iii) there are many refinements to remove redundancy from proofs [13, 12]; (iv) lemmaizing and caching can be naturally defined and implemented [1]. All these techniques can be efficiently implemented using Prolog technology [13]. Finally, (v) tableaux are suitable for human interaction. This is a major advantage in program verification, where frequently very large or not valid formulas occur. In the latter case also the use of specialised decision procedures e.g. for certain arithmetical theories is very helpful.

The basic idea of order restricted resolution is as follows:[3] assume we are given a partial ordering $<$ on first-order terms. Now admit only resolution steps wherein the resolved literal is $<$-maximal in the resolvent clause. For certain

---

[*] Supported by BMFT within the project KORSO.
[2] We assume the reader is familiar with basic expositions of semantic tableaux like [8] and of resolution like [6].
[3] In order to gain familarity with ordering restricted resolution we recommend [7] as an up-to-date account.

orderings this restriction turn out to be still complete for first-order logic. Moreover, resolution with (variants of) ordering restrictions can be used to decide certain classes of first-order logic [7] and to enhance the performance of basic resolution [18]. Hence, in the light of applications such as program verification it is extremely desirable to make ordering refinements available for tableaux as well, thus bringing together paradigms that have already been proven successful separately.

The part of resolution that corresponds to the selection of a closure substitution in tableau is the selection of a pair of parent clauses and complementary literals therein. A wrong choice does not require backtracking, since resolution is *proof confluent*, but an unnecessary clause can in turn produce many useless clauses, and if resolution proofs are unsuccessful, then usually because the system gets choked with useless clauses.

Every reasonable tableau procedure with free variables and unification must employ backtracking over the possible substitutions that close a branch. For complex problems the resulting number of proofs becomes simply too large even if techniques like caching [1], anti-lemmata or regularity conditions [13] are used to reduce the number of choices. For this reason ordering restrictions are obvious candidates for further enhancement of tableaux, as they are well understood in the case of resolution.

The paper is organized as follows: in Section 1 for convenience of reading we repeat some basic definitions; in Section 2 we introduce refutation graphs which constitute the central link between semantic trees and semantic tableaux, and prove some basic properties. In Section 3 we use refutation graphs to give a completeness proof of (order-restricted) binary resolution, while in Section 4 we do the same for (unrestricted) ground tableaux. Finally, in Section 5 we prove our main result, namely that a straightforward and natural ordering restriction of first-order ground tableau is a complete proof procedure for first-order logic.

Due to space restrictions most proofs had to be omitted. A long version of this paper with full proofs of all theorems can be obtained from the authors on request or via anonymous ftp to 129.13.31.2 (switch to binary mode and get the file pub/haehnle/Ordered-Tableaux-Long.ps.Z).

1 Semantic Trees

Throughout the paper we use a standard first-order CNF clause language without equality.

The proof of completeness of resolution via semantic trees is well known. Unfortunately, it is not easy to adapt this proof to (clausal) tableau procedures. The first obstacle is to identify the exact counterpart of clauses and resolvents in a tableau. The propositional resolution rule can be seen as a cut rule restricted to atomic formulas and clausal tableau proofs corresponds to cut free sequent proofs. Cut elimination, however, results in additional copies of certain subformulas which are spread around the proof. Hence, more than one formula in a tableau corresponds to one generated clause in a resolution proof and one has to keep track of these formulas.

Example 1.

In Figure 1 the closure of the rightmost branch corresponds to an application of the resolution rule with parent clauses $A \vee -B \vee C \vee D$ and $A \vee E \vee -D$. The resolvent $A \vee -B \vee C \vee A \vee E$ occurs as the union of the framed nodes. The parent clauses may be spread over several nodes in the tableau. The only restriction is that the nodes of the literals resolved upon are labelled with a single literal. Note that the parent clauses are not available any longer. Otherwise, a node can belong to different clauses and cause problems, especially in the case of free variable first-order tableaux.

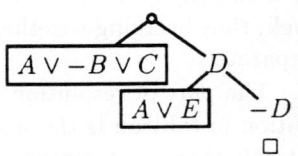

Fig. 1. The tableau is symbolized by a labelled tree. □ denotes the closing of a branch.

The classic completeness proof for resolution [11] is based on two presuppositions: (i) Only subsumed clauses and tautological clauses are discarded; (ii) The resolvent belonging to an inference node retains all interesting ground instances of the parents. As we just have seen we have to discard also clauses in a tableau procedure that are not subsumed. Thus we have to modify the proof idea for completeness based on semantic trees.

We assume the reader is familiar with the basic notions and results of computational logic, in particular with Herbrand bases and Herbrand's theorem (see, for instance [14]). In the following let B_0 be the Herbrand base of our first-order language, and let $B_0 = A_0, A_1, \ldots$ be an arbitrary, but fixed, enumeration; $\sim L$ denotes the complementary of a literal L; sometimes we will treat clauses as sets without mentioning it. $C \backslash L$ means, that each occurrence of the literal L is deleted from the clause C. Most of the following definitions may be found in [14, 7].

Definition 1. Let $a_0, a_1, \ldots$ be a fixed enumeration of a set of atoms B. The labelled, binary tree ST, which is defined as the smallest tree obeying the following conditions, is called **semantic tree** for B.

1. The root node of ST is unlabelled; its left child is labelled with a_0, its right child is labelled with $-a_0$.
2. If a node is labelled with a_i or $-a_i$, then its left child is labelled with a_{i+1}, and its right child is labelled with $-a_{i+1}$.

The semantic tree for B_0 is denoted with ST_0. Two nodes in ST with the same parent node are called **siblings**. A **path** in ST is a (finite or infinite) sequence of node labels $\langle l_i, l_{i+1}, \ldots \rangle$, such that the node belonging to l_{j+1} is the child of the node belonging to l_j for all $j \geq i$. An maximal path in ST is a **branch** of ST. Two paths p_1, p_2 are called **sibling-paths** wrt a pair of sibling nodes n_1, n_2 iff p_1 and p_2 are both of the form $\langle l_1, l_2, \ldots, l_{i-1}, l_{n_j}, l_{i+1}, \ldots \rangle$, where l_{n_j} is the label of n_j for $j = 1, 2$. With each node n of a semantic tree ST we associate its **refutation set** which consists of the labels on the path from the root to n.

Definition 2. A clause C **fails** at a node n of a semantic tree ST iff there is a substitution σ such that for each literal L in $C\sigma$, $\sim L$ is in the refutation set of n. A node n is a **failure node** for a set of clauses M iff some clause from M fails at n, but no clause from M fails at a node above n. A node n is an **inference node** iff both of its children are failure nodes. A semantic tree ST is **closed** by M iff every branch of ST contains a failure node for M.

See Figure 2 for an example illustrating these definitions.

Definition 3. An **A-ordering** on a set of atoms B is a binary relation $<_A$ such that for all $a, b, c \in B$

Irreflexivity $a \not<_A a$.
Transitivity $a <_A b$ and $b <_A c$ imply $a <_A c$.
Substitutivity $a <_A b$ implies $a\sigma <_A b\sigma$ for all substitutions σ.

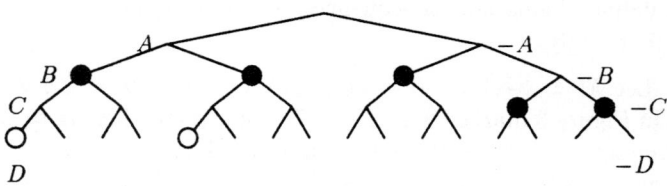

Fig. 2. The failure nodes in a semantic tree for $\{A, B, C, D\}$ for the clause set $\{-A \vee -B,\ -A \vee B,\ A \vee -B,\ A \vee B \vee -C,\ A \vee B \vee C\}$. The paths $\langle A, B, C, D \rangle$ and $\langle A, -B, C, D \rangle$ going from the node labelled with A to the nodes marked with o are sibling-paths the sequences of their labels differing only in the second position. The failure nodes are marked with •. The node labelled with A is an inference node (among others). Since all branches contain a failure node, the tree is closed.

2 Refutation Graphs[4]

Our goal is to adapt completeness proofs via semantic trees to tableau procedures. In the resolution case, completeness proofs via semantic trees are based on the successive elimination of failure nodes, yielding smaller and smaller closed semantic trees for an inconsistent clause set. At the end the tree consisting solely

---

[4] This should not be confused with a concept bearing the same name introduced by Shostak [16] in the context of clause graph resolution. Since we feel that the phrase 'refutation graph' catches exactly our intention we decided to keep the name nevertheless. Independently, Bibel [4] recently developed the concept of a so-called π-regular graph which is somewhat related to our refutation graphs.

of the root node is left. Since the only clause that fails at the root node is the empty clause, it must have been inferred by that time.

Unfortunately, this argument does not work in the case of tableaux. Instead of reducing the size of the whole closed semantic tree one has to have a closer look on the nodes of a semantic tree that establish the closure of the tree. We will collect such nodes according to their mutual relationship into more complicated graphs on the node set of the semantic tree. If such a graph meets certain requirements, it can be seen as refutation of a corresponding set of clauses. The number of nodes in this graph will play a similar role in our completeness proof as the size of the semantic tree in the classic proof.

Definition 4. Let M be a set of clauses, and G be a ground instance of a clause $C \in M$. We call a set ch of nodes of ST_0 a **chain** of G (or of C) iff

- all nodes in ch are on the same path in ST_0, and
- L is a label of a node in ch iff $\sim L \in G$.

We denote the set of all chains of M in ST_0 by $CH(M)$, and the set of all nodes occurring in a set CH of chains by $\overline{CH}$. The **empty chain** by definition corresponds to the empty clause.

Informally speaking, a chain is a minimal subset of a path that refutes a clause. We define chains not as sequences but as sets, because we will mainly use them accordingly.

Example 2. Let $M = \{-A \lor -B \lor -C, -A \lor C, -B, A \lor -C, A \lor C, -A \lor B\}$ The chains in Figure 3 marked with $1, \ldots, 7$, respectively, belong to the following clauses: 1 belongs to $-A \lor -B \lor -C$, 2 and 3 belong to $-A \lor C$, 4 belongs to $-B$, 5 belongs to $A \lor -C$, 6 belongs to $A \lor C$, and 7 belongs to $-A \lor B$. Note that different chains can belong to the same clause, because different nodes can have the same labels.

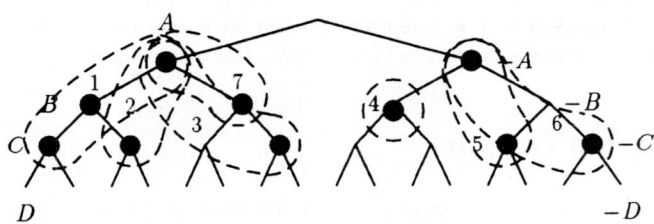

Fig. 3. Chains that belong to various clauses in the semantic tree for the clause set from Example 2.

Since all nodes of a chain are contained in the same path of a semantic tree they are ordered in a natural way. The successor relation together with the relation of being sibling nodes is used to define the graphs mentioned above.

Definition 5. We write $n_1 > n_2$, if n_2 is closer to the root. Moreover, a node n_1 of a chain ch is called a **successor node** of $n_2 \in ch$, if $n_1 > n_2$ and there is no node $n_3 \in ch$ such that $n_1 > n_3 > n_2$. A node without a successor node is called an **end node** of that chain.

Definition 6. For an arbitrary set of chains CH in ST_0, we define a binary relation $\prec_{CH}$ on the nodes of ST_0: $n_1 \prec_{CH} n_2$ iff n_1 and n_2 are sibling nodes or belong to the same chain in CH and n_2 is a successor of n_1.

A set CH of chains **closes** a subtree ST of ST_0 iff it contains the empty chain or each path of ST_0 through the root of ST contains a chain of CH whose end node is in ST.

Note that for each clause set M the semantic tree ST_0 is closed by $CH(M)$ iff ST_0 is closed by M in the sense of Definition 2.

Definition 7. Given a non-empty set of chains CH and a subtree ST of ST_0, the **graph** defined by ST and CH is $G(CH, ST) = \langle \overline{CH} \cap ST, \prec_{CH} \rangle$. A graph $G(CH, ST)$ is called a **refutation graph** iff it is finite and for all nodes $n_1 \in G(CH, ST)$, $n_2 \in ST_0$ we have that $n_1 \prec_{CH} n_2$ implies $n_2 \in G(CH, ST)$. A graph $G(CH, ST)$ is a **semi refutation graph** iff it is finite and for each node, but the root node of ST, also its sibling is contained in $G(CH, ST)$. A (semi) refutation graph $G(CH, ST)$ is **minimal** iff for no $CH_1 \subsetneq CH$ $G(CH_1, ST)$ is a (semi) refutation graph. A refutation graph $G(CH, ST)$ is **connected** if there are no $\emptyset \neq CH_1, CH_2 \subsetneq CH$, $CH_1 \neq CH_2$ such that $G(CH_1, ST)$ and $G(CH_2, ST)$ are both refutation graphs.

Note that the empty graph that corresponds to the set of chains consisting only of the empty chain is a refutation graph, and is called the **empty refutation graph**.

Example 3. Let M be as in Example 2. The chains of Example 2 do not establish a refutation graph, since sibling nodes for nodes in the chains 3 and 4 are missing. However, consider the subgraph consisting of the chains $CH_1 = \{1, 2, 5, 6, 7\}$. The corresponding refutation graph $G(CH_1, ST_0)$ is shown on top in Figure 4. Note that the clauses corresponding to chains $3, 4$ are not necessary for a refutation by resolution or tableau. We get a much simpler refutation graph $G(CH_2, ST_0)$ (depicted on bottom in Figure 4), if we take the chains $5, 6, 7$ and 8 as a chain set CH_2 with the new chain 8 stemming from the clause $-B$. One can consider chain 8 as a copy of chain 4 placed in the left subtree of ST_0 instead of the right. The next definition will deal with this copy operation more rigorously.

Definition 8. Let ch be a chain on a path p in the tree ST_0 with end node n. Let $c \notin ch$ be a node with $c < n$ and let q be the sibling-path of p with respect to c and its sibling. We denote with $copy(ch, c)$ the unique chain situated on q, whose nodes have the same labels as ch. The extension of the definition $copy(CH, c)$ to sets of chains CH is obvious.

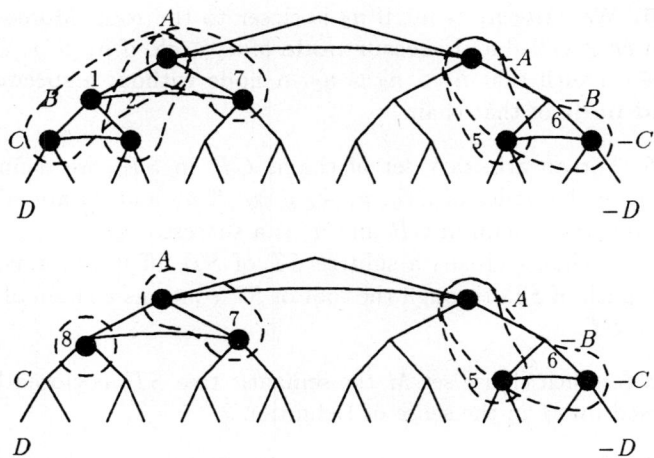

Fig. 4. The refutation graphs $G(CH_1, ST_0)$ and $G(CH_2, ST_0)$ from Example 3.

Example 4. Using the numbering of the previous examples, $copy(4, n_A) = 8$, where n_A is the node labelled with A.

Theorem 9. *Let M be a clause set.*
1. *M is unsatisfiable iff ST_0 is closed by M.*
2. *If ST_0 is closed by M, then there is a set of chains $CH \subseteq CH(M)$ such that $G(CH, ST_0)$ is a refutation graph.*
3. *If CH is a set of chains belonging to clauses in M, and $G(CH, ST_0)$ is a refutation graph, then ST_0 is closed by M.*

Proof. 1. Proved in [7].
2. The semi refutation graph $G(CH, ST_0)$ constructed in Lemma 10 is sufficient, because ST_0 contains all nodes.
3. Proved in Lemma 11 for arbitrary subtrees ST of ST_0.

Lemma 10. *Let M be a set of clauses and ST a subtree of ST_0 closed by $CH(M)$. There is a set of chains $CH \subseteq CH(M)$ such that*

- *CH closes ST*
- *CH contains only chains that have end nodes in ST*
- *no node in CH is end node of different chains*
- *$G(CH, ST)$ is a semi refutation graph*

Proof. The proof is by induction over the maximal distance d between any inference node and the root node n_0 of ST.

If $d = 0$ (that is, the root node of ST is an inference node), then both its successors are failure nodes with corresponding chains ch_1, ch_2. Let $CH = \{ch_1, ch_2\}$; all conditions are fulfilled trivially.

If $d \neq 0$ several cases must be distinguished. Consider the successors n_1, n_2 of n_0. We proof only the semi refutation graph property, all others are trivial. If n_i is the end node of a chain ch_i, then set $K_i = \{ch_i\}$. Otherwise, if n_i is the root of a closed subtree, then let K_i be the set of chains obtained by applying the induction hypothesis.

Case 1: $n_i \in \overline{K_i}$ for $i = 1, 2$ or $n_i \notin \overline{K_i}$ for $i = 1, 2$. Set $CH = K_1 \cup K_2$.

Case 2: $n_1 \in \overline{K_1}$, but $n_2 \notin \overline{K_2}$. This time we cannot take the union of K_1 and K_2 as CH, because, by $n_1 \in \overline{K_1}$, its sibling n_2 would have to be in $\overline{CH}$ which is not the case. We remedy this situation by discarding all chains ending in the subtree rooted at n_1, and use instead the semi refutation graph we have already below n_2. Formally, let $\tilde{K}_1 := copy(K_2, n_2)$. Set $CH = \tilde{K}_1 \cup K_2$. All chains in $\tilde{K}_1$ have their end nodes in the subtree rooted in n_1 and neither n_1 nor n_2 is in $G(CH, ST)$. Hence, by the induction hypothesis, for all nodes of $G(CH, ST)$, with the exception of n_0, also its sibling is in $G(CH, ST)$.

Case 3: $n_2 \in \overline{K_2}$, but $n_1 \notin \overline{K_1}$. Analogous to case 2.

Lemma 11. *If ST is a subtree of ST_0, $CH \subseteq CH(M)$, and $G(CH, ST)$ is a refutation graph, then ST_0 is closed by M.*

3 Refutation Graphs and Resolution

We intend to use refutation graphs for completeness proofs. In the present section we outline how this is done in the case of resolution. If we can prove that a refutation procedure finds and recognizes a refutation graph of an unsatisfiable clause set, as a consequence of Theorem 9, it is refutation complete. First we will redefine some notions related to semantic trees in the context of refutation graphs.

Definition 12. Let M be a clause set, CH a subset of its chains, and $G(CH, ST)$ a refutation graph. A node $n \in \overline{CH}$ is a **failure node** of $G(CH, ST)$, iff it is an end node of a chain $ch \in CH$ and no other chain of CH than ch contains a node $n_1 \geq n$.

We call the parent node of two sibling nodes that are failure nodes[5] an **inference node** of $G(CH, ST)$ and any two chains having siblings as failure nodes **inference chains**. If ch_1, ch_2 are a pair of inference chains with failure nodes n_1 and n_2, then the chain with nodes $(ch_1 \cup ch_2) \setminus \{n_1, n_2\}$ is called **resolvent chain** of ch_1 and ch_2.

In a refutation graph the sibling of a failure node is also contained in the graph. Since each refutation graph is finite, a minimal, non-empty refutation graph contains at least one inference node.

---
[5] Recall that by definition for each node in a refutation graph also its sibling node must be present.

Theorem 13. *Let $G(CH, ST)$ be a minimal refutation graph, let ST be a subtree of ST_0, and e the resolvent chain of a pair c, d of inference chains. Then the graph $G((CH\setminus\{c,d\}) \cup \{e\}, ST)$ is a refutation graph and it has lesser nodes than $G(CH, ST)$.*

To each chain there belongs a unique ground clause (the empty clause to the empty chain). Thus the well-known lifting lemma (see [14] for the versions for basic resolution and for ordered resolution) tells us that resolving of chains can be lifted to basic resolution (with factorisation). Moreover, since in our setting only maximal nodes (literals) are involved in a resolution step, the restrictions imposed by any A-ordering are obeyed, provided the ordering is compatible with the enumeration of the Herbrand base chosen for ST_0.

Example 5. Consider the set of chains CH_2 used in the lower part of Figure 4. Chains 5 and 6 constitute a pair of inference chains with failure nodes labelled with C and $-C$. $G(CH_2, ST_0)$ is a minimal refutation graph. If we resolve on chains 5 and 6 we obtain the minimal refutation graph $G(CH_3, ST_0)$ depicted in Figure 5, where $CH_3 = (CH_2\setminus\{\{-A, C\}, \{-A, -C\}\}) \cup \{\{-A\}\}$. Note that CH_2 corresponds to the set of clauses $\{-B, -A \vee B, A \vee -C, A \vee C\}$, and the application of Theorem 13 corresponds to resolving the last two clauses on C.

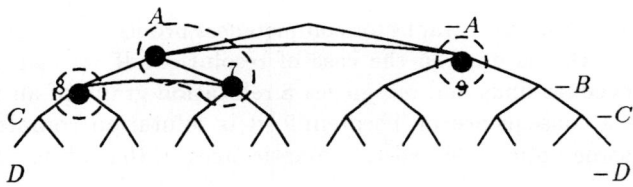

Fig. 5. The minimal refutation graph $G(CH_3, ST_0)$ from Example 5.

Robinson resolution or A-resolution can recognize a refutation graph that consists of a pair of sibling nodes (that are resolved to the empty clause). The search strategies of these procedures enrich the set of clauses. Let us think about adding all resolvents of factors of parents in a clause set S_i yielding the new clause set S_{i+1} as a single step. We know that S_{i+1} contains a refutation graph with lesser nodes. For an unsatisfiable clause set S_0 the well-founded order on sets of clauses $S < T$ iff $\alpha(S) < \alpha(T)$, where $\alpha(S) = \min\{|G(CH', ST)| : CH' \subseteq CH(S), ST \subseteq ST_0, G(CH', ST)$ a is a refutation graph$\}$ garuantees termination since under this ordering $S_i > S_{i+1}$ for all $i \geq 0$.

Resolution takes care of all graphs at each step and uses the fact that certain resolution steps decrease the number of nodes of a refutation graph. A-resolution employs that these resolution steps are among the A-resolvents. Therefore, instead of resolving each possible pair of parent clauses, A-resolution looks for certain patterns, namely for inference nodes. The corresponding pair of clauses is A-resolvable.

4 Refutation Graphs for Tableaux

We will deal with tableaux for ground clauses in a enumeration strategy for first-order logic. Thus it is convenient to modify our notion of tableaux in order to reflect the clausal form (cf. also [12] for *clausal tableaux*). The reader, who is familiar with Fitting's [8] conception of tableaux, can consider our tableaux as generated by a single extension rule that, at the same time, generalizes the β extension rule to any finite number of disjuncts, and combines it with the substitution rule.

Definition 14. Let M be a clause set. We call a finitely branching tree T a **tableau** for a clause set M, if

- the root node is labelled with T, all other nodes are labelled with a literal, and
- if $D = L_1 \vee \cdots \vee L_n$, where $\{L_1, \ldots, L_n\}$ are the literals labelling the set of direct successor nodes of some node, then there is a substitution σ and a clause $C \in M$, such that $D = C\sigma$.

T is a **ground tableau** iff all its labels are ground. A ground tableau is **closed** iff every branch contains two complementary literals. A **subtableau** is a proper subtree of a tableau (hence, the root of a subtableau is labelled with a literal).

We use the following notation for tableaux: $T = (L, T_1, \ldots, T_n)$, where L is the label of the root of T, and $T_1, \ldots, T_n$ are immediate subtableaux.

Tableau procedures for clause sets regard the chains of each clause as a part of a possible refutation graph. To find a refutation graph they look for a chain that contains a sibling node of a node of a chain already present in the tableau. If a suitable chain is found, a corresponding clause can be used to extend the current tableau and close a branch of the extended tableau. If some siblings of the used chain (clause) are not yet on the extended branch one has to find counterparts for these nodes also. One can consider constructing a tableau as walking through graphs. The next definition makes this idea precise.

Definition 15. Let seq be a finite sequence $\langle n_1, n_2, \ldots, n_m \rangle$ of nodes of a graph $G(CH, ST)$, $m > 1$. We call seq a **connection** in $G(CH, ST)$ iff $n_i \neq n_j$ for $i \neq j$ (i.e. seq is repetition free), all n_i, n_{i+1} are siblings for odd i, and all n_i, n_{i+1} are members of the same chain in CH for even i. We say that n_m is **connected** to n_1. If n_m is connected to n_1, then obviously there is a chain that contains n_m and all of whose nodes are also connected to n_1. We call the set $cc(CH, n_1)$ of chains in $G(CH, ST)$ all of whose nodes are connected to n_1 the **connected component** of n_1.

Example 6. Consider the graph $G(CH, ST)$ in Figure 6 with chains as indicated, without the shaded chain. The nodes $\langle n_1, n_2, n_3, n_4, n_7 \rangle$ form a connection from n_1 to n_7. All chains in CH but $\{n_1\}$ are connected to n_1, hence $cc(CH, n_1) = CH \backslash \{\{n_1\}\}$. On the other hand, none of the nodes in $\{n_1, \ldots, n_7\}$ is connected to n_7, thus $cc(CH, n_7) = \emptyset$. Hence, we see immediately that the connectedness relation is irreflexive and not symmetric.

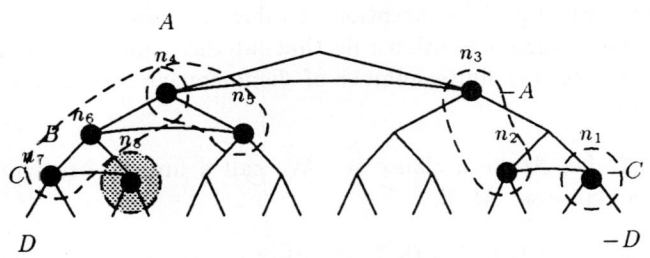

Fig. 6. Connections in a graph.

Let us adopt the convention $\sim \Box = \mathsf{T}$.

Definition 16. Let CH be an arbitrary set of chains, T a ground (sub)tableau and L a label. We say that T **represents** the pair (CH, L) iff the following holds: If CH is empty, then $T = (\sim L)$, else $T = (\sim L, T_1, \ldots, T_n)$ with subtableaux T_i representing (CH_i, L_i), where the L_i are the labels of the nodes n_i of a chain $ch \in CH$ and the $CH_i = cc(CH \backslash \{ch\}, n_i)$ are connected components of the roots of the subtrees (each of the CH_i has less chains than CH provided CH is finite).

If we fix a rule telling us how to choose the next chain $ch \in CH$, when there are several, we obtain a unique tableau representant of a given pair (CH, L). This is the case, for example, when there is an enumeration of the ground instances of a given clause set M. If the chains of CH belong to ground instances of M, one can take the chain with the lowest index.

Example 7. Assume that the graph of Figure 6 is extended to a refutation graph $G(CH, ST)$ by the shaded chain. In order to construct a tableau that represents $(CH, \Box)$, we begin by taking an arbitrary chain from CH, say $\{n_4, n_5\}$, and extend the root tableau node with the complements of its labels, see Figure 7(a). We number the branches, so that we can identify them more easily.

The next chain that is used to extend branch (1) must be taken from the set $cc(CH \backslash \{\{n_4, n_5\}\}, n_4) = \{\{n_1\}, \{n_2, n_3\}\}$. We take $\{n_1\}$ and extend (1) with C. After that only one chain, namely $\{n_2, n_3\}$, is left, and we obtain the tableau in Figure 7(b).

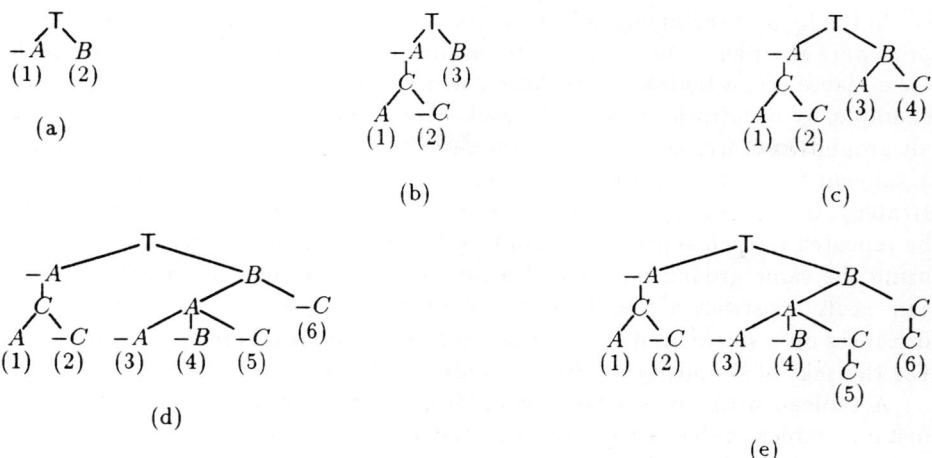

Fig. 7. Constructing a closed tableau that represents a refutation graph.

Now consider branch (3) in Figure 7(b). Observe that node n_5 is connected to every chain in $CH\setminus\{\{n_4, n_5\}\}$. Hence, $CH_{(1)} := cc(CH\setminus\{\{n_4, n_5\}\}, n_5) = CH\setminus\{\{n_4, n_5\}\}$ is the next connected component. We choose $\{n_2, n_3\}$ and extend the tableau with the corresponding labels in Figure 7(c).

We focus on branch (3) in Figure 7(c). The leaf A is the label of n_3. We must look for the chains in $CH_{(2)} := cc(CH_{(1)}\setminus\{\{n_2, n_3\}\})$ that are connected with n_3. These are $\{\{n_4, n_6, n_7\}, \{n_8\}\}$. We expand the tableau with the labels corresponding to $\{n_4, n_6, n_7\}$, and obtain Figure 7(d). The node corresponding to the leaf of (3) is n_4. If we look for the chains in $CH_{(3)} := cc(CH_{(2)}\setminus\{\{n_4, n_6, n_7\}\}, n_4)$ we see that this is the empty set, therefore, we cannot expand the branch anymore. Indeed, branch (3) in Figure 7(d) is closed with $(-A, A)$. The same is true for branches (1), (2), and (4). In each case the connected component is empty.

The node corresponding to the leaf of (5) is n_7. The connected component for n_7 in $CH_{(4)} := cc(CH_{(2)}\setminus\{\{n_4, n_6, n_7\}\}, n_7)$ is $\{n_8\}$, and we extend (5) with C. The situation in branch (6) is similar, and we finally obtain the closed tableau shown in Figure 7(e).

The following lemma states that it was not by coincidence that the tableau in the example was closed.

Lemma 17. *If T represents $(CH, \square)$ and $G(CH, ST)$ is a refutation graph, then T is closed. Moreover, if a subtableau T' of T represents a pair (CH', L), then for each node n in CH' the following holds:*

(*) Either the sibling of n is in $\overline{CH'}$ or the literal that labels n is already on the branch.

In the light of the previous lemma and Theorem 9 it is sufficient for a tableau procedure in order to be complete to ensure that, if there is a refutation graph for a clause set, a tableau representing it is found. One can guruantee this, for example, by backtracking over all possible tableaux, or by successively bringing all ground instances of all clauses on each branch. In the latter case there is a tableau for each refutation graph compatible with the chosen enumeration strategy. If unnecessary literals are put onto a branch the closing steps have to be repeated in each superfluous resulting branch. Of course, there is no need to bring the same ground instance of a literal more than once on a branch. One can easily construct a closed tableau without multiple occurrences of ground instances on a single branch if a closed tableau without this restriction is given. For the sake of simplicity we did not address this problem in Lemma 17.

A tableau with free variables as in [8] deals simultaneously with all ground instance tableaux that are obtained by the application of a ground substitution to the variables of the tableau. One has to ensure then that it can be closed, if one of its instances can be closed. This problem will be addressed in a forthcoming paper.

5 Tableaux with A-Ordering

In this section we explain how to incorporate A-orderings into a tableau refutation procedure for clause sets.

Definition 18. Let T be a ground (sub)tableau representing (CH, L) and $d(n)$ the distance of a node n from the root node. We say T is **ordered** iff either

- T is a leaf or
- all immediate subtableaux $T_1, \ldots, T_n$ of T are ordered, and there is no node $d(n) > d(n'_{\max})$ in $\overline{CH}$, where $n'_{\max}$ is the sibling of the maximal node of the chain $ch \in CH$ that determines $T_1, \ldots, T_n$.

Example 8. The tableau in Figure 7(e) is not ordered, because $d(n_5)$ is not maximal in the set $\overline{CH}$ of Example 7. On the other hand, both subtrees below the root node are ordered.

The following theorem is a consequence of Lemma 17.

Theorem 19. *If T is ordered, represents $(CH, \Box)$, and $G(CH, ST)$ is a refutation graph, then T is closed.*

Proof. Each subtableau T' of T represents a pair (CH', L). The theorem is a consequence of Lemma 17, provided that CH' contains a chain that obeys the restriction of Definition 18. In each chain set there are chains whose end nodes have maximal distance from the root. Every such chain is suitable.

One can combine this strategy with enumeration, if one selects the chain with the lowest index if several candidates are present.

Definition 20. Let $<_A$ be an A-ordering on a clause set M. An **A-ordered ground instance enumeration tableau** is a tableau constructed by the following restricted expansion rule:

Tableau branches are expanded solely by ground instances of clauses in M not already on the branch

1. that have a maximal literal complementary to a literal already on the branch or
2. to a maximal literal of another ground instance of a clause in M

Theorem 21. *If M is an unsatisfiable clause set, then there is a closed A-ordered ground instance enumeration tableau.*

Proof. If M is an unsatisfiable clause set, then let $CH \subseteq CH(M)$, and T' an ordered tableau representing $(CH, \square)$ such that $G(CH, ST_0)$ is a refutation graph. This is possible by Theorem 9. By Theorem 19, T' is closed.

The property (*) of Lemma 17 holds for T'. By definition of ordered tableaux, the sibling of the maximal literal of an expansion is either complementary to the maximal literal of a ground instance or not in the corresponding connected component. In the latter case, by (*), the complement of the maximal literal is already on the branch. Thus all the instances in T' are permitted by the restriction imposed in the A-ordered ground instance enumeration tableau procedure.

6 Conclusions and Related Work

In this paper we can only lay the ground for further investigations. We provided a technique for proving completeness of first-order enumeration ground tableaux and Robinson resolution in a uniform way. The completeness proofs can be adapted to ordering refinements for resolution and tableau as well. The outlined ground tableau procedure can be immediately implemented in tableau provers like HARP [15] or Tatzelwurm [10].

In future work we intend to investigate the combination of ordering restrictions with free variable tableaux, tableaux with equality, and tableaux as decision procedures.

There have been several approaches to involve ordering restrictions into first-order theorem proving. Soon after the introduction of basic resolution by J. A. Robinson a wide variety of refinements for resolution followed. Ordering restrictions and semantic clash were first investigated by Slagle [17]. Kowalski and Hayes [11] combined ordering restrictions with a completeness proof using semantic trees. Joyner used this refinement as a decision procedure for a certain class of first-order formulas [9]. His results, however, were not widely discussed until recently Fermüller et al. [7] revived the ideas of Joyner. Another approach to decidability stems from the inverse method due to Maslov, and was further developed by Zamov [19] and Tammet [18] in a very similar way as resolution in the work of Joyner. Both approaches contribute to [7]. Further, we mention the work of Bachmaier and Ganzinger [2]. They generalize completion procedures similar to the well known Knuth-Bendix completion to refutational first-order theorem proving with equality. A different approach to using tableaux as a decision procedure for certain first-order problems is due to Caferra [5]. It is not based on ordering restrictions, but on keeping track of models with a decidable equality language.

References

1. O. Astrachan and M. Stickel. Caching and lemmaizing in model elimination theorem provers. In D. Kapur, editor, *Proc. 11th Conference on Automated Deduction, Albany/NY, USA*, pages 224–238. Springer LNAI 607, 1992.
2. L. Bachmair and H. Ganzinger. Rewrite-based equational theorem proving with selection and simplification. Technical Report MPI-I-9-1-208, Max-Planck-Institut für Informatik, MPI at Saarbrücken, Germany, 1991.
3. W. Bibel. *Automated Theorem Proving*. Vieweg, Braunschweig, 2nd edition, 1987.
4. W. Bibel and E. Eder. Decomposition of tautologies into regular formulas and strong completeness of connection graph resolution. Technical Report AIDA-94-0?, FG Intellektik, FB Informatik, TH Darmstadt, Mar. 1994.
5. R. Caferra and N. Zabel. A tableaux method for systematic simultaneous search for refutations and models using equational problems. *Journal of Logic and Computation*, 3(1):3–26, 1993.
6. C.-L. Chang and R. C.-T. Lee. *Symbolic Logic and Mechanical Theorem Proving*. Academic Press, London, 1973.
7. C. Fermüller, A. Leitsch, T. Tammet, and N. Zamov. *Resolution Methods for the Decision Problem*. Springer LNAI 679, 1993.
8. M. C. Fitting. *First-Order Logic and Automated Theorem Proving*. Springer, New York, 1990.
9. W. H. Joyner. Resolution strategies as decision procedures. *Journal of the Association for Computing Machinery*, 23(3):398–417, 1976.
10. T. Käufl and N. Zabel. Cooperation of decision procedures in a tableau-based theorem prover. *Revue d'Intelligence Artificielle*, 4(3):99–126, 1990.
11. R. Kowalski and P. Hayes. Semantic trees in automatic theorem-proving. *Machine Intelligence*, 4:87–101, 1969.
12. R. Letz. *First-Order Calculi and Proof Procedures for Automated Deduction*. PhD thesis, TH Darmstadt, June 1993.
13. R. Letz, J. Schumann, S. Bayerl, and W. Bibel. SETHEO: A high-perfomance theorem prover. Technical report, Forschungsgruppe KI, TU München, 1991.
14. D. W. Loveland. *Automated Theorem Proving. A Logical Basis*, volume 6 of *Fundamental Studies in Computer Science*. North-Holland, 1978.
15. F. Oppacher and E. Suen. HARP: A tableau-based theorem prover. *Journal of Automated Reasoning*, 4:69–100, 1988.
16. R. E. Shostak. Refutation graphs. *Artificial Intelligence*, 7:51–64, 1976.
17. J. R. Slagle. Automatic theorem proving with renamable and semantic resolution. *Journal of the Association for Computing Machinery*, 14(4):687–697, 1967.
18. T. Tammet. The resolution program, able to decide some solvable classes. In *Proceedings COLOG-88, Talinn*, pages 300–312. Springer, LNCS 417, 1990.
19. N. Zamov. Maslov's inverse method and decidable classes. *Annals of Pure and Applied Logic*, 42:165–194, 1989.

Strongly Analytic Tableaux for Normal Modal Logics

Fabio Massacci

Dipartimento di Informatica e Sistemistica
Università di Roma "La Sapienza"
via Salaria 113, I-00198 Roma, Italy
E-mail: massacci@assi.dis.uniroma1.it

Abstract. A strong analytic tableau calculus is presentend for the most common normal modal logics. The method combines the advantages of both sequent-like tableaux and prefixed tableaux. Proper rules are used, instead of complex closure operations for the accessibility relation, while non determinism and cut rules, used by sequent-like tableaux, are totally eliminated. A strong completeness theorem without cut is also given for symmetric and euclidean logics. The system gains the same modularity of Hilbert-style formulations, where the addition or deletion of rules is the way to change logic. Since each rule has to consider only adjacent possible worlds, the calculus also gains efficiency. Moreover, the rules satisfy the strong Church Rosser property and can thus be fully parallelized. Termination properties and a general algorithm are devised. The propositional modal logics thus treated are K, D, T, KB, K4, K5, K45, KDB, D4, KD5, KD45, B, S4, S5, OM, OB, OK4, OS4, OM$^+$, OB$^+$, OK4$^+$, OS4$^+$. Other logics can be constructed with different combinations of the proposed rules, but are not presented here.

1 Introduction

The use of modal logic spreads across artificial intelligence and computer science, as a way of modelling knowledge and belief [10] or as a formalism to specify the behaviour of distributed systems [9]. Nonmonotonic versions are taking grounds too [14, 16]. Thus, devising efficient and human-oriented proof procedures becomes a compelling task in automated deduction.

The main problem, even in the case of propositional modal logics, is that traditional proof systems, such as tableaux or sequent calculus [2, 4, 5, 8, 18], are not very efficient, whereas other proposals like resolution [17] and matrix proof methods [19] seem to need a "user-friendly" interface for the presentation of proofs and proof search. The prefixed tableaux of Fitting [3, 4] appear to stand in between, yet they are more oriented towards model checking.

Beside that, a common limitation affects all these proposals: the unsatisfactory treatment of strong deducibility for euclidean (and sometimes also symmetric logics), which are now playing a major role in nonmonotonic reasoning [13, 14, 16]. At this point, sequent-like tableaux or sequent themselves treat these logics in an entirely non-deterministic way, since cut has been eliminated for K45

[8, 18] but only for weak deducibility (i.e. without premises). Other methods do not cover them at all. Apart from Hilbert-style proof procedures, the only way to cope with them seemed to be the following: revert from automated deduction to model checking and try to construct explicitly a (counter) model; a construction that requires to calculate many times the euclidean, symmetric etc. closure of the accessibility relation [1, 10].

We propose a new tableau calculus that combines the advantages of sequent-like and prefixed tableaux and allows an analytic, deterministic, effective, uniform, strongly confluent and "user friendly" approach to most normal propositional modal logics, including symmetric and euclidean ones: K, T, D, KB, K4, K5, K45, KDB, D4, KD5, KD45, KD45, T, B, S4, S5, OM, OB, OK4, OS4, OM^+, OB^+, $OK4^+$, $OS4^+$ and many others. For any of these logics, we provide a strong completeness theorem without cut, a necessary property if global assumptions are going to play the role of a knowledge base and local assumptions that of information provided by the user.

Intuitively, a tableau is a refutation, a failed attempt to build a countermodel for a given formula. The different behaviour of prefixed and sequent-like tableaux is due to the way they code information. A sequent-like tableau uses formulae, and thus "travels" across the supposed countermodel, moving formulae from one world to the next. Choosing a different accessibility relation means that different formulae may be taken along. Such a tableau considers only one world at a time and cannot look back. Hence, beside the usual nondeterminism brought by disjunction, further levels of nondeterminism are necessary to cope with alternative possible worlds or with symmetric and euclidean logics. In contrast, a prefixed tableau uses prefixes (sequence of integers) to "name" the possible worlds where a particular formula holds. It can move back but is forced to behave as a model checker and to devise a relation over prefixes that is a "carbon copy" of the accessibility relation over possible worlds. Hence a prefix must possibly access other prefixes "very far away" and the efficiency of the systems is limited.

Our proof system — we called it Single Step Tableaux (SST for short) — also uses prefixes to keep track of the worlds where formulae hold. The key difference is that the "accessibility relation" between prefixes is not used in the proof procedure, but only in the proof of correctness and completeness theorems (an idea also used for matrix proof methods, see [19]). Instead, we define rules to specify which formulae may be moved forward to the next close possible world, as sequent-like tableaux do, or backward, as they cannot do. The name "single step" has been used since the reduction of a formula, with a given prefix, makes only use of immediately accessibile prefixes, without caring of those accessible via the transitive, symmetric etc. closure.

The intuition is very close to the "visa rules", proposed for first order modal logic in Labelled Deductive Systems [6]: each logic grants a different "visa" to constants for moving from one world to another. Why can different logics not grant different visas to formulae?! In this spirit, our SST may be viewed as a particular labelled deductive system.

Moreover, going "step by step" does not reveal to be an obstacle: a "hidden" property of prefixed tableaux (a kind of proof theoretical counterpart of Generation Theorem of Segerberg [12]) is used to get completeness. Thus the method gains a lot of efficiency, since only a small part of the worlds' graph must be taken into account for each reduction. Yet, it is human oriented enough for doing proof by hand.

The SST-calculus provides other advantages: it offers the same flexibility of Hilbert-style proof methods (to change logic we just have to add or take away a rule), and may be used to obtain countermodels for non-valid formulae i.e. prefixes form a minimal spanning tree of the underlaying frame. The transitive, reflexive etc. closure of the accessibility relation may be done afterwards, if needed. Moreover, as a tableaux system, it can be used for establishing both satisfiability and validity.

Other properties of our formal system are also of great interest from the point of view of automated deduction and practical implementation. Our rules satisfy the strong Church-Rosser property and thus the calculus is strongly confluent. Hence, at any stage of the computation, the execution sequence of applicable rules may be changed in arbitrary fashion. Therefore SST can be a good basis for a flexible, i.e. compatible with many search heuristic, and highly parallelizable implementation. In practice, any formula could be assigned to a different processor, provided the processors exchange information to keep prefixes naming consistent.

The SST proof method allows us to separate the rules which treat positive and negative information and to prove some interesting properties about termination and the underlying prefixes graph structure. These properties, along with the Church-Rosser theorem, are used to devise a general and terminating algorithm (obviously in the case where global and local assumptions are finite).

The results we present may be easily used to extend other proof systems to cover the same logics. Indeed Fitting's prefixed tableaux are directly extended using the conditions over prefixes described in Table 6, in Sect. 3. Matrix proof methods and other systems may be enhanced in a similar way.

In Sect. 2 we present the SST-calculus, whereas correctness and completeness proofs are sketched in Sect. 3. Church Rosser and other termination properties are presented in Sect. 4. A general algorithm is sketched in Sect. 5. Related works and conclusions are discussed in Sect. 6.

2 The Single Step Tableaux calculus

2.1 Preliminaries

Familiarity with the usual definitions of modal language and formulae is assumed (see [12] for an introduction). In particular we use the signed version of modal formulae. Thus an *unsigned formula* is constructed from propositional letters by the usual connectives $\wedge, \vee, \rightarrow, \neg, \Box, \Diamond$ and a *signed formula* is an unsigned one prefixed by the operator T or F. Informally, signs can be interpreted as the qualifiers "is true" or "is false". In the sequel X and Y will range over unsigned formulae, Z and Q over signed ones and p over propositional letters.

Signed formulae allow us to use the uniform notation of Smullyan and Fitting that classifies signed formulae according to their signs and principal connectives, as shown in Table 1. Informally speaking, formulae with a similar semantic behaviour (see below) are classified accordingly.

Table 1. Classification of Signed Formulae

α	α_1	α_2	β	β_1	β_2	π	π_0	ν	ν_0
$T.X \wedge Y$	$T.X$	$T.Y$	$F.X \wedge Y$	$F.X$	$F.Y$	$F.\Box X$	$F.X$	$T.\Box X$	$T.X$
$F.X \vee Y$	$F.X$	$F.Y$	$T.X \vee Y$	$T.X$	$T.Y$	$T.\Diamond X$	$T.X$	$F.\Diamond X$	$F.X$
$F.X \to Y$	$T.X$	$F.Y$	$T.X \to Y$	$F.X$	$T.Y$				
$F.\neg X$	$T.X$	$T.X$	$T.\neg X$	$F.X$	$F.X$				

The semantics of modal logic is the usual one, based on *frames* i.e. pairs $\langle W, \mathcal{R} \rangle$ where W is a non empty set and $\mathcal{R}$ is a relation over W. Different modal logics may be obtained by varying $\mathcal{R}$, as shown in Table 2.

Once again, the purpose of this work is not to choose among these logics, but to develop a proof system for each alternative. Thus, to keep exposition as compact and parametric as possible, we refer to L-frames, L-models or L-tableaux etc. when referring to frames, models or tableaux etc. for a particular logic L, viz. one among those listed in Table 2.

A L-model is thus a triple $\langle W, \mathcal{R}, \Vdash \rangle$ where $\langle W, \mathcal{R} \rangle$ is a L-frame and $\Vdash$ is a relation between possible worlds and signed formulae such that,

$w \Vdash T.\top$ and $w \not\Vdash F.\bot$.

$w \Vdash \alpha$ iff $w \Vdash \alpha_1$ and $w \Vdash \alpha_2$.

$w \Vdash \beta$ iff $w \Vdash \beta_1$ or $w \Vdash \beta_2$.

$w \Vdash \nu$ iff $\forall w^* \in W : w\mathcal{R}w^* \Rightarrow w^* \Vdash \nu_0$.

$w \Vdash \pi$ iff $\exists w^* \in W : w\mathcal{R}w^*$ and $w^* \Vdash \pi_0$.

The conjugate $\overline{Q}$ of a formula Q is obtained by exchanging F with T. Clearly, for any L-model $\langle W, \mathcal{R}, \Vdash \rangle$ and any $w \in W$, $w \Vdash Q$ iff $w \not\Vdash \overline{Q}$.

Validity and satisfiability of a formula in a L-model are given in the usual way, using local and global assumptions (set of signed formulae):

Definition 1. A signed formula Q is L-valid with local assumptions U and global assumptions G i.e. $G \models_L U \Rightarrow Q$ if and only if, in every L-model $\langle W, \mathcal{R}, \Vdash \rangle$, such that for every $v \in W$ it is $v \Vdash G$, if $w \in W$ and $w \Vdash U$ then $w \Vdash Q$.

For short we write $w \Vdash S$ instead of $\forall Z \in S : w \Vdash Z$. The unsigned version may be reduced to the signed one, just by tagging all formulae in G, U, or Q itself, with the sign T.

Table 2. Conditions on Accessibility Relation

Logic	Accessibility Relation $\mathcal{R}$ on frames
K	any relation
KB	symmetric
K4	transitive
K5	euclidean
K45	transitive and euclidean
D	serial
KDB	serial and symmetric
D4	serial and transitive
KD5	serial and euclidean
KD45	serial, transitive and euclidean
T	reflexive
B	reflexive and symmetric
S4	reflexive and transitive
S5	equivalence (reflexive, symmetric and transitive)
OM	almost reflexive
OB	almost reflexive and almost symmetric
OK4	almost transitive
OS4	almost reflexive and transitive
OM$^+$	serial and almost reflexive
OB$^+$	serial, almost reflexive and almost symmetric
OS4$^+$	serial, almost reflexive and transitive

Note. $\mathcal{R}$ is serial iff $\forall w \in W \ \exists w^* \in W$ s. t. $w\mathcal{R}w^*$. Almost reflexive iff $\forall w_0, w \in W$, $w_0 \mathcal{R} w$ implies $w \mathcal{R} w$ and almost symmetric iff $\forall w_0, w, w^* \in W$, $w_0 \mathcal{R} w$ implies $w \mathcal{R} w^* \Rightarrow w^* \mathcal{R} w$. A similar condition holds for almost transitivity.

2.2 Tableaux and Prefixes

SST use *prefixed formulae* i.e. pairs $\langle \sigma : Z \rangle$ where σ is a non empty sequence of integers called *prefix* and Z is a signed formula. Intuitively σ "names" the possible world where Z holds.

In the sequel, σ or $\langle n_1 n_2 \ldots n_p \rangle$, with $p \geq 1$, will be prefixes, ε the empty sequence, $\sigma_0 \cdot \sigma_1$ the concatenation of the sequence σ_0 with the sequence σ_1 and σn a short cut for $\sigma \cdot \langle n \rangle$. The relation $\sqsubseteq$ (intuitively $\sigma_0 \sqsubseteq \sigma$ holds if and only if σ_0 is an initial part of σ) and the binary operator $\sqcap$ ($\sigma_1 \sqcap \sigma_2$ is, in practice, the maximal initial part common to both prefixes) are also used.

The definitions of branch and tableau will be the same (but the rules!) of the prefixed tableau proposed by Fitting [3, 4].

Remark. Differently from sequent-like tableaux, where many tableaux may be needed to prove a formula satisfiable, here there is *exactly one* tableau for both satisfiability and validity.

The definition of closure and termination is equally simple: a branch is *closed* if there are contradictory prefixed formulae on that branch (i.e. there is a σ such that, for some X, both $\langle \sigma : T.X \rangle$ and $\langle \sigma : F.X \rangle$ are present on the branch or either $\langle \sigma : T.\bot \rangle$ or $\langle \sigma : F.\top \rangle$ are present); a tableau is closed if every branch is closed. *Termination* occurs when no operation is possible. A branch is *open* if it is terminated and not closed and a tableau is open if at least one branch is such.

Definition 2 Tableau Proof. Given a set of global assumptions G and local assumptions U in the modal logic L, a *proof* for the signed formula Q, i.e. $G \vdash_L U \Rightarrow Q$, is the tableau starting with $\langle 1 : \overline{Q} \rangle$ and closed by using only SST-rules for the logic L with global assumptions G and local assumptions U.

Intuitively a tableau proof of Q is a failed attempt to prove that $\overline{Q}$ is satisfiable. When the tableau closes, e.g. because $\langle \sigma : T.\bot \rangle$ is present on the branch, this means that the assumption of $\overline{Q}$ leads to a contradiction in some possible world (the one "tagged" by σ) and thus Q must be valid. To check that Q is satisfiable, one just needs to start with $\langle 1 : Q \rangle$ and end with an open tableau.

2.3 SST-Rules

Table 3 shows the rules which are common to all logics viz. those which add local or global assumptions and those which reduce α, β and π formulae. In the case of π-rule some more terminology is needed: a prefix is *unrestricted* on a branch ϑ, if it is not the initial part of any other prefix already present on the branch.

Table 4 shows the different rules which may be used to reduce a ν-formula. Each rule may be added or deleted from the system, with the same flexibility of Hilbert-style axiomatization. Thus, each collection of SST rules gives us an SST calculus for a different logic, as shown in Table 5.

Remark. Each rule does not depend on a particular logic nor on some accessibility relation over prefixes: it just states which formulae may be "passed" to some close neighbours.

2.4 Examples

The following example, for the logic K4, should better clarify the key difference between SST and prefixed tableaux (as devised in [4]).

prefixed K4   K4-SST

$$\frac{\sigma : \nu}{\sigma nmk : \nu_0} \quad \Longleftrightarrow \quad \frac{\dfrac{\dfrac{\sigma : \nu}{\sigma n : \nu}}{\sigma nm : \nu}}{\sigma nmk : \nu_0}$$

Table 3. SST-rule Common to All Logic

$$\alpha: \quad \dfrac{\sigma:\alpha}{\begin{array}{c}\sigma:\alpha_1\\ \sigma:\alpha_2\end{array}} \qquad \beta: \quad \dfrac{\sigma:\beta}{\sigma:\beta_1 \ \mid \ \sigma:\beta_1}$$

$$Loc: \quad \dfrac{\vdots}{1:Z} \qquad \text{if} \quad Z \in U$$

$$Glob: \quad \dfrac{\vdots}{\sigma:Z} \qquad \text{if} \quad \sigma \text{ is present on the branch and } Z \in G$$

$$\pi: \quad \dfrac{\sigma:\pi}{\sigma n:\pi_0} \qquad \text{with} \quad \sigma n \text{ unrestricted on the branch}$$

Table 4. SST-rules for ν Formulae

$$K: \ \dfrac{\sigma:\nu}{\sigma n:\nu_0} \qquad D: \ \dfrac{\sigma:\nu}{\sigma:\pi} \qquad T: \ \dfrac{\sigma:\nu}{\sigma:\nu_0}$$

$$B: \ \dfrac{\sigma n:\nu}{\sigma:\nu_0} \qquad 4: \ \dfrac{\sigma:\nu}{\sigma n:\nu} \qquad 4^R: \ \dfrac{\sigma n:\nu}{\sigma:\nu}$$

$$T^D: \ \dfrac{\sigma n:\nu}{\sigma n:\nu_0} \qquad B^D: \ \dfrac{\sigma nm:\nu}{\sigma n:\nu_0} \qquad 4^D: \ \dfrac{\sigma n:\nu}{\sigma nm:\nu}$$

Note. σ, σn and σmn must all be already present on the branch, i.e. some π-rules must have introduced them already, and

$$\dfrac{\sigma:\nu}{\sigma:\pi} \quad \text{equals to} \quad \dfrac{\sigma:T.\Box X}{\sigma:T.\Diamond X} \quad \text{or} \quad \dfrac{\sigma:F.\Diamond X}{\sigma:F.\Box X}.$$

Table 5. Modal Logics and SST-rules

Logics	ν-SST-rules	Logics	ν-SST-rules	Logics	ν-SST-rules
K	K	T	K + T	OM	K + T^D
KB	K + B	B	K + T + B	OB	K + T^D + B^D
K4	K + 4	S4	T + 4	OK4	K + 4^D
K5	K + 4^D + 4^R	S5	T + 4 + 4^R	OS4	K + T^D + 4^D
K45	K + 4^D + 4^R	KD*	K*-rules + D	O*+	O*-rules + D

Note 3. For K4 prefixed tableaux, one has to prove that $\sigma \vartriangleright_{K4} \sigma nmk$ and that σmnk already occurs on the branch, whereas, for K4-SST, one has to prove only that σn, σnm and σnmk are already present on the branch.

Remark. The key feature is that SST-rules access only immediate neighbours of the current prefix, thus boosting efficiency without affecting completeness.

Therefore, when a formula of the type $\langle \sigma : \pi \rangle$ is reduced, and a new prefix is introduced, the ν-formulae, which need to be reduced again, are *only those with the same prefix* σ. "Far away formulae may well be left sleeping". At the same time a Generation Lemma (see Sec. 3) ensures us that, if a "far" prefix occurs on a branch, all intermediate prefixes must occur too (e.g. if σnmk occurs then σ, σn, σnm also do).

A simple proof for the logic K5 is given in Fig. 1.

(1) $1 : F.\Diamond X \to \Box \Diamond X$
(2a) $1 : T.\Diamond X$ α − rule at (1)
(2b) $1 : F.\Box \Diamond X$
(3) $11 : T.X$ π − rule at (2a)
(4) $12 : F.\Diamond X$ π − rule at (2b)
(5) $1 : F.\Diamond X$ 4^R − rule at (4) backward from $\langle 12 \rangle$ to $\langle 1 \rangle$
(6) $11 : F.X$ K − rule at (5) forward from $\langle 1 \rangle$ to $\langle 11 \rangle$
      Contradiction in $\langle 11 \rangle$ tableau closes

Fig. 1. Proof of the euclidity axiom $\Diamond X \to \Box \Diamond X$ in K5

3 Correctness and Completeness

Only at this stage the relation $\vartriangleright$ over prefixes need to be introduced. $\vartriangleright$ will be the "carbon copy" of the accessibility relation $\mathcal{R}$ over possible worlds and will allow us to prove correctness and completeness with a suitable extension of Fitting prefixed tableaux. Obviously, conditions imposed on $\vartriangleright$ depend on the logic and are specified in Table 6. In the following, we just sketch the main ideas; details can be found in [15].

To prove the correctness theorem it is necessary to merge the techniques used for sequent-like and prefixed tableaux:

1. show that properties of $\mathcal{R}$ induce the required behaviour of formulae in different worlds (e.g. when $\mathcal{R}$ is transitive if $w\mathcal{R}w^*$ and $w\|\!\!\!-\nu$ then $w^*\|\!\!\!-\nu$);

Table 6. Conditions on Prefixes

Logic	Conditions on the relation over prefixes $\sigma \triangleright \sigma^*$
K	$\sigma^* = \sigma n$
KB	$\sigma^* = \sigma n$ or $\sigma^* n = \sigma$
$K4$	$\sigma = \sigma_0 \cdot \sigma_1$ and $\sigma^* = \sigma_0 \cdot \sigma_2$ and $\sigma_0 \neq \varepsilon$ and $\sigma_1 = \varepsilon$ and $\sigma_2 \neq \varepsilon$
$K5$	$\sigma = \sigma_0 \cdot \sigma_1$ and $\sigma^* = \sigma_0 \cdot \sigma_2$ and $\sigma_0 \neq \varepsilon$ and or $\sigma_1 = \varepsilon$ and $\sigma_2 = n$ or $\sigma_1 \neq \varepsilon$ and $\sigma_2 \neq \varepsilon$ or $\|\sigma_0\| \geq 2$
$K45$	$\sigma = \sigma_0 \cdot \sigma_1$ and $\sigma^* = \sigma_0 \cdot \sigma_2$ and $\sigma_0 \neq \varepsilon$ and or $\sigma_2 \neq \varepsilon$ or $\|\sigma_0\| \geq 2$
$D - logics$	same conditions for $K - logics$
T	$\sigma^* = \sigma n$ or $\sigma^* = \sigma$
B	$\sigma^* = \sigma n$ or $\sigma^* n = \sigma$ or $\sigma^* = \sigma$
$S4$	$\sigma = \sigma_0 \cdot \sigma_1$ and $\sigma^* = \sigma_0 \cdot \sigma_2$ and $\sigma_0 \neq \varepsilon$ and $\sigma_1 = \varepsilon$
$S5$	$\sigma = \sigma_0 \cdot \sigma_1$ and $\sigma^* = \sigma \cdot \sigma_2$ and $\sigma_0 \neq \varepsilon$
$O - logics$	similar to $K - logics$ adding condition $\|\sigma_0\| \geq 2$

Note. It is always $\sigma_0 \doteq \sigma \sqcap \sigma^*$.

2. define a mapping to assign "names" (prefixes) to "things" (possible worlds) such that there is a suitable matching between $\triangleright$ and $\mathcal{R}$ (see [3, 4]);
3. prove a safe step lemma such that, if a tableau is satisfiable then the tableau obtained by the application of any SST rule is still satisfiable;
4. prove the correctness theorem i.e. show that if the tableau closes then the formula is valid.

Theorem 4 Correctness. *If Q has a proof with the SST for the logic L with global assumptions G and local assumptions U then Q is valid i.e. $G \models_L U \Rightarrow Q$.*

A simple completeness proof for prefixed tableaux, yet restricted to non euclidean modal logics, has been given by Fitting [3, 4]. The same ideas may be applied here, taking into account the fact we are dealing with SST:

1. apply a systematic procedure to the tableau;
2. if the algorithm terminates (no rule may be further applied) and the tableau does not close, then any open branch is saturated for the SST

3. if the algorithm never terminates then, by König Lemma, it is possible to prove that there is an infinite branch that will never close, even "ad infinitum" [4]. This branch can then be chosen as the saturated one;
4. prove that, given the conditions for $\triangleright$ as in Table 6, a branch saturated with the SST rules is also saturated with the prefixed rules of Fitting;
5. prove a strong model existence theorem, building a model with the saturated branch, using the property of L-Pref-SAT and a lemma that prove that the accessibility relation between prefixes has the same properties of the accessibility relation between possible worlds.
6. thus the proof of the completeness theorem is just round the corner.

The idea behind *prefixed downward saturation* for a logic L (L-Pref-SAT for short) is that all possible rules have been applied to a branch ϑ and yet no contradiction has been found. For example, to get consistency of ϑ, $\langle \sigma : Z \rangle$ and $\langle \sigma : \overline{Z} \rangle$ cannot be both in ϑ. To get saturation w. r. t. the ν-rule (ν-L-Pref-SAT), ϑ must be such that if $\langle \sigma : \nu \rangle$ is in ϑ then $\langle \sigma^* : \nu_0 \rangle$ must be in ϑ for every σ^* in ϑ such that $\sigma \triangleright \sigma^*$, according to the logic L (Table 6). Details may be found in [4, 15].

However L-Pref-SAT is a powerful propriety, and thus too "expensive": some sets are not the result of any tableau reduction. For instance the set $\{\langle 1 : T.\Box p \rangle$, $\langle 1111 : T.p \rangle\}$ is K4-Pref-SAT with $U = G = \emptyset$, but there is no K4-tableau corresponding to this set.

The following property better characterizes prefixed tableaux and SST.

Definition 5 Generated Submodel Property. A set of prefixed formulae ϑ satisfies the generated submodel property iff

1. for any prefix $\langle n_0 n_1 n_2 \ldots n_k \rangle$ present in ϑ all "ancestors" $\langle n_0 n_1 \ldots n_i \rangle$ must be in ϑ with $i = 0, 1 \ldots k$
2. there is exactly one prefix $\langle n_0 \rangle$ in ϑ such that, for every prefix σ in ϑ, it is $n_0 \sqsubseteq \sigma$

Lemma 6 Generation. *If ϑ is a branch of a tableau then it satisfies the generated submodel property.*

Remark. This may be seen as the proof theoretical counterpart of the Generation Theorem by Segerberg (see [12]): checking validity in a model reduces to checking validity in the generated submodels.

An analogous definition of L-SST-SAT can be given: change the ν-condition for Pref-SAT into a similar one for SST-rules. e.g. for K45:

K-SST-SAT if $\langle \sigma : \nu \rangle \in \vartheta$ then $\langle \sigma n : \nu_0 \rangle \in \vartheta$ for every σn present in ϑ.
4-SST-SAT if $\langle \sigma : \nu \rangle \in \vartheta$ then $\langle \sigma n : \nu \rangle \in \vartheta$ for every σn present in ϑ.
4^R-SST-SAT if $\langle \sigma n : \nu \rangle \in \vartheta$ then $\langle \sigma : \nu_0 \rangle \in \vartheta$ for every[1] σ in ϑ.

From these and the Generation Lemma the following key lemma results:

---
[1] Indeed just one.

Lemma 7 Prefix Closure. *If a set of prefixed formulae ϑ is L-SST-SAT and satisfies the generated submodel property then it is also L-Pref-SAT.*

The main idea behind the proof (see [15]) is to "let a formula travel throughout prefixes until it reaches the right point", e.g. first example of Sect. 2.

One more lemma — indeed a lemma for any logic L — needs to be proved:

Lemma 8 Matching. *Given the conditions on the relation over prefixes $\triangleright$ for the logic L (in Table 6), $\triangleright$ has the same properties of the accessibility relation over possible worlds $\mathcal{R}$ of L-frames (in Table 2).*

Now we have all the necessary machinery to state the following

Theorem 9 Strong Model Existence Theorem. *If ϑ is a set of prefixed formulae that is L-SST-SAT with the local assumptions U and global assumptions G then there is an L-model where ϑ is satisfiable, with local assumptions U and global assumptions G and a suitable L-interpretation.*

Proof. We reduce SST-SAT to Pref-SAT thanks to Prefix Closure and Generation Lemma and then construct the model using prefixes as possible worlds and $\triangleright$ as the accessibility relation i.e.

$$W \doteq \{\sigma \in \mathbb{N}^+ : \sigma \text{ is present in } \vartheta\}$$
$$\sigma \mathcal{R} \sigma^* \quad \text{iff} \quad \sigma \triangleright \sigma^*$$
$$v(\sigma, p) \doteq \begin{cases} \text{true} & \text{if } \langle \sigma : T.p \rangle \in \vartheta \\ \text{false} & \text{otherwise} \end{cases}$$

Then it is possible to prove that

1. $\forall Z \in U : 1 \Vdash Z$,
2. $\forall Z \in G : \forall \sigma \in W : \sigma \Vdash Z$,
3. $\forall \langle \sigma : Z \rangle \in \vartheta : \sigma \Vdash Z$,
4. $\langle W, \mathcal{R} \rangle$ is an L-frame.

The first two items are a consequence of the properties of L-Pref-SAT and the third must be proved by induction on the complexity of the formula Z, by L-Pref-SAT too. The fourth is a consequence of the Matching Lemma. □

And strong completeness follows:

Theorem 10 Strong Completeness. *If Q is L-valid with G as global assumptions and U as local assumptions, i.e. $G \models_L U \Rightarrow Q$, then Q has a L-SST proof.*

4 Some Properties of SST

Theorem 11 Church-Rosser. *The SST calculus satisfies the strong Church-Rosser Property i.e. it is strongly confluent up to isomorphic renaming of prefixes.*

Proof. SST-rules may be seen as particular term rewriting ones in which ν formulae are not deleted (and thus can be used more than once on the same branch). It's clear that a formula of a given type may be reduced only by the corresponding rule and that one cannot reduce a subformula before reducing its parent. Hence no superposition exists between different rules and it is possible only between different ν-rules. But ν-formulae are not deleted, and thus every critical pair is locally confluent. Moreover, SST rules are left linear (just look at Tables 3 and Table 4). Hence, by Neuman's theorem, the result follows [11]. □

To prove termination properties we devised the founding system:

$\dim(\vartheta) =$ the sum of sizes of formulae not yet processed or potentially reusable (the latter are only ν-formulae);
$\text{scope}(\vartheta) =$ the number of prefixes where a ν-formula is active (i.e. not yet processed) or absent (thus potentially introducible by some suitable ν-rule).

Proposition 12. *Given a finite SST, the iterative application of α, β, π-rules and the introduction of local assumptions terminates for any logic L.*

Proposition 13. *Given a finite SST, the iterative application of α, β, ν-rules and the introduction of local and global assumptions terminates for any logic L.*

More results may be proved about the prefixes' graph structure.

Lemma 14. *On a tableaux branch, the formulae with a given prefix are bounded by the subformulae of $G \cup U \cup \{\overline{Q}\}$.*

Lemma 15. *On a tableau branch, the number of prefixes of a given length n is finite and bounded by the number of π-subformulae of $G \cup U \cup \{\overline{Q}\}$ raised to the $n - 1$ (π and ν subformulae for D and O^+ logics).*

Proof. By induction on n. If $n = 1$ then the only prefix is $\langle 1 \rangle$. For the inductive step just note that the only way to introduce a new prefix is to reduce a π-formula and those are bounded by Lemma 14. Then a multiplication suffices.
□

Theorem 16. *If a tableau branch is infinite, it must contain a "growing chain" of prefixes and, after a finite number of steps, the chain becomes "periodical" i.e. formulae will just be repeated with a longer prefix.*

Proof. By Lemma 14 and 15 the number of formulae with prefixes of a given length is finite and thus a growing chain of prefixes must exist. By Lemma 14 a prefix can have only formulae contained in $G \cup U \cup \{\overline{Q}\}$ and thus the chain must be periodical. □

Thus, even if the underlying countermodel is potentially infinite, an algorithm may check for this periodicity and thus terminate in any case. Clearly the number of steps to realize a branch cannot be closed are bounded by the size of the powerset of subformulae of $G \cup U \cup \{\overline{Q}\}$ but, from a practical point of view, a better condition has been devised for the algorithm of Sect. 5.

5 A General Algoritm

To cope with the reuse of ν-formulae, each formula must be set to one among the following states:

Active: the formula must be processed;

Done: the formula has been processed once and needs not be processed anymore (typically α, β, π-formulae);

Quiet: the formula has been already processed (and thus it isn't anymore active) but may be awaken at any time. Typically it's a ν-formula that becomes active as soon as a π-rule creates a new neighbour prefix.

The Church-Rosser property allows us to choose a suitable ordering of rule application in order to exploit the termination properties of Sec. 4. We just sketch the main procedure and the intuition that lies behind. Therefore, to check if $\overline{Q}$ is satisfiable with the local assumptions U and global assumptions G:

1. set $\langle 1 : \overline{Q} \rangle$ and $\langle 1 : Z \rangle$ for all $Z \in U$ as the starting branch of the tableaux and set them as **Active**;
2. repeat
   (a) choose an open branch;
   (b) iteratively apply all π-rules (thus spannig the tree of possible worlds);
   (c) apply, if necessary, Glob-rule to every prefix (thus both new and old worlds will have the same global assumptions);
   (d) iteratively apply all ν, α, β-rules (thus moving information from a prefix (a world) to another);
   (e) check if any periodicity in the prefix structure has been reached and then suspend the formulae with the periodical prefix (or awake them if the structure is no longer periodical);
3. until a branch is open (with all formulae either **Quiet** or **Done**) and thus $\overline{Q}$ is SATISFIABLE and hence Q NON VALID;
4. or until all branches close, $\overline{Q}$ is NON-SAT and hence Q is VALID.

Note 17. When a ν-formula is reduced, all availables ν-rules are applied and only afterwards the ν-formula is set to **Quiet**. When, later on, a π-rule is applied, and a new prefix σn is created, the *only* ν-formulae that must be "awaken", and set back to **Active**, are those with the prefix σ!

Note 18. To suspend the search on a prefix (thus setting its π-formulae from **Active** to **Quiet**), it is enough to find a shorter prefix such that: the same formulae occur in both, the same rules may be applied to both and the same rule may introduce formulae in both. Later on, the proof search may show that prefixes were not really periodical and then π-formulae may be reset **Active** to continue the computation further.

In any case, only a limited number of prefixes must be considered at any time, and the rest may well be left in secondary memory to be retrieved only when the proof search approaches them.

Obviously, if U and G are huge and largely irrelevant, this algorithm may be inefficient but, thanks to the Church-Rosser property, we can also proceed in an incremental way etc. Anyhow, this and other considerations (e.g. discarding closed branches, pushing β-rule down as much as possible etc.) are at stake during the implementation of the system and will not be considered here.

Remark. We stress the fact that the system does not depend on the particular strategy. For example depth first strategies may well be devised or a particular ordering on formula reduction may be fixed etc.

6 Related Works and Conclusion

Sequent-like (or Fitch-style) tableaux are one of the traditional tools for deduction in modal logics and may be found in [2, 4, 5, 8], whereas full model checkers are the tableaux for K45 proposed in [10] or the TABLEAUX system presented in [1]. The latter has been implemented and of treats many logics. A dual approach may be found in [7], where the accessibility relation is made explicit by using restricted quantification, and deduction is performed in first order logic.

Other methods are, more or less, linked by the same idea: use labels (called prefixes, indexes, worlds' paths, etc.) to name the worlds where formulae are supposed to hold.

An intermediate approach is that of prefixed tableaux, proposed by Fitting [3, 4], which are closer to model checking than others. This feature allows one to use this system as a precious tool to build completeness theorems for other proof methods (see [15, 19] or here).

A comprehensive work about modal resolution is presented by Ohlbach [17] where world paths are used to identify worlds where a formula holds. Different unification procedures are used to cope with different accessibility relations. The system is also lifted to a full first order framework.

The application of matrix proof methods to modal logics is due to Wallen [19]. Prefixes and unification procedures, varying with the logic, are used to ensure connections are drawn between formulae belonging to the same world. The method is definitely effective and spans to first order modal logics but needs a human-oriented interface for the presentation of proofs and proof search (see [19] for further references).

A common limit is that systems not based on pure model checking do not handle euclidean or symmetric logic as well as our proposal. Moreover, passing from a logic to another is not simple, since information is coded in the unification procedure or in the accessibility relation.

We have proposed a new tableau calculus that brings together the advantages of sequent-like tableaux and prefixed tableaux and allows an analytic, deterministic, reasonably efficient, strongly confluent, uniform and "user friendly" approach to many normal propositional modal logics, including symmetric and euclidean ones: K,T,D, KB, K4, K5, K45, KDB, D4, KD5, KD45, KD45, T, B, S4, S5, OM, OB, OK4, OS4, OM^+, OB^+, $OK4^+$, $OS4^+$ and many others. Future work is in the direction of lifting this framework to first order logic.

Moreover, we think that other deduction methods may directly benefit from this work, especially prefixed tableaux and matrix proof methods: they just need to extend respectively ▷ and the unification procedure with the new conditions for euclidean and other logics contained in Table 6.

Acknowledgements

A warm thank goes to L. Carlucci Aiello and F. Pirri for reading drafts of this paper and for their many useful hints and suggestions. Comments from W. Bibel and anonymous referees helped to improve the presentation of this paper.

References

1. L. Catach. TABLEAUX, a general theorem prover for modal logics. *J. of Automated Reasoning*, 7, 1991.
2. F. Fitch. Tree proofs in modal logic. *J. of Symbolic Logic*, 31, 1966.
3. M. Fitting. Tableau methods of proof for modal logic. *Notre Dame J. of Formal Logic*, 13, 1972.
4. M. Fitting. *Proof Methods for Modal and Intuitionistic Logics*. Reidel, Dordrecht, 1983.
5. M. Fitting. First order modal tableaux. *J. of Automated Reasoning*, 4, 1991.
6. D. M. Gabbay. *Labelled Deductive Systems*. Oxford Univ. Press, Oxford, 1993.
7. I. Gent. Theory Matrices (for Modal Logics) Using Alphabetical Monotonicity. *Studia Logica*, 52, 1993.
8. R. Goré. Cut-free Sequent and Tableau Systems for Propositonal Normal Modal Logics. PhD thesis, Computer Laboratory, Univ. of Cambridge, 1992.
9. J. Y. Halpern and R. Fagin. Modelling knowledge and action in distributed systems. *Distributed computing*, 3(4), 1989.
10. J. Y. Halpern and Y. Moses. A guide to the modal logic of knowledge and belief. In *Proc. of IJCAI-85*, 1985.
11. G. Huet. Confluent reductions: Abstract properties and applications to term rewriting systems. *J. of the ACM*, 1980.
12. G. E. Hughes and M. J. Cresswell. *a Companion to Modal Logic*. Methuen, London, 1984.
13. K. Konolige. On relation between default and autoepistemic logic. *Artificial Intelligence*, 35(3), 1988.
14. W. Marek, G. F. Schwarz, and M. Truszczynski. Modal nonmonotonic logics: ranges, characterization, computation. In *Proc. of KR-91*, San Matteo, CA, 1991.
15. F. Massacci. Ragionamento autoepistemico automatico. Tesi di Laurea, Fac. Ingegneria, Univ. di Roma "La Sapienza", 1993. In Italian.
16. R. C. Moore. Possible world semantics for autoepistemic logic. In M. Ginsberg, editor, *Readings on Nonmonotonic Reasoning*. Morgan Kaufmann, 1990.
17. H. J. Ohlbach. A resolution calculus for modal logic. In *Proc. of CADE-88*, LNCS 310. Springer Verlag, 1988.
18. G. F. Schvarts. Gentzen style systems for K45 and K45D. In A. Meyer and M. Taitslin, editors, *Logic at Botik '89, Symposium on Logical Foundations of Computer Science*. LNCS 363. Springer-Verlag, 1989.
19. L. A. Wallen. Matrix proof methods for modal logics. In *Proc. of IJCAI-87*, 1987.

Reconstructing Proofs at the Assertion Level

Xiaorong Huang

Fachbereich Informatik, Universität des Saarlandes
Postfach 15 11 50, D-66041 Saarbrücken, Germany
huang@cs.uni-sb.de

Abstract. Most automated theorem provers suffer from the problem that they can produce proofs only in formalisms difficult to understand even for experienced mathematicians. Effort has been made to *reconstruct* natural deduction (ND) proofs from such machine generated proofs. Although the single steps in ND proofs are easy to understand, the entire proof is usually at a low level of abstraction, containing too many tedious steps. To obtain proofs similar to those found in mathematical textbooks, we propose a new formalism, called ND style proofs at the *assertion level*, where derivations are mostly justified by the application of a definition or a theorem. After characterizing the structure of compound ND proof segments allowing assertion level justification, we show that the same derivations can be achieved by domain-specific inference rules as well. Furthermore, these rules can be represented compactly in a tree structure. Finally, we describe a system called *PROVERB*, which substantially shortens ND proofs by *abstracting* them to the assertion level and then transforms them into natural language.

1 Introduction

This paper concerns the presentation of machine generated proofs. Viewing automated theorem provers as a special sort of expert systems, this problem is very similar to that of the explanation component of an expert system. In order to aid the understanding of an end-user, methods are devised to augment, to prune, or even to transform the trace of reasoning left behind by an expert system [Sho76, WS89]. Explanations produced in this way are in general *tightly bound* with the authentic movement of an expert system from the initial data to the conclusion. Although such explanations are apparently appropriate for system developers or knowledge engineers, they do not meet the requirement of a typical end-user. To solve this problem, a new, so called *reconstructive* paradigm for explanation has emerged in recent years [WT92]. The central idea of this approach is that a distinct knowledge base should be used to reconstruct a new solution based on the original one found by the expert system.

The *reconstructive* approach for explanation has been pursued in the field of automated reasoning as well, because not only the line of reasoning can be unnatural and obscure, the formalism in which the proofs are encoded is usually extremely machine oriented. Procedures have been developed to transform proofs from machine oriented formalisms into more natural formalisms

[And80, Mil83, Pfe87, Lin90]. As the target formalism, usually a variation of the *natural deduction* (ND) proof first proposed by G. Gentzen [Gen35] is chosen. Heuristics of various kinds are developed to improve the quality of the target ND proof. For instance, C. Lingenfelder utilizes the topological structures of the refutation graph both to produce more direct proofs as well as to avoid redundancy by inserting lemmas [Lin90]. Another technique for inserting lemmas is reported in [PN90].

Until now the reconstruction stops here and ND proofs are used as inputs by systems producing proofs in natural language. The first such attempt was made by D. Chester [Che76]. His system EXPOUND is usually characterized as an example of *direct translation*. Although a sophisticated linearization is applied on the input ND proofs, the steps are translated locally in a template driven way. Equipped with more advanced techniques developed in the field of natural language generation, a more coherent translation was obtained by the MUMBLE system of D. McDonald [McD83], where emphasis was laid on the generation of utterances highlighting important global structures of the proofs, as well as utterances mediating between subproofs. A more recent attempt can be found in THINKER [EP93], where different styles for explaining ND proofs are exploited. In short, it was believed that ND proofs can be adequately presented by resorting solely to *ordering*, *pruning*, and *augmentation*.

All these systems suffer from the same problem: The derivations they convey are exclusively at the level of the inference rules of the ND calculus. In contrast to informal proofs found in standard mathematical textbooks, such proofs are composed of derivations familiar from elementary logic, where the focus of attention is on syntactic manipulations rather than on the underlying semantic ideas. The main problem, we believe, lies on the lack of intermediate structures of ND proofs, which allow atomic justifications at a higher level of abstraction.

To gain more reliable experience with the *levels of justifications*, we have analyzed proofs in mathematical textbooks like [Deu71]. Based on our preliminary empirical study, justifications are provided at three levels.

- *Logic level* justifications are simply verbalizations of the ND inference rules, such as the rule of Modus Ponens.
- *Assertion level* justifications account for a derivations in terms of the application of an axiom, a definition or a theorem (collectively called an assertion). The following is an example:
  "since a is an element of the set S_1, and S_1 is a subset of S_2, *according to the definition of subset*, a is an element of S_2".
- *Proof level* justifications are at a still higher level and are comparatively rare. One example is justifying a proof segment as a whole by resorting to its similarity to a previous proof segment.

Among the three levels mentioned above, the assertion level plays a dual role in presentation. On the one hand assertion level justifications are *logically compound*, that is, mathematicians can explain such steps by providing a logic level proof segment. On the other hand, assertion level justifications are primitive with respect to presentation, since proof segments justifiable atomically at the assertion level is practically never expanded to a logic level proof segment. On account of this, while proof level structures are also very useful, the reconstruction of assertion level units in ND proofs is of paramount importance and is indispensable for the purpose of presenting proofs in a natural way.

Section 2 first defines the structure of the logic level proof segments which can be justified atomically at the assertion level. Section 3 accounts for the acquisition of domain-specific assertion level inference rules and shows how they can be organized in a tree structure. Then in section 4, we illustrate how this tree structure can be used to abstract ND proofs to the assertion level and report our experience with them in the subsequent translation into natural language. Finally, a look into the future work concludes this paper.

2 Compound Proof Segment at the Assertion Level

The existence of a hierarchy of proof units in proofs constructed by mathematicians can be accounted for by a computational model of human deductive reasoning [Hua93]. Following A. Bundy [Bun88], this theory cast theorem proving as a planing process, where a planner constructs a proof by applying *methods* (called tactics in some earlier systems [GMW79, CAB+86]) on open goals. The proof under construction is represented as a hierarchical and partially elaborated plan called a *proof tree*. The execution of each method results in the integration of a subtree constituting a proof unit with internal structure. In the light of this, the intuitive notion of the application of an assertion is technically realized either by a compound proof unit composed of applications of ND rules, or by a atomic proof unit justified by a *domain-specific* inference rule.

Fig. 1 is an example of a compound proof unit inferring $a_1 \in F_1$ from $U_1 \subset F_1$ and $a_1 \in U_1$ by applying the definition of subset encoded as

$$\forall_{S_1,S_2} S_1 \subset S_2 \Leftrightarrow \forall_x x \in S_1 \Rightarrow x \in S_2 \tag{1}$$

The leaf with the label $\mathcal{A}$ contains the *assertion* being applied.

$$\cfrac{\cfrac{\cfrac{\cfrac{\cfrac{\mathcal{A}: \ \forall_{S_1,S_2} S_1 \subset S_2 \Leftrightarrow (\forall_x x \in S_1 \Rightarrow x \in S_2)}{U_1 \subset F_1 \Leftrightarrow (\forall_x x \in U_1 \Rightarrow x \in F_1)} \forall D}{U_1 \subset F_1 \Rightarrow (\forall_x x \in U_1 \Rightarrow x \in F_1)} \Leftrightarrow D, \ U_1 \subset F_1}{\forall_x x \in U_1 \Rightarrow x \in F_1} \Rightarrow D}{a_1 \in U_1 \Rightarrow a_1 \in F_1} \forall D, \ a_1 \in U_1}{a_1 \in F_1} \Rightarrow D$$

Fig. 1. Natural Expansion 1 for Subset Definition

Actually, the procedure applying assertions by constructing a compound proof segment is specified in terms of a so called *decomposition-composition* constraint imposed on such proof segments identified in our preliminary empirical study [Hua92]. The following two definitions are necessary for the discussion of this constraint.

Definition: An inference rule of the form $\frac{\Delta \vdash F, \Delta \vdash P_1, \ldots, \Delta \vdash P_n}{\Delta \vdash Q}$ is a *decomposition* rule with respect to the formula schema F, if all applications of it, written as $\frac{\Delta \vdash F', \Delta \vdash P'_1, \ldots, \Delta \vdash P'_n}{\Delta \vdash Q'}$ satisfy the following condition: each $P'_1, \ldots, P'_n$ and Q' is

- a proper subformula of F', or
- a specialization of F' or of one of its proper subformula, or
- a negation of one of the first two cases.

Under this definition, $\wedge D, \Rightarrow D, \forall D$ are the only elementary decomposition rules in the natural deduction calculus $\mathcal{NK}$. Compare Fig. 1 for the meaning of the rules.

Definition: An inference rule of the form $\frac{\Delta \vdash P_1, \ldots, \Delta \vdash P_n}{\Delta \vdash Q}$ is called a *composition* rule if all applications of it, written as $\frac{\Delta \vdash P'_1, \ldots, \Delta \vdash P'_n}{\Delta \vdash Q'}$, satisfy the following condition: each $P'_1, \ldots P'_n$ is

- a proper subformula of Q', or
- a specialization of Q' or of one of its proper subformula, or
- a negation of one of the first two cases.

As illustrated in Fig. 1, the decomposition-composition constraint requires that a logic level proof segment applying an assertion $\mathcal{A}$ consists of a linear decomposition of $\mathcal{A}$ along the branch from $\mathcal{A}$ to the root. Other premises needed in the series of decompositions (the leaves $U_1 \subset F_1$ and $a_1 \in U_1$ in Fig. 1) can be obtained by compositions. For an example of such composition, see Fig. 2. For a precise definition of this constraint, the readers are referred to [Hua92]. In the sequel, proof segments satisfying this constraint will be referred to as the *natural expansion* of corresponding assertion level justification. This constraint is closely related to one of Johnson-Laird's *effective procedures* [JL83], aimed at accounting for spontaneous daily reasoning. Unfortunately, the psychological explanations provided by him can not be extended to predicate logic straightforwardly.

3 Assertion Level Inference Rules

In this section, we show that deductions justifiable by the application of a particular assertion $\mathcal{A}$ can be covered by a finite set of *domain-specific* inference rules at the assertion level. In the sequel, we denote this set of rules applying an assertion $\mathcal{A}$ by $Rules(\mathcal{A})$. It is this finiteness that makes this concept useful both for proof presentation, as well as for interactive proof development environments [HKK+94].

3.1 Acquisition of Assertion Level Inference Rules

There are two ways for acquiring new assertion level rules:

- learning by *chunking-and-variablization*,
- learning by *contraposition*.

Chunking-and-Variablization First, since there is evidence that input-output patterns of repeated actions will be remembered as new operators, we believe that patterns of repeated applications of an assertion may be remembered as new rules. Similar phenomena is called in other systems the learning of *macro-operators* [FHN72], or *chunking* [New90]. On account of this, domain-specific rules are also referred to as compound rules or macro-rules. We continue with our subset example to illustrate this.

Example 1 (Continued): Suppose that a reasoner has just derived $a_1 \in F_1$ from the premises $a_1 \in U_1$ and $U_1 \subset F_1$ by applying the definition of subset (1). Our assumption is that apart from merely drawing a concrete conclusion from the premises, possibly he learns the following macro-rule as well:

$$\frac{\Delta \vdash a \in U, \Delta \vdash U \subset F}{\Delta \vdash a \in F} \qquad (2)$$

where a, U and F are metavariables standing for object variables. More generally, hand in hand with deductive steps corresponding to the natural expansions with $P'_1, ..., P'_m$ as the leaves and P' as the root, the inference rule below may be acquired:

$$\frac{\Delta \vdash P_1, ..., \Delta \vdash P_m}{\Delta \vdash P} \qquad (3)$$

where $P_1, ..., P_n$ are formula schemata generalized from $P'_1, ..., P'_m$ and P is the formula schema generalized from P'. This generalization replaces constant symbols not originally occurring in $\mathcal{A}$, the assertion being applied, by new metavariables. A similar *variablization* is a standard technique employed in the context of explanation based generalization [Moo90]. Obviously, the replaced constant symbols must occur in formulas serving as premises, such as a_1, U_1 and F_1 in $a_1 \in U_1$ and $U_1 \subset F_1$ in our example.

Contraposition The second way of acquiring assertion level rules can be viewed as a generalized contraposition, described by the following schema: if r is an existing rule of the form:

$$r = \frac{\Delta \vdash p_1, ..., \Delta \vdash p_n}{\Delta \vdash q}$$

then r' below can be acquired by contraposition:

$$r' = \frac{\Delta \vdash p_1, ..., \Delta \vdash p_{i-1}, \Delta \vdash p_{i+1}, ..., \Delta \vdash p_n, \Delta \vdash \neg q}{\Delta \vdash \neg p_i}$$

For instance, after the acquisition of

$$\frac{\Delta \vdash a \in U, U \subset F}{\Delta \vdash a \in F}$$

two other rules

$$\frac{\Delta \vdash a \in U, a \notin F}{\Delta \vdash U \not\subset F} \quad \text{and} \quad \frac{\Delta \vdash a \notin F, U \subset F}{\Delta \vdash a \notin U}$$

can be derived as contrapositions (see [Hua92] for more details).

3.2 The Complete Set of Assertion Level Rules

Now let us turn to our main concern, namely the set of inference rules $Rule(\mathcal{A})$, associated with a particular assertion $\mathcal{A}$. As we have argued, rules in $Rule(\mathcal{A})$ are either generated in a chunking-and-variablization manner, or by contraposition. Therefore:

$$Rules(\mathcal{A}) = R(\mathcal{A}, \mathcal{NK} \cup Contra(\mathcal{NK})) \cup Contra(R(\mathcal{A}, \mathcal{NK} \cup Contra(\mathcal{NK}))) \quad (4)$$

where $R(\mathcal{A}, \mathcal{B})$ denotes the set of rules applying $\mathcal{A}$, which can be acquired in a chunking-and-variablization manner with respect to $\mathcal{B}$, denoting the set of logic level rules at the disposal of the reasoner for constructing logic level proof segment. In our theory, we assume the ND calculus $\mathcal{NK}$ [Gen35], together with their contrapositions, as the available rules at the logic level. $Contra(S)$ denotes the set of rules which are contrapositions of rules in the set of rules S. There are redundancies in $R(\mathcal{A}, \mathcal{NK} \cup Contra(\mathcal{NK}))$ and $Contra(R(\mathcal{A}, \mathcal{NK} \cup Contra(\mathcal{NK})))$, because many rules in the latter may have a direct derivation as well.

Example 1 (continued):
With a rule $\frac{a_1 \in U_1, U_1 \subset F_1}{a_1 \in F_1}$ already acquired from the subset definition, supported by the ND proof segment illustrated in Fig. 1, it is only natural for a human to be able to apply the following contraposition: $\frac{a_1 \in U_1, a_1 \notin F_1}{U_1 \not\subset F_1}$. This, however, has a corresponding compound proof segment of its own, given in Fig. 2.

$$\frac{\forall_{S_1,S_2} S_1 \subset S_2 \Leftrightarrow (\forall_x x \in S_1 \Rightarrow x \in S_2)}{U_1 \subset F_1 \Leftrightarrow \forall_x x \in U_1 \Rightarrow x \in F_1} , \frac{\begin{array}{c} a_1 \in U_1, a_1 \notin F_1 \\ \neg(a_1 \in U_1 \Rightarrow a_1 \in F_1) \end{array}}{\neg(\forall_x x \in U_1 \Rightarrow x \in F_1)}}{U_1 \not\subset F_1}$$

Fig. 2. Natural Expansion 2 for Subset Definition

In general, if Fig. 3(a) is the corresponding tree schema for a rule $\frac{c1,c2,b1}{b2}$, acquired, the corresponding tree schema for its contraposition $\frac{b1,\neg b2,c1}{\neg c2}$ can be constructed, using the corresponding contrapositions of the logic level rules, as depicted in Fig. 3(b).

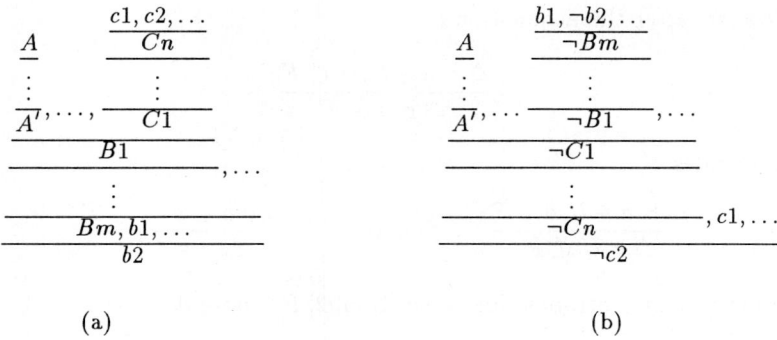

Fig. 3. Expansion for Contrapositions

The following property makes a more succinct representation of equation (4) possible [Hua91]:

$$R(\mathcal{A}, Contra(\mathcal{B}) \cup \mathcal{B}) = R(\mathcal{A}, \mathcal{B}) \cup Contra(R(\mathcal{A}, \mathcal{B}))$$

where $\mathcal{B}$ is an arbitrary set of logic level inference rules. A natural corollary is:

$$Contra(R(\mathcal{A}, Contra(\mathcal{B}) \cup \mathcal{B})) \subset R(\mathcal{A}, Contra(\mathcal{B}) \cup \mathcal{B})$$

Intuitively, this means contraposition of compound rules will bring forth no more new rules if contraposition is already applied to elementary rules. Thus

$$Rules(\mathcal{A}) = R(\mathcal{A}, \mathcal{NK}) \cup Contra(R(\mathcal{A}, \mathcal{NK})) \tag{5}$$

3.3 Tree Schemata for Assertion Level Inference Rules

As stated in (5), the application of an assertion $\mathcal{A}$ can be covered by the union of $R(\mathcal{A}, \mathcal{NK})$ and its contraposition. According to the definition of $R(\mathcal{A}, \mathcal{NK})$, each of its rules corresponds to a tree schema generalized from a natural expansion. Therefore $R(\mathcal{A}, \mathcal{NK})$ can be represented by a set of tree schemata covering all natural expansions, denoted by $Tree(\mathcal{A}, \mathcal{NK})$. Since some members in $Tree(\mathcal{A}, \mathcal{NK})$ are subtrees of others and can therefore be omitted, we show in this section that this set can be represented in a very compact way. For almost all examples, $Tree(\mathcal{A}, \mathcal{NK})$ consists usually of only one or two trees.

Example 1. (continued)

If we apply the variablization described in section 3 on the proof segment in Fig. 1 by replacing a_1, U_1 and F_1 by metavariables a, U and F, respectively, the tree schema in Fig. 4 can be obtained.

Because every subtree (with the subset definition as one of its leaves) of the tree schema in Fig. 4 is a schema of natural expansion, this tree contains a whole set of assertion level inference rules. Apart from the one listed in (2), $\dfrac{U \subseteq F}{\forall_x x \in U \Rightarrow x \in F}$ is another rule contained in this tree, for instance.

$$\cfrac{\cfrac{\cfrac{\cfrac{\mathcal{A}: \forall_{S_1,S_2} S_1 \subset S_2 \Leftrightarrow \forall_x x \in S_1 \Rightarrow x \in S_2}{U \subset F \Leftrightarrow \forall_x x \in U \Rightarrow x \in F}}{U \subset F \Rightarrow \forall_x x \in U \Rightarrow x \in F}, U \subset F}{\forall_x x \in U \Rightarrow x \in F}}{a \in U \Rightarrow a \in F}, a \in U}{a \in F}$$

Fig. 4. Tree Schema for Subset Definition

Now we are ready to examine the set of proof tree schemata designated by $Tree(\mathcal{A}, \mathcal{NK})$. We do this by defining $Tree(\mathcal{A}, \mathcal{B})$ as a restricted deductive closure of the composition and decomposition rules in $\mathcal{B}$, an arbitrary set of logic level inference rules. Technically, for all $r \in R(\mathcal{A}, \mathcal{B})$, there is a tree schema $t \in Tree(\mathcal{A}, \mathcal{B})$, such that r can be accounted for by a subtree of t. Below is a constructive definition:

i Start with the tree in Fig. 5(a), which corresponds to the rule $\frac{}{\triangle A \vdash A}$,

ii If there is a tree t in the form of Fig. 5(b), $r = \frac{\triangle A \vdash a, \triangle A \vdash p1, \ldots, \triangle A \vdash pn}{\triangle A \vdash Q} \in \mathcal{B}$ is a decomposition rule with respect to a, and if there exists a substitution σ, such that $A' = a\sigma$, then extend t to a tree t' in form of Fig. 5(c).

iii If there is a tree t in the form of Fig. 5(b), and $r = \frac{\triangle A \vdash p'1, \ldots, \triangle A \vdash p'n}{\triangle A \vdash Q} \in \mathcal{B}$ is a composition rule with respect to Q, now if there exists a substitution σ, such that $p = Q\sigma$, then extend t to a tree t' in form of Fig. 5(d).

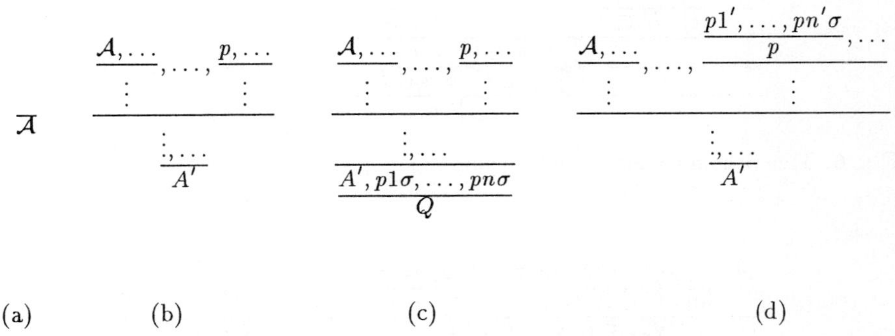

(a)　(b)　(c)　(d)

Fig. 5. Construction of Tree Schemata

Some explanations: i) initializes a tree with only one node, corresponding to the initial inference rule $\frac{}{\triangle A \vdash A}$, ii) and iii) extend existing trees by decomposing the root or the leaves. The informations contained in this set can be redundant, since some rules accounted for by one tree schema are contrapositions of rules accounted for by another tree schema.

Example 1. (Continued): We illustrate the structure of the tree schemata introduced above by continuing with the example used throughout this paper. Two trees are needed as shown in Fig. 6 and Fig. 7, since the equivalence "$\Leftrightarrow$" is understood as the shorthand of the conjunction of two implications and therefore can be decomposed in two different ways. Table 1 is a list of some of the rules in $Rule(\mathcal{A})$, with their corresponding tree schemata indicated.

No.	Inference Rule	Derivation(Tree or Contraposition)
(1)	$\dfrac{\Delta \vdash a \in U, \Delta \vdash U \subset F}{\Delta \vdash a \in F}$	Tree in Fig. 6
(2)	$\dfrac{\Delta \vdash a \notin F, \Delta \vdash U \subset F}{\Delta \vdash a \notin U}$	Contraposition of (1)
(3)	$\dfrac{\Delta \vdash a \in U, \Delta \vdash a \notin F}{\Delta \vdash U \not\subset F}$	Contraposition of (1)
(4)	$\dfrac{\Delta \vdash U \subset F}{\Delta \vdash \forall_x x \in U \Rightarrow x \in F}$	Subtree of Fig. 6, rooted at node [c]
(5)	$\dfrac{\Delta \vdash a \in U \Rightarrow a \in F}{\Delta \vdash U \subset F}$ where a does not occur in $\mathcal{A}$ Tree in Fig. 7.	
(6)	$\dfrac{\Delta \vdash \forall_x x \in U \Rightarrow x \in F}{\Delta \vdash U \subset F}$	Subtree of Fig. 7 rooted at node [c]
(7)	$\dfrac{\Delta \vdash U \not\subset F}{\Delta \vdash \neg \forall_x x \in U \Rightarrow x \in F}$	Contraposition of (5)

Table 1. Some Inference Rules for Subset Definition

$$\dfrac{\dfrac{\dfrac{\dfrac{\mathcal{A} : \forall_{S_1, S_2} S_1 \subset S_2 \Leftrightarrow \forall_x x \in S_1 \Rightarrow x \in S_2}{[a] : U \subset F \Leftrightarrow \forall_x x \in U \Rightarrow x \in F}}{[b] : U \subset F \Rightarrow \forall_x x \in U \Rightarrow x \in F}, U \subset F}{[c] : \forall_x x \in U \Rightarrow x \in F}}{[d] : a \in U \Rightarrow a \in F}, a \in U$$
$$[e] : a \in F$$

Fig. 6. Tree Schema 1 for Subset Definition

$$\dfrac{\dfrac{\dfrac{\mathcal{A} : \forall_{S_1, S_2} S_1 \subset S_2 \Leftrightarrow \forall_x x \in S_1 \Rightarrow x \in S_2}{[a] : U \subset F \Leftrightarrow \forall_x x \in U \Rightarrow x \in F}}{[b] : (\forall_x x \in U \Rightarrow x \in F) \Rightarrow U \subset F}, \dfrac{a \in U \Rightarrow a \in F}{\forall_x x \in U \Rightarrow x \in F}}{[c] : U \subset F}$$

Fig. 7. Tree Schema 2 for Subset Definition

Notice, as we argued above, every subtree containing the subset definition as a leaf corresponds to a rule of inference, if we take the other leaves as preconditions and the root as the conclusion. In other words, only subtrees rooted along the path from the leave which is the assertion being applied to the root, called the *main branch*, are of interest. Fig. 6 has five such subtrees and Fig. 7 has three, namely, the length of the main branch. Nodes along the main branch are numbered in Fig. 6 and Fig. 7 for convenience. Each such subtree represents a rule of inference, directly, and its contraposition, indirectly.

For instance, rule (1) is directly represented by Fig. 6 itself and (2), (3) are contrapositions of (1). Rule (4) is represented by a subtree rooted at node [c] in Fig. 6. Rule (5) is represented by one of the subtrees rooted at [c] in Fig. 7, which has no associated rules because of its variable condition.

4 Abstracting ND-Proofs to Assertion Level Proofs

As describe above, assertion level justifications are used both for compound logic level proof segments satisfying the composition-decomposition constraint and atomic derivations justified by an assertion level inference rule. Logically, the two kinds of derivations are equivalent. In this section an algorithm is devised which replaces as many compound proof segments in machine found ND proofs as possible, by atomic derivations justified by assertion level rules. Most importantly, the replacement is not restricted to natural expansions, but includes other logically equivalent compound segments. This procedure is the preprocessor of *PROVERB*, a system transforming natural deduction proofs into natural language [Hua94]. *PROVERB* is the explanation component for Ω-MKRP, an interactive proof development environment [HKK+94].

As argued above, in order to produce natural language proofs comparable with proofs found in typical mathematical textbooks, we should first try to replace as many complex proof units as possible by atomic assertion level steps. One straightforward solution is a strict abstraction of the input ND proof by replacing all subproofs satisfying the decomposition-composition constraint by an atomic step justified by the corresponding assertion level inference rule. This approach, however, has a severe drawback: Since automated theorem provers usually work in a manner fundamentally different to that of human beings, the input ND proofs are often quite twisted so that not many units satisfying this constraint can be found. Another approach based on the assertion level rules rather than on the constraint avoids this problem. One way to do so is to go through the entire input proof, and test for every proof node N, if N can also be justified by the application of an assertion. As candidates for such assertions all proof nodes that depends on less assumptions than N could be considered. Apparently, such a procedure nearly reproves the problem based on the input proof. Although this exhaustive procedure may find optimal proofs and its complexity is theoretically still polynomial, it is quite search intensive in the practice. A more restrictive variation is employed in our system that mainly abstracts an existing proof as it is proved, but utilizes the assertion level inference rules instead of the decomposition-composition constraint.

Algorithm: Go through the entire proof tree starting from the root, for each proof node N,

1. Choose as the set of assertions AS the definitions and theorems contributed to the proof of N, namely the leaves of the subtree rooted by N, which are definitions or theorems used.

2. Among the nodes in the subtree rooted by N, test if there exist nodes $p_1, \ldots, p_n$, from that N can be derived by applying an assertion $\mathcal{A}$ in AS. In this case, reduce the proof so that N has $p_1, \ldots, p_n$ as its only direct children and the assertion $\mathcal{A}$ as its new justification.

The applicability of a particular assertion $\mathcal{A}$ can be tested by finding a subtree in $Tree(\mathcal{A}, \mathcal{NK})$ (or one of its contrapositions), and a subtree in the input proof tree rooted by the conclusion N to be justified, so that the leaves match. To maximize the factor of abstraction, we proceed in a top-down manner and gradually search for maximal subtrees satisfying the condition above. For example, suppose we are at the proof node $[e']$ in a segment of an input proof as shown in Fig. 8 (the ND rules used as justifications are omitted). The label $Subset$ indicate this hypothesis is the definition of subset.

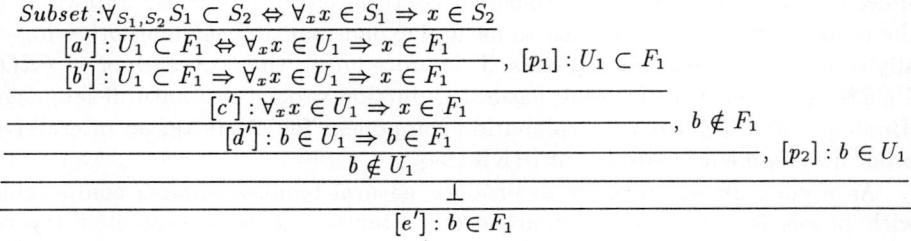

Fig. 8. A Segment of the Input Proof

Since it is recorded that the definition of subset is used as one of its hypothesis, it is tested if any assertion level rule associated with this definition can be applied. For this purpose, we search for a node in the tree schemata in Fig. 6 and Fig. 7 for a node which matches the node $[e']$ in Fig. 8. $[e]$ in Fig. 6 is found. Now we try to find two maximal subtrees rooted by $[e]$ and $[e']$ respectively which match. In this case, they are the two trees themselves. $[p_1]$ and $[p_2]$ are used as new premises of $[e']$ in the abstracted proof, and the definition of subset as the new justification. Thereby the proof segment in Fig. 8 is abstracted to the proof segment in Fig. 9 below:

$$\frac{[p_1]: U_1 \subset F_1, \quad [p_2]: b \in U_1}{[e']: b \in F_1} Subset$$

Fig. 9. The abstracted proof segment

Note that the leaf $b \notin F_1$ in Fig. 8 need not be matched, since it is a temporary assumption for the indirect proof step. The search for maximal matching subtrees is carried out in a breath-first manner, upwards from both of the roots. Note also, that not every intermediate node in the input proof segment needs to be matched and only the leaves count. The indirect proof step in the input proof segment in Fig. 8, which is apparently a detour made by the machine, is absorbed.

Searching for maximal subtrees in a breath-first manner may lose the optimal abstraction, since all intermediate nodes of the maximally matched subtree in $Tree(\mathcal{A}, \mathcal{NK})$ must be matched by a node in the input proof. However, this restriction significantly accelerates the process. The worst case of our abstraction algorithm is now only of the order $O(n^2)$, including the cost of generating tree schemata. This happens when no abstraction can be performed. For neatly written input proofs containing segments which structurally resemble tree schemata representing assertion level rules, it can even be nearly linear.

The quality of the resulting proofs depends on the input proofs in the following way:

- The algorithm works well on neatly structured ND proofs. In these cases, the reduction factor depends on the average depth of the terms in the definitions and theorems. Since mathematicians usually avoid using both too trivial and too complicated definitions and theorems, a quite stable reduction factor (about two thirds in terms of the number of the proof lines) is normally achieved.
- Most significant reduction is observed with input proofs which are essentially direct proofs, but containing machine generated detours and redundancies. At the end of this section, we show an example where a machine generated ND proof of 134 lines is shortened to a proof of 15 lines.
- The complete proof transformation procedures described in [And80, Mil83, Pfe87, Lin90] work fairly similar to a tableau prover. They tend to produce proofs which are mainly indirect, if not properly guided by heuristics. Our algorithm performs poorly on such indirect proofs, where in most of the node only $\perp$ is derived. Although such proofs are also often shortened to the half in length, the resulting proofs are still largely at the level of calculus rules and therefore still too tedious. This problem can be overcome by incorporating techniques that help to avoid indirect proofs (see [Lin90, PN90]) into the process transforming proofs in machine oriented formalisms to ND proofs. Techniques described in [Lin90] can also be adapted to be applied on ND proofs after the transformation.

Let us look at the example below, abstracted from an input proof of 134 lines, generated in the proof development environment Ω–MKRP. It is given in a linearized format, where the last column contains the justification as well as the premises. Eleven of the remaining fifteen steps are at the assertion level. The rest are justified by ND rules of more structural import: They introduce new temporary hypothesis and then discharge them (the *Hyp* and the *Choice rule* in this example). These steps are usually presented explicitly. Groups of trivial steps instantiating quantifiers or manipulating logical connectives are largely abstracted to assertion level steps. Line 7 corresponds to the proof step in Fig. 9, abstracted from the proof segment in Fig. 8. The definitions of *semigroup, group,* and *unit* are obvious and therefore omitted in the proof below. "$solution(a, b, c, F, *)$" should be read as "c is a solution of the equation $a * x = b$ in F." Notice, the proof segments replaced by assertion level steps are not necessarily a natural expansion of the latter. In contrast, they are usually proof segments produced by a automated theorem prover, which are logically equivalent to a natural expansion, but contain unnecessary detours. If we replace the assertion level steps in the proof below by their natural expansions, the result is a logic level proof of 43 lines, in contrast to the input proof of 134 lines.

Theorem: Let F be a group and U a subgroup of F, if 1_U is a unit element of U, then $1 = 1_U$.

Abstracted Proof about Unit Element of Subgroups

NNo	S;D		Formula	Reason
1.	1;	⊢	$group(F, *) \land subgroup(U, F, *) \land$ $unit(F, 1, *) \land unit(U, 1_U, *)$	(Hyp)
2.	1;	⊢	$U \subset F$	(Def-subgroup 1)
3.	1;	⊢	$1_U \in U$	(Def-unit 1)
4.	1;	⊢	$\exists_x x \in U$	(∃ 3)
5.	;5	⊢	$u \in U$	(Hyp)
6.	1;5	⊢	$u * 1_U = u$	(Def-unit 1 5)
7.	1;5	⊢	$u \in F$	**(Def-subset 2 5)**
8.	1;5	⊢	$1_U \in F$	(Def-subset 2 3)
9.	1;5	⊢	$semigroup(F, *)$	(Def-group 1)
10.	1;5	⊢	$solution(u, u, 1_U, F, *)$	(Def-solution 6 7 8 9)
11.	1;5	⊢	$u * 1 = u$	(Def-unit 1 7)
12.	1;5	⊢	$1 \in F$	(Def-unit 1)
13.	1;5	⊢	$solution(u, u, 1, F, *)$	(Def-solution 7 11 12 9)
14.	1;5	⊢	$1 = 1_U$	(Th-solution 11 10 13)
15.	1;	⊢	$1 = 1_U$	(Choice 4 14)

The appropriateness of the assertion level is supported by our experience in the verbalization of abstracted proofs using the system *PROVERB* [Hua94]. Taking as input ND style proofs at assertion level, the resulting texts are at an acceptable level of abstraction. Below is the natural language proof generated by *PROVERB*:

The Natural Language Proof

(1)Let F be a group, U be a subgroup of F, 1 be a unit element of F and 1_U be a unit element of U. (2)According to the definition of unit element, $1_U \in U$. (3)Therefore there is an X, $X \in U$. (4)Now suppose that u is such an X. (5)According to the definition of unit element, $u * 1_U = u$. (6)Since U is a subgroup of F, $U \subset F$. (7)Therefore $1_U \in F$. (8)Similarly $u \in F$, since $u \in U$. (9)Since F is a group, F is a semigroup. (10)Since $u * 1_U = u$, 1_U is a solution of the equation $u * X = u$. (11)Since 1 is a unit element of F, $u * 1 = u$. (12)Since 1 is a unit element of F, $1 \in F$. (13)Since $u \in F$, 1 is a solution of the equation $u * X = u$. (14)Since F is a group, $1_U = 1$ by the uniqueness of solution. (15)This conclusion is independent of the choice of the element u.

5 Conclusion and Future Work

This paper proposes a reconstructive approach toward the presentation of machine found proofs. It is argued that after machine found proofs are transformed into ND proofs, a reconstruction should be started anew, to obtain proofs containing justifications at a higher level of abstraction, which are intuitively understood as the application of a definition or of a theorem, collectively called an assertion. We have illustrated that compound proof segments which can be justified as the application of a certain assertion fulfill the so called decomposition-composition constraint. Furthermore, they are logically equivalent to atomic derivations justified by rules of inference at the assertion level. The complete set of such assertion level rules associated to a particular assertion can be represented in a very compact way in form of tree schemata. With the help of these tree schemata, we devised an efficient algorithm abstracting machine generated ND proofs to the assertion level. This algorithm works even better, if adequate heuristics are employed to generate well structured ND proofs.

The significance becomes more evident when it is viewed within the entire spectrum of transforming machine generated proofs into natural language. With natural deduction style proofs composed of mostly assertion level steps as an additional intermediate representation, the proofs passed to the text planner already resemble proofs produced by human mathematician, and therefore lend themselves to a natural specification of presentation strategies. Using the abstraction as a preprocessor which substantially shortens input proofs, we are able to tackle a broad class of proofs containing more than one hundred lines, and the final proofs generated are at a level of abstraction comparable with proofs found in typical mathematical text books, where authors choose a detailed style.

There is no doubt that proofs are often presented by mathematician at a even higher level of abstraction, since a loss factor of 10 to 20 is reported when using systems like AUTOMATH [dB80]. Even more radical expansion factors (about 5,000 to 10,000) are conjectured by experts for harder mathematical problems. To achieve a similar factor of reduction in the proof presentation, a much deeper understanding of the cognitive process of theorem proving is necessary. This work is only a first step toward this direction.

Acknowledgment

Thanks are due to Manfred Kerber and Daniel Nesmith, who read several drafts of this paper carefully, and to Armin Fiedler, who implemented the abstraction algorithm. I am also indebted to two CADE referees for their constructive suggestions.

References

[And80] P. B. Andrews. Transforming matings into natural deduction proofs. In *Proc. of the 5th CADE*, pages 281–292. Springer, 1980.

[Bun88] A. Bundy. The use of explicit plans to guide inductive proofs. In *Proc. of 9th CADE*, pages 111–120. Springer, 1988.

[CAB+86] R.L. Constable et al. *Implementing Mathematics with the Nuprl Proof Development System*. Prentice Hall, New Jersey, 1986.

[Che76] D. Chester. The translation of formal proofs into English. *AI*, 7:178–216, 1976.

[dB80] N. G. de Bruijn. A survey of the project AUTOMATH. In J. P. Seldin and J. R. Hindley, editors, *To H.B. Curry - Essays on Combinatory Logic, Lambda Calculus and Formalism*, pages 579–606. Academic Press, 1980.

[Deu71] P. Deussen. *Halbgruppen und Automaten*. Springer, 1971.

[EP93] A. Edgar and F J. Pelletier. Natural language explanation of natural deduction proofs. In *Proc. of the first Conference of the Pacific Association for Computational Linguistics*. Simon Fraser University, 1993.

[FHN72] R. R. Fikes, P. E. Hart, and N. J. Nilsson. Learning and executing generalized robot plans. *Artificial Intelligence*, 3:251–288, 1972.

[Gen35] G. Gentzen. Untersuchungen über das logische schließen i. *Math. Zeitschrift*, 39:176–210, 1935.

[GMW79] M. Gordon, R. Milner, and C. Wadsworth. *Edinburgh LCF: A Mechanized Logic of Computation*. LNCS 78. Springer, 1979.

[HKK+94] X. Huang et al. Ω–MKRP – a proof development environment. In these proceedings, 1994.

[Hua91] X. Huang. An extensible natural calculus for argument presentation. SEKI-Report SR-91-03, Universität Kaiserslautern, Germany, 1991.

[Hua92] X. Huang. Applications of assertions as elementary tactics in proof planning. In V. Sgurev and B. du Boulay, editors, *Artificial Intelligence V - Methodology, Systems, Applications*, pages 25–34. Elsevier, 1992.

[Hua93] X. Huang. An explanatory framework for human theorem proving. In H. J. Ohlbach, editor, *GWAI-92: Advances in Artificial Intelligence*, LNAI 671, pages 55–66. Springer, 1993.

[Hua94] X. Huang. *Human Oriented Proof Presentation: A Reconstructive Approach*. PhD thesis, Universität des Saarlandes, 1994, forthcoming.

[JL83] P. N. Johnson-Laird. *Mental Models*. Harvard Univ. Press, 1983.

[Lin90] C. Lingenfelder. *Transformation and Structuring of Computer Generated Proofs*. PhD thesis, Universität Kaiserslautern, Germany, 1990.

[McD83] D. D. McDonald. Natural language generation as a computational problem. In *Brady/Berwick: Computational Models of Discourse*. MIT Press, 1983.

[Mil83] D. A. Miller. *Proofs in Higher-Order Logic*. PhD thesis, CMU, Pittsburgh, 1983.

[Moo90] R. J. Mooney. Learning plan schemata from observation: Explanantion-based learning for plan recognition. *Cognitive Science*, 14:483–509, 1990.

[New90] A. Newell. *Unified Theories in Cognition*. Harvard University Press, 1990.

[Pfe87] F. Pfenning. *Proof Transformation in Higher-Order Logic*. PhD thesis, CMU, Pittsburgh, 1987.

[PN90] F. Pfenning and D. Nesmith. Presenting intuitive deductions via symmetric simplification. In *Proc. of 10th CADE*, pages 336–350. Springer, 1990.

[Sho76] E. H. Shortliffe. *Computer-Based Medical Consultations: MYCIN*. Elsevier, New York, 1976.

[WS89] M. R. Wick and J. R. Slagle. An explanation facility for today's expert systems. *IEEE Expert*, 4(1):26–36, 1989.

[WT92] M. R. Wick and W. B. Thompson. Reconstructive expert system explanation. *Artificial Intelligence*, 54:33–70, 1992.

Problems on the Generation of Finite Models

Jian Zhang[1]

Institute of Software, Academia Sinica
P.O.Box 8718, Beijing 100080, P.R.China

Abstract. Recently, the subject of model generation has received much attention. By *model generation* we mean the automated generation of finite models of a given set of logical formulas. In this note, we present some problems on the generation of finite models. The purpose is two-fold: (1) to offer some test problems for model generation programs; and (2) to show the potential applications of such programs. Some of the problems are easy, some are hard and even open. We also give a new result in combinatory logic, which says that the fragment $\{B, N_1\}$ does not possess the strong fixed point property.

1 Introduction

Traditionally, the major task of automated deduction (or automated reasoning) systems is proving theorems. But there are also other tasks which rely on logical reasoning. One of them is the generation of finite models, i.e. finding a finite model satisfying a given set of formulas. A trivial example is, given a set of 3 or 5 axioms for group theory, produce the multiplication tables of a 6-element group. Model generation is complementary to theorem proving: instead of proving conjectures, it produces counterexamples refuting conjectures. It is also important in its own right, e.g. in some branches of mathematics, we would like to construct an object having some desired properties. Finite model generation can be regarded as constraint satisfaction in finite domains, with emphasis on logical reasoning instead of numerical computation. And model generators can be thought of as a special class of automated reasoning programs.

Since model generation is a new subject, we feel it necessary to collect some test problems for such programs. The following problems are selected, which is partly due to their connection with Automated Deduction, and partly because they have been attacked by existing programs. Three programs have been used: FINDER (Finite Domain Enumerator [8]), Sato (an implementation of the Davis-Putnam procedure for checking satisfiability in propositional logic [14]), and FALCON (a tool for Finite ALgebra CONstruction, previously known as Mod/E [15]). In some cases, we shall give the execution times (on a Sparcstation 2) of

---
[1] Supported in part by the Natural Science Foundation of China.

the programs. Throughout this note, we adopt the convention that a model of size n has the set $\{0, 1, 2, ..., n\text{-}1\}$ as its domain.

2 The Problems

2.1 Ternary Boolean Algebra

A ternary Boolean algebra is a nonempty set with a ternary operator f and a unary operator g, satisfying the following 5 axioms:

(T1) $f(f(v,w,x), y, f(v,w,z)) = f(v,w,f(x,y,z))$
(T2) $f(y,x,x) = x$ (T4) $f(x,x,y) = x$
(T3) $f(x,y,g(y)) = x$ (T5) $f(g(y),y,x) = x$

The axioms (T4) and (T5) can be shown to depend on the remaining axioms. But the first three are independent of each other. With assistance from a theorem prover, Winker [9] established the independence of (T2) by finding a model of (T1), (T3), (T4) and (T5) which falsifies (T2). It is easy for a model generator to find the models automatically. One such model, which is different from Winker's, was given in [15].

2.2 Combinatory Logic

Combinatory logic can be defined as an equational system satisfying the combinators S and K with $a(a(a(S,x),y),z) = a(a(x,z), a(y,z))$ and $a(a(K,x),y) = x$. There are also other combinators, such as

$a(a(a(B,x),y),z) = a(x, a(y,z))$ $a(a(L,x),y) = a(x, a(y,y))$
$a(a(a(N_1,x),y),z) = a(a(a(x,y),y),z)$ $a(M,x) = a(x,x)$

In a fragment of combinatory logic (defined by a set of combinators), the strong fixed point property holds if there exists a combinator y such that for all combinators x, $a(y,x) = a(x, a(y,x))$. Using a theorem prover, Wos and McCune [13] have obtained many results on the existence (or nonexistence) of fixed point combinators in certain fragments of combinatory logic.

To prove the nonexistence of fixed point combinators, one may find a counterexample, i.e. a model in which the fixed point property does not hold. Here we give a 5-element model of $\{B, N_1\}$ which falsifies the fixed point property, answering an open problem proposed by Wos [13].

a	0	1	2	3	4
0	0	0	2	3	4
1	0	0	2	3	4
2	3	3	3	3	3
3	3	3	3	3	3
4	4	4	4	4	4

(In the model, $B = 0$, $N_1 = 1$.) FALCON-2 finds the model in about 1 second. We have also searched for finite counterexamples for the following fragments: $\{B, L\}$; $\{B, M\}$; and $\{B, S\}$. We fail to find the models when the size is restricted to be less than 6. It seems not easy to generate larger models.

2.3 Single Axioms for Group Theory

A *single axiom* for a theory is a formula from which the entire theory can be derived. With a theorem prover, one may prove that a formula is a single axiom by deriving from it the known axioms of the theory. To search for single axioms for a given theory, one usually generates a large number of candidate formulas. Many of the candidates are not single axioms. To show this, one may use a model generator to construct a model of the candidate which falsifies some known axiom (or any known theorem). In fact, FINDER has been used McCune [6] in the search for single axioms in group theory.

In [6] (and some other papers), several single axioms for groups are given. The question is: are they the simplest, or, is it possible that some of their proper instances are still single axioms? For an example, let us see the single axiom

$(G1) \quad f(x, i(f(y, f(f(f(z, i(z)), i(f(u, y))), x)))) = u$

One of its instances is (taking $x = u$)

$(G2) \quad f(u, i(f(y, f(f(f(z, i(z)), i(f(u, y))), u)))) = u$

There exists a 2-element model satisfying (G2) but falsifying (G1): $f(0,0) = f(0,1) = f(1,1) = 0$, $f(1,0) = 1$, $i(0) = i(1) = 0$. So (G2) is not a single axiom.

2.4 Equivalential Calculus

Equivalential Calculus (EC) has the following 3 axioms:

$(E1) \quad P(e(x, x))$
$(E2) \quad P(e(e(x, y), e(y, x)))$
$(E3) \quad P(e(e(x, y), e(e(y, z), e(x, z))))$

and the inference rule *condensed detachment* (CD): $P(e(x, y))$ & $P(x) \rightarrow P(y)$. It was shown [12] that (E1) is dependent on (E2) and (E3). But is it possible that one of the remaining axiom is dependent on the other? The answer is no. In fact, there is a 2-element model of (E3) and (CD) which falsifies (E2), and there is a 3-element model of (E2) and (CD) which falsifies (E3).

EC also has some short single axioms. Similarly, model generators can play some role in detecting those formulas which are not single axioms. For two testing problems, we list the following two formulas which are known not to be single axioms of EC, because (E1) is not derivable from either of them [7] [11].

$(XBB) \quad P(e(x, e(e(e(x, e(y, z)), y), z)))$
$(XAK) \quad P(e(x, e(e(e(y, z), x), z), y)))$

There is a 4-element model of (XBB) which does not satisify (E1).

2.5 Quasigroups and Latin Squares

A *quasigroup* is a set on which a binary operation $\circ$ is defined, such that, for every pair (a, b) of elements, each of the equations $a \circ x = b$ and $y \circ a = b$ has a unique solution. A quasigroup is *idempotent* iff $x \circ x = x$ holds for every element x. Mathematicians are interested in the existence of quasigroups having some additional property. Model generators can assist in such studies. In fact, various programs have solved some previously open questions [2]. To show the performance of model generators on the quasigroup problems, we note that SATO takes about 2 minutes to conclude that there is no idempotent quasigroup of order 12 satisfying the identity $((y \circ x) \circ y) \circ y = x$. In contrast to the existence problems, one may consider some enumeration problems, e.g. enumerating the number of Latin squares of order 8 [4]. (A Latin square of order n corresponds to the multiplication table of an n-element quasigroup.)

2.6 The High School Identities

The following set (HSI) of identities are true in the set $\mathbf{N}$ of positive integers, which one learns in high school:

$$x + y = y + x \qquad x + (y + z) = (x + y) + z$$
$$x * y = y * x \qquad x * (y * z) = (x * y) * z$$
$$x * 1 = x \qquad x * (y + z) = (x * y) + (x * z)$$
$$1^x = 1 \qquad x^1 = x \qquad (x^y)^z = x^{y*z}$$
$$x^{y+z} = x^y * x^z \qquad (x * y)^z = x^z * y^z$$

Tarski's High School Problem is: whether the above eleven identities serve as a basis for all the identities of $\mathbf{N}$. It was answered in the negative in 1980 by Wilkie, who showed that the following identity holds in $\mathbf{N}$, but cannot be derived from HSI: $(P^x + Q^x)^y * (R^y + S^y)^x = (P^y + Q^y)^x * (R^x + S^x)^y$ where $P = 1 + x$, $Q = 1 + x + x * x$, $R = 1 + x * x * x$, and $S = 1 + x * x + x * x * x * x$. In fact, there are finite models of HSI which do not satisfy Wilkie's identity. Such models are called *G-algebras*. An open question is: *what's the size of the smallest G-algebra?* It is known that this size is at least 7 and at most 15 [1]. The question should be a hard one for current model generation programs. The size of the model is not small, and there are 3 function symbols.

3 Concluding Remarks

We have described some problems of finite model generation. Some of them are related to previous research on automated theorem proving, while others are new. Some instances of the problems are easy, others are hard and even open. We note that model generation programs can be used to obtain independence results, to show that a formula is *not* a theorem, and to determine the existence of certain finite objects in mathematics. Combined with theorem provers, model generators may offer us a better understanding of a theory.

Due to the limit of space, many other problems are not included, such as the n-queens problem, the jobs puzzle and the zebra puzzle. We also note that, models are not necessarily finite. Among problems on the generation of infinite models, we list the following open one: does there exist an infinite Robbins algebra which is not Boolean? It is known that every finite Robbins algebra is Boolean [10]. In addition to the three programs mentioned, there are other programs which can generate finite models, such as SATCHMO [5], MGTP [3], DDPP [14] and various Constraint Logic Programming systems. We hope that more model generation programs will be developed, and more problems solved.

References

[1] Burris, S., and Lee, S., "Tarski's high school identities," *The Amer. Math. Monthly* 100 (1993) 231–236.

[2] Fujita, M. et al, "Automatic generation of some results in finite algebra," *Proc. 13th IJCAI* (1993) 52–57.

[3] Hasegawa, R. et al, "MGTP: A parallel theorem prover based on lazy model generation," *Proc. 11th CADE, LNAI* 607 (1992) 776–780.

[4] Kolesova, G. et al, "On the number of 8×8 Latin squares," *J. Combinatorial Theory* A54 (1990) 143–148.

[5] Manthey, R., and Bry, F., "SATCHMO: A theorem prover implemented in Prolog," *Proc. 9th CADE, LNCS* 310 (1988) 415–434.

[6] McCune, W., "Single axioms for groups and Abelian groups with various operations," *J. Automated Reasoning* 10 (1993) 1–13.

[7] Peterson, J. G., "The possible shortest single axioms for EC-tautologies," Report No. 105, Dept. of Mathematics, Univ. of Auckland (1977).

[8] Slaney, J., "FINDER: Finite domain enumerator. Version 3.0 notes and guide," Australian National University, (1993).

[9] Winker, S., "Generation and verification of finite models and counterexamples using an automated theorem prover answering two open questions," *J. ACM* 29 (1982) 273–284.

[10] Winker, S., "Robbins algebra: Conditions that make a near-Boolean algebra Boolean," *J. Automated Reasoning* 6 (1990) 465–489.

[11] Wos, L. et al, "A new use of an automated reasoning assistant: Open questions in equivalential calculus and the study of infinite domains," *Artificial Intelligence* 22 (1984) 303–356.

[12] Wos, L. "Meeting the challenge of fifty years of logic," *J. Automated Reasoning* 6 (1990) 213–232.

[13] Wos, L., "The kernel strategy and its use for the study of combinatory logic," *J. Automated Reasoning* 10 (1993) 287–343.

[14] Zhang, H., and Stickel, M., "Implementing the Davis-Putnam method by tries," submitted to AAAI-94.

[15] Zhang, J., "Search for models of equational theories," *Proc. 3rd Int'l Conf. for Young Computer Scientists (ICYCS-93)*, Beijing (1993) 2.60–63.

Combining symbolic computation and theorem proving: some problems of Ramanujan

Edmund Clarke Xudong Zhao
School of Computer Science
Carnegie Mellon University
Pittsburgh, PA 15213, USA

1 Introduction

One way of building more powerful theorem provers is to use techniques from symbolic computation. So far, there has been very little research in this direction. The challenge problems in this paper are taken from Chapter 2 of *Ramanujan's Notebooks* [1]. They were selected because they are non-trivial and require the use of symbolic computation techniques. The preface to Chapter 2 describes the problems as being "fairly elementary", but states that "several of the formulas are very intriguing and evince Ramanujan's ingenuity and cleverness." We suspect that several of the problems would prove quite challenging for many mathematics graduate students even with the help of a symbolic computation system.

We have developed a theorem prover based on the symbolic computation system Mathematica [8] that can prove all the challenge problems completely automatically. This theorem prover uses many of the same techniques that we incorporated in an earlier theorem prover called Analytica [2, 3]. We plan to describe the theorem prover in greater detail in a forthcoming paper.

Although decision procedures like Gosper's algorithm [4] can prove some identities involving summations, we have not found a decision procedure that can handle the problems proposed here. Moreover, decision procedures only give the final result without intermediate steps. Our theorem prover produces readable proofs in which each intermediate step is justified by an axiom or a rule of inference that can be checked by the user. This usually gives greater insight into why the theorem is true.

2 Axioms that are used in proofs

In addition to the simplification rules that are provided by the symbolic computation system, the following axioms are also needed for proving the theorems.

This research was sponsored in part by the National Science Foundation under Contract Number CCR-8722633, and also by the Defense Advanced Research Projects Agency (DOD), ARPA Order No. 4976, Amendment 20, under Contract Number F33615-87-C-1499.

All of these axioms are simple identities about summations. However, no symbolic computation system, including Mathematica, implements these identities so that they can be applied in both directions. In order to use these axioms effectively, a theorem prover (like the one we have developed) must be constructed so that cycles are avoided and termination is guaranteed.

1. $\sum_{k=i}^{j} f(k) = \sum_{k=1}^{j} f(k) - \sum_{k=1}^{i-1} f(k)$
2. $\sum_{k=1}^{-n} f(k) = -\sum_{k=-n+1}^{0} f(k)$
3. $\sum_{k=1}^{n} (f(k) + g(k)) = \sum_{k=1}^{n} f(k) + \sum_{k=1}^{n} g(k)$
4. $\sum_{k=1}^{n} c f(k) = c \sum_{k=1}^{n} f(k)$
5. $\sum_{k=1}^{n+1} f(k) = \sum_{k=1}^{n} f(k) + f(n+1)$
6. $\sum_{k=1}^{n-1} f(k) = \sum_{k=1}^{n} f(k) - f(n)$
7. $\sum_{k=1}^{2n} (-1)^k f(k) = \sum_{k=1}^{n} f(2k) - \sum_{k=1}^{n} f(2k-1)$
8. $\sum_{k=1}^{na} f(k) = \sum_{j=1}^{n} \sum_{k=0}^{a} f(nk+j)$
9. $\sum_{k=1}^{n} f(k+c) = \sum_{k=c+1}^{n+c} f(k)$

3 List of problems

The list of challenge problems is given below. All of these problems can be proved automatically by the theorem proving system we have developed. This system uses the rules for summation given in the previous section and is similar to another theorem prover that we have built called Analytica [2, 3].

Ramanujan used two abbreviations in stating the theorems. We will use these abbreviations as well.

$$\Phi(x,n) \equiv 1 + 2 \sum_{k=1}^{n} \frac{1}{-(k\,x) + k^3\,x^3}; \quad \varphi(x,n) \equiv \sum_{k=1}^{n} \frac{1}{-(k\,x) + k^3\,x^3}$$

3.1 Problems involving summation of rational functions

1. $\sum_{k=1}^{n} \frac{1}{n+k} = \frac{n}{2n+1} + \varphi(2,n)$
2. $\sum_{k=1}^{n} \frac{n-k}{n+k} = 2n\,\varphi(2,n) - \frac{n}{2n+1}$
3. $\sum_{k=1}^{2n+1} \frac{1}{n+k} = \Phi(3,n)$
4. $\left(\sum_{k=1}^{n} \frac{1}{n+k}\right) + \left(\sum_{k=0}^{n} \frac{1}{2n+2k+1}\right) = \Phi(4,n)$

5. $\left(\sum_{k=1}^{4n+1} \frac{(-1)^{k+1}}{k}\right) + \frac{1}{2}\left(\sum_{k=1}^{2n} \frac{(-1)^{k+1}}{k}\right) = \Phi(4,n)$

6. $\Phi(6,n) = \frac{2}{3}\left(\sum_{k=1}^{n} \frac{1}{n+k}\right) + \left(\sum_{k=0}^{2n} \frac{1}{2n+2k+1}\right)$

7. $2\Phi(4,n) = \Phi(2,2n) + \frac{\Phi(2,n)}{2} + \frac{1}{(4n+1)(4n+2)}$

8. $\Phi(4,n) = \frac{1}{2}\left(\sum_{k=n+1}^{2n} \frac{1}{k}\right) + \left(\sum_{k=2n+1}^{4n+1} \frac{1}{k}\right)$

9. $2\Phi(6,n) + \frac{\Phi(2,n)}{3} = \Phi(3,n) + \Phi(2,3n) + \frac{2}{(6n+1)(6n+2)(6n+3)}$

10. $\sum_{k=n+1}^{A_r} \frac{1}{k} = r + 2\left(\sum_{k=1}^{r}(r-k)\left(\sum_{j=A_{k-1}+1}^{A_k} \frac{1}{(3j)^3-3j}\right)\right) + 2r\,\varphi(3,A_0)$
where $A_0 = 1$, $A_{n+1} = 3A_n + 1$.

3.2 Problems involving infinite summations

Given that $\lim_{n\to\infty}(\sum_{k=1}^{n} \frac{1}{k} - \ln n) = \gamma$, where γ is the Euler constant, the following identities can also be proved.

1. $\Phi(2,\infty) = 2\ln(2)$

2. $\Phi(3,\infty) = \ln(3)$

3. $\Phi(4,\infty) = \frac{3}{2}\ln(2)$

4. $\Phi(6,\infty) = \frac{\ln(3)}{2} + \frac{\ln(4)}{3}$

5. $\sum_{k=1}^{\infty} \frac{1}{(2(2k-1))^3 - 2(2k-1)} = \frac{\ln(2)}{4}$

6. $\sum_{k=1}^{\infty} \frac{2(-1)^k}{(2k)^3 - 2k} = \ln(2) - 1$

7. $\sum_{k=1}^{\infty} \frac{1}{(3(2k-1))^3 - 3(2k-1)} = \frac{\ln(3)}{4} - \frac{\ln(2)}{3}$

8. $\sum_{k=1}^{\infty} \frac{2(-1)^k}{(3k)^3 - 3k} = \frac{4}{3}\ln(2) - 1$

3.3 Problems about the arctan function

The equations in this section can be proved using the standard trigonometric identities for the *arctan* function. These identities are not provided by Mathematica and may be treated as axioms.

1. $\sum_{k=1}^{2n+1} arctan(\frac{1}{n+k}) = \frac{\pi}{4} + \left(\sum_{k=1}^{n} arctan(\frac{10k}{(3k^2+2)(9k^2-1)})\right)$

2. $2\left(\sum_{k=1}^{n+1} \arctan(\frac{1}{n+k})\right) = \arctan(\frac{n+1}{n}) + \left(\sum_{k=1}^{n} \arctan(\frac{2k}{8k^4+2k^2+1})\right) +$
   $2\left(\sum_{k=1}^{n} \arctan(\frac{1}{k(4k^2+3)})\right)$

3. $\left(\sum_{k=1}^{n} \arctan(\frac{1}{n+k})\right) + \left(\sum_{k=0}^{n} \arctan(\frac{1}{2n+2k+1})\right)$
   $= \frac{\pi}{4} + \left(\sum_{k=1}^{n} \arctan(\frac{9k}{32k^4+22k^2-1})\right) + \left(\sum_{k=1}^{n} \arctan(\frac{4k}{128k^4+8k^2+1})\right)$

4. $\sum_{k=0}^{r-1} \arctan(\frac{2}{(n+2k+1)^2}) = \arctan(\frac{2r}{n^2+2nr+1})$

5. $\sum_{k=0}^{\infty} \arctan(\frac{2}{(n+2k+1)^2}) = \arctan(\frac{1}{n})$

6. $\sum_{k=1}^{\infty} \arctan(\frac{2}{(n+k)^2}) = \arctan(\frac{2n+1}{n^2+n-1})$

7. $\left(\sum_{k=1}^{\infty} (-1)^{k+1} \arctan(\frac{2}{(n+k)^2}) = \arctan(\frac{1}{n^2+n+1})\right)$

8. $\sum_{k=1}^{\infty} \arctan(\frac{1}{2(n+k)^2}) = \arctan(\frac{1}{2n+1})$

9. $\sum_{k=1}^{\infty} \arctan(\frac{2}{k^2}) = \frac{3\pi}{4}$

10. $\sum_{k=1}^{\infty} \arctan(\frac{1}{2k^2}) = \frac{\pi}{4}$

11. $\sum_{k=1}^{\infty} (-1)^{k+1} \arctan(\frac{2}{k^2}) = \frac{\pi}{4}$

12. $\sum_{k=1}^{\infty} \arctan(\frac{1}{(1+\sqrt{2}k)^2}) = \frac{\pi}{8}$

13. $\sum_{k=1}^{\infty} \arctan(\frac{8}{(2k-1+\sqrt{5})^2}) = \frac{\pi}{2}$

14. $\sum_{k=0}^{\infty} \arctan(\frac{2}{(2k+1)^2}) = \frac{\pi}{2}$

4 Outline of a typical proof

This section contains part of the proof of identity 10 in Subsection 3.1. The proof of the identity uses some elementary properties of the *harmonic numbers* $H_n = \sum_{k=1}^{n} \frac{1}{k}$ as well as the properties of summations given in Section 2. All of the simplification steps involving summation and the harmonic numbers are implemented directly by our theorem prover. None can be done by Mathematica alone. We have been extremely careful to use only very general rules. The proof has twenty steps, some of which are quite complicated. However, the time required to complete the proof is only about two minutes.

Theorem :

$$\sum_{k=n+1}^{A_r} \frac{1}{k} = r + 2\left(\sum_{k=1}^{r}(r-k)\left(\sum_{j=A_{k-1}+1}^{A_k} \frac{1}{(3j)^3 - 3j}\right)\right) + 2r\,\varphi(3, A_0)$$

Proof:
left hand side:
$$\sum_{k=n+1}^{A_r} \frac{1}{k}$$

use harmonic number notation
$$-H_n + H_{A_r}$$

right hand side:
$$r + 2\left(\sum_{k=1}^{r}(r-k)\left(\sum_{j=A_{k-1}+1}^{A_k} \frac{1}{(3j)^3 - 3j}\right)\right) + 2r\,\varphi(3, A_0)$$

simplify
$$r + 2r\sum_{k=1}^{n}\frac{1}{-3k+27k^3} + 2r\sum_{k=1}^{r}\sum_{j=1+A_{-1+k}}^{A_k}\frac{1}{-3j+27j^3} - 2\sum_{k=1}^{r}k\sum_{j=1+A_{-1+k}}^{A_k}\frac{1}{-3j+27j^3}$$

... (8 steps)

$$r - \frac{2}{3}r H_n + \sum_{j=1}^{n}\frac{1}{-1+3j} - \sum_{j=1}^{A_r}\frac{1}{-1+3j} - r\sum_{j=1}^{A_r}\frac{1}{-1+3j} + \sum_{j=1}^{n}\frac{1}{1+3j}$$
$$-\sum_{j=1}^{A_r}\frac{1}{1+3j} - r\sum_{j=1}^{A_r}\frac{1}{1+3j} + r\sum_{k=1}^{n}\frac{1}{-1+3k} + r\sum_{k=1}^{n}\frac{1}{1+3k} - \frac{2}{3}r\sum_{k=1}^{r}H_{A_k}$$
$$-\frac{2}{3}\left(H_n - H_{A_r} + \sum_{k=1}^{r}H_{A_k}\right) + \frac{2}{3}r\left(H_n - H_{A_r} + \sum_{k=1}^{r}H_{A_k}\right) + \frac{2}{3}\sum_{k=1}^{r}k H_{A_k}$$
$$-\frac{2}{3}\left(-r H_{A_r} + \sum_{k=1}^{r}k H_{A_k}\right) + \sum_{k=1}^{r}\sum_{j=1}^{A_k}\frac{1}{-1+3j} + r\sum_{k=1}^{r}\sum_{j=1}^{A_k}\frac{1}{-1+3j}$$
$$-r\left(\sum_{j=1}^{n}\frac{1}{-1+3j} - \sum_{j=1}^{A_r}\frac{1}{-1+3j} + \sum_{k=1}^{r}\sum_{j=1}^{A_k}\frac{1}{-1+3j}\right) + \sum_{k=1}^{r}\sum_{j=1}^{A_k}\frac{1}{1+3j}$$
$$+ r\sum_{k=1}^{r}\sum_{j=1}^{A_k}\frac{1}{1+3j} - r\left(\sum_{j=1}^{n}\frac{1}{1+3j} - \sum_{j=1}^{A_r}\frac{1}{1+3j} + \sum_{k=1}^{r}\sum_{j=1}^{A_k}\frac{1}{1+3j}\right)$$

... (7 stpes)

$$\frac{1}{1+3n} - H_n + H_{3n} + H_{A_r} - H_{A_{1+r}} - \left(\sum_{k=1}^{r}H_{A_k}\right) + \left(\sum_{k=2}^{1+r}H_{A_k}\right)$$

simplify range of summation
$$\frac{1}{1+3n} - H_n + H_{3n} - H_{1+3n} + H_{A_r}$$

use harmonic number notation
$$-H_n + H_{A_r}$$

both sides are the same $\square$

5 Related Work

There has been relatively little work on using symbolic computation techniques in automatic theorem proving besides our own research on Analytica [2, 3]. Suppes and Takahashi [6] have combined a resolution theorem prover with the *Reduce* system, but their prover is only able to check very small steps and does not appear to have been able to handle very complicated proofs. London and Musser [5] have also experimented with the use of Reduce for program verification, but did not consider theorems from other areas of mathematics or computer science. Bundy [7] has investigated the use of induction for finding closed forms for summations. He is able to handle some very complicated examples. However, his techniques are not applicable to the summations in this paper since they do not have closed forms.

References

[1] B.C.Berndt, *Ramanujan's Notebooks, Part I*, Springer-Verlag, 1985, pp 25-43.

[2] E.M.Clarke, X.Zhao, *Analytica – An Experiment in Combining Theorem Proving and Symbolic Computation*, Technical Report, School of Computer Science, Carnegie Mellon University, CMU-CS-92-147, Oct. 1992.

[3] E.M.Clarke, X.Zhao, *Analytica – A theorem prover for Mathematica*, The Mathematica Journal, Vol. 3, Issue 1, 1993, pp 56-71.

[4] R.W.Gosper, *Indefinite Hypergeometric sums in MACSYMA*, Proc. MACSYMA Users Conference, Berkeley CA, 1977, pp 237-252.

[5] R.L.London and D.R.Musser, *The Application of a Symbolic Mathematical System to Program Verification*, Technique Report, USC Information Science Institute.

[6] P.Suppes and S.Takahashi, *An Interactive Calculus Theorem-prover for Continuity Properties*, Journal of Symbolic Computation, No.7, 1989, pp 573-590.

[7] T.Walsh, A.Nunes, and A.Bundy, *The Use of Proof Plans to Sum Series*, Proc. of 11th International Conference on Automated Deduction, June 1992.

[8] S. Wolfram. *Mathematica: A System for Doing Mathematics by Computer*, Wolfram Research Inc., 1988.

SCOTT: Semantically Constrained Otter
System Description

John Slaney[1], Ewing Lusk[2] and William McCune[2]

[1] Australian National University, Canberra 0200, Australia
[2] Argonne National Laboratory, Argonne, Illinois 60439-4801, USA

The theorem prover SCOTT, early work on which was reported in [3], is the result of tying together the existing prover OTTER [1] and the existing model generator FINDER [4] to make a new system of significantly greater power than either of its parents. The functionality of SCOTT is broadly similar to that of OTTER, but its behaviour is sufficiently different that we regard it as a separate system.

1   OTTER

We briefly review the algorithm of OTTER, a first order theorem prover embodying the set of support strategy. It chains forward from a set of input clauses until either the search space is exhausted or the empty clause is deduced, showing that a goal has been matched. The clauses are divided into two lists: the *usable* clauses and the *set of support*. The main cycle of its proof search is as follows.

1. **Select** a given clause g from the set of support. Move g into the usable list.
2. **Generate** immediate consequences of g in combination with the usable clauses. If the empty clause is found, stop.
3. **Rewrite** the consequences if appropriate rewrite rules are in force.
4. **Filter** out unwanted consequences, such as subsumed ones or those which are too long to be worth keeping.
5. **Update** the clause database by adding the surviving consequences to the set of support.

The rules of inference available to define 'immediate consequence' for phase 2 of this loop include several varieties of resolution and hyper-resolution as well as paramodulation and demodulation (rewriting) for equational reasoning. It is clear that heuristics and other search-direction techniques may be applied at several points in the process. SCOTT gives a heuristic for use in the 'Select' phase and a restriction strategy applicable either in the 'Generate' phase or as a filter.

2   FINDER

FINDER is not a theorem prover in the ordinary sense, but searches for small models of first order theories presented to it as sets of clauses in a simple many-sorted language. The domain of interpretation being fixed first, as containing just a few objects, FINDER's problem is to seek functions defined on those objects to interpret the function symbols of the language in such a way that all of the clauses come out true. Thus it treats its search as a constraint satisfaction problem in which the ground instances of the clauses, restricted to the chosen domain, give rise to sets of constraints.

To illustrate with a trivial example, the input

```
sort      element   cardinality = 3
function *: element,element -> element.
clause    a * (b * c) = (a * b) * c.
end
```

will cause FINDER to enumerate the semigroups of order 3. Settings may be invoked, for example to make it stop after finding one model or to make it return a null result after searching unsuccessfully for two seconds or after a thousand backtracks or the like. When called as a procedure from another program such as SCOTT, FINDER may be instructed to remember a model it has generated and subsequently to test arbitrary clauses for truth or falsehood in that model. Note that a clause is regarded as having all its variables bound by implicit universal quantifiers, so it has a definite truth value in any interpretation.

3 The combined system

SCOTT itself is best seen as OTTER with some additional capabilities. It can appeal to an interpretation called the *guiding model* in which, of course, some of the clauses which occur in the proof search are true and others are false. The guiding model is discovered by a FINDER module and may be changed from time to time during the proof search in order to make more of the clauses true. Details of the guiding model are not known to the OTTER module, but the latter may send a clause to an oracle called is_good which returns the (boolean) truth value of the clause in the model.

There are two ways in which the guiding model can be used. One is the *false preference strategy*. In the selection phase, when the next given clause is chosen, OTTER normally applies a weighting function to the clauses in the set of support and chooses one of the lightest. By default, the weight is just the number of symbols in the clause, though interesting behaviour can result from more elaborate functions.[3] The false preference strategy arises from the thought that the goal is a consequence of clauses which imply it rather than of clauses which do not. Naturally, OTTER does not have access to information as to whether clauses really imply the goal (or there would be no need for a proof search) but it can tell whether they imply it in the guiding model, and this should be some approximation to real implication. That is, the guiding model is chosen so as to make the goal false (and a lot of the kept clauses true) and then some preference attaches to choosing given clauses which are false in the guiding model. This is achieved by testing each kept clause in the 'Update' phase and

---

[3] A relatively simple example is the 'deletion strategy' discussed in [2].

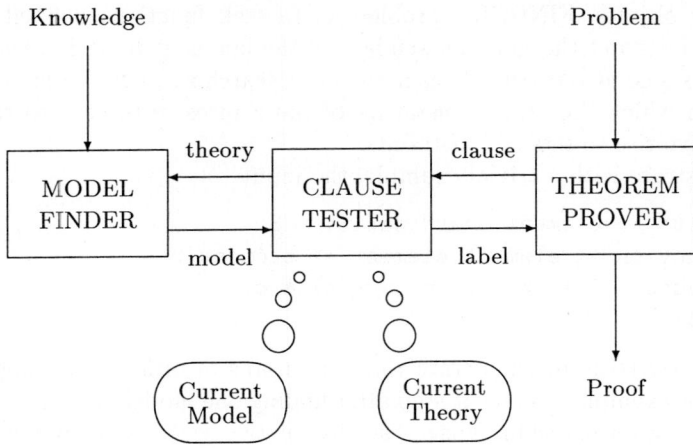

Fig. 1. Semantics in Theorem Proving

adding a constant to its weight if it is true in the guiding model. The constant is determined by an assignment in the OTTER input file.

The second use of the guiding model is for a form of rule restriction which we have dubbed *dynamic semantic resolution*. At present, we have implemented only the simple form sometimes called model resolution: in each inference, at least one of the parent clauses must be false in the guiding model. This restriction is an extension of the set of support idea: at each step, after the given clause is moved into the usable list, we can think of the clauses as divided afresh into axioms and set of support, the axioms being those usable clauses true in the guiding model (and hence a consistent set). There is never any need to resolve axioms with each other, which warrants the model resolution restriction. What makes SCOTT's implementation 'dynamic' is that the guiding model need not be given in advance but can be discovered and repeatedly changed in response to the clauses occurring in the search. In the present implementation, the restriction is applied in the 'Filter' phase, after the consequences have been deduced. It would clearly be more efficient to apply it earlier, to prevent the deductions, but the present version is sufficient for experimental purposes.

Figure 1 shows the basic structure of SCOTT. Two input files are needed. One contains the problem just as for OTTER, and the other is a FINDER file containing the language definitions, the clauses in the initial usable list (which have to be true in the guiding model for integrity in the context of OTTER's set of support algorithm) and optionally any other domain-specific knowledge which may help direct FINDER to good models. There must also be some condition such as a time limit to cause FINDER to stop the search for a model and return a null result in cases where the search fails. The two modules—prover and modeller—do not communicate directly, but exchange messages with the 'clause tester'. At any given time during the proof search, the clause tester has

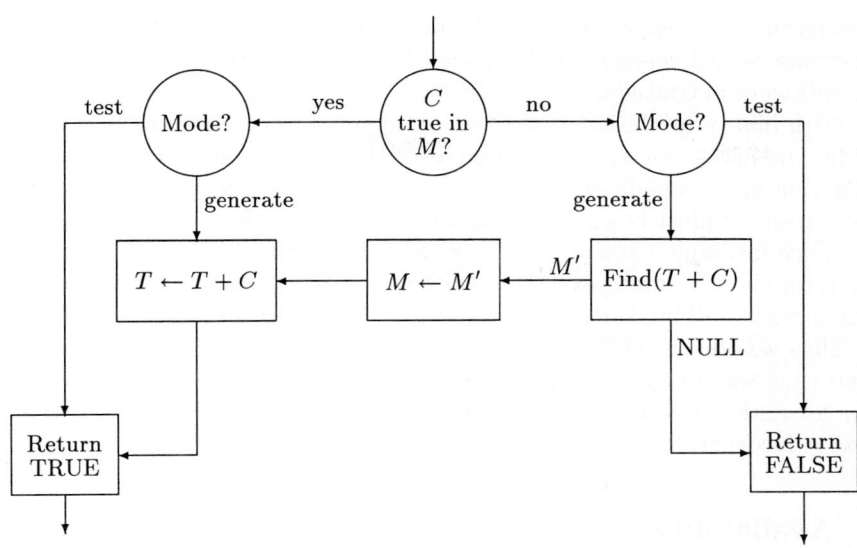

Fig. 2. Clause Tester Flowchart

in its memory a 'current theory', which is a set of clauses, and a 'current model' in which the current theory is true. After FINDER has been called initially with its input file in order to set these up, the clause tester runs for a while in 'generating' mode and then switches to 'testing' mode. The switch is governed by a setting in the FINDER file. The logic of the clause tester is shown in Figure 2. The next clause is C, the current model is M and the current theory is T. In test mode, the returned value is simply the truth value of C in M. In generate mode, if C is false in M an attempt is made to find a better M' in which all of T is true and C is also true. Then if C is true in the guiding model it is added to T. Finally, in any case, its truth value is returned.

4 Comment

SCOTT brings semantic information gleaned from the proof attempt into the service of the syntax-based theorem prover. We find it appealing that the guiding model is thus automatically adapted to the specific problem and to the particular proof search method being applied to it. There are many ways in which such information could help to guide a proof search. We have implemented two of them. The results of our experiments to date have been encouraging if somewhat mixed. We looked at some of the hard condensed detachment problems of [2] on which SCOTT is reasonably successful. As we report in [3] it generally improves on OTTER by a factor of two or so. In extreme cases, it can be over 1000 times faster than OTTER, though there are also cases where model resolution can actually cause inefficiency. The false preference strategy is generally useful, though the optimum weight to be added to true clauses has to be guessed or determined by experiment. With the correct setting, it improves OTTER's performance on a version of Luka-5, one of the hardest problems in [2], by almost two orders of magnitude.

With binary resolution as the rule of inference, its restriction by means of a guiding model obviously retains completeness. Where other rules such as hyper-resolution or paramodulation are used, the model strategy is in general incomplete. Hence it must be applied with care. The false preference strategy does not introduce incompleteness. In any case, SCOTT has OTTER as a sub-program and is capable of running exactly as OTTER, with all semantic features disabled, so in a sense nothing is lost even where incompleteness occurs.

Thus we offer SCOTT not as the solution to all known problems but as an interesting way of adding power to an already powerful prover by making some new heuristics available to it. We welcome further experimentation with the ideas it incorporates.

5 Availability

SCOTT is available by anonymous ftp from arp.anu.edu.au where it is in file pub/scott/scott-1.0.tar.Z. The sources include both OTTER and FINDER, each of which may be separately compiled and installed if desired. Upgrades and new releases of both OTTER and FINDER should be compatible with SCOTT.

References

1. W. McCune, *OTTER 2.0 Users Guide*, Technical report ANL-90/9, Argonne National Laboratory, Argonne, IL, 1990.
2. W. McCune & L. Wos, *Experiments in Automated Deduction with Condensed Detachment*, **Proc. 11th International Conference on Automated Deduction**, 1992, pp. 209–223.
3. J. Slaney, *SCOTT: A Model-Guided Theorem Prover*, **Proc. 13th International Joint Conference on Artificial Intelligence**, 1993, pp. 109–114.
4. J. Slaney, *FINDER, Finite Domain Enumerator: Version 2.0 Notes and Guide*, Technical report TR-ARP-1/92, Automated Reasoning Project, Australian National University, Canberra, 1992.
5. J. Slaney, *FINDER, Finite Domain Enumerator: Version 3.0 Notes and Guide*, Document with program sources, anonymous ftp, arp.anu.edu.au, file ARP/FINDER/finder-3.0.1.tar.Z.

PROTEIN: A *PRO*ver with a *Theory* Extension *IN*terface*

Peter Baumgartner and Ulrich Furbach

Universiät Koblenz
Institut für Informatik
Rheinau 1
56075 Koblenz, Germany

Tel.: +49–261–9119–426, +49–261–9119–433
E-mail: {peter,uli}@informatik.uni-koblenz.de

Abstract. PROTEIN (*PRO*ver with a *Theory* Extension *IN*terface) is a PTTP-based first order theorem prover over built-in theories. Besides various standard-refinements known for model elimination, PROTEIN also offers a variant of model elimination for case-based reasoning and which does not need contrapositives.

PROTEIN is a complete theorem prover for first order clause logic. It is characterized by the following features:

- PROTEIN is based on the *PTTP implementation technique* [Sti88] for model elimination [Lov69].
- PROTEIN offers alternative inference rules for *case analysis* [Lov87, BF93]. In this setting no contrapositives are needed, and hence the system is well suited as an interpreter for disjunctive logic programming.
- PROTEIN includes *theory reasoning* [Sti85, Bau92, Bau94] in a very general way. An auxiliary program can be used to derive a suitable background reasoner from a given Horn theory in a fully automatic way.
- PROTEIN includes several *calculus refinements and flags*.

The idea of the *PTTP implementation technique* ("Prolog Technology Theorem Prover") [Sti88] is to view Prolog as an "almost complete" theorem prover, which has to be extended by only a few ingredients in order to handle the non-Horn case. By this technique the WAM-technology and other benefits of optimizing Prolog compilers are accessible to theorem proving. A disadvantage of PTTP, according to Stickel ([Sti90]), is that "the high inference rate can be overwhelmed by its exponential search space" and therefore PTTP might be well suited for easy problems whereas "it is unsuitable for many difficult theorems for which more conventional theorem provers have demonstrated success".

Our system proves that PTTP can be used even for many challenging problems from the theorem proving literature if PTTP is understood only as a kernel system, which has to be augmented by additional features, like theory handling or case analysis. This is done in PROTEIN.

---

* This research was sponsored by DFG within the "Schwerpunktprogramm *Deduktion*".

The *case-analysis* style of reasoning came up with various (non-theory) calculi which do not need all contrapositives ([Lov87, Pla88]). A detailed comparison of those calculi can be found in [RL92].

In [BF93] we have made a small change to model elimination which also avoids contrapositives and has some distinguished features. This modification of model elimination is called restart model elimination; its distinguished feature are, first, that it can be very easily implemented within a PTTP-framework, and, second, the better informed search due to additional ancestor literals.

Theory reasoning was introduced by M. Stickel within the resolution calculus[Sti85]; for model elimination it is defined an investigated in [Bau92].

Technically, *theory reasoning* means to relieve a calculus from explicit reasoning in some domain (e.g. equality, partial orders, taxonomic reasoning) by taking apart the domain knowledge and treating it by special inference rules. In an implementation, this results in a universal "foreground" reasoner that calls a specialized "background" reasoner for theory reasoning. See [BFP92] for an overview.

Fortunately, the calculus' features "case-analysis reasoning" and "theory reasoning" are fairly compatible with model elimination [Bau94]. In [BF93, BF94] we have shown that "case analysis" – in the non-theory setting – requires only a small change to the calculus. PROTEIN is the respective implementation for theory reasoning.

Furthermore PROTEIN includes several *calculus refinements and flags* such as unit-lemmas, factorisation, ground reduction steps (reduction steps not affecting variables can be handled irrevocably in the proof search) and regularity[2].

Theory Reasoning with Completed Theories

Theory reasoning comes in two variants: *total* and *partial* theory reasoning. *Total* theory reasoning generalizes the idea of finding complementary literals in inferences (e.g. resolution) to a semantic level. For example, in theory resolution the foreground reasoner may select from some clauses the literal set $\{a < b, b < c, c < a\}$, pass it to the background reasoner[3] which in turn should discover that this set is contradictory.

The problem with total theory reasoning is that in general it is undecidable which literals constitute a contradictory set. As a solution, partial theory reasoning tries to break the "big" total steps into more manageable, *decidable*, smaller steps. In the example, the background reasoner might be passed $\{a < b, b < c\}$, compute the new subgoal $a < c$ and return it to the foreground reasoner. In the next step, the foreground reasoner might call the background reasoner with $\{a < c, c < a\}$ again, which detects a trivial contradiction and thus concludes this chain.

PROTEIN currently offers the following: for *total* theory reasoning the input clauses may contain a call to a Prolog-predicate. The theory then is implicitly defined via the enumeration of all answer substitutions to the Prolog-predicate. For the sake of

---

[2] Regularity means that proof attempts which repeat the same subgoal along a branch can be discarded; the regularity restriction has to be relaxed for the case-analysis variant.

[3] assume that $<$ is interpreted as a strict ordering

classification we have here a total theory-extension step with a theory-complementary set consisting of exactly one literal. As a possible application we think of *reasoning by examples* [KMS93].

The implementation of *partial* theory reasoning is currently tailored for the method of *linearizing completion* [Bau93]. Linearizing completion is a saturation technique that transforms a given Horn clause set $\mathcal{T}$ into a "completed" set, which admits (in resolution terminology) both linear and unit-resulting proofs for $\mathcal{T}$-unsatisfiable literal sets. Such a system then can be used as complete background reasoner for partial theory model elimination.

Implementation and Practical Experiments

Both PROTEIN and a tool for linearizing completion are implemented in ECLiPSe, ECRC's Prolog dialect. ECLiPSe extends Prolog by various features, with the most relevant ones for us being *sound unification* and *delayed subgoals*. While the use of sound unification is obvious, the delayed subgoal mechanism is used to implement the regularity restriction. Regularity is realized by delaying a set of constraints, where each constraint states the syntactical inequality of two path literals. A delayed constraint is checked (waked up) each time a change to one of its variable occurs.

We ran several examples known from the literature with PROTEIN and a high-performance model elimination prover. Table 1 contains the runtime results (in seconds), obtained on a SPARC 10/40. The first four columns refer to different versions of PROTEIN. Column 5 contains data for Setheo [LSBB92] in its latest version (Version 3.0).

PROTEIN was run in default mode, except where indicated in Table 1. In default mode it includes the regularity restriction and the ground-reduction refinement. Setheo was also run in its default mode, which then makes use of the following refinements and constraints: subgoal reordering, purity, anti-lemmas, regularity, tautology and subsumption.

The example referred to as *Non-obvious* is taken from the October 1986 Newsletter of the *Association of Automated Reasoning*[4]. The selected theory here consists of a transitive and symmetric relation p and a transitive relation q. In the *Graph* example a graph with a transitive and symmetric reachability relation is traversed. The *Eder*-examples are described in [LSBB92]. The *Bledsoe* examples are the first two of the five given in [Ble90]. The example referred to by $x \neq 0 \rightarrow x^2 > 0$ is to prove this theorem (x is universally quantified) from calculus. Case analysis is carried out according to the axiom $X > 0 \quad \vee \quad X = 0 \quad \vee \quad -X > 0$.

For the theory variants of PROTEIN, the background calculus was obtained completely automatical by the linearizing completion tool in a preprocessing phase. For the theories we have selected appropriate Horn-subsets of the input clauses. The runtime of the linearizing completion tool was sufficiently small and need not be mentioned. In case linearizing completion would yield an infinite inference system for background

---

[4] Entries such as MSC/MSC006-1 refer to the respective TPTP-names [SSY94]. All examples were drawn from that problem library without modification — only the theory part had to be selected by hand.

Example	ME	Restart-ME	TME	Restart-TME	Setheo
Non-obvious MSC/MSC006-1	0.3	2.7	1.6	1.1 (0.15[1])	0.5
Eder45	0.7	3.4 (1.8[1])	-	-	1.0
Eder58	22.0	110	-	-	32
Graph	10.8	∞	0.2	7.0 (0.8[2])	3.0
$x \neq 0 \rightarrow x^2 > 0$	2.4	0.7	2.2	0.6	0.8
Pelletier 48 SYN/SYN071-1	5.9	1.2 (0.6[2])	0.4	0.9 (0.1[1,2])	0.2
Pelletier 49 SYN/SYN072-1	∞	297	1.6	1.5	∞
Pelletier 55 PUZ/PUZ001-2	392	∞	21	254	3.5
Lion&Unicorn PUZ/PUZ005-1	588	∞	21	∞	47
Bledsoe 1 ANA/ANA003-4	∞	7.4[1,3]	-	-	87
Bledsoe 2 ANA/ANA004-4	∞	32[1,3]	-	-	∞
Wos 4 GRP/GRP008-1	22	20	0.3	26	13
Wos 10 GRP/GRP001-1	∞	-	14	-	850
Wos 11 GRP/GRP013-1	9.6	-	1.1	-	0.7
Wos 15 GRP/GRP035-3	384	-	47	-	478
Wos 16 GRP/GRP036-3	302	-	0.02	-	13
Wos 17 GRP/GRP037-3	∞	-	0.1	-	23

Entries: Numbers: runtimes (seconds) – ∞ no proof within reasonable time bound – "-" Not applicable –

Remarks: 1 – With selection function, 2 – With (back) factoring, 3 – With Lemmas

Fig. 1. Runtime Results for various provers: *ME* – plain model elimination version of PROTEIN; *Restart-ME* – case-analysis style reasoning; *TME* and *Restart-TME* – respective versions with theory reasoning extensions.

reasoning – in the Wos examples from group theory – a finite approximation was used. The selected theory consists here of equality (except f-substituivity) and the associativity of the group operation.

Concerning the search strategy we used iterative deepening with the costs of extension steps uniformly set to 1. The same costs are used for case analysis steps.

The runtime results suggest to us that indeed both variants — restart vs. non-restart — are valuable: every variant obtains proof not obtainable to the other variant. Furthermore, the application of theory reasoning very often helps to find a proof more quickly. If no suitable theory is at hand, the "empty" theory can be used which instantiates TME (resp.

Restart-TME) to ME (resp. Restart-ME).
In order to obtain the PROTEIN system please contact
peter@informatik.uni-koblenz.de.

References

[Bau92] P. Baumgartner. A Model Elimination Calculus with Built-in Theories. In H.-J. Ohlbach, editor, *Proceedings of the 16-th German AI-Conference (GWAI-92)*, pages 30–42. Springer, 1992. LNAI 671.

[Bau93] P. Baumgartner. Linear Completion: Combining the Linear and the Unit-Resulting Restrictions. Research Report 9/93, University of Koblenz, 1993.

[Bau94] P. Baumgartner. Refinements of Theory Model Elimination and a Variant without Contrapositives. In *Proc. ECAI'94*, 1994. (to appear).

[BF93] P. Baumgartner and U. Furbach. Model Elimination without Contrapositives and its Application to PTTP. Fachbericht Informatik 12/93, Universität Koblenz, 1993. (short version in Proc. CADE-12).

[BF94] P. Baumgartner and U. Furbach. Model Elimination without Contrapositives. In *Proc. 12th International Conference on Automated Deduction*. Springer, 1994. (in this volume).

[BFP92] P. Baumgartner, U. Furbach, and U. Petermann. A Unified Approach to Theory Reasoning. Forschungsbericht 15/92, University of Koblenz, 1992. (submitted to Journal of Automated Reasoning).

[Ble90] W. W. Bledsoe. Challenge Problems in Elementary Calculus. *Journal of Automated Reasoning*, 6:341–359, 1990.

[KMS93] Manfred Kerber, Erica Melis, and Jörg Siekmann. Reasoning with Assertions and Examples. Technical Report SEKI Report SR-93-10, Universität des Saarlandes, Fachbereich Informatik, 1993.

[Lov69] D. Loveland. A Simplified Version for the Model Elimination Theorem Proving Procedure. *JACM*, 16(3), 1969.

[Lov87] D.W. Loveland. Near-Horn Prolog. In J.-L. Lassez, editor, *Proc. of the 4th Int. Conf. on Logic Programming*, pages 456–469. The MIT Press, 1987.

[LSBB92] R. Letz, J. Schumann, S. Bayerl, and W. Bibel. SETHEO: A High-Performace Theorem Prover. *Journal of Automated Reasoning*, 8(2), 1992.

[Pla88] D. Plaisted. Non-Horn Clause Logic Programming Without Contrapositives. *Journal of Automated Reasoning*, 4:287–325, 1988.

[RL92] D. W. Reed and D. W. Loveland. A Comparison of Three Prolog Extensions. *Journal of Logic Programming*, 12:25–50, 1992.

[SSY94] G. Sutcliffe, C. Suttner, and T. Yemenis. The TPTP problem library. In *Proc. CADE-12*. Springer, 1994.

[Sti85] M.E. Stickel. Automated Deduction by Theory Resolution. *Journal of Automated Reasoning*, 1:333–355, 1985.

[Sti88] M. Stickel. A Prolog Technology Theorem Prover: Implementation by an Extended Prolog Compiler. *Journal of Automated Reasoning*, 4:353–380, 1988.

[Sti90] M. Stickel. A Prolog Technology Theorem Prover. In M.E. Stickel, editor, *Proc CADE 10, LNCS 449*, pages 673–675. Springer, 1990.

DELTA — A Bottom-up Preprocessor for Top-Down Theorem Provers
— System Abstract —

Johann M. Ph. Schumann

Institut für Informatik,
Technische Universität München
D-80290 München
email: **schumann@informatik.tu-muenchen.de**

Top-down theorem provers with depth-first search (e.g., PTTP [Sti88], METEOR [AL91], SETHEO [LSBB92]) have the general disadvantage that during the search the same goals have to be proven over and over again, thus causing a large amount of redundancy. Resolution-based bottom-up theorem provers (e.g., OTTER [McC90]), on the other hand, avoid this problem by performing backward and forward subsumption and by using elaborate storage and indexing techniques. Those provers, however, often lack the goal-orientedness of top-down provers.

In order to combine the advantages of top-down and bottom-up theorem proving, we have developed the preprocessor DELTA. DELTA processes one part of the search space (the "bottom" part) in a preprocessing phase, using bottom-up techniques (see also [Sch91]). It generates unit-clauses (e.g., by applying UR-resolution) which are added to the original formula. Then, this formula is processed by a top-down theorem prover in the usual way.

During this top-down search, the additional unit clauses are used as generalized unit lemmata. Due to the structure of the search space and the combination of advantages of both approaches (subsumption in the preprocessing phase and goal-oriented search in the subsequent top-down search), a remarkable gain of efficiency can be achieved in many cases.

DELTA uses SETHEO [LSBB92] and its logic programming facilities for generating these unit clauses. In order to obtain a high efficiency for the bottom-up phase, the unit clauses are generated level by level. This technique is similar to delta iteration as it is used in the field of data-base research.

Starting with the original formula, the following iteration step is performed (first, we only consider Horn-formulae): we let SETHEO generate all new unit clauses which can be obtained by one UR-resolution step out of the current formula. This is accomplished by adding to the formula "most general queries" $\neg p(X_1, \ldots, X_n)$ for each predicate symbol p and variables $X_1, \ldots, X_n$. For those queries, SETHEO searches for all solutions within a low bound δ_{bu} (here, normally, a depth-bound (A-literal depth) of 2 is used, but it can be varied freely). For each obtained substitution σ, we generate the unit clause "$p(\sigma X, \ldots, \sigma X_n)$". Up to now SETHEO only uses locally effective subsumption techniques, based on SETHEO's built-in constraint mechanism [GLMS94]. Although, in case of a depth bound of 2, anti-lemma constraints[1] simulate forward subsumption, all newly generated unit clauses must pass an additional subsumption test. Then, the remaining ones are added to the current formula, and the iteration step is executed again, unless until a stop condition (see below) has been reached. The original formula plus the unit clauses produced during the last iteration are used for the final top-down search.

Adding unit clauses to a formula does not affect completeness of the top-down proof procedure. Therefore, several ways of controlling the generation of unit clauses during the preprocessing phase are provided: the generation of new clauses can be restricted to specific predicate symbols, and unit clauses with an excessive size or complexity of terms can be filtered out. Furthermore, we can limit the number of generated unit clauses to a suitable value (usually around 100). These methods of control, however, are only a first step in developing more flexible filters for the generated clauses.

In the case of a Horn-formula, a proof can always be found in the final top-down step with a lower bound than needed for a pure top-down proof (assuming, however, all bottom-up unit clauses have been generated).

For non-Horn formulae, we use the same approach of bottom-up iteration. However, we generate unit clauses for p (as above) and for $\neg p$. Furthermore, we allow Model Elimination reduction steps to occur during the preprocessing phase. This increases the power of the bottom-up resolution step by introducing factorization. However, in the non-Horn case, we cannot assure that the resources needed in the final step are lower compared to those needed without DELTA. (There exists always the possibility that a leaf node in the tableau has to be closed by a reduction step almost up to the root.) Experiments showed that even with non-Horn formulae, in many cases, a gain in efficiency could be obtained, often because, there are only few reduction steps in a proof.

The following table shows the results of first experiments with well-known benchmark examples. We have used a simple prototype version of DELTA and SETHEO V3.1 as the top-down prover. This prototype has been implemented in UNIX shell, using awk [AKW88] scripts for manipulating the formula and filtering the unit clauses. As a resource bound for the preprocessing and the final top-down search, the depth of the proof tree (A-literal depth) has been used. DELTA has always been started with default parameters, except in cases where the number of newly generated clauses grew too rapidly. In that case, the maximal term complexity has been restricted to 2 († in the table). If, for Non-Horn formulae, too few unit clause have been generated, the depth bound for each iteration has been increased (‡).

As a comparison, run-times for SETHEO without bottom-up preprocessing are shown. As in the final top-down search, SETHEO has been started with its default parameters (iterative deepening with A-literal depth, and generation and usage of constraints). All run-times shown in this table are in seconds and have

---

[1] Anti-lemma constraints (for details see [LMG93]) are syntactic constraints which are generated during the search to prevent solving a subgoal more than once with an identical substitution, as long as it remains in the tableau.

been obtained on a sun sparc II. They include full compilation and assembly times (for each iteration), not just the run-times of the prover itself. However, a more efficient implementation of DELTA could further substantially reduce the preprocessing time.

Example	iteration level	gener. clauses	run-time DELTA	run-time top-down	run-time total	run-time SETHEO *only*
wos1	1	36	0.87	3.59	4.46	**1.53**
wos4	1	93	1.79	23.35	25.14	**22.93**
wos15(H)[2]	2†	256	15.58	61.82	**77.40**	807.61
wos17(H)	2†	328	8.72	9.01	**17.73**	11081.93
wos20	1†	507	3.05	25.79	**28.84**	—
wos21(H)	2†	144	3.94	20.03	**34.17**	—
wos31	4‡	114	17.2	3.65	**20.85**	—
wos33	4	28	12.97	5.97	**18.94**	—
sam(H)	3	66	12.97	5.97	**18.94**	—
LS36(H)	1	148	1.35	45.28	**46.63**	87.25
LS37a(H)	3†	257	21.67	7.52	**29.19**	—
Bledsoe-1	2	67	1.96	4.40	**6.63**	6.67
Bledsoe-2	2	68	1.90	6.87	**8.77**	114.33

The results shown are quite remarkable, compared to pure top-down theorem proving, despite the straight-forward approach and primitive prototype of DELTA. The combined system could even solve several examples for which SETHEO alone could not find a proof.

This approach of combining top-down and bottom-up theorem proving is extremely flexible and thus allows for many enhancements and future developments, leading to a dramatic increase in the power of automated theorem proving.

The prototype version of DELTA together with a manual page is available via ftp. For information, please contact the author.

References

[AKW88] A. Aho, B. Kernighan, and P. Weinberger. *The AWK Programming Language*. Eddison Wesley, 1988.

[AL91] O.L. Astrachan and D.W. Loveland. METEORs: High Performance Theorem Provers using Model Elimination. In R.S. Boyer, editor, *Automated Reasoning: Essays in Honor of Woody Bledsoe*. Kluwer Academic Publishers, 1991.

[GLMS94] Chr. Goller, R. Letz, K. Mayr, and J. Schumann. SETHEO V3.2: Recent Developments — System Abstract —. In *Proc. CADE 12*, June 1994.

[LMG93] R. Letz, K. Mayr, and C. Goller. Controlled Integrations of the Cut Rule into Connection Tableau Calculi. Technical report, Technische Universität München, 1993. submitted to JAR.

---

[2] Horn formulae are marked by a "(H)".

[LSBB92] R. Letz, J. Schumann, S. Bayerl, and W. Bibel. SETHEO: A High-Performance Theorem Prover. *Journal of Automated Reasoning*, 8(2):183–212, 1992.

[McC90] W. McCune. *Otter 2.0 Users Guide*. National Technical Information Service, U.S. Department of Commerce, Springfield, VA, 1990.

[Sch91] Johann Schumann. Combining top-down and bottom-up theorem proving: First experiments. Technical report, ICOT, Tokyo, Japan, 1991.

[Sti88] M. E. Stickel. A Prolog Technology Theorem Prover: Implementation by an Extended Prolog Compiler. *Journal of Automated Reasoning*, 4:353–380, 1988.

SETHEO V3.2: Recent Developments
– System Abstract –

Chr. Goller, R. Letz, K. Mayr, J. Schumann

Institut für Informatik,
Technische Universität München
D-80290 München
email:
setheo@informatik.tu-muenchen.de

1 Introduction

SETHEO is a theorem prover for first-order predicate logic. The original system, which has been developed within the ESPRIT project 415, is described in [LSBB92]. SETHEO is a top-down prover based on the calculus of so-called *connection tableaux* [LMG93] which generalizes *weak model elimination* [Lov78]. Proofs are found by a consecutively bounded depth-first iterative deepening search with backtracking. The search procedure is implemented as an abstract machine which extends the Warren Abstract Machine. The system is being continuously extended and enhanced with additional inference mechanisms and reduction techniques. The current version of SETHEO is V3.2. In this paper we describe the major improvements of the new system with respect to the system described in [LSBB92].

2 Additional Inference Rules

Since, due to its *cut-freeness*, the basic calculus is among the weakest proof systems concerning *proof compactness*, many proofs cannot be found, simply because no sufficiently short proofs exist in the calculus. Therefore, in [LMG93] controlled ways of integrating the *backward* cut rule are investigated. The techniques developed there are the *folding up* and the *folding down* operation. Folding up, which generalizes *C-reduction* [Sho76], represents an efficient way of supporting the basic top-down calculus with lemmata derived in a bottom-up manner. Folding down is equivalent to *factorization* (of subgoals) in model elimination and the *connection calculus*. Both mechanisms have been implemented in a very efficient manner and are now available in the system. Experiments with a number of representative examples from the field of automated deduction demonstrate that for difficult problems the system with folding up performs significantly better than the cut-free variant of connection tableaux and the one with folding down. Thus, using folding up one can easily solve problems like the intermediate value theorem which is out of scope or at least extremely difficult for current automated deduction systems, as shown below in the table of measurements.

3 Tableau Subsumption and Anti-Lemmata

In contrast to the most successful style of resolution theorem proving which is based on a formula enumeration or *saturation* procedure, such an approach is not possible in the connection tableau framework, because, unlike resolution and unlike the original tableau calculus, the connection tableau calculus is not *proof confluent*, that is, not every refutation attempt of an unsatisfiable formula can be completed successfully. This possibility of making irreversible decisions in the calculus demands a different organization of the proof process, namely, as a *tableau* enumeration instead of a formula enumeration procedure. The structure implicitly explored by the backtracking-driven tableau search procedure of SETHEO is a tree of tableaux. There are two different methodologies for reducing the search effort of tableau enumeration procedures. On the one hand, one can attempt to *refine* the tableau calculus, that is, disallow certain inference steps if they produce tableaux of a certain *structure*—the *regularity* condition described in [LSBB92] is of this type. The effect on the tableau search tree is that the respective nodes together with the dominated subtrees can be ignored, so that the branching rate of the tableau search tree decreases, whereas minimal proof lengths cannot be preserved (see [LMG93]). These *structural* methods of redundancy elimination are *local* pruning techniques in the sense that they can be performed by looking at single tableaux only.

The other approach is to improve the *proof search procedure* so that information coming from the proof search itself can be used to even eliminate proof attempts not excluded by the calculus. More specifically, these *global* methods compare *competitive* tableaux in the search tree, i.e., tableaux on different branches, and attempt to show that one tableau (together with its successors) is redundant in the presence of the other. A natural approach here is to exploit *subsumption* between tableaux, in a similar manner subsumption between clauses is used in formula saturation procedures like resolution. To this end, in [LMG93] the notion of subsumption has been generalized from clauses to literal trees, and it is shown under which conditions subsumed tableaux can be deleted. Unfortunately, a direct application of subsumption deletion is only possible in *explicit* tableau enumeration procedures. The technique is not compatible with the Prolog-like search procedure of SETHEO, since at any time only one tableau is in memory and the information about the existence of alternative (possibly subsuming) tableaux is not available. A restricted concept of subsumption deletion, however, can be achieved with the mechanism of so-called *anti-lemmata*, which avoids that the same (or more special) solutions of subgoals are computed several times.

Generation and application of anti-lemmata
Whenever a subgoal N in a tableau has been solved, the computed solution substitution σ of N is stored at the node N. If the solution of subsequent subgoals fails and the proof procedure backtracks over the node N, then σ is turned into an anti-lemma. In any alternative solution process of the subgoal N, if a substitution $\tau = \tau_1 \cdots \tau_m$ is computed such that one of the anti-lemmata stored at the node

N is more general than τ, then the proof procedure immediately backtracks. When the search state at which N was selected for solution is backtracked, then all anti-lemmata at N are deleted.

It can be shown, that whenever an anti-lemma σ for a tableau node N in a tableau T is more general than a substitution τ computed during an alternative solution attempt T' of N, then T subsumes the alternative tableau T'. Consequently, the described anti-lemma mechanism achieves a restricted form of subsumption deletion. All cases of tableau subsumption, however, cannot be captured with this technique, since when a completely failed subgoal N is deleted, then all anti-lemmata at N disappear, too. More permanent anti-lemmata require *caching* techniques as used in [Ast92]. These techniques, however, are very expensive and normally only applied in the Horn-case, whereas anti-lemmata can be used efficiently in the non-Horn-case too.

4 Constraint Technology

In [LSBB92] it has been demonstrated that the tableau refinements which forbid cases of irregularity, tautological tableau clauses, and tableau clauses which are subsumed by other input clauses can be reformulated as syntactic disequation constraints between term lists. Fortunately, the same holds for the implementation of the anti-lemma mechanism. As a consequence, in the new version a constraint handling mechanism has been integrated similar to the ones available in advanced Prolog systems like SEPIA or SIXTUS Prolog. The constraints are carried along with the current tableau in order to control the instantiations of the variables in the tableau. Whenever a constraint is violated the proof procedure backtracks. The constraint management is carried out by using *disunification*. Here maximal efficiency is obtained by keeping the constraints in a *solved form* where the left sides of the constraints are lists of variables. When variables are instantiated in the unification routine, the constraints to be checked can be accessed very quickly.

The constraint mechanism provides a uniform and highly efficient method for implementing the mentioned search pruning reductions. The integration of this technique is a natural extension of the abstract machine and it permits to detect suboptimal or unsolvable tableaux as early as possible.

5 Table of Results

In order to demonstrate the effect of the new techniques, particularly the new inference rules, we wish to present the results for running 12 well-known examples from the field of automated deduction. The problem specifications are taken from the TPTP problem library [SSY94][1]. All three versions of SETHEO

---

[1] ANA002-3 is D. Loveland's version of the intermediate value theorem which will appear in release v1.1.0 of the TPTP (note that the versions ANA002-1 and ANA002-2 in v1.0.0 are erraneous).

are using anti-lemmata. The first one is the cut-free system, the others apply factorization and c-reduction steps, respectively. As completeness bound for the iterative deepening search we have used a combined bound limiting both the depth and the inferences in the tableaux considered on each level. The values in the 'proof'-column are the numbers of proof inferences; in brackets the numbers of factorization steps and c-reduction steps are given, respectively. For a more detailed description of the experiments see [LMG93]. In order to compare our system with the performance of a theorem prover using a completely different paradigm, we have added the run times and the numbers of inference steps (binary-, hyper-resolution, factoring, etc.) of the new release 3.0 of the system Otter [McC93], started in autonomous mode. The problems were run on a SUN SPARC 10, the time is given in seconds.

Problem			SETHEO 3.2						OTTER 3.0	
			model elimination cut-free		model elimination factorization		model elimination c-reduction		autonomous mode	
Name	Horn	Size	time	proof	time	proof	time	proof	time	proof
GRP029(wos1)	yes	17	1.8	10	.2	11(2)	.1	11(1)	.5	4
GRP001(wos10)	yes	20	25.6	15	68.3	15(0)	48.5	15(0)	.4	4
GRP013(wos11)	yes	22	4.5	9	9.1	9(0)	6.5	8(0)	1.4	6
RNG004(wos22)	yes	34	3.8	14	11.8	14(0)	11.5	14(0)	8.5	7
SYN013(wos31)	no	23	$\gg 10^4$		1214.8	55(16)	12.5	55(10)	11.4	19
SYN015(wos33)	no	26	$\gg 10^4$		$\gg 10^4$		37.5	59(18)	12.4	14
MSC001(apabhp)	no	18	130.7	14	149.7	14(0)	253.9	14(1)	6.7	11
ALG002(ex5)	no	14	2.8	18	3.7	15(1)	3.6	14(1)	2.7	10
SET005(ls108)	no	16	17.8	40	8.4	23(3)	.2	21(3)	840.4	34
SET007(ls112)	no	23	28.8	56	2.7	60(8)	1.7	60(9)	$\gg 10^4$	
SET011(ls121)	no	21	5.7	27	32.4	29(0)	.3	20(3)	255.9	18
ANA002-3(ivt)	no	17	6787.6	99	9849.4	99(0)	14.3	41(13)	$\gg 10^4$	

The experiments clearly demonstrate that the proof procedure with folding up (c-reduction) performs significantly better than the other two. While the *relative loss* in efficiency is limited to a factor of 3, the *relative gain* in efficiency with respect to the other systems is often by magnitudes. Let us demonstrate the benefit of shortening proofs for the proof *search* with the intermediate value theorem (ivt). The cut-free proof we have found needs 99 inference steps and has depth 8. With the folding up mechanism the proof length can be reduced to 41 steps and, which seems even more crucial, the depth of the proof to 6. Consequently, the proof is found two levels earlier. This explains the achieved speed-up of about 500. The same holds for the problems wos31, wos33, ls108, and ls121. Interestingly, even in cases where the proof is on the same level, a speed-up can be achieved, namely, for the problems wos1 and ls112.

Additional experiments have shown that indeed techniques like anti-lemmata are needed in order to successfully compensate the redundancies caused by the new inference rules folding up and folding down. Just to give one example, the solution process of the problem wos31 needs 40 times more time if no anti-lemmata are used.

6 The SETHEO System

The SETHEO system includes an X-based graphical user interface to facilitate its usage, especially for the novice. Both the search process and the proof tree (a tableau with applied substitutions) can be displayed graphically. Features for scrolling and hiding parts of the tableau allow to represent even extremely large tableaux in a readable manner.

Since SETHEO is based on Prolog abstract machine technology, it can also be used for logic-programming purposes. A set of Prolog-style built-ins and additional features (like backtrackable global variables) are provided and can be combined with the basic theorem proving techniques of SETHEO (e.g., sound unification, access of the current path in the tableau, iterative deepening). The entire system is implemented in C as a set of independent programs. Additionally, several stand-alone preprocessing modules, like a bottom-up delta-iterator [Sch94] are available. Binaries and documentation are available via ftp from:
    flop.informatik.tu-muenchen.de in directory /fki/setheo.
For free sources and further information, send e-mail to
    setheo@informatik.tu-muenchen.de.

References

[Ast92] O. W. Astrachan and M. E. Stickel. Caching and Lemmaizing in Model Elimination Theorem Provers. *Proceedings of the 11th Conference on Automated Deduction (CADE-11)*, LNAI 607, Saratoga Springs, pages 224–238, Springer, 1992.

[Lov78] D. W. Loveland. *Automated Theorem Proving: a Logical Basis.* North–Holland, 1978.

[LSBB92] R. Letz, J. Schumann, S. Bayerl, and W. Bibel. SETHEO: A High-Performance Theorem Prover. *Journal of Automated Reasoning*, 8(2):183–212, 1992.

[LMG93] R. Letz, K. Mayr, and C. Goller. Controlled Integrations of the Cut Rule into Connection Tableau Calculi. Technical Report, Techn. Univ. Munich, 1993. Submitted to the Journal of Automated Reasoning.

[McC93] W. W. McCune. OTTER 3.0 Reference Manual and Guide DRAFT Computer Science Division, Argonne National Laboratory, 1993.

[Sho76] R. E. Shostak. Refutation Graphs. *Artificial Intelligence*, 7:51–64, 1976.

[Sch94] J. Schumann. DELTA — A Bottom-up Preprocessor for Top-Down Theorem Provers, System Abstract. Proceedings of the 12th International Conference on Automated Deduction (this volume), 1994.

[SSY94] G. Sutcliffe, C. B. Suttner, and T. Yemenis. The TPTP Problem Library. Proceedings of the 12th International Conference on Automated Deduction (this volume), 1994.

KoMeT*

W. Bibel, S. Brüning, U. Egly, and T. Rath

FG Intellektik, TH Darmstadt
Alexanderstraße 10, D-64283 Darmstadt
e-mail: {bibel,stebr,uwe,rath}@intellektik.informatik.th-darmstadt.de

Abstract. In this paper, we describe KoMeT, a theorem prover for full first order logic. KoMeT is based on the connection method. Our main goal is to develop an adequate proof procedure by integrating a variety of different proof techniques.

1 Introduction

Developing adequate proof procedures is one of the main research goals in the field of automated deduction. Roughly speaking, theorem provers based on adequate proof procedures solve simple problems faster than harder ones [8]. The hardness of a problem is related to humans or to proof methods specialized for a particular class of problems. Clearly, computational adequateness is neither an absolute measure nor can we get the ultimate adequate prover; the term is rather meant as a guiding principle for the design of the prover.

General methods like the resolution principle [24] or the connection method [6, 9] are well known. However, calculi based on these methods seem only to be adequate if they are augmented by specialized proof techniques and by means for controlling the proof search. In this paper we report our efforts to build a new general *and* more adequate theorem prover KoMeT. We first provide an overview of the system and then discuss two selected aspects concerning such specialized techniques, viz. database features and induction, in more detail.

2 Overview

The connection method [6, 9] is the basis of our work. Theoremhood is characterized by a globally unifying and spanning set of connections. A calculus based on this method is not computationally adequate in its pure form because all formulas are handled by the same general mechanism without taking their respective features into account.

Although there are refinements of such calculi like connection tableaux [18], which perform much better than a generate and test approach in order to find a spanning set of connections, there are wide classes of formulas which cannot be proved efficiently, even with high-performance theorem provers like SETHEO [19]. What is needed in order to make theorem provers based on such a general

---

* This research was partially supported by the DFG under grant Bi228/6-2

calculus more adequate is the integration and combination of different mechanisms which guide the search for a proof and prune the search space. Furthermore, extensions like induction are necessary in order to increase the applicability of the prover.

The development of KoMeT follows this approach. It is based on connection tableaux and integrates the aforementioned mechanisms. We stress the fact that the additional mechanisms are not added like in a toolbox but integrated in the underlying calculus. KoMeT is implemented in PROLOG and therefore can be modified and extended without too much effort. On the other hand we have to accept a relatively low inference rate in comparison with high-performance systems like SETHEO. Four main aspects of KoMeT are discussed in the sequel.

The first aspect concerns extensive preprocessing activities. First, a module for the transformation of the original formula into various definitional clausal normal forms [14, 22] has been integrated. Second, problem simplifications play a major role. Besides the well-known reductions of [9], we integrated a reduction originally proposed in [5]. The basic idea is to propagate possible values for variables through the given clause set. As a result, some of the resolution possibilities (connections) may no longer be possible. Thus, they are no longer considered at run time.

In order to strengthen the applicability of reductions, KoMeT was augmented by the concept of reachability [20]. A data structure, named reachability graph, tells the system which part of the given clause set fails to contribute to a proof. Besides the fact that this analysis allows to generalize some reductions of [9], the application of inference rules can be restricted yielding search spaces with lower branching degrees [26]. A further important technique to reduce the search space is DB-reduction [23] which will be discussed in more detail in Section 3.

The second aspect concerns the avoidance of derivation duplication. To this end, we augmented KoMeT by the capability to generate and apply lemmata. An unrestricted use of lemmata, however, is inefficient in most cases since, besides few useful lemmata, a huge amount of unnecessary lemmata is derived. Therefore, several criteria are included to indicate which lemmata should be stored and to decide how derived lemmata are applied. Furthermore, lemmata are generalized as much as possible and the generation of subsumed lemmata is avoided completely. It is especially for these mechanisms that KoMeT is able to prove SAM's Lemma [27] known to be an extremely hard problem for such a top down prover. An alternative approach used for avoiding proof duplication is the usage of failure "lemmata". A failure lemma for a goal indicates that this goal cannot be proved within a given depth bound (see also [2]). Recent tests have shown that the application of failure lemmata can yield drastic search space reductions.

The third aspect concerns the integration of special techniques for theory reasoning and induction. A mechanism for handling arbitrary theories, proposed in [3], has been integrated in KoMeT. It performs a completion of the theory under consideration in a preprocessing step. During the derivation, clauses from this completion are used in a restricted way. For theories having an infinite completion, the completion process is terminated upon reaching a user defined limit. If this limit is sufficiently large, completeness is not affected in most cases[2]. For the

---

[2] The completion may also be generated in a process deepening the limit iteratively.

special case of equality reasoning, two further mechanisms were implemented. The first one is based on eq-connections and eq-literals [6]. As unification algorithm for such eq-connections and eq-literals we use an algorithm based on rigid E-unification [4, 17]. The second mechanism for equality reasoning is similar to the modification method proposed in [10]. The basic idea is to encode possible applications of equality axioms into the given clause set. This can be done efficiently using the technique of flattening. Using these mechanisms, some problems occurring in verification tasks were solved without any user guidance. The integration of induction is described in Section 4.

The fourth aspect concerns the exploitation of global properties of the formula under consideration. Detecting global connection structures in order to reduce the proof length and the search space is one major matter of concern. In [15, 16] techniques are presented which may reduce the length of proofs. They are based on symmetry detection and on function introduction. By analyzing the change of terms during a derivation and by tests based on subsumption between goals and their subgoals, redundant, infinite, or failing derivations [13, 12] may be detected in advance. The integration of these techniques is currently under development.

In addition to these four main aspects, there are several further concepts which have been integrated. Among these are an interactive interface of KoMeT consisting of a comfortable user interface including proof presentation, an efficient propositional theorem prover, constraint mechanisms and bottom-up enhancements. Additionally, different search strategies may be applied. All the mechanisms listed above are further refined by investigating their behavior on numerous problems like the one given in [1] which KoMeT is able to prove in reasonable time. We plan to apply an advanced version of KoMeT in the future in various domains such as planning, program synthesis, program verification, and knowledge representation.

3 DB-Reduction and DB-Unification

A lot of theorem provers face serious problems when proving formulas which contain a large amount of facts. These problems may be overcome by using DB-reductions, first described in [7] and generalized in [23]. Using DB-reductions it is possible to merge clauses containing similar literals. For illustration consider the following clause set written in a PROLOG-like notation.

$$F_1 : \quad p(a_1).$$
$$\vdots \quad \vdots$$
$$F_5 : \quad p(a_5).$$
$$F_6 : \quad q(a_5).$$
$$R : \quad q(f(X)) : - q(X).$$
$$: - p(Y), q(f^5(Y)).$$

In this clause set, the facts $F_1, \ldots, F_5$ are merged by a DB-reduction. This yields the fact $p(V_{\mathrm{DB}})$, where substitutions of the variable V_{DB} are restricted to terms of the set $S_{\mathrm{DB}} = \{a_1, \ldots, a_5\}$. Now we start to prove this modified clause set. At first, the literal $p(Y)$ of the query is unified with the fact $p(V_{\mathrm{DB}})$. It remains to prove $q(f^5(V_{\text{DB}}))$. Five applications of rule R yield the subgoal $q(V_{\text{DB}})$, which is solved with the fact F_6. To this end V_{DB} has to be unified with the term a_5. While performing this unification, the restriction imposed on V_{DB} has to be considered, which is done by an extended unification method called DB-unification. Since a_5 is included in S_{DB}, V_{DB} obviously can be bound to a_5 and the proof is completed. Overall we needed 7 proof-steps to prove the given formula[3]. The reader may verify that without using DB-mechanisms 31 proof- and 24 backtracking-steps are needed to reach the same result if a PROLOG-like deduction is assumed. Thus, DB-reductions and DB-unification can lead to a significant speed-up. In particular, these techniques are successful for examples including a considerable amount of facts like the steamroller problem [25] or constraint satisfaction problems like the n-queens problem.

Not only facts but arbitrary clauses containing similar literals may be merged by DB-reductions. Furthermore, abstraction trees [21] rather than sets are actually used to represent possible substitutions. With this indexing technique an efficient access to unifiable terms is possible as well as a fast performance of join operations, if predicates with an arity greater than one are merged.

4 Induction

For a wide range of applications, the ability of a theorem prover to employ induction seems to be inevitable. Therefore, one main emphasis in the development of KoMeT is the integration of methods for handling induction. These methods are based on ideas sketched in [6] and worked out in [11]. Roughly speaking, the criterion triggering a proof attempt with induction is the detection of an (infinite) recursion. Using the information given by this recursion, an appropriate induction axiom is generated and it is checked whether this axiom can be applied successfully.

Up to now, the generation of suitable induction axioms still follows a rough heuristic. There are no means implemented to generate lemmata which may be needed for a proof. Even so, we successfully proved a number of examples, some by adding suitable lemmata by hand. For example, KoMeT can prove that

$$\forall n. \sum_{i=1}^{n} i^3 = \left(\frac{n(n+1)}{2}\right)^2 \quad (1)$$

holds in the standard model of the natural numbers. Besides the usual definitions for the arithmetical operators, the following lemmata are used which are given to the system by the user.

$$\forall n. \left(\frac{(n+1)(n+2)}{2}\right)^2 = \left(\frac{n(n+1)}{2}\right)^2 + (n+1)^3 \quad (2)$$

$$\forall y. [y = 0 \rightarrow P(0)] \wedge [y > 0 \rightarrow P((y-1)+1)] \rightarrow P(y) \quad (3)$$

Formula (3) merely allows to perform a case analysis. For a successful proof, $P(y)$ has to be instantiated to $\sum_{i=1}^{y} i^3 = \left(\frac{y(y+1)}{2}\right)^2$.

---

[3] Recognition of the cycle in the rule would result in even fewer steps.

References

1. K. Ammon. An automatic proof of Gödel's incompleteness theorem. *Artificial Intelligence*, 61(2):291–306, 1993. Research note.
2. O. L. Astrachan and M. E. Stickel. Caching and Lemmaizing in Model Elimination Theorem Provers. In D. Kapur, editor, *Proceedings of the Conference on Automated Deduction*, pages 224 – 238. Springer Verlag, 1992.
3. P. Baumgartner. Combining Model Elimination with Unit-Resulting Resolution. In *Proc. of the Workshop on Theorem Proving with Analytic Tableaux and Related Methods*, pages 15–18. Technical Report MPI-I-93-213, MPI Saarbrücken, 1993.
4. B. Beckert. Ein vervollständigungsbasiertes Verfahren zur Behandlung von Gleichheit im Tableaukalkül mit freien Variablen. Diplomarbeit, Karlsruhe, 1993.
5. W. Bibel and B. Buchberger. Towards a connection machine for logical inference. Future Generations Computer Systems Journal, 1(3):177–188, 1985.
6. W. Bibel. *Automated Theorem Proving*. Vieweg, second edition, 1987.
7. W. Bibel. Advanced topics in automated deduction. In R. Nossum, editor, *Advanced Topics in Artificial Intelligence*, pages 41–59, 1988. Springer, LNCS 345.
8. W. Bibel. Perspectives on automated deduction. In R. S. Boyer, editor, *Automated Reasoning: Essays in Honor of Woody Bledsoe*, pages 77–104. Kluwer Academic, Utrecht, 1991.
9. W. Bibel. *Deduction: Automated Logic*. Academic Press, London, 1993.
10. D. Brand. Proving Theorems with the Modification Method. *SIAM Journal on Computing*, 4(4):412–430, 1975.
11. M. Breu. Einbeziehung einfacher Induktionsbeweise in den Konnektionenkalkül (in german). Diplomarbeit, Technische Universität München, 1986.
12. S. Brüning. On Loop Detection in Connection Calculi. *Proceedings of the Kurt Gödel Colloquium*, pages 144–151. Springer Verlag, 1993.
13. S. Brüning. Search Space Pruning by Checking Dynamic Term Growth. *Proceedings of the International Conference on Logic Programming and Automated Reasoning*, pages 52–63. Springer Verlag, 1993.
14. E. Eder. *Relative Complexities of First Order Calculi*. Vieweg, 1992.
15. U. Egly. A First Order Resolution Calculus with Symmetries. *Proceedings of the International Conference on Logic Programming and Automated Reasoning*, pages 110–121. Springer Verlag, 1993.
16. U. Egly. On Different Concepts of Function Introduction. *Proceedings of the Kurt Gödel Colloquium*, pages 172–183. Springer Verlag, 1993.
17. J. Gallier, P. Narendran, S. Raatz, and W. Snyder. Theorem proving using equational matings and rigid E-unification. *JACM*, 39(2):377–429, 1992.
18. R. Letz. First-Order Calculi and Proof Procedures for Automated Deduction, 1993. PhD thesis.
19. R. Letz, J. Schumann, S. Bayerl, and W. Bibel. SETHEO — A High–Performance Theorem Prover for First–Order Logic. *JAR*, 8:183–212, 1992.
20. G. Neugebauer. From Horn Clauses to First Order Logic: A Graceful Ascent. Technical Report AIDA–92–21, FG Intellektik, TH Darmstadt, 1992.
21. H. J. Ohlbach. Abstraction tree indexing for terms. In *ECAI 90. Proceedings of the 9th European Conference on Artificial Intelligence*, pages 479–484, 1990.
22. D. A. Plaisted and S. Greenbaum. A Structure-Preserving Clause Form Translation. *JSC*, 2:293 – 304, 1986.
23. T. Rath. Datenbankunifikation. Technical Report AIDA–92–09, FG Intellektik, TH Darmstadt, 1992.
24. J. A. Robinson. A machine-oriented logic based on the resolution principle. *JACM*, 12(1):23–41, 1965.
25. M. E. Stickel. Schubert's steamroller problem: Formulations and solutions. *JAR*, 2:89–101, 1986.
26. G. Sutcliffe. Linear-Input Subset Analysis. In D. Kapur, editor, *Proceedings of the Conference on Automated Deduction*, pages 268 – 280. Springer Verlag, 1992.
27. L. Wos. Automated Reasoning: 33 Basic Research Problems. Prentice Hall, 1988.

Ω-MKRP: A Proof Development Environment

Xiaorong Huang   Manfred Kerber   Michael Kohlhase   Erica Melis
Dan Nesmith   Jörn Richts   Jörg Siekmann

Fachbereich Informatik, Universität des Saarlandes
Postfach 1150, 66041 Saarbrücken, Germany
Phone: [++49] (681) 302–4627, –4628, –4629
{huang|kerber|kohlhase|melis|nesmith|richts}@cs.uni-sb.de
siekmann@dfki.uni-sb.de

1 Introduction

In the following we describe the basic ideas underlying Ω-MKRP, an interactive proof development environment [6]. The requirements for this system were derived from our experiences in proving an interrelated collection of theorems of a typical textbook on semi-groups and automata [3] with the first-order theorem prover MKRP [11]. An important finding was that although current automated theorem provers have evidently reached the power to solve non-trivial problems, they do not provide sufficient assistance for proving the theorems contained in such a textbook.

On account of this, we believe that significantly more support for proof development can be provided by a system with the following two features:
- The system must provide a comfortable human-oriented problem-solving environment. In particular, a human user should be able to specify the problem to be solved in a natural way and communicate on proof search strategies with the system at an appropriate level.
- Such a system is interesting only if it relieves the user of non-trivial reasoning tasks and provides the foundation for a practicable increased reasoning power. We are convinced that this requires not only task-specific tactics but also the strong reasoning power of a general logic engine.

In section 2 and 3 we describe how these two requirements are achieved in Ω-MKRP and in section 4 we give a bird's-eye view of its problem-solving cycle.

2 A Comfortable Environment

In our experiments to prove the theorems in the above-mentioned textbook with MKRP, we recognized that the reasoning power alone of such an automated theorem prover does not provide sufficient assistance. In this section we describe the features a comfortable environment should have (see also figure 1).

One important aspect of a proof development environment is its input language. Automated theorem provers usually use variants of first-order predicate logic. When encoding the theorems of the textbook in first-order logic, we were forced to use sophisticated formulation techniques. That is, although such a language is sufficient in principle, *in practice* it is too weak to allow an adequate

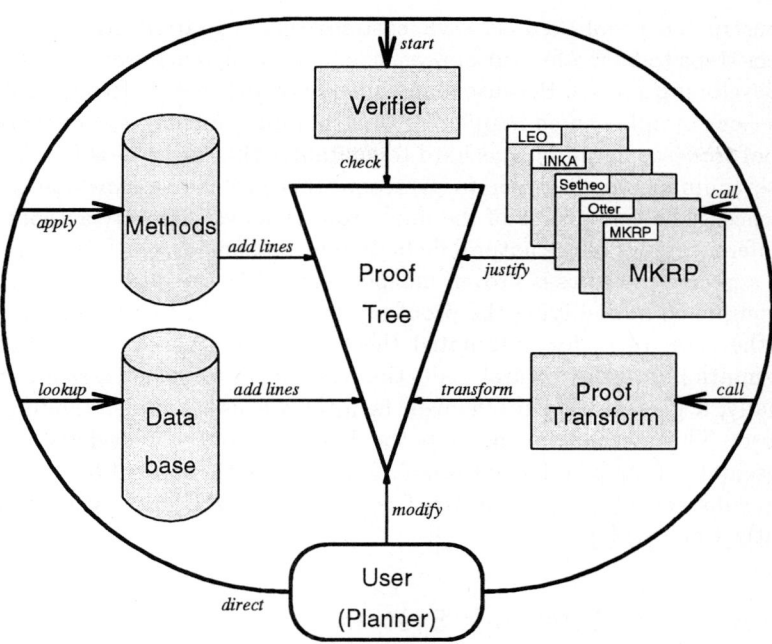

Fig. 1. Architecture of Ω-MKRP

representation. Hence, we developed an input language called $\mathcal{POST}$ (partial-function order-sorted type theory, see [7] for references), which is based on sorted higher-order logic, and thus is much closer to the mathematical language in textbooks.

In addition to the problem formulation language, the proof format is crucial for an adequate interface. As a common proof format for both the user and the system, we have chosen the well-established natural deduction formalism [4]. To attack a given problem, the user can apply natural deduction inference rules. But since this is too laborious even for trivial problems, further automatic assistance is provided.

An advanced tool for proving mathematical theorems should also enable the user to communicate his proof strategies and plans to the system, since every automated theorem prover still needs guidance through the search space for more difficult problems, and this situation will not change fundamentally in the foreseeable future. The user can guide the search for a proof by providing methods containing high-level proof plans [1], while the remaining gaps in the plan can be filled by an automatic planning component based on methods (sometimes called tactics) or by the underlying logic engines (automated theorem provers).

Since the user expects that apart from the knowledge of predicate logic, the system should also possess a large amount of mathematical definitions and theorems, we are filling a database with the mathematical knowledge contained in the textbook mentioned above [3].

It is a main task of a proof development environment to ensure the correctness of a constructed proof. In LCF style systems this is guaranteed by disallowing incorrect steps to be made. Our approach is to allow as much freedom as possible when developing a proof. Because many different components often from different origins—even implemented in different programming languages—are working on the proof tree (see figure 1), it is hard to guarantee the correctness for the results of these components. In order to overcome this problem we include a verifier which checks the correctness of the final proof. This verifier only accepts a fixed set of inference rules of the natural deduction formalism for which the correctness w. r. t. a given semantics is proven independently. This requires that the results of all components modifying the proof are formulated in this formalism. If this is not the case, (e.g., for automated theorem provers based on resolution), a transformation into the natural deduction formalism must be carried out.

Finally, a presentation mechanism facilitates a user's understanding of the final proof. This mechanism abstracts the final ND proof to a level of abstraction that resembles that of informal proofs found in mathematical textbooks. The thereby substantially shortened proof is then verbalized into natural language (currently English) [8].

3 A Powerful Inference System

MKRP [11] has become a powerful automated theorem prover. Nevertheless its weaknesses, like those of other automated theorem provers, seem to be fatal to the goal to prove a mathematical textbook. We think that a *fully* automatic theorem prover will always have these weaknesses in spite of its usefulness. So we have decided to pursue a theorem proving paradigm that combines different deduction paradigms and different theorem provers with different strengths.

Automated theorem provers can roughly be divided into two categories: those based on machine-oriented mechanisms like resolution and those based on more human-oriented mechanisms like tactics and proof planning. The former are usually based on strong general problem-solving strategies, but can hardly be adjusted for specific problems. In contrast, the latter are well suitable for describing domain-specific strategies, but lack strong general problem-solving mechanisms. To prove the problems in the textbook both are needed: the latter to construct a global proof plan and the former to fill the gaps. Therefore, our basic idea is to build a proof development environment which combines both the reasoning power of automated theorem provers as logic engines *and* that of the proof planning mechanism. We think that the combination of these complementary approaches inherits more of the advantages than of the drawbacks, because for most tasks domain-specific as well as domain-independent problem-solving know-how is required.

In Ω-MKRP we basically follow the framework proposed by Alan Bundy et al. [2] for the planning mechanism. The basic logic engines are integrated by simply considering them as primitive operators. Certain modifications of the framework are necessary in order to incorporate higher-level human proof strategies such as analogy. In particular, in our framework, methods and their tactics are defined declaratively. This offers the possibility to define meta-methods which adjust existing methods to new situations [5].

4 The Problem-Solving Cycle in Ω-MKRP

A problem-solving cycle in our system can be described briefly as follows: The user formulates his problem in $\mathcal{POST}$. In order to solve the problem he may load definitions and already-proved theorems from the database, he may invoke a method to split the original problem by hand or let a planner do this, or he may pass a subproblem to a logic engine. When the subproblem is within the capability of the logic engine, a corresponding proof will be found and translated back into the user's format, and then the user can continue with the above cycle.

There is a gap between the problem formulation language $\mathcal{POST}$ and the proof format, on the one hand, and the input and output languages of the underlying first-order theorem provers, on the other hand. In order to bridge this gap we must transform $\mathcal{POST}$ into the input languages of the underlying provers and transform the found proofs (such as a clause proof of a resolution-based prover) into natural deduction proofs. An overview of the problem solving cycle with a first-order theorem prover in Ω-MKRP is illustrated in figure 2.

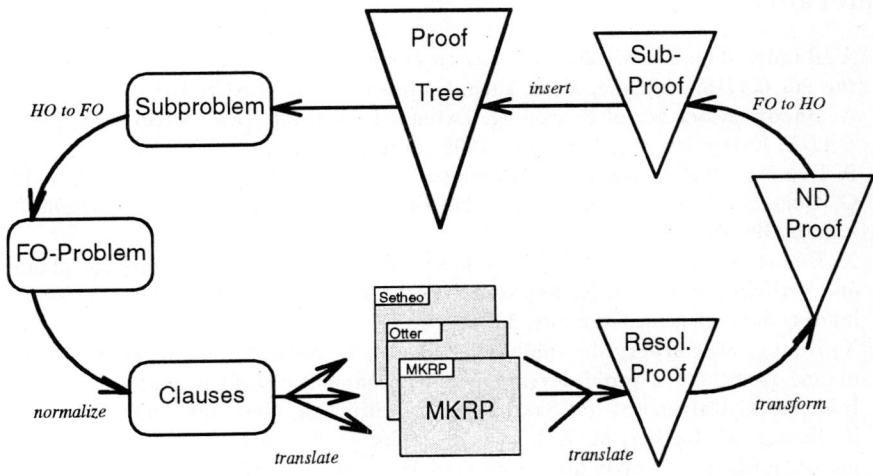

Fig. 2. Integration of first-order theorem provers in Ω-MKRP

In order to pass a subproblem to a first-order theorem prover, the problem must first be translated into first-order logic [9] and then normalized into clausal normal form. These steps are common for all selected first-order theorem provers but now a specific translation from our clause protocol into the input language of the respective theorem prover is needed, and after a successful run of the prover a specific translation from the prover's proof protocol into our resolution proof protocol. With this protocol a common representation for all theorem provers is reached and the following steps apply to all provers again. The clause proof is transformed into a first-order natural deduction proof [10], which in turn can be retranslated into a higher-order natural deduction proof.

For theorem provers using a stronger calculus than first-order logic (like induction or higher-order logic) not all transformations in this cycle are needed.

5 Current State and Future Work

The basic elements of the first feature, a comfortable environment, are already implemented. The user can define a problem, manipulate it with the inference rules and use the database of mathematical knowledge contained in [3]. The MKRP system is the first logic engine fully integrated into Ω-MKRP; OTTER and SETHEO can be called. Other systems will follow, in particular, logic engines for mathematical induction (INKA) and higher-order logic (LEO, which we are developing ourselves). The final proof can be presented in English. While the first steps towards a comfortable environment have already been made, we are now concentrating on the implementation of the proof plan mechanism.

References

1. A. Bundy. The use of explicit plans to guide inductive proofs. In *Proceedings of the 9th CADE*, Argonne, USA, 1988. Springer Verlag. LNCS 310.
2. A. Bundy. A Science of Reasoning: Extended Abstract. In *Proceedings of the 10th CADE*, Kaiserslautern, Germany, 1990. Springer Verlag. LNCS 449.
3. P. Deussen. *Halbgruppen und Automaten*. Springer Verlag, Berlin, Germany, 1971.
4. G. Gentzen. Untersuchungen über das logische Schließen II. *Mathematische Zeitschrift*, 39: 572–595, 1935.
5. X. Huang, M. Kerber, and M. Kohlhase. Methods – the basic units for planning and verifying proofs. SEKI Report SR-92-20, FB Informatik, Universität des Saarlandes, Saarbrücken, Germany, 1992.
6. X. Huang, M. Kerber, M. Kohlhase, E. Melis, D. Nesmith, J. Richts, and J. Siekmann. Ω-MKRP – a proof development environment. SEKI Report SR-92-22, FB Informatik, Universität des Saarlandes, Saarbrücken, Germany, 1992.
7. X. Huang, M. Kerber, M. Kohlhase, E. Melis, D. Nesmith, J. Richts, and J. Siekmann. KEIM: A Toolkit for Automated Deduction. In these proceedings.
8. X. Huang. *Human Oriented Proof Presentation: A Reconstructive Approach.* PhD thesis, Universität des Saarlandes, Saarbrücken, Germany, 1994, forthcoming.
9. M. Kerber. How to prove higher order theorems in first order logic. In J. Mylopoulos and R. Reiter, editors, *Proceedings of the 12th IJCAI*, pages 137–142, Sydney, 1991. Morgan Kaufman.
10. C. Lingenfelder. Structuring computer generated proofs. In N. S. Sridharan, editor, *Proceedings of the 11th IJCAI*, pages 378–383, Detroit, USA, 1989. Morgan Kaufman.
11. H. J. Ohlbach and J. H. Siekmann. The Markgraf Karl Refutation Procedure. In J.-L. Lassez and G. Plotkin, editors, *Computational Logic – Essays in Honor of Alan Robinson*, chapter 2, pages 41–112. MIT Press, 1991.

leanT^AP: Lean Tableau-Based Theorem Proving
— Extended Abstract —

Bernhard Beckert & Joachim Posegga

Universität Karlsruhe
Institut für Logik, Komplexität und Deduktionssysteme
76128 Karlsruhe, Germany
{beckert|posegga}@ira.uka.de

Abstract.
"prove((E,F),A,B,C,D) :- !, prove(E,[F|A],B,C,D).
 prove((E;F),A,B,C,D) :- !, prove(E,A,B,C,D), prove(F,A,B,C,D).
 prove(all(H,I),A,B,C,D) :- !,
        \+length(C,D), copy_term((H,I,C),(G,F,C)),
        append(A,[all(H,I)],E), prove(F,E,B,[G|C],D).
 prove(A,_,[C|D],_,_) :-
        ((A= -(B); -(A)=B)) -> (unify(B,C); prove(A,[],D,_,_)).
 prove(A,[E|F],B,C,D) :- prove(E,F,[A|B],C,D)."

implements a first-order theorem prover based on free-variable semantic tableaux. It is complete, sound, and efficient.

1 Introduction

The Prolog program listed in the abstract implements a complete and sound theorem prover for first-order logic based on free-variable semantic tableaux [6]. We call this *lean theorem proving*: the idea is to achieve maximal efficiency from minimal means. We will see that the above program is indeed very efficient—not *although* but *because* it is extremely short and compact.

Satchmo [7] can be regarded the earliest application of lean theorem proving. The core of Satchmo is about 15 lines of Prolog code, and for implementing a refutation complete version another 15 lines are required. Unfortunately, Satchmo works only for formulæ in clausal form (CNF).

Many problems become much harder when translating them to clausal form, so it seems much better to avoid CNF and to preserve position and scope of quantifiers.[1] One way to achieve this is to use a calculus based on free-variable tableaux. It is a common, but mistaken belief that tableau calculi are inefficient; we will demonstrate the contrary.

The reader is assumed to be familiar with free-variable tableaux (e.g. [6]) and the basics of Prolog. The full version of this paper [4] is available upon request

---
[1] Using a definitional CNF [5] helps at most partially: it avoids exponential growth of formulæ for the price of introducing some redundancy into the proof search. Extending the scope of quantifiers to clause level, however, cannot be avoided.

from the authors. The source code of leanT^AP (and that of a slightly improved version; see Section 4) can also be obtained free of charge.

2 The Program

The idea behind leanT^AP is to exploit the power of Prolog's inference engine as much as possible; whilst Satchmo is based upon `assert` and `retract`, we do not use these predicates at all but rely on Prolog's clause indexing scheme and backtracking mechanism. We modify Prolog's depth-first search to bounded depth-first search for gaining a complete prover.

For the sake of simplicity, our considerations are restricted to first-order formulæ in Skolemized negation normal form. This is not a serious restriction; the prover can easily be extended to full first-order logic by adding the standard tableau rules.[2] We will use Prolog syntax for first-order formulæ: atoms are Prolog terms, "-" is negation, ";" disjunction, and "," conjunction. Universal quantification is expressed as `all(X,F)`, where `X` is a Prolog variable and `F` is the scope. Thus, a first-order formula is represented by a Prolog term (e.g., `(p(0),all(N,(-p(N);p(s(N)))))` stands for $p(0) \wedge (\forall n(\neg p(n) \vee p(s(n))))$).

We use a single Prolog predicate to implement our prover:

`prove(Fml,UnExp,Lits,FreeV,VarLim)`

succeeds if there is a closed tableau for the first-order formula bound to `Fml`. The proof proceeds by considering individual branches (from left to right) of a tableau; the parameters `Fml`, `UnExp`, and `Lits` represent the current branch: `Fml` is the formula being expanded, `UnExp` holds a list of formulæ not yet expanded, and `Lits` is a list of the literals present on the current branch. `FreeV` is a list of the free variables on the branch (these are Prolog variables, which might be bound to a term). A positive integer `VarLim` is used to initiate backtracking: it is an upper bound for the length of `FreeV`.

We will briefly go through the program listed in the abstract again (using a more readable form now) and explain its behavior. The prover is started with the goal `prove(Fml,[],[],[],VarLim)`, which succeeds if `Fml` can be proven inconsistent without using more than `VarLim` free variables on each branch.

If a conjunction (α-formula[3]) "A and B" is to be expanded, then "A" is considered first and "B" is put in the list of not yet expanded formulæ:

`prove((A,B),UnExp,Lits,FreeV,VarLim) :- !,`
    `prove(A,[B|UnExp],Lits,FreeV,VarLim).`

For disjunctions (β-formulæ) we split the current branch and prove two new goals:

---

[2] Skolemization has to be carried out very carefully, since straightforwardly Skolemizing can easily hinder finding a proof [3]. The full version [4] of this paper gives more details.

[3] Due to R. Smullyan, conjunctive type formulæ are called α-formulæ in the semantic tableaux framework.

```
prove((A;B),UnExp,Lits,FreeV,VarLim) :- !,
    prove(A,UnExp,Lits,FreeV,VarLim),
    prove(B,UnExp,Lits,FreeV,VarLim).
```

Handling universally quantified formulæ (γ-formulæ) requires a little more effort. We first check the number of free variables on the branch. Backtracking is initiated if the depth bound VarLim is reached. Otherwise, we generate a "fresh" instance of the current γ-formula all(X,Fml) with copy_term. FreeV is used to avoid renaming the free variables in Fml. The original γ-formula is put at the end of UnExp[4], and the proof search is continued with the renamed instance Fml1 as the formula to be expanded next. The copy of the quantified variable, which is now free, is added to the list FreeV:

```
prove(all(X,Fml),UnExp,Lits,FreeV,VarLim) :- !,
    \+ length(FreeV,VarLim),
    copy_term((X,Fml,FreeV),(X1,Fml1,FreeV)),
    append(UnExp,[all(X,Fml)],UnExp1),
    prove(Fml1,UnExp1,Lits,[X1|FreeV],VarLim).
```

The next clause closes branches; it is the only one which is not determinate. Note, that it will only be entered with a literal as its first argument. Neg is bound to the negated literal and sound unification[5] is tried against the literals on the current branch. The clause calls itself recursively and traverses the list in its second argument; no other clause will match since UnExp is set to the empty list for this recursion.

```
prove(Lit,_,[L|Lits],_,_) :-
    (Lit = -Neg; -Lit = Neg) ->
    (unify(Neg,L); prove(Lit,[],Lits,_,_)).
```

Note, that the implication "->" introduces an implicit cut after binding Neg: this prevents generating double negation when backtracking (which would happen, if "," were used instead).

The last clause is reached if the preceding clause cannot close the current branch. We add the current formula (always a literal) to the list of literals on the branch and pick a formula waiting for expansion:

```
prove(Lit,[Next|UnExp],Lits,FreeV,VarLim) :-
    prove(Next,UnExp,[Lit|Lits],FreeV,VarLim).
```

lean$T^A P$ has two choice points: One is selecting between the last two clauses, which means closing a branch or extending it. The second choice point within the fourth clause enumerates closing substitutions during backtracking.

---

[4] Putting it at the top of the list destroys completeness: the same γ-formula would be used over and over again until the depth bound is reached (bracktracking does not change the order in which formulæ are expanded).

[5] In contrary to the built-in unification =, the predicate unify implements sound unification, i.e., unification with occurs check.

3 Performance

Although (or better: because) the prover is so small, it shows striking performance. Table 1 shows experimental results for a subset of Pelletier's problems [8].

Table 1. leanT^AP's performance for Pelletier's problem set (the runtime has been measured on a SUN SPARC 10 workstation with SICStus Prolog 2.1; "0 $msec$" means "not measurable").

No.	Limit VarLim	Branches closed	Closings tried	Time $msec$	No.	Limit VarLim	Branches closed	Closings tried	Time $msec$
17	1	14	14	0	32	3	10	10	10
18	2	1	1	9	33	1	11	11	0
19	2	4	6	0	34	??	–	–	∞
20	6	3	3	9	35	4	1	1	0
21	2	8	8	0	36	6	3	3	0
22	2	7	14	9	37	7	8	8	9
23	1	4	4	0	38	4	90	101	210
24	6	33	33	9	39	1	2	2	0
25	3	5	5	0	40	3	4	5	0
26	3	16	17	0	41	3	4	5	0
27	4	8	8	0	42	3	5	5	9
28	3	5	5	0	43	5	18	18	109
29	2	11	11	9	44	3	5	5	10
30	2	4	4	9	45	5	17	17	39
31	3	5	5	0	46	5	53	63	59

Some of the theorems, like Problem 38, are quite hard: the $_3T^AP$ prover [1], for instance, needs more than ten times as long. If leanT^AP can solve a problem, its performance is in fact comparable to compilation-based systems that search for proofs by generating Prolog programs and running them [10, 9].

Pelletier No. 34 (also called "Andrews Challenge") is a surprise: all others run really fast but for Problem 34 leanT^AP does not find a proof. One reason might be that we did not invest much time in finding the right number of free variables to allow: an iterative deepening approach (as applicable to all other problems) does not work: if VarLim is set to 4, the prover does not return after 30 minutes. It is easily possible that the right limit (which is above 12) returns a proof very quickly.

4 Conclusion & Outlook

We showed that a first-order calculus based on free-variable semantic tableaux can be efficiently implemented in Prolog with minimal means.

One could regard leanT^AP as a Prolog hack. However, we think it demonstrates more than tricky use of Prolog: it shows that semantic tableaux are an efficient calculus when implemented carefully. Besides this, the philosophy of "lean theorem proving" is interesting: We showed that it is possible to reach considerable performance by using extremely compact (and efficient) code instead of elaborate heuristics. One should not confuse "lean" with "simple": each line of a "lean" prover has to be coded with a lot of careful consideration.

There is still room for improvement without sacrificing simplicity and/or elegance of our approach: leanT^AP can easily be extended by techniques known to enhance the performance of tableau-based provers: A slightly longer version making use of "universal formulæ" [2] can, for example, solve Pelletier No. 34 in about $100 msec$ (see [4] for details).

Another possibility is to use an additional preprocessing step that translates a negation normal form into a graphical representation of a fully expanded tableau (see [9] for details). This can be implemented equivalently simply and requires only linear effort at runtime. The prover itself then reduces to two clauses, since no compound formulæ are present any more and all branches are already fully developed. The speedup will not be dramatic, but considerable. Furthermore, we can implement the compilation principle described by Posegga [9]: the idea is to translate tableau graphs into Prolog clauses that carry out the proof search at runtime. Compared with "conventional" implementations of tableau-based systems, this gains about one order of magnitude of speed. It will be subject to future research to apply this principle in the spirit of lean deduction.

References

1. B. Beckert, S. Gerberding, R. Hähnle, and W. Kernig. The tableau-based theorem prover $_3T^AP$ for multiple-valued logics. In *Proceedings, CADE 11, Albany/NY*, LNCS. Springer, 1992.
2. B. Beckert and R. Hähnle. An improved method for adding equality to free variable semantic tableaux. In *Proceedings, CADE 11, Albany/NY*, LNCS. Springer, 1992.
3. B. Beckert, R. Hähnle, and P. H. Schmitt. The even more liberalized δ-rule in free variable semantic tableaux. In *Proceedings, 3rd Kurt Gödel Colloquium, Brno, Czech Republic*, LNCS. Springer, 1993.
4. B. Beckert and J. Posegga. leanT^AP: Lean tableau-based theorem proving. To appear (available upon request from the authors), 1994.
5. E. Eder. *Relative Complexities of First-Order Calculi*. Artificial Intelligence. Vieweg Verlag, 1992.
6. M. Fitting. *First-Order Logic and Automated Theorem Proving*. Springer, 1990.
7. R. Manthey and F. Bry. SATCHMO: A theorem prover implemented in Prolog. In *Proceedings, CADE 9, Argonne*, LNCS. Springer, 1988.
8. F. J. Pelletier. Seventy-five problems for testing automatic theorem provers. *Journal of Automated Reasoning*, 2:191–216, 1986.
9. J. Posegga. Compiling proof search in semantic tableaux. In *Proceedings, ISMIS 7, Trondheim, Norway*, LNCS. Springer, 1993.
10. M. E. Stickel. A prolog technology theorem prover. In *Proceedings, CADE 9, Argonne*, LNCS. Springer, 1988.

FINDER: Finite Domain Enumerator System Description

John Slaney

Australian National University, Canberra, ACT 0200, Australia
John.Slaney@anu.edu.au

The program FINDER takes as input a set of clauses in a many-sorted first order language, together with specifications of finite cardinalities of the domains for the sorts, and generates interpretations on the given domains which satisfy all of the clauses. Thus it simply finds finite models of arbitrary first order theories. It does not treat the clauses as executable statements, so for example it imposes no order on them and there is no 'flow of control'. Hence its semantics are very declarative. The clauses are made ground by instantiating to the given domains, and these ground instances regarded as defining the constraint satisfaction problem of evaluating the function symbols of the language across the interpretation domains. Formally (see [5]) the inference mechanism it uses is ground tableau-style reasoning, driven by case analysis as is natural to such constraint satisfaction problems, with a combination of (ground) unit resolution to extend the tableau branches and a form of negative hyper-resolution to generate additional constraints during the search. Case analysis is performed on the implicit positive clauses which define the search space, the choice at each point being such as to minimise splitting of the branch. It is appropriate to see the logic as related to the Davis-Putnam algorithm. The inference process is opaque to the user.

FINDER can produce a single model, to demonstrate the existence of a solution, or can enumerate all models in its (necessarily finite) search space. It is also possible to use output from FINDER as input for another FINDER job. One use of this is to select from the models of a theory only those which do *not* have some further property such as the definability of some function.

1 Problem representation language

A FINDER input file consists of various sections. Each section specifies either sorts, function symbols, clauses or environment settings as illustrated in the example below. There may be any number of sections, in any order, except that each function symbol must be declared before it is used in a clause and each sort must be declared before it is used to specify the type of a function.

Each sort declaration is followed by a specification of its cardinality or range of cardinalities. Optionally, a sort may be enumerated by assigning a canonical name to each object.

Each function declaration is followed by a specification of its (first order) type and a (possibly null) list of characteristics. Partial functions are allowed.

Nullary function symbols (proper names) may be declared using the key word 'constant' instead of 'function'.

Clauses are written in the form

$$[\langle\text{formula}\rangle\ldots] \rightarrow [\langle\text{formula}\rangle\ldots].$$

where the arrow separates antecedents (negative literals) from consequents (positive literals). Either side of the arrow may be null, single, or multiple. In the multiple case, the antecedents are read conjunctively and the consequents disjunctively, so arbitrary clauses may be represented. A formula consists of either a variable or a function symbol applied to appropriately many formulae of the right sorts. Binary function symbols may be written in prefix or in infix notation.

There are several pre-defined expressions. The sorts `bool` and `int` are given, fairly much as expected. Truth values `true` and `false` and natural numerals are defined as constants. Every sort comes totally ordered. Since all sorts are finite, this ordering evidently does no harm and is often useful. Identity is well defined. Given the total order and discreteness, clearly any object may be 'incremented' by an integer. Hence for each sort S, '+' is available as a partial function of type $S \times \text{int} \rightarrow S$. Since partial functions are allowed, it is useful to have an existence predicate, so the pre-defined expression `E!` is made available. Where f is a formula, `E!`(f) means that f has a value (i.e. its value exists).

For full details of the problem representation language, including the various environment settings, see the documentation shipped with the program.

2 Examples

The first example is the Queens Problem. This is to enumerate all non-attacking placements of 8 queens. The problem is entirely trivial, of course, with a number as small as 8, and is given here for illustration only.

```
sort {
    rank
    cardinality = 8

    file
    cardinality = 8
}

function queen: rank -> file        % The file where there is
    bijective                       % a queen on a given rank.

clause {
    E!(a+x), queen(a+x) = queen(a)+x   ->   x = 0.
    E!(a+x), queen(a+x)+x = queen(a)   ->   x = 0.
}

end
```

Note that such properties as the bijectiveness of the `queen` function may be stipulated to FINDER directly without needing clausal representation. Note also the use of addition as a partial function. Variable `a` ranges over ranks and variable `x` over integers, so rank `a+x` may be off the edge of the board; the clauses which prohibit pairs of queens on diagonals only apply where the rank exists, hence the antecedent condition `E!(a+x)`.

The second example shows a highly nontrivial problem which FINDER solved in about a day on a Sparc-2 and which had previously been open. The problem is to find an idempotent quasigroup of order 14 satisfying the equation $(xy.x)y = x$ and equivalently $(xy.y).xy = x$, or alternatively to prove that there is no such thing. A quasigroup is just a groupoid whose 'multiplication table' is a Latin square, each value occurring in every row and in every column. A fuller account of the research program in which this experiment occurred is given in [1] and [5]. Note that the arrow may be omitted from a unit clause such as an equation. The 'verbosity' settings are given for illustration only.

```
sort element      cardinality = 14

function o: element,element -> element {
    row-bijective
    column-bijective                    % I.e. a quasigroup
}

clause {
    x o 0 > x+1  ->  false.             % To reduce isomorphisms.

    x o x = x.                          % Idempotence.
    ((x o y) o x) o y = x.              % The special equation -
    (x o y o y) o (x o y) = x.          % - and its equivalent.
}

setting {
    solutions: 1                        % Stop if one model found.
    verbosity {
        stats: full                     % Print a detailed report,
        models: full                    % and the solution if any.
    }
}

end
```

The point of the first clause is to force the cycles in the permutation constituting the leftmost column to occupy contiguous sections of the numbering. More efficient techniques are known for reducing the isomorphic subspaces searched, but they have not been shown here for reasons of clarity. This example shows that FINDER is not a mere toy but can be used (and has been used) for serious finite mathematics.

3 FINDER as module

As well as standing alone, FINDER can be linked as a module to other C programs. It can be used simply to generate models of a given theory read from a file and possibly with additional input from the parent program. It can also be used to test clauses for truth in a model which it keeps in memory and optionally to update that model from time to time in order to validate more clauses. This function of FINDER has been put to use in the theorem prover SCOTT (Semantically Constrained OTTER) where the model generator serves to tailor the guiding model to the problem in hand for such purposes as semantic resolution [2]. A library of function calls is provided to allow FINDER to be used flexibly in further applications. Again, details are supplied in the documentation shipped with the sources.

4 Availability

FINDER 3.0 is available by anonymous ftp from arp.anu.edu.au. The latest release is in file finder-3.0.⟨*patchlevel*⟩.tar.Z in directory ARP/FINDER. As of November 1993 the patchlevel is 1. The above file contains the 'C' sources for FINDER, the LaTeX sources for the full documentation, Makefile, online 'man' pages and a collection of sample input files. If you install the program, we ask that you send email to FINDER@arp.anu.edu.au letting us know that you have it, so that we may keep you informed of patches and upgrades.

References

1. M. Fujita, J. Slaney & F. Bennett, *Automatic Generation of Some Results in Finite Algebra*, **Proc. 13th IJCAI**, 1993, pp. 52–57.
2. J. Slaney, *SCOTT: A Model-Guided Theorem Prover*, **Proc. 13th IJCAI**, 1993, pp. 109–114.
3. J. Slaney, *FINDER, Finite Domain Enumerator: Version 2.0 Notes and Guide*, Technical report TR-ARP-1/92, Automated Reasoning Project, Australian National University, Canberra, 1992.
4. J. Slaney, *FINDER, Finite Domain Enumerator: Version 3.0 Notes and Guide*, Document with program sources, anonymous ftp, arp.anu.edu.au, file ARP/FINDER/finder-3.0.1.tar.Z.
5. J. Slaney, M. Stickel & M. Fujita, *Automated Reasoning and Exhaustive Search: Quasigroup Existence Problems*, **Computers and Mathematics with Applications**, Special Issue on Automated Reasoning and Its Applications, to appear 1994.

Symlog
Automated Advice in Fitch-style Proof Construction

Frederic D. Portoraro
Department of Philosophy, University of Toronto
Toronto, Ontario, M5S 1A1, Canada
frederic@ai.toronto.edu, (416) 978-3311

Abstract. *Symlog* is a system for learning symbolic logic by computer. One of *Symlog*'s components is a built-in theorem prover designed around a powerful, yet highly intuitive, set of proof construction strategies. Its role is to act as an advisor to students engaged in the construction of formal proofs in Fitch-style natural deduction systems of propositional and predicate logic.

1. Introduction

Symlog is a system for learning symbolic logic by computer and it is intended to be used by students in elementary and intermediate courses in formal logic in departments of philosophy, mathematics, computer science, and the like. The software accompanies the recently published textbook *Logic with Symlog*[1] [4] and both text and software are fully integrated with each other to cover all areas of introductory sentential and predicate logic: symbolizations, truth-tables, truth-trees, natural deduction and models. Although for each major component in the textbook there is corresponding software module, here we will restrict our discussion to *Symlog*'s natural deduction 'environments' as they will be of more interest to CADE-12 attendees and the readers of the proceedings. *Symlog*'s deduction environments provide the student with interactive proof construction, immediate feedback, the automatic checking of proofs and, not least of all, proof construction expert advice at the press of a key.

2. *Symlog*'s Derivation Systems

Symlog supports four systems of Fitch-style natural deduction: SD, SD+, PD and PD+ ([4],[2]). The system SD of sentential logic consists of a set of intelim—introduction and elimination—rules of deduction, two per each of the standard propositional connectives. SD+ extends SD by the addition of 'derived' rules of inference such as Commutation, Distribution, De Morgan, and the like. PD adds to SD the generalization and instantiation rules for the universal and existential quantifiers; PD+ is to PD what SD+ is to SD, with the addition of the rule of quantifier negation (generalized De Morgan). For pedagogical reasons, *Symlog*'s theorem prover only deals with the 'core' systems, SD and PD.
Proofs in any of these systems are constructed with *Symlog*'s interactive proof editor. The full-screen editor allows considerable freedom since proofs can be built in any order,

---

[1]The *Symlog* software, and in particular its underlying theorem prover, was designed by Frederic D. Portoraro. The textbook, *Logic with Symlog*, was co-authored by Frederic D. Portoraro and Robert E. Tully.

e.g., from the top to the bottom of the screen, or from the bottom-up (as done when employing a goal-analysis strategy). The editor keeps track of many mundane details (e.g., when deleting a line in a proof all the sentences in the proof which appeal to the line being deleted are marked as invalid and an error message is associated with each offending line as a reminder). Theorems are entered in standard logic notation and can be checked at the press of a key. 'Theorems' do not have to be justified right-away but can be left unproven until later on. The logic rules, besides justifying the theorems being proved, can be used as commands to cause new theorems to be generated in terms of previously stated theorems. The rules, by applying a goal-analysis strategy, can also be used to produce theorem or subproof 'templates' whose form suggests the theorems that must be proved in order to obtain the current goal. For instance, Figure 1 shows how the rule ¬E (negation elimination) is applied in line 9 to derive the sentence R via indirect proof; to this move, *Symlog* responds by opening a subproof with ¬R at its assumption and α and ¬α as contradictory candidates. *Symlog* produces α and ¬α inside the subproof because it does not know what two contradictory sentences the student has in mind (however, *Symlog*'s advisor can make recommendations: If asked, it would suggest ¬R and ¬¬R. See Figure 2).

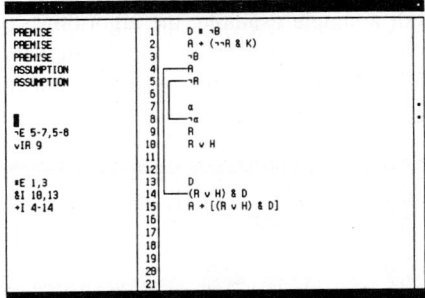 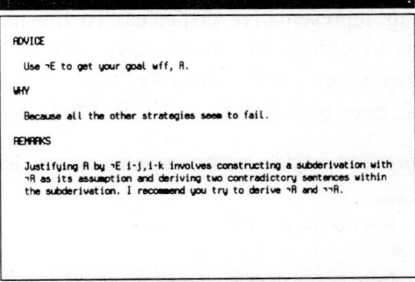

Figure 1   Figure 2

3. The Strategies and Automated Advice

Long and complex proofs can only be sensibly attacked with the help of strategies. The *Symlog* textbook provides a powerful, yet intuitive, set of strategies for proof construction which the student is urged to learn well. In fact, this very same set of strategies has been codified and implemented into *Symlog*'s automated theorem prover which does proofs in the very same way the student is being taught, namely using Fitch's method of subordinate proofs. It is also worth remarking that the theorem prover manipulates expressions in standard logic notation. Although the prover is part of *Symlog*, its full power is not directly accessible to the students but only that component which guides the construction of its proofs—the strategic component. The prover appears to the user in the form of on-line help (*Symlog*'s so-called Proof Advisor) which the student can summon at the press of a key. More specifically, the user simply places the screen cursor on the sentence he wants to derive (but does not know how) and, at the press of a key, invokes the prover which 'will look at the proof as developed thus far by the student and be able, if he has a reasonable idea, to give him help on continuing and completing the proof he has already begun' [7]. Here are, in a highly distilled form, the strategies in question (the reader is refered to [4] for more details):

1. *Basis case:* Reiteration, ¬E (sure-contradiction form)
2. *Elimination strategies:* ∃E, &E, →E, ≡E, ∨E (immediate case), ∀E
3. *Introduction strategies:* ¬I, &I, ∨I, →I, ≡I, ∀I, ∃I
4. *Cases strategy:* ∨E
5. *Catch-all case:* ¬E

The strategies are meant to be applied in the order given. The reiteration strategy says that if you need to prove α but α has already been derived then there is nothing to prove; sure-contradiction says that if two contradictory sentences have already been derived then you should quickly close your proof of α by reductio ad absurdum. Modus ponens (→E) and existential instantiation (∃E) are examples of elimination strategies, e.g., get α by modus ponens if β → $α_u$ occurs in an accessible sentence (even as a subwff) and β → $α_u$ but also β can both be proved (here $α_u$ is a wff which unifies with the goal, α). 'Immediate case' says that ∨E should be deployed if one of the disjuncts unifies with the goal. A typical example of an introduction strategy, ∀I, is to prove (∀x)α by proving one of its representative instances. To illustrate with a simple example, the algorithm says that to obtain Fa from, say, Hc & [Gab → (∀x)Fx] one should apply universal instantiation, ∀E, on (∀x)Fx; this latter sentence should be obtained by modus ponens, →E, once both Gab & (∀x)Fx and Gab have been derived.

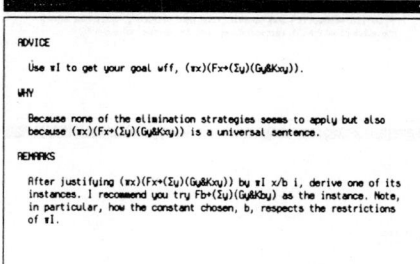 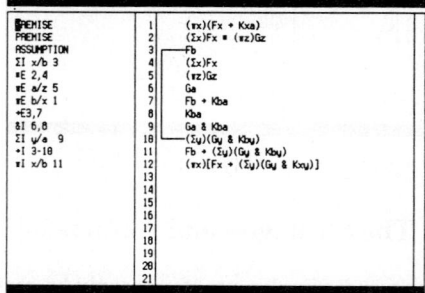

Figure 3   Figure 4

As an illustration of the effectiveness of the implementation of the strategies outlined above, Figure 3 shows[2] the automated advice that *Symlog* generates when a student asks for a hint in order to prove (∀x)[Fx → (∃y)(Gy & Kxy)] from (∀x)(Fx → Kxa) and (∃x)Fx ≡ (∀z)Gz. As seen in the figure, the actual advice consists of (1) a recommendation as to what rule of deduction to apply next, (2) why it makes sense to do so, and (3) remarks dealing mostly with the logistics of deploying the rule being recommended. If asked repeatedly for advice (something students are discouraged to do) the derivation shown in Figure 4 would eventually result. Another example of the strategies' effectiveness is the

---

[2]Due to hardware restrictions in most PCs, *Symlog* displays ∃ and ∀ as Σ and Π, respectively.

fact that the answers to the proof exercises in the *Symlog* textbook were actually solved by the theorem prover.

4. Other Design Considerations

Many researchers tend to classify a proof generated by a theorem prover as 'better' than another proof (generated by some other theorem prover) if it is shorter. The criterion for 'better proof' when designing *Symlog*'s theorem prover was, mostly for pedagogical reasons, quite different: A derivation is not considered 'better' than another if it is shorter but if it is, instead, 'more natural.' Here, 'more natural' means that the proof produced is more faithful—closer—to the set of intuitive strategies being learnt by the students. Roughly speaking, proving a conditional $\alpha \rightarrow \beta$ by conditional proof is more natural than by, say, indirect proof.

The overriding reason for designing *Symlog*'s theorem prover was, of course, to greatly augment the pedagogical value of its interactive proof subsystem. There was another, hopefully more far-reaching, reason for the prover's implementation. Some researchers have expressed concerns about the insufficient amount of recognition that the field of automated resoning has received and point to the lack of a wider exposure of its achievements as one of the reasons [8]. Although *Symlog* is strictly an educational system, to the eyes of the thousands of students engaged in the interactive construction of proofs with it, *Symlog*'s proof advisor provides a living example of a workable and useful implementation of an automated reasoning assistant. Moreover, since many of the students taking introductory logic courses do *not* ultimately specialize in fields like automated reasoning (or even mathematics or computer science, for that mater), using a program like *Symlog*'s theorem prover means exposing the value of automated deduction[3] to a wide audience comprised of individuals specializing in a great variety of fields ranging from immunology to chemical engineering.

5. Related and Future Work

Symlog is by no means the first computer program to implement theorem-proving techniques to teach logic. Actually, the original inspiration for the design of *Symlog* was Patrick Suppes' well-known EXCHECK/VALID system [6] but, as opposed to it, *Symlog*'s emphasis is on (Fitch-style) propositional and predicate logic proof construction and, at the time of writing, it does not include set theory. A system closer in spirit to *Symlog*, however, is the CMU Proof Tutor of Sieg and Scheines [5]. Their system also includes a theorem prover for Fitch-style propositional logic and Scheines informs me that work is well under way to extend it to predicate logic and to integrate it with the rest of the Proof Tutor; the idea is, of course, to have the prover act as an advisor à la *Symlog*.

Symlog's built-in theorem proving capabilities are not—and were not intended to be—as powerful as those of other stand-alone research-oriented theorem provers such as Pelletier's THINKER [3] or Li Dafa's ANDP [1]. Nevertheless, *Symlog* illustrates how a relatively simple—yet effective enough for its purposes—inference engine, when

---

[3]This value increases without bounds as the due date for an assignment or a test gets near.

coupled with a fairly good interactive proof editor and a friendly user interface, can be of great value to students of logic.[4]

The literally thousands of students that use *Symlog* every year likely constitute the world's largest user base for an interactive theorem proving system. Among other things, this translates into a demand to increase the system's sophistication. Although not yet generally available at the time of writing, a new deduction environment is close to completion which allows users to construct proofs in Fitch-style Peano arithmetic.

6. Acknowledgements

I would like to thank my colleague Robert E. Tully for reading a preliminary version of this paper, and to Alasdair Urquhart for his advice on logic-related technical issues. I am indebted to Richard Scheines and Andrew McCafferty for their fruitful discussions on the prover's implementation, and to the CADE-12 referees for their valuable comments on this paper.

References

[1] Dafa, L. A Natural Deduction Automated Theorem Proving System, *Automated Deduction—CADE-11, Lecture Notes in Artificial Intelligence*, D. Kapur editor, vol 607, pp 668-672, Springer-Verlag (1992).

[2] Fitch, F. B. *Symbolic Logic: An Introduction*, Ronald (1952).

[3] Pelletier, F. J. *Further Developments in THINKER, an Automated Theorem Prover*, Technical Report TR-ARP-16/87, Australian National University (1987).

[4] Portoraro, F. D. and Tully, R. E. *Logic with Symlog: Learning Symbolic Logic by Computer*, Prentice-Hall (1994).

[5] Sieg, W. and Scheines, R. Searching for Proofs (in Sentential Logic), *Philosophy and the Computer*, L. Burkholder editor, pp 137-159, Westview (1992).

[6] Suppes, P. et al. Part I. Interactive Theorem-Proving in CAI Courses, *University-Level Computer-Assisted Instruction at Stanford: 1968-1980*, P. Suppes editor, IMSSS, Stanford University (1981).

[7] Suppes, P. The Next Generation of Interactive Theorem Provers, *7th International Conference on Automated Deduction, Lecture Notes in Computer Science*, R. E. Shostak editor, vol 170, pp 303-315, Springer-Verlag (1984).

[8] Wos, L. The Impossibility of the Automation of Logical Reasoning, *Automated Deduction—CADE-11, Lecture Notes in Artificial Intelligence*, D. Kapur editor, vol 607, pp 1-3, Springer-Verlag (1992).

---

[4]The reader may want to keep in mind that *Symlog*'s target machine is the PC, whose memory limitations are well-known. Without going into any RAM management details, *Symlog*'s proof advisor has typically no more than 75K to make all its moves.

KEIM: A Toolkit for Automated Deduction

Xiaorong Huang   Manfred Kerber   Michael Kohlhase   Erica Melis
Dan Nesmith   Jörn Richts   Jörg Siekmann

Fachbereich Informatik, Universität des Saarlandes
66041 Saarbrücken, Germany
`keim@cs.uni-sb.de`
Telephone: (49) 681-302-4627

Abstract. KEIM is a collection of software modules, written in Common Lisp with CLOS, designed to be used in the implementation of automated reasoning systems. KEIM is intended to be used by those who want to build or use deduction systems (such as resolution theorem provers) without having to write the entire framework. KEIM is also suitable for embedding a reasoning component into another Common Lisp program. It offers a range of datatypes implementing a logical language of type theory (higher order logic), in which first order logic can be easily embedded. KEIM's datatypes and algorithms include: types; terms (symbols, applications, abstractions); unification and substitutions; proofs, including resolution and natural deduction styles.

1 Motivation

Though automated reasoning systems are among the earliest AI programs, the methods developed and implemented by the theorem-proving community are little-used outside of it. This phenomenon may be explained to some extent by the computational complexity of the programs involved (often NP-complete or undecidable), but an even larger share of the blame may be assigned to the *cognitive* complexity involved in the implementation of the programs themselves. It is easy to describe a proof process such as resolution, but actually writing a fairly-efficient resolution prover is far from trivial. In addition, a prover requires subcomponents, such as formula parsing and pretty-printing, which add to the magnitude of the job. The work and experience required to build a good theorem prover from scratch can be daunting for an outsider. Especially when the theorem prover is not the main object of study, but rather intended to be used as a component in some larger system, the foreseen difficulties will discourage many from beginning.

One may of course decide to use a prover that is already available. This has the advantage that its reliability is relatively assured and, being off-the-shelf, requires no implementation. Unfortunately, it is rare that the needs of a new application exactly fit the strengths of an existing theorem prover. Even if that were the case, one would probably have to build some kind of bridge between the two programs in order to exchange data, because the basic data structures used (terms, formulas, etc.), not to mention input syntax, are probably incompatible. One may try to modify the source code directly, but this is a very difficult task for the nonexpert.

In addition, most theorem provers are *sui generis*; they are designed to investigate a particular paradigm or approach and are not intended to be useful for all types of reasoning problems. This limits their applicability among a wide audience. And trying to get two provers to cooperate without greatly changing at least one of them is not a task for the faint of heart.

Because of these difficulties, those who wish to apply techniques developed by the theorem-proving community face the choice of either learning this 'black art' themselves by developing their own prover from scratch, or jury-rigging available provers to get some kind of result. Hardly an encouraging prospect. Even automated reasoning experts may wish to make a theoretical study of just a minor aspect (say, a comparison of term indexing schemes), and not want to go to the trouble of implementing the whole environment normally required.

We feel that what is needed to make theorem-proving technology widely available in a useful way is a framework that provides the essential tools (data structures and algorithms) to allow a theorem prover to be assembled by a non-expert. Such a framework must be modular, to allow data structure or algorithm variants to be swapped in or out, and extensible, to permit customization, as well as the addition of new modules, with relatively modest effort.

2 An Open Architecture

Despite the diversity of theorem provers currently in use and in development, there are many aspects that they share. They must support basic data structures such as terms, formulas and, often, more complex objects such as clauses and substitutions. There must be a way to parse user input into a usable form, and to pretty-print the internal format in a human-readable way. Unification and/or matching are also common components.

There are well-known algorithms and techniques for each of these areas. Their current implementations, however, are not suited for generic use, often relying on varying idiosyncratic data structures which cannot be reconciled. It is certainly necessary to continue research in the optimization of techniques such as unification, but for many applications, choosing one of the currently-known variants is good enough. Most users of theorem-proving technology do not want to reinvent the wheel, and even the best-known algorithms may be difficult to implement correctly and efficiently.

KEIM provides a framework, through documented interface functions, which allows such techniques to be implemented in a generic way, so that later improvements or customizations can be carried out without requiring changes to other modules. Just as a toolbox holds several similar tools which do roughly the same thing, KEIM will contain differing implementations of data structures and algorithms. These will be provided in a modular form that will allow unnecessary components to be left out and improved components to be swapped in. KEIM is like a cafeteria of theorem-proving tools, providing wholesome options that can be appetizingly combined.

This modularity of KEIM is essential for three reasons. First, the state of knowledge in theorem proving is always expanding; there will always be new ideas and techniques worthy of sharing. Second, the resources of any one group are limited. There is no way a single group can be expert in all the aspects of theorem proving that KEIM should offer. The cooperation and contributions of others must be possible if KEIM is to be truly useful. We wish KEIM to be a toolbox, not a toybox with only sketchy or incomplete implementations of certain techniques. Last but not least, it will be the rare case that a user will want to use exactly what is provided without any customizations. This is especially true when a theorem prover is to be embedded in an existing system, with its own data structures. Tools that solve the wrong problem would be of little use.

3 The KEIM Toolbox

KEIM version 1.2 [3] is implemented in Common Lisp, using the Common Lisp Object System (CLOS) [5]. CLOS allows great flexibility in the integration of new classes of objects. The generic function paradigm allows one to specialize the behavior of a function on a new type of object without changing its behavior on existing objects and without having to rewrite or copy existing and unrelated code, thus making it well-qualified for the implementation of modular, extensible toolboxes.

KEIM offers a range of datatypes implementing a logical language of type theory (higher order logic) called $\mathcal{POST}$ [3], in which first order logic can be easily embedded. KEIM's datatypes and algorithms include: types; terms (symbols, applications, abstractions); unification and substitutions; proofs, including resolution and natural deduction styles.

KEIM also provides functionality for the pretty-printing, error handling, formula parsing and user interface facilities which form a large part of the code of any theorem prover. These facilities are all easily customizable.

KEIM serves as the basis for the Ω-MKRP [2] proof development environment (successor to the MKRP project), and is developed as part of the German Deduction Effort, which is sponsored by the Deutsche Forschungsgemeinschaft as "Schwerpunkt Deduktion".

Cooperation with other research groups is underway, and a KEIM implementation of the ACID [1] term indexing software has been made available.

3.1 A Scenario

Suppose a user wishes to build a small resolution prover. She must first decide what logical language the prover will allow by selecting the KEIM modules that contain the corresponding CLOS class definitions. She may want to make some minor adjustments—to the pretty-printing functions, for example. Another change may be to specialize a class, *e.g.*, she may want to add a slot to the clause class which counts the number of times it was used. This will then require adding a method to the generic function that actually does the resolution, so that the method updates the slot.

Modules will be chosen for the desired types of resolution, factoring, etc. If the prover is to be interactive, commands can be defined in Lisp using the KEIM primitives. Some Lisp ability will be required to sew things together. The modules are loaded in the proper order into a Lisp environment, and the prover is ready for action. An example of a simple tableau prover and resolution prover written in KEIM are described in [4].

4 Summary and Future Directions

KEIM is a software project that intends to offer, through its software library, a way for the general AI community, as well as the theorem-proving community, to take advantage of the many developments that have been made in automated reasoning. KEIM will provide standard implementations of many techniques and algorithms, making it useful not only for building reasoning systems, but also for pedagogical purposes. KEIM is available for anonymous FTP from various locations; send e-mail to keim@cs.uni-sb.de for instructions.

We intend to extend KEIM in both breadth and depth, that is, both to improve the current implementations, as well as to add new variants (*e.g.*, various unification algorithms, equality-handling mechanisms). Another goal is to make KEIM even easier for nonexperts to use by providing a better user interface. Because of the amount of KEIM code, it can be intimidating for those who are just starting. We want to make getting started with KEIM a painless process.

We intend to explore cooperation with other groups who have expertise in particular areas, and welcome collaboration and suggestions. Currently we are setting up a KEIM user group consisting of those who do some implementation on the basis of KEIM.

KEIM's extensibility and customizability is intended to make it an open architecture for (Lisp-based) reasoning systems. We hope that KEIM, by providing the building blocks of a reasoning system, will allow others to concentrate on the research areas which are of principal interest to them.

References

1. P. Graf: Path Indexing for Term Retrieval. Technical Report MPI-I-92-237, Max-Planck-Institut für Informatik, Saarbrücken, Germany, 1992.
2. X. Huang et al.: Ω-MKRP, A Proof Development Environment. In these proceedings.
3. D. Nesmith, editor: KEIM-Manual version 1.2. Universität des Saarlandes, Im Stadtwald, Saarbrücken, Germany, 1994.
4. J. Richts and D. Nesmith: Implementing Simple Theorem Provers in KEIM: Case Studies. To appear as a SEKI Report, Universität des Saarlandes, Im Stadtwald, Saarbrücken, Germany.
5. G. Steele: Common Lisp, second edition. Digital Press, Boston, 1990.

Elf: A Meta-Language for Deductive Systems
(System Description)

Frank Pfenning[*]

Department of Computer Science,
Carnegie Mellon University,
Pittsburgh, PA 15213, U.S.A.

1 Overview

Elf is a uniform meta-language for the formalization of the theory of programming languages and logics. It provides means for

1. specifying the abstract syntax and semantics of an object language in a natural and direct way;
2. implementing related algorithms (*e.g.*, for type inference, evaluation, or proof search); and
3. representing proofs of meta-theorems about an object language, its semantics, and its implementation.

Its conceptual basis are *deductive systems* which are used pervasively in the study of logic and the theory of programming languages. Logics and type systems for programming languages, for example, are often specified via inference rules. Structured operational semantics and natural semantics also employ deductive systems, and other means for semantic specification (for example, by rewrite rules) can be easily cast into this framework. Many meta-theorems can be proved by induction over the structure of derivations.

Elf's formal foundation is the logical framework LF [5] in which systems of natural deduction can be concisely represented. LF employs the *judgments-as-types* encoding technique for the representation of derivations in a type theory with dependent types. In addition, Elf provides sophisticated type reconstruction and a constraint logic programming interpretation for LF. The latter allows the execution of algorithms when expressed as deductive systems. Proofs of meta-theorems can be represented concisely as higher-level judgments relating derivations.

The most complete reference describing the Elf language is [10]. Gentler introductions can be found in [12] and [6]. Elf has also been used in a graduate course on the theory of programming languages. A draft of the course notes may be available from the author upon request. Below we provide a brief overview of how specification, implementation, and meta-theory tasks are supported in the Elf language. The subsequent sections list some case studies and describe the implementation of Elf.

Object Language Specification. LF generalizes first-order terms by allowing objects from a dependently typed λ-calculus to represent object language expressions. This allows variables in the object language to be represented by variables in the meta-language, using the technique of *higher-order abstract syntax*. Common operations

---

[*] Internet address: fp@cs.cmu.edu

(*e.g.*, renaming of bound variables or substitution) and side-conditions on inference rules (*e.g.*, occurrence restrictions on variables) are thus built into the framework and do not need to be coded up anew for each object language. Another important advantage of this technique is that it greatly simplifies the representation of meta-theoretic proofs.

For semantic specification LF uses the *judgments-as-types* representation technique: a judgment J is represented by a type A and a derivation $\mathcal{D}$ of J by an object M of type A. Such an encoding is *adequate* if there is a bijection between canonical LF objects of type A and derivations $\mathcal{D}$ of J. For adequate encodings, checking the validity of a proposed derivation $\mathcal{D}$ for a judgment J is reduced to type-checking the representation M of $\mathcal{D}$ in the LF type theory. As type-checking for LF is decidable, this yields an effective procedure for checking derivations.

In combination with higher-order abstract syntax, this technique also allows direct representation of *parametric* (sometimes called *generic*) and *hypothetical judgments* without cumbersome, explicit side-conditions on inference rules.

Object Language Implementation. Once a language and its semantics have been represented in LF, one would often like to program and execute related algorithms. In other framework implementations, these algorithms are typically written in a different (meta-)meta-language such as ML, for example using tactics and tacticals. This has the disadvantage that it is difficult, if not impossible, to reason formally about these algorithms.

Thus we pursue another approach in which deductive systems are given an operational interpretation. This has been inspired by the logic programming language λProlog [8] which gives an operational interpretation to hereditary Harrop formulas. Similarly, Elf gives an operational interpretation to types. While λProlog aspires to be a general purpose programming language and thus includes non-logical features such as cut, primitives for input and output, *etc.*, Elf remains pure. Consequently, not all algorithms can be expressed faithfully or implemented efficiently, and Elf should be considered a prototyping and experimentation tool. On the other hand the purity and simplicity of the language enables us to represent proofs of some properties of Elf programs within Elf itself.

Our experience with Elf has also shown that in many cases specifications themselves are executable. A good example of this phenomenon is natural semantics, which is usually structured so that goal-oriented search for a derivation as performed by the Elf interpreter corresponds directly to evaluation in the object language. We would like to emphasize, however, that this is usually not the case for logics: an encoding of natural deduction for a logic does not automatically give rise to a search procedure. Theorem provers for object logics have to be programmed explicitly. Standard techniques (*e.g.* iterative deepening, or tactics and tacticals) are readily implementable in Elf. For truly interactive theorem proving, some ML programming is also necessary due to the lack of input and output primitives in Elf.

Meta-Theory. Many proofs in the theory of programming languages proceed by induction over the structure of terms or derivations. It is well-known that such proofs give rise to primitive recursive functions. Unfortunately, the presence of induction principles or primitive recursive function objects conflicts with higher-order abstract syntax and the central encoding techniques for parametric and hypothetical judgments.

Instead, we continue to follow ideas from logic programming by representing the functions which could be extracted from meta-proofs as higher-level judgments relating derivations. This captures the computational contents of meta-proofs, *i.e.*, they can be executed. While type-checking in Elf guarantees local consistency of this kind of representation, it cannot guarantee that a judgment represents a total function. Thus, while we can implement, partially verify, and execute the meta-theory of deductive systems, at present Elf cannot guarantee the validity of a meta-proof. We are currently implementing a *schema-checker* that would verify that higher-level judgments follow a schema akin to primitive recursion, thus, in combination with the type-checker, verifying meta-proofs. Some preliminary ideas related to schema-checking can be found in [13].

2 Case Studies

A number of case studies have been carried out using Elf, most of which are distributed with the Elf source. Each example deals with a language, some aspects of implementation, and some meta-theory. Among these examples are various logics and logical interpretations, following the ideas laid out in [5]. We have also investigated the theory of logic programming and uniform proofs in this context. We have further implemented a small functional language with polymorphism and recursion and proved various properties such as type preservation [6]. For a proof of compiler correctness for essentially the same language, see [4]. Penny Anderson [1] has implemented a constructive logic, an extraction procedure for functional programs, and some aspects of the correctness proof for this procedure. We have also implemented a proof of the Church-Rosser theorem for the untyped λ-calculus [9]. Other unpublished experiments include type reconstruction for the polymorphic λ-calculus, a proof of the equivalence of Cartesian Closed Categories and the simply-typed λ-calculus (A. Filinski), an implementation of Monads (W. Gehrke), and a correctness proof for CPS conversion (O. Danvy).

3 The Elf Implementation

Elf is implemented in Standard ML of New Jersey; only a few minor changes would be required for other implementations of SML. The implementation is highly modular, taking advantage of the module system of SML. The core of the implementation is a λ-calculus with *type : type* which is general enough to encompass the Calculus of Constructions and the LF type theory. Precisely which quantifications are allowed is specified in a separate module, giving rise to various concrete type theories.[1] Building upon this core, we have implemented a constraint solving algorithm [10, 11] for the full core calculus and a pre-unification algorithm [3] for LF. The main constraint solver simplifies equations between typed λ-terms. For reasons discussed in [7], the pre-unification algorithm is currently not in use, although with the ML module system it is easy to construct a system which employs it.

Based on the constraint solver, we have implemented an algorithm for type reconstruction, again for the full core calculus. Because of the undecidability of the general

---
[1] The current implementation is not general enough to encompass all Pure Type Systems.

reconstruction problem [2], the algorithm will either report a principal type, a type error, or an indication that the source term contained insufficient type information for unambiguous reconstruction. It is always possible to add enough types to the source so that the typing becomes unambiguous.

Type reconstruction for Elf is practical: parsing and checking the types of the meta-proof of the Church-Rosser theorem for the untyped λ-calculus, for example, takes about 5 seconds on a SparcStation IPX. This proof, described in [9], consists of 1852 words (374 lines) of Elf source code. After type reconstruction, the proof consists of 3922 words (439 dense lines). On many examples, the fully typed source expressions will be 3–5 times larger than the actual input.

The logic programming module is specific to Elf. It implements an interpreter using an interactive top-level similar to that of Prolog. Ignoring some of the more advanced features, one may think of it as a typed version of a Prolog interpreter which maintains derivations of queries in the form of λ-terms. Queries can be of the form

$$?-\ M\ :\ A.\quad or\quad ?-\ A.$$

The type A represents a judgment, the object M its derivation. If M is given, this represents a type-checking query; if M is a free variable, the query triggers search for an object of type A, *i.e.*, an object representing a derivation of the judgment represented by A. In the second form, the derivation M need not be explicitly constructed, which can be significantly more efficient than the first form. Queries in either form may contain free variables that may be instantiated during constraint simplification or by resolution. Search proceeds in a depth-first fashion as in Prolog and is thus predictable though incomplete as a theorem prover. Unlike λProlog's higher-order unification, the constraint solver will never make non-deterministic choices; all choices are made during subgoal and clause selection. As a consequence, constraints may remain even after execution, where each solution to the constraints (possibly none) yields an answer to the original query.

Even though Elf does not compile programs or queries, it applies some standard optimizations known from logic programming interpreters and a few which are specific to Elf. Among the general optimizations are clause indexing, elimination of some unnecessary occurs-checks, and the use of efficient global symbol tables. Among the specific optimization are avoidance of redundant unifications and elimination of unnecessary proof objects. Rationales for the various implementation design decisions and an empirical analysis of the runtime behavior of Elf programs can be found in [7].

Together with the implementation we distribute some Emacs Lisp files that allow Elf to run as an inferior process to GNU Emacs. This includes a recently completed incremental type-checker which allows the programmer to check individual declarations, thus tightening the usual feedback loop of editing, type-checking a whole file, locating the source of a possible type error, *etc.* It can also be used to query types of subterms or to obtain a menu of possible constructors for objects valid in a given subterm location.

Elf can be retrieved by anonymous ftp from alonzo.tip.cs.cmu.edu, directory /afs/cs/user/fp/public. Copies of papers related to Elf can be found in the elf-papers subdirectory. Please see the README files for further information. There is also a mailing list with announcements regarding Elf; please send mail to elf-request@cs.cmu.edu to join the list.

Acknowledgments. The author would like to acknowledge the contributions of Spiro Michaylov, Ekkehard Rohwedder, Conal Elliott, and Ken Cline to the Elf implementation. This work was supported in part by NSF Grant CCR-9303383 and by the U.S. Air Force under Contract F33615-90-C-1465, ARPA Order No. 7597.

References

1. Penny Anderson. *Program Derivation by Proof Transformation.* PhD thesis, Carnegie Mellon University, October 1993. Available as Technical Report CMU-CS-93-206.
2. Gilles Dowek. The undecidability of typability in the lambda-pi-calculus. In M. Bezem and J.F. Groote, editors, *Proceedings of the International Conference on Typed Lambda Calculi and Applications, TLCA '93*, pages 139–145, Utrecht, The Netherlands, March 1993. Springer-Verlag LNCS 664.
3. Conal Elliott. Higher-order unification with dependent types. In N. Dershowitz, editor, *Rewriting Techniques and Applications*, pages 121–136, Chapel Hill, North Carolina, April 1989. Springer-Verlag LNCS 355.
4. John Hannan and Frank Pfenning. Compiler verification in LF. In Andre Scedrov, editor, *Seventh Annual IEEE Symposium on Logic in Computer Science*, pages 407–418, Santa Cruz, California, June 1992. IEEE Computer Society Press.
5. Robert Harper, Furio Honsell, and Gordon Plotkin. A framework for defining logics. *Journal of the Association for Computing Machinery*, 40(1):143–184, January 1993.
6. Spiro Michaylov and Frank Pfenning. Natural semantics and some of its meta-theory in Elf. In L.-H. Eriksson, L. Hallnäs, and P. Schroeder-Heister, editors, *Proceedings of the Second International Workshop on Extensions of Logic Programming*, pages 299–344, Stockholm, Sweden, January 1991. Springer-Verlag LNAI 596.
7. Spiro Michaylov and Frank Pfenning. An empirical study of the runtime behavior of higher-order logic programs. In D. Miller, editor, *Proceedings of the Workshop on λProlog Programming Language*, pages 257–271, Philadelphia, Pennsylvania, July 1992. University of Pennsylvania. Available as Technical Report MS-CIS-92-86.
8. Gopalan Nadathur and Dale Miller. An overview of λProlog. In Robert A. Kowalski and Kenneth A. Bowen, editors, *Logic Programming: Proceedings of the Fifth International Conference and Symposium, Volume 1*, pages 810–827, Cambridge, Massachusetts, August 1988. MIT Press.
9. Frank Pfenning. A proof of the Church-Rosser theorem and its representation in a logical framework. *Journal of Automated Reasoning*, 199? To appear.
10. Frank Pfenning. Logic programming in the LF logical framework. In Gérard Huet and Gordon Plotkin, editors, *Logical Frameworks*, pages 149–181. Cambridge University Press, 1991.
11. Frank Pfenning. Unification and anti-unification in the Calculus of Constructions. In *Sixth Annual IEEE Symposium on Logic in Computer Science*, pages 74–85, Amsterdam, The Netherlands, July 1991.
12. Frank Pfenning. Dependent types in logic programming. In Frank Pfenning, editor, *Types in Logic Programming*, chapter 10, pages 285–311. MIT Press, Cambridge, Massachusetts, 1992.
13. Frank Pfenning and Ekkehard Rohwedder. Implementing the meta-theory of deductive systems. In D. Kapur, editor, *Proceedings of the 11th International Conference on Automated Deduction*, pages 537–551, Saratoga Springs, New York, June 1992. Springer-Verlag LNAI 607.

EUODHILOS-II on top of GNU Epoch

Takeshi Ohtani, Hajime Sawamura, Toshiro Minami

Institute for Social Information Science
FUJITSU LABORATORIES LTD.
140, Miyamoto, Numazu, Shizuoka 410-03, Japan
Email:{ohtani, hajime, minami}@iias.flab.fujitsu.co.jp

1 Introduction

Many reasoning systems supporting both the specification of logics and the construction of formal proofs for various logics have been built. The systems are classified into the following three categories according to their approaches:

(1) Encoding logics into a formal system: Isabelle [8], LF [5],
(2) Representing logics so as to directly reflect their proof theoretic nature: EUODHILOS [9][10], ICLE [2], mural [6], PROOF DESIGNER [1],
(3) Implementing logics in functional or logical programming languages: HOL [4], λProlog [3].

The second approach to a general reasoning system is fairly different from other two. It allows users to specify logics in an easier and more direct way than others which require them to learn a programming language or metalogic for encoding a logic. Moreover, it provides reasoning facilities and a unique reasoning-oriented interface to make proof construction more flexible and easier.

EUODHILOS-II is the successor to EUODHILOS, built on the UNIX system by redesigning it so as to be highly portable and much more automated. It inherits the following two main features from EUODHILOS:

- Ease and expressiveness of defining logics:
  Users can easily define various logical systems in a way that preserves the original styles of a logic.
- Flexibility of proof construction:
  EUODHILOS-II provides a proof environment that is natural for humans.

In this report, we will mainly describe how to specify logics in EUODHILOS-II, due to the space limitation.

2 Specifying Logics

Logics are stipulated by language and derivation systems. The language system is for describing the language of terms, formulas and so on, which are used in proving. The derivation system consists of axioms, inference rules and rewriting rules.

Language System. The syntax definition of EUODHILOS-II is described in terms of the augmented context-free grammar with which logicians and computer scientists are familiar. It consists of three components: *ROOT*,

META_VARIABLES and *PRODUCTIONS*. The *ROOT* part specifies a starting symbol of context-free grammar. The *META_VARIABLES* part defines lexical information with regular expressions. The symbols defined by this part are used as metavariables. Metavariables are place-holders that are to be substituted with the same syntactic category. Use of metavariables enables us to prove theorem schemata and to yield partially instantiated proofs. The *PRODUCTIONS* part consists of production rules. This is written in BNF like notation. Bound variables and their scope are the most basic concepts for logics. They can be defined within the production rule. A bound variable is expressed as a symbol that has "@" at the head of a nonterminal symbol. Its scope is enclosed by "[" and "]". The bound variable information is not used in parsing, but only in checking side conditions for axioms or inference rules. Most logical languages can be described in this manner without introducing any extra apparatus.

The syntax definition in Figure 1 is an example of a simple version of First-Order Predicate Logic, whose predicate and function symbols are at most unary.

```
%ROOT Formula
%META_VARIABLES
   Predicate = "[A-Z][A-Z0-9]*" ;
   Variable = "[x-z][0-9]*" ;
   Function = "[f-h][0-9]*" ;
%PRODUCTIONS
   Term ::= Variable ;
   Term ::= Function "(" Term ")" ;
   Formula ::= Predicate ;
   Formula ::= Predicate "(" Term ")" ;
   Formula ::= "⊥" ;
   Formula ::= "¬" Formula ;
   Formula ::= "(" Formula ")" ;
   Formula ::= Formula "∧" Formula ;
   Formula ::= Formula "∨" Formula ;
   Formula ::= Formula "⊃" Formula ;
   Formula ::= "∀" @Variable "." [ Formula ] ;
   Formula ::= "∃" @Variable "." [ Formula ] ;
```

Fig. 1. Syntax definition of a simple First-Order Predicate Logic.

Derivation System. The derivation system consists of axioms, inference rules and rewriting rules. Each axiom is defined by an axiom scheme, together with its side conditions. For the specification of inference rules, there are two important issues to be considered: side conditions and dependency. An inference rule is specified by a conclusion, its premises and their assumptions in natural deduction style as follows:

$$\frac{[Assumption_{11}]\cdots[Assumption_{1n_1}] \quad \cdots \quad [Assumption_{m1}]\cdots[Assumption_{mn_m}]}{\vdots \qquad \qquad \vdots \qquad \qquad \vdots \\ Premise_1 \qquad \cdots \qquad Premise_m}{Conclusion}$$

The sequent style rules are also expressed in this style. Axioms and inference rules may have the following four types of side conditions:

(1) (FREE-FOR t x F) $\cdots$ t is free for x in F (Substitution Condition),
(2) (NOT-FREE x F) $\cdots$ x is not a free variable in F (Variable Occurrence Condition),
(3) (NOT-FREE-IN-ASSM x F) $\cdots$ x does not freely occur in any assumption of F (Eigenvariable Condition),
(4) (SYNTAX-CAT F C) $\cdots$ F is syntactically in the class C which is a nonterminal defined in the syntax definition part (Syntactical Restriction),

where NOT-FREE-IN-ASSM is used only for inference rules. The side conditions of usual logics are covered with the four types. Since these are almost the same terms as in ordinary logic textbooks, users can define inference rules without any difficulty. Figure 2 is a screen image of the inference rule editor in which the $\exists$ elimination rule is defined.

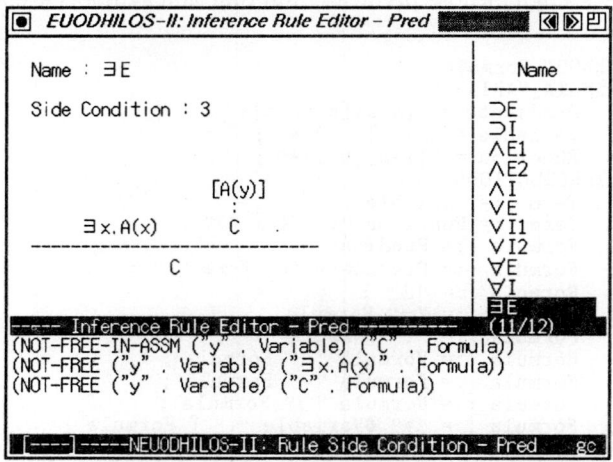

Fig. 2. Inference Rule Editor: in the upper right window, the list of inference rules are displayed. In the upper left window, the inference rule corresponding to the highlighted name is displayed. In the lower window, the side conditions for the rule are displayed.

Substitution is common to most of logics, and it deserves a subtle treatment in reasoning. The substituted expression $A(t/x)$ does not always mean that all occurrences of x in A are replaced by t. For example, in getting a conclusion $\exists x.A(x)$ by applying the $\exists$ introduction rule to $A(t/x)$, the substitution $A(t/x)$ is partial. It is likely that there are many conclusions according to the occurrence of a term. EUODHILOS-II then generates possible candidates for conclusions, so that users can simply choose one of them.

In natural deduction formalism, it is crucial to maintain the dependency among a conclusion, premises and assumptions. An expression is sometimes used as the assumption of different premises. Numbering each assumptions, EUODHILOS-II identifies assumptions with the same numbers as the same ones and assumptions with different numbers as different ones.

3 Constructing Proofs

Users can interactively construct proofs in natural deduction style on the proof editor of EUODHILOS-II. The proof editor has the following effective features:

(1) Connection of Proof Fragments by Matching/Unification:
    In order to prove larger proofs, it may be a good way for users to construct some small proof fragments with metavariables, and then to connect them to a desired proof. EUODHILOS-II automatically finds proper substitutions to connect them by pattern matching and unification.
(2) Use of Lemmas and Derived Rules:
    Lemmas and derived rules with metavariables proved by users, can be used as axioms and inference rules.
(3) Mixed Reasoning of Forward and Backward:
    Users can construct proof fragments by both forward and backward reasoning. In backward reasoning, since side conditions cannot be checked at each proof step, the consistency for a proof tree is checked whenever users want.
(4) Proof Search Strategy:
    Obvious and boring subproofs often appear in the proofs of major theorems. The tactics/tacticals provide a promising method to help it. EUODHILOS-II incorporates simple tactics/tacticals for the mingled reasoning with forward and backward.
(5) Automatic Candidate Generation for Rewriting:
    Rewriting rules are automatically applied to an expression the number of times that the user indicates, and the resulting expression is shown to the user.
(6) Proof Representation:
    At each stage of proving, EUODHILOS-II shows an abridged proof in the indented form of a relation among a conclusion, premises and assumptions. Besides it, EUODHILOS-II provides a tree-form representation for abridged proofs. Moreover, in order to show a full proof tree, EUODHILOS-II generates LaTeX macro codes to draw the proof tree.

4 Implementation

EUODHILOS-II is implemented on the GNU Epoch editor in the Emacs Lisp language. GNU Emacs produced by the GNU project is familiar to many users as a text editor, which exceeds in the ease to extend. Moreover it has the various functions for such text processing as syntactic analysis and searching with regular expressions. Epoch is an extension of GNU Emacs for the X Window System, which can manipulate multiple windows. Operations are keyboard-oriented and the usual editing commands are available in the comment, syntax definition and side condition editors. Users can customize EUODHILOS-II as they wish. The size of the source code of EUODHILOS-II is about 300K bytes.

References

[1] M. Bedau and J. Moor. PROOF DESIGNER: A Programmable Prover's Workbench. In *Philosophy and the Computer*, pages 218–228. Westview Press, 1992.

[2] M. Dawson. Using the ωp Generic Logic Environment. Technical Report, Dept. of Computing, Imperial College, 1989.

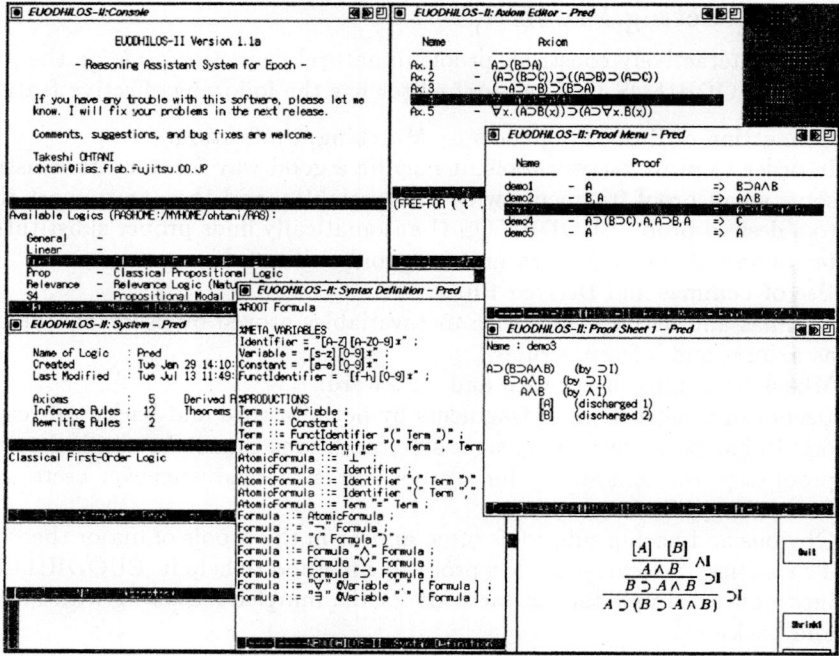

Fig. 3. Screen Layout: the upper left window is the main panel of EUODHILOS-II, in which a list of predefined logics is shown. Besides it, the console panel, the syntax definition editor, the axiom editor, the proof menu and the proof sheet are appeared. The proof tree in the lower right window is a previewer for LaTeX whose code is generated by EUODHILOS-II.

[3] A. Felty and D. Miller. Specifying Theorem Provers in a Higher-Order Logic Programming Language. In *9th International Conference on Automated Deduction, LNCS 310*, pages 61–80. Springer, 1988.

[4] M.J. Gordon. HOL – A Proof Generating System for Higher-Order Logic. In *VLSI Specification, Verification and Synthesis*, pages 73–128. Kluwer Academic Publishers, 1988.

[5] R. Harper, F. Honsell, and G. Plotkin. A Framework for Defining Logics. In *Symposium on Logic in Computer Science*, pages 194–204. IEEE, 1987.

[6] C.B. Jones, K.D. Jones, P.A. Lindsay, and R. Moore. *MURAL: A Formal Development Support System*. Springer, 1990.

[7] T. Ohtani. Reasoning Assistant System EUODHILOS-II: Operation Manual. Research Report, ISIS, FUJITSU LABORATORIES LTD., 1993. in preparation.

[8] L.C. Paulson and T. Nipkow. Isabelle Tutorial and User's Manual. Technical Report 189, University of Cambridge, Computer Laboratory, 1990.

[9] H. Sawamura, T. Minami, K. Yokota and K. Ohashi. A Logic Programming Approach to Specifying Logics and Constructing Proofs. In *Proc. of the Seventh International Conference on Logic Programming*, pages 405–424, MIT Press, 1990.

[10] H. Sawamura, T. Minami and K. Ohashi. EUODHILOS: A General Reasoning System for a Variety of Logics. In *Proc. of International Conference on Logic Programming and Automated Reasoning, LNAI 624*, pages 501–503. Springer, 1992.

Pi: an Interactive Derivation Editor for the Calculus of Partial Inductive Definitions

Lars-Henrik Eriksson

Swedish Institute of Computer Science[1]
Box 1263, S-164 28 KISTA, SWEDEN
Phone: +46 8 752 15 09
E-mail: lhe@sics.se

Abstract. Pi is a system for the interactive construction and editing of formal derivations in the calculus of finitary partial inductive definitions. This calculus can be used as a logical framework where object logics are specified, turning Pi into a derivation system for a particular object logic. Noteworthy features of Pi include: a graphic user interface where derivations are presented in tree form, direct manipulation of the derivation tree structure by selection using a mouse, and the ability to edit existing derivations by cutting and pasting as well as by changing the formulae occurring in a derivation. A simple facility for automatic theorem proving has been designed.

1. Introduction

Pi is a program for the interactive construction of formal derivations, being developed at the Swedish Institute of Computer Science. The main design goals for the Pi system are:

- To support the calculus of finitary partial inductive definitions – a formal system that can be used as a logical framework.
- To provide a highly user-friendly user interface.
- To provide functions for editing existing derivations, in order to facilitate interactive development of derivations.

The program is intended to serve the double purpose of being a vehicle for experimentation with the use of the calculus of finitary partial inductive definitions as a logical framework and as a tool for interactively proving theorems in object logics. As a tool, it has primarily been used for formal verification. Since automatic theorem proving facilities have not yet been implemented only smaller proofs have been done so far – up to a few hundred inferences – although work on substantially larger proofs is in progress.

Much of the inspiration for the design of the Pi user interface comes from the IPE (Interactive Proof Editor) system from the University of Edinburgh [6].

The author's Ph.D thesis [1] gives a thorough description of the calculus of finitary partial inductive definitions, of using it as a logical framework and of the principles of the Pi system. [2] summarises the calculus and describes its use as a logical framework.

Pi is written in SICStus Prolog and runs under X windows. It consists of about 5000 lines of Prolog code. The program is perhaps also of interest as a "production-size" Prolog program which makes virtually no use of meta-logical predicates and which uses cut only in situations where it is safe from a declarative point of view. The program is available from the author.

---
[1] This paper was written while the author was visiting the Wilhelm-Schickard-Institut für Informatik, Universität Tübingen, Germany.

2. The Meta Logic

The meta logic of the Pi system is the calculus of *finitary partial inductive definitions* [1,2] – a finitary variant of partial inductive definitions [3]. This calculus is a quite general logical formalism, where inference rules are parametrised by a finitary partial inductive definition (or simply a *definition*). In contrast to ordinary inductive definitions, the operator associated with a partial inductive definition may be non-monotonic [3].

Briefly, the calculus can be described as an intuitionistic sequent calculus including only connectives and inference rules for conjunction, implication, and universal quantification. The calculus also includes the usual structural inference rules – except cut. (Although cut is sometimes an admissible rule of the calculus.) A definition comprises a set of *clauses*, each of which "defines" some atomic formula in terms of an arbitrary formula of the calculus.

Additionally, there are two inference rules that depends on a particular definition. The inference rule for applying the definition to the succedent of a sequent ("definition-right") basically infers the succedent formula from its definition. The inference rule for applying the definition to the antecedent of a sequent ("definition-left"), implements the dual principle of *definitional reflection* [7].

The parametrisation of the inference rules makes the calculus suitable as a logical framework. A logic can be represented by providing a particular definition, which in a sense makes the inference rules of the calculus behave as inference rules of the represented (*object*) logic. The logics that have actually been represented so far include ordinary first-order logic, several modal logics, three-valued logic, naive set theory and the type system of the Edinburgh logical framework. To distinguish the connectives of the meta-logic from the corresponding connectives of object logics, they are not written with the usual symbols.

For representing first-order logic, the missing connectives (disjunction, negation, and existential quantification) can all be defined using a suitable finitary partial inductive definition. With that definition, the two definition inference rules will behave as the sequent calculus rules for these logical connectives. (In practise, conjunction, implication and universal quantification are defined anew as synonyms to the meta level connectives, to provide a complete set of connectives and inference rules at the object level.)

To provide a sufficiently expressive language for the representation of object logics, the atomic formulae of the calculus are terms of typed lambda calculus with a type system based on the simple theory of types, similar to that of λProlog [5]. The formulation of the definition inference rules requires full unification of higher-order terms – pre-unification is not enough. In practise, though, higher-order pattern unification [4] is usually sufficient.

3. The User Interface

To use Pi, the user must prepare a file with the particular partial inductive definition to be used. This file also includes relevant syntax and type declarations. In order to display formulae of an object logic as close as possible to that of the standard formulation of that logic, Pi provides a flexible syntax and can display several special symbols commonly used in logic and computer science. After starting the Pi session, the definition file is loaded and its contents determines the behaviour of the definition inference rules as well as the way formulae are parsed and printed. In the examples of the rest of this paper, we assume that a definition of first order logic has been loaded.

The construction of the derivation of a sequent proceeds in a top-down, goal-oriented fashion. The user specifies the sequent to be derived – the *endsequent* – using a built-in text editor. This creates a derivation tree with the endsequent as its only node. The user then requests the insertion of an inference step to derive this sequent. That inference step will be added to the derivation tree, having (in general) some premises which must also be derived. The user then adds inferences steps to derive each premise, and so on until no premises remain. The user is not bound to any specific order of deriving open premises, but can do this in any order desired.

Interaction with Pi takes place using a window, with a menu or text entry area at the top and the rest of the window displaying (part of) the derivation in progress. The following picture shows a typical situation[2]:

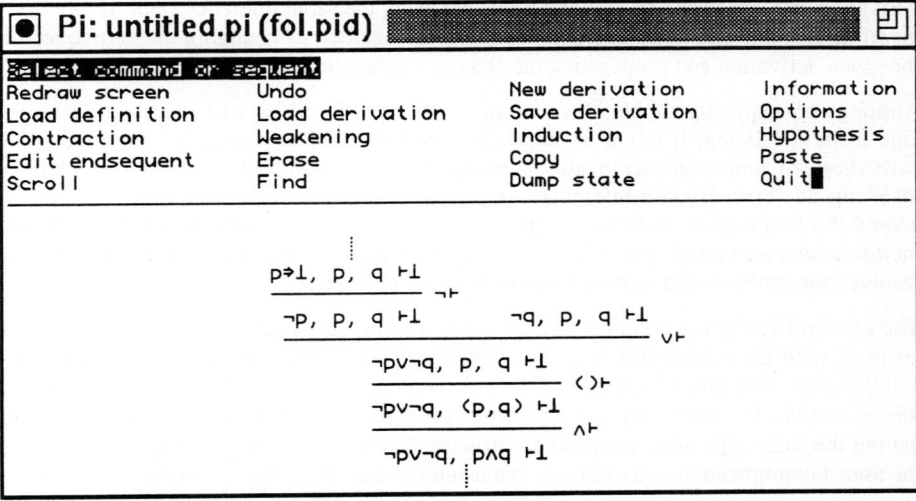

This window shows a derivation tree. Pi will attempt to arrange the derivation within the window in such a way as to make as much as possible of the derivation fit in the space available. In this case the full derivation tree is too large to fit on the screen, so dotted lines show where the derivation continues, above or below the visible part. By selecting one of the dotted lines with the mouse, the display is scrolled in the corresponding direction.

To present derivations entirely within the object logic, is it possible to suppress the display of those parts of a derivation dealing with the meta-logical connectives.

Commands are given by selecting items in the menu at the top using a mouse or other pointing device. Logical (non-structural) inference steps are added to the derivation tree by selecting the appropriate formula of a goal sequent. The form of the selected formula uniquely determines the inference rule to be inserted. E.g. selecting the formula $\neg q$ in the sequent $\neg q, p, q \vdash \bot$ causes an inference step for negation in the antecedent to be inserted.

In some cases, the user must choose between variants of a rule, or provide additional information – such as the formula to be substituted for the quantified variable when using the rule for universal quantification in the antecedent. Choices are presented in the menu area and selected using the mouse. Formulae are entered using the built-in text editor.

---

[2] Except that the size of the window is not typical, but quite small in order to fit on the page

4. Derivation Editing

A central feature of Pi is the ability to edit derivations. By "derivation editing" we mean the process of modifying an existing derivation (except by simply adding new inference steps to leaves of the derivation tree or by removing entire subderivations). There are four editing operations available to the user:

- Copying and pasting subderivations
- Changing the endsequent of the whole derivation
- Inserting an inference step into the middle of the derivation tree
- Changing an existing inference step in the middle of the derivation tree

The derivation editing functions are implemented using two basic operations: changing the derivation tree structure and deriving a sequent using the same inference steps as those of a given derivation. The latter operation can also be regarded as changing the endsequent of the given derivation and propagating the changes throughout the entire derivation.

Although conceptually simple, there are two practical difficulties with the latter operation. One difficulty is that it might be impossible to derive the new sequent using the given derivation. At some place an invalid inference step might result. Pi solves this problem by inserting an "error pseudo-inference" step at the place of the invalid inference step. It serves the function of keeping the previous subderivation so that the user can do any modifications necessary to rectify the error. In most cases Pi will notice when the user has resolved the problem and automatically remove the error step.

The other difficulty is to know which antecedent formulae of the new sequent that should be used with each inference step. Pi attempts to resolve this problem by heuristically matching the antecedent formulae of the new sequent with those of the old endsequent of the derivation. Presently, the heuristic will first match identical formulae, then formulae having the same type and outermost constructor. This matching is generally transparent to the user, but might sometimes not give the result the user might have expected.

Suppose the new sequent is $a, f(y,z), r \Rightarrow s \vdash c$ and the old endsequent of the derivation is $r \Rightarrow s, d, f(y,g(3)), a \Rightarrow b \vdash c$. The third antecedent formula of the former sequent will match the first antecedent formula of the latter sequent, since they are identical. The second formula of the former sequent will match the third formula of the latter sequent, since they have the same outermost function symbol.

For an example of an editing operation – the insertion of a step into a derivation – suppose that we want to derive $\vdash \neg\neg(p \vee \neg p)$ in intuitionistic logic. Forgetting that a contraction is needed, we create the following (incomplete) derivation:

$$\frac{\dfrac{\dfrac{\dfrac{\dfrac{\vdash p \vee \neg p \qquad \bot \vdash \bot}{p \vee \neg p \Rightarrow \bot \vdash \bot} \Rightarrow \vdash}{\neg(p \vee \neg p) \vdash \bot} \neg \vdash}{\vdash \neg(p \vee \neg p) \Rightarrow \bot} \vdash \Rightarrow}{\vdash \neg\neg(p \vee \neg p)} \vdash \neg$$

At this point we realise that a contraction step is needed at the sequent $\neg(p \vee \neg p) \vdash \bot$. By asking Pi to insert a contraction of the antecedent formula into the derivation, the derivation is changed to:

$$\cfrac{\cfrac{\cfrac{\cfrac{\cfrac{\neg(p\lor\neg p)\vdash p\lor\neg p \qquad \bot,\ \neg(p\lor\neg p)\vdash\bot}{p\lor\neg p\Rightarrow\bot,\ \neg(p\lor\neg p)\vdash\bot}\Rightarrow\vdash}{\neg(p\lor\neg p),\ \neg(p\lor\neg p)\vdash\bot}\neg\vdash}{\neg(p\lor\neg p)\vdash\bot}\text{Con}}{\vdash\neg(p\lor\neg p)\Rightarrow\bot}\vdash\Rightarrow}{\vdash\neg\neg(p\lor\neg p)}\vdash\neg$$

5. Automatic Theorem Proving

Although Pi is intended for interactive theorem proving, simple facilities for automatic theorem proving can be of great help to the user by automatically performing trivial derivation construction tasks. At the time of this writing, such facilities have been designed but not yet implemented.

The automatic theorem proving facilities consist of two parts. The first part is the possibility of using incomplete derivations as derived inference rules. This is by itself not automatic theorem proving, since no search is involved. However, the used of derived rules can greatly speed up the derivation construction process.

The second part is a mechanism that takes a set of derived rules and attempts to apply each in turn to a sequent to be derived. If some derived rule is applicable, it is inserted into the derivation. The process is then repeated in a depth-first fashion on the premises of the newly inserted rule. This is an automatic theorem proving process, although quite simple.

For the purposes of assisting interactive work, we expect these facilities to be sufficient. They also have the advantage that it is easy to construct the sets of derived rules.

6. References

1. Eriksson, L.-H.: Finitary Partial Inductive Definitions and General Logic, Ph.D. thesis, Department of Computer and Systems Sciences, Royal Institute of Technology, Stockholm 1993.
2. Eriksson, L.-H.: Finitary Partial Inductive Definitions as a General Logic, In: Dyckhoff, R. (ed.), Extensions to Logic Programming – ELP 93, Springer Lecture Notes in Artificial Intelligence, 1994.
3. Hallnäs, L.: Partial Inductive Definitions, Theoretical Computer Science, vol. 87, no. 1, 1991.
4. Miller, D.: A Logic Programming Language with Lambda-Abstraction, Function Variables and Simple Unification, In: Schroeder-Heister, P. (ed.), Extensions of Logic Programming, Springer Lecture Notes in Artificial Intelligence 475, 1991.
5. Nadathur, G. and Pfenning, F.: Types in Higher-Order Logic Programming, In: Pfenning, F. (ed.), Types in Logic Programming, The MIT Press, 1992.
6. Ritchie, B.: The Design and Implementation of an Interactive Proof Editor, Ph.D. thesis CST-57-88, Department of Computer Science, University of Edinburgh, 1988.
7. Schroeder-Heister, P.: Rules of Definitional Reflection, Proceedings of the Eighth Annual IEEE Symposium of Logic in Computer Science (LICS), pp. 222 – 232, 1993.

Mollusc
A General Proof-Development Shell for Sequent-Based Logics

Bradley L. Richards 1, Ina Kraan 2, Alan Smaill* 3, and Geraint A. Wiggins** 4

1 Artificial Intelligence Laboratory, Swiss Federal Institute of Technology, Lausanne,
bradley@lia.di.epfl.ch
2 Department of Computer Science, University of Zurich, inak@ifi.unizh.ch
3 Department of Artificial Intelligence, University of Edinburgh,
geraint@aisb.ed.ac.uk
4 Department of Artificial Intelligence, University of Edinburgh,
smaill@aisb.ed.ac.uk

Abstract. This article describes an interactive proof development shell, *Mollusc* [Richards 93], which can be used to construct and edit proofs in sequent-based logics. Conceptually, *Mollusc* may be thought of as a logic-independent successor to *Oyster* [Bundy et al 90]. However, where *Oyster* was tied to a variant of Martin-Löf type theory, *Mollusc* can be used with any sequent-based logic for which a suitable definition is provided. Although developed in a research environment, *Mollusc* should also be suitable for use in classroom exercises. In addition to proof editing facilities, *Mollusc* supports the definition of new logics, includes a proof-planner interface, and provides for automated proof construction through a tactic language and interpreter.

1 Introduction

Mollusc is a general proof-development shell, which was designed as a logic-independent successor to *Oyster*. The *Oyster* system is an interactive proof checker for a variant of Martin-Löf type theory, developed as a rational reconstruction of Nuprl [Constable *et al* 86]. It is used principally to execute proof plans developed by the proof planner *CIAM* [Bundy *et al* 90]. Recently, there has been a strong interest in applying *CIAM* to a variety of different logics, and this created a need for proof-checking support in these logics. Rather than creating a new proof-checking system for each logic, we decided to create a single shell which would work with a variety of logics, and which would assist the user in defining new logics. In addition to checking proof plans, the shell would also be meant for interactive proof development; thus, it had to include a friendly user-interface with on-line help.

---

* supported by SERC grant GR/H/23610
** supported by ESPRIT Basic Research Project #6810, "Compulog II"

These requirements formed the specification of the *Mollusc* proof development shell, which has been developed and implemented over the past two years. *Mollusc* consists of two components: the proof development shell itself and a logic-definition utility.

The proof development shell allows the user to choose a logic to work in, and then supports the user in exploring, constructing, and editing proofs in that logic. *Mollusc* supports backward inference only, i.e., moving from goals toward axioms. It provides a library mechanism and a means of tracking dependencies among proofs and definitions, as well as a tactic mechanism for automating the construction of a proof. *Mollusc* also provides a generic interface to proof planners, and can directly execute proof plans represented as tactics. The tactic language is a superset of that used by *Oyster*, and includes tacticals such as *complete, try, repeat, then, or*, as used in the LCF system [Gordon et al 79].

The logic-definition utility allows the user to formally specify the syntax of a new logic, and automatically produces both a parser and a set of "access functions" which allow the user to compose and decompose logical expressions in a declarative manner. The parser and access functions can then be used to write inference rules for the logic. Unlike the approaches taken in, e.g., [Huet & Plotkin 91, Paulson 89], where a variety of logics are represented in a single meta-logic, *Mollusc* does not imlpement a meta-logic, and generates a distinct theorem-proving system for each logic. The uniform presentation, however, does allow efficient reuse of inference procedures.

Mollusc has been used to implement propositional logic, many-sorted first-order predicate logic with equality, Martin-Löf type theory [Martin-Löf 79], Edinburgh Logical Framework [Harper et al 87], and a decidability logic for logic program synthesis [Wiggins 92]. *Mollusc* has been used most extensively in many-sorted first-order logic, where it was used to execute proof plans generated by *CIAM* to verify synthesized logic programs [Kraan 94]. The proof trees for such verifications were up to 0.75 megabytes in size; the tactic execution times for proofs of that size were under 40 seconds of CPU time on a Sparc station 10 using Quintus Prolog Release 3.1.4.

The rest of this article is organized as follows. Section 2 describes the proof development shell itself. Section 3 discusses the logic-definition utility. Section 4 discusses extensions that are planned for future releases. Finally, Section 5 describes how to obtain a copy of *Mollusc*.

2 The *Mollusc* Proof Development Shell

Mollusc provides an interactive environment for creating and editing proofs. Users can develop proofs manually or use the methods *Mollusc* provides to partially or completely automate the proof. *Mollusc* supports both scripts and tactics. A script is a file containing a series of Mollusc commands; this can be useful, for example, to set up the Mollusc environment to work with a particular set of proofs and definitions. Tactics, on the other hand, are programs representing a particular sequence of proof steps to be followed.

Although a complete description of *Mollusc* is beyond the scope of this paper, the paragraphs below describe three of the most important capabilities it provides: library support, proof manipulation, and tactics.

Library support. *Mollusc* maintains a library for each logic. A library may be distributed; normally, for example, there will be one or more central, shared repositories of definitions and lemmas, plus each user's local library containing work in progress. A user's start-up script file lists the shared libraries which should be searched, plus the user's local library. *Mollusc* loads proofs or definitions from all libraries listed, but saves work in the user's own library.

A library contains directories for proofs, tactics, and each type of definition in the logic. These items are all manipulated using the same basic set of commands (e.g., *load, save*). In addition, the user can create a file defining the dependencies of items in the library. For example, if the definition of *plus* should only be loaded after the definition of natural numbers, this can be specified in the dependency file. *Mollusc* will then automatically load all required definitions in the proper order.

Proof manipulation. The ultimate purpose of *Mollusc* is to allow the user to interactively create, edit, and display proofs. *Mollusc* provides an extensive set of navigation and display commands, allowing the user to move about a proof tree and inspect it. The display format is defined in a user-alterable file, so that the output format can be tailored to a particular logic. New display commands can also be added; this is useful, for example, if the logic has some additional feature such as extract terms.

Inferences are carried out in three ways. First, the user may explicitly invoke an inference rule. *Mollusc* can advise the user on possible inferences if the inference rules are written declaratively. Second, the user can create and execute tactics, as described below. Finally, the user can set an "autotactic", a tactic that is applied after every successful proof step. This helps eliminate many of the tedious steps in a proof, such as well-formedness goals.

Tactics. Tactics are essentially programs that describe how to perform a particular set of inferences. The tactic language is a superset of the language used in *Oyster*. At one end of the spectrum, the user might write a tactic that performs some common sequence of proof steps. At the other extreme, a tactic may include all steps necessary to completely recreate a proof. The user has control over the "grafting" of tactics. When a tactic is executed, all proof steps are normally represented explicitly in the proof tree, just as though they had been done manually. If a tactic is "grafted", then all intermediate steps are removed from the proof tree, and the tactic execution looks like a monolithic inference rule. Grafting can substantially shorten and simplify the proof tree.

Tactics may be created in three ways. First, of course, the user may write the tactic manually. Second, a tactic may be extracted from an existing proof or portion of a proof. Finally, *Mollusc* provides an interface to proof planners (e.g., $C\mathcal{I}^AM$ [Bundy et al 90]), so that a proof planner can create a tactic for *Mollusc* to execute.

3 The Logic Definition Utility

When starting *Mollusc*, the first thing the user does is choose a logic to work in. *Mollusc* then loads a parser, a set of inference rules, a substitution algorithm, and a pretty printer for sequents in the logic. When a user wants to create a new logic, all of these must be created. While they can be written manually, much of the work is tedious and repetitive. To make the process of creating a logic easier, *Mollusc* provides standard templates for some files (e.g., the pretty printer) and a *define_logic* utility to help create a parser.

The *define_logic* utility allows the user to specify the syntax of the logic in a BNF-like language. The utility uses this specification to create a parser for the logic. It can also create a set of "access functions" which compose and decompose terms in the logic. These access functions make the process of writing declarative inference rules considerably easier.

4 Limitations and Planned Extensions

As with any new system, there are a number of areas where *Mollusc* could be enhanced. The major improvements being considered are:

- **Complete logic independence.** *Mollusc* currently assumes that target logics are sequent-based. This assumption could be lifted, making the shell completely logic independent. The principal difficulty is that *Mollusc* currently manipulates and displays hypothesis lists directly. In a completely logic-independent shell, utility predicates for these functions would have to be provided by each external logic definition.
- **Custom input parser.** *Mollusc* currently allows Prolog to parse all input items. This means that syntax errors result in uninformative messages. It also requires the user to end all entries with a period, and enclose descriptive items in single quotation marks. Providing *Mollusc* with its own input parser would eliminate these problems.
- **Custom display predicates for proof trees.** Currently, the logic designer can create custom display predicates for sequents, but not for the entire proof tree.
- **Automatic dependency determination.** In principle, *Mollusc* should determine dependencies automatically, by observing when definitions are used in the course of a proof. Adding this capability would simplify or even eliminate the manually constructed dependency file.

5 Obtaining *Mollusc*

Mollusc is available from the Mathematical Reasoning Group at the Department of Artificial Intelligence, University of Edinburgh. The contact person is Alan Smaill, smaill@aisb.ed.ac.uk.

Mollusc is written in Quintus Prolog, and runs under versions 3.1 or later. While it has not been ported to other Prolog implementations, this should not be difficult as long as the target Prolog supports modules. As of this writing, one user is porting *Mollusc* to SICSTUS Prolog.

References

[Bundy et al 90] A. Bundy, F. van Harmelen, C. Horn, and A. Smaill. The Oyster-Clam system. In M.E. Stickel, editor, *10th International Conference on Automated Deduction*, pages 647–648. Springer-Verlag, 1990. Lecture Notes in Artificial Intelligence No. 449. Also available from Edinburgh as DAI Research Paper 507.

[Constable et al 86] R.L. Constable, S.F. Allen, H.M. Bromley, et al. *Implementing Mathematics with the Nuprl Proof Development System*. Prentice Hall, 1986.

[Gordon et al 79] M.J. Gordon, A.J. Milner, and C.P. Wadsworth. *Edinburgh LCF - A mechanised logic of computation*, volume 78 of *Lecture Notes in Computer Science*. Springer Verlag, 1979.

[Harper et al 87] R. Harper, F. Honsell, and G. Plotkin. A framework for defining logics. In *Proc. of the Second Symposium on Logic in Computer Science*, 1987.

[Huet & Plotkin 91] G. Huet and G.D. Plotkin. *Logical Frameworks*. CUP, 1991.

[Kraan 94] I. Kraan. *Proof Planning for Logic Program Synthesis*. Unpublished PhD thesis, Department of Artificial Intelligence, University of Edinburgh, 1994. Submitted February 1994.

[Martin-Löf 79] Per Martin-Löf. Constructive mathematics and computer programming. In *6th International Congress for Logic, Methodology and Philosophy of Science*, pages 153–175, Hanover, August 1979. Published by North Holland, Amsterdam. 1982.

[Paulson 89] L. Paulson. The foundation of a generic theorem prover. *Journal of Automated Reasoning*, 5:363–397, 1989.

[Richards 93] B. L. Richards. Mollusc user's guide version 1.1. DAI Technical paper 23, University of Edinburgh, September 1993.

[Wiggins 92] G. A. Wiggins. Synthesis and transformation of logic programs in the Whelk proof development system. In K. R. Apt, editor, *Proceedings of JICSLP-92*, pages 351–368. M.I.T. Press, Cambridge, MA, 1992.

KITP-93: An Automated Inference System for Program Analysis

T. C. Wang and Allen Goldberg

Kestrel Institute, 3260 Hillview Avenue, Palo Alto CA 94304

KITP is an automated inference system developed for supporting the formal design, verification, and validation of computer programs. It has evolved from an automated verification system, RVF [5]. The latest version of KITP, KITP-93, features a typed formulation and deduction, meta-level reasoning, and resolution-based proving enhanced by conditional term-rewriting. This report describes KITP-93 from a user's perspective.

1 Logical Framework

KITP-93 employs a classical higher-order language at the user level, and a typed clausal language for the inference. It accepts, in general, any well-formed higher-order formula. Internally, all input axioms and lemmas are transformed into clausal normal form, and stored as rules in the knowledge-base (KB) of the system. Only these KB rules are directly accessed by the deductive components of the system. For example, if the user inputs the following statement,

1. $\forall(s)(valid\text{-}key(s) \Rightarrow (stringp(s) \land \forall(k:char)(k \in s \Rightarrow (k \geq \#2 \land k \leq \#7))))$,

the system will transform the statement into three KB rules ($s : seq(char)$ in 1a is transformed from $stringp(s)$ by rewriting):

1a. $(s : seq(char))/valid\text{-}key(s)/true$
1b. $(\#2 \leq k) \lor \neg(k \in s) \lor \neg valid\text{-}key(s))/(k : char)$
1c. $(k \leq \#7) \lor \neg(k \in s) \lor \neg valid\text{-}key(s))/(k : char)$

Formulas in clausal normal forms are divided into two classes: typing rules and typed clauses. A typing rule has a form $t: \tau/H/R$. It specifies that a term t has a type τ if H and R hold. $t:\tau$ is called a type description (TD). R is a conjunction of TDs. H is a conjunction of ordinary literals. For example, a rule $(x:nat)/(x \geq 0)/(x:int)$ states that x is of type nat if x is of type int and $x \geq 0$ holds. A *typed clause* has a form K/R, where K is an ordinary clause and R is a conjunction of TDs, which means that K holds under R. 1a is a typing rule; 1b and 1c are typed clauses.

The inference supported by this framework is predicate calculus with equality and a restricted form of higher-order reasoning. It employs two kinds of resolution. One is kernel resolution, which is resolution of two typed clauses K_1/R_1 and K_2/R_2 upon literals of K_1 and K_2. The other is TP-resolution, which is resolution made by unifying a TD of a typed clause (e.g., a member of R_1 in K_1/R_1) and the head of a typing rule (e.g., $t:\tau$ in $t:\tau/H/R$). As will be described later, kernel resolution is used to replace ordinary resolution; TP-resolution is used for type-checking and TP-deduction.

The ability to perform higher-order reasoning is due to a special feature of the clausal language: predicate and function symbols can be denoted by higher-order variables. This feature is similar to the use of meta variables in Bundy's meta-level inference system [1]. For instance, the variable f contained in KB rule 12 (see next page) denotes an arbitrary unary function. With this and some other KB rules, KITP-93 proves the conjecture,

$$\forall (f: map(\alpha, \beta), a: set(\alpha), b: set(\alpha))(image(f, a \cup b) = image(f, a) \cup image(f, b)).$$

Even through KITP-93 uses a typed higher-order language, it employs an ordinary Skolemization algorithm and a form of first-order unification in deduction. Its ability in general higher-order theorem proving has not been well developed. Its logical framework is essentially a predicate calculus expanded with specialized notions and inference rules for type relations. The purpose of this specialization is for improving the efficiency of inference and the clarity of representation. There is no requirement that every variable must be typed. In particular, this framework assumes the set of standard (untyped) equality axioms, and uses (untyped) paramodulation for equality-oriented deduction.

2 Knowledge Base

KITP-93 contains a large KB (300+ rules), which is built on data type theory for integers, reals, characters, strings, sets, sequences, tuples and maps. Natural number is introduced as a subtype (*nat*) of integer (*int*). A difficult problem in proving theorems in a real programming environment is the possible search explosion caused by reasoning about large KBs. To attack this problem, we have developed a KB management mechanism to control the use of KB rules. KB rules are classified into different rule types. Among them, a *typing-rule* rule will be used only for type-checking and TP-deduction, a *reduction* rule only for term-rewriting, a *forward-implication* rule only by the forward inference procedure (FIP), and a *backward-implication* rule only by the backward inference procedure (BIP). An *any-rule* rule can be used by both FIP and BIP. A *manual-rule* rule will not be used unless it is explicitly invoked by a particular inference task. The rule type of a KB rule is assigned by the system based on both the user's annotation on its parent formula, and its syntactic feature. For example, if the user annotates the formula 1, given earlier, with *forward-implication,* then 1b and 1c are *forward-implication* rules, and 1a is a *typing-rule* rule. Many KB rules are classified as *reduction* rules in order to promote the use of term-rewriting. The user may modify the KB or define a new KB. The following are some examples of KB rules. (Those without annotations are *any-rule* rules by default).

2. $(x: nat)/(x \geq 0)/(x: int)$ [*typing-rule*]
3. $(x: int)/true/(x: nat)$ [*typing-rule*]
4. $(x \geq 0)/(x: nat)$
5. $stringp(x) = (x: seq(char))$ [*reduction*]
6. $(x: string) = (x: seq(char))$ [*reduction*]

7. $(conc(x,y): seq(char))/true/(x: seq(char))(y: seq(char))$ [*typing-rule*]
8. $(x \in conc(p_1, p_2)) \vee \neg(x \in p_1)/(x: char)(p_1: seq(char))(p_2: seq(char))$
9. $(x \in conc(p_1, p_2)) \vee \neg(x \in p_2)/(x: char)(p_1: seq(char))(p_2: seq(char))$
10. $(x \in p_1) \vee (x \in p_2) \vee \neg(x \in conc(p_1, p_2))/(x: char)(p_1: seq(char))(p_2: seq(char))$
11. $(y: \alpha)/x = y/(x: \alpha)$ [*manual-rule*]
12. $image(f, s \text{ with } e) = image(f, s) \text{ with } f(e)/(f: map(\alpha, \beta))(s: set(\alpha))(e: \alpha)$

3 Proof Objects

KITP-93 provides inference service through an interface construct called proof-object. A proof-object is a record of classified information about an inference task. The basic usage of a proof-object is to introduce a proof-obligation to the prover. But it can also be used to introduce user-directions and to help incremental development of proofs. A proof-object produced by a batch-mode verification/analysis procedure usually contains no other information except a conjecture to be proved. However, the proof-obligation given in a simple proof-object may not always be discharged automatically and efficiently. Sometimes, a more complex proof-object needs to be used to guide the prover. Such a proof-object can be created by the user from a scratch or by editing a proof-object produced by an application procedure. For instance, in order to assist in proving a conjecture $h_1 \wedge ... \wedge h_n \Rightarrow C$, one can decompose the hypothesis into a set of proof-rules, $h_1, ..., h_n$, and annotate each of them with a specific rule type and constraints. Other information that can be provided by a proof-object includes local bindings (constants with a lexical scope of the object body), KB rules to be disabled or enabled, additional lemmas, hypotheses, and conjectures, values of control parameters, and proof-strategies, etc. The following figure contains an example of proof-object.

Proof valid-key-prop
  local-bindings {valid-key}
  proof-rules   valid-key-def-1 : *forward-implication* (definition)
      $\forall(s)(valid\text{-}key(s) \Rightarrow stringp(s) \wedge \forall(k: char)(k \text{ in } s \Rightarrow (k \geq \#2 \wedge k \leq \#7)))$
    valid-key-def-1 : *backward-implication* (definition)
      $\forall(s)(stringp(s) \wedge \forall(k: char)(k \text{ in } s \Rightarrow (k \geq \#2 \wedge ch \leq \#7)) \Rightarrow valid\text{-}key(s))$
  proof-conjectures
    conjecture   valid-key-prop [*any-rule*]
      $\forall(a) \wedge \forall(b)(valid\text{-}key(a) \wedge valid\text{-}key(b) \Rightarrow valid\text{-}key(conc(a, b)))$
    forward-step-limit {1}
    backward-step-limit {8}
    max-induced-vars {1}
    end-conjecture

---

4 Typed Deduction with Conditional Term Rewriting

The KITP-93 prover incorporates natural deduction, term-rewriting, partial evaluation, unit resolution and paramodulation, set of support strategy, and hierarchical deduction. Except for the use of typed deduction, the basic architecture of this inference engine is similar to the prover documented in [5]. Here we discuss some of the advanced features recently added to the prover, namely goal-oriented deduction enhanced by conditional term rewriting, type-checking, and TP-deduction.

Two models of term rewriting are defined. The basic mode of term rewriting uses existing facts to verify the conditions of rewriting rules. It is applied exhaustively to each input and all formulas generated by other inference procedures of the prover. The advanced mode is employed by BIP. For BIP, if the (instantiated) condition of a rewriting rule, which is applied to the current goal G, can not be established immediately, then a resolvent will be produced by combining the instantiated condition and the result of rewriting G. Thus, the condition of a rewriting rule can be handled similarly to the subgoals inherited from a backward-implication rule. Consider applying a rewriting rule $p \Rightarrow if(p,q,r) = q/(p\colon boolean)$ to a goal, $\neg evenp(if(n = 0, ans, foo(n-1, n \times ans))) \vee G_1$. If the (instantiated) condition p (i.e., $n = 0$) is not contained in the current KB, then a candidate goal $\neg evenp(ans) \vee n \neq 0 \vee G_1$ will be produced.

The type restriction attached to a rewriting rule is verified by the type-checking procedure. Given a type restriction R, the procedure will try to prove the consistency of R with the entire set T of typing rules contained in the KB. For example, to apply a rewrite rule $(x < y) = (x + 1 \leq y)/(x\colon int)(y\colon int)$ to an expression $t_1 < t_2$, the expression will not be rewritten into $t_1 + 1 \leq t_2$ unless $T \models (t_1 : int) \wedge (t_2 : int)$ has been proved (the proof will fail if, for example, the operator $<$ is overloaded, and t_1 or t_2 is of type $real$).

BIP is based on typed hierarchical deduction, which differs from the untyped one described in [4] mainly in two aspects. First, the typed deduction uses kernel resolution to carry out hierarchical deduction. Let K/R be the current goal and assume that it is not a kernel-empty clause (i.e. $K \neq box$). Then the subgoal to be focused on must be a literal chosen from K, and the rules to be used must chosen from the set of typed clauses, but not from T. Moreover, for each resolvent K'/R' produced by kernel resolution, type-checking is applied immediately to the type restriction R' in order to determine if R' is satisfiable in T (in the sense that the result of the TP-deduction procedure described below is non-empty); and is discarded if it is not.

Second, if a goal is a kernel-empty clause box/R, then the TP-deduction procedure will be applied to R. This procedure plays a role, in essential, similar to Milner's type inference algorithm [3]. It deduces from R and T a set of relations by a sequence of TP-resolution. Each relation is a form $H \wedge Q$, where H is a conjunction of ordinary literals and Q is a conjunction of variable TDs (such as $(x\colon nat) \wedge (y\colon \alpha)$). For example, assuming that R is $((a + x)\colon nat)$, and T contains $(a\colon int)$, $(x + y\colon int)/true/(x\colon int)(y\colon int)$, and typing rules 2 and 3 given earlier, then the procedure will deduce a condition $a + x \geq 0 \wedge (x\colon int)$. From this condition, the BIP will deduce a resolvent $\neg(a + x \geq 0)/(x\colon int)$.

TP-deduction is implemented by a variant of SLD-resolution extended to abductive reasoning. Type-checking for both rewriting and BIP is implemented by a simplified version of TP-deduction. A reasonable restriction to the structure of the underlying typing theory has been made, which guarantees the finite termination and an efficient implementation of these procedures. The results are cached for reuse. The deterministic nature (and the efficiency) of these special inference procedures helps improve the performance of the prover. For example, proving the conjecture *valid-key-prop*, the prover produced a kernel-empty resolvent (labeled) -21. By applying the TP-deduction procedure to this resolvent, the BIP produced immediately one and only one resolvent (labeled) -22. ($k!$ is a Skolem symbol.)

$-21.\ box/(k!(conc(a,b)): char)(a: seq(char))(b: seq(char))(conc(a,b): seq(char))$
$-22: valid\text{-}key(conc(a,b)) \lor \neg valid\text{-}key(b) \lor \neg valid\text{-}key(a)/true$

However, to deduce this resolvent from -21 with untyped deduction, the prover must take at least three steps and use three distinct rule clauses (typing rules): $1a$, 7, and one from the proof-rule valid-key-def-1.

5 Conclusion

Our goal is to produce a powerful inference system capable of dealing with a large number of KB rules, and conjectures with diverse features. To achieve this goal, we have built KITP-93 with a logical framework that allows convenient user interaction and easy incorporation of existing inference techniques. We have developed a management mechanism for supporting controlled use of KB rules, high-level user interaction, and incremental development of proofs. We have designed an inference engine by incorporating a variety of efficient inference techniques, and emphasizing the role of term-rewriting, goal-oriented deduction, and decision procedures, as well as interactive proof utilities. KITP-93 has been incorporated as an inference server by a number of formal environments. Significantly, it has been used successfully by a large industrial user in the control flow analysis of Ada procedures. A review of the use of KITP in solving real world problems is included in [2]. Besides proving theorems, other inference services that KITP-93 provides include disproving a non-theorem, simplifying program fragments, and deducing antecedents.

References

1. Bundy, A., Sterling, L.: Meta-level inference: Two applications. *J. Automatic Reasoning,* **4**(1) (1988) 15-18
2. Jüllig, R. K.: Applying formal software synthesis. *IEEE Software,* **10**(3) (1993) 11-22
3. Milner, R.: A theory of type polymorphism in programming. *J. Comput. System Science,* **17** (1978) 348-375
4. Wang, T. C., Bledsoe, W. W.: Hierarchical Deduction. *J. Automatic Reasoning,* **3**(1) (1987) 35-71
5. Wang, T. C., Goldberg, A.: RVF: an automated formal verification system. *Proceedings CADE-11* (ed. D. Kapur), LNCS **607** (1992) 735-739

SPIKE: a system for sufficient completeness and parameterized inductive proofs

ADEL BOUHOULA

CRIN & INRIA-Lorraine
BP 239, 54506 Vandoeuvre-lès-Nancy, France
email: bouhoula@loria.fr

Abstract. The system SPIKE is an automated system for theorem proving in theories presented by a set of Horn clauses with equality. SPIKE is written in Caml Light and contains more than 10000 lines of code. It runs on SUN4 workstations under Unix with a graphical user-friendly interface in Xwindow system realized in Tk. It has been designed to provide users with facilities to direct and monitor proofs easily. The main novelty is the use of a new inference rule, which permit us to prove and disprove automatically inductive properties in parameterized conditional specifications [Bou93]. The motivation for this is that theorem proving in parameterized specifications allows for shorter and more structured proofs. Moreover, a generic proof can be given only once and reused for each instantiation of the parameters. Our procedure also extends our previous work [BR93a] to non-free constructors. Based on computer experiments, the method appears to be more practical and efficient than inductive proofs in non-parameterized specifications. We have also implemented a new procedure for testing sufficient completeness for parameterized conditional specifications [Bou93]. Moreover, SPIKE offers facilities to check and complete definitions.

1 Related systems

In recent years, many tools using inductive reasoning have been developed. The first method applies explicit induction arguments on the structure of terms. NQTHM [BM79] was developed on this principle. It was for many years the only significant automated theorem proving system for induction. However NQTHM requires interaction with the user even for simple proof tasks. Many of the heuristics in NQTHM have been rationally reconstructed in the prover CLAM [Bun90]. RRL [ZKK88] is another theorem proving system that supports a *cover set* method which is closely related to Boyer and Moore's approach. Although this system is efficient, no procedure is known for deriving covering sets. A second method involves a proof by consistency [Mus80]. It is generally accepted that such techniques may be very inefficient since the completion procedure often diverges. Recently, new methods have been proposed that do not rely on completion [KR90, Red90, BKR92a, BR93a]. The system SPIKE has been developed in this framework. It incorporates many optimizations such as powerful simplification techniques. SPIKE has proved several interesting theorems in a completely automatic way, that is, without interaction with the user and without ad-hoc heuristics. It has also proved the challenging Gilbreath card trick with only 2 easy lemmas which are given in the beginning of the proof. This example was treated by B. Boyer in NQTHM and H. Zhang in RRL. Unlike us, they require a lot of lemmas, some of them being non-obvious.

To our knowledge, SPIKE is the only one that can disprove non-trivial inductive theorems in conditional theories without any interaction. Note also that NQTHM, CLAM and

RRL were not designed to refute false conjectures. For an inexperienced user, a serious weakness of the NQTHM, CLAM and RRL systems is that it does not provide much useful information when it fails. In particular, it is not clear from the generated output whether the conjecture being proved is false or a proof of the conjecture is likely to need additional lemmas. Unlike the latter, SPIKE garantes when it fails that one of the initial conjectures is not an inductive theorem provided that the axioms are boolean and ground convergent with completely defined functions.

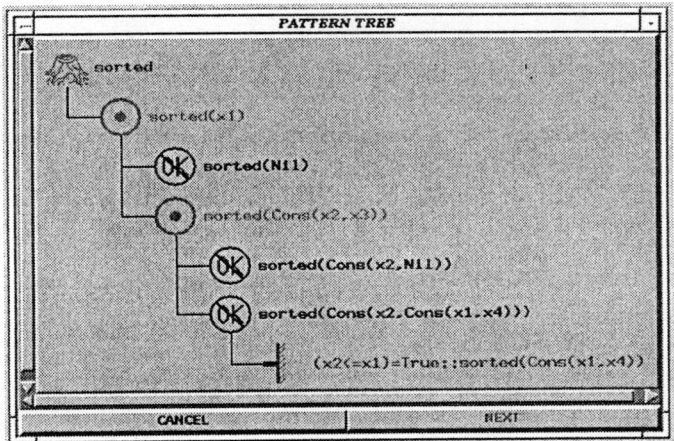

Fig. 1. The function *sorted* is not sufficiently complete

2 Parameterized Conditional Specifications

A parameterized conditional specification is a pair $PS = (PAR, BODY)$ of specifications: $PAR = (\Sigma_{PAR}, E_{PAR})$ and $BODY = (\Sigma_{BODY}, E_{BODY})$, where E_{PAR} is the set of parameter constraints consisting of equational clauses in Σ_{PAR}, and E_{BODY} is the set of axioms of the parameterized specifications, which are conditional rules over $\Sigma = \Sigma_{PAR} \cup \Sigma_{BODY}$; where Σ_{PAR} and Σ_{BODY} are signatures. A signature $\Sigma = (S, F)$ consists of a set S of sort symbols and a set F of function symbols with arities in S^*. We assume that we have a partition of F_{BODY} in two subsets, the first one, C_{BODY}, contains the constructor symbols and the second, D_{BODY}, is the set of defined operators.

Example 1. Consider the following parameterized specification: $S_{PAR} = \{bool, elem\}$, $F_{PAR} = \{true :\rightarrow bool, false :\rightarrow bool, \leq: elem \times elem \rightarrow bool\}$, E_{PAR} contains the following *constraints*: $\{true \neq false, x \leq x = true, x \leq y = true \vee x \leq y = false, x \leq y = true \vee y \leq x = true, x \leq y = false \vee y \leq z = false \vee x \leq z = true\}$. $S = S_{PAR} \cup S_{BODY}$ where $S_{BODY} = \{nat, list\}$, $F = F_{PAR} \cup C_{BODY} \cup D_{BODY}$ where $C_{BODY} = \{0 :\rightarrow nat, s : nat \rightarrow nat, nil :\rightarrow list, cons : elem \times list \rightarrow list\}$ and $D_{BODY} = \{length : list \rightarrow nat, sorted : list \rightarrow bool, insert : elem \times list \rightarrow list, isort : list \rightarrow list\}$. Suppose that

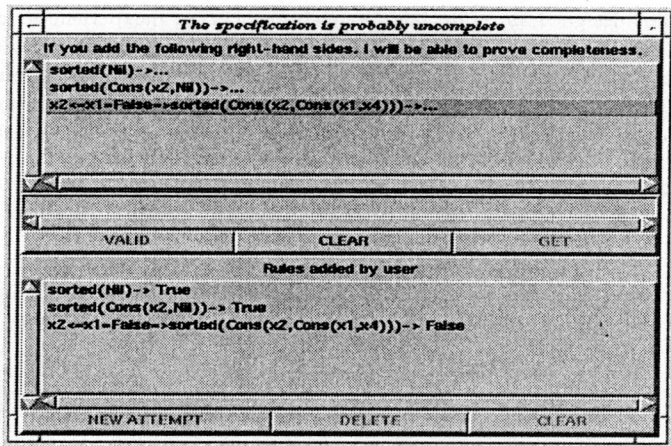

Fig. 2. Suggestions

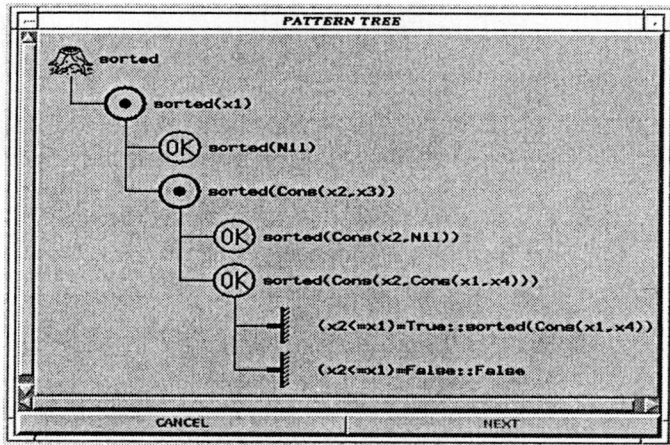

Fig. 3. The function *sorted* is now sufficiently complete

length, insert and *isort* (sort by insertion) are defined as usual and let the predicate *sorted* defined by the rule: $x \leq y = True \Rightarrow sorted(cons(x, cons(y, z))) \rightarrow sorted(cons(y, z))$.

3 Sufficient Completeness

SPIKE checks automatically if an operator f in a specification is sufficiently complete. The program builds a pattern tree for f. The leaves of the tree give a partition of the possible arguments for f. If all the leaves are reducible, the answer is affirmative [Bou93]. If one of the leaves is not reducible then SPIKE suggests a new rule for completing the specification. This rule is not entirely determined but rather a possible schema for it is proposed, namely: *(condition, left-hand-side)*. Once the user has chosen the new rule, usually by simply giving its right-hand side, SPIKE replays the test. Note that *sorted* in example 1 is not sufficiently complete. Here we describe a session with SPIKE to give an idea about the interaction with the user if the specification is not sufficiently complete (see figure 1 and figure 2). We therefore add three rules and try again (see figure 3).

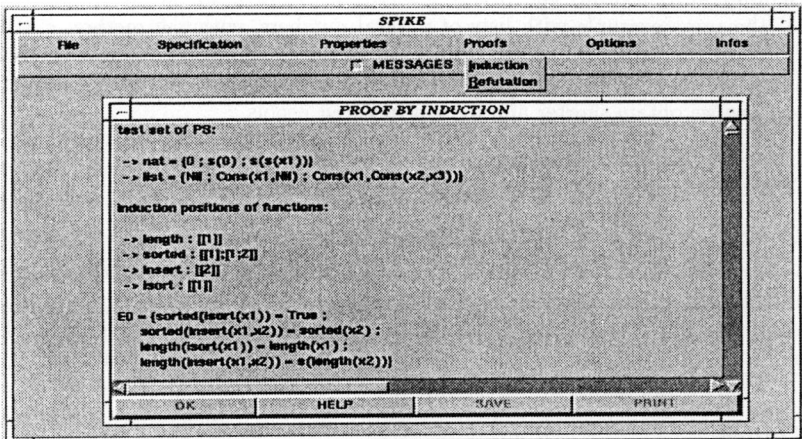

Fig. 4. Example

4 Implicit Induction

SPIKE *first* instantiates conjectures by induction schemes called *test sets* and *second* simplifies them by axioms, other conjectures or induction hypotheses. Every cycle generates new lemmas that are processed in the same way as the initial conjectures. The main characteristics of our method are: the well-founded ordering on which our induction is based is exactly the termination ordering used to orient the axioms into rules; it gives automatically induction schemes through algorithms for computing test sets; conjectures are processed in a non-hierarchical list, new lemmas to be proved are simply added to this list; using conjectures for mutual simplification simulates *simultaneous induction*. We can also handle non-orientable equations by new simplification techniques and therefore, we avoid divergence on many problems which defeat other systems or which need much interaction with the user. Finally, note that our strategy is refutationally complete for a class of rewrite systems that can specify numerous interesting examples. This class contains the boolean ground convergent rewrite systems with completely defined functions. In other words, with our procedure every false conjecture will be disproved in finite time.

Our implementation is based on the method presented in [Bou93]. Here is an overview of the algorithm. The main data structures are: the list E_{BODY} of axioms, that are conditional rules built with the constructor discipline, the list E of conjectures (clauses) to be checked, the list E_{PAR} of parameter constraints, that are equational clauses in Σ_{PAR} and finally, the set H of induction hypotheses (initialized by $\emptyset$). The first step in a proof session is to check if all defined functions are completely defined. The second step is to compute test set for PS and also induction positions. After these preliminary tasks, the proof is started.

Consider example 1 with E_0 the set of conjectures to be proved (see figure 4). SPIKE can prove these conjectures in a completely automatic way, using 104 steps. Note that a single lemma (automatically generated) is sufficient to prove the initial conjectures. Now consider the same example with lists of natural numbers, with the method of [BR93a], SPIKE proves the same conjectures using 246 steps. In addition, 40 lemmas are generated automatically during the proof. This example illustrates that with parameterized specifications we obtain shorter and more structured proofs.

5 Conclusion

We have briefly described the main functionalities of our prover SPIKE. This system has been successfully tested with difficult problems [Bou94]. The proofs are obtained with a minimal interaction with the user thanks to the mutual simplification of the conjectures and the application of new simplification rules involving non-orientable equations. Experiments illustrate that parameterized specifications allows shorter and structured proofs.

References

[Bun90]   A. Bundy, F. van Harmelen, C. Horn and A. Smaill. The Oyster-Clam system. 10th International Conference on Automated Deduction. LNAI No. 44. pages 647–648. 1990.

[BKR92a] A. Bouhoula, E. Kounalis and M. Rusinowitch. Automated mathematical induction. Technical Report 1636, INRIA, 1992. Submitted.

[BR93a]   A. Bouhoula and M. Rusinowitch. Automatic Case Analysis in Proof by Induction. In *Proceedings of the 13th International Joint Conference on Artificial Intelligence*, volume 1, page 88–94. Chambéry France, 1993.

[Bou93]   A. Bouhoula. Parameterized Specifications: Sufficient Completeness and Implicit Induction. Technical Report 2129, INRIA, 1993. Submitted.

[Bou94]   A. Bouhoula. *Preuves Automatiques par Récurrence dans les Théories Conditionnelles*. PhD thesis, Thèse de l'université de Nancy I, 1994.

[BM79]   R. S. Boyer and J. S. Moore. *A Computational Logic*. Academic Press, New York, 1979.

[KR90]   E. Kounalis and M. Rusinowitch. Mechanizing inductive reasoning. In *Proceedings of the AAAI Conference, Boston*, pages 240–245. July 1990.

[Mus80]   D. R. Musser. On proving inductive properties of abstract data types. In *Proceedings 7th ACM Symp. on Principles of Programming Languages*, pages 154–162. Association for Computing Machinery, 1980.

[Red90]   U. S. Reddy. Term rewriting induction. In M. E. Stickel, editor, *Proceedings 10th ICADE, Kaiserslautern (Germany)*, volume 449 of *LNCS*, pages 162–177. Springer-Verlag, 1990.

[ZKK88]   H. Zhang, D. Kapur, and M. S. Krishnamoorthy. A mechanizable induction principle for equational specifications. In *Proceedings 9th ICADE. Argonne (Ill., USA)*, volume 310 of *LNCS*, pages 162–181. Springer-Verlag, 1988.

Distributed Theorem Proving by *Peers*

Maria Paola Bonacina[1]* and William W. McCune[2]

[1] Department of Computer Science
University of Iowa
Iowa City, IA 52242-1419, USA
bonacina@cs.uiowa.edu

[2] Mathematics and Computer Science Division
Argonne National Laboratory
Argonne, IL 60439-4801, USA
mccune@mcs.anl.gov

1 Overview

Peers is a prototype for parallel theorem proving in a distributed environment. It implements a number of strategies for refutational, contraction-based theorem proving in equational theories, whose signatures may include associative-commutative (AC) function symbols. "Contraction-based" strategies is a more general term for "simplification-based", "rewriting-based" or "(Knuth-Bendix) completion-based" strategies.

Such strategies are parallelized in Peers according to the *Clause-Diffusion* methodology for distributed deduction [2, 4]. This is a *coarse-grain* approach to parallel theorem proving, where *concurrent, asynchronous deduction processes* work cooperatively to solve the given theorem proving problem. In Clause-Diffusion, the deduction processes are rather independent and *loosely-coupled*: each process executes a theorem proving strategy, has its own data base of clauses and develops its own derivation. The processes are all "peers" (hence the name of the prototype) in the sense that there is no master-slave type of organization between them and there are no specialized processes, e.g. a process that performs only normalization of terms. In Clause-Diffusion, the processes may execute the same strategy or different strategies. In the current version of Peers, all processes execute the same strategy.

The work is subdivided among the processes by dynamically partitioning the data base of clauses. Whenever a clause is generated, a *distributed allocation algorithm* is used to decide which process is its owner. Each process executes the allocation algorithm for every clause it generates: all allocation decisions are taken locally. A partition of the clauses induces a subdivision of the inferences by stipulating that each process is responsible for the inference steps on its clauses, called its *residents*. This "data-driven" distribution of the theorem proving task serves two purposes. The first one is to subdivide the inferences, in such a way that each parallel process has less work than a single sequential process would. The second one is to balance the work-load of the parallel processes. As a consequence of this partition, each process has direct access to just a subset of the data base. Since finding a proof requires in general global knowledge, the processes need to communicate clauses to each other by *message-passing*. Thus, a derivation by a process includes both inference steps and communication steps and the collection of these derivations forms the *distributed derivation*. As soon as one of the processes finds a proof, the distributed derivation halts successfully.

The Clause-Diffusion methodology does not assume a specific parallel architecture. Given the coarse-grain type of parallelism in Clause-Diffusion, a purely distributed environment, such as a network of computers or a multiprocessor with distributed memory, is a natural choice. However, Clause-Diffusion may also take advantage of a shared memory, by implementing send/receive of messages as read/write in a shared part of the memory. In Peers, the creation of parallel processes and the communication are realized by using the p4 library of functions for parallel programming [6]. Since p4 works on a variety of parallel machines, neither the method (Clause-Diffusion) nor the software (p4 and C) pose significant architecture-related restrictions. Indeed, we developed Peers on a network of workstations, but we also ported it to a

---

* Supported in part by the GE Foundation Faculty Fellowship to the University of Iowa.

shared-memory Sequent. Networks of workstations remain our basic target environment, because they are widely available. In such clusters, each workstation hosts a deductive process and the deductive processes communicate via message-passing over the network.

Peers is one of the few coarse-grain, contraction-based, parallel theorem provers written so far (two other such provers are e.g. [1, 5]). The current version is for equational theories only, but Peers can be extended to larger logics without modifying its basic architecture. To our knowledge, Peers is the first parallel prover to feature built-in treatment of AC operators. Peers is small, portable and easy to use: it is not interactive, but the user can drive the behaviour of the prover by setting *flags* (boolean options) and *parameters* (integer options) in the input file. We refer to [2, 4] for more details on Clause-Diffusion. In the following, we describe the design of Peers, some of its strategies, and we discuss a selection of experiments.

2 The Structure of Peers

The Peers program expects an input file, which contains the input clauses and options. One of the processes, $peer\_0$, reads the input file, pre-processes the input clauses according to the strategy and broadcasts the input clauses, the symbol table for the input signature, the options and the same initialization information to all the processes. After this initialization phase, all "peers" become active and the distributed derivation starts.

During the distributed computation, each deduction process may be in one of three states: answering a message from another process, performing local work, i.e. inferences in its local data base, or being idle. Each peer executes a basic loop, called the *work-loop*, where it tests for pending messages, selects the pending message of highest priority and serves it by executing the appropriate action. If no message is pending, the process executes a unit of local work. If no such work is available, the process becomes idle and remains idle until it receives a message. After processing a message or a unit of local work, the peer repeats the selection. It is necessary to break the local work into units, because a theorem proving process may generate an infinite derivation by performing local inferences only, and thus it would indefinitely postpone handling the messages. Also, all messages from the outside have higher priority than local work, on the ground that communication is often the bottleneck in distributed computation and thus should have the highest priority.

If a peer finds a proof or raises an exception, e.g. running out of memory, it broadcasts a *halt message* and halts. All other peers will halt upon receiving a *halt message*. The Clause-Diffusion methodology is fault-tolerant: if a deductive process runs out of memory and halts, the other processes may continue without hindrance for the soundness and completeness of the derivation [2, 5]. The choice of halting all peers when one raises an exception was motivated by the intent of keeping this prototype simple. In addition to successful termination and exceptional termination, Peers terminates if all the processes are idle and all messages ever sent have been received. In order to detect this condition, Peers implements the *Dijkstra-Pnueli global termination detection algorithm*: this is a distributed algorithm that circulates a token, carrying status information, among the processes, and establishes termination if the token has circulated at least twice without detecting any activity (see e.g. [9], pages 48–49 and 182–185, for a definition and proof of correctness of this algorithm).

Given this organization, specific Clause-Diffusion strategies can be implemented by determining the types of work (messages and units of local work), with their priorities and corresponding actions.

3 The "Types of Work"

In the current version of Peers, the main types of work (e.g. excluding the messages used only during the initialization phase and the messages for termination) are two types of messages, *new settlers* and *inference messages*, and a type for the unit of local work, called *expansion work* from *expansion inference rules*, e.g. paramodulation and resolution.

New settlers and *inference messages* are the two basic types of messages in the Clause-Diffusion methodology. The purpose of *new settlers* is to subdivide the work among the processes by dynamically distributing the clauses. Whenever a new clause c is generated by a process, say $peer_i$, c is subject first to *forward contraction*, that is contraction, e.g. simplification and subsumption, with respect to the existing clauses in the data base of $peer_i$. If c is not deleted, $peer_i$ executes the allocation algorithm to decide which process c should be a resident of. If the result is that c belongs to $peer_i$, c is used for *backward contraction*, that is, c is applied to contract the existing clauses in the data base of $peer_i$. Then it is stored in the data base as a resident of $peer_i$. If the allocation algorithm decides that c should be a resident of another process $peer_j$, $peer_i$ sends c to $peer_j$ as a new settler message. Upon reception of the message, $peer_j$ performs forward and backward contraction on c and then stores it as one of its residents.

Each process performs the expansion inferences between its residents. If each process executes only steps between its residents, the strategy would not be complete. The purpose of *inference messages* is to preserve completeness by making inferences between residents at different nodes possible. We refer to [2, 3] for a treatment of completeness of Clause-Diffusion. When a process $peer_i$ selects its resident c for *expansion work*, it also broadcasts c as an inference message. Appropriate book-keeping is in place to avoid sending a clause as an inference message more than once. Any other process $peer_j$ will receive c and perform expansion between c and its residents and (backward) contraction on c.

Peers features two mechanisms for organizing expansion inferences, one of which can be selected by setting appropriate flags. In the first mechanism, similar to that of Otter [8], a unit of local work consists in selecting a *given clause* c and performing all expansion inferences between c and the other clauses in the local data base. In the second mechanism, a unit of local work consists in selecting a pair of clauses (c, d) and performing all expansion inferences between c and d. The second mechanism was designed for AC theories: if paramodulation is done modulo AC, there are generally so many AC-paramodulants that generating all the AC-paramodulants between the given equation and all the other equations is a too large amount of work to be a single unit of expansion work. In both mechanisms, the inferences are subdivided based on the ownership of clauses: for instance, for paramodulation, each process executes only paramodulations into its residents, i.e. it paramodulates either a resident or a received inference message into a resident. Therefore, distinct peers perform different selections of steps in different orders, so that their computations are different.

4 Some Strategies in Peers

Peers has 23 flags and 13 parameters and different settings of some of these options define different strategies. An important parameter is the one that controls the choice of the allocation algorithm. Peers currently has three allocation algorithms. In the *"rotate"* allocation algorithm, each peer keeps track of the most recently used destination q, including itself, and simply picks $q + 1 \mod n$, where n is the number of processes. The *"syntax"* allocation algorithm maps a clause to a process based on the syntax of the clause. Each symbol in the signature is associated to an integer code, its key in the symbol table. A clause is assigned to the process whose number is the sum of the codes of all the symbols in the clause modulo the number of processes. This algorithm has the nice property that it sends all identical copies of a clause to the same destination, where all but one can be deleted by subsumption. The *"select-min"* allocation algorithm chooses the peer which is estimated to have minimum work-load. For the purpose of this allocation criterion, each process needs to have some information on the work-load of all the other processes. Currently, the number of residents at a process is used as measure of its work-load. A process $peer_i$ may estimate the number of residents of another process $peer_j$ based on the inference messages that $peer_i$ receives from $peer_j$. Intuitively, the higher are the identifiers of inference messages from $peer_j$, the higher is the number of residents at $peer_j$. Thus, process $peer_i$ saves the identifier of the most recently received inference message from $peer_j$. This number is regarded by $peer_i$ as an esteem of the work-load of $peer_j$.

Another option which is relevant to strategies definition, is the parameter that controls the treatment of backward contracted clauses. In many sequential contraction-based theorem provers, the reduced form of a backward-contracted clause is regarded as a new clause. This approach is available in Peers. It has the advantage of being a uniform, simple treatment of all backward-contracted clauses. However, in the distributed case, one may want to restrict the extent to which backward contraction of clauses causes the clauses to be re-allocated. Accordingly, another possibility in Peers is to treat backward-contracted residents as new clauses, but for re-allocation: if c, resident of $peer_i$, is reduced to c' by backward contraction at $peer_i$, c' is regarded as a new settler settling down at $peer_i$. A third possibility is to regard backward-contracted clauses as new clauses, except that they are not re-allocated and they do not get an entirely new identifier (just a new "birth-time"). This mechanims allows to recognize, based on common identifier, different reduced forms of a common ancestor and delete all of them but one (see [2] for details of this mechanism).

5 Experiments

In the following we give the performances of Peers on a few problems. N-Peers is Peers with n nodes. The reported run time (in seconds) of n-Peers is the CPU time of the first Peer to succeed. The other Peers run till either they receive a halting message or also find a proof, whichever happens first. The nodes are workstations Sun Sparc 2, communicating over the Ethernet. The workstations used for our experiments were not isolated from the rest of the network and were possibly simultaneously used by other users.

Problem	1-Peers	2-Peers	4-Peers	6-Peers	8-Peers
x3	96.45	50.29	43.28	30.66	7.51
r2	40.04	16.51	18.74	34.97	22.31
sa1	15.99	7.30	16.06	12.96	9.65
sa2	24.28	20.09	12.76	81.05	20.34

Problem $x3$ is proving that $x^3 = x$ implies commutativity in ring theory. Problem $r2$ is a problem in Robbins algebra. It consists in deriving Huntington's axiom $(-(-x + y) + (-(-x + (-y))) = x)$ from Robbins' axiom $(-(-(x + y) + (-(x + (-y)))) = x)$ and the hypothesis that there exists an element C such that $C + C = C$ [10]. Problems $sa1$ and $sa2$ are "single axioms problems" in group theory, where the goal is to prove that a given single axiom is sufficient to axiomatize group theory. All problems are solved by contraction-based strategies (Knuth-Bendix completion-based strategies), working modulo AC for $x3$ and $r2$. In some cases, e.g. $x3$, the run-time decreases somewhat regularly, albeit not linearly, as the number of nodes increases. In others, e.g. $sa1$, the run-time first improves and then gets worse with 6 and 8 nodes. The latter observation may be explained in part by redundancy, e.g. duplication of clauses among the processes, and the non-determinism of the distributed prover.

The following table shows the run-times for five runs on the problem $x3$. In each run a different strategy was selected. Strategies a and c use the *"rotate"* allocation algorithm, b and d use *"syntax"* and e uses *"select-min"*. Strategies a and b treat residents reduced by backward contraction as new clauses, while c, d and e, only update their birth-times:

Problem	1-Peers	2-Peers	4-Peers	6-Peers	8-Peers
x3a	96.28	53.58	46.87	54.04	25.95
x3b	96.45	50.29	43.28	30.66	7.51
x3c	96.06	51.37	44.06	43.52	28.06
x3d	95.86	49.16	44.52	31.65	8.60
x3e	96.36	87.64	38.34	24.93	31.02

Peers displays a considerable unstability of performances. This phenomenon, which appeared also in [5] and in radically different approaches to parallelization, e.g. [7], has not yet been studied systematically. We did not compare Peers with highly optimized sequential provers such as Otter, because Peers is a prototype, a tool to understand the problems in distributed deduction. We feel that Clause-Diffusion has the potential of achieving significant speed-ups on some non-trivial class of problems. Much more work is needed, on both the method and its implementation, in order to obtain more stable and more satisfactory performances.

Acknowledgements

Peers was written when the first author was visiting the Argonne National Laboratory in January/February 1993, supported by the Argonne National Laboratory and the Università degli Studi di Milano. The work on Peers continued when the first author was at INRIA-Lorraine and CRIN in the Spring of 1993, supported by INRIA-Lorraine and CRIN and the Università degli Studi di Milano.

References

1. J.Avenhaus and J.Denzinger, Distributing Equational Theorem Proving, in C.Kirchner (ed.), *Proceedings of the Fifth Conference on Rewriting Techniques and Applications*, Montréal, Canada, June 1993, Springer Verlag, Lecture Notes in Computer Science 690, 62–76, 1993.
2. M.P.Bonacina, Distributed Automated Deduction, Ph.D. Thesis, Department of Computer Science, State University of New York at Stony Brook, December 1992.
3. M.P.Bonacina and J.Hsiang, On fairness in distributed deduction, in P.Enjalbert, A.Finkel and K.W.Wagner (eds.), *Proceedings of the Tenth Symposium on Theoretical Aspects of Computer Science*, Würzburg, Germany, February 1993, Springer Verlag, Lecture Notes in Computer Science 665, 141–152, 1993.
4. M.P.Bonacina and J.Hsiang, The Clause-Diffusion methodology for distributed deduction, submitted for publication.
5. M.P.Bonacina and J.Hsiang, Distributed Deduction by Clause-Diffusion: the Aquarius Prover, in A.Miola (ed.), *Proceedings of the Third International Symposium on Design and Implementation of Symbolic Computation Systems*, Gmunden, Austria, September 1993, Springer Verlag, Lecture Notes in Computer Science 722, 272–287, 1993.
6. R.Butler and E.L.Lusk, User's Guide to the p4 Programming System, Technical Report ANL-92/17, Argonne National Laboratory, Argonne, Illinois, October 1992.
7. E.L.Lusk and W.W.McCune, Experiments with ROO: a Parallel Automated Deduction System, in B.Fronhöfer and G.Wrightson (eds.), *Parallelization in Inference Systems*, Springer Verlag, Lecture Notes in Artificial Intelligence 590, 139–162, 1992.
8. W.W.McCune, OTTER 2.0 Users Guide, Technical Report ANL-90/9, Argonne National Laboratory, Argonne, Illinois, March 1990.
9. S.Taylor, *Parallel Logic Programming Techniques*, Prentice Hall.
10. L.Wos, Searching for Open Questions, Newsletter of the Association for Automated Reasoning, No. 15, May 1990.

Author index

Anderson, Penny, 575

Bachmair, Leo, 435
Baker, Siani, 177
Basin, David A., 466
Baumgartner, Peter, 87, 769
Beckert, Bernhard, 678, 793
Bibel, W., 783
Bonacina, Maria Paola, 841
Bouhoula, Adel, 836
Bourely, Christophe, 72
Boyer, Robert S., 237
Bronsard, Francois, 102
Bruijn, N.G. de, 237
Brüning, Stefan, 222, 783
Bündgen, Reinhard, 693

Caferra, Ricardo, 72
Chazarain, Jacques, 118
Chou, Shang-Ching, 401
Chu, Heng, 192
Clarke, Edmund, 758
Constable, Robert L., 529

Domenjoud, Eric, 267

Egly, U., 783
Eriksson, Lars-Henrik, 821

Farmer, William M., 356
Felty, Amy, 605
Fribourg, Laurent, 311
Furbach, Ulrich, 87, 769

Gallagher, J.P., 207
Ganzinger, Harald, 435
Gao, Xiao-Shan, 401
Goldberg, Allen, 831
Goller, Ch., 778
Goubault, Jean, 499
Graf, Peter, 514
Gramlich, Bernhard, 162
Guttman, Joshua D., 356

Hähnle, Reiner, 708
Hasker, Robert W., 102
Hermann, Miki, 560
Hines, Larry M., 416
Howe, Douglas, 605
Huang, Xiaorong, 738, 788, 807
Huet, Gérard, 237
Hutter, Dieter, 29

Iwanuma, Kouji, 296

Jackson, Paul, 590
Johann, Patricia, 620

Kerber, Manfred, 371, 788, 807
Klay, Francis, 267
Klingenbeck, Stefan, 708
Kohlhase, Michael, 371, 620, 788, 807
Kolaitis, Phokion G., 560
Kounalis, Emmanuel, 118
Kraan, Ina, 826

Letz, R., 778
Lowry, Michael, 341
Lusk, Ewing, 764
Lysne, Olav, 133

Martin, Ursula, 432
Massacci, Fabio, 723
Mayr, K., 778
McCune, William W., 764, 841
McPhee, Nicholas Freitag, 401
Melis, Erica, 788, 807
Middeldorp, Aart, 451
Minami, Toshiro, 816
Müller, Olaf, 650

Nadel, Mark E., 356
Nesmith, Dan, 788, 807
Nieuwenhuis, Robert, 545

Ohtani, Takeshi, 816

Paulson, Lawrence C., 148

Peixoto, Marcos Veloso, 311
Peltier, Nicolas, 72
Pfenning, Frank, 811
Plaisted, David A., 57, 192
Platek, Richard, 431
Portoraro, Frederic D., 802
Posegga, Joachim, 793
Prehofer, Christian, 635
Pressburger, Thomas, 341
Protzen, Martin, 42

Quantz, J. Joachim, 326

Rath, T., 783
Reddy, Uday S., 102
Richards, Bradley L., 826
Richts, Jörn, 788, 807
Ringeissen, Christophe, 267
Royer, Veronique, 326
Rubio, Albert, 545

Salzer, Gernot, 282
Sawamura, Hajime, 816
Schumann, Johann M. Ph., 774, 778
Segre, Alberto Maria, 484
Siekmann, Jörg, 788, 807
Slaney, John, 1, 764, 798
Smaill, Alan, 826
Socher-Ambrosius, Rolf, 665
Stickel, Mark, 341
Sturgill, David B., 484
Sutcliffe, Geoff, 252
Suttner, Christian, 252

Thayer, F. Javier, 356
Trybulec, Andrzej, 237

Underwood, Ian, 341

Vigneron, Laurent, 530

Waal, D.A. de, 207
Waldinger, Richard, 341
Walsh, Toby, 14, 466
Wang, Dongming, 386
Wang, T.C., 831
Weber, Franz, 650

Wiggins, Geraint A., 826
Wirth, Claus-Peter, 162

Yemenis, Theodor, 252

Zantema, Hans, 451
Zhang, Jian, 753
Zhao, Xudong, 758

Springer-Verlag and the Environment

We at Springer-Verlag firmly believe that an international science publisher has a special obligation to the environment, and our corporate policies consistently reflect this conviction.

We also expect our business partners – paper mills, printers, packaging manufacturers, etc. – to commit themselves to using environmentally friendly materials and production processes.

The paper in this book is made from low- or no-chlorine pulp and is acid free, in conformance with international standards for paper permanency.

Lecture Notes in Artificial Intelligence (LNAI)

Vol. 619: D. Pearce, H. Wansing (Eds.), Nonclassical Logics and Information Processing. Proceedings, 1990. VII, 171 pages. 1992.

Vol. 622: F. Schmalhofer, G. Strube, Th. Wetter (Eds.), Contemporary Knowledge Engineering and Cognition. Proceedings, 1991. XII, 258 pages. 1992.

Vol. 624: A. Voronkov (Ed.), Logic Programming and Automated Reasoning. Proceedings, 1992. XIV, 509 pages. 1992.

Vol. 627: J. Pustejovsky, S. Bergler (Eds.), Lexical Semantics and Knowledge Representation. Proceedings, 1991. XII, 381 pages. 1992.

Vol. 633: D. Pearce, G. Wagner (Eds.), Logics in AI. Proceedings. VIII, 410 pages. 1992.

Vol. 636: G. Comyn, N. E. Fuchs, M. J. Ratcliffe (Eds.), Logic Programming in Action. Proceedings, 1992. X, 324 pages. 1992.

Vol. 638: A. F. Rocha, Neural Nets. A Theory for Brains and Machines. XV, 393 pages. 1992.

Vol. 642: K. P. Jantke (Ed.), Analogical and Inductive Inference. Proceedings, 1992. VIII, 319 pages. 1992.

Vol. 659: G. Brewka, K. P. Jantke, P. H. Schmitt (Eds.), Nonmonotonic and Inductive Logic. Proceedings, 1991. VIII, 332 pages. 1993.

Vol. 660: E. Lamma, P. Mello (Eds.), Extensions of Logic Programming. Proceedings, 1992. VIII, 417 pages. 1993.

Vol. 667: P. B. Brazdil (Ed.), Machine Learning: ECML – 93. Proceedings, 1993. XII, 471 pages. 1993.

Vol. 671: H. J. Ohlbach (Ed.), GWAI-92: Advances in Artificial Intelligence. Proceedings, 1992. XI, 397 pages. 1993.

Vol. 679: C. Fermüller, A. Leitsch, T. Tammet, N. Zamov, Resolution Methods for the Decision Problem. VIII, 205 pages. 1993.

Vol. 681: H. Wansing, The Logic of Information Structures. IX, 163 pages. 1993.

Vol. 689: J. Komorowski, Z. W. Raś (Eds.), Methodologies for Intelligent Systems. Proceedings, 1993. XI, 653 pages. 1993.

Vol. 695: E. P. Klement, W. Slany (Eds.), Fuzzy Logic in Artificial Intelligence. Proceedings, 1993. VIII, 192 pages. 1993.

Vol. 698: A. Voronkov (Ed.), Logic Programming and Automated Reasoning. Proceedings, 1993. XIII, 386 pages. 1993.

Vol. 699: G.W. Mineau, B. Moulin, J.F. Sowa (Eds.), Conceptual Graphs for Knowledge Representation. Proceedings, 1993. IX, 451 pages. 1993.

Vol. 723: N. Aussenac, G. Boy, B. Gaines, M. Linster, J.-G. Ganascia, Y. Kodratoff (Eds.), Knowledge Acquisition for Knowledge-Based Systems. Proceedings, 1993. XIII, 446 pages. 1993.

Vol. 727: M. Filgueiras, L. Damas (Eds.), Progress in Artificial Intelligence. Proceedings, 1993. X, 362 pages. 1993.

Vol. 728: P. Torasso (Ed.), Advances in Artificial Intelligence. Proceedings, 1993. XI, 336 pages. 1993.

Vol. 743: S. Doshita, K. Furukawa, K. P. Jantke, T. Nishida (Eds.), Algorithmic Learning Theory. Proceedings, 1992. X, 260 pages. 1993.

Vol. 744: K. P. Jantke, T. Yokomori, S. Kobayashi, E. Tomita (Eds.), Algorithmic Learning Theory. Proceedings, 1993. XI, 423 pages. 1993.

Vol. 745: V. Roberto (Ed.), Intelligent Perceptual Systems. VIII, 378 pages. 1993.

Vol. 746: A. S. Tanguiane, Artificial Perception and Music Recognition. XV, 210 pages. 1993.

Vol. 754: H. D. Pfeiffer, T. E. Nagle (Eds.), Conceptual Structures: Theory and Implementation. Proceedings, 1992. IX, 327 pages. 1993.

Vol. 764: G. Wagner, Vivid Logic. XII, 148 pages. 1994.

Vol. 766: P. R. Van Loocke, The Dynamics of Concepts. XI, 340 pages. 1994.

Vol. 770: P. Haddawy, Representing Plans Under Uncertainty. X, 129 pages. 1994.

Vol. 784: F. Bergadano, L. De Raedt (Eds.), Machine Learning: ECML-94. Proceedings, 1994. XI, 439 pages. 1994.

Vol. 795: W. A. Hunt, Jr., FM8501: A Verified Microprocessor. XIII, 333 pages. 1994.

Vol. 798: R. Dyckhoff (Ed.), Extensions of Logic Programming. Proceedings, 1993. VIII, 360 pages. 1994.

Vol. 799: M. P. Singh, Multiagent Systems: Intentions, Know-How, and Communications. XXIII, 168 pages. 1994.

Vol. 804: D. Hernández, Qualitative Representation of Spatial Knowledge. IX, 202 pages. 1994.

Vol. 808: M. Masuch, L. Pólos (Eds.), Knowledge Representation and Reasoning Under Uncertainty. VII, 237 pages. 1994.

Vol. 810: G. Lakemeyer, B. Nebel (Eds.), Foundations of Knowledge Representation and Reasoning. VIII, 355 pages. 1994.

Vol. 814: A. Bundy (Ed.), Automated Deduction — CADE-12. Proceedings, 1994. XVI, 848 pages. 1994.

Lecture Notes in Computer Science

Vol. 780: J. J. Joyce, C.-J. H. Seger (Eds.), Higher Order Logic Theorem Proving and Its Applications. Proceedings, 1993. X, 518 pages. 1994.

Vol. 781: G. Cohen, S. Litsyn, A. Lobstein, G. Zémor (Eds.), Algebraic Coding. Proceedings, 1993. XII, 326 pages. 1994.

Vol. 782: J. Gutknecht (Ed.), Programming Languages and System Architectures. Proceedings, 1994. X, 344 pages. 1994.

Vol. 783: C. G. Günther (Ed.), Mobile Communications. Proceedings, 1994. XVI, 564 pages. 1994.

Vol. 784: F. Bergadano, L. De Raedt (Eds.), Machine Learning: ECML-94. Proceedings, 1994. XI, 439 pages. 1994. (Subseries LNAI).

Vol. 785: H. Ehrig, F. Orejas (Eds.), Recent Trends in Data Type Specification. Proceedings, 1992. VIII, 350 pages. 1994.

Vol. 786: P. A. Fritzson (Ed.), Compiler Construction. Proceedings, 1994. XI, 451 pages. 1994.

Vol. 787: S. Tison (Ed.), Trees in Algebra and Programming – CAAP '94. Proceedings, 1994. X, 351 pages. 1994.

Vol. 788: D. Sannella (Ed.), Programming Languages and Systems – ESOP '94. Proceedings, 1994. VIII, 516 pages. 1994.

Vol. 789: M. Hagiya, J. C. Mitchell (Eds.), Theoretical Aspects of Computer Software. Proceedings, 1994. XI, 887 pages. 1994.

Vol. 790: J. van Leeuwen (Ed.), Graph-Theoretic Concepts in Computer Science. Proceedings, 1993. IX, 431 pages. 1994.

Vol. 791: R. Guerraoui, O. Nierstrasz, M. Riveill (Eds.), Object-Based Distributed Programming. Proceedings, 1993. VII, 262 pages. 1994.

Vol. 792: N. D. Jones, M. Hagiya, M. Sato (Eds.), Logic, Language and Computation. XII, 269 pages. 1994.

Vol. 793: T. A. Gulliver, N. P. Secord (Eds.), Information Theory and Applications. Proceedings, 1993. XI, 394 pages. 1994.

Vol. 795: W. A. Hunt, Jr., FM8501: A Verified Microprocessor. XIII, 333 pages. 1994. (Subseries LNAI).

Vol. 796: W. Gentzsch, U. Harms (Eds.), High-Performance Computing and Networking. Proceedings, 1994, Vol. I. XXI, 453 pages. 1994.

Vol. 797: W. Gentzsch, U. Harms (Eds.), High-Performance Computing and Networking. Proceedings, 1994, Vol. II. XXII, 519 pages. 1994.

Vol. 798: R. Dyckhoff (Ed.), Extensions of Logic Programming. Proceedings, 1993. VIII, 360 pages. 1994. (Subseries LNAI).

Vol. 799: M. P. Singh, Multiagent Systems: Intentions, Know-How, and Communications. XXIII, 168 pages. 1994. (Subseries LNAI).

Vol. 800: J.-O. Eklundh (Ed.), Computer Vision – ECCV '94. Proceedings 1994, Vol. I. XVIII, 603 pages. 1994.

Vol. 801: J.-O. Eklundh (Ed.), Computer Vision – ECCV '94. Proceedings 1994, Vol. II. XV, 485 pages. 1994.

Vol. 802: S. Brookes, M. Main, A. Melton, M. Mislove, D. Schmidt (Eds.), Mathematical Foundations of Programming Semantics. Proceedings, 1993. IX, 647 pages. 1994.

Vol. 803: J. W. de Bakker, W.-P. de Roever, G. Rozenberg (Eds.), A Decade of Concurrency. Proceedings, 1993. VII, 683 pages. 1994.

Vol. 804: D. Hernández, Qualitative Representation of Spatial Knowledge. IX, 202 pages. 1994. (Subseries LNAI).

Vol. 805: M. Cosnard, A. Ferreira, J. Peters (Eds.), Parallel and Distributed Computing. Proceedings, 1994. X, 280 pages. 1994.

Vol. 806: H. Barendregt, T. Nipkow (Eds.), Types for Proofs and Programs. VIII, 383 pages. 1994.

Vol. 807: M. Crochemore, D. Gusfield (Eds.), Combinatorial Pattern Matching. Proceedings, 1994. VIII, 326 pages. 1994.

Vol. 808: M. Masuch, L. Pólos (Eds.), Knowledge Representation and Reasoning Under Uncertainty. VII, 237 pages. 1994. (Subseries LNAI).

Vol. 809: R. Anderson (Ed.), Fast Software Encryption. Proceedings, 1993. IX, 223 pages. 1994.

Vol. 810: G. Lakemeyer, B. Nebel (Eds.), Foundations of Knowledge Representation and Reasoning. VIII, 355 pages. 1994. (Subseries LNAI).

Vol. 811: G. Wijers, S. Brinkkemper, T. Wasserman (Eds.), Advanced Information Systems Engineering. Proceedings, 1994. XI, 420 pages. 1994.

Vol. 812: J. Karhumäki, H. Maurer, G. Rozenberg (Eds.), Results and Trends in Theoretical Computer Science. Proceedings, 1994. X, 445 pages. 1994.

Vol. 813: A. Nerode, Y. N. Matiyasevich (Eds.), Logical Foundations of Computer Science. Proceedings, 1994. IX, 392 pages. 1994.

Vol. 814: A. Bundy (Eds.), Automated Deduction—CADE-12. Proceedings, 1994. XVI, 848 pages. 1994. (Subseries LNAI).

Vol. 815: R. Valette (Eds.), Application and Theory of Petri Nets 1994. Proceedings. IX, 587 pages. 1994.